INTERNATIONAL PLUMBING CODE®

CODE AND COMMENTARY

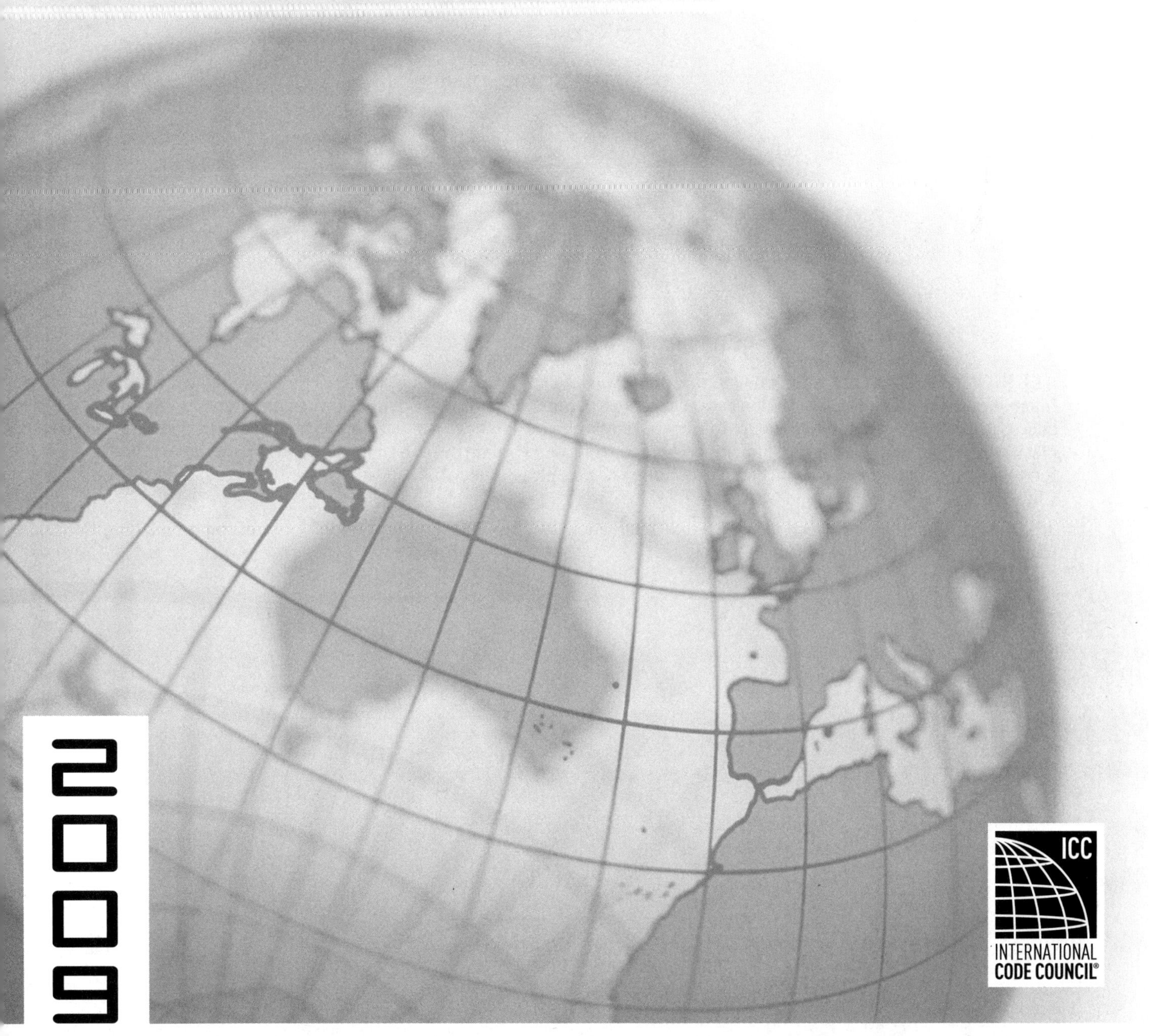

2009

INTERNATIONAL CODE COUNCIL®

2009 International Plumbing Code®—Code and Commentary

First Printing: June 2010
Second Printing: September 2011

ISBN: 978-1-58001-904-0

35924-T013310

PRINTED IN THE U.S.A.

PREFACE

Significant changes in the plumbing industry, as well as in manufacturing technology, had become so commanding that a radically new approach to the design and installation of plumbing systems seemed an imperative. The reply to that imperative was the *International Plumbing Code*, a document emphasizing prescriptive and performance-related provisions.

As a follow-up to the *International Plumbing Code*, we offer a companion document, the *International Plumbing Code Commentary*. The basic appeal of the Commentary is thus: it provides in a small package and at reasonable cost thorough coverage of many issues likely to be dealt with when using the *International Plumbing Code* — and then supplements that coverage with historical and technical background. Reference lists, information sources and bibliographies are also included.

Throughout all of this, strenuous effort has been made to keep the vast quantity of material accessible and its method of presentation useful. With a comprehensive yet concise summary of each section, the Commentary provides a convenient reference for plumbing regulations. In the chapters that follow, discussions focus on the full meaning and implications of the code text. Guidelines suggest the most effective method of application, and the consequences of not adhering to the code text. Illustrations are provided to aid understanding; they do not necessarily illustrate the only methods of achieving code compliance.

The format of the Commentary includes the full text of each section, table and figure in the code, followed immediately by the commentary applicable to that text. Each section's narrative includes a statement of its objective and intent, and usually includes a discussion about why the requirement commands the conditions set forth. Code text and commentary text are easily distinguished from each other. All code text is shown as it appears in the *International Plumbing Code*, and all commentary is indented below the code text and begins with the symbol ❖.

Readers should note that the Commentary is to be used in conjunction with the *International Plumbing Code* and not as a substitute for the code. **The Commentary is advisory only;** the code official alone possesses the authority and responsibility for interpreting the code.

Comments and recommendations are encouraged, for through your input, we can improve future editions. Please direct your comments to the Codes and Standards Development Department at the Chicago District Office.

TABLE OF CONTENTS

Chapter 1: Scope and Administration

General Comments

The law of building regulation is grounded on the police power of the state. In terms of how it is used, this is the power of the state to legislate for the general welfare of its citizens. This power enables passage of such laws as a plumbing code. It is from the police power delegated by the state legislature that local governments are able to enact building regulations. If the state legislature has limited this power in any way, the municipality may not exceed these limitations. Although the municipality may not further delegate its police power (e.g., by delegating the burden of determining code compliance to the building owner, contractor or architect), it may turn over the administration of building regulations to a municipal official, such as a code official, provided that he or she is given sufficient criteria to clearly establish the basis for decisions as to whether or not a proposed building, including its plumbing systems, conforms to the code.

Chapter 1 is largely concerned with maintaining "due process of law" in enforcing the performance criteria contained in the code. Only through careful observation of the administrative provisions can the code official reasonably hope to demonstrate that "equal protection under the law" has been provided. Although it is generally assumed that the administrative and enforcement sections of a code are geared toward the code official, this is not entirely true. The provisions also establish the rights and privileges of the design professional, contractor and building owner. The position of the code official is merely to review the proposed and completed work and determine whether a plumbing installation conforms to the code requirements. The design professional is responsible for the design of a safe, sanitary plumbing system.

The contractor is responsible for installing the system in strict accordance with the plans.

During the course of the construction of a plumbing system, the code official reviews the activity to make sure that the spirit and intent of the law are being met and that the plumbing system provides adequate protection of public health. As a public servant, the code official enforces the code in an unbiased, proper manner. Every individual is guaranteed equal enforcement of the code. Furthermore, design professionals, contractors and building owners have the right of due process for any requirement in the code.

Purpose

A plumbing code, as with any other code, is intended for adoption as a legally enforceable document to safeguard health, safety, property and public welfare. A plumbing code cannot be effective without adequate provisions for its administration and enforcement. The official charged with the administration and enforcement of plumbing regulations has a great responsibility, and with this responsibility goes authority. No matter how detailed the plumbing code may be, the code official must, to some extent, exercise judgment in determining compliance. The code official has the responsibility for establishing that the homes in which the citizens of the community reside and the buildings in which they work are designed and constructed to be reasonably free from hazards associated with the presence and use of plumbing appliances, appurtenances, fixtures and systems. The code is intended to establish a minimum acceptable level of safety.

PART 1—SCOPE AND APPLICATION

SECTION 101 GENERAL

101.1 Title. These regulations shall be known as the *International Plumbing Code* of [NAME OF JURISDICTION] hereinafter referred to as "this code."

❖ This section sets forth the scope and intent of the code as it applies to new and existing structures. The adopted regulations are identified by inserting the name of the adopting jurisdiction into the code.

101.2 Scope. The provisions of this code shall apply to the erection, installation, alteration, repairs, relocation, replacement, addition to, use or maintenance of plumbing systems within this jurisdiction. This code shall also regulate nonflammable medical gas, inhalation anesthetic, vacuum piping, nonmedical oxygen systems and sanitary and condensate vacuum collection systems. The installation of fuel gas distribution piping and equipment, fuel-gas-fired water heaters and water heater venting systems shall be regulated by the *International Fuel Gas Code*. Provisions in the appendices shall not apply unless specifically adopted.

Exception: Detached one- and two-family dwellings and multiple single-family dwellings (townhouses) not more than three stories high with separate means of egress and their accessory structures shall comply with the *International Residential Code.*

❖ This section describes the types of plumbing system construction-related activities to which the code is intended to apply. The applicability of the code encompasses the initial design of plumbing systems, the installation and construction phases and the maintenance of operating systems. Section 101.2 excludes nothing plumbing related and does not limit applicability of the code to any device, fixture, system and associated equipment that could fall under, or is construed to fall under, the definition of "Plumbing" (see the definition of "Plumbing" in Chapter 2). The code is intended to govern plumbing systems provided for use by and for the general safety and well-being of occupants of a building. The code intends to regulate any and all plumbing-related appliances, systems and associated equipment that can affect the health, safety and welfare of building occupants insofar as they are affected by the installation, operation and maintenance of such appliances and systems. Plumbing systems include the associated equipment by definition of "Plumbing system" in Chapter 2.

In addition, nonflammable medical gas, inhalation anesthetic, vacuum piping, nonmedical oxygen systems and sanitary and condensate vacuum collection systems are regulated by the code. Other than the actual connections to the potable water system, the code does not regulate hydronic piping for space heating or cooling, lawn sprinkler (irrigation) systems or automatic fire sprinkler and standpipe systems. These are not considered to be plumbing systems because they have nothing to do with supplying potable water or the transport of liquid wastes and water-borne solid wastes. Flammable/combustible liquid piping, hydronic piping, fire suppression system piping and nonplumbing-related systems are typically addressed in the *International Building Code*® (IBC®), the *International Fire Code*® (IFC®) and the *International Mechanical Code*® (IMC®). Most hydronic heating and cooling systems and fire suppression systems have one or more connections to the plumbing system. Such connections involve direct connections to the water supply system and indirect connections to the drainage system. For example, an automatic fire sprinkler system may be supplied by the public potable water supply and may have one or more drains and test apparatus that discharge water to the building drainage system. In the case of irrigation systems, for example, the plumbing system terminates at the backflow prevention device that joins the potable water supply to the irrigation piping system. All interfaces between plumbing systems and nonplumbing systems are subject to the requirements of the code.

This section references the *International Fuel Gas Code*® (IFGC®) for all fuel-gas-related regulations. This is the result of an agreement between the International Code Council® (ICC®) and the American Gas Association (AGA) to develop the IFGC.

The exception is actually a distinct requirement that sends the user to the *International Residential Code*® (IRC®) for one- and two-family dwellings that are less than four stories in height and townhouses that are less than four stories in height, as these structures are within the scope of the IRC. It is the intent of the International Codes® that the code regulate plumbing in all structures that do not fall within the scope of the IRC. Structures falling within the scope of the IRC are to be regulated by the IRC.

101.3 Intent. The purpose of this code is to provide minimum standards to safeguard life or limb, health, property and public welfare by regulating and controlling the design, construction, installation, quality of materials, location, operation and maintenance or use of plumbing equipment and systems.

❖ The intent of the code is to set forth requirements that establish the minimum acceptable level to safeguard life or limb, health, property and public welfare. Intent becomes important in the application of sections such as Sections 102, 104.2, 105.2 and 108, as well as any enforcement-oriented interpretive action or judgment. As with any code, the written text is subject to interpretation. Interpretations should not be affected by economics or the potential impact on any party. The only consideration should be protection of the public health, safety and welfare.

101.4 Severability. If any section, subsection, sentence, clause or phrase of this code is for any reason held to be unconstitutional, such decision shall not affect the validity of the remaining portions of this code.

❖ Only invalid sections of the code (as established by the court of jurisdiction) can be set aside. This is essential to safeguard the application of the code text to situations in which a provision of the code is declared illegal or unconstitutional. This section preserves the legislative action that put the legal provisions in place.

SECTION 102
APPLICABILITY

102.1 General. Where there is a conflict between a general requirement and a specific requirement, the specific requirement shall govern. Where, in any specific case, different sections of this code specify different materials, methods of construction or other requirements, the most restrictive shall govern.

❖ Specific requirements of the code override or take precedence over general requirements. For example, while the code specifies the types of piping materials that can be used for vent systems, the specific requirements for chemical waste systems require that the vent piping be suitable for the service intended and be approved by the code official.

102.2 Existing installations. Plumbing systems lawfully in existence at the time of the adoption of this code shall be permitted to have their use and maintenance continued if the use, maintenance or repair is in accordance with the original design and no hazard to life, health or property is created by such plumbing system.

❖ An existing plumbing system is generally considered to be "grandfathered in" with code adoption if the system meets a minimum level of safety. Frequently the criteria for this level are the regulations (or code) under which the existing building was originally constructed. If there are no previous code criteria to apply, the code official is to apply those provisions that are reasonably applicable to existing buildings. A specific level of safety is dictated by provisions dealing with hazard abatement in existing buildings and maintenance provisions, as contained in the code and the *International Property Maintenance Code*® (IPMC®), the *International Existing Building Code*® (IEBC®) and the IFC.

102.3 Maintenance. All plumbing systems, materials and appurtenances, both existing and new, and all parts thereof, shall be maintained in proper operating condition in accordance with the original design in a safe and sanitary condition. All devices or safeguards required by this code shall be maintained in compliance with the code edition under which they were installed.

The owner or the owner's designated agent shall be responsible for maintenance of plumbing systems. To determine compliance with this provision, the code official shall have the authority to require any plumbing system to be reinspected.

❖ All plumbing systems and equipment are subject to deterioration resulting from aging, wear, accumulation of dirt and debris, corrosion and other factors. Maintenance is necessary to keep plumbing systems and equipment in proper operating condition. Required safety devices and controls must be maintained to continue providing the protection that they afford. Existing equipment and systems could have safety devices or other measures that were necessary because of the nature of the equipment, and such safeguards may have been required by a code that predates the current code. Safeguards required by previous or present codes must be maintained for the life of the equipment or system.

The maintenance of plumbing systems as prescribed in this section is the responsibility of the owner of the property. The owner may authorize another party to be responsible for the property, in which case that party is responsible for the maintenance of the plumbing systems involved.

The reinspection authority of the code official is needed to ensure compliance with the maintenance requirements in this section.

102.4 Additions, alterations or repairs. Additions, alterations, renovations or repairs to any plumbing system shall conform to that required for a new plumbing system without requiring the existing plumbing system to comply with all the requirements of this code. Additions, alterations or repairs shall not cause an existing system to become unsafe, insanitary or overloaded.

Minor additions, alterations, renovations and repairs to existing plumbing systems shall meet the provisions for new construction, unless such work is done in the same manner and arrangement as was in the existing system, is not hazardous and is *approved.*

❖ Simply stated, new work must comply with current code requirements. Any alteration or addition to an existing system involves some new work, and therefore is subject to the requirements of the code. Additions or alterations to an existing system can place additional loads or different demands on the system, which could necessitate changing all or part of the existing system. For example, the addition of plumbing fixtures to an existing system may necessitate an increase in drain piping size and water distribution piping size. Additions and alterations must not cause an existing system to be any less in compliance with the code than it was before the changes.

Repair of an existing nonconforming plumbing system is permitted without having to completely replace the nonconforming portion. This typically occurs when repairing a fixture or piping. Although some types of fixtures or piping arrangements are no longer permitted, existing fixtures or piping can be repaired and remain in service if a health hazard or insanitary condition is not maintained or created. This section distinguishes between alterations (subject to applicable provisions of the code) and ordinary repairs (maintenance activities not requiring a permit). The intent of this section is to allow the continued use of existing plumbing systems and equipment that may or may not be designed and constructed as required for new installations.

Existing plumbing systems and equipment will normally require repair and component replacement to remain operational. This section permits repair and component replacements without requiring the redesign, alteration or replacement of the entire system. In other words, the plumbing system is allowed to stay as it was if it is not hazardous. It is important to note that the word "minor" in this section is intended to modify "additions," "alterations," "renovations" and "repairs." It is not the intent of this section to waive code requirements for the replacement of all or major portions of systems under the guise of repair. Any work other than minor repairs or replacement of minor portions of a system must be considered as new work subject to all applicable provisions of the code. Repairs and minor component replacements are permitted in a manner that is consistent with the existing system if those repairs or replacements are approved by the code official; are not hazardous; do not cause the system or equipment to be any less in compliance with the code than before; and are, to the extent practicable, in compliance with the provisions of the code applicable to new work.

102.5 Change in occupancy. It shall be unlawful to make any change in the *occupancy* of any structure that will subject the structure to any special provision of this code applicable to the new *occupancy* without approval of the code official. The code official shall certify that such structure meets the intent of the provisions of law governing building construction for the proposed new *occupancy* and that such change of *occupancy* does not result in any hazard to the public health, safety or welfare.

❖ When a building undergoes a change of occupancy, the plumbing system must be evaluated to determine what effect the change has on the plumbing system. For example, if a mercantile building was converted to a restaurant, additional plumbing fixtures would be required for the public based on the increased occupant load. If an existing system serves an occupancy that is different from the one it served when the code went into effect, the plumbing system must comply with the applicable code requirements for the newer occupancy. Depending on the nature of the previous occupancy, changing a building's occupancy classification could result in the requirement for additional or different types of plumbing fixtures.

102.6 Historic buildings. The provisions of this code relating to the construction, alteration, repair, enlargement, restoration, relocation or moving of buildings or structures shall not be mandatory for existing buildings or structures identified and classified by the state or local jurisdiction as historic buildings when such buildings or structures are judged by the code official to be safe and in the public interest of health, safety and welfare regarding any proposed construction, alteration, repair, enlargement, restoration, relocation or moving of buildings.

❖ This section gives the code official the widest possible flexibility in enforcing the code when the building in question has historic value. This flexibility, however, does not come without conditions. The most important criterion for application of this section is that the building must be specifically classified as being of historic significance by a qualified party or agency. Usually this is done by a state or local authority after considerable scrutiny. Most, if not all, states have such authorities, as do many local jurisdictions. Agencies with such authority typically exist at the state or local government level.

102.7 Moved buildings. Except as determined by Section 102.2, plumbing systems that are a part of buildings or structures moved into or within the jurisdiction shall comply with the provisions of this code for new installations.

❖ Buildings that have been relocated are subject to the requirements of the code as if they were new construction. Placing a building where one did not previously exist is analogous to constructing a new building. It is the intent of this section to require alteration of the existing plumbing systems to the extent necessary to bring them into compliance with the provisions of the code applicable to new construction or make the existing plumbing system comply with Section 102.2.

102.8 Referenced codes and standards. The codes and standards referenced in this code shall be those that are listed in Chapter 13 and such codes and standards shall be considered as part of the requirements of this code to the prescribed extent of each such reference. Where differences occur between provisions of this code and the referenced standards, the provisions of this code shall be the minimum requirements.

❖ The code references many standards promulgated and published by other organizations. A complete list of referenced standards appears in Chapter 13. The wording of this provision, "shall be those that are listed in Chapter 13," was carefully chosen to establish the edition of the standard that is enforceable under the code.

Although a standard is referenced, its full scope and content is not necessarily applicable. The standard is applicable only to the extent indicated in the text in which the standard is specifically referenced. The code takes precedence when the requirements of the standard conflict with the requirements of the code. For example, Section 412.1 requires all floor drains to conform to ASME A112.6.3 or CSA B79. These standards regulate the design and construction of a floor drain and permit both fixed and removable strainers. Section 412.2 requires all floor drains to have removable strainers. Although a floor drain without a removable cover may conform to ASME A112.6.3 or CSA B79, removable covers would be required because the code takes precedence over the standard. A referenced standard or a portion thereof is an enforceable extension of the code as if the content of the standard were included in the body of the code. For example, Section 410.1 references ARI 1010 for water coolers. The use and application of referenced standards are limited to those portions of the standards that are specifically identified. Although it is the intention of the code to be in harmony with referenced standards, the code text governs if conflicts occur.

102.9 Requirements not covered by code. Any requirements necessary for the strength, stability or proper operation of an existing or proposed plumbing system, or for the public safety, health and general welfare, not specifically covered by this code shall be determined by the code official.

❖ Evolving technology in our society will inevitably result in a situation in which the code is comparatively silent on an identified hazard. The reasonable application of the code to such hazardous, unforeseen conditions is addressed in this section. Clearly this section and the code official's judicious and reasonable application of it is necessary. The purpose of this section, however, is not to impose requirements that may be preferred when the code provides alternative methods or is not silent on the circumstances. Additionally, this section can be used to implement the general performance-oriented language of the code to specific enforcement situations.

102.10 Other laws. The provisions of this code shall not be deemed to nullify any provisions of local, state or federal law.

❖ Other laws enacted by the local, state or federal government may be applicable to an installation that is also governed by a requirement in the code. In such circumstances, the requirements of the code are in addition to those other laws, even though the building official may not be responsible for the enforcement of those laws. For example, the local health department might require the pots/pans sink waste lines in a restaurant to be indirectly connected to the drainage system whereas the plumbing code allows these sinks to be either directly or indirectly connected.

102.11 Application of references. Reference to chapter section numbers, or to provisions not specifically identified by number, shall be construed to refer to such chapter, section or provision of this code.

❖ In a situation where the code makes reference to a chapter or section number or to another code provision without specifically identifying its location in the code, assume that the referenced section, chapter or provision is in this code and not in a referenced code or standard.

PART 2 ADMINISTRATION AND ENFORCEMENT

SECTION 103 DEPARTMENT OF PLUMBING INSPECTION

103.1 General. The department of plumbing inspection is hereby created and the executive official in charge thereof shall be known as the code official.

❖ This section describes the plumbing inspection department and the duties of its employees. The executive official in charge of the plumbing department is named the "code official" by this section. In actuality, the person who is in charge of the department may hold a different title such as building commissioner, plumbing inspector, construction official, etc. For the purpose of the code, the person is referred to as the "code official" on being appointed.

103.2 Appointment. The code official shall be appointed by the chief appointing authority of the jurisdiction.

❖ This section establishes the code official as an appointed position from which he or she cannot be removed, except for cause subject to a due process review.

103.3 Deputies. In accordance with the prescribed procedures of this jurisdiction and with the concurrence of the appointing authority, the code official shall have the authority to appoint a deputy code official, other related technical officers, inspectors and other employees. Such employees shall have powers as delegated by the code official.

❖ This section gives the code official the authority to appoint other individuals to assist with the administration and enforcement of the code. These individuals have authority and responsibility as designated by the code official.

103.4 Liability. The code official, member of the board of appeals or employee charged with the enforcement of this code, while acting for the jurisdiction in good faith and without malice in the discharge of the duties required by this code or other pertinent law or ordinance, shall not thereby be rendered liable personally, and is hereby relieved from all personal liability for any damage accruing to persons or property as a result of any act or by reason of an act or omission in the discharge of official duties.

Any suit instituted against any officer or employee because of an act performed by that officer or employee in the lawful discharge of duties and under the provisions of this code shall be defended by the legal representative of the jurisdiction until the final termination of the proceedings. The code official or any subordinate shall not be liable for costs in any action, suit or proceeding that is instituted in pursuance of the provisions of this code.

❖ The code official is not intended to be held liable for actions performed in accordance with the code in a reasonable and lawful manner. The responsibility of the code official in this regard is subject to local, state and federal laws that may supersede this provision. This section further establishes that the code official (or subordinates) is not liable for costs in any legal action instituted in response to the performance of lawful duties. These costs are to be assumed by the state or municipality. The best way to be certain that the code official's action is a "lawful duty" is to always cite the applicable code section on which the enforcement action is based.

SECTION 104 DUTIES AND POWERS OF THE CODE OFFICIAL

104.1 General. The code official is hereby authorized and directed to enforce the provisions of this code. The code official shall have the authority to render interpretations of this code and to adopt policies and procedures in order to clarify the application of its provisions. Such interpretations, policies and procedures shall be in compliance with the intent and purpose of this code. Such policies and procedures shall not have the effect of waiving requirements specifically provided for in this code.

❖ The duty of the code official is to enforce the code, and he or she is the "authority having jurisdiction" for all matters relating to the code and its enforcement. It is the duty of the code official to interpret the code and to determine compliance. Code compliance will not always be easy to determine and will require judgment and expertise. In exercising this authority, however, the code official cannot set aside or ignore any provision of the code.

104.2 Applications and permits. The code official shall receive applications, review construction documents and issue permits for the installation and alteration of plumbing systems, inspect the premises for which such permits have been issued, and enforce compliance with the provisions of this code.

❖ The code enforcement process is normally initiated with an application for a permit. The code official is responsible for processing the applications and issuing permits for the installation, replacement, addition to or modification of plumbing systems in accordance with the code.

104.3 Inspections. The code official shall make all the required inspections, or shall accept reports of inspection by *approved* agencies or individuals. All reports of such inspections shall be in writing and be certified by a responsible officer of such *approved* agency or by the responsible individual. The code official is authorized to engage such expert opinion as deemed necessary to report on unusual technical issues that arise, subject to the approval of the appointing authority.

❖ The code official is required to make inspections as necessary to determine compliance with the code or to accept written reports of inspections by an approved agency. The inspection of the work in progress or accomplished is another significant element in determining code compliance. Even though a department may not have the resources to inspect every aspect of all work, the required inspections are those that are dictated by administrative rules and procedures based on many parameters, including available inspection resources. To expand the available inspection resources, the code official may approve an inspection agency that, in his or her opinion, possesses the proper qualifications. When unusual, extraordinary or complex technical issues arise relative to plumbing installations or the safety of an existing plumbing system, the code official has the authority to seek the opinion and advice of experts. A technical report from an expert can be used to assist the code official in the approval process.

104.4 Right of entry. Whenever it is necessary to make an inspection to enforce the provisions of this code, or whenever the code official has reasonable cause to believe that there exists in any building or upon any premises any conditions or violations of this code that make the building or premises unsafe, insanitary, dangerous or hazardous, the code official shall have the authority to enter the building or premises at all reasonable times to inspect or to perform the duties imposed upon the code official by this code. If such building or premises is occupied, the code official shall present credentials to the occupant and request entry. If such building or premises is unoccupied, the code official shall first make a reasonable effort to locate the owner or other person having charge or control of the building or premises and request entry. If entry is refused, the code official shall have recourse to every remedy provided by law to secure entry.

When the code official shall have first obtained a proper inspection warrant or other remedy provided by law to secure entry, no owner or occupant or person having charge, care or control of any building or premises shall fail or neglect, after proper request is made as herein provided, to promptly permit entry therein by the code official for the purpose of inspection and examination pursuant to this code.

❖ The first part of this section establishes the right of the code official to enter the premises to make the permit inspections required by Section 107. Permit application forms typically include a statement signed by the applicant (who is the owner or owner's agent) granting the code official the authority to enter specific areas to enforce code provisions related to the permit. The right to enter other structures or premises is more limited. First, to protect the right of privacy, the owner or occupant must grant the code official permission before the interior of the property can be inspected. Permission is not required for inspections that can be accomplished from within the public right-of-way. Second, access may be denied by the owner or occupant. Unless the inspector has reasonable cause to believe that a violation exists, access may be unattainable. Third, code officials must present proper identification (see commentary, Section 104.5) and request admittance during reasonable hours—usually the normal business hours of the establishment. Fourth, inspections must be aimed at securing or determining compliance with the provisions and intent of the regulations that are specifically within the established scope of the code official's authority.

Searches to gather information for the purpose of enforcing the other codes, ordinances or regulations are considered unreasonable and are prohibited by the Fourth Amendment to the U.S. Constitution. "Reasonable cause" in the context of this section must be distinguished from "probable cause," which is required to gain access to property in criminal cases.The burden of proof establishing reasonable cause may vary among jurisdictions. Usually, an inspector must show that the property is subject to inspection under the provisions of the code; that the interests of the public health, safety and welfare outweigh the individual's right to maintain privacy; and that such an inspection is required solely to determine compliance with the provisions of the code.

Many jurisdictions do not recognize the concept of an administrative warrant and may require the code official to prove probable cause in order to gain access upon refusal. This burden of proof is usually more substantial, often requiring the code official to stipulate in advance why access is needed (usually access is restricted to gathering evidence for seeking an indictment or making an arrest); what specific items or information is sought; its relevance to the case against the individual subject; how knowledge of the relevance of the information or items sought was obtained; and how the evidence sought will be used. In all such cases, the right to privacy must always be weighed against the right of the code official to conduct an inspection to establish public health, safety and welfare. Such important and complex constitutional issues should be discussed with the jurisdiction's legal counsel. Jurisdictions should establish procedures for securing the necessary court orders when an inspection is deemed necessary following a refusal.

The last paragraph in this section requires the owner or occupant to permit entry for inspection if a proper warrant or other documentation required by law has been obtained.

104.5 Identification. The code official shall carry proper identification when inspecting structures or premises in the performance of duties under this code.

❖ This section requires the code official (including by definition all authorized designees) to carry identification in the course of conducting the duties of the position. This identification removes any question concerning the purpose and authority of the inspector.

104.6 Notices and orders. The code official shall issue all necessary notices or orders to ensure compliance with this code.

❖ An important element of code enforcement is the necessary advisement of deficiencies and correction, which is accomplished through notices and orders. The code official is required to issue orders to abate illegal or unsafe conditions. Sections 108.7, 108.7.1, 108.7.2 and 108.7.3 contain additional information for these notices.

104.7 Department records. The code official shall keep official records of applications received, permits and certificates issued, fees collected, reports of inspections, and notices and orders issued. Such records shall be retained in the official records for the period required for the retention of public records.

❖ In keeping with the need for an efficiently conducted business practice, the code official must keep records pertaining to permit applications, permits, fees collected, inspections, notices and orders issued. Such documentation provides a valuable resource if questions arise regarding the department's actions with respect to a building. It requires that other documents be kept for the length of time mandated by a jurisdiction's, or its state's, laws or administrative rules for retaining public records.

SECTION 105 APPROVAL

105.1 Modifications. Whenever there are practical difficulties involved in carrying out the provisions of this code, the code official shall have the authority to grant modifications for individual cases, upon application of the owner or owner's representative, provided the code official shall first find that special individual reason makes the strict letter of this code impractical and the modification conforms to the intent and purpose of this code and that such modification does not lessen health, life and fire safety requirements. The details of action granting modifications shall be recorded and entered in the files of the plumbing inspection department.

❖ The code official may amend or make exceptions to the code as needed where strict compliance is impractical. Only the code official has authority to grant modifications. Consideration of a particular difficulty is to be based on the application of the owner and a demonstration that the intent of the code is accomplished. This section is not intended to permit setting aside or ignoring a code provision: rather, it is intended to provide acceptance of equivalent protection. Such modifications do not, however, extend to actions that are necessary to correct violations of the code. In other words, a code violation or the expense of correcting one cannot constitute a practical difficulty.

105.2 Alternative materials, methods and equipment. The provisions of this code are not intended to prevent the installation of any material or to prohibit any method of construction not specifically prescribed by this code, provided that any such alternative has been *approved*. An alternative material or method of construction shall be *approved* where the code official finds that the proposed alternative material, method or equipment complies with the intent of the provisions of this code and is at least the equivalent of that prescribed in this code.

❖ The code is not intended to inhibit innovative ideas or technological advances. A comprehensive regulatory document such as a plumbing code cannot envision and then address all future innovations in the industry. As a result, a performance code must be applicable to and provide a basis for the approval of an increasing number of newly developed, innovative materials, systems and methods for which no code text or referenced standards yet exist. The fact that a material, product or method of construction is not addressed in the code is not an indication that the material, product or method is prohibited. The code official is expected to apply sound technical judgment in accepting materials, systems or methods which, although not anticipated by the drafters of the current code text, can be demonstrated to offer equivalent performance. By virtue of its text, the code regulates new and innovative construction practices while addressing the relative safety of building occupants. The code official is responsible for determining whether a requested alternative provides an equivalent level of protection of the public health, safety and welfare as required by the code.

The most common application of an alternative approval occurs with the proposed use of new material. For example, if a new piping material is produced, the manufacturer may gain approval for use by submitting adequate technical data indicating it is equivalent in quality, strength, effectiveness, fire resistance, durability and safety to the piping material listed as acceptable in the code. At the same time, the manufacturer may submit a proposed code change to recognize the new piping material. If the code official rejects the request for an alternative approval, the applicant may appeal the decision, as regulated by Section 109.

105.2.1 Research reports. Supporting data, where necessary to assist in the approval of materials or assemblies not specifically provided for in this code, shall consist of valid research reports from *approved* sources.

❖ When an alternative material or method is proposed for construction, it is incumbent upon the code official to determine whether this alternative is, in fact, an equivalent to the methods prescribed by the code. Reports providing evidence of this equivalency are required to be supplied by an approved source, meaning a source that the code official finds to be reliable and accurate. The ICC Evaluation Service is one example of an agency that provides research reports for alternative materials and methods.

105.3 Required testing. Whenever there is insufficient evidence of compliance with the provisions of this code, or evidence that a material or method does not conform to the requirements of this code, or in order to substantiate claims for alternate materials or methods, the code official shall have the authority to require tests as evidence of compliance to be made at no expense to the jurisdiction.

❖ To provide the basis on which the code official can make a decision regarding an alternative material or type of equipment, sufficient technical data, test reports and documentation must be provided for evaluation. If evidence satisfactory to the code official proves that the alternative equipment, material or construction method is equivalent to that required by the code, the code official is obligated to approve it for use. Any such approval cannot have the effect of waiving any requirements of the code. The burden of proof of equivalence lies with the applicant who proposes the use of alternative equipment, materials or methods.

105.3.1 Test methods. Test methods shall be as specified in this code or by other recognized test standards. In the absence of recognized and accepted test methods, the code official shall approve the testing procedures.

❖ The code official must require the submission of any appropriate information and data to assist in the determination of equivalency before a permit can be issued. The type of information required includes test data in accordance with the referenced standards, evidence of compliance with the referenced standard specifications, and design calculations. An evaluation report issued by an authoritative agency, such as ICC Evaluation Service, is particularly useful in providing the code official with the technical basis for evaluation and approval of new and innovative plumbing materials and components. The use of authoritative research reports can greatly assist the code official by reducing the time-consuming engineering analysis necessary to review materials and products. Failure to adequately substantiate a request for the use of an alternative is a valid reason for the code official to deny a request.

105.3.2 Testing agency. All tests shall be performed by an *approved* agency.

❖ The testing agency must be approved by the code official. The testing agency should have technical expertise, test equipment and quality assurance to properly conduct and report the necessary testing.

105.3.3 Test reports. Reports of tests shall be retained by the code official for the period required for retention of public records.

❖ Test reports substantiating the modification must be retained in accordance with public record laws. The attorney of the jurisdiction could be asked to verify the specific time period in applicable laws of the jurisdiction.

105.4 Alternative engineered design. The design, documentation, inspection, testing and approval of an *alternative engineered design* plumbing system shall comply with Sections 105.4.1 through 105.4.6.

❖ This section permits an engineer or architect to design a plumbing system that may not comply with all of the provisions found in Chapters 3 through 13. The design must be approved by the code official and must conform to accepted engineering principles. The engineered plumbing system must provide the level of protection of the public health, safety and welfare intended by the code.

105.4.1 Design criteria. An *alternative engineered design* shall conform to the intent of the provisions of this code and shall provide an equivalent level of quality, strength, effectiveness, fire resistance, durability and safety. Material, equipment or components shall be designed and installed in accordance with the manufacturer's installation instructions.

❖ Although an engineered plumbing system may not comply with all of the minimum requirements set forth in Chapters 3 through 13, it must comply with the intent of these provisions. This section permits use of standard engineering principles in the design of an innovative system as long as there is no sacrifice of quality, strength, effectiveness, fire resistance, durability and safety. This section further reinforces the intent of Section 105.2 for the acceptance of alternative materials and equipment. The requirement for compliance with the manufacturer's installation instructions is generally intended to address entire engineered systems, such as the single-stack plumbing system. The manufacturer or appropriate industry association provides criteria contained in design and installation handbooks. The manufacturer's instructions must be followed for all innovative fittings or products regulated by this section.

105.4.2 Submittal. The registered design professional shall indicate on the permit application that the plumbing system is an *alternative engineered design*. The permit and permanent permit records shall indicate that an *alternative engineered design* was part of the *approved* installation.

❖ The permit and permanent permit records must indicate that an alternative engineered design is part of the proposed plumbing design. This is essential information to have on file to maintain a complete legal record of the plumbing system. When future permits are applied for regarding alterations or modifications, appropriate measures can then be taken to determine that the future work will not adversely affect the system design.

105.4.3 Technical data. The registered design professional shall submit sufficient technical data to substantiate the proposed *alternative engineered design* and to prove that the performance meets the intent of this code.

❖ The appropriate information and data must be submitted to the code official to assist in the approval of the alternative engineered design. This is not an option. Acceptable data to substantiate the performance of the proposed plumbing system or components include results of tests performed by an approved third-party testing agency, design calculations or an evaluation report issued by an authoritative agency, such as ICC Evaluation Service.

105.4.4 Construction documents. The registered design professional shall submit to the code official two complete sets of signed and sealed construction documents for the *alternative engineered design*. The construction documents shall include floor plans and a riser diagram of the work. Where appropriate, the construction documents shall indicate the direction of flow, all pipe sizes, grade of horizontal piping, loading, and location of fixtures and appliances.

❖ This section is used in conjunction with Section 106.3.1. The option for the code official to waive the requirements for filing construction documents, however, is not applicable. The required detailing of such documents is needed to provide the code official with the necessary information to review and approve the plans.

105.4.5 Design approval. Where the code official determines that the *alternative engineered design* conforms to the intent of this code, the plumbing system shall be *approved*. If the *alternative engineered design* is not *approved*, the code official shall notify the registered design professional in writing, stating the reasons thereof.

❖ The code official is responsible for determining whether the requested alternative engineered design provides the equivalent level of protection of public health, safety and welfare as required in the code. The code official's response to the design professional must be in writing, stating the reason for either accepting or rejecting the request. If the code official rejects the request for the alternative engineered system, the registered design professional may appeal the decision, as regulated by Section 109.

105.4.6 Inspection and testing. The *alternative engineered design* shall be tested and inspected in accordance with the requirements of Sections 107 and 312.

❖ As is the case for all plumbing installations, the code official must inspect the alternative engineered plumbing to verify that the work is in compliance with the construction documents. Section 107 requires the code official to witness the testing of the plumbing system before it is placed in service to verify that it is free from leaks or other defects, in accordance with Section 312.

105.5 Approved materials and equipment. Materials, equipment and devices *approved* by the code official shall be constructed and installed in accordance with such approval.

❖ The code is a compilation of criteria with which materials, equipment, devices and systems must comply to be suitable for a particular application. The code official has a duty to evaluate such materials, equipment, devices and systems for code compliance and when compliance is determined, approve the same for use. The materials, equipment, devices and systems must be constructed and installed in compliance with, and all conditions and limitations considered as a basis for, that approval. For example, the manufacturer's instructions and recommendations are to be followed if the approval of the material was based, even in part, on those instructions and recommendations. The approval authority given to the code official is a significant responsibility and is a key to code compliance. The approval process is first technical and then administrative and must be approached as such. For example, if data to determine code compliance is required, such data should be in the form of test reports or engineering analysis and not simply taken from a sales brochure.

105.5.1 Material and equipment reuse. Materials, equipment and devices shall not be reused unless such elements have been reconditioned, tested, placed in good and proper working condition and *approved*.

❖ The code criteria for materials and equipment have changed over the years. Evaluation of testing and materials technology has permitted the development of new criteria, which the old materials may not satisfy. As a result, used materials must be evaluated in the same manner as new materials. Used (previously installed) equipment must be equivalent to that required by the code if it is to be used in a new installation.

SECTION 106
PERMITS

106.1 When required. Any owner, authorized agent or contractor who desires to construct, enlarge, alter, repair, move, demolish or change the *occupancy* of a building or structure, or to erect, install, enlarge, alter, repair, remove, convert or replace any plumbing system, the installation of which is regulated by this code, or to cause any such work to be done, shall first make application to the code official and obtain the required permit for the work.

❖ This section contains the administrative rules governing the issuance, suspension, revocation or modification of plumbing permits. It also establishes how and by whom the application for a plumbing permit is to be made, how it is to be processed and what information it must contain or have attached to it. In general, a permit is required for all activities that are regulated by the code, and these activities cannot begin until the permit is issued.

A plumbing permit is required for the installation, replacement, alteration or modification of all plumbing systems and components that are in the scope of applicability of the code. Replacement of an existing fixture, piece of equipment or related piping is treated no differently than a new installation in new building construction. The purpose of a permit is to cause the work to be inspected to determine compliance with the intent of the code.

106.2 Exempt work. The following work shall be exempt from the requirement for a permit:

1. The stopping of leaks in drains, water, soil, waste or vent pipe provided, however, that if any concealed trap, drainpipe, water, soil, waste or vent pipe becomes defective and it becomes necessary to remove and replace the same with new material, such work shall be considered as new work and a permit shall be obtained and inspection made as provided in this code.
2. The clearing of stoppages or the repairing of leaks in pipes, valves or fixtures, and the removal and reinstallation of water closets, provided such repairs do not involve or require the replacement or rearrangement of valves, pipes or fixtures.

Exemption from the permit requirements of this code shall not be deemed to grant authorization for any work to be done in violation of the provisions of this code or any other laws or ordinances of this jurisdiction.

❖ Repair work, maintenance procedures and similar work are exempt from the permit requirement provided that such work does not involve the replacement of any system components other than minor parts.

Literally, the only plumbing activities that are exempt from the requirement for a permit are drain cleaning and rodding and the repair or replacement of faucets, fill valves, flushometers and similar fixture trim and fittings. Any plumbing work that results in the alteration or disassembly of existing drain, waste, vent and water distribution piping would require a permit. Additionally, any plumbing work that involves the alteration of a plumbing fixture would require a permit.

106.3 Application for permit. Each application for a permit, with the required fee, shall be filed with the code official on a form furnished for that purpose and shall contain a general description of the proposed work and its location. The application shall be signed by the owner or an authorized agent. The permit application shall indicate the proposed *occupancy* of all parts of the building and of that portion of the site or lot, if any, not covered by the building or structure and shall contain such other information required by the code official.

❖ This section limits persons who may apply for a permit to the building owner or an authorized agent. An owner's authorized agent could be anyone who is given written permission to act in the owner's interest for the purpose of obtaining a permit, such as an architect, an engineer, a contractor, a tenant or other. Permit forms generally have sufficient space to write a very brief description of the work to be accomplished, which is acceptable for small jobs. For larger projects, the description will be augmented by construction documents.

106.3.1 Construction documents. Construction documents, engineering calculations, diagrams and other such data shall be submitted in two or more sets with each application for a permit. The code official shall require construction documents, computations and specifications to be prepared and designed by a registered design professional when required by state law. Construction documents shall be drawn to scale and shall be of sufficient clarity to indicate the location, nature and extent of the work proposed and show in detail that the work conforms to the provisions of this code. Construction documents for buildings more than two stories in height shall indicate where penetrations will be made for pipes, fittings and components and shall indicate the materials and methods for maintaining required structural safety, fire-resistance rating and fireblocking.

> **Exception:** The code official shall have the authority to waive the submission of construction documents, calculations or other data if the nature of the work applied for is such that reviewing of construction documents is not necessary to determine compliance with this code.

❖ A detailed description of the work for which application is made must be submitted. When the work is of a "minor nature," either in scope or needed description, the code official may use judgement in determining the need for a detailed description of the work. An example of minor work that may not involve a detailed description is the replacement of an existing fixture in a plumbing system or the replacement or repair of a defective portion of a piping system.

These provisions are intended to reflect the minimum scope of information needed to determine code compliance. A statement on the construction documents such as, "All plumbing work must comply with the 2009 IPC," is not an acceptable substitute for showing the required information.

This section also requires the code official to determine compliance with any state professional registration laws as they apply to the preparation of construction documents.

106.3.2 Preliminary inspection. Before a permit is issued, the code official is authorized to inspect and evaluate the systems, equipment, buildings, devices, premises and spaces or areas to be used.

❖ Some projects might require a preliminary inspection by the code official prior to a permit being issued. This is especially useful for remodel and addition projects where the conditions of the existing building plumbing systems are unknown or are of questionable condition. This section authorizes the code official to make such inspections.

106.3.3 Time limitation of application. An application for a permit for any proposed work shall be deemed to have been abandoned 180 days after the date of filing, unless such application has been pursued in good faith or a permit has been issued; except that the code official shall have the authority to grant one or more extensions of time for additional periods not exceeding 180 days each. The extension shall be requested in writing and justifiable cause demonstrated.

❖ Once an application for a permit has been submitted for proposed work, a time limit of 180 days is established for issuance of the permit. This prevents the code official from having to hold on to incomplete or delayed applications for an indefinite amount of time. The code official can grant extensions for this time period if provided with a written request with justifiable reasons for the extension request.

106.4 By whom application is made. Application for a permit shall be made by the person or agent to install all or part of any plumbing system. The applicant shall meet all qualifications established by statute, or by rules promulgated by this code, or by ordinance or by resolution. The full name and address of the applicant shall be stated in the application.

❖ This section specifies that the permit applicant is to meet all of the qualifications that may be established by the governmental enforcing agency. For example, some governmental enforcing agencies may require the permit applicant to be a registered design professional or licensee.

106.5 Permit issuance. The application, construction documents and other data filed by an applicant for permit shall be reviewed by the code official. If the code official finds that the proposed work conforms to the requirements of this code and all laws and ordinances applicable thereto, and that the fees specified in Section 106.6 have been paid, a permit shall be issued to the applicant.

❖ This section requires the code official to review all submittals for a permit for compliance with the code and to verify that the project will be carried out in accordance with any other applicable laws. This may involve interagency communication and cooperation so that all laws are being obeyed. Once the code official verifies this, a permit may be issued upon payment of the required fees.

106.5.1 Approved construction documents. When the code official issues the permit where construction documents are required, the construction documents shall be endorsed in writing and stamped "APPROVED." Such *approved* construction documents shall not be changed, modified or altered without authorization from the code official. All work shall be done in accordance with the *approved* construction documents.

The code official shall have the authority to issue a permit for the construction of a part of a plumbing system before the entire construction documents for the whole system have been submitted or *approved*, provided adequate information and detailed statements have been filed complying with all pertinent requirements of this code. The holders of such permit shall proceed at their own risk without assurance that the permit for the entire plumbing system will be granted.

❖ Construction documents that reflect compliance with code requirements form an integral part of the permit process. Successful prosecution of the work depends on these documents. This section requires the code official to stamp the complying construction documents as being "approved." Once approved, no further revisions to the documents may be made without the express authorization of the code official in order to maintain code compliance.

In the interest of saving time and coordinating construction phases, it is common practice for contractors to seek permits solely applicable to the installation of site work, such as water services and sewers. This practice allows the project to proceed before the final construction documents are completed, thus minimizing delays in the construction process. This also allows the builder to perform site work while the weather permits.

The holder of a partial permit must realize that a permit for the remainder of the plumbing system or structure may not be granted for various reasons. Issuance of a partial permit in no way guarantees issuance of a permit for the entire scope of the project.

106.5.2 Validity. The issuance of a permit or approval of construction documents shall not be construed to be a permit for, or an approval of, any violation of any of the provisions of this code or any other ordinance of the jurisdiction. No permit presuming to give authority to violate or cancel the provisions of this code shall be valid.

The issuance of a permit based upon construction documents and other data shall not prevent the code official from thereafter requiring the correction of errors in said construction documents and other data or from preventing building operations being carried on thereunder when in violation of this code or of other ordinances of this jurisdiction.

❖ An important code section, this section states the fundamental premise that the permit is only "a license to proceed with the work." It is not a license to "violate, cancel or set aside any provisions of the code." This is important because it means that despite any errors in the approval process, the permit applicant is responsible for code compliance.

106.5.3 Expiration. Every permit issued by the code official under the provisions of this code shall expire by limitation and become null and void if the work authorized by such permit is not commenced within 180 days from the date of such permit, or if the work authorized by such permit is suspended or abandoned at any time after the work is commenced for a period of 180 days. Before such work can be recommenced, a new permit shall be first obtained and the fee therefor shall be one-half the amount required for a new permit for such work, provided no changes have been made or will be made in the original construction documents for such work, and provided further that such suspension or abandonment has not exceeded 1 year.

❖ The permit becomes invalid under two distinct situations, both based on a 6-month period. The first situation is when no work has started six months from issuance of the permit. The second situation is when there is no continuation of authorized work for six months. The person who was issued the permit should be notified in writing that it is invalid and what steps must be taken to restart the work.

This section also provides the administrative authority with a means of offsetting the costs associated with expired permits by charging a nominal fee for permit reissuance. If, however, the nature or scope of the work to be resumed is different from that covered by the original permit, the permit process essentially starts from "scratch" and full fees are charged. The same procedure would also apply if the work has not commenced within one year of the date of permit issuance or if work has to be suspended for a year or more.

106.5.4 Extensions. Any permittee holding an unexpired permit shall have the right to apply for an extension of the time within which the permittee will commence work under that permit when work is unable to be commenced within the time required by this section for good and satisfactory reasons. The code official shall extend the time for action by the permittee for a period not exceeding 180 days if there is reasonable cause. No permit shall be extended more than once. The fee for an extension shall be one-half the amount required for a new permit for such work.

❖ Although it is typical for a project to begin immediately following issuance of a permit, there are occasions when an unforeseen delay may occur. This section intends to afford the permit holder the opportunity to apply for and receive a single 180-day extension within which to begin a project under a still-valid permit (i.e., less than 180 days old). The applicant must, however, provide the code official with an adequate explanation for the delay in starting a project, which could include such things as the need to obtain approvals or permits from other agencies having jurisdiction. This section requires the code official to determine what constitutes "good and satisfactory" reasons for any delay and further allows the jurisdiction to offset its administrative costs for extending the permit by charging one-half the fee for a new permit for the extension.

106.5.5 Suspension or revocation of permit. The code official shall have the authority to suspend or revoke a permit issued under the provisions of this code wherever the permit is issued in error or on the basis of incorrect, inaccurate or incomplete information, or in violation of any ordinance or regulation or any of the provisions of this code.

❖ A permit is in reality a license to proceed with the work. The code official, however, must revoke all permits shown to be based, all or in part, on any false statement or misrepresentation of fact. An applicant may subsequently reapply for a permit with the appropriate corrections or modifications made to the application and construction documents.

106.5.6 Retention of construction documents. One set of *approved* construction documents shall be retained by the code official for a period of not less than 180 days from date of completion of the permitted work, or as required by state or local laws.

One set of *approved* construction documents shall be returned to the applicant, and said set shall be kept on the site of the building or work at all times during which the work authorized thereby is in progress.

❖ Once the code official has stamped or endorsed as approved the construction documents on which the permit is based (see commentary, Section 106.3.1), one set of approved construction documents must be kept on the construction site to serve as the basis for all subsequent inspections. To avoid confusion, the construction documents on the site must be precisely the documents that were approved and stamped. This is because inspections are based on the approved documents. Additionally, the contractor cannot determine compliance with the approved construction documents unless those documents are readily available. Unless the approved construction documents are available, the inspection should be postponed and work on the project halted.

106.5.7 Previous approvals. This code shall not require changes in the construction documents, construction or designated *occupancy* of a structure for which a lawful permit has been heretofore issued or otherwise lawfully authorized, and the construction of which has been pursued in good faith within 180 days after the effective date of this code and has not been abandoned.

❖ This section provides the code official with a useful tool to protect the continuity of permits issued under previous codes or code editions, as long as such permits are being actively executed subsequent to the effective date of the ordinance adopting the newer code.

106.5.8 Posting of permit. The permit or a copy shall be kept on the site of the work until the completion of the project.

❖ This section requires the permit (or a copy of the permit) to be on the work site until the project is completed. Having the permit at the jobsite provides project information and evidence to anyone needing to know if the project has been duly authorized.

106.6 Fees. A permit shall not be issued until the fees prescribed in Section 106.6.2 have been paid, and an amendment to a permit shall not be released until the additional fee, if any, due to an increase of the plumbing systems, has been paid.

❖ All fees are to be paid prior to permit issuance. This requirement establishes that the permit applicant intends to proceed with the work, as well as facilitates payment.

106.6.1 Work commencing before permit issuance. Any person who commences any work on a plumbing system before obtaining the necessary permits shall be subject to 100 percent of the usual permit fee in addition to the required permit fees.

❖ This section is intended to serve as a deterrent to proceeding with work on a plumbing system without a permit (except as provided in Section 106.2). As a punitive measure, it doubles the cost of the permit fee. This section does not, however, intend to penalize a contractor called upon to do emergency work after hours, provided that he or she makes prompt notification to the code official the next business day and obtains the requisite permit for the work done and has the required inspections performed.

106.6.2 Fee schedule. The fees for all plumbing work shall be as indicated in the following schedule:

[JURISDICTION TO INSERT APPROPRIATE SCHEDULE]

❖ A published fee schedule must be established for plans examination, permits and inspections. Ideally, the department should generate revenues that cover operating costs and expenses. The permit fee schedule is an integral part of this process.

106.6.3 Fee refunds. The code official shall authorize the refunding of fees as follows:

1. The full amount of any fee paid hereunder that was erroneously paid or collected.
2. Not more than [SPECIFY PERCENTAGE] percent of the permit fee paid when no work has been done under a permit issued in accordance with this code.
3. Not more than [SPECIFY PERCENTAGE] percent of the plan review fee paid when an application for a permit for which a plan review fee has been paid is withdrawn or canceled before any plan review effort has been expended.

The code official shall not authorize the refunding of any fee paid except upon written application filed by the original permittee not later than 180 days after the date of fee payment.

❖ This section allows for a partial refund of fees resulting from the revocation, abandonment or discontinuance of a plumbing project for which a permit has been issued and fees have been collected. The incomplete work for which the excess fees are to be refunded refers to the work that would have been required by the department had the permit not been terminated. The refund of fees should be related to the cost of enforcement services not provided because of termination of the project.

SECTION 107 INSPECTIONS AND TESTING

107.1 General. The code official is authorized to conduct such inspections as are deemed necessary to determine compliance with the provisions of this code. Construction or work for which a permit is required shall be subject to inspection by the code official, and such construction or work shall remain accessible and exposed for inspection purposes until *approved*. Approval as a result of an inspection shall not be construed to be an approval of a violation of the provisions of this code or of other ordinances of the jurisdiction. Inspections presuming to give authority to violate or cancel the provisions of this code or of other ordinances of the jurisdiction shall not be valid. It shall be the duty of the permit applicant to cause the work to remain accessible and exposed for inspection purposes. Neither the code official nor the jurisdiction shall be liable for expense entailed in the removal or replacement of any material required to allow inspection.

❖ The inspection function is one of the more important aspects of building department operations. This section authorizes the code official to inspect the work for which a permit has been issued and requires that the work to be inspected remain accessible to the code official until inspected and approved. Any expense incurred in removing or replacing material that conceals an item to be inspected is not the responsibility of the code official or the jurisdiction. As with the issuance of permits (see Section 106.5.2), an approval as a result of an inspection is not a license to violate the code. Approval of work that contains a violation of the code does not relieve the applicant of the responsibility for complying with the code.

107.2 Required inspections and testing. The code official, upon notification from the permit holder or the permit holder's agent, shall make the following inspections and such other inspections as necessary, and shall either release that portion of the construction or shall notify the permit holder or an agent of any violations that must be corrected. The holder of the permit shall be responsible for the scheduling of such inspections.

1. Underground inspection shall be made after trenches or ditches are excavated and bedded, piping installed, and before any backfill is put in place.
2. Rough-in inspection shall be made after the roof, framing, fireblocking, firestopping, draftstopping and bracing is in place and all sanitary, storm and water distribution piping is roughed-in, and prior to the installation of wall or ceiling membranes.
3. Final inspection shall be made after the building is complete, all plumbing fixtures are in place and properly connected, and the structure is ready for occupancy.

❖ This section requires that all portions of the plumbing system be inspected before and after fixtures are installed. The code official has the authority to require noncomplying plumbing to be brought into compliance and reinspected.

Inspections are necessary to determine that an installation conforms to all code requirements. Because the majority of a plumbing system is hidden within the building enclosure, periodic inspections are necessary before portions of the system are concealed. The code official is required to determine that plumbing systems and equipment are installed in accordance with the approved construction documents and the applicable code requirements. All inspections that are necessary to provide such verification must be conducted. Generally, the administrative rules of a department may list the required interim inspections. Construction that occurs in steps or phases may necessitate multiple inspections; therefore, an exact number of required inspections cannot be specified. Where violations are noted and corrections are required, re-inspections may be necessary. As time permits, frequent inspections of some jobsites, especially where the work is complex, can be beneficial in detecting code-compliance or other potential problems before they develop or become more difficult to correct.

It is the responsibility of the contractor, the builder, the owner or other authorized party to arrange for the required inspections and to coordinate them to prevent work from being concealed before it is inspected.

1. Inspection of underground plumbing is especially important because once it is covered, it is the most challenging part of a plumbing system in which to detect a leak. If repairs are necessary, underground repairs are proportionately more expensive because of the need for heavy

equipment and the more labor-intensive nature of working below grade level (see Section 306 for trenching, excavation and backfill requirements).

2. A rough-in inspection is a visual observation of all parts of the plumbing system that will eventually be concealed in the building structure. Rough-in inspections also include verification that the applicable test pressures are applied to the system and that leaks do not exist. The inspection must be made before any of the system is covered by building finish materials or hidden by future work.

   A rough-in inspection may be completed in one visit or as a series of inspections. This is administratively determined by the local inspections department and is typically dependent on the size of the job.

3. A final inspection may be done as a series of inspections or in one visit, similar to a rough-in inspection. A final inspection is required prior to the approval of plumbing work and installations. For the construction of a new building, final approval is required prior to the issuance of the certificate of occupancy as specified in the IBC. To verify that all previously issued correction orders have been complied with and to determine whether subsequent violations exist, a final inspection must be made. All violations observed during the final inspection must be noted and the permit holder must be advised.

   The final inspection follows the completion of the work or installation. Typically, the final inspection is an inspection of all that was installed after the rough-in inspection and not concealed in the building construction. Subsequent re-inspections are necessary if the final inspection generates a notice of violation.

107.2.1 Other inspections. In addition to the inspections specified above, the code official is authorized to make or require other inspections of any construction work to ascertain compliance with the provisions of this code and other laws that are enforced.

❖ Any item regulated by the code is subject to inspection by the code official to determine compliance with the applicable code provision, and no list can include all types of work in a given building. Also, other inspections before, during or after the rough-in could be necessary. This section gives the code official the authority to inspect any regulated work.

107.2.2 Inspection requests. It shall be the duty of the holder of the permit or their duly authorized agent to notify the code official when work is ready for inspection. It shall be the duty of the permit holder to provide *access* to and means for inspections of such work that are required by this code.

❖ This section clarifies that it is the responsibility of the permit holder to arrange for the required inspections when the completed work is ready. It also establishes his or her responsibility for keeping the work open for inspection and providing all means needed to accomplish the inspections.

107.2.3 Approval required. Work shall not be done beyond the point indicated in each successive inspection without first obtaining the approval of the code official. The code official, upon notification, shall make the requested inspections and shall either indicate the portion of the construction that is satisfactory as completed, or notify the permit holder or his or her agent wherein the same fails to comply with this code. Any portions that do not comply shall be corrected and such portion shall not be covered or concealed until authorized by the code official.

❖ This section establishes that work cannot progress beyond the point of a required inspection without the code official's approval. Upon making the inspection, the code official must either approve the completed work or notify the permit holder or other responsible party of that which does not comply with the code. Approvals and notices of noncompliance must be in writing, as required by Section 104.3, to avoid any misunderstanding as to what is required. Any work not approved cannot be concealed until it has been corrected and approved by the code official.

107.2.4 Approved agencies. The code official is authorized to accept reports of *approved* inspection agencies, provided that such agencies satisfy the requirements as to qualifications and reliability.

❖ The determination as to whether to accept an agency test report rests with the code official and the reporting agency must be acceptable to the code official.

107.2.5 Evaluation and follow-up inspection services. Prior to the approval of a closed, prefabricated plumbing system and the issuance of a plumbing permit, the code official shall require the submittal of an evaluation report on each prefabricated plumbing system indicating the complete details of the plumbing system, including a description of the system and its components, the basis upon which the plumbing system is being evaluated, test results and similar information, and other data as necessary for the code official to determine conformance to this code.

❖ As an alternative to a physical inspection in the plant or location where prefabricated components are fabricated (such as modular homes, prefabricated structures, etc.), the code official has the option of accepting an evaluation report from an approved agency detailing such inspections. These evaluation reports can serve as the basis for code compliance.

107.2.5.1 Evaluation service. The code official shall designate the evaluation service of an *approved* agency as the evaluation agency, and review such agency's evaluation report for adequacy and conformance to this code.

❖ The code official is required to review all submitted reports for conformity to the applicable code requirements. If, in the judgment of the code official, the submitted reports are acceptable, he or she should document the basis for the approval.

107.2.5.2 Follow-up inspection. Except where ready *access* is provided to all plumbing systems, service equipment and accessories for complete inspection at the site without disassembly or dismantling, the code official shall conduct the frequency of in-plant inspections necessary to ensure conformance to the *approved* evaluation report or shall designate an independent, *approved* inspection agency to conduct such inspections. The inspection agency shall furnish the code official with the follow-up inspection manual and a report of inspections upon request, and the plumbing system shall have an identifying label permanently affixed to the system indicating that factory inspections have been performed.

❖ The owner is required to provide special inspections of fabricated assemblies at the fabrication plant. The code official or an approved inspection agency must conduct periodic in-plant inspections to ensure conformance to the approved evaluation report described in Section 107.2.5.1. Such inspections would not be required where the plumbing systems can be inspected completely at the job site.

107.2.5.3 Test and inspection records. All required test and inspection records shall be available to the code official at all times during the fabrication of the plumbing system and the erection of the building, or such records as the code official designates shall be filed.

❖ All testing and inspection records related to a fabricated assembly must be filed with the code official so he or she can maintain a complete and legal record of the assembly and erection of the building.

107.3 Special inspections. Special inspections of *alternative engineered design* plumbing systems shall be conducted in accordance with Sections 107.3.1 and 107.3.2.

❖ This section establishes that the design professional has to periodically inspect the alternative engineered design, keep records of such inspections and submit a final report to the code official certifying that all work conforms to the construction documents. Because of the unusual nature and possible complexity of alternative engineered plumbing systems, it is necessary for the system designer to be involved in the inspection process.

107.3.1 Periodic inspection. The registered design professional or designated inspector shall periodically inspect and observe the *alternative engineered design* to determine that the installation is in accordance with the *approved* construction documents. All discrepancies shall be brought to the immediate attention of the plumbing contractor for correction. Records shall be kept of all inspections.

❖ The registered design professional must periodically inspect the engineered plumbing system during installation to determine that the system conforms to the approved construction documents.This is an important step because the design professional can identify any deviations from the approved plans in the early stages of the plumbing work. The design professional must then advise the plumbing contractor of any problems so that corrective measures can be taken before needless costs are incurred, and labor and materials are wasted.

The design professional must compile a complete legal record of the project, which must include all inspections made, discrepancies found and resolutions of discrepancies. It is the responsibility of the design professional to document and submit inspection records and written certification in accordance with Section 107.3.2.

107.3.2 Written report. The registered design professional shall submit a final report in writing to the code official upon completion of the installation, certifying that the *alternative engineered design* conforms to the *approved* construction documents. A notice of approval for the plumbing system shall not be issued until a written certification has been submitted.

❖ After all work is completed, the design professional is required to inspect the entire alternative engineered plumbing system. The details of that inspection, including verification of compliance with the approved construction documents, must be submitted in writing to the code official before final approval can be granted.

107.4 Testing. Plumbing work and systems shall be tested as required in Section 312 and in accordance with Sections 107.4.1 through 107.4.3. Tests shall be made by the permit holder and observed by the code official.

❖ Visual inspection is not all that is required in the determination of a plumbing system's compliance with the code. This section establishes where and how testing is to be performed to disclose leaks and defects.

107.4.1 New, altered, extended or repaired systems. New plumbing systems and parts of existing systems that have been altered, extended or repaired shall be tested as prescribed herein to disclose leaks and defects, except that testing is not required in the following cases:

1. In any case that does not include addition to, replacement, alteration or relocation of any water supply, drainage or vent piping.
2. In any case where plumbing equipment is set up temporarily for exhibition purposes.

❖ Every plumbing system must be tested before it is placed into service. Testing is necessary to make sure that the system is free from leaks or other defects. Testing is also required, to the extent practicable, for portions of existing systems that have been repaired, altered or extended.

107.4.2 Equipment, material and labor for tests. All equipment, material and labor required for testing a plumbing system or part thereof shall be furnished by the permit holder.

❖ The permit holder is responsible for performing tests, as well as for supplying all of the labor, equipment and apparatus necessary to conduct such tests. The code official observes but never performs the test.

107.4.3 Reinspection and testing. Where any work or installation does not pass any initial test or inspection, the necessary corrections shall be made to comply with this code. The work or installation shall then be resubmitted to the code official for inspection and testing.

❖ If a system or portion thereof does not pass the initial test or inspection, all violations must be corrected and the system must be reinspected.

To encourage code compliance and cover the expense of the code official's time, many jurisdictions charge fees for inspections that are required subsequent to the first reinspection.

107.5 Approval. After the prescribed tests and inspections indicate that the work complies in all respects with this code, a notice of approval shall be issued by the code official.

❖ After the code official has performed the required inspections and observed the required equipment and system tests (or has received written reports of the results of such tests), he or she must determine whether the installation or work is in compliance with all applicable sections of the code. The code official must issue a written notice of approval if it has been determined that the subject plumbing work or installation is in apparent compliance with the code. The notice of approval is given to the permit holder, and a copy of the notice is retained on file by the code official.

107.5.1 Revocation. The code official is authorized to, in writing, suspend or revoke a notice of approval issued under the provisions of this code wherever the notice is issued in error, or on the basis of incorrect information supplied, or where it is determined that the building or structure, premise or portion thereof is in violation of any ordinance or regulation or any of the provisions of this code.

❖ This section is needed to give the code official the authority to revoke a notice of approval for the reasons indicated in the code text. The code official can suspend the notice until all of the code violations are corrected.

107.6 Temporary connection. The code official shall have the authority to authorize the temporary connection of the building or system to the utility source for the purpose of testing plumbing systems or for use under a temporary certificate of occupancy.

❖ The typical procedure for a local jurisdiction is to withhold the issuance of the certificate of occupancy until approvals have been received from each code official responsible for inspection of the structure. The code official is permitted to issue a temporary authorization to make connections to the public utility system prior to the completion of all work. The certification is intended to acknowledge that, because of seasonal limitations, time constraints, the need for testing or partial operation of equipment, some building systems may be connected even though the building is not suitable for final occupancy. The intent of this section is that a request for temporary occupancy or the connection and use of plumbing equipment or systems should not be denied when the requesting permit holder has demonstrated to the code official's satisfaction that the public health, safety and welfare will not be endangered.

The code official should view the issuance of a "temporary authorization or certificate of occupancy" as substantial an act as the issuance of the final certificate. Indeed, the issuance of a temporary certificate of occupancy offers a greater potential for conflict because once the building or structure is occupied, it is very difficult to remove the occupants through legal means.

107.7 Connection of service utilities. A person shall not make connections from a utility, source of energy, fuel, power, water system or *sewer* system to any building or system that is regulated by this code for which a permit is required until authorized by the code official.

❖ This section establishes the authority of the code official to approve utility connections to a building such as water, sewer, electricity, gas and steam, and to require their disconnection when such approval has not been granted. For the protection of building occupants, including workers, such systems must have had final inspection approvals, except as allowed by Section 110.3 for temporary connections.

SECTION 108
VIOLATIONS

108.1 Unlawful acts. It shall be unlawful for any person, firm or corporation to erect, construct, alter, repair, remove, demolish or utilize any plumbing system, or cause same to be done, in conflict with or in violation of any of the provisions of this code.

❖ This section describes the citing, recording and subsequent actions pursuant to observed code violations. Violations of the code are prohibited; this is the basis for all citations and correction notices.

108.2 Notice of violation. The code official shall serve a notice of violation or order to the person responsible for the erection, installation, alteration, extension, repair, removal or demolition of plumbing work in violation of the provisions of this code, or in violation of a detail statement or the *approved* construction documents thereunder, or in violation of a permit or certificate issued under the provisions of this code. Such order shall direct the discontinuance of the illegal action or condition and the abatement of the violation.

❖ The code official is required to notify the person responsible for the erection or use of a building found to be in violation of the code. The section that is allegedly being violated must be cited so that the responsible party can respond to the notice.

108.3 Prosecution of violation. If the notice of violation is not complied with promptly, the code official shall request the legal counsel of the jurisdiction to institute the appropriate proceeding at law or in equity to restrain, correct or abate such violation, or to require the removal or termination of the unlawful occupancy of the structure in violation of the provisions of this code or of the order or direction made pursuant thereto.

❖ The code official must pursue, through the use of legal counsel of the jurisdiction, legal means to correct the violation. This is not optional.

Any extensions of time for voluntary correction of the violations must be for a reasonable, bona fide cause, or the code official may be subject to criticism for "arbitrary and capricious" actions. In general, it is better to have a standard time limitation for correction of violations. Departures from this standard must be for a clear and reasonable purpose, usually stated in writing by the violator.

108.4 Violation penalties. Any person who shall violate a provision of this code or shall fail to comply with any of the requirements thereof or who shall erect, install, alter or repair plumbing work in violation of the *approved* construction documents or directive of the code official, or of a permit or certificate issued under the provisions of this code, shall be guilty of a [SPECIFY OFFENSE], punishable by a fine of not more than [AMOUNT] dollars or by imprisonment not exceeding [NUMBER OF DAYS], or both such fine and imprisonment. Each day that a violation continues after due notice has been served shall be deemed a separate offense.

❖ A standard fine or other penalty as deemed appropriate by the jurisdiction is prescribed in this section. Additionally, this section identifies a principle that "each day that a violation continues shall be deemed a separate offense" for the purpose of applying the prescribed penalty in order to facilitate prompt resolution.

108.5 Stop work orders. Upon notice from the code official, work on any plumbing system that is being done contrary to the provisions of this code or in a dangerous or unsafe manner shall immediately cease. Such notice shall be in writing and shall be given to the owner of the property, or to the owner's agent, or to the person doing the work. The notice shall state the conditions under which work is authorized to resume. Where an emergency exists, the code official shall not be required to give a written notice prior to stopping the work. Any person who shall continue any work in or about the structure after having been served with a stop work order, except such work as that person is directed to perform to remove a violation or unsafe condition, shall be liable to a fine of not less than [AMOUNT] dollars or more than [AMOUNT] dollars.

❖ Upon receipt of a violation notice from the code official, the owner of the property, the owner's agent or the person doing the work must immediately cease all construction activities identified in the notice, except as expressly permitted to correct the violation. A stop work order can prevent a violation from becoming worse and more difficult or expensive to correct. However, it can result in inconvenience and monetary loss to the affected parties; therefore, justification must be evident and judgment must be exercised before such an order is issued.

A stop work order may be issued where work is proceeding without a permit. Hazardous conditions could develop where the code official is unaware of the nature of the work and a permit has not been issued. The issuance of a stop work order on a plumbing system often results from work done by the plumbing contractor that affects a nonplumbing component. For example, if a plumbing contractor cuts a structural element to install piping or fixtures, the structure may be weakened enough to cause a partial or complete structural failure. As determined by the adopting jurisdiction, a penalty may be assessed for failure to comply with this section, and it is to be inserted in the blanks provided.

108.6 Abatement of violation. The imposition of the penalties herein prescribed shall not preclude the legal officer of the jurisdiction from instituting appropriate action to prevent unlawful construction or to restrain, correct or abate a violation, or to prevent illegal occupancy of a building, structure or premises, or to stop an illegal act, conduct, business or utilization of the plumbing on or about any premises.

❖ Despite the assessment of a penalty in the form of a fine or imprisonment against a violator, the violation itself must still be corrected. Failure to make the necessary corrections will result in the violator being subject to additional penalties as described in the preceding section.

108.7 Unsafe plumbing. Any plumbing regulated by this code that is unsafe or that constitutes a fire or health hazard, insanitary condition, or is otherwise dangerous to human life is hereby declared unsafe. Any use of plumbing regulated by this code constituting a hazard to safety, health or public welfare by reason of inadequate maintenance, dilapidation, obsolescence, fire hazard, disaster, damage or abandonment is hereby declared an unsafe use. Any such unsafe equipment is hereby declared to be a public nuisance and shall be abated by repair, rehabilitation, demolition or removal.

❖ Unsafe conditions include those that constitute a health hazard, fire hazard, explosion hazard, shock hazard, asphyxiation hazard, physical injury hazard or are otherwise dangerous to human life and property.

In the course of performing duties, the code official may identify a hazardous condition. Such condition must be declared in violation of the code and, therefore, must be abated. For example, while inspecting a plumbing installation, the code official may notice a leaking soil pipe that is causing structural decay, a missing cleanout plug or a defective water-heater relief valve. Even though the defective plumbing may be unrelated to the installation that the code official was called to inspect, he or she is obligated to address the hazard and have it corrected.

108.7.1 Authority to condemn equipment. Whenever the code official determines that any plumbing, or portion thereof, regulated by this code has become hazardous to life, health or property or has become insanitary, the code official shall order in writing that such plumbing either be removed or restored to a safe or sanitary condition. A time limit for compliance with such order shall be specified in the written notice. No person shall use or maintain defective plumbing after receiving such notice.

When such plumbing is to be disconnected, written notice as prescribed in Section 108.2 shall be given. In cases of immediate danger to life or property, such disconnection shall be made immediately without such notice.

❖ When a plumbing system or plumbing equipment is determined to be unsafe, the code official is required to notify the owner or agent of the building as the first step in correcting the difficulty. Such notice is to describe the repairs and improvements necessary to correct the deficiency or require removal or replacement of the unsafe equipment or system. All such notices must specify a time frame in which the corrective actions must occur. Additionally, such notice should require the immediate response of the owner or agent. If he or she is not available, public notice of such declaration should suffice for the purposes of complying with this section. The code official may also determine that disconnection of the system is necessary to correct an unsafe condition and must give written notice to that effect (see commentary, Section 108.2) unless immediate disconnection is essential for public health and safety reasons (see commentary, Section 108.7.2).

108.7.2 Authority to disconnect service utilities. The code official shall have the authority to authorize disconnection of utility service to the building, structure or system regulated by the technical codes in case of an emergency, where necessary, to eliminate an immediate danger to life or property. Where possible, the owner and occupant of the building, structure or service system shall be notified of the decision to disconnect utility service prior to taking such action. If not notified prior to disconnecting, the owner or occupant of the building, structure or service systems shall be notified in writing, as soon as practical thereafter.

❖ The code official should have the authority to order disconnection of any plumbing supplied to a building, structure or equipment regulated by the code when it is determined that the equipment or any portion thereof has become an immediate danger. Written notice of an order to disconnect service and the causes therefore, should be given to the owner and the occupant of the building, structure or premises. However, disconnection should be done without such notice in cases of immediate danger to life or property.

108.7.3 Connection after order to disconnect. No person shall make connections from any energy, fuel, power supply or water distribution system or supply energy, fuel or water to any equipment regulated by this code that has been disconnected or ordered to be disconnected by the code official or the use of which has been ordered to be discontinued by the code official until the code official authorizes the reconnection and use of such equipment.

When any plumbing is maintained in violation of this code, and in violation of any notice issued pursuant to the provisions of this section, the code official shall institute any appropriate action to prevent, restrain, correct or abate the violation.

❖ Once the reason for discontinuation of use or disconnection of the plumbing system no longer exists, only the code official can authorize resumption of use or reconnection of the system after it is demonstrated to his or her satisfaction that all repairs or other work are in compliance with applicable sections of the code. This section also requires him or her to take action to abate code violations (see commentary, Section 108.2).

SECTION 109
MEANS OF APPEAL

109.1 Application for appeal. Any person shall have the right to appeal a decision of the code official to the board of appeals. An application for appeal shall be based on a claim that the true intent of this code or the rules legally adopted thereunder have been incorrectly interpreted, the provisions of this code do not fully apply, or an equally good or better form of construction is proposed. The application shall be filed on a form obtained from the code official within 20 days after the notice was served.

❖ This section holds that any aggrieved party with a material interest in the decision of the code official may challenge such a decision before a board of appeals. This provides a forum, other than the court of jurisdiction, in which to review the code official's actions.

This section literally allows any person to appeal a decision of the code official. In practice, this section has been interpreted to permit appeals only by aggrieved parties with a material or definitive interest in the decision of the code official. An aggrieved party may not appeal a code requirement per se. The intent of the appeal process is not to waive or set aside a code requirement; rather, it is to provide a means of reviewing a code official's decision on an interpretation or application of the code or to review the equivalency of protection to the code requirements.

109.2 Membership of board. The board of appeals shall consist of five members appointed by the chief appointing authority as follows: one for 5 years, one for 4 years, one for 3 years, one for 2 years and one for 1 year. Thereafter, each new member shall serve for 5 years or until a successor has been appointed.

❖ The board of appeals is to consist of five members appointed by the "chief appointing authority"—typically, the mayor or city manager. One member is to be appointed for five years, one for four, one for three, one for two and one for one year. This method of appointment allows for a smooth transition of board of appeals members, allowing continuity of action over the years.

109.2.1 Qualifications. The board of appeals shall consist of five individuals, one from each of the following professions or disciplines:

1. Registered design professional who is a registered architect; or a builder or superintendent of building construction with at least 10 years' experience, 5 years of which shall have been in responsible charge of work.
2. Registered design professional with structural engineering or architectural experience.
3. Registered design professional with mechanical and plumbing engineering experience; or a mechanical and plumbing contractor with at least 10 years' experience, 5

years of which shall have been in responsible charge of work.

4. Registered design professional with electrical engineering experience; or an electrical contractor with at least 10 years' experience, 5 years of which shall have been in responsible charge of work.
5. Registered design professional with fire protection engineering experience; or a fire protection contractor with at least 10 years' experience, 5 years of which shall have been in responsible charge of work.

❖ The board of appeals consists of five persons with the qualifications and experience indicated in this section. One must be a registered design professional (see Item 2) with structural or architectural experience. The others must be registered design professionals, construction superintendents or contractors with experience in various areas of building construction. These requirements are important in that technical people rule on technical matters. The board of appeals is not the place for policy or political deliberations. It is intended that these matters be decided purely on their technical merits, with due regard for state-of-the-art construction technology.

109.2.2 Alternate members. The chief appointing authority shall appoint two alternate members who shall be called by the board chairman to hear appeals during the absence or disqualification of a member. Alternate members shall possess the qualifications required for board membership, and shall be appointed for 5 years or until a successor has been appointed.

❖ This section authorizes the chief appointing authority to appoint two alternate members who are to be available if the principal members of the board are absent or disqualified. Alternate members must possess the same qualifications as the principal members and are appointed for a term of five years or until such time that a successor is appointed.

109.2.3 Chairman. The board shall annually select one of its members to serve as chairman.

❖ It is customary to determine chairmanship annually so that a regular opportunity is available to evaluate and either reappoint the current chairman or appoint a new one.

109.2.4 Disqualification of member. A member shall not hear an appeal in which that member has any personal, professional or financial interest.

❖ All members must disqualify themselves regarding any appeal in which they have a personal, professional or financial interest.

109.2.5 Secretary. The chief administrative officer shall designate a qualified clerk to serve as secretary to the board. The secretary shall file a detailed record of all proceedings in the office of the chief administrative officer.

❖ The chief administrative officer is to designate a qualified clerk to serve as secretary to the board. The secretary is required to record the proceedings using detailed records.

109.2.6 Compensation of members. Compensation of members shall be determined by law.

❖ Members of the board of appeals need not be compensated unless required by the local municipality or jurisdiction.

109.3 Notice of meeting. The board shall meet upon notice from the chairman, within 10 days of the filing of an appeal or at stated periodic meetings.

❖ The board must meet within 10 days of the filing of an appeal or at regularly scheduled meetings.

109.4 Open hearing. All hearings before the board shall be open to the public. The appellant, the appellant's representative, the code official and any person whose interests are affected shall be given an opportunity to be heard.

❖ All hearings before the board must be open to the public. The appellant, the appellant's representative, the code official and any person whose interests are affected must be heard.

109.4.1 Procedure. The board shall adopt and make available to the public through the secretary procedures under which a hearing will be conducted. The procedures shall not require compliance with strict rules of evidence, but shall mandate that only relevant information be received.

❖ The board is required to establish and make available to the public written procedures detailing how hearings are to be conducted. Additionally, this section provides that although strict rules of evidence are not applicable, the information presented must be deemed relevant.

109.5 Postponed hearing. When five members are not present to hear an appeal, either the appellant or the appellant's representative shall have the right to request a postponement of the hearing.

❖ When all five members of the board are not present, either the appellant or the appellant's representative may request a postponement of the hearing.

109.6 Board decision. The board shall modify or reverse the decision of the code official by a concurring vote of three members.

❖ A concurring vote of three members of the board is needed to modify or reverse the decision of the code official.

109.6.1 Resolution. The decision of the board shall be by resolution. Certified copies shall be furnished to the appellant and to the code official.

❖ A formal decision in the form of a resolution is required to provide an official record. Copies of this resolution are to be furnished to both the appellant and the code official. The code official is bound by the action of the board of appeals unless he or she thinks that the board of appeals has acted improperly. In such cases, relief through the court having jurisdiction may be sought by corporate council.

109.6.2 Administration. The code official shall take immediate action in accordance with the decision of the board.

❖ To avoid any undue hindrance in the progress of construction, the code official is required to act without delay based on the board's decision. This action may be to enforce the decision or to seek legislative relief if the board's action can be demonstrated to be inappropriate.

109.7 Court review. Any person, whether or not a previous party of the appeal, shall have the right to apply to the appropriate court for a writ of certiorari to correct errors of law. Application for review shall be made in the manner and time required by law following the filing of the decision in the office of the chief administrative officer.

❖ This section allows any person to request a review by the court of jurisdiction of perceived errors of law. Application for such review must be made after the decision of the board is filed with the chief administrative officer. This helps to establish the observance of due process for all concerned.

SECTION 110 TEMPORARY EQUIPMENT, SYSTEMS AND USES

110.1 General. The code official is authorized to issue a permit for temporary equipment, systems and uses. Such permits shall be limited as to time of service, but shall not be permitted for more than 180 days. The code official is authorized to grant extensions for demonstrated cause.

❖ The code official is permitted to issue temporary authorization to make connections to a public utility system prior to completion of all work. This acknowledges that, because of seasonal limitations, time constraints, or the need for testing or partial operations of equipment, some building systems may be safely connected even though the building is not suitable for final occupancy. The temporary connection and utilization of connected equipment should be approved when the requesting permit holder has demonstrated to the code official's satisfaction that public health, safety and welfare will not be endangered.

110.2 Conformance. Temporary equipment, systems and uses shall conform to the structural strength, fire safety, means of egress, accessibility, light, ventilation and sanitary requirements of this code as necessary to ensure the public health, safety and general welfare.

❖ Even though a utility connection may be temporary, the only way to make sure that the public health, safety and general welfare are protected is for those temporary connections to comply with the code.

110.3 Temporary utilities. The code official is authorized to give permission to temporarily supply utilities before an installation has been fully completed and the final certificate of completion has been issued. The part covered by the temporary certificate shall comply with the requirements specified for temporary lighting, heat or power in the code.

❖ Commonly, the utilities on many construction sites are installed and energized long before all aspects of the system are completed. This section would allow such temporary systems to continue provided that they comply with the applicable safety provisions of the code.

110.4 Termination of approval. The code official is authorized to terminate such permit for temporary equipment, systems or uses and to order the temporary equipment, systems or uses to be discontinued.

❖ This section provides the code official with the necessary authority to terminate the permit for temporary equipment, systems and uses if conditions of the permit have been violated or if temporary equipment or systems pose an imminent hazard to the public. This enables the code official to act quickly when time is of the essence in order to protect public health, safety and welfare.

Bibliography

The following resource materials are referenced in this chapter or are relevant to the subject matter addressed in this chapter.

ARI 1010-02, *Self-contained, Mechanically-refrigerated Drinking-water Coolers*. Arlington, VA: Air-Conditioning & Refrigeration Institute, 2002.

ASME A112.6.3-01, *Floor and Trench Drains*. New York, NY: American Society of Mechanical Engineers, 2001.

IBC-09, *International Building Code*. Washington, DC: International Code Council, Inc., 2009.

IEBC-09, *International Existing Building Code*. Washington, DC: International Code Council, Inc., 2009.

IFC-09, *International Fire Code*. Washington, DC: International Code Council, Inc., 2009.

IFGC-09, *International Fuel Gas Code*. Washington, DC: International Code Council, Inc., 2009.

IMC-09, *International Mechanical Code*. Washington, DC: International Code Council, Inc., 2009.

IPMC-09, *International Property Maintenance Code*. Washington, DC: International Code Council, Inc., 2009.

IRC-09, *International Residential Code*. Washington, DC: International Code Council, Inc., 2009.

Legal Aspects of Code Administration. Washington, DC: International Code Council.

Chapter 2: Definitions

General Comments

The words or terms defined in this chapter are deemed to be of prime importance in both specifying the subject matter of code provisions and giving meaning to certain terms used throughout the code for administrative or enforcement purposes.

Purpose

Codes, by their very nature, are technical documents. As such, literally every word, term and punctuation mark can add to or change the meaning of the intended result.

This is even more so with a performance code where the desired result often takes on more importance than the specific words.

Furthermore, the code, with its broad scope of applicability, includes terms inherent in a variety of construction disciplines. These terms can often have multiple meanings depending on the context or discipline being used at the time.

For these reasons, it is necessary to maintain a consensus on the specific meaning of terms contained in the code. Chapter 2 performs this function by stating clearly what specific terms mean for the purpose of the code.

SECTION 201 GENERAL

201.1 Scope. Unless otherwise expressly stated, the following words and terms shall, for the purposes of this code, have the meanings shown in this chapter.

❖ This section contains language and provisions that are supplemental regarding the use of Chapter 2. The subsections give guidance to the use of the defined words relevant to tense, gender, etc. Finally, this chapter provides the means to resolve those terms not defined.

201.2 Interchangeability. Words stated in the present tense include the future; words stated in the masculine gender include the feminine and neuter; the singular number includes the plural and the plural the singular.

❖ Although the definitions contained in Chapter 2 are to be taken literally, gender and tense are considered to be interchangeable. This is so that any grammatical inconsistencies within the code do not hinder the understanding or enforcement of the requirements.

201.3 Terms defined in other codes. Where terms are not defined in this code and are defined in the *International Building Code*, *International Fire Code*, *International Fuel Gas Code* or the *International Mechanical Code*, such terms shall have the meanings ascribed to them as in those codes.

❖ When a word or term appears in the code that is not defined in this chapter, other references may be used to find its definition, such as the other *International Codes*® (I-Codes®) in the ICC family of codes, which are coordinated to prevent conflict between documents.

201.4 Terms not defined. Where terms are not defined through the methods authorized by this section, such terms shall have ordinarily accepted meanings such as the context implies.

❖ Another resource for defining words or terms not defined herein or in other codes is their "ordinarily accepted meanings." The intent of this statement is that a dictionary definition could suffice, provided that such definition refers to the context.

Some of the construction terms used throughout the code may not be defined in Chapter 2 or in a dictionary. In such a case, one would first turn to the definitions contained in the referenced standards (see Chapter 13) and then to published textbooks on the subject in question.

SECTION 202 GENERAL DEFINITIONS

ACCEPTED ENGINEERING PRACTICE. That which conforms to accepted principles, tests or standards of nationally recognized technical or scientific authorities.

❖ The code makes frequent reference to a variety of consensus standards. Where a requirement is not based on a specific standard but on a body of knowledge of a particular engineering or construction discipline, this term is used.

It is also understood that where the code is silent on a subject, any rules that could be applicable in the interest of public safety must be in accordance with that body of knowledge called "accepted engineering practice." The code official can enforce this concept through the use of Section 102.9, titled "Requirements not covered by code." Care and caution need to be used by the code official when enforcing the "accepted engineering practice" concept to be sure that the concept is sound and safe.

ACCESS (TO). That which enables a fixture, appliance or equipment to be reached by ready *access* or by a means that first requires the removal or movement of a panel, door or similar obstruction (see "Ready *access*").

❖ Access to plumbing fixtures, appliances and appurtenances is necessary to facilitate inspection, observation, maintenance, adjustment, repair or replacement. Access to something means that it can be easily physically reached without having to remove a permanent portion of the structure. It is acceptable, for example, to install a device in a space that would require removal of lay-in suspended ceiling panels to gain access. A fixture, appliance or device would not be considered accessible if it were necessary to remove or open any portion of a structure other than panels, doors, covers or similar obstructions intended to be removed or opened normally, without special tools or knowledge. See the definition of "Ready access."

Access can be described as the capability of being reached or approached for inspection, observation, maintenance, adjustment, repair or replacement. Achieving access could require the removal or opening of a panel, door or similar obstruction and might require overcoming an obstacle such as elevation. When access is being provided to an emergency shutoff valve, the designer should be aware that more damage can occur during the time it takes to actually "access" the valve.

ACCESS COVER. A removable plate, usually secured by bolts or screws, to permit *access* to a pipe or pipe fitting for the purposes of inspection, repair or cleaning.

❖ A removable access cover is normally installed to conceal a plumbing fitting that must be inspected, serviced, repaired or cleaned through a finished material or surface. Where the floor or paved surface is subject to pedestrian or vehicular traffic, the access cover must be structurally capable of supporting the load applied to it. Most access covers are secured with either fasteners, such as bolts or screws, or a simple latching mechanism. In special installations where security or safety is a concern, the method for securing the access cover could become more extensive and complicated. These special installations should be limited in number wherever possible because of the concern over the time it takes to access the fitting behind the cover.

Access covers placed in finished walls are often set flush with the face of the wall. Some access covers have provisions that allow for insertion of a portion of the actual wall finish material to help conceal the access cover. Where the access cover is in a floor surface, it must be installed flush so that it does not create a tripping hazard.

ADAPTER FITTING. An *approved* connecting device that suitably and properly joins or adjusts pipes and fittings which do not otherwise fit together.

❖ Where approved by the code official, these fittings are used to join pipes that are of different diameters, different materials or both. Adapter fittings are used to convert from one joining method to another. Refer to Sections 605 and 705 for specific information related to joints.

AIR ADMITTANCE VALVE. One-way valve designed to allow air to enter the plumbing drainage system when negative pressures develop in the piping system. The device shall close by gravity and seal the vent terminal at zero differential pressure (no flow conditions) and under positive internal pressures. The purpose of an air admittance valve is to provide a method of allowing air to enter the plumbing drainage system without the use of a vent extended to open air and to prevent *sewer* gases from escaping into a building.

❖ An air admittance valve (AAV) is a device used as an alternate method for the venting of fixture traps. AAV's reduce the amount of vent piping required to terminate outdoors. An AAV is designed to open and admit air into the drainage system when negative drain system pressures occur. An AAV closes by gravity and seals the vent terminal when the drain system pressure is equal to or exceeds atmospheric pressure. The seal prevents sewer gas from entering the building [see Commentary Figures 202(1) and 202(2)].

It is important to note that a mechanical vent is not considered an air admittance valve [see Commentary Figure 202(3)]. A mechanical vent is a device that operates in a manner similar to that of an air admittance valve. The mechanical vent opens to admit air under negative pressure conditions and closes under zero and positive pressure conditions. It cannot, however, be classified as an air admittance valve because it closes against the force of gravity by a lightweight wire spring. The use of a spring to overcome the forces of gravity to return the seal to the closed position introduces a higher probability of device failure than that of an air admittance valve.

AIR BREAK (Drainage System). A piping arrangement in which a drain from a fixture, appliance or device discharges indirectly into another fixture, receptacle or interceptor at a point below the *flood level rim* and above the trap seal.

❖ An air break is an indirect drainage method in which waste discharges to the drainage system through piping that terminates below the flood level rim and above the trap seal of an approved receptor. An air break is commonly used to protect mechanical equipment from sewage backup in the event that stoppage occurs. It also protects the drainage system from adverse pressure conditions caused by pumped discharge [see Commentary Figure 202(4)].

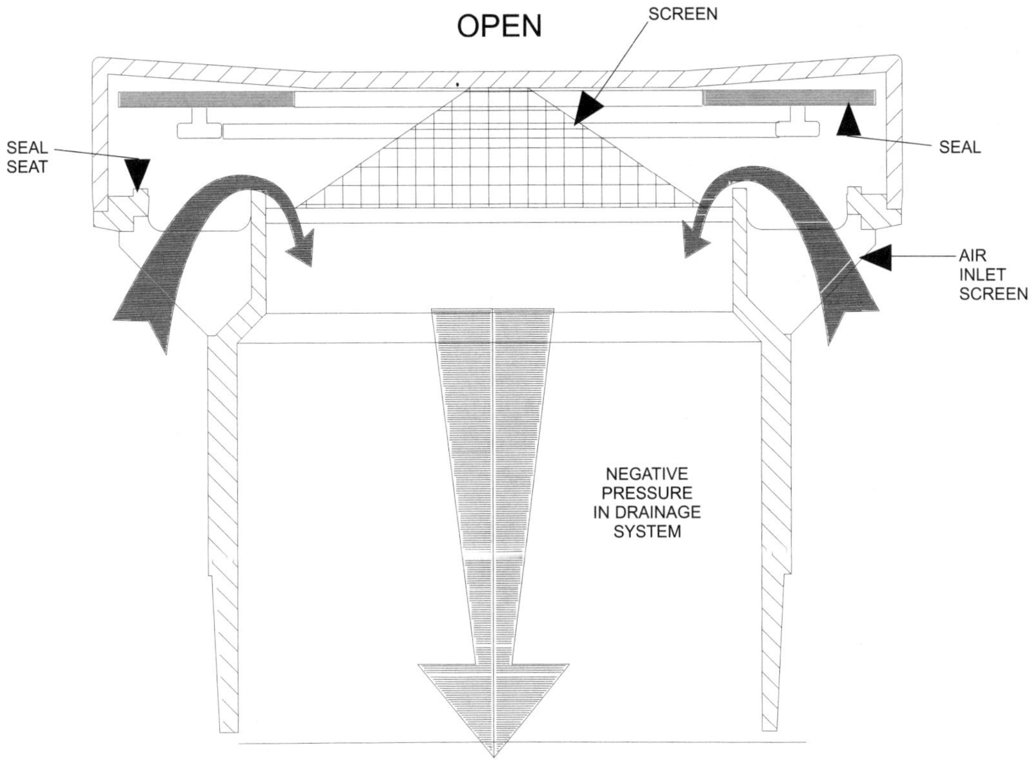

Figure 202(1)
AIR ADMITTANCE VALVE OPENS

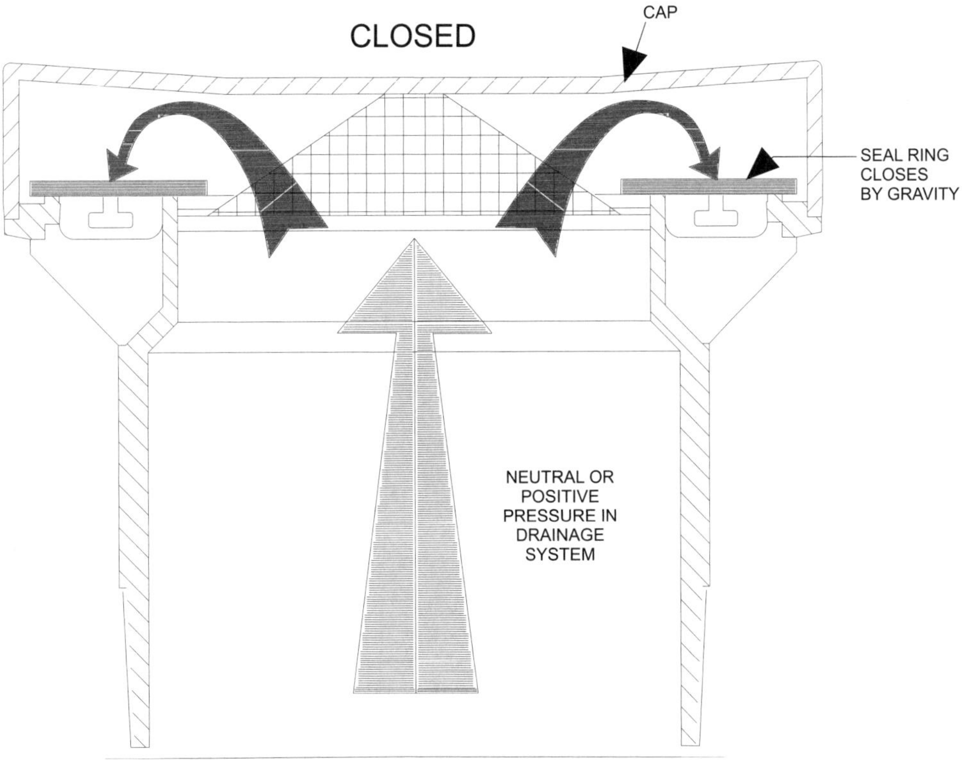

Figure 202(2)
AIR ADMITTANCE VALVE CLOSES

AIR GAP (Drainage System). The unobstructed vertical distance through the free atmosphere between the outlet of the waste pipe and the *flood level rim* of the receptacle into which the waste pipe is discharging.

❖ An air gap (drainage system) is a type of indirect waste arrangement where the waste is discharged to the drainage system through piping that terminates at a specified distance above the flood level rim of an approved receptor.

An air gap is commonly installed in drain lines serving equipment that is used in the preparation, storage and service of food; the conveyance of potable water; and the sterilization of medical equipment. The air gap serves as an impossible barrier for sewage to overcome in the event that stoppage occurs in the receptor drain, because sewage backup would overflow the receptor drain flood level before it came in contact with the drain line above [see Commentary Figure 202(4)].

AIR GAP (Water Distribution System). The unobstructed vertical distance through the free atmosphere between the lowest opening from any pipe or faucet supplying water to a tank, plumbing fixture or other device and the *flood level rim* of the receptacle.

❖ An air gap is the most reliable and effective means of backflow protection. It is simply the vertical airspace

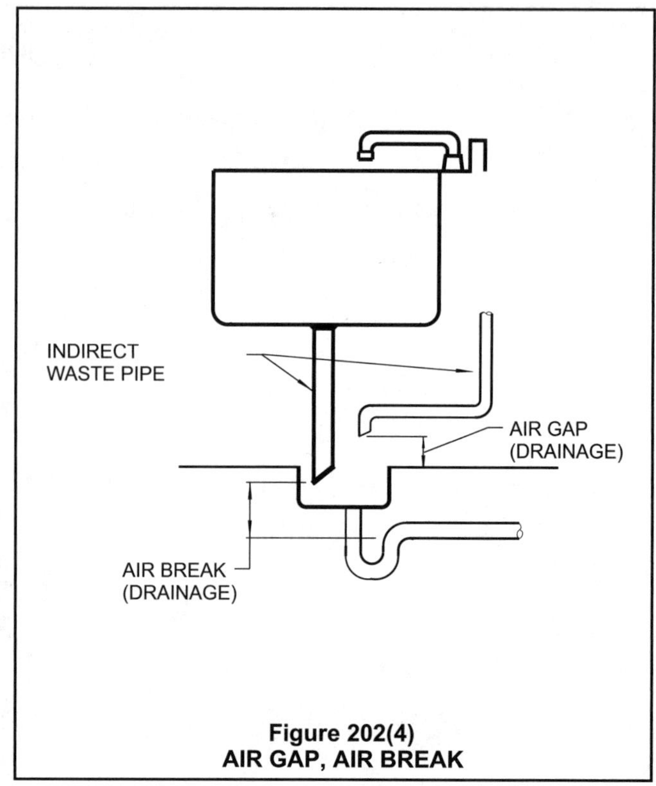

Figure 202(4)
AIR GAP, AIR BREAK

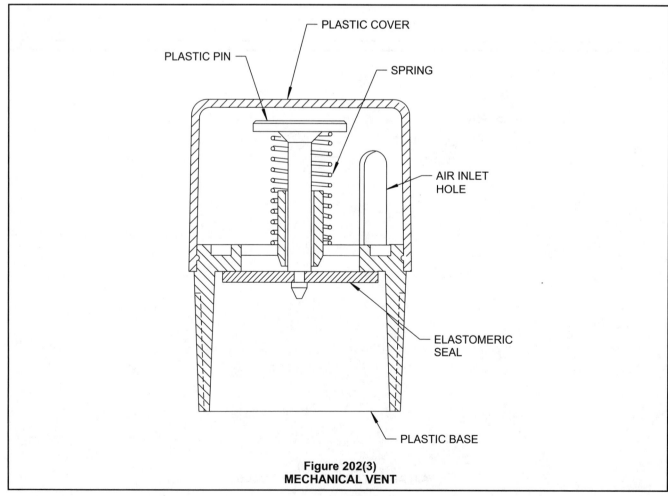

Figure 202(3)
MECHANICAL VENT

between the potable water supply outlet and the possible source of contamination. This air gap prevents possible contamination of the potable water supply by preventing the supply outlet, such as a faucet, from backsiphoning waste water, such as from a basin. Many manufacturers have developed air gap fittings that provide a rigid connection to the drainage system while maintaining the minimum level of protection from contamination.

ALTERNATIVE ENGINEERED DESIGN. A plumbing system that performs in accordance with the intent of Chapters 3 through 12 and provides an equivalent level of performance for the protection of public health, safety and welfare. The system design is not specifically regulated by Chapters 3 through 12.

❖ Plumbing systems designed by accepted engineering practice (as previously defined in this section) to provide the same performance as achieved by conventional plumbing design methods are considered as alternative engineered designs. An alternative engineered design could include water supply and distribution, sanitary and storm drainage, venting, traps, fixtures and connections, devices and appurtenances. Alternative engineered designs must be at least equivalent to that prescribed by Chapters 3 through 12 (see commentary, Sections 105.4 and 918).

ANCHORS. See "Supports."

❖ Anchors are the principle components of a support device or system that affix the support device or system to the building structure. Perhaps the most common use of the term "anchor" in the building industry is for indicating a "concrete anchor" which is an imbedded (either integrally cast-in-place or drilled-and-inserted) fastener. However, the term can also include nails, screws, bolts or adhesives used in conjunction with fastening the support device or system to the structure.

ANTISIPHON. A term applied to valves or mechanical devices that eliminate siphonage.

❖ This term refers to the function of certain valves and devices used to break or prevent the siphon effect that can be created in plumbing systems.

APPROVED. Acceptable to the code official or other authority having jurisdiction.

❖ As related to the process of acceptance of building installations, including materials, equipment and construction systems, this definition identifies where ultimate authority rests. Whenever this term is used, it intends that only the enforcing authority can accept a specific installation or component as complying with the code. It should be noted that the research reports prepared and published by ICC Evaluation Service can be used by code officials to aid in their review and approval of the material or method described in the report. Publishing a report does not indicate any form of "approval" for the material or method described in the report.

APPROVED AGENCY. An established and recognized agency approved by the code official and that is regularly engaged in conducting tests or furnishing inspection services.

❖ An established and recognized agency approved by the code official that is regularly engaged in conducting tests or furnishing inspection services.

The word "approved" means "as approved by the code official." The basis for approval of an agency for a particular activity could include the capacity and capability of the agency to perform the work in accordance with Section 303.

AREA DRAIN. A receptacle designed to collect surface or storm water from an open area.

❖ An area drain is a storm water and surface drainage collection point installed in outdoor open areas and paved areas, such as driveways, patios, outside stairwells, window wells and parking lots. These devices are placed at points of lower elevation to allow water to flow toward them. Additionally, this location causes debris, ice, etc., to flow toward them. Therefore, attention must be given to their design and installation to prevent them from becoming inoperative, often at the time when they are most needed.

ASPIRATOR. A fitting or device supplied with water or other fluid under positive pressure that passes through an integral orifice or constriction, causing a vacuum. Aspirators are also referred to as suction apparatus, and are similar in operation to an ejector.

❖ Within the context of this code, an aspirator is a device that uses a flowing liquid to create a reduced pressure condition. A simple example of an aspirator is a laboratory filter pump. Where water is forced through a smooth tube having a zone with an abrupt restricted area, the flow velocity will necessarily increase in the restricted area, thereby creating a reduced pressure condition at the restriction. Highly effective aspirators can produce vacuums approaching zero atmospheres (perfect vacuum).

Since an aspirator has no moving parts and only requires a flow of liquid to function, they are an economic method for drawing liquids (or gases) into the flow stream. Common examples of aspirators can be found in many food service facilities to "inject" cleaning chemicals for the dishwasher or for mop bucket filling. The flow of water through the aspirator draws the chemical into the flow stream without the need for pumps. Where an aspirator device is connected to a potable water supply, a backflow protection device must be installed between the water supply and the aspirator due to the connection of a contaminate to the aspirator. Most manufacturers of chemical dispensing equipment that are designed to use internal aspirators will also provide integral backflow protection within the equipment.

BACKFLOW. Pressure created by any means in the water distribution system, which by being in excess of the pressure in the water supply mains causes a potential backflow condition.

❖ Backflow is the reversal of flow from its intended direction. In the context of this code, backflow refers to a reversed flow condition that occurs in a closed water piping system caused by an event or device within the system. Since movement of liquids in a pipe is caused by a difference in pressure, in order for flow to reverse, the event or device must cause a change in pressure that is either greater than or less than the differential pressure conditions that cause flow in a normal direction (see "Backflow preventer").

Backpressure, low head. A pressure less than or equal to 4.33 psi (29.88 kPa) or the pressure exerted by a 10-foot (3048 mm) column of water.

❖ Low-head backpressure is created where the source of backpressure comes from a hose elevated to a level that produces 10 feet (3048 mm) or less of water column at the outlet. A hose connection, such as a wall hydrant or faucet, is where "low head backpressure" is typically experienced (also see the commentary to "Backpressure").

Backsiphonage. The backflow of potentially contaminated water into the potable water system as a result of the pressure in the potable water system falling below atmospheric pressure of the plumbing fixtures, pools, tanks or vats connected to the potable water distribution piping.

❖ Backsiphonage occurs where the pressure within the potable water distribution system drops below atmospheric (subatmospheric) pressure or negative gauge pressure. It, like backpressure, allows the normal direction of flow to be reversed. A siphon can result in an unprotected cross connection, causing contamination or pollution of the potable water supply.The main difference between backsiphonage and backpressure is the pressure in the system. Water distribution systems having a pressure less than zero gauge have the potential to create a siphon, resulting in backsiphonage. Where backpressure typically occurs in a closed system, backsiphonage can occur in an open system or a system that is open to the atmosphere.

Drainage. A reversal of flow in the drainage system.

❖ Backflow drainage is the flow of waste in the drainage system opposite the direction intended, regardless of the cause. Backflow is generally the result of drainage system stoppages, overloads or a higher pressure on the drainage pipe side of the trap. Drainage backflow can result in interior flooding, contamination of fixtures and contamination of the potable water system.

Water supply system. The flow of water or other liquids, mixtures or substances into the distribution pipes of a potable water supply from any source except the intended source.

❖ The flow of water or other liquids, mixtures or substances into the distribution pipes of a potable water supply from any source except the intended source.

BACKFLOW CONNECTION. Any arrangement whereby backflow is possible.

❖ Backflow is a possibility in most, if not all, piping systems. The theory of backflow prevention is based on the possibility, not on the probability or likelihood, of an event.

BACKFLOW PREVENTER. A device or means to prevent backflow.

❖ Section 608.13 identifies the types of backflow preventers permitted by the code. Air gaps and barometric loops are means of backflow prevention, but are not backflow prevention devices. Backflow preventers are designed for different applications, depending on the pressures and types of hazards involved. Selection of the device or method depends upon the degree of hazard posed by the cross connection. There are six basic types of backflow preventers that can be used to correct cross connections: air gaps; barometric loops; vacuum breakers, both atmospheric pressure and spillproof type; double check valves with intermediate atmospheric vent; double check-valve assemblies and reduced pressure principle devices.

BACKWATER VALVE. A device or valve installed in the *building drain* or *sewer* pipe where a *sewer* is subject to backflow, and which prevents drainage or waste from backing up into a lower level or fixtures and causing a flooding condition.

❖ Backwater valves are a type of check valve designed for installation in drainage piping. The valves have a lower invert downstream of the flapper to help prevent solids from interfering with valve closure.

BASE FLOOD ELEVATION. A reference point, determined in accordance with the building code, based on the depth or peak elevation of flooding, including wave height, which has a 1 percent (100-year flood) or greater chance of occurring in any given year.

❖ The base flood elevation is the height to which floodwaters are predicted to rise during passage or occurrence of a base flood. The Federal Emergency Management Agency (FEMA) uses commonly accepted computer models that estimate hydrologic and hydraulic conditions to determine the 1-percent annual chance (base) flood. Along rivers and streams, statistical methods and computer models may have been used to estimate runoff and develop flood elevations. The models take into consideration watershed characteristics and the shape and nature of the flood plain, including natural ground contours and the presence of buildings, bridges and culverts. Along coastal areas, base flood elevations can be developed using models that take into account offshore bathymetry, historical storms and typical wind patterns. In many coastal areas, the base flood elevation includes wave heights.

BATHROOM GROUP. A group of fixtures consisting of a water closet, lavatory, bathtub or shower, including or excluding a bidet, an *emergency floor drain* or both. Such fixtures are located together on the same floor level.

❖ This term is used in Table 709.1 and Section 909.1. Special consideration is given to a bathroom group when sizing drain, waste and vent piping because of the probability of fixture usage. A bathroom group has historically referred to a single water closet, a lavatory and a bathtub or shower that are all located within a single room. Such an arrangement would normally allow only one occupant in the room; therefore, the likelihood of simultaneous fixture discharge is remote. For the fixture usage theory of the bathroom group to be valid, the fixtures in the group must be located within the same bathroom. A bidet has been added to the group of fixtures considered as a bathroom group. A bathroom group may or may not contain a bidet, an emergency floor drain or both, and in any case, the drainage fixture unit (dfu) value is the same (see commentary, Table 709.1).

BEDPAN STEAMER OR BOILER. A fixture utilized for scalding bedpans or urinals by direct application of steam or boiling water.

❖ A bedpan steamer or boiler is a special health care fixture designed to sterilize bedpans by the direct application of steam or boiling water. Bedpan washers and sterilizers combine the functions of a human waste elimination fixture and a sterilizing fixture.

As with all hose connections to the potable water supply, the potential for backflow is present, and the flushing devices and hand-held hose apparatus must be protected against backflow in accordance with Section 608.

BEDPAN WASHER AND STERILIZER. A fixture designed to wash bedpans and to flush the contents into the sanitary drainage system. Included are fixtures of this type that provide for disinfecting utensils by scalding with steam or *hot water*.

❖ See the definition of "Bedpan steamer or boiler."

BEDPAN WASHER HOSE. A device supplied with hot and cold water and located adjacent to a water closet or clinical sink to be utilized for cleansing bedpans.

❖ A bedpan washer hose consists of a flexible hose connected on one end to a source of water and having a spray head with a shut off valve one the other end. The flexibility of the hose allows the user to flush all inside areas fo a bedpan without having to change the position of the bedpan being held above a water closet or clinical sink.

BRANCH. Any part of the piping system except a riser, main or *stack*.

❖ Risers, stacks and mains (including a building drain) are the principal arteries of water and waste plumbing systems. Pipes connecting to these arteries are called branches. Pipes connecting to those branches are also called branches, except that a fixture drain pipe is not a fixture branch (see commentary for "Fixture drain"). A branch could be the same size or smaller than the pipe to which it connects. Branches in plumbing systems are analogous to branches of trees.

BRANCH INTERVAL. A vertical measurement of distance, 8 feet (2438 mm) or more in *developed length*, between the connections of horizontal branches to a drainage *stack*. Measurements are taken down the *stack* from the highest horizontal *branch* connection.

❖ One branch interval corresponds to a length of a drainage stack that has two horizontal branch connections at least 8 feet (2438 mm) apart. In typical multistory construction practice with 8 foot (2438 mm) ceilings, horizontal branch drains connect just below each floor level to a vertical stack; therefore, a branch interval will correspond to each floor in the building, below the top floor. If the lowest floor fixtures drain into a horizontal branch drain (and not the building drain) the number of branch intervals is the number of floors minus one, assuming that each floor has fixtures that discharge to the stack. The numbering of branch intervals starts at the uppermost horizontal branch connection to the stack and proceeds down the stack.

Note that the number of branch intervals does not correspond to the number of stories in a building. Because stories are not limited in height, multiple horizontal branches could enter the stack within a story and some floors might not have any horizontal branch connections. See Commentary Figures 202(5) and 202(10) illustrating where branch intervals do and do not exist.

Branch intervals are a design factor for drainage pipe sizing and venting design. Drain, waste and vent (DWV) system design must consider the nature of waste and airflow in a stack and the effects that branch connections have on such flow [see commentary, Table 710.1(2) and Sections 903.2 and 914].

BRANCH VENT. A vent connecting one or more individual vents with a vent *stack* or *stack* vent.

❖ Commentary Figure 202(7) illustrates a branch vent.

BUILDING. Any structure occupied or intended for supporting or sheltering any *occupancy*.

❖ Where this term is used in the code, it means a structure used to provide shelter or support for some activity or occupancy.

BUILDING DRAIN. That part of the lowest piping of a drainage system that receives the discharge from soil, waste and other drainage pipes inside and that extends 30 inches (762 mm) in *developed length* of pipe beyond the exterior walls of the building and conveys the drainage to the *building sewer*.

❖ Building drains are usually considered to be the main drain of a drainage system within a structure. Building drains are horizontal (including vertical offsets) and are the portion of the drainage system that is at the lowest elevation in the structure. All horizontal drains above the elevation of the building drain are horizontal branches. The building drain terminates at the point where it exits the building [see commentary, Table 710.1(1)].

Combined. A *building drain* that conveys both sewage and storm water or other drainage.

❖ See the definitions of "Building drain" and "Building sewer, combined."

Sanitary. A *building drain* that conveys sewage only.

❖ The sanitary building drain is the main drain of the sanitary drainage system and does not conduct storm drainage [see Commentary Figure 202(8)].

Storm. A *building drain* that conveys storm water or other drainage, but not sewage.

❖ The storm building drain is the main drain of the storm drainage system and does not conduct sewage (see the definition of "Building drain").

BUILDING SEWER. That part of the drainage system that extends from the end of the *building drain* and conveys the discharge to a *public sewer*, *private sewer*, individual sewage disposal system or other point of disposal.

❖ The building sewer is the extension of the building drain and is located entirely outside of the building exterior walls [see commentary, Table 710.1(1) and Commentary Figure 202(8)].

Combined. A *building sewer* that conveys both sewage and storm water or other drainage.

❖ Combination storm and sanitary sewers are no longer constructed because of the burden placed on sewage treatment plants. More than 1,100 cities housing 42 million people nationwide, however, are served by existing combination storm and sanitary sewers. Storm water does not require treatment; therefore, combining it with sanitary drainage unnecessarily increases the volume of influent to treatment facilities (see commentary, Section 1101.3). Controlled-flow storm drainage systems should be considered in all combined systems to minimize the impact on municipal sewage disposal plants.

Sanitary. A *building sewer* that conveys sewage only.

❖ The sanitary sewer extends from the point of connection to the sanitary building drain to the point of connection to the public sewer system or the private sewage disposal system. Most jurisdictions arbitrarily set a distance such as 3 or 5 feet (914 or 1524 mm) outside of the building foundation as the point at which the building drain becomes the building sewer [see Figure 202(8)].

Storm. A *building sewer* that conveys storm water or other drainage, but not sewage.

❖ The storm sewer extends from the point of connection to the storm building drain to the point of connection to the public storm sewer or other point of disposal (see the definition of "Sewer, sanitary"). Storm sewers are prohibited by Section 1101.3 from being combined with a sanitary sewer.

BUILDING SUBDRAIN. That portion of a drainage system that does not drain by gravity into the *building sewer*.

❖ Building subdrains must discharge to sumps or receivers equipped with pumps or ejectors to lift the influent to the elevation of the gravity sewer.

BUILDING TRAP. A device, fitting or assembly of fittings installed in the *building drain* to prevent circulation of air between the drainage system of the building and the *building sewer*.

❖ A building trap (house trap) is a "running" trap that prevents the circulation of sewer gases between the building drain and the building sewer (see commentary, Section 1002.6 and Commentary Figure 1002.6). The installation of a building trap should not occur unless required by the code official because of the condition of the public sanitary sewer.

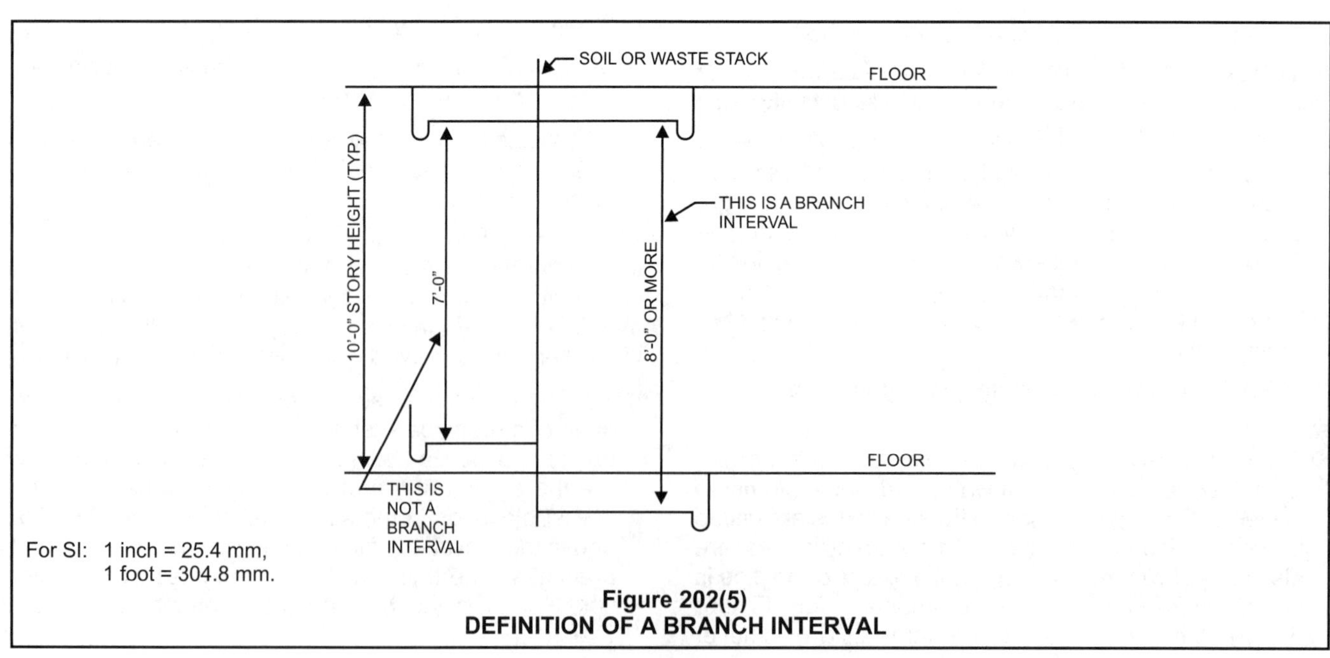

Figure 202(5)
DEFINITION OF A BRANCH INTERVAL

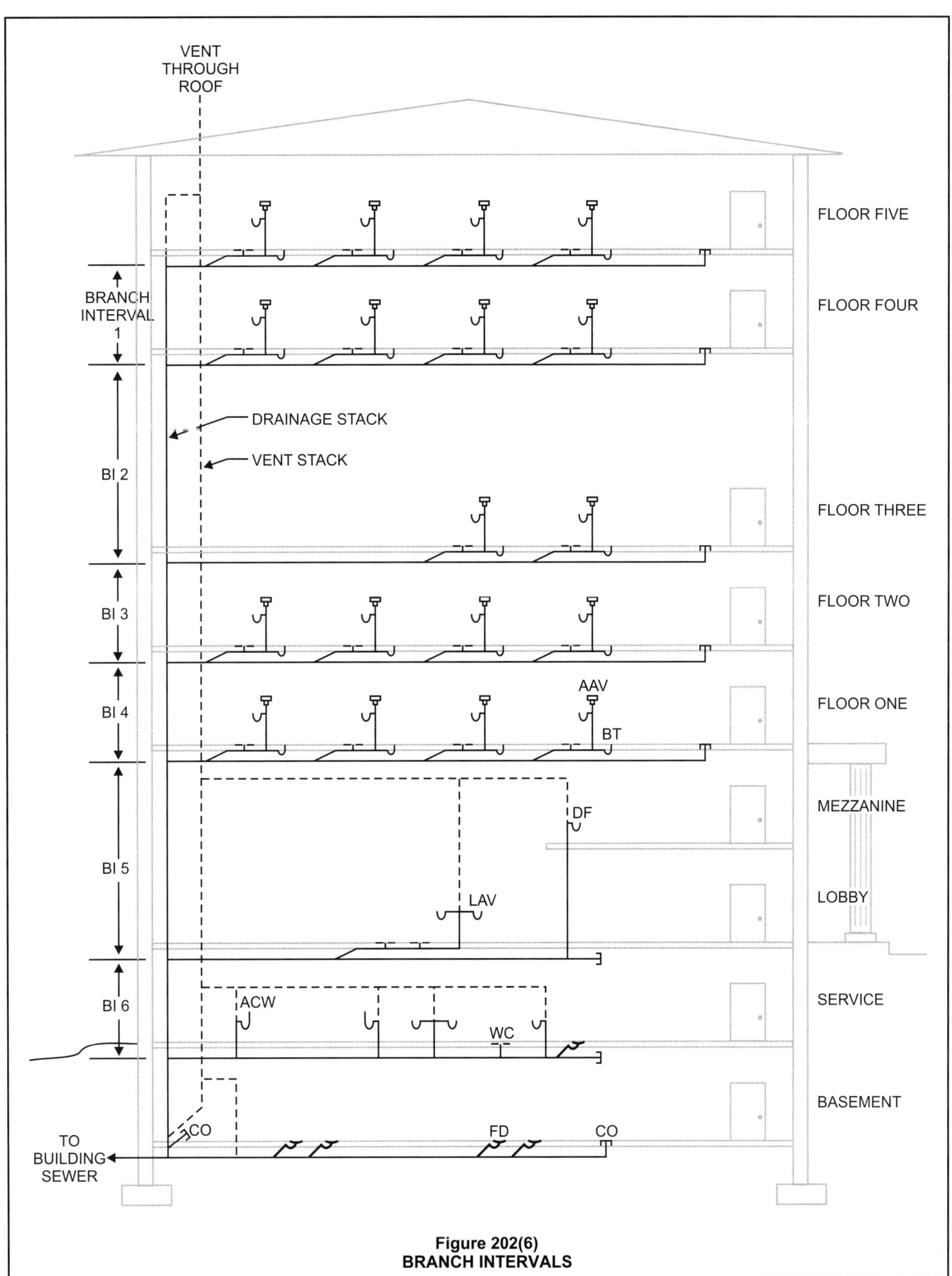

Figure 202(6)
BRANCH INTERVALS

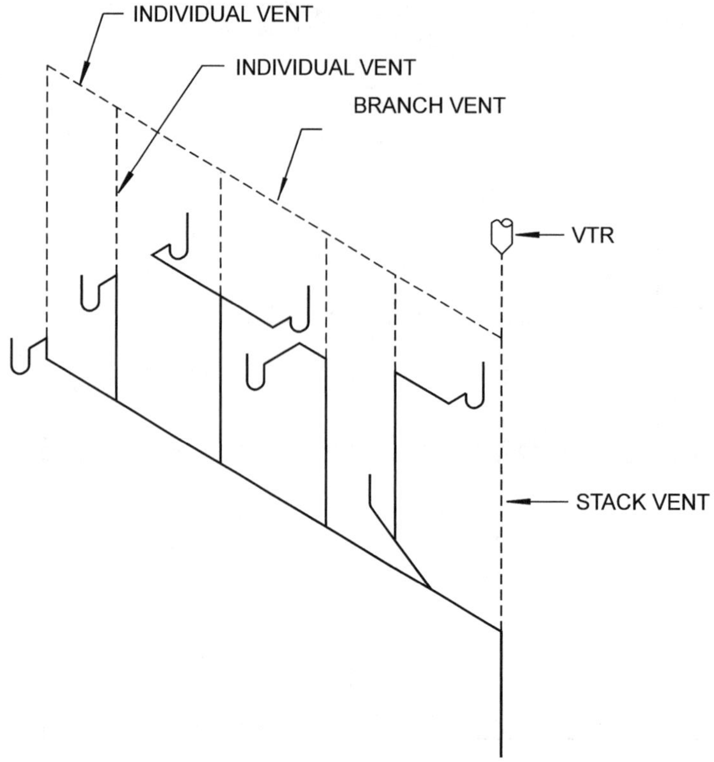

Figure 202(7)
BRANCH VENT

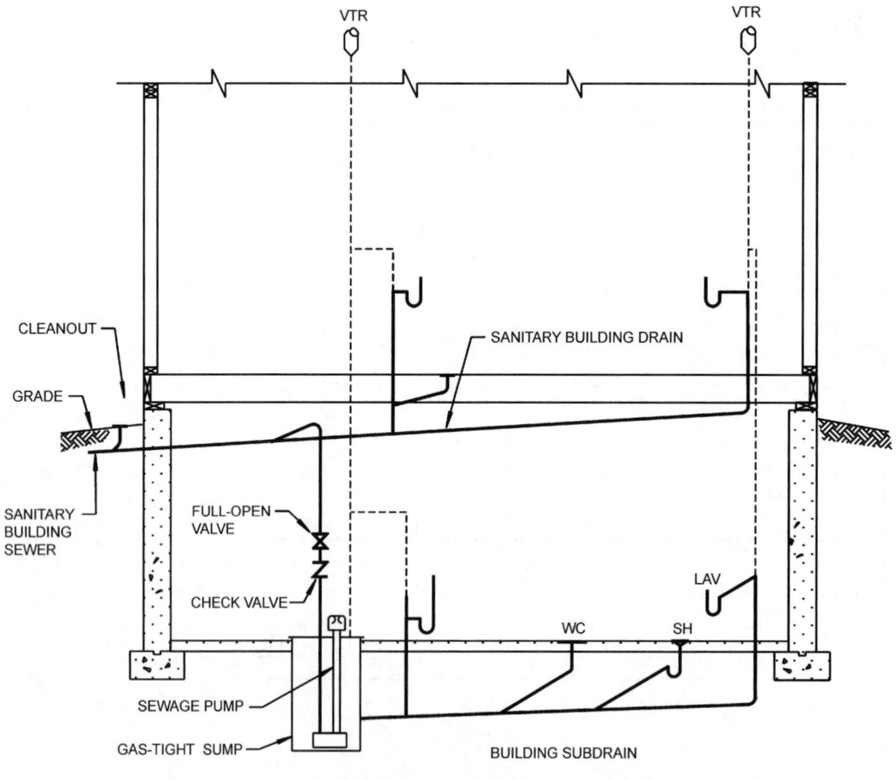

Figure 202(8)
DRAINAGE SYSTEM COMPONENTS

CIRCUIT VENT. A vent that connects to a horizontal drainage *branch* and vents two traps to a maximum of eight traps or trapped fixtures connected into a battery.

❖ A circuit vent is a means of venting multiple fixtures using only one or two vents. The horizontal drainage branch actually serves as a wet vent for the fixtures located downstream of the circuit vent connection (see Commentary Figure 911.1).

CISTERN. A small covered tank for storing water for a home or farm. Generally, this tank stores rainwater to be utilized for purposes other than in the potable water supply, and such tank is placed underground in most cases.

❖ A cistern is a reservoir, usually placed underground, used for the storage of rainwater. Once common, cisterns are rarely used today because of the availability of potable water supplies from wells and public water distribution systems. Cisterns were used as the primary or supplemental source of water for drinking, cooking, bathing and washing. When cisterns are used, care needs to be taken to prevent contamination during construction, maintenance and use.

CLEANOUT. An *access* opening in the drainage system utilized for the removal of obstructions. Types of cleanouts include a removable plug or cap, and a removable fixture or fixture trap.

❖ Cleanouts are broadly defined as any access opening point into the drainage system for the purpose of removing a clog or other obstruction in the drain line. A cleanout is intended to provide convenient access to the piping interior without significant disassembly of the plumbing installation. A "P" trap with a removable U-bend is typically acceptable as a cleanout for the fixture drain or branch to which the trap discharges. Although a water closet connection (flange) is permitted to serve as a cleanout, such use will require removal and reinstallation of the water closet. The installation of a water closet is an activity that requires a permit, whereas the clearance of drain stoppages does not (see commentary, Sections 106.1 and 106.2 and Commentary Figure 708.1). Removal of "fixed" objects should occur only as a last resort when no other cleanouts are available.

CODE. These regulations, subsequent amendments thereto, or any emergency rule or regulation that the administrative authority having jurisdiction has lawfully adopted.

❖ The adopted regulations are generally referred to as "the code" and include not only the code, but any adopted modifications to the code and all other associated building codes, related rules and regulations promulgated and enacted by the jurisdiction.

CODE OFFICIAL. The officer or other designated authority charged with the administration and enforcement of this code, or a duly authorized representative.

❖ The statutory power to enforce the code is normally vested in a building department (or the like) of a state, county or municipality whose designated enforcement officer is termed the "code official" (see commentary, Section 103).

COMBINATION FIXTURE. A fixture combining one sink and laundry tray or a two- or three-compartment sink or laundry tray in one unit.

❖ These fixtures have more than one waste outlet and, depending on their design, are allowed to be served by a single trap (see commentary, Section 1002.1, Exceptions 2 and 3). Combination fixtures include two- and three-well sinks, such as pots-and-pans sinks, laundry sinks, bar sinks and culinary sinks.

COMBINATION WASTE AND VENT SYSTEM. A specially designed system of waste piping embodying the horizontal wet venting of one or more sinks or floor drains by means of a common waste and vent pipe adequately sized to provide free movement of air above the flow line of the drain.

❖ A designed system of waste piping in which the horizontal wet venting of floor drains, drinking fountains, sinks and lavatories is accomplished by means of a common waste and vent pipe that is oversized to provide free movement of air above the flow line of the drain. See Section 912.1 which expands the applicability of this system to lavatories and drinking fountains.

COMBINED BUILDING DRAIN. See "*Building drain, combined.*"

❖ See the commentary for "Building drain, combined."

COMBINED BUILDING SEWER. See "*Building sewer, combined.*"

❖ See the commentary for "Building sewer, combined."

COMMON VENT. A vent connecting at the junction of two fixture drains or to a fixture *branch* and serving as a vent for both fixtures.

❖ Where two fixtures are connected to the same drainage pipe, a common vent could be used. This common vent is sized and classified as an individual vent. Any two fixtures may be common vented to either a vertical or horizontal drainage pipe.

The most typical form of common venting is where two fixtures are connected at the same level. Two fixtures connecting at different levels but within the same story are still considered to be common vented. When one of the fixtures connected is at a different level, the vertical drain between the fixtures is oversized because it functions as both a drain for the upper fixture and a vent for the lower fixture.

Additional information on common venting is found in the commentary to Section 908 and Commentary Figure 908.1.

CONCEALED FOULING SURFACE. Any surface of a plumbing fixture which is not readily visible and is not scoured or cleansed with each fixture operation.

❖ The concealed fouling surface is any part of the plumbing fixture that is not readily visible and cleanable by the occupants. This is of concern because of the possible contamination of these surfaces by waste and the resulting growth of bacteria. The amount of concealed fouling surface is to be kept to a minimum. This is controlled in the standards for each type of fixture. As the amount of water used with each operation of the fixture is reduced for conservation purposes, the possibility for contamination of these concealed fouling surfaces increases. This concern needs to be addressed in the related fixture standards.

CONDUCTOR. A pipe inside the building that conveys storm water from the roof to a storm or combined *building drain*.

❖ A conductor is an interior vertical drain pipe (including horizontal offsets) that generally serves roof drains (also see the definition of "Leader").

CONSTRUCTION DOCUMENTS. All of the written, graphic and pictorial documents prepared or assembled for describing the design, location and physical characteristics of the elements of the project necessary for obtaining a building permit. The construction drawings shall be drawn to an appropriate scale.

❖ For the code official to determine that proposed construction is in compliance with code requirements, sufficient information must be submitted for review. This typically consists of the drawings (floor plans, elevations, sections, details, etc.), specifications and product information describing the proposed work. This term replaces the less descriptive "plans and specifications" used in earlier editions of the codes.

CONTAMINATION. An impairment of the quality of the potable water that creates an actual hazard to the public health through poisoning or through the spread of disease by sewage, industrial fluids or waste.

❖ The Environmental Protection Agency (EPA) sets maximum levels for various chemicals and bacteria in drinking water. When an unacceptable level of one or more contaminants is present, the water is considered to be nonpotable. One of the primary purposes of the code is to protect potable water supplies from contamination. Contamination consists of either hazardous chemicals or raw sewage and is considered by the code to represent a high hazard when present in the water supply. Refer to Table 608.1 for the required potable protection requirements related to high-hazard situations.

The purpose of the code and other related regulations, such as those from the EPA, is to reduce, if not totally eliminate, the risk to the general public of poisoning or health impairment through sewage, industrial fluids or waste entering the potable water supply. It should be noted that these contaminants would have a harmful effect on the occupants of a building should such water be consumed. Contamination is quite different from pollution in that polluted water is still considered to be nonpotable, but would not kill or harm if ingested. Refer to the definition of "Pollution."

CRITICAL LEVEL (C-L). An elevation (height) reference point that determines the minimum height at which a backflow preventer or vacuum breaker is installed above the *flood level rim* of the fixture or receptor served by the device. The critical level is the elevation level below which there is a potential for backflow to occur. If the critical level marking is not indicated on the device, the bottom of the device shall constitute the critical level.

❖ The critical level determines the installation height of vacuum breaker backflow-preventer devices. This is referred to as critical because whether or not the vacuum breaker will function as intended is dependent on that level (see commentary, Section 608.13.6). The critical level (C-L) will be either marked on the device or stated in the manufacturer's installation instructions.

CROSS CONNECTION. Any physical connection or arrangement between two otherwise separate piping systems, one of which contains potable water and the other either water of unknown or questionable safety or steam, gas or chemical, whereby there exists the possibility for flow from one system to the other, with the direction of flow depending on the pressure differential between the two systems (see "Backflow").

❖ Cross connections are the links through which it is possible for contaminating materials to enter a potable water supply. The contaminant enters the potable water supply when the pressure of the polluted source exceeds the pressure of the potable source.The action may be called "backsiphonage" or "backpressure." Many serious outbreaks of illness and disease have been traced to cross connections. The intent of the code is to eliminate cross connections or prevent backflow where cross connections cannot be eliminated. Also refer to the definitions and commentary for the terms "Backflow," "Backsiphonage" and "Backpressure."

DEAD END. A *branch* leading from a soil, waste or vent pipe; a *building drain*; or a *building sewer*, and terminating at a *developed length* of 2 feet (610 mm) or more by means of a plug, cap or other closed fitting.

❖ A dead end refers to horizontal piping that does not conduct waste flow, but is connected to piping that does. A dead end will collect solid waste as a result of both the normal flow depth in the drain pipe to which it is connected (see commentary, Section 704.3) and as a result of drainage system stoppages.

DEPTH OF TRAP SEAL. The depth of liquid that would have to be removed from a full trap before air could pass through the trap.

❖ This is the distance measured from the bottom of the "dip" to the top of the "crown of the trap" (crown weir) [see Commentary Figure 202(9)]. The definition prior to this edition of the code referred to water. But with the advent of liquid seal traps for nonwater urinal designs, this definition changed water to liquid.

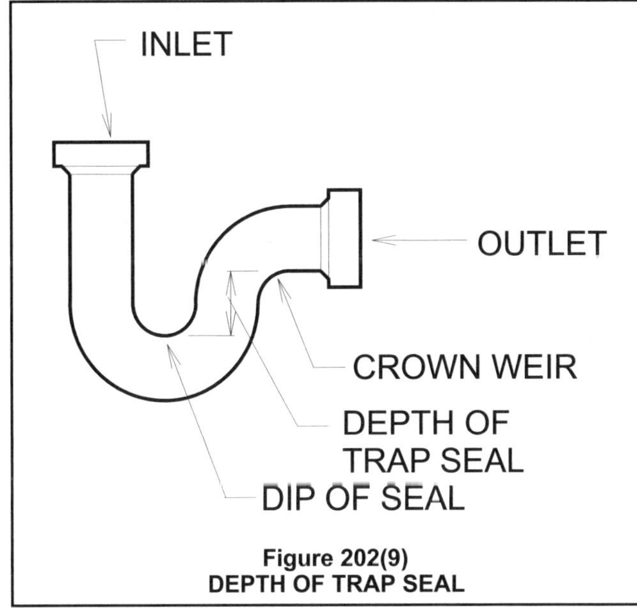

Figure 202(9)
DEPTH OF TRAP SEAL

DESIGN FLOOD ELEVATION. The elevation of the "design flood," including wave height, relative to the datum specified on the community's legally designated flood hazard map.

❖ The design flood elevation is the height to which floodwaters are predicted to rise during the passage or occurrence of the design flood. The datum specified on the flood hazard map is important because it may differ from that used locally for other purposes. Communities adopt the Flood Insurance Rate Maps as prepared by the Federal Emergency Management agency (FEMA) or another flood hazard map that shows at least the same flood hazard areas. FEMA uses commonly accepted computer models that estimate hydrologic and hydraulic conditions to determine the 1-percent annual chance (base) flood. Along rivers and streams, statistical methods and computer models may have been used to estimate the runoff and to develop flood elevations. The models take into consideration watershed characteristics and the shape and nature of the floodplain, including natural ground contours and the presence of buildings, bridges and culverts. Along coastal areas, base flood elevations may be developed using models that take into account offshore bathymetry, historical storms and typical wind patterns. In many coastal areas, the base flood elevation includes wave heights.

DEVELOPED LENGTH. The length of a pipeline measured along the centerline of the pipe and fittings.

❖ This term identifies a concept necessary for computing the actual length of piping in a plumbing system. The developed length is measured along the actual flow path of piping and includes the piping lengths in all offsets and changes in direction. Several code requirements are dependent on the actual length of plumbing piping (see commentary, Sections 708.3.3, 802.2, 906.1 and 916.3).

DISCHARGE PIPE. A pipe that conveys the discharges from plumbing fixtures or appliances.

❖ This type of pipe typically refers to waste piping that either empties into a sump or conveys the discharge of a pump (see commentary, Sections 712 and 1111).

DRAIN. Any pipe that carries wastewater or water-borne wastes in a building drainage system.

❖ A drain is any pipe in a plumbing system that carries sanitary waste, clear-water waste or storm water. In the case of wet vents, common vents, waste stack vents, circuit vents and combination drain and vents, the drain may also be conducting airflow.

DRAINAGE FITTINGS. Type of fitting or fittings utilized in the drainage system. Drainage fittings are similar to cast-iron fittings, except that instead of having a bell and spigot, drainage fittings are recessed and tapped to eliminate ridges on the inside of the installed pipe.

❖ The second sentence of the definition above is outdated. Fittings for galvanized steel drain piping were called drainage fittings. Drainage fittings were made substantially different from standard cast pipe fittings so that restrictions to drainage flow were minimized. These special fittings were made of cast iron with recessed internal threads to enable connection with Schedule 40 galvanized pipe. The threads on the fittings were cut in a manner that a horizontal pipe would be sloped $^{1}/_{4}$ inch per foot (20.8 mm/m) to cause gravity flow. A drain piping system using galvanized pipe and these special threaded cast fittings was called a Durham system, named after prominent manufacturer's product line of threaded drainage fittings. Thus, all threaded drainage pattern fittings were coined "Durham" fittings. These special drainage fittings are no longer used in modern construction. Today, the term "drainage fitting" applies to any fitting which has been made to a recognized (ASTM) drainage pattern for use in a gravity flow drain system.

DRAINAGE FIXTURE UNIT

Drainage (dfu). A measure of the probable discharge into the drainage system by various types of plumbing fixtures. The drainage fixture-unit value for a particular fixture depends on its volume rate of drainage discharge, on the time duration of a single drainage operation and on the average time between successive operations.

❖ The conventional method of designing a sanitary drainage system is based on dfu load values. The fixture unit approach takes into consideration the probability of load on a drainage system. The dfu is an arbitrary loading factor assigned to each fixture relative to its impact on the drainage system. The dfu values are determined based on:

- Average rate of water discharge by a fixture;
- Duration of a single operation; and
- Frequency of use or interval between each operation.

Because dfu values have a built-in probability factor, they cannot be directly translated into flow rates or discharge rates.

A dfu is not the same as a water supply fixture unit (wsfu) in Table E103.3(2), Appendix E.

DRAINAGE SYSTEM. Piping within a *public* or *private* premise that conveys sewage, rainwater or other liquid wastes to a point of disposal. A drainage system does not include the mains of a *public sewer* system or a private or public sewage treatment or disposal plant.

❖ The drainage system is the network of piping starting from the waste outlets of fixtures or receptors and ending at the public sewer system or other point of disposal [see Commentary Figure 202(8)].

Building gravity. A drainage system that drains by gravity into the *building sewer*.

❖ Plumbing systems are designed to drain by gravity wherever possible. Where gravity drainage is not possible, pumps or ejectors must be used to lift the waste to higher elevations.

Sanitary. A drainage system that carries sewage and excludes storm, surface and ground water.

❖ See the commentary to Section 701.

Storm. A drainage system that carries rainwater, surface water, subsurface water and similar liquid wastes.

❖ The storm drainage system conducts water from roof drains, area drains, subsurface drains and clear-water waste sources to an approved location or to the public storm sewer (see Section 1101). The storm drainage system is independent of the sanitary drainage system.

EFFECTIVE OPENING. The minimum cross-sectional area at the point of water supply discharge, measured or expressed in terms of the diameter of a circle or, if the opening is not circular, the diameter of a circle of equivalent cross-sectional area. For faucets and similar fittings, the *effective opening* shall be measured at the smallest orifice in the fitting body or in the supply piping to the fitting.

❖ The effective opening is used to determine the minimum air gap required between a potable water supply outlet or opening and the flood level rim of a receptacle, fixture or other potential source of contamination (see Table 608.15.1).

EMERGENCY FLOOR DRAIN. A floor drain that does not receive the discharge of any drain or indirect waste pipe, and that protects against damage from accidental spills, fixture overflows and leakage.

❖ A floor drain installed in a bathroom group is an example of a emergency drain.

ESSENTIALLY NONTOXIC TRANSFER FLUIDS. Fluids having a Gosselin rating of 1, including propylene glycol; mineral oil; polydimethylsiloxane; hydrochlorofluorocarbon, chlorofluorocarbon and carbon refrigerants; and FDA-approved boiler water additives for steam boilers.

❖ See the commentary to Section 608.16.3.

ESSENTIALLY TOXIC TRANSFER FLUIDS. Soil, waste or gray water and fluids having a Gosselin rating of 2 or more including ethylene glycol, hydrocarbon oils, ammonia refrigerants and hydrazine.

❖ Transfer fluids are liquids or gases that transfer heat to or remove heat from another fluid, such as water. The exchange of heat energy takes place through a heat exchanger material that separates the transfer fluid from the fluid being heated or cooled (see commentary, Section 608.16.3).

EXISTING INSTALLATIONS. Any plumbing system regulated by this code that was legally installed prior to the effective date of this code, or for which a permit to install has been issued.

❖ Existing installations is a term that applies to all plumbing work that has been legally installed, and for which a permit to install has been issued prior to the effective date of the code. Plumbing that has been illegally installed prior to the effective date of the code is not considered existing and is subject to all code requirements for new installations. Sections that address existing plumbing systems and installations include Sections 102 and 108.

FAUCET. A valve end of a water pipe through which water is drawn from or held within the pipe.

❖ A faucet is a type of valve or terminal fitting that controls the volume and directional flow of water. In the context of the code, faucets are installed in the potable water distribution system. Faucets terminate to the atmosphere by means of a spout or similar opening and are considered as outlets for the potable water distribution system.

FILL VALVE. A water supply valve, opened or closed by means of a float or similar device, utilized to supply water to a tank. An antisiphon fill valve contains an antisiphon device in the form of an *approved air gap* or vacuum breaker that is an integral part of the fill valve unit and that is positioned on the discharge side of the water supply control valve.

❖ A fill valve is a float-actuated valve used in flush tanks. The fill valve serves to control the water supply for refilling the flush tank and restoring the trap seal in the water closet bowl. All flush tanks must be equipped with an antisiphon type installed 1 inch (25 mm) above the overflow pipe.

FIXTURE. See "Plumbing fixture."

❖ See the commentary for "Plumbing fixture."

FIXTURE BRANCH. A drain serving two or more fixtures that discharges to another drain or to a *stack.*

❖ See Commentary Figure 202(10).

FIXTURE DRAIN. The drain from the trap of a fixture to a junction with any other drain pipe.

❖ A fixture drain is the section of pipe that connects the outlet weir of a trap to a stack, fixture branch or any other drain [see Commentary Figures 202(10) and 202(11)]. A fixture drain conveys the discharge of only a single trap and can consist of both horizontal and vertical sections of pipe. A fixture drain ends at the point where the discharge from any other fixture is introduced. Although commonly referred to as a "trap arm," a fixture drain can extend beyond the limits of a trap arm. Many code sections depend on the definition of fixture drain including Sections 906.1, 906.2, 907.1, 908.2, 908.3, 909.1, 909.2 and 911.1.

FIXTURE FITTING

❖ Fixture fittings include faucets, spouts, terminal fittings, diverters, supply stops, wastes, traps and other "trim" items that attach to, or in the case of supply fittings are accessible from, a plumbing fixture. These items are necessary to allow the fixture to function as intended.

Supply fitting. A fitting that controls the volume and/or directional flow of water and is either attached to or accessible from a fixture, or is used with an open or atmospheric discharge.

❖ These fittings are associated with the supply and control of flow of water to a fixture. Supply fittings can also include fittings that serve the same purpose but are not mounted directly to the fixture.

Waste fitting. A combination of components that conveys the sanitary waste from the outlet of a fixture to the connection to the sanitary drainage system.

❖ These fittings include the fixture traps that are not an integral part of the fixture (see commentary, Section 1002) and the parts used to connect piping to the sanitary drainage system (see Section 424.1.2).

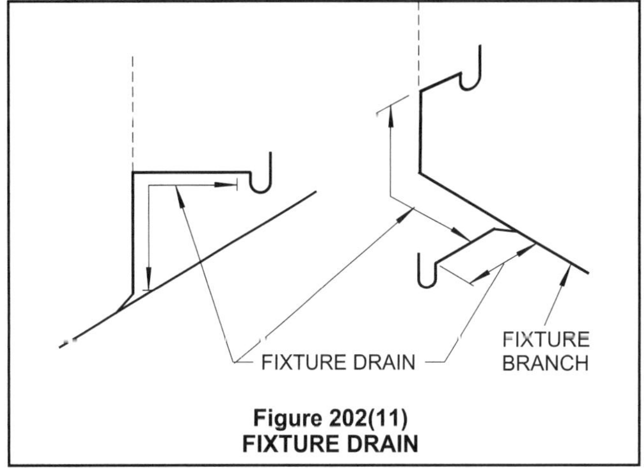

Figure 202(11)
FIXTURE DRAIN

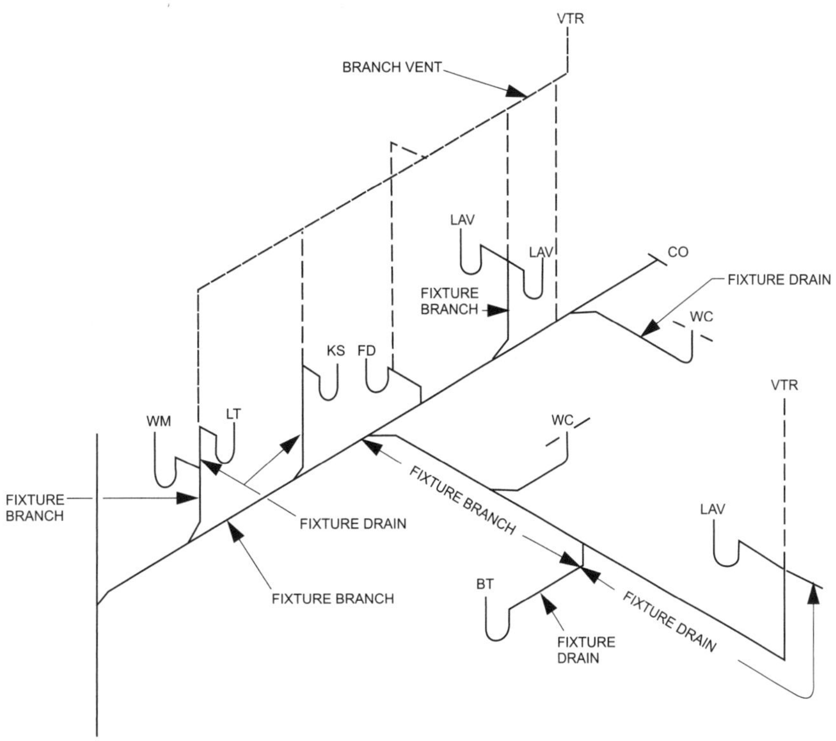

Figure 202(10)
FIXTURE DRAINS AND FIXTURE BRANCHES

FIXTURE SUPPLY. The water supply pipe connecting a fixture to a *branch* water supply pipe or directly to a main water supply pipe.

❖ A fixture supply is a water supply pipe that serves a single fixture (see commentary, Section 604.5).

FLOOD HAZARD AREA. The greater of the following two areas:

1. The area within a flood plain subject to a 1-percent or greater chance of flooding in any given year.
2. The area designated as a *flood hazard area* on a community's flood hazard map or as otherwise legally designated.

❖ The Flood Insurance Rate Maps that are prepared by the Federal Emergency Management Agency (FEMA) show areas that are predicted to be flooded under conditions of the 1-percent annual chance (base) flood. Some states and local jurisdictions develop and adopt maps of flood hazard areas that are more extensive than the areas shown on FEMA's maps. For the purposes of the code, the flood hazard area within which the requirements are to be applied is the greater of the two delineated areas.

FLOOD LEVEL RIM. The edge of the receptacle from which water overflows.

❖ The flood level rim corresponds to the highest elevation that liquid can be contained in a receptacle without spilling over the side. The flood level rim is the point above which liquid will flow over and out of a fixture or receptor. For example, the flood level rim of a sink is the top edge of the basin, and the flood level rim of a floor drain is the floor surface [see Commentary Figure 608.15.1(2)].

FLOW CONTROL (Vented). A device installed upstream from the interceptor having an orifice that controls the rate of flow through the interceptor and an air intake (vent) downstream from the orifice that allows air to be drawn into the flow stream.

❖ Referenced standard ASME A112.14.3 and Section 1003.4.2 use this term in connection with grease interceptors. This device is necessary to control the flow rate of waste through the interceptor and to entrain air to assist in separation of grease and oily waste from the water. Grease interceptors depend on having the necessary waste retention time to allow the fats, oils and grease to separate. If the influent flow rate is too high, the waste will pass through the interceptor too quickly, not allowing sufficient time for separation to occur. Entraining air in the waste influent will cause air bubbles to attach to grease globules, making them more buoyant and thereby enhancing the separation process (see Commentary Figure 1003.3.4.2).

FLOW PRESSURE. The pressure in the water supply pipe near the faucet or water outlet while the faucet or water outlet is wide open and flowing.

❖ Each fixture requiring water has a minimum flow pressure that is necessary for the fixture to operate properly and to help protect the potable water supply from contamination. Flow pressure is a factor used in the sizing of the water supply system. The flow pressure of a plumbing system is less than the static pressure because energy is lost in putting the fluid in motion (see Table 604.3).

FLUSH TANK. A tank designed with a fill valve and flush valve to flush the contents of the bowl or usable portion of the fixture.

❖ The flush tank for a water closet or urinal must be able to supply water in sufficient quantity to flush the contents of the fixture, cleanse the fixture bowl and refill the fixture trap. The definition states clearly that the flush valve is considered to be part of the flush tank.

FLUSHOMETER TANK. A device integrated within an air accumulator vessel that is designed to discharge a predetermined quantity of water to fixtures for flushing purposes.

❖ These devices are a hybrid between flushometer valves and gravity flush tanks. A flushometer tank uses a compression tank that holds water under a pressure equivalent to the water supply pressure. The flushing action created by the flushometer tank is similar to that of the flushometer valve. The amount of water used is limited, as it is with a gravity flush tank.

FLUSHOMETER VALVE. A valve attached to a pressurized water supply pipe and so designed that when activated it opens the line for direct flow into the fixture at a rate and quantity to operate the fixture properly, and then gradually closes to reseal fixture traps and avoid water hammer.

❖ These flushing devices are a type of metering valve and are used primarily in public occupancy applications because of the more powerful flushing action.

GREASE INTERCEPTOR. A plumbing appurtenance that is installed in a sanitary drainage system to intercept oily and greasy wastes from a wastewater discharge. Such device has the ability to intercept free-floating fats and oils.

❖ The interceptor is used to separate and retain grease, thereby preventing it from being discharged into the normal gravity system, which would result in rapid clogging of the waste drainage system.

The primary objection to its use is that it requires maintenance to function properly. All items that require maintenance are subject to failure if maintenance is not performed regularly. In the case of the grease interceptor, this maintenance is often ignored because the task of cleaning it is unpleasant.

Grease interceptors, because of their allowable flow capacity, often serve more than one fixture [see Commentary Figure 1003.3(1)]. Note that the code no longer uses the term "grease trap."

GREASE-LADEN WASTE. Effluent discharge that is produced from food processing, food preparation or other sources where grease, fats and oils enter automatic dishwater prerinse stations, sinks or other appurtenances.

❖ The effluent discharge from food preparation and processing of animal products will usually produce grease-laden waste. Additionally, many industrial processes that use animal fats or petroleum products will produce grease-laden waste. This waste has two main deleterious effects on sewer systems: frequent stoppages resulting from congealed grease and overloading of the sewage treatment system.

GREASE REMOVAL DEVICE, AUTOMATIC (GRD). A plumbing appurtenance that is installed in the sanitary drainage system to intercept free-floating fats, oils and grease from wastewater discharge. Such a device operates on a time- or event-controlled basis and has the ability to remove free-floating fats, oils and grease automatically without intervention from the user except for maintenance.

❖ Referenced standard ASME A112.14.4 and Section 1003.3.4 use this term in connection with grease interceptors. An automatic grease removal device serves the same function as a grease interceptor with the distinction that the unpleasant aspect of grease removal is performed automatically. There are two main types of automatic GRD; pump out and skimmer type. The pump out type has an internal sensor that senses a build up of collected grease, liquefies it and pumps it out to an external container for proper environmental disposal. The skimmer type has a timer-activated skimmer mechanism to mechanically collect and transport the collected grease to an external container. Both types usually have internal heaters to keep the grease in liquid form for easy pumping or skimming.

Although the solids interceptor section of the GRD still requires manual maintenance to remove waste stream particulates such as food scraps, broken glass and plastics, the task is relatively easy, thus having the likelihood of being performed regularly [see Commentary Figures 202(12) through 202(14)].

GRIDDED WATER DISTRIBUTION SYSTEM. A water distribution system where every water distribution pipe is interconnected so as to provide two or more paths to each fixture supply pipe.

❖ A gridded water distribution system is more hydraulically efficient than a parallel or branch line layout but not necessarily more cost efficient. The design balances pressure and equalizes flow so that fixtures farthest from the water service entry point will operate under the same pressure and flow conditions as those closest to the water service entry point. Commentary Figure 202(15) illustrates various distribution layouts.

HANGERS. See "Supports."

❖ Hangers include a variety of devices ranging from perforated metal straps to specialized or proprietary manufactured components (see also the definition of "Supports").

HORIZONTAL BRANCH DRAIN. A drainage *branch* pipe extending laterally from a soil or waste *stack* or *building drain*, with or without vertical sections or branches, that receives the discharge from two or more fixture drains or branches and conducts the discharge to the soil or waste *stack* or to the *building drain*.

❖ See Figure 202(10) and Sections 909 and 911.

HORIZONTAL PIPE. Any pipe or fitting that makes an angle of less than 45 degrees (0.79 rad) with the horizontal.

❖ This definition is needed for the application of many code provisions that apply only to horizontal piping, such as those contained in Chapters 3, 7, 9 and 11.

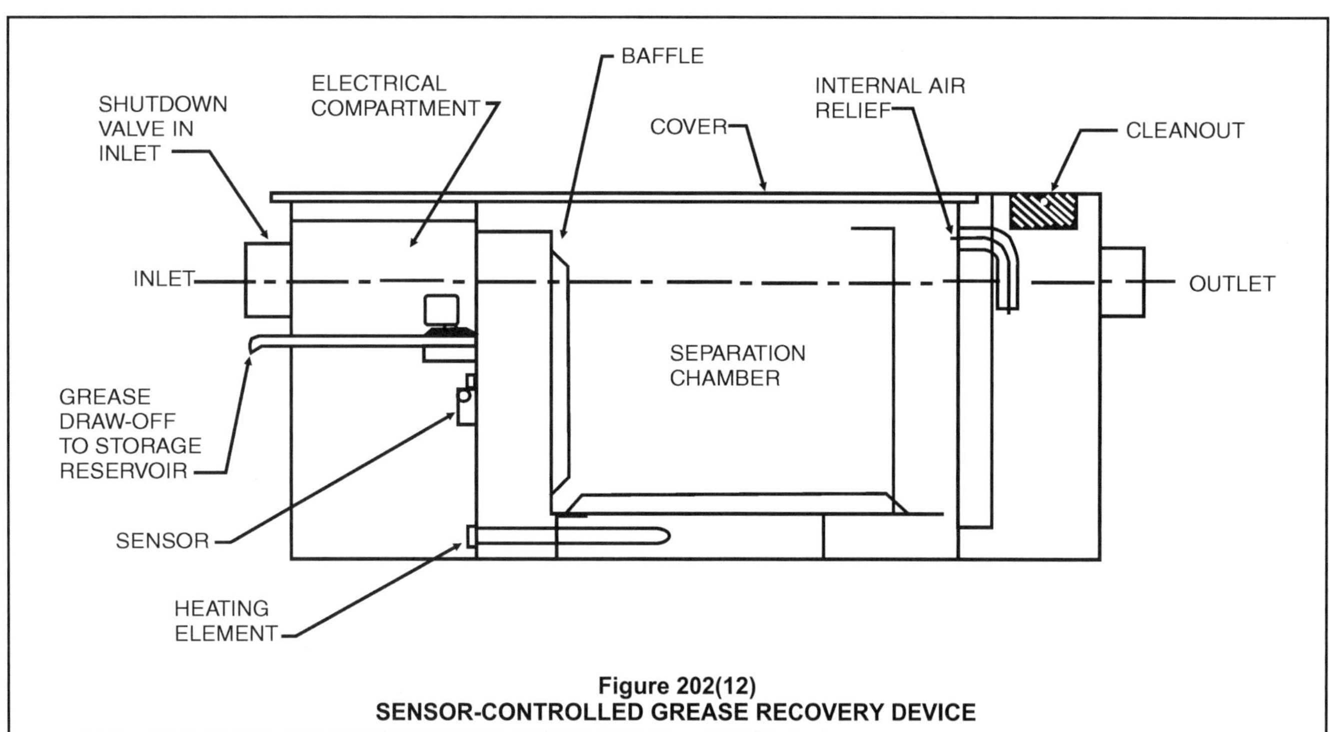

Figure 202(12)
SENSOR-CONTROLLED GREASE RECOVERY DEVICE

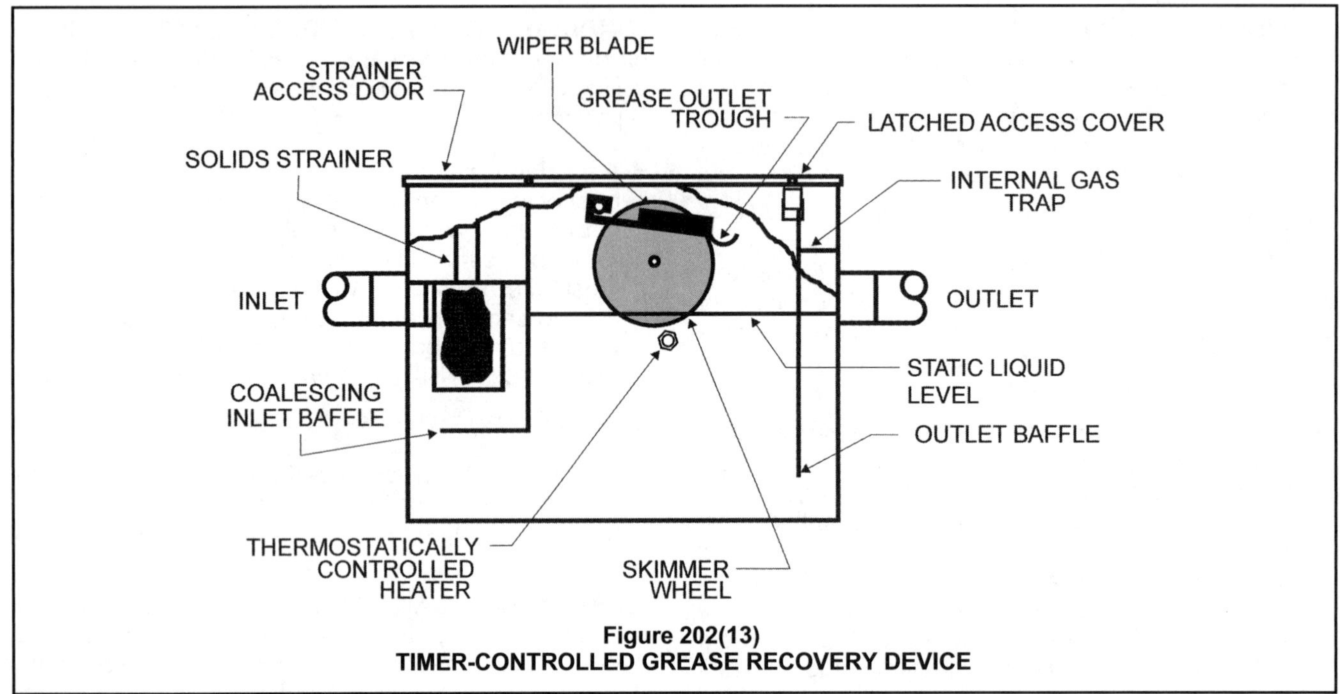

Figure 202(13)
TIMER-CONTROLLED GREASE RECOVERY DEVICE

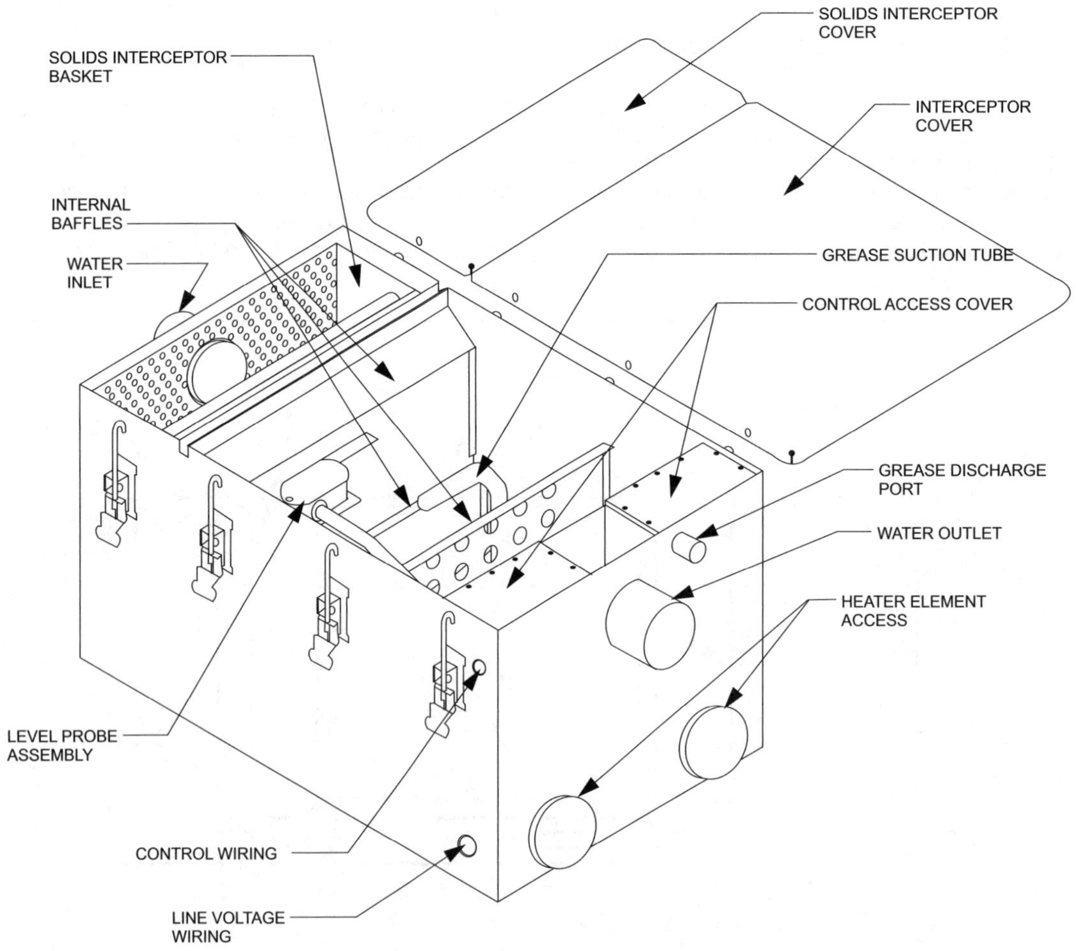

202(14)
GREASE REMOVAL DEVICE, AUTOMATIC

HOT WATER. Water at a temperature greater than or equal to 110°F (43°C).

❖ The code does not specify a maximum temperature for hot water; however, it should be limited to minimize the risk of burn injury and thermal stress to plumbing system components. The designer, the installer or the owner has the choice of determining the temperature. Underwriters Laboratories Standard UL 174 requires that the temperature-regulating thermostat or control be set at the factory at a maximum of 125°F (51.7°C). The minimum temperature is set to define what is considered as hot water.

Water temperature below the minimum temperature of 110°F (43°C) is considered to be either tempered or cold. Care must be used when installing a system that has temperature capability of scalding.

HOUSE TRAP. See "Building trap."

❖ A house trap (building trap) is a "running" trap that isolates the building drain from the building sewer (see commentary, Section 1002.6 and Commentary Figure 1002.6). A house trap should not be installed unless required by the code official because of the condition of the public sanitary sewer.

INDIRECT WASTE PIPE. A waste pipe that does not connect directly with the drainage system, but that discharges into the drainage system through an *air break* or *air gap* into a trap, fixture, receptor or interceptor.

❖ Often referred to as "open site" or "safe waste," indirect waste pipe installations are intended to protect fixtures, equipment and systems from the backflow or backsiphonage of waste (see commentary, Sections 801.1 and 802.1).

INDIVIDUAL SEWAGE DISPOSAL SYSTEM. A system for disposal of domestic sewage by means of a septic tank, cesspool or mechanical treatment, designed for utilization apart from a public *sewer* to serve a single establishment or building.

❖ When installed in accordance with the *International Private Sewage Disposal Code®* (IPSDC®) and approved by the administrative authority, these systems are acceptable as a point of disposal for sewage when a public sewer system is not available.

INDIVIDUAL VENT. A pipe installed to vent a fixture trap and that connects with the vent system above the fixture served or terminates in the open air.

❖ An individual vent is the simplest and most common method of venting a single plumbing fixture (see commentary, Section 907.1).

INDIVIDUAL WATER SUPPLY. A water supply that serves one or more families, and that is not an *approved* public water supply.

❖ This type of water supply includes wells, springs, streams and cisterns. Surface bodies of water and land cisterns can be used when properly treated and approved (see commentary, Sections 602.3 and 608.17).

INTERCEPTOR. A device designed and installed to separate and retain for removal, by automatic or manual means, deleterious, hazardous or undesirable matter from normal wastes, while permitting normal sewage or wastes to discharge into the drainage system by gravity.

❖ These devices are often necessary for the initial treatment of sewage before it is discharged to an approved point of disposal. Interceptors are intended to eliminate or reduce the amounts of substances in sewage that can be detrimental to drainage and sewage treatment systems. When the interceptor is designed for a

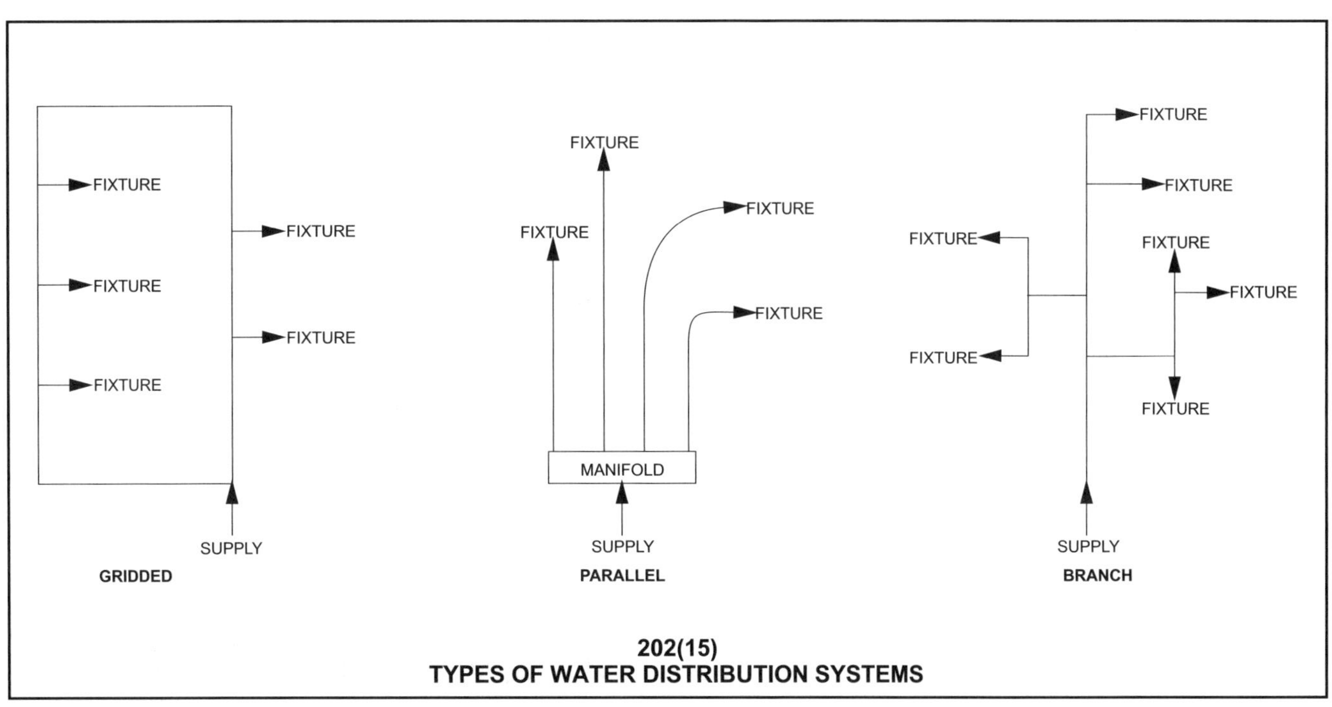

202(15)
TYPES OF WATER DISTRIBUTION SYSTEMS

specific use, it often bears that name, such as a grease interceptor.

JOINT

❖ For a piping system, either supply or drainage, to be installed in a structure, there must be a series of pipe sections and pipe fittings. The connection of these two types of components is a joint in its most general definition. The four types of joints listed below are special types used for a particular installation. These joints are very specific and should be made using the proper components and procedures.

Expansion. A loop, return bend or return offset that provides for the expansion and contraction in a piping system and is utilized in tall buildings or where there is a rapid change of temperature, as in power plants, steam rooms and similar occupancies.

❖ See the commentary to Sections 305.3 and 308.8.

Flexible. Any joint between two pipes that permits one pipe to be deflected or moved without movement or deflection of the other pipe.

❖ See the commentary to Section 305.3.

Mechanical. See "Mechanical joint."

❖ See the commentary for "Mechanical joint."

Slip. A type of joint made by means of a washer or a special type of packing compound in which one pipe is slipped into the end of an adjacent pipe.

❖ A slip joint is a joint commonly found in both fixture tailpieces and "P" traps (see commentary, Section 405.8 and Commentary Figure 1002.2).

LEAD-FREE PIPE AND FITTINGS. Containing not more than 8.0-percent lead.

❖ Ingestion of lead by humans has come under strict control and regulation. The fact that lead can cause harmful effects, particularly in young children, has resulted in an effort to reduce, if not totally eliminate, the amount of lead present in various components of construction. When the EPA enacted the lead solder ban in June 1986, it limited the lead content of pipe and fittings to a maximum of 8 percent. This limitation applies to the various alloys of brass used in producing faucets, fittings and valves. The amount of lead found in brass pipe and brass nipples is minimal, less than 1 percent.

The addition of lead to brass makes the material softer and easier to cut and shape. All commercially available faucets, fittings and valves have less than 8 percent lead. Most of the brass used has a lead content between 1 and 6 percent.

LEAD-FREE SOLDER AND FLUX. Containing not more than 0.2-percent lead.

❖ A soldered joint is the most common method of joining copper pipe and tubing. ASTM B 32 covers many grades of solder, including tin-silver solders. Previously, the grades used in the plumbing industry were typically 40-60 (40 percent tin and 60 percent lead), 50-50 (50 percent tin and 50 percent lead) and 95-5 (95 percent tin and 5 percent antimony). Lead-based solders, however, are no longer permitted for joining copper tubing used for the distribution of potable water. Lead-bearing solders are still acceptable for joining drainage and vent piping, but their use needs to be carefully controlled so that the installer of the potable water tubing does not use lead-bearing solders by accident.

In addition to 95-5 (95 percent tin and 5 percent antimony), which is considered a viable alternative to lead solders in many applications, manufacturers have developed many replacement solder alloys for the plumbing industry (see commentary, Section 705.9.3).

LEADER. An exterior drainage pipe for conveying storm water from roof or gutter drains to an *approved* means of disposal.

❖ Otherwise known as a downspout, this type of drain serves the storm drainage system and typically connects to hanging or built-in gutters and scuppers (see the definition of "Conductor").

LOCAL VENT STACK. A vertical pipe to which connections are made from the fixture side of traps and through which vapor or foul air is removed from the fixture or device utilized on bedpan washers.

❖ Local vents are designed to conduct moisture vapor and foul air to the exterior of the building. Local vents connect to a fixture on the house side (fixture side) of the fixture trap; therefore, such vents must be entirely independent of the sanitary drainage system vents.

MACERATING TOILET SYSTEMS. An assembly consisting of a water closet and sump with a macerating pump that is designed to collect, grind and pump wastes from the water closet and up to two other fixtures connected to the sump.

❖ This assembly includes a water closet and a sump facilitating the connection of two other fixtures and is normally used where fixtures are installed below the building sewer invert. Section 712.4.1 specifically requires these assemblies to be installed according to the manufacturers' instructions and ASME A112.3.4 or CSA B45.9. The assembly includes a grinder pump that reduces the size of solids, allowing the use of discharge piping that is smaller than conventional the size.

MAIN. The principal pipe artery to which branches are connected.

❖ This is a generic item that applies to the primary (and typically the largest) water supply system piping or drainage system piping to which all other piping is connected. The main is the trunk line of branch system piping designs.

MANIFOLD. See "Plumbing appurtenance."

❖ A manifold is a multiple-opening header to which one or more branch lines connect (see the definition of "Plumbing appurtenance" and Section 604.10).

MECHANICAL JOINT. A connection between pipes, fittings, or pipes and fittings that is not screwed, caulked, threaded, soldered, solvent cemented, brazed or welded. A joint in which compression is applied along the centerline of the pieces being joined. In some applications, the joint is part of a coupling, fitting or adapter.

❖ Mechanical joints form a seal by compressing a ferrule or seal around the outside circumference of the pipe or tube or by compressing the tubing around a fitting inserted into the tubing. Mechanical joints are often proprietary and must be installed in accordance with the manufacturers' instructions.

MEDICAL GAS SYSTEM. The complete system to convey medical gases for direct patient application from central supply systems (bulk tanks, manifolds and medical air compressors), with pressure and operating controls, alarm warning systems, related components and piping networks extending to station outlet valves at patient use points.

❖ Nonflammable medical gases are distributed through piping systems from central supply systems, such as cylinders or medical air compressors, to station outlets throughout the medical facility (i.e., patient rooms, operating/delivery rooms or emergency treatment facilities). Medical gases include, but are not limited to, oxygen, nitrogen, nitrous oxide, carbon dioxide and compressed air.

Because more than one medical gas is usually delivered to an outlet station, care must be exercised so that the correct gas is being used for a given procedure. Accidental cross connection can result in injury or death of a patient. All piping, supply systems and outlet stations must be clearly labeled with the name of the gas being supplied. In addition, color-coded or noninterchangeable fittings, keyed to the specific gas, are usually incorporated to further safeguard against cross connections.

The delivery of medical gases is critical to the operation of a health care facility, and provisions must be made to ensure continuous operation of the system. Secondary (backup) cylinders and air compressors are required, and the compressors should be connected to the emergency electrical power grid.

Detailed guidance for the design, installation and maintenance of a medical gas system is provided in NFPA 99C.

MEDICAL VACUUM SYSTEMS. A system consisting of central-vacuum-producing equipment with pressure and operating controls, shutoff valves, alarm-warning systems, gauges and a network of piping extending to and terminating with suitable station inlets at locations where patient suction may be required.

❖ Central medical vacuum systems, including surgical and bedside vacuum systems and waste gas scavenging systems, have become a standard feature in most health care facilities. Inlet stations are typically installed in patient rooms, operating/delivery rooms, nurseries, emergency rooms and other patient care facilities. A vacuum system may also be provided in the laboratory, but is not usually connected to the medical vacuum system.

Each vacuum inlet station requires a collection bottle for collecting fluids and a trap bottle to prevent any overflow fluids from entering and clogging the suction line.This dual bottle system helps prevent the spread of contaminants within the medical facility.

Because the medical vacuum system has become so critical to patient care, continuous service from the system must be assured. At least two vacuum pumps should be installed, each sized to meet all of the facility's needs if the other fails.The pumps should also be connected to the facility's emergency electrical power grid.

Detailed guidance for the design, installation and maintenance of a medical vacuum system is contained in NFPA 99C.

NONPOTABLE WATER. Water not safe for drinking, personal or culinary utilization.

❖ Myriad substances can make water unfit for consumption, washing or bathing. Water not meeting the parameters for quality as defined in the EPA Clean Drinking Water Act is considered to be nonpotable.The code has numerous requirements for preventing the cross connection of nonpotable water to the potable water supply (see commentary, Section 608).

NUISANCE. Public nuisance as known in common law or in equity jurisprudence; whatever is dangerous to human life or detrimental to health; whatever structure or premises is not sufficiently ventilated, sewered, drained, cleaned or lighted, with respect to its intended *occupancy*; and whatever renders the air, or human food, drink or water supply unwholesome.

❖ The meaning of the term "nuisance" can be interpretive. Although there are certainly situations wherein a health hazard is clearly present, other conditions that represent an unreasonable interference with "the enjoyment of life and property" require a subjective decision on the part of the administrative authority. An indirect waste discharge onto a floor adjacent to a floor receptor could be a slipping hazard and therefore judged to be a nuisance. Additionally, any insanitary condition caused by a plumbing installation or lack of a plumbing installation could be considered a nuisance.

OCCUPANCY. The purpose for which a building or portion thereof is utilized or occupied.

❖ The occupancy classification of a building is an indication of the level of hazard to which occupants are exposed as a function of the building's use. Occupancy in terms of an occupancy group classification is one of the primary considerations in the development and application of many code requirements that are designed to offset the hazards specific to each occupancy group designation.

OFFSET. A combination of *approved* bends that makes two changes in direction bringing one section of the pipe out of line but into a line parallel with the other section.

❖ Offsets are necessary to route piping around or through structural elements such as beams, joists, trusses and columns. An offset always involves at least two changes in direction to keep the piping heading in the same direction. Any other arrangement would be considered a bend or a simple change in direction.

OPEN AIR. Outside the structure.

❖ This term clarifies the intent of the code with regard to the location of vent terminals. "Open air" refers to the atmosphere that is exterior to the envelope or outside of a building.

PLUMBING. The practice, materials and fixtures utilized in the installation, maintenance, extension and alteration of all piping, fixtures, plumbing appliances and plumbing appurtenances, within or adjacent to any structure, in connection with sanitary drainage or storm drainage facilities; venting systems; and public or private water supply systems.

❖ The term refers to the collective system of piping, fixtures, fittings, components or devices and appurtenances that transport potable water and liquid and solid wastes associated with cleaning, washing, bathing, food preparation, drinking and the elimination of bodily wastes.

The word "plumbing" comes from the Latin word for the element lead (*plumbum*), because lead was used extensively in the construction of piping systems. The practice of installing piping systems and the materials used in such systems became known as plumbing.

Piping, fittings, devices, faucets, vessels, containers and receptacles that are used to supply, distribute, receive or transport potable water, liquid wastes or solid wastes suspended in water are considered as plumbing. Plumbing also includes appliances, equipment and systems such as water conditioners, water heaters, storm drainage systems, water coolers, water filters and waste treatment systems.

PLUMBING APPLIANCE. Any one of a special class of plumbing fixtures intended to perform a special function. Included are fixtures having the operation or control dependent on one or more energized components, such as motors, controls, heating elements, or pressure- or temperature-sensing elements.

Such fixtures are manually adjusted or controlled by the owner or operator, or are operated automatically through one or more of the following actions: a time cycle, a temperature range, a pressure range, a measured volume or weight.

❖ Examples of plumbing appliances include water heaters, hot water dispensers, garbage disposals, dishwashers, clothes washers, water purifiers and water softeners.

PLUMBING APPURTENANCE. A manufactured device, prefabricated assembly or an on-the-job assembly of component parts that is an adjunct to the basic piping system and plumbing fixtures. An appurtenance demands no additional water supply and does not add any discharge load to a fixture or to the drainage system.

❖ Examples of plumbing appurtenances include water closet seats, hand-held showers, manifolds, backflow preventers, water-hammer arrestors, strainers and filters.

PLUMBING FIXTURE. A receptacle or device that is either permanently or temporarily connected to the water distribution system of the premises and demands a supply of water therefrom; discharges wastewater, liquid-borne waste materials or sewage either directly or indirectly to the drainage system of the premises; or requires both a water supply connection and a discharge to the drainage system of the premises.

❖ Plumbing fixtures include water closets, urinals, bidets, lavatories, sinks, showers, bathtubs, floor drains and drinking fountains.

A separate class of plumbing fixtures defined as plumbing appliances includes clothes washing machines, dishwashers, water heaters, water softeners, hot water dispensers, garbage disposals, water purifiers and water coolers.

PLUMBING SYSTEM. Includes the water supply and distribution pipes; plumbing fixtures and traps; water-treating or water-using equipment; soil, waste and vent pipes; and sanitary and storm sewers and building drains; in addition to their respective connections, devices and appurtenances within a structure or premises.

❖ The plumbing system includes all piping, fixtures and components that transport potable water and convey liquid and liquid-borne solid wastes [see Commentary Figure 202(16) and the definition of "Plumbing"]. Also included would be devices that treat the water prior to its use.

POLLUTION. An impairment of the quality of the potable water to a degree that does not create a hazard to the public health but that does adversely and unreasonably affect the aesthetic qualities of such potable water for domestic use.

❖ Many books have been written trying to define the term "pollution." For use in the code, pollution is anything that reduces the quality of the potable water supply so that it is undesirable for consumption or use. A pollutant in the water supply causes the water to look, smell or taste bad, but does not make drinking the water harmful. This does not change the fact that polluted water is still considered to be nonpotable. Pollutants are thus not considered potentially harmful, whereas a contaminant would be considered potentially harmful. Refer to the definition of the term "Contamination" to find additional information and the differences between pollution and contamination.

The methods for determining potable and nonpotable water supplies are very closely regulated by the EPA. For the purposes of protecting a water supply, the code considers a polluted source to represent a low hazard when selecting a backflow preventer. Refer to Table 608.1 for potable protection requirements related to low-hazard situations.

It should be noted that most forms of pollution are the direct result of some action by humans on their environment and thus, humans are responsible for preventing this pollution.

POTABLE WATER. Water free from impurities present in amounts sufficient to cause disease or harmful physiological effects and conforming to the bacteriological and chemical quality requirements of the Public Health Service Drinking Water Standards or the regulations of the public health authority having jurisdiction.

❖ The EPA Clean Drinking Water Act defines the quality requirements for water to be classified as potable. The word "potable" means fit to drink.

PRIVATE. In the classification of plumbing fixtures, "*private*" applies to fixtures in residences and apartments, and to fixtures in nonpublic toilet rooms of hotels and motels and similar installations in buildings where the plumbing fixtures are intended for utilization by a family or an individual.

❖ Because there are code requirements specific to public plumbing fixtures and facilities, the distinction must be made between public and private.

PUBLIC OR PUBLIC UTILIZATION. In the classification of plumbing fixtures, "*public*" applies to fixtures in general toilet rooms of schools, gymnasiums, hotels, airports, bus and railroad stations, public buildings, bars, public comfort stations, office buildings, stadiums, stores, restaurants and other installations where a number of fixtures are installed so that their utilization is similarly unrestricted.

❖ Public or public utilization is a descriptor for toilet facilities that are available to the general public. Employee plumbing fixtures are considered to be public fixtures unless their use is restricted to an individual (see the definition of "Private").

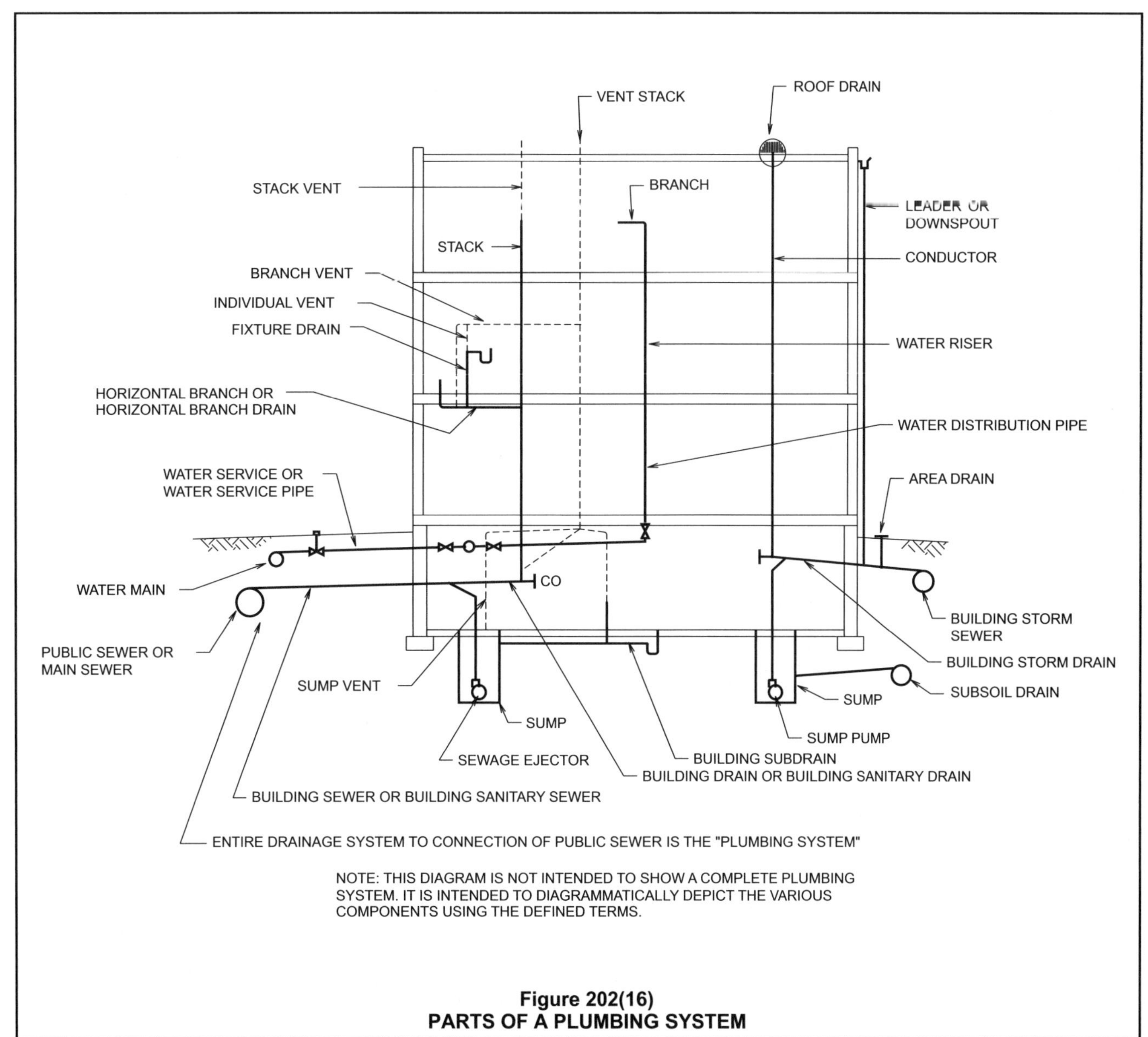

Figure 202(16)
PARTS OF A PLUMBING SYSTEM

PUBLIC WATER MAIN. A water supply pipe for public utilization controlled by public authority.

❖ See the definition of and commentary to "Water main."

QUICK-CLOSING VALVE. A valve or faucet that closes automatically when released manually or that is controlled by a mechanical means for fast-action closing.

❖ A quick-closing valve would be any type of solenoid-actuated valve; spring-loaded, self-closing faucets or any other device capable of instantaneously reducing water flow from full flow to no flow. This rapid change from a full-flow to no-flow condition can cause water-hammer-related damage. Although most faucets can be manually closed fast enough to produce water hammer, it is not the intent of the code to consider manually closed valves or faucets as "quick closing."

Typical quick-closing valves include spring-loaded faucets and drinking fountain control valves that abruptly close when manually released; electrically actuated valves such as those found in dishwashing machines, clothes washing machines, boiler makeup water feeders, self-washing air cleaners or grease extractors; and process equipment and electronically controlled "touchless" faucets. Some self-closing valves and faucets are not considered as quick closing. For example, typical metering faucets and typical flushometer valves are slowly, automatically closed by a diaphragm/bleed orifice or pilot valve assembly. Additionally, the typical float-actuated valve is automatic, yet slow in closing.

READY ACCESS. That which enables a fixture, appliance or equipment to be directly reached without requiring the removal or movement of any panel, door or similar obstruction and without the use of a portable ladder, step stool or similar device.

❖ Ready access can be described as the capability of being quickly reached or approached for the purpose of operation, inspection, observation or emergency action. Plumbing installations requiring ready access must not be covered or concealed by a door, panel or similar obstruction, nor are they permitted to be obstructed by physical obstacles, including elevation.

REDUCED PRESSURE PRINCIPLE BACKFLOW PREVENTER. A backflow prevention device consisting of two independently acting check valves, internally force-loaded to a normally closed position and separated by an intermediate chamber (or zone) in which there is an automatic relief means of venting to the atmosphere, internally loaded to a normally open position between two tightly closing shutoff valves and with a means for testing for tightness of the checks and opening of the relief means.

❖ This device is considered to be one of the most reliable mechanical devices (second only to an air gap) for the prevention of backflow. These devices use two spring-loaded check valves with a relief valve in between that monitors the system pressure upstream and downstream of the device. The zone between the check valves is maintained at a pressure that is less than the water supply pressure (see commentary, Section 608.13.2).

REGISTERED DESIGN PROFESSIONAL. An individual who is registered or licensed to practice professional architecture or engineering as defined by the statutory requirements of the professional registration laws of the state or jurisdiction in which the project is to be constructed.

❖ Legal qualifications for engineers and architects are established by the state having jurisdiction. Licensing and registration of engineers and architects is accomplished by written or oral examinations offered by states or by reciprocity (licensing in other states).

RELIEF VALVE

❖ In general, a relief valve is a safety device designed to prevent the development of a potentially dangerous condition in a closed piping system. This is normally associated with a closed system that has a heat source (see commentary, Section 504.4).

Pressure relief valve. A pressure-actuated valve held closed by a spring or other means and designed to relieve pressure automatically at the pressure at which such valve is set.

❖ This valve is designed to be actuated by excessive pressures to prevent rupture or explosion of tanks, vessels or piping. The degree of valve opening is directly proportional to the pressure acting on the valve disk.

Temperature and pressure relief (T&P) valve. A combination relief valve designed to function as both a temperature relief and a pressure relief valve.

❖ These devices combine the components of pressure and temperature relief valves and are designed to be actuated by excessive pressures, excessive temperatures or both.

Temperature relief valve. A temperature-actuated valve designed to discharge automatically at the temperature at which such valve is set.

❖ This valve is designed to be actuated by excessive temperatures to prevent dangerously high water temperatures that can cause tank, vessel or pipe explosions.

RELIEF VENT. A vent whose primary function is to provide circulation of air between drainage and vent systems.

❖ A relief vent is required for some circuit venting configurations and for stacks of more than 10 branch intervals. The relief vent acts to relieve pressures that develop in stacks as a result of the flow of waste (see commentary, Sections 911.4, 914 and 916.2).

RIM. An unobstructed open edge of a fixture.

❖ The rim of a fixture or receptor is the reference point for the measurement of installation dimensions, such as for critical levels, air gaps, air breaks and horizontal vent piping (see the definition of "Flood level rim").

RISER. See "Water pipe, riser."

❖ See the commentary to "Water pipe, riser."

ROOF DRAIN. A drain installed to receive water collecting on the surface of a roof and to discharge such water into a leader or a conductor.

❖ Roof drains act as a receptor for storm water flowing across the surface of a building roof. Roof drains are equipped with inlet strainers that are designed to prevent the entry of leaves, debris and roof ballast materials.

ROUGH-IN. Parts of the plumbing system that are installed prior to the installation of fixtures. This includes drainage, water supply, vent piping and the necessary fixture supports and any fixtures that are built into the structure.

❖ Rough-in refers to the stage of construction prior to the installation of any materials that would conceal plumbing installations in the building structure. This term generally refers to the installation of drain, waste, vent and water distribution piping that will be concealed in floor, ceiling and wall cavities or underground or under slab. At this stage of construction, the plumbing systems are tested for leaks and the entire plumbing "rough-in" is inspected for code compliance.

SELF-CLOSING FAUCET. A faucet containing a valve that automatically closes upon deactivation of the opening means.

❖ Use of self-closing faucets is increasing, primarily in public restroom facilities, as public awareness of water conservation efforts increases. The flow of water automatically shuts off when the faucet handle is released, resulting in less water consumption per use.

SEPARATOR. See "Interceptor."

❖ The term is used interchangeably with "interceptor" (see the definition of "Interceptor").

SEWAGE. Any liquid waste containing animal or vegetable matter in suspension or solution, including liquids containing chemicals in solution.

❖ Sewage is the general term referring to the discharge from all plumbing fixtures and primarily includes human bodily wastes and the wastes associated with cleaning, washing, bathing and food preparation.

SEWAGE EJECTORS. A device for lifting sewage by entraining the sewage in a high-velocity jet of steam, air or water.

❖ Ejectors are used to lift sewage from sumps serving subdrains. Ejectors are distinct from sewage pumps that use motor-driven impellers to lift the sewage [see Commentary Figure 712.1(1)].

SEWER

❖ This is the general term used to describe piping that conducts waste from structures to the point of disposal. It also refers to piping within a public system that transports sewage to treatment facilities.

Building sewer. See "Building sewer."

❖ See the commentary to "Building sewer."

Public sewer. A common *sewer* directly controlled by public authority.

❖ A public sewer is the piping system used to collect sewage from building sewers and to transport it to treatment facilities (see commentary, Section 701.2).

Sanitary sewer. A *sewer* that carries sewage and excludes storm, surface and ground water.

❖ See the definition of "Building sewer, sanitary."

Storm sewer. A *sewer* that conveys rainwater, surface water, subsurface water and similar liquid wastes.

❖ This is the general term used to describe the piping that conducts waste other than sewage from structures to the point of disposal.

SLOPE. The fall (pitch) of a line of pipe in reference to a horizontal plane. In drainage, the slope is expressed as the fall in units vertical per units horizontal (percent) for a length of pipe.

❖ Commonly referred to as pitch, fall or grade, slope is the gradual change in elevation of horizontal piping. Slope is necessary to cause waste and sewage to flow by gravity (see commentary, Section 704.1).

SOIL PIPE. A pipe that conveys sewage containing fecal matter to the *building drain* or *building sewer*.

❖ In the plumbing trade, soil pipe is a common name for cast-iron drainage pipe. The term, however, is not specific to cast-iron pipe and refers to any drain pipe that conveys the discharge of water closets or urinals or any other fixture that receives human waste.

SPILLPROOF VACUUM BREAKER. An assembly consisting of one check valve force-loaded closed and an air-inlet vent valve force-loaded open to atmosphere, positioned downstream of the check valve, and located between and including two tightly closing shutoff valves and a test cock.

❖ Spillproof vacuum breakers are designed primarily for indoor use where leakage or spillage caused by a loss of system pressure could cause problems such as water damage or health risks, but the performance requirements of a pressure-type vacuum breaker are still necessary.

STACK. A general term for any vertical line of soil, waste, vent or inside conductor piping that extends through at least one story with or without offsets.

❖ Traditionally, the term "stack" has been used to describe vertical piping [90 degrees (1.57 rad) from the horizontal] that extends through one or more stories of a building and constitutes a main to which branch piping connects. Vertical vent branch piping that extends to the open air or through two or more stories is considered a stack.

STACK VENT. The extension of a soil or waste *stack* above the highest horizontal drain connected to the *stack*.

❖ A stack vent is the dry extension of a soil or waste stack. Generally, a stack vent extends to the open air and can serve as a vent to which branch vents connect.

STACK VENTING. A method of venting a fixture or fixtures through the soil or waste *stack*.

❖ This is a method of wet venting where a uniformly oversized vertical drain serves as the vent for one or more fixtures.The fixtures being vented are independently connected to the stack. This venting method is limited to use with waste fixtures.

STERILIZER

❖ A sterilizer is a plumbing appliance that uses heat to disinfect instruments, utensils and other equipment used in health care facilities.

Boiling type. A boiling-type sterilizer is a fixture of a nonpressure type utilized for boiling instruments, utensils or other equipment for disinfection. These devices are portable or are connected to the plumbing system.

❖ The boiling of water is the oldest and most widely used form of microbial destruction. However, temperatures of water boiling at atmospheric pressure might not be high enough [212°F (100°C) or less] to kill all forms of bacteria. Thus, these types of sterilizers are not used widely where serious bacteria elimination is required.

Instrument. A device for the sterilization of various instruments.

❖ Usually this refers to a small benchtop pressure-type, self-contained sterilizer used to disinfect hand instruments used in dental or minor surgery offices.

Pressure (autoclave). A pressure vessel fixture designed to utilize steam under pressure for sterilizing.

❖ Pressure sterilizers (autoclave) achieve temperatures well above 212°F (100°C) for complete bacteria elimination. The size of these devices can range from bench-top size to room-size units. The basic principle of operation is similar to that of a household pressure cooker. Water in a closed vessel having an air space above the water is heated. Steam generated from the water rises into the air space creating a pressure in the vessel. The increased pressure in the vessel raises the boiling point of the water. Thus, a steam temperature of greater than 212°F (100°C) is generated. For example, water heated in a closed vessel under a pressure of 10 psi (69.9 kPa) above atmospheric pressure will generate steam at a temperature of 240°F (115.6°C).

Pressure instrument washer sterilizer. A pressure instrument washer sterilizer is a pressure vessel fixture designed to both wash and sterilize instruments during the operating cycle of the fixture.

❖ This device is a combination washer and a pressure-type sterilizer, all in one cabinet or unit (similar to a dishwasher). It removes debris from instruments and specialized surgical equipment (e.g., endoscopes).

Utensil. A device for the sterilization of utensils as utilized in health care services.

❖ This type of sterilizer is used for devices that are larger than small handheld instruments and includes trays, pans, buckets, saws and drills.

Water. A water sterilizer is a device for sterilizing water and storing sterile water.

❖ Sterilized water is required for many dental, medical and scientific procedures to prevent bacterial contamination. Water sterilizers are generally of the distillation-type that condenses wet steam to produce sterile water.

STERILIZER VENT. A separate pipe or *stack*, indirectly connected to the building drainage system at the lower terminal, that receives the vapors from nonpressure sterilizers, or the exhaust vapors from pressure sterilizers, and conducts the vapors directly to the open air. Also called vapor, steam, atmospheric or exhaust vent.

❖ Sterilizer vents are independent of the sanitary venting system and serve only sterilizer appliances. Furthermore, sterilizer vents cannot serve more than one type of sterilizer (see commentary, Section 713.8).

STORM DRAIN. See "Drainage system, storm."

❖ See the commentary for "Drainage system, storm."

STRUCTURE. That which is built or constructed or a portion thereof.

❖ This definition is intentionally broad to include within its scope, and therefore the scope of the code (see commentary, Section 101.2), everything that is built as an improvement to real property. The phrase "or a portion thereof" is included so that those words do not have to be inserted at each location in the code where a provision applies to a portion of a structure.

SUBSOIL DRAIN. A drain that collects subsurface water or seepage water and conveys such water to a place of disposal.

❖ These drains are generally installed adjacent to the foundation of buildings to alleviate problems caused by subsurface (ground) water. Additionally, any other piping that collects either subsurface water or seepage would be termed a "subsoil drain."

SUMP. A tank or pit that receives sewage or liquid waste, located below the normal grade of the gravity system and that must be emptied by mechanical means.

❖ This term refers to the receiving tank or vessel used to collect and store waste from drainage systems that are incapable of draining by gravity. A sump generally houses an ejector or pump used to evacuate the contents of the sump. A sump can refer to a receiver of either waste water or storm water.

SUMP PUMP. An automatic water pump powered by an electric motor for the removal of drainage, except raw sewage, from a sump, pit or low point.

❖ These pumps are used for removing waste from sumps other than sewage sumps. Sump pumps handle storm water only.

SUMP VENT. A vent from pneumatic sewage ejectors, or similar equipment, that terminates separately to the open air.

❖ A sump vent is needed to neutralize pressures that develop in closed sumps as a result of either the rise and fall of waste levels or the operation of pneumatic ejectors (see commentary, Sections 916.5, 916.5.1 and 916.5.2).

SUPPORTS. Devices for supporting and securing pipe, fixtures and equipment.

❖ Types of supports include hangers, anchors, braces, brackets, strapping and any other material or device used to support plumbing system piping and components (see commentary, Section 308).

SWIMMING POOL. Any structure, basin, chamber or tank containing an artificial body of water for swimming, diving or recreational bathing having a depth of 2 feet (610 mm) or more at any point.

❖ Swimming pools, for the application of the code, are man-made structures that are at least 2 feet (610 mm) deep and are designed to hold water. Bodies of water such as ponds, streams, lakes or retention ponds are not considered to be swimming pools.

Swimming pools are generally filled with potable water. The water is then filtered to remove contaminants and debris and is chemically treated to control bacterial growth, plant life growth and acidity/alkalinity levels. Swimming pools are primarily used for recreational purposes and include private and public swimming pools, indoor and outdoor swimming pools, hot tubs and spas. See related sections in the applicable building or swimming pool code.

Bathtubs and therapeutic whirlpools used for hygienic or medical purposes are not considered to be swimming pools. The water discharged from a swimming pool is considered to be clear-water waste.

TEMPERED WATER. Water having a temperature range between 85°F (29°C) and 110°F (43°C).

❖ Tempered water (warm water) is at a temperature suitable for bathing, handwashing and similar purposes and is defined as a range between 85°F (29°C) and 110°F (43°C) (see commentary, Section 607.1).

THIRD-PARTY CERTIFICATION AGENCY. An *approved* agency operating a product or material certification system that incorporates initial product testing, assessment and surveillance of a manufacturer's quality control system.

❖ These agencies evaluate products, verify compliance with specific standards or criteria and conduct periodic inspections of the manufacturer's facilities to determine ongoing compliance with manufacturing processes, standards or criteria (see Section 303.4).

THIRD-PARTY CERTIFIED. Certification obtained by the manufacturer indicating that the function and performance characteristics of a product or material have been determined by testing and ongoing surveillance by an *approved third-party certification agency*. Assertion of certification is in the form of identification in accordance with the requirements of the *third-party certification agency*.

❖ A product that is certified by a third-party certification agency (see commentary, "Third-party certification agency") to comply with requisite standards or criteria, usually identified by a certification mark attached to or permanently marked on it. Additionally, the agency usually provides a list of the products it certifies, which provides information relative to the certification (see Section 303.4).

THIRD-PARTY TESTED. Procedure by which an *approved* testing laboratory provides documentation that a product, material or system conforms to specified requirements.

❖ A product that has been tested by an approved agency (see the definition of and commentary for "Approved agency") to determine compliance with specific product standards or criteria (see Section 303.4).

TRAP. A fitting or device that provides a liquid seal to prevent the emission of *sewer* gases without materially affecting the flow of sewage or wastewater through the trap.

❖ The sole purpose of a trap is to isolate the interior of occupiable spaces from the sanitary drainage and vent system. A trap "traps" liquid waste and retains it to form a seal or barrier through which gases and vapors in the drainage and vent system cannot pass under normal operating conditions. Traps are either separate devices or integral with fixtures. Other than a grease trap or trap that is integral with a fixture, the only type of trap permitted by the code is the "P" type (see commentary, Sections 1002.1, 1002.3, 1002.4 and 1002.5).

TRAP SEAL. The vertical distance between the weir and the top of the dip of the trap.

❖ The trap seal is a measurement of the quantity of liquid held in the dip of a trap. Trap seals are required to have a minimum depth (see commentary, Section 1002.4).

UNSTABLE GROUND. Earth that does not provide a uniform bearing for the barrel of the *sewer* pipe between the joints at the bottom of the pipe trench.

❖ Any earth that will be used to support underground sewer piping that is not capable of providing adequate, uniform bearing so that the sewer piping and its related fittings are not damaged is defined as unstable ground. This term can relate to either earth that exists as is or any earth disturbed by construction that is not properly replaced and compacted.

Typically, when unstable ground is encountered during the course of construction, the portion of the ground that is unstable is either removed and replaced, or is made stable through various engineering methods, such as dewatering or chemical injections.

VACUUM. Any pressure less than that exerted by the atmosphere.

❖ Commonly referred to as "negative pressure" and "suction," vacuum is measured in inches of mercury or inches of water column. Any pressure below atmospheric pressure is a partial vacuum. A perfect vacuum is a pressure of zero pounds per square inch (psi) (0 kPa) absolute. Atmospheric pressure at sea level is approximately 14.7 psi (101 kPa), which converts to approximately 29.92 inches (760 mm) of mercury or approximately 34 feet (10 363 mm) of water. The phenomenon of siphonage depends on vacuum (negative pressures).

VACUUM BREAKER. A type of backflow preventer installed on openings subject to normal atmospheric pressure that prevents backflow by admitting atmospheric pressure through ports to the discharge side of the device.

❖ A vacuum breaker is a device used as a backflow preventer. Whenever this device senses a system pressure that is less than atmospheric pressure (e.g., a partial vacuum), a port opens to allow atmospheric pressure to enter the device to relieve (break) the vacuum condition in the piping.

VENT PIPE. See "Vent system."

❖ The term "vent pipe" refers to any pipe in the vent system that conducts air to the drainage system, from the drainage system or both.

VENT STACK. A vertical vent pipe installed primarily for the purpose of providing circulation of air to and from any part of the drainage system.

❖ A dry vent that extends from the base of a waste or soil stack and typically runs parallel and adjacent to the waste or soil stack it serves. The vent stack can connect either to the horizontal drain downstream of the drain stack or directly to the drainage stack (see commentary, Section 903.2).

VENT SYSTEM. A pipe or pipes installed to provide a flow of air to or from a drainage system, or to provide a circulation of air within such system to protect trap seals from siphonage and backpressure.

❖ A piping arrangement that is designed to maintain pressure fluctuation at fixture traps to within plus or minus 1 inch of water column (249 Pa) (see commentary, Section 901.2).

VERTICAL PIPE. Any pipe or fitting that makes an angle of 45 degrees (0.79 rad) or more with the horizontal.

❖ A vertical pipe is any pipe that is not defined as a horizontal pipe (see the definition of "Horizontal pipe"). Vertical pipes are pipes that are angled 45 degrees (0.79 rad) or more above the horizontal.

WALL-HUNG WATER CLOSET. A wall-mounted water closet installed in such a way that the fixture does not touch the floor.

❖ A wall-hung water closet is designed for installation above the floor. The water closet discharges horizontally through the wall (see Commentary Figure 405.4.3).

WASTE. The discharge from any fixture, appliance, area or appurtenance that does not contain fecal matter.

❖ Any plumbing fixture discharge other than that from a water closet, urinal, clinical sink or bedpan washer is defined as waste. This material does not contain human fecal matter.

WASTE PIPE. A pipe that conveys only waste.

❖ Waste pipe is any waste-conducting pipe not falling under the definition of soil pipe. Waste pipes do not convey human fecal matter.

WATER-HAMMER ARRESTOR. A device utilized to absorb the pressure surge (water hammer) that occurs when water flow is suddenly stopped in a water supply system.

❖ Water-hammer arrestors are factory-built mechanical devices that use expandable elements or piston-type compression chambers to dissipate the energy of rapidly decelerating water (see commentary, Section 604.9). Site-built air chambers do not qualify as acceptable water-hammer arresters.

WATER HEATER. Any heating appliance or equipment that heats potable water and supplies such water to the potable *hot water* distribution system.

❖ The types of water heaters include storage (tank) type, circulating type and instantaneous type. Point-of-use water heaters typically supply hot water to a single fixture or outlet and are located in close proximity to it.

WATER MAIN. A water supply pipe or system of pipes, installed and maintained by a city, township, county, public utility company or other public entity, on public property, in the street or in an *approved* dedicated easement of public or community use.

❖ A water main is owned and operated by municipalities, rural water districts, privately owned water purveyors and other such entities.

WATER OUTLET. A discharge opening through which water is supplied to a fixture, into the atmosphere (except into an open tank that is part of the water supply system), to a boiler or heating system, or to any devices or equipment requiring water to operate but which are not part of the plumbing system.

❖ Literally, a water outlet is any point where water is withdrawn from the potable water distribution system. An outlet, such as a faucet or spigot, discharges water through the atmosphere into a fixture or receptacle. Outlets are not direct connections to other systems except into an open tank that is part of the water supply system (see commentary, Section 608.15).

WATER PIPE

Riser. A water supply pipe that extends one full story or more to convey water to branches or to a group of fixtures.

❖ A riser extends vertically to supply water to upper stories.

Water distribution pipe. A pipe within the structure or on the premises that conveys water from the water service pipe, or from the meter when the meter is at the structure, to the points of utilization.

❖ Water distribution pipe refers to all potable water piping that is downstream of the water meter or termination of the water service.

Water service pipe. The pipe from the water main or other source of potable water supply, or from the meter when the meter is at the public right of way, to the water distribution system of the building served.

❖ Water pipe is any code-accepted pipe material used to convey potable water.

WATER SUPPLY SYSTEM. The water service pipe, water distribution pipes, and the necessary connecting pipes, fittings, control valves and all appurtenances in or adjacent to the structure or premises.

❖ The water supply system includes all piping and components that convey potable water from the public main or other source to the points of water usage.

WELL

❖ A well is a vertical shaft that extends down into the earth for the purpose of extracting water from underground formations or strata. Wells can be constructed by drilling, boring or digging in the earth or created by driving a pipe (casing) into the earth. Wells are the source of water for many public water supply systems and are also used to serve private (individual) water supply systems. Depending on the location, some water from wells may be potable without requiring any treatment, whereas some must be treated to make it potable.The following four methods are the most common or typical forms of well construction and are self-defining.

Bored. A well constructed by boring a hole in the ground with an auger and installing a casing.

❖ See the commentary for "Well."

Drilled. A well constructed by making a hole in the ground with a drilling machine of any type and installing casing and screen.

❖ See the commentary for "Well."

Driven. A well constructed by driving a pipe in the ground. The drive pipe is usually fitted with a well point and screen.

❖ See the commentary for "Well."

Dug. A well constructed by excavating a large-diameter shaft and installing a casing.

❖ See the commentary for "Well."

WHIRLPOOL BATHTUB. A plumbing appliance consisting of a bathtub fixture that is equipped and fitted with a circulating piping system designed to accept, circulate and discharge bathtub water upon each use.

❖ Whirlpool bathtubs are equipped with circulator pumps, suction fittings, piping and related appurtenances designed to agitate water in the tub for relaxation or therapeutic purposes (see commentary, Section 421). In recent years, consumers have embraced multihead luxury showers which can discharge large volumes of water during a showering event. There is an direct cost of the water and sewer charges and a indirect cost of high-capacity water heating systems and larger piping for high flows. Those facts, coupled with not being a conservative water practice, have led some manufacturers to also offer "whirlpool tub/showers." These multipiece units utilize a bathtub with a vertical panel having multiple spray heads and jets. The bathtub serves as the water reservoir for a whirlpool type pump to convey water to the shower heads and jets in a closed system arrangement. An in-line heater maintains the desired water temperature. Once the tub is filled with a sufficient volume of water, the shower can be used as long as desired without further introduction of water resulting in much less water use than a shower where all the water discharged goes to the drain system.

YOKE VENT. A pipe connecting upward from a soil or waste *stack* to a vent *stack* for the purpose of preventing pressure changes in the stacks.

❖ This vent derives its name from the manner in which it connects a drainage stack to the attendant vent stack. Its purpose is to prevent pressure changes in the stack.

Bibliography

The following resource materials are referenced in this chapter or are relevant to the subject matter addressed in this chapter.

ANSI Z21.22-99 (R2003), *Relief Valves for Hot Water Supply Systems*. New York, NY: American National Standards Institute, 1999.

ASME A112.14.3-00, Grease Interceptors. New York, NY: American Society of Mechanical Engineers, 2000.

ASPE-98, *Data Book Volume 1*. Westlake Village, CA: American Society of Plumbing Engineers, 1998.

ASSE 1013-99, *Performance Requirements for Reduced Pressure Principle Backflow Preventers and Reduced Pressure Fire Protection Principle Backflow Preventers*. Westlake, OH: American Society of Sanitary Engineering, 1999.

ASTM B 32-03, *Specification for Solder, Metal*. West Conshohocken, PA: American Society for Testing and Materials, 2003.

IECC-09, *International Energy Conservation Code*. Washington, DC: International Code Council, Inc., 2009.

IFGC-09, *International Fuel Gas Code*. Washington, DC: International Code Council, Inc., 2009.

IMC-09, *International Mechanical Code*. Washington, DC: International Code Council, Inc., 2009.

IPSDC-09, *International Private Sewage Disposal Code*. Washington, DC: International Code Council, Inc., 2009.

IPMC-09, *International Property Maintenance Code*. Washington, DC: International Code Council, Inc., 2009.

IZC-09, *International Zoning Code*. Washington, DC: International Code Council, Inc., 2009.

NFPA 99C-02, *Gas and Vacuum Systems*. Quincy, MA: National Fire Protection Association, 2002.

UL 174-04, UL *Standard for Safety Household Electric Storage Tank Water Heaters*. Northbrook, IL: Underwriters Laboratories, Inc., 2004.

Chapter 3: General Regulations

General Comments

The content of Chapter 3 is often referred to as "miscellaneous," rather than general regulations. Chapter 3 received that label because it is the only chapter in the code whose requirements do not interrelate. If a requirement cannot be located in another chapter, it should be located in this chapter.

Some nonplumbing regulations merely reference other codes that have the specific requirements. The requirements provide a cross reference to the appropriate document, recognizing that it affects the plumbing system but the details are not specifically contained in the code [Sections 307, 309, 310 and 313 reference other *International Codes*® (I-Codes®)].

The jurisdictional requirements specify that the water and sewer must connect to the public system when a public system is provided (Sections 602.1 and 701.2 are more specific on this issue).

Nonplumbing requirements include surface requirements for walls and floors in a toilet room, light and ventilation, floodproofing and rodentproofing.

Purpose

Chapter 3 contains safety requirements for the installation of plumbing and nonplumbing requirements for toilet rooms, including requirements for the identification of pipe, pipe fittings, traps, fixtures, materials and devices used in plumbing systems.

The safety requirements provide protection for the building's structural members, stress and strain of pipe and sleeving. The building's structural stability is protected by the regulations for cutting and notching of structural members. Additional protection for the building occupants includes requirements to maintain the plumbing in a safe and sanitary condition, as well as privacy for those occupants.

SECTION 301 GENERAL

301.1 Scope. The provisions of this chapter shall govern the general regulations regarding the installation of plumbing not specific to other chapters.

❖ The requirements included in Chapter 3 are not interrelated, as is typical with other chapters. Many regulations are not specific plumbing requirements, but relate to the overall plumbing system.

301.2 System installation. Plumbing shall be installed with due regard to preservation of the strength of structural members and prevention of damage to walls and other surfaces through fixture usage.

❖ Plumbing components and materials are to be installed in accordance with the installation requirements of the applicable standard listed in the code.

Where a standard is not provided, the manufacturer's instructions must be followed. For example, because there are very few standards available that regulate the installation of valves, the manufacturer's instructions must be used to install these components.

301.3 Connections to the sanitary drainage system. All plumbing fixtures, drains, appurtenances and appliances used to receive or discharge liquid wastes or sewage shall be directly connected to the sanitary drainage system of the building or premises, in accordance with the requirements of this code. This section shall not be construed to prevent the indirect waste systems required by Chapter 8.

❖ Plumbing fixtures, drains, appurtenances and appliances that either receive or discharge liquid waste or sewage are to be directly connected to the sanitary drainage system unless required by Chapter 8 to be indirectly connected. Typically, food-handling equipment, clear water waste and health care-related fixtures and equipment must discharge to the drainage system through an indirect waste pipe by means of an air gap or air break. Fixtures not required by Chapter 8 to be indirectly connected must be directly connected to the plumbing system in accordance with Chapter 7 (see Section 802).

301.4 Connections to water supply. Every plumbing fixture, device or appliance requiring or using water for its proper operation shall be directly or indirectly connected to the water supply system in accordance with the provisions of this code.

❖ Fixtures that supply water or recycled gray water (see Appendix C) for the occupant's use are required to have either a direct or indirect connection to the potable water supply system. Chapter 6 contains specific requirements governing connections to water supply and distribution systems.

Indirect connections include faucets or fixture fittings discharging into fixtures such as tubs and lavatories. Direct connections occur at water closets and urinals. Water closets and urinals can be supplied with gray flushing water through a direct connection to the fixture if the code official approves the use of a gray water recycling system as an alternate and the system meets the requirements of Appendix C.

301.5 Pipe, tube and fitting sizes. Unless otherwise specified, the pipe, tube and fitting sizes specified in this code are expressed in nominal or standard sizes as designated in the referenced material standards.

❖ Pipe, tube and fitting sizes referenced in the code refer to the inside diameter (ID) of the pipe, tube or fitting. The ID measurement in the text is expressed in both English and metric units (inches and millimeters). Systeme International d'Unites (SI) metric unit conversions are indicated at the bottom of each table.

301.6 Prohibited locations. Plumbing systems shall not be located in an elevator shaft or in an elevator equipment room.

Exception: Floor drains, sumps and sump pumps shall be permitted at the base of the shaft, provided that they are indirectly connected to the plumbing system and comply with Section 1003.4.

❖ Plumbing components within the elevator pit, such as floor drains, sumps and sump pumps that are indirectly connected to the building drainage system, are allowed within elevator shafts. An indirect connection is required to prevent waste or gases from backing up into the elevator shaft. Other plumbing systems are strictly prohibited in these spaces because of inaccessibility for repairs and the possible water damage that could be caused to the elevator equipment if a leak developed in the piping system. The exception references Section 1003.4 to alert the reader to the fact that if the elevator is a hydraulic type, an oil separator is required in the sump drain line before it discharges into the building drain or other point of disposal.

301.7 Conflicts. In instances where conflicts occur between this code and the manufacturer's installation instructions, the more restrictive provisions shall apply.

❖ A conflict refers to instances where the code and manufacturer's instructions differ. The code official must evaluate each circumstance of perceived conflict and identify the requirements that provide the greatest level of protection for life and property.

SECTION 302
EXCLUSION OF MATERIALS DETRIMENTAL TO THE SEWER SYSTEM

302.1 Detrimental or dangerous materials. Ashes, cinders or rags; flammable, poisonous or explosive liquids or gases; oil, grease or any other insoluble material capable of obstructing, damaging or overloading the building drainage or *sewer* system, or capable of interfering with the normal operation of the sewage treatment processes, shall not be deposited, by any means, into such systems.

❖ This section prohibits the disposal of detrimental or dangerous materials into the sewer system. Such materials can cause the pipes to clog or accelerate the clogging of pipes, which prevents the proper disposal of sewage waste. Section 1003 contains design and installation details for the use of interceptors, grease traps and separators to remove oil, grease, sand and other detrimental substances.

Discharge of materials that are flammable or combustible into the public sewer system is prohibited because an accumulation of these types of materials poses a fire and explosion hazard. Insoluble chemicals that are not processed before disposal could react with other discharged chemicals to cause damage to the piping and components of the drainage, sewer and waste treatment systems. Section 803.2 provides details for using approved dilution or neutralizing devices to process harmful chemicals prior to disposal.

302.2 Industrial wastes. Waste products from manufacturing or industrial operations shall not be introduced into the *public sewer* until it has been determined by the code official or other authority having jurisdiction that the introduction thereof will not damage the *public sewer* system or interfere with the functioning of the sewage treatment plant.

❖ Harmful or hazardous industrial waste must be treated before it is discharged to the sewer. This can require the complete removal or neutralization of certain chemicals or substances.

SECTION 303
MATERIALS

303.1 Identification. Each length of pipe and each pipe fitting, trap, fixture, material and device utilized in a plumbing system shall bear the identification of the manufacturer.

❖ The manufacturer is given the option of determining the type of marking for the material. If there is no applicable standard or the applicable standard does not require that a material be identified, identification of the manufacturer is still required by the code. Where the code indicates compliance with an approved standard, the manufacturer must comply with the requirements for marking in accordance with the applicable standard.

303.2 Installation of materials. All materials used shall be installed in strict accordance with the standards under which the materials are accepted and *approved*. In the absence of such installation procedures, the manufacturer's installation instructions shall be followed. Where the requirements of referenced standards or manufacturer's installation instructions do not conform to minimum provisions of this code, the provisions of this code shall apply.

❖ Plumbing components and materials are to be installed in accordance with the installation requirements of the applicable standard listed in the code. Where a standard is not provided, the manufacturer's instructions must be followed. For example, because there are very few standards available that regulate the installation of valves, the manufacturer's instructions must be used to install these components.

303.3 Plastic pipe, fittings and components. All plastic pipe, fittings and components shall be third-party certified as conforming to NSF 14.

❖ Plastic piping, fittings and plastic pipe-related components, including solvent cements, primers, tapes, lubricants and seals used in plumbing systems, must be tested and certified as conforming to NSF 14.This includes all water service, water distribution, drainage piping and fittings and plastic piping system components, including but not limited to pipes, fittings, valves, joining materials, gaskets and appurtenances. This section does not apply to components that only include plastic parts such as brass valves with a plastic stem, or to fixture fittings such as fixture stop valves. Plastic piping systems, fittings and related components intended for use in the potable water supply system must comply with NSF 61 in addition to NSF 14.

303.4 Third-party testing and certification. All plumbing products and materials shall comply with the referenced standards, specifications and performance criteria of this code and shall be identified in accordance with Section 303.1. When required by Table 303.4, plumbing products and materials shall either be tested by an *approved* third-party testing agency or certified by an *approved third-party certification agency*.

❖ This section requires that all plumbing products and materials comply with the referenced standards. However, the provisions contained in Section 105.2 regarding the evaluation and approval of alternative materials, methods and equipment are still applicable (see commentary, Section 105.2). Additionally, the code has been revised to include specific requirements for third-party certification and testing of plumbing products. Table 303.4 lists these requirements. "Third-party certified" indicates that the minimum level of quality required by the applicable standard is maintained and the product is often referred to as "listed." "Third-party tested" indicates a product that has been tested by an approved testing laboratory and found to be in compliance with the standard. Although the code does not specifically state the identification or marking requirements, except for the manufacturer's identification, the applicable referenced standard states the minimum information required. The identification or marking requirements typically include the name of the manufacturer, product name or serial number, installation specifications, applicable tests and standards, testing agency and labeling agency. Commentary Figure 303.4 shows an example of a typical label for PVC plastic pipe.

TABLE 303.4. See below.

❖ See commentary to Section 303.4.

SECTION 304
RODENTPROOFING

304.1 General. Plumbing systems shall be designed and installed in accordance with Sections 304.2 through 304.4 to prevent rodents from entering structures.

❖ Rodents are known to be carriers of diseases and their presence around human beings presents serious health risks to humans. To prevent the spread of disease, Sections 304.2 through 304.4 require plumbing systems to be installed in a manner that will reduce the potential for rodent entry into structures.

304.2 Strainer plates. All strainer plates on drain inlets shall be designed and installed so that all openings are not greater than $^1/_2$ inch (12.7 mm) in least dimension.

❖ Rodents often travel and live within sanitary sewer systems. The limitation for opening size in strainer plates for floor and shower drains as well as receptor strainers provides two forms of protection. If rodents are in the sewer system, strainer plates prevent them from entering the building through the floor or shower drain. If rodents are within the structure itself, the strainer plate prevents rodent access to the drainage system.

TABLE 303.4
PRODUCTS AND MATERIALS REQUIRING THIRD-PARTY TESTING AND THIRD-PARTY CERTIFICATION

PRODUCT OR MATERIAL	THIRD-PARTY CERTIFIED	THIRD-PARTY TESTED
Potable water supply system components and potable water fixture fittings	Required	—
Sanitary drainage and vent system components	Plastic pipe, fittings and pipe-related components	All others
Waste fixture fittings	Plastic pipe, fittings and pipe-related components	All others
Storm drainage system components	Plastic pipe, fittings and pipe-related components	All others
Plumbing fixtures	—	Required
Plumbing appliances	Required	—
Backflow prevention devices	Required	—
Water distribution system safety devices	Required	—
Special waste system components	—	Required
Subsoil drainage system components	—	Required

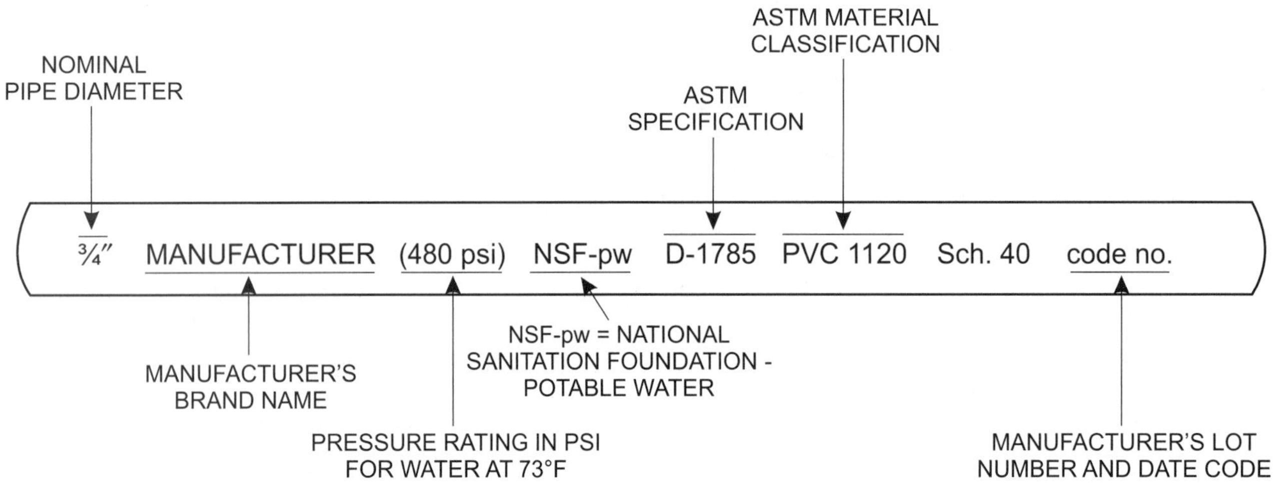

ASTM MATERIAL CLASSIFICATION

NOMINAL PIPE DIAMETER

ASTM SPECIFICATION

¾″ MANUFACTURER (480 psi) NSF-pw D-1785 PVC 1120 Sch. 40 code no.

MANUFACTURER'S BRAND NAME

NSF-pw = NATIONAL SANITATION FOUNDATION - POTABLE WATER

PRESSURE RATING IN PSI FOR WATER AT 73°F

MANUFACTURER'S LOT NUMBER AND DATE CODE

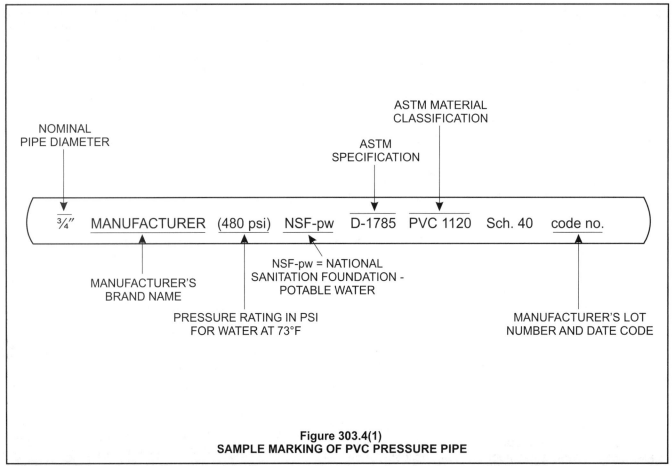

Figure 303.4(1)
SAMPLE MARKING OF PVC PRESSURE PIPE

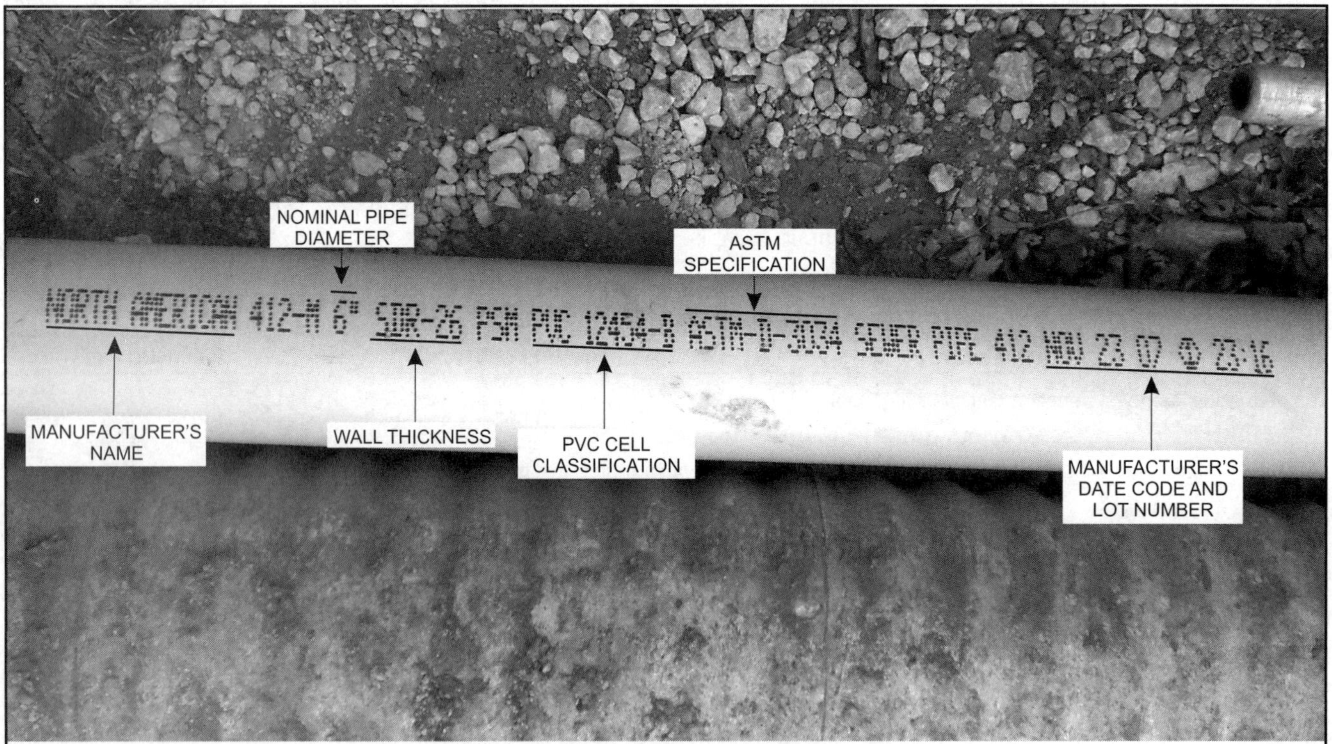

NOMINAL PIPE DIAMETER

ASTM SPECIFICATION

NORTH AMERICAN 412-M 6" SDR-26 PSM PVC 12454-B ASTM-D-3034 SEWER PIPE 412 NOV 23 07 23:16

MANUFACTURER'S NAME

WALL THICKNESS

PVC CELL CLASSIFICATION

MANUFACTURER'S DATE CODE AND LOT NUMBER

Figure 303.4(2)
MARKING OF PVC GRAVITY SEWER PIPE

304.3 Meter boxes. Meter boxes shall be constructed in such a manner that rodents are prevented from entering a structure by way of the water service pipes connecting the meter box and the structure.

❖ The water service pipe may be tunneled into the building from the meter box. When such an installation occurs, the annular space around the pipe must be protected to prevent rodents from getting into the building. This can be accomplished by a barrier to block their entry, such as a corrosion-resistant heavy wire screen or metal plate that is securely fastened in place.

304.4 Openings for pipes. In or on structures where openings have been made in walls, floors or ceilings for the passage of pipes, such openings shall be closed and protected by the installation of *approved* metal collars that are securely fastened to the adjoining structure.

❖ Annular spaces created by the passage of pipes through walls, floors or ceilings must be protected to prevent rodents from entering the building. This protection can be achieved by placing a metal collar around penetrating pipes. The collar material must be durable for the exposure conditions and of sufficient strength to prevent rodents from chewing through it and entering the building.

SECTION 305 PROTECTION OF PIPES AND PLUMBING SYSTEM COMPONENTS

305.1 Corrosion. Pipes passing through concrete or cinder walls and floors or other corrosive material shall be protected against external corrosion by a protective sheathing or wrapping or other means that will withstand any reaction from the lime and acid of concrete, cinder or other corrosive material. Sheathing or wrapping shall allow for movement including expansion and contraction of piping. Minimum wall thickness of material shall be 0.025 inch (0.64 mm).

❖ Metallic pipes made from brass, copper, cast iron and steel are subject to corrosion when exposed to lime and acid of concrete, cinder, some types of soil or other corrosive material. A protective sheathing having a minimum thickness of 0.025 inch (0.64 mm) is to be placed over the piping before installation in these corrosive environments. The protective sheathing should expand and contract to prevent rubbing and abrasion. Typical protective coatings include coal tar wrapped with paper, epoxy or plastic coatings and plastic sheathing.

305.2 Breakage. Pipes passing through or under walls shall be protected from breakage.

❖ Prior to the 1995 code edition, this section included the text of Section 305.5. The current text of this section was and should still be considered as the introductory/intent statement for Section 305.5. By itself, the text of this section provides no guidance as to what type of protection is required is ambiguous. See commentary for Section 305.5.

305.3 Stress and strain. Piping in a plumbing system shall be installed so as to prevent strains and stresses that exceed the structural strength of the pipe. Where necessary, provisions shall be made to protect piping from damage resulting from expansion, contraction and structural settlement.

❖ A plumbing system must not be damaged by stresses, strains or movement of the building components. In piping systems, provisions must be made for the expansion and contraction of the pipes themselves. Each piping material has a different rate of expansion and contraction that must be considered when designing the restraint system for the piping and the structure that surrounds the piping system.

Changes in temperature can cause distortion of the pipe material; for example, heat causes the material to expand. The greatest amounts of expansion and contraction in piping will occur along the length of the pipe. Hot water piping can experience significant movements in short runs of piping. Though the amount of expansion per unit length is low, large movements can occur in long lengths of pipe. Commentary Table 305.3 contains the expansion rate for various pipe materials.

Table 305.3
THERMAL EXPANSION RATES FOR PIPING MATERIALS

PIPING MATERIAL	RATE OF THERMAL EXPANSION in/in/°F
Brass-red	0.000009
Copper	0.00001
Cast iron	0.0000056
Carbon steel	0.000005
Ductile iron	0.0000067
Stainless steel	0.0000115
Borosilicate (glass)	0.0000018
ABS	0.00005
CPVC	0.000035
HDPE	0.00011
PE	0.00008
PEX	0.00093
PP	0.000065
PVC	0.00004
PVDF	0.000096

Source: Facility Piping Systems Handbook 2nd Edition, Frankel, McGraw Hill, 2002.

The most common method to absorb thermal expansion in plumbing piping systems is the installation of one or more offsets in the piping. The typical offset piping arrangements used are the "L" bend (1-elbow change in direction), the "Z" bend (2-elbow offset) and the "U" bend (4-elbow offset) [see commentary Figures 305.3(1) through 305.3(3)]. The length, L, in each of these figures is determined from the following equation:

$$L = [1.5 \times (E/S) \times D_o \times (G)]^{1/2}$$

where,

E = Modulus of elasticity at pipe temperature (psi)

S = Maximum allowable stress for the pipe material at the highest in-service pipe temperature (psi)

D_o = Outside diameter of pipe (inches)

G = Change in length of piping due to temperature change (inches). Thermal expansion rate of the pipe material from Commentary Table 305.3 x Change in temperature of the pipe (installation day to highest in-service temperature expected) x Length of pipe in feet between points where the pipe is restricted from growing in length x 12 inches per foot.

The values for E and S are obtained from piping manufacturers, engineering publications or the material standards for the pipe products.

Sample Problem 1: A 160-foot-long straight run of 1 inch copper tube size CPVC pipe is to be installed to convey hot water at 140°F. The ends of the piping will be attached to equipment. The temperature at the time of installation will be 50°F. Determine the required offset length, L, to accommodate the thermal expansion that will occur when the system is operating.

Problem Approach

Obtain the thermal expansion rate from Commentary Table 305.3, calculate the amount of thermal expansion of the piping run, obtain the modulus of elasticity and allowable stress values from the piping manufacturer and calculate the required offset length, L.

Solution

For CPVC piping material, the thermal expansion rate listed in Commentary Table 305.3 is 0.000035 inches/inch/°F. The change in temperature of the pipe from installation to service condition is 140°F minus 50°F which equals 90°F. The change in length of the pipe, G, is calculated as follows:

G = 0.000035 inches/inch/°F x 90°F x 160 feet x 12 inches per foot

G = 6.1 inches

The pipe manufacturer provides a modulus of elasticity of 3.23 x 10^5 psi and an allowable stress of 1000

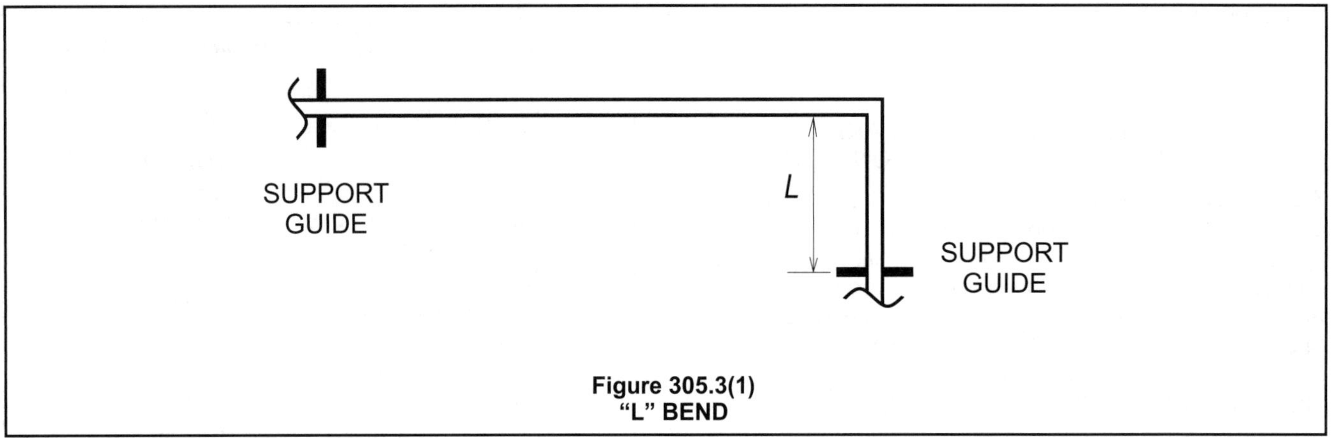

Figure 305.3(1)
"L" BEND

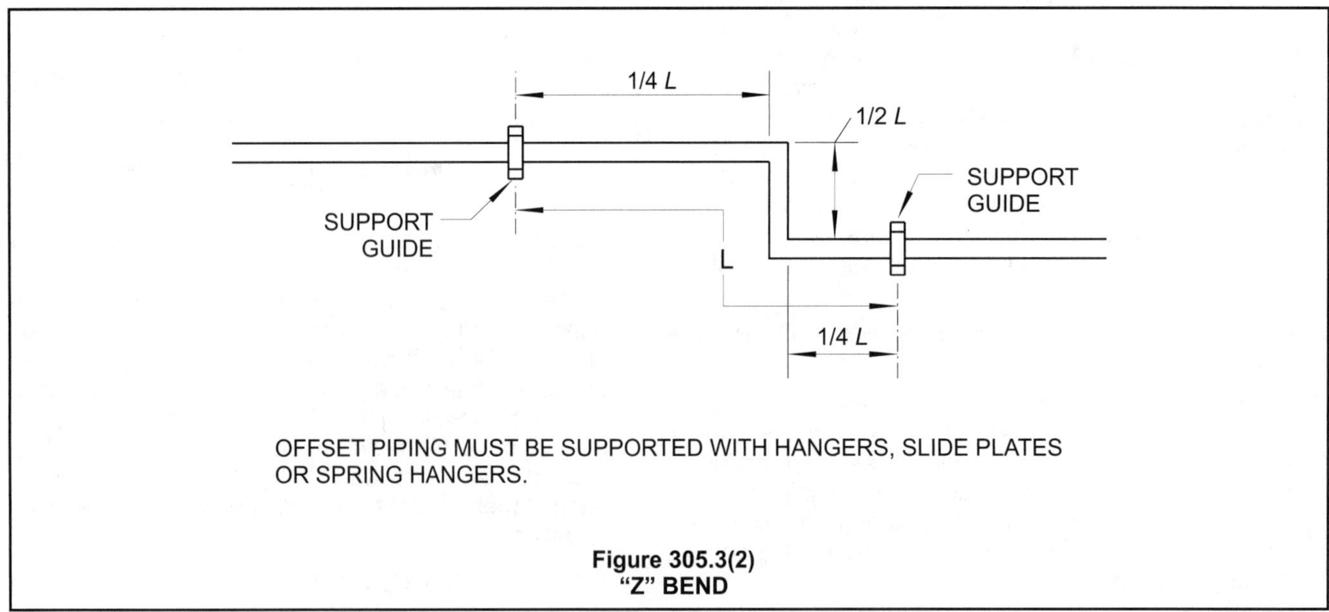

Figure 305.3(2)
"Z" BEND

psi, both at the service temperature of 140°F. The outside diameter of the pipe, D_o, is 1.125 inches. Therefore,

$L = [1.5 \times (E/S) \times D_o \times G]^{1/2}$

$L = [1.5 \times (3.23 \times 10^5 /1000) \times 1.125 \times 6.1]^{1/2}$

L = 57.7 inches

Expansion joints can also be used to absorb thermal expansion. These products are pre-engineered or can be custom engineered for any application. The code does not regulate the design or installation of expansion joints.

The change in dimension due to temperature change is not limited to the length of the pipe but affects the diameter as well. As the temperature of a pipe increases, the pipe will "swell" to a larger diameter. If the thermal expansion is restrained such as by casting of the pipe in a concrete wall, significant stresses can develop in the pipe wall.

305.4 Sleeves. Annular spaces between sleeves and pipes shall be filled or tightly caulked in an *approved* manner. Annular spaces between sleeves and pipes in fire-resistance-rated assemblies shall be filled or tightly caulked in accordance with the *International Building Code*.

❖ The annular space created between a sleeve and a pipe in an assembly that is not fire-resistance rated must be filled with a material that prevents structural loading of the pipe and rodent infiltration (see commentary, Sections 304.4 and 305.3). This material is typically a caulk with coal tar or an asphaltic compound with some degree of flexibility to prevent structural loading of the piping. It should be noted that the materials used to fill the annular space must be compatible with the pipe as well as the sleeve material [see Commentary Figure 307.3(2)].

Annular spaces created by penetrations in fire-resistance-rated assemblies must also be filled and must maintain the fire-resistance rating of the assembly. The fire-resistance rating can be protected by providing a through-penetration system, which can consist of caulks, intumescent materials and sleeves installed around the penetrating pipe. All through-penetration systems must be tested in accordance with ASTM E 814 as provided in the *International Building Code*® (IBC®) [see Commentary Figures 307.3(3) and 307.3(4)].

305.5 Pipes through or under footings or foundation walls. Any pipe that passes under a footing or through a foundation wall shall be provided with a relieving arch, or a pipe sleeve pipe shall be built into the foundation wall. The sleeve shall be two pipe sizes greater than the pipe passing through the wall.

❖ Piping installed within or under a footing or foundation wall must be structurally protected from any transferred loading from the footing or foundation wall. This protection may be provided by a relieving arch or a pipe sleeve.

When a sleeve is used, it must be sized to be two pipe sizes larger than the penetrating pipe. For example, a 4-inch (102 mm) penetrating pipe would require a 6-inch (152 mm) sleeve. This space will allow for any differential movement of the pipe. By providing structural protection to the piping system, the piping will not be subjected to undue stresses that could cause it to rupture and leak [see commentary, Figures 305.5(1) and 305.5(2)].

305.6 Freezing. Water, soil and waste pipes shall not be installed outside of a building, in attics or crawl spaces, concealed in outside walls, or in any other place subjected to freezing temperatures unless adequate provision is made to protect such pipes from freezing by insulation or heat or both. Exterior water supply system piping shall be installed not less than 6 inches (152 mm) below the frost line and not less than 12 inches (305 mm) below grade.

❖ Water, soil and waste pipes must be protected from freezing. Although vent piping is not required to be protected from freezing, it may be prudent to insulate these pipes in extreme freezing climates to prevent condensation freezing and blocking of vents. Where a water or drain pipe is installed in an exterior wall or ceiling of a space intended for occupancy (in other words, a heated space), some degree of freeze protection can be achieved by making sure that the ther-

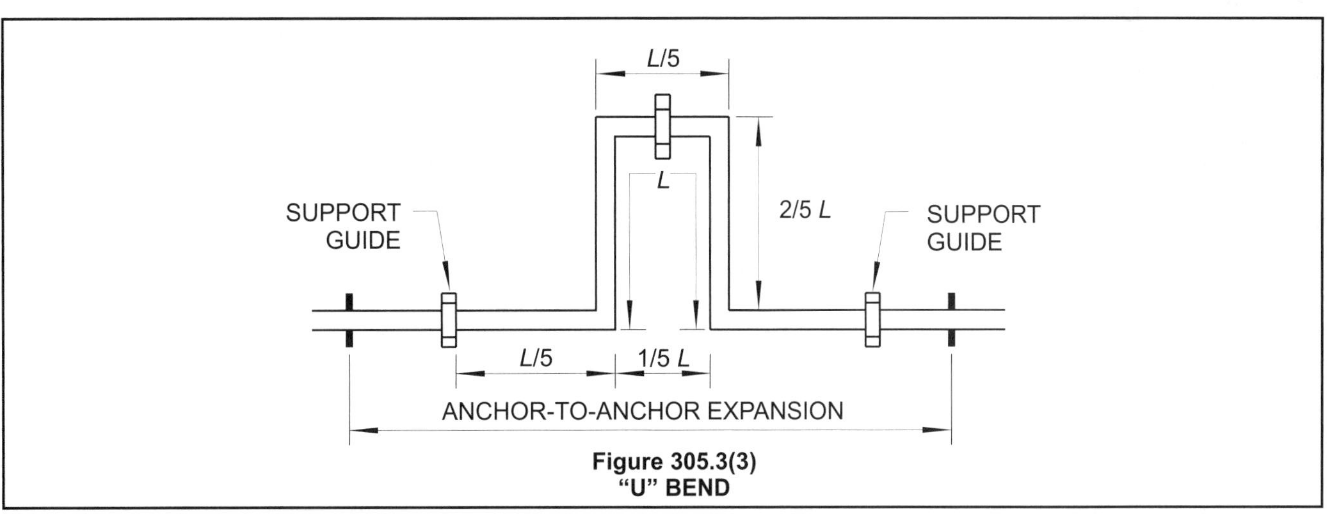

Figure 305.3(3)
"U" BEND

mal insulation for the wall (or ceiling) is installed between the outdoor surface and the piping [see commentary, Figures 305.6(1) and 305.6(2)]. Whether or not this arrangement will prevent freezing temperatures at the piping location depends on the climatic conditions, the thickness of insulation (between the outdoor surface and the piping) and the room temperature.

Note that there must always be a heat source along with an appropriate insulation thickness in order to protect pipes from freezing conditions. Insulation by itself (without a heat source) cannot protect a pipe from freezing; insulation only slows the rate of heat loss. The closer a pipe is to the heat source and the less insulation there is between the pipe and the heat source; the warmer the pipe will be. Commentary Figure

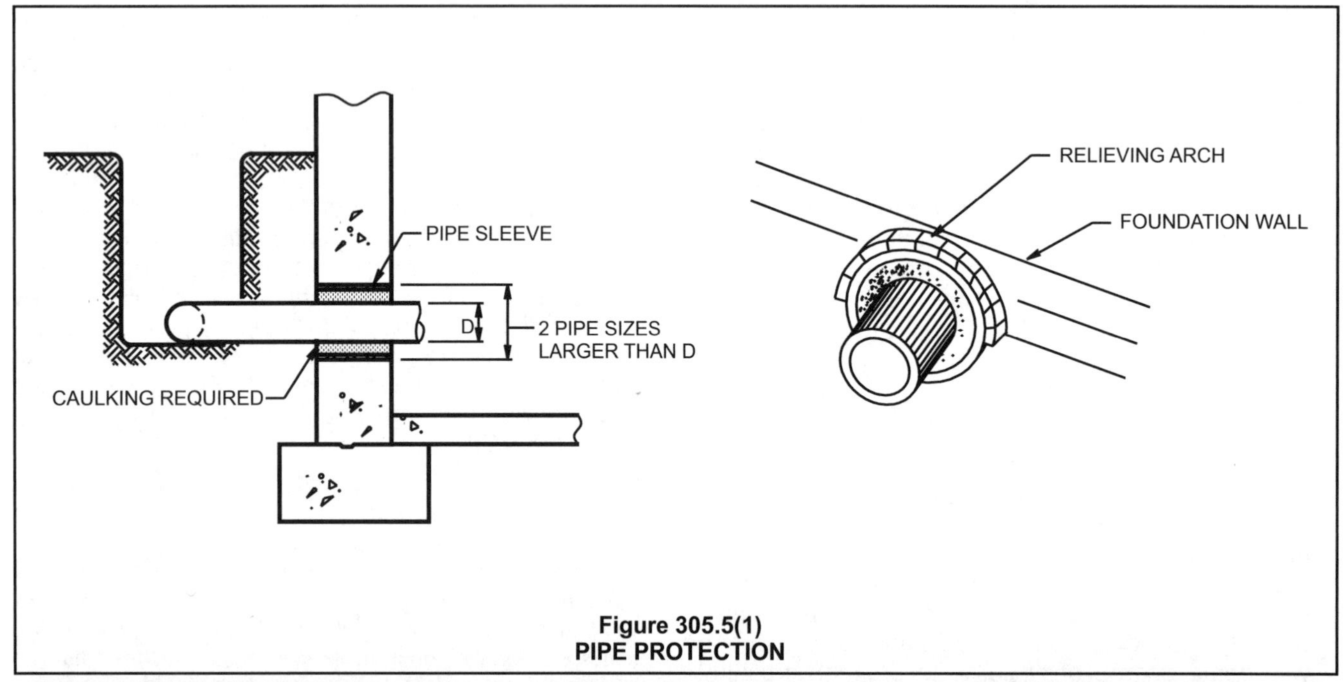

Figure 305.5(1)
PIPE PROTECTION

Figure 305.5(2)
BUILDING DRAIN PENETRATION THROUGH FOUNDATION WALL

305.6(3) shows piping in an exterior wall of a building. The water distribution piping has been covered with foam tube insulation. On the left side of the photo, note the close proximity of the horizontal water pipe to the exterior sheathing. While the insulation on the pipe between the pipe and the sheathing does help protect the pipe from outdoor temperatures, the insulation on the pipe between the pipe and the room will insulate the pipe from the heat coming from the room. In all likelihood, the wall insulation will be installed over the insulated pipe resulting in further insulation of the pipe from the room heat. On the right side of this photo,

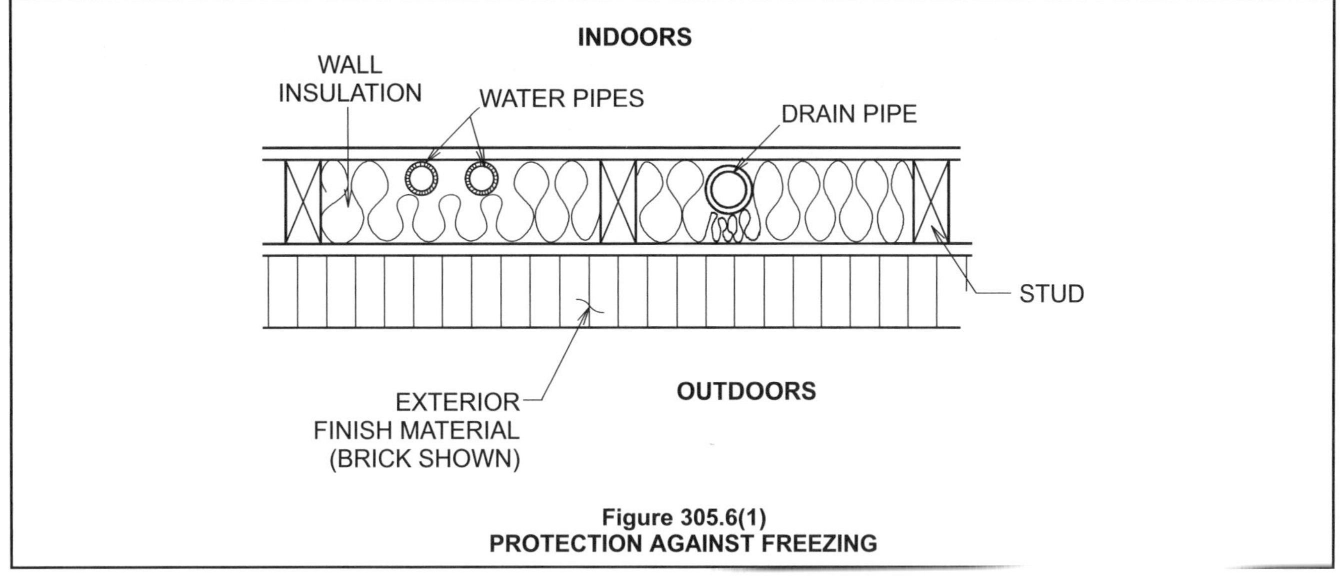

Figure 305.6(1)
PROTECTION AGAINST FREEZING

Figure 305.6(2)
WATER DISTRIBUTION PIPING IN EXTERIOR WALL

note the multitude of water distribution pipes behind the fixture location (probably a laundry tub). Again, while the pipe insulation between the pipes and the sheathing is beneficial for freeze protection of the pipes, the insulation on the room side of the pipes is detrimental because it insulates the pipes from the room heat. The congestion in this area will undoubtedly cause difficulty for installing wall insulation, especially between the piping and the sheathing. In the center of this photo, note that the trap for the clothes washer stand pipe is in close proximity to the sheathing so much so that it will be difficult to install any insulation between the trap and the sheathing. The arrangement and insulation of pipes shown in this figure would most likely be suitable only for a building located where freezing climatic conditions are unlikely or are of limited duration.

Where piping is not directly adjacent to heated spaces in a building, electric resistance heat tapes or cables can be used to supply heat to the piping. Heat tapes should not be used on piping in concealed spaces as the tapes can burn out and require replacement. Plastic piping requires self-limiting type heat tape to prevent overheating of the pipe.

The code does not provide any guidelines for insulating methods to ensure that piping will not be exposed to freezing temperatures. The code official is responsible for determining whether water and drain piping is adequately protected for the climatic conditions expected in the geographic area.

Freezeproof Hose Faucets

Hose bibbs and wall hydrants located on the exterior wall must be protected when installed in areas subject to freezing temperature. This can be accomplished by installing devices, such as freezeproof hose bibbs, that locate the valve seat within the heated space and allow residual water within the hydrant to drain after the valve is closed. A freezeproof hose bibb cannot be installed with the valve seat located in an unheated garage or storage room. The valve seat must extend through to the heated side of the exterior wall. Valve casings on these devices are available in various lengths to accommodate a variety of wall thicknesses [see commentary, Figure 305.6(4)].

Freezer Floor Drains

Where floor drains are located in areas (e.g., walk-in freezers or frozen food warehouses) maintained at temperatures at or below 32°F (0°C), the traps for the floor drains could freeze and damage the piping. Therefore, floor drains in these areas must not have traps and must be indirectly connected in accordance with Section 802.1.2.

Figure 305.6(3)
PIPING IN EXTERIOR WALL

Underground Water Service Piping

Water service piping that is installed exterior to a building and underground must be buried at least 12 inches (305 mm) below grade or 6 inches (152 mm) below the established frost penetration depth for the geographic area [see commentary, Figure 305.6(5)], whichever is the greater depth. For example, if the frost depth for the region is 24 inches (610 mm) below grade, the water service pipe must be installed so that the top surface of the pipe is at least 30 inches (762 mm) below grade. If the frost depth is 5 inches (127 mm) below grade, the water service pipe must be installed such that the top surface of the pipe is at least 12 inches (305 mm) below grade. Where frost pene-

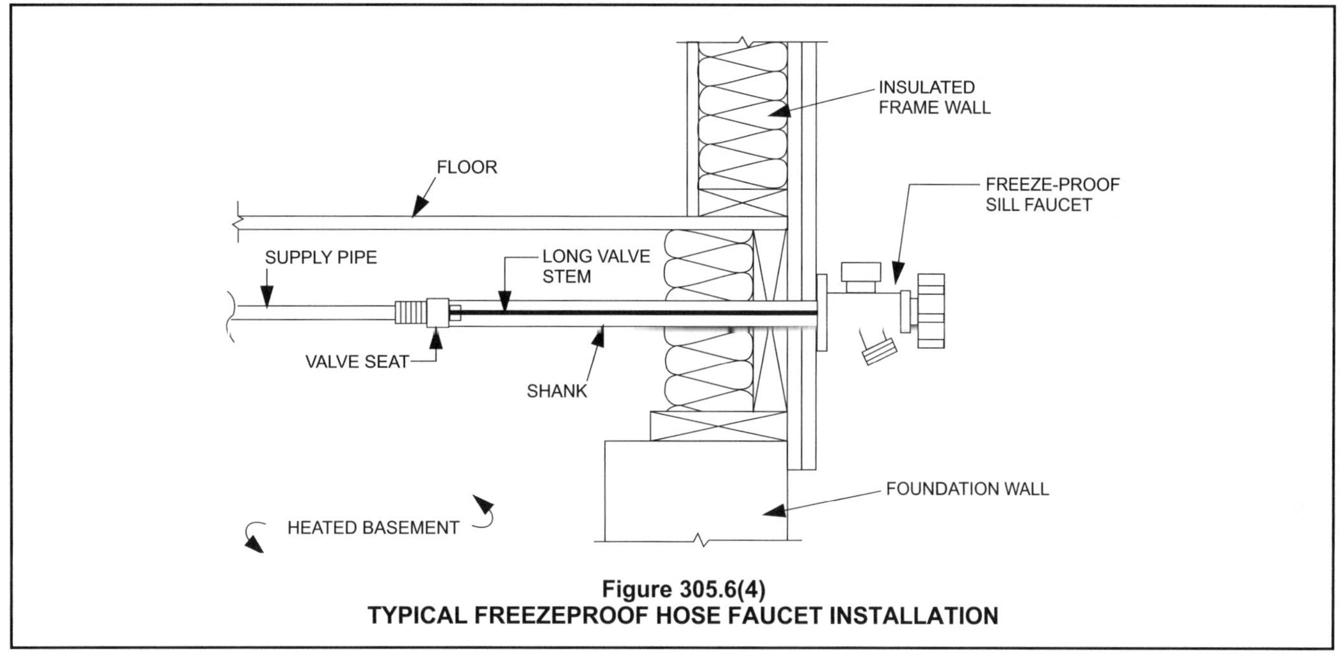

Figure 305.6(4)
TYPICAL FREEZEPROOF HOSE FAUCET INSTALLATION

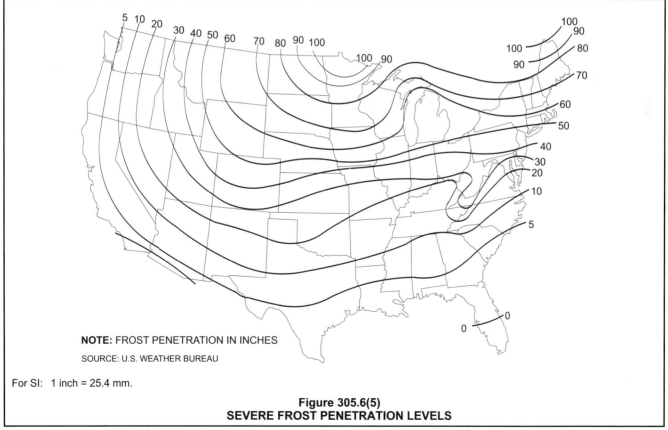

NOTE: FROST PENETRATION IN INCHES

SOURCE: U.S. WEATHER BUREAU

For SI: 1 inch = 25.4 mm.

Figure 305.6(5)
SEVERE FROST PENETRATION LEVELS

tration depth controls the burial depth, the 6 inch buffer protects the pipe from forces caused by the freezing and expansion of moisture in the above soil (i.e., "frost heave") [see commentary Figure 305.6(6)]. The minimum burial depth of 12 inches protects the pipe from the most common accidental damage: shallow hand digging for landscape plantings.

305.6.1 Sewer depth. Building sewers that connect to private sewage disposal systems shall be a minimum of [NUMBER] inches (mm) below finished grade at the point of septic tank connection. Building sewers shall be a minimum of [NUMBER] inches (mm) below grade.

❖ A building sewer is not subject to the same constraints as a water pipe because a sewer carries heat to the surrounding soil as the drainage flows through it. A building sewer also has an intermittent flow. As a result, the frost depth is higher in the immediate area of the building sewer; therefore, the burial depth of a building sewer does not have to be below the recorded frost depth. A typical burial depth for a building sewer ranges from 6 to 18 inches (152 to 457 mm) [see Commentary, Figure 305.6(1)].

The installation depth of the sewer pipe connecting to a septic tank could be different to facilitate the proper operation of the private sewage disposal system. The connection at the septic tank is typically 18 inches (457 mm) below grade; however, the code official is to determine the depths necessary for the locality.

305.7 Waterproofing of openings. Joints at the roof and around vent pipes, shall be made water-tight by the use of lead, copper, galvanized steel, aluminum, plastic or other *approved* flashings or flashing material. Exterior wall openings shall be made water-tight.

❖ Where a pipe penetrates a roof or an exterior wall, the annular space around the pipe must be sealed, or the pipe must be provided with flashing to prevent the entry of moisture. If this annular space is improperly sealed or not sealed at all, moisture from precipitation can enter the structure and damage the surrounding structure and finishes.

Because of direct exposure to rain and, in colder climates, exposure to snow accumulation, piping that penetrates a roof must be provided with flashing. Many types of flashings have been developed to prevent leakage around penetrations. In a typical installation, the flashing surrounds the pipe and rises above the roof surface, and is made water tight by wrapping over the top of the pipe or applying an elastomeric seal around the pipe. The installation of materials, such as lead, that wrap over the top of the pipe must not reduce the required cross-sectional area of the vent terminal (see Commentary Figure 305.7).

Several of the flashing materials listed in Section 305.7 have been used for many years and are recog-

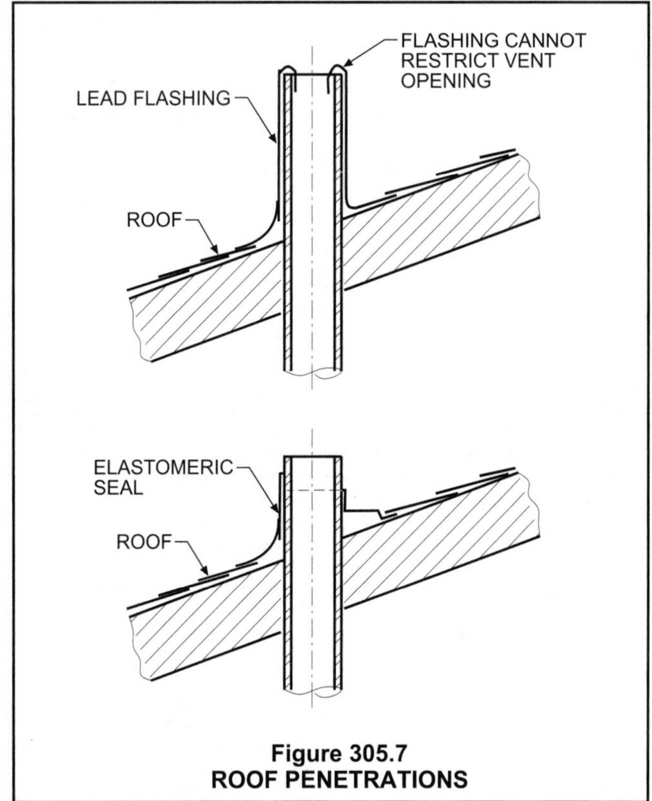

Figure 305.7
ROOF PENETRATIONS

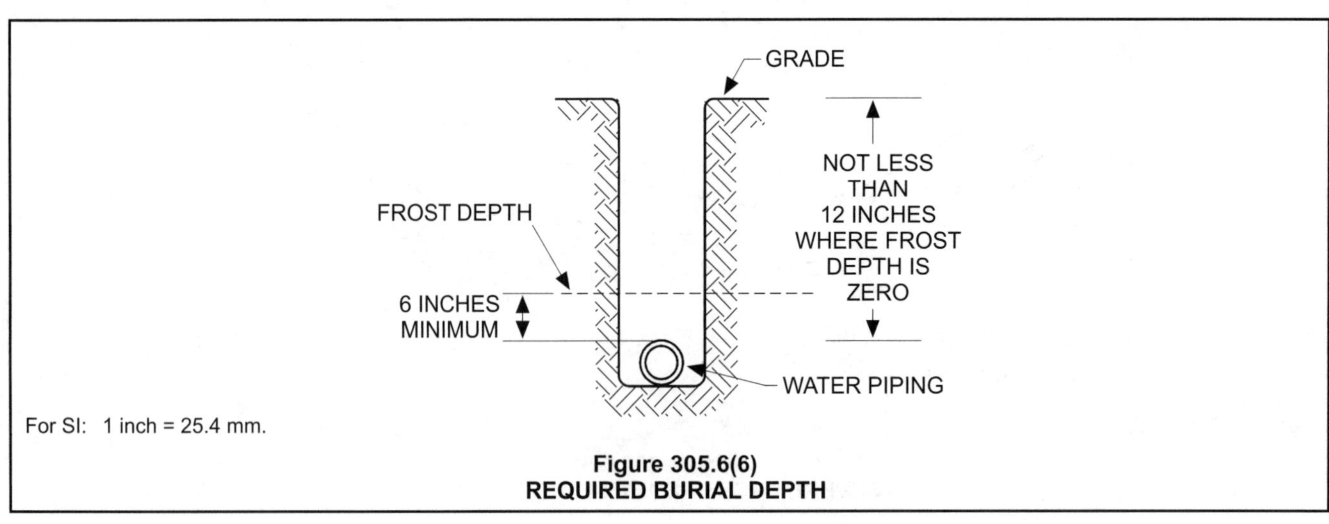

Figure 305.6(6)
REQUIRED BURIAL DEPTH

nized as effective moisture penetration protection. This section also provides for the use of new and innovative materials that are not specifically listed when approved by the code official. Whenever an alternative material is proposed for use as a flashing, the material must be proven to be corrosion resistant and durable against the effects of exposure to the elements, including but not limited to resistance to impact, freeze-thaw and ultraviolet exposure and aging. Another consideration in the selection of the flashing material is that it must be compatible with the roof-covering material, the assembly penetrated and the pipe that it will be in contact with.

Because exterior walls are typically not subjected to the same severity of exposure as a roof, the space around pipes that penetrate exterior walls need not be sealed with flashing; rather, it must only be made watertight. Openings in exterior walls, therefore, can be sealed with an approved sealant, mechanical device, flashing or other approved method that will prevent penetration into the structure.

305.8 Protection against physical damage. In concealed locations where piping, other than cast-iron or galvanized steel, is installed through holes or notches in studs, joists, rafters or similar members less than $1^1/_2$ inches (38 mm) from the nearest edge of the member, the pipe shall be protected by steel shield plates. Such shield plates shall have a thickness of not less than 0.0575 inch (1.463 mm) (No. 16 gage). Such plates shall cover the area of the pipe where the member is notched or bored, and shall extend a minimum of 2 inches (51 mm) above sole plates and below top plates.

❖ This section is intended to minimize the possibility of drainage and water pipe damage caused by nails, screws or other fasteners. Because nails and screws sometimes miss the stud, rafter, joist or sole or top plates, the shield must protect the pipe through the full width of the member on each side and must extend not less than 2 inches (51 mm) above or below the sole or top wall plates. Commentary Figure 305.8(1) shows typical shield plates. Commentary Figure 305.8(2) shows shield plates improperly installed because the plate(s) do not extend 2 inches (51 mm) below the bottom member of the double top plate. Cast-iron and galvanized steel pipe each have wall thicknesses greater than the required thicknesses for the shield plate, which makes them inherently resistant to nail and screw penetrations.

The shield plate minimum thickness specified represents the "low end" of the thickness tolerance range for No. 16 gage galvanized sheet metal in accordance with Sheet Metal and Air Conditioning Contractors National Association (SMACNA) HVAC duct construction standards. The "low end" thickness dimension is stated so that if a field measurement of shield plate thickness is necessary, the absolute minimum thickness will allow all possible thicknesses of No. 16 gage shield plates to be acceptable. No. 16 gage has been proven to be the minimum thickness to prevent pene-

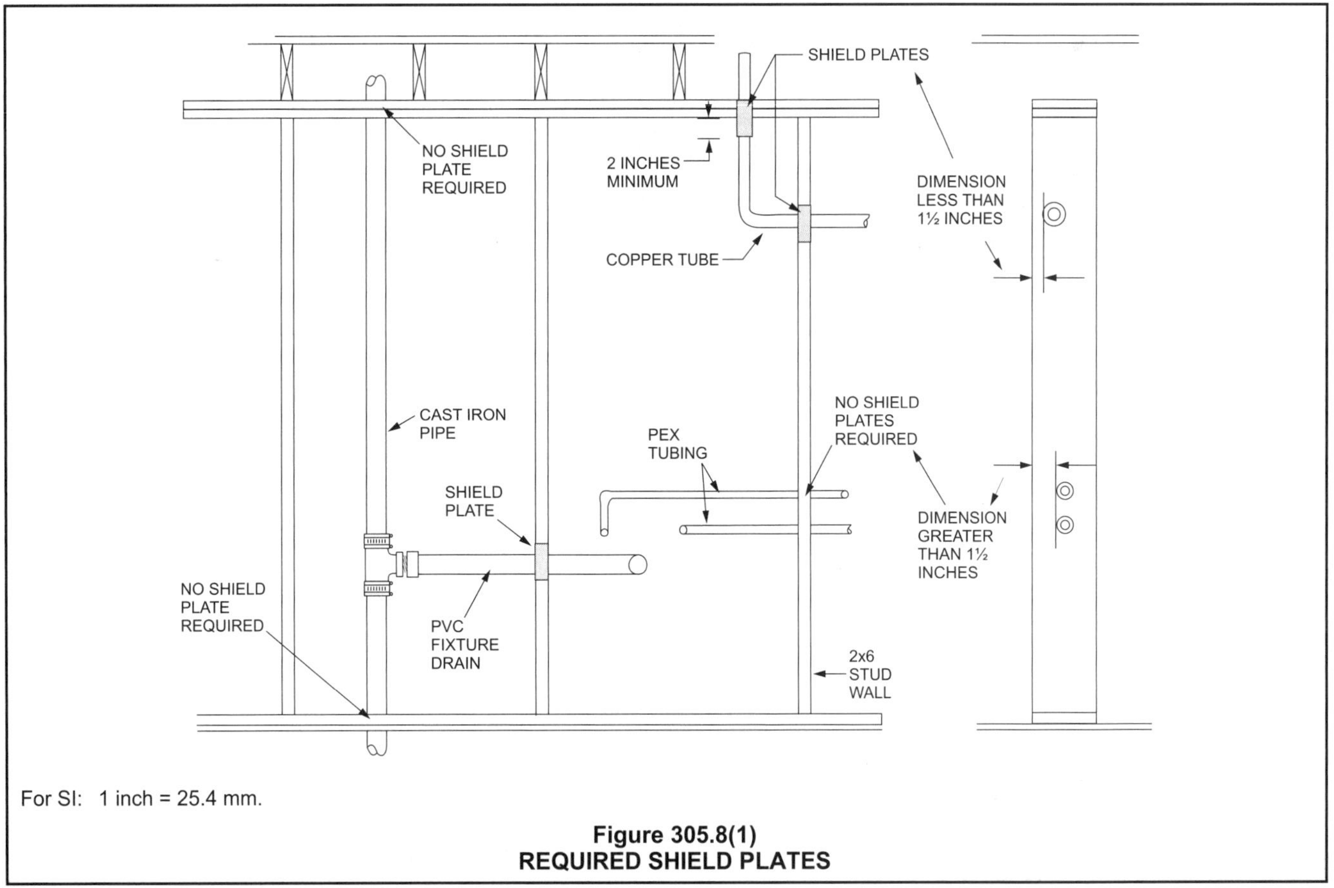

Figure 305.8(1)
REQUIRED SHIELD PLATES

tration by commonly used fasteners without causing noticeable "bulge" appearance problems in the wall sheathing.

305.9 Protection of components of plumbing system. Components of a plumbing system installed along alleyways, driveways, parking garages or other locations exposed to damage shall be recessed into the wall or otherwise protected in an *approved* manner.

❖ Plumbing system components such as water supply pipes, vents, stacks, leaders or downspouts that are subject to possible motor vehicle impact must be adequately protected (see Commentary Figure 305.9). The method of protection must be approved by the code official.

SECTION 306
TRENCHING, EXCAVATION AND BACKFILL

306.1 Support of piping. Buried piping shall be supported throughout its entire length.

❖ Piping that is installed below grade must be continuously supported to prevent a number of possible problems including, but not limited to, reverse slope (of gravity flow pipelines), flexible connector misalignment/failure and pipe breakage.

306.2 Trenching and bedding. Where trenches are excavated such that the bottom of the trench forms the bed for the pipe, solid and continuous load-bearing support shall be provided between joints. Bell holes, hub holes and coupling holes shall be provided at points where the pipe is joined. Such pipe shall not be supported on blocks to grade. In instances where the materials manufacturer's installation instructions are more restrictive than those prescribed by the code, the material shall be installed in accordance with the more restrictive requirement.

❖ A trench must be wide enough to allow for proper alignment of the piping system. Piping is best supported when it rests directly on the bottom of a solid, continuous trench for its entire length, as required by Section 306.1 (see Commentary Figure 306.2). Sections 306.2.1 through 306.2.3 contain specific requirements for excavating and rebedding the trench below the installation level of the pipe.

If bell or hub pipe is installed, the bottom of the

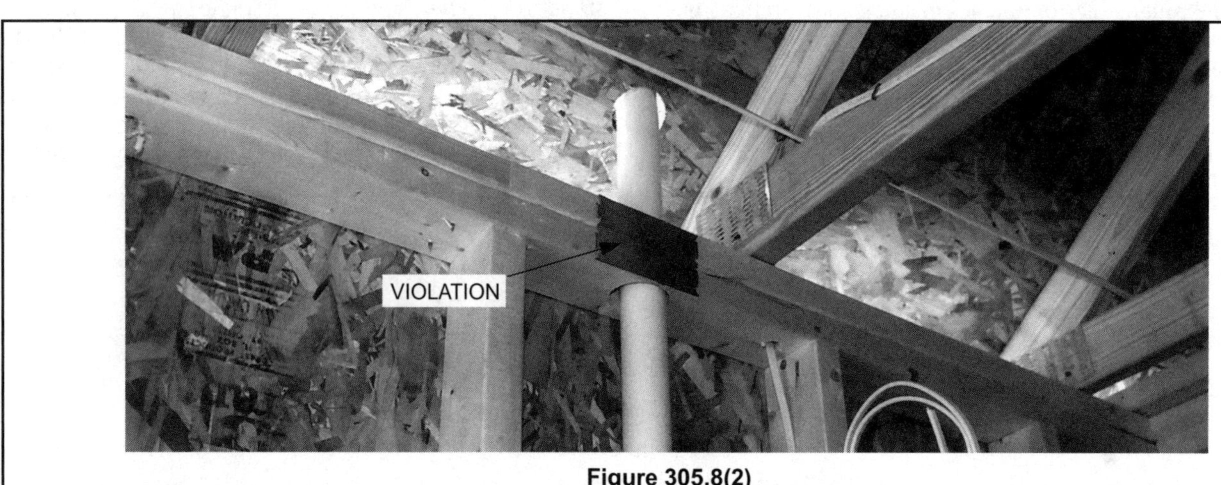

Figure 305.8(2)
PIPE PROTECTION VIOLATION

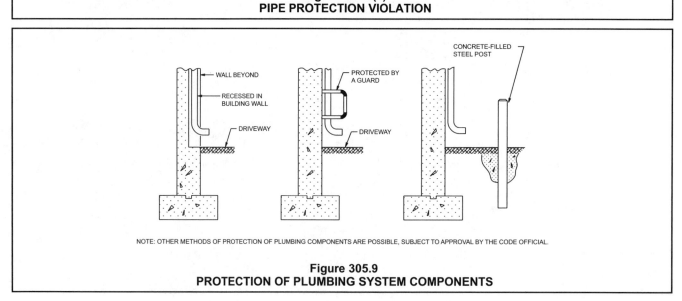

NOTE: OTHER METHODS OF PROTECTION OF PLUMBING COMPONENTS ARE POSSIBLE, SUBJECT TO APPROVAL BY THE CODE OFFICIAL.

Figure 305.9
PROTECTION OF PLUMBING SYSTEM COMPONENTS

trench must be dug out around the hub to maintain continuous support of the pipe (see Commentary Figure 306.2). For additional information on burial techniques, consult installation standards such as ASTM D 2321 (plastic sewer) and ASTM D 2774 (plastic pressure pipe).The standards include criteria such as the size of backfill particles, classification and compaction density requirements of various embedment materials and the deflection requirements of different pipe materials. They also address the minimum backfill cover required before heavy equipment can traverse the pipe trench. When the manufacturer's installation instructions are more restrictive than the requirements of this section, the manufacturer's instructions govern. The rationale for this is that the manufacturer has tested the assembly for various characteristics, including proper support of the piping system for various areas where the piping may be installed.

306.2.1 Overexcavation. Where trenches are excavated below the installation level of the pipe such that the bottom of the trench does not form the bed for the pipe, the trench shall be backfilled to the installation level of the bottom of the pipe with sand or fine gravel placed in layers of 6 inches (152 mm) maximum depth and such backfill shall be compacted after each placement.

❖ Trench overexcavation occurs frequently because it is sometimes difficult to maintain an even grade line. The removal of large rocks, construction debris or abandoned piping can create voids below the desired grade line. Where over-excavation occurs, the trench bottom must be built up with compacted layers of sand or fine gravel (see commentary, Figure 306.2.1). The layers of fill material must not exceed 6 inches (152 mm) in thickness so that compaction is effective throughout the fill layer. Compaction is necessary to prevent future settling and to decrease the permeability of the fill material so that any ground water flow does not wash out the fill material.

306.2.2 Rock removal. Where rock is encountered in trenching, the rock shall be removed to a minimum of 3 inches (76 mm) below the installation level of the bottom of the pipe, and the trench shall be backfilled to the installation level of the bottom of the pipe with sand tamped in place so as to provide uniform load-bearing support for the pipe between joints. The pipe, including the joints, shall not rest on rock at any point.

❖ Piping must not be supported intermittently by hard surfaces, such as rock or concrete as this creates "point loads" on the pipe wall. Bedding provides a continuous "smooth" cradled support to insure that the pipe is not exposed to concentrated "point" loads which could ultimately cause pipe failure.

306.2.3 Soft load-bearing materials. If soft materials of poor load-bearing quality are found at the bottom of the trench, stabilization shall be achieved by overexcavating a minimum of two pipe diameters and backfilling to the installation level of the bottom of the pipe with fine gravel, crushed stone or a concrete foundation. The concrete foundation shall be bedded with sand tamped into place so as to provide uniform load-bearing support for the pipe between joints.

❖ If soil at the bottom of an excavated trench is of poor load-bearing quality, such as peat or muck, the trench must be overexcavated by two pipe diameters and the area backfilled with material that will properly support the piping (see commentary for Section 306.2.1). For example, where a 4-inch (102 mm) pipe is to be installed in a trench where there is soggy soil, the trench

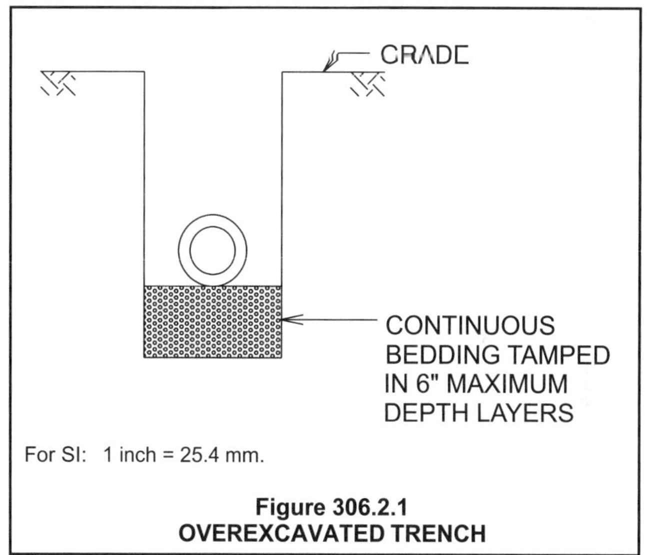

Figure 306.2.1
OVEREXCAVATED TRENCH

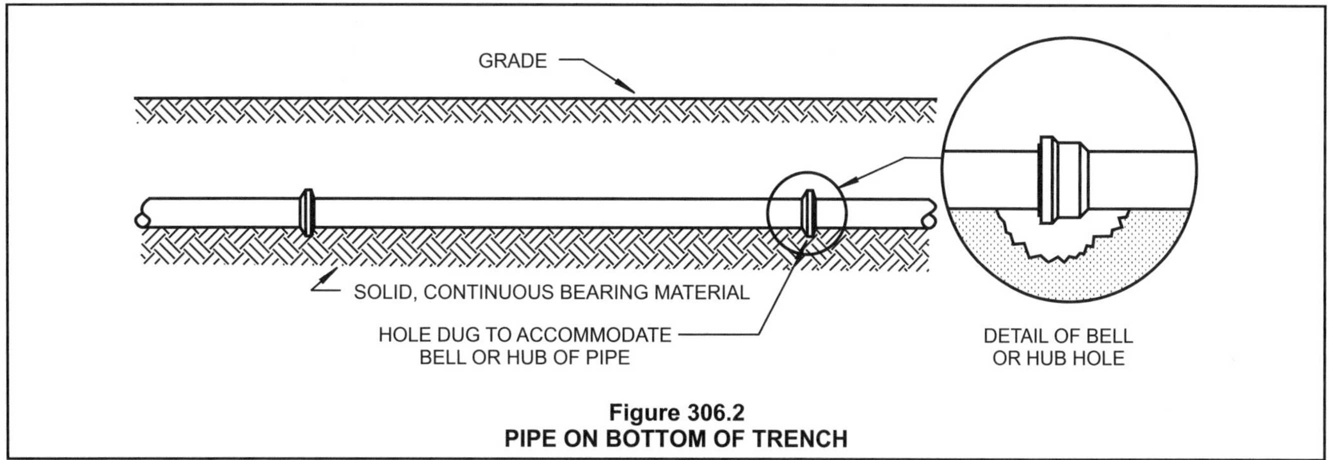

Figure 306.2
PIPE ON BOTTOM OF TRENCH

must be overexcavated at least 8 inches (203 mm) below the level where the pipe is to be installed. The overexcavated area must be backfilled with crushed stone, or fine gravel.

In some cases, the top of abandoned concrete foundations might be found at the bottom of the trench or new concrete might have to be placed to span "bottomless" soft earth locations. In either case, the pipe must be isolated from the concrete with a layer of tamped or compacted sand (two pipe diameters deep) to protect the pipe from the hard and often uneven concrete surface.

306.3 Backfilling. Backfill shall be free from discarded construction material and debris. Loose earth free from rocks, broken concrete and frozen chunks shall be placed in the trench in 6-inch (152 mm) layers and tamped in place until the crown of the pipe is covered by 12 inches (305 mm) of tamped earth. The backfill under and beside the pipe shall be compacted for pipe support. Backfill shall be brought up evenly on both sides of the pipe so that the pipe remains aligned. In instances where the manufacturer's installation instructions for materials are more restrictive than those prescribed by the code, the material shall be installed in accordance with the more restrictive requirement.

❖ This section provides a general approach for backfilling piping that has been properly bedded in a trench. Proper backfilling technique is important to ensure that the pipe remains aligned in its bedded position, especially where there will be loads from additional backfill above the 12 inches (305 mm) of tamped backfill cover over the pipe. Note that the last line of text in this section requires that the installer consider "manufacturer's installation instructions" to determine if more restrictive requirements exist. Different types of soil conditions, deep trenches or additional loads anticipated at grade level (such as rolling trucks) could require different backfilling methods (see commentary to Section 306.2).

It is not the intent of this section to require that all piping be buried with at least 12 inches (305 mm) of cover. Sections 305.6 and 305.6.1 indicate the required burial depths for some types of piping systems. Piping in trenches under a slab-on-ground buildings need not be buried any more than what is necessary to provide for a full thickness of the concrete slab above. Piping in trenches in a crawl space need not have any bedding or backfill installed just as long as the piping is supported at the same required intervals as if the piping was suspended from the structure above.

Discarded construction materials such as broken concrete block, brick, and lumber scraps must be cleared out of the backfill material to prevent the pipe from being misaligned or the pipe becoming "point loaded" from above when the backfill is compacted. Large rocks, frozen clumps of soil and broken concrete must also be kept out of the initial backfill (haunching) because these materials do not easily flow into the voids next to the pipe. The backfill for the sides of the pipe must readily flow and be easily placed to completely fill the open trench space next to the pipe. An even distribution of backfill along both sides of the pipe is required to prevent the pipe from being pushed sideways. Backfilling above the pipe is to be accomplished by placing 6 inches (152 mm) of backfill material in the trench, tamping that material firm and then placing another 6 inches (152 mm) of material in the trench and tamping that layer of material firm. The process is repeated until the tamped material above the top of the pipe is at least 12 inches thick (305 mm) (see commentary Figure 306.3) or the trench is filled to grade. Any remaining backfill can then be placed into the trench in any layer thickness that can be adequately compacted with the available equipment or placed in the trench without compacting and left to naturally settle over time.

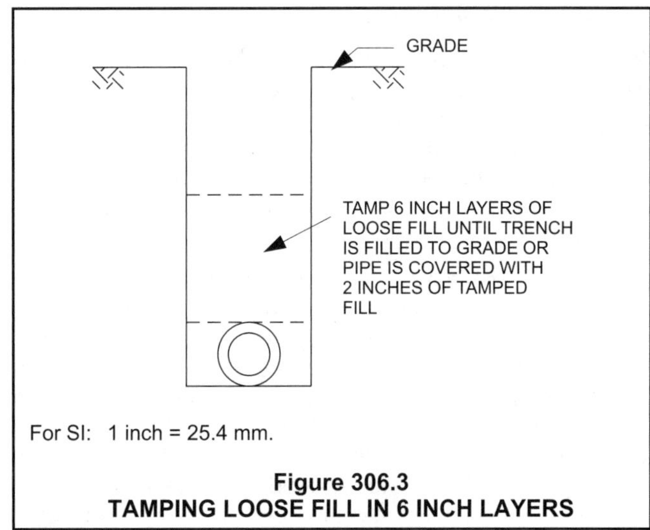

Figure 306.3
TAMPING LOOSE FILL IN 6 INCH LAYERS

306.4 Tunneling. Where pipe is to be installed by tunneling, jacking or a combination of both, the pipe shall be protected from damage during installation and from subsequent uneven loading. Where earth tunnels are used, adequate supporting structures shall be provided to prevent future settling or caving.

❖ When installing pipe under a paved area or sidewalk, tunneling, jacking or a combination of both is required. The pipe must be protected by proper backfilling in the tunneled area to prevent caving or settlement around the piping system. The length of tunneling should be limited to only that required to clear the obstacle above. Tunnels of any length are difficult to refill with proper backfill and the longer the tunnel, the more difficult it is to prevent settlement or caving. If an uneven loading condition exists, the piping system could eventually fail.

SECTION 307
STRUCTURAL SAFETY

307.1 General. In the process of installing or repairing any part of a plumbing and drainage installation, the finished floors, walls, ceilings, tile work or any other part of the building or premises that must be changed or replaced shall be left in a safe structural condition in accordance with the requirements of the *International Building Code*.

❖ Quite often, the installation of a plumbing system affects other systems in the building. Structural components, for example, may have to be modified to allow for a plumbing system to be installed in accordance with code requirements. Any change to a building system or component that is necessary for the proper installation of the plumbing system must result in the modified building system or component still complying with the code requirements regulating that particular system. Structural components must meet building code minimums.

307.2 Cutting, notching or bored holes. A framing member shall not be cut, notched or bored in excess of limitations specified in the *International Building Code*.

❖ The code recognizes that it is occasionally necessary to modify framing members by boring holes, notching or cutting in the course of installing plumbing piping or plumbing fixture supports. While the original intent of this section was to address modifications to solid sawn wood members (commonly called "dimensional lumber"), modern building construction includes a mix of "engineered" wood members and assemblies, cold-formed steel light-framed members, structural steel members and log members. To maintain a building's structural integrity, the load carrying capacity of these members must not be reduced. If a structural member is modified beyond what is specifically permitted by the IBC, the member must either be replaced with an unmodified member or a structural analysis of that member must be performed by a registered design professional (and approved by the code official) to verify that the modified member will have the required structural capacity for the loading conditions. As either solution could be costly, the plumbing installer should carefully consider the specified limitations for modification of all framing members before such alterations are performed.

Bearing Walls versus Nonbearing Walls

The terms "bearing" and "nonbearing" are used in the prescriptive allowances for the alteration of studs. In order to apply the allowances in the least restrictive manner, the plumbing installer must understand how the roof and floor loads of a building are supported by the wall framing. Exterior walls are always considered to be (load) bearing walls. Where floor joists, trusses or I-joists are supported by an interior wall, the wall is a (load) bearing wall. Roof trusses might also require support by interior walls and as such, those walls would be (load) bearing. However, not all walls located under the bottom chord of a roof truss are intended to support roof trusses. Where a wall only supports itself and no other framing member, it is non-(load) bearing. If the plumbing installer has doubt about whether a wall is bearing or nonbearing, he or she should either consult with the party responsible for the framing design or assume that the wall is (load) bearing.

Sawn Wood Members

The IBC offers prescriptive allowances for cutting, notching and boring of sawn wood studs, joists and rafters. These allowances apply to wood-constructed buildings that have been designed in accordance with the allowable stress design method, the load resistance design method or conventional light-frame construction design method. However, in some rare cases, some portions of or entire sawn wood constructed structures might be specially designed such that standard limitations for cutting, notching or boring of members are restricted further or prohibited altogether. The plumbing contractor should always ask the general contractor or framing contractor where, if any, such special designs are located in the building.

IBC Section 2308.8.2 covers notches and holes in wood floor joists.

IBC Section 2308.9.10 covers notching and cutting of wood studs.

IBC Section 2308.9.11 covers boring holes in wood studs.

IBC Section 2308.10.4.2 covers notching and holes in wood rafters and ceiling joists.

IBC Section 2308.9.8 covers holes and notches in wood wall sole and top plates.

Commentary Figure 307.2(1) shows the IBC limitations for modifications to wood studs. Commentary Figure 307.2(2) shows the IBC limitations for modifications to wood joists and rafters.

In Commentary Figure 307.2(3), the bearing wall studs have $2^1/_8$ inch (54 mm) holes drilled in 2 by 4 studs for passage of the $1^1/_2$-inch (38 mm) drain and vent pipes. The holes exceed 40 percent of the $3^1/_2$ inch (89 mm) actual depth of the stud and therefore, according to the IBC, the modified studs should have been evaluated by a design professional for structural adequacy. However, in this situation, the code official allowed the use of metal stud brackets (often called "stud shoes") to reinforce the studs at the oversized hole locations. Note that the IBC is silent on the use of metal stud shoes to reinforce wood studs where holes are larger than the IBC limitations. Although it may be common practice for stud shoes to be installed in these situations, the code official would have to approve their use. Commentary Figure 307.2(3) also shows notching on the face of the studs for $^3/_4$-inch (19 mm) copper water pipe. The allowable notch depth for a bearing stud is 25 percent of the stud depth or $^7/_8$ inch (22 mm). The notches in these studs are deeper than $^7/_8$ inch and thus are in violation of the notching rules.

Commentary Figures 307.2(4) through 307.2(8) show various violations of the cutting, notching and boring limitations for sawn wood members.

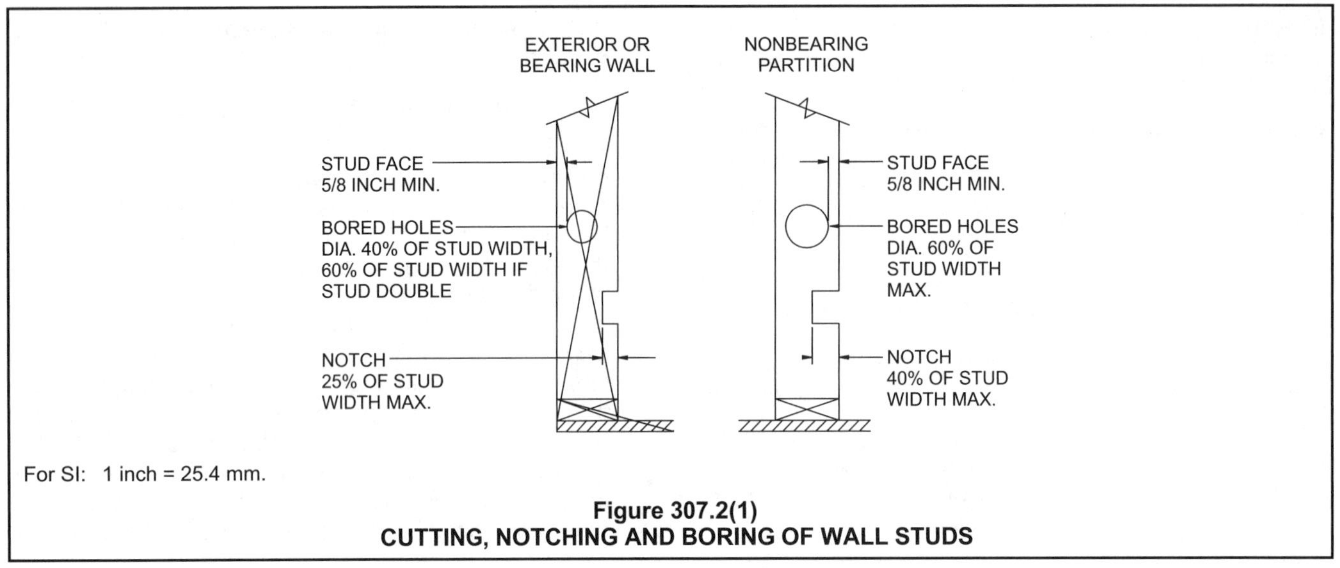

Figure 307.2(1)
CUTTING, NOTCHING AND BORING OF WALL STUDS

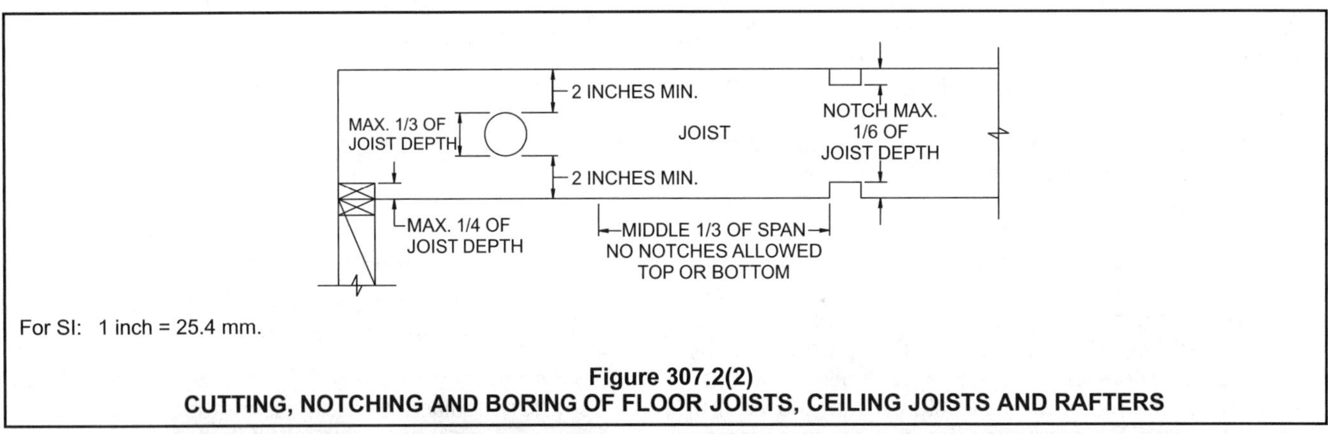

Figure 307.2(2)
CUTTING, NOTCHING AND BORING OF FLOOR JOISTS, CEILING JOISTS AND RAFTERS

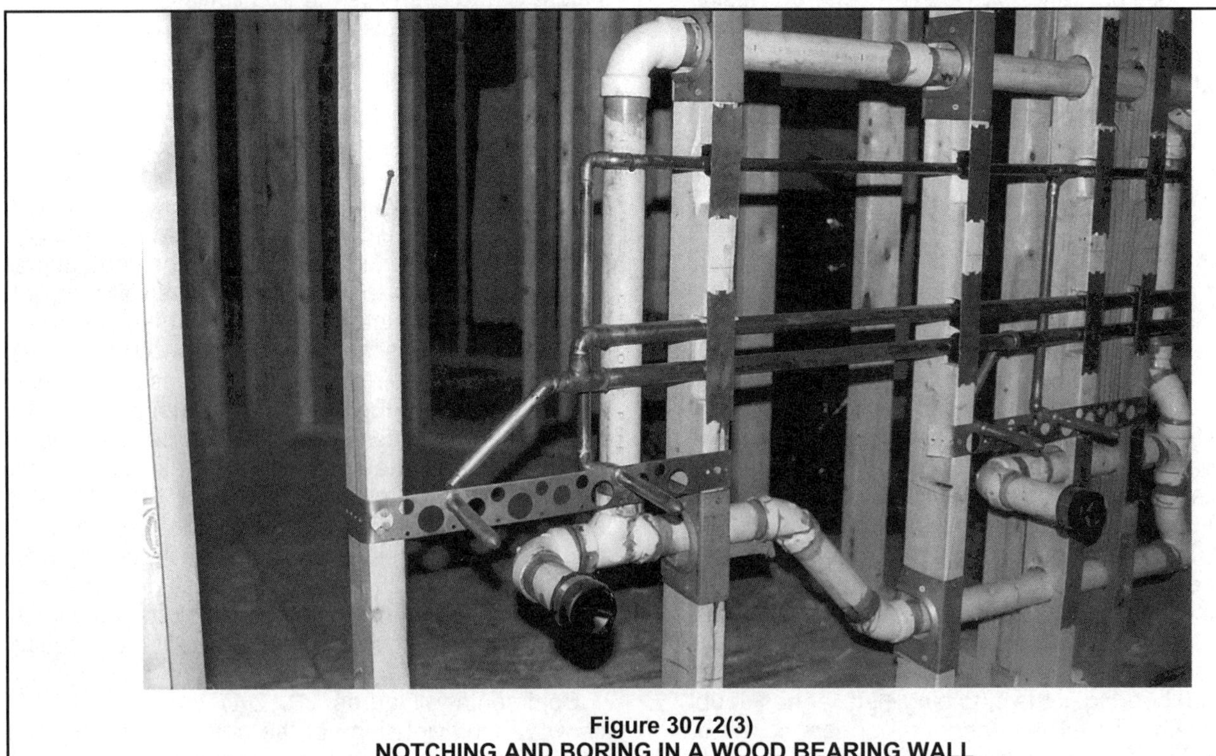

Figure 307.2(3)
NOTCHING AND BORING IN A WOOD BEARING WALL

Figure 307.2(4)
VIOLATION: NOTCH IN FLOOR JOIST TOO DEEP

Figure 307.2(5)
VIOLATION: HOLES TOO CLOSE TO BOTTOM OF JOIST

Figure 307.2(6)
VIOLATION: NOTCHES TOO DEEP IN WALL STUDS

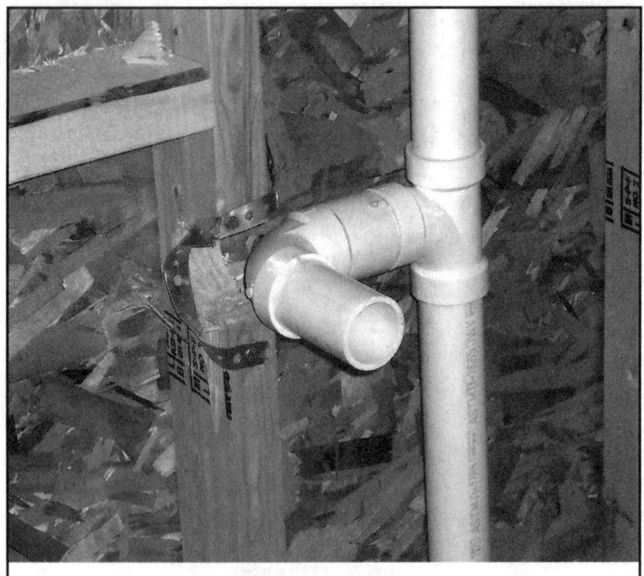

Figure 307.2(7)
VIOLATION: EXTERIOR
WALL STUD NOTCHED TOO DEEP

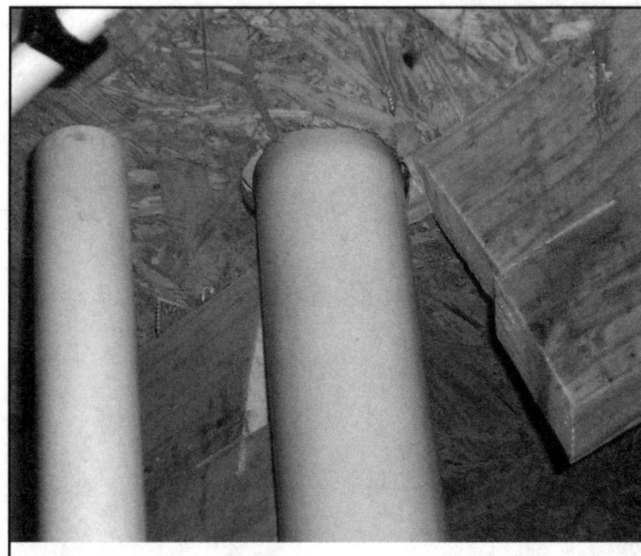

Figure 307.2(8)
VIOLATION: FLOOR JOIST COMPLETELY SEVERED

Figure 307.2(9)
VIOLATION: EXTERIOR WALL STUD COMPLETELY SEVERED

Sawn Wood Wall Plate Reinforcement

It is common for piping to be routed through holes or notches in the top plate of a (load) bearing wall (exterior or interior). Where this occurs, IBC Section 2308.9.8 requires that the top wall plate be reinforced with a metal tie (reinforcing strap). The IBC is silent as to what size of hole can exist without a metal tie being installed. Where the plate is completely severed, the IBC is also silent as to whether a metal tie is required on both faces of the wall. The IBC is also silent as to how wide a notch or missing section of top plate can be. The ICC Plan Review Department provided the following guidance based upon common sense and general knowledge of the types of loads carried by the top plate of a prescriptively designed (load) bearing wall:

1. Where a hole in the top plate leaves less than 1 inch (25 mm) between the edge of the hole and a face of the wall, that face of the top plate requires a metal tie to be installed. If the hole leaves less than 1 inch (25 mm) between the edge of the hole and both faces of the wall, both faces of the top plate require a metal tie to be installed.
2. Where the top plate is notched 50 percent or less of the width of the top plate, a metal tie is only required to be installed on the notched face of the top plate. If the notch exceeds 50 percent of the width of the top plate (including complete severing of the top plate), both faces of the top plate require a metal tie.
3. If a metal tie is required to be installed on the face of a wall but that face has structural sheathing installed, the metal tie is not required to be installed on that wall face.
4. The length of a notch (or the length of the section of plate completely removed) should not exceed the length allowed by the metal tie manufacturer's installation instructions. Typically, metal tie manufacturers offer ties for a maximum notch length of 12 inches (305 mm).
5. The final determination on the extent of the reinforcement required for a top wall plate that has been notched or bored is the responsibility of the code official.

The metal tie must be at least $1^1/_2$ inch (38 mm) wide by 0.058 inch (1.47 mm) No.16 galvanized gage thick. Each end of the tie must be fastened to the top plate with six 16d nails [see Commentary Figure 307.2(10)].

Note that the installation of reinforcing ties on both faces of a wall presents a complication. As each tie requires six 16d [diameter of 0.162 inches (4 mm)] nails [$3^1/_2$ inches (89 mm) long] at each end of the tie, nail-

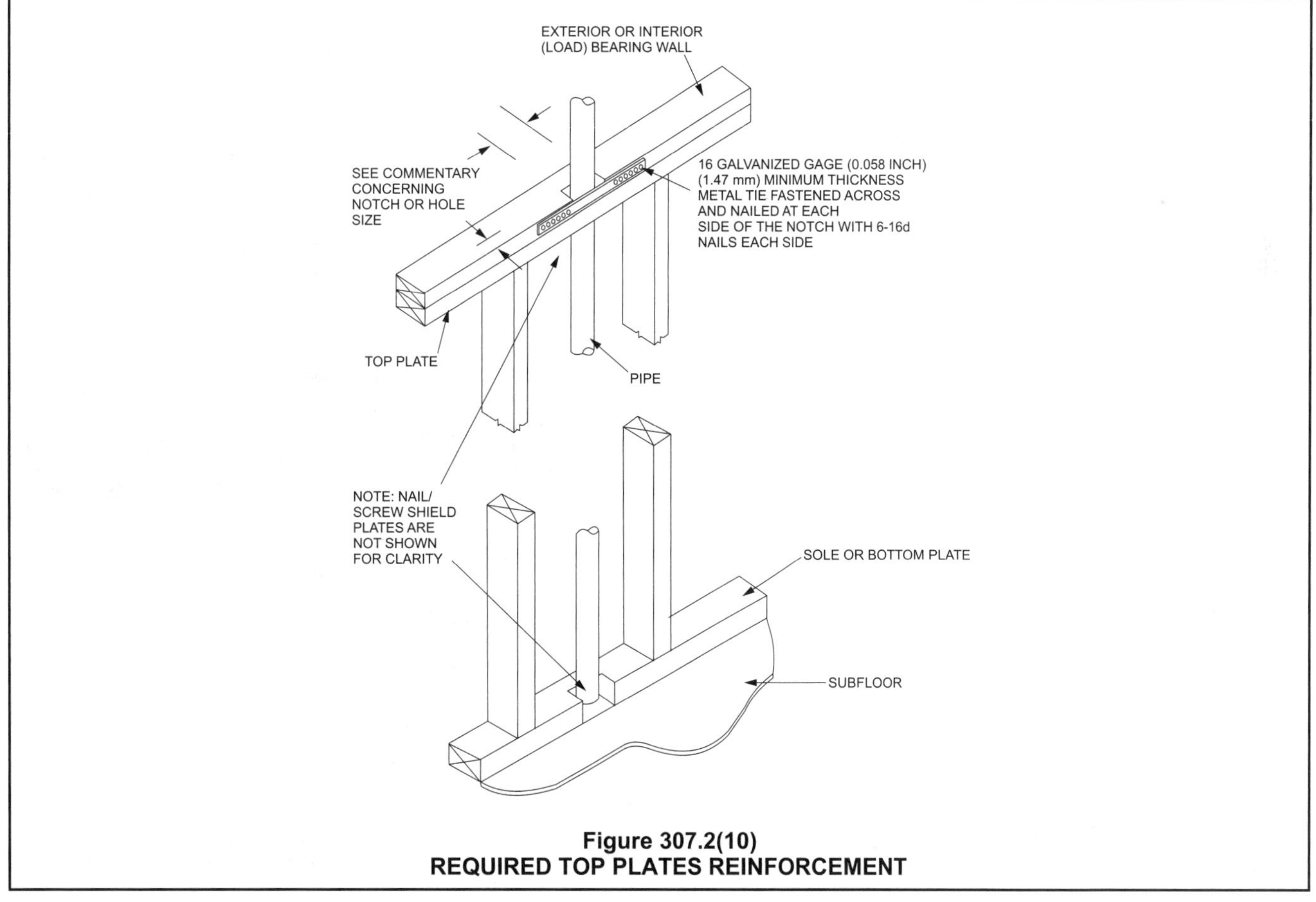

Figure 307.2(10)
REQUIRED TOP PLATES REINFORCEMENT

ing a tie on the same plate member on both faces of the wall will split the wood member. From this, one could conclude that where a single top plate is used, it would be prudent to limit the depth of a notch or size of a bored hole in the plate such that only one metal tie is necessary. Where the top plate is a double (member) plate, one face of the wall could have the tie installed on top member of the top plate and the other face of the wall have the tie installed on the lower member of the top plate. Some tie manufacturers might allow for the use of shorter (and smaller diameter) nails so as to permit nailing from both sides of the same top plate member without splitting the wood. Refer to the tie manufacturer's installation instructions for alternate fastener sizes. While shorter, smaller diameter, nails might slightly decrease the load carrying capacity of the tie, the consensus is that it is better to have the ties securely nailed to the plate member (with the tie manufacturer's prescribed alternate fasteners) rather than the ties having little to no load carrying capacity due to the plate member being split by 16d nailing on both faces of the plate member. Screws must not be used to fasten the ties to the plates unless specified by the tie manufacturer.

Where copper or plastic piping passes though wall frame members (including plates) within $1^1/_2$ inches (38 mm) of the edge of the plate (face of the wall), steel shield plates are required to be installed to protect the pipes from fastener penetration (see Section 305.8). The installation of the steel shield plates and the top plate reinforcing ties must be coordinated. If the top plate is a single member top plate (or a double member top plate requiring a tie on the lower member of the top plate) then, in order to prevent a bulge in the wall sheathing material, the edges of the wall plate should be notched for the combined thicknesses of the shield plate, the tie, and the heads of the tie fastening nails. The shield plate would then be installed first with the tie installed over the shield plate. Although some tie/shield plate manufacturers offer wide shield plates that have the required height dimension to cover a double member top plate plus two inches (50 mm) below the plate, these plates might not be suitable for use as top plate reinforcing ties. Use of a single shield plate to serve as the top plate reinforcing tie would require alternate approval by the code official in accordance with Section 105.2.

In summary, the plumbing designer or installer must carefully consider the impact of routing piping though the top plates of walls with respect to the requirements for metal ties (reinforcing straps). Commentary Figure 307.2(11) shows the severed top plates of an interior (load) bearing wall. Note that the close proximity of other pipes penetrating the plates would prevent the installation of the nails required for fastening of the required reinforcing ties. As always, thoughtful planning of a piping installation will allow the installer to avoid most, if not all, problematic drilling, cutting and notching. Good advice to plumbers would be: "Think before you pick up that saw or drill."

Figure 307.2(11)
VIOLATION: WALL TOP PLATES NOT REINFORCED

Engineered Wood Members

Engineered wood members are manufactured in many ways. The most common products used are laminated veneer lumber (LVL) beams [see Commentary Figure 307.2(12)], trusses of dimensional lumber, and I-joists of oriented strand board (OSB) webs with dimensional lumber top and bottom flanges (chords). Because most LVL manufacturers either prohibit the boring of holes or severely limit the size, location and number of holes in LVL members, as a general rule, holes should never be bored in LVL members. Truss members must not be altered in any manner.

The top and bottom chords of I-joist members must never be cut or notched. Although most I-joist manufacturers allow holes to be cut in the web of the member, the maximum size and location of holes must be in accordance with the specific manufacturer's product instructions. The cutting, notching and boring limi-

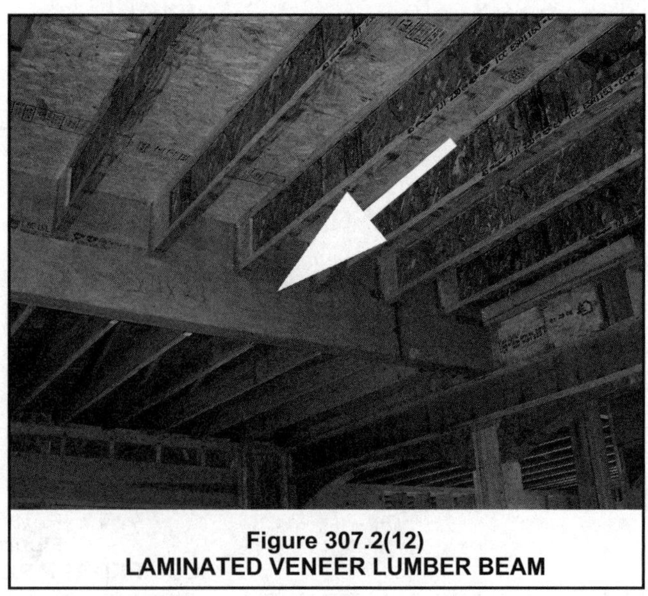

Figure 307.2(12)
LAMINATED VENEER LUMBER BEAM

tations indicated for sawn lumber [Commentary Figure 307.2(1)] do not apply to I-joists. The installer and inspector must have access to the product instructions in order to determine if hole sizes and locations are acceptable. A review of several I-joist manufacturers' installation instructions does provide some generalizations that can help the plumbing installer generally plan piping routes as well as alert the inspector that a violation might exist. The generalizations are:

1. The closer the hole is located to the bearing locations (supports) of the I-joist, the smaller the allowable size of the hole becomes. Typically, holes near the bearing ends are limited to $1^1/_2$ inches (38 mm) in diameter with a spacing (hole edge to hole edge) of not less than 12 inches (305 mm).

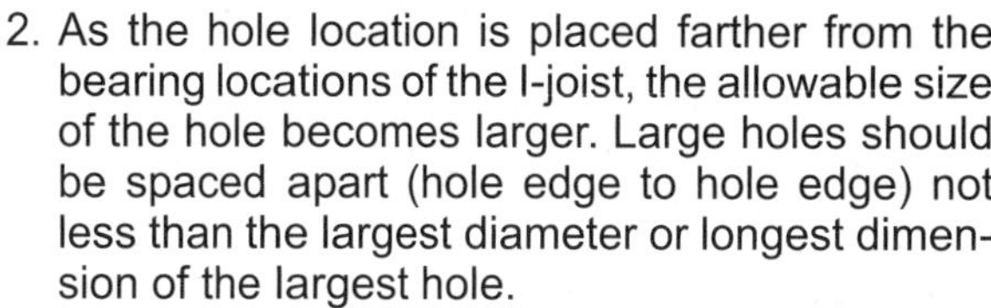

2. As the hole location is placed farther from the bearing locations of the I-joist, the allowable size of the hole becomes larger. Large holes should be spaced apart (hole edge to hole edge) not less than the largest diameter or longest dimension of the largest hole.

3. Holes should be cut round, square or rectangular. The edge of any hole should not be any closer than $^1/_4$ inch (6.4 mm) to a top or bottom chord of the I-joist.

As these rules are only generalizations, always refer to the I-joist manufacturer's installation instructions for specific hole limitations. Commentary Figures 307.2(13) through 307.2(17) show prohibited boring, cutting and notching of trusses and I-joists.

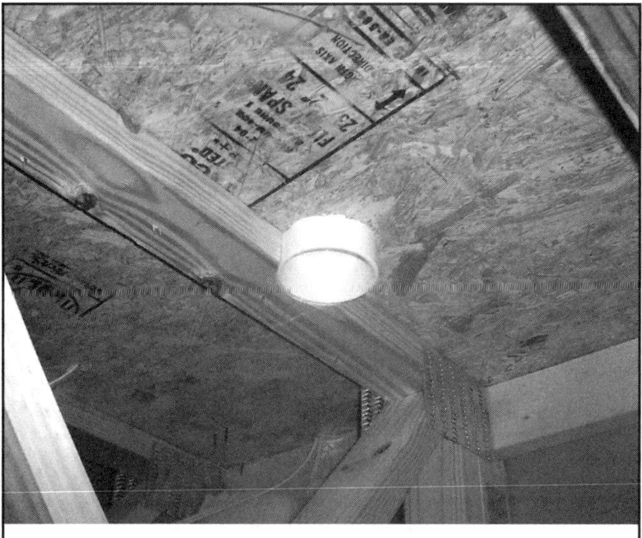

Figure 307.2(13)
VIOLATION: TOP CHORD OF TRUSS NOTCHED

Figure 307.2(14)
VIOLATION: HOLE IN BOTTOM CHORD OF ROOF TRUSS

Figure 307.2(15)
VIOLATION: TOP CHORD OF I-JOIST COMPLETELY SEVERED

Figure 307.2(16)
VIOLATION: I-JOIST WEB HOLES IRREGULAR SHAPE, TOO LARGE, TOO CLOSE TOGETHER

Cold-formed Steel Light-framed Members

Cold-formed steel light-framed members of trusses must not be altered in any manner. Flanges and lips of stud, floor joist, ceiling joist and rafter members must not be cut, notched, bored or punched [see Commentary Figure 307.2(18) for identification of flanges, lips and webs]. Holes in the webs of studs, floor joists, ceiling joists and rafter members are prohibited except where:

1. The holes are punched at the factory,
2. The design drawings for the building specifically indicate the location and size of holes, or
3. The prescriptive design method was used for the building framing design.

Because the IBC limits the use of the prescriptive design method to only one- and two-family detached dwellings or townhouses, up to three stories in height, a building outside that scope cannot be designed using the prescriptive method. [In most jurisdictions, the *International Residential Code®* (IRC®) is adopted as the controlling code for one- and two-family detached dwellings or townhouses, up to three stories in height. Where the IRC is the adopted code for these types of structures, refer to the IRC for all structural and plumbing requirements]. The IBC references the standard AISI A230 for the prescriptive design method for cold-formed steel-light frame construction. This standard has prescriptive allowances for holes and their location in the webs of studs, floor joists, ceiling joists and rafters [see commentary, Figures 307.2(19) and 307.2(20)].

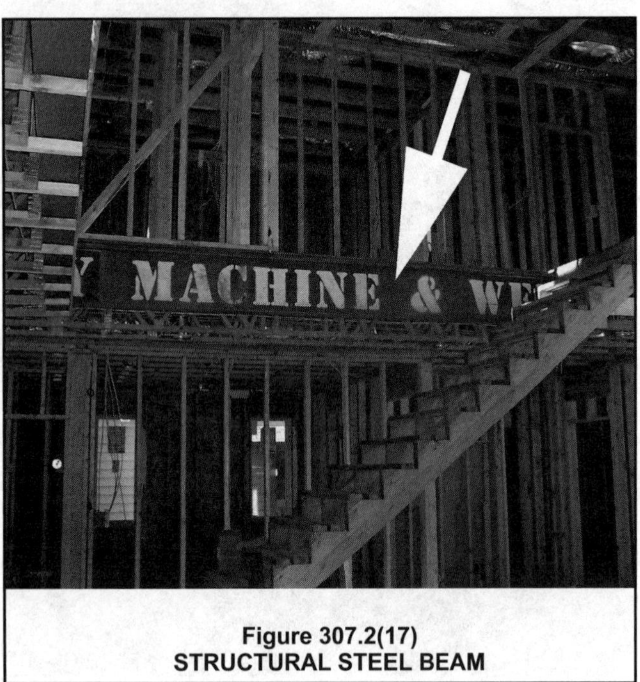

Figure 307.2(17)
STRUCTURAL STEEL BEAM

Structural Steel Members and Structural Steel Trusses

Any modifications to structural steel members [see commentary Figure 307.2(19)] or structural steel trusses must be approved by a registered design professional.

Log Members

The IBC references the standard ICC-400 for the construction of log structures. ICC-400 Section 302.2.4 covers the limitations for notching and boring of holes in logs used in structural applications. Commentary Figure 307.2(21) illustrates the notching and boring limitations.

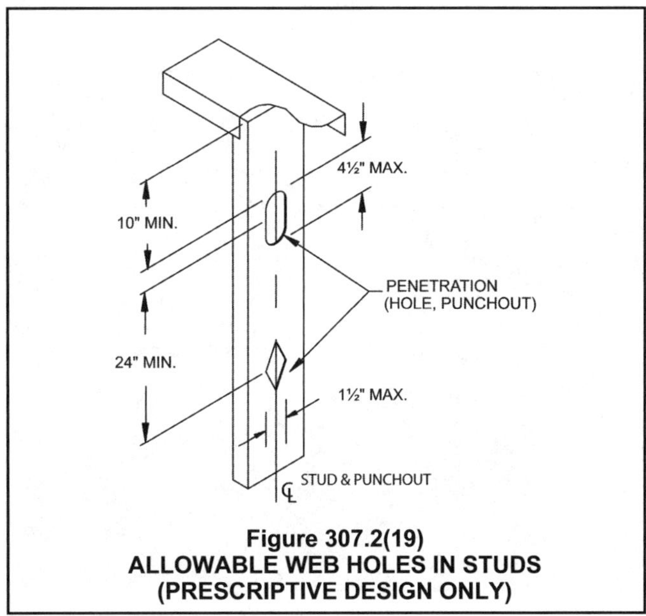

Figure 307.2(19)
ALLOWABLE WEB HOLES IN STUDS (PRESCRIPTIVE DESIGN ONLY)

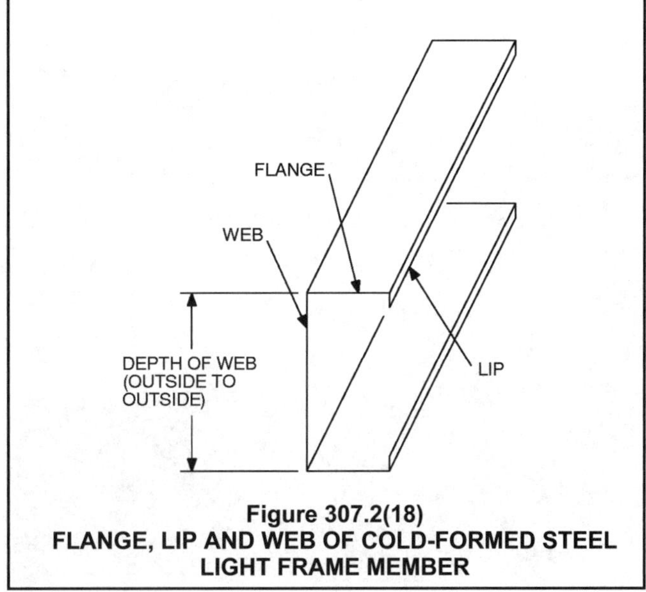

Figure 307.2(18)
FLANGE, LIP AND WEB OF COLD-FORMED STEEL LIGHT FRAME MEMBER

307.3 Penetrations of floor/ceiling assemblies and fire-resistance-rated assemblies. Penetrations of floor/ceiling assemblies and assemblies required to have a fire-resistance rating shall be protected in accordance with the *International Building Code*.

❖ This section states the requirement for the protection of penetrations and openings necessary for plumbing pipe in building assemblies must be fire-resistance rated. In addition to maintaining the structural integrity of walls, floors, ceilings and roofs when these elements are penetrated by the plumbing pipe, the penetrations must also be protected to prevent the passage of fire and other products of combustion through the fire-resistance-rated assembly because of the pipe penetrating the assembly. These penetrations must be protected in accordance with the IBC [see IBC Sections 711 and 712 and IPC Commentary Figures 307.3(1) through 307.3(4)].

[B] 307.4 Alterations to trusses. Truss members and components shall not be cut, drilled, notched, spliced or otherwise altered in any way without written concurrence and approval of a registered design professional. Alterations resulting in the addition of loads to any member (e.g., HVAC equipment, water heater) shall not be permitted without verification that the truss is capable of supporting such additional loading.

❖ Trusses are engineered products, and any alteration to specific elements may prevent the truss from performing as intended. The code does not permit drilling and notching of truss members and components without documentation that the design has taken this possibility into account. Additionally, trusses are designed to resist specific loads; the loading on the truss or on any of its members should not be changed without approval of a registered design professional (see Commentary Figure 307.4).

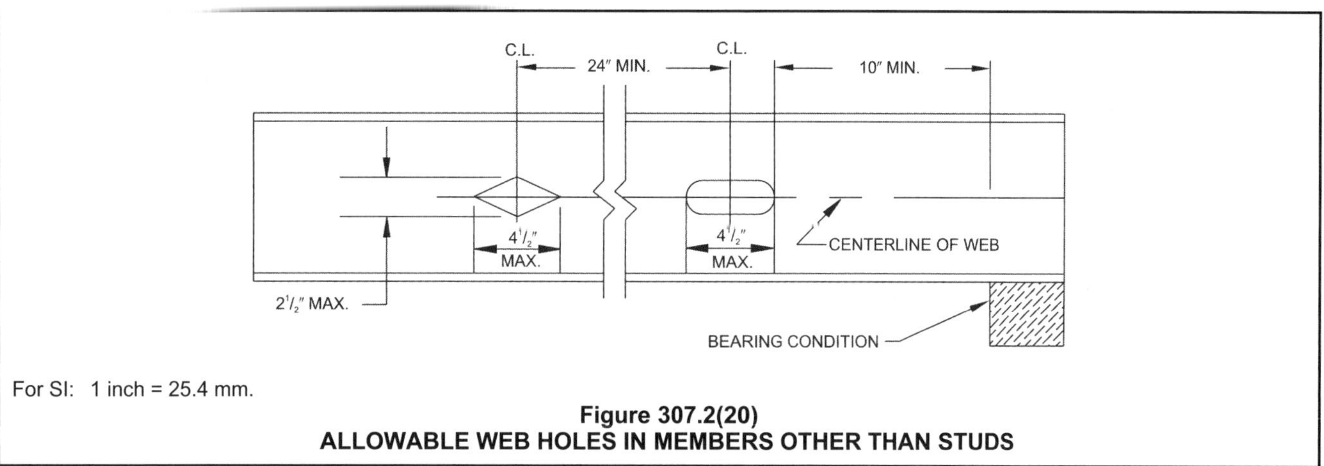

For SI: 1 inch = 25.4 mm.

Figure 307.2(20)
ALLOWABLE WEB HOLES IN MEMBERS OTHER THAN STUDS

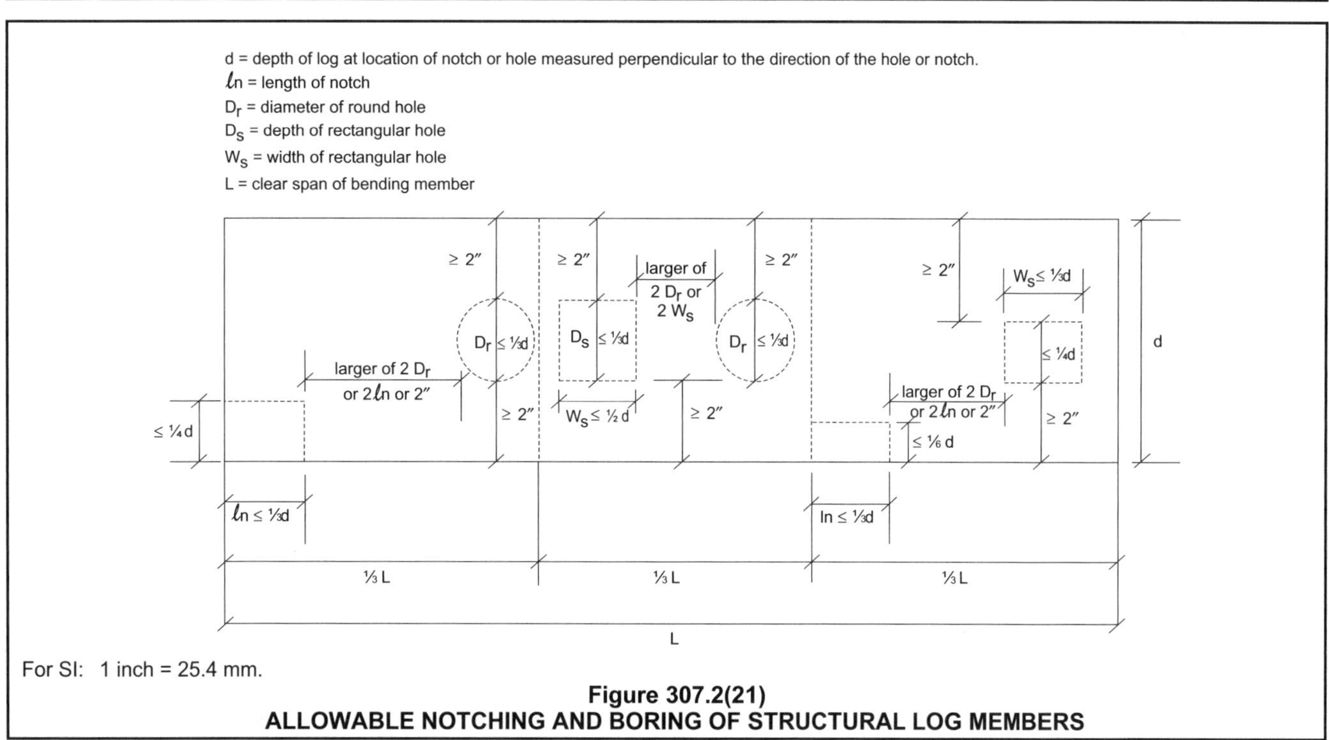

For SI: 1 inch = 25.4 mm.

Figure 307.2(21)
ALLOWABLE NOTCHING AND BORING OF STRUCTURAL LOG MEMBERS

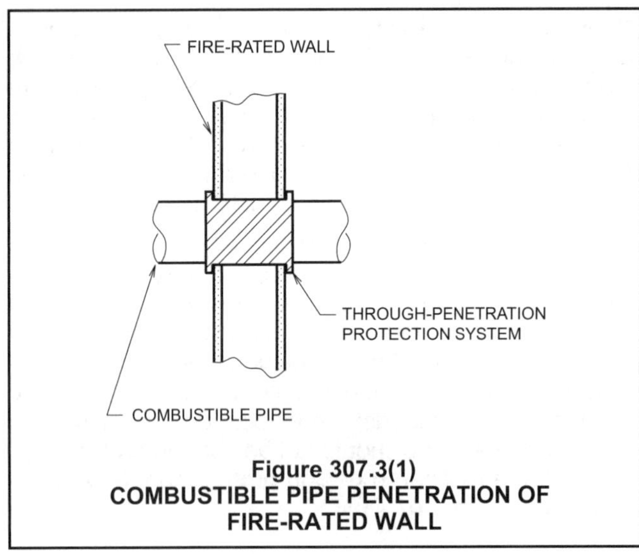

Figure 307.3(1)
COMBUSTIBLE PIPE PENETRATION OF FIRE-RATED WALL

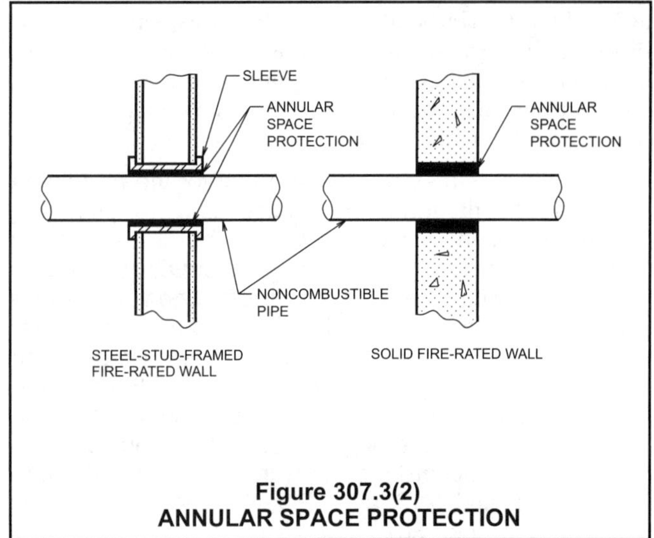

Figure 307.3(2)
ANNULAR SPACE PROTECTION

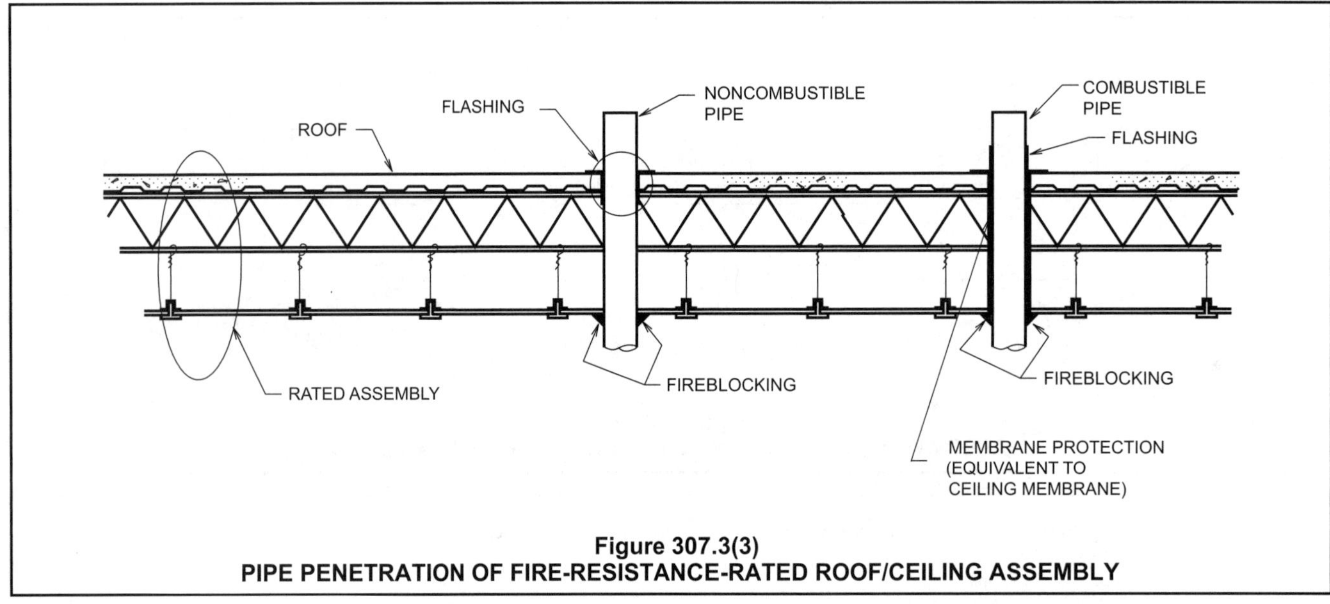

Figure 307.3(3)
PIPE PENETRATION OF FIRE-RESISTANCE-RATED ROOF/CEILING ASSEMBLY

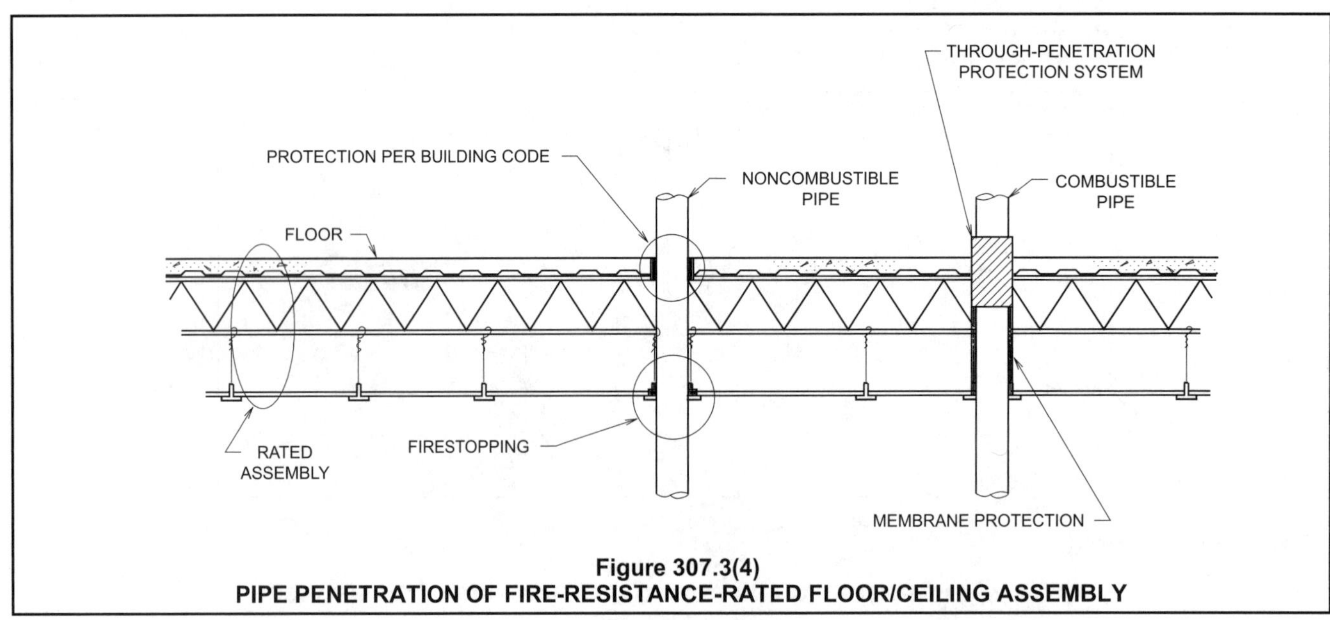

Figure 307.3(4)
PIPE PENETRATION OF FIRE-RESISTANCE-RATED FLOOR/CEILING ASSEMBLY

307.5 Trench location. Trenches installed parallel to footings shall not extend below the 45-degree (0.79 rad) bearing plane of the footing or wall.

❖ A footing requires a minimum load-bearing area to distribute the weight of the building. This load-bearing distribution plane extends downward at approximately a 45-degree (0.79 rad) angle from the base of the footing. Water and sewer piping must not be installed below this load-bearing plane. Excavation for the installation of pipe below the plane could affect the load capacity of the footing, or cause the excavation to collapse (see commentary, Figure 307.5).

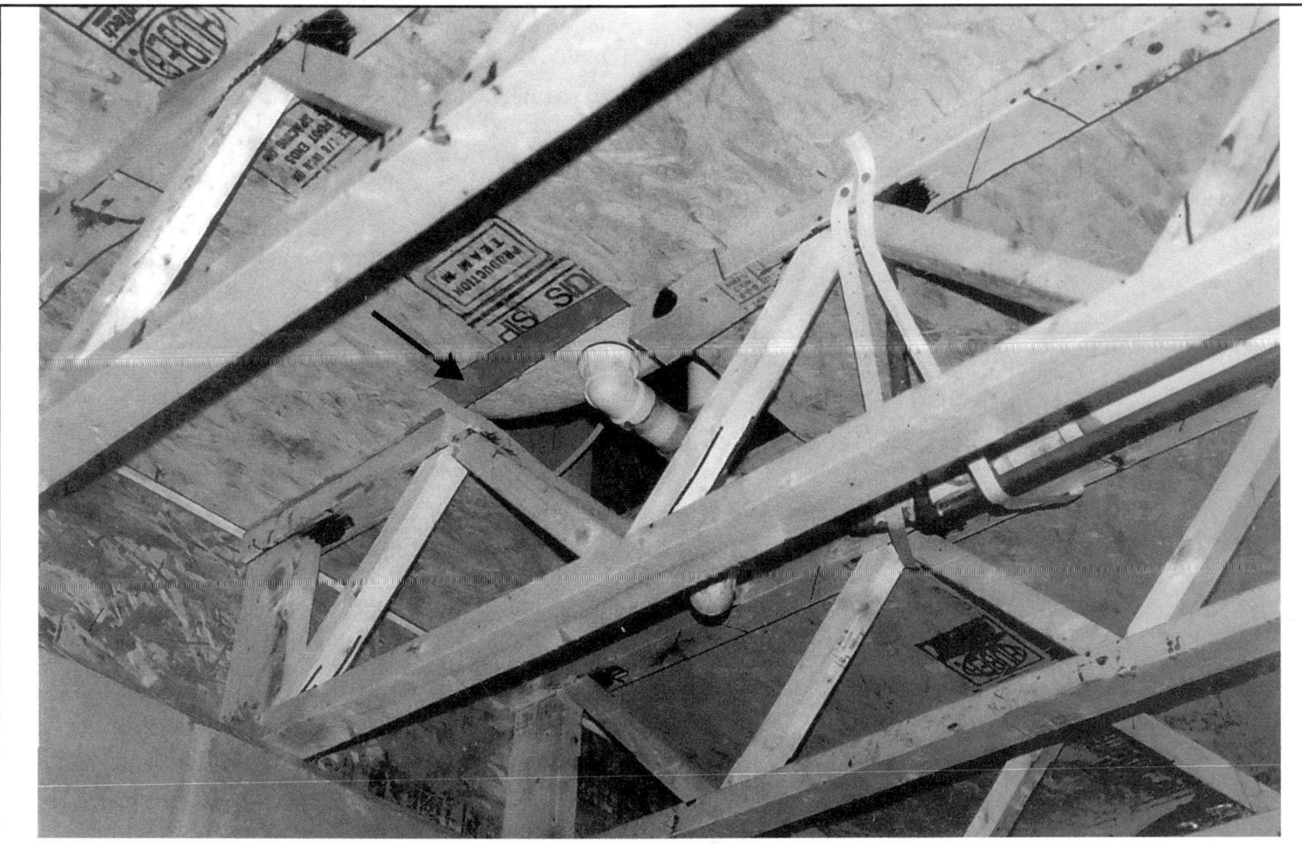

Figure 307.4
VIOLATION: TOP CHORD OF FLOOR TRUSS REMOVED

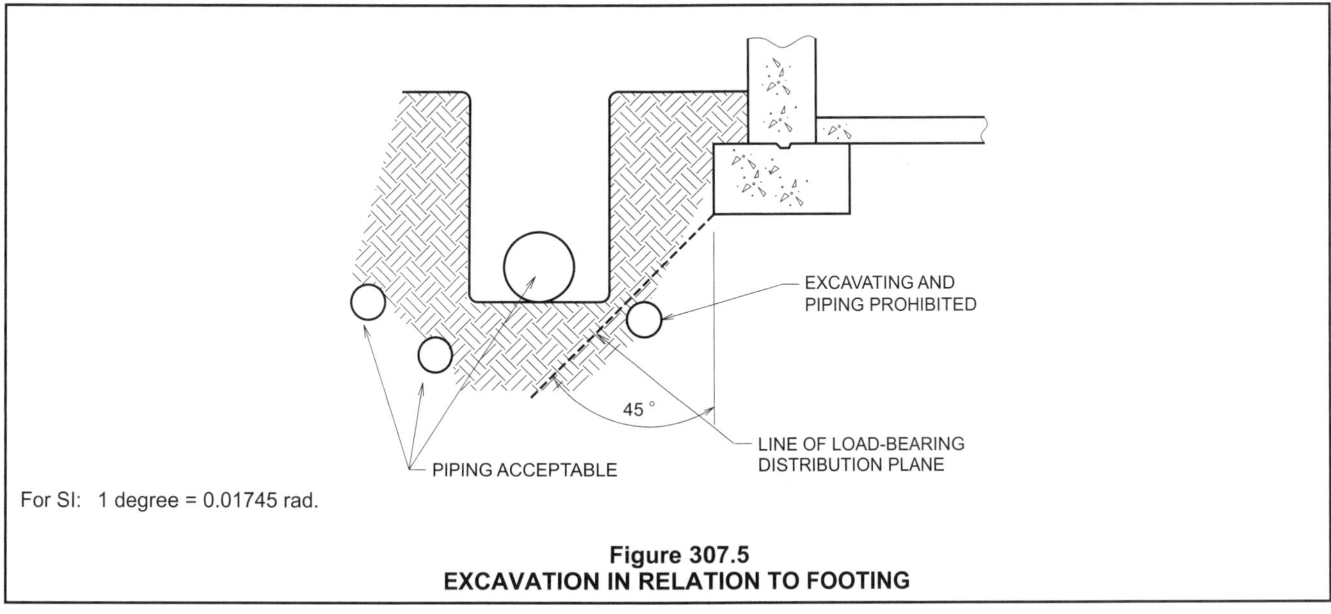

For SI: 1 degree = 0.01745 rad.

Figure 307.5
EXCAVATION IN RELATION TO FOOTING

307.6 Piping materials exposed within plenums. All piping materials exposed within plenums shall comply with the provisions of the *International Mechanical Code*.

❖ Materials exposed within a plenum must be noncombustible or have a flame spread index of 25 or less and a smoke-developed index of 50 or less when tested in accordance with ASTM E 84. Typically, the plenum space is often used to accommodate components of the plumbing system and other building systems such as electrical, fire protection and mechanical. The code permits exposure of limited types of combustibles used for these systems within the plenum. A flame spread index of 25 or less and a smoke-developed index of 50 or less are viewed as acceptable values for combustibles in a plenum because of the minimal risk associated with those materials.

SECTION 308
PIPING SUPPORT

308.1 General. All plumbing piping shall be supported in accordance with this section.

❖ There are many variables to consider during the design and installation of hangers and supports. Consideration must be given to the type and size of the pipe material; the weight of the pipe, fittings and contents; the structural element available for attachment; earthquake motions; vibration-sensitive applications and thermal expansion and contraction. Other load conditions to consider include the weight of valves and appurtenances; insulating materials; hanger and support components; and wind, snow and ice loads.

308.2 Piping seismic supports. Where earthquake loads are applicable in accordance with the building code, plumbing piping supports shall be designed and installed for the seismic forces in accordance with the *International Building Code*.

❖ While the IBC prescribes the level of seismic forces to be considered in the design of buildings in seismic zones, it does not provide prescriptive methods for designing, sizing or installing piping supports. A registered design professional should be consulted to provide the required seismic support designs and installation instructions.

308.3 Materials. Hangers, anchors and supports shall support the piping and the contents of the piping. Hangers and strapping material shall be of *approved* material that will not promote galvanic action.

❖ The hangers and supports for plumbing system piping must be capable of supporting the load imposed by the piping system and not be detrimental to the pipe they support. When the hanger or strapping material is not compatible with the piping material it supports, corrosion caused by galvanic action can occur. This happens when dissimilar metals are in contact with one another and sufficient moisture is present to carry an electrical current. Such corrosion can deteriorate the hanger, anchor or piping to the point of failure.

Where metallic pipe is installed, the hangers or supports must be of similar material to prevent any corrosive galvanic action. For example, if the water distribution system is copper tubing or copper pipe, copper hangers, straps and clamps are required or such supports must be of a material or clad by a material (such as plastic) that will not react with copper. Some support devices have special plastic coatings or are made entirely of plastic to prevent contact of dissimilar metals and are acceptable alternatives (see commentary, Figures 308.3(1) and 308.3(2).

Figure 308.3(1)
PLASTIC HANGER STRAP

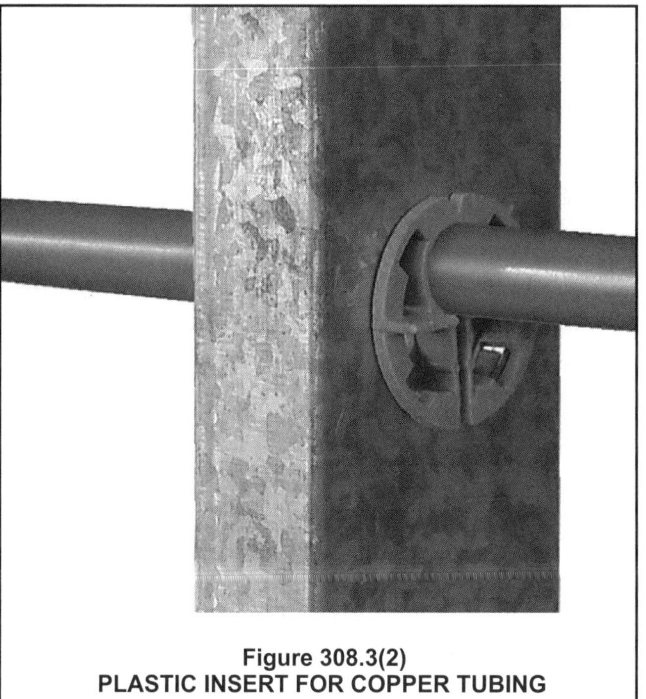

Figure 308.3(2)
PLASTIC INSERT FOR COPPER TUBING THROUGH METAL STUD
(Photo Courtesy of Holdrite-Hubbard Enterprises)

308.4 Structural attachment. Hangers and anchors shall be attached to the building construction in an *approved* manner.

❖ The selection of an appropriate hanger or support depends on the structural element available for hanger attachment. The integrity of all hangers and anchors depends on the method of attachment to the building construction. The means of attachment must provide the required strength and must not adversely affect or overstress the construction to which it is attached. Ultimately, a support is only as strong as its means of attachment and the member to which it is attached (see commentary, Figure 308.4).

Figure 308.4
POOR HANGER ATTACHMENT

308.5 Interval of support. Pipe shall be supported in accordance with Table 308.5.

Exception: The interval of support for piping systems designed to provide for expansion/contraction shall conform to the engineered design in accordance with Section 105.4.

❖ Piping must be supported at distances not to exceed those indicated in Table 308.5. The exception allows for design of special piping support systems in order to accommodate the movements caused by thermal expansion of piping.

TABLE 308.5
HANGER SPACING

PIPING MATERIAL	MAXIMUM HORIZONTAL SPACING (feet)	MAXIMUM VERTICAL SPACING (feet)
ABS pipe	4	10[b]
Aluminum tubing	10	15
Brass pipe	10	10
Cast-iron pipe	5[a]	15
Copper or copper-alloy pipe	12	10
Copper or copper-alloy tubing, $1^1/_4$-inch diameter and smaller	6	10
Copper or copper-alloy tubing, $1^1/_2$-inch diameter and larger	10	10
Cross-linked polyethylene (PEX) pipe	2.67 (32 inches)	10[b]
Cross-linked polyethylene/aluminum/cross-linked polyethylene (PEX-AL-PEX) pipe	2.67 (32 inches)	4
CPVC pipe or tubing, 1 inch and smaller	3	10[b]
CPVC pipe or tubing, $1^1/_4$ inches and larger	4	10[b]
Steel pipe	12	15
Lead pipe	Continuous	4
Polyethylene/aluminum/polyethylene (PE-AL-PE) pipe	2.67 (32 inches)	4
Polypropylene (PP) pipe or tubing 1 inch and smaller	2.67 (32 inches)	10[b]
Polypropylene (PP) pipe or tubing, $1^1/_4$ inches and larger	4	10[b]
PVC pipe	4	10[b]
Stainless steel drainage systems	10	10[b]

For SI: 1 inch = 25.4 mm, 1 foot = 304.8 mm.

a. The maximum horizontal spacing of cast-iron pipe hangers shall be increased to 10 feet where 10-foot lengths of pipe are installed.

b. Midstory guide for sizes 2 inches and smaller.

❖ The horizontal support intervals listed in the table are based on the limiting mid-span deflection (sag) of pipes. The maximum amount of sag occurs at a point equidistant from the supports when the pipe is full of water. The table is designed to limit sag (and the resulting pipe stress) as well as provide for the maintaining of slopes required for gravity drainage.

Because any given length of pipe becomes more resistant to mid-span deflection as the pipe diameter becomes larger, the spacing in the table is calculated for the pipe diameters at the low end of the pipe size ranges stated [see commentary, Figures 308.5(1) and 308.5(2)]. As this is a generalized table intended to cover a wide range of manufacturers, the installation instructions from the specific manufacturer should also be consulted to determine if support spacing is required to more be restrictive.

The spacing intervals indicated will result in the minimum number of supports allowed. Extra supports should be used where necessary to accommodate items such as fittings, connections, changes in direction or the connection to a vertical pipe.

Hubless cast-iron pipe, because of the nonrigid nature of the joints, must be supported at sufficiently close intervals to maintain alignment and prevent sagging or grade reversal. For example, approved anchors, spaced at least every 5 feet (1524 mm) for 5-foot (1524 mm) lengths of pipe or every 10 feet (3048 mm) for 10-foot (3048 mm) lengths of pipe, must be installed in accordance with the manufacturer's instructions.

The support requirements for polybutylene pipe and tube were removed in the 2009 code edition as polybutylene is no longer for water service or water distribution.

308.6 Sway bracing. Rigid support sway bracing shall be provided at changes in direction greater than 45 degrees (0.79 rad) for pipe sizes 4 inches (102 mm) and larger.

❖ This section contains lateral support requirements for pipe 4 inches (102 mm) and larger in diameter.

The flow of waste in drainage piping subjects the piping to the forces that result from the momentum of the waste.

These forces are a product of the mass and velocity

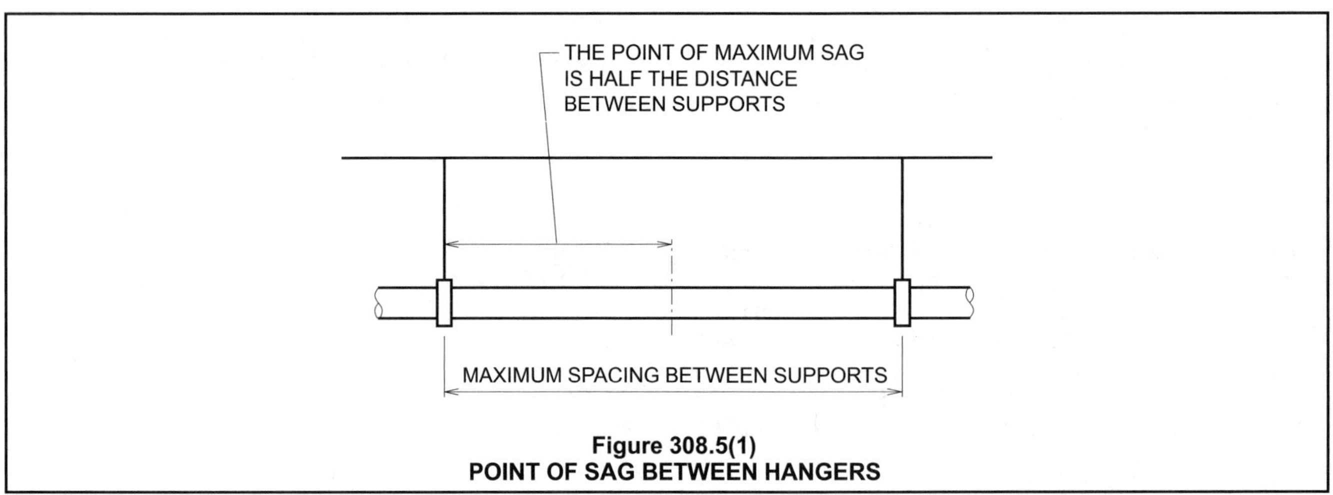

Figure 308.5(1)
POINT OF SAG BETWEEN HANGERS

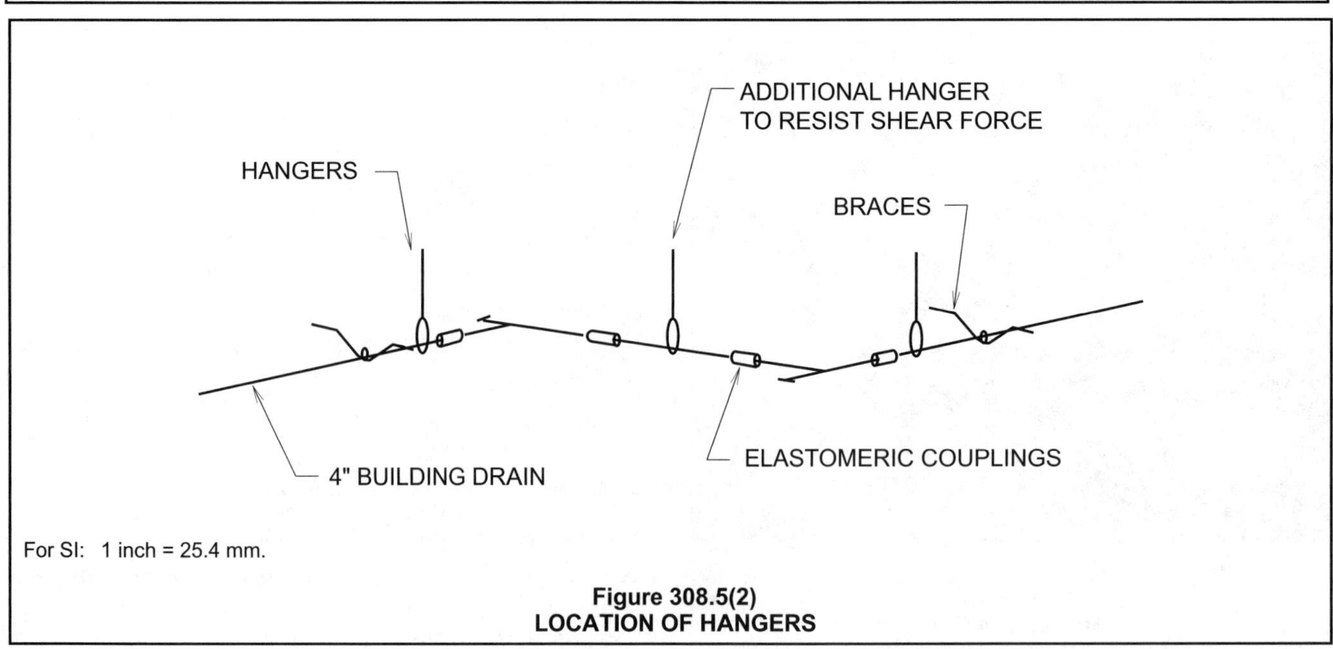

Figure 308.5(2)
LOCATION OF HANGERS

of the flow together with the change in direction of the pipe and can be quite large, especially in piping that is 4 inches (102 mm) and larger. Hangers alone might not be sufficient to resist such forces; therefore, rigid bracing is required at all changes of direction greater than 45 degrees (0.79 rad) for piping that is 4 inches (102 mm) or larger in size. Without adequate bracing, the piping could be damaged or joint failure could occur. Bracing is intended to restrict or eliminate lateral movement of both horizontal and vertical piping (see commentary, Section 308.7.1 and commentary, Figure 308.6).

308.7 Anchorage. Anchorage shall be provided to restrain drainage piping from axial movement.

❖ This section requires a method of resisting axial movement of piping systems to prevent joint separation, regardless of the type of fittings or connections used. In particular, mechanical couplings using an elastomeric seal (typically hubless piping systems) have a limited ability to resist axial movement (pulling apart); therefore, pipe restraints must be installed to prevent joint separation. The required anchorage locations of axial restraints are specified in Section 308.7.1. Such joints also have a limited ability to resist shear forces. The hanger and support system must, therefore, prevent the couplings and connections from being subjected to shear forces that can damage the joint (see commentary, Figure 308.7). Although mechanical couplings using elastomeric seals are most often the type of joint to experience damage caused by axial movement, this requirement pertains to all portions of drainage piping systems.

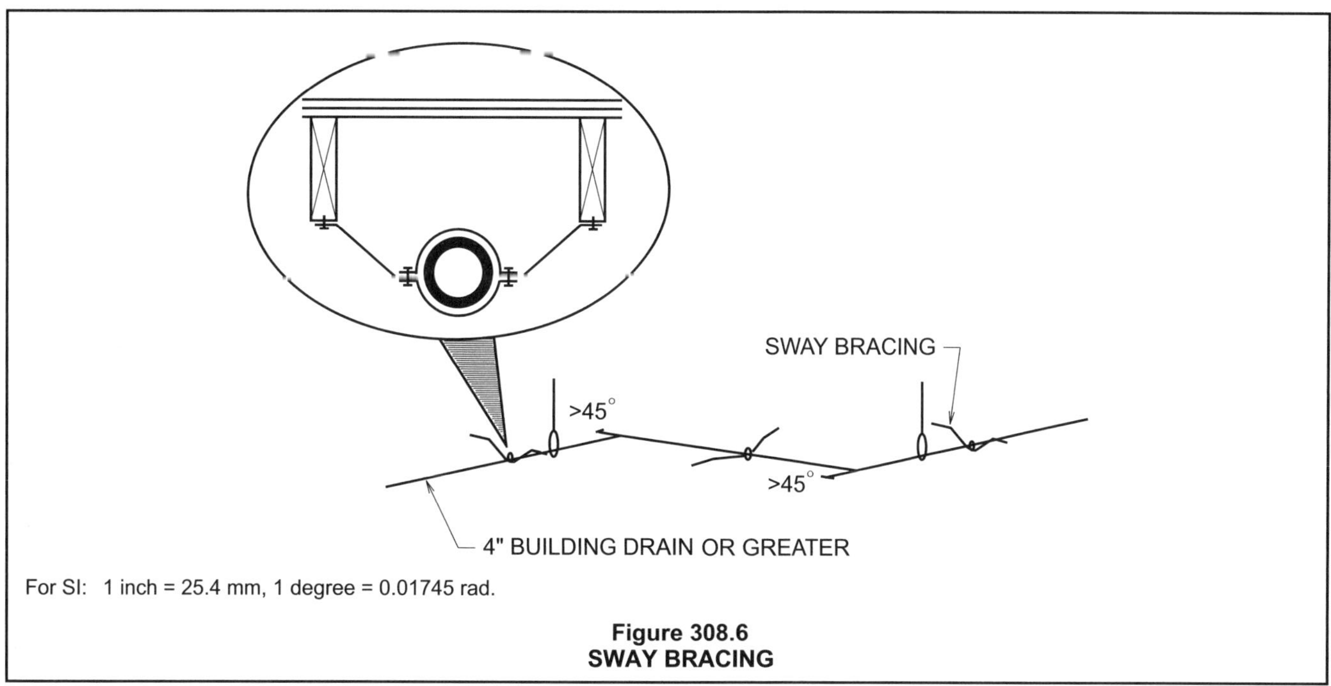

Figure 308.6
SWAY BRACING

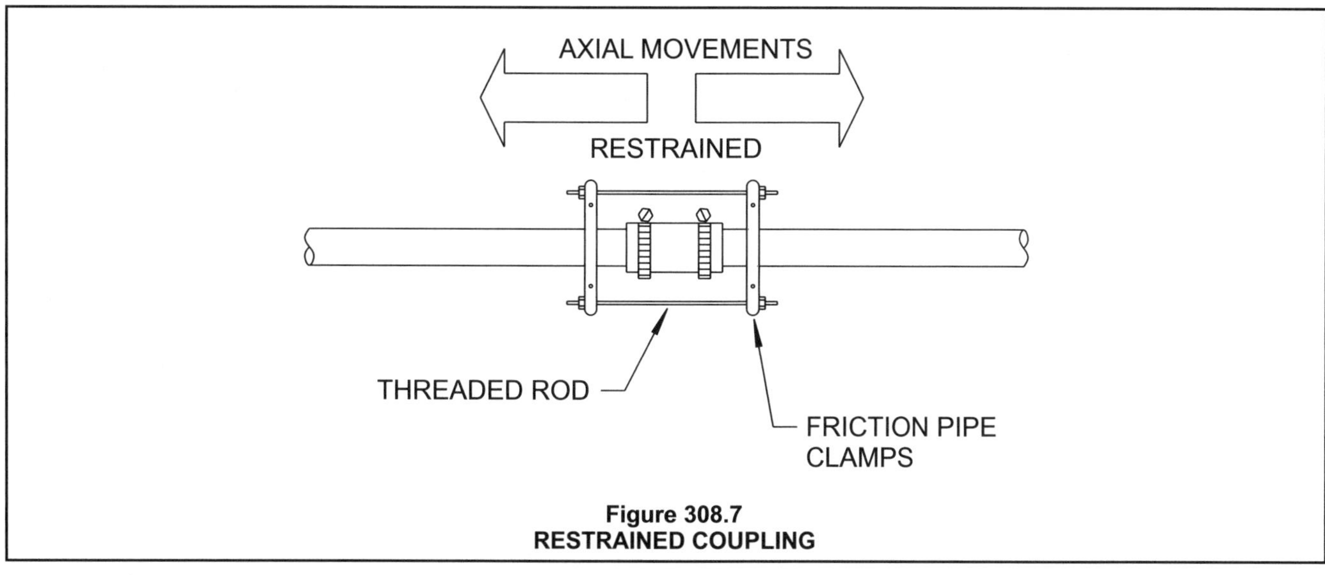

Figure 308.7
RESTRAINED COUPLING

308.7.1 Location. For pipe sizes greater than 4 inches (102 mm), restraints shall be provided for drain pipes at all changes in direction and at all changes in diameter greater than two pipe sizes. Braces, blocks, rodding and other suitable methods as specified by the coupling manufacturer shall be utilized.

❖ In drain pipes that are large in diameter, particularly those having a diameter of 4 inches (102 mm) or more, the flow of drainage can create sizable forces in the system resulting from the mass and velocity of the waste. The force could have the effect of pushing apart mechanical and other types of couplings when there is a change in direction or diameter. To resist such forces in the piping, additional anchorage and supports would be necessary to hold the pipe and couplings in place and prevent joint separation caused by this dynamic force (see commentary, Figure 308.7.1).

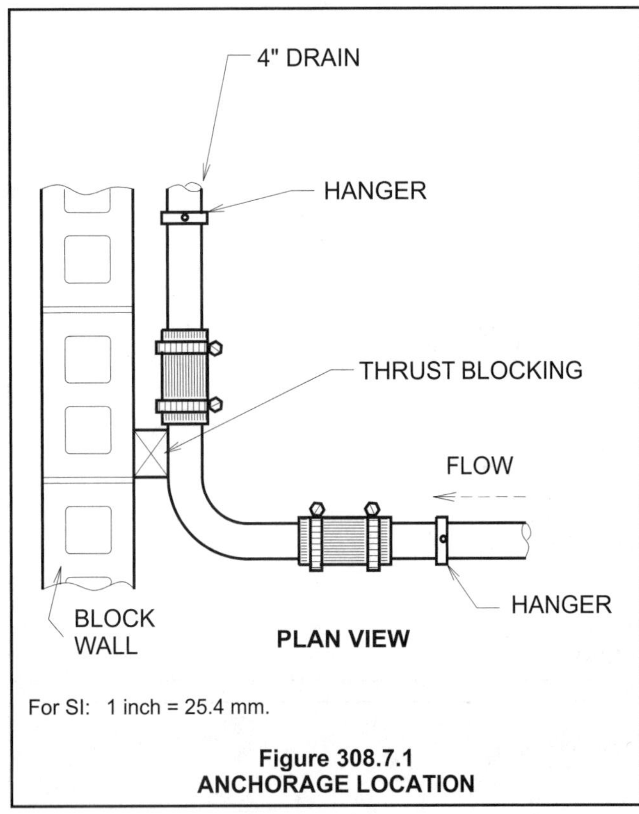

Figure 308.7.1
ANCHORAGE LOCATION

308.8 Expansion joint fittings. Expansion joint fittings shall be used only where necessary to provide for expansion and contraction of the pipes. Expansion joint fittings shall be of the typical material suitable for use with the type of piping in which such fittings are installed.

❖ Plumbing and drainage pipe materials experience changes in length caused by the changing temperature of the pipe material, typically as a result of a change in water temperature or ambient air temperature. Repeated expansion and contraction of the pipe material can induce stress into the pipe and its attached fittings. To relieve this stress and thus prevent damage to the pipe, its fittings and the surrounding building components, properly designed expansion joints are installed in the piping runs. The designer of the piping system must determine whether an expansion joint in the pipe run is required based on the length of the run, the pipe material and diameter, the expected temperature range of the material inside the pipe and the exact configuration of the installed pipe and fittings.

Where installed, expansion joints must be of a material compatible with the pipe. The installer of the expansion joint must follow the manufacturer's instructions so that the finished joint is leak free while still providing the required expansion. Special attention needs to be given to the number and type of pipe hangers used when a pipe run is equipped with an expansion joint, so that the hangers do not restrict the movement of either the pipe or the expansion joint while providing the required support to both.

308.9 Parallel water distribution systems. Piping bundles for manifold systems shall be supported in accordance with Table 308.5. Support at changes in direction shall be in accordance with the manufacturer's installation instructions. Hot and cold water piping shall not be grouped in the same bundle.

❖ Parallel water distribution systems require different methods of pipe support than conventional piping systems. A parallel water distribution system consists of a central manifold that feeds independent supply lines to each fixture. Plastic pipe is a common material used for both the independent supply lines from, and the supply line to, the manifold. To reduce labor costs, supply lines that make up the parallel water distribution system have common routes and are generally bundled together with plastic tie wraps. A bundle of plastic supply lines is treated as a single unit and supported by hangers in accordance with the spacing requirements found in Table 308.5.

In the case of semirigid plastic pipe, changes in direction must be supported to protect the piping and maintain the position of piping at the bend. Plastic pipe manufacturers provide specific pipe bending instructions and offer special devices to protect pipe bends from flattening and bending stresses that can damage the pipe [see commentary, Figures 308.9(1) through 308.9(4)].

Bundles must be held together loosely by plastic ties to allow for the expansion and contraction of the individual pipes. To prevent thermal transfer between hot and cold water piping, they must be segregated to avoid contact with each other.

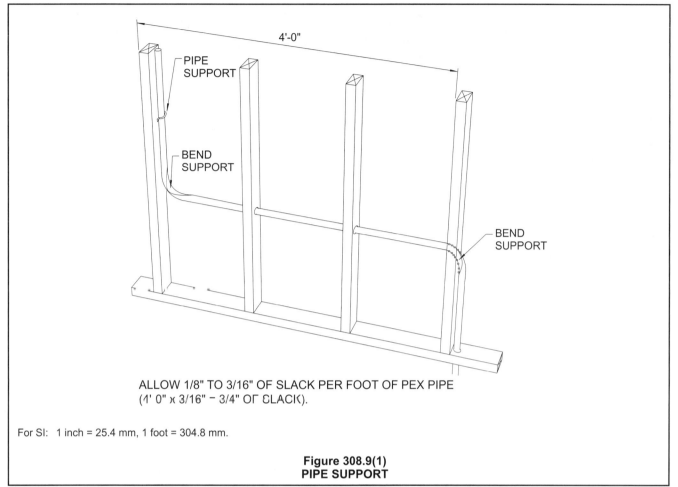

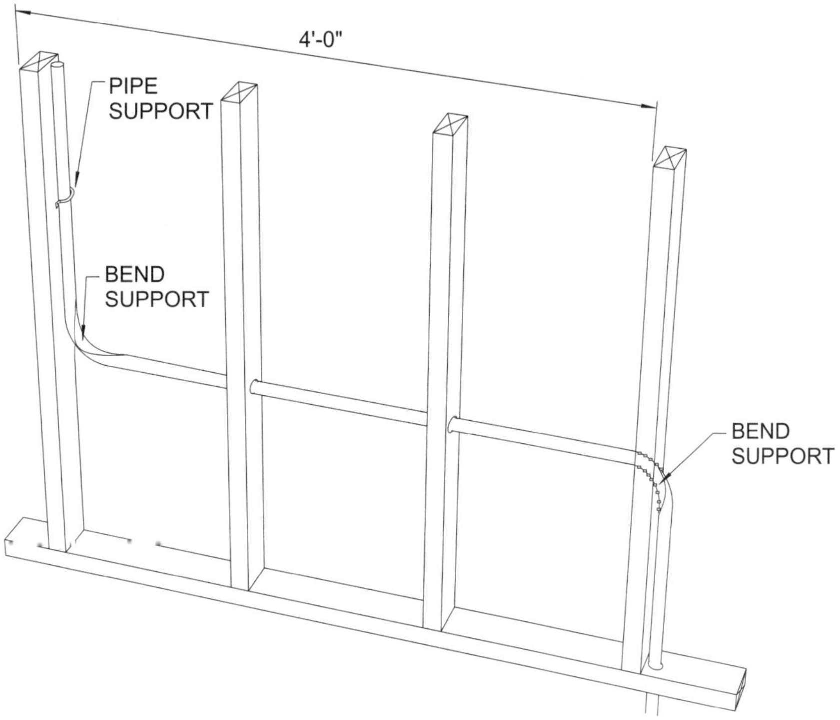

4'-0"

PIPE SUPPORT

BEND SUPPORT

BEND SUPPORT

ALLOW 1/8" TO 3/16" OF SLACK PER FOOT OF PEX PIPE
(4' 0" x 3/16" = 3/4" OF SLACK).

For SI: 1 inch = 25.4 mm, 1 foot = 304.8 mm.

Figure 308.9(1)
PIPE SUPPORT

Figure 308.9(2)
PIPE BEND SUPPORT MAINTAINING BEND RADIUS
(Photo courtesy of Holdrite-Hubbard Enterprises)

Figure 308.9(3)
PIPE BEND SUPPORT MAINTAINING BEND RADIUS
(Photo courtesy of Holdrite-Hubbard Enterprises)

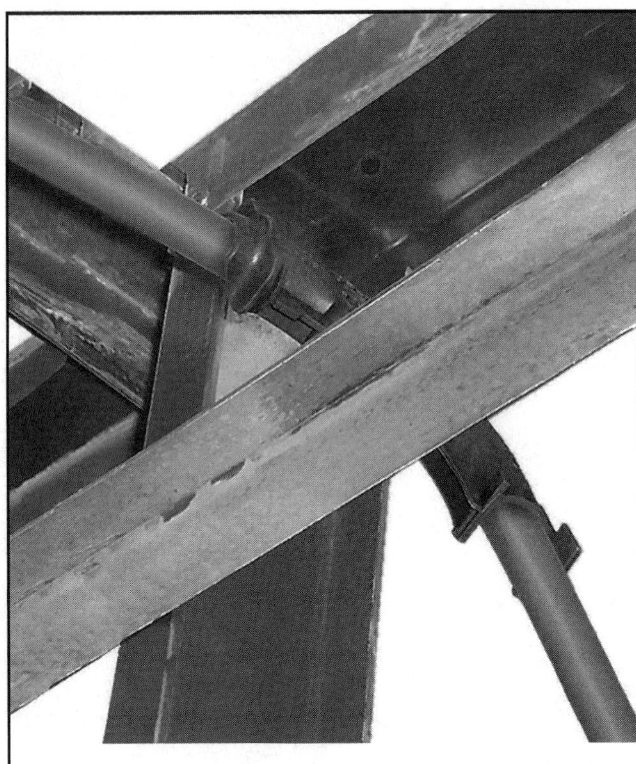

Figure 308.9(4)
PIPE BEND SUPPORT MAINTAINING BEND RADIUS
(Photo courtesy of Holdrite-Hubbard Enterprises)

SECTION 309
FLOOD HAZARD RESISTANCE

309.1 General. Plumbing systems and equipment in structures erected in flood hazard areas shall be constructed in accordance with the requirements of this section and the *International Building Code*.

❖ Buildings and structures erected in flood hazard areas must be designed to resist the forces caused by flooding, including flotation, collapse and lateral movement from flowing water, standing water or wave action. This section sets a broad requirement that plumbing systems and equipment are to be constructed according to the provisions of the IBC. IBC Section 1612 references ASCE 24, *Flood Resistant Design and Construction*. Where possible, plumbing fixtures and equipment are to be elevated above the design flood elevation or they must be specifically designed and intended for placement in locations that may be inundated. For additional guidance refer to FEMA 348, *Protecting Building Utilities from Flood Damage*.

[B] 309.2 Flood hazard. For structures located in flood hazard areas, the following systems and equipment shall be located at or above the *design flood elevation*.

Exception: The following systems are permitted to be located below the *design flood elevation* provided that the systems are designed and installed to prevent water from entering or accumulating within their components and the systems are constructed to resist hydrostatic and hydrodynamic loads and stresses, including the effects of buoyancy, during the occurrence of flooding to the *design flood elevation*.

1. All water service pipes.
2. Pump seals in individual water supply systems where the pump is located below the *design flood elevation*.
3. Covers on potable water wells shall be sealed, except where the top of the casing well or pipe sleeve is elevated to at least 1 foot (305 mm) above the *design flood elevation*.
4. All sanitary drainage piping.
5. All storm drainage piping.
6. Manhole covers shall be sealed, except where elevated to or above the *design flood elevation*.
7. All other plumbing fixtures, faucets, fixture fittings, piping systems and equipment.
8. Water heaters.
9. Vents and vent systems.

❖ Although most plumbing facilities and system elements within a building or structure must be located at or above the design flood elevation, certain plumbing system elements and equipment are expected to extend below that elevation. In such instances, the installation must be designed and installed to withstand anticipated flood forces. The flood forces may include buoyancy as well as dynamic loads imposed by moving water and/or waves and the impact of floating debris. This section specifically addresses those plumbing systems and equipment that are to withstand flood forces. It also identifies those systems and equipment that are to be sealed to prevent infiltration of floodwaters into the systems or loss of contents out of systems during the conditions of flooding.

1. Potable water service pipes located below elevated buildings, and those buried below ground, are to be water tight to prevent floodwater from entering the pipes. Water service pipes could develop leaks when subjected to the additional stress from floodwater. This means that the entire piping system subject to floodwater contact must be designed accordingly; for example, using hangers that are capable of resisting not only the normal downward load of the pipe and its contents, but also lateral or upward forces on the pipe caused by floodwaters.
2. Pump seals in individual water supply systems are not normally considered water tight from the "outside." This item requires that these be capable of keeping out the floodwater.
3. To prevent contamination of the below-ground water supply, private well heads that are subject to submersion below floodwater must be water

tight. This should also include any attachment to the well head, such as the electrical connection.

4. Sanitary drainage piping should be designed so that the additional stress of the floodwater does not cause it to fail and thus contaminate the floodwater.
5. Storm drainage systems are not normally connected to any other system, such as the sanitary drainage system. The potential exists that the storm drainage system could carry water from the flood to an area currently not flooded. This item is trying to contain the potential damage from floodwater to the areas immediately subject to that water.
6. As noted in Item 5 above, the concern is potential for floodwater to get into another area or system which may cause backup into buildings. Therefore, there is a requirement for sealing these potential avenues for floodwater to travel.
7. Item 7 is intended to cover any other item not specifically addressed in the six items above. It requires both the code official and the designer to verify that all plumbing systems subject to contact with floodwater have been properly designed to withstand the forces of that water.
8. Water heaters and mechanical equipment must be protected from water or elevated above the design flood elevation. Exposure to water can cause deterioration of most system components and serious malfunctions.
9. The venting system is no different from any other plumbing system component that must be protected from flood levels. Floodwaters can cause damage if the venting system is not properly anchored, supported, sealed, installed and elevated above the design flood elevation or designed to prevent water from entering or accumulating within such system.

[B] 309.3 Flood hazard areas subject to high-velocity wave action. Structures located in flood hazard areas subject to high-velocity wave action shall meet the requirements of Section 309.2. The plumbing systems, pipes and fixtures shall not be mounted on or penetrate through walls intended to break away under flood loads.

❖ Breakaway walls are required if elevated buildings have enclosed areas beneath the elevated floors. Breakaway walls are designed and constructed to fail under flood loads in order to avoid transferring the loads and damaging the primary structural support of the building. When plumbing systems, pipes and fixtures are mounted on or penetrate through the breakaway walls, they can prevent the walls from breaking away, thus transferring additional load to the building. Also, damage to the mechanical systems is certain to occur when the walls fail as intended.

SECTION 310
WASHROOM AND TOILET ROOM REQUIREMENTS

310.1 Light and ventilation. Washrooms and toilet rooms shall be illuminated and ventilated in accordance with the *International Building Code* and *International Mechanical Code.*

❖ The IBC requires every bathroom or toilet room to have natural or artificial light. Minimum lighting is necessary to permit not only proper use of the facilities, but also maintenance and proper housekeeping.

Natural or mechanical ventilation requirements for bathrooms and toilet rooms are specified in the IBC and the *International Mechanical Code*® (IMC®). Minimum ventilation is necessary to remove odors and moisture to maintain a healthful environment.

310.2 Location of fixtures and piping. Piping, fixtures or equipment shall not be located in such a manner as to interfere with the normal operation of windows, doors or other means of egress openings.

❖ Fixtures and piping located in washrooms or toilet rooms cannot adversely affect the operation of building components such as windows and doors. The location of the fixtures and piping should not interfere with or block an occupant's egress to or from that space (see commentary, Figure 310.2). Obstructions caused by fixtures or piping can seriously hamper egress in an emergency condition or cause injury to the occupants.

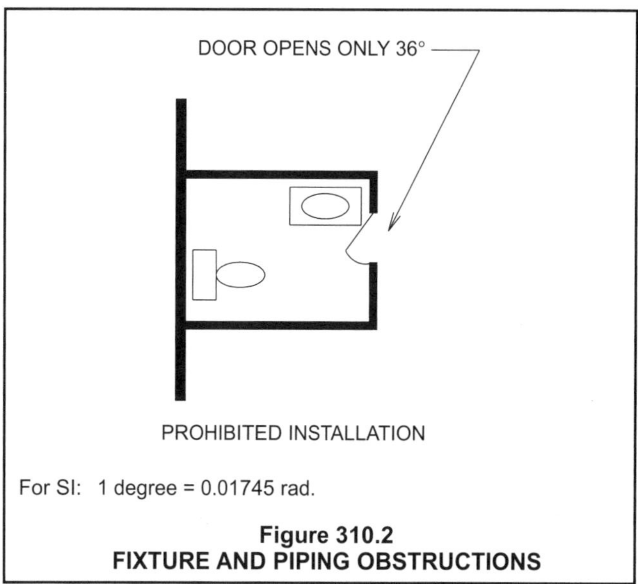

For SI: 1 degree = 0.01745 rad.

Figure 310.2
FIXTURE AND PIPING OBSTRUCTIONS

310.3 Interior finish. Interior finish surfaces of toilet rooms shall comply with the *International Building Code.*

❖ In other than dwelling units, the interior finish of toilet and bathing room floors and walls must have a smooth, hard, nonabsorbent surface. The water-resistant surface for floors must extend upward onto the walls at least 6 inches (152 mm). Walls within 2 feet

(610 mm) of urinals and water closets must have a water-resistant surface to a height of 4 feet (1219 mm) above the floor (see IBC Section 1210). These types of finishes will allow thorough cleaning of the surfaces (see commentary, Section 419.3).

310.4 Water closet compartment. Each water closet utilized by the *public* or employees shall occupy a separate compartment with walls or partitions and a door enclosing the fixtures to ensure privacy.

Exceptions:

1. Water closet compartments shall not be required in a single-occupant toilet room with a lockable door.
2. Toilet rooms located in day care and child-care facilities and containing two or more water closets shall be permitted to have one water closet without an enclosing compartment.
3. This provision is not applicable to toilet areas located within Group I-3 housing areas.

❖ Psychological studies have proven that lack of privacy places a burden on an individual's physical ability to use bathroom facilities. This is caused by uneasiness, inhibition or indignation. A partitioned compartment can provide the necessary privacy. Also, a single-occupant toilet room with a lockable door can provide the required privacy as stated in the first exception. The second exception, allowing an unenclosed water closet in a child care facility, recognizes the need for people to be able to assist children. The third exception was added in the 2009 code edition to allow for security personal to better monitor for illicit activities of inmates in Group I-3 occupancies.

310.5 Urinal partitions. Each urinal utilized by the *public* or employees shall occupy a separate area with walls or partitions to provide privacy. The walls or partitions shall begin at a height not more than 12 inches (305 mm) from and extend not less than 60 inches (1524 mm) above the finished floor surface. The walls or partitions shall extend from the wall surface at each side of the urinal a minimum of 18 inches (457 mm) or to a point not less than 6 inches (152 mm) beyond the outermost front lip of the urinal measured from the finished back wall surface, whichever is greater.

Exceptions:

1. Urinal partitions shall not be required in a single occupant or family/assisted-use toilet room with a lockable door.
2. Toilet rooms located in day-care and child-care facilities and containing two or more urinals shall be permitted to have one urinal without partitions.

❖ Users of urinals in a public toilet room need to be afforded a level of privacy for the reasons stated in the commentary for Section 310.4. In addition, partitions aid in protecting adjacent users from splashing. Because of the likelihood of splashing, the partition surfaces must be designed so that they can be easily cleaned and sanitized during normal restroom maintenance. Exception 1 assumes that in a single-occupant toilet room or a family/assisted use toilet room, privacy is afforded by locking the door. Exception 2 allows day care and child care centers to have one urinal out of two or more urinals in a toilet room to be without partitions. This affords care-providers the ability to assist or supervise certain children with use of the urinal (e.g., those just learning to use a urinal or those with difficult clothing arrangements). Where urinal partitions are provided, occupants are more inclined to use a urinal rather than using a water closet for urination. This will result in less soiling of water closet seats and surrounding surfaces [see Commentary Figures 310.5(1) and (2)]. Note that where partitions are used, the distance from the center of the urinal to the face of the partition must be at least 15 inches (381 mm) to be in compliance with Section 405.3.1.The thickness of partitions cannot encroach into the required spaces for fixtures.

Figure 310.5(1)
URINAL PARTITIONS
(Photo courtesy of Falcon Waterfree Technologies)

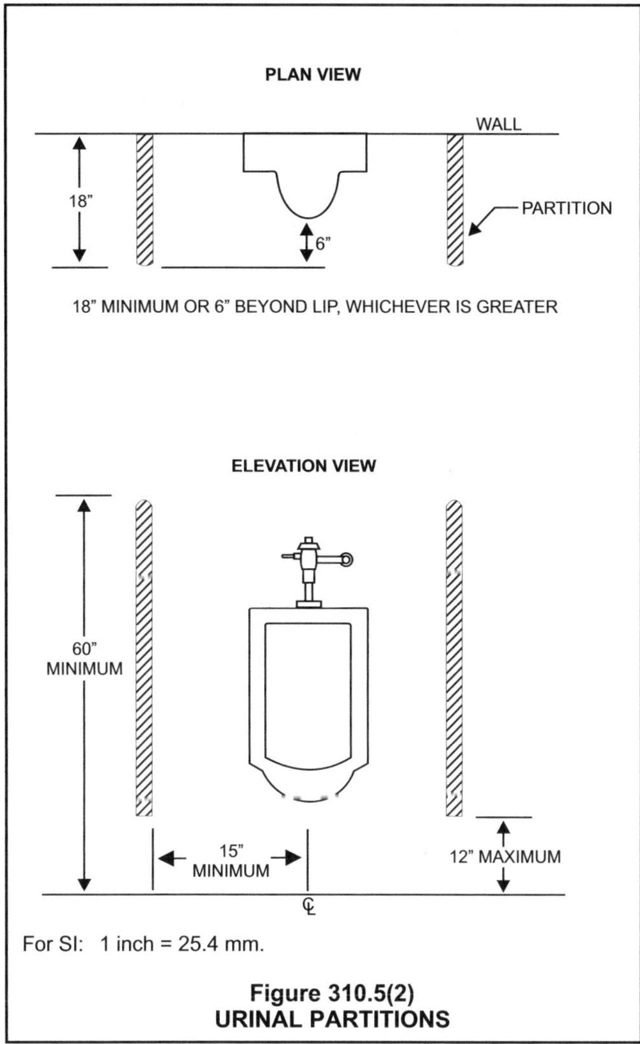

Figure 310.5(2)
URINAL PARTITIONS

Table 311.1
MINIMUM NUMBER OF TOILET FACILITIES

NUMBER OF EMPLOYEES	MINIMUM NUMBER OF TOILET FACILITIES
	If serviced once per week[a]
1 - 10	1
11 - 20	2
21 - 30	3
31 - 40	4
Over 40	1 additional facility for each 10 additional employees
	If serviced more than once per week[a]
1 - 15	1
16 - 35	2
36 - 55	3
56 - 75	4
76 - 95	5
Over 95	1 additional facility for each 20 additional employees

a. "Servicing" refers to the emptying of waste and the cleaning of the toilet facility. A camp equipped with flush toilets shall meet the standard for "minimum number of toilet facilities if serviced more than once per week."

(Source: ANSI Z4.3-95)

SECTION 311
TOILET FACILITIES FOR WORKERS

311.1 General. Toilet facilities shall be provided for construction workers and such facilities shall be maintained in a sanitary condition. Construction worker toilet facilities of the nonsewer type shall conform to ANSI Z4.3.

❖ Construction employees must have plumbing facilities available to them during the construction of a building. These facilities may be either permanent or portable. Portable facilities are regulated by ANSI Z4.3, which specifies various construction requirements. Commentary Table 311.1 has been provided to show the minimum number of fixtures required by ANSI Z4.3.

SECTION 312
TESTS AND INSPECTIONS

312.1 Required tests. The permit holder shall make the applicable tests prescribed in Sections 312.2 through 312.10 to determine compliance with the provisions of this code. The permit holder shall give reasonable advance notice to the code official when the plumbing work is ready for tests. The equipment, material, power and labor necessary for the inspection and test shall be furnished by the permit holder and the permit holder shall be responsible for determining that the work will withstand the test pressure prescribed in the following tests. All plumbing system piping shall be tested with either water or, for piping systems other than plastic, by air. After the plumbing fixtures have been set and their traps filled with water, the entire drainage system shall be submitted to final tests. The code official shall require the removal of any cleanouts if necessary to ascertain whether the pressure has reached all parts of the system.

❖ This section states that testing and inspection is required on plumbing systems and that the responsibility for this rests with the permit holder. This testing verifies that the plumbing system complies with the code and operates as intended. Although visual inspection plays a vital part in determining the integrity of the plumbing system, it is impossible for the code official to identify potential leaks and defects without testing. Because there is no guarantee that a plumbing system that passes visual inspection and testing will continue to operate properly and remain free from leaks and defects, plumbing systems must also be maintained in a safe and sanitary condition.

Plumbing work should remain uncovered and unconcealed until it can be approved for two very important reasons. First, the plumbing work can be visually inspected for defects and code violations can be addressed. Second, leaks and defects exposed during testing can be easily located and repaired or corrected. The sequence for inspection and testing should be established to prevent code violations and defects from going unnoticed, thus eliminating the potential for covering up nonconforming work (see the commentary to Section 107.1 for other aspects of testing and inspection).

This section lists the specific testing and inspections required for plumbing systems. It also states that the permit holder is responsible for this testing and for supplying the labor, power, material, equipment and apparatus necessary to conduct it. The code official observes the testing but never performs or directs it. The permit holder is responsible not only for giving the code official reasonable advance notice of the tests, but also for the costs associated with this work.

The permit holder is the person responsible for determining that the plumbing system is capable of withstanding the test pressure prescribed in the section for the specific test. Visual inspection is not all that is in the determination of a plumbing system's compliance with the code. This section establishes where and how testing is to be performed to disclose leaks and defects.

Every plumbing system must be tested before it is placed into service to determine that it is free from leaks or other defects. Testing is required, to the extent practicable, for portions of existing systems that have been repaired, altered or extended. This section states that the permit holder has the option of using either water or, for piping systems other than plastic, testing by air.

There is an inherent danger in testing plastic piping with compressed air. When testing with water (hydrostatic testing), structural failure of a pipe or joint will result in the instantaneous release of pressure because water is not compressible and there is little stored potential energy in the system. However, when a compressed gas is used, there will be much energy stored in the compressed gas and pipe/fitting failure could result in pieces and shards of plastic being propelled at high velocity toward personnel. To determine that the test is being properly performed on the entire system, the code official can require the removal of cleanout plugs to determine that the test medium has reached every portion of the plumbing system.

This section also states that the entire system is to be tested after the plumbing fixtures have been set and the traps filled with water.

Although not specifically referenced in this section, Section 107 addresses other aspects of testing and inspection related to plumbing systems.

312.1.1 Test gauges. Gauges used for testing shall be as follows:

1. Tests requiring a pressure of 10 pounds per square inch (psi) (69 kPa) or less shall utilize a testing gauge having increments of 0.10 psi (0.69 kPa) or less.
2. Tests requiring a pressure of greater than 10 psi (69 kPa) but less than or equal to 100 psi (689 kPa) shall utilize a testing gauge having increments of 1 psi (6.9 kPa) or less.
3. Tests requiring a pressure of greater than 100 psi (689 kPa) shall utilize a testing gauge having increments of 2 psi (14 kPa) or less.

❖ It is common to encounter test gauges that read in 1 or 2 pounds per square inch (psi) increments while attempting to perform a DWV inspection that only requires a 5 psi (34.5 kPa) test. A small drop in pressure will be difficult to notice on a gauge with 1 or 2 psig (6.895 or 13.79 kPa) increments and therefore, they are not appropriate for low test pressures.

In each case, the user needs to ensure that the proper pressure gauge was selected with respect to indicating range and design. Most standard dial-type pressure gauges use a bourdon tube sensing element generally made of a copper alloy (brass) or stainless steel. The construction is simple and operation does not require any additional power source. The C-shaped or spirally wound bourdon tube flexes when pressure is applied, producing a rotational movement, which in turn causes the pointer to indicate the measured pressure. Such gauges are the most accurate when reading in the midrange of their scale; therefore, the test gauge should be chosen to reflect the test pressure in the midrange of the scale. A shutoff valve should be installed between the measuring point and the pressure gauge with another valve to depressurize the gage only. This will allow an exchange of the pressure gauge and checks on the zero setting while the system remains under test pressure or operating pressure.

312.2 Drainage and vent water test. A water test shall be applied to the drainage system either in its entirety or in sections. If applied to the entire system, all openings in the piping shall be tightly closed, except the highest opening, and the system shall be filled with water to the point of overflow. If the system is tested in sections, each opening shall be tightly plugged except the highest openings of the section under test, and each section shall be filled with water, but no section shall be tested with less than a 10-foot (3048 mm) head of water. In testing successive sections, at least the upper 10 feet (3048 mm) of the next preceding section shall be tested so that no joint or pipe in the building, except the uppermost 10 feet (3048 mm) of the system, shall have been submitted to a test of less than a 10-foot (3048 mm) head of water. This pressure shall be held for at least 15 minutes. The system shall then be tight at all points.

❖ This section discusses the procedure for performing a water test on portions of or the entire drain, waste and vent (DWV) system. A rough-in test is commonly performed on the entire DWV system. In multistory buildings buildings, this could be a dangerous practice because the lower floors are exposed to much higher head pressure than that for which the system was designed. This excess pressure could be detrimental to the plumbing system. Testing should be conducted in stages so that no one section of piping is exposed to excessive pressures.

This section requires that, except for the uppermost 10 feet (3048 mm) of the system, no section be tested with less than a 10-foot (3048 mm) head of water. This 10-foot (3048 mm) head of water is equivalent to 4.33 psi (30 kPa) and must be held tight for at least 15 minutes. The water test allows system defects to be found more readily than in air tests but introduces the risks of water damage to the structure, freeze damage to traps and piping system damage caused by excessive head pressures and excessive weight. This portion of the testing could be performed with air in compliance with Section 312.3.

312.3 Drainage and vent air test. An air test shall be made by forcing air into the system until there is a uniform gauge pressure of 5 psi (34.5 kPa) or sufficient to balance a 10-inch (254 mm) column of mercury. This pressure shall be held for a test period of at least 15 minutes. Any adjustments to the test pressure required because of changes in ambient temperature or the seating of gaskets shall be made prior to the beginning of the test period.

❖ This section allows testing to be performed using compressed air in the system. The system must be filled with air until there is either a uniform gauge pressure of 5 psi (34.5 kPa) or sufficient pressure to balance a 10-inch (254 mm) column of mercury. This pressure must then be held for a period of at least 15 minutes without the introduction of additional air.

Air tests are easy to apply and involve no risk of freeze damage. They do, however, have disadvantages. Defects in the system are more difficult to locate using an air test. System temperature changes can also cause misleading changes in pressure, which can cause false indications. Additionally, some gasket materials will provide proper seals only after they are given a brief period to seat themselves under pressure. Adjustments for changes in pressure resulting from ambient temperature fluctuations or the seating of gaskets must be made prior to the start of the 15-minute test period.

Although this section does not specifically prohibit the testing of DWV systems with air, testing with a compressed gas (air) is inherently more hazardous to personnel than hydrostatic testing because of the energy release that can occur from a system rupture or failure. Note that Section 312.1 appears to prohibit the testing of all plastic piping with compressed air, including DWV piping. The 5 psi (34.5 kPa) testing pressure stated in this section is often exceeded in the field by the persons performing the test to make the location of leaks easier. Increasing the test pressure beyond the safe limits of the material can result in a dangerous situation and should not be allowed. In cases where compressed air is to be used to test the piping for leaks, a positive means of controlling pressure should be used, such as a relief valve or a pressure-limiting switch.

312.4 Drainage and vent final test. The final test of the completed drainage and vent systems shall be visual and in sufficient detail to determine compliance with the provisions of this code. Where a smoke test is utilized, it shall be made by filling all traps with water and then introducing into the entire system a pungent, thick smoke produced by one or more smoke machines. When the smoke appears at *stack* openings on the roof, the *stack* openings shall be closed and a pressure equivalent to a 1-inch water column (248.8 Pa) shall be held for a test period of not less than 15 minutes.

❖ In preparation for the final drainage and vent test, the plumbing fixtures must be fully assembled, set, connected to the drain system (if required), and the fixture traps filled with liquid. Venting applications requiring an air admittance valve must have the valve installed. The items checked in the final visual test of the drainage and vent system is up to the discretion of the code official.

A smoke test of the completed drainage and vent system could be required by the code official where there is justification for such testing. Smoke testing requires that the outlet of the drain, waste and vent system be plugged and a thick smoke be pumped into the DWV system so that smoke circulates throughout the entire system. After all vent openings are capped off, the system is then continuously pressurized to 1 inch water column (248.8 Pa) for a period of 15 minutes. During the test period, smoke must not be visually detected inside of the building in order for the system to pass the test [see commentary, Figures 312.4(1) and 312.4(2) which illustrate smoke being used during a rough-in test].

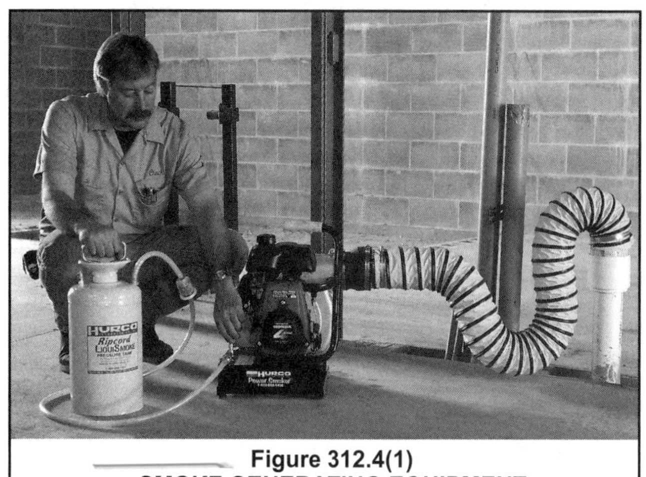

Figure 312.4(1)
SMOKE GENERATING EQUIPMENT
(Photo courtesy of Hurco Technologies, Inc.)

Figure 312.4(2)
SMOKE INDICATING PRESENCE OF LEAK
(Photo courtesy of Hurco Technologies, Inc.)

312.5 Water supply system test. Upon completion of a section of or the entire water supply system, the system, or portion completed, shall be tested and proved tight under a water pressure not less than the working pressure of the system; or, for piping systems other than plastic, by an air test of not less than 50 psi (344 kPa). This pressure shall be held for at least 15 minutes. The water utilized for tests shall be obtained from a potable source of supply. The required tests shall be performed in accordance with this section and Section 107.

❖ The water supply system consists of piping from the water meter at the curb, the stop valve at the curb or the well system to the ends of the water distribution piping in the building. Where only portions of the water supply system have been installed or worked upon, only those portions are required to be tested.

The term "working pressure" is not defined in the code. It is understood to be the pressure in water supply system that will exist under normal operating conditions. The working pressure can be different depending on where the pressure is measured in a system. For example, the working pressure in a water service line from the curb stop valve to the entry point into the building might be 120 psi (827 kPa). As this pressure is greater than 80 psi (551 kPa), in accordance with Section 604.8, a pressure reducing valve must be installed to limit the pressure in the building's water distribution system to not greater than 80 psi. Therefore, the working pressure in the water distribution system portion of the water supply system is 80 psi or less.

The phrase "the system shall be proved tight," means that by visual inspection, no evidence of leakage from the piping system is observed. Evidence of leakage is typically determined by attaching a pressure gauge to the system, pressurizing the system to the test pressure and without further addition of test water or air to the system, verifying after 15 minutes that the pressure gauge indication has not changed from the reading taken at the beginning of the test. Note the critical test gage requirements of Section 312.1.1. Where minor repairs or modifications are made to a water supply system and the system is tested with water, the pressure gauge method is not necessary because evidence of leakage could be simply determined by observing each piping connection for the presence of leaking water.

If air is used to test a water supply system the test pressure need only be 50 psi (344 kPa). Where testing with compressed air, it is advisable to allow the system under test pressure to thermally stabilize before the start of the observation period. Warm air from a compressor introduced into cold piping will cool and result in a decrease in the test pressure, falsely indicating a leak. Gaskets and o-rings in shower mixing valves are intended to seal against water pressure and typically do not initially seat or seal well when first pressurized with air. In most cases, leaks from these locations will eventually cease after a "seating in" period of time. Adjustments for changes in pressure resulting from ambient temperature fluctuations or the seating of gaskets must be made prior to the start of the 15-minute test period.

Air testing of water supply systems of plastic material is prohibited by the code and many plastic piping manufacturers because of the risk of personal injury caused by flying shards of plastic should the piping rupture. Compressed air stores potential energy not unlike a compressed coil spring stores energy. If the piping ruptures, the stored potential energy becomes kinetic energy that can propel plastic pieces with great force and velocity.

Section 107 specifies three general construction stages that require inspection: (1) underground rough-in, (2) above-ground rough-in and (3) final (completion of plumbing work). Depending on the size and required sequence of a project, multiple inspections may be necessary for each stage. The intent of this code section is to ensure that all underground and above-ground water supply system piping is tested prior to concealment so that any leaks can be readily located and repaired. The final inspection and "test" (Section 312.4) provides for visual inspection to find leaks in exposed water supply system connections such as water heater piping, stop valves, supply tube connections and fixture/faucet assemblies.

312.6 Gravity sewer test. Gravity *sewer* tests shall consist of plugging the end of the *building sewer* at the point of connection with the *public sewer*, filling the *building sewer* with water, testing with not less than a 10-foot (3048 mm) head of water and maintaining such pressure for 15 minutes.

❖ The building sewer must be subjected to the same testing as required for the DWV system, except that the air test is not an option for building sewers. This section states specific points of procedure for conducting this test, including plugging the sewer at its point of connection with the public sewer, the minimum test water head and the time duration of the test.

312.7 Forced sewer test. Forced *sewer* tests shall consist of plugging the end of the *building sewer* at the point of connection with the *public sewer* and applying a pressure of 5 psi (34.5 kPa) greater than the pump rating, and maintaining such pressure for 15 minutes.

❖ Although not specifically defined in the code, the term "forced sewer" refers to any sewer system that cannot drain solely because of the force of gravity. The forced sewer has its contents placed under pressure for some length of travel. This length of travel is often a short vertical rise to allow the sewage to again travel using the forces of gravity. This type of arrangement normally involves the use of a sewage pump or sewage ejector (refer to Section 712 for requirements related to sumps and ejectors).

This section requires that the forced sewer be tested in a manner similar to that required for the gravity sewer, except that this test is to be conducted with an applied test pressure of 5 psi (34.5 kPa) greater than the highest pressure that the pump can generate. Many times, this pressure is called the "shut off" pressure because for many pumps, the highest pressure is when the flow is zero. The pump pressure curve or data table must be reviewed to determine the maximum pump pressure.

Although this section does not specify the test medium (air or water), the intent is that forced sewers be tested using the same methods and precautions as gravity sewer testing except that the pressure will be according to the maximum pump pressure.

312.8 Storm drainage system test. *Storm drain* systems within a building shall be tested by water or air in accordance with Section 312.2 or 312.3.

❖ Piping used for drainage of roof water that is located inside a building has the same potential for leaks and defects as building sanitary drainage system. Because of this potential, this section requires that piping used for storm drains, conductors or downspouts and related branch lines that are located inside a building be tested. This testing is to be performed in a manner similar to the methods that would be used to test building drain, waste and vent systems.

312.9 Shower liner test. Where shower floors and receptors are made water-tight by the application of materials required by Section 417.5.2, the completed liner installation shall be tested. The pipe from the shower drain shall be plugged water tight for the test. The floor and receptor area shall be filled with potable water to a depth of not less than 2 inches (51 mm) measured at the threshold. Where a threshold of at least 2 inches (51 mm) high does not exist, a temporary threshold shall be constructed to retain the test water in the lined floor or receptor area to a level not less than 2 inches (51 mm) deep measured at the threshold. The water shall be retained for a test period of not less than 15 minutes, and there shall not be evidence of leakage.

❖ Although Section 417.5.2 always required field-fabricated shower floors and receptors to be made "water tight" (i.e., free of any leaks when exposed to water) by the application of liner materials, the code did not actually require testing to prove that the installation was leak free. Many jurisdictions have required testing of field-fabricated shower pans for many years due to the potential for water damage if the installation of liner material is not performed properly. The resulting water damage, increased difficulty in making repairs and additionally expense involved associated with discovering a leak after a building has been completed have prompted the addition of this testing requirement section for the 2009 edition of the code.

This section clearly spells out how to perform the test. The evidence of leakage could be either the lowering of the water level from the full threshold level or water drips/seepage outside of the receptor area (see commentary, Figure 312.9).

312.10 Inspection and testing of backflow prevention assemblies. Inspection and testing shall comply with Sections 312.10.1 and 312.10.2.

❖ Backflow prevention assemblies are critical components that protect the potable water system from pollution or contamination sources. As such, these assemblies must be inspected and tested after installation at the project site to ensure that the devices are providing the level of protection intended.

312.10.1 Inspections. Annual inspections shall be made of all backflow prevention assemblies and air gaps to determine whether they are operable.

❖ Backflow prevention assemblies, including air gaps, must be inspected annually. Air gaps must be inspected to determine that proper clearances are maintained in accordance with Table 608.15.1. Additionally, backflow prevention assemblies are to be inspected for damage or evidence of tampering that could cause the device to malfunction.

Because these devices must be periodically inspected, they cannot be located in concealed spaces. Depending on the type of device and the installed location, they might be located behind access panels. Backflow devices must not be allowed to be installed in a pit or other below grade locations as flooding could prevent the device from properly operating.

312.10.2 Testing. Reduced pressure principle backflow preventer assemblies, double check-valve assemblies, pressure vacuum breaker assemblies, reduced pressure detector fire protection backflow prevention assemblies, double check detector fire protection backflow prevention assemblies, hose connection backflow preventers, and spillproof vacuum breakers shall be tested at the time of installation, immediately after repairs or relocation and at least annually. The testing procedure shall be performed in accordance with one of the following standards:

ASSE 5013, ASSE 5015, ASSE 5020, ASSE 5047, ASSE 5048, ASSE 5052, ASSE 5056, CSA B64.10 or CSA B64.10.1.

❖ The listed standards contain the required testing procedures for backflow preventer assemblies. The test procedures are intended to simulate possible fault conditions of an assembly. Backflow prevention assemblies are to be tested after completion of new installations, repairs and relocations and at least annually thereafter. Although not specifically required, there should be some method of recording when these assemblies were tested. This record should be maintained near the assembly or at a central maintenance office.

SECTION 313
EQUIPMENT EFFICIENCIES

313.1 General. Equipment efficiencies shall be in accordance with the *International Energy Conservation Code.*

❖ The *International Energy Conservation Code®* (IECC®) requires minimum efficiencies for appliances and hot water storage tanks.

SECTION 314
CONDENSATE DISPOSAL

[M] 314.1 Fuel-burning appliances. Liquid combustion by-products of condensing appliances shall be collected and discharged to an *approved* plumbing fixture or disposal area in accordance with the manufacturer's installation instructions. Condensate piping shall be of *approved* corrosion-resistant material and shall not be smaller than the drain connection on the appliance. Such piping shall maintain a minimum horizontal slope in the direction of discharge of not less than one-eighth unit vertical in 12 units horizontal (1-percent slope).

❖ Condensation must be collected from appliances in accordance with the manufacturers' installation instructions and this section. High-efficiency condensing-type appliances (Category IV) and some mid-efficiency appliances produce liquid as a combustion by-product. This condensate is collected at various points in the appliance's heat exchangers and venting system and must be disposed of. Because of impurities in the fuel and combustion air, the condensate can be acidic, thus corrosive, to many materials. For example, such condensate can contain hydrochloric and also sulfuric acids. Such drain piping is part of the mechanical system and, as such, is also covered in the IMC. Although condensate piping is not plumbing, it is covered in the code because of the possibility of plumbers installing such piping.

[M] 314.2 Evaporators and cooling coils. Condensate drain systems shall be provided for equipment and appliances containing evaporators or cooling coils. Condensate drain systems shall be designed, constructed and installed in accordance with Sections 314.2.1 through 314.2.4.

❖ Appliances and equipment containing evaporators or cooling coils, including refrigeration, dehumidification and comfort cooling equipment, can produce conden-

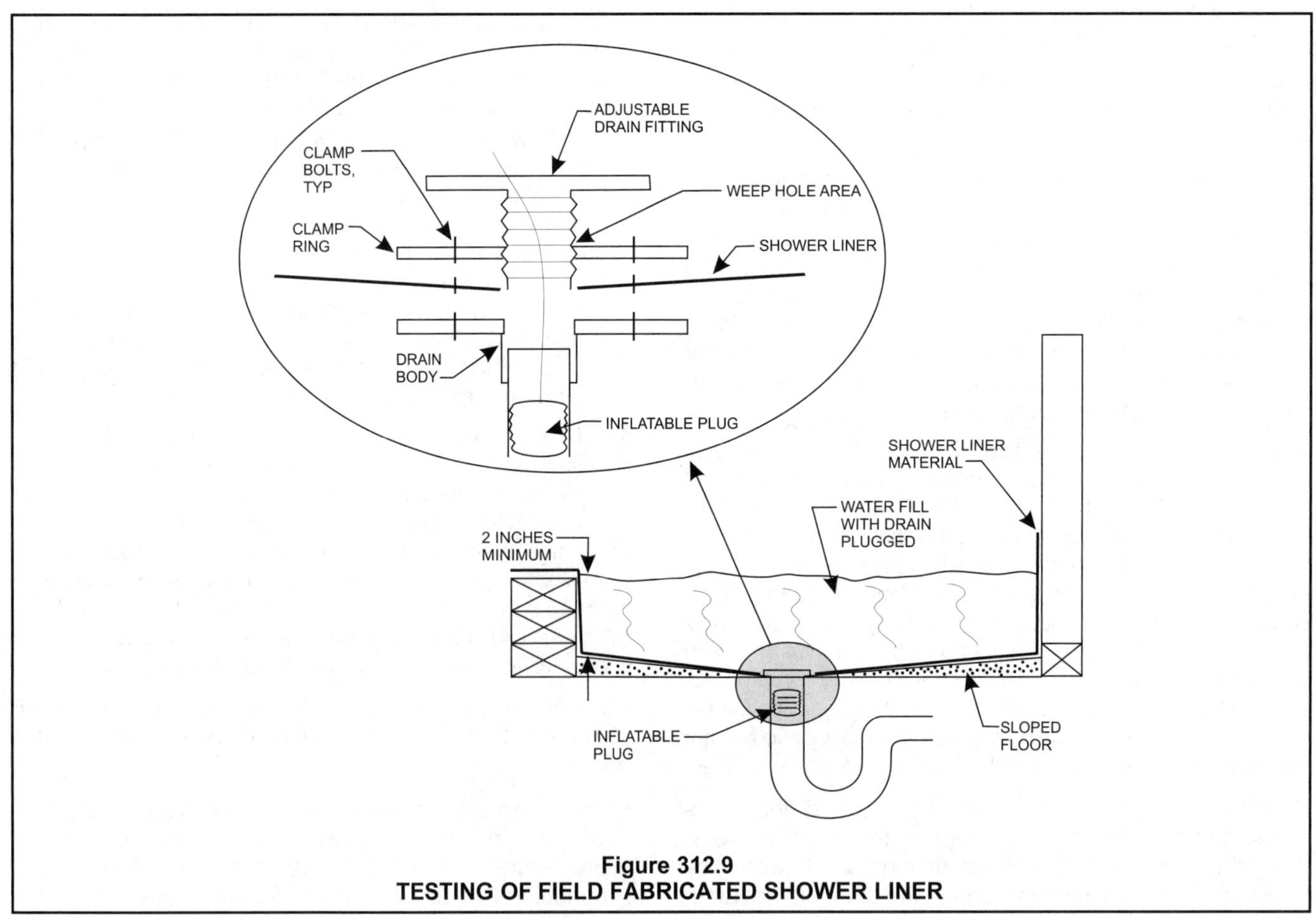

Figure 312.9
TESTING OF FIELD FABRICATED SHOWER LINER

sate from the water vapor in the atmosphere. A drainage system is necessary to dispose of the condensate and prevent damage to the structure.

[M] 314.2.1 Condensate disposal. Condensate from all cooling coils and evaporators shall be conveyed from the drain pan outlet to an *approved* place of disposal. Such piping shall maintain a minimum horizontal slope in the direction of discharge of not less than one-eighth unit vertical in 12 units horizontal (1-percent slope). Condensate shall not discharge into a street, alley or other areas so as to cause a nuisance.

❖ The "approved place of disposal" specified in this section will vary from jurisdiction to jurisdiction. While condensate is typically "clear" water, it can gather dust, bacteria and mold spores which can result in a solids buildup in the waste stream. A slope of at least 1 percent creates a minimum flow velocity to aid in transport of these solids towards the point of disposal.

Some jurisdictions strictly forbid condensate from being drained into the sanitary sewer system primarily because of overloading the system, but could allow it depending on the quantity of effluent produced. Common points of disposal are to grade in areas where a nuisance will not be created, onto rooftops where the condensate will evaporate or drain to grade, to seepage pits where soil condition allows or to a storm drain. Section 802.1.5 prohibits condensate drains from being directly connected to a sanitary sewer because of the possibility of sewer gases or sewage entering the equipment or system.

Additional consideration must be given to the disposal of condensate from fuel-burning appliances (see commentary, Section 314.1).

[M] 314.2.2 Drain pipe materials and sizes. Components of the condensate disposal system shall be cast iron, galvanized steel, copper, cross-linked polyethylene, polybutylene, polyethylene, ABS, CPVC or PVC pipe or tubing. All components shall be selected for the pressure and temperature rating of the installation. Joints and connections shall be made in accordance with the applicable provisions of Chapter 7 relative to the material type. Condensate waste and drain line size shall be not less than $^3/_4$-inch (19 mm) internal diameter and shall not decrease in size from the drain pan connection to the place of condensate disposal. Where the drain pipes from more than one unit are manifolded together for condensate drainage, the pipe or tubing shall be sized in accordance with Table 314.2.2.

❖ Condensate drains must be constructed of a material listed in this section. The previous editions of this section required that horizontal sections of the condensate piping be installed in uniform alignment at a uniform slope. This was to prevent dips and sags that could impede flow and also prevented the use of flexible piping that could not maintain a uniform slope. Section 314.2.1 requires a slope of at least 1 percent [$^1/_8$ inch per foot (10 mm per meter)]. Condensate drains must have the minimum prescribed slope to promote gravity drainage and lessen the possibility of clogging. Greater slopes are allowed, and in all cases, the slope must be maintained in the direction of flow such that no dips or reverse slopes occur.

TABLE [M] 314.2.2. See below.

❖ While the code is clear on a minimum condensate pipe size, designers previously had no approved method for sizing condensate lines connecting multiple units. This table provides minimum sizing for those condensate piping systems connecting multiple units.

[M] 314.2.3 Auxiliary and secondary drain systems. In addition to the requirements of Section 314.2.1, where damage to any building components could occur as a result of overflow from the equipment primary condensate removal system, one of the following auxiliary protection methods shall be provided for each cooling coil or fuel-fired appliance that produces condensate:

1. An auxiliary drain pan with a separate drain shall be provided under the coils on which condensation will occur. The auxiliary pan drain shall discharge to a conspicuous point of disposal to alert occupants in the event of a stoppage of the primary drain. The pan shall have a minimum depth of $1^1/_2$ inches (38 mm), shall not be less than 3 inches (76 mm) larger than the unit or the coil dimensions in width and length and shall be constructed of corrosion-resistant material. Galvanized sheet metal pans shall have a minimum thickness of not less than 0.0236-inch (0.6010 mm) (No. 24 gage) galvanized sheet metal. Nonmetallic pans shall have a minimum thickness of not less than 0.0625 inch (1.6 mm).

TABLE [M] 314.2.2
CONDENSATE DRAIN SIZING

EQUIPMENT CAPACITY	MINIMUM CONDENSATE PIPE DIAMETER
Up to 20 tons of refrigeration	$^3/_4$ inch
Over 20 tons to 40 tons of refrigeration	1 inch
Over 40 tons to 90 tons of refrigeration	$1^1/_4$ inch
Over 90 tons to 125 tons of refrigeration	$1^1/_2$ inch
Over 125 tons to 250 tons of refrigeration	2 inch

For SI: 1 inch = 25.4 mm, 1 ton of capacity = 3.517 kW.

2. A separate overflow drain line shall be connected to the drain pan provided with the equipment. Such overflow drain shall discharge to a conspicuous point of disposal to alert occupants in the event of a stoppage of the primary drain. The overflow drain line shall connect to the drain pan at a higher level than the primary drain connection.

3. An auxiliary drain pan without a separate drain line shall be provided under the coils on which condensate will occur. Such pan shall be equipped with a water-level detection device conforming to UL 508 that will shut off the equipment served prior to overflow of the pan. The auxiliary drain pan shall be constructed in accordance with Item 1 of this section.

4. A water-level detection device conforming to UL 508 shall be provided that will shut off the equipment served in the event that the primary drain is blocked. The device shall be installed in the primary drain line, the overflow drain line, or in the equipment-supplied drain pan, located at a point higher than the primary drain line connection and below the overflow rim of such pan.

Exception: Fuel-fired appliances that automatically shut down operation in the event of a stoppage in the condensate drainage system.

❖ An auxiliary or secondary drain system is required for equipment locations where condensate overflow would cause damage to a building or its contents. These systems are intended to catch condensate spilling from the primary drain or drain pan if the primary drain becomes clogged. These methods protect the building from structural damage. This section applies to cooling and evaporator coils and condensing type fuel-fired appliances.

Condensate drains are notorious for clogging because of debris (lint and dust) from air-handling systems and the natural affinity to produce slime growth in drain pans and pipes.

Note that four distinct options are provided for the auxiliary/secondary system. Option 1 is a redundant drain pan and drain piping system that is completely independent of the primary system. Option 2 is the simplest method because it involves only an overflow drain connected to the overflow opening on the primary pan. Such an opening is typically found on factory-built primary pans. The overflow drain must be completely independent of the primary drain. Option 3 involves an auxiliary pan without a drain and depends on a float switch or sensor to detect water in the pan and shut off the equipment producing the condensate [see commentary, Figure 314.2.3(3)]. Option 4 involves a float-switch assembly designed to be threaded into the overflow tapping provided on the primary pan or installed at some point in a drain line. Like Option 3, the float-switch assembly will detect overflow from the primary pan and shut off the equipment [see commentary, Figures 314.2.3(1) and (2)].

The exception is for fuel-fired appliances having integral controls to prevent operation of the appliance if condensate is not draining. The standards for some appliances, such as gas-fired furnaces, will permit either of two performance modes for an appliance with a failed condensate drainage system: (1) the appliance will shut down or, (2) the appliance will continue to operate provided that such operation does not in any way present a hazard.

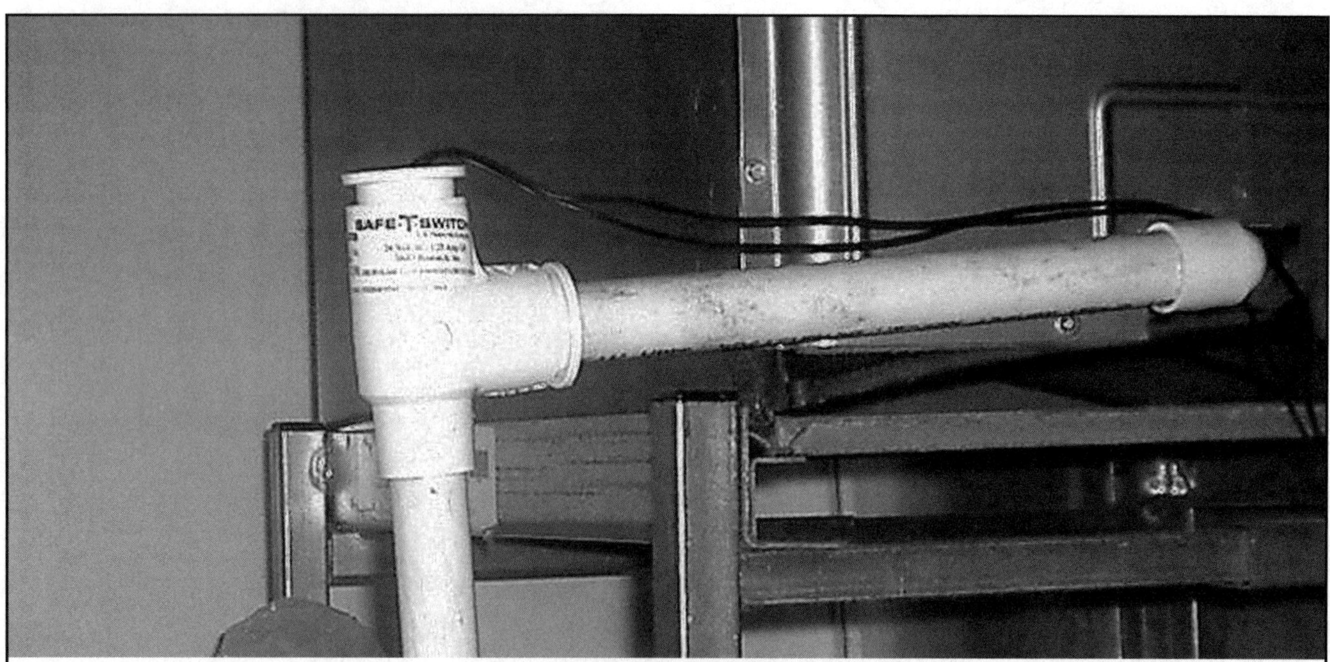

Figure 314.2.3(1)
CONDENSATE OVERFLOW SWITCH
(Photo courtesy of SMD Research, Inc.)

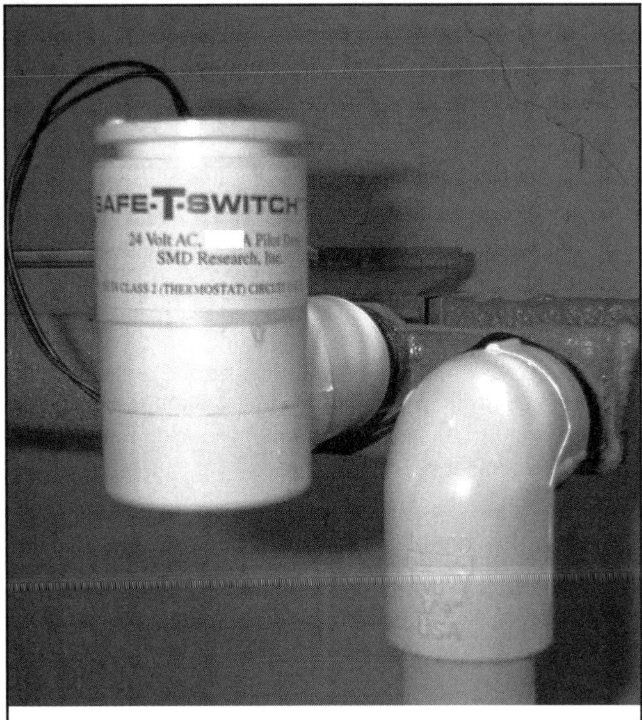

Figure 314.2.3(2)
CONDENSATE OVERFLOW SWITCH
(Photo courtesy of SMD Research, Inc.)

Figure 314.2.3(3)
PRIMARY OR AUXILIARY DRAIN PAN FLOAT SWITCH
(Photo courtesy of SMD Research, Inc.)

[M] 314.2.3.1 Water-level monitoring devices. On down-flow units and all other coils that do not have a secondary drain or provisions to install a secondary or auxiliary drain pan, a water-level monitoring device shall be installed inside the primary drain pan. This device shall shut off the equipment served in the event that the primary drain becomes restricted. Devices installed in the drain line shall not be permitted.

❖ Some appliances (downflow type) and coil arrangements cannot be fitted with a secondary drain or installed with an auxiliary drain pan. This is particularly true for the typical rooftop HVAC unit where drain pan overflow could damage ducts and duct insulation. An internal water monitoring device must be used in these situations so that the device shuts off the unit prior to overflow of the primary drain pan. If the device was installed external to the appliance, a drain blockage in or near the appliance might occur upstream of the device thereby, rendering it useless.

[M] 314.2.3.2 Appliance, equipment and insulation in pans. Where appliances, equipment or insulation are subject to water damage when auxiliary drain pans fill such portions of the appliances, equipment and insulation shall be installed above the *flood level rim* of the pan. Supports located inside of the pan to support the appliance or equipment shall be water resistant and *approved*.

❖ If a condensate overflow situation occurs, appliances sitting directly on the bottom of the pan will be exposed to the condensate water, even though the pan may have a working drain. Most appliances are not suitable for constantly damp or intermittent wet environments. Sheet metal enclosures will quickly rust under these conditions. Exterior and interior insulation will be degraded by the water providing ideal conditions for growth of mold in concealed locations. This section requires that any such appliance be elevated above the flood rim level of the pan below so that the appliance is not directly exposed to wet conditions. The code is silent on what type of material is to be used for support other than requiring it to be water resistant and approved by the code official (see commentary, Figure 314.2.3.2).

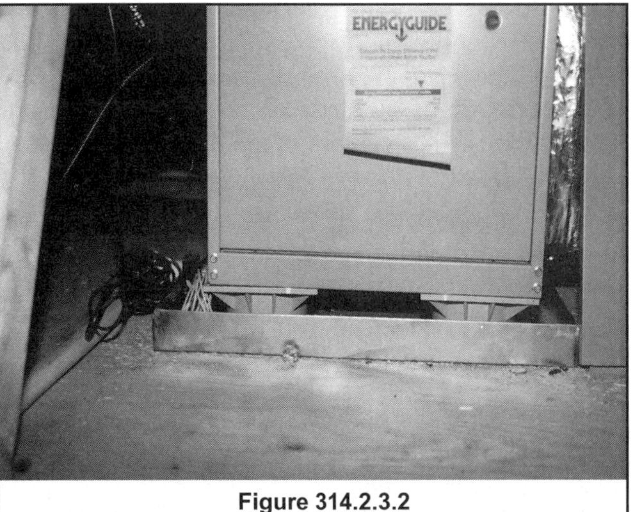

Figure 314.2.3.2
FURNACE ELEVATED ABOVE
TOP OF CONDENSATE PAN

[M] 314.2.4 Traps. Condensate drains shall be trapped as required by the equipment or appliance manufacturer.

❖ The appliance or equipment manufacturer determines the need for a trap and often specifies its depth and configuration. The traps addressed in this section are unrelated to plumbing traps and serve a different purpose. Condensate drain traps do not directly connect in any way to the plumbing DWV system of a building. Condensate drain traps are installed to prevent air from being pushed or pulled through the drain piping. Such airflow can impede condensate flow, causing overflow or abnormal water depth in drain pans. Some drain pans, such as those under pull-through cooling coils, might not drain without a trap to block airflow in the drain piping.

Bibliography

The following resource materials are referenced in this chapter or are relevant to the subject matter addressed in this chapter.

AISI S230-07, *Standard for Cold-formed Steel Framing-Prescriptive Method for One- and Two-family Dwellings*. Washington, DC: American Iron and Steel Institute, 2007.

ANSI Z4.3-95, *Minimum Requirements for Nonsewered Waste-Disposal Systems*. New York, NY: American National Standards Institute, Inc., 1995.

ASCE 24-05, *Flood Resistant Design and Construction*. Reston, VA: American Society of Civil Engineers, 2005.

ASTM E 814-02, *Test Method of Fire Tests of Through-penetration Fire-stops*. West Conshohocken, PA: American Society for Testing and Materials, 2002.

ASTM D 2321-05, *Practice for Underground Installation of Thermoplastic Pipe for Sewers and Other Gravity Flow Applications*. West Conshohocken, PA: American Society for Testing and Materials, 2005.

ASTM D 2774-05, *Standard Practice for Underground Installation of Thermoplastic Pressure Piping*. West Conshohocken, PA: American Society for Testing and Materials, 2005.

Cast Iron Soil Pipe and Fittings Handbook, Cast Iron Soil Pipe Institute, 2006.

FEMA 348, *Protecting Building Utilities from Flood Damage: Principles and Practices for the Design and Construction of Flood Resistant Building Utility Systems*. Washington, DC: Federal Emergency Management Agency, 1999.

IBC-09, *International Building Code*. Washington, DC: International Code Council, Inc., 2009.

IECC-09, *International Energy Conservation Code*. Washington, DC: International Code Council, Inc., 2009.

IMC-09, *International Mechanical Code*. Washington, DC: International Code Council, Inc., 2009.

IRC-09, *International Residential Code*. Washington, DC: International Code Council, Inc., 2009.

ICC-400, *Standard on the Design and Construction of Log Structures*. Washington, DC: International Code Council, Inc., 2009.

NSF 14-03, *Plastic Piping System Components and Related Materials*. Ann Arbor, MI: National Sanitation Foundation, 2003.

NSF 61-03e, *Drinking Water System Components—Health Effects*. Ann Arbor, MI: National Sanitation Foundation, 2003.

Chapter 4: Fixtures, Faucets and Fixture Fittings

General Comments

Determining the exact number of plumbing fixtures needed by the maximum number of possible users in any given building continues to be a challenging problem for the plumbing industry. Various methods have been used to evaluate the number and type of plumbing fixtures needed for an occupancy. Studies of office buildings have produced design guidelines based on occupancy times, arrival rates and patterns of fixture use that offer insight into the number of required plumbing fixtures for a desired level of service. The Building Technology Research Laboratory at the Stevens Institute of Technology conducted a study based on the Queueing Theory which involves probabilities for waiting times and fixture use times based upon a preferred level of service. The study resulted in design guidelines for the quantities of fixtures needed in public toilet facilities. For residential-type buildings and health care facilities, plumbing fixture numbers are based on the minimum need, resulting in at least one water closet and one lavatory for each dwelling unit, guestroom or hospital room.

Studies completed by the U.S. military have been used for dormitories and prisons to determine the number of fixtures required based on a simultaneous need in a regimented society. This assumes that everyone rises at approximately the same time and has a limited amount of time to shower and use the water closet and lavatory.

The National Restaurant Association conducted a study based on the difference in use between a restaurant and a nightclub. It should be noted that the study did not take into account today's fast-food-style restaurant, nor did it allow for restaurants located along heavily traveled routes, such as those at highway rest stops or oases.

Fixture requirements for factory and industrial uses are based on the same requirements as for storage facilities. This method establishes a realistic minimum requirement for factory occupancies. The reasonableness in the number of plumbing fixtures was established through a limited study of factory projects in Henrico County, Virginia.

The fixture needs for the remaining occupancies were determined based on empirical data, experience and tradition. There are no exact studies providing values that are definitively supportable; the values have been modified periodically based on general observations. Various studies will most likely continue to be performed in order to develop a completely rational method for determining the minimum number of plumbing fixtures for any occupancy.

Purpose

The purpose of Chapter 4 is to provide a building with the necessary number of plumbing fixtures of a specific type and quality. The fixtures must be properly installed to be usable by the individuals occupying the building. The quality and design of every fixture must conform to the applicable referenced standard.

SECTION 401 GENERAL

401.1 Scope. This chapter shall govern the materials, design and installation of plumbing fixtures, faucets and fixture fittings in accordance with the type of *occupancy*, and shall provide for the minimum number of fixtures for various types of occupancies.

❖ This section contains the scoping requirements for Chapter 4. Compliance with this chapter will result in a building or structure having adequate plumbing fixtures for the sanitary, hygienic, cleaning, washing and food preparation needs of the occupants. Many of the requirements of this chapter have been duplicated in the *International Building Code*® (IBC®), Chapter 29, for the convenience of the architect, who is often responsible for the layout and number of plumbing fixtures when creating the floor plan of a building. In fact, Chapter 29 of the IBC is under the complete control of the plumbing code.

401.2 Prohibited fixtures and connections. Water closets having a concealed trap seal or an unventilated space or having walls that are not thoroughly washed at each discharge in accordance with ASME A112.19.2M shall be prohibited. Any water closet that permits siphonage of the contents of the bowl back into the tank shall be prohibited. Trough urinals shall be prohibited.

❖ Water closet fixture designs that do not adequately clean the inside bowl surfaces with flushing water can harbor bacteria, resulting in an insanitary condition. The inspection and testing requirements of ASME 112.19.2M ensure that water closet designs have the necessary scouring action as well as a visible trap seal. Water closets must not be located in unventilated spaces as stagnant air conditions in toilet rooms are known to have caused widespread public disease. Tank-type water closets must be designed so that the bowl contents cannot be siphoned into the tank as this can expose the occupants to water-borne diseases. Trough urinals are not allowed to be installed as they do not offer the necessary privacy to the users and urine stream splashing creates insanitary conditions for adjacent users.

401.3 Water conservation. The maximum water flow rates and flush volume for plumbing fixtures and fixture fittings shall comply with Section 604.4.

❖ The impetus for the plumbing code to begin regulating the amount of water used in the operation certain plumbing fixtures was the Federal Energy Policy Act of 1992 (EPAct). The process of transporting of raw water to treatment plants, the treatment of water, the conveyance of treated water to the users, the transportation of wastewater to treatment plants and the treatment of wastewater consume tremendous amounts of energy. As the amount of water processed through this never-ending cycle becomes greater, the energy consumed increases proportionally. The production of energy for this purpose (primarily electricity) depletes the planet's nonrenewable resources of coal, gas, oil and fissionable materials. The purpose of the Energy Act of 1992 was to put controls in place to reduce the consumption rate of nonrenewable resources used for power generation. Among the many restrictions required by EPAct, certain plumbing fixtures were targeted for a reduction in the rate of water use. These flow and volume restrictions are reflected in Table 604.4.

Water is a resource that all life forms on this planet must have to survive. While there is an abundance of water on the Earth, the total amount of water is finite and is not necessarily located where water is needed. As the number of people on this planet continues to increase and as more countries become industrialized, the demand for potable water continues to rise. Many regions of the world are quickly depleting the once plentiful fresh water resources. Water conservation practices are now a major factor in building designs as there is already a real crisis of exhausting fresh water resources in some regions of the world.

SECTION 402
FIXTURE MATERIALS

402.1 Quality of fixtures. Plumbing fixtures shall be constructed of *approved* materials, with smooth, impervious surfaces, free from defects and concealed fouling surfaces, and shall conform to standards cited in this code. All porcelain enameled surfaces on plumbing fixtures shall be acid resistant.

❖ This section is provided to direct plumbing fixture standard creators as to the core requirements that all plumbing fixtures must meet. Most mass-produced plumbing fixtures for installation in accordance with the code have the design and quality regulated by the various standards as specified in this code. Where a plumbing fixture that doesn't have a standard is desired to be used, the information in this section assists the code official in evaluating the product in accordance with Section 105.2.

402.2 Materials for specialty fixtures. Materials for specialty fixtures not otherwise covered in this code shall be of stainless steel, soapstone, chemical stoneware or plastic, or shall be lined with lead, copper-base alloy, nickel-copper alloy, corrosion-resistant steel or other material especially suited to the application for which the fixture is intended.

❖ Most commercially available fixtures are designed and made in compliance with nationally recognized standards that are referenced in the code. Where a fixture is produced to a fixture standard not referenced by the code or no fixture standard exists, the code official must evaluate and approve the fixture in accordance with Section 105.2, regardless of whether the fixture is made of the materials listed in this section. For example, consider a kitchen sink which is one of the materials listed in this section but is in compliance with only an unrecognized fixture standard. The faucet hole(s) and drain outlet dimensions could be such that fittings complying with the referenced standards in this code will not fit properly. The stainless steel material used to make the sink might be of a grade that will quickly corrode or be stained in normal service. Therefore, the code official might deem the fixture not suitable since he or she is charged with the responsibility to insure that the sink will provide for reasonable durability, sanitary conditions and ease of maintenance.

Shower bases and group wash-up sinks made of terrazzo are fixtures that are not covered by standards referenced in this code and, as such, must be specifically approved by the code official.

Other specialty fixtures include lavatory bowls made of exotic metals, glass and antique wash basins; bathtubs of hollowed-out blocks of granite or other natural materials such as wood; and custom-fabricated copper kitchen sinks. If these products are not made to a referenced standard listed in this code, these products must be approved by the code official.

402.3 Sheet copper. Sheet copper for general applications shall conform to ASTM B 152 and shall not weigh less than 12 ounces per square foot (3.7 kg/m^2).

❖ In the past, general-use sheet copper was predominantly used as a lining in wooden plumbing fixtures. Today, the material is rarely used for those purposes. The ASTM standard regulates cold-rolled tempered sheet copper, hot-rolled tempered sheet copper and annealed copper sheet. Note that Section 425.3.3 permits the use of 10-ounce-per-square-foot (0.026 kg/m^2) sheet copper to line flush tanks, and Section 902.2 permits the use of 8-ounce-per-square-foot (0.021 kg/m^2) sheet copper for vent flashings.

402.4 Sheet lead. Sheet lead for pans shall not weigh less than 4 pounds per square foot (19.5 kg/m^2) coated with an asphalt paint or other *approved* coating.

❖ In the past, sheet lead was frequently used a shower pan liner and for other lining applications. Today, sheet lead is rarely used in the plumbing industry. The approximate thickness of 4-pound-per-square-foot (20 kg/m^2) sheet lead is $^{1}/_{16}$ inch (1.6 mm).

SECTION 403 MINIMUM PLUMBING FACILITIES

403.1 Minimum number of fixtures. Plumbing fixtures shall be provided for the type of *occupancy* and in the minimum number shown in Table 403.1. Types of occupancies not shown in Table 403.1 shall be considered individually by the code official. The number of occupants shall be determined by the *International Building Code*. *Occupancy* classification shall be determined in accordance with the *International Building Code*.

❖ This section requires that the type and number of plumbing fixtures be based on the use of an occupancy and the occupant load as determined for the means of egress for a building or space. Occupancy classifications (groups) are determined from Chapter 3 of the IBC and occupant load is determined from Section 1004 of the IBC. Table 403.1 shows ratios indicating the maximum number of occupants that can be served by one fixture of each type. For example, "1 per 75" means one fixture will serve up to 75 occupants. These "ratios" are applied to the occupant load of the building or space in order to determine the total number of each required fixture type.

The "description" of plumbing fixture use determines which set of fixture ratios should be used for calculating plumbing fixture requirements. In most cases, the description of plumbing fixture use will match the IBC occupancy classification. However, there are situations where a description of fixture use different than the IBC occupancy classification might be a more reasonable alternative approach to determining the number of required plumbing fixtures. These alternative methods require approval by the code official in accordance with Section 105.2 of this code. This approach might be useful for certain educational and business facilities as illustrated in the following discussions.

Educational Facilities

Consider an educational facility (Grades 1-12) with a gymnasium with a stage. The IBC states that assembly areas that are accessory to Group E occupancies are not to be considered as a separate occupancy. Therefore, the number of water closets should be determined from section 3 of Table 403.1, which requires one water closet per 50 occupants. Because this gymnasium has a stage, a greater occupant density than for a gymnasium-only space must be chosen as nonfixed chairs will most likely be set up for viewing on-stage activities. Therefore, for egress purposes, the gymnasium is considered to be an assembly area with an occupant density of 7 square feet per person. The resulting occupant load number applied with the required Group E occupancy plumbing fixture ratio can result in an excessive number of plumbing fixtures for the building. An alternative approach might be to consider using the plumbing fixture ratios in the fourth row of Table 403.1 for a gymnasium space since these ratios reflect how the space is intended to be used with respect to the occupancy density chosen for the space. In other words, fixture ratios might be chosen to "agree" with the actual use of the space.

A question that is often asked about Group E occupancies is whether the concept of nonsimultaneous occupancy could be considered in order to further reduce the number of required plumbing fixtures. For example, where a school auditorium is occupied by only the students within the school, the students are either in the classroom or the auditorium but not in both places at the same time. Therefore, why should each area require a number of fixtures as if both areas were occupied simultaneously? The code does not recognize a nonsimultaneous use concept for any building, as simultaneous use could easily occur. For example, while students are in the classrooms, a school auditorium could be used temporarily for blood drives, regional science fairs and town meetings. However, this is not to say that in some circumstances, a nonsimultaneous use could, in some way, be completely guaranteed such that the local code official might entertain a proposed reduction in the required number of fixtures for the building.

Business Facilities

Consider a barber college where the IBC classifies the entire building as a Group B occupancy. The building has several large assembly rooms where the intent is to have training sessions for large groups of students. Clearly, these areas are used for assembly and, therefore, the use of the fixture ratios in section 1, row 4 of Table 403.1 could be proposed to the code official.

The choice of an occupancy use for the purposes of determining plumbing fixtures does not affect the occupancy group chosen for IBC purposes, that being for egress. In other words, using a previous example, the school gymnasium with stage space may be chosen to be an "assembly use" for plumbing fixture requirements, but the entire building is still a Group B occupancy for the purposes of egress as far as the IBC is concerned.

Note that 2009 IBC Section 1004.1.1 has an exception which allows for the designer to present an "actual" occupancy load number to the code official for approval instead of the calculated load for the occupancy square footage. While this exception would allow a smaller occupancy load to be chosen for the purposes of determining the required number of plumbing fixtures, code officials must carefully consider and appropriately justify the reduction. The code official should consider the potential for the occupancy to be loaded with more persons than stated by the designer, future use by different tenants of the same occupancy classification and the difficulty of enforcement of the maximum occupancy load based upon the limited number of fixtures provided.

TABLE 403.1. See page 4-6.

❖ The brief wording in the "descriptions" column of Table 403.1 is not intended to be complete or all-inclusive of all uses for a particular occupancy. However, the descriptions do identify a majority of the types of uses encountered in the design of most buildings. Because the descriptions "narrow in" on what row of ratios to use, users of this table will benefit by using the description column as the entry point into the table. The number, classification and occupancy columns are intended only for reference purposes. The requirements for type and number of plumbing fixtures are driven by the actual use of a building or space, which is usually consistent with the occupancy group classification.

Some use description rows have a different ratio for male and female water closets, and in some cases, lavatories. The smaller ratio for female fixtures provides an "equality of fixture availability" in those particular occupancies. The occupancies described by these use descriptions have historically had long lines of females waiting to use toilet facilities while male facilities had no lines. The reasons for this include the following:

1. For a variety of social and physical reasons, women generally take a longer period of time to use the facilities;
2. Women outnumber men in the general population and this becomes especially evident in large groups of people; and
3. Women, in general, tend to use the facilities more frequently.

The term "potty parity" was coined in a general sense to mean that a sufficient number of female plumbing fixtures were available such that women did not have to wait any longer than men (in a queue or line) to use an equivalent type of fixture as men. In a specific sense, "potty parity" also indicates the ratio of the number of male to the number of female fixtures of the same type. For example, if 24 male and 48 female water closets are installed in a building, then the potty parity is stated to be "1 to 2."

The code is silent concerning the installation of child-size water closets. In a child care facility having children age 6 years old and under, the provision of child-size water closets (and sinks mounted at child height) could be beneficial to a facility's operation. Generally, for children over 6 years old, standard-sized fixtures are suitable because most children have learned to use the same size fixtures in a home setting. Note that the accessibility standard, ICC/ANSI A117.1 (as referenced by Chapter 11 of the IBC), has specific requirements for accessible child-size water closets.

Some school facilities are designed to have a single-occupant toilet room that can be accessed only from within the classroom. The code is silent on whether the fixtures in this limited access toilet room can be counted toward the required number of plumbing fixtures for the building.

The intent of table note d concerning outdoor seating and entertainment areas is to require that outdoor patios, decks, balconies, beer gardens and similar areas, whether those areas have seating or not, be included for calculating the number of plumbing fixtures. While seating does provide a definitive occupant load for these types of areas, there are many situations where these areas are not provided with seating in order to accommodate as many customers as possible. Therefore, the occupant load must be based upon standing space density. This occupancy load number must be applied to the ratios in the appropriate use description row in Table 403.1. For example, a restaurant has a fixed seating area with fixed booths and an outdoor entertainment area that has no seating or tables. The fixed seating area load is simply a count of the number of seats according to Section 1004.7 of the IBC. The restaurant fixture ratios (section 1, row 3 of Table 403.1) are then applied to this load to determine the number of plumbing fixtures for the fixed seating area. The outdoor area requires a different approach, as it is a standing space area, not fixed seating. The occupant load is calculated from the area using IBC Table 1004.1.1. Even though the building is a restaurant, this outdoor entertainment area might be used, not as a restaurant, but as a nightclub/dance floor. Therefore, the fixture ratios in row 2 of Table 403.1 (for nightclub/dance halls) are applied to the standing occupant load. Using row 3 fixture ratios for this area would be inappropriate. The use of proper occupant loads and use ratios will ensure that an adequate number of fixtures are provided where the occupant load is substantially increased by the addition of outdoor seating or standing areas.

Choosing the proper use description for a restaurant having a bar or for a nightclub serving food can be challenging. Some restaurant operations evolve into a nightclub/dance hall in the late evening. Bars with dance floors often have kitchens for preparing food to serve to patrons at the bar. The code is silent on how to determine which use description (restaurant or night club) is the best "fit" for the demand on the toilet facilities. The answer to this question lies in determining what use is primary. Is the bar's main purpose to support the restaurant operation? Or is the serving of food in support of the bar operation? In situations where there is no clear answer, perhaps considering the occupancy as a mixed use arrangement is a practical answer.

The drinking fountain ratio indicated for an occupancy use description does not necessarily imply that every occupancy must have a drinking fountain within that occupancy. Just as toilet facilities are only required to be provided for an occupancy and available when the occupancy is in use, the provision of a drinking fountain follows similar logic. For example, consider a strip center with multiple tenant spaces, each tenant space having the required number of plumbing fixtures for and within each space. The drinking fountain for that space (and all the adjacent spaces) could be located exterior to all the tenant spaces in a common area. The calculation for the required number of drinking fountains to be used by multiple occupancies is performed in the same manner as for other plumbing fixtures. See the Sample Problems in this commentary section.

The "service sinks" required by Table 403.1 are intended to be of a type suitable for janitorial and building maintenance purposes. Service sinks include mop sinks/basins, utility tub/sinks, janitor sinks, slop sinks and laundry trays. In all occupancy classifications, except Groups I-2, R-2 and R-3, only one service sink is required for the entire building, except that hospitals are required to have a service sink on each floor. The one service sink per building must be available from all portions of the building. The intent of the code is that employees and tenants of the building be able to access the required sink. This does not mean that the service sink is required to be in an open area of the building for use by any occupant (e.g., visitors). The service sink can be located in a locked janitor's closet for which all employees have access by key or door lock code to gain entry. However, the required service sink cannot be located in areas that are access-controlled by a single tenant. For example, if the service sink is located on floor one within a tenant's space, it would be possible for the tenant of floor one to lock out (exclude) other tenants from accessing the required service sink. Therefore, that service sink within one tenant-controlled space cannot be the required service sink for the building.

Although the IBC classifies parking garages as a Group S occupancy, there is no entry in Table 403.1 for parking structures. It is not the intent of the code to require public toilet facilities in parking garages. A parking garage is similar to a parking lot in that it is not normally the final destination of a person parking a car. The only reason a person is in a parking garage is to park their car and then leave. A parking lot, having no enclosing structure, is not required to be provided with toilet facilities for the occupants that are parking their vehicles, so it stands to reason that a parking garage would not be required to be provided with toilet facilities. When people drive into a parking garage, their normal expectation is that the toilet facilities will be located in the building or buildings served by the parking garage. If public toilet facilities were provided in a parking garage, significant maintenance, vandalism and security issues would probably result in the eventual closure of such facilities. Not requiring public toilet facilities in a parking garage does not provide justification for not providing toilet facilities for parking garage employees (such as ticket kiosk attendants). These employees must be provided with toilet facilities in accordance with Section 403.3. However, these facilities are not required to be located in the parking garage but only within the travel limitations of Section 403.3.2.

Note e was added in the 2009 edition to clarify the requirements for drinking fountain accessibility in accordance with Chapter 11 of the IBC. In order to comply with both the IPC and the IBC, an occupancy that is required to have a minimum of one drinking fountain is required to have either a unit that has two nozzles, one high and one low, or two separate units, one with a high spout and one with a low spout. The accessibility standard ICC A117.1, as referenced by the IBC, specifies the elevation range for the spouts for both the high and low spout drinking fountains. See commentary for Section 404 for additional details concerning wheelchair approach to drinking fountains.

Note f was added in the 2009 edition to allow occupancies with 15 or fewer occupants not to have a drinking fountain. The requirement for drinking fountains in all occupancies put an undue hardship on small establishments where the floor space is usually limited. Adding to this problem is the IBC requirement stating that where one drinking fountain is provided, two drinking fountains must be installed, one high (for standing persons) and one low (for wheelchair-seated persons). One required drinking fountain in a small establishment turned into an expensive, space consuming fixture that, according to the proponents of the change, is rarely is used.

A change for the 2009 edition of the code altered the women's water closet requirements for Groups A-4 and A-5 occupancies to increase the initial threshold value from 1500 to 1520. This change was implemented to make the result of dividing 1520 by 40 a whole number so as to avoid the need to deal with the rounding up of multiple fractional fixture results.

Other changes to the 2009 edition of the code were (1) a category was added for employees in a Group I-3 occupancy and (2) a category was added for congregate living facilities (Group R-3 occupancy) with 16 or fewer persons.

TABLE 403.1
MINIMUM NUMBER OF REQUIRED PLUMBING FIXTURES[a]
(See Sections 403.2 and 403.3)

NO.	CLASSIFICATION	OCCUPANCY	DESCRIPTION	WATER CLOSETS (URINALS SEE SECTION 419.2)		LAVATORIES		BATHTUBS/ SHOWERS	DRINKING FOUNTAIN[e, f] (SEE SECTION 410.1)	OTHER
				MALE	FEMALE	MALE	FEMALE			
1	Assembly	A-1[d]	Theaters and other buildings for the performing arts and motion pictures	1 per 125	1 per 65	1 per 200		—	1 per 500	1 service sink
		A-2[d]	Nightclubs, bars, taverns, dance halls and buildings for similar purposes	1 per 40	1 per 40	1 per 75		—	1 per 500	1 service sink
			Restaurants, banquet halls and food courts	1 per 75	1 per 75	1 per 200		—	1 per 500	1 service sink
		A-3[d]	Auditoriums without permanent seating, art galleries, exhibition halls, museums, lecture halls, libraries, arcades and gymnasiums	1 per 125	1 per 65	1 per 200		—	1 per 500	1 service sink
			Passenger terminals and transportation facilities	1 per 500	1 per 500	1 per 750		—	1 per 1,000	1 service sink
			Places of worship and other religious services.	1 per 150	1 per 75	1 per 200		—	1 per 1,000	1 service sink
		A-4	Coliseums, arenas, skating rinks, pools and tennis courts for indoor sporting events and activities	1 per 75 for the first 1,500 and 1 per 120 for the remainder exceeding 1,500	1 per 40 for the first 1,520 and 1 per 60 for the remainder exceeding 1,520	1 per 200		—	1 per 1,000	1 service sink
		A-5	Stadiums, amusement parks, bleachers and grandstands for outdoor sporting events and activities	1 per 75 for the first 1,500 and 1 per 120 for the remainder exceeding 1,500	1 per 40 for the first 1,520 and 1 per 60 for the remainder exceeding 1,520	1 per 200		—	1 per 1,000	1 service sink
2	Business	B	Buildings for the transaction of business, professional services, other services involving merchandise, office buildings, banks, light industrial and similar uses	1 per 25 for the first 50 and 1 per 50 for the remainder exceeding 50		1 per 40 for the first 80 and 1 per 80 for the remainder exceeding 80		—	1 per 100	1 service sink
3	Educational	E	Educational facilities	1 per 50		1 per 50		—	1 per 100	1 service sink
4	Factory and industrial	F-1 and F-2	Structures in which occupants are engaged in work fabricating, assembly or processing of products or materials	1 per 100		1 per 100		(see Section 411)	1 per 400	1 service sink

(continued)

TABLE 403.1—continued
MINIMUM NUMBER OF REQUIRED PLUMBING FIXTURES[a]
(See Sections 403.2 and 403.3)

NO.	CLASSIFICATION	OCCUPANCY	DESCRIPTION	WATER CLOSETS (URINALS SEE SECTION 419.2)		LAVATORIES		BATHTUBS/ SHOWERS	DRINKING FOUNTAIN[e, f] (SEE SECTION 410.1)	OTHER
				MALE	FEMALE	MALE	FEMALE			
5	Institutional	I-1	Residential care	1 per 10		1 per 10		1 per 8	1 per 100	1 service sink
		I-2	Hospitals, ambulatory nursing home patients[b]	1 per room[c]		1 per room[c]		1 per 15	1 per 100	1 service sink per floor
			Employees, other than residential care[b]	1 per 25		1 per 35		—	1 per 100	—
			Visitors, other than residential care	1 per 75		1 per 100		—	1 per 500	—
		I-3	Prisons[b]	1 per cell		1 per cell		1 per 15	1 per 100	1 service sink
			Reformitories, detention centers, and correctional centers[b]	1 per 15		1 per 15		1 per 15	1 per 100	1 service sink
			Employees[b]	1 per 25		1 per 35		—	1 per 100	—
		I-4	Adult day care and child care	1 per 15		1 per 15		1	1 per 100	1 service sink
6	Mercantile	M	Retail stores, service stations, shops, salesrooms, markets and shopping centers	1 per 500		1 per 750		—	1 per 1,000	1 service sink
7	Residential	R-1	Hotels, motels, boarding houses (transient)	1 per sleeping unit		1 per sleeping unit		1 per sleeping unit	—	1 service sink
		R-2	Dormitories, fraternities, sororities and boarding houses (not transient)	1 per 10		1 per 10		1 per 8	1 per 100	1 service sink
		R-2	Apartment house	1 per dwelling unit		1 per dwelling unit		1 per dwelling unit	—	1 kitchen sink per dwelling unit; 1 automatic clothes washer connection per 20 dwelling units
		R-3	One- and two-family dwellings	1 per dwelling unit		1 per dwelling unit		1 per dwelling unit	—	1 kitchen sink per dwelling unit; 1 automatic clothes washer connection per dwelling unit
		R-3	Congregate living facilities with 16 or fewer persons	1 per 10		1 per 10		1 per 8	1 per 100	1 service sink
		R-4	Residential care/assisted living facilities	1 per 10		1 per 10		1 per 8	1 per 100	1 service sink

(continued)

TABLE 403.1—continued
MINIMUM NUMBER OF REQUIRED PLUMBING FIXTURES[a]
(See Sections 403.2 and 403.3)

NO.	CLASSIFICATION	OCCUPANCY	DESCRIPTION	WATER CLOSETS (URINALS SEE SECTION 419.2) MALE	WATER CLOSETS (URINALS SEE SECTION 419.2) FEMALE	LAVATORIES MALE	LAVATORIES FEMALE	BATHTUBS/ SHOWERS	DRINKING FOUNTAIN[e, f] (SEE SECTION 410.1)	OTHER
8	Storage	S-1 S-2	Structures for the storage of goods, warehouses, storehouse and freight depots. Low and Moderate Hazard.	1 per 100		1 per 100		See Section 411	1 per 1,000	1 service sink

a. The fixtures shown are based on one fixture being the minimum required for the number of persons indicated or any fraction of the number of persons indicated. The number of occupants shall be determined by the *International Building Code.*

b. Toilet facilities for employees shall be separate from facilities for inmates or patients.

c. A single-occupant toilet room with one water closet and one lavatory serving not more than two adjacent patient sleeping units shall be permitted where such room is provided with direct access from each patient sleeping unit and with provisions for privacy.

d. The occupant load for seasonal outdoor seating and entertainment areas shall be included when determining the minimum number of facilities required.

e. The minimum number of required drinking fountains shall comply with Table 403.1 and Chapter 11 of the *International Building Code.*

f. Drinking fountains are not required for an occupant load of 15 or fewer.

403.1.1 Fixture calculations. To determine the occupant load of each sex, the total occupant load shall be divided in half. To determine the required number of fixtures, the fixture ratio or ratios for each fixture type shall be applied to the occupant load of each sex in accordance with Table 403.1. Fractional numbers resulting from applying the fixture ratios of Table 403.1 shall be rounded up to the next whole number. For calculations involving multiple *occupancies*, such fractional numbers for each *occupancy* shall first be summed and then rounded up to the next whole number.

Exception: The total occupant load shall not be required to be divided in half where *approved* statistical data indicates a distribution of the sexes of other than 50 percent of each sex.

❖ In previous editions of the code, a common question was whether the gender distribution was to be applied before applying the fixture ratios or if the gender distribution was to be applied after the fixture ratios were applied. In most cases, it did not matter because the results were identical. However, where "graduated" fixture ratios are involved, different results could be obtained. A business occupancy has graduated fixture ratios for water closets and lavatories. The following example illustrates how two different results can be obtained:

Assume a total business occupant load of 130.

A. Applying the water closet fixture ratio to the total occupant load and then dividing in half results in the following:

For the first 50 occupants 50 x 1/25 = 2

For the remaining 80 occupants: 80 x 1/50 = 1.6

Total number of water closets = 3.6 → round up to 4

Therefore, each gender requires two water closets.

B. Applying the gender distribution first and then applying the fixture ratio results in the following:

Number of occupants of each sex: 130/2 = 65

For the first 50 occupants: 50 x 1/25 = 2

For the remaining 15 occupants: 15 x 1/50 = 0.3

Water closets for each sex = 2.3 → round up to 3

Therefore, each gender requires three water closets.

Both answers were correct as the code did not recognize this anomaly. Thus, for the 2009 edition of the code, this section was revised to require that the gender distribution be applied first before the fixture ratios are applied.

This section requires that the occupant load be divided by two in order to determine the quantities of males and females. This is known as a 50-50 gender distribution meaning that 50 percent of the occupants are males and 50 percent of occupants are female. While a 50-50 distribution is used most of the time, there might be situations where this assumption is inaccurate. The exception to this section allows for a different gender distribution if statistical data is submitted and approved by the code official for an occupancy. Example occupancies that might require a modified distribution would be an all-women's health club, a males-only boarding school, a convent or a monastery.

Once the appropriate use description row in Table 403.1 is chosen, the ratios for each fixture type and for each gender are used to determine the minimum number of plumbing fixtures for a building or space.

Although not specifically indicated in this code section, prior to performing any plumbing fixture calculations for an occupancy, the exceptions of Section 403.2 must be reviewed for applicability. For example, a small restaurant could have two employees and 12 seats. If the fixture calculations were performed without consideration of Section 403.2, this restaurant would be required to have a toilet facility for each gender. However, because the occupancy load is less than 15, a single toilet facility is all that is required.

The following sample problems illustrate calculation methods for various situations.

Sample Problem 1: A mixed-use building has a business-use occupant load of 538, a library-use occupant load of 115 and storage-use occupant load of 82. The occupancies are not separate tenant spaces. Determine the total number of required plumbing fixtures for the building.

Problem Approach

Check Section 403.2 for applicability and if not applicable, proceed with calculations.

Apply the gender distribution for each occupancy use, calculate the number of fixtures required by each gender, sum the gender totals for each type of fixture from all occupancies and round up the number of each type of fixture for each gender to a whole number. Note that drinking fountains are not gender specific. Therefore, the total number of occupants for the occupancy is applied to the drinking fountain fixture ratio.

Solution—Part I

For the business-use occupancy, find the required ratios in Table 403.1, section 2. Because a gender distribution was not specified, the number of occupants must be divided by two to obtain the number of occupants for each gender (refer to Section 403.1.1). Therefore, in this business-use occupancy, there will be

538/2 = 269 occupants of each gender

1. Calculate the number of water closets for 269 males:

   For the first 50 males, the ratio of 1 per 25 is applied:

   50 x 1/25 = 2 water closets

   For the remaining number of males, the ratio of 1 per 50 is applied:

   (269 - 50) x 1/50 = 219 x 1/50 = 4.38 water closets

   The required number of water closets for males is:

   2 + 4.38 = 6.38 water closets for males

   Because the gender distribution is equal and the water closet ratio for females is the same as for males, the number of water closets for females is also 6.38.

2. Calculate the number of lavatories for 269 males:

   For the first 80 males, the ratio of 1 per 40 is applied:

   80 x 1/40 = 2 lavatories

   For the remaining males, the ratio of 1 per 80 is applied:

   (269 - 80) x 1/80 = 189 x 1/80 = 2.36 lavatories.

   The required number of lavatories for males is:

   2 + 2.36 = 4.36 lavatories for males

Because the gender distribution is equal and the lavatory ratio for females is the same as for males, the number of lavatories for females is also 4.36.

3. Calculate the number of drinking fountains for 538 occupants:

   The ratio of 1 per 100 is applied:

   538 x 1/100 = 5.38 drinking fountains for male and female use

Solution—Part II

For the library-use occupancy, find the required ratios in Table 403.1 in section 1. Because a gender distribution was not specified, the number of occupants must be divided by two to obtain the number of occupants for each gender (refer to Section 403.1.1). Therefore, in this library-use occupancy, there will be

115/2 = 57.5 occupants of each gender

1. Calculate the number of water closets for males:

   The ratio of 1 per 125 is applied:

   57.5 x 1/125 = 0.46 water closets for males

2. Calculate the number of water closets for females:

   The ratio of 1 per 65 is applied:

   57.5 x 1/65 = 0.88 water closets for females

3. Calculate the number of lavatories for 57.5 males:

   The ratio of 1 per 200 is applied:

   57.5 x 1/200 = 0.29 lavatories for males

   Because the gender distribution is equal and the lavatory ratio for females is the same as for males, the number of lavatories for females is also 0.29.

3. Calculate the number of drinking fountains for 115 occupants:

   The ratio of 1 per 500 is applied:

   115 x 1/500 = 0.23 drinking fountains of both males and females.

Solution—Part III

For the storage-use occupancy, find the required ratios in Table 403.1 in section 8. Because a gender distribution was not specified, the number of occupants must be divided by two to obtain the number of occupants for each gender (refer to Section 403.1.1). Therefore, in this storage-use occupancy, there will be

82/2 = 41 occupants of each gender

1. Calculate the number of water closets for 41 males:

   The ratio of 1 per 100 is applied:

   41 x 1/100 = 0.41 water closets for males

   Because the gender distribution is equal and the water closet ratio for females is the same as for males, the number of water closets for females is also 0.41.

2. Calculate the number of lavatories for 41 males:

   The ratio of 1 per 100 is applied:

   41 x 1/100 = 0.41 lavatories for males

   Because the gender distribution is equal and the lavatory ratio for females is the same as for males, the number of lavatories for females is also 0.41.

3. Calculate the number of drinking fountains for 82 occupants:

   The ratio of 1 per 1000 is applied:

   82 x 1/1000 = 0.082 drinking fountains for males and females.

Solution—Summary

See Commentary Table 403.1(1) for the solution summary for Sample Problem 1. Note that if the raw fixture numbers were rounded up prior to summation for the building, the calculated minimum fixture requirements would have been increased by one water closet for two drinking fountains.

Sample Problem 2: A mixed-use building has a business-use occupant load of 940 and a restaurant-use occupant load of 360. The occupancies are separate tenant spaces. The toilet facilities are to be located in each tenant space. For the business occupancy, bottle water coolers are to be substituted for drinking fountains to the maximum allowable extent. Determine the total number of required plumbing fixtures for the building.

Problem Approach

Check Section 403.2 for applicability and if not applicable, proceed with calculations.

Note that the plumbing fixtures required by each occupancy will be located in each tenant space. Thus, the required number of plumbing fixtures for each tenant space is calculated as if the space stood alone, in other words, in its own building. Drinking fountain substitution is calculated last.

Solution—Part I

For the business-use occupancy, find the required ratios in Table 403.1 in section 2. Because a gender distribution was not specified, the number of occupants must be divided by two to obtain the number of occupants for each gender (refer to Section 403.1.1). Therefore, in this business-use occupancy, there will be

940/2 = 470 occupants of each gender

1. Calculate the number of water closets for 470 males:

   For the first 50 males, the ratio of 1 per 25 is applied:

   50 x 1/25 = 2 water closets

   For the remaining number of males, the ratio of 1 per 50 is applied:

   (470 - 50) x 1/50 = 420 x 1/50 = 8.4 water closets

   The required number of water closets for males is:

   2 + 8.4 = 10.4 water closets for males

   Because the gender distribution is equal and the water closet ratio for females is the same as for males, the number of water closets for females is also 10.4.

3. Calculate the number of lavatories for 470 males:

   For the first 80 males, the ratio of 1 per 40 is applied:

   80 x 1/40 = 2 lavatories

   For the remaining males, the ratio of 1 per 80 is applied:

   (470 - 80) x 1/80 = 390 x 1/80 = 4.88 lavatories.

   The required number of lavatories for males is:

   2 + 4.88= 6.88 lavatories for males

   Because the gender distribution is equal and the lavatory ratio for females is the same as for males, the number of lavatories for females is also 6.88.

3. Calculate the number of drinking fountains for 940 occupants:

   The ratio of 1 per 100 is applied:

   940 x 1/100 = 9.4 drinking fountains for males and females

Soultion—Part II

For the restaurant-use occupancy, find the required ratios in Table 403.1 in section 1. Because a gender distribution was not specified, the number of occupants must be divided by two to obtain the number of occupants for each gender (refer to Section 403.1.1). Therefore, in this restaurant-use occupancy, there will be

360/2 = 180 occupants of each gender

1. Calculate the number of water closets for males:

   For the first 25 males, the ratio of 1 per 25 is applied:

   25 x 1/25 = 1 water closet

   For the remaining 155:

   The ratio of 1 per 75 is applied:

   155 x 1/75 = 2.07 water closets for males

   Therefore, 3.07 male water closets are needed.

Because the gender distribution is equal and the water closet ratio for females is the same as for males, the number of water closets for females is also 3.07.

2. Calculate the number of lavatories for 180 males:

   For the first 40 males, the ratio of 1 per 40 is applied:

   40 x 1/40 = 1 lavatory

   For the remaining 140 the ratio of 1 per 200 is applied:

   140 x 1/200 = 0.7 lavatories for males are required.

   Therefore, 1.7 male lavatories are needed.

   Because the gender distribution is equal and the lavatory ratio for females is the same as for males, the number of lavatories for females is also 1.7.

3. Calculate the number of drinking fountains for 360 occupants:

   The ratio of 1 per 500 is applied:

   360 x 1/500 = 0.72 drinking fountains of both males and females.

Solution—Part III

The number of required drinking fountains is allowed to be reduced by up to 50 percent by substitution with bottled water coolers or bottled water dispensers (see Section 410.1 and related commentary). In restaurants, drinking fountains are not required if drinking water is served. Chapter 11 of the IBC requires that any drinking fountains that are provided be accessible. Where only one drinking fountain is being provided, either two installed at different heights or one combination unit with two bowls (each with a spout) at different heights, must be installed. (The spout height requirements are specified in ICC A117.1 and are repeated for convenience in the commentary for Section 404).

Bottled water cooler (or dispenser) substitution might not be appropriate for applications where located in areas that are completely open to the public. For example, a bottled water cooler to substitute for one of three required drinking fountains that are to be placed outdoors at a large strip center would most likely be subject to vandalism. However, a bottled water cooler substituted for one of three required drinking fountains in a large office building (a more con-

Table 403.1(1)
SOLUTION SUMMARY FOR SAMPLE PROBLEM 1

OCCUPANCY		WATER CLOSETS				LAVATORIES					
USE	LOAD	RATIO	MALE	RATIO	FEMALE	RATIO	MALE	FEMALE	DF RATIO	DRINKING FOUNTAINS	SERVICE SINK
Business	538	1 per 25 for the first 50 and 1 per 50 for the remainder exceeding 50	6.38	1 per 25 for the first 50 and 1 per 50 for the remainder exceeding 50	6.38	1 per 40 for the first 80 and 1 per 80 for the remainder exceeding 80	4.36	4.36	1 per 100	5.38	Note 1
Libraries, Halls, Museums, etc.	115	1/125	0.46	1/65	0.88	1/200	0.29	0.29	1 per 500	0.23	
Storage	82	1/100	0.41	1/100	0.41	1/100	0.41	0.41	1/1000	0.082	
Subtotals			7.25		7.67		5.06	5.06		5.69	
Required Totals			8		8		6	6		6	

Notes:
1. The code requires only one service sink per building if all occupancies have access to the service sink at all times.

trolled environment) would be appropriate. Where the plumbing code requires two drinking fountains, the accessibility regulations of IBC Chapter 11 require two drinking fountains, one with a high spout and one with a low spout. The following example illustrates how the accessibility requirements for drinking fountains relate to the plumbing code requirements for drinking fountains and bottled water cooler substitutions.

Consider the business occupancy requiring ten drinking fountains. Bottled water dispensers can substitute for five of the ten required drinking fountains. Of the five drinking fountains to be installed, two are mounted at the high spout level, two are mounted at the low spout level and the third can be mounted at either the high or low spout level. A bottled water cooler is not a drinking fountain. While the IBC provides accessibility requirements for drinking fountains, it does not address bottled water coolers.

Solution—Summary

See Commentary Table 403.1(2) for the solution summary for Sample Problem 2. Note that if the raw fixture numbers were rounded up prior to summation for the building, the calculated minimum fixture requirements would have increased by one water closet for each gender, one lavatory for each gender and two drinking fountains.

Sample Problem 3: A stadium has an occupant load of 5,200. Urinals are to be substituted for male water closets to the maximum extent allowed. Determine the total number of required plumbing fixtures for the stadium.

Problem Approach

Check Section 403.2 for applicability and if not applicable, proceed with calculations.

Apply the gender distribution, calculate the number of fixtures required by each gender, sum the gender totals for each type of fixture and round up the number of each type of fixture for each gender to a whole number. Apply the urinal substitution (Section 419.2) to reduce the number of male water closets.

Solution

For the stadium-use occupancy, find the required ratios in Table 403.1 in row 8. Because a gender distribution was not specified, the number of occupants must be divided by two to obtain the number of occupants for each gender (refer to Section 403.1.1). Therefore, in this stadium-use occupancy, there will be

5200/2 = 2600 occupants of each gender

1. Calculate the number of water closets for 2600 males:

   For the first 1500 males, the ratio of 1 per 75 is applied:

   1500 x 1/75 = 20 water closets

   For the remaining number of males, the ratio of 1 per 120 is applied:

   (2600 - 1500) x 1/120 = 1100 x 1/120 = 9.17 water closets

Table 403.1(2)
SOLUTION SUMMARY OF SAMPLE PROBLEM 2

OCCUPANCY		WATER CLOSETS				LAVATORIES			DF RATIO	DRINKING FOUNTAINS	SERVICE SINK
USE	LOAD	RATIO	MALE	RATIO	FEMALE	RATIO	MALE	FEMALE			
Business	940	1 per 25 for the first 50 and 1 per 50 for the remainder exceeding 50	10.4	1 per 25 for the first 50 and 1 per 50 for the remainder exceeding 50	10.4	1 per 40 for the first 80 and 1 per 80 for the remainder exceeding 80	6.88	6.88	1 per 100	9.4	Note 1
Business Required Totals			11		11		7	7		10	
Restaurants, Banquet Halls, and Food Courts	360	1 per 75	2.4	1 per 75	2.4	1 per 200	1.8	1.8	1 per 500	0.72	
Restaurant Required Total			3		3		2	2		1	
Building Required Totals			14		14		9	9		11	

Notes:

1. The code only requires one service sink per building if all occupancies have access to it at all times. If this building has a central area (such as janitorial closet for the building) that both tenants can access at all times, only one service sink is required for the building. But if the service sink is

The required number of water closets for males is:

20+ 9.17 = 29.17 → round to 30 water closets for males

2. Calculate the number of water closets for 2600 females:

For the first 1520 females, the ratio of 1 per 40 is applied:

1520 x 1/40 = 38 water closets

For the remaining number of females, the ratio of 1 per 60 is applied:

(2600 - 1520) x 1/60 = 1080 x 1/60 = 18 water closets

The required number of water closets for females is:

38 + 18 = 56 water closets for females.

3. Calculate the number of lavatories for 2600 males:

The ratio of 1 per 200 is applied:

2600 x 1/200 = 13 lavatories for males

4. Calculate the number of lavatories for 2600 females:

The ratio of 1 per 150 is applied:

2600 x 1/150 = 17.3 lavatories → round to 18 lavatories for females

5. Calculate the number of drinking fountains for 5200 occupants:

The ratio of 1 per 1000 is applied:

5200 x 1/1000 = 5.2 → round to 6 drinking fountains for males and females

6. Calculate the number of urinals to be substituted for male water closets. Because a stadium is in the assembly classification, 67 percent of the water closets can be replaced with urinals.

The ratio of 67/100 is applied:

30 x 67/100 = 20.1 urinals → Round down to 20 urinals because rounding up would result in the number of urinals being in excess of 67 percent of the required number of water closets.

Solution—Summary

See Commentary Table 403.1(3) for the solution summary for Sample Problem 3.

403.1.2 Family or assisted-use toilet and bath fixtures. Fixtures located within family or assisted-use toilet and bathing rooms required by Section 1109.2.1 of the *International Building Code* are permitted to be included in the number of required fixtures for either the male or female occupants in assembly and mercantile *occupancies*.

❖ A family or assisted-use toilet room is required by the IBC in assembly or mercantile occupancies having an total of six or more required male and female water closets for the assembly or mercantile occupancy. The term "total" means the combined number of required male and female water closets before any urinal substitutions. A family or assisted-use bathing room is required by the IBC in recreational facilities having separate sex bathing rooms except where the separate sex bathing rooms only have one shower or fixture. Family or assisted-use toilet (and bathing room) facilities are intended for use by individuals who need the assistance of a family member or care-provider to properly use plumbing fixtures in a private and dignified manner. Fixtures located in family or assisted-use toilet or bathing rooms are allowed to reduce the number of re-

Table 403.1(3)
SOLUTION SUMMARY FOR SAMPLE PROBLEM 3

OCCUPANCY		WATER CLOSETS				LAVATORIES				DRINKING FOUNTAINS	
USE	LOAD	MALE RATIO	MALE	FEMALE RATIO	FEMALE	MALE RATIO	MALE	FEMALE RATIO	FEMALE	RATIO	ALL GENDERS
Stadiums	5200	1 per 75 for the first 1500 and 1 per 120 for the remainder exceeding 1500	29.17	1 per 40 for the first 1520 and 1 per 60 for the remainder exceeding 1520	56	1 per 200	13	1 per 150	17.3	1 per 1000	5.3
Required Totals			30		56		13		18		6

quired fixtures for the occupancy in either the male or female standard toilet rooms, but not both.

Sample Problem 4: A theater has an occupant load of 520. Urinals are to be substituted for male water closets to the maximum extent allowed. Determine the total number of plumbing fixtures that will be in building.

Problem Approach

Apply the gender distribution, calculate the number of fixtures required by each gender, sum the gender totals for each type of fixture and round up the number of each type of fixture for each gender to a whole number. Consider IBC Section 1109.2.1requirement for a family or assisted-use toilet room. Apply the urinal substitution (Section 419.2) to reduce the number of male water closets.

Solution

Calculation method is similar to Sample Problem 1—Part II.

Solution—Summary

See Commentary Tables 403.1(4) and 403.1(5) for the solution summary for Sample Problem 4.

Because a theater is a use description of assembly and the total of required water closets is 7, one family or assisted-use toilet facility is required. The designer has the choice of reducing the number of fixtures in either the male or female toilet rooms by one lavatory and one water closet. In this case, the designer chose to reduce the number of required male fixtures. It may be more logical to reduce the male fixture count because of the greater need for fixtures in the female facilities, but the code is silent in this regard.

The code is silent as to whether the fixtures in a voluntarily added family or assisted-use toilet facility can be counted toward the required number of fixtures in any occupancy. For example, in Sample Problem 3, perhaps the building owner chooses to have three family or assisted-use toilet facilities in addition to the required number of fixtures in the public toilet facilities. If the fixtures in these "extra" family or assisted-use toilet facilities are allowed to count toward the required number of fixtures, which gender toilet facility should have fixture counts reduced? It is the author's opinion that voluntarily added family or assisted-use toilet facilities should not count toward the required number of fixtures, no matter the occupancy, because the code does not specify a maximum reduction, and any reductions seem to go against the intent of the code to provide ample gender-specific toilet facilities.

Multiple Single Occupant Toilet Rooms

Where the calculations require multiple fixtures for a gender, the code is silent as to whether those required fixtures can be provided in multiple, single-occupant toilet rooms. For example, if two water closets for each gender are required, could four single-occupant toilet facilities (2 labeled male, 2 labeled female), be provided? While the arrangement of multiple fixtures in one toilet room is cost efficient and space saving, some business owners might perceive a "marketing advantage" for their business by having all single-occupant toilet facilities. For example, families with young children and the elderly might be more attracted to a restaurant that has all single-occupant toilet facilities so that a family member can privately assist another family member using the facility. Or simply, some customers might prefer the extra privacy that a single-occupant toilet room offers.

The provision of the required number of plumbing fixtures in multiple single-occupant toilet facilities should be discouraged because the ratios provided in Table 403.1 never anticipated such arrangements. Users of single-occupant toilet rooms will generally take longer as they have the security of a locking door. Users in multiuser toilet facility environments are "encouraged" to be quick due to the bustle of activity in the more open environment (people waiting outside stalls or standing in line at urinals). In other words, (4)

Table 403.1(4)
SOLUTION SUMMARY FOR SAMPLE PROBLEM 4

OCCUPANCY		WATER CLOSETS				LAVATORIES				DRINKING FOUNTAINS	
USE	LOAD	MALE RATIO	MALE	FEMALE RATIO	FEMALE	MALE RATIO	MALE	FEMALE RATIO	FEMALE	RATIO	ALL GENDERS
Theater	520	1 per 125	2.08	1 per 65	4	1 per 200	1.3	1 per 200	1.3	1 per 500	1.04
Required Totals			3		4		2		2		2

Table 403.1(5)
BUILDING PLUMBING FIXTURE TOTALS FOR SAMPLE PROBLEM 4

Theater Building Totals with Urinal Substitution	MALE TOILET FACILITY			FEMALE TOILET FACILITY		DRINKING FOUNTAINS	SERVICE SINK	FAMILY OR ASSISTED-USE	
	URINAL	WATER CLOSETS	LAVATORIES	WATER CLOSETS	LAVATORIES			WATER CLOSETS	LAVATORIES
	1	1	1	4	2	2	1	1	1

single-use toilet facilities (marked 2 male and 2 female) do not provide the level of use efficiency as does a multiuser public toilet facility. Even though the code doesn't discuss efficiency with regard to toilet facilities, the desired outcome of Table 403.1 is to prevent long waiting lines and, thus, efficiency is a significant factor.

Note that where multiple single-occupancy toilet rooms are provided, Chapter 11 of the IBC requires that at least 50 percent of the single-occupancy toilet rooms within a cluster of toilet rooms be of accessible design. If the single-occupant toilet rooms are not clustered but are located individually throughout a building or tenant space, all toilet rooms are required to be of accessible design. For example, four single-occupant toilet rooms, two labeled "men" and two labeled "women" are clustered together. Only one of each gender of those toilet rooms must be of accessible design. However, if the single occupant toilet rooms are not grouped together in the building, all of the single-occupant toilet rooms must be of accessible design.

403.2 Separate facilities. Where plumbing fixtures are required, separate facilities shall be provided for each sex.

Exceptions:

1. Separate facilities shall not be required for dwelling units and sleeping units.
2. Separate facilities shall not be required in structures or tenant spaces with a total occupant load, including both employees and customers, of 15 or less.
3. Separate facilities shall not be required in mercantile *occupancies* in which the maximum occupant load is 50 or less.

❖ Separate facilities are required for males and females. Exception 1 is redundant as Table 403.1 already indicates the types of occupancies where one single-occupancy toilet room is sufficient. Exception 2 allows one single-occupant toilet room to be provided in small establishments because the space requirement and installation cost is considered to be a significant hardship for the minimal floor area that results in a 15-person occupant load. Exception 3 allows one single-occupant toilet room to be provided in what are considered to be small floor area mercantile establishments so as to not create a significant hardship in both space and installation cost. There is no specific documentation as to why 15 and 50 persons, respectively, were chosen as the thresholds for these exceptions other than what is perceived to be reasonable for such small spaces.

The separate facilities requirement for males and females addresses two main concerns: privacy and safety. Users of toilet facilities often experience embarrassment if members of the opposite sex are in the same room. This increased embarrassment can lead to difficulty or prevention of waste elimination for many users. While some of these inhibitions are often temporarily "given up" under special conditions such as in co-ed college dormitories or where one gender's facilities are inadequate for the demand, these inhibitions quickly return for most people as it is innate to the human species to desire privacy during waste elimination. In a public environment, safety for female users of toilet facilities is of paramount concern. While a female restroom placard is no barrier to those intent on harming female occupants, a female user confronted by (or even hearing) a male in the female-only toilet facility will immediately recognize the potential threat and take action. A toilet facility intended to be used by both sexes simultaneously does not offer the same level of immediate situational awareness for females, and therefore, it is not perceived to be as safe as a facility intended for one gender only. Also, some individuals may be inhibited to use a facility that is used by the opposite sex because of notions of cleanliness and perceived appropriateness.

403.3 Required public toilet facilities. Customers, patrons and visitors shall be provided with *public* toilet facilities in structures and tenant spaces intended for public utilization. The number of plumbing fixtures located within the required toilet facilities shall be provided in accordance with Section 403 for all users. Employees shall be provided with toilet facilities in all *occupancies*. Employee toilet facilities shall be either separate or combined employee and *public* toilet facilities.

❖ Toilet facilities must be available for all public establishments that are used by persons engaged in activities involved with the purpose of the establishment. Public establishments include but are not limited to restaurants, nightclubs, theaters, offices, retail shops, stadiums, libraries and churches. Persons, engaged in the activities of the establishment include but are not limited to, buyers of merchandise, recipients of services, viewers of displays, receivers of information materials, employees and those persons in attendance with those engaging in the activities. The code is silent about whether the toilet facilities are for use by persons not engaged in the activities involved with the purpose of the establishment, such as passers-by, street maintenance workers or vagrants. However, because the number of plumbing fixtures for any establishment is based only upon either the square footage or number of seats of the establishment's space, the intent of the code is to serve only the people involved with the activities of the establishment. The photo of the front door of the restaurant shown in Commentary Figure 403.3 might first appear to be in conflict with the requirements of this section; however, this sign is actually a promotion of the intent of this code section: "If you are engaged with the activities involved with the purpose of this establishment, toilet facilities are available for you."

The quantities of plumbing fixtures required for an establishment are specified in Section 403 of the code. Employees of the establishment must also have access to the toilet facilities for the establishment whether those facilities are separate from or combined with the facilities for the other users. Except for the specific requirements of Group I-2 occupancies, I-3 occupancies and footnote b of Table 403.1, employee toilet facilities are not required to be separate from other toilet facilities.

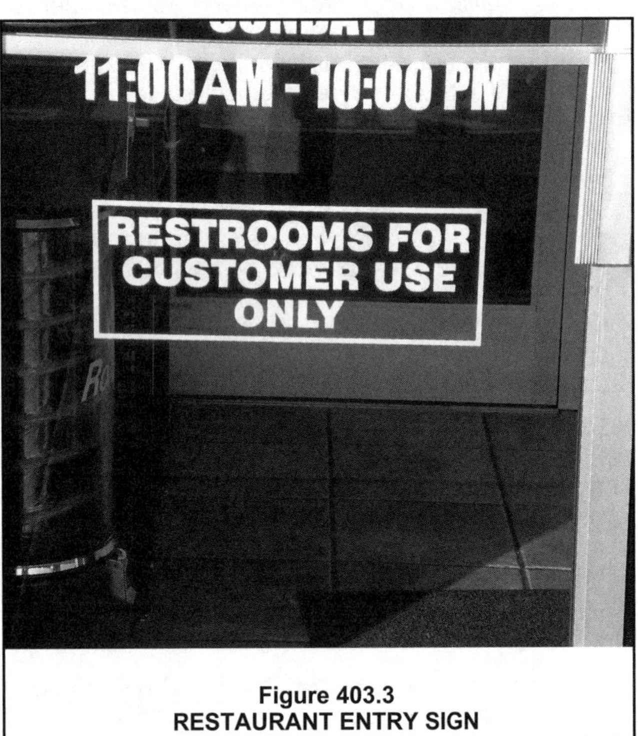

Figure 403.3
RESTAURANT ENTRY SIGN

403.3.1 Access. The route to the *public* toilet facilities required by Section 403.3 shall not pass through kitchens, storage rooms or closets. Access to the required facilities shall be from within the building or from the exterior of the building. All routes shall comply with the accessibility requirements of the *International Building Code*. The public shall have access to the required toilet facilities at all times that the building is occupied.

❖ It is inappropriate and dangerous to locate toilet facilities intended for use by persons who are not employees of the establishment, such that the path of travel is through kitchens, storage rooms or closets. This provision avoids placing occupants from the general public in an unfamiliar area of the building where they could be confused about the path of egress. Especially during an emergency situation, public users of the toilet facility will instinctively want to try to exit the building along the same path of travel used to access the toilet facilities, which could be through an area where the emergency condition exists, such as a kitchen [see Commentary Figure 403.3.1(2)]. Note that IBC Section 1210.5 prohibits toilet rooms from directly opening into a room used for the preparation of food to be served to the public.

Prior to the 2009 code edition, this section was not specific about how the required toilet facilities should be accessed for a building (other than within the travel limitations of Sections 403.3.2 and 403.3.3). It has been common practice in many areas that the required toilet facilities were located on the outside of a building, such as is sometimes found in automotive service stations and convenience store buildings. Outlet mall complexes sometimes elect to have a central toilet facility building separate from the tenant spaces. While the code is silent as to whether the required toilet facilities are allowed to be in another building not under the direct control of the owner or tenant of the building from where the toilet user travels to the toilet facilities, the intent of the code is that public toilet facilities will be available (and open for use) when anyone is using a building or tenant space. However, it is unreasonable for toilet facilities in a building under different ownership or in a tenant space under different control to serve as the required toilet facilities for a building or tenant space.

Walk-up and drive-through service windows are unique applications with regard to requirement for toilet facilities. Where these establishments do not allow customers to enter the building, the code is silent on whether toilet facilities for the public are required. However, if facilities were required, how would one calculate the occupant loading to determine the required number of fixtures? Examples are ice cream shops with only a walk-up window, coffee shops with only a drive-up window and 24-hour fast food drive-throughs that close the building entrances to walk-in customers during late night/early morning hours. Where outdoor seating for customers is made available by these types of businesses, it is the author's opinion that public toilet facilities for these walk-up/drive-through establishments should be available [see Commentary Figure 403.3.1(1)]. If such outdoor seating areas can be "closed for public use" such as by a perimeter fence and locked entry gates, the seating area is not available for use, and no requirement would exist to provide public toilet facilities.

Variations of the above service window applications are dry cleaners, pizza pick-ups and banks that pro-

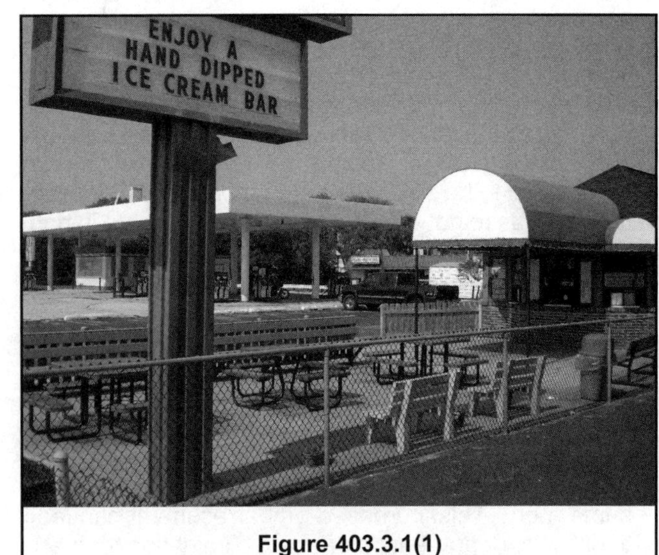

Figure 403.3.1(1)
WALK-UP ICE CREAM SHOP
WITH OUTDOOR SEATING

vide service or products to customers from indoor counters or kiosks. Because the public must walk into these buildings for service, public toilet facilities are required for the public. Obviously, these establishments are required to have toilet facilities for the employees; however, those facilities might not be properly located for public access. For example, if the location of the toilet facilities in a dry cleaning establishment requires that the public walk a convoluted path around machinery and hanging clothes racks, the user could become confused about how to safely exit the toilet facilities and return to where he or she started from. Public toilet facilities must be provided along an obvious and safe path of travel. Note that this path of travel must also meet the accessible route requirements for disabled persons. In the pizza pick-up situation, because the public is not allowed to pass though a kitchen, the toilet facilities would have to be located on the customer side of the pick-up counter. Banks (as well as check cashing establishments) do not have code restrictions for the areas that the public can pass through; however, the operators of these establishments might wish to restrict public access to "backroom" areas for security reasons. These self-imposed restrictions do not justify denying the public access the toilet facilities located in restricted areas. If the toilet facilities are located in restricted areas, public toilet facilities must be provided outside of those self-imposed restricted areas.

403.3.2 Location of toilet facilities in occupancies other than covered malls. In occupancies other than covered mall buildings, the required *public* and employee toilet facilities shall be located not more than one story above or below the space required to be provided with toilet facilities, and the path of travel to such facilities shall not exceed a distance of 500 feet (152 m).

Exception: The location and maximum travel distances to required employee facilities in factory and industrial *occupancies* are permitted to exceed that required by this section, provided that the location and maximum travel distance are *approved*.

❖ To prevent health problems for building occupants related to difficulty in accessing a needed toilet facility in a timely fashion, the code clearly provides two conditions that must be met: (1) the required toilet facilities must be located within a certain travel distance and (2) the building occupant must not be required to travel beyond the next adjacent story above or below his or her location.

The distribution of plumbing fixtures within a building must satisfy the maximum travel limitations for occupants in the building. Sections 403.3.2 and 403.3.3 limit the horizontal and vertical distances that an occupant has to travel to use a toilet facility. A 20-story high-rise office building cannot have all of the required plumbing fixtures located on the 10th floor as the path of travel must neither exceed a travel distance of 500

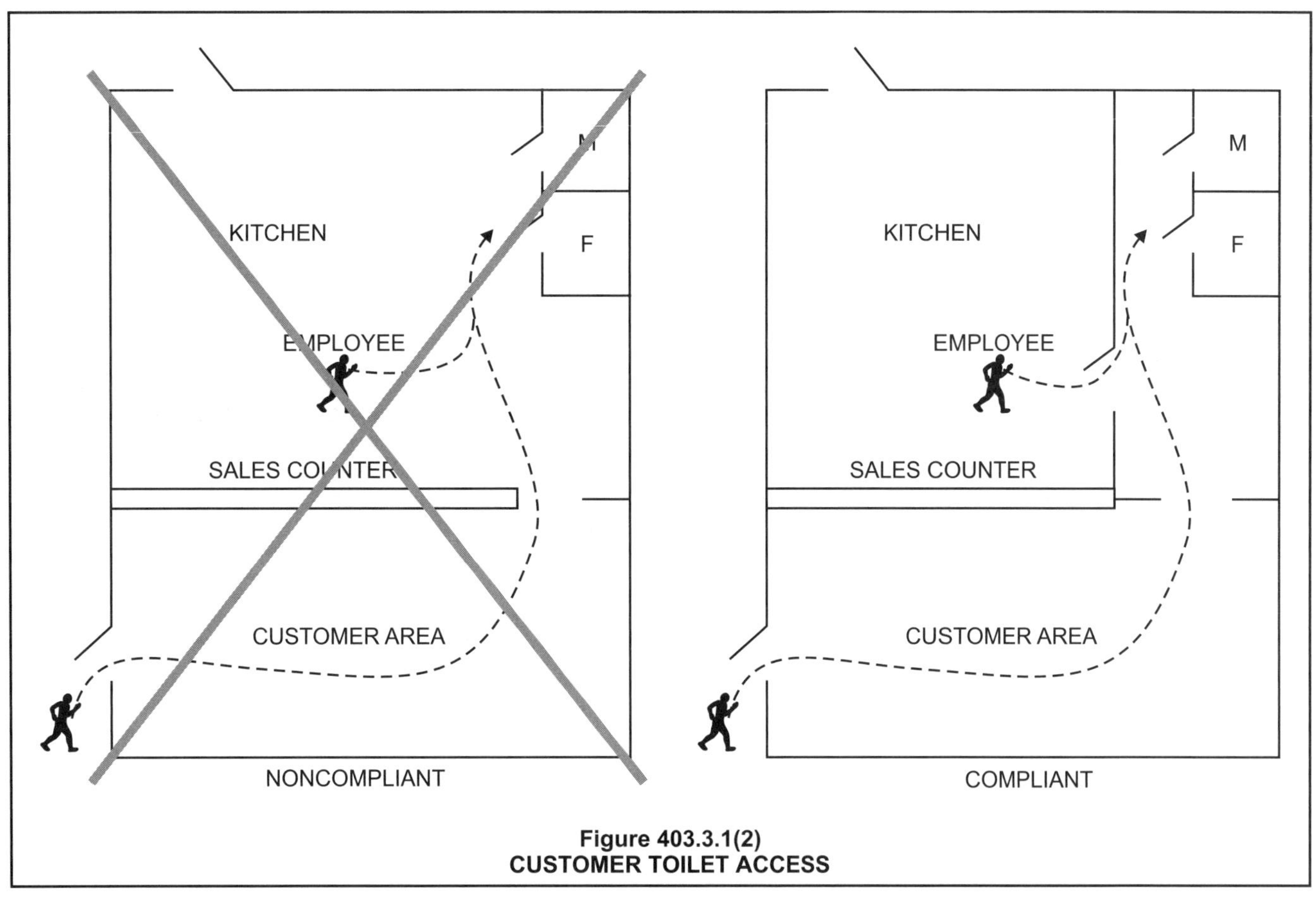

Figure 403.3.1(2)
CUSTOMER TOILET ACCESS

feet (152 m) nor require travel beyond the adjacent story above or below. While this travel path limitation can result in toilet facilities not being located on every floor of a multistory building, the intent of the code is for a proportional number of fixtures to be available for the intended occupant use. For example, occupants on the 20th, 19th and 18th floors could reasonably be expected to utilize toilet facilities located only on the 19th floor (as long as the 500 foot travel distance limitation is not exceeded). However, that toilet facility cannot have just one fixture of each type to serve 3 floors of occupants. The distribution of fixtures must be proportioned according to the number of occupants that will be using the toilet facility. To illustrate, assume that each of the 20 floors in this high-rise business occupancy building has an occupant load of 100. Applying the business occupancy water closet fixture ratio to the occupant load for the entire building results in 40 male water closets required for the building. Three floors of occupants will require (300 floor occupants/2,000 total occupants) x 40 w/c = 6 male water closets on the 19th floor.

The exception in this section provides for the unique nature of the workplace in factory and industrial occupancies where the designer is better equipped to determine the location of the employee toilet facilities.

403.3.3 Location of toilet facilities in covered malls. In covered mall buildings, the required *public* and employee toilet facilities shall be located not more than one story above or below the space required to be provided with toilet facilities, and the path of travel to such facilities shall not exceed a distance of 300 feet (91 440 mm). In covered mall buildings, the required facilities shall be based on total square footage, and facilities shall be installed in each individual store or in a central toilet area located in accordance with this section. The maximum travel distance to central toilet facilities in covered mall buildings shall be measured from the main entrance of any store or tenant space. In covered mall buildings, where employees' toilet facilities are not provided in the individual store, the maximum travel distance shall be measured from the employees' work area of the store or tenant space.

❖ In covered malls, the path of travel to required toilet facilities must not exceed a distance of 300 feet (91 400 mm) and is measured from the main entrance of any store or tenant space. Note, however, that if the tenant space does not have toilets for the employees, the starting point for measuring the travel path is the employee's work area. As for other occupancies, the facilities must be located no more than one story above or below the occupant's location. This shorter travel distance addresses the fact that covered malls are frequently very congested and occupants are unfamiliar with their surroundings. It is also logical to reduce the travel time for employees for security reasons. The minimum number of required plumbing facilities is based on total square footage of the mall, including tenant spaces. This section does not prohibit the installation of separate employee toilet facilities in individual tenant spaces; however, such facilities are not deductible from the total number of common facilities required.

403.3.4 Pay facilities. Where pay facilities are installed, such facilities shall be in excess of the required minimum facilities. Required facilities shall be free of charge.

❖ Pay facilities have been included in some public areas to prevent vagrants from loitering in the public bathrooms. They may also be installed to offer customers facilities with greater amenities where the customer desires such upgrades. This section does not prevent pay facilities from being installed; rather, it requires that such facilities be in excess of those required by Section 403.

403.4 Signage. Required *public* facilities shall be designated by a legible sign for each sex. Signs shall be readily visible and located near the entrance to each toilet facility.

❖ Public facilities must be designated by a legible sign for each sex and located near the entrance to such facilities. The code is silent on the specifics of the type of designation, making it dependent on the approval of the building official. Such signage should be universally understood by all potential users of the facilities.

403.4.1 Directional signage. Directional signage indicating the route to the *public* facilities shall be posted in accordance with Section 3107 of the *International Building Code*. Such signage shall be located in a corridor or aisle, at the entrance to the facilities for customers and visitors.

❖ This section strengthens the intent of Section 403.3 which is to make sure that public toilet facilities are provided for spaces intended for pubic utilization. Where the public facilities are not clearly visible from within the public areas or the facilities are locked, patrons and visitors are sometimes told that public toilet facilities are not available or that they are for employee use only. Mandated directional signs will inform the public that public restrooms do exist and where they are located. It is difficult to tell a customer or patron that there are no public facilities when there is an obvious sign advertising their existence. This code requirement provides to the code official the means for issuing a citation to the building owner if the signs are found to be removed in an attempt by the building owner or tenant to limit public access to toilet facilities. The location of the sign is to be near the entrance to the facilities in either an aisle or corridor where it will be readily seen. The code is silent on specific details of the signage such as wording, size, color or mounting height. Note that IBC Section 1110 covering signage for accessibility requires that directional signage be provided at the location of separate-sex toilet facilities indicating the location of family or assisted-use toilet facilities, if such facilities are required.

SECTION 404
ACCESSIBLE PLUMBING FACILITIES

404.1 Where required. Accessible plumbing facilities and fixtures shall be provided in accordance with the *International Building Code*.

❖ In almost all occupancies, toilet rooms, bathing rooms and kitchens as well as the plumbing fixtures contained within are required to be constructed in such a manner that persons with physical disabilities will have "access" to those facilities in a manner similar to persons without disabilities. A portion of the plumbing fixtures in each facility must be designed and installed so that persons with physical disabilities have the "ability" to use fixtures of a similar type to those used by persons not having disabilities. One- and two-family detached dwellings are exempt from the requirements to provide for accessibility. The requirement for "accessibility" is mandated by several federal and state laws, as well as by the IBC, and thus the defined term "accessible" has a very specific intent where it is used across all of the I-Codes [except for the *International Residential Code*® (IRC®)]. For example, in Section 403.3 of this code, the term accessible is used to describe the route to public toilet facilities.

In nearly all jurisdictions in the United States using the I-Codes, the federal requirements for accessibility are generally deemed to be accomplished if a building, site or facility complies with the applicable provisions of Chapter 11 of IBC and ICC/ANSI A117.1. Chapter 11 of the IBC regulates the extent (scoping) of required accessible features to be provided in various occupancies, and ICC/ANSI A117.1 regulates the (technical) characteristics of those accessible features.

While the adoption of the IBC is nearly 100 percent for all ICC member jurisdictions, the reader must be made aware that there are jurisdictions that choose to delete, replace or modify IBC Chapter 11 in favor of using other (or slightly modified) accessibility requirements or standards. Users of the code are cautioned to thoroughly review each jurisdiction's (and project's) specific requirements for accessibility to determine if Chapter 11 of the IBC and ICC A117.1 can be used for design and installation of accessible building features, or whether additional or alternate standards apply.

Prior to the 2000 IPC, Section 404 contained numerous subsections with accessibility requirements and details that were a duplication of the requirements for accessible plumbing facilities stated in the building code as well as a duplication of the accessible plumbing fixture details as required by CABO A117.1 (the precursor to ICC A117.1). Beginning with the 2000 IPC, those subsections were removed from the code for two reasons: (1) the "scoping" requirements for accessibility were deemed to be strictly a building code issue and (2) there was not an automatic administrative mechanism to update the "technical" details upon future revisions of the A117.1 standard.

In a perfect world, every architect and designer would have a complete understanding of the accessibility requirements and would provide a detailed drawing showing the exact arrangement and location of each required accessible plumbing fixture and every dimension needed for the plumbing contractor to install those fixtures. However, not every project is given the deserved attention to accessibility details. In some cases, the information that is provided is based on ancient accessibility guidelines and not the latest requirements of IBC Chapter 11 and the ICC A117.1 standard. It is the responsibility of the plumbing contractor to catch these errors or assume the role of designer of accessible plumbing facilities? No; however, an uninformed plumbing contractor will be an unhappy contractor if rework is required to correct a plumbing installation. It is better to be informed in order to be able to ask appropriate questions to insure that the plumbing for accessible fixtures is installed correctly. Thus, the following information is provided to assist those using the code to understand how the plumbing-related features required by IBC Chapter 11 are to be installed in accordance with ICC A117.1-2003, a standard that is not referenced by the IPC. The reader must note that this commentary on the installation of accessible plumbing features does not focus on other required elements for providing accessibility, such as routes, required quantities, location of rooms and defining what level of accessibility is required for a building or space. Chapter 11 of the IBC along with ICC A117.1 provides the complete package of information for the designer of accessible spaces.

Dimension Reference Points and Dimension Ranges

The dimensions to floor and wall surfaces in all of the commentary figures for Section 404 (which are dimensionally identical to those in the ICC A117.1 standard) are to *finished* wall and floor surfaces and not to the face of a framed wall or top of a subfloor, respectively. Although this seems like a minor detail, this is often the cause of frequent plumbing fixture and plumbing trim location errors made by designers and plumbing contractors. Many of the dimensions shown in the standard (and in the commentary figures) are given as a dimension range [e.g., 16 to 18 inches (405 to 455 mm) or 34 inches (432 mm) maximum]. Improper understanding of the purpose and use of these dimension ranges could result in violations at final inspection.

The ranges on dimensions are not intended to give the designer or installer the initial "latitude" on where to locate the item. The ranges are also not intended to absorb errors due to misinterpreting specific measurement reference points or to permit a general carelessness in making measurements. The ranges are intended to allow for normal variances associated with field construction practices so that upon final inspection, a measurement of the location of the item has a very high probability of being within the required dimension range. The following examples illustrate some of the most commonly occurring problems.

Example A

Consider the required 16 to 18 inch (405 to 455mm) dimension range for the centerline of a water closet for accessible design to the finished surface of the adjacent sidewall. Suppose the designer chooses to show 18 inches (457 mm) on the drawing or the contractor chooses 18 inches (457 mm) (from the dimension range shown in the standard) because "that's the dimension that we have used for years." The 18-inch (457 mm) dimension complies with the accessibility standard. The contractor necessarily makes adjustments to this dimension to be able to measure to reference points that he or she has available to work from such as a string line indicating where the face of a framed wall will be located or an existing block wall that will eventually be "furred out," insulated and covered with finish materials (e.g., gypsum board, cement board with tile, etc.). Even though the contractor uses "correct" nominal thicknesses of materials to adjust the 18 inch (457 mm) dimension in order to locate the drain piping, small variations in measurement accuracy, actual material thicknesses, straightness of materials, placement of the future wall, the "plumbness" of the wall, the final mounting position of the fixture, and the perceived center of the water closet bowl could cause the final inspection measurement to exceed 18 inches (457 mm) some percentage of the time.

The lesson to be learned from this example is that the "target" dimension to be used for the initial layout should be the value that is midway between the extremes of the dimension ranges shown in the following commentary figures.

Example B

Consider the 34 inch (864 mm) maximum dimension required for the height of a carrier-mounted lavatory above a finished floor. If the contractor uses the 34-inch (864 mm) dimension as the basis for figuring out a dimension to an unfinished floor surface that he or she must use as a reference point, for similar reasons as given in Example A, the final inspection measurement could exceed 34 inches (864 mm) some percentage of the time.

The lesson to be learned from this example is that the "target" dimension for maximum (or minimum) dimensions to be used for the initial layout should be a value that is slightly less than the maximum (or slightly more than the minimum). How much less (or more) must be based on an understanding of the possible variations that could be encountered in each situation.

Why is compliance to the dimensions in the ICC A117.1 standard so critical? Although the standard indicates that the dimensions "are subject to conventional industry tolerances," there has never been a consensus on what those tolerances should be. The standard also states "dimensions not stated as maximum or minimum are absolute". This seems to indicate that there is "no tolerance" for installations that have measurements that fall outside of the indicated dimension limits. Because of past litigation that has occurred in many jurisdictions, it has become commonplace for some code officials to expect plumbing fixtures for all levels of accessibility to be installed exactly to the dimensions shown on the project drawings or in the ICC A117.1 standard, if the project drawings lack the critical dimensions. The accessibility details on the project drawings should show the same dimensional limits as ICC A117.1. The plumbing contractor should interpret the dimensional limits appropriately to provide for the highest probability that the fixtures, upon final inspection, will be located within the dimensional limits.

Clear Floor Space and Approach Direction

The operation of a wheelchair requires ample floor area for turning and positioning. This required floor area is termed "clear floor space." The size and orientation of clear floor space around plumbing fixtures can impact the location and trim features required for the fixtures.

The direction of approach by a user of a wheelchair can also impact the location and features required for the fixtures. Fixtures are intended to be approached by either the forward or parallel methods. A forward approach is where the user approaches in a manner that is perpendicular to the wall that the back of the fixture is against. A parallel approach is where the user approaches in a manner that is parallel to the wall that the back of the fixture is against.

Required Clearances around Grab Bars

The information on grab bar locations is provided in this commentary to illustrate potential interferences with plumbing components. The first potential for interference is between the required blocking in the walls for the grab bars and piping installed (or to be installed) in the walls. This is especially critical where framed walls are only $3^1/_2$ inches (89 mm) thick as 2 inch (51 mm) nominal thickness blocking is typically installed for grab bar attachments. The second potential for interference is between the grab bars and plumbing components that are located within the finished area. The following provides the clearance requirements and examples of typical plumbing components located in finished areas.

For water closet grab bar installations, the minimum clearance between the top of a grab bar and any element protruding from the wall above is 12 inches (305 mm). A minimum clearance of $1^1/_2$ inches (38 mm) is required below, and at the ends of grab bars. For example, if a flushometer valve location is required to be below the rear-mounted grab bar of a water closet, the top of the flushometer valve is required to be located at least $1^1/_2$ inches (38 mm) below the bottom surface of the grab bar. A greater clearance might be required to enable proper access for repairing the valve without removal of the grab bar.

For grab bars in bathtubs and showers, the minimum clearance between the grab bar top and bottom surfaces and ends and other objects such as shower controls/shower fittings is $1^1/_2$ inches (38 mm). For example, the outermost point of a shower control handle must not pass any closer than $1^1/_2$ inches (38 mm) to a grab bar surface.

Three Designs for Accessibility

There are three designs for accessibility that can be required: (1) Accessible, (2) Type B and (3) Type A. The Accessible design provides the greatest level of accessibility. Accessible design is used for public facilities and for some sleeping units and dwelling units. The standards in ICC A117.1 for the Accessible design are intended to be equivalent to the accessibility requirements of the Americans with Disabilities Act (ADA). The Type B design provides the most flexible level of accessibility. The standards in ICC A117.1 for the Type B design are intended to comply with the requirements of the Federal Fair Housing Act. The Type A design provides a level of accessibility that is a compromise between the Accessible and Type B designs. Type A and B designs are used only for sleeping units and dwelling units. The architect or designer of a building is responsible for determining which of the designs for accessibility are required in a building as well as the locations of accessible toilet and bathing facilities and the quantities of plumbing fixtures required in those facilities.

Installation of Plumbing Fixtures in the Accessible Design

Water Closets for Wheelchair Access (Accessible Design)

Toilet and bathing rooms in the Accessible design must have accessible plumbing features installed in accordance with Sections 603 through 610 of ICC A117.1. Sections 603 and 604 of ICC A117.1 cover water closets (and grab bar locations). Commentary Figures 404(1) through 404(5) indicate the required dimensions for locating a water closet and the minimum clear floor space and grab bar locations for a water closet. The required dimension from the centerline of the water closet to the side wall having the grab bar must be from 16 to 18 inches (405 to 455 mm) to the finished wall surface. For reasons stated previously in this commentary, the recommended "target" dimension for installation is 17 inches (432 mm).

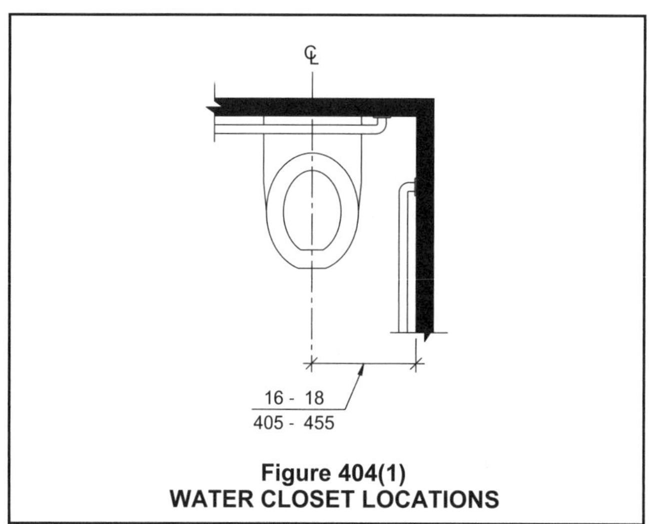

Figure 404(1)
WATER CLOSET LOCATIONS

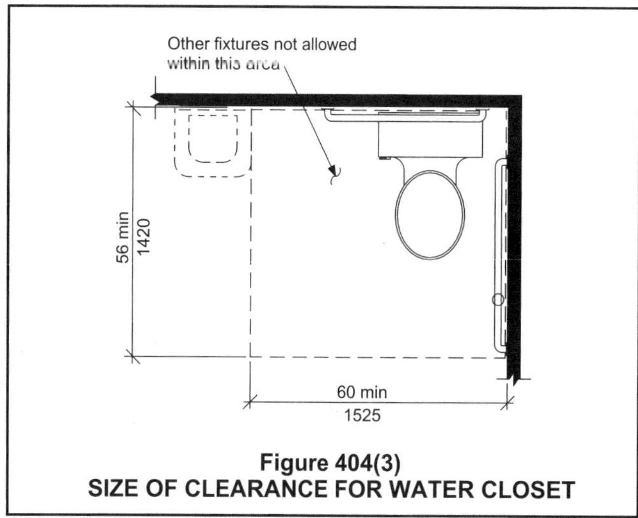

Figure 404(3)
SIZE OF CLEARANCE FOR WATER CLOSET

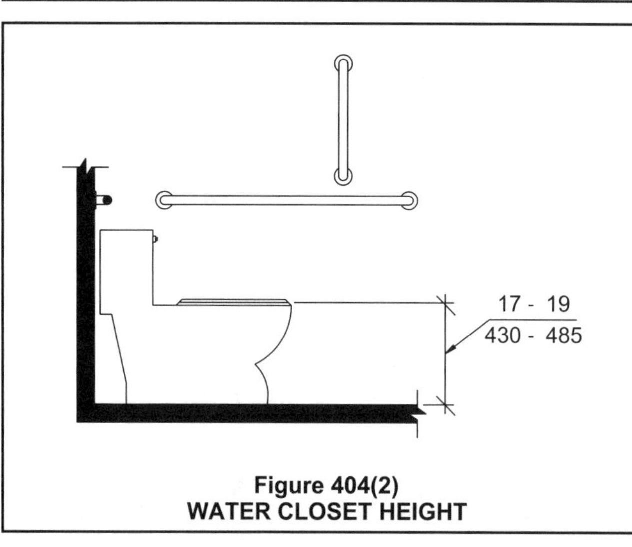

Figure 404(2)
WATER CLOSET HEIGHT

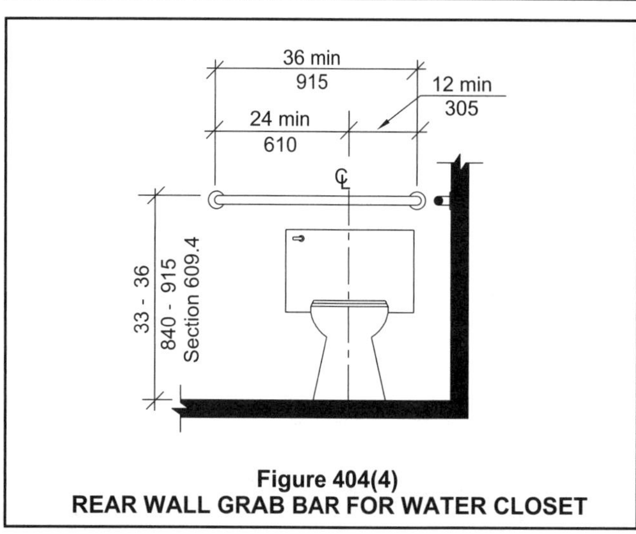

Figure 404(4)
REAR WALL GRAB BAR FOR WATER CLOSET

The heights of the water closets shown in the commentary figures are to the top of the seat and not to the top of the bowl rim or to the top of a lid (cover) for the seat. (Note that Section 420 of this code requires that any public or employee water closet must be of the elongated type and have an open-front seat.) A water closet in a single-occupant toilet room accessed only through private office and not for public use is not required to have a seat height of 17 to 19 inches [430 to 480mm] (see IBC Section 1109.2, Exception 1.2).

Even though the commentary figures depict a flush tank-type water closet, floor or wall-mount water closet bowls with flushometer valves might be used in these applications. As previously indicated, the ICC A117.1 standard specifies the minimum allowable clearances above, below and at the ends of grab bars. The standard intends that the area bound by the top of the clearance zone above the bar and the bottom of the clearance zone below the bar, extending in a direction that is horizontal and perpendicular away from the wall surface, is not to be blocked. Some operators of healthcare facilities require flushometer valves be located at or near the maximum allowable flush handle height elevation of 48 inches (1219 mm). This is to protect a seated user from abrasions that could be caused by the flushometer valve body located at lower elevations. This higher mounting height location will create an obstruction in front of the rear grab bar. In these situations, the ICC A117.1 standard allows for the rear grab bar to be supplied in two sections or the one piece bar shifted towards the open side of the water closet to eliminate interference with the flush valve.

ICC A117.1 requires that a hand-actuated flush handle (for either a flush tank or flushometer) be on the "open side" of the water closet location. For example, where facing the water closet, if the wall is on the right, then the flush lever must be on the left. If the wall is on the left, the flush lever must be on the right. For flushometer valves, attention to this right/left detail at the rough-in stage will eliminate last minute corrective actions in order to pass final inspection. The required flush handle height is 15 to 48 inches (381 to 1219 mm) above the finished floor. This height range eliminates the possibility of a foot-operated flush device. Where flush tank water closets are utilized, right-hand flush lever configurations might be a special order item with a long delivery time.

Flush valves and flush tank controls are required to meet accessible (operational) requirements. Flushometer and water closet tank manufacturers are responsible for determining which flushometers and water closet tank flush controls in their product lines comply with the requirements. Packaging, product labeling or specification sheets can be used to verify compliance. The indication for compliance might be the words "ADA-Compliant," "Compliant with (ICC) ANSI A117.1" or the International Symbol for Accessibility [see Commentary Figure 404(6)].

Commentary Figures 404(7) and 404(8) indicate the required dimensions for an accessible child size water closet installation, should one be specified by the designer. The centerline-to-side-wall dimension range of 12 to 18 inches (305 to 457mm) for an accessible child water closet encompasses the standard ac-

Figure 404(6)
INTERNATIONAL SYMBOL FOR ACCESSIBILITY

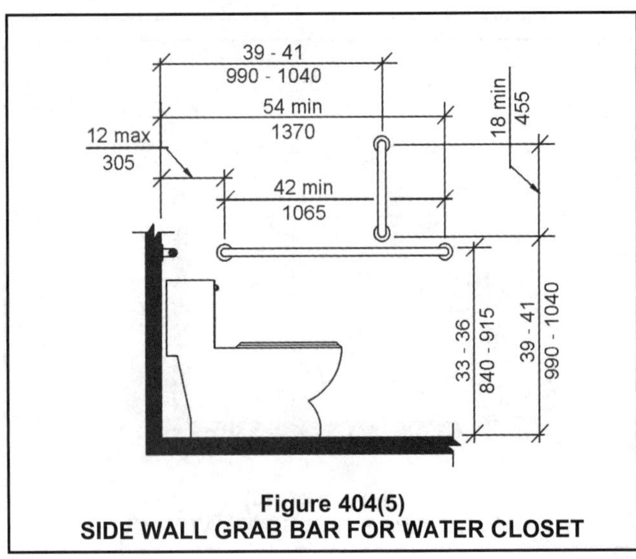

Figure 404(5)
SIDE WALL GRAB BAR FOR WATER CLOSET

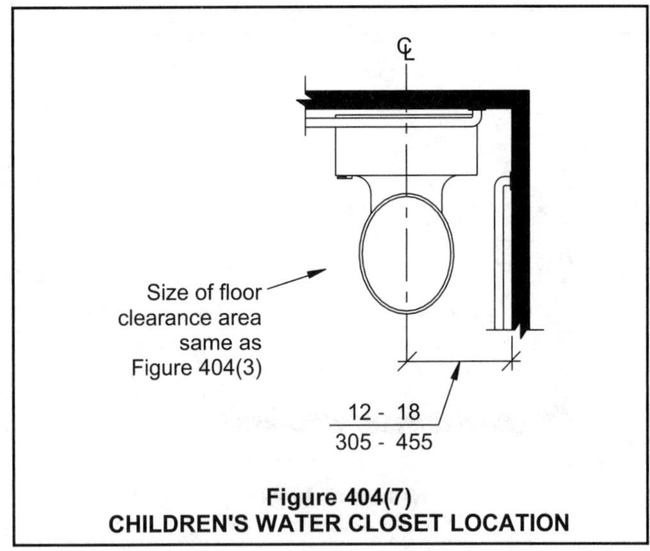

Figure 404(7)
CHILDREN'S WATER CLOSET LOCATION

cessible water closet centerline-to-side-wall dimension range of 16 to 18 inches (406 to 457 mm). If the building's use has a reasonable probability of requiring a future change such that a standard accessible adult water closet would be needed for that location; prudent design planning would be to use a target dimension of 17 inches (432 mm) for locating the accessible child water closet. Therefore, a future change-over to a standard accessible water closet would not require substantial plumbing rework. The required floor clearance area for the accessible child water closet is the same as for the standard accessible water closet as indicated in Commentary Figure 404(3). However, the side wall grab bar height is lower than for a standard accessible water closet [see Commentary Figure 404(8)]. The flush handle height for child water closet is required to be not more than 36 inches (914 mm) above the finished floor.

Water Closets for Ambulatory Access (Accessible Design)

Water closets for ambulatory access have a wall or partition on both sides of the water closet such that the compartment width is 36 inches (914 mm). This enables an ambulatory disabled user to use both side grab bars to assist in lowering onto and rising from the water closet seat. ICC A117.1 allows for the water closet centerline to finished side wall (or partition) dimension to range from 17 to 19 inches (430 to 485 mm). For the reasons stated previously in this commentary, the recommended "target" dimension for installation is 18 inches (457 mm). The required flush handle height is 15 to 48 inches (381 to 1219 mm) above the finished floor. This height range eliminates the possibility of a foot-operated flush device.

Note that the IBC requires a water closet for ambulatory access where there are 6 or more water closets in the same toilet room (or bathing room). This requirement does not increase the number of water closets but only requires one of the water closets to have ambulatory access.

Figure 404(8)
CHILDREN'S WATER CLOSET HEIGHT AND SIDE WALL GRAB BARS

Urinals for Wheelchair Access (Accessible Design)

Section 605 of ICC A117.1 specifies that urinals either be of the stall type (where the drain is very near or at floor level) or of a wall-hung type having the "rim" located at a maximum dimension of 17 inches (430 mm) above the finished floor [see Commentary Figures 404(9) and 404(10)]. The 17 inch (430 mm) dimension is a maximum height; a lesser dimension is acceptable. The "rim" of the urinal is understood to be that part of the fixture which is the highest point of the "bowl" that is the farthest point away from the wall mounting surface.

Flush handles can be located on either side of the centerline of a urinal. The required flush handle height is 15 to 48 inches (381 to 1219 mm) above the finished floor. This height range eliminates the possibility of a foot-operated flush device.

A critical point that is often overlooked by designers and contractors is the requirement for the distance between urinal partitions for the wheelchair access urinal to be at least 36 inches (914 mm). The centerline of the urinal must be at least 18 inches (457 mm) from the face of each partition.

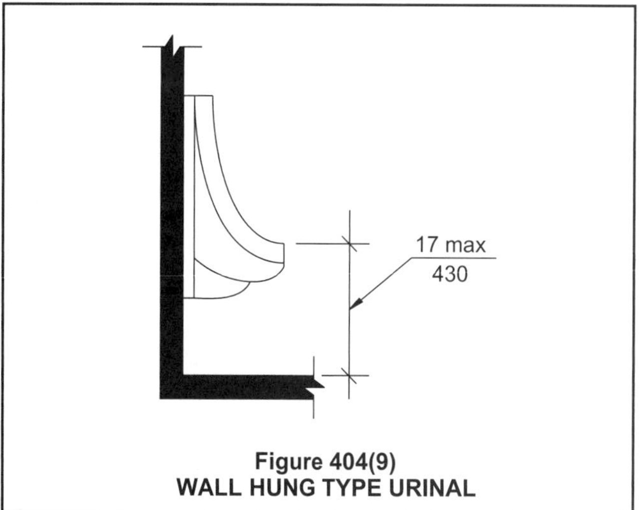

Figure 404(9)
WALL HUNG TYPE URINAL

Figure 404(10)
ACCESSIBLE HEIGHT URINAL ON LEFT

Note that where there is only one urinal in a toilet or bathing room, the urinal is not required to be accessible (see IBC Section 1109.2, Exception 4). However, it might be prudent to install the single urinal at the accessible height to allow use by wheelchair seated users as well as children.

Bathtubs (Accessible Design)

Section 607 of ICC A117.1 covers the requirements for bathtubs and grab bar locations. Commentary Figures 404(11) and 404(12) illustrate the two possible types of bathtub arrangements, one without a permanent seat and one with a permanent seat. The bathtub without a permanent seat requires a clear floor space in front of the tub of 30 inches (760 mm) by the length of the tub. The drain of the tub can be located on either end. For bathtubs with a permanent seat, the seat end of the tub must be located on the end where the 12 inch (305 mm) extended area of the required clear floor space is located [see Commentary Figure 404(12)]. The drain end of the bathtub unit must be located on the end that is opposite of the 12-inch (305 mm) extended area of required clear floor space.

Commentary Figures 404(13) and (14) show the grab bar locations for bathtubs with and without permanent seats, respectively. This information is especially critical where the building walls are only $3^1/_2$ inch (89 mm) (rough dimension) deep as 2-inch (50 mm) nominal thickness blocking is typically installed for grab bar attachments. Where a prefabricated Acces-

Figure 404(11)
BATHTUB WITHOUT PERMANENT SEAT

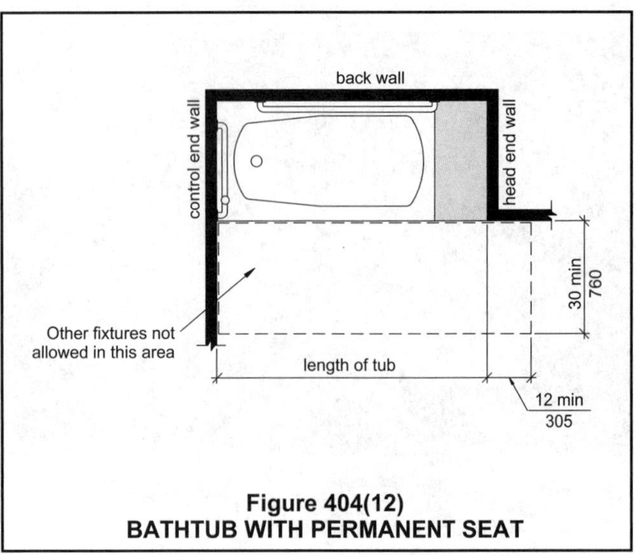

Figure 404(12)
BATHTUB WITH PERMANENT SEAT

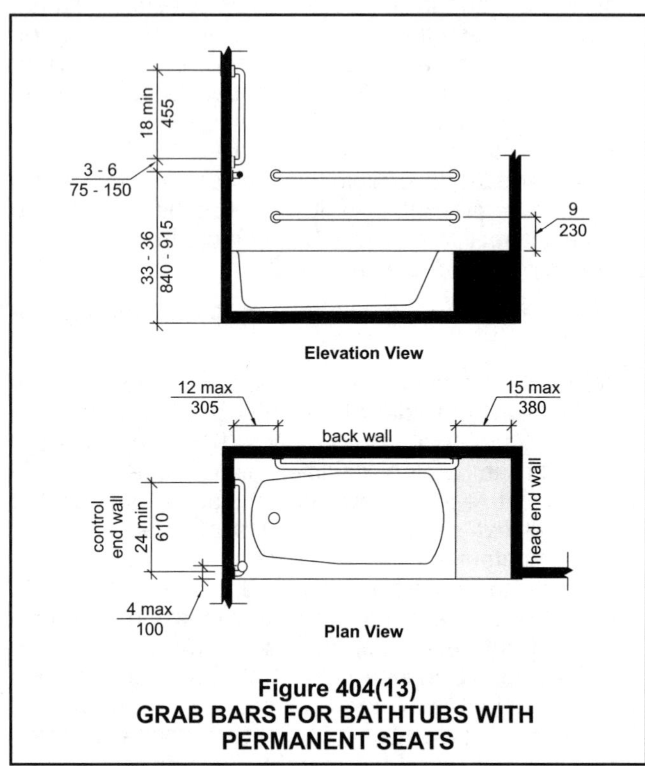

Figure 404(13)
GRAB BARS FOR BATHTUBS WITH PERMANENT SEATS

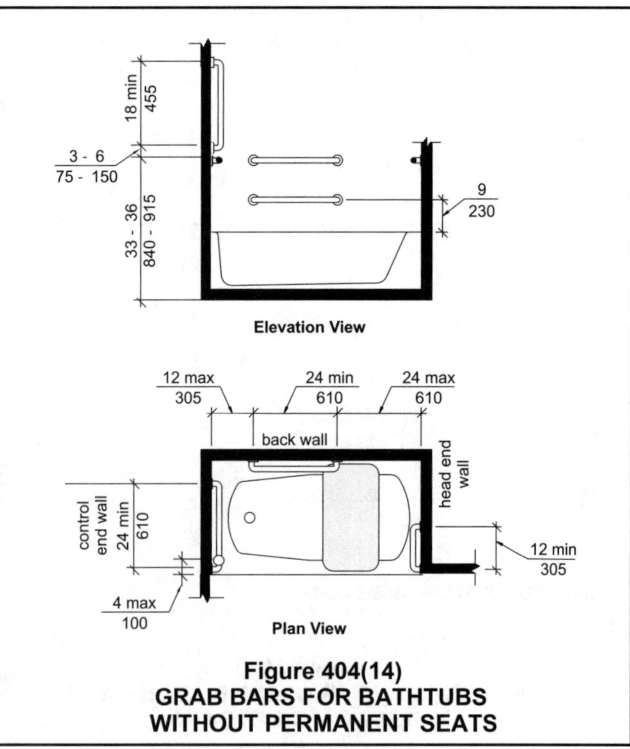

Figure 404(14)
GRAB BARS FOR BATHTUBS WITHOUT PERMANENT SEATS

sible design bathtub module already having factory-installed grab bars is being installed, blocking for the grab bars is typically not required to be installed in the building wall framing.

Bathtubs must have the water control located in the hatched area as shown in Commentary Figure 404(15). The depicted zone should not be interpreted as requiring that the entire control face plate (escutcheon) be located within the zone. For a single handle control, locating the centerline of the control handle (and a diverter spout) on the centerline of the tub meets the intent of ICC A117.1. Where multiple handle controls are installed, the centerline of the control handle farthest from the open side of the tub must be located not farther away than the centerline of the tub. Note that any control centerline should probably not be located very near to or on the top edge of the indicated hatched area due to the potential for interference of the control handle with the required grab bar location. The minimum allowable clearance of the path of the control handle to the grab bar is $1^1/_2$ inches (38 mm). The location of the control should also provide ample space for service access to the control without requiring removal of the grab bar.

Commentary Figures 404(11) through (15) and the text of ICC A117.1 do not specify a minimum or maximum width for the bathtub [e.g., 30 inches (762 mm), 36 inches (914 mm), 42 inches (1067 mm)]. The ICC A117.1 originally based the control location requirements on a 30 inch (762 mm) wide bathtub so that floor plans requiring accessible bathtubs would not require a floor space larger than that required for the minimum required bathtub size for a nonaccessible unit. This does not mean that a bathtub of a width greater than 30 inches (762 mm) cannot be used. However, it is advisable to consider limiting the most remote vertical edge of the designated control zone at a dimension that does not exceed 18 inches (457 mm) from the open side of the bathtub. This limitation is suggested in order to enable the control to still be within an easy side reach range for wheelchair-seated users who are parked outside the bathtub.

The maximum height of the bathtub wall on the open side of the tub (i.e., the "apron" side) is controlled by the type of seat used for the installation. Where a removable seat is used, the apron side tub wall cannot be higher than the underside of the removable seat located with its top at 19 inches (483 mm) above the finished floor. Where a permanent seat is used, the apron side tub wall must be not higher than 19 inches (483 mm) above the finished floor. While most standard bathtubs have apron side wall heights less than 17 inches (432 mm), some models of soaker tubs, whirlpool tubs or free-standing tubs might have heights that exceed the allowable dimension.

Showers (Accessible Design)

Section 608 of ICC A117.1 covers the requirements for showers. There are three types of shower designs: (1) transfer type, (2) standard roll-in type, and (3) alternate roll-in type.

Commentary Figure 404(16) illustrates the dimensional and clear floor space requirements for a transfer-type shower. The seat of the shower compartment must be arranged such that there is a clear floor space to position the wheel chair and align the wheelchair seat with the shower seat for a side transfer. This point is made to draw attention to the necessity for locating the control on the wall opposite of the seat wall. Reversing the arrangement in some circumstances might not provide adequate wheelchair "rear parking area" [the 12-inch (305 mm) area past the opening of the shower] to enable alignment of the wheelchair seat to the shower seat for a proper and safe transfer of the user.

end wall

Figure 404(15)
LOCATION OF BATHTUB CONTROLS

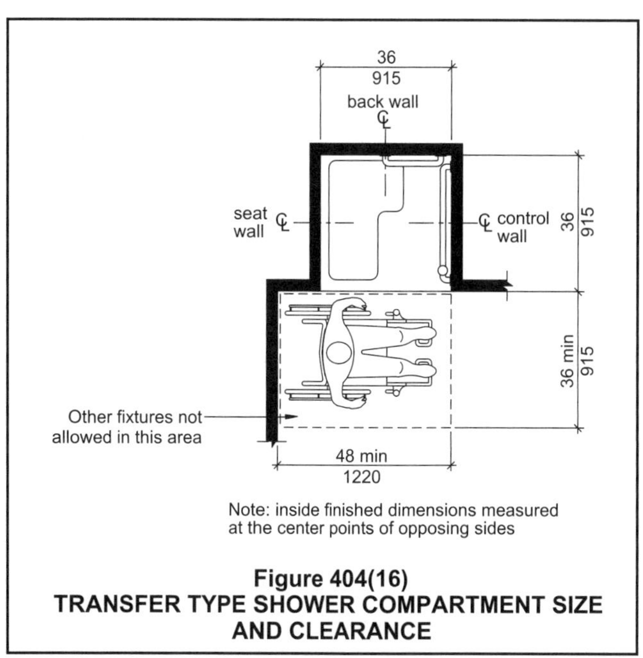

Figure 404(16)
TRANSFER TYPE SHOWER COMPARTMENT SIZE AND CLEARANCE

Commentary Figures 404(17) and (18) are provided to show the required grab bar installation so that the plumbing contractor can avoid installing piping in locations where blocking might be required to be installed in the building framing.

Commentary Figure 404(18) also indicates the required zone (matched) to locate the control for a transfer shower. Refer to the previous commentary for bathtubs in this section for general information on control placement in the designated control zone.

Commentary Figure 404(19) shows the minimum dimensions for a standard roll-in shower. Note that an accessible lavatory can be located in the 30 inch x 60 inch (762 by 1524 mm) clear floor space as shown in the figure. As long as the control for the shower is located on the back wall, the lavatory is allowed to be at either end of the clear floor space. However, if the roll-in shower does not have a seat and the shower control is to be located on a side wall, the lavatory must be located on the end of the clear floor space that is opposite of the shower side wall having the control. This allows ample clearance for ambulatory users to operate the control prior to entering the shower.

Commentary Figure 404(20) is provided to show the required grab bar installation so that the plumbing contractor can avoid installing piping in locations where blocking might be required to be installed in the building framing.

Commentary Figure 404(21) indicates the allowable control location zones (matched) for a standard roll-in shower. The installer must determine if a seat

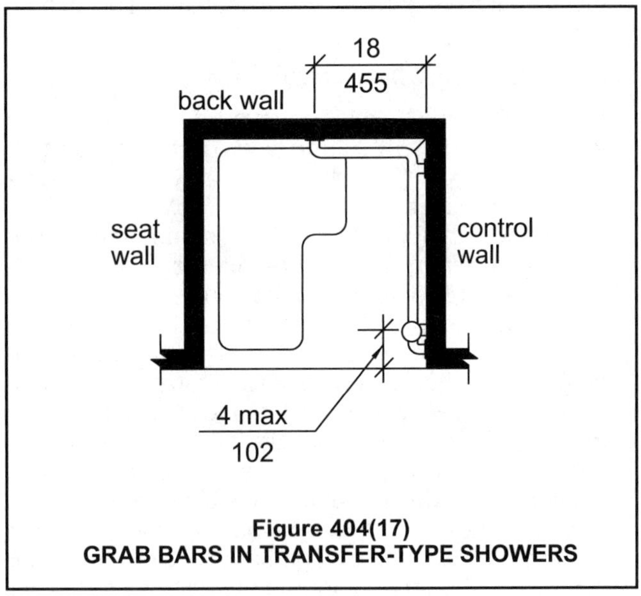

Figure 404(17)
GRAB BARS IN TRANSFER-TYPE SHOWERS

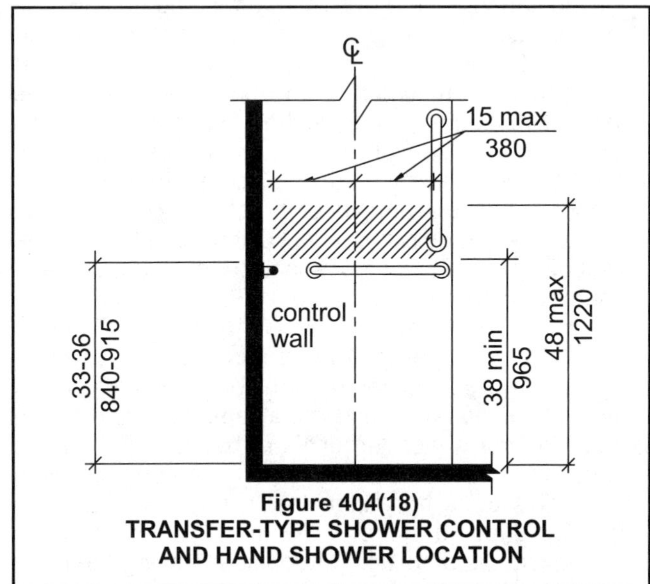

Figure 404(18)
TRANSFER-TYPE SHOWER CONTROL AND HAND SHOWER LOCATION

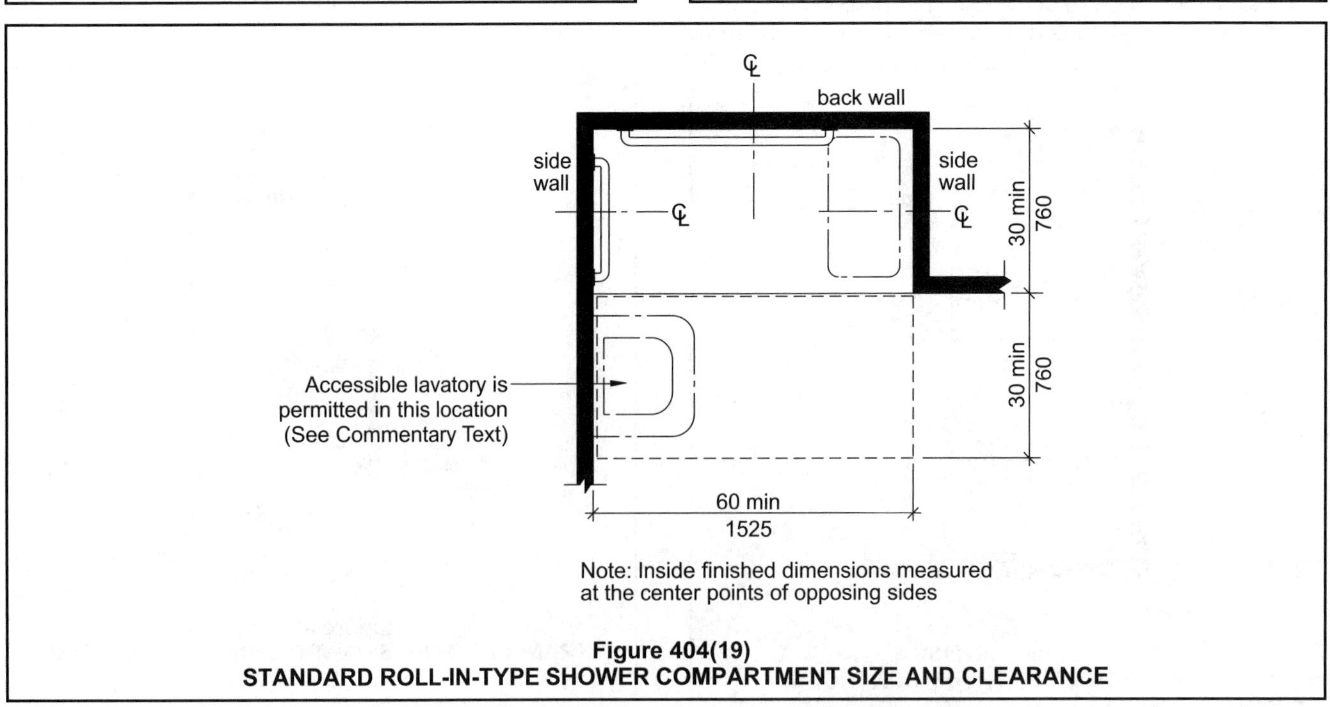

Figure 404(19)
STANDARD ROLL-IN-TYPE SHOWER COMPARTMENT SIZE AND CLEARANCE

(or the blocking for the seat) will be installed in order to properly locate the control. If a seat is to be installed (or blocking is to be installed for a future seat), the control must be located on the back wall in the narrow zone adjacent to the seat. Not all standard roll-in showers are required to have seats or blocking for seat installation. The designer of the space is responsible for making the determination on whether a seat is needed or desired. Refer to the previous commentary for bathtubs in this section for general information on control placement in the designated control zone.

Commentary Figure 404(22) shows the minimum dimensions for an alternate roll-in shower. ICC A117.1 requires that this type of shower arrangement have a seat or at least have the blocking in the wall so that a seat can be installed at some future point in time.

Commentary Figure 404(23) is provided to show the required grab bar installation so that the plumbing contractor can avoid installing piping in locations where blocking might be required to be installed in the building framing.

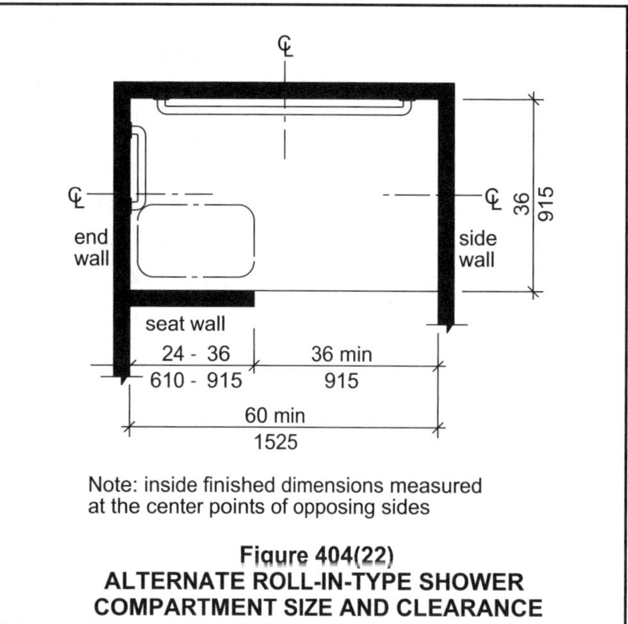

Figure 404(22)
ALTERNATE ROLL-IN-TYPE SHOWER COMPARTMENT SIZE AND CLEARANCE

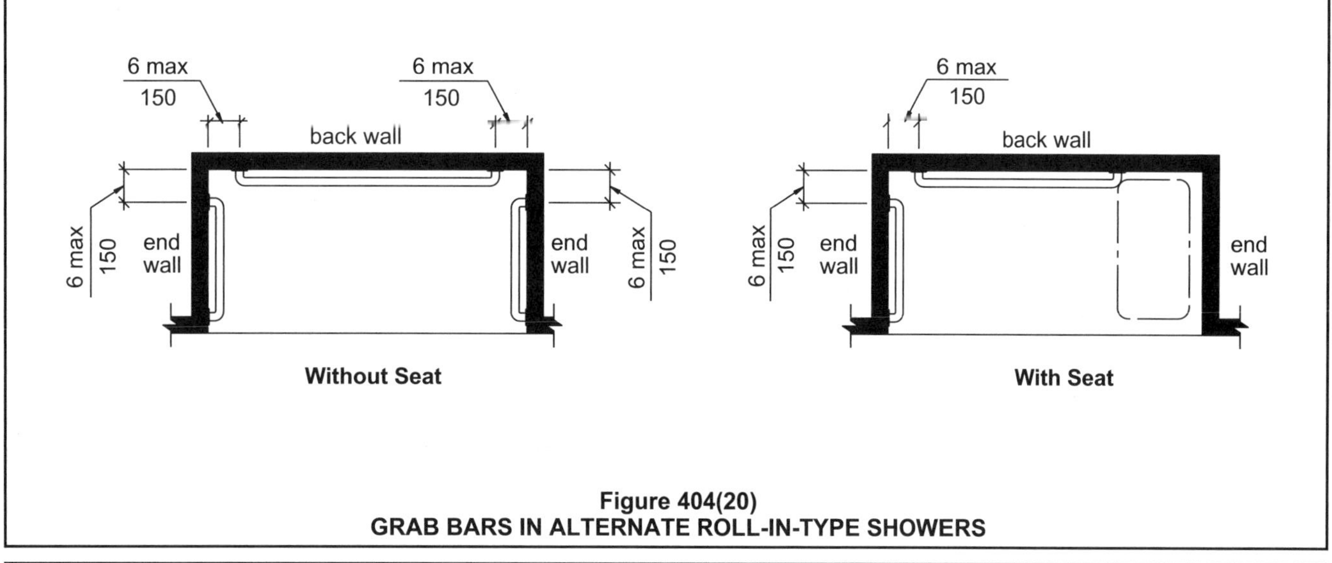

Figure 404(20)
GRAB BARS IN ALTERNATE ROLL-IN-TYPE SHOWERS

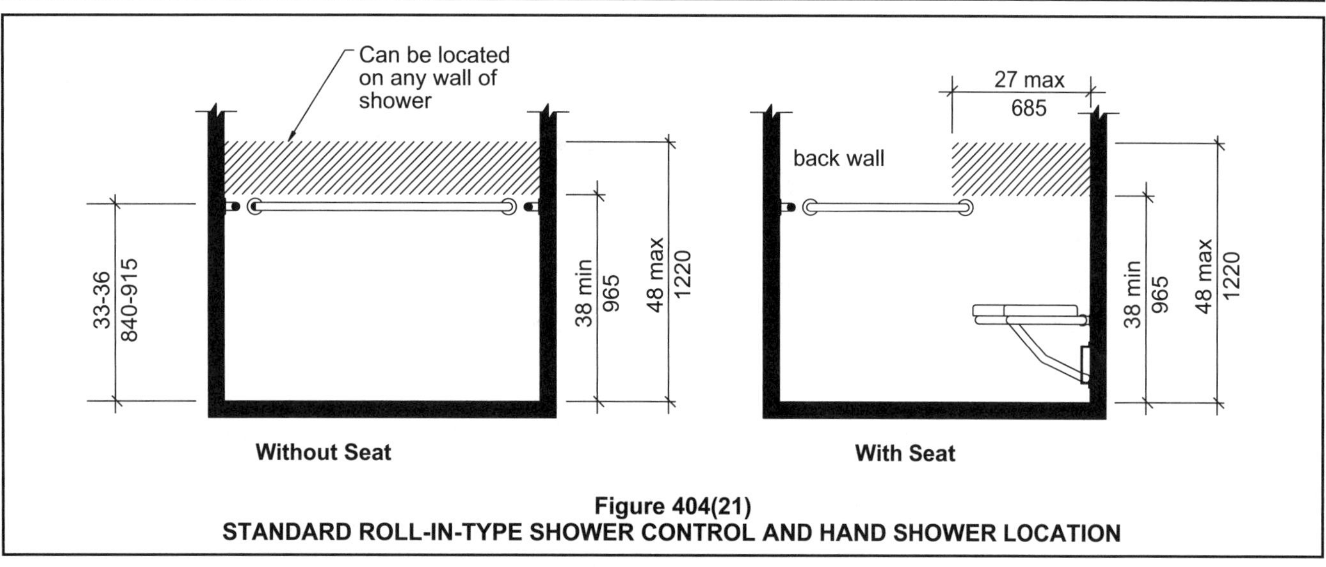

Figure 404(21)
STANDARD ROLL-IN-TYPE SHOWER CONTROL AND HAND SHOWER LOCATION

The diagrams in Commentary Figure 404(24) illustrate where controls and hand shower locations are required to be located, depending on the depth of the shower. If the shower depth is the minimum dimension of 36 inches (915 mm), the controls and hands however can be located in either of the positions shown in diagrams "a" and "b." The shower-seated user is able to reach the controls in either position. If the depth of the shower is greater than 36 inches (914 mm), the controls and hand shower must be located according to diagram "a" because the shower-seated user is not able to reach controls on a wall that is farther away than 36 inches (914 mm). Even though a seat might not ever be installed in the shower, because blocking for a seat is required to be installed in the wall, the control and hands however locations are dictated by the potential seat location. Refer to the previous commentary under bathtubs in this section for general information on control placement in the designated control zone.

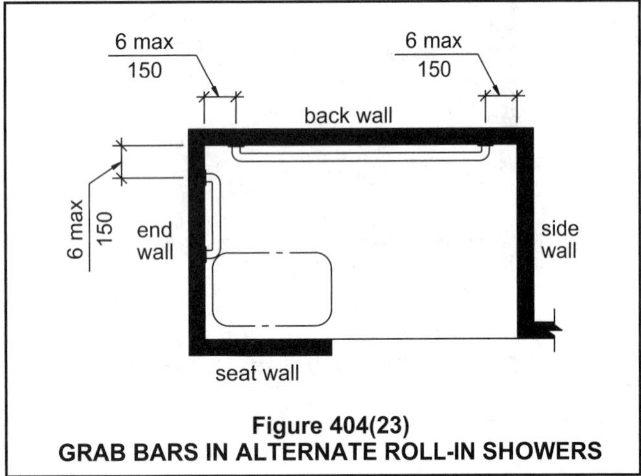

Figure 404(23)
GRAB BARS IN ALTERNATE ROLL-IN SHOWERS

Hand Showers (Accessible Design)

All bathtubs and showers for the Accessible design must be provided with a hand shower that can also be placed ina vertically adjustable sliding mount so that the user is not always required to hold the hand shower for use. This vertical adjustability feature also allows stationary positioning of the showerhead for both standing and seated users. Although the ICC A117.1 does not indicate a required length for this bar, many hand shower manufacturers commonly offer

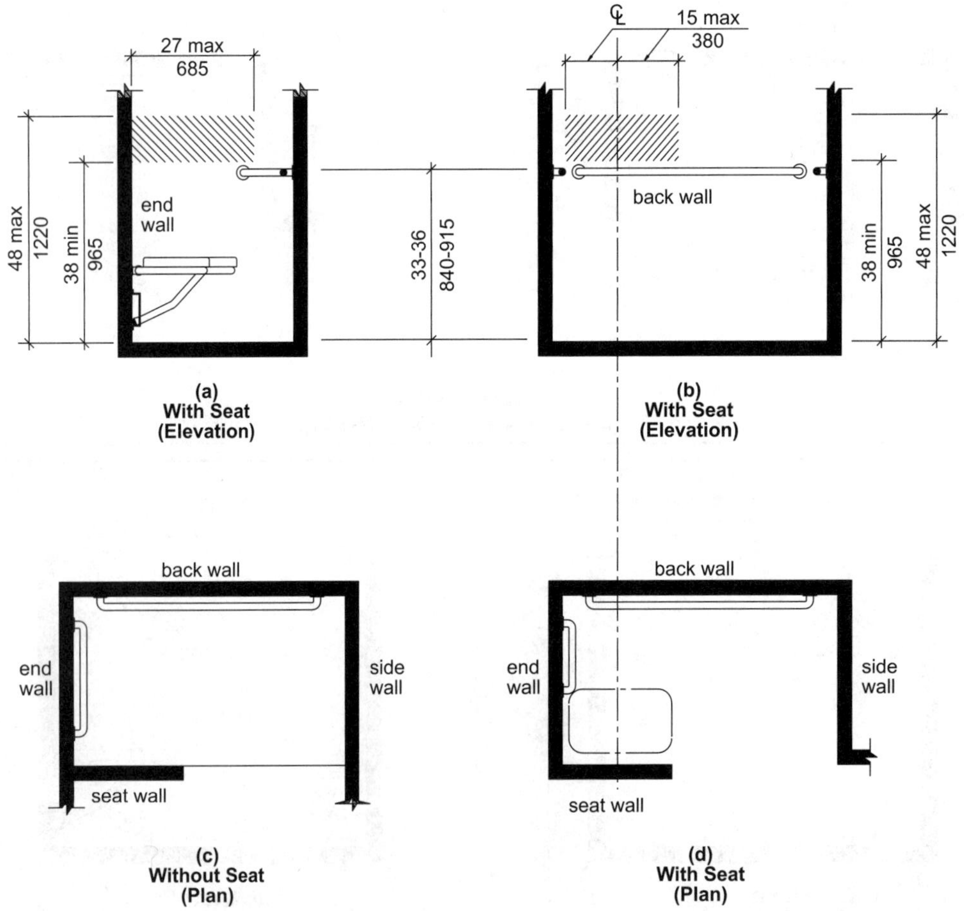

(a) With Seat (Elevation)

(b) With Seat (Elevation)

(c) Without Seat (Plan)

(d) With Seat (Plan)

Figure 404(24)
ALTERNATE ROLL-IN-TYPE SHOWER CONTROL AND HAND SHOWER LOCATION

hand shower sliding mount bar assemblies in 24 inch (610 mm) and 36 inch (914 mm) long versions that meet the accessible operational requirements [i.e., no grasping, no twisting and no force greater than 5 pounds (22.2 N) required to operate].

Although the current standard does not require the bar for the hand shower mount to meet the requirements for grab bars [i.e., resist a 250 pound (1112 N) force in any direction], many accessible design product offerings are made to meet these requirements. If the vertical sliding bar unit is capable of or even appears to be capable of meeting the force requirements, the installer is advised to make sure that the necessary blocking for solid attachment of the bar to the structure is installed at the rough-in stage. Because the installer of this vertical bar is often not the same person who installs the grab bars or folding seats, the required blocking for the vertical bar might not be installed during the framing stage of the project. Thus, at the plumbing fixture "set out" stage, having nothing substantial to which to mount the bar, an installer might be tempted to use expansion-type wall anchors which might not have the necessary pull-out resistance. Careful coordination between the designer, plumbing contractor and framing contractor is necessary to avoid missing this small yet important detail. The point of this comment is to prevent a vertical bar installation that looks to be adequate for supporting the user, but could fail upon use and possibly injure the user.

The intent of ICC A117.1 is that the vertical bar of the sliding mount be located such that the hand shower (handle) and the height adjustment lever for the mount can be positioned at an elevation corresponding to the elevation of the designated control zone. The hand shower does not have to be located in the control zone but can be to the side of the control zone on the same wall. Because all of the indicated shower control zones are indicated be above the elevation of the horizontal grab bar location, the lower end of the vertical slide bar can be mounted near the bottom elevation of the control zone to meet the handshower position requirement [see Commentary Figure 404(25)].

The ICC A117.1 is not as specific as to the location of the hand shower and mount adjustment device for bathtub arrangements. The lower end of the vertical slide bar cannot be mounted at the elevation within the control zone because the control zone is below the horizontal grab bar elevation. The vertical slide bar crossing over the grab bar would block user access to the grab bar. Therefore, the lower end of the vertical slide bar must be located above the horizontal grab bar. Even though this might put the mounted hand shower slightly out of reach for some users sitting on a bathtub seat, it is an acceptable compromise in order to provide for proper access to the horizontal grab bar. Typically, if the user wants to have the hand shower free of the mount, they will move the hand shower to the bottom of the tub while still seated in the wheelchair parked outside the tub. Once seated on the bathtub seat, the hose for the hand shower is very near to the control zone so that the user is able to "snag" the hose to get the hand shower to his or her body. Although not specified by the standard, to enable the wheelchair-seated user to reach the hand shower mount, the vertical line of the vertical slide bar should probably not be beyond the farthest vertical edge of the designated control zone.

The hand shower hose length must be at least 59 inches [1499 mm]. ICC A117.1 does not specify a required location for the hose connect supply elbow. Because this item is not a "control", it is not required to be located in the designated control zone. However, one must consider where a hand shower user will be seated in order to provide the user with ample useable hose length for showering of all body parts. For example, for a bathtub arrangement, locating the supply elbow near the traditional shower arm location might result in limited usability of the hand shower for the seated user. Another example would be locating the supply elbow too close to the grab bar surfaces such that the fittings or the hose, being relatively stiff and immovable, at the connection to the fitting presents an interference for use of the grab bars. Refer to previous commentary concerning required clearances around

Figure 404(25)
CONTROLS AND HAND SHOWER IN ACCESSIBLE SHOWER
(Photo courtesy of B&I Contractors, Inc.)

grab bars. The designer or installer has the responsibility to intelligently locate the supply fitting for the hand shower hose connection.

Hand shower heads must have a control that provides a "leaky" shut-off of the water flow. In other words, the "pause" control on the hand shower head must not completely shut off the water flow so as to alert the user that the main water control is still open, to prevent the water in the hose from cooling off and to reduce the potential for cross flow between the hot and cold supply to the control valve. Note that Section 424 of this code has requirements for hand showers and and the requirements for backflow protection for hand showers.

ICC A117.1 standard provides an exception for deleting the hand shower requirement and supplying a fixed showerhead. This exception "allows" for the code official to grant alternate approval (to those designers who request this variance) without being in violation of the accessibility standard. The sole purpose for the exception is to provide an alternative to hand showers in locations where vandalism of hand shower and hose assemblies would be problematic. Jails and public use areas (e.g., beach/pool bath houses) are examples of locations having a high potential for vandalism. ICC A117.1 does not address the specific type, location or required height for this fixed shower head. As this variance is for the protection against vandalism, the "fixed" showerhead is probably intended to be one of robust construction that has a fixed angle for the flow stream. The positioning of a fixed shower head is dependent on the nozzle angle of the showerhead to be used. Most major shower head manufacturers offer institutional/vandal-proof shower heads in 15 degree and 30 degree versions. Therefore, by estimating a head/neck height of a user sitting on the seat of the shower (or seated in a wheelchair) and the location of the seat (or wheelchair), the installer should be able to "accurately" locate a reasonable position for the particular angle of fixed showerhead to be installed.

The requirements of ICC A117.1 do not directly address other possible arrangements for providing a hand shower. For example, some designs of shower controls use either an integral or separate transfer valve for controlling the water flow from the main water control to a fixed shower head, a body spray array or a hand shower. Should the transfer valve be located in the designated control zone indicated in Commentary Figure 404(15)? The answer is "yes" because it is a control. If a stationary showerhead for standing users is supplied in addition to a hand shower, is it necessary to provide the vertical bar for the adjustable mount for the hand shower? The answer is "yes" because according ICC A117.1, a hand shower must be mounted on a vertical bar that allows adjustment of the height of the hand shower.

Hand shower (and slide bar) manufacturers are responsible for determining which hand showers (and slide bars) in their product lines comply with the accessible design (operational) requirements. Packaging, product labeling or specification sheets can be used to verify compliance. The indication for compliance might be the words "ADA-Compliant," "Compliant with (ICC) ANSI A117.1" or the International Symbol for Accessibility [see Commentary Figure 404(6)].

A hand shower connected to a "push-pull" type diverter valve assembly located on a shower head arm (or supply elbow) does not comply with the requirements of ICC A117.1 for the following reasons: (1) these types diverter valves require significant grasping or twisting to operate and (2) the mounting height would be well outside of the designated control zone (e.g., they are usually intended to be attached to a shower head arm located for a standing person's height).

Lavatories and Sinks (Accessible Design)

Section 606 of ICC A117.1 covers the requirements for lavatories and sinks for the Accessible design. In the context of ICC A117.1, the term "sink" includes a sink in a kitchen, a wet bar or a break/lunch room. All lavatories and most sinks for the Accessible design must be capable of being used by a person in a wheelchair using a "forward approach" to the fixture. In other words, the feet and legs of the user seated in a wheelchair must be provided with a clearance of specific dimensions under the plumbing fixture. Where a double bowl (kitchen) sink is installed, knee and toe clearances are required only under one of the bowls. There are a few limited exceptions for sink applications that allow for a parallel approach by a person in a wheelchair. Kitchen sinks (not in a space having a cooktop or range) and wet bar sinks are permitted to have a parallel approach. Although a parallel approach eliminates the need for providing knee/toe clearances and pipe protection under those fixtures, proper location with respect to the minimum clear floor space requirements is essential. The center of the sink bowl or the bowl of a double bowl sink for parallel approach can be at any location that is not less than 24 inches (610 mm) from at least one end of the clear floor space length dimension.

Commentary Figure 404(26) shows the maximum allowable height for a lavatory or a sink measured from the finished floor to the rim of the lavatory or sink; or the counter surface in which the sink is mounted, whichever is higher. This requirement applies to sinks with forward or parallel approach. The "rim" is understood to be that part of the fixture which is the highest point of the "bowl" that is the closest to the user. The dimension is a maximum; the rim height could be located slightly lower to provide some reasonable installation tolerance for practicality. However, the installed rim height cannot be too low because of the 27-inch (685 mm) minimum knee clearance and the 8-inches (203 mm) minimum from the front edge clearance under the sink as shown in diagram "a" of Commentary Figure 404(27). Most wall-mount lavatories marketed for the Accessible design have "underside" bowl dimensions that easily accommodate a slightly lower rim mounting height [e.g., $33^1/_2$ inch (851 mm)] without violating the minimum required knee clearance under the lavatory [see Commentary Figure 404(28)]. However, where lavatories and sinks for the Accessible design are constructed using countertops with sink bowls attached, the designer/installer must carefully review the bowl projection below the countertop to ascertain whether adequate knee clearance will be provided at the chosen installation height of the bowl rim or countertop.

Commentary Figures 404(27) and 404(29) show the required knee and toe clearances for forward approach. This information is provided so that the plumbing installer realizes that plumbing components such as stops, traps, supply piping and any exterior surfaces of chase coverings for piping cannot be located in zones required for knee and toe clearances. One possible situation where piping might have to be located near to knee or toe clearance areas is where drain or supply piping cannot be installed in the wall on which the lavatory is mounted because of firewall restrictions, freeze protection or retrofit considerations. Supply and drain piping can extend up from the floor in these applications but only where placement does not violate the required clearance zones. Any piping under lavatories or sinks must be configured so that user contact is prevented (e.g., by the installation of a removable panel), or if piping (including stops, traps and other plumbing devices) is exposed, those items must have a smooth covering installed so as to protect the user from those surfaces.

Commentary Figure 404(30) shows an arrangement for a forward approach kitchen sink for Accessible design in an employee break room. A standard countertop height for kitchens and employee break

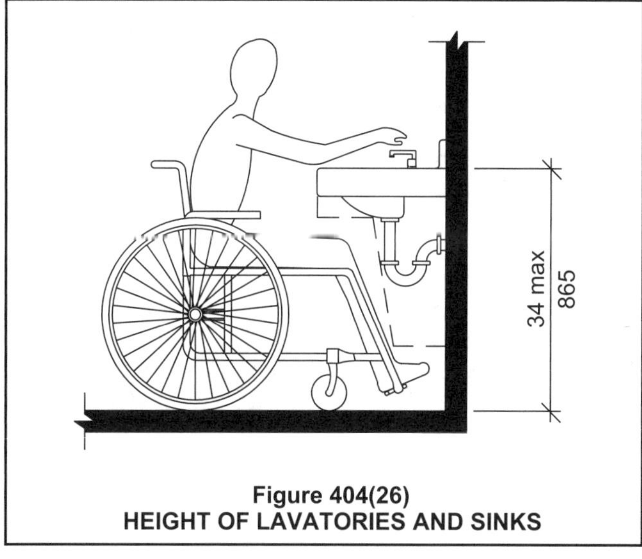

Figure 404(26)
HEIGHT OF LAVATORIES AND SINKS

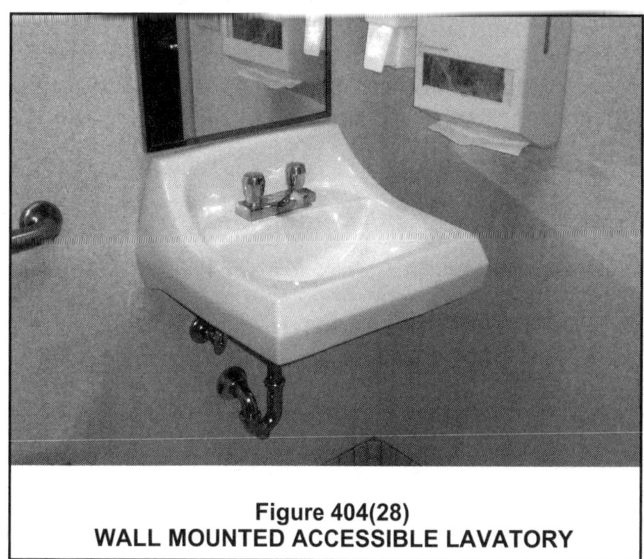

Figure 404(28)
WALL MOUNTED ACCESSIBLE LAVATORY

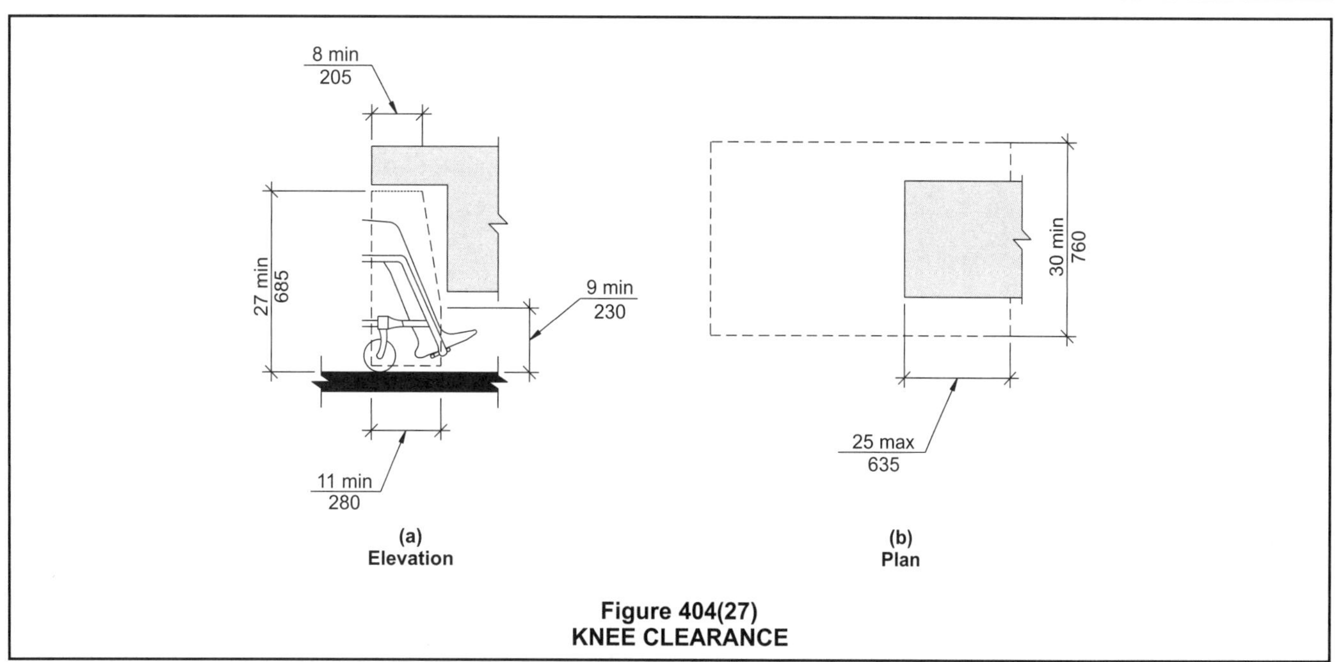

Figure 404(27)
KNEE CLEARANCE

rooms is approximately 36 inches (914 mm) above the floor. However, Accessible design requires that the portion of the countertop having the sink must have a height not greater than 34 inches (864 mm) from the top of the countertop (or bowl rim, whichever is higher) to the finished floor.

For a toilet or bathing room that can be accessed only through a private office, not for common or public use and intended for use by a single occupant, the lavatory height, knee and toe clearances do not apply (see IBC Section 1109.2, Exception 1.3). However, ICC A117.1 requires that the toilet and bathing room space and door width be sized so that the room can be easily converted to an Accessible design. Although not required by the standard, it would be prudent to locate the lavatory supply and drain piping such that if a conversion would occur, the required plumbing rework would be minimal.

Where a lavatory or sink will be used primarily by children of ages 6 though 12, ICC A117.1 Section 606.2, Exception 2, allows the 27-inch (685 mm) minimum knee clearance limitation to be reduced to 24 inches (610 mm) and limits the bowl rim or countertop height to a maximum of 31 inches (785 mm) above the finished floor.

Drinking Fountains (Accessible Design)

Most occupancies require the installation of drinking fountains in common or public areas of the building (see Table 403.1 and footnote e of the table). Where drinking fountains are provided, Chapter 11 of the IBC requires those drinking fountains to serve wheelchair-seated persons and standing persons.

For wheelchair-seated drinking fountain users, drinking fountains must be located with respect to the approach method to the drinking fountain. Drinking fountains are required to have a forward approach direction and require a clear floor space of 30 inches wide by 48 inches deep (762 by 1219 mm). The center of the drinking fountain can be at any location that is not less than 15 inches (381 mm) from at least one edge of the clear floor space width dimension. The required knee and toe clearances for a forward approach drinking fountain are the same as for forward approach lavatories and sinks [see commentary, Figures 404(28) and (29)]. The spout outlet of a wheelchair-accessible drinking fountain must not be greater than 36 inches (914 mm) above the finished floor. The spout height can be slightly lower (to provide a reasonable installation tolerance) as long as the minimum knee and toe clearances below the unit can still be achieved. A review of most drinking fountain manufacturer's drawings for drinking fountains show

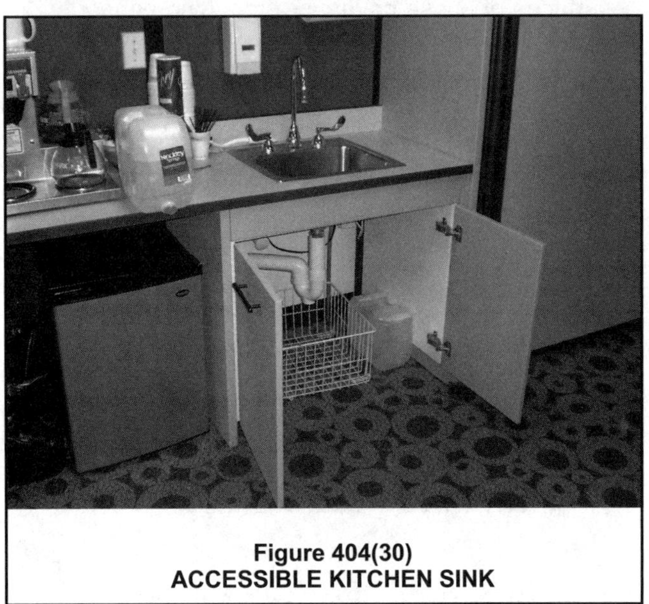

Figure 404(30)
ACCESSIBLE KITCHEN SINK

9
230
6 max
150
(a)
Elevation

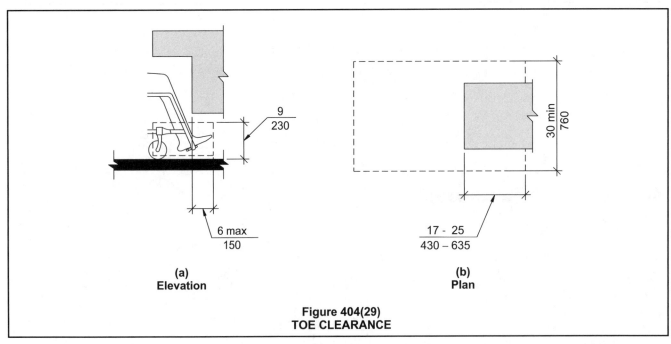

30 min
760
17 - 25
430 – 635
(b)
Plan

Figure 404(29)
TOE CLEARANCE

plumbing rough-in and mounting dimensions based upon the minimum knee clearance height of 27 inches (685 mm) above the finished floor. Although this guarantees that the spout outlet height will be well below the 36-inch (914 mm) maximum height required by ICC A117.1, careful consideration must be given to the thickness of floor coverings that might cause the knee clearance height to be violated. As stated previously, requiring accessible fixtures to be installed at maximum or minimum dimensions can lead to a code violation at final inspection [see commentary, Figure 404(31)].

Figure 404(31)
ACCESSIBLE DRINKING FOUNTAIN

A parallel approach to a drinking fountain is only allowed for drinking fountains that are a replacement for an existing drinking fountain that required a parallel approach. The center of the drinking fountain can be at any location within the clear floor space length that is not less than 24 inches (610 mm) from at least one end of the clear floor space length dimension.

Drinking fountains to be used by standing persons must have the spout outlet installed at a height not less than 38 inches (965 mm) and not greater than 43 inches (1090 mm) above the finished floor.

Where drinking fountains are to be used primarily by wheelchair-seated children, a spout height not greater than 30 inches (760 mm) above the finished floor is allowed to be used but only where the drinking fountain is provided with the clear floor space for the parallel approach method. The lower spout height for this application does not allow for the required knee and toe clearance for a forward approach to the drinking fountain.

Drinking fountains must comply with the accessible (operational) requirements of ICC A117.1 and also meet the requirements of Section 410 of this code. Drinking fountain manufacturers are responsible for determining which drinking fountains in their product lines comply with the accessible (operational) requirements. Packaging, product labeling or specification sheets can be used to verify compliance. The indication for compliance might be the words "ADA-Compliant," "Compliant with (ICC) ANSI A117.1" or the International Symbol for Accessibility.

Enhanced Reach Range Lavatories (Accessible Design).

The IBC requires an enhanced reach range lavatory in toilet and bathing rooms having 6 or more lavatories. This requirement does not increase the required number of lavatories; it only requires one of the lavatories to have enhanced reach range characteristics. It is a lavatory for use by those of short stature having short arm lengths. The requirements of Section 606.5 of ICC A117.1 are that the faucet controls and soap dispenser controls be located not farther than 11 inches (279 mm) from the front edge of the lavatory. This also means that water or soap streams, from their respectively spouts, must also be positioned at a dimension not greater than 11-inches (279 mm) from the front edge of the lavatory. As this is a relatively new requirement, wall-mount lavatory manufacturers have not yet created special lavatory designs that meet this requirement. However, using certain currently available electronic sensor-controlled faucets and soap dispensers with certain Accessible design compliant wall-mount lavatories, the 11-inch (279 mm) requirement could be achieved. Where a lavatory is constructed using a countertop with a separately mounted lavatory bowl, accessible design-compliant faucets and soap dispensers can be located along the sides of the bowl to provide for the 11-inch (279 mm) reach range. The enhanced reach range lavatory is not required to be mounted at the accessible design height [34 inches (864 mm)] maximum). Although not discussed in ICC A117.1, it is suggested by proponents of enhanced reach range lavatories that these types of lavatories have a rim height of not greater than 28 inches (711 mm) above the finished floor.

Faucets (Accessible Design)

Faucet manufacturers are responsible for determining which faucets in their product lines comply with the accessible (operational) requirements. Packaging, product labeling or specification sheets can be used to verify compliance. The indication for compliance might be the words "ADA-Compliant," "Compliant with (ICC) ANSI A117.1" or the International Symbol for Accessibility [see commentary, Figure 404(6)]. Note that metering faucets are required to stay open for at least 10 seconds.

Installation of Plumbing Fixtures in the Type A Design

The plumbing fixture installation requirements for the Type A design are covered by Section 1003.11 of ICC A117.1. Type A design could be called the "adaptable design" as all of the "prep work" must be done to allow future changes to be easily accomplished in order for the access to plumbing fixtures to be similar to that provided in the Accessible design. Type A-designed facilities have door widths, clear floor spaces and other built-in features that are identical to what is required for the Accessible design but the original finish construction has a residential look and feel. This residential look and feel is important in order to market Type A design units to persons who do not require or immediately want some of the obvious accessible design features in their living environment. Because all buildings covered by the IBC are required to be designed for accessibility, the Type A designed unit satisfies the intent for accommodating someone who requires an Accessible design unit and also satisfies the builder's, contractor's or owner's need to be able sell, rent or lease the unit to someone who does not want to live in an Accessible designed unit.

Water Closets for Wheelchair Access (Type A Design)

The only difference from the Accessible design for water closets is that the low end for the dimension range for the top of the water closet seat to the floor is 15 inches (381 mm) as opposed to 17 inches (432 mm). This allows for a more "standard height" water closet to be installed providing for a more aesthetically-pleasing "residential like" appearance. Grab bars are also not required to be initially installed but the blocking for the bars must be in place. From a clear floor space perspective, a lavatory is allowed to be located against the same rear wall for the water closet, with the side of the lavatory located no closer than 18 inches (455 mm) to the centerline of the water closet. A Type A design allows for a lavatory to be located on a the rear wall alongside the water closet if the following conditions exits (1) the side of the lavatory is located not less than 18 inches (455 mm) t the centerline of the water closet and (2) the clear floor space dimension for the water closet is not less than 66 inches (1671 mm), measured perpendicular from the rear wall, by a width of not less than 15 inches (381 mm), measured to either side of the centerline of the water closet.

Urinals for Wheelchair Access (Type A Design)

There is no requirement for urinals for the Type A design. However, there is nothing to prohibit the installation of a urinal in a Type A design unit other than the urinal must not be a substitute for an accessible water closet. The urinal does not have to be installed in accordance with the accessible design dimensional requirements because the water closet serves as the required accessible fixture. However, it would be prudent to mount the urinal at or below the accessible design height so that wheelchair users and children could benefit from the urinal installation.

Bathtubs (Type A Design)

Bathtubs are required to comply with the Accessible design except that countertops and cabinetry (with floor covering/finish under such cabinetry) are allowed to be installed at the control end of the bathtub's clear floor space. The countertops and cabinetry must be removable. The 12 inch (305 mm) extended area beyond the head end of the bathtub's clear floor space remains a requirement.

Showers (Type A Design)

The requirements for showers are the same as for the Accessible design for showers. Countertops and cabinetry (with floor covering/finish under such cabinetry) are allowed to be installed at the control end of the shower's clear floor space provided that countertops and cabinetry are removable. Although not specifically stated in ICC A117.1, it should be intuitively obvious that this countertop/cabinet allowance must not be used where a transfer shower is installed. It is intended to be for standard roll-in shower designs only. The 12-inch (305 mm) extended area of the shower's clear floor space beyond the seat end of the shower (whether the seat is actually installed or not) remains a requirement.

Hand Showers (Type A Design)

The hand shower requirements for Type A design are the same as for the Accessible design.

Lavatories (Type A Design)

The requirements for lavatories for Type A design are the same as for lavatories for Accessible design. However, the Type A design allows cabinetry to be installed under the lavatory provided that: (1) the cabinetry can be removed without removal of the lavatory (i.e., the lavatory or lavatory countertop is wall-mounted), (2) the current floor finish material extends up under such cabinetry, and 3) the wall behind and surrounding the cabinet is finished. These provisions allow for the installation to initially have a more "residential-like" appearance but have the capability to be easily converted allowing forward approach to the lavatory. Refer to the previous discussion of water closets for the Type A design for an allowance for locating a lavatory adjacent to a water closet.

Sinks (Type A Design)

For a multibowl sink, the knee and toe clearances according to Commentary Figures 404(27) and 404(29) are only required under one bowl. This allows ample area for the required drain piping connections under an adjacent bowl that ordinarily might not be easily accommodated if the knee and toe clearance were required under all bowls of a multibowl sink. If the kitchen sink is of a single bowl design, the knee and toe clearances must be provided under the entire sink (bowl).

Cabinetry is also allowed to be installed under kitchen sink provided that: (1) the cabinetry can be removed without removal of the sink or countertop, (2) the current floor finish must extend under such cabinetry, and (3) the wall behind and the sides of the cabinetry (adjacent to the cabinet to be removed) must be finished before the cabinetry is initially installed under the kitchen sink.

Although not required by ICC A117.1, plumbing piping should not interfere with the removal of cabinetry as the intent of the removable cabinet feature is that the cabinetry can be easily and economically removed to convert a Type A design sink installation to an Accessible design sink installation. Providing this ease of cabinet removal requirement might be problematic if plumbing piping is required to extend up from the floor, outside the wall cavity. Therefore, prior planning and coordination with the cabinet installer is recommended. There are no requirements for the removable cabinetry to be removed "intact"; it could be destructively dismantled. However, prudent design of the removable cabinetry would allow for nondestructive removal so that reinstallation is possible to return the room environment back to a more standard-looking "residential-like" environment.

Plumbing components such as stops, traps or supply piping, must not be in required knee and toe clearance areas, and where located near to knee and toe clearances, those components must have a smooth covering installed to protect the user. It is the intent of ICC A117.1 that these coverings be installed upon initial installation of the plumbing installation so as to make the installation compliant without the cabinet in place. In other words, no components would be required to be added to the installation to comply with forward approach requirements once the cabinetry.

Drinking Fountains (Type A Design)

There are no requirements for drinking fountains in Type A design.

Enhanced Reach Range Lavatories (Type A Design)

There are no requirements for this type of lavatory in Type A design.

Faucets (Type A Design)

Faucets must be located and comply with the requirements for Accessible design.

Installation of Plumbing Fixtures in the Type B Design

The plumbing fixture installation requirements for Type B design are covered by Section 1004.11 of ICC A117.1. Although the Type B design could be considered to provide a lesser degree of accessibility than the designs previously discussed, the Type B design should be viewed as a design that provides for basic accessibility requirements rather than a downgraded Type A design or Accessible design. The Type B design can either be of an Option A or an Option B configuration.

Because the origins of the Type B design different from those of the Type A design or the Accessible design, some of the requirements of Type B design might appear to be counterintuitive as compared to the other designs. For example, while a water closet in an Accessible design can be as close as 16 inches (406) to an adjacent finished wall at final inspection, the Type B design limits the water closet location to no closer than 18 inches (455 mm) to the finished wall. The reader is cautioned not to assume that all configurations of Type B designs have the same dimensional requirements or allowances as Type A or Accessible designs.

Water Closets for Wheelchair Access (Type B, Option A Design)

The water closet centerline is required to be located18 inches (457 mm) from the finished surface of an adjacent wall. Because this dimension is not given as a range or indicated as a minimum or maximum, the code official must determine what tolerance, if any, is acceptable for this dimension (see previous commentary under "Dimension reference points and dimension ranges").

Where a bathtub or lavatory is adjacent to a water closet, the centerline dimension of the water closet to the outermost edge of these fixtures is dependent upon the "approach direction" to the water closet and the clear floor space around the water closet. Refer to Section 1004.11.3.1.2 and Figure 1004.11.3.1.2 in ICC A117.1 for more details.

Type B, Option A design has no requirements for water closet seat height or flush handle height or location. In most cases, blocking is required for grab bar installation in accordance with the dimensional requirements for the Accessible design. Blocking is also sometimes required in the rear wall behind the water closet for a rear wall-mounted side grab bars [see Commentary Figure 404(32) as well as Section 1004.11.2 of ICC A117.1 for more details].

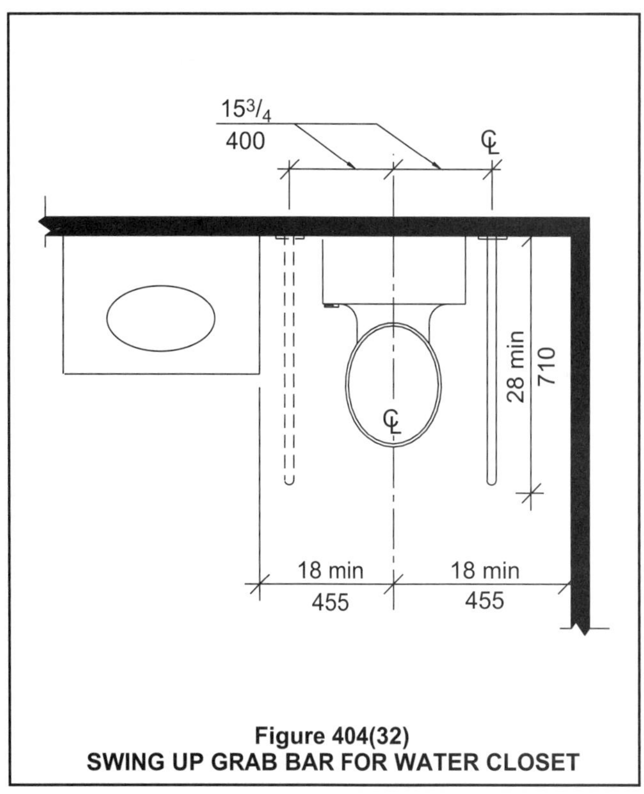

Figure 404(32)
SWING UP GRAB BAR FOR WATER CLOSET

Urinals for Wheelchair Access (Type B, Option A Design)

There are no requirements for urinals in the Type B design. A urinal could be installed in a Type B design, but it must not be a substitute for the accessible water closet. The urinal does not have to be installed in accordance with the Accessible design dimensional requirements because the water closet serves as the required accessible fixture. However, it would be prudent to install the urinal at or below the Accessible design height so that wheelchair users (as well as children) could benefit from the urinal installation.

Bathtubs (Type B, Option A Design)

For Type B, Option A design, Commentary Figures 404(33) and 404(34) show the required clear floor space for the two possible approaches to the bathtub. Diagram "a" of Commentary Figure 404(33) shows the required clear floor space where there is either no fixture located in the clear floor space or a lavatory in compliance with the Accessible design is located in the clear floor space. Diagram "b" shows the required clear floor space where there is a countertop (with or without a vanity below) having a lavatory bowl. The lavatory in diagram "b" must have the required clear space for a parallel approach to the lavatory [see commentary for "Lavatories (Type B, Option A Design)"]. The "jog" in the control end wall is the result of providing at least the required minimum length of clear floor space of 48 inches (1219 mm) plus the depth of the lavatory countertop.

Commentary Figure 404(34) shows the required clear floor space where there is a forward approach to the bathtub and a water closet is adjacent to the bathtub [see the previous commentary for "Water Closets for Wheelchair Access (Type B, Option A Design)"].

ICC A117.1 does not require the controls for the

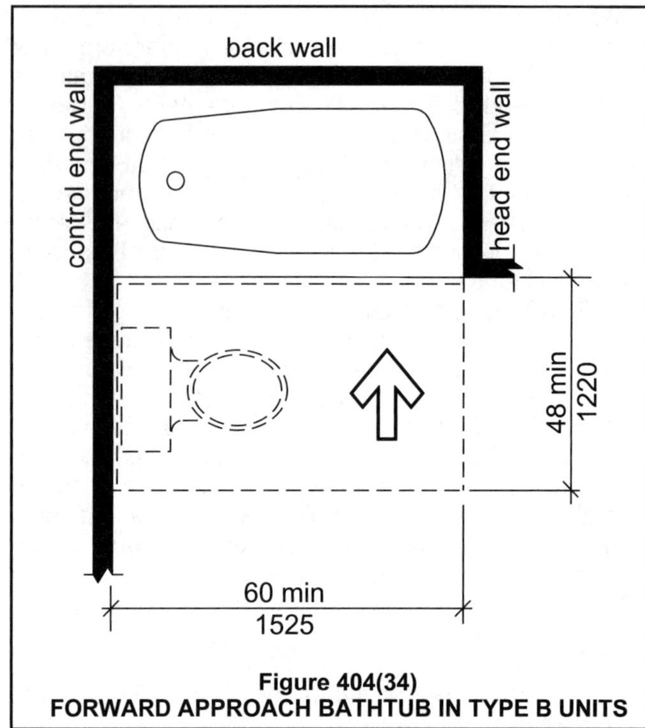

Figure 404(34)
FORWARD APPROACH BATHTUB IN TYPE B UNITS

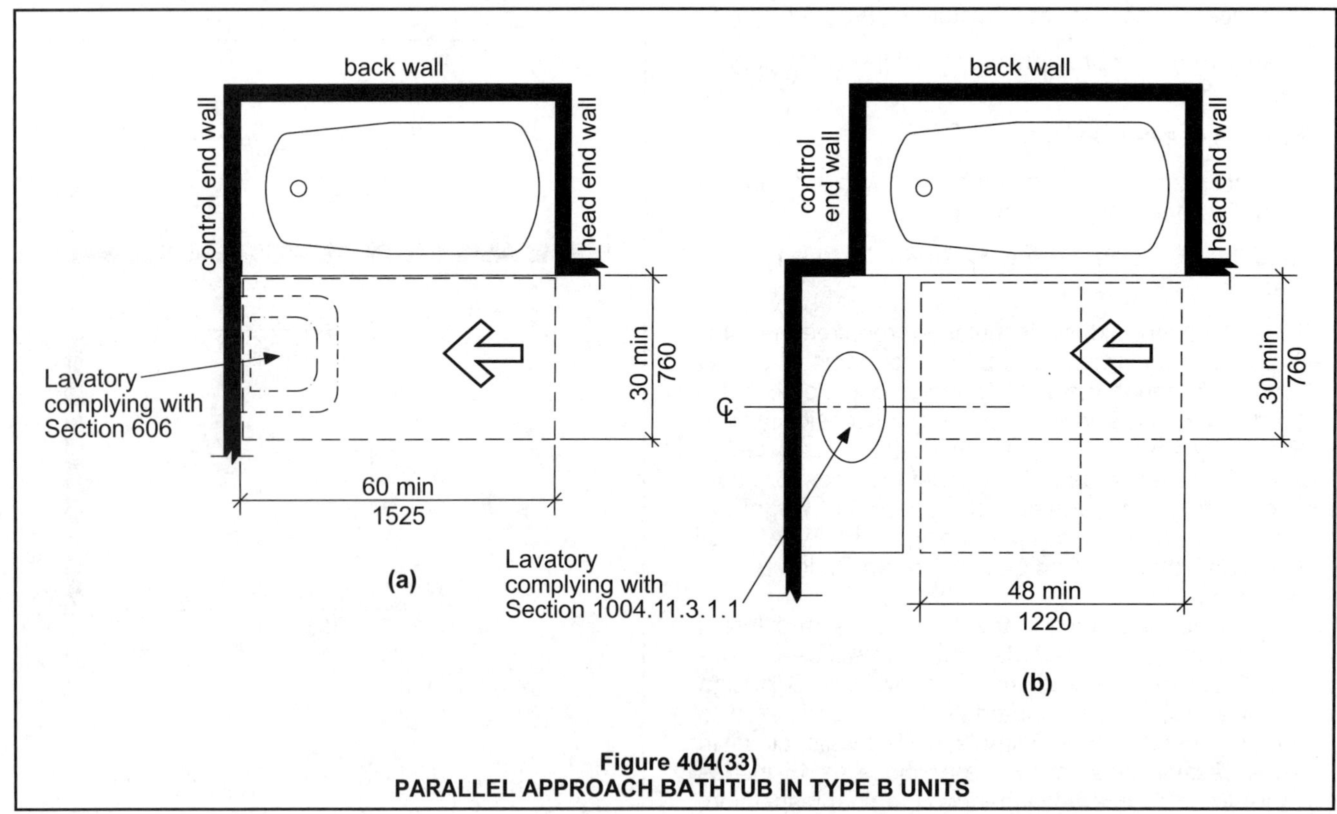

Figure 404(33)
PARALLEL APPROACH BATHTUB IN TYPE B UNITS

bathtub to be mounted in any particular location or be compliant to the operational limitations for accessible controls.

Grab bar blocking is required to be installed in walls in accordance with the dimensional requirements for the Accessible design.

Showers (Type B, Option A Design)

If the only bathing fixture in the Type B, option A designed unit is a shower, it must at least the size of an Accessible design transfer-type shower as shown in Commentary Figure 404(16) with the following modifications:

1. The clear floor space is required to be only 30 inches (760 mm) wide.
2. The blocking for the seat is not required if the shower is larger in size than both of the minimum shower dimensions.

Controls for showers are not required to be mounted in any particular location or be compliant with the operational limitations for accessible controls.

Grab bar blocking is required to be installed in accordance with the dimensional requirements for the Accessible design.

Hand showers (Type B, Option A Design)

Hand showers are not required in either showers or bathtubs. Standard showerheads are acceptable. If a hand shower is installed, there are no requirements for supplying or installing the hand shower in accordance with previously discussed accessible designs or that the hand shower be compliant with the operational limitations for accessible controls.

Lavatories (Type B, Option A Design)

There are three possible arrangements for lavatories in Type B, option A units.

1. Where there is at least a 30 inch x 48 inch (762 by 1219 mm) clear floor space for parallel approach in front of the lavatory, a lavatory of any height, with or without permanent cabinetry below, can be installed [see commentary, Figure 404(33) diagram "b"]. The lavatory bowl center can be at any location that is not less than 24 inches (610 mm) from at least one end of the clear floor space length dimension.
2. Where there is at least a 30 inch x 60 inch (762 by 1219 mm) clear floor space for forward approach to the lavatory, a wall mounted lavatory that complies with all of the requirements for an Accessible design lavatory can be installed [see Commentary Figure 404(33), diagram "a"]. The lavatory bowl center can be at any location that is not less than 15 inches (381 mm) from at least one end of the clear floor space width dimension.
3. Where removable cabinetry is planned for below a lavatory, a clear floor space of at least 30 inch x 60 inch (762 by 1219 mm) for forward approach (after the cabinet is removed) must be provided. The removable cabinet must not support the lavatory/countertop and the wall/floor must be finished under/behind the cabinet. The lavatory bowl center can be at any location not less than 15 inches (381 mm) from either end of the clear floor space width dimension. The lavatory/countertop must comply with the height, knee/toe clearances and pipe protection requirements required for an Accessible design lavatory. The intent of ICC A117.1 is for the pipe protection to be installed at the time of the initial installation such that only cabinet removal is required to comply with forward approach lavatory requirements.

Drinking Fountains (Type B, Option A Design)

There is no requirement for drinking fountains in Type B, option A units.

Sinks (Type B, Option A Design)

There are two possible arrangements for kitchen sinks.

1. Where the approach is parallel to the sink and knee/toe clearances are not provided under the sink, the sink can be any height. A clear floor space of at least 30 inches (762 mm) x 48 inches (1219 mm) for parallel approach must be provided in front of the sink. The center of the bowl (or one of the bowls of a two bowl sink) can be at any location that is not less than 24 inches (610 mm) from at least one end of the clear floor space length dimension.
2. A sink that complies with all of the requirements for Accessible design can be installed. A clear floor space of at least 30 inches x 48 inches (762 by 1219 mm) for forward approach must be provided in front of the bowl (or one of the bowls of a two bowl sink). The center of the bowl (or one of the bowls of a two-bowl sink) can be at any location not less than 15 inches (381 mm) from either edge of the clear floor space width dimension. The sink is also permitted to have parallel approach configuration having a clear floor space of at least 30 inches x 48 inches (762 by 1219 mm). The center of the bowl (or one of the bowls of a two- bowl sink) can be at any location that is not less than 24 inches (610 mm) from at least one end of the clear floor space length dimension.

Enhanced Reach Range Lavatories (Type B, Option A Design)

There are no requirements for this type of lavatory in Type B option A design.

Faucets (Type B, Option A Design)

Except for a faucet on a kitchen sink which is compliant with the Accessible design [see previous commentary under "Sinks (Type B, Option A Design)" item 2], faucets are not required to be mounted in any particular location or to be compliant with the operational limitations required for accessible controls.

Water Closets for Wheelchair Access (Type B, Option B Design)

A water closet centerline can be not closer than 18 inches (457 mm) to an adjacent wall and to a bathtub or lavatory where those fixtures are on a side opposite the wheelchair approach direction to the water closet. Where the bathtub or lavatory is not on a side opposite of the approach direction to the water closet, the dimension of the water closet centerline to those fixtures must be not less than 15 inches (381 mm). The purpose of the minimum required 18-inch (457mm) dimension on the side of the water closet opposite the approach direction is to allow for the future installation of a grab bar on that side [see commentary, Figure 404(33)].

There are no requirements for seat height or the location of the flush handle for water closets.

Rear wall grab bar blocking is required to be installed in accordance with the dimensional requirements for the Accessible design. Where the required 18-inch (457 mm) clearance side of the water closet is adjacent to a bathtub or lavatory, blocking is required in the rear wall for a rear wall-mounted side grab bar.

Urinals for Wheelchair Access (Type B, Option B Design)

There is no discussion of urinals for the Type B design. However, there is nothing to prohibit the installation of a urinal in a Type B design unit other than the urinal must not be a substitute for an accessible water closet. The urinal does not have to be installed in accordance with the Accessible design dimensional requirements because the water closet serves as the required accessible fixture. However, it would be prudent to mount the urinal at the accessible design height so that male wheelchair users and children could benefit from the urinal installation.

Bathtubs (Type B, Option B Design)

The possible bathtub arrangements are based upon either a parallel approach or forward approach. For parallel approaches to the bathtub, diagram "a" of Commentary Figure 404(34) allows for a wall mounted lavatory to be located in the clear floor area. The lavatory must be centered on the 30 inch (760 mm) dimension of the clear floor space and meet the requirements for Type A design lavatories, including the allowance for cabinetry below along with the same ease of removal requirements. Diagram "b" of Commentary Figure 404(34) allows for a lavatory in a wall-supported countertop at the end of the indicated required clear floor area in front of the bathtub as long as the lavatory is provided with a parallel approach and a 30 inch by 48 inch (762 by 1219 mm) clear floor space, centered on and directly in front of the lavatory. In other words, the center of the lavatory is required to be at least 24 inches (610 mm) from the face of the bathtub but not less than 24 inches (610 mm) from the other end of the clear floor space in front of the lavatory.

For a forward approach to the bathtub, Commentary Figure 404(35) indicates the clear floor area required. A water closet is allowed to be located in the clear floor space and is required to have the centerline located at least 18 inches (457 mm) from the face of the bathtub.

Controls for bathtubs are not required to be mounted in any particular location or to be compliant with the operational limitations required for accessible controls.

Note that grab bar blocking is required to be installed in walls that will exist in those locations in accordance with the dimensional requirements for the Accessible design.

Showers (Type B, Option B Design)

If the only bathing fixture in the dwelling unit is a shower, then it must be at least the size of a transfer-type shower as shown in Commentary Figure 404(14) with the following modifications:

1. The clear floor area is only required to be 30 inches (760 mm) wide.
2. A seat is not required; blocking for a seat must be installed if the shower is of the minimum size and a wall exists in that location.
3. Seat blocking is not required if the shower is larger in size than both of the minimum shower dimensions.

Controls for showers are not required to be mounted in any particular location or be compliant with the operational limitations required for accessible controls.

Grab bar blocking is required to be installed in accordance with the dimensional requirements for the Accessible design.

Handshowers (Type B, Option B Design)

Hand showers are not required in either showers or bathtubs. Standard showerheads are acceptable. If a hand shower is installed, there are no requirements for installing the hand shower in accordance with previously discussed accessible designs or that the hand shower be compliant with the operational limitations for accessible controls.

Lavatories (Type B, Option B Design)

Cabinetry is allowed to be installed under the countertop as long as it meets the ease of removal requirements as indicated for Type A design lavatories. Alternatively, a wall-mount lavatory similar to the one allowed in diagram "a" of Commentary Figure 404(34) can be installed, but it is required to have the centerline located at least 24 inches (610 mm) from the side of the bathtub to enable a parallel approach to the lavatory.

Sinks (Type B, Option B Design)

Refer to previous commentary for "Sinks (Type B, Option A Design)."

Drinking Fountains (Type B, Option B Design)

There is no requirement for drinking fountains in Type B, option B design.

Enhanced reach range lavatories (Type B, Option B Design)

There are no requirements for this type of lavatory in the Type B, option B design.

Faucets (Type B, Option B Design)

Faucets are not required to be mounted in any particular location or be compliant with the operational limitations for accessible controls.

SECTION 405
INSTALLATION OF FIXTURES

405.1 Water supply protection. The supply lines and fittings for every plumbing fixture shall be installed so as to prevent backflow.

❖ This section identifies minimal installation requirements for water supply protection. Backflow prevention is perhaps the most important aspect of a plumbing code (see commentary, Section 608). All plumbing fixtures, plumbing appliances and water distribution system openings, outlets and connections are potentially capable of contaminating the potable water supply.

405.2 Access for cleaning. Plumbing fixtures shall be installed so as to afford easy *access* for cleaning both the fixture and the area around the fixture.

❖ Poorly designed facilities lead to challenges for housekeeping professions. For proper sanitation, every plumbing fixture must be capable of being cleaned. The installation of any fixture must not result in concealed spaces that do not facilitate proper cleaning.

405.3 Setting. Fixtures shall be set level and in proper alignment with reference to adjacent walls.

❖ The operational design of plumbing fixtures and the connections to those fixtures require that the fixture be installed with certain surfaces either level or plumb. This level fixture setting is required regardless of any out-of-level or out-of-plumb condition of the building structure that the fixture attaches to or rests upon. For example, a floor surface around a water closet connection may be out of level, but when the water closet is installed, the base of the water closet must be either shimmed or grouted such that the fixture is in level condition. Where a water closet is in an out-of-level condition, the integral trap may not provide the required water seal depth, and the flushing action within the fixture may be compromised. Where a wall-mounted lavatory is mounted to an out-of-plumb wall, water in the basin may not completely drain and the trap below may not have the required water seal depth due to also being in an out-of-plumb condition. The lavatory must be installed so as to be level, regardless of the wall condition.

Proper alignment to adjacent walls is a subjective requirement but the intent is that the installed fixture appear to be "square" and "in alignment" with the walls that are in close proximity to the fixture. An example would be the clearance behind a water closet flush tank to the wall behind it. The gap should appear to be even. Another example is a one-piece lavatory countertop with bowl on a vanity. The back splash of the counter top should appear to be even against the wall so as to prevent large gaps that are difficult to caulk and maintain a seal against possible splashing. While the code is not specific on the precision of the alignments, the intent is to ensure that fixtures are installed with attention to details that could affect the operation and future sanitary condition of the fixture.

405.3.1 Water closets, urinals, lavatories and bidets. A water closet, urinal, lavatory or bidet shall not be set closer than 15 inches (381 mm) from its center to any side wall, partition, vanity or other obstruction, or closer than 30 inches (762 mm) center-to-center between adjacent fixtures. There shall be at least a 21-inch (533 mm) clearance in front of the water closet, urinal, lavatory or bidet to any wall, fixture or door. Water closet compartments shall not be less than 30 inches (762 mm) wide and 60 inches (1524 mm) deep (see Figure 405.3.1).

❖ Code Figure 405.3.1 attempts to illustrate the minimum clearances stated in this section but due to the

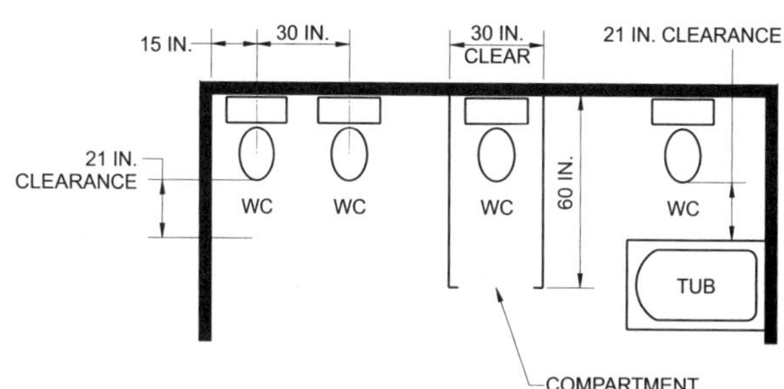

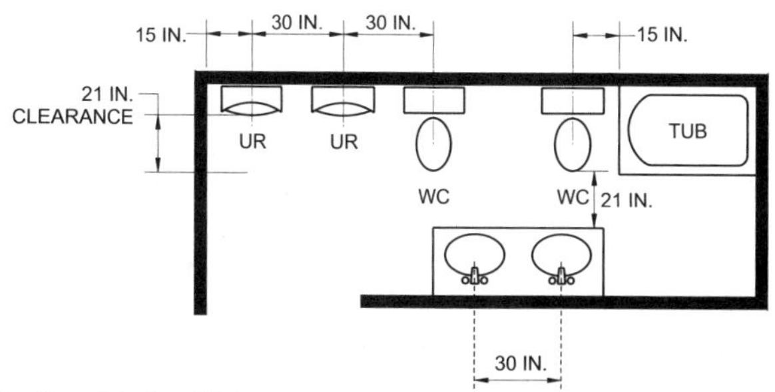

For SI: 1 inch = 25.4 mm.

FIGURE 405.3.1
FIXTURE CLEARANCE

endless possibilities of fixture arrangements, the figure understandably falls short of providing a clear understanding of the requirements. In the context of this section only, bathtubs and showers are considered to be but "obstructions" for which the minimum clearance dimensions from the indicated fixtures must be observed. Required minimum clearances in front of bathtubs and showers are not addressed in the IPC.

The requirements for placement and installation of plumbing fixtures for accessibility in toilet and bathing facilities "override" this section's dimensional requirements for "nonaccessible" plumbing fixture installation (see commentary for Section 404). A critical point that is often overlooked by designers and contractors is the accessibility requirement that the distance between urinal partitions for the wheelchair access urinal be at least 36 inches (914 mm). The centerline of the wheelchair accessible urinal must be at least 18 inches (457 mm) from the face of each partition and not the minimum of 15 inches (381 mm).

Even though the code section is silent on exactly how the dimensions indicated in this section are to be measured, the intent is that the dimensions are to be measured to finished surfaces of walls (including tile or other coverings) and to imaginary vertical planes located against the most exterior feature of a fixture or obstruction. The code is also silent on whether minor protrusions from walls such as paper product dispensers, hand dryers, cove moldings, diaper changing units, display cases, vending machines and grab bars impact the measurements discussed in this section.

The centers of the fixtures discussed in this section are not specific points on the fixtures, but a point which is "derived." For example, a centerline of a lavatory bowl is typically ascertained by calculating the "halfway" dimension between the widest outside dimensions of the bowl. Given the difficulty in locating an exact center of a fixture and a lack of coordination between trades, establishing an inspection tolerance for the spacing and clearance dimensions required by this code section is next to impossible. However, code officials will necessarily impart their own acceptance criteria (i.e., tolerances) during an inspection in order to accept or reject a clearance dimension. Therefore, designers of toilet facilities are cautioned against creating layouts which require building to the exact minimum clearance dimensions. A fixture layout having greater-than-the-minimum-clearance dimensions is inherently easier to build and results in far less inspection difficulties.

Note that all of the minimum dimensions stated in this code section might be overridden by the accessibility requirements of Chapter 11 of the IBC. See the commentary for Section 404 for more information.

Both layouts in Code Figure 405.3.1 show prohibited arrangements for a public toilet facility because, where multiple water closets exist, they are required to be in compartments. For a private bathroom, the arrangements would be unlikely to have more than one water closet in an unenclosed room. The intent of the Code Figure 405.3.1 and the commentary figures to follow is to show the required minimum clearance requirements for each plumbing fixture but not necessarily how a typical restroom or bathroom layout should look. Good design practices for toilet facilities and bathrooms usually provide clearances in excess of the minimum requirements in order to provide for appealing aesthetics and user friendliness.

The 2-inch (533 mm) minimum clearance from the front edge of a fixture to any wall, fixture or door applies only to water closets, urinals, lavatories and bidets. This dimension has been determined to be the necessary space required by a user to properly and safely use these fixtures. Because a door has a swing path, a possible interpretation of this required clearance might be to consider the position of the door at any point in the swing path. However, this section is not concerned about the operability or use of other elements of the built environment and the user of the fixture will position the door as needed to access and use the fixture. Therefore, it is the author's opinion that the intent of this code section is for the clearance to the door to be measured with the door closed.

Note that the egress requirements of the IBC require that egress doors not be prevented from fully opening (see Commentary Figure 310.2).

Code Figure 405.3.1 indicates that where two fixtures face each other, the minimum clearance is 21-inches (533 mm) and not 42 inches (1067 mm). In other words, the code does not prohibit these "user spaces" from overlapping.

If there are two fixtures facing each other, the 21 inch (533 mm) clearance in front of each fixture can overlap; however, the minimum egress width requirements in the IBC would most likely prevent a complete overlap of the frontal minimum clearances. Accessibility requirements for fixtures that are required to be accessible will require greater frontal approach dimensions. Door swing paths, in some accessible situations, are not allowed to overlap the accessible frontal approach dimensions.

The 15-inch (381 mm) minimum clearance dimension from the center of a fixture to any side wall, partition, vanity or other obstruction applies only to water closets, urinals, lavatories and bidets. Note that the requirement does not include "to any fixture" as the 30-inch minimum (762 mm) center-to-center for fixtures, discussed below, controls the spacing between fixtures. Fixtures that are required to be accessible require greater clearances from centerline to side wall, partition, vanity or other obstructions (see commentary, Section 404).

The 30-inch (762 mm) minimum center-to-center dimension is only between the centers of water closets, urinals, lavatories and bidets. Public restrooms, having the requirement for partitions and compartments separating fixtures, cannot have fixtures located at the minimum of 30 inches (762 mm) center-to-center. The controlling aspect is the 15-inch (381 mm) minimum dimension from the center of the fixture to the required partition or compartment wall, not the center of the adjacent fixture. In other words, the thicknesses of partitions must be considered, thereby necessitating dimensions greater than 30 inches (762 mm) center to center.

Lavatories, both private and public, that are part of vanities require special attention. As indicated in the previous paragraph, a vanity, an obstruction in the context of this section, can be no closer than 15 inches (381 mm) to the centerline of an adjacent fixture. The center of the lavatory bowl, which is in the lavatory top that rests on the vanity, must be no closer than 30 inches (762 mm) to the centerline of an adjacent fixture. If the center of the lavatory is less than 15 inches (381 mm) from the side edge of the lavatory top, then the 30-inch (762 mm) center-to-center minimum dimension controls [see commentary, Figure 405.3.1(1)] If the center of the lavatory is greater than 15 inches (381 mm), then the 15 inches (381 mm) from the centerline of the adjacent fixture to the nearest obstruction (i.e., the edge of the lavatory top, not necessarily the vanity cabinet) controls. Note that locating the vanity cabinet at 15 inches (381 mm) from the centerline of the adjacent fixture would not necessarily provide the required minimum clearance as most lavatory tops overhang the edge of the vanity cabinet to some extent [e.g., $^{3}/_{4}$ inch (19.1 mm)]. For example, a 48 inch (1219 mm) wide vanity cabinet is provided with a 49 inch (1245 mm) wide cultured marble integral lavatory bowl countertop. Theoretically, this countertop could overhang along one side as much as 1 inch (25.4 mm).

Improper location of lavatories and lavatory vanities narrower than 30 inches (762 mm), offset-to-one-side bowls in lavatory countertops and corner lavatory arrangements can unknowingly cause code violations. Where there is a strong emphasis on minimizing bathroom spaces, designers might be tempted to reduce the space for the lavatory by choosing a narrow vanity width or providing a wall mount or pedestal lavatory of a narrow dimension with the intent of locating this fixture as close as possible to a side wall. This can be a violation. Bowls in the lavatory countertops are often located offset to one side in order to provide clearance for vanity cabinet drawers beneath. Again, a violation can be created, as there is not sufficient clearance from the center of the lavatory (bowl) to the adjacent sidewall or to an adjacent fixture to provide for proper use the lavatory.

The code does not offer any specific insight concerning corner sinks. One possible method for determining the required clearances would be to establish a vertical plane perpendicular to a horizontal centerline line extending from the corner though the center of the lavatory bowl and positioning this plane at the front edge of the fixture. This "user width" could be determined by measuring horizontally within the established plane from the centerline of the lavatory to a point where the plane insects the adjacent wall. The intent of this section appears to be to provide a "user width" of at least 30 inches (762 mm). If the measured dimensions left and right of the lavatory centerline are 15 inches (381 mm) or more, then the "user width" requirement of this section is satisfied. The front clearance measurement for this type of fixture could also

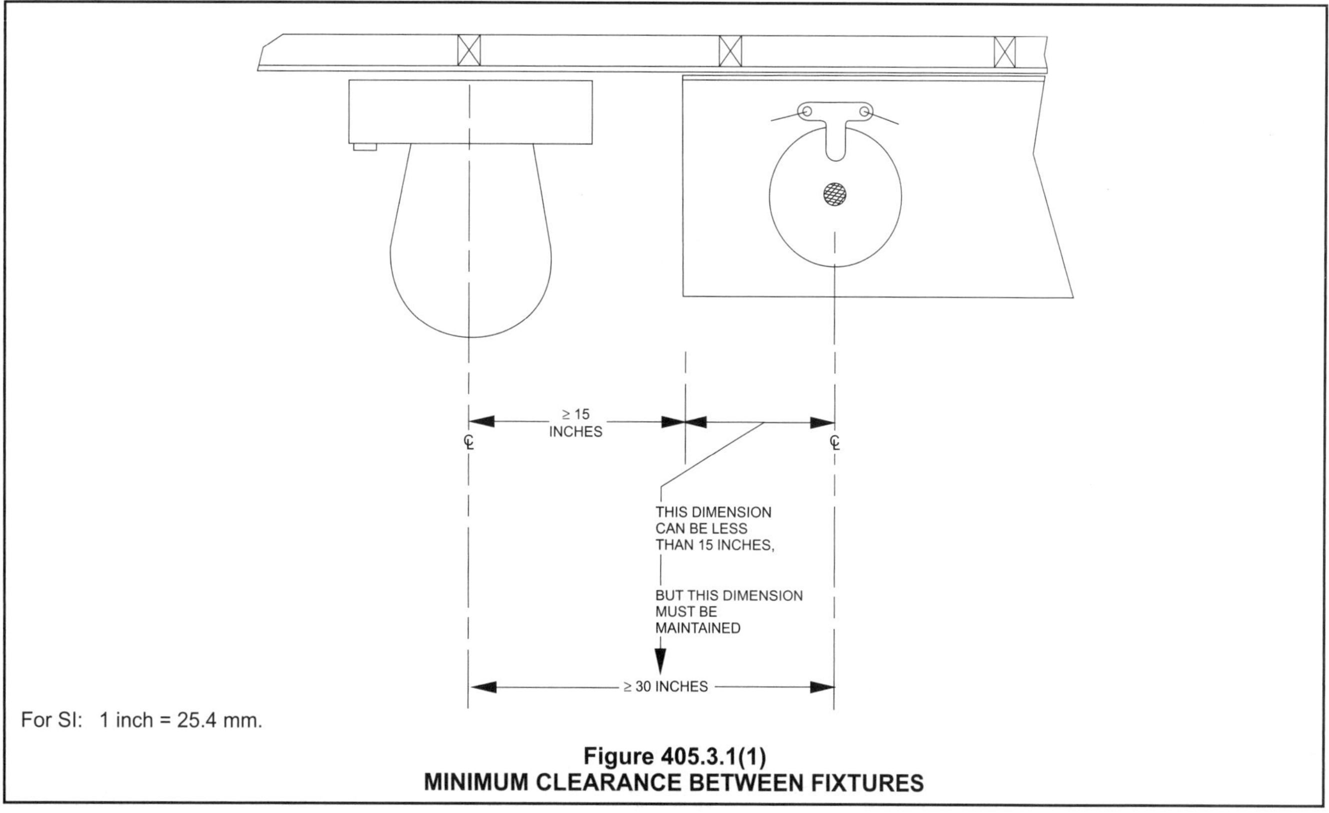

Figure 405.3.1(1)
MINIMUM CLEARANCE BETWEEN FIXTURES

utilize the imaginary front edge plane as a reference point. The intent of this section appears to be to provide a "user depth" of at least 21 inches (533 mm). Therefore, for a corner lavatory, the "user space" of 30 inches (762 mm) wide by 21 inches (533 mm) deep must not be encroached upon by other fixtures, closed doors or walls [see commentary, Figure 405.3.1(2)].

405.3.2 Public lavatories. In employee and *public* toilet rooms, the required lavatory shall be located in the same room as the required water closet.

❖ In employee and public toilet rooms, the required lavatory must be located in the same room as the required water closet or in an adjacent room connected by openings without doors. This provides the user with necessary sanitary facilities to promote proper hygiene and to prevent the spread of infectious diseases. In some public toilet arrangements, the water closets and urinals are located in a separate room from the lavatories, thereby requiring the user to engage door hardware to access the lavatories after using a urinal or water closet. Such an arrangement would be a violation of this section as the intent is that the lavatories be available to allow the user to wash his or her hands prior contacting any door handle or push plate for exiting the toilet facility. The typical airport terminal toilet room has the water closet and urinal in one room and the lavatories in an adjacent room. Such rooms are connected through doorways without doors and this scenario meet the intent of this section.

405.4 Floor and wall drainage connections. Connections between the drain and floor outlet plumbing fixtures shall be made with a floor flange. The flange shall be attached to the drain and anchored to the structure. Connections between the drain and wall-hung water closets shall be made with an *approved* extension nipple or horn adaptor. The water closet shall be bolted to the hanger with corrosion-resistant bolts or screws. Joints shall be sealed with an *approved* elastomeric gasket, flange-to-fixture connection complying with ASME A112.4.3 or an *approved* setting compound.

❖ The setting compound most commonly used is a wax ring made of a synthetic wax. In the past, putty was commonly used, but this practice is now rare as putty dries out and cracks. Elastomeric gaskets are also used between fixture outlets (horns) and flanges (see Commentary Figure 405.4). Corrosion-resistant bolts or screws are required in this application so that there is not a reduction of strength of the fastener over time due to corrosion. Failure of the bolting of a wall-hung water closet could result in serious personal injury and fixture damage. The flange-to-fixture connection complying with ASME A112.4.3 is a special plastic fitting that is only suitable for floor-mounted, bottom-discharge water closets. The device seals to the inside of the drain pipe with an elastomeric gasket and seals to the outlet of the water closet bowl with an elastomeric gasket. Where this device is used, the floor flange only serves as an attachment point for the closet bowl bolts.

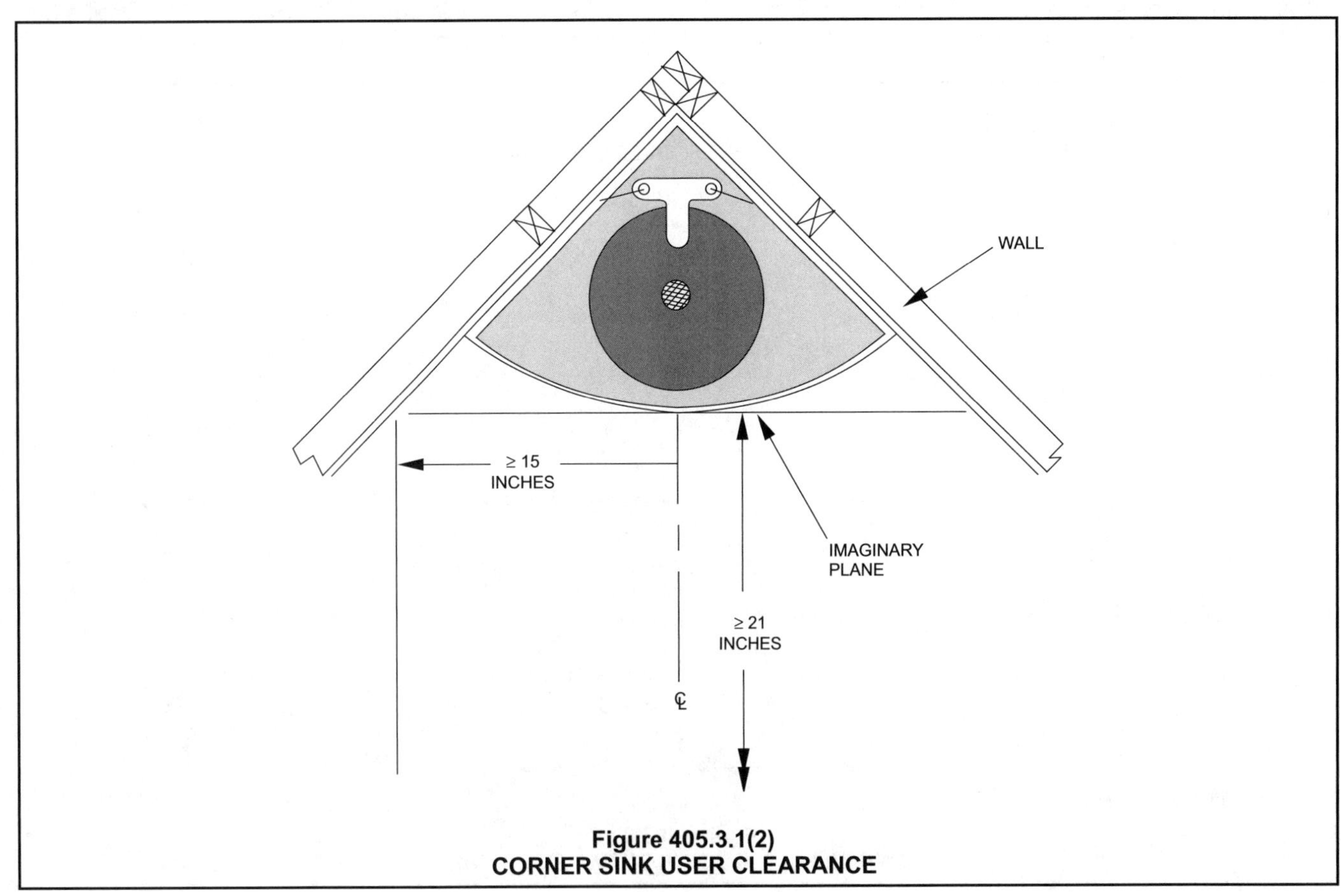

Figure 405.3.1(2)
CORNER SINK USER CLEARANCE

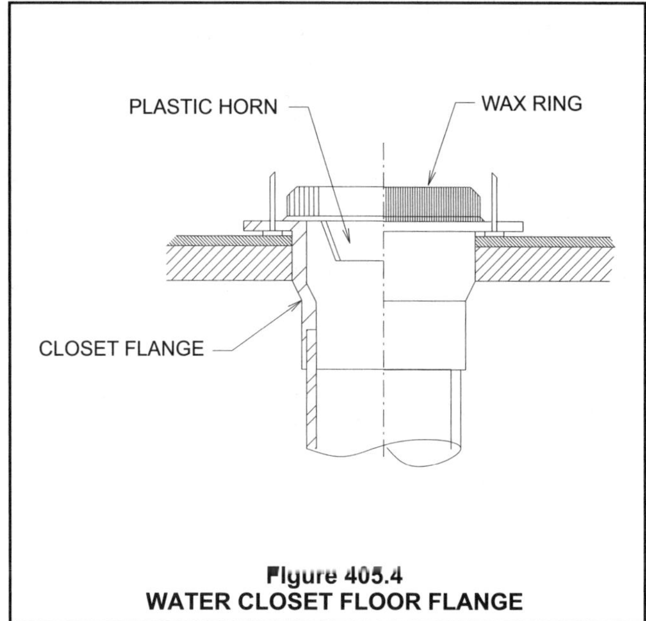

Figure 405.4
WATER CLOSET FLOOR FLANGE

405.4.1 Floor flanges. Floor flanges for water closets or similar fixtures shall not be less than 0.125 inch (3.2 mm) thick for brass, 0.25 inch (6.4 mm) thick for plastic, and 0.25 inch (6.4 mm) thick and not less than a 2-inch (51 mm) caulking depth for cast-iron or galvanized malleable iron.

Floor flanges of hard lead shall weigh not less than 1 pound, 9 ounces (0.7 kg) and shall be composed of lead alloy with not less than 7.75-percent antimony by weight. Closet screws and bolts shall be of brass. Flanges shall be secured to the building structure with corrosion-resistant screws or bolts.

❖ Closet bolts must be made of brass because of its strength and corrosion resistance and because they alone secure the water closet in place. Closet bolts are commonly available in other materials such as nylon and brass-plated steel; however, the code is specific on the use of brass for closet bolts.

Closet flanges must be firmly secured to the building structure because they receive the closet bolts that hold down the water closet. This ensures that the water closet is adequately supported and no load is transferred to any piping material attached to the flange. The fasteners securing the closet flange to the floor must also be corrosion resistant so that there is not a reduction of strength of the fasteners over time. Failure of these fasteners can allow the water closet to shift around and break down the wax ring seal resulting in undetectable water leakage under the water closet. Since the secure mounting of the water closet depends on the secure connection to the flange with brass closet bolts, it also depends on the secure mounting of the closet flange to the structure. The use of nails or drywall screws for installing closet flanges on wood subfloors is not code compliant and can result in structural damage soon after the initial installation.

With respect to securing closet flanges to concrete slabs, self-tapping, hex head, alloy carbon steel concrete screws are frequently used. The code does not define corrosion-resistant materials; therefore, it is unclear if alloy carbon steel has sufficient corrosion-resistance to satisfy the intent of this code section or provide for a long service life. Other concrete attachment methods include drop-in type lead or plastic anchors with flat head brass or stainless steel sheet metal screws. Solid brass and stainless steel screw materials have historically performed well in this application.

Occasionally, in light-frame construction, the required location for a water closet flange interferes with a structural member below the subfloor. Typically, if the structural member is "dead center" of where the water closet flange is to be located, the only two options are to either move the location of the water closet flange or move/modify the structural member. In some of these interference situations, only a small portion of the required hole in the subfloor is blocked by the structural member. Since the code does not prohibit the use of an "offset closet flange" in these situations, they can be used just as long as the structural member does not require notching in excess of the limitations. See the entire commentary for Section 307, Structural Safety, for further information.

405.4.2 Securing floor outlet fixtures. Floor outlet fixtures shall be secured to the floor or floor flanges by screws or bolts of corrosion-resistant material.

❖ Floor outlet fixtures, such as water closets and bidets, are held in place by bolts or screws that attach to the floor structure or to a drain pipe floor flange, which is attached to the floor structure. These bolts and screws could be exposed to moisture and, therefore, must be corrosion resistant. The floor outlet fixture screws and bolts are usually constructed of brass (see commentary, Section 405.4.1).

405.4.3 Securing wall-hung water closet bowls. Wall-hung water closet bowls shall be supported by a concealed metal carrier that is attached to the building structural members so that strain is not transmitted to the closet connector or any other part of the plumbing system. The carrier shall conform to ASME A112.6.1M or ASME A112.6.2.

❖ A wall-hung water closet differs greatly in structural design from a floor-mounted water closet because the structural load is transferred to the wall; the floor provides no support. The hanger or carrier for wall-hung water closets must distribute the structural load to the building structural elements without transferring any load to the plumbing piping system. These off-the-floor water closet support assemblies consist of hardware to mount and connect the water closet, the fitting and conduit component means for conveying waste, the structural anchor assembly carrier and necessary gasketing. The design of these assemblies must comply with Section 706.3 (see commentary, Section 706.3).

The referenced standard ASME A112.6.1 or ASME A112.6.2 contains design and test requirements for the loading of the hanger (see commentary, Figure 405.4.3).

405.5 Water-tight joints. Joints formed where fixtures come in contact with walls or floors shall be sealed.

❖ This section addresses the surface connection of the fixture to the floor or wall. The point of contact does not form a joint with any piping material; however, it is still an important area to seal where a fixture is installed to prevent a concealed fouling surface.

The contact joint is sealed with a substance such as plaster-of-Paris, grout or silicone caulking. The exact material selected to seal the joint should be capable of withstanding any anticipated movement of the fixture during its normal use. The seal prevents the accumulation of moisture, dirt or vermin that could lead to the growth of bacteria and structural deterioration in the concealed space [see commentary, Figures 405.5(1), 405.5(2) and 405.5(3)].

Figure 405.5(2)
VIOLATION: LAVATORY NOT ATTACHED OR SEALED TO WALL

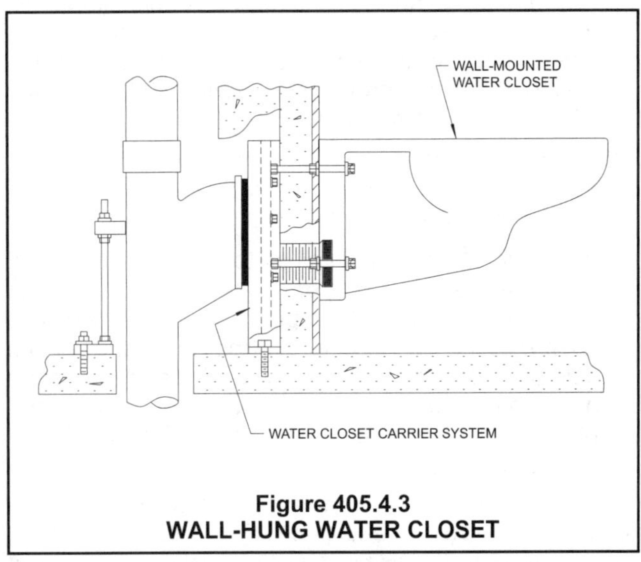

Figure 405.4.3
WALL-HUNG WATER CLOSET

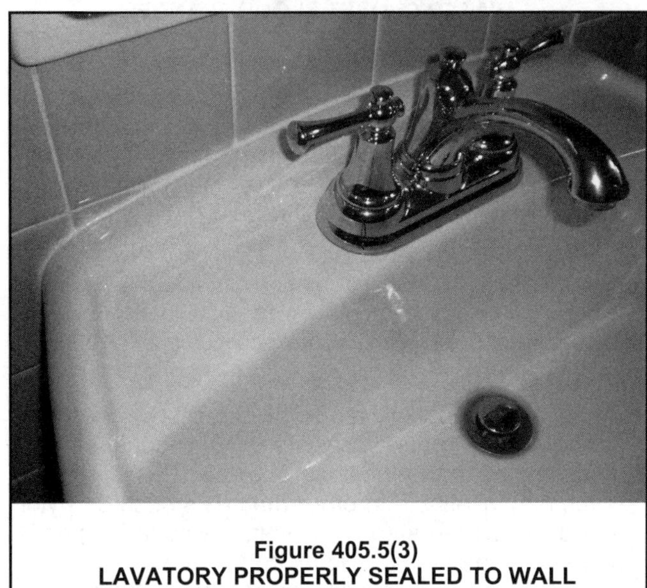

Figure 405.5(3)
LAVATORY PROPERLY SEALED TO WALL

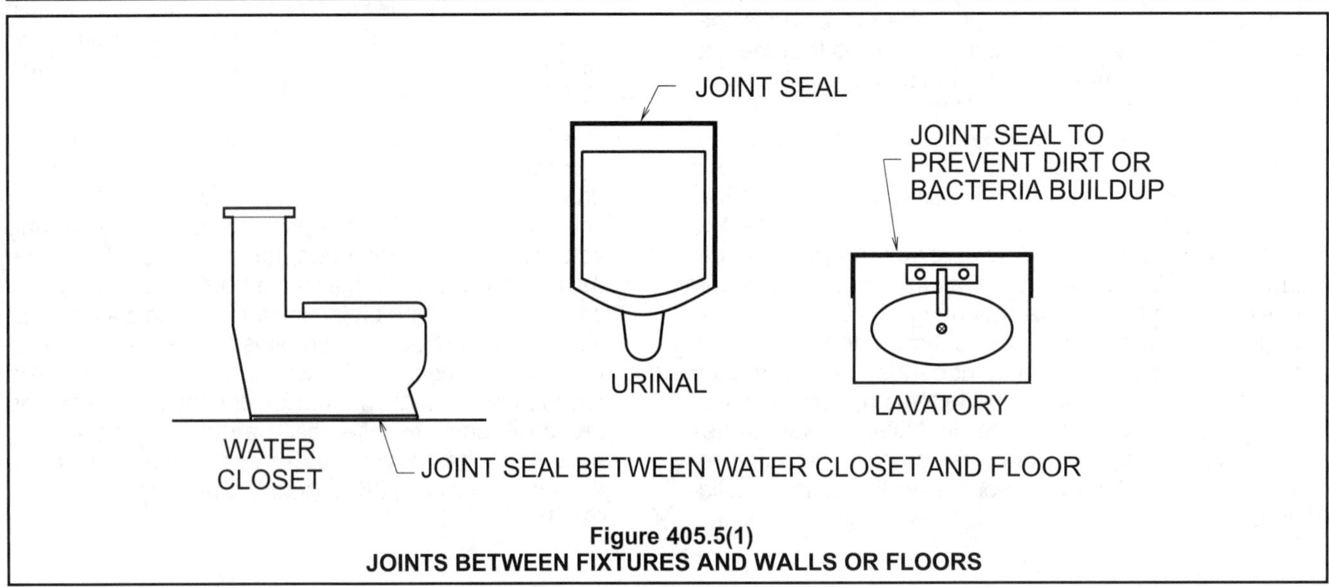

Figure 405.5(1)
JOINTS BETWEEN FIXTURES AND WALLS OR FLOORS

405.6 Plumbing in mental health centers. In mental health centers, pipes or traps shall not be exposed, and fixtures shall be bolted through walls.

❖ In certain mental health centers, patients are considered a danger to themselves; thus, special care is required to secure the plumbing fixtures and piping in these facilities.

405.7 Design of overflows. Where any fixture is provided with an overflow, the waste shall be designed and installed so that standing water in the fixture will not rise in the overflow when the stopper is closed, and no water will remain in the overflow when the fixture is empty.

❖ An overflow is provided for certain fixtures to prevent them from flooding (overflowing) the area when the stopper is closed; therefore, the overflow must not be affected by the stopper, thus permitting water to flow through the overflow to the drainage pipe.

Common fixtures having an overflow are bathtubs and lavatories. The overflow is a precaution provided for fixtures used without close observation, such as a bathtub. A person preparing to use a bathtub will start the water and usually continue to do something else while the bathtub fills. If the person does not return in time or is delayed, the overflow drains the excess water, preventing the bathroom from flooding (see Commentary Figure 405.7). Note that many faucets are capable of supplying water at rates that can exceed the flow-rate capacity of some overflows. In such cases, this does not prevent the fixture from overflowing.

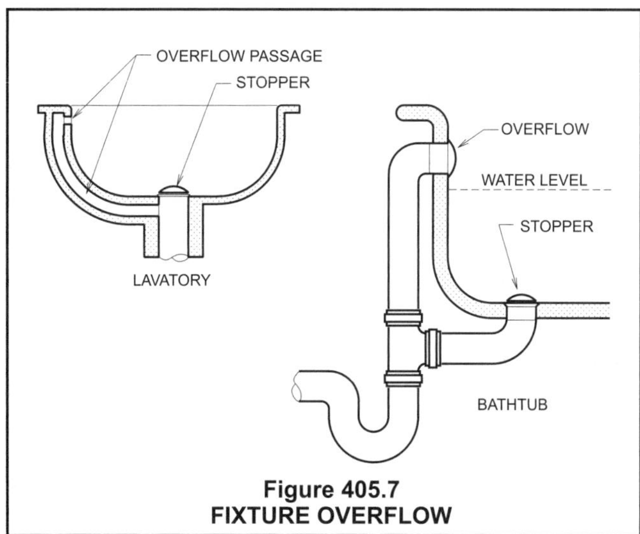

Figure 405.7
FIXTURE OVERFLOW

405.7.1 Connection of overflows. The overflow from any fixture shall discharge into the drainage system on the inlet or fixture side of the trap.

> **Exception:** The overflow from a flush tank serving a water closet or urinal shall discharge into the fixture served.

❖ The overflow must also be protected by the trap seal, thus preventing sewer gases from the drainage system from entering the building.

The exception recognizes that flush tank overflows discharge to the fixture served by the flush tank. Flush tanks normally have an open standpipe that allows any overflow to discharge directly into the water closet or urinal.

405.8 Slip joint connections. Slip joints shall be made with an *approved* elastomeric gasket and shall only be installed on the trap outlet, trap inlet and within the trap seal. Fixtures with concealed slip-joint connections shall be provided with an *access* panel or utility space at least 12 inches (305 mm) in its smallest dimension or other *approved* arrangement so as to provide *access* to the slip joint connections for inspection and repair.

❖ An access panel is required only for slip-joint connections using an elastomeric gasket to seal the joint. Fixtures such as bathtubs and bidets commonly use slip joints in concealed spaces that would require access in accordance with this section. Access is not required should all the joints be soldered, solvent cemented or screwed joints. For example, Schedule 40 ABS and PVC tub waste and overflow assemblies are now widely available and allow all joints to be solvent cemented so that an access panel is not required.

405.9 Design and installation of plumbing fixtures. Integral fixture fitting mounting surfaces on manufactured plumbing fixtures or plumbing fixtures constructed on site, shall meet the design requirements of ASME A112.19.2M or ASME A112.19.3M.

❖ The design and installation of site-built plumbing fixtures and the installation of fixture fittings are addressed through ASME A112.19.2M and ASME A112.19.3M. The air gap requirements for fixture fittings must be designed and installed so that the backflow protection is maintained without compromising the integrity of the design of the plumbing fixture. All openings and outlets must be protected by an air gap between the fixture fitting and the fixture flood level rim as specified in Table 608.15.1.

SECTION 406
AUTOMATIC CLOTHES WASHERS

406.1 Approval. Domestic automatic clothes washers shall conform to ASSE 1007.

❖ The performance requirements (typically noted in a referenced standard) and water and waste connection provisions for automatic clothes washers are contained in this section. The most important aspect of ASSE 1007 is its requirement for protection of potable water against backflow. Domestic automatic clothes washers conforming to ASSE 1007 have the potable water supply protected by an integral air gap or backflow preventer. Because the machine has its own built-in backflow protection, an additional backflow preventer installed in accordance with Section 608 is not required (see commentary, Section 406.2).

406.2 Water connection. The water supply to an automatic clothes washer shall be protected against backflow by an *air gap* installed integrally within the machine conforming to ASSE 1007 or with the installation of a backflow preventer in accordance with Section 608.

❖ Any appliance or piece of equipment with a water supply connection is a potential source of backflow contamination to the potable water supply. An automatic clothes washer is no exception. As stated in Section 406.1, conformance to ASSE 1007 will verify that integral backflow protection is present in domestic clothes washers (see commentary, Section 406.1). It should be noted that some commercial clothes washers are not equipped with integral backflow prevention, and the water supply to these units therefore must be protected with a backflow preventer in accordance with Section 608. Prior to installing the automatic clothes washer, the installer should verify whether the appliance conforms to ASSE 1007 and that an integral backflow protection device is present and functional.

406.3 Waste connection. The waste from an automatic clothes washer shall discharge through an *air break* into a standpipe in accordance with Section 802.4 or into a laundry sink. The trap and *fixture drain* for an automatic clothes washer standpipe shall be a minimum of 2 inches (51 mm) in diameter. The automatic clothes washer *fixture drain* shall connect to a *branch* drain or drainage *stack* a minimum of 3 inches (76 mm) in diameter. Automatic clothes washers that discharge by gravity shall be permitted to drain to a waste receptor or an *approved* trench drain.

❖ The minimum required size of a trap and fixture drain for a clothes washer is 2 inches (51 mm). In the 2006 IPC, this section was changed to require that a clothes washer fixture drain be connected to a 3-inch (76 mm) drain or stack at the point where other fixture flow merged with the automatic clothes washer (ACW) flow. An ACW fixture drain of 2 inches (51 mm) has no length limit, but, at the point where the ACW fixture drain becomes a branch drain, an increase in size is required. Investigations by the proponent of this change revealed that modern automatic clothes washers can pump waste at flows approaching 21 gallons (79.5 liters) per minute. These higher flows caused trap siphonage and sudsing in other fixture drains downstream of the washing machine fixture drain connection where the drain piping remained a 2-inch (51 mm) size after the connection. Increasing the fixture branch drain size or stack size to 3 inches (76 mm) eliminated these problems in all cases.

Where a clothes washer discharges to a laundry sink, the high rate of discharge does not affect the drain size for the laundry sink. The capacity of the laundry sink allows a slower discharge flow rate to the fixture drain. A clothes washer must discharge to the drainage system through an air break. A direct connection would allow sewage to back up into the appliance in the event of a blockage in the drain pipe. A clothes washer is usually discharged to an individually trapped standpipe or a laundry sink. The use of a laundry sink as a waste receptor is common practice. For commercial washers that discharge by gravity, trench drains are a suitable discharge point. Some large commercial machines drain by gravity by opening a solenoid-actuated valve located at the low point of the drum. Such machines could not use a standpipe (see Section 412 and commentary, Figure 406.3).

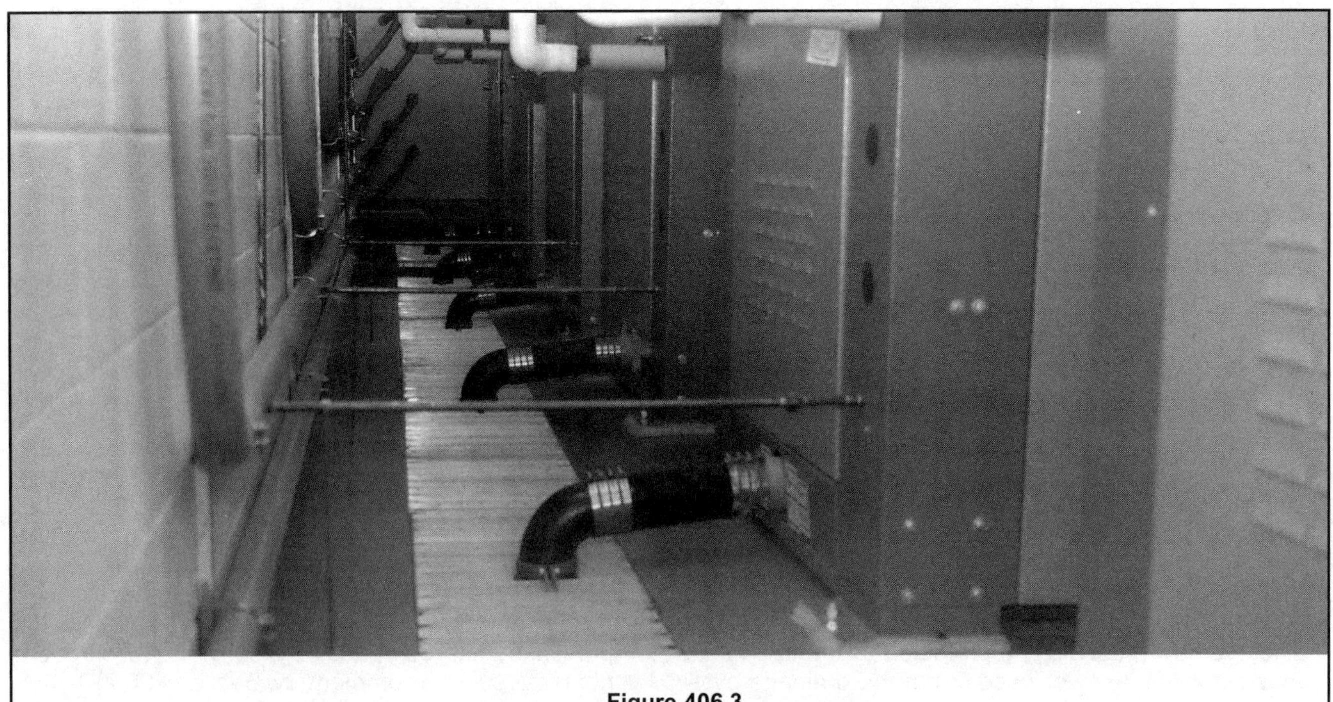

Figure 406.3
CLOTHES WASHER TRENCH DRAIN

SECTION 407 BATHTUBS

407.1 Approval. Bathtubs shall conform to ANSI Z124.1, ASME A112.19.1M, ASME A112.19.4M, ASME A112.19.9M, CSA B45.2, CSA B45.3 or CSA B45.5.

❖ Bathtubs are available in many materials, including enameled cast iron, porcelain-enameled formed steel or plastic. Plastic bathtubs can be constructed of a variety of synthetic materials, including acrylonitrile butadiene styrene (ABS), polyvinyl chloride (PVC), gel-coated fiberglass-reinforced plastic, acrylic, cultured marble cast-filled fiberglass, polyester and cultured marble acrylic. Because they can be equipped with a shower, bathtubs manufactured in accordance with ASME A112.19.4 must be tested for slip-resistance in accordance with ASTM F 462, reducing the possibility of an accidental fall. A slip-resistant surface is required for enameled cast iron and optional for porcelain-enameled formed steel and plastic bathtubs.

Plastic bathtubs are subjected to an ignition test. Plastic bathtubs must be impregnated with fire-retardant chemicals, reducing their potential fuel contribution during a fire and protecting them against accidental fire from a plumber's torch. However, it should be noted that bathtubs need not be subjected to flame spread and smoke-density tests such as ASTM E 84.

Bathtubs are supplied in a variety of shapes and sizes designed for greater comfort in accommodating the human body. Some bathtubs are larger to allow for use by more than one individual at the same time. These larger bathtubs can require special structural assemblies to support the load. A bathtub can have a water capacity in excess of 300 gallons (1136 L). The weight of the water, the bathtub unit itself, plus two individuals can add up to more than 2,900 pounds (1317 kg). This can cause structural failures because live loading for a typical residential installation is 40 pounds per square foot (195 kg/m^2); thus, for a 3-foot by 6-foot (914 mm by 1829 mm) tub, the live load is assumed to be only 720 pounds (340 kg).

407.2 Bathtub waste outlets. Bathtubs shall have waste outlets a minimum of $1^1/_2$ inches (38 mm) in diameter. The waste outlet shall be equipped with an *approved* stopper.

❖ This section requires that waste outlet drain assemblies for bathtubs be not less than $1^1/_2$ inches (38 mm) in diameter. While all bathtubs will necessarily have a drain opening in the floor of the tub for a waste outlet assembly to be installed. Bathtubs are not required by the code to have an opening for a waste overflow assembly. However, an overflow opening might be required by the standard to which the product is made to comply or might be provided at the manufacturer's discretion.

A device for blocking the flow of tub water from the drain must be provided. The device could be as simple as a "pull out" type rubber plug that tightly fits in the opening of the waste assembly or other approved methods such as trip lever waste units, cable-actuated closure devices or metal push-pull or pop up devices.

It should be noted that a tub waste (and overflow) assembly using slip joints must be located in areas that can be accessed in accordance with Section 405.8.

407.3 Glazing. Windows and doors within a bathtub enclosure shall conform to the safety glazing requirements of the *International Building Code*.

❖ Glass (clear or otherwise), light transmitting ceramic or light transmitting plastic used in doors, windows and enclosures of hot tubs, whirlpools, saunas, steam rooms, bathtubs and showers where the bottom edge is less than 60 inches (1524 mm) above the floor surface of the fixture or floor of the enclosed fixture must comply with the safety glazing requirements of IBC Chapter 24. Due to the likely presence of water on the standing or walking surfaces within these types of enclosed areas, there is high probability of user slips and falls such that the user might thrust his or her arm or leg through the panel. Materials that shatter into large, sharp chards would be extremely dangerous if installed immediately adjacent to these areas. A glazing material is considered "safe" if it complies with the impact testing requirements of the specific standards for the type of product. Glass block masonry walls are covered by Chapter 21 of the IBC and not the safety glazing requirements of IBC Chapter 24.

Partitions and windows outside of the enclosure need not be of safety glazing, except as might be required by other sections of the IBC. Exterior windows in exterior walls that serve as part of the tub enclosure must be of safety glazing unless the bottom exposed edge of the glass is at least 60 inches (1524 mm) above the tub floor surface. Note that the height dimension is taken from the tub floor and not from the finished floor outside the tub area or from a deck that might surround a tub. A sitting ledge or other horizontal surface provided around a tub next to large glass windows does not always decrease the hazard of someone falling into the windows; therefore, safety glazing would likely be required. If the width of the sitting ledge is increased such that it becomes more of a walking surface, it might not be considered to be part of the tub enclosure; however, the glazing would then have to be evaluated in accordance with the other safety glazing requirements of the IBC. Good judgment must be used in applying this section while keeping the code intent in mind.

Note that with regard to bathtub and shower enclosures, the requirement for safety glazing is not dependent on the area (size) of the glazing panels or panes. The intent is to require safety glazing wherever it is possible for the user to fall into or thrust a limb into a glass panel.

407.4 Bathtub enclosure. Doors within a bathtub enclosure shall conform to ASME A112.19.15.

❖ A bathtub with a pressure-sealed door allows the bather to enter the fixture at approximately the same level as the floor. The door can be opened to allow the bather easier access to the bathtub. Once the bather is inside the fixture, the door is closed and the water is turned on. Sensor switches, which note the presence of water entering the vessel, activate an air pump that pressurizes the door seal, thereby keeping the water within the bathing vessel. Upon completion of bathing, the door can be opened only when the bathwater is drained from the vessel, thereby allowing the door to open.

SECTION 408 BIDETS

408.1 Approval. Bidets shall conform to ASME A112.19.2M, ASME A112.19.9M or CSA B45.1.

❖ This section of the code establishes the standards for material requirements and water connection provisions for bidets.

The code references standards for both vitreous china and nonvitreous ceramic plumbing fixtures. The standards require that the bidet meet material, dimensional and performance requirements.

The bidet is a small bathing fixture used by both sexes. It is not designed for the elimination of human waste. Although used primarily to clean the perineal area, it can also be used for bathing other parts of the body, including feet. A bidet is normally equipped with a water spray, designed for external body cleaning, that directs a jet of water upward to the body.

408.2 Water connection. The water supply to a bidet shall be protected against backflow by an *air gap* or backflow preventer in accordance with Section 608.13.1, 608.13.2, 608.13.3, 608.13.5, 608.13.6 or 608.13.8.

❖ Most bidets present the hazard of a submerged outlet because the typical design uses a water supply nozzle located below the flood level rim of the fixture. This outlet is considered to be subject to high-hazard backsiphonage. The water supply to a bidet must be protected from backflow in accordance with the provisions of Section 608 (see commentary, Section 608).

408.3 Bidet water temperature. The discharge water temperature from a bidet fitting shall be limited to a maximum temperature of 110°F (43°C) by a water temperature limiting device conforming to ASSE 1070.

❖ Prior to this edition of the code, users of bidets were at significant risk for scalding of sensitive areas of the body, especially if use of other fixtures caused water pressure fluctuations resulting in loss of cold water flow. A temperature-limiting device meeting ASSE 1070 is required to prevent scalding (see Commentary Table 424.4).

SECTION 409 DISHWASHING MACHINES

409.1 Approval. Domestic dishwashing machines shall conform to ASSE 1006. Commercial dishwashing machines shall conform to ASSE 1004 and NSF 3.

❖ The referenced standards for both domestic and commercial dishwashing machines are contained in this section, as are water supply and waste connection requirements. The most important aspect of the ASSE 1006 standard for residential or domestic-type dishwashing machines is the protection of the potable water supply against backflow. If the unit conforms to the standard, the water supply is adequately protected against backflow by an internal integral backflow prevention device. Any additional precautions as specified in Section 608 are unnecessary. Prior to installing the residential or domestic-type dishwashing machine, the installer should verify that the appliance does in fact conform to ASSE 1006 and that the internal integral backflow prevention device is present and functional.

The water supply to a commercial dishwashing machine must be protected against backflow by an air gap or a backflow preventer. An air gap, installed in accordance with ASSE 1004, must be twice the diameter of the water supply inlet and not less than 1 inch (25 mm). Where a backflow preventer (vacuum breaker) is installed, it must comply with ASSE 1001 and ASSE 1004 for installation requirements. A commercial dishwashing machine must also conform to NSF 3, which establishes minimum public health and sanitation requirements for the materials, design, construction and performance of spray-type commercial dishwashing machines and their related components. This standard references ASSE 1004 for water supply protection and gives provisions for the location of water supply piping and valves in relation to the machine tank or hood.

409.2 Water connection. The water supply to a dishwashing machine shall be protected against backflow by an *air gap* or backflow preventer in accordance with Section 608.

❖ To protect the water supply to a residential or domestic-type dishwashing machine from backflow, the machine must be equipped with integral backflow protection as required by ASSE 1006, or the potable water supply must have either a backflow preventer in accordance with Section 608 or an air gap (see commentary, Section 409.1).

The water supply to a commercial dishwashing machine conforming to ASSE 1004 is required by the standard to be protected by an air gap or a vacuum breaker conforming to ASSE 1001. The air gap must be located on the outside of the machine and protected against suds, spray, splash or flooding. The vacuum breaker must be installed on the discharge side of the control valve, a minimum of 6 inches (152 mm) above the flood level rim.

409.3 Waste connection. The waste connection of a dishwashing machine shall comply with Section 802.1.6 or 802.1.7, as applicable.

❖ Requirements for waste connections are described in Sections 802.1.6 and 802.1.7.

SECTION 410 DRINKING FOUNTAINS

410.1 Approval. Drinking fountains shall conform to ASME A112.19.1M, ASME A112.19.2M or ASME A112.19.9M and water coolers shall conform to ARI 1010. Drinking fountains and water coolers shall conform to NSF 61, Section 9. Where water is served in restaurants, drinking fountains shall not be required. In other occupancies, where drinking fountains are required, water coolers or bottled water dispensers shall be permitted to be substituted for not more than 50 percent of the required drinking fountains.

❖ There have been numerous changes to this section throughout many editions of the *International Plumbing Code* and the legacy codes which have ultimately led to some confusion as to what equipment constitutes "meeting the intent" of providing a drinking fountain. As far back as the 1968 BOCA legacy code, the term "water cooler" is found in this section's text. In later editions, the term "bottled water cooler" shows up in the text "as a substitution for, or an equivalent to" a drinking fountain. The 1999 Supplement to the IPC changed the bottled water term "cooler" to "dispenser" because it was felt that the intent of the code was to only supply water, not chilled or cooled water. The allowed substitution of all drinking fountains with bottled water dispensers continued until the 2003 IPC which required that bottled water dispensers be allowed to substitute for only up to 50 percent of the required number of drinking fountains. That change was necessary to prevent a situation of no drinking water should the bottle run dry or the bottled water service be discontinued. Requiring drinking fountains connected to the potable water distribution system ensures that there would always be a source of drinking water available to building occupants. A change for the 2006 IPC reintroduced the term "water cooler" in association with "bottled water dispenser" so as to specifically allow "bottled water coolers" in addition to bottled water dispensers as an allowed substitute. Thus, from this historical perspective, the code's intent for the term "water cooler" appears to mean a unit that dispenses chilled bottled water and the term "bottled water dispenser" means a unit that dispenses room temperature bottled water.

Each of the ASME standards referenced in this section are primarily concerned with basic dimensional requirements and quality tests for many types of plumbing fixtures made of vitreous china, nonvitreous ceramic, and porcelain enameled cast iron. Only two of these standards have some reference to drinking fountains. Those standards simply require that the bubbler head mounting base be above the rim (flood rim level) of the bowl and provide a few dimensions for the basic wall mount (with rear boss), semi-recessed, and fully recessed drinking fountains that were commonly installed in buildings many years ago.

Drinking fountains are not required to provide chilled water. Whether designers, contractors and building officials realize this or not, most modern drinking fountains that supply just "cold water" as supplied from a potable water system do not comply with any standards listed in this section (other than NSF 61 Section 9). Where a "drinking fountain" supplies chilled water, this unit is called a "bubbler-type pressure water cooler" (as defined by the ARI 1010 standard). It is assumed that even though the code requirement is for drinking fountains, bubble-type pressure water coolers are equivalent. The use of the term "water cooler" in "bubbler-type pressure water cooler" is not to be confused with "bottled water cooler" with respect to the substitution language in this code section. Although both are "water coolers" that must comply with ARI 1010, the term "water coolers" used in conjunction with the limitation on substitution for drinking fountains means only "bottled water coolers" as previously explained.

IBC Chapter 11 for accessible design of building spaces specifies the configuration of the drinking fountains that are to be installed. Where the required number of drinking fountains (from Code Table 403.1) to be installed is equal to one, IBC Chapter 11 requires two drinking fountains, one installed at a height for standing persons and one installed for use by wheelchair-seated persons. (An exception in the IBC allows a single "combination unit" that has 2 spouts at different heights, each with a separate bowl.) Only the IPC allows for substitution of the required number of drinking fountains (from Code Table 403.1) by up to 50 percent with bottled water dispensers (or bottled water coolers). Chapter 11 of the IBC is specific in requiring that whatever is the number of drinking fountains to be installed, they are to be of accessible design for standing persons and wheelchair-seated persons. The IBC does not prohibit bottled water dispensers, bottled water coolers or any other type of freestanding unit from being substituted for accessible drinking fountains.

This code section does not require restaurants to have drinking fountains as long as "water is served". Even though the language of this exception has evolved, the intent of the code has always been that a restaurant provide drinking water free of charge in order to take advantage of this exception. Where a restaurant provides purchased beverages in disposable cups, drinking water must be still be free of charge even though the restaurant incurs a cost for the disposable cups.

The required location for drinking fountains is often questioned but the IPC continues to be silent on this subject. Drinking fountains are not required to be near public toilet facilities or necessarily within the same travel path requirements for public toilets. Because all installed drinking fountains are required to be of accessible design (either standing or wheelchair-seated), they must be on an accessible route (see IBC Chapter 11). Drinking fountains must also be available to all of the occupants that which create the demand for the drinking fountains. For example, one drinking fountain serving several tenant spaces cannot be located in one of the tenant spaces as the occupants of the other tenant space might not have unlimited access. One drinking fountain in the public area of a building (not within any tenant space) can serve the needs of all tenant spaces, as long as one drinking fountain will serve the total occupant load. Could all required drinking fountains for a building be located on one floor of a multifloor building? Perhaps, but this is not often done in order to provide a higher level of convenience for the occupants. In multistory buildings, the requirement for toilet facilities to be not more than one floor above or below usually creates an automatic opportunity for the placement of drinking fountains. In large, single-floor buildings, the locations for required drinking fountains are usually distributed equidistant from each other and with respect to where most people might usually travel. Locating a drinking fountain in areas that are traveled infrequently usually results in vandalism of the unit.

The commentary in Section 405 discusses drinking fountain height and clear floor area for wheelchair-seated persons. Because the IBC references ICC A117.1 and that standard requires a specific height for drinking fountains for standing persons, that requirement is repeated here: For a drinking fountain to be used by standing persons, the spout outlet of a drinking fountain must be not less than 38 inches (965 mm) and not greater than 43 inches (1090 mm) above the finished floor.

410.2 Prohibited location. Drinking fountains, water coolers and bottled water dispensers shall not be installed in *public* restrooms.

❖ Installation of drinking fountains in public toilet rooms is prohibited, thereby preventing the contamination of drinking water. Biological aerosols present in these rooms can contaminate exposed surfaces. The drinking fountain must be protected against surface contamination.

SECTION 411
EMERGENCY SHOWERS AND EYEWASH STATIONS

411.1 Approval. Emergency showers and eyewash stations shall conform to ISEA Z358.1.

❖ Water and waste connection requirements for specialized plumbing fixtures, such as emergency showers and eyewash stations, are provided in this section.

The code does not require the installation of emergency eyewash stations and showers. This section has been provided to address them where they are installed. Usually, governmental agencies regulating worker and building occupant safety establish the circumstances under which these fixtures must be installed.

Emergency showers and eyewash stations are installed in areas where an individual may come in contact with substances that are immediately harmful to the body. These emergency devices provide an instantaneous deluge of water to help neutralize any adverse reactions. Emergency showers and eyewash stations are located where people are exposed to chemicals, acids, nuclear products and fire. The water supply requirements for an emergency eyewash station depend on the manufacturer's requirements for the specific equipment installed (see Commentary Figure 411.1).

Although cold-flushing fluid temperatures provide immediate cooling after chemical contact, prolonged exposure to cold fluids affects the ability to maintain adequate body temperature and can result in the premature cessation of first aid treatment. The ISEA standard referenced (which is also ANSI Z358.1) covers minimum performance and use requirements for emergency shower, eyewash equipment, eye/face wash equipment, handheld drench hoses and combination shower and eyewash or eye/face equipment.

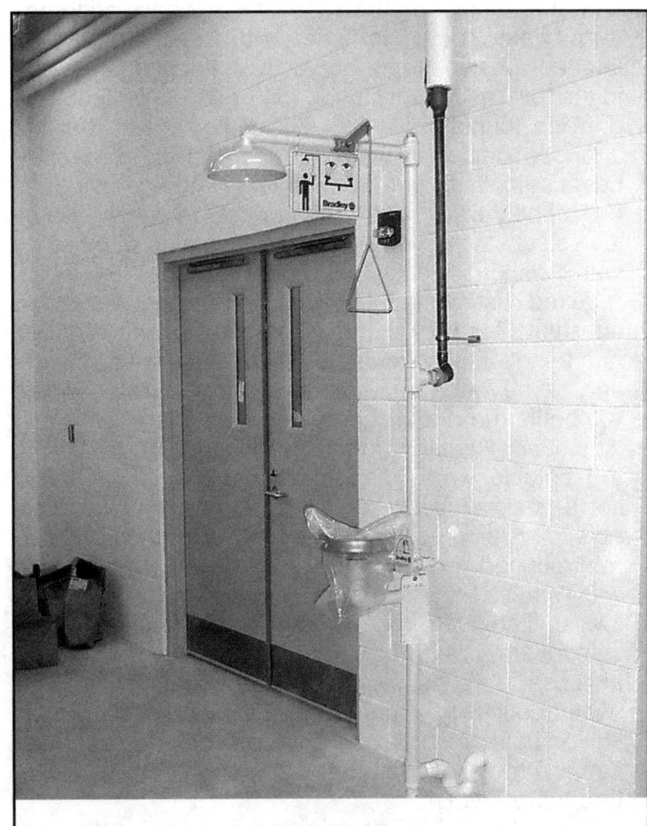

Figure 411.1
EMERGENCY SHOWER AND EYEWASH STATION
(Photo Courtesy of B&I Contractors, Inc.)

The mandatory portion of the standard requires that the delivered flushing fluid temperature be tepid and defines tepid as being lukewarm; no temperature range is given. However, the voluntary (i.e., not enforceable) section of this standard contains a suggested temperature range. Ultimately, each installer of such equipment should consult a medical advisor for specific temperature settings based on the types of chemical exposure anticipated.

It is unclear whether emergency showers under this section are considered to be individual showers according to this code. Individual showers would require a mixing valve conforming to ASSE 1016 or CSA B125 (see Section 424.3), however, just any ASSE 1016 valve may not be suitable for the application because of the large flows and instantaneous temperature control needed. With respect to eyewash stations, the code and the ISEA/ANSI standard are silent on what type of temperature control is necessary.

411.2 Waste connection. Waste connections shall not be required for emergency showers and eyewash stations.

❖ The designer should take into account, where possible, the potential discharge volume from an emergency shower or eyewash with respect to cleanup procedures and property damage. Consideration should be given to the proper disposal of possible hazardous waste flushed from the user of emergency showers and eyewash stations.

SECTION 412
FLOOR AND TRENCH DRAINS

412.1 Approval. Floor drains shall conform to ASME A112.3.1, ASME A112.6.3 or CSA B79. Trench drains shall comply with ASME A112.6.3.

❖ Except as required by Section 412.4, floor drains and trench drains are typically installed at the discretion of the design professional where an accumulation of waer on a floor or other horizontal area could create a personnel hazard or could cause damage to building finishes or structures.

ASME A112.6.3 specifies the design and quality criteria for floor drains and trench drains for buildings that are typically nonresidential. Materials covered by this standard are cast iron, copper alloy, leaded nickel bronze, ABS and PVC. Note that stainless steel is not included in the list of materials. ASME A112.3.1 specifies performance and installation requirements for floor drains of stainless steel material. Even though this standard also covers trench drains of stainless steel, the code references this standard only for floor drains. Trench drains must comply with ASME A112.6.3 which covers the design and quality criteria for drains made of cast iron, copper alloy, leaded nickel bronze, ABS and PVC. Because stainless steel is not in this list of materials, trench drains of stainless steel must be approved by the code official under Section 105.2 of this code.

CSA B79 specifies the requirements for residential floor drains constructed of cast iron, ductile iron, bronze, nickel bronze, white bronze, aluminum alloy, plastic and stainless steel.

Field-constructed floor drains and trench drains made of concrete are not covered by the above standards and therefore must be approved by the code official under Section 105.2, "Alternate materials, methods and equipment."

A floor drain is for draining liquids from indoor floor areas within a building. Floor drains are prohibited from connecting to a storm drain (see Section 1104.3) because building occupants typically assume that floor drains can be used for the disposal of any liquid. Therefore, a floor drain must be connected through a trap to a sanitary sewer system or a private sewage disposal system. Because Section 1101.3 prohibits storm water from being drained into a sanitary sewer, a floor drain must not be used for storm water disposal. For example, a rain water conductor must not discharge to a floor drain.

Trench drains are used in a variety of indoor and outdoor applications (see Commentary Figure 412.1). They are frequently used to intercept storm water at the toe of wide outdoor sloped areas such as driveways and parking lots. Where used to collect and convey only storm water, trench drain outlets are not re-

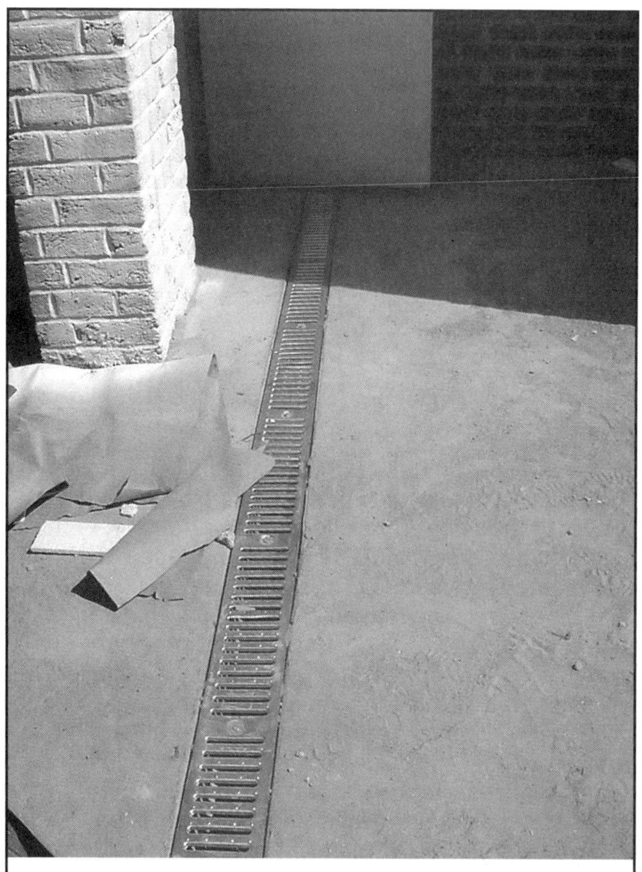

Figure 412.1
TRENCH DRAIN
(Photo Courtesy of B&I Contractors, Inc.)

quired to be connected to a trap prior to connection to a storm water system. Trench drains are also used indoors where equipment and large areas are frequently washed down for maintenance and sanitization. A trench drain in this application must be connected through a trap to a sanitary sewer system or a private sewage disposal system.

Trench drains are also often located just inside the bay entrances of covered or indoor vehicle repair facilities. The primary intent of a trench drain in this application is to collect rainwater and ice/snow melt runoff from vehicles that are moved into a repair garage area. However, trench drains in these applications must be connected through a trap that is connected to an oil separator. The reason for this is the significant potential for the trench drain to receive grease-laden floor wash water and accidental or intentional spills of vehicle fluids. See Section 1003.4 concerning the requirement for an oil separator.

In vehicle repair facilities, multiple trench drain assemblies, each with an outlet, must not be connected together prior to connecting to a trap. Each outlet from a trench drain is considered to be a separate fixture and must have a trap installed and located in compliance with Section 1002.1. Because all traps must be vented, and trench drains are often used in large open areas without walls, a combination drain and vent system is typically used to connect the traps and carry the flow to the desired point of discharge.

The code does not provide a specific drainage fixture unit value for trench drains. Therefore, Table 709.2 can be used to determine the dfu value.

Commentary Figure 406.3 shows a trench drain serving as the waste discharge point for a commercial laundry washer.

412.2 Floor drains. Floor drains shall have removeable strainers. The floor drain shall be constructed so that the drain is capable of being cleaned. *Access* shall be provided to the drain inlet. Ready *access* shall be provided to floor drains.

> **Exception:** Floor drains serving refrigerated display cases shall be provided with *access*.

❖ The strainer of the floor drain is usually flush with the floor surface to avoid creating a tripping hazard. However, if the floor drain is located in a remote, typically untraveled area such as the corner or edge of a room, a dome-type strainer might be more suitable. All strainers must be removable to enable cleaning of the drain body and rodding of drain. Because floor drains are purposely located in areas where drainage is a necessity, either due to an emergency or normal maintenance, it is imperative that the drain be able to be serviced quickly and easily. Therefore, floor drains require ready access and must not be located under appliances, equipment or shelving, within wall cavities or covered by objects that might obscure their location.

The exception allows for floor drains located under refrigerated display cases because locating the drain exterior to the case footprint would most likely create a tripping hazard.

Some floor drain designs have an integral trap [see Commentary Figures 412.2(1) and 412.2(2)]. If a trap is not integral to the design, the drain must be connected to a trap (see commentary for Sections 1002.1 and 1002.4).

A floor drain equipped with a combination strainer/funnel extending 1 inch (25 mm) or more above the finished floor surface is considered to be a waste receptor, even though it may also function as a floor drain (see Section 802.3).

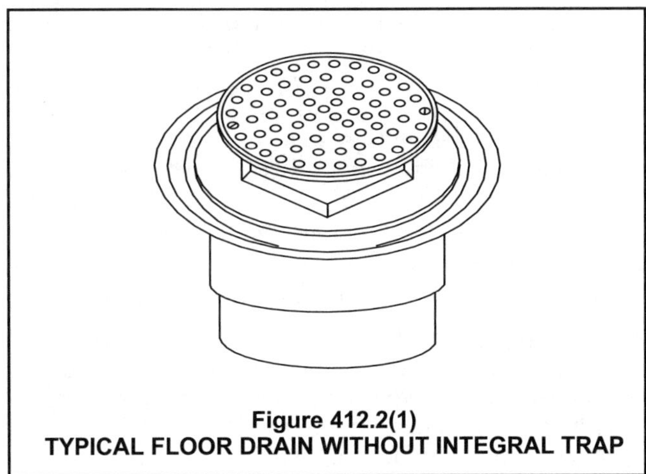

Figure 412.2(1)
TYPICAL FLOOR DRAIN WITHOUT INTEGRAL TRAP

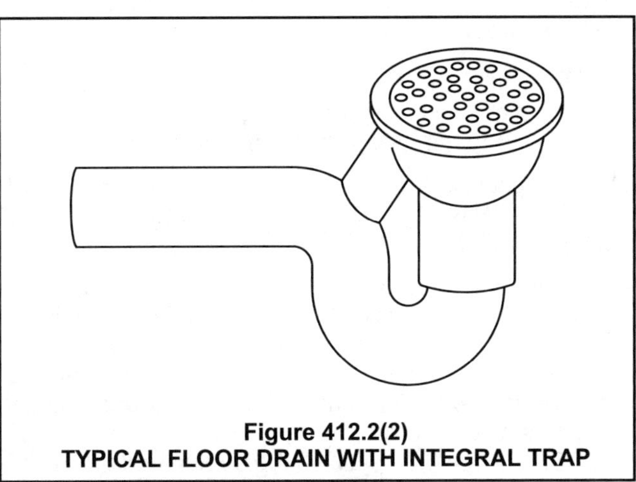

Figure 412.2(2)
TYPICAL FLOOR DRAIN WITH INTEGRAL TRAP

412.3 Size of floor drains. Floor drains shall have a minimum 2-inch-diameter (51 mm) drain outlet.

❖ This section emphasizes the requirements of Table 709.1, which provides the minimum size of the drain outlet for the particular dfu value of the floor drain fixture. It should be noted that Section 412.4 requires a minimum floor drain size of 3 inches (76 mm) in public coin-operated laundries and in central washing facilities of multiple-family dwellings.

412.4 Public laundries and central washing facilities. In public coin-operated laundries and in the central washing facilities of multiple-family dwellings, the rooms containing automatic clothes washers shall be provided with floor drains located to readily drain the entire floor area. Such drains shall have a minimum outlet of not less than 3 inches (76 mm) in diameter.

❖ A floor drain is required in public coin-operated laundries and in central washing facilities of multiple-family dwellings that contain automatic clothes washers to protect against damage from accidental spills, fixture overflows and leakage. A 3-inch (76 mm) floor drain is required because of the potential for a large volume of water discharging to the floor.

This plumbing fixture is considered to be an emergency floor drain that has a dfu value of zero. Such fixtures do not add to the load used to compute drainage pipe sizing because their sole purpose is to serve only in the event of an emergency.

It should be noted that, under normal circumstances, an emergency floor drain does not receive any waste and, therefore, is required to be provided with a means to protect the trap seal from loss by evaporation in accordance with Section 1002.4.

SECTION 413
FOOD WASTE GRINDER UNITS

413.1 Approval. Domestic food waste grinders shall conform to ASSE 1008. Commercial food waste grinders shall conform to ASSE 1009. Food waste grinders shall not increase the *drainage fixture unit* load on the sanitary drainage system.

❖ This section includes standards for both domestic and commercial food waste grinder units, along with waste outlet provisions for both types of units. Water supply requirements for food waste grinders are also provided.

Food waste grinders are designed to reduce most foods, including bones, into small-size particles that will flow through the drainage lines. Discharging fibrous or stringy food such as corn husks, celery, banana skins and onions in a food waste grinder is not recommended. Rather than reduce in size, the fibers tend to bypass the grinding hammers and slip down the drain, creating stoppages in the drainage piping.

Domestic food waste grinders are supplied with water from a sink supply faucet and do not increase the dfu load of a kitchen sink. Domestic food waste grinders complying with ASSE 1008 are not intended for installation in food-handling establishments.

413.2 Domestic food waste grinder waste outlets. Domestic food waste grinders shall be connected to a drain of not less than $1^1/_2$ inches (38 mm) in diameter.

❖ A domestic food waste grinder is installed in conjunction with a kitchen sink. The drain pipe size required is consistent with that for a kitchen sink.

413.3 Commercial food waste grinder waste outlets. Commercial food waste grinders shall be connected to a drain not less than $1^1/_2$ inches (38 mm) in diameter. Commercial food waste grinders shall be connected and trapped separately from any other fixtures or sink compartments.

❖ Historically, the minimum size of a commercial food waste grinder drain has been 2 inches (51 mm). In 2009, this section was changed to allow for a $1^1/_2$ inch (38 mm) drain size as smaller commercial size grinders with $1^1/_2$ inch (38 mm) outlets have become available. To help prevent contamination of other fixtures, commercial food waste grinders must be trapped and independently connected to the drainage system (see Commentary Figure 413.3). Although this section does not specifically state that an indirect connection

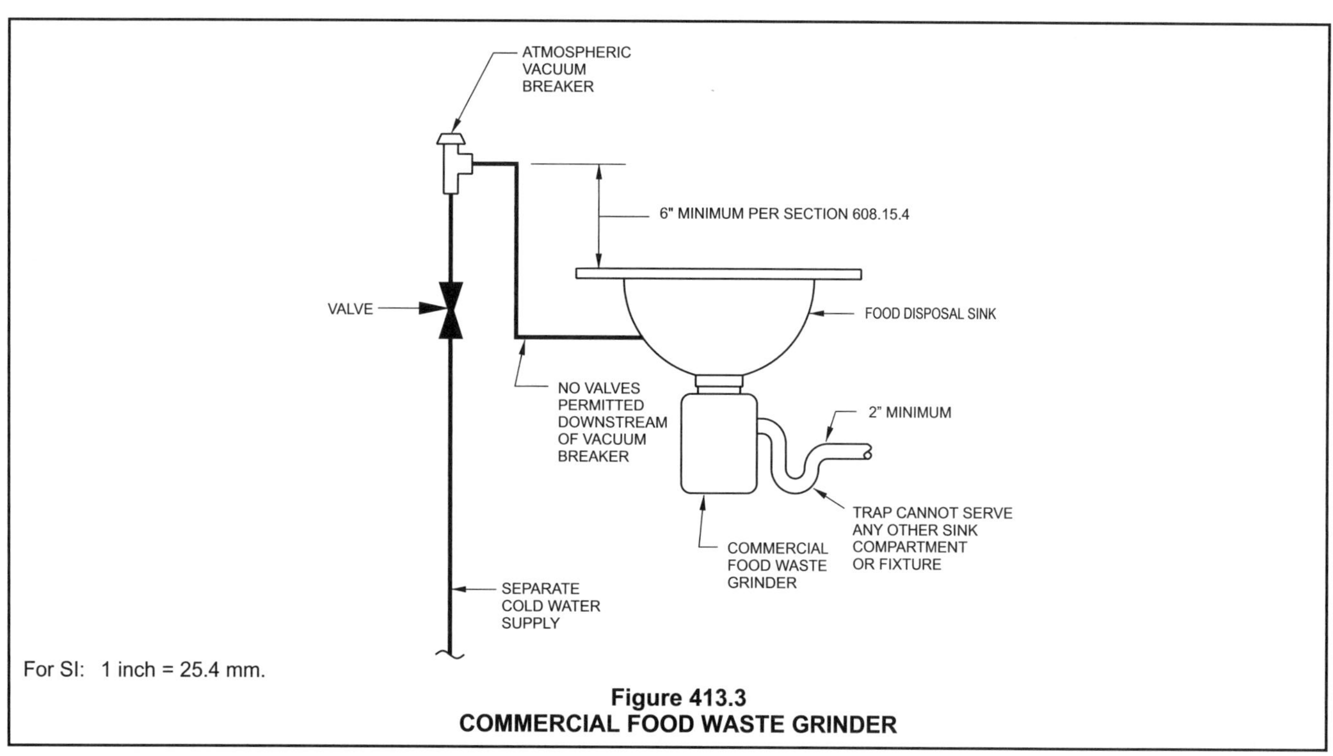

Figure 413.3
COMMERCIAL FOOD WASTE GRINDER

is not allowed, the intent of the code is that the discharge be directly connected to the sanitary drain system (see Section 802.1, last sentence). Food waste grinder discharge is often of a high velocity nature such that an indirect connection would most likely result in significant splatter, clogging of a waste receptor, offensive odors, a potential health hazard and a maintenance problem. This would be analogous to providing an indirect connection for the discharge of a water closet. Obviously, water closets must not be indirectly connected (see Section 802.1).

413.4 Water supply required. All food waste grinders shall be provided with a supply of cold water. The water supply shall be protected against backflow by an *air gap* or backflow preventer in accordance with Section 608.

❖ A supply of cold water is necessary to act as the transport vehicle and lubricant during the grinding operation. The water flushes the grinding chamber and carries the waste into the drain. If a waste grinder is used without running water, a drain blockage will result. While hot or warm water could be considered instead of cold water, most grinder manufacturers suggest cold water in order to keep greases more solid for best grinding action. The water supply serving a commercial food waste grinder must be protected from backflow by an air gap or backflow preventer in accordance with Section 608 (see commentary, Figure 413.3 and commentary, Section 608). In a residential kitchen sink application, the kitchen faucet spout mounted above the sink provides the necessary air gap for protection of the potable water supply.

SECTION 414 GARBAGE CAN WASHERS

414.1 Water connection. The water supply to a garbage can washer shall be protected against backflow by an *air gap* or a backflow preventer in accordance with Section 608.13.1, 608.13.2, 608.13.3, 608.13.5, 608.13.6 or 608.13.8.

❖ Water and waste connection requirements for garbage can washers are stated in this section. There are no referenced material or performance standards for these types of plumbing fixtures. These fixtures are typically floor sinks or receptors with a grated cover and a single-orifice spray nozzle protruding through the center of the grate. The spray nozzle directs a stream of water upward to clean the interior of an overturned garbage can.

A garbage can washer is a potential source of water supply contamination. The potable water supply to a garbage can washer is considered to be a submerged inlet subject to high-hazard backsiphonage and must be protected in accordance with the provisions of Section 608 (see commentary, Sections 608.13.1, 608.13.2, 608.13.3, 608.13.5, 608.13.6 and 608.13.8).

414.2 Waste connection. Garbage can washers shall be trapped separately. The receptacle receiving the waste from the washer shall have a removable basket or strainer to prevent the discharge of large particles into the drainage system.

❖ A removable basket on the drainage receptor is equivalent to a small waste interceptor. The basket collects the large garbage particles, preventing them from passing into the drainage system and creating a possible stoppage in the downstream drainage piping (see Commentary Figure 414.2).

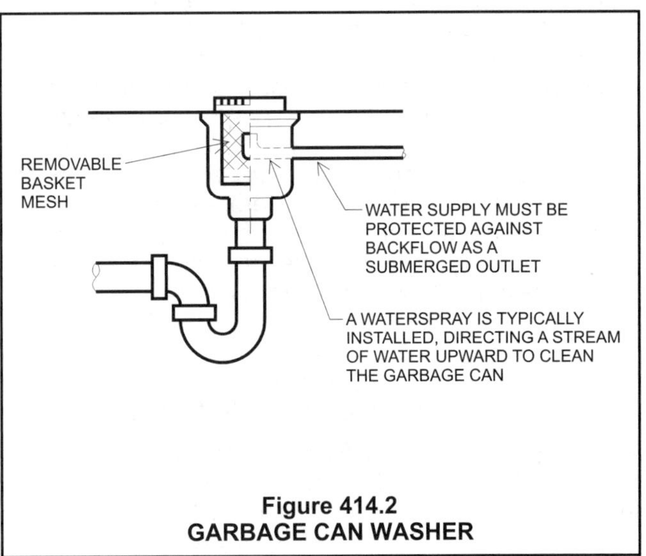

Figure 414.2
GARBAGE CAN WASHER

SECTION 415 LAUNDRY TRAYS

415.1 Approval. Laundry trays shall conform to ANSI Z124.6, ASME A112.19.1M, ASME A112.19.3M, ASME A112.19.9M, CSA B45.2 or CSA B45.4.

❖ Material standards, along with waste outlet requirements for laundry trays, are provided in this section. A laundry tray is a sink typically used in a laundry room of a residential occupancy. The referenced standards regulate enameled cast-iron, plastic, stainless steel and nonvitreous ceramic laundry trays. Another common material used to construct laundry trays is soapstone. This type of laundry tray requires an alternative approval in accordance with Section 105.2 (see commentary, Section 402.2). It should be noted that concrete laundry trays are not permitted by this section because they do not provide a smooth, impervious sanitary surface. Besides serving as a sink in a laundry room, a laundry tray is commonly used to receive the discharge of an automatic clothes washer (see Section 406.3).

415.2 Waste outlet. Each compartment of a laundry tray shall be provided with a waste outlet a minimum of $1^1/_2$ inches (38 mm) in diameter and a strainer or crossbar to restrict the clear opening of the waste outlet.

❖ The outlet must be equipped with either a strainer or crossbar to catch debris or materials that could be discharged into the sink and cause drain clogging.

SECTION 416 LAVATORIES

416.1 Approval. Lavatories shall conform to ANSI Z124.3, ASME A112.19.1M, ASME A112.19.2M, ASME A112.19.3M, ASME A112.19.4M, ASME A112.19.9M, CSA B45.1, CSA B45.2, CSA B45.3 or CSA B45.4. Group wash-up equipment shall conform to the requirements of Section 402. Every 20 inches (508 mm) of rim space shall be considered as one lavatory.

❖ This section cites standards for lavatories, including cultured marble types, and their waste outlet requirements. This section also regulates group wash-up equipment.

Lavatory means wash basin. It is derived from the Latin word "*lavatorium*," which means washing vessel, and the French word "*laver*," meaning to wash. Lavatories are available in enameled cast iron, vitreous china, stainless steel, porcelain-enameled formed steel, plastic and nonvitreous ceramic. They also come in a variety of shapes and sizes such as wall-mounted, hanger-mounted, pedestal, rimmed, above-counter basins and under mounted types.

Standards previously required lavatories to have an overflow; however, that is not currently the case. The provision of an overflow and its location is an option of the manufacturer. The reason for eliminating the overflow requirement was the lack of use, which resulted in the growth of bacteria and microorganisms.

Where a lavatory has an overflow, the standards require the cross-sectional area of the overflow to be a minimum of $1^1/_8$ square inches (726 mm^2). This is designed to prevent too small an overflow, which would promote rapid bacterial growth. Additionally, the overflow must be capable of preventing the lavatory from overflowing for a minimum of 5 minutes when tested from the onset of water flowing into the overflow opening until the water begins to overflow the fixture rim.

A group wash-up fixture, shown in Commentary Figure 416.1, is a single large basin accommodating more than one person. A group wash-up fixture has to comply with Sections 402.1 and 402.2 and be approved by the code official because it is generally not addressed by any of the referenced material or fixture standards. These fixtures are installed in areas subject to large volumes of traffic, such as facilities in sports arenas and schools.

The rim space calculation is used to determine the minimum number of required fixtures in accordance with Table 403.1 and the maximum permitted water consumption listed in Table 604.4. This calculation is not intended for use in determining the dfu value of the fixture, which is based on the trap size in accordance with Table 709.2.

416.2 Cultured marble lavatories. Cultured marble vanity tops with an integral lavatory shall conform to ANSI Z124.3 or CSA B45.5.

❖ Vanity tops with an integral lavatory that is all plastic must conform to ANSI Z124.3 or CSA B45.5. Lavatory units can be constructed of a variety of synthetic materials, including ABS, PVC, gel-coated fiberglass reinforced plastic, acrylic, cultured marble cast-filled fiberglass, polyester and cultured marble acrylic.

Cultured marble vanity tops are impregnated with fire-retardant chemicals, reducing the fuel contribution of the unit during a fire. They are also designed to resist the effect of a burning cigarette left unattended on the vanity top.

416.3 Lavatory waste outlets. Lavatories shall have waste outlets not less than $1^1/_4$ inches (32 mm) in diameter. A strainer, pop-up stopper, crossbar or other device shall be provided to restrict the clear opening of the waste outlet.

❖ The sizing requirements of Table 709.1 for a lavatory are emphasized. Each lavatory opening must have a restriction, preventing items such as rings, toothbrushes and cosmetic devices from inadvertently entering the drain.

416.4 Moveable lavatory systems. Moveable lavatory systems shall comply with ASME A112.19.12.

❖ Moveable and pivoting lavatories allow the user to adjust the location of a lavatory upward and downward,

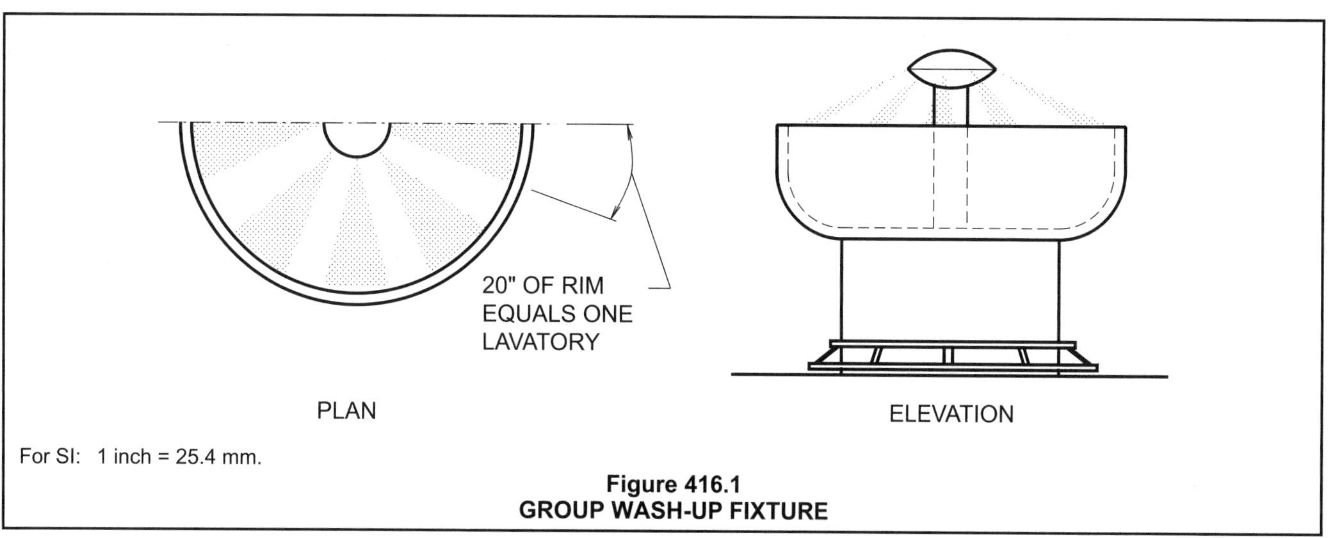

Figure 416.1
GROUP WASH-UP FIXTURE

from side to side, and front to back to maximize the ease of use of the fixture. Some products also pivot to allow greater ease of use. Commentary Figure 416.4 shows an adjustable shampoo bowl that complies with ASME A112.19.12.

Figure 416.4
ADJUSTABLE HEIGHT SHAMPOO BOWL
(Photo Courtesy of Accessible Systems LLC)

416.5 Tempered water for public hand-washing facilities. *Tempered water* shall be delivered from *public* hand-washing facilities. *Tempered water* shall be delivered through an *approved* water-temperature limiting device that conforms to ASSE 1070 or CSA B125.3.

❖ This section requires that the temperature of water delivered from hand-washing faucets be limited to that meeting the definition of tempered water. Hot water for public hand-washing facilities is rarely provided by a dedicated water heater set at a temperature that prevents scalding. In most cases, the hot water is provided from a central water heater that provides hot water for all purposes within the building. In occupancies that require high temperature water for dishwashing, cleaning and other commercial applications, public users might be abruptly exposed to this excessively hot water without warning. Installation of a temperature limiting device conforming to ASSE 1070 or CSA B125.3 will prevent scalding.

An ASSE 1016 tempering valve cannot be used in this application because the testing protocol for ASSE 1016 does not require the valve to perform the temperature control function at a flow less than 2.25 gpm. The ASSE 1070 and CSA B 125.3 compliant tempering valves require temperature regulation at the manufacturer's stated minimum flow rate (usually 0.5 gpm). Because public hand-washing facilities are required to have flow regulated to not more than 0.5 gpm, an ASSE 1070 or CSA B125.3 compliant device is required.

Note that hand-washing facilities not associated with use of public toilet facilities are also required to have tempered water control. Examples include hand washing sinks in commercial kitchens, restaurant bar hand sinks, medical examination room hand-washing sinks and dental office hand-washing sinks. See definition of "Public" and "Private."

SECTION 417
SHOWERS

417.1 Approval. Prefabricated showers and shower compartments shall conform to ANSI Z124.2, ASME A112.19.9M or CSA B45.5. Shower valves for individual showers shall conform to the requirements of Section 424.3.

❖ This section contains requirements governing prefabricated showers and field-constructed shower compartments. Material standards are cited, as are water supply riser, waste outlet and sizing limitations for such compartments. Provisions are also established for the shower floor or receptor wall area, along with specific support and pan requirements. As with bathtubs, a cross reference to the IBC is provided for safety glazing requirements.

The referenced standards for prefabricated shower compartments regulate plastic and nonvitreous ceramic units. Prefabricated units could include the entire shower compartment or the shower floor or base. The plastic units or floors can be constructed of a variety of synthetic materials, including ABS, PVC, gel-coated fiberglass-reinforced plastic, acrylic, cultured marble cast-filled fiberglass, polyester and cultured marble acrylic. The units are impregnated with fire-retardant chemicals to reduce the fuel contribution during a fire and to protect against an accidental fire from a plumber's torch.

417.2 Water supply riser. Water supply risers from the shower valve to the shower head outlet, whether exposed or concealed, shall be attached to the structure. The attachment to the structure shall be made by the use of support devices designed for use with the specific piping material or by fittings anchored with screws.

❖ The shower riser pipe must be firmly attached to a building structural element to protect the piping from leaks caused by stress fractures or joint failures. If the riser is not firmly attached, the movement of the shower head may result in movement of the riser piping or possible failure of the piping.

Previous editions of the code called for attachment of the water supply riser to the structure in an "approved manner" which was much too ambiguous. As a result, steel nails bent over piping, roofing nails, copper-coated J-nails or hanger strap were often used to restrain the riser, resulting in undesirable long-term effects. This section now requires fittings (such as a drop ear elbow) to be screwed to the structure or use of support devices specifically designed for riser support or pipe support [see commentary, Figures 417.2(1) and (2)].

Where a shower riser pipe is of the exposed type, the manufacturer typically provides a clamp or other means of support that matches the finish of the exposed riser. The clamp or other support must be firmly secured to a structural member and attached to the riser.

417.3 Shower waste outlet. Waste outlets serving showers shall be at least $1^1/_2$ inches (38 mm) in diameter and, for other than waste outlets in bathtubs, shall have removable strainers not less than 3 inches (76 mm) in diameter with strainer openings not less than $^1/_4$ inch (6.4 mm) in minimum dimension. Where each shower space is not provided with an individual waste outlet, the waste outlet shall be located and the floor pitched so that waste from one shower does not flow over the floor area serving another shower. Waste outlets shall be fastened to the waste pipe in an *approved* manner.

❖ In gang or multiple showers, the shower room floor must slope toward the shower drains to prevent waste water from flowing from one shower area through another shower area. It would be undesirable and unhygienic for a bather to stand in waste water from someone else's shower (see commentary, Figure 417.3).

417.4 Shower compartments. All shower compartments shall have a minimum of 900 square inches (0.58 m^2) of interior cross-sectional area. Shower compartments shall not be less than 30 inches (762 mm) in minimum dimension measured from the finished interior dimension of the compartment, exclusive of fixture valves, showerheads, soap dishes, and safety grab bars or rails. Except as required in Section 404, the minimum required area and dimension shall be measured from the finished interior dimension at a height equal to the top of the threshold and at a point tangent to its centerline and shall be continued to a height not less than 70 inches (1778 mm) above the shower drain outlet.

Exception: Shower compartments having not less than 25 inches (635 mm) in minimum dimension measured from the finished interior dimension of the compartment, provided that the shower compartment has a minimum of 1,300 square inches (.838 m^2) of cross-sectional area.

❖ A minimum of 900 square inches (0.58 m^2) of cross-sectional area is required to facilitate an average-size adult cleaning the lower body extremities while bending over. A smaller-size shower would not provide adequate space. The 30-inch (762 mm) minimum dimension is based on this movement of the body while cleaning oneself.

The exception allows for shower compartments with one dimension being as small as 25 inches (635 mm) so long as the compartment has at least 1300 square inches (0.838 m^2) cross-sectional area. This is especially useful in remodeling situations when old bath tubs are removed and the space turned into a standup shower.

Figure 417.2(2)
SPECIAL BRACKET FOR SHOWER RISER
(Photo Courtesy of Holdrite-Hubbard Enterprises)

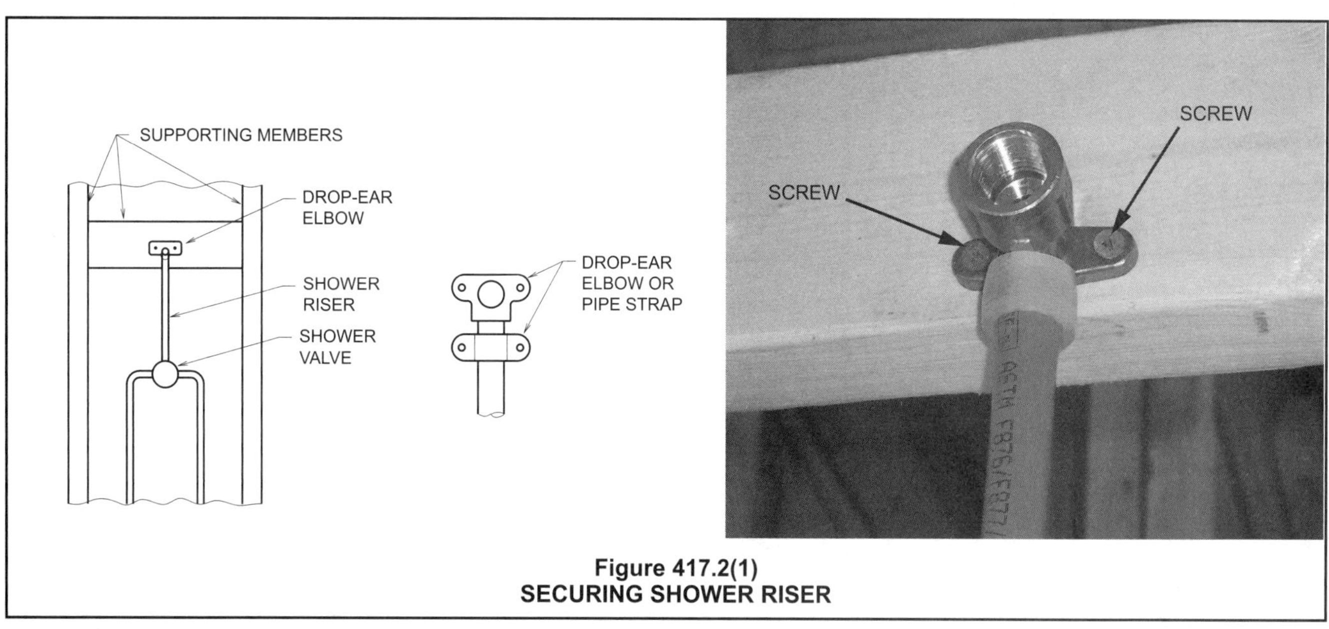

Figure 417.2(1)
SECURING SHOWER RISER

417.4.1 Wall area. The wall area above built-in tubs with installed shower heads and in shower compartments shall be constructed of smooth, noncorrosive and nonabsorbent waterproof materials to a height not less than 6 feet (1829 mm) above the room floor level, and not less than 70 inches (1778 mm) where measured from the compartment floor at the drain. Such walls shall form a water-tight joint with each other and with either the tub, receptor or shower floor.

❖ The shower walls must extend to a height of 6 feet (1829 mm) above the floor level of the room [or 70 inches (1778 mm) above the shower floor] and be constructed of smooth, corrosion-resistant and non-absorbent materials to protect the building materials from water damage. The height of the showerhead above the floor is not regulated by the code. The standard design practice is to locate the showerhead between 70 and 80 inches (1778 and 2032 mm) above the floor (see commentary, Section 424.3 for an explanation of the shower valve requirements). Where ceramic tile is installed, it must be attached to a waterproof backing surface. The older method, which is still in use, is called a mud pack. A mud pack job consisted of a backing of wire mesh and special grout. The common method of backing ceramic tile is with gypsum or cement wallboard. The IBC requires gypsum wallboard to be water resistant in this application. The water-resistant material is required only in the shower compartment area. It is not required in the remainder of the bathroom, even though it is a common practice to use water-resistant gypsum wallboard because of the frequent high moisture levels in the room (see commentary, Figure 417.4).

417.4.2 Access. The shower compartment access and egress opening shall have a minimum clear and unobstructed finished width of 22 inches (559 mm). Shower compartments required to be designed in conformance to accessibility provisions shall comply with Section 404.1.

❖ Many injuries in the home are related to accidents occurring in the tub or shower. The minimum opening requirements for access to showers allow the user to have the room necessary to safely step in and out of

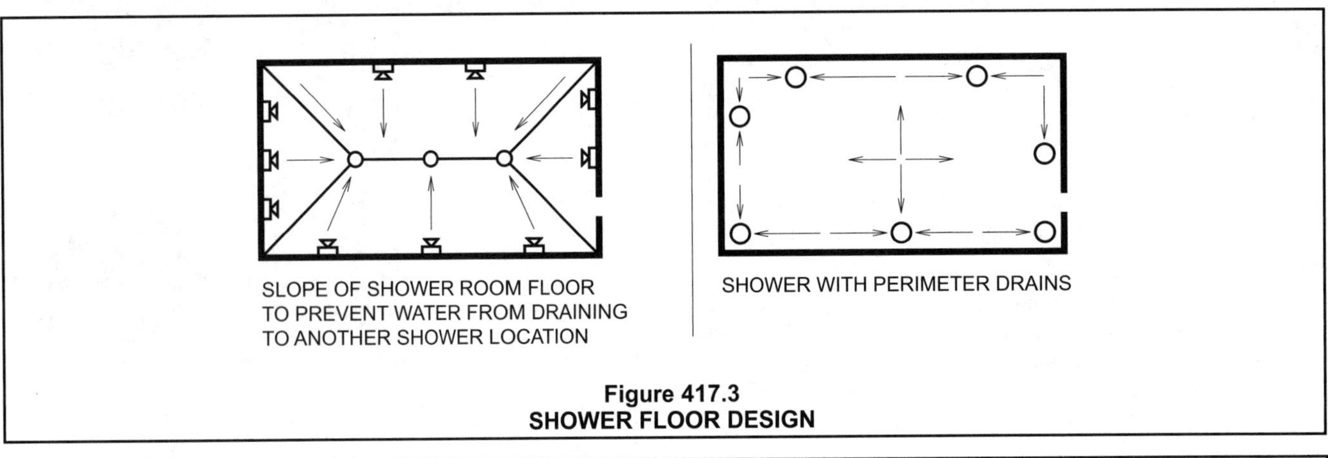

Figure 417.3
SHOWER FLOOR DESIGN

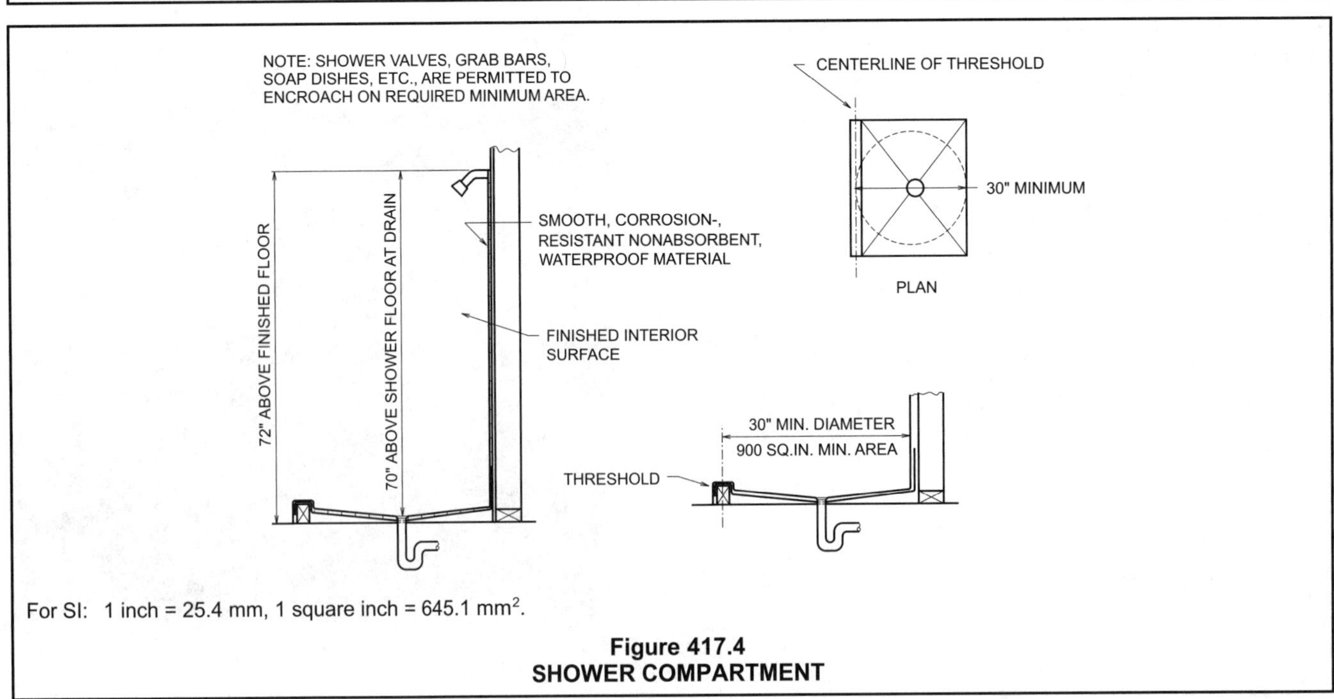

For SI: 1 inch = 25.4 mm, 1 square inch = 645.1 mm^2.

Figure 417.4
SHOWER COMPARTMENT

the shower area without having to twist or turn through a narrow opening. The 22 inches (559 mm) is based on the approximate shoulder width of an average adult. Prior to this edition of the IPC, no minimum opening dimension was required. Therefore, any minimum size opening could not be disapproved by the code official, even though the opening was unreasonably narrow. The minimum opening dimension also takes into account that besides being more user friendly, it allows for (1) better access for servicing valves, shower head and drain, (2) emergency egress and (3) emergency response and rescue access, should the need arise.

417.5 Shower floors or receptors. Floor surfaces shall be constructed of impervious, noncorrosive, nonabsorbent and waterproof materials.

❖ The floor surfaces of the shower compartment must be made water tight with smooth, corrosion-resistant, nonabsorbent, waterproof materials. Joints created between the floor and walls of the shower must be properly flashed or caulked with sealants to avoid moisture penetration. The code does not currently require a specialized slip-resistant floor surface.

417.5.1 Support. Floors or receptors under shower compartments shall be laid on, and supported by, a smooth and structurally sound base.

❖ An important requirement for shower floors is proper structural support. The support base for the shower floor must be designed for applicable dead and live loads.

417.5.2 Shower lining. Floors under shower compartments, except where prefabricated receptors have been provided, shall be lined and made water tight utilizing material complying with Sections 417.5.2.1 through 417.5.2.5. Such liners shall turn up on all sides at least 2 inches (51 mm) above the finished threshold level. Liners shall be recessed and fastened to an *approved* backing so as not to occupy the space required for wall covering, and shall not be nailed or perforated at any point less than 1 inch (25 mm) above the finished threshold. Liners shall be pitched one-fourth unit vertical in 12 units horizontal (2-percent slope) and shall be sloped toward the fixture drains and be securely fastened to the waste outlet at the seepage entrance, making a water-tight joint between the liner and the outlet. The completed liner shall be tested in accordance with Section 312.9.

Exceptions:

1. Floor surfaces under shower heads provided for rinsing laid directly on the ground are not required to comply with this section.
2. Where a sheet-applied, load-bearing, bonded, waterproof membrane is installed as the shower lining, the membrane shall not be required to be recessed.

❖ Where a shower is built in place and not prefabricated, a shower pan is required under the shower's finished floor. Various materials are used for shower pans, including sheet lead, sheet copper, polyethylene sheet, chlorinated polyethylene sheet, pre-formed ABS, and bonded waterproof membranes. The shower pan is designed to collect water that leaks through the shower floor. The shower pan must connect to the shower drain with a flashing flange or clamping device and weep holes or seepage openings to allow the leaking water to enter the drainage system. The flashing flange or clamping device is used to clamp down the waterproof membrane and make a water-tight joint (see commentary, Figure 417.5.2).

A standard practice with lead and copper shower pans is to coat the finished shower pan with tar. If a plastic material or bonded waterproof membrane is installed, the shower pan must not be tarred because it may adversely affect or react with the plastic or membrane material. Care must be taken during the installation of a shower pan to prevent damage from occurring to the pan itself, which could result in a shower pan leak.

The exception permits installing rinsing showers such as those used at beach facilities for sand removal without a shower pan liner. Note that this does not negate the requirements for shower floors in Section 417.5.

For the 2009 edition, sheet-applied, load-bearing, bonded, waterproof membrane systems were in-

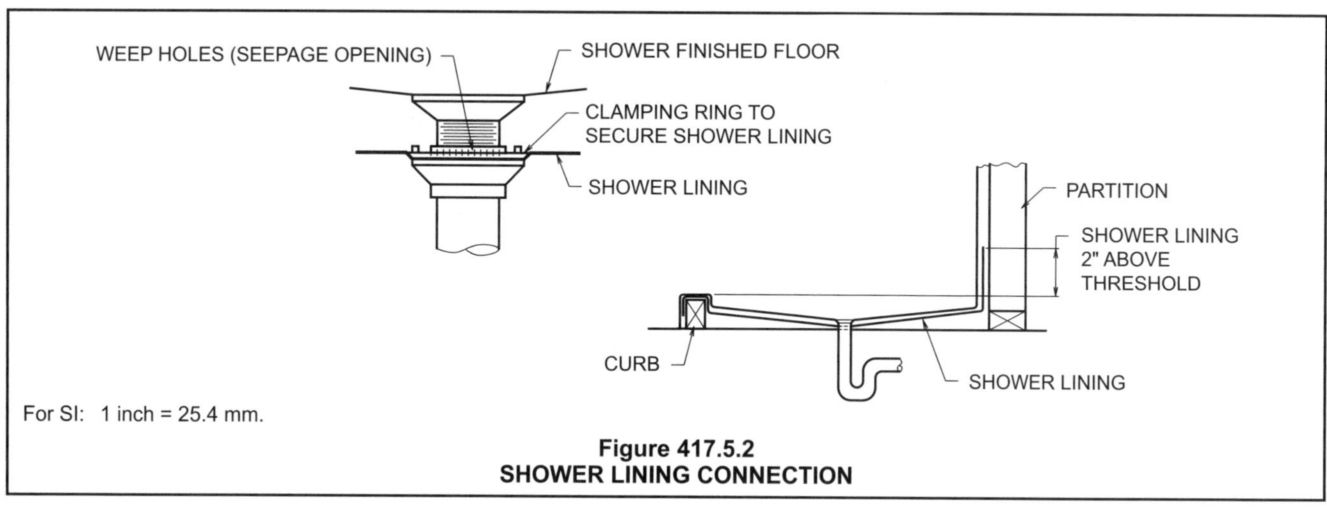

Figure 417.5.2
SHOWER LINING CONNECTION

cluded as an acceptable shower liner material. Because these membranes are very thin, Exception 2 was added as there is no need to recess this type of liner material.

417.5.2.1 PVC sheets. Plasticized polyvinyl chloride (PVC) sheets shall be a minimum of 0.040 inch (1.02 mm) thick, and shall meet the requirements of ASTM D 4551. Sheets shall be joined by solvent welding in accordance with the manufacturer's installation instructions.

❖ The PVC sheet is 0.040 inch (1.02 mm), commonly referred to as 40 mil, flexible plastic membrane that is fitted into a shower compartment on top of the subfloor to serve as the liner required by Section 417.5.2 to protect the building from water damage. The sheet is cut at the job site to accommodate the shower drain and line the subfloor, threshold and rough sidewalls of the shower compartment to a point at least 2 inches (51 mm) above the finished threshold level. The PVC sheet should be fastened to solid sidewalls and should not be nailed or perforated at any point below the flood level rim of the shower compartment.

417.5.2.2 Chlorinated polyethylene (CPE) sheets. Nonplasticized chlorinated polyethylene sheet shall be a minimum 0.040 inch (1.02 mm) thick, and shall meet the requirements of ASTM D 4068. The liner shall be joined in accordance with the manufacturer's installation instructions.

❖ This section contains the minimum requirements for chlorinated polyethylene (CPE) sheets used as shower pan liners. The industry standard addressing this product is ASTM D 4068. CPE sheets are manufactured as one piece, impervious to water and resistant to permeation by water vapor, which provides a high degree of certainty that secondary damage from leakage will not occur.

417.5.2.3 Sheet lead. Sheet lead shall not weigh less than 4 pounds per square foot (19.5 kg/m^2) coated with an asphalt paint or other *approved* coating. The lead sheet shall be insulated from conducting substances other than the connecting drain by 15-pound (6.80 kg) asphalt felt or its equivalent. Sheet lead shall be joined by burning.

❖ Various materials are used for shower pans, including sheet lead. Shower pans are designed to collect the water that leaks through the shower floor. The bottom layer is fitted to the formed subbase and each succeeding layer thoroughly burned (hot mopped) to the one below.

417.5.2.4 Sheet copper. Sheet copper shall conform to ASTM B 152 and shall not weigh less than 12 ounces per square foot (3.7 kg/m^2). The copper sheet shall be insulated from conducting substances other than the connecting drain by 15-pound (6.80 kg) asphalt felt or its equivalent. Sheet copper shall be joined by brazing or soldering.

❖ The industry standard addressing this product is ASTM B 152. There are 13 different types of coppers listed in this standard; each has a unique property making it suitable for the specific application. Annealed copper sheets are typically used for pan liners.

417.5.2.5 Sheet-applied, load-bearing, bonded, waterproof membranes. Sheet-applied, load-bearing, bonded, waterproof membranes shall meet requirements of ANSI A118.10 and shall be applied in accordance with the manufacturer's installation instructions.

❖ Sheet applied, load bearing, bonded, waterproof membranes are used as part of a system fo waterproofing for a shower floor when installing tile. The installation of this material must be in compliance with the manufacturer's installation instructions. The membrane is a finely woven cloth of synthetic fibers that is impregnated with a proprietary waterproofing system. The membrane has a stiffness similar to that of heavy brown paper and a roughness similar to that of burlap cloth. It can be cut with scissors as well as folded with a minimum radius to fit well into corners. Installation involves spreading a thin layer of special mortar base adhesive, laying the membrane on the mastic layer and then troweling (embedding) the membrane into the mastic. seams in the membrane are made by overlapping the material and using mastic to adhere the over lapping materials.

417.6 Glazing. Windows and doors within a shower enclosure shall conform to the safety glazing requirements of the *International Building Code*.

❖ If a window is installed in the shower compartment, the IBC requires the window to be of safety glazing. The compartment size is based on the 6-foot (1829 mm) dimension above the room floor level or the 70-inch (1778 mm) dimension above the shower compartment floor for nonabsorbent wall surfacing material.

Glass doors enclosing the shower compartment must also be of safety glazing to provide protection. If a slip or fall occurs during a shower, the occupant could be severely injured by broken glass. Safety glazing abates the hazard associated with broken glass (see commentary, Section 407.3).

SECTION 418
SINKS

418.1 Approval. Sinks shall conform to ANSI Z124.6, ASME A112.19.1M, ASME A112.19.2M, ASME A112.19.3M, ASME A112.19.4M, ASME A112.19.9M, CSA B45.1, CSA B45.2, CSA B45.3 or CSA B45.4.

❖ Standards containing material and performance requirements are provided in this section of the code for plumbing fixtures known as sinks. This section also contains waste outlet requirements for sinks. Sinks are considered a different item than lavatories by the code, even though the terms can be used interchangeably.

Sinks are available in enameled cast-iron, vitreous china, stainless steel, porcelain-enameled formed steel, plastic and nonvitreous ceramic materials. Soapstone sinks are not regulated by any standard; therefore, those sinks must be approved as an alternative in accordance with Section 105.2.

Although the most common classification is kitchen sinks, this section also regulates service sinks, bar sinks, mop sinks and wash sinks.

418.2 Sink waste outlets. Sinks shall be provided with waste outlets a minimum of $1^1/_2$ inches (38 mm) in diameter. A strainer or crossbar shall be provided to restrict the clear opening of the waste outlet.

❖ A food waste grinder has a standard opening of $3^1/_2$ inches (89 mm). Most kitchen sinks have this size opening, permitting the installation of a food waste grinder. The standard kitchen sink basket strainer is also $3^1/_2$ inches (89 mm).

418.3 Moveable sink systems. Moveable sink systems shall comply with ASME A112.19.12.

❖ See the commentary to Section 416.4.

SECTION 419
URINALS

419.1 Approval. Urinals shall conform to ANSI Z124.9, ASME A112.19.2M, ASME A112.19.19, CSA B45.1 or CSA B45.5. Urinals shall conform to the water consumption requirements of Section 604.4. Water-supplied urinals shall conform to the hydraulic performance requirements of ASME A112.19.6, CSA B45.1 or CSA B45.5.

❖ A urinal is a fixture that has been designed specifically for urination; a water closet is primarily designed for defecation. Although urinals are normally associated with men, there have been urinals designed for women. Female urinals have not been popular because of clothing styles, the need for privacy and uncertainty regarding the proper use of the fixture.

Although it appears that the main feature in the design of the urinal was to have the user avoid contact with the fixture surface that can be contaminated, it was actually designed for practicality and speed. The installation of this fixture is the reason why there is frequently no line for the men's room, while a long line gathers for the women's room.

The design and construction of the urinal is regulated by the referenced standards. Trough urinals are prohibited because they do not conform to the referenced standards and are considered unhygienic.

There are four types of water-supplied urinals: stall, blowout, siphon jet and wash down. A stall urinal is floor mounted, with other urinals being predominantly wall hung. Both blowout and siphon-jet urinals flush completely by siphonic action; the contents of the bowl are completely evacuated during the flushing cycle, and the trap is refilled.

ANSI Z124.9 covers water supplied and waterless urinals made of plastic. ASME A112.19.19 covers waterless urinals made of vitreous china. ASME A112.19.2M and CSA B45.1 cover water-supplied urinals of vitreous china. CSA B45.5 covers water-supplied urinals of plastic materials.

The hydraulic performance requirements for water-supplied urinals were previously covered by ASME A112.19.6; however, ASME revised standard A112.19.2 to cover the hydraulic performance of both water closets and urinals, abandoning the ASME A112.19.6 standard. While CSA B45.1 and B45.5 provide for material performance for ceramic and plastic urinals, respectively, CSA standard B45.0, section 15.8.7, covers the hydraulic testing for urinals.

Like a stall urinal, a washout urinal does not typically have an integral trap. Some manufacturers, however, design washout urinals with an integral trap. Neither stall nor washout urinals flush with a siphon action. The flush cycle is accomplished through a combination of water exchange and dilution. The water consumption limitation for a urinal is 1.0 gallon (3.785 L) per flushing cycle (see commentary, Section 604.4).

As water conservation has become more important in recent years, manufacturers have responded with waterless urinal designs. Waterless urinals are a popular consideration for sustainable "green" building projects. One design uses a special liquid that is less dense than the urine and the liquid creates the trap seal. Urine passes through the liquid seal and into the drain without affecting the liquid seal. The liquid is held in a special cartridge that is easily replaced for maintenance. Table 709.1 assigns a DFU value of 0.5 to "nonwater supplied" urinals. Because of the obvious conservation of energy associated with supplying potable water and treating wastewater, they are also popular in areas having water shortages [see commentary Figures 419.1(1) and (2)].

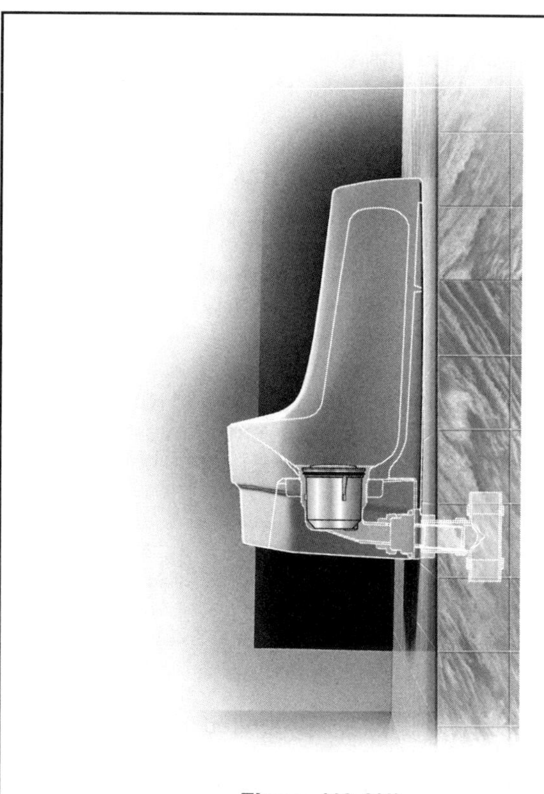

Figure 419.1(1)
WATERLESS URINAL
(Photo Courtesy of Falcon Waterfree Technologies)

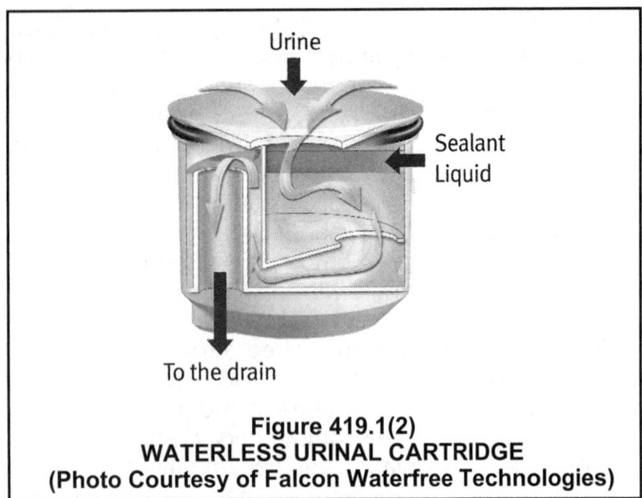

Figure 419.1(2)
WATERLESS URINAL CARTRIDGE
(Photo Courtesy of Falcon Waterfree Technologies)

419.2 Substitution for water closets. In each bathroom or toilet room, urinals shall not be substituted for more than 67 percent of the required water closets in assembly and educational *occupancies*. Urinals shall not be substituted for more than 50 percent of the required water closets in all other *occupancies*.

❖ This section allows urinals to be substituted for a maximum of 67 percent of the required number of water closets in assembly and educational occupancies and a maximum of 50 percent in all other occupancies. For example, if five water closets are required in an educational occupancy men's room, the design professional may choose to install two water closets and three urinals, or three water closets and two urinals. If four urinals are installed, however, two water closets are also required. Experience has shown that water closet shortages result with a 67 percent substitution allowance in occupancies other than E and A; thus, a 50 percent maximum allowance was introduced for those occupancies.

[B] 419.3 Surrounding material. Wall and floor space to a point 2 feet (610 mm) in front of a urinal lip and 4 feet (1219 mm) above the floor and at least 2 feet (610 mm) to each side of the urinal shall be waterproofed with a smooth, readily cleanable, nonabsorbent material.

❖ Wall and floor space to a point 2 feet (610 mm) in front of a urinal lip and 4 feet (1219 mm) above the floor and at least 2 feet (610 mm) to each side of the urinal must be waterproofed with a smooth, readily cleanable, nonabsorbent material.

One of the major problems with the design of a urinal is the splash factor. When a forceful stream of urine hits the wall of the urinal or the water seal, there is a splash that could bounce out of the fixture. To prevent the splash from contaminating the building material surrounding the urinal, smooth, waterproof, nonabsorbent material is required on the surfaces most likely to be splashed. The finished material will allow for easy cleaning of the area, thus preventing any bacterial growth (see commentary, Figure 419.3). Note that there are requirements for urinal partitions in Section 310.5.

Section 1210.2 of the IBC also stipulates that the material be "hard." The intent of this section and the IBC section is that the surrounding material should not be of a type that is easily damaged such that the material could become absorbent and thus, not readily cleanable (and sanitized). Painted drywall, even where coated with highly durable paints, does not meet the intent of these sections. The typical materials used for this application are ceramic tile, fiberglass reinforced plastic (FRP) panels, phenolic panels, laminated plastic panels, painted steel and stainless steel. Painted concrete block and poured concrete walls that have had holes and cracks filled with block filler material have also been used.

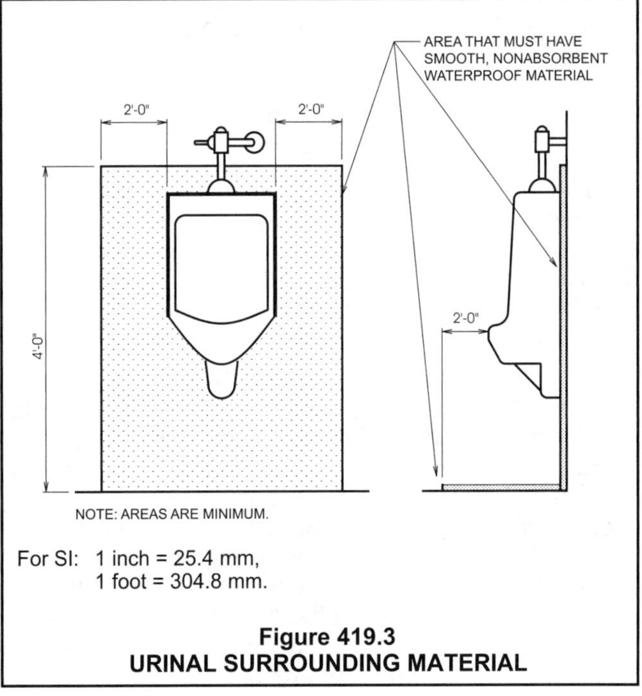

For SI: 1 inch = 25.4 mm,
1 foot = 304.8 mm.

Figure 419.3
URINAL SURROUNDING MATERIAL

SECTION 420
WATER CLOSETS

420.1 Approval. Water closets shall conform to the water consumption requirements of Section 604.4 and shall conform to ANSI Z124.4, ASME A112.19.2M, CSA B45.1, CSA B45.4 or CSA B45.5. Water closets shall conform to the hydraulic performance requirements of ASME A112.19.6. Water closet tanks shall conform to ANSI Z124.4, ASME A112.19.2, ASME A112.19.9M, CSA B45.1, CSA B45.4 or CSA B45.5. Electro-hydraulic water closets shall comply with ASME A112.19.13.

❖ The most recognizable and most used plumbing fixture is the water closet. Material and performance requirements are included in the referenced standards for water closets.

A water closet is referred to by many other names, including toilet and commode. The term "water closet" remains the technical expression. The term "water closet" originates from the first attempt to bring plumbing indoors. The human elimination process was done in a closet which contained a cistern. The closet was shaped like a chair to accommodate a person in the squatting or sitting position. The expression was soon associated with the room because it was the size of a closet. The term, however, applies today as the name of the fixture itself.

The most common word used in the United States to identify a water closet is "*toilet*." The word "toilet" is derived from the French word "toilette," meaning dressing. The word "toilet" was considered more socially acceptable than the term "water closet." The word "toilet" more accurately describes the room in which water closets are installed.

The water closet is the primary fixture responsible for the modernization of plumbing. The fixture effectively disposes of human waste in a sanitary manner.

Water closets must be of the water-conservation type. The applicable fixture standard permits a maximum average of 1.6 gallons (6.1 L) of water per flushing cycle. No cycle during the testing can exceed 2.0 gallons (7.6 L) per flush (see Section 604.4).The exceptions to Section 604.4 give the cases where use of a water closet with an increased water usage per flush, up to a maximum of 3.5 gallons (13 L) per flushing cycle is permitted. To determine conformance, a representative model of the water closet must be tested for its flushing efficiency. The flushing test uses polypropylene balls and polyethylene granule discs. There is also a test for the ability to remove an ink stain on the inside surface of the water closet bowl, and a dye test measuring the exchange rate of water in the bowl. Another test measures the capacity of the drain line attached to the water closet to carry away the solids from the bowl following flushing.

There are three styles of water closets: close coupled, one piece and flushometer valve.The close-coupled water closet has a bowl and separate gravity-type tank or flushometer tank that is supported by the bowl. A one-piece water closet is constructed with the gravity-type tank or flushometer tank and the bowl as one integral unit. A flushometer valve water closet is a bowl with a flushometer valve.

Water closet bowls come in six styles: blowout, siphon jet, reverse trap, wash down, siphon vortex and siphon wash. The following provides a brief description of each style:

Blowout: A blowout water closet must have a flushometer valve because the flushing relies on the higher pressure of the entering water. The water pressure forces the bowl contents through the trapway with little reliance on siphonic action. A blowout is the noisiest of the water closets.

Siphon jet: The siphon-jet bowl has water entering the bottom of the bowl directed into the trapway. Water enters through the rim of the bowl creating the siphon action started by the trapway opening. The water rises in the bowl until the siphon action draws the contents out of the bowl.

Reverse trap: The reverse trap is similar to the siphon-jet bowl. The difference between the two fixtures is the water surface area and the depth of the trap seal. The reverse-trap bowl is smaller in area with less of a trap seal.

Washdown: The washdown bowl has the trapway in the front of the bowl with a small water surface area. The siphon action is started by the water entering the rim. There is typically a flush opening directed into the trapway.

Siphon vortex: The siphon vortex is a one-piece water closet. The water entering the rim creates a swirling vortex in the center of the bowl. There is usually no flush opening in the trapway. The siphon action is created only by the water entering the rim of the bowl.

Siphon wash: The siphon wash is designed for maximum water conservation. The trap seal depth and water surface area are the smallest of all the designs. The flushing action is similar to the washdown principle.

Commentary Figure 420.1(1) shows a cutaway of four of the styles of water closet bowls. The top left is a siphon jet and the top right is a reverse trap. The bottom left is a blowout and the bottom right is a wash down [see Commentary Figure 420.1(2)].

The IPC does not specify water closet bowl rim heights (above finished floors) for any application [see commentary, Figure 420.1(3). Chapter 11 of the IBC, by reference to ICC/ANSI A117.1 covering accessible design, does specify water closet seat heights (above finished floors) for accessible adult and accessible child water closet applications. An Accessible adult water closet must have a seat height in the range of 17 to 19 inches (432 to 483 mm). An accessible child water closet must have a seat height in the range of 11 to 17 inches (279 to 432 mm). However, where a jurisdiction adopts an accessibility design standard other than ICC/ANSI A117.1, the requirements for accessible seat height or bowl rim height may be different.

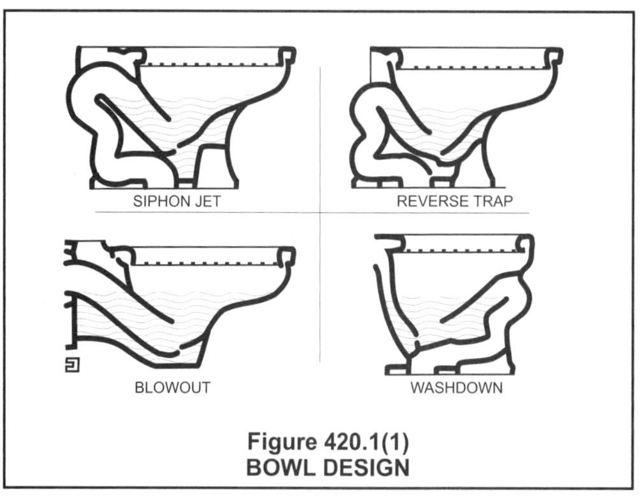

Figure 420.1(1)
BOWL DESIGN

Where wall-mounted, non-accessble, water closets are being installed, the following bowl rim height (above finished floors) guidelines can be used for setting water closet carriers:

Standard bowl rim height:
14 to 15 inches (356 to 381 mm)

Adult "comfort" bowl rim height:
15 to 16 inches (381 to 406 mm)

Child (2-6 years) bowl rim height:
$9^1/_2$ to $10^1/_2$ inches (241 to 267 mm)

Juvenile (6-12 years) bowl rim height:
13 to 14 inches (330 to 356 mm)

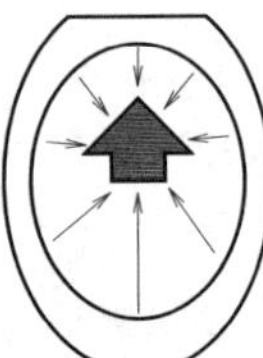

SIPHON VORTEX
WATER ENTERS THROUGH PUNCHINGS AROUND THE RIM, CREATING A WHIRLING VORTEX THAT DRAWS WASTE DOWN AND WASHES THE BOWL.

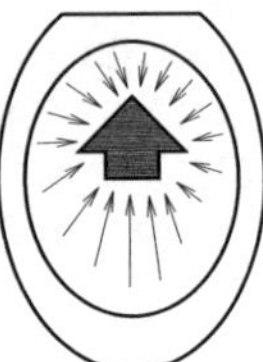

SIPHON JET
WATER ENTERS THROUGH THE RIM AND SIDE JETS, FILLING THE TRAPWAY AND CREATING A SIPHON ACTION THAT CAUSES RAPID, SANITARY WITHDRAWAL OF WASTE AND WATER FROM THE BOWL.

BLOWOUT
A STRONG JET ACTION EVACUATES CONTENTS RAPIDLY. AVAILABLE ONLY IN VALVE-OPERATED WATER CLOSETS.

Figure 420.1(2)
FLUSHING ACTION FOR CERTAIN WATER CLOSETS

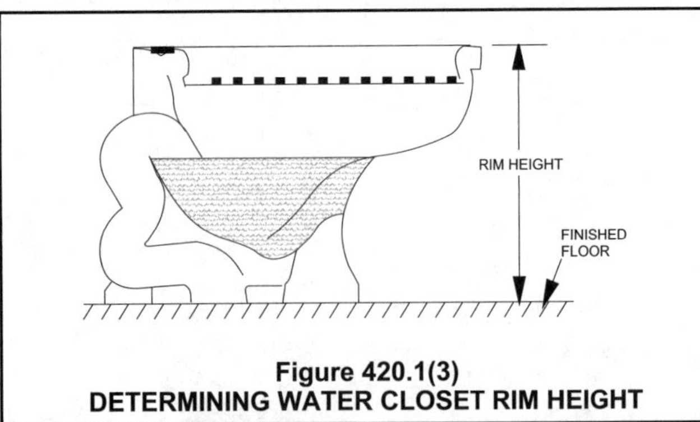

Figure 420.1(3)
DETERMINING WATER CLOSET RIM HEIGHT

420.2 Water closets for public or employee toilet facilities. Water closet bowls for *public* or employee toilet facilities shall be of the elongated type.

❖ An elongated water closet bowl is required for public or employee use because it gives the user an extended area in which to eliminate human waste. The concept is to reduce the possibility of the user missing the bowl and soiling the water closet seat and surroundings.

The elongated water closet bowl also has a larger water surface area, which helps keep the inside surface of the bowl clean.

An elongated water closet bowl is 2 inches (51 mm) longer than a regular bowl. The bowl extension is toward the front of the fixture.

Elongated bowls are not required for various nonpublic or nonemployee uses, including hotel guest-rooms and private facilities.

420.3 Water closet seats. Water closets shall be equipped with seats of smooth, nonabsorbent material. All seats of water closets provided for *public* or employee toilet facilities shall be of the hinged open-front type. Integral water closet seats shall be of the same material as the fixture. Water closet seats shall be sized for the water closet bowl type.

❖ A water closet seat must be designed to facilitate cleaning and prevent the accumulation of dirt or bacterial growth. The seat must also be designed for the bowl on which it is installed.

The requirement for a hinged open-front seat was originally to aid women. The open seat, however, is equally beneficial to men. The open seat allows women to wipe the perineal area after using the water closet, while it also eliminates an area that could be contaminated with urine when used by men. The open-front design also eliminates the user's genital contact with the seat.

420.4 Water closet connections. A 4-inch by 3-inch (102 mm by 76 mm) closet bend shall be acceptable. Where a 3-inch (76 mm) bend is utilized on water closets, a 4-inch by 3-inch (102 mm by 76 mm) flange shall be installed to receive the fixture horn.

❖ A 4-inch by 3-inch (102 mm by 76 mm) closet bend is not considered a reduction in drainage pipe size (see Commentary Figure 420.4 and commentary, Section 704.2).

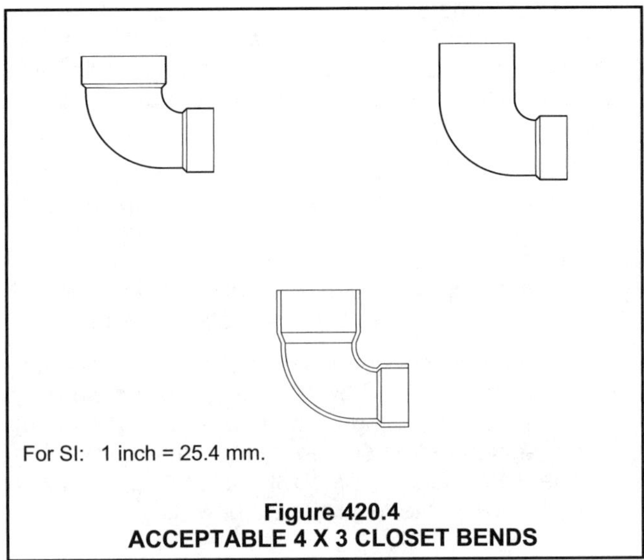

For SI: 1 inch = 25.4 mm.

Figure 420.4
ACCEPTABLE 4 X 3 CLOSET BENDS

SECTION 421 WHIRLPOOL BATHTUBS

421.1 Approval. Whirlpool bathtubs shall comply with ASME A112.19.7M or with CSA B45.5 and CSA B45 (Supplement 1).

❖ This section of the code provides material and performance requirements for bathtubs in which there is continuous movement of water. Installation requirements along with drain and suction fitting requirements are also provided.

ASME A112.19.7, CSA B45.5 and CAN/CSA B45 (Supplement 1) were developed because of an increase in the use of whirlpool bathtubs. These standards regulate fixture design and construction, installation instructions and field-testing requirements. Whirlpool bathtubs must conform to these standards.

421.2 Installation. Whirlpool bathtubs shall be installed and tested in accordance with the manufacturer's installation instructions. The pump shall be located above the weir of the fixture trap.

❖ The referenced standard requires the manufacturer to indicate that each whirlpool bathtub must be filled with water and tested during the rough-in inspection. This is necessary to determine whether there are any leaks in the factory-piped whirlpool lines.

The pump, being a mechanical device, must be accessible to facilitate any repairs, maintenance or replacement. The pump (circulation) must be located above the tub trap weir to verify that the pump body drains each time the tub is drained.

421.3 Drain. The pump drain and circulation piping shall be sloped to drain the water in the volute and the circulation piping when the whirlpool bathtub is empty.

❖ The whirlpool tub, suction lines and pump must be designed to drain completely after each use. This improves the sanitation of a whirlpool bathtub. Wastewater left standing in the pump body or other waterways will promote the growth of bacteria.

421.4 Suction fittings. Suction fittings for whirlpool bathtubs shall comply with ASME A112.19.8M.

❖ ASME A112.19.8M was developed to regulate suction fittings on whirlpool bathtubs. ASME A112.19.8M contains performance criteria to establish a standard for suction fittings. The purpose of the standard is to prevent the use of a fitting that may create a potential hazard for hair or body entrapment.

421.5 Access to pump. *Access* shall be provided to circulation pumps in accordance with the fixture or pump manufacturer's installation instructions. Where the manufacturer's instructions do not specify the location and minimum size of field-fabricated *access* openings, a 12-inch by 12-inch (305 mm by 305 mm) minimum sized opening shall be installed to provide *access* to the circulation pump. Where pumps are located more than 2 feet (609 mm) from the *access* opening, an 18-inch by 18-inch (457 mm by 457 mm) minimum sized opening shall be installed. A door or panel shall be permitted to close the opening. In all cases, the *access* opening shall be unobstructed and of the size necessary to permit the removal and replacement of the circulation pump.

❖ A whirlpool pump must be accessible for removal for repairs/replacement. The opening is needed for access to the mounting bolts, pipe unions and electrical connections to physically remove the pump without damage to the pump or to the surrounding finished walls or ceiling. The larger opening is required when the pump is located deeper in the enclosure because the service personnel will have to position more of their body in the opening to reach and remove the pump. Note that the whirlpool tub manufacturer may dictate specific access requirements and such requirements prevail over what is specified in this section [see commentary, Figures 421.5(1) and (2)].

421.6 Whirlpool enclosure. Doors within a whirlpool enclosure shall conform to ASME A112.19.15.

❖ See the commentary to Section 407.4.

SECTION 422 HEALTH CARE FIXTURES AND EQUIPMENT

422.1 Scope. This section shall govern those aspects of health care plumbing systems that differ from plumbing systems in other structures. Health care plumbing systems shall conform to the requirements of this section in addition to the other requirements of this code. The provisions of this section shall apply to the special devices and equipment installed and maintained in the following *occupancies*: nursing homes, homes for the aged, orphanages, infirmaries, first aid stations, psychiatric facilities, clinics, professional offices of dentists and doctors, mortuaries, educational facilities, surgery, dentistry, research and testing laboratories, establishments manufacturing pharmaceutical drugs and medicines, and other structures with similar apparatus and equipment classified as plumbing.

❖ In health care facilities, there are plumbing fixtures that have special requirements because of their unique application or use. This section outlines occupancies where special health care plumbing devices and equipment may be installed. Although these fixtures are typically found in hospitals, they might be located in other health care facilities.

Note that the list of occupancies is not all inclusive. It is intended only to be representative of the occupancies in which special plumbing fixtures are regulated by this chapter. Occupancies other than those listed can be equipped with some of these special fixtures. When this occurs, the requirement shown here would apply to those occupancies.

422.2 Approval. All special plumbing fixtures, equipment, devices and apparatus shall be of an *approved* type.

❖ Because many health care plumbing fixtures are not regulated by any fixture standards, each must be evaluated and approved by the code official. The code official should consider backflow prevention and the ability to maintain sanitary conditions of the special fixtures.

Figure 421.5(1)
NONCOMPLIANT WHIRLPOOL BATHTUB PUMP ACCESS

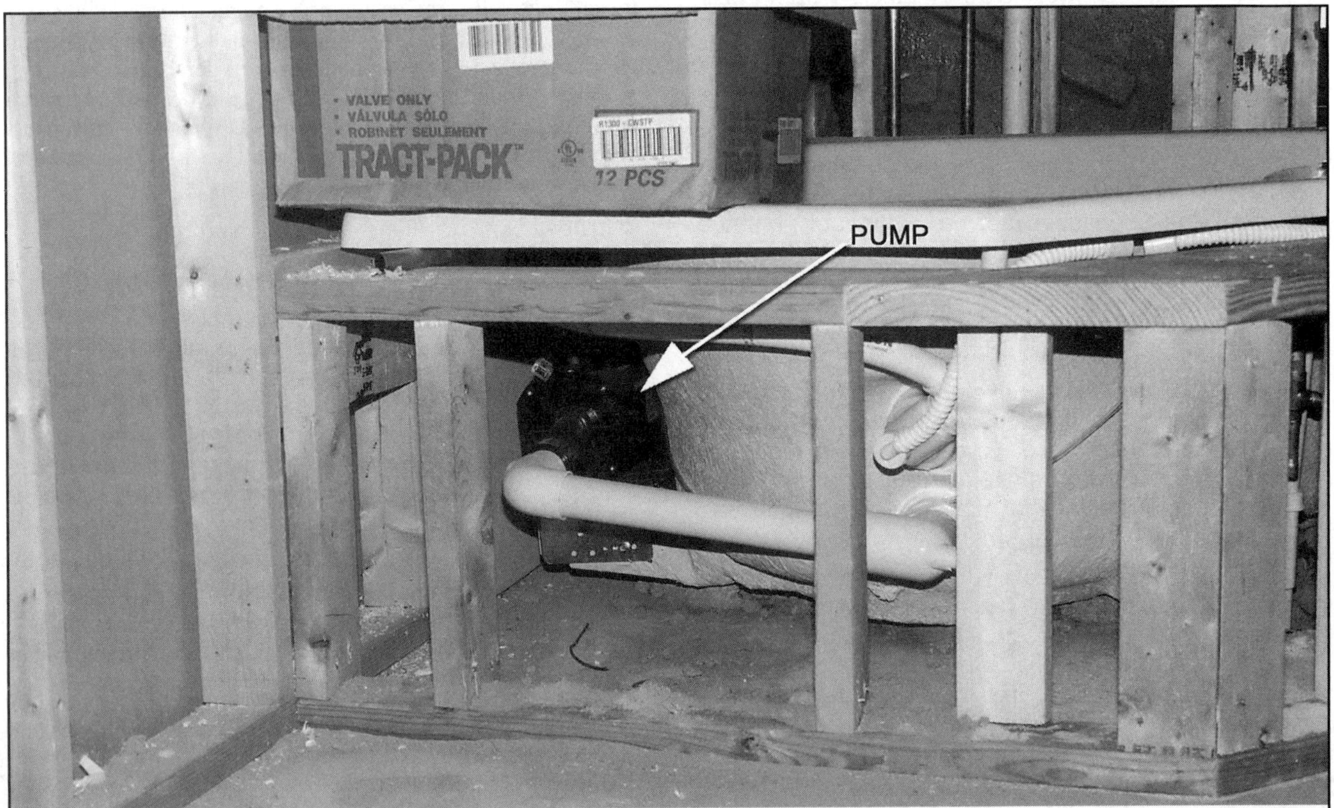

Figure 421.5(2)
NONCOMPLIANT WHIRLPOOL BATHTUB PUMP ACCESS

422.3 Protection. All devices, appurtenances, appliances and apparatus intended to serve some special function, such as sterilization, distillation, processing, cooling, or storage of ice or foods, and that connect to either the water supply or drainage system, shall be provided with protection against backflow, flooding, fouling, contamination of the water supply system and stoppage of the drain.

❖ This section requires protection of fixtures, appliances and equipment against contamination resulting from backflow. The water supply system must also be protected against contamination resulting from backflow from such fixtures, appliances and equipment.

This section combines the intent of Sections 608, 609 and 802.

422.4 Materials. Fixtures designed for therapy, special cleansing or disposal of waste materials, combinations of such purposes, or any other special purpose, shall be of smooth, impervious, corrosion-resistant materials and, where subjected to temperatures in excess of 180°F (82°C), shall be capable of withstanding, without damage, higher temperatures.

❖ This section requires the materials used for special plumbing fixtures to be cleanable and capable of resisting staining, corrosion and the accumulation of waste in any pores or other minute openings in the finish of the material. The material used for the fixture must also be capable of withstanding the temperatures to which it will be exposed during normal service conditions.

422.5 Access. *Access* shall be provided to concealed piping in connection with special fixtures where such piping contains steam traps, valves, relief valves, check valves, vacuum breakers or other similar items that require periodic inspection, servicing, maintenance or repair. *Access* shall be provided to concealed piping that requires periodic inspection, maintenance or repair.

❖ Any plumbing fixture, appliance or device having mechanical components must be accessible to facilitate any needed repair or maintenance. A mechanical component is always prone to failure and, therefore, access is required.

422.6 Clinical sink. A clinical sink shall have an integral trap in which the upper portion of a visible trap seal provides a water surface. The fixture shall be designed so as to permit complete removal of the contents by siphonic or blowout action and to reseal the trap. A flushing rim shall provide water to cleanse the interior surface. The fixture shall have the flushing and cleansing characteristics of a water closet.

❖ A clinical sink is the main special fixture used in health care facilities. It is designed to both wash a bedpan and remove its contents; therefore, the other name for a clinical sink is a bedpan washer.

A clinical sink is like a water closet in that it has a flushing rim, which washes the interior walls of the fixture and discharges urine and fecal matter by siphonic action.

A hose or washing arm is provided with a clinical sink to wash the contents of the bedpan into the fixture.

In some installations, a water closet functions as both a clinical sink and a water closet. The common use of the dual-purpose fixture reduces the number of plumbing fixtures required in a health care facility.

422.7 Prohibited usage of clinical sinks and service sinks. A clinical sink serving a soiled utility room shall not be considered as a substitute for, or be utilized as, a service sink. A service sink shall not be utilized for the disposal of urine, fecal matter or other human waste.

❖ A service sink is not designed for the elimination of human waste, nor does it have a flushing rim to clean the fixture with each use. By not completely eliminating the waste material via a flushing method, a service sink could allow bacteria to grow and infect the surrounding areas. A service sink and a clinical sink, therefore, must not be interchanged.

422.8 Ice prohibited in soiled utility room. Machines for manufacturing ice, or any device for the handling or storage of ice, shall not be located in a soiled utility room.

❖ In rooms where fecal matter is eliminated, biological aerosols are released into the air, making contact with exposed surfaces. The feces of an unhealthy person could also contain pathogens that may release into the air.

To avoid the possible contamination of fluid that is to be ingested, locating ice machines in the same room as a clinical sink is prohibited. For the same reason, installation of drinking fountains in toilet rooms is prohibited.

422.9 Sterilizer equipment requirements. The approval and installation of all sterilizers shall conform to the requirements of the *International Mechanical Code*.

❖ The *International Mechanical Code*® (IMC®) regulates the installation of appliances, including sterilizers. The appliance must bear the label of an approved agency and be installed in accordance with the manufacturer's instructions. If a sterilizer is gas fired, the *International Fuel Gas Code*® (IFGC®) specifies the requirements for gas piping, combustion air and venting.

422.9.1 Sterilizer piping. *Access* for the purposes of inspection and maintenance shall be provided to all sterilizer piping and devices necessary for the operation of sterilizers.

❖ The term "Access" is defined in Chapter 2. Access to piping and devices associated with sterilizers is essential to allow for periodic inspection, maintenance and repair.

422.9.2 Steam supply. Steam supplies to sterilizers, including those connected by pipes from overhead mains or branches, shall be drained to prevent any moisture from reaching the sterilizer. The condensate drainage from the steam supply shall be discharged by gravity.

❖ Because condensate could contain contaminants that interfere with the operation of the sterilizer, it must not be allowed to enter the sterilizer. The condensate must be drained by gravity, which is the most dependable method. Drainage of the condensate also reduces the likelihood of it entering the sterilizer.

422.9.3 Steam condensate return. Steam condensate returns from sterilizers shall be a gravity return system.

❖ Steam condensate returns from sterilizers are required to drain by gravity to reduce the likelihood of condensate backing up into a sterilizer, which could violate the sterile conditions.

422.9.4 Condensers. Pressure sterilizers shall be equipped with a means of condensing and cooling the exhaust steam vapors. Nonpressure sterilizers shall be equipped with a device that will automatically control the vapor, confining the vapors within the vessel.

❖ To avoid the release of steam into the space that houses the sterilizer, units that operate above atmospheric pressures must condense the exhaust steam vapors. For the same reason, sterilizers that operate at approximately atmospheric pressure must confine the steam vapors within the vessel.

422.10 Special elevations. Control valves, vacuum outlets and devices protruding from a wall of an operating, emergency, recovery, examining or delivery room, or in a corridor or other location where patients are transported on a wheeled stretcher, shall be located at an elevation that prevents bumping the patient or stretcher against the device.

❖ Many of the control devices used in health care plumbing systems protrude from the wall. They must be located in a way to prevent any accidental contact by a patient in a wheelchair or stretcher. Besides protecting an individual from hitting any control device, it also protects the device from any possible damage.

SECTION 423 SPECIALTY PLUMBING FIXTURES

423.1 Water connections. Baptisteries, ornamental and lily pools, aquariums, ornamental fountain basins, swimming pools, and similar constructions, where provided with water supplies, shall be protected against backflow in accordance with Section 608.

❖ This section addresses the myriad of specialty and decorative items that are not commonly thought of as plumbing fixtures. The most important requirement for these types of construction is to prevent water supply contamination. Note that Section 423.2 requires that specialty fixtures be approved by the code official. Requirements for specialty fixture materials are located in Section 402.2. Baptismal pools, ornamental ponds and fountains, swimming pools, specialty food service equipment, dispensers, pedicure chairs and ice makers are examples of items and equipment that, although not classified as plumbing fixtures, require a water supply connection or waste connection.

The majority of these "special fixtures" are not governed by standards; therefore, extraordinary precautions must be taken to verify that the potable water supply is not endangered. The code official must evaluate each installation independently to secure compliance with Section 608 (see commentary, Section 608).

A pedicure chair (having integral foot bath tub) that is permanently connected to the water supply and drain system is a good example of a specialty plumbing fixture. While these factory assembled units have water and waste connections and are conceptually similar to some appliances such as a whirlpool tub, they might not be designed or tested to any plumbing code referenced standard. Therefore, the code official must review the equipment information prior to approving such fixtures for installation so he or she can advise the installer on special requirements for protecting the potable water supply from contamination.

423.2 Approval. Specialties requiring water and waste connections shall be submitted for approval.

❖ Fixtures or devices not conforming to referenced standards must be approved by the code official as an alternative material in accordance with Section 105.2. Water supplies serving such fixtures or devices must be protected against backflow to prevent contamination of the potable water supply (see Section 423.1).

SECTION 424 FAUCETS AND OTHER FIXTURE FITTINGS

424.1 Approval. Faucets and fixture fittings shall conform to ASME A112.18.1/CSA B125.1. Faucets and fixture fittings that supply drinking water for human ingestion shall conform to the requirements of NSF 61, Section 9. Flexible water connectors exposed to continuous pressure shall conform to the requirements of Section 605.6.

❖ Prior to the 2009 code edition, the ASME and CSA standards organizations joined forces to produce a "harmonized standard" for faucet and fixture fittings thus eliminating the need to indicate separate standards for each organization. The indicated standard regulates plumbing supply fittings located between the supply line stop and the terminal fitting, inclusive. The terminal fitting is the fitting that has an open or atmospheric discharge, e.g., a faucet spout or hose bibb. This standard covers automatic compensating valves for individual wall-mounted showering systems, bath and shower supply fittings, bidet supply fittings, clothes washer supply fittings, drinking fountain supply fittings, humidifier supply stops, kitchen sink and lavatory supply fittings, laundry tub supply fittings, lawn and sediment faucets, metering and self-closing supply fittings and supply stops. The standard does not apply to pipes, tubes or the fittings for pipes and tubes.

The main focus of this standard is to ensure that mounting dimensions and connection interchangeability exist among manufacturers of these types of plumbing products. The standard also covers all of the quality requirements and uniform test methods including, but not limited to, strength (pressure and structural), finish durability, operational life, backflow protection, force/torque to operate and maximum leakage amounts for the fittings.

The second line of this code section requires that faucet and fittings that supply drinking water for human ingestion conform to the requirements of NSF 61, Section 9. This is also a requirement of the ASME A112.18.1/CSA B125.1 standard. It is often assumed that because a potable water distribution system conveys potable water, water discharged from any point in the system is of drinking water quality. However, this is not the case, as the following explanation of the Safe Drinking Water Act and NSF 61 will clarify.

In 1974, the Safe Drinking Water Act (SDWA) was enacted by the United States Federal government to minimize chemical and bacterial contamination of drinking water. The Act authorized the U.S. Environmental Protection Agency to set national health-based standards for drinking water to protect the public from naturally occurring and man-made contaminants. In 1986, an amendment to the SDWA addressed concerns for lead contamination of drinking water by requiring that all pipes, solders, pipe fittings and plumbing fixtures, intended to provide water for human consumption and sold after July 1986, be lead-free. Lead-free in this context is defined to mean that solder and flux cannot contain more than 0.2 percent lead and that pipes, fittings and fixtures cannot contain more than 8.0 percent lead (see commentary for Sections 605.2 and 605.14.3).

Because studies indicated that plumbing (supply) fittings and fixtures within the last liter volume of the water discharge point were significant contributors of contaminants (especially lead) in dispensed water, the 1986 amendment to the SDWA also required that by August 1997, a set of voluntary consensus standards and testing protocols be developed for these products. In preparation for meeting the August 1997 deadline, the initial standard, NSF 61, *Drinking Water System Components—Health Effects*, was voluntarily developed in 1988 by a consortium of NSF International, the American Water Works Association and the Association of State Drinking Water Administrators with support from the U.S.E.P.A. In 1994, Section 9 of NSF 61 was completed in order to comply with the August 1997 deadline.

The reader might question why plumbing (supply) fittings and fixtures within the last liter volume of the discharge point are required to comply with more stringent standards beyond that of a maximum lead content of 8 percent. Because the majority of plumbing (supply) fittings and fixtures are made from copper and brass alloys, the casting, forging, machining and finishing processes to produce those products can cause the lead in the alloys to be much more "leachable" than the lead in pipe and fitting products. If water remains at rest int he plumbing (supply) fittings and fixtures for a long period of time such as might happen overnight, the next draw or water from the plumbing (supply) fittings or fixtures often contains a lead concentration much higher than the maximum allowable lead concentration established by the U.S.E.P.A. for safe drinking water.

NSF 61 Section 9 only covers plumbing (supply) fittings and fixtures intended by the manufacturer to supply water for human consumption. Examples are kitchen faucets, lavatory faucets, bar faucets, drinking fountains, residential icemakers, hot/cold water dispenser and commercial kitchen pot fillers. NSF 61 section 9 specifically excludes certain "end point" devices such as bath/shower valves, utility faucets, laboratory faucets, bidet controls, any faucets with hose thread fittings at the discharge point, nonlavatory hand wash stations, and faucets that are self-closing, metering or electronically activated. In summary, any end point device where the water is specifically intended to be drawn for consumption by humans must be certified to NSF 61, Section 9.

The last line of this code section refers the reader to another section that specifically covers flexible water connectors as the requirements of this section do not apply to that type of product.

424.1.1 Faucets and supply fittings. Faucets and supply fittings shall conform to the water consumption requirements of Section 604.4.

❖ Faucets are the end point devices that allow the user to turn the water flow on and off, regulate the flow and, in some cases, regulate the temperature at the point of use. A standard lavatory faucet is an example of an end point device that must have a water discharge rate of not greater than the maximum stated in Section 604.4. Faucets that are required to have the flow rate limited utilize a removable flow-restricting device at the outlet of the faucet, commonly called an aerator. Not all of these devices "aerate" (entrain air in) the flow stream.

The original use of aerators was to reduce the degree of flow stream splashing as it impacted the basin or objects being washed. Entrained air "softens" the impact of the flow stream. However, as water conservation was mandated (see Section 604.4), aerators took on the more important role of limiting flow from faucets and remain today as the key water flow limiting device for most faucets.

As stated previously, not all "aerators" entrain air into the flow stream. A common example is a laminar flow device found in many medical and dental faucet applications. In these environments, bacteria-laden room air entrained into the flow stream by an aerator is detrimental to infection control efforts. Since laminar flow devices do not entrain air into the flow stream, bacterial infection is reduced. Further towards the effort of infection control, some laminar flow devices now include minute amounts of silver (Ag) which has been proven to prevent bacteria growth within the laminar flow device during periods of nonuse.

In applications where water flow from faucets is significantly reduced such as in public lavatories (limited to 0.5 gallons per minute), an aerated or laminar flow stream does not provide enough force to allow the user to effectively remove soap and dirt from the hands. A nonaerated spray pattern device is used which produces numerous small but powerful streams for better cleaning [see commentary, Figures 424.1.1 (1) and (2)].

Supply fittings, as referenced in this section, are those devices downstream of faucets or control valves. Examples are shower heads and hand held showers. Water conservation in these supply fittings are typically accomplished with internal flow restricting devices.

424.1.2 Waste fittings. Waste fittings shall conform to ASME A112.18.2/CSA B125.2, ASTM F 409 or to one of the standards listed in Tables 702.1 and 702.4 for above-ground drainage and vent pipe and fittings.

❖ Prior to the 2009 code edition, the ASME and CSA standards organizations joined forces to produce a "harmonized standard" for plumbing waste fittings thus eliminating the need to indicate separate standards for each of these organizations. The indicated standard covers plumbing waste fittings of sizes 2 inches (51 mm) nominal and smaller and includes traps, tailpieces, trap wall adapters, diversion tees, twin waste elbows overflow elbows, drain elbows, and strainer bodies of various materials including plastic.

ASTM F 409 covers waste fittings made of plastic materials only.

The main focus of these standards is to ensure that mounting dimensions and connection interchangeability exist among manufacturers of these types of plumbing products. The standard also covers all of the quality requirements and uniform test methods including, but not limited to, strength (pressure and structural) and finish durability.

424.2 Hand showers. Hand-held showers shall conform to ASME A112.18.1 or CSA B125.1. Hand-held showers shall provide backflow protection in accordance with ASME A112.18.1 or CSA B125.1 or shall be protected against backflow by a device complying with ASME A112.18.3.

❖ Because it is possible for a hand-held shower to be submerged in the bathtub or shower compartment base, a cross connection to wastewater could be created. The hand-held shower must have adequate protection against backflow. The backflow protection may be integral to the hand-held shower or may be provided by an external backflow device that meets ASME A112.18.3. These types of showers are commonly installed in accessible shower enclosures. Their use, however, has increased in popularity in showers within a dwelling unit. The referenced standards specify the requirements for backflow protection of the hand-held shower.

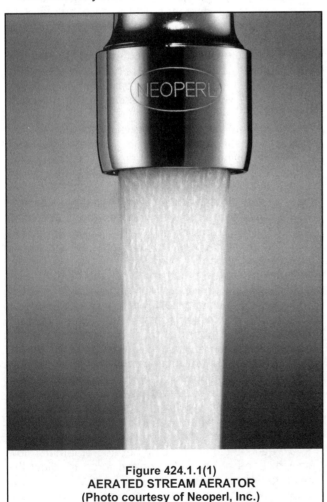

Figure 424.1.1(1)
AERATED STREAM AERATOR
(Photo courtesy of Neoperl, Inc.)

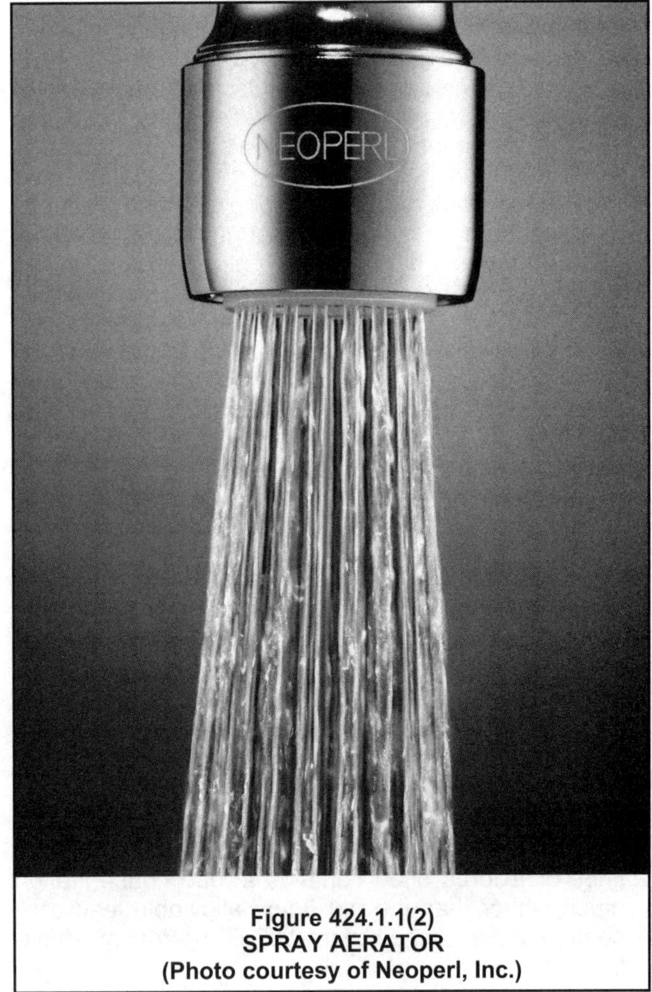

Figure 424.1.1(2)
SPRAY AERATOR
(Photo courtesy of Neoperl, Inc.)

A hand-held shower, used in a hand-held mode, is not considered to be the type of shower that requires a means to protect the user from scalding. The reason for this comes from the justifications that eventually led to the code requirement for anti-scald shower valves (see commentary for Section 424.3). Therefore, if the hand-held shower is mounted for use in a hands-free mode with the user in a standing position, a hand-held shower is no different than any other fixed-type shower head and must be supplied though a anti-scald shower valve in accordance with Section 424.3.

However, where a hand-held shower is used in a hand-held mode, most users, when subjected to a high temperature surge of water spray, will either drop or simply redirect the hand-held shower. If the user is standing, the potential for a fall injury is minimal. If the user is sitting, fall injuries cannot occur. Therefore, where a hand-held shower is used only in a hand-held mode, the water discharged from it is not required to be supplied through an anti-scald shower valve.

424.3 Individual shower valves. Individual shower and tub-shower combination valves shall be balanced-pressure, thermostatic or combination balanced-pressure/thermostatic valves that conform to the requirements of ASSE 1016 or ASME A112.18.1/CSA B125.1 and shall be installed at the point of use. Shower and tub-shower combination valves required by this section shall be equipped with a means to limit the maximum setting of the valve to 120°F (49°C), which shall be field adjusted in accordance with the manufacturer's instructions. In-line thermostatic valves shall not be utilized for compliance with this section.

❖ Commentary Table 424.3 shows exposure times at various water temperatures to cause third-degree burns to human skin. Industry safety experts have chosen 120°F (49°C) to be the maximum safe water for bathing purposes. In a showering application where the shower head is fixed, a user subjected to a surge of high temperature water usually reacts by abruptly moving away from the hot spray. This abrupt movement often causes falls resulting in the stunned user not being able to escape the scalding water stream. Children and some elderly persons often have delayed or no physical reaction to the scalding water such that they continue to stand in the stream. Therefore, showers and combination tub/showers must have a shower control valve that is capable of protecting an individual from being exposed to water temperatures in excess of 120°F (49°C). The control valve must be installed at the point of use. In other words, the person in the shower must have access to the control handle(s) of the valve.

There are two types of events that the control valve must be designed to protect against: (1) extreme temperature fluctuations from the user set temperature caused by changes in hot or cold water distribution line pressures and (2) extreme temperature conditions caused by the user either purposely or accidentally adjusting the control valve to deliver the hottest water available from the hot water distribution system. Where water inlet pressures or temperatures fluctuate during shower use, control valves complying with ASME A112.18.1/CSA B125.1 and ASSE 1016 must automatically and rapidly adjust to maintain the water discharge temperature to within ±3.6°F (±2°C) of the user selected temperature.

The standards also require that shower control valves have a maximum temperature limit device that is not adjustable by the user at the point of use. The high-limit stop is typically an adjustable set screw or cam that is manually set to limit the travel of the control valve handle. The high-limit stop must be field adjusted at the time of installation to limit the delivered water temperature to a maximum of 120°F (49°C). Most control valve manufacturer's installation instructions require that a thermometer be used to verify the maximum discharge temperature from the valve.

Table 424.3
TEMPERATURE BURN CHART

TIME AND TEMPERATURE RELATIONSHIP TO SERIOUS BURNS		
	Time required for a third-degree burn to occur	
Water temperature	Adults (skin thickness of 2.5 mm)	Children (skin thickness of .56 mm)
155°F 68°C	1 second	0.5 second
148°F 64°C	2 seconds	1 second
140°F 60°C	5 seconds	1 second
133°F 56°C	15 seconds	4 seconds
127°F 53°C	1 minute	10 seconds
124°F 51°C	3 minutes	1.5 minutes
120°F 49°C	5 minutes	2.5 minutes
100°F 38°C	Safe temperature for bathing	Safe temperature for bathing

For SI: °C = [(°F) - 32]/1.8.

The three types of shower control valves available are pressure balancing, thermostatic and combination thermostatic/pressure-balancing. A pressure-balancing valve senses changes in pressure of the hot and cold water supplies (up to 50 percent pressure change). If the pressure on one side changes, the valve reacts so that the flow from each side of the valve is adjusted to maintain the user's temperature selection. The thermostatic valve senses the discharge water temperature and adjusts flows of supply water to the valve to maintain the user set temperature. A thermostatic valve provides some limited protection against hot and cold supply pressure fluctuations (up to 20 percent pressure change). A combination thermostatic/pressure-balancing valve adjusts to changes in supply pressures (up to 50 percent pressure change) and senses discharge temperature.

Pressure-balancing control valves are used in a majority of shower applications. One slight disadvantage of the pressure balanced type control valve is that any change in the temperature setting of the water heating system will affect the maximum discharge temperature available from the control valve. If the system temperature is lowered, some users might complain that they are not receiving hot enough water. If the system temperature is raised, the maximum water temperature available from the valve might be in excess of 120°F (49°C). Seasonal fluctuations in the cold water supply temperature can also affect the maximum discharge temperature. Thus, pressure balancing type control valves might require periodic readjustment of the high-limit setting.

Although this type of valve might significantly reduce the need for readjustments to the high limit setting, due to its inherent limited protection against supply side pressure changes, a thermostatic valve might not be suitable for applications where large pressure fluctuations (greater than 20 percent change in pressure) are expected.

A combination pressure balanced/thermostatic control valve offers the greatest degree of protection for the shower user. As the name implies, this type of valve offers the best features of both the pressure balanced type valve and the thermostatic type valve designs.

This section does not apply to control valves that supply water only for filling tubs (see Section 424.5) or emergency showers (see Section 411). Inline-type thermostatic and pressure balancing valves do not satisfy the requirements of this section because they are not the final control valve that the user adjusts at the point of use and they do not have a high limit setting feature.

424.4 Multiple (gang) showers. Multiple (gang) showers supplied with a single-tempered water supply pipe shall have the water supply for such showers controlled by an *approved* automatic temperature control mixing valve that conforms to ASSE 1069 or CSA B125, or each shower head shall be individually controlled by a balanced-pressure, thermostatic or combination balanced-pressure/thermostatic valve that conforms to ASSE 1016 or CSA B125 and is installed at the point of use. Such valves shall be equipped with a means to limit the maximum setting of the valve to 120°F (49°C), which shall be field adjusted in accordance with the manufacturer's instructions.

❖ In a shower facility that has multiple shower areas where all shower heads are supplied with tempered water from a single source (a master valve), the tempered water supply must be controlled by an automatic temperature control mixing valve conforming to either of the referenced standards. Master thermostatic mixing valves must be sized to meet the peak demand of showers located downstream of the valve so that there is sufficient water volume during peak usage, and more importantly, so the master thermostatic mixing valve works properly. Regardless of shower control valve operation or supply pressure fluctuations, the user will be protected from scalding injury because the temperature of the water supplied to the shower will not exceed 120°F (49°C) (see commentary, Figure 424.4).

In lieu of using a master thermostatic valve, each shower head must have a control valve at the point of use that is capable of protecting an individual from being scalded when taking a shower.

The water heater thermostat cannot be used instead of a master thermostatic valve or individually protected showers to limit water temperature to 120°F (49°C).

424.5 Bathtub and whirlpool bathtub valves. The *hot water* supplied to bathtubs and whirlpool bathtubs shall be limited to a maximum temperature of 120°F (49°C) by a water-temperature limiting device that conforms to ASSE 1070 or CSA B125.3, except where such protection is otherwise provided by a combination tub/shower valve in accordance with Section 424.3.

❖ This section addresses faucets used only for the filling of whirlpool tubs or bathtubs. The water discharged from these "tub filler" faucets must be limited to a maximum temperature of 120°F (49°C) to prevent skin burns (see commentary, Section 424.3). It is common for a user to fill a bathtub in advance, with water hotter than he or she could stand, in an attempt to compensate for the cooling effect caused by the tub walls or a planned delay in entering the tub. These users typically avoid burns by testing the temperature of the water in the tub and add cold water to temper the hot water before getting into the tub. However, there have been cases in which the user was sitting on the side of the tub, reached to adjust the control for cold water, accidentally slipped into the tub and received third degree-burns. In other cases, children placed in A tub being filled have been scalded because the water became hotter some time after an adult tested the flowing water temperature at the beginning of tub filling. In some cases, an unsupervised child has adjusted a faucet and cause scalding. While it would seem that hot water supply temperatures might be simply controlled by setting the water heater thermostat to 120°F (49°C), experience has indicated that this is an unreliable means for limiting the temperature of hot water in any type of building because such thermostats are very inaccurate and are too easy for laypersons to adjust. The adjustment to a higher temperature setting is usually in response to users' complaints that there is not enough hot water during periods of high demand. Adjusting a water heater to a higher temperature effectively creates more useable hot water. This is because less, higher temperature, hot water flow is required for any particular mixed hot water use and, thus, there is effectively more hot water available to the users. Because the code does not regulate the temperature to which water heater thermostats must be set and the thermostats are not tamper-proof or have the necessary repeatable accuracy, water temperature control devices unrelated to the water heater unit must be used to positively limit the discharge water temperature to 120°F (49°C).

The temperature limiting device required by this section is a control valve that might require periodic adjustment, maintenance or replacement. Obviously, these valves should be located so that they can be accessed. Access panels in finished walls floors or ceilings might be required in order to provide access.

This code section allows for a combination tub/shower valve (complying with ASSE 1016 or

Table 424.4
MAXIMUM TEMPERATURES ALLOWED AT USE LOCATIONS

PLUMBING FIXTURE	MAXIMUM DISCHARGE TEMPERATURE	REFERENCE STANDARD
1. Bidet	≤ 110°F	ASSE 1070
2. Public hand-washing facilities	> 85°F and < 110°F	ASSE 1070
3. Individual shower valves	≤ 120°F	ASSE 1016 OR CSA B125
4. Multiple (gang) showers	≤ 120°F	ASSE 1069 OR CSA B125
5. Bathtub/whirlpool bathtubs	≤ 120°F	ASSE 1070, Optional combination tub/shower valve conforming to ASSE 1016 or CSA B125

For SI: °C = [(°F) - 32]/1.8.

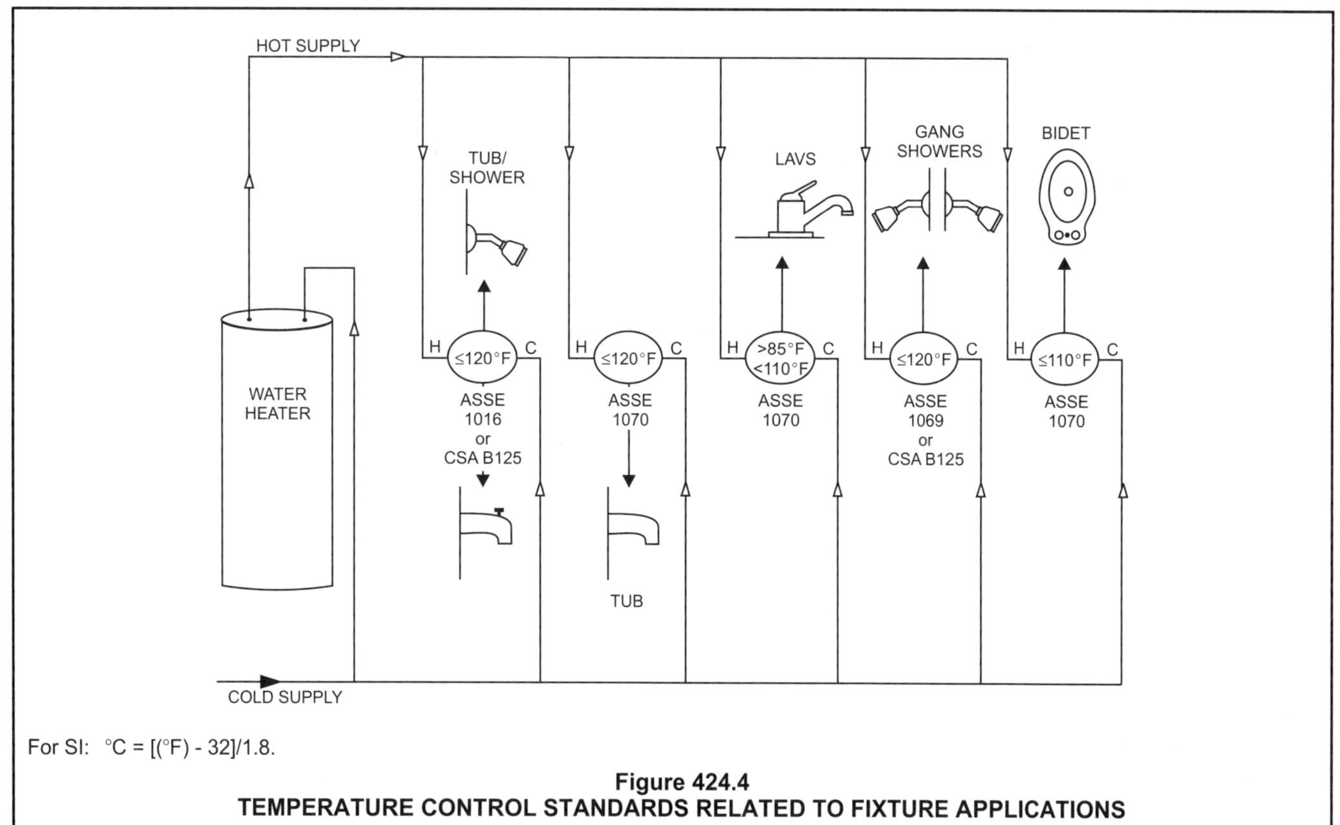

For SI: °C = [(°F) - 32]/1.8.

Figure 424.4
TEMPERATURE CONTROL STANDARDS RELATED TO FIXTURE APPLICATIONS

ASME A112.18.1/CSA B125.1), with or without a fixed shower head, to be used for filling a tub, including a large whirlpool tub. However, in actual practice, the installation of a combination tub/shower valve as a tub filler for a large tub (such as a garden or whirlpool tub) is rarely done, for several reasons: (1) faucet manufacturers generally don't provide or offer "extra long" wall mount tub spouts to enable the discharge of the spout to be located over the tub well so as to not splash onto the tub deck, and (2) the placement of the water control on the wall makes it difficult for a user sitting in a large tub to add more water if needed. This "allowance" for a combination tub/shower valve for a tub application might lead the reader to assume that any device meeting ASSE 1016 or ASME A112.18.1/CSA B125.1 is also acceptable for meeting the requirements of this section but this is not the case. Combination tub/shower valves are the only ASSE 1016 or ASME A112.18.1/CSA B125.1 device that can be used to meet the requirements of this section.

Why can't ASSE 1016 valves (other than combination tub/shower valves) be used for the ASSE 1070 or CSA B125.3 device required by this section? ASSE 1016 or CSA B125.3 devices, other than combination tub/shower valves, have a limited useful "maximum flow" rating. In other words, for flows in excess of 2.5 gpm (9.5 lpm), the pressure drop through these devices severely limits the user's ability to fill a large tub in short order. ASSE 1070 and CSA B125.3 devices have a much higher usable flow rate, typical of what is needed to rapidly fill a large bathtub or whirlpool tub. The IPC "allowance" to use a combination tub/shower valve complying with Section 424.3 for a bathtub filler faucet is a modest compromise directed at the most commonly installed, minimum size [30 inches (762 mm) x 60 inches (1524 mm) x 16 inches (406 mm) deep], standard tub found in nearly every private bathroom. The filling time of this standard tub size, using a combination tub/shower valve through a standard tub spout is deemed "reasonable" based upon years of user experience with filling tubs through anti-scald type combination tub/shower valves. However, where tub sizes are larger, higher faucet flow rates are required to fill those tubs within an acceptable length of time. Requiring the use of an ASSE 1070 device with the higher flow through rating assures that larger tubs can be filled quickly so that the tub water temperature has less time to "cool off" (since these devices are limiting the discharge temperature to 120°F (49°C).

Where a hand-held shower is connected to a tub filler faucet, the hand shower is not required to be provided with hot water supplied through an ASSE 1016 or ASME A112.18.1/CSA B125.1 device as is required for a standup-type shower application (see commentary, Section 424.2). The hand-held shower also does not "reclassify" a tub filler faucet as a "combination tub/shower valve" such that it would be required to comply with ASSE 1016 or ASME A112.18.1/CSA B125.1.

424.6 Hose-connected outlets. Faucets and fixture fittings with hose-connected outlets shall conform to ASME A112.18.3M or CSA B125.

❖ Typically, this is a deck-mounted fitting with a moveable spout or outlet connected to the body of the faucet through a flexible hose. Applications could include kitchen sinks, bar and lavatory sinks or shampoo bowls. The standard governing requirements for backflow protection devices in plumbing fixture fittings are found in ASME A112.18.3M or CSA B125. These standards establish the performance requirements for backflow prevention devices that provide protection consistent with the level of risk associated with the application.

424.7 Temperature-actuated, flow reduction valves for individual fixture fittings. Temperature-actuated, flow reduction devices, where installed for individual fixture fittings, shall conform to ASSE 1062. Such valves shall not be used alone as a substitute for the balanced pressure, thermostatic or combination shower valves required in Section 424.3.

❖ Temperature-actuated flow reduction (TAFR) devices were developed primarily for existing shower applications where the existing shower control was not of a compensating design i.e., a balanced pressure, thermostatic or combination balanced pressure/thermostatic design. Buildings constructed prior to the code requiring compensating shower controls typically have "two handle" (one for hot, one for cold) shower controls. There is a high potential for these types of showers to discharge scalding water due to changes in system pressures (e.g., flushing of a nearby water closet), or changes in hot water supply temperature (e.g., a boiler firing interval). Building owners, concerned about providing safe showering temperature for the building occupants, install this device between the shower head and shower arm pipe in order to protect the user. When the water temperature reaches 115° (46°C), the device severely restricts flow to not greater than 0.25 gpm (0.95 L/m). After the water cools below 115°F (46°C), flow resumes. Where compensating type shower controls are installed, these devices could be installed as an extra level of protection. However, these devices cannot serve as a substitute for installing a compensating type shower control where a new shower control is being installed.

424.8 Transfer valves. Deck-mounted bath/shower transfer valves containing an integral atmospheric vacuum breaker shall conform to the requirements of ASME A112.18.7.

❖ Deck-mounted bath/shower transfer valves with integral backflow protection must comply with ASME A112.18.7. With the introduction of hand-held showers and other bathing appliances mounted on the deck of bathtubs and other fixtures, the need for backflow and backsiphonage protection became evident. This is because of the accessories that are being furnished typically incorporate a hose or pull out features that could be submerged. ASME A112.18.7 combines safety features such as a vent to the atmosphere and check members to protect against backflow or backsiphonage.

SECTION 425 FLUSHING DEVICES FOR WATER CLOSETS AND URINALS

425.1 Flushing devices required. Each water closet, urinal, clinical sink and any plumbing fixture that depends on trap siphonage to discharge the fixture contents to the drainage system shall be provided with a flushometer valve, flushometer tank or a flush tank designed and installed to supply water in quantity and rate of flow to flush the contents of the fixture, cleanse the fixture and refill the fixture trap.

❖ This section contains the applicability requirements for flushing devices recognized by the code. Material and performance requirements are specified in referenced standards, as are provisions for flush tanks.

There are three classifications of devices that are required to flush a water closet, urinal or other self-siphoning type of plumbing fixture: a flush tank, a flushometer valve and a flushometer tank. The flush tank discharges to the fixture by gravity, using only the head pressure created by the height of the tank. A flushometer valve is a controlling device discharging to the fixture at the full line pressure of the water supply. A flushometer tank is a hydropneumatic tank that creates a flushing action like a flushometer valve and fills like a gravity tank.

425.1.1 Separate for each fixture. A flushing device shall not serve more than one fixture.

❖ Each fixture must have its own flushing device. Individual control valves are necessary in an effort to conserve water.

425.2 Flushometer valves and tanks. Flushometer valves and tanks shall comply with ASSE 1037. Vacuum breakers on flushometer valves shall conform to the performance requirements of ASSE 1001 or CAN/CSA B64.1.1. *Access* shall be provided to vacuum breakers. Flushometer valves shall be of the water-conservation type and shall not be utilized where the water pressure is lower than the minimum required for normal operation. When operated, the valve shall automatically complete the cycle of operation, opening fully and closing positively under the water supply pressure. Each flushometer valve shall be provided with a means for regulating the flow through the valve. The trap seal to the fixture shall be automatically refilled after each valve flushing cycle.

❖ There is a variety of styles of flushometer valves and tanks, all of which must be equipped with a vacuum breaker to prevent backflow. The flushometer valve operation is either a diaphragm type or a piston type. The majority of flushometer valves are diaphragm type because of its reliability. The working components of a flushometer valve are either exposed or concealed. When concealed, the valve must be accessible through an access opening, cover or plate.

Flushometers are either manually or automatically operated. Common manual valves are activated by a hand lever, push button or foot valve. Automatic valves are activated electrically by infrared sensors or other devices that sense the presence of an individual using the fixture. A segment diaphragm separates the valve into an upper and lower chamber with the pressure being the same on both sides of the diaphragm, equalized through the bypass in the diaphragm. The greater surface area on the top of the diaphragm holds the valve closed. The slightest touch of the handle grip in any direction pushes the plunger inward, which tilts the relief valve, releasing the pressure in the upper chamber. Then, the pressure below raises the diaphragm and working parts as a unit, allowing the water that flushes the bowl to go down through the barrel of the valve. While this is occurring, a small amount travels up through the bypass, gradually filling the upper chamber to close the valve.

The newest category of flushing device is a flushometer tank. The flushing device incorporates a hydropneumatic tank to flush the fixture at full line pressure. The flushing of the fixture is similar to a flushometer valve; however, the water demand is the same as a flush tank. Rather than using the water supply pressure from the piping system, a flushometer tank stores the necessary water for flushing in the hydropneumatic accumulator tank. The water in the tank is stored under a pressure equivalent to the water supply pressure connected to the tank.

425.3 Flush tanks. Flush tanks equipped for manual flushing shall be controlled by a device designed to refill the tank after each discharge and to shut off completely the water flow to the tank when the tank is filled to operational capacity. The trap seal to the fixture shall be automatically refilled after each flushing. The water supply to flush tanks equipped for automatic flushing shall be controlled with a timing device or sensor control devices.

❖ The flush tank is designed to permit a large volume of water to discharge at once, creating the flushing action in the fixture. The tank then refills while simultaneously refilling the trap seal of the fixture. Most flush tanks have a float device (fill valve) to open and close the water supply mechanism. When the tank empties, the float lowers, opening the fill valve. As the float rises, the water supply closes at the predetermined height in the tank.

425.3.1 Fill valves. All flush tanks shall be equipped with an antisiphon fill valve conforming to ASSE 1002 or CSA B125.3. The fill valve backflow preventer shall be located at least 1 inch (25 mm) above the full opening of the overflow pipe.

❖ The fill valve is the refilling component that controls the flow of water into the flush tank. Because the fill valve connects the potable water to the flush tank, it must have adequate protection against backflow. The referenced standards require an antisiphon device in the fill valve. An antisiphon device is a type of vacuum breaker that must be located a minimum of 1 inch (25 mm) above the overflow pipe to prevent its submersion. The 1-inch (25 mm) minimum is equivalent to the vacuum breaker critical level.

425.3.2 Overflows in flush tanks. Flush tanks shall be provided with overflows discharging to the water closet or urinal connected thereto and shall be sized to prevent flooding the tank at the maximum rate at which the tanks are supplied with water according to the manufacturer's design conditions. The opening of the overflow pipe shall be located above the flood level rim of the water closet or urinal or above a secondary overflow in the flush tank.

❖ The overflow in the flush tank prevents flooding of the tank and submersion of the fill valve. The fill valve antisiphon device must be located above water at all times to maintain the protection against backflow. The overflow discharges into the water closet or urinal, ultimately discharging to the drainage system.

When the flush tank is a separate unit (two-piece water closets), the overflow is always above the flood level rim of the water closet or urinal. If the water closet or urinal drain is obstructed, the water discharges onto the floor before submerging the fill valve antisiphon device. In certain styles of one-piece water closets, the overflow opening is below the water closet flood level rim; therefore, openings are provided in the tank to serve as a secondary overflow to prevent the flooding of the fill valve.

425.3.3 Sheet copper. Sheet copper utilized for flush tank linings shall conform to ASTM B 152 and shall not weigh less than 10 ounces per square foot (0.03 kg/m^2).

❖ Years ago, flush tanks for urinals and water closets were made of wood and hung on the wall high above the fixture. Since wood was not impervious to water, sheet copper, formed and soldered, was used to line the wood tank. Today, perhaps only historical building renovations would require application of this code section.

425.3.4 Access required. All parts in a flush tank shall be accessible for repair and replacement.

❖ A flush tank contains components that are subject to wear that could cause improper operation of the flush tank. It should not be necessary to replace a complete flush tank assembly where only a component in the flush tank needs repair or replacement.

425.4 Flush pipes and fittings. Flush pipes and fittings shall be of nonferrous material and shall conform to ASME A112.19.5 or CSA B125.

❖ Flush pipes and fittings connect a water closet tank to the bowl or a flush tank to a urinal. The ASME standard specifies only the dimensional requirements for the flush pipe and fittings. For the material requirement, this section stipulates only that it must be nonferrous material, such as brass, copper alloy or plastic. These materials are intended to extend the life of the pipe and fittings.

SECTION 426 MANUAL FOOD AND BEVERAGE DISPENSING EQUIPMENT

426.1 Approval. Manual food and beverage dispensing equipment shall conform to the requirements of NSF 18.

❖ NSF 18 establishes minimum food protection and sanitation requirements for the materials, design, construction and performance requirements for manual food and beverage dispensing equipment. This standard does not apply to vending machines, dispensing freezers and bulk milk dispensing equipment covered by the scope of other National Sanitation Foundation (NSF) standards.

SECTION 427 FLOOR SINKS

427.1 Approval. Sanitary floor sinks shall conform to the requirements of ASME A112.6.7.

❖ Floor sinks must comply with ASME A112.6.7, which covers the material requirements for cast-iron and PVC sanitary floor sinks. All floor sinks must have a removable strainer or accessible grate.

Bibliography

The following resource materials are referenced in this chapter or are relevant to the subject matter addressed in this chapter.

ANSI Z124.1-95, *Plastic Bathtub Units*. New York, NY: American National Standards Institute, 1995.

ANSI Z124.2-95, *Plastic Shower Receptors and Shower Stalls*. New York, NY: American National Standards Institute, 1995.

ANSI Z124.3-95, *Plastic Lavatories*. New York, NY: American National Standards Institute, 1995.

ANSI Z124.4-96, *Plastic Water Closet Bowls and Tanks*. New York, NY: American National Standards Institute, 1996.

ANSI Z124.6-97, *Plastic Sinks*.New York, NY: American National Standards Institute, 1997.

ANSI Z358.1-04, *Emergency Eyewash and Shower Equipment*. New York, NY: American National Standards Institute, 2004.

ANSI Z124.9-94, *Plastic Urinal Fixtures*. New York, NY: American National Standards Institute, 1994.

ARI 1010-02, *Self-contained, Mechanically-refrigerated Drinking-water Coolers*. Arlington, VA: Air-conditioning & Refrigeration Institute, 2002.

ASME A112.3.1-2007, *Stainless Steel Drainage Systems for Sanitary DWV, Storm, and Vacuum Applica-*

tions, Above- and Below Ground. New York, NY: American Society of Mechanical Engineers, 2007.

ASME A112.4.3-1999, *Plastic Fittings for Connecting Water Closets to the Sanitary Drainage System*. New York, NY: American Society of Mechanical Engineers, 1999.

ASME A112.6.1M-1997(R2002), *Floor Affixed Supports for Off-the-Floor Plumbing Fixtures for Public Use*. New York, NY: American Society of Mechanical Engineers, 2002.

ASME A112.6.2-2000, *Framing-affixed Supports for Off-the-floor Water Closets with Concealed Tanks*. New York, NY: American Society of Mechanical Engineers, 2000.

ASME A112.6.3-2002, *Performance Requirements for Backflow Protection Devices and Systems in Plumbing Fixture Fittings*. New York, NY: American Society of Mechanical Engineers, 2002.

ASME A112.6.7-2001, *Enameled and Epoxy Coated Cast Iron and PVC Plastic Sanitary Floor Sinks*. New York, NY: American Society of Mechanical Engineers, 2001.

ASME A112.18.1-2003, *Plumbing Fixture Fittings*. New York, NY: American Society of Mechanical Engineers, 2003.

ASME A112.18.3-2003, *Plumbing Fixture Fittings*. New York, NY: American Society of Mechanical Engineers, 2003.

ASME A112.18.7-1999, *Deck-mounted Bath/Shower Transfer Valves with Integral Backflow Protection*. New York, NY: American Society of Mechanical Engineers, 1999.

ASME A112.19.1 M-1994 (R1999), *Enameled Cast-iron Plumbing Fixtures*. New York, NY: American Society of Mechanical Engineers, 1999.

ASME A112.19.2M-2003, *Vitreous China Plumbing Fixtures—with 1996 Errata*. New York, NY: American Society of Mechanical Engineers, 2003.

ASME A112.19.3M-2000, *Stainless Steel Plumbing Fixtures (Designed for Residential Use)*. New York, NY: American Society of Mechanical Engineers, 2000.

ASME A112.19.4M-1 994 (R1999), *Porcelain Enameled Formed Steel Plumbing Fixtures*. New York, NY: American Society of Mechanical Engineers, 1999.

ASME A112.19.5-1999, *Trim for Water-closet Bowls, Tanks, and Urinals*. New York, NY: American Society of Mechanical Engineers, 1999.

ASME A112.19.6-1995, *Hydraulic Performance Requirements for Water Closets and Urinals*. New York, NY: American Society of Mechanical Engineers, 1995.

ASME A112.19.7M-1995, *Whirlpool Bathtub Appliances*. New York, NY: American Society of Mechanical Engineers, 1995.

ASME A112.19.8M-1987 (R1996), *Suction Fittings forUse in Swimming Pools, Wading Pools, Spas, HotTubs, and Whirlpool Bathtub Appliances*. New York, NY: American Society of Mechanical Engineers, 1996.

ASME A112.19.9M-1991(R2002), *Nonvitreous Ceramic Plumbing Fixtures*. New York, NY: American Society of Mechanical Engineers, 2002.

ASPE-95, *A Critique of Queueing Theory Approaches to Plumbing Design*. Westlake, CA: American Society of Plumbing Engineers, 1995.

ASSE 1001-02, *Performance Requirements for Pipe-applied Atmospheric-type Vacuum Breakers*. Westlake, OH: American Society of Sanitary Engineering, 2002.

ASSE 1002-99, *Performance Requirements for Water Closet Flush Tank Fill Valves*. Westlake, OH: American Society of Sanitary Engineering, 1999.

ASSE 1004-90, *Performance Requirements for Commercial Dishwashing Machines*. Westlake, OH: American Society of Sanitary Engineering, 1990.

ASSE 1006-89, *Performance Requirements for Residential Use (Household) Dishwashers*. Westlake, OH: American Society of Sanitary Engineering, 1989.

ASSE 1007-92, *Performance Requirements for Home Laundry Equipment*. Westlake, OH: American Society of Sanitary Engineering, 1992.

ASSE 1008-89, *Performance Requirements for Household Food Waste Disposer Units*. Westlake, OH: American Society of Sanitary Engineering, 1989.

ASSE 1009-90, *Performance Requirements for Commercial Food Waste Grinder Units*. Westlake, OH: American Society of Sanitary Engineering, 1990.

ASSE 1014-90, *Performance Requirements for Hand-held Showers*. Westlake, OH: American Society of Sanitary Engineering, 1990.

ASSE 1016-96, *Performance Requirements for Individual Thermostatic, Pressure Balancing and Combination Control Valves for Bathing Facilities*. Westlake, OH: American Society of Sanitary Engineering, 1996.

ASSE 1017-99, *Performance Requirements for Temperature Actuated Mixing Valves for Hot Water Distribution Systems*. Westlake, OH: American Society of Sanitary Engineering, 1999.

ASSE 1037-90, *Performance Requirements for Pressurized Flushing Devices for Plumbing Fixtures*. Westlake, OH: American Society of Sanitary Engineering, 1990.

ASSE 1062-97, *Performance Requirements for Temperature Actuated, Flow Reduction Valves to Individual*

Fixture Fittings. Westlake, OH: American Society of Sanitary Engineering, 1997.

ASSE 1070-04, *Performance Requirements for Water Temperature Limiting Devices*. Westlake, OH: American Society of Sanitary Engineering, 2004.

ASTM B 152/B152M-00, *Specification for Copper Sheet, Strip Plate and Rolled Bar*. West Conshohocken, PA: Society for Testing and Materials, 2000.

ASTM D 4068-01, *Specification for Chlorinated Polyethylene (CPE) Sheeting for Concealed Water-containment Membrane*. West Conshohocken, PA: American Society for Testing and Materials, 2001.

ASTM D 4551-96(2001), *Specification for Poly (Vinyl Chloride) (PVC) Plastic Flexible Concealed Water-Containment Membrane*. West Conshohocken, PA: American Society for Testing and Materials, 2001.

ASTM E 84-04, *Test Method for Surface Burning Characteristics of Building Materials*. West Conshohocken, PA: American Society for Testing and Materials, 2004.

ASTM F 409-02, *Specification for Thermoplastic Accessible and Replaceable Plastic Tube and Tubular Fittings*. West Conshohocken, PA: American Society for Testing and Materials, 2002.

CSA B45.1-02, *Ceramic Plumbing Fixtures*. Rexdale (Toronto), Ontario, Canada: Canadian Standards Association, 2002.

CSA B45.2-02, *Enameled Cast-iron Plumbing Fixtures*. Rexdale (Toronto), Ontario, Canada: Canadian Standards Association, 2002.

CSA B45.3-02, *Porcelain Enameled Steel Plumbing Fixtures*. Rexdale (Toronto), Ontario, Canada: Canadian Standards Association, 2002.

CSA B45.4-02, *Stainless-steel Plumbing Fixtures*. Rexdale (Toronto), Ontario, Canada: Canadian Standards Association, 2002.

CSA B45.5-02, *Plastic Plumbing Fixtures*. Rexdale (Toronto), Ontario, Canada: Canadian Standards Association, 2002.

CSA B45.10-02, *Hydromassage Bathtubs*. Rexdale (Toronto), Ontario, Canada: Canadian Standards Association, 2002.

CSA B79-94(2000), *Floor, Area and Shower Drains, and Cleanouts for Residential Construction*. Rexdale (Toronto), Ontario, Canada: Canadian Standards Association, 2000.

CSA B125-01, *Plumbing Fittings*. Rexdale (Toronto), Ontario, Canada: Canadian Standards Association, 2001.

CSA B125.1-05, *Plumbing Supply Fittings*. Rexdale (Toronto), Ontario, Canada: Canadian Standards Association, 2005.

CSA B125.3-05, *Plumbing Fittings*. Rexdale (Toronto), Ontario, Canada: Canadian Standards Association, 2005.

FS TT-P-1 536a-75, *Federal Specification for Plumbing Fixture Setting Compound*. Washington, DC: Federal Specification, 1975.

IBC-09, *International Building Code*. Washington, DC: International Code Council, 2009.

ICC A117.1-2003, *Accessible and Usable Buildings and Facilities*. Washington, DC: International Code Council, 2003.

IFGC-09, *International Fuel Gas Code*. Washington, DC: International Code Council, 2009.

IMC-09, *International Mechanical Code*. Washington, DC: International Code Council, 2009.

IRC-09, *International Residential Code*. Washington, DC: International Code Council, 2009.

NSF 3-2003, *Commercial Spray-type Dishwashing Machines*. Ann Arbor, MI: National Sanitation Foundation, 2003.

NSF 2004, *Manual Food and Beverage Dispensing Equipment*. Ann Arbor, MI: National Sanitation Foundation, 2004.

NSF 61-2003e, *Drinking Water System Components—Health Effects*. Ann Arbor, MI: National Sanitation Foundation, 2003

Chapter 5: Water Heaters

General Comments

A water heater is any appliance that heats potable water and supplies it to the potable hot water distribution system of the building. Some water heaters might also provide hot water for space heating purposes in addition to supplying hot water to the water distribution system. This code does not apply to appliances, equipment and devices that provide hot water for the sole purpose of space heating (hydronic heating systems) because (1) by definition in the code, a water heater supplies water to the potable water distribution system and (2) Section 607.1 of the code requires a supply of hot or tempered water to all plumbing fixtures and equipment used for bathing, washing, culinary purposes, cleansing, laundry or building maintenance within occupied structures. Hydronic heating systems are regulated by the *International Mechanical Code*® (IMC®).

Historically, the main focus of this chapter has been to regulate the installation and safety features for automatic storage tank water heaters and unfired hot water storage tanks. Other water heating methods and equipment such as solar, electric heat pump, ground source thermal transfer, boiler water recirculation, steam heat transfer, solid-fuel-fired technology and the necessary controls/safety devices for those methods are not specifically addressed in the code but many are covered in the IMC.

Questions often arise as to how the regulations in this chapter are intended to apply to these alternative water heating methods. Where the code doesn't specifically address those questions, the manufacturer's installation instructions must serve as the basis for installation of these products (see Section 502.1). For example, Section 504.7 requires that a pan be installed under a water heater if leakage from the heater could cause damage. If the water heater is a tankless water heater, is a pan required? In all probability, the proponents that originally created this section did not envision tankless water heaters or any other alternative water heating methods when drafting the wording. Therefore, in this example, it is not likely that the intent of the code is to require a pan under a tankless water heater. However, since the application of the code in every situation depends upon how the code official interprets the code requirements stated, the designer or installer of these alternative water heating methods is advised to consult with the authority having jurisdiction before work is performed.

Wherever hot water is stored in a tank, there is a potential for explosion. There are documented cases where improperly installed (or poorly maintained) water heaters have exploded with such force that they have been propelled through floor and roof assemblies and over 100 feet (30 480 mm) in the air. Thus, the intent of this chapter is to regulate the materials, design and installation of water heaters and related safety devices in order to protect property and life.

Purpose

Chapter 5 regulates the design, approval and installation of water heaters and related safety devices. The intent is to minimize the hazards associated with the installation and operation of water heaters.

SECTION 501 GENERAL

501.1 Scope. The provisions of this chapter shall govern the materials, design and installation of water heaters and the related safety devices and appurtenances.

❖ This section defines the scope of the chapter and contains requirements for water heaters used for space heating, the installation of drain valves, the location of water heaters, third-party certification of water heaters, tankless water heaters, marking of storage tanks and temperature controls for hot water supply systems.

The materials, design and installation of water heaters and related safety equipment and appurtenances are regulated by this chapter. The requirements in this chapter are intended to establish that water heaters and the necessary controls and safety devices are designed and installed so that they perform in an acceptable manner.

501.2 Water heater as space heater. Where a combination potable water heating and space heating system requires water for space heating at temperatures higher than 140°F (60°C), a master thermostatic mixing valve complying with ASSE 1017 shall be provided to limit the water supplied to the potable *hot water* distribution system to a temperature of 140°F (60°C) or less. The potability of the water shall be maintained throughout the system.

❖ Where a water heater has a dual purpose of supplying hot water and serving as a heat source for a hot water space heating system, the maximum outlet water temperature for the potable hot water distribution system is limited to 140°F (60°C). A master thermostatic mixing valve conforming to ASSE 1017 must be installed to limit the water temperature to 140°F (60°C) or less. These valves are used extensively in applications for domestic service to mix hot and cold water to reduce high service water temperature to the building distribution system. These devices are not intended for final temperature control at fixtures and appliances.

A water heater used as part of a space heating system must be protected from any conditions that can cause contamination of the potable water supply system. A typical installation might be an under-floor radiant heating system. Because the water heater is part of the potable water system, materials used in the heating system must be approved for use in a potable water system, and all connections must be protected against contamination. In the summer months when the heating system is inactive, a method to prevent stagnation of the water should be employed. Small orifices in the isolation valves to permit a small amount of water to circulate through the system could be provided for this purpose. Chemicals of any type must not be added to the heating system, because this would directly contaminate the potable water supply. Protection of the potable water supply must be in accordance with Section 608 (see Commentary Figure 501.2).

501.3 Drain valves. Drain valves for emptying shall be installed at the bottom of each tank-type water heater and *hot water* storage tank. Drain valves shall conform to ASSE 1005.

❖ A water heater must be drainable to facilitate service, sediment removal, repair or replacement. Drain valves must be constructed and tested in accordance with ASSE 1005, which requires the valve inlet to be $^3/_4$-inch (19 mm) nominal iron pipe size, the outlet to be equipped with a standard $^3/_4$-inch (19 mm) male garden hose thread and a straight-through waterway of not less than $^1/_2$-inch (13 mm) diameter.

501.4 Location. Water heaters and storage tanks shall be located and connected so as to provide *access* for observation, maintenance, servicing and replacement.

❖ Because water heaters and the potable water connections require routine inspections, maintenance, repairs and possible replacement, access is required. Additionally, access recommendations or requirements are usually stated in the manufacturers' installation instructions. Thus, the provisions stated herein are intended to supplement the manufacturers' installation instructions.

The intent is to provide access to all components that require observation, inspection, adjustment, servicing, repair and replacement. Access is also necessary to conduct operating procedures such as start-up or shutdown.

The code defines "Access" as being able to be reached but which may first require the removal of a panel, door or similar obstruction. Access to a water heater is not achieved if it is necessary to remove any portion of the structure's permanent finish materials, such as drywall, plaster, paneling, built-in furniture or cabinets, or any other similar permanently affixed building component.

There is not always sufficient room for installation of a water heater in spaces such as basements, alcoves, utility rooms and furnace rooms. In an effort to save space or simplify an installation, water heaters are of-

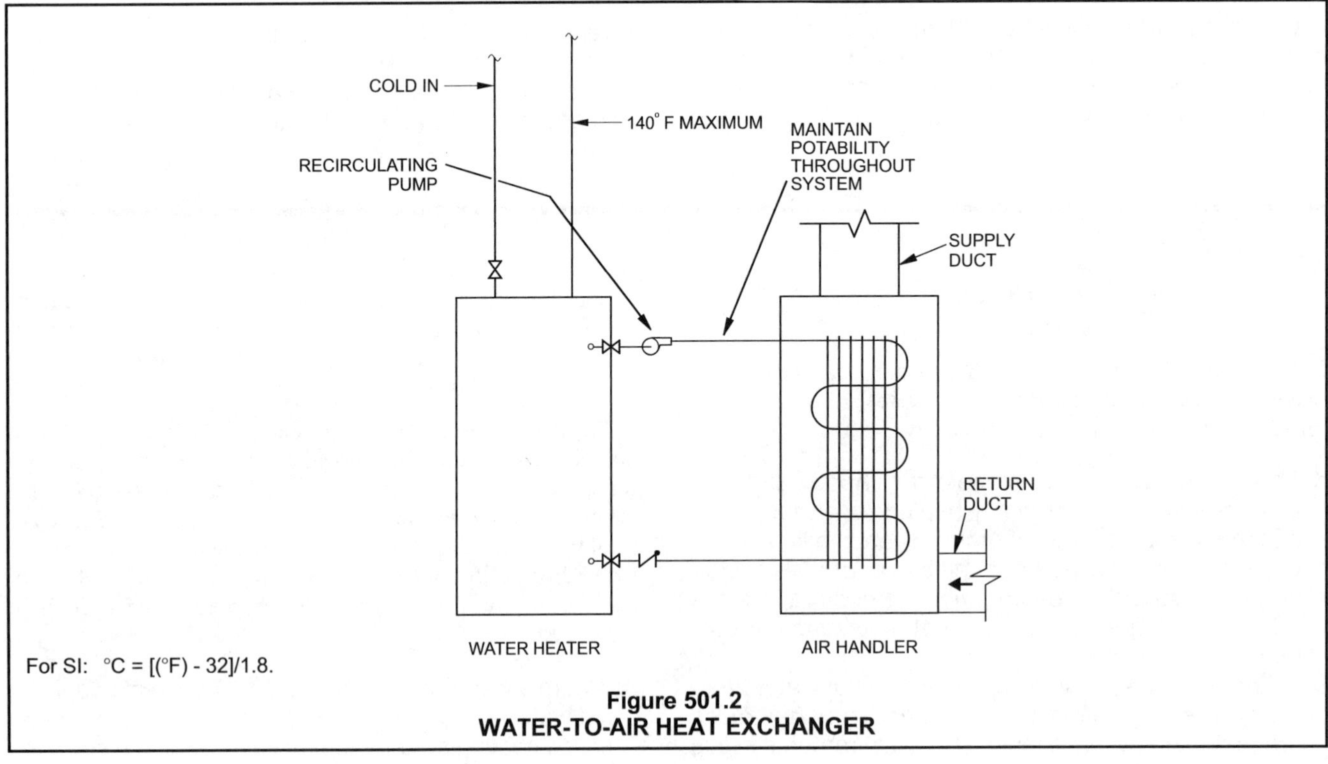

For SI: °C = [(°F) - 32]/1.8.

Figure 501.2
WATER-TO-AIR HEAT EXCHANGER

ten installed in crawl spaces, attics or similar remote locations. Access to and servicing of water heaters could be difficult; therefore, a walkway, work space or platform is necessary to provide a large enough path of travel to, and adequate space for, such activities in accordance with Section 502.3.

501.5 Water heater labeling. All water heaters shall be third-party certified.

❖ See the commentary for Section 303.4 for a general description of third-party certification of products. The code does not specify the standards that water heaters must comply with; it simply requires that all water heaters be third-party certified. Other I-Codes require compliance with the following standards as applicable to the type of water heating unit.

The *International Fuel Gas Code*® (IFGC®) requires that fuel gas fired water heaters comply with one of the following:

ANSI Z21.10.1 Gas-fired storage tank water heaters with Btu input of 75,000 or less

ANSI Z21.10.3 Gas-fired storage tank water heaters with Btu input greater than 75,000 and gas-fired circulating and (gas-fired) instantaneous water heaters.

The IMC requires that oil-fired water heaters comply with the following:

UL 732 Oil-fired storage tank water heaters with a Btu input of not greater than 200,000, having a storage capacity not more than 120 gallons, and producing water temperatures not higher than 200°F.

The IMC requires that electric water heaters comply with one of the following as applicable:

UL 174 Household electric storage water heaters from 1 to 120 gallons having an input of 12kW or less.

UL 1453 Commercial electric booster heaters and storage water heaters that meet at least one of the following criteria:

- Have a capacity over 120 gallons
- Have an input greater than 12 kW
- Have the capability of being set to discharge water temperature higher than 185 degrees F

Note that the IMC does not indicate a standard for hot water dispensers or electric instantaneous water heaters. This does not mean that these devices cannot be installed just because the codes do not specify a standard that is appropriate for a particular type of water heating unit. The code simply requires that water heaters be third-party certified so that any type of water heating unit that is not covered by the standards indicated by the IMC or the IFGC will be built and third-party certified to a nationally recognized standard. A few examples are:

UL 499 *Electric Heating Appliances*. This standard covers hot water dispensers of less than 1 gallon capacity and all electric instantaneous water heaters.

UL 1995 *Heating and Cooling Equipment*. This standard covers electric heat pump water heaters.

The certification mark is the primary, if not the only, indicator to the installer, the inspector and the end user that the water heater has been evaluated and tested by an approved agency and has been determined to perform in an acceptable manner where installed and operated in accordance with the manufacturer's instructions.

The approved agency can require the manufacturer to alter, add or delete information in the installation instructions as necessary to achieve compliance with the applicable standard. Because water heaters must be installed in accordance with the manufacturer's instructions, these instructions must be available to the code official at the time of inspection to verify that the water heater has been properly installed.

501.6 Water temperature control in piping from tankless heaters. The temperature of water from tankless water heaters shall be a maximum of 140°F (60°C) when intended for domestic uses. This provision shall not supersede the requirement for protective shower valves in accordance with Section 424.3.

❖ The intent of this section is to prevent excessively high water temperatures from reaching plumbing fixtures. Tankless heaters do not have a storage capacity and are often called "instantaneous" because the water is heated at the same rate it is used. Instantaneous heaters can discharge an uncertain range of temperatures at any given time, depending on the use. Therefore, some form of temperature control is necessary to protect the user against exposure to excessively hot water discharged from domestic fixtures, such as lavatories, kitchen sinks, tubs and laundry trays. This could be accomplished by installation of a tempering valve [set at 140°F (60°C)] on the outlet of the water heater or by equipping the heater with a temperature-limiting device or thermostat which has a maximum setting of 140°F (60°C).

Where a tankless water heater is used to supply hot water to a shower or a tub/shower combination, the maximum shower or tub/shower control valve outlet temperature must be limited to 120°F (49°C) in accordance with Section 424.3.

Tankless water heaters have become increasingly popular in recent years for heating potable water. But, tankless water heating is not new; it was invented in the 1870s. They were the first devices to provide a "quick" supply of hot water as compared to previous methods of placing pots over a fire and waiting for the water to come to a simmer or boil. These tankless units worked well for applications where there was a demand for an "instant" supply of small volumes of hot water. For larger "quick" hot water demands, storage tank water heaters were invented and soon became the most economical market solution for meeting most hot water demands. Tankless water heating soon became obsolete in many parts of the world, including North America. However, because users in recent years are demanding continuous unlimited streams of hot water for the simultaneous operation of hot water-consuming appliances/fixtures as well as for providing large volumes of hot water for multihead luxury showers, tankless water heating has been reborn in North America and abroad. The desire to save floor space and to conserve energy by reducing standby losses are also major factors in this trend.

The sizing of any type of water heating equipment for domestic applications is not an exact science. The difficulty arises because user behavior patterns are not 100 percent predictable. There have been many studies of hot water demand for a variety of residential, institutional and commercial applications. At best, these studies have concluded that without significant oversizing of the hot water producing equipment, users will inevitably run out of hot water as a result of changes in their normal patterns of hot water use. For instance, a family might have an established pattern of bathing, clothes washing and dishwashing that allows the water heater to provide just for those hot water draws without a noticeable decrease in delivery temperature. If the family's living pattern changes because of schedule changes, visiting relatives, a child being born or family members simultaneously being home for shorter intervals, hot water demand will undoubtedly increase, possibly resulting in the family running out of hot water. Customer satisfaction surveys in one study revealed that in general, single-family home occupants are satisfied with their hot water availability even if they run out of hot water up to 12 times per year. Most residential users realize that there is a limit to hot water availability and if they run out of hot water frequently, they adjust their living patterns to accommodate the current water heating capacity. Only in severe cases will users opt for changing out the water heating unit to a larger capacity to accommodate their family's pattern of higher demand. Clearly, user expectations have been molded by generations of use patterns that accommodated water heating sizing based on rules-of-thumb/best-methods-of-practice developed over the years.

Tankless water heaters are rated at so many gallons per minute (gpm) at so many °F of water temperature rise. There are many different models of electric and fuel-gas powered tankless water heaters, each having a specific rating. If the largest tankless water heater cannot accommodate the total demand, additional tankless units are added to meet the total demand. For electric units, each can be installed to serve a specific fixture or group of fixtures that can be accommodated by that unit. Where additional fuel-gas units are required to meet the total demand, each unit could be connected to serve only specific fixture groups within the building. However, fuel-gas powered unit controls are often sophisticated enough so that multiple units can be connected to a common header and the controls interconnected so as to cause additional units to "come on line" when hot water demand increases beyond the capability of the first unit.

The first challenge in tankless water heater sizing is to determine the demand in gallons per minute. Early studies by Roy B. Hunter in the area of overall water demand established the concept of "water supply fixture units" (w.s.f.u.).This unit of measure combined the probability and frequency of the use of common fixtures and related w.s.f.u. to an average demand in gallons per minute. The study was based on a variety of buildings having large total demands (hundreds of fixtures and thousands of gallons flow). Demand curves were developed and from those curves, Appendix E Tables E103.3(2) and E103.3(3) were extrapolated. Unfortunately, because of the size of water demands in the studies as well as the variety of occupancies examined, the accuracy of these tables for estimating precise demands for small numbers of fixtures is questionable. Nonetheless, these tables provide a base of factual information from which to, at least, begin to estimate hot water demand for any particular building.

Because water supply fixture units already include the probability and frequency of fixture use, one could simply add up the hot water w.s.f.u. for all fixtures within a building and then look up in Table E103.3(3) the gallons per minute that the total w.s.f.u. represents.

Example

A two bedroom house has one shower, one bathtub with shower, two lavatories, one clothes washer, one dishwasher, one kitchen sink, and one laundry tray. Determine the required gallon-per-minute rating for a tankless water heater.

From Table E103.3(2), the hot water w.s.f.u. load = 1+1+0.5+0.5+1+1.4+1+1=7.4 w.s.f.u. Using Table E103.3(3), this load equates to 12.2 gpm of hot water needed at any particular moment. This is not a maximum peak demand, i.e., all these fixtures/appliances operating at once, but a flow that might have a reasonable probability of occurring.

The second challenge in tankless water heater sizing is to determine what temperature rise is required for that flow. "Temperature rise" refers to the change in water temperature across the heater (i.e., the difference between inlet and outlet temperatures).

One of the largest savings in energy used to heat water can be realized by not heating water to any higher temperature than is necessary for the end use. For instance, with a storage tank water heater set at 120°F (49°C) , the hot water for a shower must be mixed with cold water to achieve a comfortable shower temperature of approximately 105°F (41°C). Water is "overheated" for the intended use at 105°F

(41°C) and thus energy is wasted. The ideal situation is to heat the water as it is being used and only to a temperature level that just meets the end use temperature requirement.

A comfortable temperature for hand washing, showers and bathtubs is generally accepted to be 105°F (41°C). Clothes washing temperatures can be variable because some use cold water only to wash clothes and some wash in warm water and rinse in cold. The kitchen sink and laundry tray (sink) might have needs for higher temperature water but generally, these sinks are user adjusted to discharge water that is comfortable to the touch. Dishwashers usually need at least 120°F (49°C) water to operate properly. An assumption is made to consider 105°F (41°C) water as sufficient for most needs, but, as will be explained later, higher temperature water will be available from a tankless water heater for any application under conditions that the user can establish by adjusting his/her uses patterns.

The manufacturer's ratings of tankless water heaters are given in table form [see Commentary Table 501.6(1)] or in graph form [see Commentary Figure 501.6(1)].

The term "temperature rise" means that the water temperature at the outlet of the heater will be so many degrees hotter than the incoming water temperature. If the incoming water temperature is cooler or warmer, the tankless water heater will increase the temperature of the incoming water by the same amount. Thus, as incoming cold water temperature varies throughout the year, the outlet water temperature will also vary accordingly.

The minimum temperature of the incoming water must be determined. For this example, a minimum temperature of 55°F (13°C) is assumed. Therefore, a tankless water heater having a temperature rise of 105 - 55 = 50°F at 12.2 gpm is required. For the electric tankless water heater ratings given in Commentary Figure 501.6(2), approximately five Model B units would be required to meet the demand because each unit is capable of only 2.6 gpm at a 50°F temperature rise.

However, in actual practice, smaller numbers of tankless water heater units or units with smaller ratings are being used with good success. The tankless water heater industry promotes sizing techniques that use the number of showers that will be operating simultaneously to guide the system designer toward a particular model. Although a particular building might have many fixtures that require hot water, not all those fixtures will be used at the same time. Because users tend to complain the most about not having enough hot water when showers are taken, that fixture has been selected to be the driver for selection of tankless water heaters. In the example, an assumption could be made that both showers would be used simultaneously.

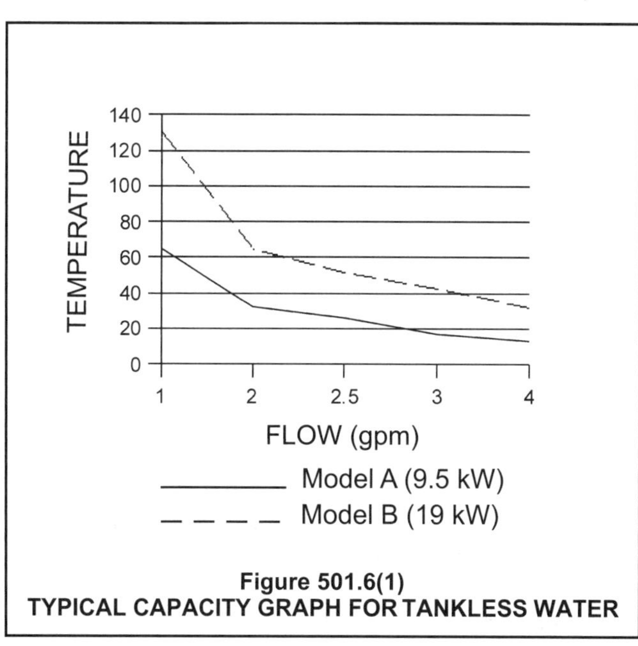

Figure 501.6(1)
TYPICAL CAPACITY GRAPH FOR TANKLESS WATER

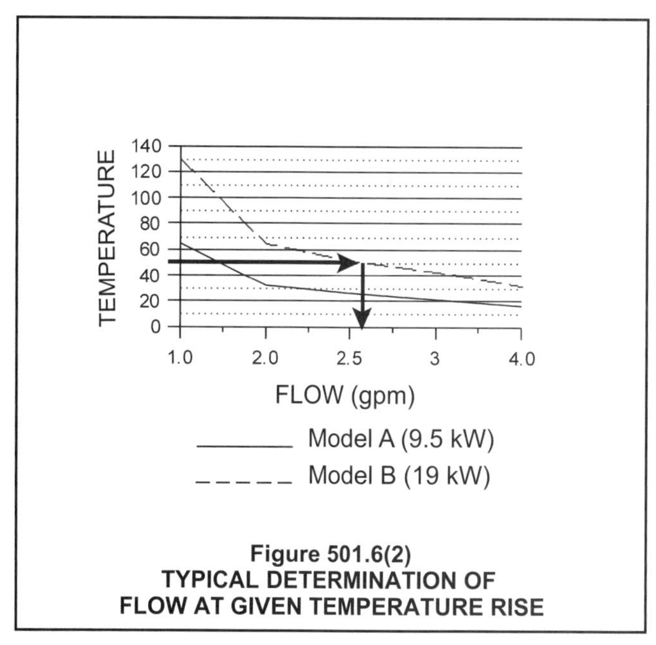

Figure 501.6(2)
TYPICAL DETERMINATION OF FLOW AT GIVEN TEMPERATURE RISE

Table 501.6(1)
TYPICAL CAPACITY TABLE FOR TANKLESS WATER HEATERS

FLOW	TEMPERATURE RISE, °F		
	1.0 gpm	**2.0 gpm**	**4.0 gpm**
Model A (9.5 kW)	65	32	16
Model B (19 kW)	130	65	32

According to one manufacturer, an appropriately sized tankless unit for this house having two showers operating at the same time would be one Model B. However, in reviewing the ratings table for Model B, it is rated for only 2.6 gpm at 50°F temperature rise. Even though this is a large discrepancy from the originally calculated 12.2 gallons per minute requirement, single Model B units are used successfully in applications like the example.

If the users make some accommodations, these small tankless water heaters can provide minimum acceptable flows of hot water. For instance, when two showers are being taken at the same time, no cold water can be mixed into the shower hot water stream. The flow from each shower head may have to be restricted so that no more than 1.25 gallons per minute flows from each head. Both shower mixing valves are adjusted to full open hot positions. This results in a shower water temperature of noT more than 105°F. If anyone else opens another hot water faucet in the house, the pressure will decrease at the showers and the temperature of the hot water at all outlets will be cooler.

User satisfaction where only one Model B is installed has to do with a change in user expectations and patterns when using hot water supplied by a tankless water heater. Obviously, if the user does not want to change his or her expectations for hot water use, more and/or larger tankless water heater units will be needed to lessen the effects of multiple simultaneous uses. As tankless water heaters are designed to provide a specific temperature rise at a given maximum flow rate (because passing higher flows of water through that same unit does not allow the water to be appreciably heated), there is a pressure loss associated with flows in excess of a unit's usable flow rating. For instance, in a house with a tankless water heater, it might be possible to have a nearly "no flow of hot water" condition if too many faucets are opened simultaneously. Tankless water heater users need to be informed that if such a condition develops, they must close some of the open faucets to restore flow at the remaining open faucets.

Hot-water-using appliances requiring water temperatures above 105°F are easily accommodated by (1) slightly closing the hot water supply valve at the appliance so that the desired temperature of water is achieved, and (2) operating no other faucets or hot water-using appliances when the "primary" appliance is demanding hot water. Any fixture can offer water hotter than 105°F by (1) allowing no other hot water to be used elsewhere and (2) reducing the hot water flow until the desired water temperature is achieved.

Depending on how systems were originally designed, the users of tankless water heater systems might be required to adjust their expectations concerning when and to what extent simultaneous hot water demands can be expected before temperature (and flow) of the hot water is affected. This adjustment of expectations is no different than for most users of storage tank water heater systems that implicitly understand that there is a limit to how much hot water can be used before the temperature of the hot water becomes unacceptable for the purposes intended and how much time it takes for the system to recover sufficiently in order to, again, provide an acceptable supply and temperature level of hot water.

501.7 Pressure marking of storage tanks. Storage tanks and water heaters installed for domestic hot water shall have the maximum allowable working pressure clearly and indelibly stamped in the metal or marked on a plate welded thereto or otherwise permanently attached. Such markings shall be in an accessible position outside of the tank so as to make inspection or reinspection readily possible.

❖ Water heaters, like all pressure vessels, must be able to withstand the working pressures to which they will be subjected. The maximum working pressure of the water heater must be known so that a properly rated relief valve not exceeding the manufacturer's rated working pressure can be installed. The working pressure must be marked in such a way that it is permanent, not susceptible to damage and legible for the life of the water heater. Therefore, if the relief valve requires replacement, a properly sized valve can be reinstalled. The marking on the water heater must be located so that it is readily accessible during inspections and servicing.

501.8 Temperature controls. All hot water supply systems shall be equipped with automatic temperature controls capable of adjustments from the lowest to the highest acceptable temperature settings for the intended temperature operating range.

❖ The intent of this section is to prohibit manually controlled water-heating systems that might cause dangerous water temperatures to be generated. Automatic controls provide for reliable, consistent operation of the unit and most often have integrated safety devices to protect the equipment from unsafe operating conditions. Where systems consist of circulating-type heaters with separate hot water storage tanks, automatic controls connected to a remote temperature sensor in the storage tank(s) are required for control of the stored water temperature.

The user must be provided with a means for adjustment of the temperature of the hot water generated so that an appropriate water temperature for the intended purposes can be selected. Other than the requirement in Section 501.6 for tankless water heaters, the code does not specify a minimum or maximum temperature setting for a water heater temperature control because of the variety of applications requiring "hot" water. A few examples are (1) 180°F of (82°C) hot water might be necessary for the operation of a commercial dishwasher to provide for sterilization of dishware, (2) 120°F (49°C) hot water might be required for the potable water distribution system of a hotel not having a kitchen and (3) 140°F (60°C) hot water might be required to provide for space heating in addition to supplying water to potable water distribution system. The user and installer of the water heating system are responsible for selecting the appropriate temperature setting for the water heater.

Although the need for adjustability of the hot water temperature setting of a water heater is understandable, this adjustability feature has drawbacks. Where a water heater serves fixtures in addition to the appliances or equipment that require a hot water temperature greater than 120°F (49°C), the potential for skin burns exists at those fixtures. Commentary Table 424.3 is a temperature burn chart illustrating the amount exposure time at various hot water temperatures for third-degree burns of human skin. Because most water heater temperature controls can be easily adjusted by nearly anyone, the code requires certain fixtures to have temperature limiting devices to protect users from high temperature hot water. These fixtures include showers, bathtubs, whirlpool tubs, bidets and faucets for public handwashing. The requirements for temperature limiting devices for these fixtures are located in Chapter 4. All other fixtures such as sinks, lavatories in sleeping and dwelling units, bar sinks and laundry trays (tubs) are not required to have temperature limiting devices and as such, could be locations where burn injury could occur.

A water heater temperature setting that is too low could possibly cause health problems, specifically Legionaires' disease. The disease was realized and named after an outbreak of pneumonia-like illness that was contracted by 221 persons attending an American Legion convention at a hotel in Philadelphia in July 1976. Two-thirds of the patients required hospitalization of which 34 died. Although this type of bacteria had been in existence and causing disease for many years before that, this large-scale tragic event caused significant research to be performed to fully understand the origin and transmission of what is now called Legionella. Legionella are warm water organisms that are found primarily in surface waters of lakes, ponds, rivers and streams. Because Legionella are protected by being inside of host organisms, typical potable water supply disinfection methods do not eradicate the bacteria. Therefore, Legionella can be found in many potable water systems of buildings. The potential for Legionella to cause disease dramatically increases where conditions for growth (amplification) exist and actions/activities cause the bacteria to be inhaled (aspirated). Because the optimum temperature range for Legionella growth in potable water is 86 to 104°F (30°C to 40°C), water heater temperature settings in this range should be avoided. Temperatures of 122°F (50°C) for 2 hours or 140°F (60°C) for 2 minutes results in 90 percent destruction of Legionella bacteria. A temperature of approximately 158°F (70°C) results in instaneous 100 percent destruction of the bacteria. Although not a requirement of the code, many hot water system designers require stored hot water to be at a temperature not less than 140°F (60°C) so as to reduce the potential for Legionella to exist and grow in hot water storage tanks. Regardless of the water heater temperature setting chosen, this code requires tempering valves to be installed at locations where the risk of burn injury is especially high (see Chapter 4).

SECTION 502 INSTALLATION

502.1 General. Water heaters shall be installed in accordance with the manufacturer's installation instructions. Oil-fired water heaters shall conform to the requirements of this code and the *International Mechanical Code*. Electric water heaters shall conform to the requirements of this code and provisions of NFPA 70. Gas-fired water heaters shall conform to the requirements of the *International Fuel Gas Code*.

❖ The installation of a water heater relies heavily on the manufacturer's installation instructions as the instructions are an integral part of the third-party certification for these appliances. However, in some cases, the referenced *International Codes*® (I-Codes®) contain more restrictive requirements which must be followed as well. Fuel-gas-fired water heater installation requirements are covered by the IFGC. The IMC covers installation requirements for oil-fired and electric water heaters.

Water heaters are appliances and, as such, the IMC or the IFGC contains additional requirements concerning them. For example, installation guidance for fuel pipe sizing and connections, appliance location and clearances, determination of the adequacy of combustion air, combustion air opening size and location and venting requirements are contained in the IFGC. NFPA 70 contains additional requirements for the installation of electric water heaters, such as the size of the electrical supply and disconnect provisions. In many situations, the installation requirements between the code and the manufacturer's installation instructions are identical. However, where there is a conflict between the code and the manufacturers' installation instructions, the requirements that provide the greatest level of safety apply.

Even if the water heating appliance appears to be in compliance with the manufacturer's instructions, the installation cannot be considered complete nor can it be approved until all associated components, connections and systems that serve the water heater are also in compliance with the applicable code provisions. For example, a fuel-fired water heater installation cannot be approved if the water heater is connected to a deteriorated, undersized or otherwise unsafe chimney or vent that does not comply with the IMC or the IFGC. Likewise, the same installation could not be approved if the existing fuel piping has insufficient capacity to supply the gas demand of the water heater, or the existing electrical supply conductors are inadequate or unsafe.

502.1.1 Elevation and protection. Elevation of water heater ignition sources and mechanical damage protection requirements for water heaters shall be in accordance with the *International Mechanical Code* and the *International Fuel Gas Code*.

❖ An ignition source could be an open flame such as a pilot light, an electrical switch, open resistance heating coils or an electrical igniter unit (such as hot surface or spark ignition types). Certain areas of a building are known to have a high potential for volatile liquids, such as gasoline and paint thinners, to be spilled or leak from their containers. Because the vapors of these liquids are heavier than air, they collect and concentrate just above floor level, posing an explosion hazard. Testing has indicated that locating any possible ignition sources at some elevation above the floor in these areas provides a degree of safety due to the time delay for the vapors from a spilled combustible liquid to build up to ignitable concentrations at the elevation of the ignition source. However, given large enough spills and enough time, any ignition source in an enclosed area could cause an explosion to occur.

Section 304.3 of the IMC requires that where equipment and appliances having an ignition source are located in hazardous locations, repair garages, private garages, motor fuel dispensing facilities or parking garages, the ignition source must be located at least 18 inches (457 mm) above the floor of where the equipment rests. Equipment and appliances having an ignition source are prohibited in Group H occupancies or control areas where open use, handling or dispensing of combustible, flammable or explosive materials occurs. Note that rooms or spaces that are not part of the living space of a dwelling unit and that communicate directly with a private garage are considered to be part of the private garage. For example, a storage room on the same level of the garage having a door into the garage is in “direct communication” (in terms of the exchange of air) with the garage.

Most electric water heater thermostats have enclosed contacts but they are not sealed gas tight. Therefore, if an electric water heater has a thermostat located less than 18 inches (457 mm) from the bottom of the unit, the IMC requires that the unit be elevated so that the ignition source (thermostat) is at least 18 inches (457 mm) above the garage floor. Electric water heaters having all switching controls located above 18 inches (457 mm) from the bottom of the water heater are not required to be elevated.

Sections 305.2 and 305.3 of the IFGC mirror the same ignition source requirements as the IMC except where the appliance is listed as flammable vapor ignition resistant (FVIR). Note that an FVIR-listed appliance is still prohibited from be located in a Group H area. Although FVIR technology has been proven to be reliable, some jurisdictions might still require elevation of FVIR fuel-gas-fired water heaters. Therefore, always check with the authority having jurisdiction regarding elevation requirements for FVIR listed appliances. (See IFGC commentary for more information on FVIR appliances.)

A frequently asked question is how this section should be applied to a replacement water heater that originally was not required by the code at that time to be elevated or have protection against mechanical damage. The code makes no distinction between an installation in new construction and an installation in existing construction. Where installers are confronted with obstacles that appear to prevent strict compliance with the code, they should consult with the code official before performing the replacement work.

502.2 Rooms used as a plenum. Water heaters using solid, liquid or gas fuel shall not be installed in a room containing air-handling machinery when such room is used as a plenum.

❖ Installation of fuel-fired water heaters is not permitted in rooms that are used as a plenum because of the potential for toxic products of combustion to be spread throughout the building in the event that the water heater or venting system malfunctions. Also, the negative and positive pressures developed within a plenum space can seriously affect the operation of a fuel-fired water heater located there.

For example, if a natural-draft fuel-fired water heater was installed in a room used as a return air plenum, the negative pressure of the plenum could overcome the draft of the water heater's venting system, and flue gases could be drawn into the plenum space. The improper operation of a water heater will greatly accelerate the deterioration of the water heater and its vent.

502.3 Water heaters installed in attics. Attics containing a water heater shall be provided with an opening and unobstructed passageway large enough to allow removal of the water heater. The passageway shall not be less than 30 inches (762 mm) high and 22 inches (559 mm) wide and not more than 20 feet (6096 mm) in length when measured along the centerline of the passageway from the opening to the water heater. The passageway shall have continuous solid flooring not less than 24 inches (610 mm) wide. A level service space at least 30 inches (762 mm) deep and 30 inches (762 mm) wide shall be present at the front or service side of the water heater. The clear *access* opening dimensions shall be a minimum of 20 inches by 30 inches (508 mm by 762 mm) where such dimensions are large enough to allow removal of the water heater.

❖ There is not always sufficient room for installations of water heaters in spaces such as basements, alcoves, utility rooms and furnace rooms. In an effort to save space or simplify an installation, water heaters are often installed in attics. Access to water heaters in attics is often difficult due to the lack of walking surfaces other than the edges of exposed ceiling joists. The intent of this section is to require a suitable access opening, passageway and work space that will provide for safe access by service personnel and to provide for water heater replacement without the need for removal of permanent portions of the building structure.

It is not uncommon for water heaters that are located in attics to be placed into position long before finish wall, ceiling and door/access opening trim materials are installed. A water heater of a size that “just fits” through the rough opening between ceiling joists or a water heater that was set prior to the gable ends of the attic space being completed might not be removable after the finish materials are in place. Coordination is required between the building designer and supplier/installer of the water heater to ensure that water heaters can be replaced without compromising the integrity of the building structure or requiring significant repair work to the building.

502.4 Seismic supports. Where earthquake loads are applicable in accordance with the *International Building Code*, water heater supports shall be designed and installed for the seismic forces in accordance with the *International Building Code*.

❖ Chapter 16 of the *International Building Code*® (IBC®) contains seismic maps that designate which areas of the country must design for earthquake loads. Section 308.2 of the code requires that piping supports be designed to resist seismic loads in designated zones. Similarly, this section requires that water heater supports be designed to resist the same seismic loads in accordance with the IBC. Failure of water heater supports or dislocation of a water heater has been shown to be a threat to health and safety of the building occupants. the requirements of this section are the minimum required criteria in view of life safety considerations.

502.5 Clearances for maintenance and replacement. Appliances shall be provided with *access* for inspection, service, repair and replacement without disabling the function of a fire-resistance-rated assembly or removing permanent construction, other appliances or any other piping or ducts not connected to the appliance being inspected, serviced, repaired or replaced. A level working space at least 30 inches deep and 30 inches wide (762 mm by 762 mm) shall be provided in front of the control side to service an appliance.

❖ A water heater (an appliance) typically requires some inspection, service or repair over the life of the unit and eventually, the water heater requires replacement. Because previous editions of the code and many water heater manufacturers' installation instructions did not require a working space in front (control side) of the unit, some water heaters would be installed without consideration for providing ample access to the controls. With no prohibitions against installing piping, ductwork, or other equipment not associated with the water heater such that access to the water heater was blocked, the code official had no code substantiation to require proper access to the water heater. Where the replacement of a "blocked" water heater was necessary, removal and replacement of those items as well as any permanent construction often resulted in improper reinstallation of those items because the persons performing water heater replacements are often unfamiliar with those other systems. Improper reinstallation (or worse, no reinstallation) of those removed items could lead to dangerous conditions Piping, ducts, vents and wiring serving other appliances and systems should not have to be disturbed in order to access any water heater..

The required size of the space in front (control side) of a water heater allows for ample kneeling area and provides a level area to set up a step ladder to access the top of tall water heaters, especially if the water heater is elevated above the floor. As this section prohibits removal of permanent construction even for replacement of the unit, building designers need to consider the size of the water heater with respect to its path of removal to the exterior of the building. During construction of a building, a water heater might be set before all of the walls, doorways and ceilings are completed resulting in no possible way to move the old water heater out of building and move a new unit of similar size into position without removing permanent parts of the building. The removal and reinstallation of permanent portions of the building could cause problems with structural integrity, fire safety or operation of other building systems.

SECTION 503
CONNECTIONS

503.1 Cold water line valve. The cold water *branch* line from the main water supply line to each hot water storage tank or water heater shall be provided with a valve, located near the equipment and serving only the hot water storage tank or water heater. The valve shall not interfere or cause a disruption of the cold water supply to the remainder of the cold water system. The valve shall be provided with *access* on the same floor level as the water heater served.

❖ This section requires that a valve located near the water heater be installed in the cold water branch line from the main water supply. This section also provides requirements for separate water heater and storage tank hot water systems.

Valves are needed to isolate each hot water storage tank or water heater from the water distribution system to facilitate service, repair and replacement and to allow for emergency shutoff in the event of a failure leak. The shutoff valve must be adjacent to the hot water storage tank or water heater, conspicuously located and within reach to permit it to be easily located and operated in the event of an emergency. The valve is to serve only the hot water storage tank or water heater, thereby not disrupting the flow of the cold water supply to other portions of the water distribution system when the hot water storage tank or water heater is taken out of service [see Commentary Figure 503.1(1)]. In accordance with Section 606.1, Item 8, the valve must be a full-open valve located on the water supply pipe to every water heater (see commentary, Section 606.1).

Where plastic cold and hot water piping connections are made to a draft hood equipped gas- or oil-fired water heater, the Plastic Pipe and Fittings Association recommends maintaining at least a 6-inch (152 mm) clearance between the piping and the draft hood/vent pipe. Metallic pipe or metallic flexible connectors should be used in these locations [see Commentary Figure 503.1(2)].

This section does not prohibit the installation of a shutoff valve on the hot water line from the water heater. Some installations might have a water heater located below the system being served, and placement of a valve in the hot water line enables heater servicing without a drain down of the hot water distribution system. The code is silent on what type of valve should be used if a valve is installed at this location.

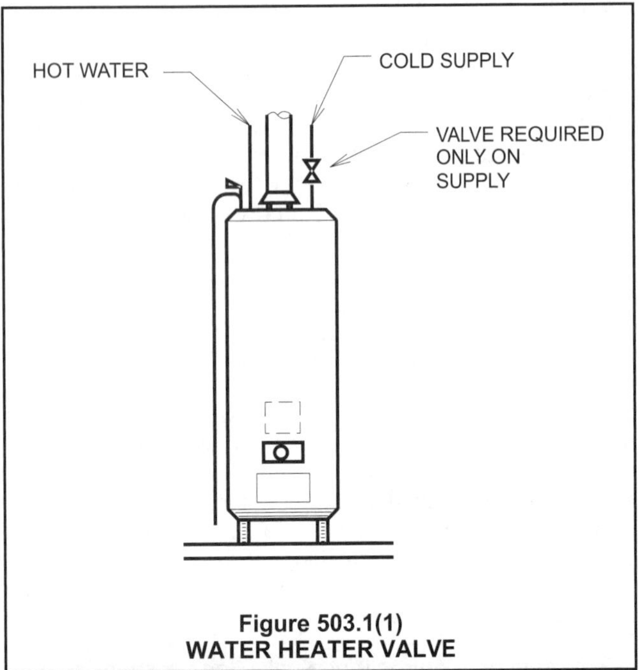

Figure 503.1(1)
WATER HEATER VALVE

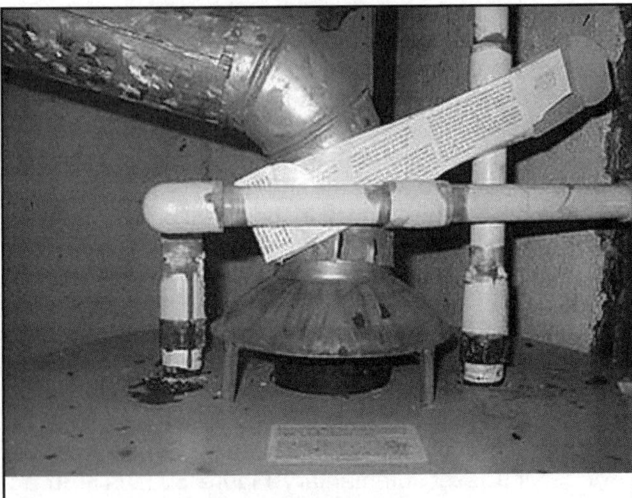

Figure 503.1(2)
PLASTIC PIPE TOO CLOSE TO DRAFT HOOD OF GAS-FIRED WATER HEATER

503.2 Water circulation. The method of connecting a circulating water heater to the tank shall provide proper circulation of water through the water heater. The pipe or tubes required for the installation of appliances that will draw from the water heater or storage tank shall comply with the provisions of this code for material and installation.

❖ Occupancies such as restaurants, motels, hotels and laundries typically have high demands for hot water. In such structures, it is not uncommon to have a separate water heater and storage tank. When a separate water heater and storage tank are installed, they must be connected to provide proper circulation of water through the water heater (see Commentary Figure 503.2).

Gravity and forced circulation are two methods of providing circulation between the water heater and the storage tank. Gravity circulation is achieved by locating the water heating equipment below the storage tank. As water is heated, the hot water will rise out of the heat source, through the piping and into the storage tank. As the water in the storage tank cools, it will travel by gravity through the pipe back to the heat source. Forced circulation uses a pump or a circulator to send the water between the heat source and the storage tank. When there is high demand for hot water, forced circulation is used to meet the demand because it allows for a much faster recovery time (i.e., it can replace the hot water used faster than the slower gravity system).

The manufacturer's installation instructions must be followed for proper operation of the system. The manufacturer will specify the direction of flow, the pipe and pump size, necessary valves, safety flow switches and pressure relief valves necessary for the efficient operation of the system and to prevent damage to the water heater.

The materials used in connecting the water heater

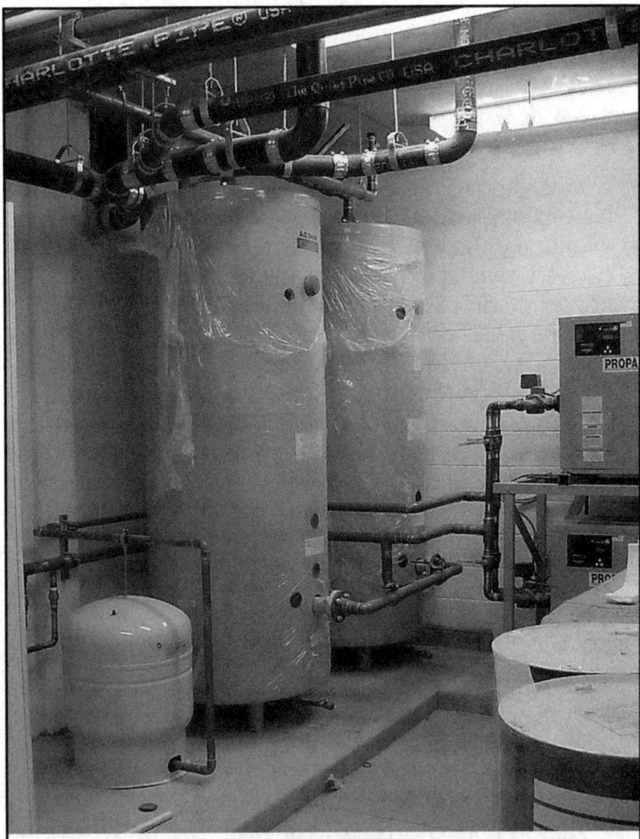

Figure 503.2
CIRCULATING TYPE WATER HEATERS CONNECTED TO STORAGE TANKS
(Photo courtesy of B&I Contractors, Inc.)

to the storage tank and for the hot water distribution system must meet the minimum quality standards prescribed in Section 605 (see commentary, Section 605).

SECTION 504
SAFETY DEVICES

504.1 Antisiphon devices. An *approved* means, such as a cold water "dip" tube with a hole at the top or a vacuum relief valve installed in the cold water supply line above the top of the heater or tank, shall be provided to prevent siphoning of any storage water heater or tank.

❖ This section establishes the requirements for the approval and installation of safety devices on water heaters.

Water heaters are designed to operate only when they are full of water. If some or all of the water is siphoned out of the tank during an interruption in the cold water supply, damage to the water heater from overheating could occur. Typically, water heater designs include a cold water "dip" tube, which directs the incoming cold water to the bottom of the tank. At the top of the tank, a hole is provided in the dip tube to prevent water from being siphoned from the tank through the tube. ANSI Z21.10.1 and UL 174 both require that the hole in the dip tube be located within 6 inches (152 mm) of the top of the tank.

If the water heater does not have an antisiphoning hole in the dip tube, a vacuum relief valve or other approved means must be installed to break any siphoning effect that might occur.

504.2 Vacuum relief valve. Bottom fed water heaters and bottom fed tanks connected to water heaters shall have a vacuum relief valve installed. The vacuum relief valve shall comply with ANSI Z21.22.

❖ A vacuum-relief valve must be installed in bottom-fed water heaters and bottom-fed tanks connected to water heaters. The vacuum relief valve is intended to prevent a possible reduction within the tank to below atmospheric pressure (i.e., a partial vacuum). Many tanks are not designed to resist external pressures exceeding internal pressures; therefore, a vacuum relief valve is necessary to prevent atmospheric pressure from possibly collapsing or otherwise damaging the tank or internal flueways.

The vacuum relief valve operates by automatically venting the system to the atmosphere when a partial vacuum is created, thereby permitting air to enter and maintaining the internal pressure (see Commentary Figure 504.2). Bottom-fed tanks would be susceptible to the creation of a vacuum by tank drainage resulting from negative supply pipe pressure or leakage.

Vacuum relief valves must be constructed and tested in accordance with ANSI Z21.22. This essentially requires the valve to be third-party certified as an indication to the installer, the inspector and the end user that it has been tested and evaluated by an approved agency and has been determined to perform in an acceptable manner. The presence of a certification mark is part of the information that the code official is to consider in the approval of the vacuum relief valve and the water heater.

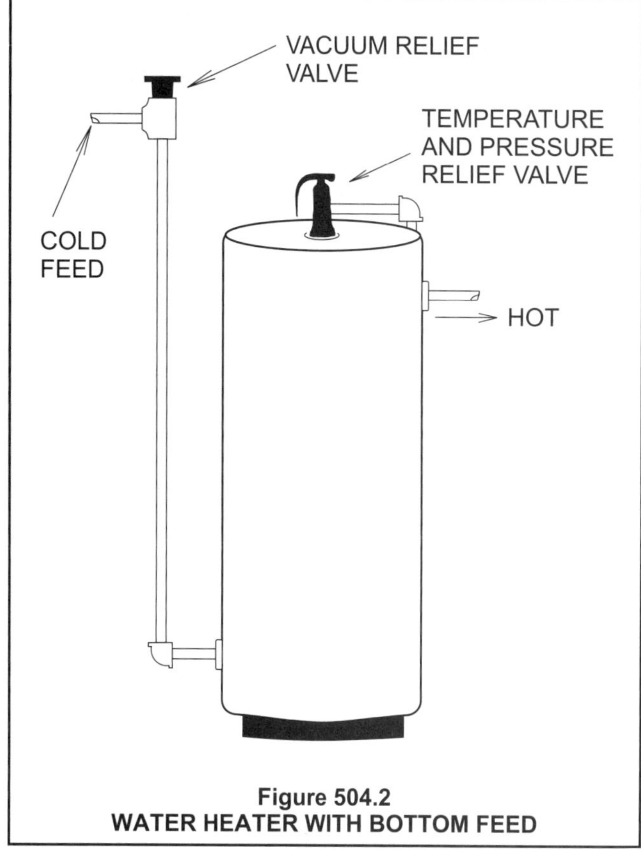

Figure 504.2
WATER HEATER WITH BOTTOM FEED

504.3 Shutdown. A means for disconnecting an electric hot water supply system from its energy supply shall be provided in accordance with NFPA 70. A separate valve shall be provided to shut off the energy fuel supply to all other types of hot water supply systems.

❖ Just as the IMC and the IFGC require a shutoff valve (i.e., a disconnect means) for the energy supply, NFPA 70 requires electric water heaters have an electric disconnect switch. The shutoff valves for fuel-fired water heaters and disconnects for electric water heaters are necessary to allow for service, repairs and temporary and emergency shutdown (see Commentary Figure 504.3).

504.4 Relief valve. All storage water heaters operating above atmospheric pressure shall be provided with an *approved*, self-closing (levered) pressure relief valve and temperature relief valve or combination thereof. The relief valve shall conform to ANSI Z21.22. The relief valve shall not be used as a means of controlling thermal expansion.

❖ Every storage tank-type water heater must have both temperature and pressure relief protection to prevent tank failure. In many applications, the temperature and pressure relief valve safety functions are accomplished by a single combination-type pressure and temperature relief valve.

Where water is heated in a vessel open to atmospheric pressure (at sea level elevation), the water cannot be heated to a temperature any greater than 212°F (100°C) because water converts to steam at this temperature. However, if water is heated in a vessel that is under a pressure greater than atmospheric pressure, water temperatures greater than 212°F (100°C) are possible because the pressure in the vessel prevents the conversion of water into steam. Water at a temperature greater than its atmospheric pressure boiling point is said to be superheated. If the pressure in a vessel containing superheated water is rapidly lowered such as would occur when the vessel ruptures, the water will instantly and violently boil to create steam. The conversion of water to steam is an expansive process; that is, steam created from water occupies significantly more volume. Every gallon (3.8 L) of water converted to steam (at atmospheric pressure) will occupy approximately 224 cubic feet (6.3 cubic meters).

Similar conditions in a storage tank-type water heater under normal water distribution system pressures could occur if the heating control failed to cease heating the water at the time the thermostat setting was achieved and the temperature and pressure relief valve was inoperable. The sudden release of energy (the conversion of water into steam) resulting from tank rupture can have devastating results. The sudden conversion to steam (flashing) acts as propellant not unlike the fuel of a rocket engine. Exploding water heaters have been known to be propelled through several stories of a building, knock down concrete walls and be propelled through roofs to land over a 100 yards away. Human deaths and injuries have been attributed to water heater explosions.

Figure 504.3
ELECTRIC DISCONNECT ADJACENT TO WATER HEATER
(Photo courtesy of B&I Contractors, Inc.)

A temperature relief valve prevents the water temperature in the tank from exceeding 210°F (99°C). If the water temperature should near 210°F (99°C), the valve opens to relieve any pressure buildup in the tank and to allow cooler water to enter the tank to lower the tank water temperature. With a properly installed and operating temperature relief valve, the water in the tank can never reach a temperature where steam could be generated under any pressure condition.

Many water heater tanks are designed for a working pressure of not greater than one-half of the design pressure for the tank. A pressure relief valve is required to prevent the tank pressure from exceeding the working pressure of the tank. Although the tank is designed for a pressure twice that of the working pressure, other factors such as variations in manufacturing, in-service corrosion and weld joint fatigue due to pressure cycling necessitate limiting the operating pressure to not greater than the "working pressure" tank rating. Although pressure relief valves might be available with various pressure relief settings, at no time should a water heater tank be protected by a relief valve having a pressure relief setting greater than the working pressure rating of the tank. However, installing a pressure relief valve having a relief setting lower than the working pressure of the tank is not prohibited.

Obviously, the design, construction and quality of pressure and temperature relief valves must be controlled to ensure that these devices reliably perform the intended safety functions. The ANSI Z21.22 standard prescribes the necessary features, construction and testing of temperature and pressure relief valves. Note that because these valves are water distribution system safety devices, they are required to be third-party certified in accordance with Table 303.4 of this code.

Note that this section does not address the relief valve requirements for instantaneous and circulating water heaters. However, where these types of gas-fired water heaters are used, the applicable standards for these products require over temperature protection as part of the burner control system and overpressure protection by installation of a pressure relief valve installed on one of the water connections to unit. The UL 499 standard for electric instantaneous water heaters requires that over-temperature protection be built into the temperature control system but does not require over-pressure protection (see commentary, Figure 504.4).

As stated previously, when water is heated, the water will expand in volume. If the water being heated is in a closed vessel, the increase in volume will result in an increase in pressure. A storage tank-type water heater connected to a water distribution system could be a "closed vessel" such that when the water is heated, the system pressure will rise (see commentary, Section 607.3.2). Where thermal expansion-caused pressures must be controlled, the water heater pressure relief valve must not be used as the method to control the pressure increases. Relief valves are not designed to act as operating controls. A relief valve required to open frequently could become damaged or inoperative due to wear, corrosion or the valve might leak because of debris in the seat area. In any case, the intended safety function of the valve could be compromised.

Figure 504.4
COMBINATION TEMPERATURE- AND PRESSURE-RELIEF VALVE
(Photo courtesy of CASH-ACME, Inc.)

504.4.1 Installation. Such valves shall be installed in the shell of the water heater tank. Temperature relief valves shall be so located in the tank as to be actuated by the water in the top 6 inches (152 mm) of the tank served. For installations with separate storage tanks, the valves shall be installed on the tank and there shall not be any type of valve installed between the water heater and the storage tank. There shall not be a check valve or shutoff valve between a relief valve and the heater or tank served.

❖ Storage water heaters typically have factory-installed openings that are properly located to receive a relief valve device. Such openings must be used for that purpose. ANSI Z21.10.1 and UL 174 both require installations of a temperature- and pressure-relief valve in a location specified by the manufacturer or for the water heater to be equipped with a temperature- and pressure relief-valve. Where there is no tapping in the tank for a relief valve, special tee fittings and extended-element relief valves are required.

The highest water temperature in a tank-type water heater is in the top of the tank. The thermal-sensing element of the relief valve must be in contact with the hottest water [i.e., within 6 inches (152 mm) of the top of the tank] to accurately sense the water temperature and to provide the earliest response. Otherwise, the tank temperature can become considerably hotter than the 210°F (99°C) temperature relief setting because of a thermal lag that could exist between the relief valve and the tank (see Commentary Figure 504.4.1). Any type of valve installed on either side of a relief valve (inlet or outlet) could render the relief valve inoperative, thereby creating a potentially dangerous condition.

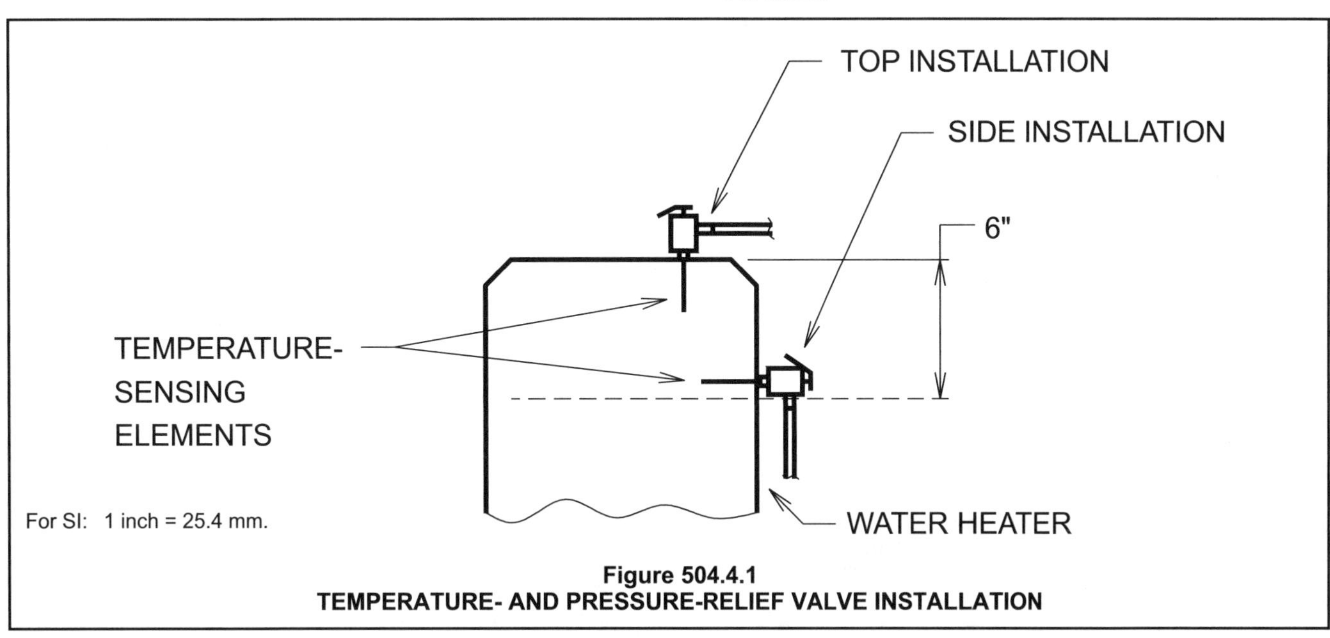

Figure 504.4.1
TEMPERATURE- AND PRESSURE-RELIEF VALVE INSTALLATION

504.5 Relief valve approval. Temperature and pressure relief valves, or combinations thereof, and energy cutoff devices shall bear the label of an *approved* agency and shall have a temperature setting of not more than 210°F (99°C) and a pressure setting not exceeding the tank or water heater manufacturer's rated working pressure or 150 psi (1035 kPa), whichever is less. The relieving capacity of each pressure relief valve and each temperature relief valve shall equal or exceed the heat input to the water heater or storage tank.

❖ Relief valves must be third-party certified by an approved agency as complying with ANSI Z21.22. The certification mark is the primary, if not the only, indication to the installer, the inspector and the end user that the relief valve has been tested and evaluated by an approved agency and has been determined to perform in an acceptable manner when installed and operated in accordance with its application, installation and operation instructions. The presence of a certification mark is part of the information that the code official is to consider in the approval of the relief valve and the water heater.

Water heater temperature relief valves are set to open at a temperature not higher than 210°F (99°C). A relief valve device must be able to dissipate energy at a rate equal to or greater than the energy input rate of the water heater. Relief valve sizing information is provided by both water heater and relief device manufacturers (see Commentary Figure 504.5).

Water heaters must be marked with the maximum working pressure, and the relief valve must be set to open at this pressure or 150 pounds per square inch (psi) (1035 kPa), whichever is less. Marking requirements for water heaters will also specify the minimum required relief-valve capacity. A relief valve must be able to dissipate energy at a rate equal to or greater than the rate of energy input to the water heater. An undersized (insufficient capacity) valve would be unable to prevent the heater pressure from exceeding the maximum capacity and a dangerous vessel failure could result. Relief-valve capacity is expressed in terms of British thermal units per hour (Btu/h).

A relief valve is designed to open in direct proportion to the water pressure and temperature acting on its closure disk. The higher the temperature or pressure, the greater the force and the more the valve opens.

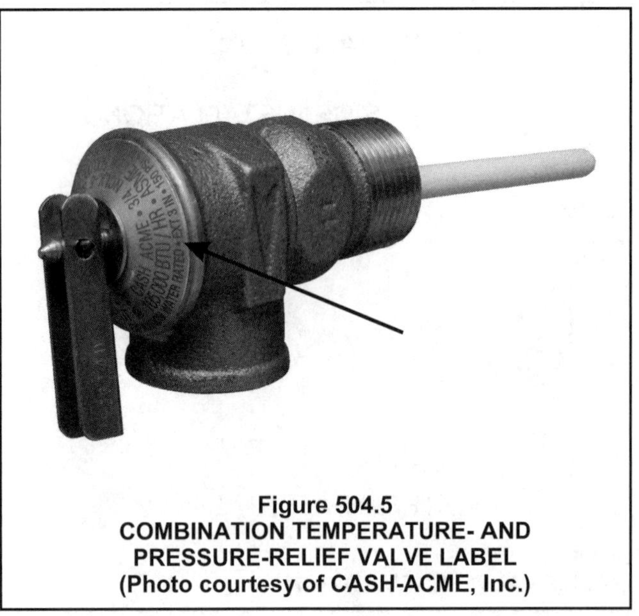

Figure 504.5
COMBINATION TEMPERATURE- AND PRESSURE-RELIEF VALVE LABEL
(Photo courtesy of CASH-ACME, Inc.)

504.6 Requirements for discharge piping. The discharge piping serving a pressure relief valve, temperature relief valve or combination thereof shall:

1. Not be directly connected to the drainage system.
2. Discharge through an *air gap* located in the same room as the water heater.
3. Not be smaller than the diameter of the outlet of the valve served and shall discharge full size to the *air gap*.
4. Serve a single relief device and shall not connect to piping serving any other relief device or equipment.
5. Discharge to the floor, to the pan serving the water heater or storage tank, to a waste receptor or to the outdoors.
6. Discharge in a manner that does not cause personal injury or structural damage.
7. Discharge to a termination point that is readily observable by the building occupants.
8. Not be trapped.
9. Be installed so as to flow by gravity.
10. Not terminate more than 6 inches (152 mm) above the floor or waste receptor.
11. Not have a threaded connection at the end of such piping.
12. Not have valves or tee fittings.
13. Be constructed of those materials listed in Section 605.4 or materials tested, rated and *approved* for such use in accordance with ASME A112.4.1.

❖ Because the discharge pipe from a water heater temperature and pressure relief valve is an extension of the potable water distribution system, the outlet of the discharge pipe must be treated as a potable water supply system outlet. Although Section 608.15.1 of the code already requires protection of potable water supply outlets by requiring an air gap, installers of water heaters might not realize that the termination point of the relief valve discharge pipe must also provide for an air gap.

The requirement for an air gap at the end of the discharge pipe automatically provides for an indirect connection, should it be desired for the discharge to be captured by any type of drainage system. However, as water heater installers might not realize that the discharge pipe is not a waste pipe; the requirement for indirect connection is reiterated in this section.

A water heater temperature and pressure relief valve is an important safety device for the water heater as well as for the water distribution system to which the heater it is connected (see commentary, Section

504.4). If the water heater is operating properly, the pressure in the water distribution system is not greater than is required by the code and the relief valve is in proper operating condition; there should not be any water discharge from the relief valve. Therefore, any water discharge from a relief valve is an indication of a water distribution system problem, a water heater problem or a relief valve malfunction. Because all of these problems are safety related and should be rectified as soon as possible, the discharge from the relief valve is an important "tell tale" indicator to alert the occupants of the building to call for service. Therefore, it is important to terminate the relief valve discharge pipe within the same room as the water heater since most building occupants will usually associate evidence of leaking water at or near a water heater location as a potential water damage problem that deserves immediate attention [see Commentary Figure 504.6(1)].

An even more compelling reason for the relief valve discharge pipe to terminate at an air gap in the same room as the water heater is that a continuous routing of the discharge pipe from the relief valve through concealed spaces to a remote termination point can pose an extreme risk. Because the satisfactory condition of the discharge pipe is just as important as a properly operating relief valve, the condition of the discharge pipe must be observable by all who might need to perform service to the relief valve. Concealed discharge piping cannot be readily observed to verify that (1) it has not been damaged (i.e., flattened, kinked, bent or cut); (2) the proper slope for gravity drainage exists; (3) the pipe has not been tapped into, valved off or capped; and (4) the pipe length and number of elbows in the pipe are not in excess of the valve manufacturer's installation instructions. And because relief valve discharges are often just small amounts or continuous trickles of water, where freezing conditions might exist in spaces containing concealed piping or at a remote termination point, these small flows could easily freeze resulting in blockage of the pipe. The code requirement for the relief valve discharge pipe to terminate at an air gap in the same room as the water heater eliminates all uncertainties about the condition of the discharge pipe and its termination point. Note that excessive developed length of a discharge pipe will cause excessive back pressure on the relief valve thus reducing its relief capacity. The valve manufacturer's instructions will limit the number of fittings and pipe length.

Even though discharge of water from a relief valve should be a very rare occurrence, when discharge does occur, the flow of water must have a way to escape the area without causing damage to the building structure such as wood members, drywall and insulation. Note that the code is not concerned with damage to an occupant's property within the building. For example, if a water heater is located in a garage with a floor that slopes to the exterior through a garage door opening, and the perimeter walls of the garage at the floor level are of concrete block construction (without insulation or drywall applied), the water from the discharge pipe could flow to the outdoors without the

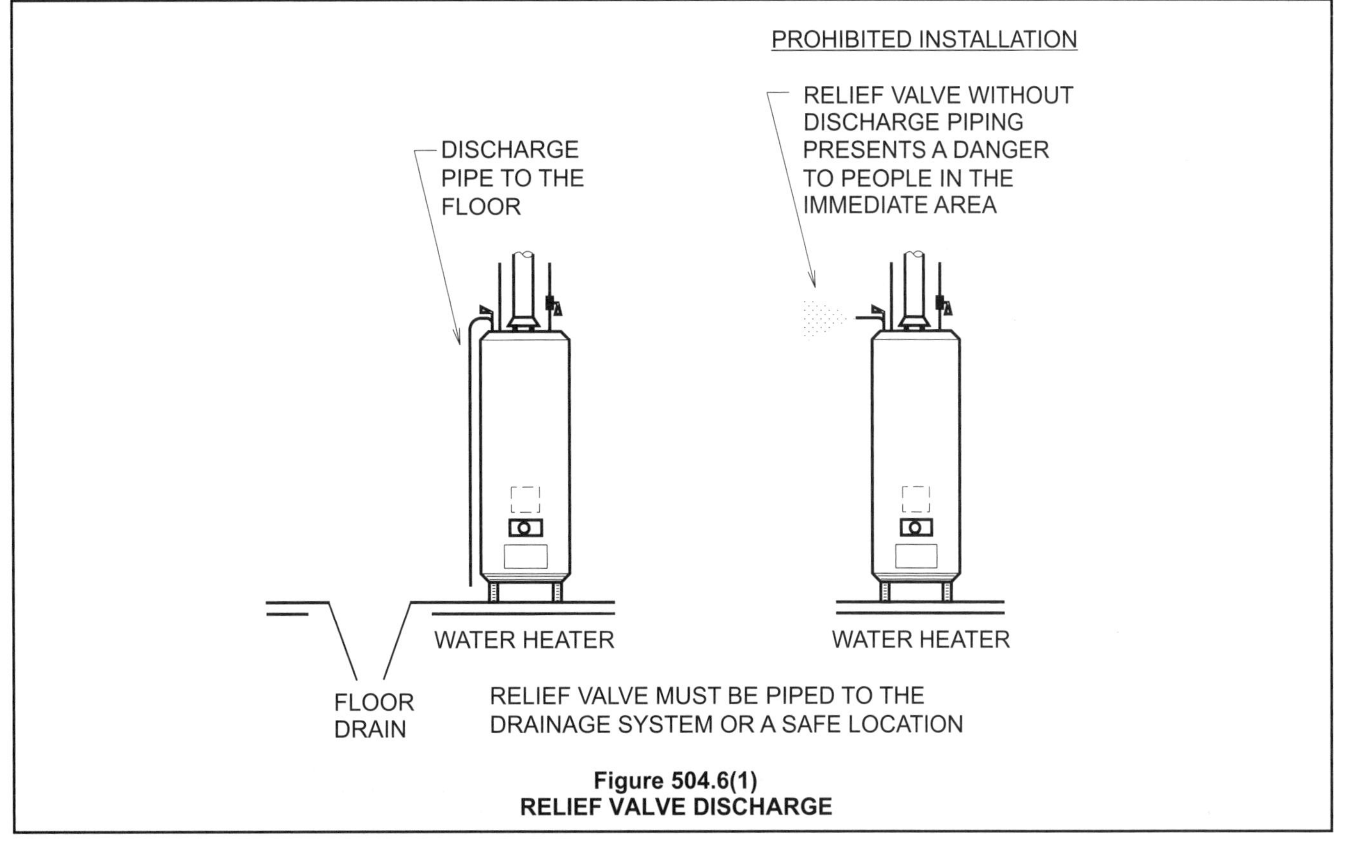

Figure 504.6(1)
RELIEF VALVE DISCHARGE

need for a floor drain or waste receptor. The code is not concerned about damage to an occupant's property that might be stored on the floor of the garage. However, if the perimeter walls of the garage are of wood construction (with or without insulation or drywall) that rests directly upon the floor of the garage, the potential for damage to the building structure exists. In this situation, the discharge of water from the relief valve pipe must be captured and conveyed to a drain system or to the outdoors. Although it might be argued that the slope of the garage floor would convey the water out the garage door to the outdoors before contacting the wall, concrete floors are rarely perfectly sloped and an occupant's property within the space might cause diversion of the water towards the walls, thus causing damage to the structure.

The discharge pipe of a temperature and pressure relief valve must not be of a size that is smaller than the relief valve outlet size at any point from the outlet to the termination of the pipe. For example, a reduction in the size of the pipe would from $^{3}/_{4}$ inch (19 mm) nominal to $^{1}/_{2}$ inch (12.7 mm) nominal size would impede the discharge flow, and reduce the discharge capacity of the relief valve. The discharge pipe must not have valves or tees installed. Obviously, a valve in the pipe that could be inadvertently closed would obstruct the discharge pipe and a tee fitting would allow connection of piping from other sources, both of which are dangerous conditions. The code is silent concerning the use of types of relief valve discharge piping that requires the use of insert fittings. Common examples are PEX and PEX-AL-PEX tubing. Whether or not these insert fittings constitute enough of a restriction in flow to cause a reduction in maximum discharge capacity of the relief valve is unknown. The authority having jurisdiction must make the decision on whether the use of insert fittings in relief valve discharge piping violates the intent of "no reduction in size" of the discharge pipe.

The relief valve discharge pipe must terminate not more than 6 inches (152 mm) above the floor surface, a waste receptor (which includes a floor drain) in order to prevent hot water discharge from being directed onto a building occupant [see Commentary Figure 504.6(1)]. The relief valve discharge pipe is also allowed to discharge to the pan serving the water heater or hot water storage tank [see Commentary Figure 504.6(3)]. Note that a pan receiving the discharge from a relief valve could be easily flooded if a relief valve opens fully. Obviously, for termination points to the floor, the floor must be a suitable location for water discharge as previously discussed. Otherwise, a floor drain or waste receptor must be provided to capture and direct discharges from the pipe. In multilevel buildings where water heaters are in the same location on each floor, it is common to provide an indirect piping system to capture any discharge and direct the flow to a single discharge location such as a floor drain on the first level [see commentary, Figure 504.6(2)] or to the outdoors. The code is silent as to the size/shape of waste receptors and the gravity drain line size for capturing and conveying the discharged water from relief valves.

Discharge of a relief valve to laundry trays/tubs, kitchen/utility sinks and shower floors are not suitable locations as this violates Item number 6 of this code section. Persons using the fixture could be injured by hot water and steam that could come from the pipe.

The relief valve discharge pipe must drain by gravity and must not be trapped or have sags in the piping that would act as traps. Traps (or piping sags) would restrict the discharge flow and also cause water to be retained in the pipe. This retained water might cause a build up, scale/corrosion in the pipe or, if the water heater is located in an area subject to freezing such as a garage or attic space, the water might freeze. Both situations have the potential to block the discharge flow from the relief valve which would create a dangerous condition.

A threaded end on the termination of a relief valve discharge pipe is prohibited as this would be an invita-

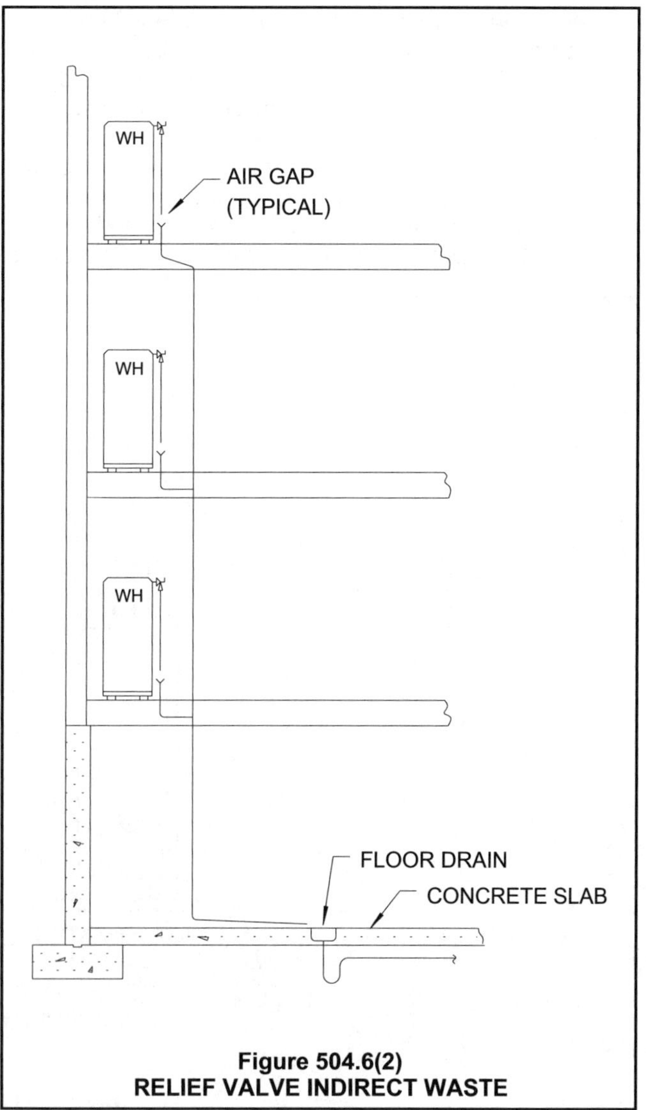

Figure 504.6(2)
RELIEF VALVE INDIRECT WASTE

tion for someone to screw on a cap or valve to stop water from dripping from the pipe. What might be perceived as a repair would, in reality, create a very dangerous condition [see commentary, Figure 504.6(4)].

Relief valve discharge piping must be one of the piping materials listed in Table 605.4. Although the pipe materials in this table must have a pressure rating of at least 100 psi (690 kPa) at 180°F (82°C) to serve as water distribution piping all pipe materials, all of these materials will most likely have the capability to survive limited exposure to higher temperatures at minimal (nearly atmospheric) pressure. Because a water heater relief valve is required to open at a water temperature of 210°F (99°C) and due to the short amount of time that this high temperature condition would usually exist (e.g., 15 minutes or less), any of the water distribution piping materials listed in Table 605.4 are deemed suitable for this application.

The code does not prohibit the use of the flexible types of water distribution piping listed in Table 605.4 for the relief valve discharge. However, Item 6 of this section would require that the end of the relief valve discharge pipe of flexible material be attached in some manner to direct the water flow to a "safe" location as well as prevent the end of the pipe from moving around so as to not expose a building occupant to a hot water discharge.

504.7 Required pan. Where water heaters or hot water storage tanks are installed in locations where leakage of the tanks or connections will cause damage, the tank or water heater shall be installed in a galvanized steel pan having a material thickness of not less than 0.0236 inch (0.6010 mm) (No. 24 gage), or other pans *approved* for such use.

❖ Where leaks from a water heater or storage tank or the associated connections could cause damage to the building structure, the water heater or hot water storage tank must be installed in a pan. See Commentary Figures 504.7(1). The pan must be made of galvanized steel of at least the thickness stated or be made of other materials as approved by the code official. Even though prefabricated aluminum and plastic pans are commonly available for many diameters of water heaters and are widely used by installers, these pans must be approved for use by the code official because they are not of galvanized steel. Note that the water heater manufacturer's installation instructions must also be reviewed for pan requirements because some manufacturers require the use of only a metal pan.

The 2009 code edition changed the required thickness of galvanized sheet metal to coincide with the minimum thickness tolerance for 24 gauge galvanized sheet metal so that a field measurement of thickness would allow for all thicknesses within the tolerance range of 24 gauge to be acceptable.

Figure 504.6(3)
T&P RELIEF VALVE PERMITTED TO DISCHARGE TO THE PAN

Figure 504.6(4)
CAPPED TANK TAPPING FOR RELIEF VALVE AND PLUGGED RELIEF VALVE OPENING

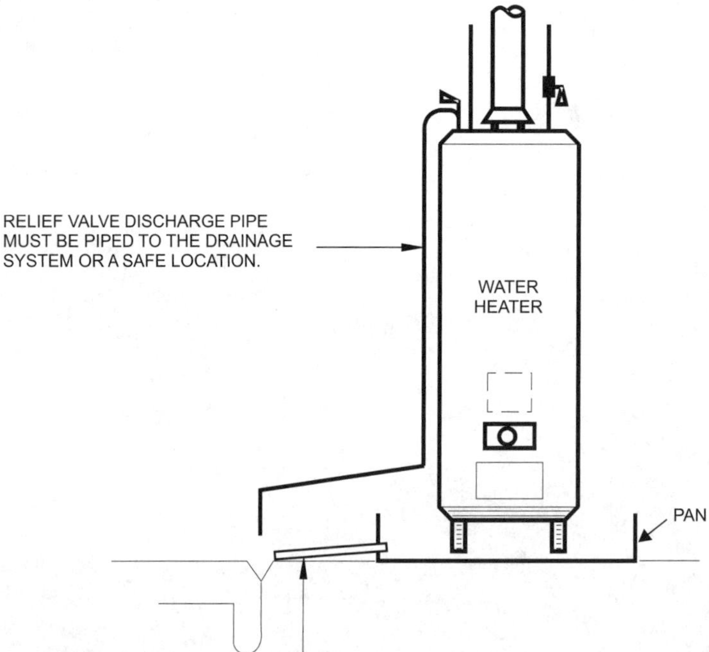

RELIEF VALVE DISCHARGE PIPE MUST BE PIPED TO THE DRAINAGE SYSTEM OR A SAFE LOCATION.

WATER HEATER

PAN

THE PAN DRAIN MUST DISCHARGE TO WASTE RECEPTOR, FLOOR DRAIN OR TO THE EXTERIOR OF THE BUILDING

Figure 504.7(1)
WATER HEATER IN PAN

504.7.1 Pan size and drain. The pan shall be not less than $1^1/_2$ inches (38 mm) deep and shall be of sufficient size and shape to receive all dripping or condensate from the tank or water heater. The pan shall be drained by an indirect waste pipe having a minimum diameter of $^3/_4$ inch (19 mm). Piping for safety pan drains shall be of those materials listed in Table 605.4.

❖ The pan is required to capture water leakage from the water heater or hot water storage tank and convey the water to the pan drain. Therefore, the pan must be slightly larger than the footprint of the tank. Generally, storage tank water heater manufacturers require that the pan minimum dimension be at least 2 inches (52 mm) greater than the diameter of the tank. The depth of the pan must be at least $1^1/_2$ inches (38 mm) to provide for the drain connection fitting which is typically installed in the side wall of the pan. The pan drain must be at least $^3/_4$ inch (19 mm) nominal pipe size and must be of one of the materials listed in Table 605.4. The pan drain pipe must not be directly connected to the sanitary drain system. Just as drain pipes downstream of the required air gaps at the relief valve discharge pipe terminations can be combined together [see Commentary Figure 504.6(2)], pan drains can also be combined together in a similar manner to terminate at a single location at a floor drain, a waste receptor or to the outdoors.

Note that the pan drain pipe material must be of one of the materials listed for piping material suitable for water distribution system use. The concern is that a hot water storage tank or the water connections to the tank could leak hot water for extended periods of time until discovered. If the pan drain pipe material was not made of a material suitable for hot water, the pipe might sag between supports causing sluggish draining and clogging which could cause the drain pan to overflow.

A frequently asked question is how this section should be applied to an existing water heater location that did not originally have a code requirement for installation of pan. The water heater might be in a basement where the building drain is above the elevation of the bottom of the water heater or the water heater might be in the middle of a slab-on-grade building, without a nearby floor drain and not adjacent to an outside wall.

As stated in the commentary for Section 502.1.1, the code is silent on this topic, thus requiring the code official to determine to what degree the installation must comply with the code. Where water heater replacement installers are confronted with obstacles that prevent strict compliance to this code section, they should consult with the authority having jurisdiction before performing the replacement work.

504.7.2 Pan drain termination. The pan drain shall extend full-size and terminate over a suitably located indirect waste receptor or floor drain or extend to the exterior of the building and terminate not less than 6 inches (152 mm) and not more than 24 inches (610 mm) above the adjacent ground surface.

❖ Although a pan drain is not required to follow the same rules as for gravity flow sanitary drainage, it should be obvious that any reductions in pan drain pipe size would restrict flow and be a potential clogging point especially where pans have collected years of dust and possible rust flakes from corroding steel tanks. Since the drainage from the pan is basically clear water, the drain can terminate outdoors or can terminate at a waste receptor or floor drain. Indirect connections at waste receptors or floor drains can be through either an air break or an air gap since the only concern is to prevent a possible sewage backup into the pan. Where the pan drain terminates outdoors, a termination point of at least 6 inches (152 mm) above the ground protects the open end from being blocked by careless landscaping work and vegetation growth which might also allow for insects to build nests in the opening. A maximum height of 24 inches (610 mm) provides an allowance for snow and ice build up on the ground but is low enough so as to not create a safety concern for passersby.

SECTION 505 INSULATION

[E] 505.1 Unfired vessel insulation. Unfired hot water storage tanks shall be insulated to R-12.5 (h · ft^2 · °F)/Btu (R-2.2 m^2 · K/W).

❖ Previous editions of this code section specified an allowable heat loss from the unfired storage tank which required a thermodynamic problem to be solved. The code now prescribes an *R*-value for the insulation to be installed, making the insulation requirement straight forward.

This section requires that unfired hot water storage tanks be insulated with a minimum required *R*-value of 12.5. External insulation is necessary for unfired hot water storage tanks to conserve energy. Such tanks are used in conjunction with circulating-type water heaters, hot water supply boilers and various heat exchangers. Unlike storage-type water heaters, unfired storage tanks can be purchased without an insulation jacket or covering.

The previous editions of this code required that insulation for these tanks be designed for limiting the heat loss rate to a maximum value. This performance approach required a designer or installer to perform thermal calculations in order to determine a required insulation thickness. The revised text specifies the minimum required *R*-value so that calculations are not required and enforcement of this section only requires verification of the insulation *R*-value.

Bibliography

The following resource materials are referenced in this chapter or are relevant to the subject matter addressed in this chapter.

ANSI Z21.10.1-00, *Gas Water Heaters—Volume I: Storage Water Heaters with Input Ratings of 75,000 Btu per Hour or Less*. New York, NY: American National Standards Institute, 2000.

ANSI Z21.10.3-98, *Gas Water Heaters—Volume III: Storage, with Input Ratings Above 75,000 Btu per Hour, Circulating and Instantaneous Water Heaters—with Z21*. 10.3a-99 Addendum. New York, NY: American National Standards Institute, 1998.

ANSI Z21.22-99 (R2003), *Relief Valves and Automatic Gas Shutoff Devices for Hot Water Supply Systems*. New York, NY: American National Standards Institute, 2003.

ASME A112.4.1-1993 (R2002), *Water Heater Relief Valve Drain Tubes*. New York, NY: American Society of Mechanical Engineers, 2002.

ASSE 1005-99, *Performance Requirements for Water Heater Drain Valves—with 1986 Revisions*. Westlake, OH: American Society of Sanitary Engineering, 1999.

ASSE 1016-96, *Performance Requirements for Individual Thermostatic, Pressure Balancing and Combination Control Valves for Bathing Facilities*. Westlake, OH: American Society of Sanitary Engineering, 1996.

ASSE 1017-03, *Performance Requirements for Temperature Actuated Mixing Valves for Hot Water Distribution Systems*. Westlake, OH: American Society of Sanitary Engineering, 2003.

IBC-09, *International Building Code*. Washington, DC: International Code Council, Inc., 2009.

IFGC-09, *International Fuel Gas Code*. Washington, DC: International Code Council, Inc., 2009.

IMC-09, *International Mechanical Code*. Washington, DC: International Code Council, Inc., 2009.

Legionella 2003: Update and Statement by the Association of Water Technologies, Association of Water Technologies, Rockville, MD.

NFPA 70-08, *National Electric Code*. Quincy, MA: National Fire Protection Association, 2008.

UL 174-98, *Household Electric Storage Tank Water Heaters—with Revisions through December 1998*. Northbrook, IL: Underwriters Laboratories Inc., 1998.

UL 499-05, *Electric Heating Appliances*. Northbrook, IL: Underwriters Laboratories, Inc., 2005.

UL 732-95, *Oil-Fired Storage Tank Water Heaters—with Revisions through January 1999*. Northbrook, IL: Underwriters Laboratories Inc., 1995.

UL 1453-95, *Electric Booster and Commercial Storage Tank Water Heaters—with Revisions through September 1998*. Northbrook, IL: Underwriters Laboratories Inc., 1995.

Chapter 6: Water Supply and Distribution

General Comments

Potable water is viewed by the U.S. government as a precious commodity. The U.S. Environmental Protection Agency's (EPA) Office of Drinking Water specifies the quality requirements for water to be classified as potable. Water quality requirements regarding allowable levels of contaminants are becoming increasingly restrictive.

With this emphasis on water quality, plumbing equipment manufacturers have become increasingly concerned with the potential for introducing contaminants into the water through material selection. The banning of lead solders was only the start of evaluating material performance. Manufacturers are paying greater attention to the selection of materials that are exposed to potable water. All materials used in the water supply system must be evaluated for their possible effect on the potable water with which they are in contact.

There is also a great concern for the quality of ground water, which is often used as a source of drinking water. The EPA has enacted regulations aimed at protecting the quality of ground water.

Regulations for underground storage tanks were designed to protect against leakage that can contaminate ground water. Regulations are also in place for cleanup of hazardous material spills. All of these regulations are intended to protect the environment and the natural resources that sustain us.

Purpose

Chapter 6 regulates the supply of potable water from both public and individual sources to every fixture and outlet so that it remains potable and uncontaminated through cross connection. Chapter 6 also regulates the design of the water distribution system, which will allow fixtures to function properly. The unique requirements of the water supply for health care facilities are addressed separately. It is critical that the potable water supply system remain free of actual or potential sanitary hazards by providing protection against backflow.

SECTION 601 GENERAL

601.1 Scope. This chapter shall govern the materials, design and installation of water supply systems, both hot and cold, for utilization in connection with human occupancy and habitation and shall govern the installation of individual water supply systems.

❖ Requirements for the delivery of potable water from the source to the point of use are contained in this chapter. The requirements apply to both public water sources and individual water supplies, including wells, springs, streams and cisterns.

601.2 Solar energy utilization. Solar energy systems used for heating potable water or using an independent medium for heating potable water shall comply with the applicable requirements of this code. The use of solar energy shall not compromise the requirements for cross connection or protection of the potable water supply system required by this code.

❖ Solar heating systems consist of two basic types: direct connection and indirect connection. In a direct connection system, the heat transfer fluid is potable water. This type of system has potential safety problems that must be addressed in its design, inspection and maintenance. See the *International Mechanical Code*® (IMC®) for information on the design and installation of solar heating systems. The main concerns over contamination of the potable water source are from piping and joint materials used in the system and from the inadvertent introduction of nonpotable or toxic transfer fluids at a future date. The use of direct connection systems is typically limited to solar water-heating systems where the potable water is heated directly by the solar collector and circulated through the system, and is suited for use only in areas where the water in the collectors is not subject to freezing.

In an indirect connection system, a freeze-protected heat transfer fluid is circulated through a closed loop to a heat exchanger. The heat is then transferred indirectly to the potable water. The fluids in these systems are not potable and are occasionally toxic. The type of heat exchanger used in these systems depends on the exact nature of the transfer fluid used. As required by Section 608.16.3, a single-wall heat exchanger may be used in systems where the fluid is nontoxic, and a double-wall heat exchanger must be used in systems using toxic fluids.

601.3 Existing piping used for grounding. Existing metallic water service piping used for electrical grounding shall not be replaced with nonmetallic pipe or tubing until other *approved* means of grounding is provided.

❖ In some older existing buildings, equipment is grounded by connecting a grounding conductor to the interior metal water piping system. More commonly, a metal water pipe is used as a junction point for multiple grounding electrode conductors. The continuity of this electrical grounding is essential to maintaining the safety of the electrical system. The replacement of a portion of that metal piping system with nonmetallic piping could interrupt the continuity of that electrical grounding system and create a potentially hazardous situation. When the replacement of any portion of an existing metal water piping system with nonmetallic piping is proposed, the method of maintaining electrical system continuity in the existing building must be approved by the code official.

Bonding of underground metallic water service piping is mandated in NFPA 70 and the *International Residential Code®* (IRC®) as part of the grounding electrode system.

601.4 Tests. The potable water distribution system shall be tested in accordance with Section 312.5.

❖ Requirements for water supply system testing are described in Section 312.

SECTION 602 WATER REQUIRED

602.1 General. Every structure equipped with plumbing fixtures and utilized for human occupancy or habitation shall be provided with a potable supply of water in the amounts and at the pressures specified in this chapter.

❖ This section establishes the requirement that potable water must be provided to every building intended to be occupied. The requirements for hot and tempered water in this section are similar to those stated in Section 607.1. Although tempered water is suitable for handwashing and similar purposes typically required in workplaces, it is not satisfactory to meet the various sanitation needs such as cleaning, washing and bodily cleansing of the occupants of dwellings (see commentary, Section 607.1 and the definition of "Tempered water" in Section 202).

602.2 Potable water required. Only potable water shall be supplied to plumbing fixtures that provide water for drinking, bathing or culinary purposes, or for the processing of food, medical or pharmaceutical products. Unless otherwise provided in this code, potable water shall be supplied to all plumbing fixtures.

❖ Potable water is required for all plumbing fixtures unless nonpotable water, otherwise known as gray water, is used for purposes indicated in Appendix C, provided the appendix is made enforceable through specific adoption or is used as a basis for alternative approval. Some common uses of nonpotable water are irrigation and flushing water closets and urinals.

Although the use of nonpotable water is not common today, it may be used more in the future as a means of conserving water resources. Present-day uses of a nonpotable water supply to plumbing fixtures are typically limited to factory and industrial buildings. Refer to Appendix C for gray-water recycling systems.

602.3 Individual water supply. Where a potable public water supply is not available, individual sources of potable water supply shall be utilized.

❖ An individual water supply must be potable. Where a public system is unavailable, an individual source must be provided in conformance to the requirements of this section.

602.3.1 Sources. Dependent on geological and soil conditions and the amount of rainfall, individual water supplies are of the following types: drilled well, driven well, dug well, bored well, spring, stream or cistern. Surface bodies of water and land cisterns shall not be sources of individual water supply unless properly treated by *approved* means to prevent contamination.

❖ The sources for an individual water supply include wells, springs, cisterns, streams and surface water impoundments. Wells are probably the most common source of individual water supply used today and consist of a vertical shaft that extends down into the earth to a water-bearing stratum. Wells may be bored, drilled, driven or dug. See Section 202 for the definitions of each well type and a cistern.

Surface water impoundments always have to be treated if used as a supply of potable water. This may include some form of chlorination or ozone treatment.

602.3.2 Minimum quantity. The combined capacity of the source and storage in an individual water supply system shall supply the fixtures with water at rates and pressures as required by this chapter.

❖ The amount and quality of water supplied to a building are critical to both the health of the occupants and the safe and efficient use of the plumbing fixtures and drainage system.

The design professional must determine the water demand rate for the building in accordance with the requirements of Section 604 and in the same manner as though the water were supplied from a public source. The private water supply system must be capable of supplying the required demand.

602.3.3 Water quality. Water from an individual water supply shall be *approved* as potable by the authority having jurisdiction prior to connection to the plumbing system.

❖ Water from a private supply must be analyzed by an approved or EPA-certified test laboratory to determine that it is potable.

602.3.4 Disinfection of system. After construction or major repair, the individual water supply system shall be purged of deleterious matter and disinfected in accordance with Section 610.

❖ A private water supply system must be disinfected after it has been constructed or after major repair or alteration. The disinfection procedure must be as prescribed in Section 610 for all potable water systems (see commentary, Section 610).

602.3.5 Pumps. Pumps shall be rated for the transport of potable water. Pumps in an individual water supply system shall be constructed and installed so as to prevent contamination from entering a potable water supply through the pump units. Pumps shall be sealed to the well casing or covered with a water-tight seal. Pumps shall be designed to maintain a prime and installed such that ready *access* is provided to the pump parts of the entire assembly for repairs.

❖ A pump must be designed and installed to provide continuous service and is to be accessible to facilitate service, repair or replacement without requiring the removal or movement of any panel, door or similar obstruction and without the use of a portable ladder, step stool or similar device. The pump and its components that are in direct contact with the potable water must be rated for use in potable water systems. This is generally determined by the pump's compliance with referenced consensus standards related to materials used in potable water systems.

602.3.5.1 Pump enclosure. The pump room or enclosure around a well pump shall be drained and protected from freezing by heating or other *approved* means. Where pumps are installed in basements, such pumps shall be mounted on a block or shelf not less than 18 inches (457 mm) above the basement floor. Well pits shall be prohibited.

❖ A well pit, being below grade, has the potential for collecting rainwater, surface water or ground water and, subsequently, contaminating the well. The pump must be located to prevent the contamination of the well. When located in a basement, the pump must be elevated in the event that the basement becomes partially flooded. Pumps located outside a building must be properly protected from freezing during the winter months. Heat is typically provided by a small heating device controlled by a thermostat.

SECTION 603 WATER SERVICE

603.1 Size of water service pipe. The water service pipe shall be sized to supply water to the structure in the quantities and at the pressures required in this code. The minimum diameter of water service pipe shall be $^3/_4$ inch (19.1 mm).

❖ The water service pipe transports water from the source of supply to the building, either from a public main or an individual water supply. It is typically an underground pipe; however, in certain climates it might be located above ground.

The minimum size of $^3/_4$ inch (19.1 mm) is based on the nominal pipe or tube size of material. The sizes are regulated by the pipe material standard. Hence, a $^3/_4$-inch (19.1 mm) copper water service tube has a different inside diameter than a $^3/_4$-inch (19.1 mm) galvanized steel pipe (see commentary, Section 301.5).

For sizing considerations, see the commentary to Section 604. Additionally, two methods of sizing the water service have been provided in Appendix E. Table 605.3 lists various types of piping materials that are acceptable for use as water service pipe or tubing. Though the pipe sizing methodology in Appendix E is valid for all of these types of piping materials, the methodology cannot be fully applied without using the friction loss chart for the specific piping or tubing material. At this time, Appendix E contains only six friction loss charts. Therefore, to size a pipe not included in Appendix E, use a friction loss chart that closely resembles the pipe material used or the friction loss for that piping material has to be obtained.

One of the most widely used methods for determining the friction losses of piping material is the Hazen-Williams Formula, which is:

$$f = 0.2083\,(100/C)^{1.85}\,[(q)^{1.85}/(d)^{4.8655}]$$

where:

f = Friction head, in feet of liquid per 100 feet of pipe.

d = Inside diameter of pipe, in inches.

q = Fluid flow, in gpm.

C = Surface roughness constant.

Typical *C* factors used in design, which take into account some increase in roughness as pipe ages, are as follows:

- Asbestos-cement—140
- Cast iron—100
- Cement-Mortar Lines Ductile Iron Pipe—140
- Concrete—100
- Copper—150
- Steel—120
- Galvanized steel—120
- Polyethlyene—150
- Polyvinyl chloride (PVC)—150
- Fiber-reinforced plastic (FRP)—150

As an example, cross-linked polyethylene plastic tubing (PEX) is listed in Table 605.3 as an acceptable type of piping material for the water service, but Appendix E does not contain the friction loss chart for PEX. Using the Hazen-Williams Formula for SDR 9 PEX, the friction losses can be determined, thus allowing the methodology in Appendix E to be used.

The following example has been included to establish how Appendix E, Section E103.3—Segmented loss method—can be used to size the water service.

Example: Determine the minimum size of the water service pipe and the pressure available at the water service entrance for the arrangement shown in Commentary Figure 603.1(1).

A. To size the water service pipe:

Step 1: Use Commentary Table 603.1 to convert the total demand for the building from water supply fixture units (wsfu) to a flow rate in gallons per minute (gpm).

30 wsfu = 23.3 gpm

Step 2: Use Commentary Figure 603.1(2) to determine the smallest pipe size that will convey 23.3 gpm at a flow velocity of no greater than 10 feet per second.

Commentary Figure 603.1(2) shows that approximately a 1-inch-diameter pipe will deliver 23.3 gpm at approximately 9 feet per second.

B. To determine the available pressure at the water meter, "m":

Step 1: Use Commentary Figure 603.1(2) to determine the pressure loss through a 1-inch Type L copper pipe with a delivery rate of 23.3 gpm.

Commentary Figure 603.1(2) shows that a 1-inch water service pipe, 100 feet in length, at 23.3 gpm, has a pressure drop of 10 psi.

Step 2: Calculate the pressure loss:

Pressure available as the water meter = Pressure at the water main–Pressure drop through water service pipe

Pressure remaining = 90 psi - 1 psi = 80 psi

For SI: 1 gallon per minute = 0.06309 L/S,
1 pound per square inch = 6.895 kPa,
1 foot per second = 0.3048 m/s,
1 inch = 25.4 mm, 1 foot = 304.8 mm.

ASSUME THE FOLLOWING:
- PIPE MATERIAL IS TYPE "L" COPPER
- WATER CLOSETS PREDOMINANTLY FLUSH TANKS
- TOTAL DEMAND FOR BUILDING IS 30 WSFU
- MAINTAIN VELOCITY AT OR BELOW 10 FT/SEC
- WATER SERVICE HAS NO CHANGE IN ELEVATION
100 FT
M
PUBLIC WATER MAIN
90 PSI

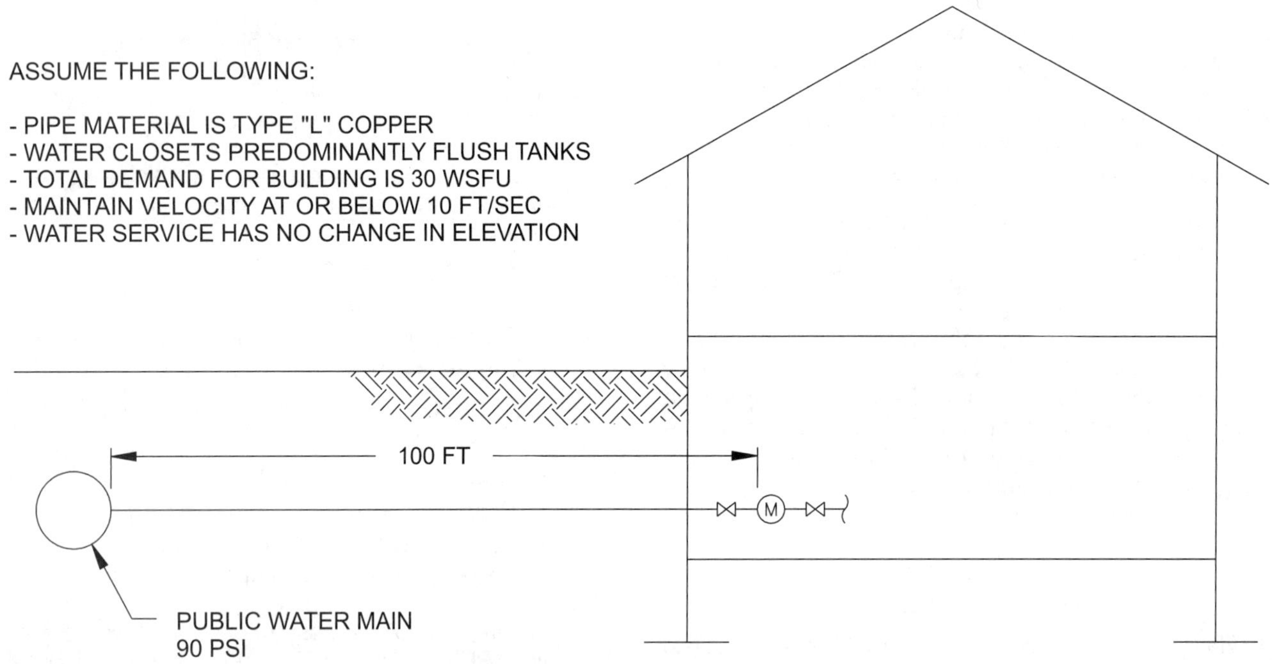

For SI: 1 foot = 304.8 mm, 1 pound per square inch = 6.895 kPa, 1 foot per second = 0.3048 m/s.

Figure 603.1(1)
RESIDENTIAL WATER SERVICE SCHEMATIC

Table 603.1
TABLE FOR ESTIMATING DEMAND

SUPPLY SYSTEMS PREDOMINANTLY FOR FLUSH TANKS		
Load	**Demand**	
(Water supply fixture units)	**(Gallons per minute)**	**(Cubic feet per minute)**
1	3.0	0.4104
2	5.0	0.6684
3	6.5	0.86892
4	8.0	1.06944
5	9.4	1.2566
6	10.7	1.4308
7	11.8	1.57743
8	12.8	1.71111
9	13.7	1.83142
10	14.6	1.95174
11	15.4	2.05868
12	16.0	2.13889
13	16.5	2.20573
14	17.0	2.27257
15	17.5	2.33941
16	18.0	2.40624
17	18.4	2.45972
18	18.8	2.51319
19	19.2	2.56667
20	19.6	2.62014
25	21.5	2.87413
30	23.3	3.11476
35	24.9	3.32865
40	26.3	3.5158
45	27.7	3.70295

For SI: 1 gallon per minute = 3.785 L/m, 1 cubic foot per minute = 0.0004719 m^3/s.

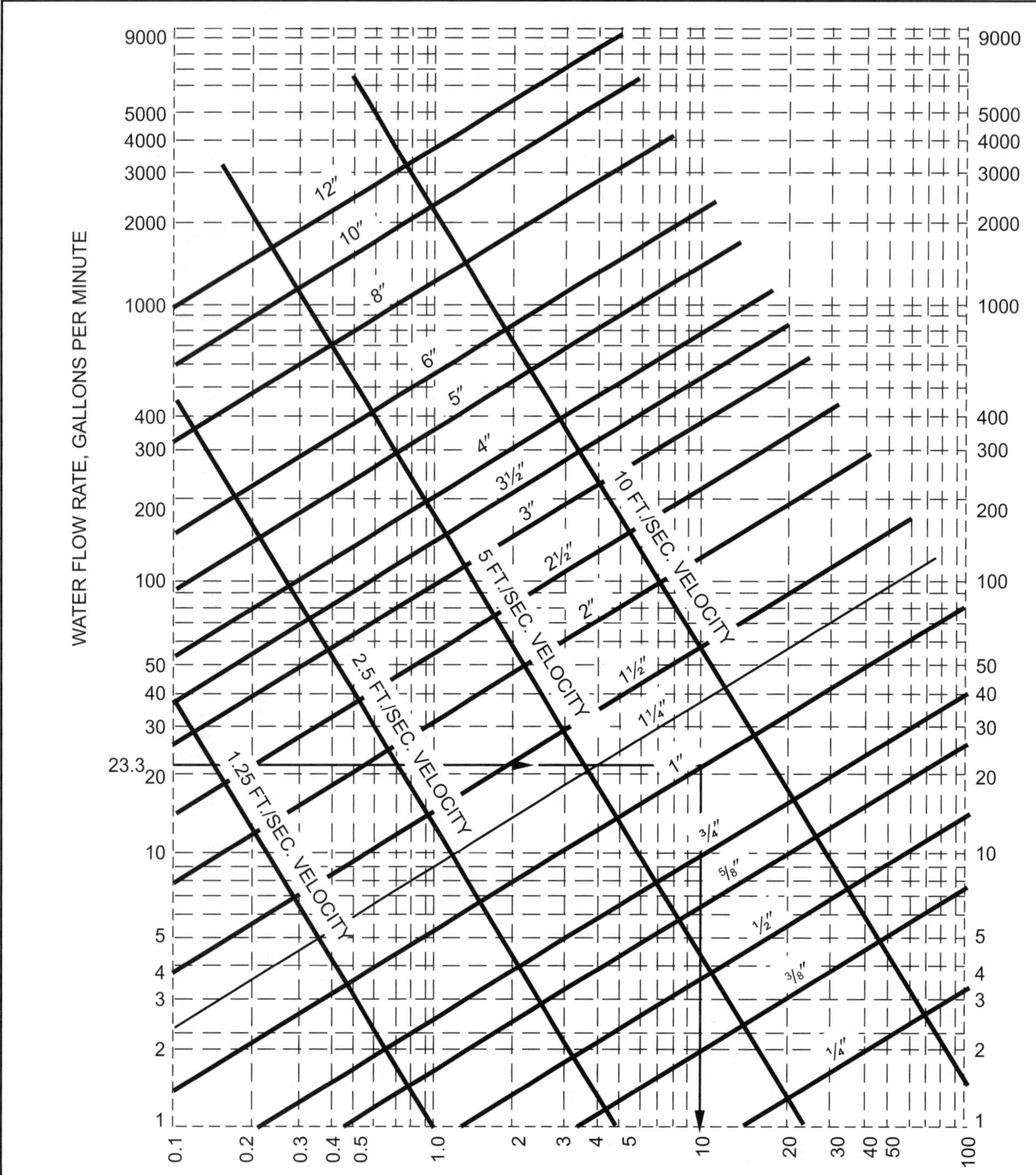

For SI: 1 inch = 25.4 mm, 1 foot = 304.8 mm, 1 gallon per minute = 3.785 L/m, 1 pound per square inch = 6.895 kPa, 1 foot per second = 0.3048 m/s.

Figure 603.1(2)
FRICTION LOSS IN SMOOTH PIPE[a]
(Type L, ASTM B 88 Copper Tubing)

[a]This chart applies to smooth new copper tubing with recessed (Streamline) soldered joints and to the actual sizes of type indicated on the diagram.
(Note: This is a reprint of Figure E103.3(3) in Appendix E.)

603.2 Separation of water service and building sewer. Water service pipe and the *building sewer* shall be separated by 5 feet (1524 mm) of undisturbed or compacted earth.

Exceptions:

1. The required separation distance shall not apply where the bottom of the water service pipe within 5 feet (1524 mm) of the *sewer* is a minimum of 12 inches (305 mm) above the top of the highest point of the *sewer* and the pipe materials conform to Table 702.3.
2. Water service pipe is permitted to be located in the same trench with a *building sewer*, provided such *sewer* is constructed of materials listed in Table 702.2.
3. The required separation distance shall not apply where a water service pipe crosses a *sewer* pipe, provided that the water service pipe is sleeved to at least 5 feet (1524 mm) horizontally from the *sewer* pipe centerline on both sides of such crossing with pipe materials listed in Table 605.3, 702.2 or 702.3.

❖ The requirements for separating the water service from the building sewer are intended to reduce the possibility of the sewer contaminating the potable water supply, assuming failure of the water supply and building sewer pipes.

Contamination can occur when there is a leak in the building sewer located near the water service pipe.The soil then becomes contaminated around the water pipe, and, if the water service pipe subsequently leaks, contamination of the potable water supply could occur.

The simplest means of reducing the possibility of potable water contamination is installing the building sewer and water service pipe in two trenches separated horizontally by undisturbed or compacted earth [see Commentary Figure 603.2(1)]. Any contamination will tend to saturate the excavated soil before settling into the undisturbed, compacted earth.

The first exception permits the water service to be located within 5 feet (1524 mm) horizontally of a building sewer if it is located 12 inches (305 mm) above the highest point of the sewer and the sewer is constructed of acrylonitrile butadiene styrene (ABS) plastic pipe, cast-iron pipe, copper or copper-alloy tubing, polyvinyl chloride (PVC) plastic pipe or stainless steel conforming to one of the standards listed in Table 702.3 for building sewer pipe.The 12-inch (305 mm) vertical separation is based on the soil saturation profile if a leak occurs in the building sewer.The soil contamination would not rise 12 inches (305 mm) unless affected by a ground-water condition. This exception also applies to conditions where a water service pipe crosses over a building sewer [see Commentary Figure 603.2(2)].

The second exception permits the water service to be located in the same trench if the materials conform to Table 702.2 for underground building drainage and vent pipe. This exception results in greater flexibility for the installation of water and sewer pipes.

The third exception provides for the installation where the water service pipe crosses a sewer pipe (over or under). Sometimes this installation is unavoidable, especially with existing installations. The water service must be sleeved to a point at least 5 feet (1524 mm) horizontally from the sewer pipe centerline, on both sides of the crossing. Simply stated, the water service must be run in a continuous conduit constructed of piping listed in Table 605.3 for water service pipe, Table 702.2 for underground building drainage and vent pipe or Table 702.3 for building sewer pipe for a minimum distance of 10 feet (3048 mm) to reduce the possibility of soil contamination to the potable water supply. Note that sealing the ends of the conduit containing the water service pipe is not necessary or required.

603.2.1 Water service near sources of pollution. Potable water service pipes shall not be located in, under or above cesspools, septic tanks, septic tank drainage fields or seepage pits (see Section 605.1 for soil and groundwater conditions).

❖ Because septic systems rely on the soil for treatment, locating the water service pipe in the area is prohibited. Commentary Table 603.2.1 references the minimum horizontal separation distances between water source elements and components of soil absorption systems found in Sections 406 and 802 of the *International Private Sewage Disposal Code*® (IPSDC®).

Table 603.2.1
MINIMUM HORIZONTAL SEPARATION DISTANCES BETWEEN WATER SOURCE ELEMENTS AND SOIL ABSORPTION SYSTEM COMPONENTS

ELEMENT	ABSORPTION FIELD	SEPTIC OR TREATMENT TANK
	DISTANCE (feet)	
Cistern	50	25
Lot line	5	2
Spring	100	50
Water main	50	—[a]
Water service	10	5
Water well	50	25

For SI: 1 foot = 304.8 mm.
a. No Limitation is listed in Table 802.8 of the IPSDC.

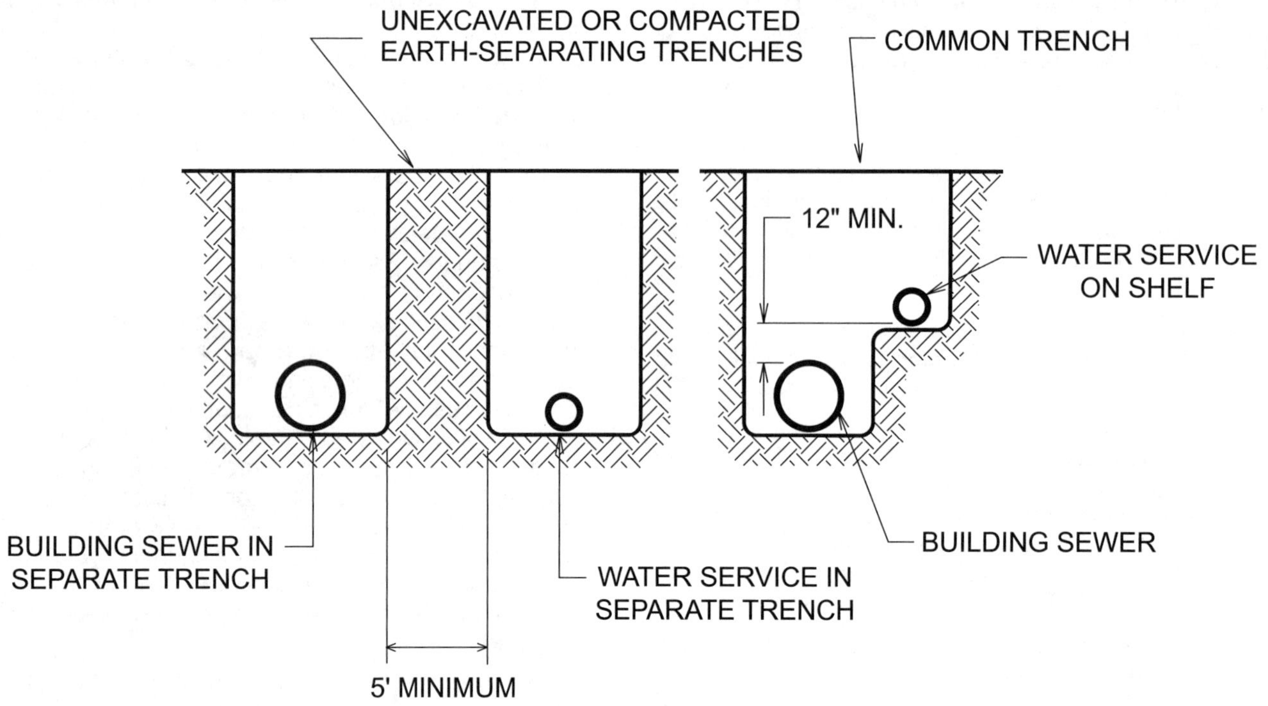

For SI: 1 inch = 25.4 mm, 1 foot = 304.8 mm.

Figure 603.2(1)
SEPARATION OF WATER SERVICE AND BUILDING SEWER

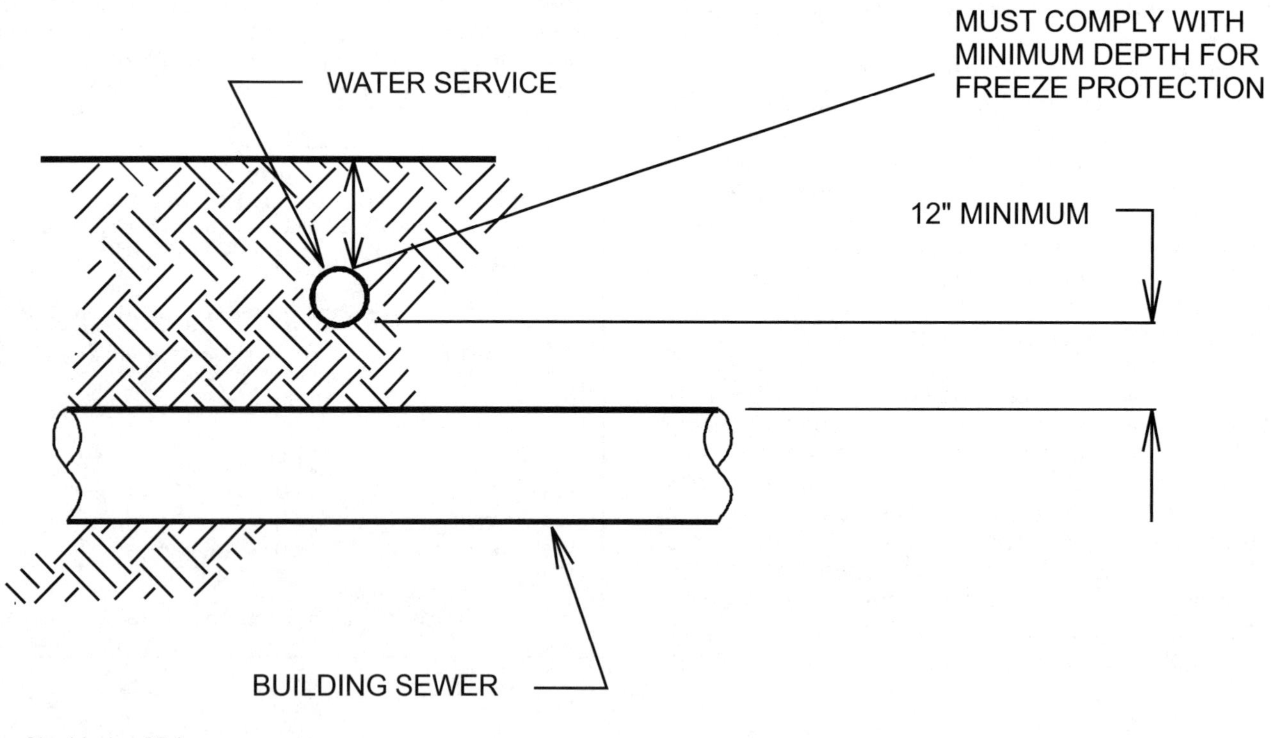

For SI: 1 inch = 25.4 mm.

Figure 603.2(2)
WATER SERVICE CROSSING OVER BUILDING SEWER

SECTION 604 DESIGN OF BUILDING WATER DISTRIBUTION SYSTEM

604.1 General. The design of the water distribution system shall conform to *accepted engineering practice*. Methods utilized to determine pipe sizes shall be *approved*.

❖ The water distribution system must be capable of supplying a sufficient volume of potable water at pressures adequate to enable plumbing fixtures, devices and appurtenances to function properly and without undue noise. This section is intended to allow the option to use any sizing method available if it conforms to accepted engineering practice, without containing or making reference to a specific water pipe sizing method. Appendix E contains one method of sizing water piping systems.

There are two basic reasons why the code does not contain or make reference to a specific water pipe sizing method. First, it is impractical to specify precise, detailed rules for sizing because of the numerous variable parameters encountered in hydraulic design. Second, it is impossible to create an exact model to predict the demand for a building and determine the probability of use and flow characteristics of various fixtures within various occupancies.

It is the sole responsibility of the code official to determine that the water supply system has been designed to meet the applicable sections of the code. The code official is not the water supply system designer. This section requires that the water distribution system conform to accepted engineering practice and that methods used to determine the pipe sizes be approved by the code official. This language places the responsibility for the selection of the design method with the system designer, while providing the code official with the authority to accept or reject the design methodology proposed for use. This text is also intended to allow the system designer the option to use any pipe sizing method available if the design produced by the method used complies with the intent of the code and is in accordance with generally accepted engineering practice.

In addition to the approval of the proposed water pipe sizing design method, the code official is responsible for determining that the construction documents contain pipe sizes, sizing calculations indicating the method used and any other information necessary to establish the adequacy of the proposed water supply system as required by Section 106.3.1 (see commentary, Section 106.3.1).

Water pipe sizing methods are ultimately based on hydraulics, which is the study of the principles and laws governing the behavior of liquids. The factors that influence the sizing of the water supply system and must be considered in the system design are pressure, velocity limitations, materials, characteristics of the water source and demand.

Pressure

Pressure is one of the key variables for sizing the water supply system. Pressure must be adequate to overcome pressure losses that occur in the system from friction and elevation, so that plumbing fixtures operate properly.

Several of the requirements contained in Section 604 deal directly with the requirements for water supply system pressure. Table 604.3 lists the minimum flow pressures (also known as residual pressures) required at the fixture for proper operation with water flowing.

System pressure losses resulting from elevation must be considered in the system design. The difference in elevation between the water supply and the highest water supply outlet has a significant impact on the sizing of the water supply system.

The pressure caused by the height of water can be converted from a measurement of feet of head to pounds per square inch.

$$P = 0.433 \times h$$

where:

P = Static head or pressure produced by the height of water (psi).

h = Difference in elevation (feet).

The difference in elevation between the water supply source and the highest water supply outlet will result in an increase or decrease in available pressure. This pressure adjustment is often referred to as static head.

The difference in elevation usually results in a loss in the available pressure when the water supply outlet is located above the water supply source, as shown in Commentary Figure 604.1(1). The loss is the result of the pressure required to overcome the weight of water in order to deliver water to the fixture outlet. The pressure created by the height of water is treated as a pressure loss and is subtracted from the available pressure of the water source.

Elevation difference, however, does not always result in pressure losses. Where the highest water supply outlet is located below the water supply source, as shown in Commentary Figure 604.1(2), the pressure created by the height of water is treated as a gain and is added to the available pressure of the water source.

Pressure loss caused by pipe and fittings (friction) is another item to be considered in the system design. Pressure loss is produced when water particles rub against the inside of the pipe or the flow direction changes. The amount of pressure loss caused by friction is affected by the velocity of the water and the roughness, developed length and diameter of the pipe.

For flow rate, the effects of friction from pipe diameter decrease as the diameter of the pipe increases.

The effects of friction increase as velocity, coefficient of friction and length increase.

Fittings impose even greater losses than the pipe itself. To account for these losses, the fittings are generally converted to equivalent lengths of pipes. The larger the fitting, the larger the equivalent length becomes.

In determining the total equivalent length, Tables E103.3(4), E103.3(5) and E103.3(6) of Appendix E can be used.

Another method of calculating the equivalent length of fittings is to assume 50 percent of the developed length of pipe or:

Equivalent developed length of pipe and fittings = Actual developed length of pipe x 1.5

Sizing methods require the determination of the "most hydraulically remote" fixture to compute the pressure loss caused by pipe and fittings. The most hydraulically remote fixture represents the most downstream fixture along the circuit of piping requiring the most available pressure to operate properly. This is usually the highest, most distant fixture from the supply source. Consideration must be given to pressure demands and losses, such as friction in pipe, fittings and equipment, elevation and the residual pressure required by Code Table 604.3 for the fixture to operate properly.

Pressure caused by equipment, such as backflow preventers, check valves, water meters, strainers, filters, instantaneous or tankless water heaters and softeners, can also impose friction losses. This equipment can impart much greater pressure losses than the piping; therefore, the manufacturers of such equipment should be consulted to obtain pressure loss data and charts.

Velocity Limitations

The velocities within the water supply and distribution system must be high enough to minimize the deposit of suspended material while avoiding the destructive effects of erosion-corrosion, cavitation and pressure surges.

In accordance with good engineering practice, the maximum velocity in water supply pipe should be limited to no more than 8 feet per second (2.4 m/s). The effects of friction increase with increased velocity and are even more dramatic in smaller-diameter pipe.

The abrasive or erosive effects of water are also increased with increased velocity, depending on pipe material and characteristics of the water supply. Erosion is caused by entrained air bubbles, sand and other suspended solid matter scouring the inner surface of the pipe.

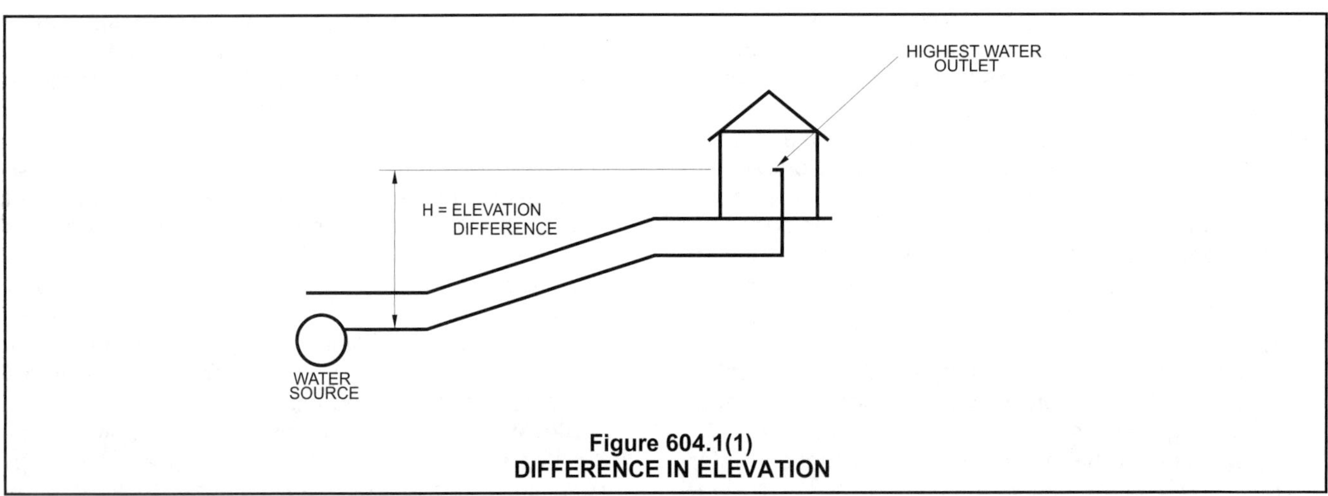

Figure 604.1(1)
DIFFERENCE IN ELEVATION

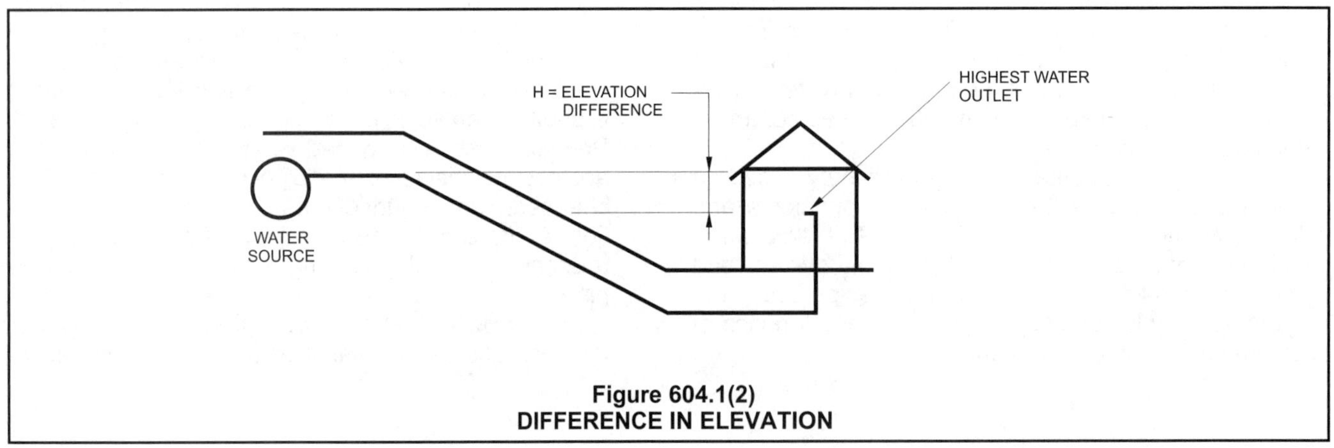

Figure 604.1(2)
DIFFERENCE IN ELEVATION

Cavitation is caused by sharp changes in the direction of flow at high velocities, as shown in Commentary Figure 604.1(3). As a result, the pressure within the cavitation zone is reduced to the vapor pressure of water. Under such conditions, vapor bubbles form (boiling occurs) and then collapse (condense). The noise produced by cavitation sounds like gravel bouncing in the piping or popping balloons.

The term "water hammer" is used to describe the noise and vibrations produced by destructive forces known as hydraulic shock, which develops in a piping system when an instantaneous change in velocity of flowing water occurs or when water flowing at a given velocity is abruptly stopped (see commentary, Section 604.9).

Materials

Commentary Table 604.1 is a list of maximum velocities for various materials as recommended by manufacturers. The table accounts for noise, shock damage and the effects of erosion-corrosion.

The selection of materials has an influence in the sizing of the water supply system. Each pipe material has a capacity different from other materials having the same nominal pipe size. This is caused by varying surface roughness and actual inside diameter. Additionally, every material has a different tolerance to corrosion and scaling.

Characteristics of the Water Source

Knowledge of local water conditions is necessary to properly design the water supply system. The corrosivity and scale-forming tendency of the water supply can have a marked effect on the sizing of the water supply system. A chemical analysis may be necessary to determine dissolved solid content, pH range, carbon dioxide content and the presence of dissolved oxygen.

Supply piping conveying water containing dissolved mineral salts, such as sulfates, bicarbonates, chlorides of calcium, sodium and magnesium, can form a scale or buildup on the inner walls of the pipe. This phenomenon, known as caking, can significantly affect the capacity of a pipe over time. Caking occurs more rapidly in hot water distribution piping because the solubility of the minerals increases with higher temperatures.

Corrosion occurs in metallic piping conveying dissolved gases such as oxygen and carbon dioxide. The rate of corrosion tends to increase with increasing velocity. Generally, corrosion occurs in areas of turbulence, such as joints or fittings. The turbulence interferes with the formation of a protective film that normally forms on the inside of the pipe. This is especially important for the preservation of copper pipe and tubing.

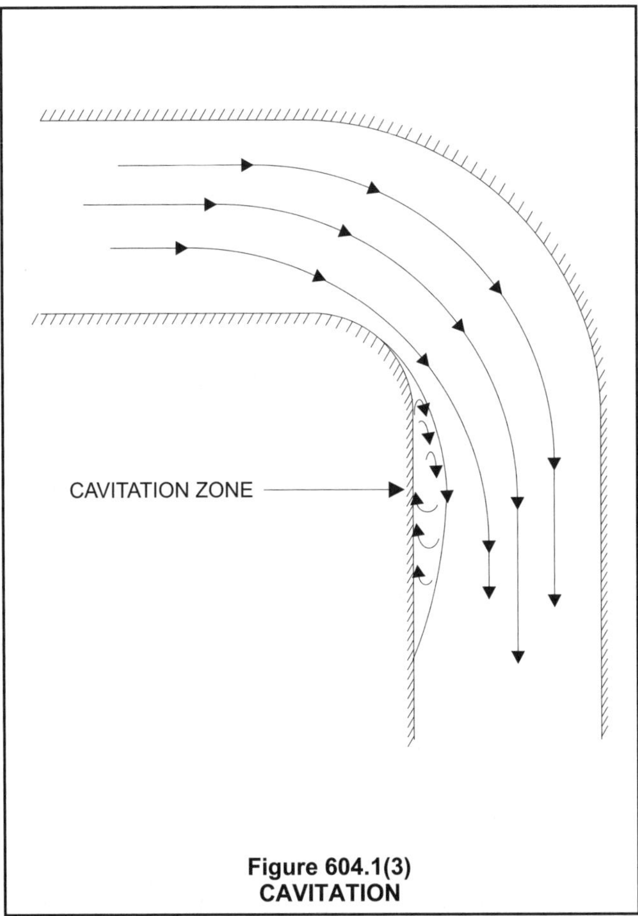

Figure 604.1(3)
CAVITATION

Table 604.1
MAXIMUM FLOW VELOCITY

MATERIAL	PIPE OR TUBE SIZE (inches)	MAXIMUM VELOCITY (feet per second)
All pipe and tube material	$^1/_2$ and smaller	5
Brass pipe	$^5/_8$ and larger	8
Copper or copper-alloy pipe and tubing (hot water systems)	$^5/_8$ and larger	5
Copper or copper-alloy pipe and tubing (cold water systems)	$^5/_8$ and larger	8
Chlorinated polyvinyl (CPVC) plastic pipe	$^5/_8$ - 1 $1^1/_4$ and larger	8 10
Cross-linked polyethylene (PEX) plastic tubing	$^5/_8$ and larger	8
Cross-linked polyethylene/aluminum/cross-linked polyethylene (PEX-AL-PEX) pipe	$^5/_8$ and larger	8
Galvanized steel pipe	$^5/_8$ - 1 2 - 4 4 and larger	8 10 12
PE-AL-PE	$^5/_8$ and larger	8

For SI: 1 inch = 25.4 mm, 1 foot per second = 0.3048 m/s.

Various pH levels create environments that can accelerate the effects of corrosion and caking, resulting in the "aging" of pipe.

Over time, caking and corrosion reduce both the cross-sectional area and the carrying capacity of the pipe. The rate of aging varies among pipe materials.

Estimating Demand

Estimating demand is the most difficult and most important aspect of water pipe sizing. When designing the water supply system, the designer must make a determination of peak demand of water usage.

Peak demand is the maximum amount (flow rate) of water at any given time during the day. It is unrealistic to calculate the maximum flow assuming that all fixtures will operate at the same time. Determining an accurate estimate of peak demand is difficult because of the intermittent operation and irregular frequency of fixture use. Peak demand occurs at different times of day, depending on the occupancy. Typical times of peak demand are shown in Commentary Table 604.3.

The water distribution system piping must be sized such that under conditions of peak demand, the capacities at the fixture supply pipe outlet are not less than those shown in Code Table 604.3 (see commentary, Section 604.3).

In determining the maximum probable flow at peak conditions, the loading effects of each fixture must be considered. Loading effects are determined by:

- Rate of water required.
- Duration of each use.
- Frequency of each use.

The various water pipe sizing methods try to model the peak demand (maximum probable flow) by considering the flow and concurrent use of fixtures based on the duration and frequency of use. Most of the sizing methods are derived from the National Bureau of Standards Report BMS 79, *Water-Distribution Systems for Buildings*, by Dr. Roy B. Hunter, published in 1941.

When estimating peak demand, sizing methods use water supply fixture units (wsfu), which is a numerical factor on an arbitrary scale assigned to intermittently used fixtures to calculate their load-producing effects on the water supply system.

The use of fixture units makes it possible to simplify the difficult task of calculating the load-producing characteristics of any fixture to a common basis. Units of different kinds of fixtures can be applied to a single basic probability curve found in the various sizing methods. Note that fixture units generally cannot be applied to constant-use fixtures, such as hose bibbs, or continuous supply systems, such as lawn sprinklers and air conditioners. The demand for constant-use fixtures should be calculated separately and added to the demand for fixtures used intermittently. It should be noted that lawn sprinkler systems can be considered as a nonsimultaneous demand when sizing the water supply system. For a lawn sprinkler system to be considered separately and not included in the determination of peak demand, the system should have an automatic control system programmed for off-hour watering (for instance, 2:00 a.m. to 6:00 a.m.). Otherwise, the demand from the largest zone in the system must be added to the peak demand for the building [lawn sprinkler heads typically range from 0.25 to 6.5 gpm (0.95 to 24.6 L/m)].

Finally, automatic fire sprinkler systems can impact the size of the water supply system served. NFPA 13R, NFPA 13D and limited-area sprinkler systems are permitted to be served by a common water supply that also serves the domestic supply system. In most cases, however, the sprinkler system dictates the sizing of supply lines. Check the *International Building Code*® (IBC®) to determine which sprinkler systems are acceptable or where one is required.

604.2 System interconnection. At the points of interconnection between the hot and cold water supply piping systems and the individual fixtures, appliances or devices, provisions shall be made to prevent flow between such piping systems.

❖ When a plumbing fixture, device or appliance is provided with both a hot and a cold water supply connection (other than terminal devices that blend hot and cold water, such as tempering valves, pressure-balanced mixing valves, etc., to achieve a desired temperature), the water must be supplied in such a way that there will be no possibility for the flow of cold water to the hot water piping or vice versa. This is required so the hot water supply does not become cooled or tempered below the temperature required (see the definition of "Hot water" in Section 202) by the introduction of cold water, and that when a fixture is operated to provide cold water; hot water does not get delivered. Another concern is that if there is a continuous flow of cold water into the hot water supply, this could lead to increased operation of the water heater in an attempt to maintain the water temperature, resulting in an increased use of energy and a shortened heater life.

This section might also be applicable to "no return pipe" hot water recirculation system arrangements which have become popular in recent years. Although originally intended as a retrofit method of providing for hot water recirculation where the installation of hot water return piping was impossible or too costly, these systems are also being installed in new construction due to the labor and materials savings. Although system component designs vary between manufacturers, the system concept utilizes a pump in the hot water line to cause hot water flow through a special valve and into the cold water line near a fixture that is most remote from the water heater. When the valve senses that the temperature of the hot water flow becomes warm [approximately 90°F (32°C)], the valve automatically closes to stop flow so that the cold water line does not become filled with hot water. The valve also prevents flow of hot water to the cold supply line while cold water is flowing from the faucet. The minimal flow rate through the valve limits the amounts of warm water flowing into the cold water line such that when the user opens only the cold water control of the faucet, warm water flows from the faucet for only a few seconds before cooler flow begins to flow. Typically, most users rarely sense the short period of warm water flow from the cold only operation of the faucet. Even if they did have their hand under the faucet at the moment the cold water valve was opened, the temperature of the flow is well within safe "tempered water" temperatures. Thus, the user is protected from the potential of being inadvertently scalded by hot water flow from cold water operation of the faucet. Because the recirculation flow is minimal, the operational cost of these systems is generally much less than a system utilizing a dedicated hot water return pipe. Thus, the energy "waste" for this type of hot water recirculation system is no greater than for a system using a dedicated hot water return pipe. Therefore, with respect to the apparent intents of this section for scald protection and preventing hot water storage depletion (and the resultant energy waste), these "no return pipe" systems do not appear to be a violation of this section.

However, if this section is read and interpreted in a literal manner, these "no return pipe" recirculation systems constitute an interconnection between hot and cold systems that does allow flow to occur between the systems and, therefore, these systems would violate this section. Ultimately, the code official must decide if these "no return pipe" systems comply with or violate this section based upon his or her interpretation of this section.

604.3 Water distribution system design criteria. The water distribution system shall be designed, and pipe sizes shall be selected such that under conditions of peak demand, the capacities at the fixture supply pipe outlets shall not be less than shown in Table 604.3. The minimum flow rate and flow pressure provided to fixtures and appliances not listed in Table 604.3 shall be in accordance with the manufacturer's installation instructions.

❖ Each fixture has a given flow rate and minimum water supply pressure needed for proper operation. The minimum values are specified in Table 604.3. Fixtures not listed in the table must be provided with flow rates and pressures as specified by the manufacturer.

The minimum flow rate and pressure requirements must be satisfied during peak demand, which is the maximum amount (flow rate) of water required at any given time during the day.

To determine peak demand, there should be knowledge of the approximate time it occurs. The peak demand does not necessarily occur just once during the day. The time that peak demand typically occurs for certain types of buildings is listed in Commentary Table 604.3.

Table 604.3
PEAK DEMAND PERIODS

BUILDING	TIME OF PEAK DEMAND
Football stadium	Halftime
Office building	Midmorning/after lunch
Residential	Early morning
Theater	Intermission
Restaurant	During meals
Factory	Shift time
Schools	Class break
Mercantile	Any time
Night clubs/bars	10:00 p.m. - 11:00 p.m.

TABLE 604.3
WATER DISTRIBUTION SYSTEM DESIGN CRITERIA REQUIRED CAPACITY AT FIXTURE SUPPLY PIPE OUTLETS

FIXTURE SUPPLY OUTLET SERVING	FLOW RATE[a] (gpm)	FLOW PRESSURE (psi)
Bathtub, balanced-pressure, thermostatic or combination balanced-pressure/thermostatic mixing valve	4	20
Bidet, thermostatic mixing valve	2	20
Combination fixture	4	8
Dishwasher, residential	2.75	8
Drinking fountain	0.75	8
Laundry tray	4	8
Lavatory	2	8
Shower	3	8
Shower, balanced-pressure, thermostatic or combination balanced-pressure/thermostatic mixing valve	3	20
Sillcock, hose bibb	5	8
Sink, residential	2.5	8
Sink, service	3	8
Urinal, valve	12	25
Water closet, blow out, flushometer valve	25	45
Water closet, flushometer tank	1.6	20
Water closet, siphonic, flushometer valve	25	35
Water closet, tank, close coupled	3	20
Water closet, tank, one piece	6	20

For SI: 1 pound per square inch = 6.895 kPa,
1 gallon per minute = 3.785 L/m.

a. For additional requirements for flow rates and quantities, see Section 604.4.

❖ This table specifies the design parameters required for a water distribution system. These minimum values must be satisfied in the design of the system for proper functioning of each plumbing fixture. The design professional must use the table when calculating pipe sizes, as required in Section 604.1 (see commentary, Section 604.1).

For the 2009 edition, flow pressures for bathtub control valves, bidet control valve, urinal flushometers, water closets (except for one piece) were all increased to be in alignment with the requirements of the latest standards for those products. The flow rates for urinal and blowout water closet flushometers were decreased reflecting the requirements of the latest product standards.

604.4 Maximum flow and water consumption. The maximum water consumption flow rates and quantities for all plumbing fixtures and fixture fittings shall be in accordance with Table 604.4.

Exceptions:

1. Blowout design water closets having a maximum water consumption of $3^1/_2$ gallons (13 L) per flushing cycle.
2. Vegetable sprays.
3. Clinical sinks having a maximum water consumption of $4^1/_2$ gallons (17 L) per flushing cycle.
4. Service sinks.
5. Emergency showers.

❖ This section intends to conserve water resources by limiting the consumption of plumbing fixtures and the flow rates of outlets.

Water conservation is important to conserve not only our limited water supplies but also the energy required to pump, transport, treat, condition and heat water. Water conservation also lessens the burden on water treatment facilities and private sewage disposal systems. The relationship between water usage and energy usage is evidenced by the fact that the federal government water conservation requirements are part of the Energy Policy Act (EPAct).

The exceptions permit specified fixtures to exceed the limitations of Table 604.4, but in the interest of conservation, the designer could specify compliance with Table 604.4 for any of those exempted applications.

Federal legislation mandates the design and use of urinals that have flush consumption restricted to 1.0 gallon (3.8 L) per flushing cycle and water closets that have flush consumption restricted to 1.6 gallons (6.1 L) per flushing cycle. Since the advent of 1.6 gpf water closets, high efficiency water closets (HET) have been perfected which employ large flush valve diameters and tray way improvements to achieve flush volumes as low as 1.28 gpf. Exception 1 of Section 604.4 allows blowout-type water closets to exceed 1.6 gallons per flush.

Water closets falling under the exception to this section are regulated by a maximum consumption of 3.5 gallons (13 L) per flushing cycle. This maximum cannot be exceeded under any circumstance.

Exception 1 exempts blowout water closets from the requirements of Table 604.4 because such fixtures depend on high-volume and high-velocity water flow to evacuate the contents of the bowl; therefore, such fixture designs may not be able to function at lower consumption rates. Blowout fixtures are commonly installed in assembly occupancies because they are considered to be less susceptible to the abuse and vandalism typically associated with public toilet facilities.

Exception 2 exempts vegetable sprays because they are not included in EPAct.

Exception 3 exempts clinical sinks because they are specialty fixtures used in institutional occupancies for the disposal of human waste and the cleaning of bedpans.

Exception 4 exempts service sinks because such fixtures are used for maintenance and janitorial operations. Because sink faucets are usually used to fill mop buckets and similar containers or the sink itself, there is nothing to be gained by limiting the flow rate of such faucets. The same logic is extended to mop sinks, pots-and-pans sinks and bathtubs. Such fixtures are filled with water or are used to fill other containers, and limiting the flow rate of the faucet will not affect the amount of water needed by the user of the fixture, but will affect the filling time with no apparent benefit.

Exception 5 exempts emergency showers because they are used only in the event of an emergency. Emergency showers are provided for body and eye washing in occupancies where people are exposed to chemical hazards. When addressing life safety, water consumption is not a concern.

There has been much speculation and a considerable amount of research in the area of 1.6-gallon per-flush (6.1 L/flush) water closets. Research has shown that the latest generation of these fixtures performs as well as traditional 3.5-gallon-per-flush (13 L/flush) fixtures and that such fixtures can conform to the hydraulic performance requirements of ASME A112.19.2 and ASME A112.19.6. The occasional need for double-flushing has been a common complaint from users; however, the water savings potential is significant even when this occurs.

The primary concern involving 1.6-gallon-per-flush (6.1 L/flush) water closets has been the occurrence of stoppages in buildings having oversized drain piping and in buildings that do not have gray-water discharging fixtures such as showers, bathtubs, laundry sinks and standpipes. Where water closets and lavatories are the only load on a building drain or sewer, there is a concern that there will be insufficient waste-water flow to transport solid waste.

These problems have not been proven to exist except in isolated examples where other factors, such as fixture abuse and improperly sloped piping, are involved.

Although the oversizing of drain piping is poor practice in most applications, it is especially undesirable in systems with low-consumption water closets. Because water-conserving plumbing will remain and evolve in the future, it is foreseeable that drain and vent sizing methodology could change in response to the reduced fixture discharge rates.

TABLE 604.4
MAXIMUM FLOW RATES AND CONSUMPTION FOR PLUMBING FIXTURES AND FIXTURE FITTINGS

PLUMBING FIXTURE OR FIXTURE FITTING	MAXIMUM FLOW RATE OR QUANTITY[b]
Lavatory, private	2.2 gpm at 60 psi
Lavatory, public (metering)	0.25 gallon per metering cycle
Lavatory, public (other than metering)	0.5 gpm at 60 psi
Shower head[a]	2.5 gpm at 80 psi
Sink faucet	2.2 gpm at 60 psi
Urinal	1.0 gallon per flushing cycle
Water closet	1.6 gallons per flushing cycle

For SI: 1 gallon = 3.785 L, 1 gallon per minute = 3.785 L/m, 1 pound per square inch = 6.895 kPa.

a. A hand-held shower spray is a shower head.

b. Consumption tolerances shall be determined from referenced standards.

❖ Consumption tolerances are determined from referenced standards. The fixtures and faucets listed in the table are those that either function in a cyclical manner or are allowed to consume water continuously for the duration of fixture use. Water is conserved by limiting both the flow rate for manually controlled fixtures and the volume-per-cycle usage for cyclically operating fixtures.

To determine maximum flow rates and consumption for fixtures designed for multiple use, such as group hand-washing basins, the number of equivalent lavatories is computed based on linear inches of rim space of the fixture as referenced in Section 416.1. By determining the equivalent number of lavatories for any given group fixture, the cumulative maximum flow rate for the fixture can be computed using the table values. The maximum flow rates for nonmetering showers, lavatory and sink faucets depend on supply pressure; therefore, a reference pressure is stated for each item in the table. Where the actual supply pressure is less than the pressure in the table, the actual flow rate of the faucet or shower head will also be less than the maximum flow rate in the table.

604.5 Size of fixture supply. The minimum size of a fixture supply pipe shall be as shown in Table 604.5. The fixture supply pipe shall not terminate more than 30 inches (762 mm) from the point of connection to the fixture. A reduced-size flexible water connector installed between the supply pipe and the fixture shall be of an *approved* type. The supply pipe shall extend to the floor or wall adjacent to the fixture. The minimum size of individual distribution lines utilized in gridded or parallel water distribution systems shall be as shown in Table 604.5.

❖ Each dimension listed in the table is the minimum size that the supply pipe can be from the point of source to the fixture. The pipe can be reduced in size for the last 30 inches (762 mm). This is common practice when connecting faucets to sinks or lavatories. To simplify installation, a flexible or semirigid connector is typically used between the faucet or fixture and its water supply pipe (see Commentary Figure 604.5). Flexible fixture connectors are considered fixture fittings. As such, they must conform to ASME A112.18.6 for flexible water connectors (see Section 605.6).

TABLE 604.5
MINIMUM SIZES OF FIXTURE WATER SUPPLY PIPES

FIXTURE	MINIMUM PIPE SIZE (inch)
Bathtubs[a] (60″ × 32″ and smaller)	$^1/_2$
Bathtubs[a] (larger than 60″ × 32″)	$^1/_2$
Bidet	$^3/_8$
Combination sink and tray	$^1/_2$
Dishwasher, domestic[a]	$^1/_2$
Drinking fountain	$^3/_8$
Hose bibbs	$^1/_2$
Kitchen sink[a]	$^1/_2$
Laundry, 1, 2 or 3 compartments[a]	$^1/_2$
Lavatory	$^3/_8$
Shower, single head[a]	$^1/_2$
Sinks, flushing rim	$^3/_4$
Sinks, service	$^1/_2$
Urinal, flush tank	$^1/_2$
Urinal, flush valve	$^3/_4$
Wall hydrant	$^1/_2$
Water closet, flush tank	$^3/_8$
Water closet, flush valve	1
Water closet, flushometer tank	$^3/_8$
Water closet, one piece[a]	$^1/_2$

For SI: 1 inch = 25.4 mm, 1 foot = 304.8 mm, 1 pound per square inch = 6.895 kPa.

a. Where the developed length of the distribution line is 60 feet or less, and the available pressure at the meter is a minimum of 35 psi, the minimum size of an individual distribution line supplied from a manifold and installed as part of a parallel water distribution system shall be one nominal tube size smaller than the sizes indicated.

❖ This table assists the piping system designer by specifying the minimum pipe size permitted for a given fixture. Note that the table lists the fixture pipe sizes in traditional plumbing trade sizing designations. The sizes specified are based on the demand of the fixture and determined by the methods reviewed in the commentary to Section 604.1. Note a permits the minimum size of a supply pipe to be reduced if it is an individual distribution pipe served by a manifold because of the difference in hydraulics between conventional (series) and manifold (parallel) piping systems. It should be noted that the pressure drop caused by the required flow rate passing through 60 feet (18 288 mm) of $^3/_8$-inch (9.5 mm) pipe is approximately 20 psi (138 kPa). Although Note a states that the required minimum available pressure must be 35 psi (508 kPa) at the meter, this pressure may not be adequate to meet the minimum flow pressures required for fixtures by Table 604.3. For instances where the available pressure is insufficient, Sections 604.3 and 604.7 require larger pipe sizes or a method of increasing pressure.

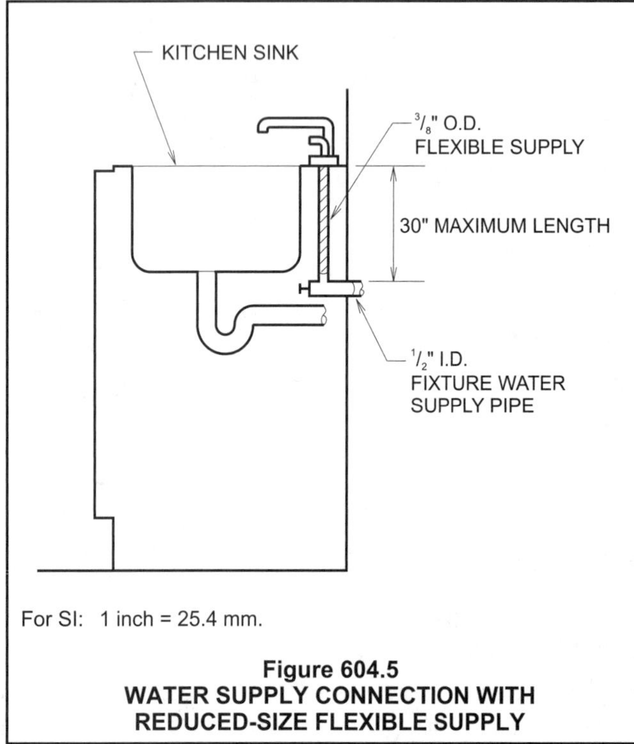

For SI: 1 inch = 25.4 mm.

Figure 604.5
WATER SUPPLY CONNECTION WITH REDUCED-SIZE FLEXIBLE SUPPLY

604.6 Variable street pressures. Where street water main pressures fluctuate, the building water distribution system shall be designed for the minimum pressure available.

❖ In the design of the water distribution system, the lowest pressure must be determined and used in the calculations to guarantee a continuous, adequate supply of water. In many areas of the country, the pressure in the public water main will vary throughout the day and seasonally.

The lowest pressure in a public main usually occurs in the summer because of lawn sprinkling and supplying makeup water for air-conditioning cooling towers. The lowest pressure in well systems generally occurs during the pump on-off cycles. The minimum pressure can usually be determined by a review of the records of pressures at different times of the day and year kept by the local water authority. The pressure can also be checked from nearby buildings or from fire department hydrant inspections. Future demands placed on a public main as a result of large community growth or expansion will result in a decrease of available pressure as additional loads are placed on the system. Anticipated losses caused by growth and expansion must be considered in determining minimum pressure.

604.7 Inadequate water pressure. Wherever water pressure from the street main or other source of supply is insufficient to provide flow pressures at fixture outlets as required under Table 604.3, a water pressure booster system conforming to Section 606.5 shall be installed on the building water supply system.

❖ Section 606.5 regulates the various methods for increasing the pressure of the water distribution system (see commentary, Section 606.5).

604.8 Water-pressure reducing valve or regulator. Where water pressure within a building exceeds 80 psi (552 kPa) static, an *approved* water-pressure reducing valve conforming to ASSE 1003 with strainer shall be installed to reduce the pressure in the building water distribution piping to 80 psi (552 kPa) static or less.

Exception: Service lines to sill cocks and outside hydrants, and main supply risers where pressure from the mains is reduced to 80 psi (552 kPa) or less at individual fixtures.

❖ The piping system, components and fixtures in a water distribution system are all designed to withstand a given pressure. The code establishes 80 psi (552 kPa) as the maximum working pressure of the system. When the pressure from the public water main or the individual water supply exceeds 80 psi (552 kPa), a pressure-reducing valve must be installed at the point where the water service enters the building.

In multistory buildings, it is sometimes advantageous to use higher pressure to supply water to the upper floors. For this type of system, the pressure is reduced at the branches to prevent the fixtures and appliances from being exposed to excessive pressures. The design professional must also account for the higher pressures in the piping for this system.

604.8.1 Valve design. The pressure-reducing valve shall be designed to remain open to permit uninterrupted water flow in case of valve failure.

❖ Water-pressure reducing valves must be designed so that if the pressure-reducing component of the valve fails to respond to the adjusted water pressure, the valve will remain in an open position to allow a continuous flow of water to the building. A valve without this fail-safe feature could close completely in event of a failure, thus depriving the building and its occupants of water service.

604.8.2 Repair and removal. All water-pressure reducing valves, regulators and strainers shall be so constructed and installed as to permit repair or removal of parts without breaking a pipeline or removing the valve and strainer from the pipeline.

❖ Water-pressure reducing valves must be designed and installed so that in the event of a valve malfunction, repairs can be made without requiring the removal of the valve from the water supply piping. Additionally, the design and installation of the strainer for the valve must allow for periodic inspection and removal for maintenance of the strainer without removal of the valve. The inability to service a valve and strainer in place could result in a prolonged shutdown of water service to the building. This is consistent with the requirements of ASSE 1003.

604.9 Water hammer. The flow velocity of the water distribution system shall be controlled to reduce the possibility of water hammer. A water-hammer arrestor shall be installed where quick-closing valves are utilized. Water-hammer arrestors shall be installed in accordance with the manufacturer's specifications. Water-hammer arrestors shall conform to ASSE 1010.

❖ Water hammer is one of the harmful effects created by hydraulic shock and is one symptom of a dangerous condition. Hydraulic shock occurs when fluid flowing through a pipe is subjected to sudden and rapid changes in velocity.

Water hammer is a phenomenon that occurs from the dissipation of the kinetic energy of flowing water. The dissipation of energy occurs in the form of a pressure surge that produces a hydraulic shock wave traveling in excess of 3,000 miles per hour (1341 m/s). The pressure surge is sometimes accompanied by a banging noise from the piping system that gives "water hammer" its name.

Any time flowing water is abruptly stopped, there is water hammer. For example, when a valve is closed, there is water hammer. To prevent any adverse effect from water hammer, the code requires the intensity of the water hammer to be controlled.

Water hammer is more than simply an annoyance. Although noise is generally associated with the occurrence of water hammer, it can occur without audible sound. Quick valve closure always creates some form of shock, with or without noise. Therefore, the absence of noise does not indicate that water hammer or shock is nonexistent in a water distribution system. The shock wave is accompanied by a pressure surge that expands the wall of the pipe. The pressure surge can result in damage to the piping system, fixture or water heater. The shock wave is generally smaller in magnitude in plastic piping systems because the material absorbs some of the energy.

The intensity of water hammer is affected by the velocity of the flowing water, the rate at which the water flow is stopped and the diameter and material of the pipe. The general expression for kinetic energy (KE) (flowing water) is:

$$KE = \frac{1}{2} mv^2$$

where:

m = Mass.

v = Velocity of flow.

As evidenced, the critical factor is the velocity of flow. The intent of the code is to protect the entire water distribution system and its components from the possible destructive forces that result from the rapid deceleration of water flow. The code requires that the system be designed for flow velocities that minimize the occurrence and magnitude of water hammer and further requires installation of water-hammer arrestor devices where quick-closing valves are used (see commentary, Section 604.1).

A quick-closing valve is any solenoid or spring loaded valve, self-closing faucet or fixture fitting capable of instantaneously reducing water from full flow to no flow (see the definition of "Quick-closing valve" in Section 202). Mechanical water-hammer arrestors (shock arrestors) help alleviate water-hammer intensity by absorbing the energy [see Commentary Figure 604.9(1)]. When installed in areas where velocity cannot be adequately controlled, mechanical arrestors should be located where they produce the maximum effect of controlling the intensity. The manufacturer typically supplies guidelines for locating the arrestor.

Because many water-hammer arrestors are now permanently charged and factory sealed, they require no maintenance and should last the life of the system. For this reason, the code official could consider allowing such devices to be installed in a concealed location without access.

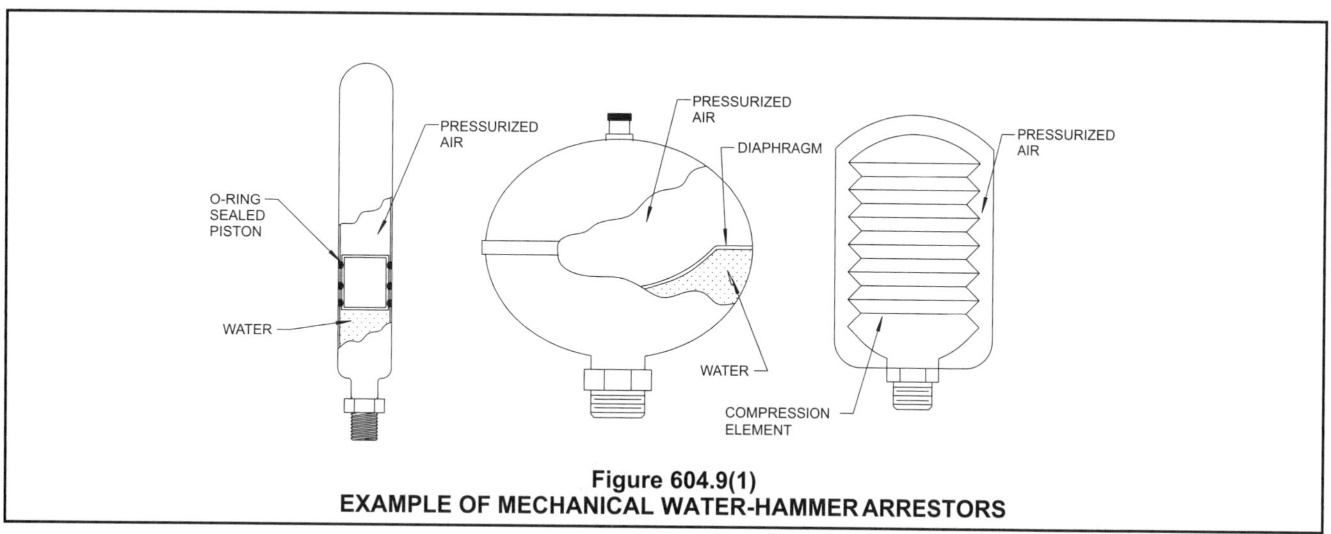

Figure 604.9(1)
EXAMPLE OF MECHANICAL WATER-HAMMER ARRESTORS

Note the distinction between water-hammer arrestors and simple air chambers. An air chamber, as shown in Commentary Figure 604.9(2), typically constructed of a capped section of vertical pipe, is ineffective in controlling water hammer. Not too far in the past, an air chamber located at the supply to every fixture was assumed to help control water hammer; however, air chambers become waterlogged in a short time. Air chambers allow a direct interface between the trapped air and the water, which allows the air to be absorbed or displaced.

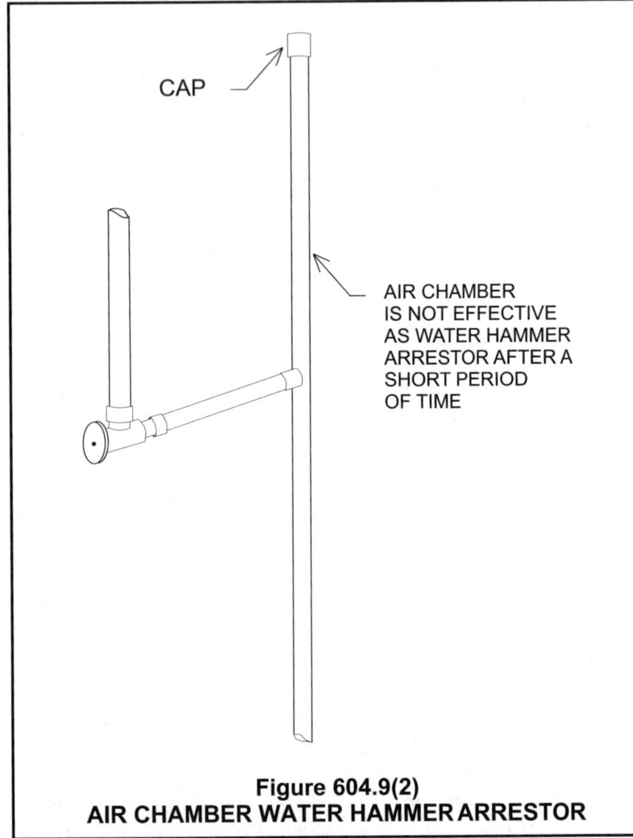

Figure 604.9(2)
AIR CHAMBER WATER HAMMER ARRESTOR

604.10 Gridded and parallel water distribution system manifolds. Hot water and cold water manifolds installed with gridded or parallel connected individual distribution lines to each fixture or fixture fitting shall be designed in accordance with Sections 604.10.1 through 604.10.3.

❖ Traditionally, water distribution systems in occupancies have been designed as branch systems. Gridded and parallel distribution systems differ from branch systems and use individual supply pipes that extend to each fixture or outlet from a central supply point. The central supply point is a multiple-outlet manifold to which the distribution lines connect. Manifold systems could be field-fabricated assemblies using conventional fittings or specialized fitting systems, or they could be constructed of factory-built units [see Commentary Figures 604.10(1) and 604.10(2)].

Such systems offer the advantages of simple design, small-size distribution lines and water conservation. In traditional branch distribution systems, the water contained in larger-diameter piping is often wasted by letting the water flow until it reaches the desired temperature. Where semirigid (bendable) tubing is used, gridded or parallel distribution systems can be installed with few, if any, fittings or connections in concealed spaces, thereby improving the overall dependability of the system.

Table 604.10.1 dictates the minimum size of manifolds to be installed in parallel distribution systems. Table 604.3 dictates the minimum flow rates and flow pressures that must be provided at fixture outlets. Note that the maximum allowable flow rates for showers, lavatories and sinks are specified in Table 604.4.

604.10.1 Manifold sizing. Hot water and cold water manifolds shall be sized in accordance with Table 604.10.1. The total gallons per minute is the demand of all outlets supplied.

❖ Manifolds serving gridded or parallel distribution systems must be sized in accordance with Table 604.10.1, based on both the total flow rate demand supplied by the manifold and the chosen design flow velocity of the system. Table 604.10.1 should be used by the designer to size manifold systems while considering the minimum and maximum flow rates required by Sections 604.3 and 604.4. The velocity rates given in Table 604.10.1 are not required minimum and maximum velocities for the design of the manifold system. As with all piping, the design flow velocity should not exceed that recommended by the pipe manufacturer. Generally, lower design flow velocities will prolong the life of the system and minimize noise (see Commentary Table 604.1).

TABLE 604.10.1
MANIFOLD SIZING

NOMINAL SIZE INTERNAL DIAMETER (inches)	MAXIMUM DEMAND (gpm)	
	Velocity at 4 feet per second	Velocity at 8 feet per second
$^1/_2$	2	5
$^3/_4$	6	11
1	10	20
$1^1/_4$	15	31
$1^1/_2$	22	44

For SI: 1 inch = 25.4 mm, 1 gallon per minute = 3.785 L/m,
1 foot per second = 0.305 m/s.

❖ The table provides sizing criteria for manifolds serving gridded or parallel distribution systems. Such systems are generally used in small buildings; therefore, the range of sizes is limited to those most likely to be encountered in such structures. The velocity columns represent the most common design velocities for the types of materials typically used in these systems. Again, these do not represent required minimum and maximum velocities that must be used in the design of the manifold system.

Figure 604.10(1)
FIELD-BUILT MANIFOLD FOR PARALLEL WATER DISTRIBUTION

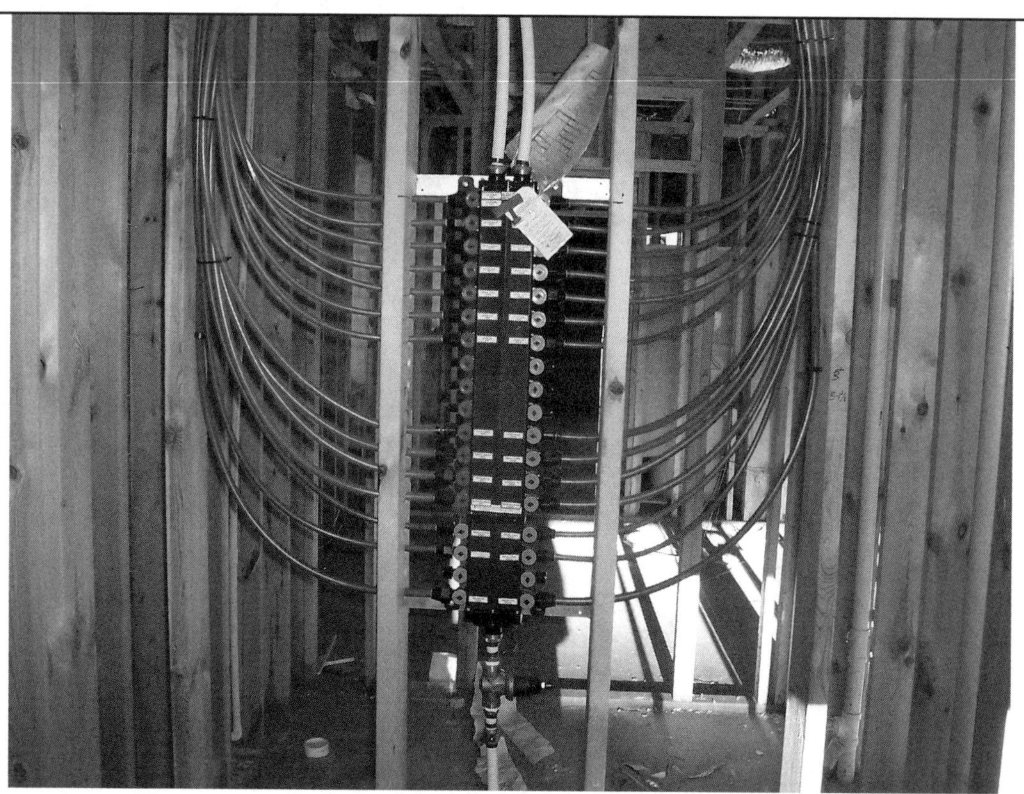

Figure 604.10(2)
FACTORY-BUILT MANIFOLD FOR PARALLEL WATER DISTRIBUTION

604.10.2 Valves. Individual fixture shutoff valves installed at the manifold shall be identified as to the fixture being supplied.

❖ Where parallel water distribution system manifolds are used, this section does not override the requirements of Item 1 in Section 606.2, specifying requirements for individual fixture supply shutoff valves. In most instances, the manifold valves are considered as branch control valves that could be used in isolating sections of the manifold system from the main water distribution supply. In the instance where a designer supplies one individual fixture directly from the manifold, such as a single-hose bibb opening, the manifold valve could be used to meet the requirements of Section 606.2 only if that valve is identified in accordance with Sections 604.10.2 and 606.4 and approved by the code official. The control function of valves located at the manifold would not be obvious; therefore, the valves must be labeled.

604.10.3 Access. *Access* shall be provided to manifolds with integral factory- or field-installed valves.

❖ Manifolds with valves, including those with field-installed valves, must be able to be accessed once the building is completed. Previous editions of the code required all manifolds to have access, even those manifolds that did not have valves. This was determined to be overly restrictive and unnecessary.

604.11 Individual pressure balancing in-line valves for individual fixture fittings. Where individual pressure balancing in-line valves for individual fixture fittings are installed, such valves shall comply with ASSE 1066. Such valves shall be installed in an accessible location and shall not be utilized alone as a substitute for the balanced pressure, thermostatic or combination shower valves required in Section 424.3.

❖ These valves must comply with ASSE 1066, which applies to automatic pressure balancing in-line valves used to equalize incoming hot- and cold-water line pressures for the purpose of minimizing mixed water-temperature variations resulting from pressure fluctuations when used in conjunction with a noncompensating mixing valve or two-handle valve set. They are not designed to limit the maximum outlet temperature at the point of use. These devices are intended for use in individual plumbing fixture fittings, such as bath faucets, sinks and lavatory faucets.

SECTION 605
MATERIALS, JOINTS AND CONNECTIONS

605.1 Soil and ground water. The installation of a water service or water distribution pipe shall be prohibited in soil and ground water contaminated with solvents, fuels, organic compounds or other detrimental materials causing permeation, corrosion, degradation or structural failure of the piping material. Where detrimental conditions are suspected, a chemical analysis of the soil and ground water conditions shall be required to ascertain the acceptability of the water service or water distribution piping material for the specific installation. Where detrimental conditions exist, *approved* alternative materials or routing shall be required.

❖ When pipe is buried, the surrounding soil conditions might cause the pipe to degrade or corrode at an accelerated rate. The soil can be evaluated, or knowledge of the historical effects from the soil would be used to determine whether additional requirements are necessary to protect the piping material.

Most soils are corrosive to galvanized steel pipe. The pipe is either coated or wrapped with coal tar, or an elastomeric or epoxy coating is applied to the exterior of the pipe.

Some soils affect copper tubing, and soil with cinders contains acid produced by combining water with sulfur compounds which will attack unprotected copper pipe; therefore, copper tubing in these instances must be coated with a protective layer. Another example of degradation is thermoplastic pipe in soil having heavy concentrations of hydrocarbons.

605.2 Lead content of water supply pipe and fittings. Pipe and pipe fittings, including valves and faucets, utilized in the water supply system shall have a maximum of 8-percent lead content.

❖ When the EPA enacted the lead solder ban in June 1986, it also limited the lead content of pipe and fittings to a maximum of 8 percent. This limitation would apply to the various qualities of brass used in producing faucets, fittings and valves. The amount of lead found in brass pipe and brass nipples is minimal, less than 1 percent.

The addition of lead to brass makes the material softer and easier to cut and shape. Commercially available faucets, fittings and valves have less than 8-percent lead. Most of the brass used has a lead content between 1 and 6 percent.

Pipe, fittings and pipe-related components must comply with NSF 61, and mechanical plumbing devices installed within the water distribution system and intended to dispense water for human consumption must comply with Section 9 of NSF 61. As a result, water supply pipe, faucets and fittings are limited to lead concentration far below the 8 percent required by this section (see Sections 605.3, 605.4, 605.5 and 424.1).

605.3 Water service pipe. Water service pipe shall conform to NSF 61 and shall conform to one of the standards listed in Table 605.3. All water service pipe or tubing, installed underground and outside of the structure, shall have a minimum working pressure rating of 160 psi (1100 kPa) at 73.4°F (23°C). Where the water pressure exceeds 160 psi (1100 kPa), piping material shall have a minimum rated working pressure equal to the highest available pressure. Water service piping materials not third-party certified for water distribution shall terminate at or before the full open valve located at the entrance to the structure. All ductile iron water service piping shall be cement mortar lined in accordance with AWWA C104.

❖ Piping material that comes in contact with the potable water source is required to conform to the requirements of NSF 61. The intent is to control the potential adverse health effects produced by indirect additives, products and materials that come in contact with potable water. Plastic water pipe tested and labeled in accordance with NSF 14 as potable water pipe (see commentary, Section 303.3) also conforms to NSF 61 because NSF 14 makes reference to and requires compliance with NSF 61.

Not all water service piping materials are third-party certified for water distribution (in other words, comply with Table 605.4). For example, PVC pipe is not suitable for water distribution pipe in a structure because it cannot tolerate exposure to hot water temperatures. Therefore, these types of piping must terminate at or before the main water shut off valve located where the water service pipe enters the structure [see Commentary Figure 605.3(1)].

Certain plastic piping materials, including chlorinated polyvinyl (CPVC) and cross-linked PEX, are permitted for both water service and water distribution. For all such materials, a transition or termination would not be required.

TABLE 605.3. See page 6-22.

❖ This table lists common types of pipe that are acceptable for use as water service pipe. Other types of pipe not listed in this table might also be suitable for water service pipe application but would require alternate material approval in accordance with Section 105.2.

ABS plastic pipe. ABS pipe is rarely manufactured for water service. ASTM D 1527 is the referenced standard for Schedule 40 and 80 ABS pipe. The schedule number indicates the wall thickness: the higher the schedule number, the thicker the pipe wall. ASTM D 2282 is the referenced standard for SDR-PR pipe. "SDR" is an abbreviation for standard pipe dimension ratio, which is the ratio of pipe diameter to wall thickness. The "PR" designation indicates that the pipe is pressure rated. The pipe is black, must be rated for distribution of potable water and is marked at intervals of not more than 5 feet (1524 mm) with the following information:

- The manufacturer's name,
- The designation ASTM D 1527 or ASTM D 2282 SDR,
- Certification for conformance to NSF 61,
- Nominal pipe size,
- ABS 40 or 80 psi rating "ABS 1210,"
- Coextruded ABS pipe "ABS Coex."

ASTM D 2468 is the standard for ABS plastic pipe fittings, Schedule 40. Fittings meeting the requirements of ASTM D 2468 are permitted with Schedule 40 pipe, and any fitting having a male thread must have an internal diameter not larger than Schedule 80 pipe of the same size.

Asbestos-cement pipe. This material was predominantly used for street water main piping under streets. It was also used for large-size water service piping as it is manufactured only in 4-inch through 36-inch (102 mm through 914 mm) diameters.

Asbestos fibers are used to bond portland cement when manufactured into pipe. Because the asbestos is bonded in the pipe, it has not been considered a health hazard. However, very little asbestos-cement pipe is being produced because there is a hazard of breathing asbestos during production and cutting. ASTM C 296 is the referenced standard for asbestos-cement pipe.

Brass pipe. This piping material is a red brass composed of approximately 85-percent copper and 15-percent zinc. The pipe is available in both standard

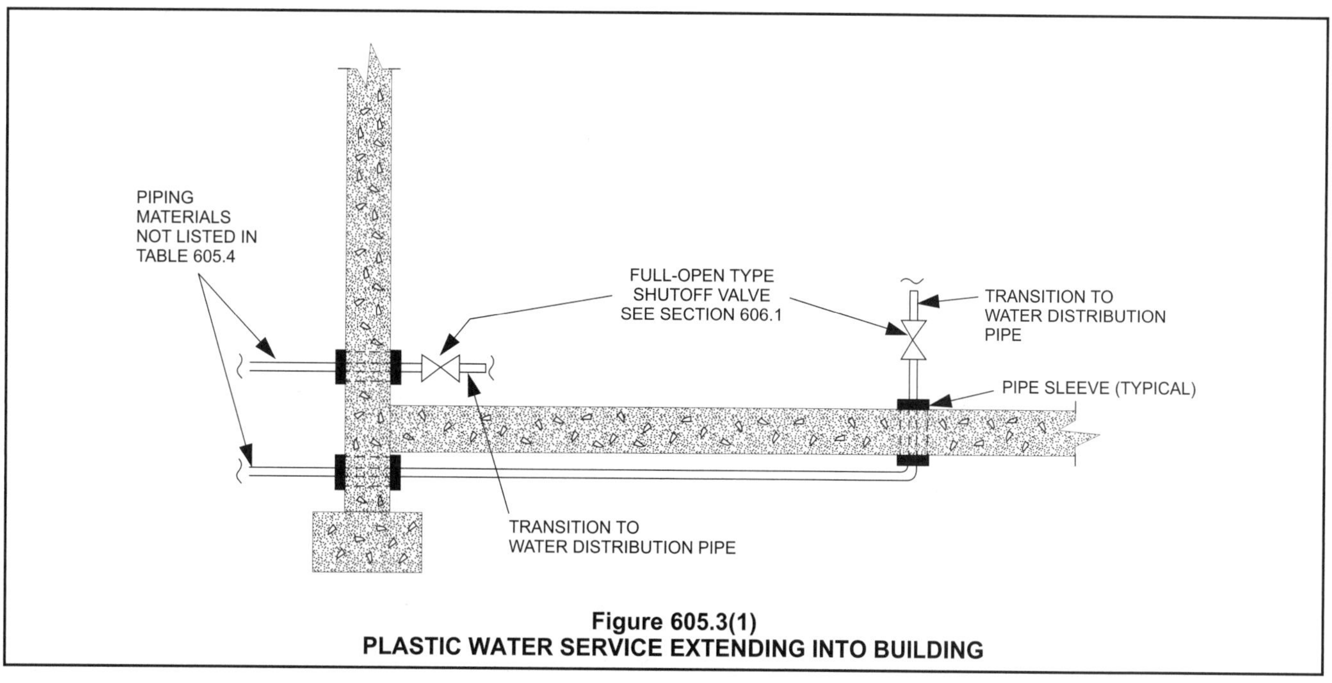

Figure 605.3(1)
PLASTIC WATER SERVICE EXTENDING INTO BUILDING

and extra-strong weight. Both weights have the same outside diameter as standard-weight steel pipe, but extra strong has a thicker pipe wall.

Standard-weight pipe is often called by the expression "Schedule 40," and extra-strong pipe is named "Schedule 80." Brass pipe internal dimensions are very similar to steel pipe internal dimensions. ASTM B 43 is the referenced standard for brass pipe, available in diameters from $^1/_8$ inch (3.2 mm) to 12 inches (305 mm) and manufactured in 12-foot (3658 mm) lengths. The pipe must be rated for distribution of potable water and is marked with the following information:

- The manufacturer's name,
- The designation ASTM B 43,
- Certification for conformance to NSF 61,
- Nominal pipe size,
- Schedule 40 standard weight or Schedule 80 extra-strong weight.

Copper or copper-alloy pipe. There are two types of copper pipe: threaded and threadless. "Threaded" means that the copper pipe has sufficient wall thickness to be threaded and used with screwed fittings. "Threadless" means that the copper pipe wall thickness is too thick to accommodate threads, and therefore the pipe must be brazed to fittings. Both pipes are seamless and have a chemical composition of 99.9-percent copper.

ASTM B 302 is the standard for threadless copper pipe available in diameters from $^1/_4$ inch (6.4 mm) to 12 inches (305 mm) and manufactured in 20-foot (6096 mm) lengths.

Threadless copper pipe has the same outside diameter as threaded copper pipe, steel pipe and brass pipe. The wall thickness of the pipe is less than that of threaded copper pipe; therefore, the inside diameter is greater. Threadless copper pipe is intended to be joined by brazing and is continuously marked in a gray color. The designation "TP" for threadless pipe is included in the marking.

ASTM B 42 is the standard for threaded copper pipe available in $^1/_8$ inch to 10 inches (3 mm to 254 mm) diameters. "Regular" and "extra strong" weights are available, both in 12-foot (3658 mm) lengths.

Section 303.4 requires identification of copper pipe used for potable water with the mark of an approved third-party agency and notation that it conforms to the requirements of NSF 61 (see commentary, Section 303.4).The pipe must be rated for distribution of potable water and be marked with the following information:

- The manufacturer's name,
- The designation ASTM B 302 TP or ASTM B 42,
- Certification for conformance to NSF 61,
- Nominal pipe size,
- Schedule 40 standard weight or Schedule 80 extra-strong weight.

TABLE 605.3
WATER SERVICE PIPE

MATERIAL	STANDARD
Acrylonitrile butadiene styrene (ABS) plastic pipe	ASTM D 1527; ASTM D 2282
Asbestos-cement pipe	ASTM C 296
Brass pipe	ASTM B 43
Chlorinated polyvinyl chloride (CPVC) plastic pipe	ASTM D 2846; ASTM F 441; ASTM F 442; CSA B137.6
Copper or copper-alloy pipe	ASTM B 42; ASTM B 302
Copper or copper-alloy tubing (Type K, WK, L, WL, M or WM)	ASTM B 75; ASTM B 88; ASTM B 251; ASTM B 447
Cross-linked polyethylene (PEX) plastic tubing	ASTM F 876; ASTM F 877; CSA B137.5
Cross-linked polyethylene/aluminum/cross-linked polyethylene (PEX-AL-PEX) pipe	ASTM F 1281; ASTM F 2262; CAN/CSA B137.10M
Cross-linked polyethylene/aluminum/high-density polyethylene (PEX-AL-HDPE)	ASTM F 1986
Ductile iron water pipe	AWWA C115; AWWA C151
Galvanized steel pipe	ASTM A 53
Polyethylene (PE) plastic pipe	ASTM D 2239; ASTM D 3035; CSA B137.1
Polyethylene (PE) plastic tubing	ASTM D 2737; CSA B137.1
Polyethylene/aluminum/polethylene (PE-AL-PE) pipe	ASTM F 1282; CAN/CSA B137.9
Polypropylene (PP) plastic pipe or tubing	ASTM F 2389; CSA B137.11
Polyvinyl chloride (PVC) plastic pipe	ASTM D 1785; ASTM D 2241; ASTM D 2672; CSA B137.3
Stainless steel pipe (Type 304/304L)	ASTM A 312; ASTM A 778
Stainless steel pipe (Type 316/316L)	ASTM A 312; ASTM A 778

Copper or copper-alloy tubing (Type K, WK, L, WL, M or WM). The tubing is manufactured in two different tempers, which are identified as drawn (called hard copper with the designation "H") and annealed (called soft copper with the designation "O"). Hard (drawn) copper tubing is available only in straight lengths, whereas soft (annealed) copper tubing is available in both straight lengths and coils.

Annealed copper may be formed by bending.

Bending temper drawn copper is identified on the tube with the designation "BT." Only bending temper and annealed copper tubing can be joined by flared connections.

Seamless copper water tubing is available in Type K, L or M. The tubing type indicates the wall thickness. Type K has the thickest wall, followed by Types L and M, respectively. ASTM B 75 is the standard for annealed seamless copper tubing. ASTM B 88 and ASTM B 251 are the standards for both drawn and annealed seamless copper water tubing. The outside diameter of the tubing is the same for all three types; only the inside diameter varies with wall thickness. Welded copper tubing has the designation "WK," "WL" or "WM." The "W" indicates that the tubing is welded seam. ASTM B 447 is the standard for welded copper tubing. The tube dimensions correspond to Type K, L or M seamless tubing.

The types of copper tubing are identified with a continuous colored marking for ease of identification. These colored markings are as follows: Type K (WK) is green, Type L (WL) is blue and Type M (WM) is red. All are available in diameters from $^1/_4$ inch (6.4 mm) to 12 inches (305 mm). Straight lengths are manufactured in 12-foot (3658 mm), 18-foot (5486 mm) and 20-foot (6096 mm) lengths and annealed copper coils are manufactured in 40-foot (12 192 mm), 60-foot (18 288 mm) and 100-foot (30 480 mm) lengths.

Section 303.4 requires identification of copper tube used for potable water with the mark of an approved third-party agency (see Commentary Section 303.4). The tube must be rated for distribution of potable water and is marked at intervals of not more than 3 feet (914 mm) with the following information:

- The manufacturer's name,
- The standard designation (ASTM Number),
- NSF 61 certification mark,
- Nominal size (inches, CTS),
- Color marking green, blue or red.

CPVC plastic pipe. CPVC raw material used to produce pipe is a polyvinyl chloride that has been chlorinated to improve the material characteristics. The resulting pipe is more resistant to temperature extremes than PVC pipe.

CPVC plastic pipe may be either a white or a milky white (cream) color. The pipe listed is continuously marked with the manufacturer's name, the ASTM standard and CPVC 4120 or CPVC 41, followed by two additional numbers. Section 303.3 requires identification of all plastic pipe and fittings with the mark of an approved agency establishing conformance to NSF 14 (see Commentary Section 303.4).

The first two digits in designation CPVC 4120 identifies the grade of the material used to produce the pipe. CPVC 4120 pipe is made with Grade 23447 plastic material. The first digit indicates that the base resin is of CPVC. The remaining four digits indicate the impact strength, tensile strength, modulus of elasticity and deflection temperature.

The last two digits in CPVC 4120 indicate the hydrostatic design stress in hundreds of psi. CPVC 4120 has a hydrostatic design stress of 2,000 psi (13 790 kPa).The hydrostatic design stress is not an indication of the pressure rating of the pipe; however, the pressure rating of the pipe at 73°F (23°C) may be computed using the hydrostatic design stress by the following expression:

$$P = \frac{2S}{(O.D./t) - 1} = \frac{2S}{R - 1}$$

where:

P	=	Pressure rating (psi).
S	=	Hydrostatic design stress (psi).
$O.D.$	=	Average outside diameter (inches).
t	=	Minimum wall thickness (inches).
R	=	Standard dimension ratio (SDR based on outside diameter).

ASTM D 2846 is the standard for SDR 11 CPVC plastic pipe and CPVC socket fittings. This standard includes CPVC products with outside diameters equal to those of copper tube sizes in the range of $^1/_4$ to 2 inches (6 mm to 50 mm). SDR means standard dimension ratio, which is a ratio of the average outside diameter to the minimum wall thickness.

ASTM F 441 is the standard for Schedule 40 and 80 plastic pipe. The dimensions of the pipe are the same as Schedule 40 and 80 steel pipe.

The standard for SDR 13.5, 17, 21, 26 and 32.5 CPVC pipe is ASTM F 442. The pipe is designated "PR," indicating that it is pressure rated. Section 303.4 requires identification of CPVC pipe used for potable water with the mark of an approved third-party agency. It also must conform to the requirements of NSF 61 (see Commentary Section 303.4). The pipe must be rated for distribution of potable water and is marked at intervals of not more than 3 feet (914 mm) with the following information:

- The manufacturer's name,
- The designation ASTM D 2846, ASTM D 441/442 or CSA B 137.6,
- Certification for conformance to NSF 61,
- Nominal pipe size,
- Type of plastic material (CPVC 4120),
- Schedule size and pressure rating (40 or 80 at 200 psi or SDR 11 at 180 psi).

Ductile-iron water pipe. Ductile iron pipe is a cast iron pipe that is made to have higher ductility, resulting in the name "ductile iron." It is not as brittle as cast-iron pipe. Section 605.3 requires that the inside surface of ductile-iron water pipe be cement-mortar lined in accordance with AWWA C104 (see Commentary Table 605.3). This is necessary to provide a protective barrier between the potable water supply and the ductile-iron pipe to prevent impurities and contaminants from leaching into the water supply.

Table 605.3
DUCTILE IRON PIPE LINING THICKNESS

MINIMUM LINING THICKNESS (inches)	PIPE DIAMETER (inches)
$^{1}/_{4}$	3-12
$^{3}/_{32}$	14-24
$^{1}/_{8}$	30-54

For SI: 1 inch = 25.4 mm.

AWWA C151 and AWWA C115 are the referenced standards for ductile-iron water pipe. The pipe is manufactured with a bell hub end in lengths of 18 and 20 feet (5486 and 6096 mm). The pipe is available in diameters of 3 through 64 inches (76 through 1626 mm), and all required markings must be on or near the bell of the pipe with the following information:

- The manufacturer's name or mark,
- The country where cast,
- The weight, class or nominal thickness,
- The year in which the pipe is produced,
- The letters "DI" or the word "Ductile."

Galvanized steel pipe. A wide variety of pipe is classified as galvanized steel pipe. The grade of steel and the method of manufacturing can differ greatly.

The common pipe used in plumbing application is either continuous welded or electric-resistance welded (ERW). Continuous-welded pipe is heated and formed, whereas ERW pipe is cold rolled and welded.

Galvanized steel pipe is available in different wall thicknesses, the most common being standard weight, extra strong and double extra strong. These thicknesses are also referred to as Schedules 40, 80 and 160, respectively. As the schedule number increases, the wall thickness of the pipe increases. The outside diameter, however, remains the same for each pipe size. Most galvanized steel pipe is electrodeposited zinc coated to produce a galvanized protective coating, both inside and outside. New galvanized pipe is a shiny silver, and, as the pipe ages, the coating oxidizes to protect the zinc. As a result, the color turns to a dull gray.

Galvanized steel pipe is sometimes called "galvanized pipe" or mistakenly called "wrought-iron pipe." Wrought-iron pipe has not been manufactured in the United States in more than 30 years. All pipe referred to as "wrought-iron pipe" is steel pipe.

The steel pipe industry claims to be responsible for nominal pipe dimensions, meaning the inside pipe diameter of the pipe (standard weight) is approximately the size by which it is identified. The dimensions date back to 1862 when Robert Briggs of Pascal Iron Works wrote the first pipe standards in the United States. It is believed that the dimensions were based on the pipe Briggs was manufacturing. That is why a $^{3}/_{4}$-inch (19.1 mm) galvanized steel pipe has an outside diameter of 1.05 inches (27 mm) and an average inside diameter of 0.824 inches (21 mm) for standard-weight pipe. The pipe is available in diameters of $^{1}/_{8}$ inch through 26 inches (3.2 mm through 660 mm) and various lengths of 12 feet (3658 mm) to 22 feet (6706 mm). Galvanized steel pipe is marked with the following information:

- The manufacturer's name or mark,
- The type of pipe (i.e., A, B, XS, XXS),
- The specification number,
- The length of pipe.

PB plastic pipe and tubing. Polybutylene pipe and tubing was the subject of much media attention decades ago. This was a result primarily of the settlement of a multimillion-dollar lawsuit and the announcement by the primary producer that it stopped manufacturing polybutylene resin for water supply piping. While some old stock of polybutylene may still be in warehouses, polybutylene pipe or tubing for potable water service or water distribution has not been produced in the United States for over 30 years. However, many homes built as late as the 1980's have had this product installed in water distribution systems. PB tubing for water service has been found installed as late as the 1990's as contractors and supply houses continued to deplete this obsolete stock. Because polybutylene pipe is no longer listed in the tables as an approved material for water service or water distribution piping, this material must not be installed in new construction, remodeling or repair work.

For reference and identification purposes only, PB pipe and tubing were marked "ASTM D 2662, D 2666 or D 3309." These materials were colored blue, black or gray.

PE plastic pipe and tubing. Polyethylene is an inert polyolefin material. Because the material is highly resistant to chemical attack, solvent-cement joining of polyethylene pipe and tubing is not possible.

Polyethylene pipe is blue or black when supplied for water service. The pipe is continuously marked with the manufacturer's name, the ASTM or CSA standard and the grade of polyethylene.

Various grades of polyethylene resin can be used to produce pipe or tubing. In accordance with the ASTM standard, the material is identified by the term "PE," followed by four digits. The first digit indicates the type of material based on density. The second digit is the category identifying the extrusion flow rate. The last two digits show the hydrostatic design stress in hundreds of psi.

The hydrostatic design stress is not an indication of the pressure rating of the pipe. The pressure rating of the pipe at 73°F (23°C), however, can be computed by the same formula as provided previously for CPVC pipe and tubing.

ASTM D 2239 is the referenced standard for SDIR-PR polyethylene pipe. The SDIRs of the pipe are 5.3, 7, 9, 11.5 and 15. The grades of polyethylene permitted include PE 1404, PE 2305, PE 2306, PE 3306, PE 3406 and PE 3408.

ASTM D 2737 is for polyethylene tubing with an SDR of 7.3,9 and 11. The SDR for tubing is based on the average outside diameter.

Cross-linked PEX plastic tubing. Cross-linked PEX tubing has been used extensively in Europe for decades for hot and cold potable water distribution systems. The material is available in nominal pipe sizes (NPS) of $^{1}/_{4}$-inch (6.4 mm) through 2 inches (51 mm) and a standard dimension ratio (SDR 9). Figure 605.3(3) shows a sample marking of PEX tubing.

Unlike most other pipe materials, PEX can be produced by several different manufacturing processes, such as the Engel, Silane, Monosil, Point a'Mousson, Daoplas and electron beam methods. Although each manufacturing process can yield tubing complying with the referenced standards, each produces a tube that possesses unique characteristics.

The cross-linked molecular structuring gives the pipe additional resistance to rupture over a wider range of temperatures and pressures than that of other polyolefin plastics [PE and polypropylene (PP)]. Because of PEX pipe's unique molecular structure and resistance to heat, heat fusion is not permitted as a joining method. Because PEX is a member of the polyolefin plastic family, it is resistant to solvents. Therefore, the pipe cannot be joined by solvent cementing.

PEX pipe is flexible, allowing it to be bent. The tubing can be bent using either hot or cold bending methods.

For hot bending, a hot-air gun with a diffuser nozzle is used. The pipe cannot be exposed to an open flame. The hot air meeting the pipe surface must not exceed 338°F (170°C), and the heat-up time must not exceed 5 minutes. The pipe is bent using conventional methods and is allowed to cool to room temperature before removal from the bending tool. The minimum hot-bending radius is two and one-half times the outside diameter.

PEX can be bent at room temperature (cold bending) to a minimum radius of six times the outside diameter.

Mechanical connectors and fittings for PEX pipe are proprietary and must be used only with the pipe for which they have been designed and tested. Fitting systems currently available include compression fittings with inserts and ferrules or O-rings, insert fittings using a metallic lock (crimp) ring and insert fittings with compression collars using an expander tool. It is important to consult the manufacturer's installation instructions to identify the fittings authorized for use with PEX piping (see commentary, Section 605.17.2).

Cross-linked PEX-AL-PEX pipe. PEX-AL-PEX is a composite pipe made of an aluminum tube laminated to interior and exterior layers of cross-linked polyethylene. The layers are bonded together with an adhesive. The cross-linked molecular structuring gives the pipe additional resistance to rupture than that of polyethylene (see the commentary for cross-linked polyethylene tubing in this section). Therefore, the pipe is suitable for hot and cold water distribution and is pressure rated for 125 psi at 180°F (862 kPa at 82°C).

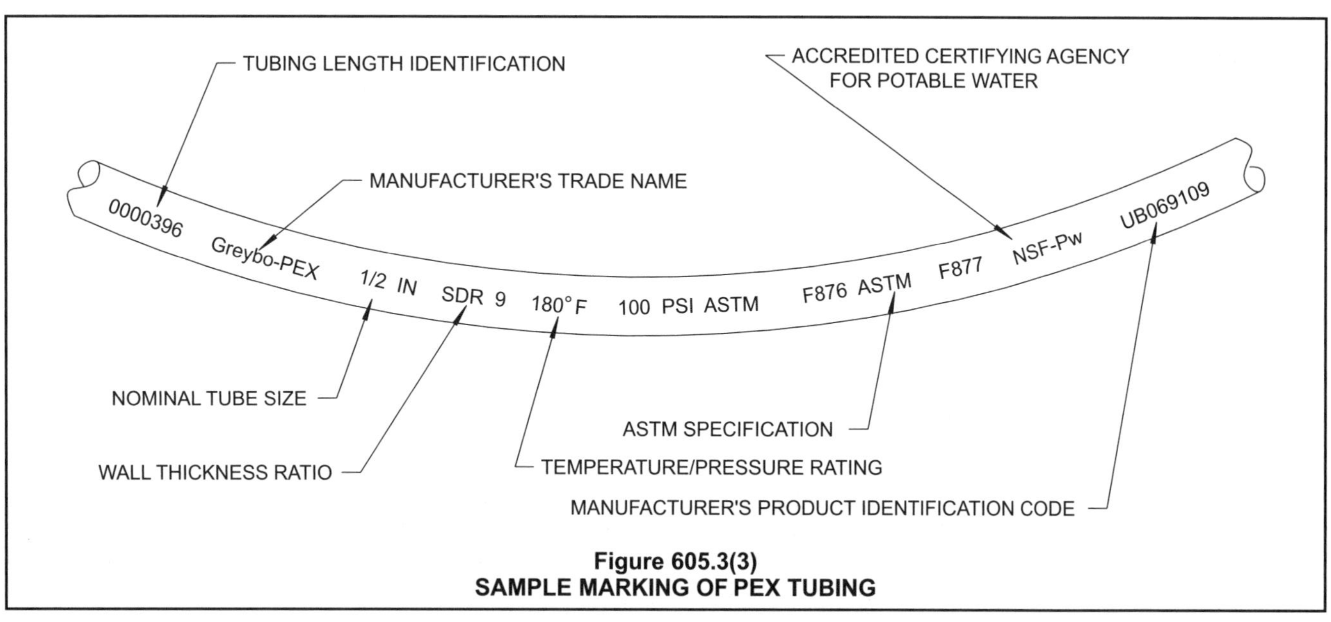

Figure 605.3(3)
SAMPLE MARKING OF PEX TUBING

Although it is partially plastic, the PEX-AL-PEX pipe resembles metal tubing in that it can be bent by hand or with a suitable bending device while maintaining its shape without fittings or supports. It is often called "form-stable" PEX piping. The minimum bending radius specified by manufacturers is five times the outside diameter.

Mechanical joints are the only method currently available to join PEX-AL-PEX pipe. A number of proprietary mechanical-compression-type connectors have been developed for use with this composite pipe to permit transition to other pipes and fittings. Such fittings must be installed in accordance with the manufacturers' instructions (see Commentary Section 605.20.3).

Polyethylene/aluminum/polyethylene (PE-AL-PE) pipe. PE-AL-PE is identical to PEX-AL-PEX composite pipe except for the physical properties of the polyethylene. Polyethylene displays the same resistance to temperature and pressure as cross-linked polyethylene does; therefore, it is suitable for hot and cold water distribution and is pressure rated for 125 psi at 180°F (862 kPa at 82°C) (see the commentary to PEX-AL-PEX in this section).

PVC plastic pipe. PVC water service pipe is a different material than PVC drainage pipe, although both materials are white. The pipe must be continuously marked with the manufacturer's name, the ASTM or CSA standard and the grade of the PVC material.

A number of grades of PVC material are used to produce pipe. In accordance with the ASTM standards, the compounds are identified as PVC 12454-B, 12454-C and 14333-D. The first digit indicates the base resin, with the following four digits identifying the impact strength, tensile strength, modulus of elasticity and deflection temperature. The letter suffix indicates the material's chemical resistance.

The compounds were previously identified by type and grade: 12454-B was Type 1, Grade 1; 12454-C was Type 1, Grade 2 and 14333-D was Type 2, Grade 1. The marking on the pipe lists the material grade by the term "PVC," followed by four digits. The first two digits use the previous type and grade numbers to identify the compound. The last two digits indicate the hydrostatic design stress in hundreds of psi.

The hydrostatic design stress is not an indication of the pressure rating of the pipe. The pressure rating of the pipe at 73°F (23°C), however, can be computed by the same formula as provided previously for CPVC pipe and tubing.

ASTM D 1785 is the standard for Schedule 40, 80 and 160 PVC plastic pipe. The pipe dimensions are the same as for Schedule 40, 80 and 160 steel pipe.

ASTM D 2241 is the standard for SDR-PR PVC plastic pipe. "SDR" means standard dimension ratio, which is the ratio between the average outside diameter of the pipe and the minimum wall thickness. "PR" indicates that the pipe is pressure rated. This pipe is available in SDRs of 13.5, 17, 21, 26, 32.5, 41 and 64.

ASTM D 2672 is the standard for bell-end PVC plastic pipe conforming to either ASTM D 1785 or ASTM D 2241.

Stainless steel pipe. Stainless steel is a family of iron-based alloys that must contain at least 10 percent or more chromium. The presence of chromium creates an invisible surface film that resists oxidation and makes the material corrosion resistant. Stainless steel is produced in an electric arc furnace using argon-oxygen degassing melting technology. This method provides for closely controlled chemistry in the final product. Most semifinished product receives a final annealing and pickling operation. The fact that stainless steel has a great resistance to corrosion means that using stainless will result in a very long pipe life compared to carbon steel pipe.The benefits of stainless include corrosion resistance, fire and heat resistance, hygiene (easy cleaning ability), aesthetic appearance, strength-to-weight advantage, ease of fabrication, impact resistance and long-term value.

Wrought austenitic stainless steels unite a useful combination of outstanding corrosion and heat resistance with good mechanical properties over a wide temperature range. This type of pipe is typically used for chemicals, petroleum, pollution abatement, food processing, transportation equipment, water service and water distribution. There are several grades of stainless steel, such as ferritic, martensitic, austenitic, duplex and precipitation hardened. All grades other than the austenitic (300 series) stainless steels exhibit some degree of magnetism; the austenitic grades are nonmagnetic. Some austenitic welds may show some slight degree of magnetism because of the desired requirement to have a small amount of residual ferrite in the finished weld. Thus, one should not limit the selection of wrought austenitic stainless steels to those included herein, but rather should ask producers for the best grade—standard or specialty—for a particular use.

ASTM A 312 and ASTM A 778 are the standards for water service and water distribution pipe. The material is suitable for both applications. Types 304/304L and 316/316L are included in those standards. Type 304/304L is the low-carbon grade, which provides the designer higher minimum design properties, while offering the benefits of the low-carbon grades of material suitable for welding. Type 316/316L stainless steels are characterized by addition of molybdenum and are more corrosion resistant than Type 304/304L.

Standard wrought austenitic stainless steels can be supplied in a variety of standard finishes, which include dull and smooth, pickled, ground finishes up to 500 grit, electropolished, rouge polished and tumbled.

Annealed tubing can be formed to minimum bend radii of two to six times outside diameter (o.d.). Tube ends can be flanged, flared or beaded. In addition to mechanical fastening, standard wrought austenitic stainless steels may be joined by brazing, soldering or welding.

Brazing can be readily accomplished, but should be well above the heat sensitizing range. Brazing without fluxes can be done in vacuum or in atmospheres of dry hydrogen, argon or helium.

Stainless steel can also be soldered using, for example, tin-lead filler metals. Fluxes are required to break down the protective oxide film and permit wetting. Chloride-free or phosphate fluxes should be used. Inert gas shielding is necessary to prevent porosity and oxidation of the weld metal.

605.3.1 Dual check-valve-type backflow preventer. Where a dual check-valve backflow preventer is installed on the water supply system, it shall comply with ASSE 1024 or CSA B64.6.

❖ The purpose of the dual check-valve-backflow preventer is to keep polluted water from flowing back into the potable water system when pressure is temporarily higher in the polluted part of the system than in the potable water piping. This standard focuses on dual check valves, which fulfill "low-hazard" protective needs. These devices are generally suitable for protection by containment at residential supply service lines, where pollutants could enter the potable water. Additionally, dual check-valve-type devices are generally suitable for cold water service under continuous or intermittent pressure conditions.

605.4 Water distribution pipe. Water distribution pipe shall conform to NSF 61 and shall conform to one of the standards listed in Table 605.4. All hot water distribution pipe and tubing shall have a minimum pressure rating of 100 psi (690 kPa) at 180°F (82°C).

❖ Piping material that comes in contact with potable water is required to conform to NSF 61 (see Commentary Section 605.3). Pipe used for the distribution of hot water must be capable of withstanding high temperatures. The maximum reasonable temperature of a hot water distribution system is 180°F (82°C). This section requires the water distribution piping be suitable for an internal pressure of 100 psi (690 kPa) at that temperature. This requirement applies to both cold and hot water distribution piping and fittings.

CPVC, PEX and PB plastic pipe or tubing is often identified as rated for 180°F at 100 psi (82°C at 690 kPa). This type of pipe can withstand temperatures in excess of 212°F (100°C). As the temperature increases, however, the pressure rating decreases.

605.5 Fittings. Pipe fittings shall be *approved* for installation with the piping material installed and shall comply with the applicable standards listed in Table 605.5. All pipe fittings utilized in water supply systems shall also comply with NSF 61. Ductile and gray iron pipe fittings shall be cement mortar lined in accordance with AWWA C104.

❖ Pipe (and tube) fittings must be appropriate for both the type of piping material and the service application. Even though a Schedule 40 PVC pipe fitting will fit Schedule 80 PVC pipe, the Schedule 40 fitting does not have the same pressure rating as the Schedule 80 pipe. Even though an insert fitting with crimp ring might appear to fit the PEX tubing of many manufacturers, only the fittings recommended by the manufacturer are to be used (see Section 605.17.2). Schedule 40 ABS pipe and fittings are dimensionally interchangeable with PVC Schedule 40 fittings; however, the code does not allow PVC and ABS to be solvent cemented together.

In previous code editions, this section prohibited fittings that had ledges, shoulders or reductions capable of retarding flow. While an appropriate requirement for

TABLE 605.4
WATER DISTRIBUTION PIPE

MATERIAL	STANDARD
Brass pipe	ASTM B 43
Chlorinated polyvinyl chloride (CPVC) plastic pipe and tubing	ASTM D 2846; ASTM F 441; ASTM F 442; CSA B137.6
Copper or copper-alloy pipe	ASTM B 42; ASTM B 302
Copper or copper-alloy tubing (Type K, WK, L, WL, M or WM)	ASTM B 75; ASTM B 88; ASTM B 251; ASTM B 447
Cross-linked polyethylene (PEX) plastic tubing	ASTM F 876; ASTM F 877; CSA B137.5
Cross-linked polyethylene/aluminum/cross-linked polyethylene (PEX-AL-PEX) pipe	ASTM F 1281; ASTM F 2262; CAN/CSA B137.10M
Cross-linked polyethylene/aluminum/high-density polyethylene (PEX-AL-HDPE)	ASTM F 1986
Ductile iron pipe	AWWA C151/A21.51; AWWA C115/A21.15
Galvanized steel pipe	ASTM A 53
Polyethylene/aluminum/polyethylene (PE-AL-PE) composite pipe	ASTM F 1282
Polypropylene (PP) plastic pipe or tubing	ASTM F 2389; CSA B137.11
Stainless steel pipe (Type 304/304L)	ASTM A 312; ASTM A 778
Stainless steel pipe (Type 316/316L)	ASTM A 312; ASTM A 778

drain waste and vent piping flows, this requirement conflicted with the basic design of water systems in that appropriate fitting losses are included in the pressure loss calculations. Therefore, the prohibition statement was removed for the 2009 edition.

Compliance to NSF 61 for all water supply system components is paramount for protection of the public against contamination of the water supply (see Commentary Section 605.3 or 605.4).

605.5.1 Mechanically formed tee fittings. Mechanically extracted outlets shall have a height not less than three times the thickness of the branch tube wall.

❖ Mechanically formed copper tee fittings are formed in a continuous operation consisting of drilling a pilot hole and drawing out the tube surface to form a collar having a height not less than three times the thickness of the tube wall. Tees formed in accordance with Section 605.5.1 lack the quality control that is imposed in a manufacturer's plant to meet applicable standards. Tolerances on tees mechanically formed in the field do not always meet standards. Because of the short collars produced by this method, soft soldering is inadequate to ensure the joint strength. Joints must be brazed, as required by Section 605.5.1.2.

605.5.1.1 Full flow assurance. Branch tubes shall not restrict the flow in the run tube. A dimple/depth stop shall be formed in the branch tube to ensure that penetration into the collar is of the correct depth. For inspection purposes, a second dimple shall be placed $^{1}/_{4}$ inch (6.4 mm) above the first dimple. Dimples shall be aligned with the tube run.

❖ Notching and dimpling, as required by Section 605.5.1.1, are of utmost importance to ensure that the branch copper tube does not enter the run of pipe having the mechanically extracted outlets. Inspection of the dimple and its location after installation will ensure proper penetration of the branch tube because notching and dimpling are both done with the same operation.

605.5.1.2 Brazed joints. Mechanically formed tee fittings shall be brazed in accordance with Section 605.14.1.

❖ See the commentary to Section 605.14.1.

605.6 Flexible water connectors. Flexible water connectors exposed to continuous pressure shall conform to ASME

TABLE 605.5
PIPE FITTINGS

MATERIAL	STANDARD
Acrylonitrile butadiene styrene (ABS) plastic	ASTM D 2468
Cast-iron	ASME B16.4; ASME B16.12
Chlorinated polyvinyl chloride (CPVC) plastic	ASSE 1061; ASTM D 2846; ASTM F 437; ASTM F 438; ASTM F 439; CSA B137.6
Copper or copper alloy	ASSE 1061; ASME B16.15; ASME B16.18; ASME B16.22; ASME B16.23; ASME B16.26; ASME B16.29
Cross-linked polyethylene/aluminum/high-density polyethylene (PEX-AL-HDPE)	ASTM F 1986
Fittings for cross-linked polyethylene (PEX) plastic tubing	ASSE 1061; ASTM F 877; ASTM F 1807; ASTM F 1960; ASTM F 2080; ASTM F 2098; ASTM F 2159; ASTM F 2434; CSA B137.5
Gray iron and ductile iron	AWWA C110; AWWA C153
Insert fittings for polyethylene/aluminum/polyethylene (PE-AL-PE) and cross-linked polyethylene/aluminum/cross-linked polyethylene (PEX-AL-PEX)	ASTM F 1974; ASTM F1281; ASTM F1282; CAN/CSA B137.9; CAN/CSA B137.10
Malleable iron	ASME B16.3
Metal (brass) insert fittings for polyethylene/aluminum/polyethylene (PE-AL-PE) and cross-linked polyethylene/aluminum/cross-linked polyethylene (PEX-AL-PEX)	ASTM F 1974
Polyethylene (PE) plastic pipe	ASTM D 2609; ASTM D 2683; ASTM D 3261; ASTM F 1055; CSA B137.1
Polypropylene (PP) plastic pipe or tubing	ASTM F 2389; CSA B137.11
Polyvinyl chloride (PVC) plastic	ASTM D 2464; ASTM D 2466; ASTM D 2467; CSA B137.2; CSA B137.3
Stainless steel (Type 304/304L)	ASTM A 312; ASTM A 778
Stainless steel (Type 316/316L)	ASTM A 312; ASTM A 778
Steel	ASME B16.9; ASME B16.11; ASME B16.28

A112.18.6. *Access* shall be provided to all flexible water connectors.

❖ASTM A 112.18.6 is the standard for flexible water connectors used in potable water systems under continuous pressure. Plastic materials coming into contact with potable water must comply with the applicable sections of NSF 14. Flexible connectors for delivery of drinking water must comply with NSF 61. Solder and fluxes containing lead in excess of 0.2 percent must not be used. Metal alloys must not exceed 8 percent lead content. Copper alloy components used with the assembled connector must contain not less than 58 percent copper.

Flexible water connectors are designed to function at water pressure up to 125 psi (862 kPa) and supply temperature from 40°F to 180°F (4°C to 82°C).

605.7 Valves. All valves shall be of an *approved* type and compatible with the type of piping material installed in the system. Ball valves, gate valves, globe valves and plug valves intended to supply drinking water shall meet the requirements of NSF 61.

❖There are very few standards regulating valves used in plumbing systems. The code official must evaluate the various valves for their suitability in a plumbing system. The valves should operate easily, serve the intended function and have a reasonable service life.

Valves must be the same material as or a compatible material to the pipe to avoid chemical or corrosive action. Valves used in the water supply system that come into contact with the potable water source must conform to the requirements of NSF 61.

605.8 Manufactured pipe nipples. Manufactured pipe nipples shall conform to one of the standards listed in Table 605.8.

❖The referenced table specifies the acceptable materials for manufactured pipe nipples.

Pipe nipples are short pieces of threaded pipe used in piping assemblies. Long ago nipples were cut and prepared by either the supply house or the and threaded contractor. Today, the majority of pipe nipples are cut and threaded in a manufacturing facility and boxed for wholesale distribution.

TABLE 605.8
MANUFACTURED PIPE NIPPLES

MATERIAL	STANDARD
Brass-, copper-, chromium-plated	ASTM B 687
Steel	ASTM A 733

❖This table specifies acceptable materials for pipe fittings, along with the appropriate standards with which the various fittings must comply.

ASTM A733 specifies the piping material for steel pipe nipples. The pipe must conform to ASTM A 53. Steel nipples are available in pipe sizes $^1/_8$ inch to 12 inches (3.2 mm to 305 mm) and in lengths up to 24 inches (610 mm).

ASTM B 687 requires brass and chrome-plated (brass) nipples to conform to ASTM B 43 and copper nipples to conform to ASTM B 42. Both types of nipples are available in pipe sizes $^1/_4$ inch to 8 inches (6.4 mm to 203 mm) and in lengths up to 12 inches (305 mm).

See the commentary to Section 605.8.

605.9 Prohibited joints and connections. The following types of joints and connections shall be prohibited:

1. Cement or concrete joints.
2. Joints made with fittings not *approved* for the specific installation.
3. Solvent-cement joints between different types of plastic pipe.
4. Saddle-type fittings.

❖Cement and concrete joints are not effective in sealing a pipe joint and such materials are inflexible and susceptible to cracking and displacement.

Fittings that are not approved for a specific installation can cause premature failure of the joint because of an incompatibility of materials used in the fitting or a contamination of the water supply from the use of improper materials in direct contact with it.

A solvent-cement joint is a homogeneous chemical bond between a pipe and fitting. The bond is accomplished because the chemical composition of both joint surfaces is the same. If the joint surface materials are different, the joint may not form a proper chemical bond, and the strength and integrity of the joint will be adversely affected. For example, solvent cementing is prohibited between ABS and PVC; PVC and CPVC or ABS and CPVC (see commentary, Section 605.10.2).

A saddle-type fitting, which is also known as a saddle tee, is made where a hole is drilled into the water supply pipe and a saddle clamp with a rubber gasket is set in place over the hole. A back-up plate is then attached, and the assembly is tightened in place on the supply pipe. These fittings are not an approved type because they reduce the flow in the piping system as a result of the hole drilled for the saddle being typically restricted to one-half the diameter of the supply pipe the fitting is attached to. Saddle-tap valves are considered to be a saddle-type fitting. Because of the gasket and the means of attaching the valve to the pipe, such valves are considered to be less dependable and less permanent.

Additionally, if these fittings are subject to impact, they may shift and further retard the flow of water through the fitting and/or leak.

605.10 ABS plastic. Joints between ABS plastic pipe or fittings shall comply with Sections 605.10.1 through 605.10.3.

❖Acceptable joints for ABS plastic pipe are identified in the referenced sections.

605.10.1 Mechanical joints. Mechanical joints on water pipes shall be made with an elastomeric seal conforming to ASTM D 3139. Mechanical joints shall only be installed in underground systems, unless otherwise *approved*. Joints shall be installed only in accordance with the manufacturer's instructions.

❖ Mechanical joints are predominantly used in underground applications (e.g., bell or hub compression gasket joints). Mechanical joints are typically of the unrestrained type. Above-ground mechanical joints are limited primarily to rehabilitation installations and connections to existing systems when it is impractical to make another type of joint.

The referenced standard regulates only the elastomeric material used to make the seal in a mechanical joint, not the entire coupling or fitting assembly.

605.10.2 Solvent cementing. Joint surfaces shall be clean and free from moisture. Solvent cement that conforms to ASTM D 2235 shall be applied to all joint surfaces. The joint shall be made while the cement is wet. Joints shall be made in accordance with ASTM D 2235. Solvent-cement joints shall be permitted above or below ground.

❖ The difference between ABS, CPVC and PVC solvent-cement joints is based on the type of solvent cement and the fact that CPVC and PVC joint surfaces must first be primed. These joining surfaces must be softened by the use of a primer solvent so that penetration and dissolving can be achieved when using solvent cement. As the solvent dissipates, the cement layer and dissolved surfaces will harden with a corresponding increase in joint strength. This section clarifies that solvent cementing is an acceptable joining method for both above-ground and below-ground installations. Although the most common joining method, solvent cementing is also one of the most misunderstood methods of joining pipe. ASTM D 2235 contains an appendix that has the recommended procedure for making joints.

The pipe and fittings should be approximately the same temperature prior to solvent cementing. The pipe end and socket fitting must be clean, dry and free from grease, oil and other foreign materials.

The solvent cement must be uniformly applied to the socket of the fitting and to the end of the pipe to the depth of the socket. Excessive solvent cement must be avoided to prevent a bead from forming on the inside of the pipe joint.

The joint is made by twisting the pipe into the fitting socket immediately after applying the cement. The pipe must be held in place for a few seconds until the solvent cement begins to set. The pipe will tend to back out unless it is restrained.

A small bead of solvent cement forms between the pipe exterior and fitting. The bead must be removed with a clean cloth to avoid the possibility of weakening the pipe wall.

A solvent-cement joint should remain undisturbed for a period of 24 hours before being exposed to the working pressure. There should be no rough handling of the pipe for at least 1 hour after the joint is made.

Because each type of solvent cement is specifically designed for a given piping material, all-purpose solvent cement or universal solvent cement cannot be used to join ABS pipe or fittings unless it conforms to ASTM D 2235. Solvent cement for each plastic pipe material requires a unique mixture or combination of solvents and dissolved plastic resins to conform to the standard. Although all-purpose cement could dissolve the surfaces of ABS, it might be too aggressive or not aggressive enough to make a joint strong enough to meet the minimum requirements of ASTM D 2235. Solvent cement must be identified on the can as complying with ASTM D 2235.

605.10.3 Threaded joints. Threads shall conform to ASME B1.20.1. Schedule 80 or heavier pipe shall be permitted to be threaded with dies specifically designed for plastic pipe. *Approved* thread lubricant or tape shall be applied on the male threads only.

❖ Pipe thread dimensions can be historically traced to Robert Briggs, already mentioned in the commentary to Table 605.3 for galvanized steel pipe. The ANSI standard identifies pipe dimensions as "NPT." NPT is a coded designation: N stands for USA Standard, P indicates pipe and T means that the threads are tapered.

This section prohibits cutting threads into Schedule 40 ABS pipe. Threaded joints in Schedule 40 ABS pipe must be accomplished with a factory-fabricated solvent cemented adapter fitting made for that purpose. This section also recognizes that Schedule 80 pipe has sufficient pipe-wall thickness and strength to permit field threading.

Thread lubricant (pipe-joint compound) must be specifically designed for ABS plastic pipe. It is not the same lubricant that is commonly used for metallic pipe, such as steel. Thread lubricant and tape are designed for two purposes: to lubricate the joint for proper thread mating and to fill in small imperfections on the surfaces of the threads.

605.11 Asbestos-cement. Joints between asbestos-cement pipe or fittings shall be made with a sleeve coupling of the same composition as the pipe, sealed with an elastomeric ring conforming to ASTM D 1869.

❖ The only joining method recognized for asbestos-cement pipe is a mechanical coupling, which makes the seal with an elastomeric O-ring. ASTM D 1869 regulates the material and dimensional requirements for the O-ring (see Commentary Figure 605.11).

605.12 Brass. Joints between brass pipe or fittings shall comply with Sections 605.12.1 through 605.12.4.

❖ Acceptable joints for brass pipe or fittings are identified in the referenced sections.

605.12.1 Brazed joints. All joint surfaces shall be cleaned. An *approved* flux shall be applied where required. The joint shall be brazed with a filler metal conforming to AWS A5.8.

❖ Although brazed joints are similar to soldered joints, brazed joints require a higher joining temperature [in excess of 1,000°F (538°C)]. Because of a brazed joint's inherent resistance to elevated temperatures, it is much stronger than a soldered joint. Brazing is often referred to as "silver soldering." Silver soldering is more accurately described as silver brazing and uses silver-bearing alloys primarily composed of silver along with small amounts of copper and zinc. Silver soldering (brazing) typically requires temperatures in excess of 1,000°F (538°C), and such solders are classified as "hard" solders.

There has always been confusion over the distinction between "silver solder" and "silver-bearing solder." Silver solders are unique and can be further subdivided into soft and hard categories, which are determined by the percentages of silver and the other component elements of the particular alloy. The distinction must be made that silver-bearing solders [with a melting point below 600°F (315°C)] are used in soft-soldered joints, whereas silver solders [with a melting point above 1,000°F (538°C)] are used in silver-brazed joints.

The two common series of filler metals used for brazing are BCuP and BAg. BCuP contains phosphorus and copper alloys. BAg contains 30- to 60-percent silver with zinc and copper alloys. Some of the BAg filler metals also contain cadmium.

The joint surfaces to be brazed must be cleaned, typically by polishing the pipe with an emery cloth. The socket fitting is cleaned with a specially designed brush.

Brazing flux is applied to the joint surfaces. Certain types of BCuP filler metals do not require flux for certain types of joints. The manufacturer's recommendations should be consulted to determine when flux is necessary.

The joints are typically heated with an oxyacetylene torch, though an air acetylene torch can be used on a smaller-diameter pipe. The pipe and fitting must be at approximately the same temperature when making the joint. The filler metal flows to the heat with the aid of the flux. The pipe should first be heated above the fitting. When the flux becomes clear, the heat should be directed to the base of the fitting socket. When the flux becomes clear and quiet, the filler metal is applied to the joint. The heat draws the filler metal into the socket (see Commentary Figure 605.12.1). Any remaining flux residue should be removed from the pipe and joint surfaces.

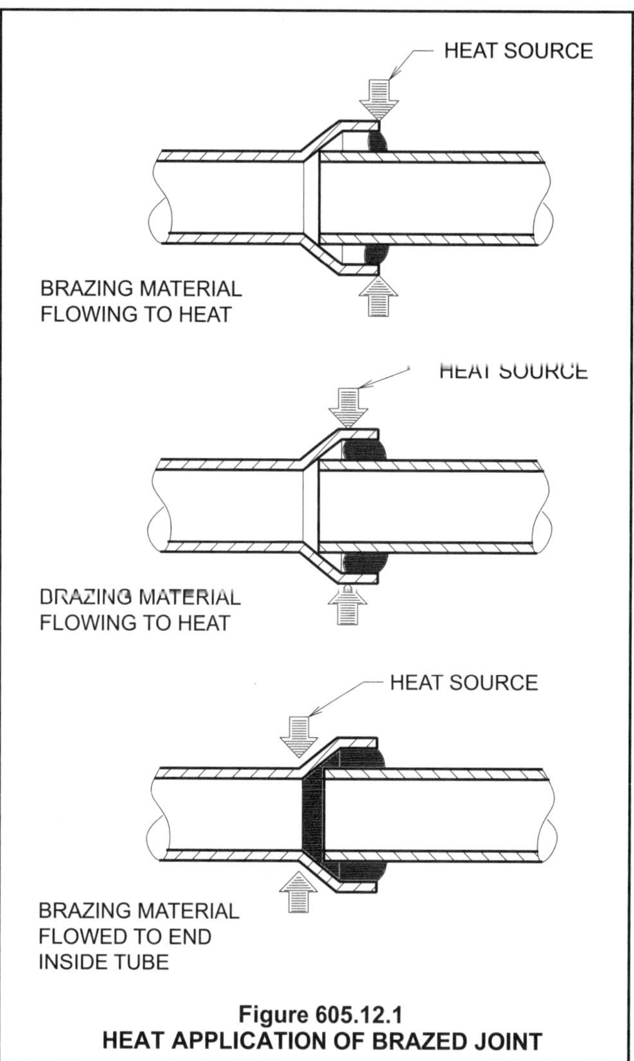

Figure 605.12.1
HEAT APPLICATION OF BRAZED JOINT

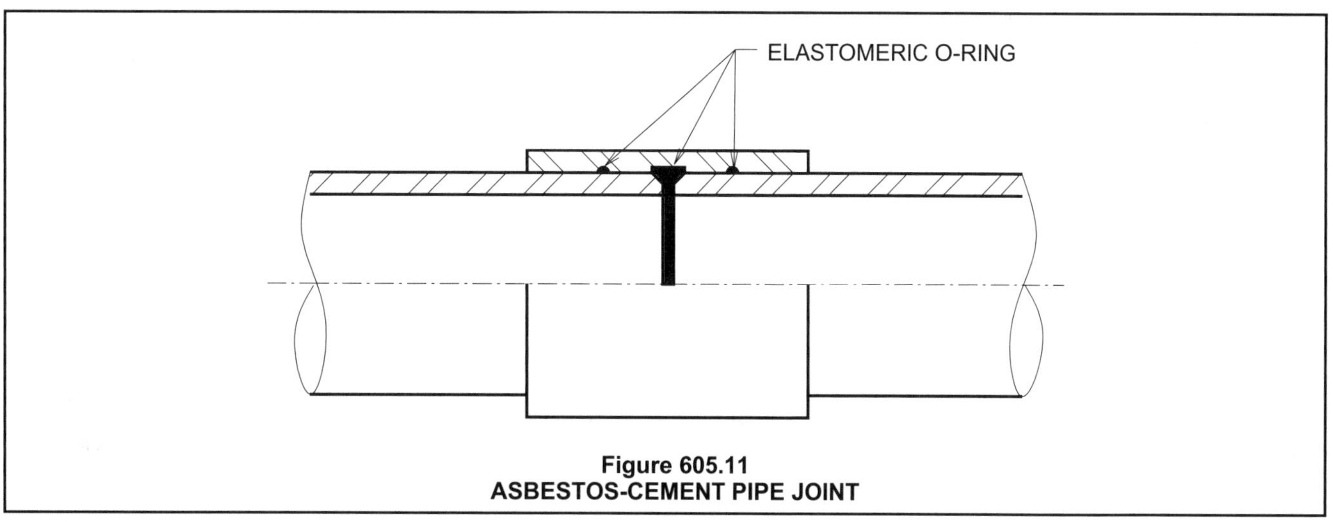

Figure 605.11
ASBESTOS-CEMENT PIPE JOINT

605.12.2 Mechanical joints. Mechanical joints shall be installed in accordance with the manufacturer's instructions.

❖ Mechanical joints are unique in that many acceptable types are available. Most mechanical joints use an elastomeric material to form a seal. Compression fittings that use a compression ring, which forms a metal-to-metal seal, are also considered as mechanical joints. In all cases, the joint must be carefully installed in accordance with the manufacturer's instructions.

605.12.3 Threaded joints. Threads shall conform to ASME B1.20.1. Pipe-joint compound or tape shall be applied on the male threads only.

❖ Pipe-joint compound is typically designed for metallic pipe. The pipe-joint compound used for brass pipe will also be applied for copper pipe and steel pipe (see Commentary Section 605.10.3).

605.12.4 Welded joints. All joint surfaces shall be cleaned. The joint shall be welded with an *approved* filler metal.

❖ A welded joint is similar to a brazed joint. The primary differences are the temperature at which the joint is made, the type of filler metals used and the fact that welding reaches the melting point of the base metal, whereas brazing temperatures are well below the melting point of the pipe (see Commentary Section 605.12.1).

605.13 Gray iron and ductile iron joints. Joints for gray and ductile iron pipe and fittings shall comply with AWWA C111 and shall be installed in accordance with the manufacturer's installation instructions.

❖ This section requires rubber-gasket joints to connect ductile-iron water pipe. Although not specifically identified, the following are considered to be rubber-gasket joints:

Mechanical joints: a bottled joint of the stuffing type including a bell, an exterior flange with bolt openings, a socket, a sealing gasket, a follower gland with bolt holes and bolts.

Push-on joint: a single rubber-gasket joint consisting of a plain-end pipe forced into a socket and compressing the gasket around the outside of the pipe.

Flange joint: consists of two flanged end pipes joined together by bolts that compress a ring gasket between the flange faces.

Modified push-on and mechanical joints: variations to the above-described joints that comply with the performance requirements of AWWA C111.

605.14 Copper pipe. Joints between copper or copper-alloy pipe or fittings shall comply with Sections 605.14.1 through 605.14.5.

❖ Acceptable joints for copper or copper-alloy pipe or fittings are identified in the referenced sections.

605.14.1 Brazed joints. All joint surfaces shall be cleaned. An *approved* flux shall be applied where required. The joint shall be brazed with a filler metal conforming to AWS A5.8.

❖ The installation technique and materials used to make a brazed joint for copper or copper-alloy pipe are identical to that of brass pipe and copper tubing (see Commentary Section 605.12.1).

605.14.2 Mechanical joints. Mechanical joints shall be installed in accordance with the manufacturer's instructions.

❖ Mechanical joints for copper or copper-alloy pipe are similar to those used to join brass pipe (see commentary, Section 605.12.2). Because the different mechanical joint couplings available are often unique to each manufacturer, the couplings must be installed in accordance with the manufacturer's instructions.

605.14.3 Soldered joints. Solder joints shall be made in accordance with the methods of ASTM B 828. All cut tube ends shall be reamed to the full inside diameter of the tube end. All joint surfaces shall be cleaned. A flux conforming to ASTM B 813 shall be applied. The joint shall be soldered with a solder conforming to ASTM B 32. The joining of water supply piping shall be made with lead-free solder and fluxes. "Lead free" shall mean a chemical composition equal to or less than 0.2-percent lead.

❖ A soldered joint is the most common method of joining copper pipe and tubing.

Pipe must be cut square for proper alignment, adequate surface area for joining and an interior free from obstructions. When the pipe is cut, an edge (burr) is left protruding into the pipe. The pipe must be properly reamed to remove the burr. Chamfering is required to bevel the outer edge of a cut pipe end. Undercutting (excessive reaming) can reduce pipe wall thickness (see Commentary Figure 605.14.3).

ASTM B 32 covers many grades of solder, including tin-silver solders. Previously, the grades used in the plumbing industry were typically 40-60 (40 percent tin and 60 percent lead), 50-50 (50 percent tin and 50 percent lead) and 95-5 (95 percent tin and 5 percent antimony). Lead-based solders, however, are no longer permitted for joining copper tubing where used for the distribution of potable water.

Lead-based solders and fluxes have been implicated as a cause for an increased concentration of lead in drinking water. Potable water is exposed to the lead in the solder that commonly coats the inside surface of the pipe or fitting near the joint. The water can dissolve lead from the solder. The chemically combined or free lead in potable water can be ingested by drinking the water or by consuming food or beverages prepared with the water. In an effort to reduce the amount of lead in drinking water, lead-bearing solders and fluxes have been banned from use in joining water supply piping. After the code implemented requirements for lead-free solders and fluxes, the federal government followed suit. In June 1986, the EPA banned lead-bearing solders and fluxes throughout the country. Additionally, all solders and fluxes used on potable water lines must be tested and listed for compliance with NSF 61, which places further restrictions on the concentrations of lead permitted in these materials.

In addition to 95-5 (95 percent tin and 5 percent antimony), which is considered a viable alternative to lead solders in many applications, manufacturers have developed replacement solders for the plumbing industry.

Lead-bearing solder can sometimes be visually identified because the solder tends to darken with age. Replacement solders, tin-antimony solders and tin-silver solders tend to stay bright. Test kits are available that can be used to identify lead-bearing solders with absolute certainty.

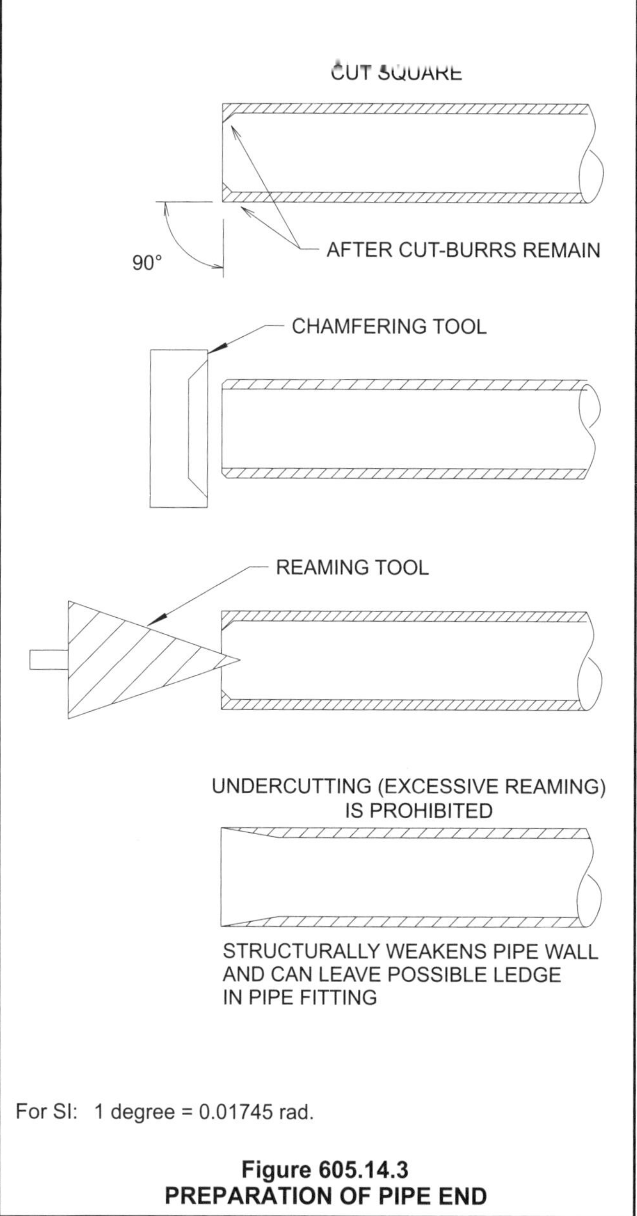

For SI: 1 degree = 0.01745 rad.

Figure 605.14.3
PREPARATION OF PIPE END

ASTM B 828 governs the procedures for making capillary joints by soldering copper and copper alloy tubing and fittings. The joint surfaces for a soldered joint must be cleaned and shined bright, typically accomplished with emery cloth and specially designed socket brushes. Once the copper is shined bright, the joint surfaces should not be touched by the human hand. The natural body oils on the hand will affect the making of the joint.

Flux must be applied to the joint surfaces. Flux is a chemically active material that removes and excludes oxides from the joint area during heating and allows the melted solder to spread out on the surfaces to be joined. The flux material must conform to ASTM B 813, which establishes minimum performance criteria, such as corrosivity solubility, as well as a maximum lead content not to exceed 0.2 percent. The lead content is consistent with the 1986 amendments to the EPA Safe Drinking Water Act. The solder will flow by capillary action toward the heat. As such, the temperatures of the joint surfaces play a major role in making a properly soldered joint. The fitting and the pipe should be approximately the same temperature when the solder is applied. Heat causes the copper atoms to move further apart from each other and the solder atoms enter the spaces between the copper atoms, creating a strong bond when solidified.

If the temperatures are incorrect, either too high or uneven, the solder will run down the inside of the pipe and fitting or on the outside of the pipe. The result is a solder-lined pipe.

Care must also be exercised using wrought fittings because there is less of a ledge to prevent the flow of solder into the fitting socket. The same care must also be exercised when making a swage joint. A swage joint is where the end of one pipe is expanded to form a socket fitting. A swaging tool, similar to a hammer-type flaring tool, is used to expand the pipe. Because it is similar to a flared joint, only annealed and bending-tempered drawn copper can be swaged.

605.14.4 Threaded joints. Threads shall conform to ASME B1.20.1. Pipe-joint compound or tape shall be applied on the male threads only.

❖ Pipe-joint compound (also referred to as "pipe dope") is typically designed for metallic pipe. Pipe-joint compound used for copper pipe will also be used for brass pipe and steel pipe (see Commentary Section 605.10.3). Pipe-joint compound and tape must not be applied to female threads because this would allow the compound to enter and contaminate the piping system.

605.14.5 Welded joints. All joint surfaces shall be cleaned. The joint shall be welded with an *approved* filler metal.

❖ The installation technique and materials used to make a welded joint for copper or copper-alloy pipe are similar to those for brass pipe (see Commentary Section 605.12.4).

605.15 Copper tubing. Joints between copper or copper-alloy tubing or fittings shall comply with Sections 605.15.1 through 605.15.4.

❖ Acceptable joints for copper or copper-alloy tubing or fittings are identified in the referenced sections.

605.15.1 Brazed joints. All joint surfaces shall be cleaned. An *approved* flux shall be applied where required. The joint shall be brazed with a filler metal conforming to AWS A5.8.

❖ The brazing of a joint for copper or copper-alloy tubing is identical to that of copper, copper-alloy and brass pipe (see Commentary Section 605.12.1).

605.15.2 Flared joints. Flared joints for water pipe shall be made by a tool designed for that operation.

❖ Because the pipe end is expanded in a flared joint, only annealed and bending-tempered drawn copper tubing may be flared. There are two types of flaring tools: hammered and screw yoke and block. The hammer-type flaring tool is inserted into the pipe and hammered. With the screw yoke and block, the end of the pipe is firmly held by the block assembly. The screw yoke is attached and a flaring surface is pressed into the pipe end, creating the flare.

The flared surface of the pipe is compressed against the mating surface of the fitting by the flare nut to form a water-tight seal. Commonly, pipe-joint compound is placed on flare-fitting threads to lubricate the assembly [see Commentary Figure 605.15.2(1)].

A flared joint is restricted to soft (annealed) copper tubing because hard-drawn pipe is subject to splitting when flared. The joint is readily dismantled and is, in effect, a type of union connection [see Commentary Figure 605.15.2(2)].

605.15.3 Mechanical joints. Mechanical joints shall be installed in accordance with the manufacturer's instructions.

❖ Mechanical joints for copper tubing include a large array of fittings. The most common mechanical joint is a compression fitting. This type of fitting uses a compression ring (ferrule) to form the seal. The ring compresses around the pipe as the fitting is tightened. Commonly, pipe-joint compound is used on the fitting threads to lubricate the assembly.

605.15.4 Soldered joints. Solder joints shall be made in accordance with the methods of ASTM B 828. All cut tube ends shall be reamed to the full inside diameter of the tube end. All joint surfaces shall be cleaned. A flux conforming to ASTM B 813 shall be applied. The joint shall be soldered with a solder conforming to ASTM B 32. The joining of water supply piping shall be made with lead-free solders and fluxes. "Lead free" shall mean a chemical composition equal to or less than 0.2-percent lead.

❖ The requirements and procedures for a soldered joint for copper and copper-alloy tubing are identical to those for copper and copper-alloy pipe (see Commentary Section 605.14.3).

605.16 CPVC plastic. Joints between CPVC plastic pipe or fittings shall comply with Sections 605.16.1 through 605.16.3.

❖ Acceptable joints for CPVC plastic pipe or fittings are identified in the referenced sections.

605.16.1 Mechanical joints. Mechanical joints shall be installed in accordance with the manufacturer's instructions.

❖ Many mechanical joints use an elastomeric material to form a seal. Because mechanical joints are typically unique to each particular manufacturer, they must be installed in accordance with the manufacturer's instructions (see Commentary Section 605.10.1).

605.16.2 Solvent cementing. Joint surfaces shall be clean and free from moisture, and an *approved* primer shall be applied. Solvent cement, orange in color and conforming to ASTM F 493, shall be applied to all joint surfaces. The joint shall be made while the cement is wet, and in accordance with ASTM D 2846 or ASTM F 493. Solvent-cement joints shall be permitted above or below ground.

> **Exception:** A primer is not required where all of the following conditions apply:
>
> 1. The solvent cement used is third-party certified as conforming to ASTM F 493.
> 2. The solvent cement used is yellow in color.

Figure 605.15.2(1)
TYPICAL FLARED TUBE JOINT

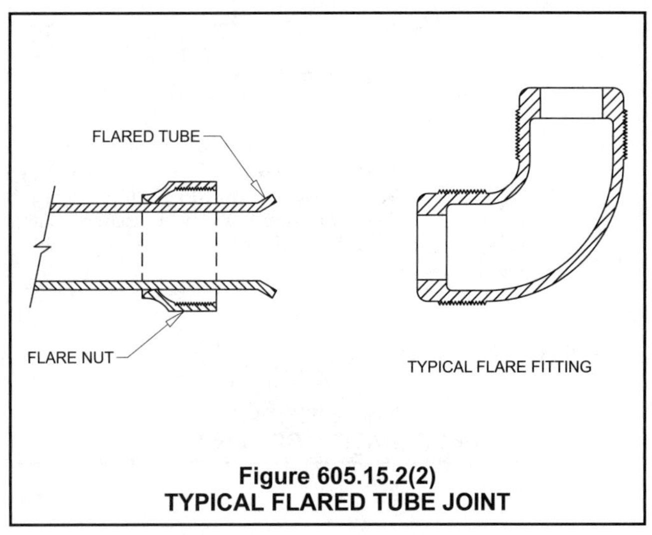

Figure 605.15.2(2)
TYPICAL FLARED TUBE JOINT

3. The solvent cement is used only for joining $^1/_2$ inch (12.7 mm) through 2 inch (51 mm) diameter CPVC pipe and fittings.

4. The CPVC pipe and fittings are manufactured in accordance with ASTM D 2846.

❖ The difference between ABS and CPVC solvent-cement joints is the type of solvent cement and the fact that a primer must be applied to all CPVC joint surfaces prior to applying the solvent cement. The primer breaks down the glossy exterior surface of the pipe to permit fusion of the pipe material with solvent cement.

Failure to apply a primer to a CPVC solvent-cemented joint will result in an improper joint and may affect the pipe manufacturer's product warranty. Note exception for yellow "one step" cement. If a clear primer has been used, close examination is required to determine that the primer has been applied. Primer removes the shiny finish on the pipe, leaving a dull-colored surface.

The solvent cement for CPVC must conform to ASTM F 493 and must be orange in color to indicate that the proper solvent cement has been used (see commentary, Section 605.10.2).

A primer is not required where all of the following conditions apply: the solvent cement conforms to the requirements of ASTM F 493 and is yellow, the solvent cement joins pipe and fittings sized 2 inches (51 mm) or less and the pipe and fittings are manufactured in accordance with ASTM D 2846. ASTM D 2846 indicates that CPVC solvent cements that meet the requirements of ASTM F 493 can be used in accordance with the manufacturer's recommendations without a primer or cleaner. CPVC solvent cement that is used without a primer must be yellow to distinguish it from the traditional cement that requires a primer (see commentary, Section 605.10.2).

605.16.3 Threaded joints. Threads shall conform to ASME B1.20.1. Schedule 80 or heavier pipe shall be permitted to be threaded with dies specifically designed for plastic pipe, but the pressure rating of the pipe shall be reduced by 50 percent. Thread by socket molded fittings shall be permitted. *Approved* thread lubricant or tape shall be applied on the male threads only.

❖ The requirements regarding CPVC pipe are similar to those of ABS plastic pipe (see Section 605.10.3). The key difference is that when CPVC plastic pipe is field threaded (Schedule 80 or heavier), the pressure rating of the pipe must be reduced by one-half because when the pipe is tested for a specific pressure, the sample of the pipe subjected to testing had a full wall thickness. Therefore, reducing pipe-wall thickness with the threading affects the pressure rating of the pipe.

The use of molded fittings for the joining of plastic pipe does not result in the 50-percent reduction in the pressure rating of the pipe because the molded fittings are fabricated so that the wall thickness of the material is maintained at the threads.

605.17 Cross-linked polyethylene plastic. Joints between cross-linked polyethylene plastic tubing or fittings shall comply with Sections 605.17.1 and 605.17.2.

❖ Acceptable joints for cross-linked PEX plastic tubing or fittings are identified in the referenced sections.

605.17.1 Flared joints. Flared pipe ends shall be made by a tool designed for that operation.

❖ Flared joints must be made with tools specific to the design of the connector system. The assembly method must be in accordance with the manufacturer's instructions specific to the type of connectors being used. Flare configurations will vary depending on the tool employed. Using a cold flaring tool, which expands the tubing outside diameter mechanically, secures the flared end between the fitting. The flared surface serves as the sealing area between the tubing and fitting.

605.17.2 Mechanical joints. Mechanical joints shall be installed in accordance with the manufacturer's instructions. Fittings for cross-linked polyethylene (PEX) plastic tubing shall comply with the applicable standards listed in Table 605.5 and shall be installed in accordance with the manufacturer's instructions. PEX tubing shall be factory marked with the appropriate standards for the fittings that the PEX manufacturer specifies for use with the tubing.

❖ ASTM F 877, ASTM F 1807, ASTM F 1960, ASTM F 2098, ASTM F 2159, ASTM F 2434 and ASTM F 2080 are the standards for insert fittings, crimp rings and manifolds for cross-linked PEX plastic tubing systems. They are intended for use in 100 psi (690 kPa) cold and hot water distribution systems operating at temperatures up to and including 180°F (82°C). Insert fittings must be installed in accordance with the manufacturers' instructions. One type of system has insert fittings joined to PEX tubing by the compression of a copper crimp ring around the outer circumference of the tubing, forcing the tubing material into annular spaces formed by ribs on the fittings.

The crimping procedure involves sliding the crimp ring onto the tubing, then inserting the ribbed end of the fitting into the end of the tubing until the tubing contacts the shoulder of the fitting or tube stop. The crimp ring must then be positioned on the tubing so the edge of the crimp ring is $^1/_8$ to $^1/_4$ inch (3.2 to 6.4 mm) from the end of the tube. The jaws of the crimping tool must be centered over the crimp ring, and the tool must be held so that the crimping jaws are approximately perpendicular to the axis of the barb. The jaws of the crimping tool must be closed around the crimp ring, compressing the crimp ring onto the tubing. The crimp ring must not be crimped more than once.

Another type of system uses a special expander tool inserted into the end of the PEX tubing and the tube is expanded. The tool is removed and the insert fitting is quickly inserted before the tubing shrinks back to its original diameter. Depending on the proprietary design, a PEX or metal ring or collar is forced over the tube, locking it to the insert fitting (see Commentary Figure 605.17.2).

Insert fittings and crimp rings are not considered to be interchangeable between different manufacturers of PEX tubing. Therefore, in the 2009 edition of the code, it is required that the PEX fittings specified by each manufacturer as indicated by the markings on the tube be used.

Figure 605.17.2
PEX TUBING CONNECTION

605.18 Steel. Joints between galvanized steel pipe or fittings shall comply with Sections 605.18.1 and 605.18.2.

❖ Acceptable joints for galvanized steel pipe or fittings are identified in the referenced sections.

605.18.1 Threaded joints. Threads shall conform to ASME B1.20.1. Pipe-joint compound or tape shall be applied on the male threads only.

❖ Pipe-joint compound is typically designed for metallic pipe. The pipe-joint compound used for steel pipe will also be used for copper pipe and brass pipe (see Commentary Section 605.10.3). The application of pipe- joint compound to only male threads prevents the compound from entering the piping system.

605.18.2 Mechanical joints. Joints shall be made with an *approved* elastomeric seal. Mechanical joints shall be installed in accordance with the manufacturer's instructions.

❖ A wide variety of mechanical joints are available to join steel pipe, including rolled groove, cut groove and plain-end types.

A union, often mistakenly identified as a mechanical joint, qualifies as a pipe fitting in accordance with Section 605.5.

605.19 Polyethylene plastic. Joints between polyethylene plastic pipe and tubing or fittings shall comply with Sections 605.19.1 through 605.19.4.

❖ Acceptable joints for PE plastic pipe and tubing or fittings are identified in the referenced sections. Sections 605.19.1 and 605.19.3 are also applicable to cross-linked polyethylene (PEX), PEX-AL-PEX composite pipe and PE-AL-PE composite pipe.

605.19.1 Flared joints. Flared joints shall be permitted where so indicated by the pipe manufacturer. Flared joints shall be made by a tool designed for that operation.

❖ Flared joints for PE pipe are acceptable only when specified as such by the pipe manufacturer.

605.19.2 Heat-fusion joints. Joint surfaces shall be clean and free from moisture. All joint surfaces shall be heated to melt temperature and joined. The joint shall be undisturbed until cool. Joints shall be made in accordance with ASTM D 2657.

❖ The method of heat fusion of polyethylene pipe or tubing is similar to that of polyethylene (see Commentary Section 705.6.1).

605.19.3 Mechanical joints. Mechanical joints shall be installed in accordance with the manufacturer's instructions.

❖ This section addresses mechanical joining methods for polyethylene piping. This section also applies to the joining of cross-linked PEX, cross-linked PEX-AL-PEX and PE-AL-PE piping (see commentary, Table 605.4).

Mechanical joints for these materials include insert-type fittings, metallic lock-ring fittings, compression fittings and crimp-type fittings. Each fitting is intended for a specific pipe material and cannot be interchanged unless specified by the pipe manufacturer. In all cases, the fittings connecting any type of polyethylene pipe must be assembled and installed in accordance with the manufacturer's instructions.

605.19.4 Installation. Polyethylene pipe shall be cut square, with a cutter designed for plastic pipe. Except where joined by heat fusion, pipe ends shall be chamfered to remove sharp edges. Kinked pipe shall not be installed. The minimum pipe bending radius shall not be less than 30 pipe diameters, or the minimum coil radius, whichever is greater. Piping shall not be bent beyond straightening of the curvature of the coil. Bends shall not be permitted within 10 pipe diameters of any fitting or valve. Stiffener inserts installed with compression-type couplings and fittings shall not extend beyond the clamp or nut of the coupling or fitting.

❖ Any properly made pipe joint requires that the pipe ends be cut square for proper alignment, proper insertion depth and adequate surface area for joining (see Commentary Figure 605.14.3). Because polyethylene plastic pipe and tubing can be kinked and stress-weakened, the minimum pipe bending radius is specified to not exceed the structural capacity of the material. The minimum bending radius must not be less than the radius of the pipe coil as it came from the manufacturer, or the minimum radius must not be less than 30 pipe diameters, whichever of these two criteria is the largest (see Commentary Figure 605.19.4). The pipe can be straightened as it is taken from the coil, but it cannot be bent in a direction opposite of the bending direction of the coil (see Commentary Figure 605.19.4). To avoid stress on connected fittings or valves, the location of pipe bends is restricted to specific distances from such components.

605.20 Polypropylene (PP) plastic. Joints between PP plastic pipe and fittings shall comply with Section 605.20.1 or 605.20.2.

❖ Acceptable joints for PP plastic pipe and tubing or fittings are identified in the referenced sections.

605.20.1 Heat-fusion joints. Heat-fusion joints for polypropylene pipe and tubing joints shall be installed with socket-type heat-fused polypropylene fittings, butt-fusion polypropylene fittings or electrofusion polypropylene fittings. Joint surfaces shall be clean and free from moisture. The joint shall be undisturbed until cool. Joints shall be made in accordance with ASTM F 2389.

❖ The method of heat fusion of polypylene pipe or tubing is similar to that for polyethylene (see Commentary Section 705.16.1).

605.20.2 Mechanical and compression sleeve joints. Mechanical and compression sleeve joints shall be installed in accordance with the manufacturer's instructions.

❖ This section addresses mechanical joining methods for polypropylene piping. Mechanical joints include insert-type fittings, metallic lock-ring fittings, compression fittings and crimp-type fittings.

Although an insert-type fitting reduces the inside diameter of the pipe, the reduction may be calculated as part of the design considerations of the water distribution system.

605.21 Polyethylene/aluminum/polyethylene (PE-AL-PE) and cross-linked polyethylene/aluminum/cross-linked polyethylene (PEX-AL-PEX). Joints between PE-AL-PE and PEX-AL-PEX pipe and fittings shall comply with Section 605.21.1.

❖ See commentary for Section 605.21.1.

605.21.1 Mechanical joints. Mechanical joints shall be installed in accordance with the manufacturer's instructions. Fittings for PE-AL-PE and PEX-AL-PEX as described in ASTM F 1974, ASTM F 1281, ASTM F 1282, CAN/CSA B137.9 and CAN/CSA B137.10 shall be installed in accordance with the manufacturer's instructions.

❖ The ASTM standards listed are for insert fittings with either crimp rings or stainless steel clamps. Manufacturer's instructions must be followed for installation of the fittings to the pipe.

605.22 PVC plastic. Joints between PVC plastic pipe or fittings shall comply with Sections 605.22.1 through 605.22.3.

❖ Acceptable joints for polyvinyl chloride (PVC) plastic pipe or fittings are identified in the referenced sections.

605.22.1 Mechanical joints. Mechanical joints on water pipe shall be made with an elastomeric seal conforming to ASTM D 3139. Mechanical joints shall not be installed in above-ground systems unless otherwise *approved*. Joints shall be installed in accordance with the manufacturer's instructions.

❖ PVC plastic hub pipe is connected using an elastomeric compression gasket. (See Commentary Section 605.10.1).

605.22.2 Solvent cementing. Joint surfaces shall be clean and free from moisture. A purple primer that conforms to ASTM F 656 shall be applied. Solvent cement not purple in color and conforming to ASTM D 2564 or CSA-B137.3 shall be applied

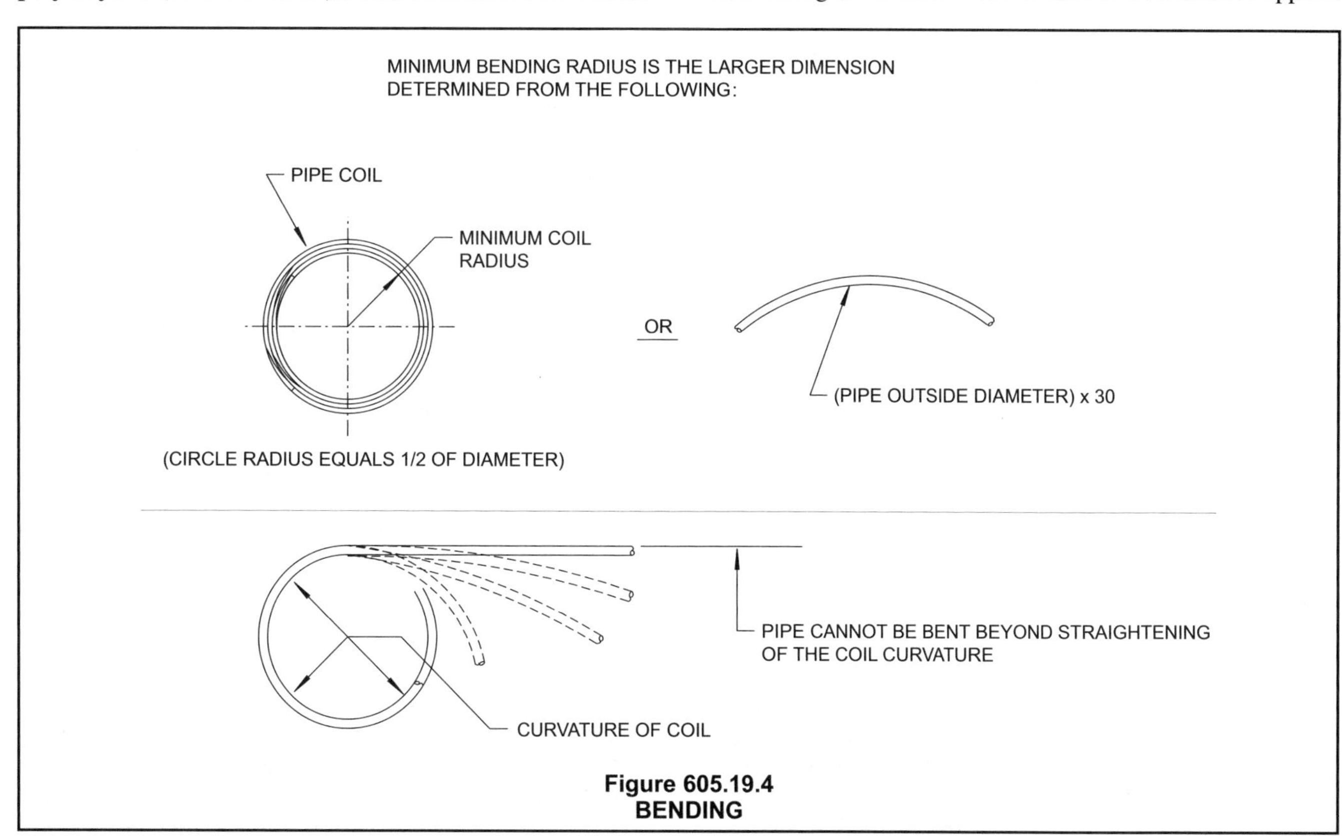

Figure 605.19.4
BENDING

to all joint surfaces. The joint shall be made while the cement is wet and shall be in accordance with ASTM D 2855. Solvent-cement joints shall be permitted above or below ground.

❖ Like CPVC, PVC plastic pipe also requires a primer prior to solvent cementing. Although ASTM F 656 only recommends that PVC primer be purple, this is a code requirement. The purple coloring is intended to allow the installer and the inspector to verify that the required primer has been applied.

To distinguish the difference between primer and solvent cement, the solvent cement must be a color other than purple. Solvent cement for PVC is typically clear (see Commentary Sections 605.10.2 and 605.16.2). A PVC primer is first applied to both the pipe and the socket fitting, then PVC solvent cement is applied to both. The joint is made with a slight twisting until the pipe reaches the full depth of the socket fitting. It is then held in place until the solvent cement begins to set. The solvent-cementing must be done in accordance with the referenced standard.

To minimize the health and fire hazards associated with solvent cements and primers, the manufacturers' instructions for the use and handling of these materials must be followed, in addition to the requirements in ASTM D 2855.

605.22.3 Threaded joints. Threads shall conform to ASME B1.20.1. Schedule 80 or heavier pipe shall be permitted to be threaded with dies specifically designed for plastic pipe, but the pressure rating of the pipe shall be reduced by 50 percent. Thread by socket molded fittings shall be permitted. *Approved* thread lubricant or tape shall be applied on the male threads only.

❖ The requirements regarding threaded joints for PVC plastic pipe are identical to those for CPVC plastic pipe (see Commentary Section 605.16.3).

605.23 Stainless steel. Joints between stainless steel pipe and fittings shall comply with Sections 605.23.1 and 605.23.2.

❖ Acceptable joints for stainless steel pipe or fittings are identified in the referenced sections.

605.23.1 Mechanical joints. Mechanical joints shall be installed in accordance with the manufacturer's instructions.

❖ Mechanical joints such as compression, grooved couplings, hydraulic pressed fittings or flange design can be used to join stainless steel pipe and fittings and are specified by the manufacturer. Such joints must be assembled and installed in compliance with the manufacturer's instructions.

605.23.2 Welded joints. All joint surfaces shall be cleaned. The joint shall be welded autogenously or with an *approved* filler metal as referenced in ASTM A 312.

❖ An important part of successful welding of the austenitic grades requires proper selection of alloy (for both the base metal and the filler rod) and corrects welding procedures. Two important objectives in making weld joints in austenitic stainless steel are (1) preservation of corrosion resistance, and (2) prevention of cracking. Welding naturally produces a temperature gradient in the metal being welded, ranging from the melting temperature of the fused weld metal to ambient temperature at some distance from the weld.

The two basic methods for welding stainless steels are fusion welding and resistance welding. In fusion welding, heat is comes from an electric arc struck between an electrode and the metal to be welded. In resistance welding, bonding is the result of heat and pressure. Heat is produced by the resistance to the flow of electric current through the parts to be welded, and pressure is applied by the electrodes.

Welds and the surrounding area should be thoroughly cleaned to avoid impairment of corrosion resistance. Weld spatter, flux or scale may become focal points for corrosive attack if not properly removed, especially in aggressive environments.

605.24 Joints between different materials. Joints between different piping materials shall be made with a mechanical joint of the compression or mechanical-sealing type, or as permitted in Sections 605.24.1, 605.24.2 and 605.24.3. Connectors or adapters shall have an elastomeric seal conforming to ASTM D 1869 or ASTM F 477. Joints shall be installed in accordance with the manufacturer's instructions.

❖ Mechanical joints can be used to join dissimilar piping materials. Because many piping materials have the same outside diameters, many fittings are designed by manufacturers for use with a number of different piping materials [see Commentary Figures 605.24.1(1) and (2)].

605.24.1 Copper or copper-alloy tubing to galvanized steel pipe. Joints between copper or copper-alloy tubing and galvanized steel pipe shall be made with a brass fitting or dielectric fitting or a dielectric union conforming to ASSE 1079. The copper tubing shall be soldered to the fitting in an *approved* manner, and the fitting shall be screwed to the threaded pipe.

❖ When copper or copper-alloy tubing is joined with galvanized steel pipe, protecting against galvanic corrosion is required. Galvanic corrosion occurs when two different metals come in contact in the presence of an electrolyte, such as water. Galvanic corrosion accelerates the natural corrosion process that occurs in all metals, and, because certain metals corrode faster than others, they have been placed in a hierarchy in order of rate of corrosion (see Commentary Table 605.24.1). The more reactive metal at a connection is called the "anode" (galvanized steel pipe), and the less reactive metal is called the "cathode" (copper tubing). When the copper tubing and galvanized steel pipe are coupled, the galvanized steel pipe will tend to dissolve in the electrolyte, thereby generating electric current flow between the metals.

This section prescribes three methods of protection against galvanic corrosion: (1) a brass fitting such as a short section of brass pipe or a brass valve body, (2) a dielectric fitting such as a plastic-lined short pipe section, or (3) a dielectric union either flange type or insulated nut type. While the code text seems to imply that all of these types must conform to ASSE 1079, only the union type is covered by the referenced standard.

A variety of dielectric fitting designs is available. Commentary Figures 605.24.1(1) and 605.24.1(2) illustrate two of the most commonly used types. Other dielectric fittings of varying designs are available. Dielectric fittings provide an electrical barrier to prevent the passage of electric current either between dissimilar metals or between the piping and the electrolyte. Because the code does not define the term "dielectric fitting" or prescribe a standard for gauging the design or performance of a dielectric fitting, the code official must evaluate and approve the type of dielectric fitting and its application.

The joint between the copper tube and the threaded galvanized steel pipe can also be made with a cast-brass fitting. A wrought-copper fitting is not permitted for this transition. A cast-brass fitting is easily recognized by its yellowish color. The brass fitting serves as a buffer unlike the dielectric fitting, which provides an electrically insulating barrier. The brass fitting greatly retards the rate of corrosion. Because brass, copper and galvanized steel are dissimilar metals, there will be a minimal amount of corrosion among such materials.

Because plastic is usually the material that provides the electrical barrier in a dielectric fitting, special care

Table 605.24.1
CORROSIVE HIERARCHY OF METALS
(Galvanic Series)

Anode end (least noble, the reactive or corroding end)
- Magnesium
- Magnesium alloys
- Zinc
- Galvanized steel
- Aluminum
- Aluminum alloys
- Carbon steel
- Cast iron
- Stainless steel (active)
- Tin
- Lead
- Nickel (active)
- Brass
- Copper
- Bronze
- Nickel-copper alloys
- Nickel (passive)
- Titanium
- Silver
- Graphite
- Gold
- Platinum

Cathode end (most noble, the protected or less-reactive end)

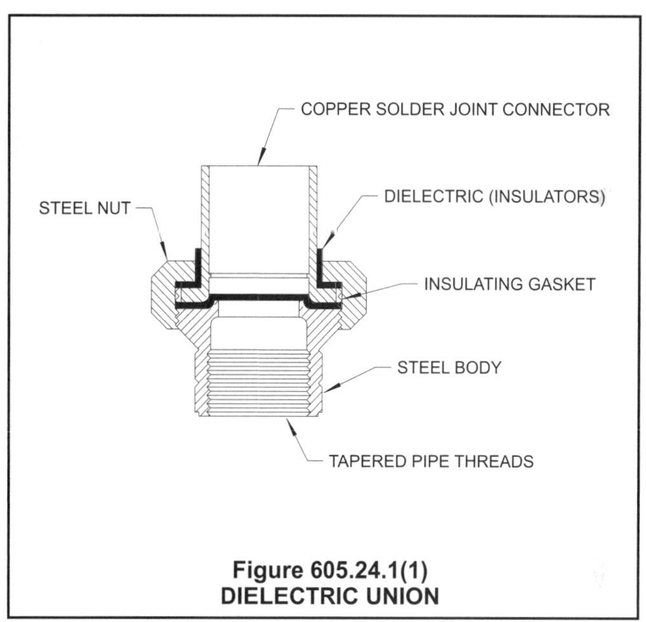

Figure 605.24.1(1)
DIELECTRIC UNION

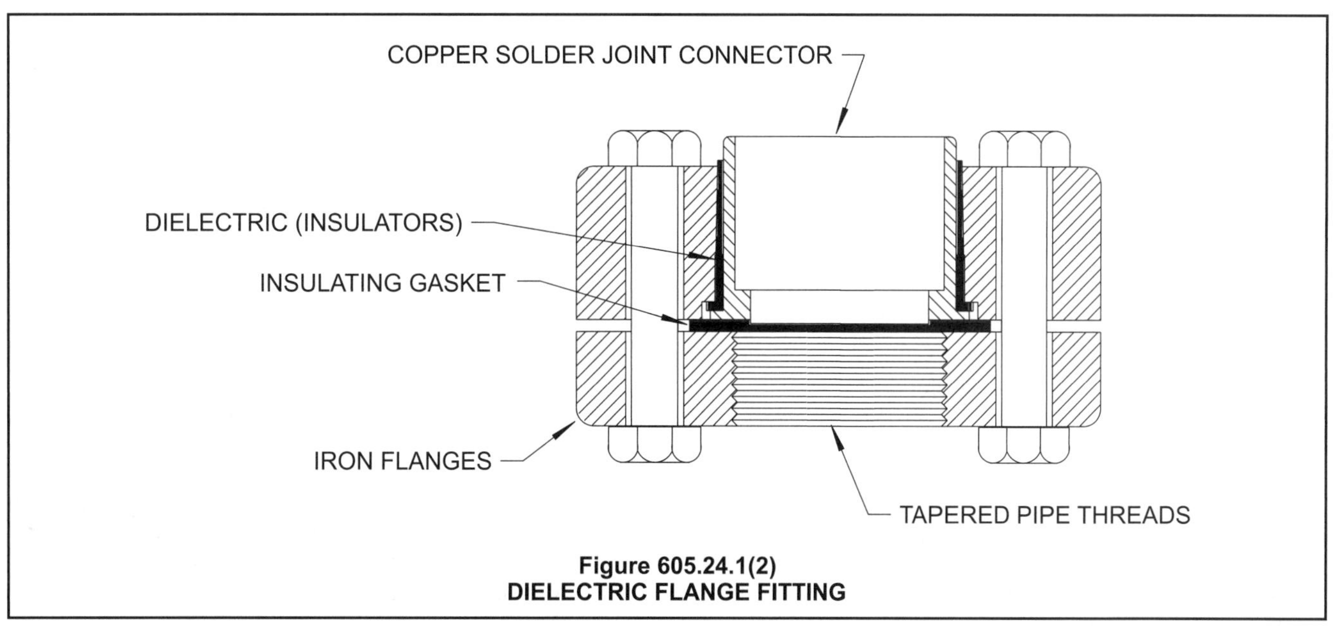

Figure 605.24.1(2)
DIELECTRIC FLANGE FITTING

is necessary when applying heat to the joint to avoid damaging the plastic sleeve. This is usually accomplished by sliding the plastic sleeve and steel nut away from the area to be heated. The soldered joint must conform to the requirements of Section 605.15.4. Commentary Figure 605.24.1(3) illustrates the result of connecting copper piping directly to a steel water heater tank. Welds on the top of this water heater tank began to leak in less than 6 years of service.

Figure 605.24.1(3)
CORROSION DUE TO ABSENCE OF DIELECTRIC FITTINGS

605.24.2 Plastic pipe or tubing to other piping material. Joints between different grades of plastic pipe or between plastic pipe and other piping material shall be made with an *approved* adapter fitting.

❖ The joining of differing types of plastic pipe and plastic pipe to other piping materials such as copper, brass and galvanized steel is permitted only through the use of adapter fittings. The adapter fittings must be evaluated individually and approved by the code official. Some key items to review when evaluating adapter fittings are joint tightness and compatibility of the adapter to the dissimilar materials being joined.

605.24.3 Stainless steel. Joints between stainless steel and different piping materials shall be made with a mechanical joint of the compression or mechanical sealing type or a dielectric fitting or a dielectric union conforming to ASSE 1079.

❖ Where two dissimilar metals are in contact in the presence of an electrolyte, a battery effect is created, current flows and one of the metals corrodes. Any metal in this sense will tend to have corrosion accelerated when it is coupled in the presence of an electrolyte. A very important factor to consider in evaluating the potential for galvanic corrosion is the relative surface area of the two different metals in contact.

In theory all metals immersed in a electrolyte, such as water, have a voltage potential. The relative activity of the metal is determined by voltage potential, with those metals with negative voltages being most active and more likely to corrode. For example, magnesium and iron have a negative voltage and corrode very rapidly when exposed to a moist atmosphere. Copper and stainless steel have a positive voltage and are less likely to corrode.

Dielectric sleeves may be used with flange bolting or as gaskets for connections between different types of materials to prevent such corrosion. The plastic or rubber gasket engages the thread of the fitting attached, thereby insulating the metal from the water. This prevents corrosion on either part.

While the code text seems to imply that all of these types must conform to ASSE 1079, only the union type is covered by the referenced standard [see Commentary Figures 605.24.1(1) and 605.24.1(2)].

SECTION 606 INSTALLATION OF THE BUILDING WATER DISTRIBUTION SYSTEM

606.1 Location of full-open valves. Full-open valves shall be installed in the following locations:

1. On the building water service pipe from the public water supply near the curb.
2. On the water distribution supply pipe at the entrance into the structure.
3. On the discharge side of every water meter.
4. On the base of every water riser pipe in occupancies other than multiple-family residential *occupancies* that are two stories or less in height and in one- and two-family residential *occupancies*.
5. On the top of every water down-feed pipe in *occupancies* other than one- and two-family residential *occupancies*.
6. On the entrance to every water supply pipe to a dwelling unit, except where supplying a single fixture equipped with individual stops.
7. On the water supply pipe to a gravity or pressurized water tank.
8. On the water supply pipe to every water heater.

❖ Valves are needed to isolate components and portions of a water distribution system to facilitate service, repair and replacement and to allow emergency shutoff in the event of system failure. As defined in legacy codes, a full-open valve is a shutoff valve that in the full-open position has a straight-through flow passageway with a diameter of not less than one nominal pipe size smaller than the nominal pipe size of the connecting pipe, or with an area not less than 85 percent of the cross-sectional area of the connecting pipe. Unlike shutoff valves and stops, full-open valves offer very little resistance to flow and, therefore, have a minimal effect on supply pressure. Full-open valves are also referred to as "water service valves" and include gate, ball and butterfly valves.

A curb valve, commonly called a "corporation cock," is typically installed by the water utility company and is usually a lubricated rotor (plug)-type valve [see Commentary Figure 606.1(1)].

The building entrance water supply valve is usually located before the water meter where the meter is located inside the building.

The location of valves before and after the water meter allows meter replacement without having to drain down the building piping system [see Commentary Figure 606.1(2)].

A water pipe must extend completely through a story before being considered a riser. One- and two-family dwellings are exempt from the water riser valve requirement as are multiple-family dwellings that are not over two stories in height. A downfeed pipe is supplied by a pipe at the top of the riser. The direction of flow is down rather than up [see Commentary Figure 606.1(3)].

The dwelling unit valve must be located at the point where the water distribution pipe enters a dwelling unit. The valve must be able to be accessed from within the dwelling unit and is intended to isolate the water distribution system to that dwelling unit in the event of a leak at the fixtures, a water pipe break within the dwelling unit or the need for repairs [see Commentary Figures 606.1(4), (5) and (6)].

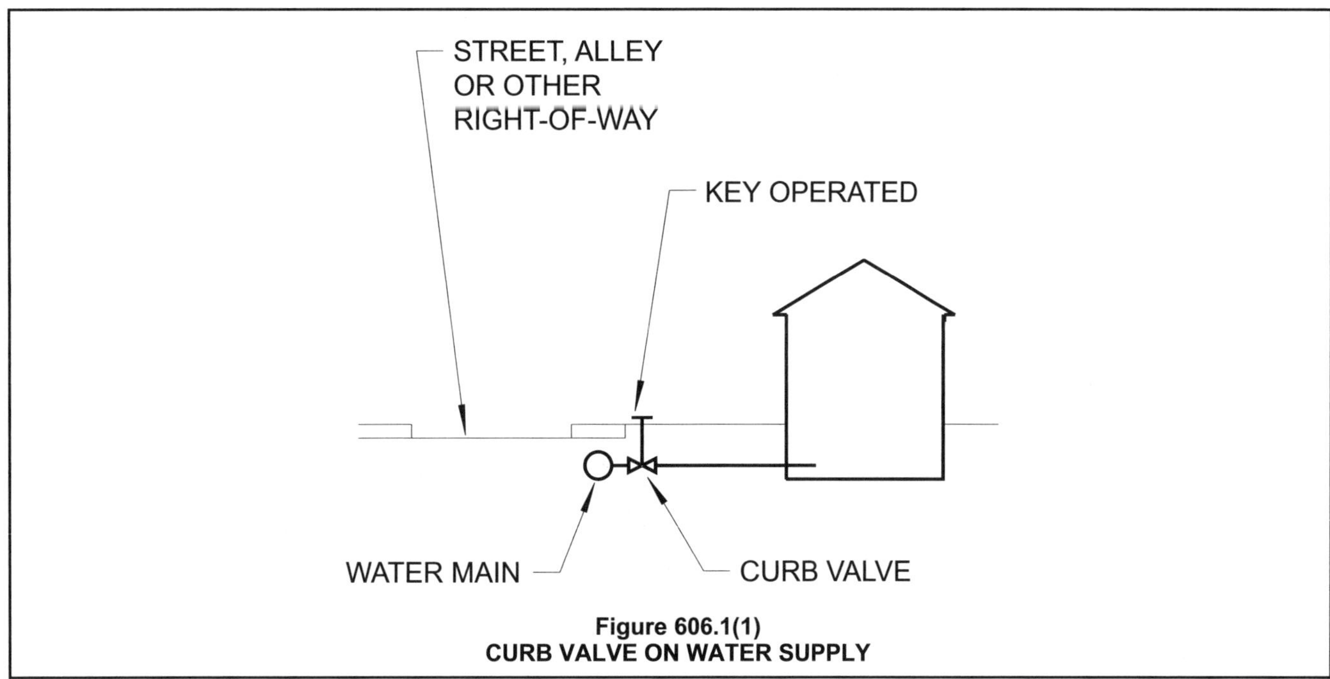

Figure 606.1(1)
CURB VALVE ON WATER SUPPLY

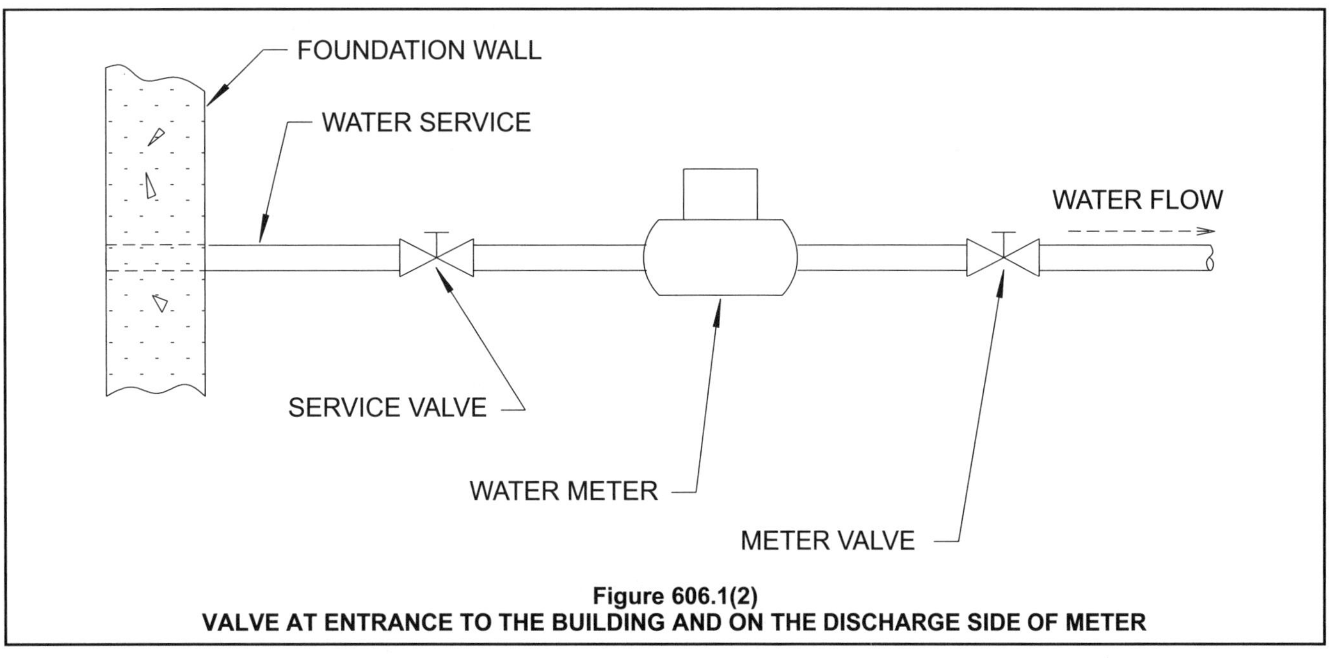

Figure 606.1(2)
VALVE AT ENTRANCE TO THE BUILDING AND ON THE DISCHARGE SIDE OF METER

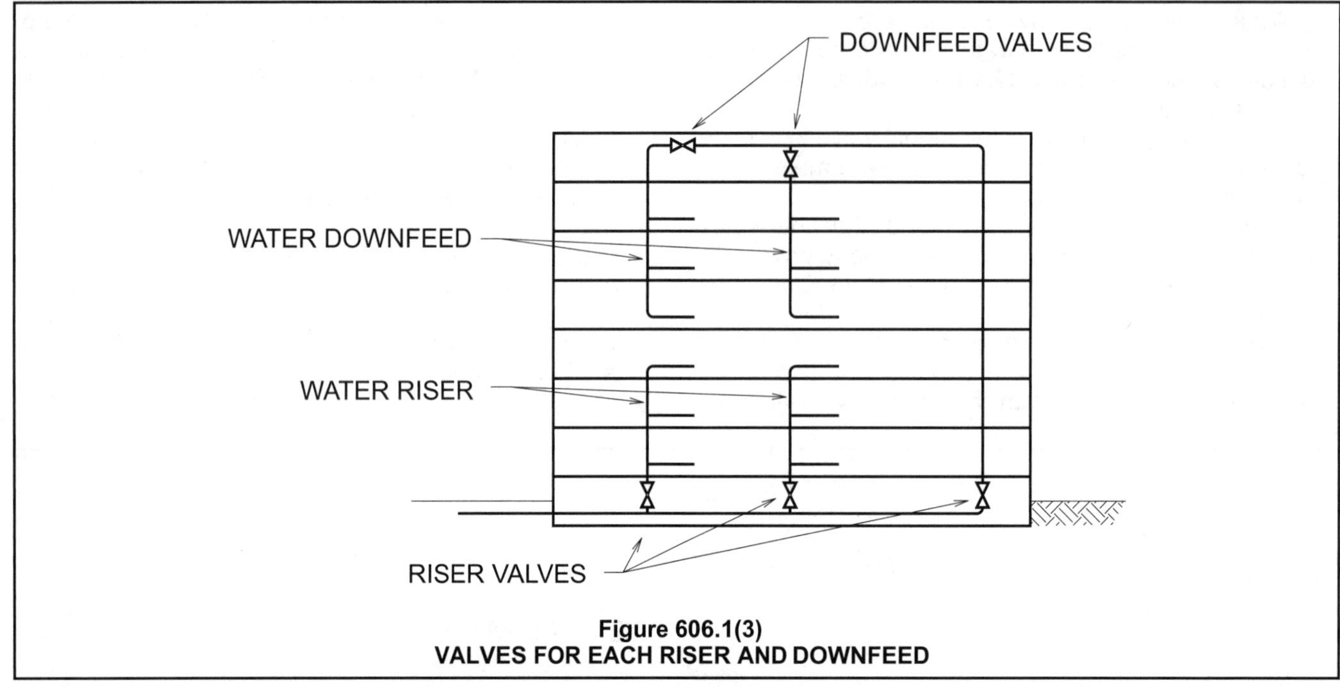

Figure 606.1(3)
VALVES FOR EACH RISER AND DOWNFEED

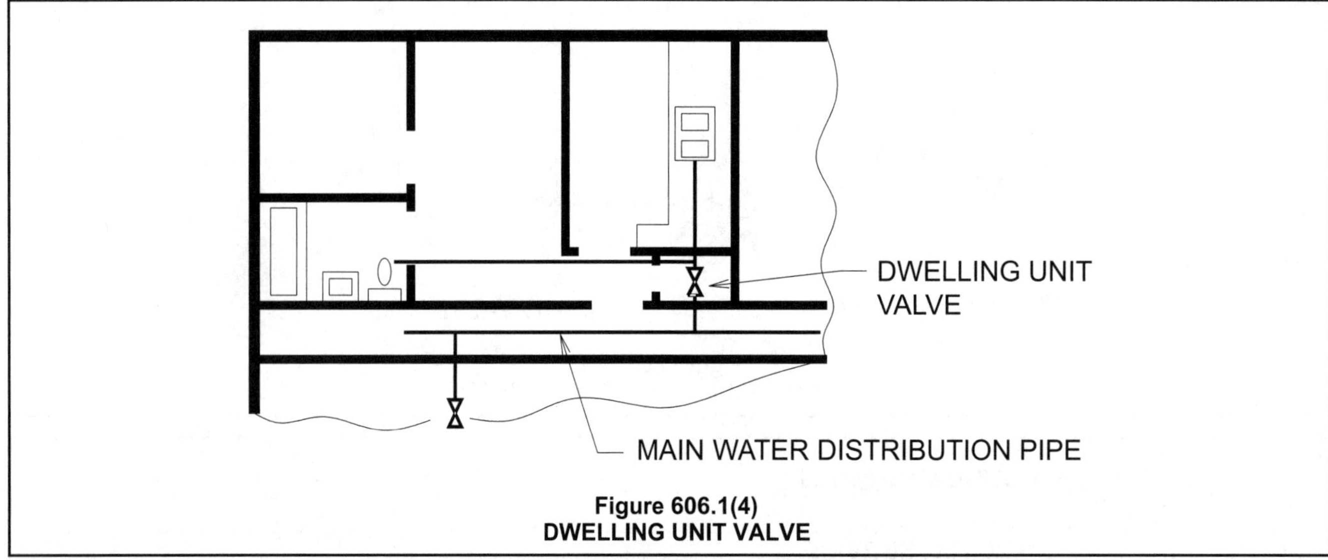

Figure 606.1(4)
DWELLING UNIT VALVE

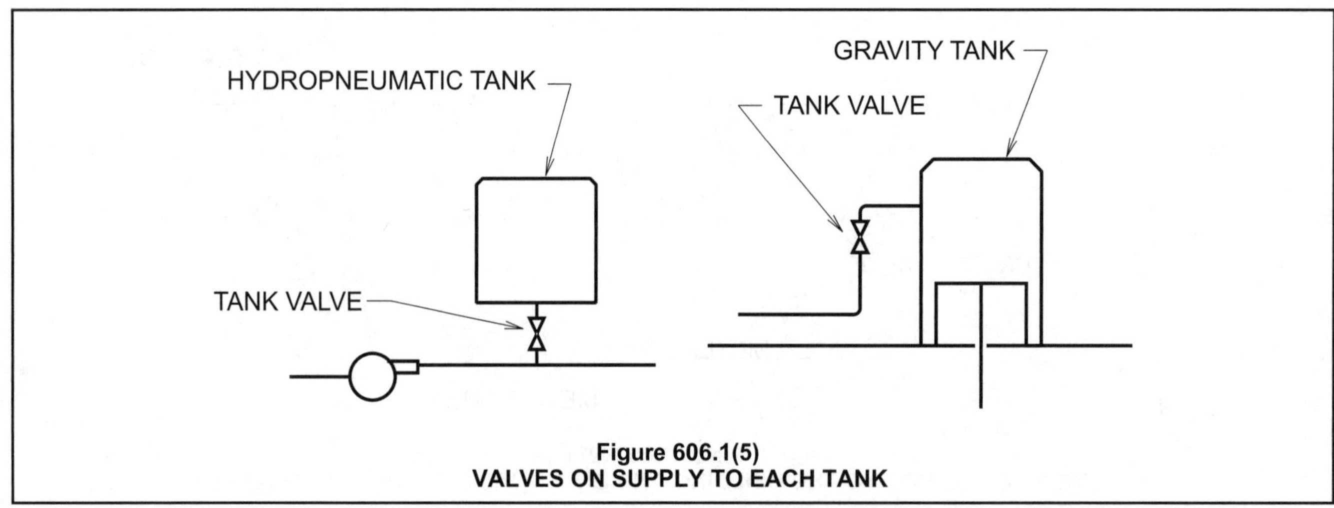

Figure 606.1(5)
VALVES ON SUPPLY TO EACH TANK

606.2 Location of shutoff valves. Shutoff valves shall be installed in the following locations:

1. On the fixture supply to each plumbing fixture other than bathtubs and showers in one- and two-family residential *occupancies*, and other than in individual sleeping units that are provided with unit shutoff valves in hotels, motels, boarding houses and similar *occupancies*.
2. On the water supply pipe to each sillcock.
3. On the water supply pipe to each appliance or mechanical equipment.

❖ A shutoff valve, unlike a full-open valve, has no requirements for the cross-sectional area of flow. Therefore, there is a greater pressure drop through a shutoff valve when compared to a full-open valve. Shutoff valves are commonly referred to as "stops" and include globe valves and straight and angle stops. Substitution of full-open valves for shutoff valves is allowed, but not vice versa. This section does not specify the exact location of the required shutoff valves; therefore, the valves can be either immediately adjacent to the fixture, appliance or equipment, or remotely located and identified in accordance with Section 606.4 [see Commentary Figures 606.2(1) and 606.2(2) and Commentary Section 604.10.2].

Shutoff valves must be installed on the fixture supply to each plumbing fixture other than bathtubs and showers in residential occupancies and individual sleeping units provided with unit shutoff valves in hotels, motels, boarding houses and similar occupancies. Bathtubs and showers typically do not have shutoff valves because they would not be accessible for maintenance. Some tub and shower control valves are available with integral stops (shutoff valves) that can be accessed by removing the escutcheon plate. The necessity for individual shutoff valves can easily be

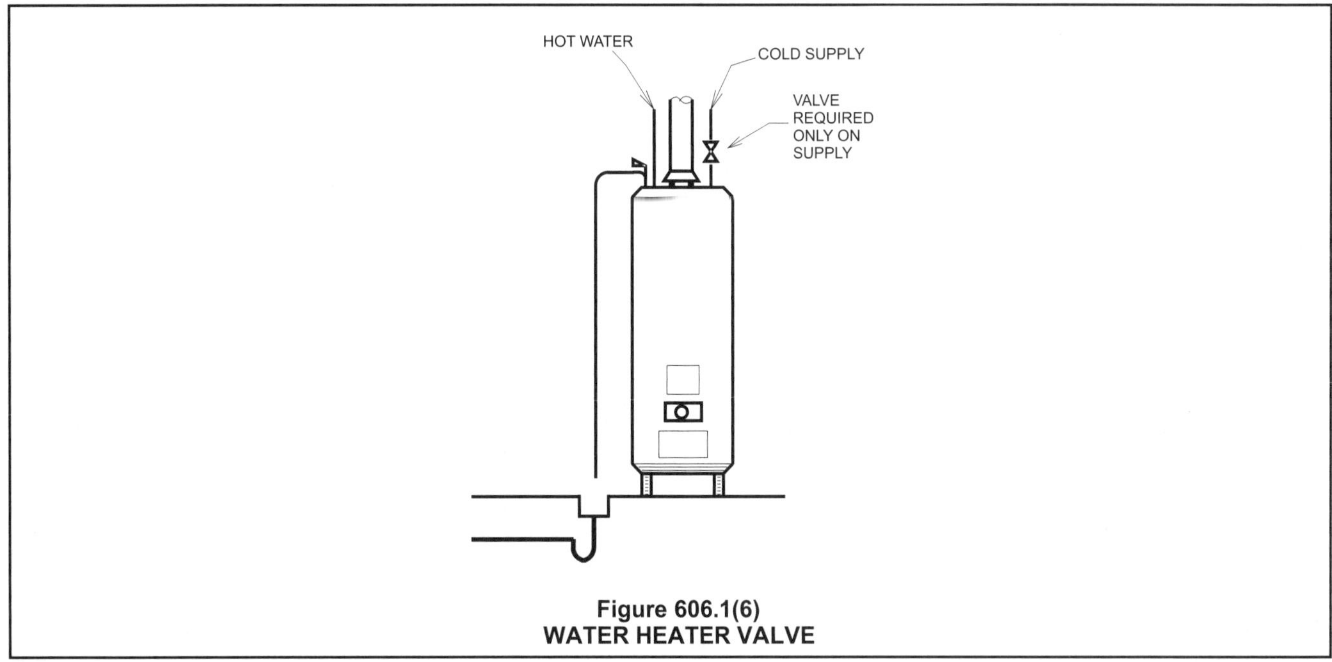

Figure 606.1(6)
WATER HEATER VALVE

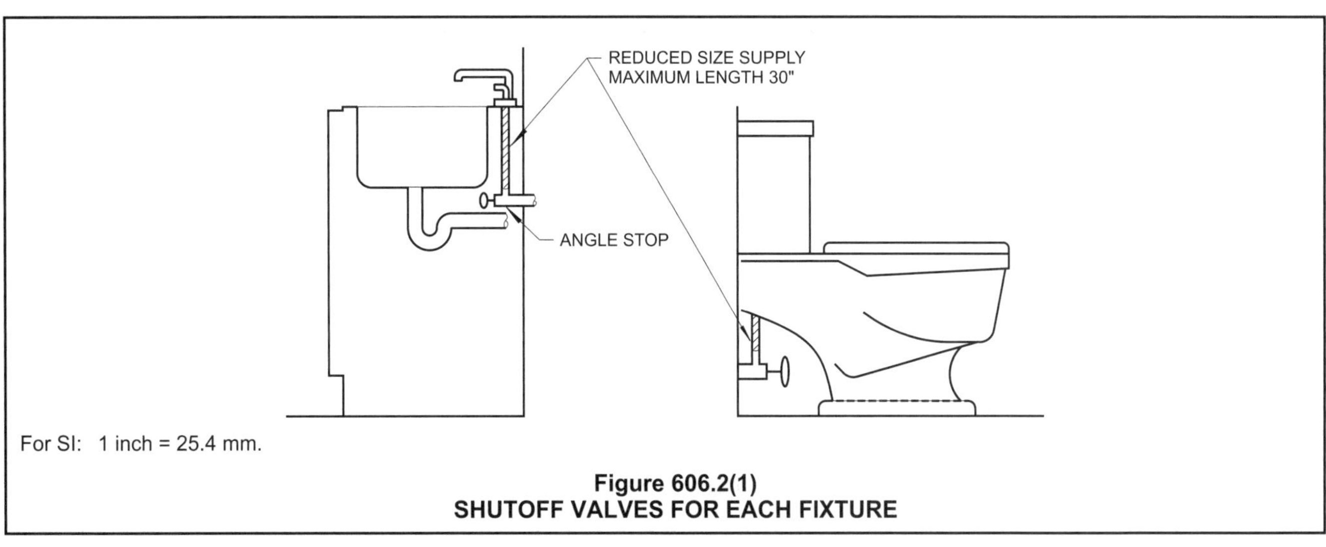

For SI: 1 inch = 25.4 mm.

Figure 606.2(1)
SHUTOFF VALVES FOR EACH FIXTURE

determined when considering the scenario of a damaged sill cock. Sill cocks are frequent locations for vandalism, abuse and freeze damage (even frostproof hose bibbs are susceptible to freezing when the hose is connected). Without shutoff valves at these locations, extended periods can lapse where the water service to the building is shut down until repairs can be made. It is therefore unreasonable to rely upon the main shutoff valve to isolate these outlets.

The prevailing argument against requiring individual shutoff valves in dwelling units has been that these devices are potential sources of leakage and maintenance, outweighing the benefits of their convenience.

606.3 Access to valves. *Access* shall be provided to all full-open valves and shutoff valves.

❖ To serve their intended purpose, full-open and shutoff valves must be able to be accessed for operation (see the definition of "Access" in Chapter 2).

606.4 Valve identification. Service and hose bibb valves shall be identified. All other valves installed in locations that are not adjacent to the fixture or appliance shall be identified, indicating the fixture or appliance served.

❖ In all cases, water service valves as required by Section 606.1 and shutoff valves serving sill faucets, yard hydrants and wall hydrants must have a tag or similar method of identification showing the valve's purpose. All other valves must be identified if the valve's purpose is not obvious because of the proximity of the fixture or appliance served by the valve (see Commentary Section 604.10.2). Identification is often helpful and sometimes critical in locating valves in an emergency in which quick action is necessary. Identification of isolation valves is even more important in dwelling units because individual fixture valves are not required.

606.5 Water pressure booster systems. Water pressure booster systems shall be provided as required by Sections 606.5.1 through 606.5.10.

❖ Requirements for water pressure booster systems are described in Sections 606.5.1 through 606.5.10.

606.5.1 Water pressure booster systems required. Where the water pressure in the public water main or individual water supply system is insufficient to supply the minimum pressures and quantities specified in this code, the supply shall be supplemented by an elevated water tank, a hydropneumatic pressure booster system or a water pressure booster pump installed in accordance with Section 606.5.5.

❖ When the pressure of the public system is inadequate for meeting the minimum requirements specified in Table 604.3, a pressure boosting system must be installed to increase the pressure. Such systems are generally required for multistory buildings of a height approaching that of the water towers that supply the public system. Some form of pumping equipment is necessary to increase the pressure. The pumping equipment may be used alone or in conjunction with a compression tank or an elevated tank. The use of elevated (gravity) tanks is declining in favor of sophisticated in-line pumping systems. The components of a booster pump system include a booster pump and control devices to maintain (one or more) preselected values of pressure.

606.5.2 Support. All water supply tanks shall be supported in accordance with the *International Building Code*.

❖ The weight of a full water tank must be considered when calculating the dead loads in the building's structural design.

606.5.3 Covers. All water supply tanks shall be covered to keep out unauthorized persons, dirt and vermin. The covers of gravity tanks shall be vented with a return bend vent pipe with an area not less than the area of the down-feed riser pipe, and the vent shall be screened with a corrosion-resistant screen of not less than 16 by 20 mesh per inch (630 by 787 mesh per m).

❖ Because the water tank contains potable water, it must be protected against contamination. The vent is necessary to prevent pressures from developing in the tank resulting from water being withdrawn or pumped into the tank and to allow for thermal expansion.

606.5.4 Overflows for water supply tanks. Each gravity or suction water supply tank shall be provided with an overflow

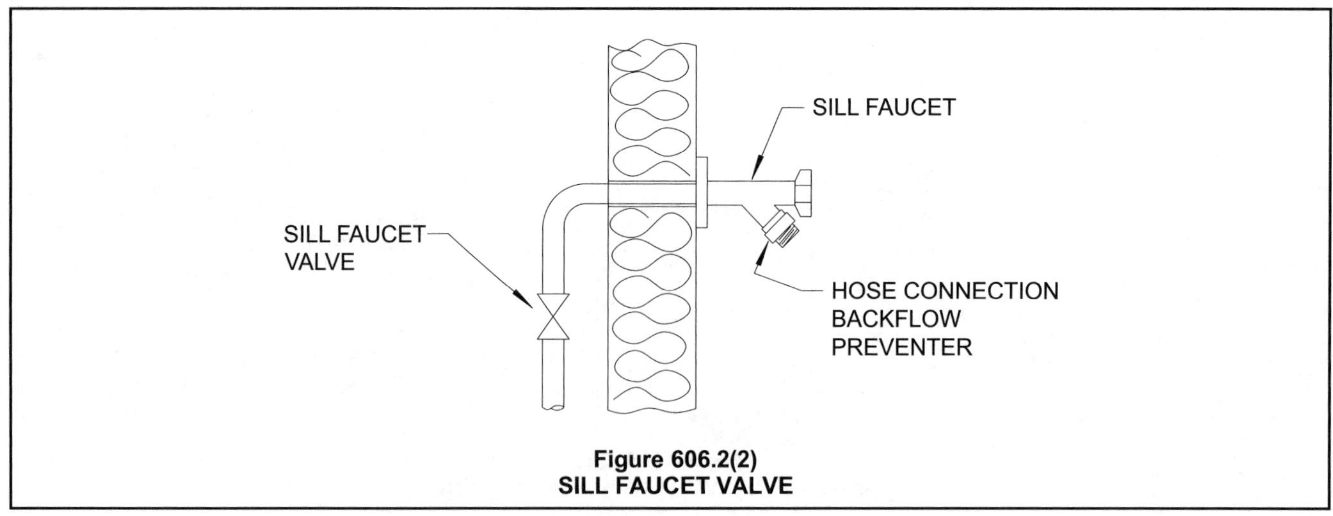

Figure 606.2(2)
SILL FAUCET VALVE

with a diameter not less than that shown in Table 606.5.4. The overflow outlet shall discharge at a point not less than 6 inches (152 mm) above the roof or roof drain; floor or floor drain; or over an open water-supplied fixture. The overflow outlet shall be covered with a corrosion-resistant screen of not less than 16 by 20 mesh per inch (630 by 787 mesh per m) and by $^1/_4$-inch (6.4 mm) hardware cloth or shall terminate in a horizontal angle seat check valve. Drainage from overflow pipes shall be directed so as not to freeze on roof walks.

❖ Each gravity tank must have an overflow that prevents the tank from being exposed to pressure. Such tanks are not designed to be pressurized. The overflow is installed as an indirect waste to protect the tank from backflow or backsiphonage and is sized using Table 606.5.4. The screen and swing check valve are intended to keep out insects and small animals, again to protect the potable water from contamination.

An elevated tank system is made up of the following components:

- A gravity tank that stores water at atmospheric pressure,
- Pumps that fill the tank by pumping water to the tank from the source,
- Controls that turn the pump on and off when the water inside the tank reaches a preset level,
- Alarms that alert operating personnel that a malfunction exists, and
- Safety devices that operate when a malfunction occurs, thus avoiding a potential accident.

TABLE 606.5.4
SIZES FOR OVERFLOW PIPES FOR WATER SUPPLY TANKS

MAXIMUM CAPACITY OF WATER SUPPLY LINE TO TANK (gpm)	DIAMETER OF OVERFLOW PIPE (inches)
0 - 50	2
50 - 150	$2^1/_2$
150 - 200	3
200 - 400	4
400 - 700	5
700 - 1,000	6
Over 1,000	8

For SI: 1 inch = 25.4 mm, 1 gallon per minute = 3.785 L/m.

❖ The minimum size of the overflow is based on discharging the water at a rate that is commensurate with the filling rate. Gravity tanks are not designed for pressurization.

606.5.5 Low-pressure cutoff required on booster pumps. A low-pressure cutoff shall be installed on all booster pumps in a water pressure booster system to prevent creation of a vacuum or negative pressure on the suction side of the pump when a positive pressure of 10 psi (68.94 kPa) or less occurs on the suction side of the pump.

❖ The creation of low pressures on the upstream (suction) side of pumps increases the possibility of backflow in the portion of the supply system that precedes the pumping equipment. Additionally, when there is inadequate pressure and volume of water on the suction side of the pump, cavitation can occur in the pump housing. Cavitation, which is the formation of partial vacuums within a flowing liquid (e.g., air bubbles), can rapidly deteriorate a pump positive suction head (NPSH). Requirements vary with impeller design and must be checked for each pump. A low-pressure cutoff prevents the pump from operating when the supply pressure at the pump inlet is 10 psi (69 kPa) or less.

606.5.6 Potable water inlet control and location. Potable water inlets to gravity tanks shall be controlled by a fill valve or other automatic supply valve installed so as to prevent the tank from overflowing. The inlet shall be terminated so as to provide an *air gap* not less than 4 inches (102 mm) above the overflow.

❖ Several types of devices, such as float switches, float valves and electrode controllers are available for controlling the flow of water into a tank. The water level in gravity tanks must be maintained by an automatic valve, usually a float-operated device such as a fill valve. An air gap must be installed in the potable water supply to the tank with the highest point of the tank overflow considered as the flood level rim for the purpose of measuring the required air gap.

606.5.7 Tank drain pipes. A valved pipe shall be provided at the lowest point of each tank to permit emptying of the tank. The tank drain pipe shall discharge as required for overflow pipes and shall not be smaller in size than specified in Table 606.5.7.

❖ A valve is required on every tank so that it may be emptied for service, repair or replacement of the tank.

TABLE 606.5.7
SIZE OF DRAIN PIPES FOR WATER TANKS

TANK CAPACITY (gallons)	DRAIN PIPE (inches)
Up to 750	1
751 to 1,500	$1^1/_2$
1,501 to 3,000	2
3,001 to 5,000	$2^1/_2$
5,000 to 7,500	3
Over 7,500	4

For SI: 1 inch = 25.4 mm, 1 gallon = 3.785 L.

❖ The valve at the base of the water tank is sized to empty the tank quickly.

606.5.8 Prohibited location of potable supply tanks. Potable water gravity tanks or manholes of potable water pressure tanks shall not be located directly under any soil or waste piping or any source of contamination.

❖ Any opening on a water tank must be protected against contamination. If drainage piping is located above a tank, there is a possibility of a leak from the piping or from a fixture served by such piping dripping onto the tank. Similar occurrences in the past have led to major outbreaks of disease.

606.5.9 Pressure tanks, vacuum relief. All water pressure tanks shall be provided with a vacuum relief valve at the top of the tank that will operate up to a maximum water pressure of 200 psi (1380 kPa) and up to a maximum temperature of 200°F (93°C). The minimum size of such vacuum relief valve shall be $^{1}/_{2}$ inch (12.7 mm).

> **Exception:** This section shall not apply to pressurized captive air diaphragm/bladder tanks.

❖ This section addresses pressure tanks that are part of a water pressure booster system. There is always the possibility of a pressure tank being subjected to a vacuum, especially elevated tanks supplying down-feed piping. Many tanks are not designed for a vacuum condition; therefore, a vacuum relief valve is necessary to prevent atmospheric pressure from collapsing or otherwise damaging the tank. A vacuum relief valve is a type of check valve having a shielded inlet.

606.5.10 Pressure relief for tanks. Every pressure tank in a hydropneumatic pressure booster system shall be protected with a pressure relief valve. The pressure relief valve shall be set at a maximum pressure equal to the rating of the tank. The relief valve shall be installed on the supply pipe to the tank or on the tank. The relief valve shall discharge by gravity to a safe place of disposal.

❖ Hydropneumatic tanks are rated for a given pressure. When connected in line with pumping equipment, there is a possibility of exceeding the pressure rating of the tank. The pressure relief valve installed on the tank prevents the buildup of excessive pressure, thereby preventing the possibility of tank rupture or failure.

606.6 Water supply system test. Upon completion of a section of or the entire water supply system, the system, or portion completed, shall be tested in accordance with Section 312.

❖ It is advantageous to conduct a water test before the system is concealed. Any leaks are easily repaired at this time without damaging or replacing existing materials. The test pressure required by Section 312.5 using water must be equal to or greater than the working pressure of the water supply system that will serve the building water service and water distribution system. When using air, the test pressure must not be less than 50 psi (344 kPa) (see Section 312.5).

SECTION 607
HOT WATER SUPPLY SYSTEM

607.1 Where required. In residential *occupancies*, *hot water* shall be supplied to all plumbing fixtures and equipment utilized for bathing, washing, culinary purposes, cleansing, laundry or building maintenance. In nonresidential *occupancies*, *hot water* shall be supplied for culinary purposes, cleansing, laundry or building maintenance purposes. In nonresidential *occupancies*, *hot water* or *tempered water* shall be supplied for bathing and washing purposes. *Tempered water* shall be supplied through a water temperature limiting device that conforms to ASSE 1070 and shall limit the *tempered water* to a maximum of 110°F (43°C). This provision shall not supersede the requirement for protective shower valves in accordance with Section 424.3.

❖ This section establishes the requirement for hot water and regulates the installation of hot water supply systems. The code defines hot water as water at a temperature of 110°F (43°C) or greater. Hot water is necessary for the convenience and the comfort of building occupants.

In nonresidential occupancies, tempered water [water ranging in temperature between 85°F and 110°F (29°C and 43°C) (see the definition of "Tempered water" in Section 202)] must be supplied at the hand-washing facilities (see Section 416.5). The device supplying the tempered water must conform to ASSE 1070 and must limit the temperature to 110°F (43°C). For example, in a restaurant, tempered water must be supplied to a hand-washing sink instead of hot water.

In all nonresidential locations other than at hand-washing facilities, in any occupancy, either hot or tempered water may be supplied. The use of an ASSE 1070 device does not negate the requirements for control valves referenced in Sections 424.3 and 424.4.

607.2 Hot water supply temperature maintenance. Where the *developed length* of hot water piping from the source of hot water supply to the farthest fixture exceeds 100 feet (30 480 mm), the hot water supply system shall be provided with a method of maintaining the temperature in accordance with the *International Energy Conservation Code.*

❖ A significant amount of water can be wasted by a fixture user while waiting for hot water to arrive at the fixture. In addition to being an inconvenience, the costs for the water, the energy to heat the water, and the disposal of the water are literally going down the drain. There will be less waste and inconvenience at the fixture where the source of hot water is located near to the fixture.

This code section requires that hot water be within 100 feet (30 480 mm) of developed pipe length of the fixture requiring hot water. Where the developed length of piping is greater than 100 feet (30 480 mm), the distance must be reduced to be not greater than 100 feet (30 480 mm). The following example illustrates the impact of this 100-foot (30 480 mm) maximum length. A 100-foot (30 480 mm) length of $^{1}/_{2}$ inch (12.7 mm) pipe contains approximately 1.2 gallons (4.5 liters) of water. A public lavatory is limited to a flow rate of 0.5 gpm (1.9 lpm). Therefore, if the entire 100-foot (30 480 mm) length of pipe is at ambient temperature, the fixture user would have to wait approximately 2.4 minutes for hot water to arrive at the fixture. Under these conditions, most public lavatory users would have given up long before hot water arrived. In reality, where fixtures are being used frequently and the hot water piping system is well insulated, the wait time (and water waste) is substantially reduced.

The code is silent on the methods that can be used to ensure that a source of hot water is within the required distance limit. For example, if a lavatory is located 125 feet (38 100 mm) of developed piping length from a water heater, several methods could be used to reduce the distance to not greater than 100 feet (30 480 mm): The most common methods are (1) locate the water heater closer to the fixture, (2) install a hot water recirculation system such that the piping of that system provides a connection point for hot water that is closer to the fixture or (3) install a pipe heating system on the piping beginning at the source of the hot water to a point that is closer to the fixture.

In small buildings where all the fixtures are clustered together in one location, it is often possible to locate a central water heater within the specified distance. Where a building is large and the fixtures requiring hot water are "spread out," multiple water heating units can be strategically located in service groups of fixtures. Single "point of use" water heaters are frequently installed to provide hot water to just one fixture. In larger buildings where all of the water heating appliances are located in a mechanical room, it is often impossible to locate the mechanical room within 100 feet (30 480 mm) of developed piping length to all fixtures requiring hot water. Where water heating units cannot be located near to the fixtures, pumped hot water recirculation piping systems are typically installed to convey the hot water to a point near to all fixtures.

A hot water recirculating system circulates hot water from the water heater or hot water storage tank through a loop of piping that returns the water to the water heater or hot water storage tank. A continuous flow of hot water in the piping loop maintains a specified water temperature at the point of connection for the piping to each fixture. Ideally, the location of the recirculation piping would pass in very close proximity to all fixtures requiring hot water. This code section only requires that the recirculation piping be within 100 feet (30 480 mm) of developed piping length to each fixture (see Commentary Figure 607.2). There are also "no-return pipe" hot water recirculation systems available.

Electric cable or steam tracing lines can also be used to maintain water temperatures in hot water piping.

Operating cost for the methods previously discussed is a concern for building designers, especially if they are designing for maximum energy efficiency. Recirculating pumps, heating cables and steam tracing all require energy. There is also the concern about heat loss from the piping which requires additional energy for maintaining the temperature of the hot water in the system. While the costs of that energy may be insignificant as compared to the costs of the wasted potable water and its disposal, the *International Energy Conservation Code®* (IECC®) regulates some aspects of hot water temperature maintenance systems.

The IECC does not prescribe methods for providing hot water to within 100 feet (30 480 mm) of developed pipe length of fixtures. The IECC does require that automatic-circulating hot water system pumps or heat trace be automatically de-energized or be configured to be conveniently manually de-energized during peri-

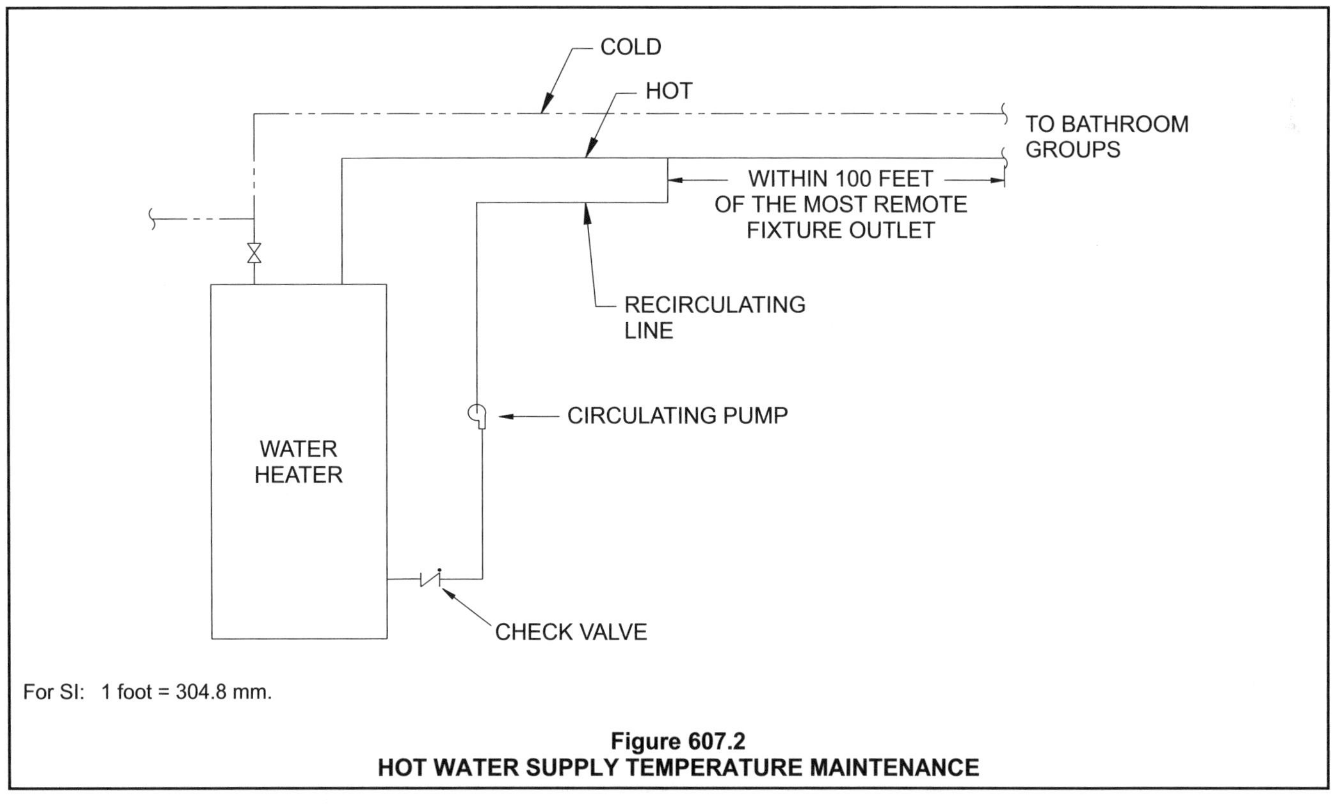

Figure 607.2
HOT WATER SUPPLY TEMPERATURE MAINTENANCE

ods of substantially reduced demand or no demand for hot water at the fixtures (see IPC Section 607.2.1). See Chapters 4 (for residential) and 5 (for commercial) of the IECC for further details.

607.2.1 Piping insulation. Circulating hot water system piping shall be insulated in accordance with the *International Energy Conservation Code.*

❖ Insulation of the hot water distribution system is required only when a return circulation system is installed. In circulating systems, the hot water is exposed to heat loss throughout the entire distribution system as long as water is circulating. Therefore, recirculating loops constructed of uninsulated piping can result in significant heat losses and energy waste. Even noncirculating systems require insulation on the first 8 feet (2438 mm) of outlet and inlet piping in commercial applications. See Chapters 4 (for residential) and 5 (for commercial) of the IECC for further details.

[E] 607.2.2 Hot water system controls. Automatic circulating hot water system pumps or heat trace shall be arranged to be conveniently turned off, automatically or manually, when the hot water system is not in operation.

❖ Energy is required to operate pumps on a circulating line as well as heat tracing systems. Because hot water circulation is not required at all times in most buildings, there must be a means of shutting off the circulating pump to save energy.

607.2.3 Recirculating pump. Where a thermostatic mixing valve is used in a system with a hot water recirculating pump, the *hot water* or *tempered water* return line shall be routed to the cold water inlet pipe of the water heater and the cold water inlet pipe or the hot water return connection of the thermostatic mixing valve.

❖ Where hot water systems have a master thermostatic mixing valve and a circulating pump downstream of the mixing valve to maintain temperatures at remote areas of the system, the hot water or tempered water return line from the mixing valve must be split and routed to the cold water inlet pipe of the water heater and the cold water inlet of the mixing valve or hot water return connection of the mixing valve [see Commentary Figure 607.2.3(1)]. If the hot water return pipe is routed only to the water heater [see Commentary Figure 607.2.3(2)], the mixed water system temperature will rise when there is no fixture use in the system. Where the hot water return pipe is routed only to the mixing valve [see Commentary Figure 607.2.3(3)], the mixed water system temperature will drop to ambient temperature when there is no fixture use in the system.

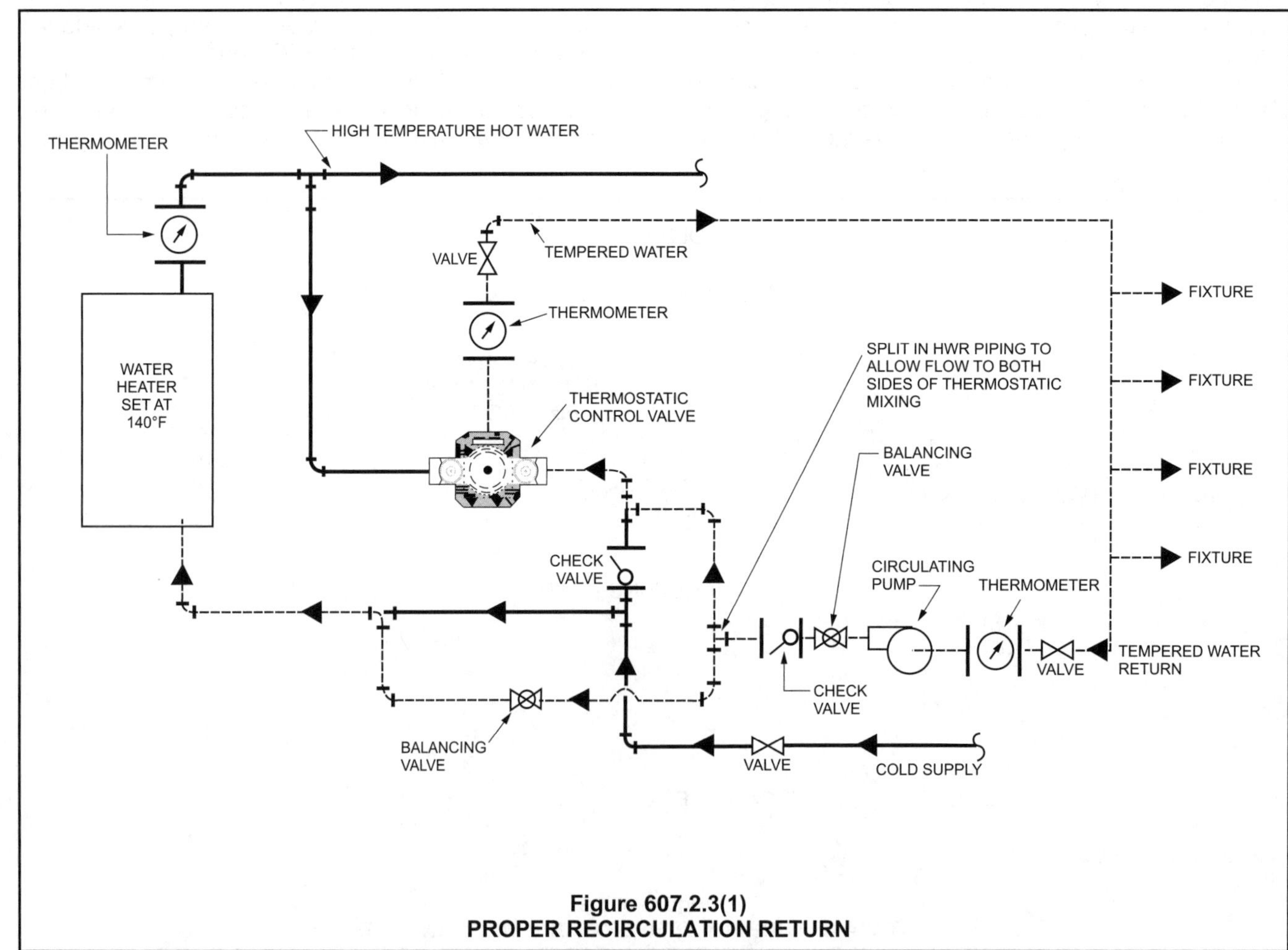

Figure 607.2.3(1)
PROPER RECIRCULATION RETURN

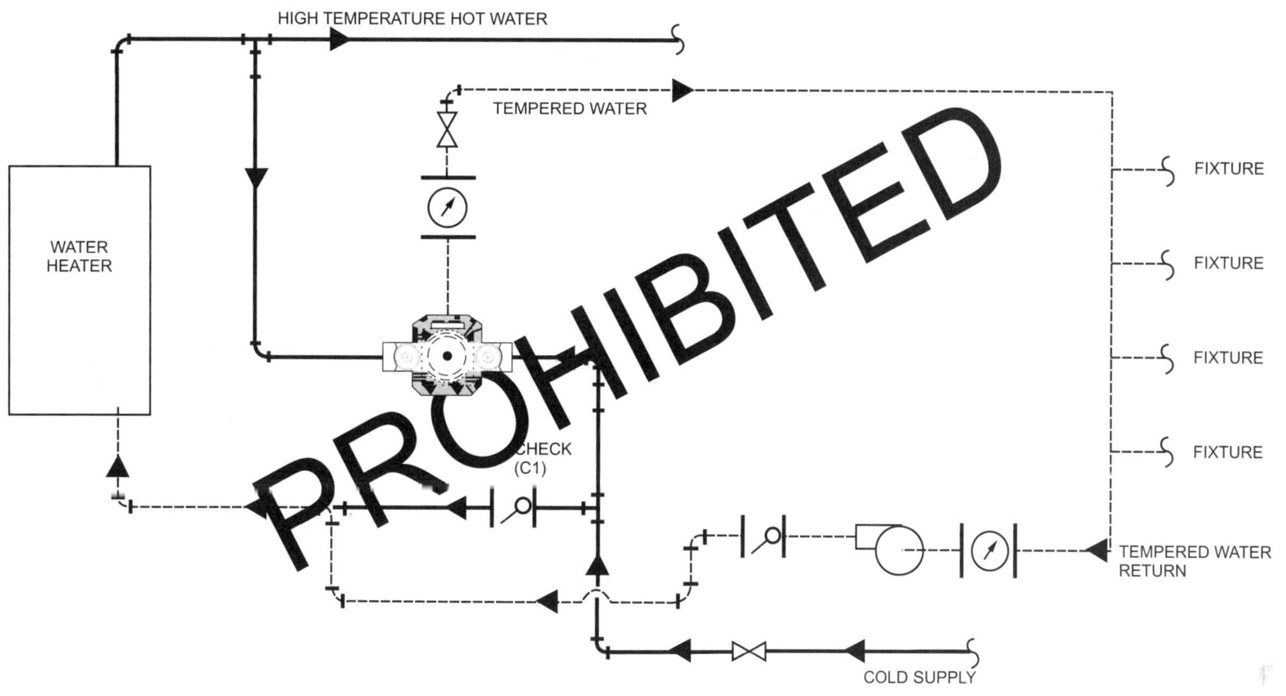

HIGH TEMPERATURE HOT WATER
TEMPERED WATER
WATER HEATER
FIXTURE
FIXTURE
FIXTURE
FIXTURE
CHECK (C1)
TEMPERED WATER RETURN
COLD SUPPLY
PROHIBITED

Figure 607.2.3(2)
PROHIBITED RECIRCULATION RETURN

HIGH TEMPERATURE HOT WATER
TEMPERED WATER
WATER HEATER
FIXTURE
FIXTURE
FIXTURE
FIXTURE
TEMPERED WATER RETURN
COLD SUPPLY
PROHIBITED

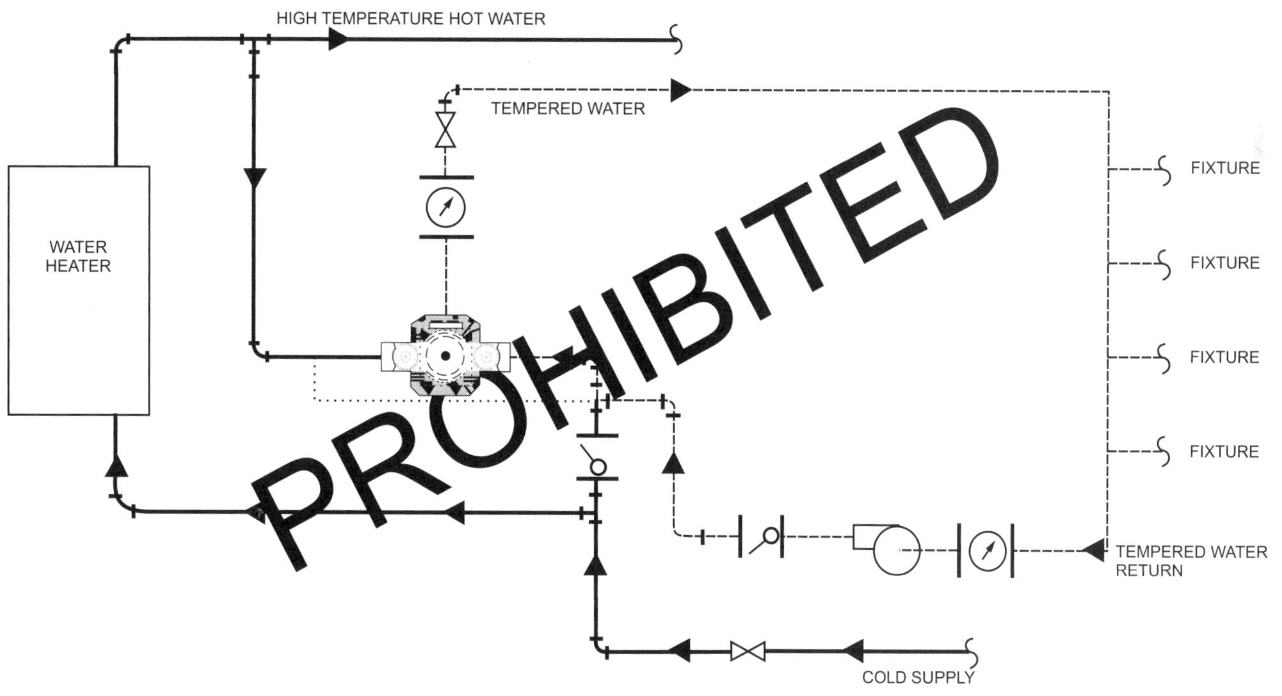

Figure 607.2.3(3)
PROHIBITED RECIRCULATION RETURN

607.3 Thermal expansion control. A means of controlling increased pressure caused by thermal expansion shall be provided where required in accordance with Sections 607.3.1 and 607.3.2.

❖ Liquids expand in volume when heated. If the liquid being heated completely fills a vessel of fixed volume, the pressure in the vessel will increase in proportion to the rise in temperature of the liquid. As long as sufficient heat is applied to cause a rise in liquid temperature, the pressure in the vessel will continue to rise until liquid is allowed to escape from the vessel (relieving the pressure) or the vessel ruptures due to overpressurization. Because a hot water distribution system is necessarily connected to a water heating device, there is the potential for thermal expansion to cause a rise in system pressure. The intent of this section is to require the control of any pressure increase caused by thermal expansion of water being heated.

Uncontrolled pressure increases due to thermal expansion can have a detrimental effect on all components in a water distribution system. Faucets, typically designed for long life operation at pressures not greater that 80 psig (552 kPa), will wear out rapidly and leak at higher pressures. Other system components, such as supply shutoff valves and flexible water supply connectors, may leak or fail at pressures exceeding 125 psig (862 kPa). Leaks might develop in water distribution piping that was tested to only the "working pressure" of the system. Water heater tank welds may fatigue and fail due to frequent pressure fluctuations caused by uncontrolled thermal expansion.

See the commentaries for Sections 607.3.1 and 607.3.2 for explanations of the circumstances that require control of pressure increases caused by thermal expansion of water being heated in a closed system.

607.3.1 Pressure-reducing valve. For water service system sizes up to and including 2 inches (51 mm), a device for controlling pressure shall be installed where, because of thermal expansion, the pressure on the downstream side of a pressure-reducing valve exceeds the pressure-reducing valve setting.

❖ Prior to the 2003 edition, this section required that pressure-reducing valves be fitted with an integral (pressure) bypass check valve (or other device) to prevent pressure (caused by thermal expansion of water being heated) downstream of the valve from exceeding the main supply (upstream) pressure. For example, if a 100 psi (689 kPa) main water supply line was fitted with a pressure-reducing valve set at 60 psi (413 kPa), the pressure downstream of the pressure-reducing valve was allowed to increase up to 100 psi (689 kPa) before the integral bypass check valve would open to relieve water (and the pressure) to the upstream side of the valve. While an integral bypass check valve (or other devices) did "control" thermal expansion to the extent that the downstream pressure could never become higher than the upstream pressure at the valve, it was realized that this was not enough thermal expansion "control" in all circumstances. For example, if the main supply (upstream) pressure was 200 psig (1378 kPa), the pressure downstream of the reducing valve could rise to 200 psi (1378 kPa) before the integral (pressure) bypass valve (or other device) would open. But because water heaters are required to have pressure relief safety valves set to not higher than 150 psi (1033 kPa), a thermal expansion-induced pressure increase over 150 psi (1033 kPa) would be relieved by the water heater T + P valve. Repeated opening of water heater T + P valves can cause the valves to become worn, corroded and ineffective (a safety concern) or leak constantly, de-

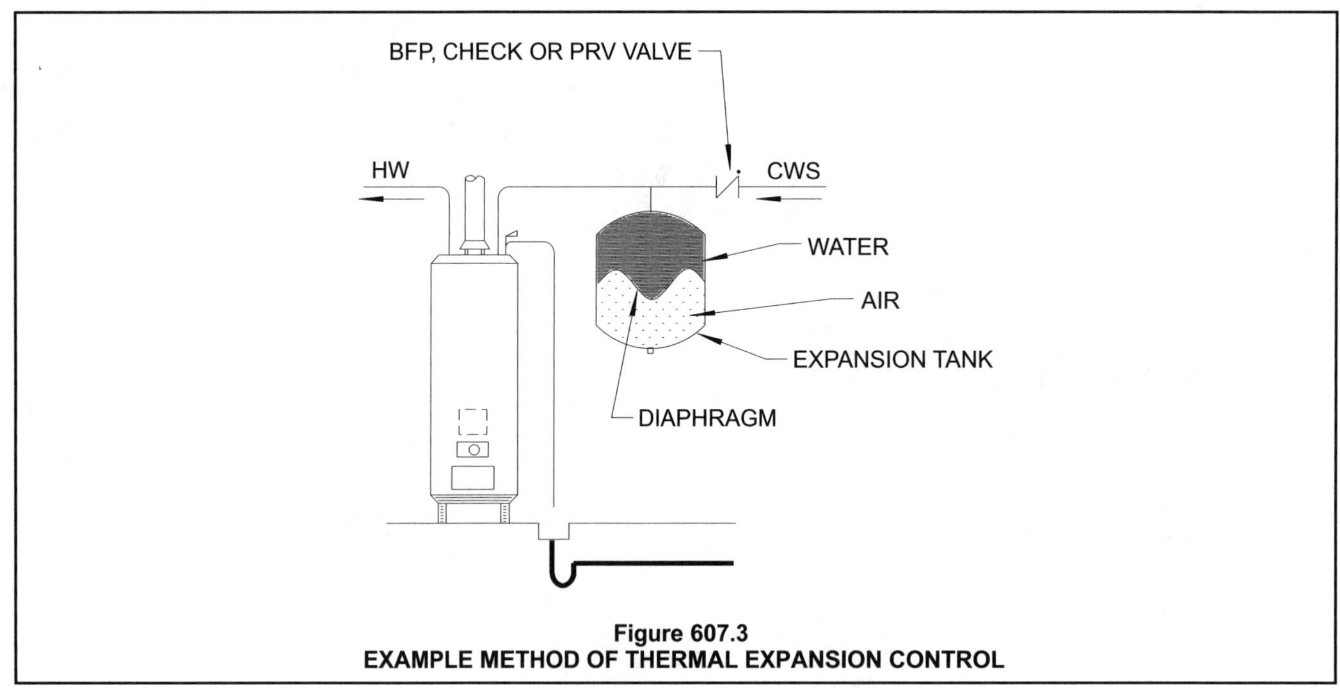

Figure 607.3
EXAMPLE METHOD OF THERMAL EXPANSION CONTROL

pleting available hot water as well as wasting significant quantities of water.

Beginning with the 2003 edition of the code, an attempt was made to change the intent of this section from a pressure-reducing valve (PRV) requirement to a requirement that the pressure downstream of a pressure-reducing valve be not greater than the valve setting. This change eliminated any benefit to having an integral (pressure) bypass check valve on a pressure-reducing valve and, thus, the last sentence of this section in the 2000 code (requiring an integral bypass or other device on a pressure-reducing valve) was eliminated. Considering that the purpose for installing a PRV was that the water main pressure exceeded the maximum allowable pressure of 80 psi (551 kPa), the bypass offers no protection against excess pressures above 80 psi (551 kPa). The change did not remove the text, "For water service system sizes up to and including 2 inches (51 mm)," as this was original text that reflected the maximum size of PRVs (available at that time) that had an integral bypass device. Regardless of the size of the incoming water service line, any pressure increases due to the thermal expansion of water being heated must be controlled to prevent the pressure downstream of a pressure-reducing valve from becoming greater than the setting of the pressure-reducing valve.

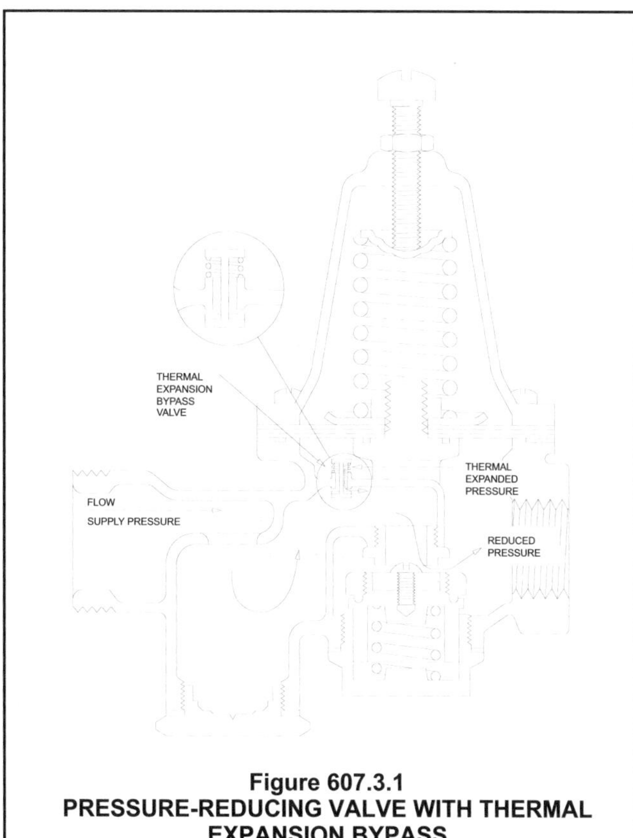

Figure 607.3.1
PRESSURE-REDUCING VALVE WITH THERMAL EXPANSION BYPASS

607.3.2 Backflow prevention device or check valve. Where a backflow prevention device, check valve or other device is installed on a water supply system utilizing storage water heating equipment such that thermal expansion causes an increase in pressure, a device for controlling pressure shall be installed.

❖ This section has the same intent as Section 607.3.1 but identifies devices other than pressure-reducing valves that could create a "closed system" where an increase in pressure could occur due to the thermal expansion of water being heated. This begs the question: "Is there any water distribution system where control of pressure due to thermal expansion is not required?" The following examples illustrate the few cases where pressure control is not required or necessary:

Case A: Consider a building where the main water supply line pressure is always below 80 psig (therefore, no pressure reducing valve is installed) and there are no backflow preventers, check valves or other one-way flow devices in the system or in the connections to the main water supply line (the public water main). When water in the water distribution system is heated, the increased volume is easily absorbed by the massive volume of the public water main and thus, no pressure increase in the water distribution system can be detected. Since there is not a pressure increase, pressure control is not required.

Case B: Consider a building where a private water source (e.g., a well system with an accumulator tank) supplies the water distribution system and no one-way flow devices (check valves) are between the accumulator tank and the water distribution system. When water in the water distribution system is heated, the increased volume is easily absorbed by the relatively large volume of the accumulator tank (also known as a pressure tank) and thus, there is no pressure increase in the water distribution system. Therefore, no thermal expansion pressure control is required.

With respect to Case A, low pressure (80 psig or less) public water mains are becoming a rarity as fire service demands, new building construction demands and better water main pipe materials all contribute toward new and replacement public water mains being designed for pressures much higher than 80 psig. Therefore, most buildings will require a pressure-reducing valve and, thus, a device to control pressures due to thermal expansion of water being heated is required.

Regardless of the public water main pressure (and the need for a pressure-reducing valve), it is becoming commonplace for water purveyors to install a backflow prevention device at the water meter or to require the building owner to provide a backflow prevention device in the water service piping to the building. The water purveyor requires these backflow prevention devices in order to comply with federal regulations and maintain certification as a safe drinking water provider. As such, most water providers have necessarily established backflow prevention programs in order to ensure that the public water system is protected from possible backflow contamination from buildings connected to the water system. At a minimum, most water purveyors install a dual check valve (in accordance with ASSE 1024 or CSA B64.6) on the upstream side of the water meter for a building. However, certain building occupancies such as restaurants and factories are required by the water purveyor to install backflow devices having a higher degree of protection. The installation of any backflow device will require the control of water distribution system pressures caused by thermal expansion of water being heated.

Note that existing buildings are typically not exempt from having to comply with a water purveyor's demands to install backflow protection. As water purveyors ramp up their backflow prevention programs, existing buildings (having no or limited backflow protection) are identified by their backflow hazard potential and a schedule for compliance is established. For example, the first year of a program might target all existing high-hazard factory and industrial occupancies. The next year might target all restaurants and smaller "medium hazard" commercial establishments. The following year might be all residential sprinkler systems. In some cases, periodic replacement of worn out, inaccurate, outdated water meters often triggers the automatic installation of a backflow prevention device (dual check valve) with the new water meter. While most water providers will inform the building owner, in writing, that a backflow device was installed with the new water meter, these written letters are often tossed aside without due respect for the need to install thermal expansion pressure control devices.

Beyond the requirements for backflow prevention devices, the installation of check valves (a one-way flow device) is often necessary for proper operation of a plumbing system. A check valve is sometimes installed in the cold supply pipe to a water heater to prevent hot water migration into the cold water distribution piping. Some hot water recirculation systems require the installation of check valves that could result in a closed system. Water pressure booster pumps might also have internal check valves that could create a closed system.

Note that this section identifies only "storage water heating equipment" as the generator of hot water. While this appears to imply that thermal expansion-induced pressure increases are only created by storage tank water heaters, this is not the case. A question that is frequently asked is, "Does a water distribution system with tankless water heater require control of pressure increases caused by thermal expansion?" The answer is provided in the following two system examples:

System A: Water distribution system with a tankless, on-demand water heater (electric or gas) without a storage tank. In this system, water is heated only when a hot water outlet is open. Because the water distribution system is now an "open" system (i.e., a hot water outlet is open), any thermal expansion is relieved through the open hot water outlet and no pressure increase can occur. When the outlet is closed, the system becomes a closed system but, since the flow has ceased, the tankless water heater stops heating the water. Therefore, methods for controlling pressure increases due to thermal expansion are not required.

System B: Water distribution system with tankless, on-demand water heater (electric or gas) with a storage tank. The water service line has a backflow preventer (such as a dual check valve at the water meter). In this system, an unfired hot water storage tank may be required to provide for hot water recirculation to lessen the "wait time" for hot water to arrive at any fixture. Or the storage tank might be required to provide for large simultaneous hot water demands (that the tankless water heater cannot produce without the storage). For either reason, as the hot water storage tank cools off, a circulation pump is switched on to cause flow out of the storage tank, through the tankless water heater and back into the storage tank until the water in the storage tank is at an acceptable temperature level. Because there will be times where no hot water outlet is opened during the time that the tankless water heater is heating water, the water distribution system is a closed system. Therefore, a means for controlling pressure increases due to the thermal expansion must be provided.

607.4 Flow of hot water to fixtures. Fixture fittings, faucets and diverters shall be installed and adjusted so that the flow of hot water from the fittings corresponds to the left-hand side of the fixture fitting.

> **Exception:** Shower and tub/shower mixing valves conforming to ASSE 1016 or ASME A112.18.1/CSA B125.1, where the flow of hot water corresponds to the markings on the device.

❖ One of the oldest expressions in plumbing is: "Hot on the left, cold on the right." The code mandates that "hot" correspond to the left side of the fixture fitting for safety reasons. It has become an accepted practice to equate the left side of the faucet with hot water. The intent is to protect an individual from potential scalding when turning on what is believed to be cold water and get scalded by hot water.

Some manufacturers design a fixture fitting that can be reversed inside the faucet if a plumber inadvertently (or intentionally in the case of back-to-back faucets) pipes the water supplies in reverse. This type of installation would be acceptable. The exception provides for single-handle shower valves that comply with ASSE 1016 or CSA B125 and that have identifiable hot designations on the control valve. These devices operate by rotating a single handle or knob away from the "off" position to initiate water flow. The handle is moved farther in the same direction to increase the temperature of flowing water. Because the rotating operation is limited to one direction, the user cannot mistakenly activate water flow at the maximum hot-water setting without first rotating the handle or knob through the cold-water setting. Also, ASSE 1016 requires control valves to have settings that indicate clearly the direction or means of adjustment to change temperature.

SECTION 608
PROTECTION OF POTABLE WATER SUPPLY

608.1 General. A potable water supply system shall be designed, installed and maintained in such a manner so as to prevent contamination from nonpotable liquids, solids or gases being introduced into the potable water supply through cross-connections or any other piping connections to the system. Backflow preventer applications shall conform to Table 608.1, except as specifically stated in Sections 608.2 through 608.16.10.

❖ This section contains the requirements for protecting the potable water supply from contamination. The most important aspect of a plumbing code is the protection of the potable water system. History is filled with local and widespread occurrences of sickness and disease caused by not safeguarding the water supply. It is imperative that the potable water supply be maintained in a safe-for-drinking condition at all times and at all actual or potential connections and outlets.

To understand the importance of protecting the potable water supply, identifying potential threats and determining the appropriate backflow prevention method, an understanding of how backflow prevention can occur is necessary.

Backflow depends on atmospheric pressure, water pressures and variations in the water supply at any moment. The following is a discussion on pressure and backflow.

Pressure

Atmospheric pressure (psi): This is pressure exerted by the weight of the atmosphere above the earth. Atmospheric pressure at sea level is 14.7 psi (101 kPa); however, it varies with altitude [see Commentary Figure 608.1(1)].

Gauge pressure (psig): The pressure read on a gauge. The reference point for the gauge is atmo-

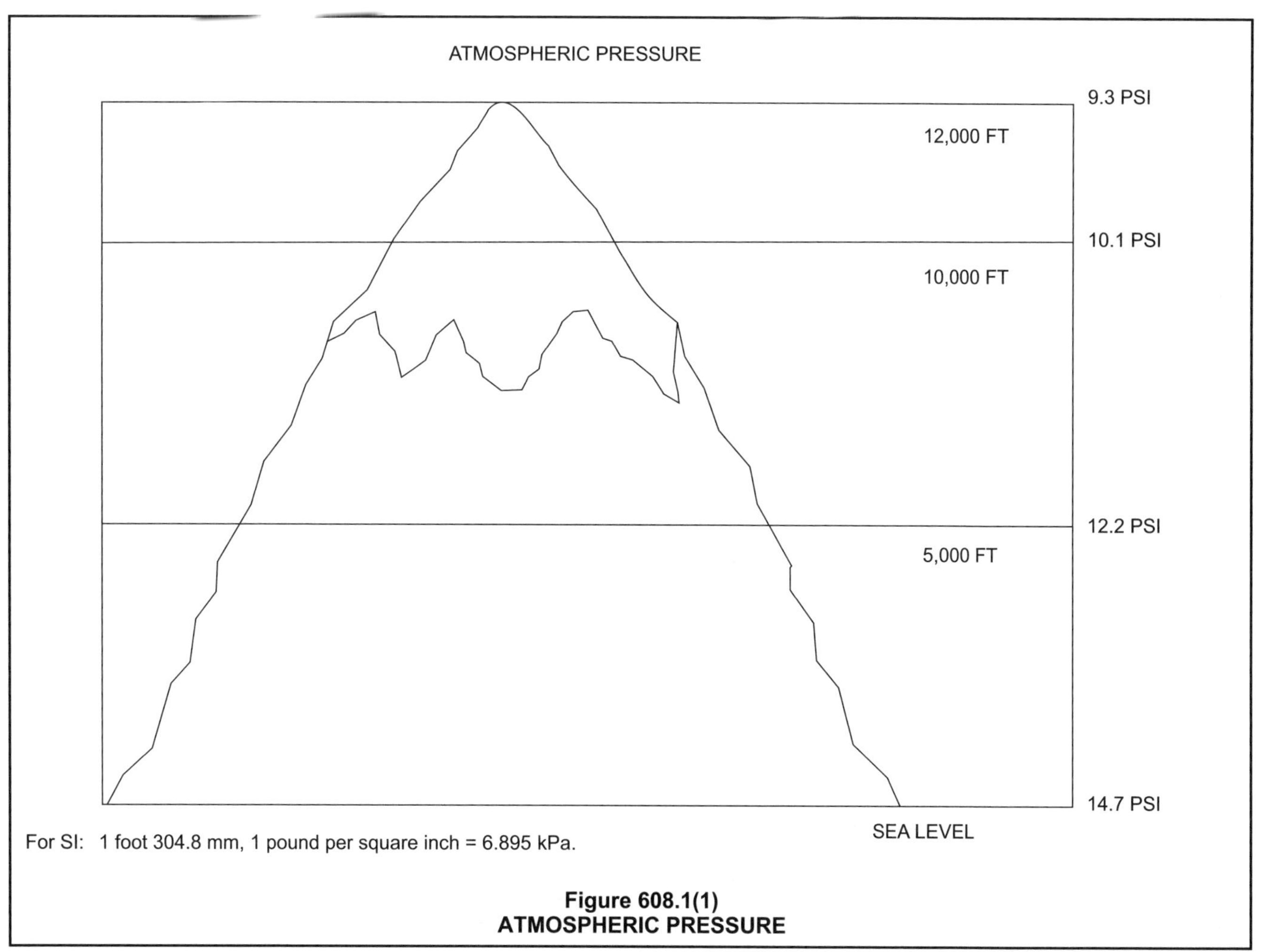

For SI: 1 foot 304.8 mm, 1 pound per square inch = 6.895 kPa.

Figure 608.1(1)
ATMOSPHERIC PRESSURE

spheric pressure equal to zero. In other words, gauge pressure is the amount of pressure above or below atmospheric pressure.

Absolute pressure (psia): The sum of atmospheric pressure and gauge pressure. For example, if a gauge reads 20 psig (138 kPa), the absolute pressure would be determined as follows:

$P_{ABSOLUTE}$ = 14.7 psi (atmospheric) + 20 psi (gauge reading)

$P_{ABSOLUTE}$ = 34.7 psia

For SI: 1 pound per square inch = 6.895 kPa.

Water pressure: This is a function of the weight and height of a column of water. A cubic foot of water [see Commentary Figure 608.1(2)] weighs 62.4 pounds (28.3 kg); the pressure exerted at the base of the cube is 62.4 pounds per square foot (psf) (0.433 lb/in.^{2}). Therefore, the height of a column of water determines the pressure that is exerted at its base [see Commentary Figure 608.1(3)].

The diameter of the pipe or vessel has no bearing on this pressure. For example, a column of water 100 feet (30 480 mm) high and 20 feet (6096 mm) in diameter has the same pressure at its base as a 100-foot-high (30 480 mm) column of water, 1 inch (25 mm) in diameter [see Commentary Figure 608.1(4)]. Columns of water with a free surface (exposed to atmosphere) will have atmospheric pressure in addition to pressure caused by the weight of water exerted at its base [see Commentary Figure 608.1(5)].

Base pressure = 14.7 psi + 43.3 psig = 58.0 psia

Water pressure is sometimes referred to as "head pressure," "head" or "feet of head."

1 foot of head = 0.433 psig
100 feet of head = 43.3 psig

For SI: 1 pound per square inch = 6.895 kPa.

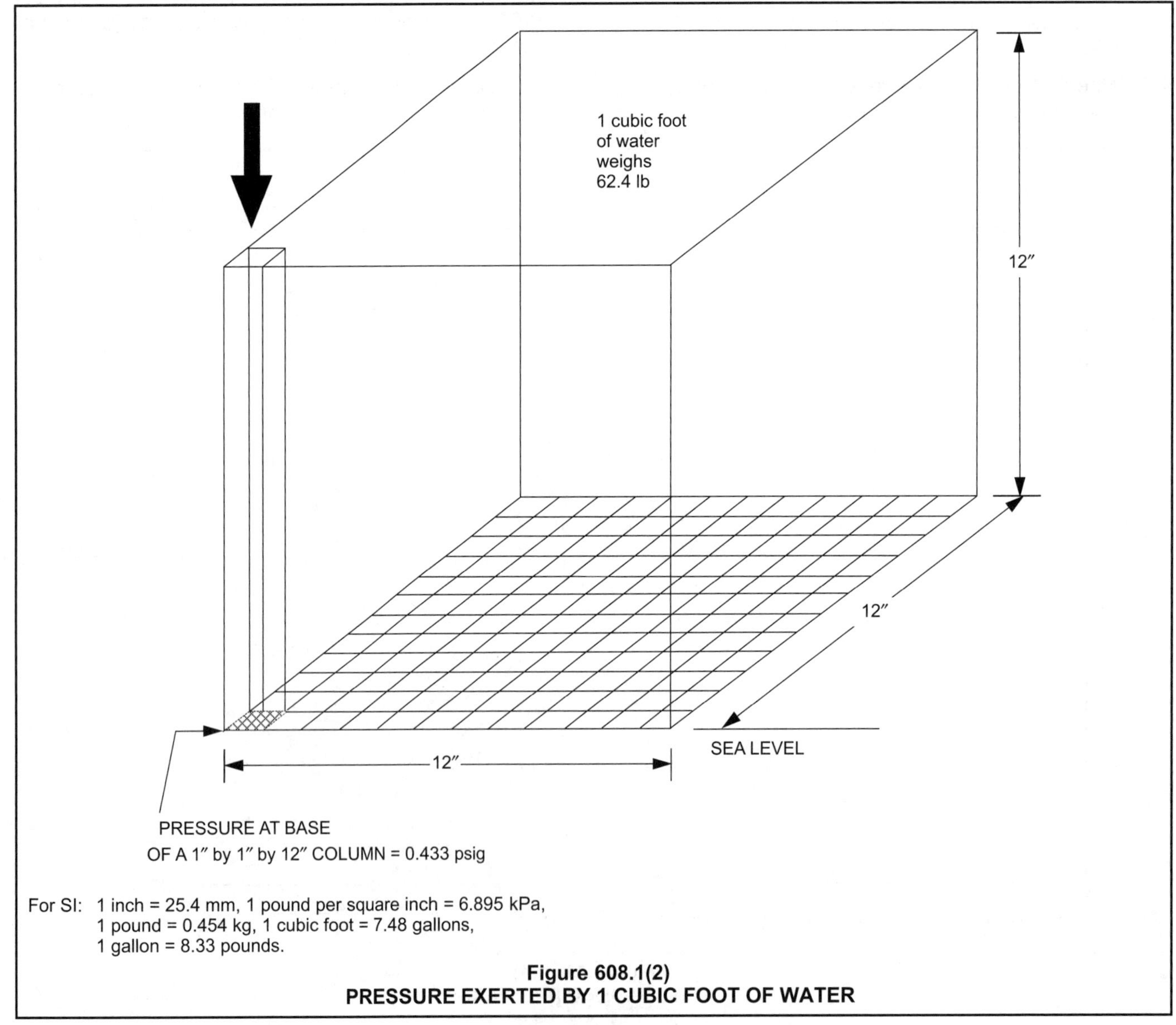

For SI: 1 inch = 25.4 mm, 1 pound per square inch = 6.895 kPa, 1 pound = 0.454 kg, 1 cubic foot = 7.48 gallons, 1 gallon = 8.33 pounds.

Figure 608.1(2)
PRESSURE EXERTED BY 1 CUBIC FOOT OF WATER

Backflow

Backflow is the flow of liquids in potable water distribution piping in reverse of their intended path. There are two types of pressure conditions that cause backflow: backsiphonage and backpressure.

Backsiphonage: This the backflow of water caused by system pressure falling below atmospheric pressure. Atmosphere supplies the force that reverses flow. Any gauge reading of pressure below atmospheric pressure will be negative. Commentary Figures 608.1(6) and 608.1(7) illustrate the effect atmospheric pressure has on water. An open tube is inserted into a container of water in Commentary Figure 608.1(6). Atmospheric pressure acts equally on the surface of the water within the tube and on the outside of the tube.

If, as shown in Commentary Figure 608.1(7), the tube is sealed and a vacuum pump is used to evacuate all the air from the tube, a vacuum with a pressure of zero psia (0 kPa) is created within the tube.

Because liquids will flow toward the point of lowest pressure, water will rise to 33.9 feet (10 333 mm). In other words, the weight of atmosphere at sea level balances the weight of a column of water 33.9 feet (10 333 mm) in height.

Backsiphonage occurs in the water distribution system when supply pressure falls below atmospheric pressure. The following are practical examples of conditions that can cause backsiphonage.

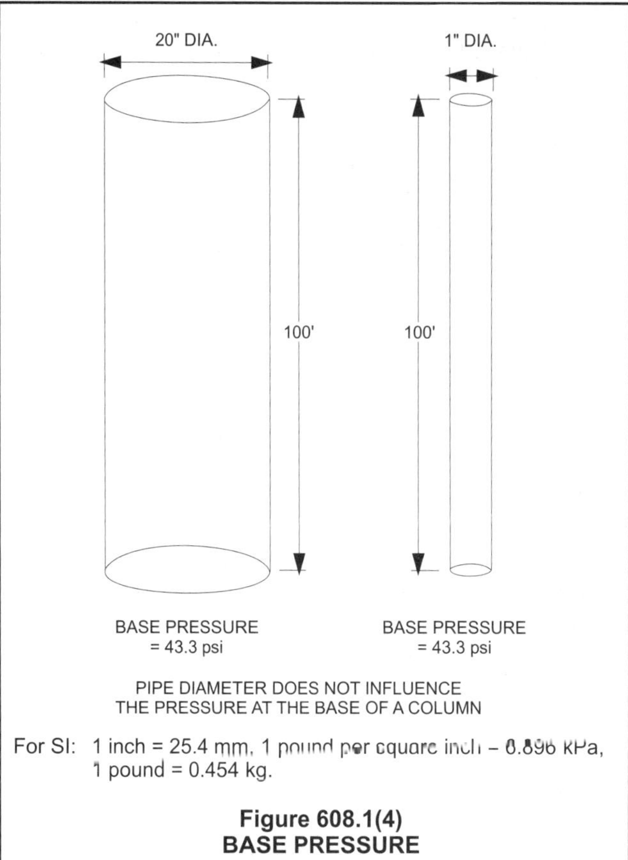

For SI: 1 inch = 25.4 mm, 1 pound per square inch = 6.896 kPa, 1 pound = 0.454 kg.

Figure 608.1(4)
BASE PRESSURE

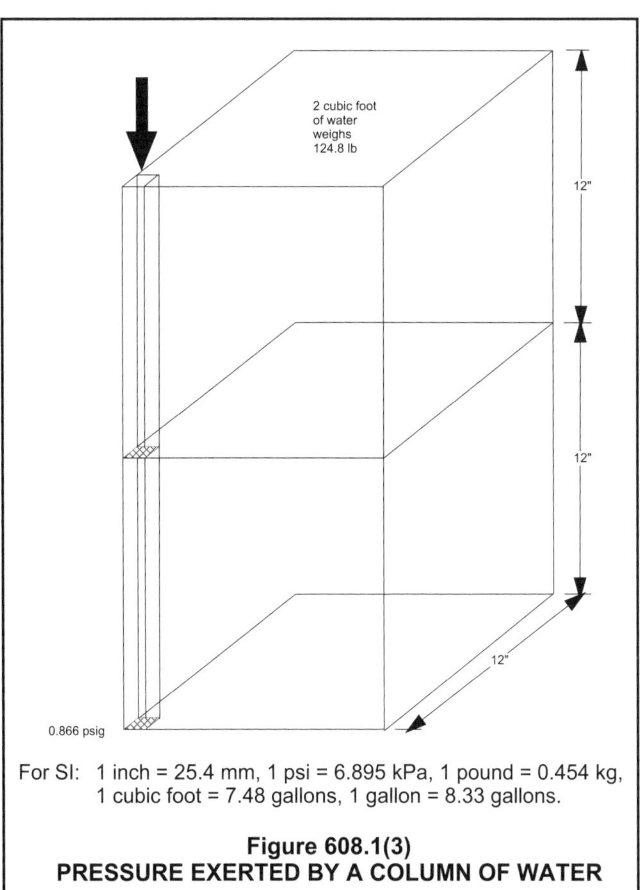

For SI: 1 inch = 25.4 mm, 1 psi = 6.895 kPa, 1 pound = 0.454 kg, 1 cubic foot = 7.48 gallons, 1 gallon = 8.33 gallons.

Figure 608.1(3)
PRESSURE EXERTED BY A COLUMN OF WATER

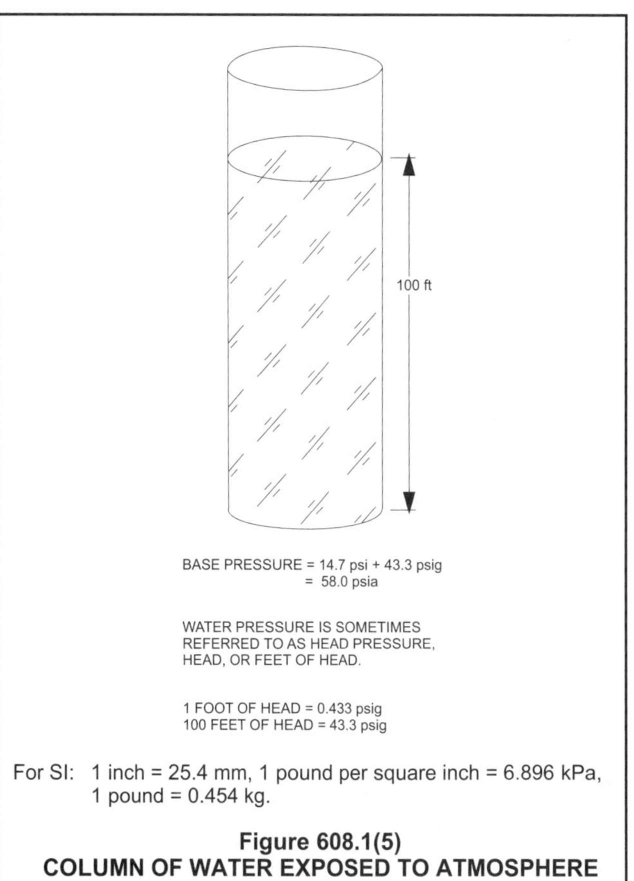

For SI: 1 inch = 25.4 mm, 1 pound per square inch = 6.896 kPa, 1 pound = 0.454 kg.

Figure 608.1(5)
COLUMN OF WATER EXPOSED TO ATMOSPHERE

A simple siphon occurs when pressure is reduced by a difference in water levels at two separate points within a continuous fluid system [see Commentary Figure 608.1(8)].

Aspiration or venturi principle is achieved when reduced pressure is created within a fluid system as a result of fluid motion. Intentional or unintentional constriction of flow through an opening or pipe increases velocity and decreases pressure. If this point of reduction is connected to a source of pollution, backsiphonage of the pollutant can occur [see Commentary Figure 608.1(9)].

Supply pressure can be reduced at the pump intake or suction side. Insufficient water pressure or undersized piping supplying the pump can cause a negative pressure in the system, allowing contaminated water to be drawn into the system [see Commentary Figure 608.1(10)].

Backpressure: This is the pressure created in a nonpotable system in excess of the water supply mains causing backflow. Backpressure can be created by mechanical means, such as a pump; by static head pressure including an elevated tank or by thermal expansion from a heat source, such as a water heater.

TABLE 608.1. See page 6-58.

❖ This table is to be used to assist in the determination of which type of backflow prevention devices should be

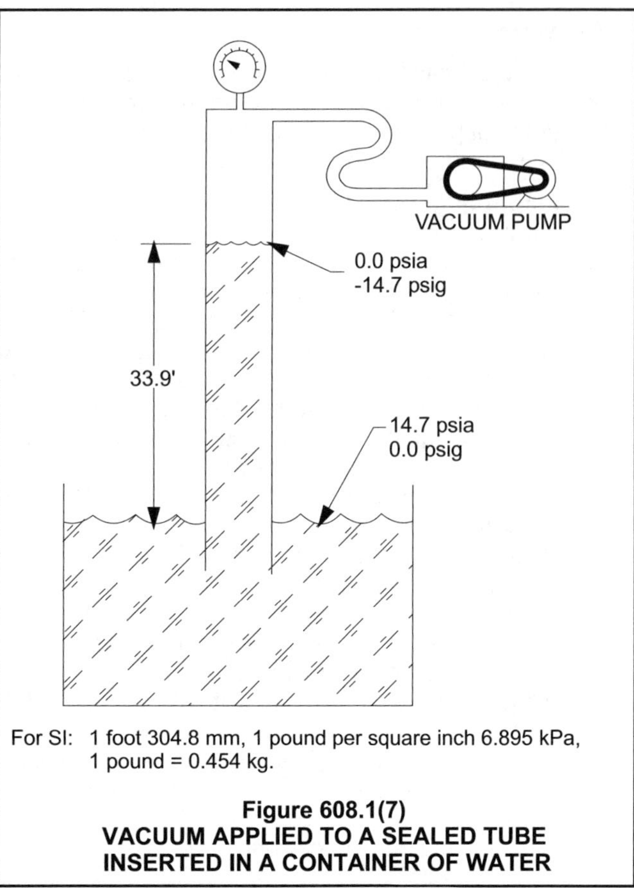

For SI: 1 foot 304.8 mm, 1 pound per square inch 6.895 kPa, 1 pound = 0.454 kg.

Figure 608.1(7)
VACUUM APPLIED TO A SEALED TUBE INSERTED IN A CONTAINER OF WATER

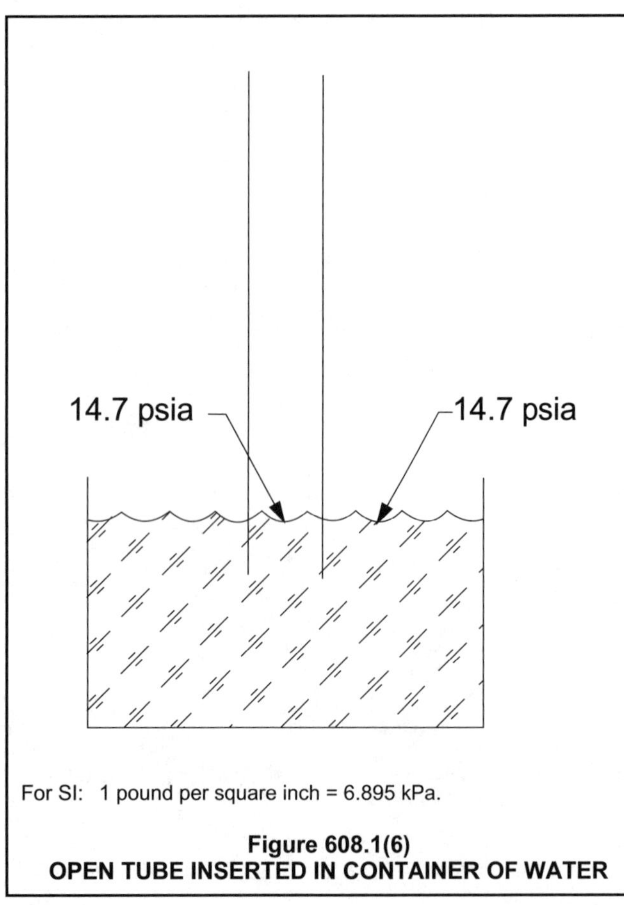

For SI: 1 pound per square inch = 6.895 kPa.

Figure 608.1(6)
OPEN TUBE INSERTED IN CONTAINER OF WATER

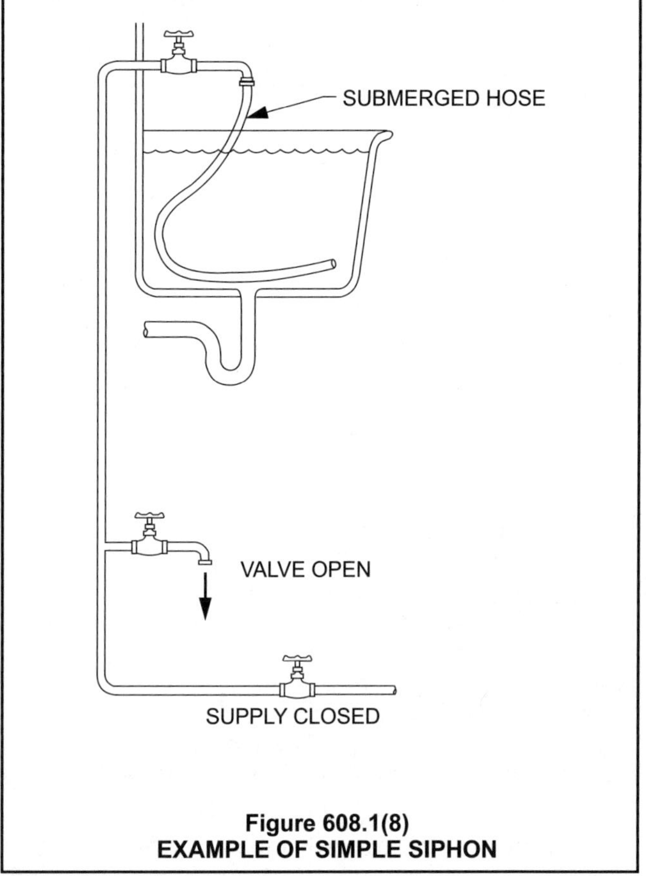

Figure 608.1(8)
EXAMPLE OF SIMPLE SIPHON

installed in a water distribution system. The "Degree of Hazard" is the level of risk to which the water distribution system is subjected, and the "Application" lists the type of backflow for which the device is rated. This table contains general information about various types of backflow preventers. There are instances throughout Section 608 where specific provisions override the general information in Table 608.1. For example, Section 608.16.6 states that potable water connections subject to backpressure must be protected by a reduced pressure principle backflow preventer. However, Table 608.1 lists several devices rated for backpressure. Therefore, to avoid the apparent conflict, the rule of thumb is the "specific" code provision overrides the general classifications given in the table.

With the 2009 edition of this code, the CSA B64.3.1 standard for backflow preventers for carbonated beverage machines was removed from the table. See commentary for Section 608.16.1.

608.2 Plumbing fixtures. The supply lines and fittings for every plumbing fixture shall be installed so as to prevent backflow. Plumbing fixture fittings shall provide backflow protection in accordance with ASME A112.18.1.

❖ To prevent the potable water supply system from being contaminated, plumbing fixtures must be installed in a manner that will prevent backflow. Backflow devices must be in accordance with ASME A112.18.1.

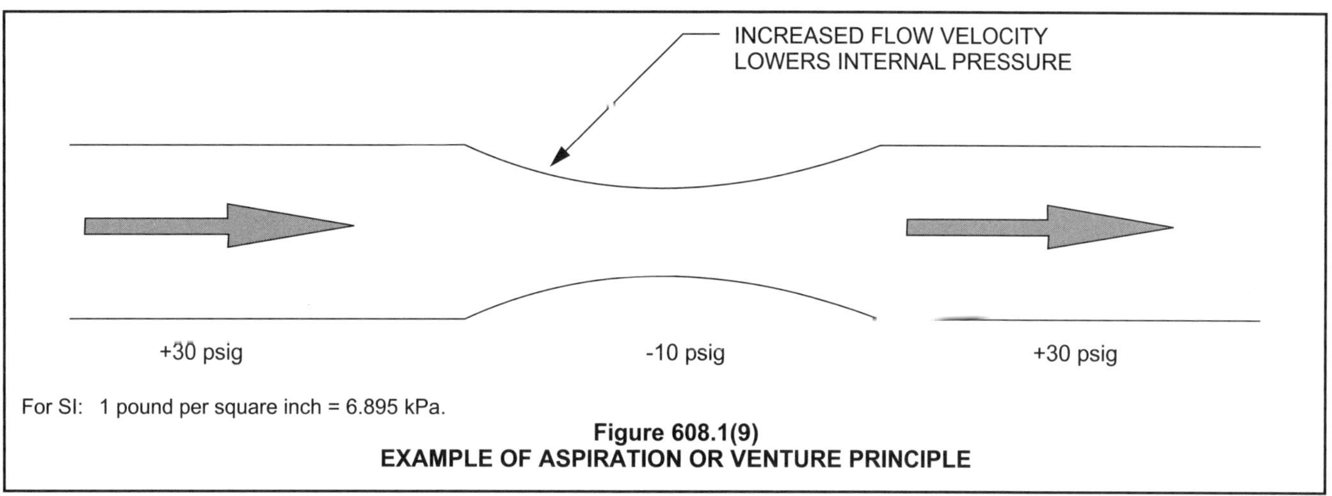

For SI: 1 pound per square inch = 6.895 kPa.

Figure 608.1(9)
EXAMPLE OF ASPIRATION OR VENTURE PRINCIPLE

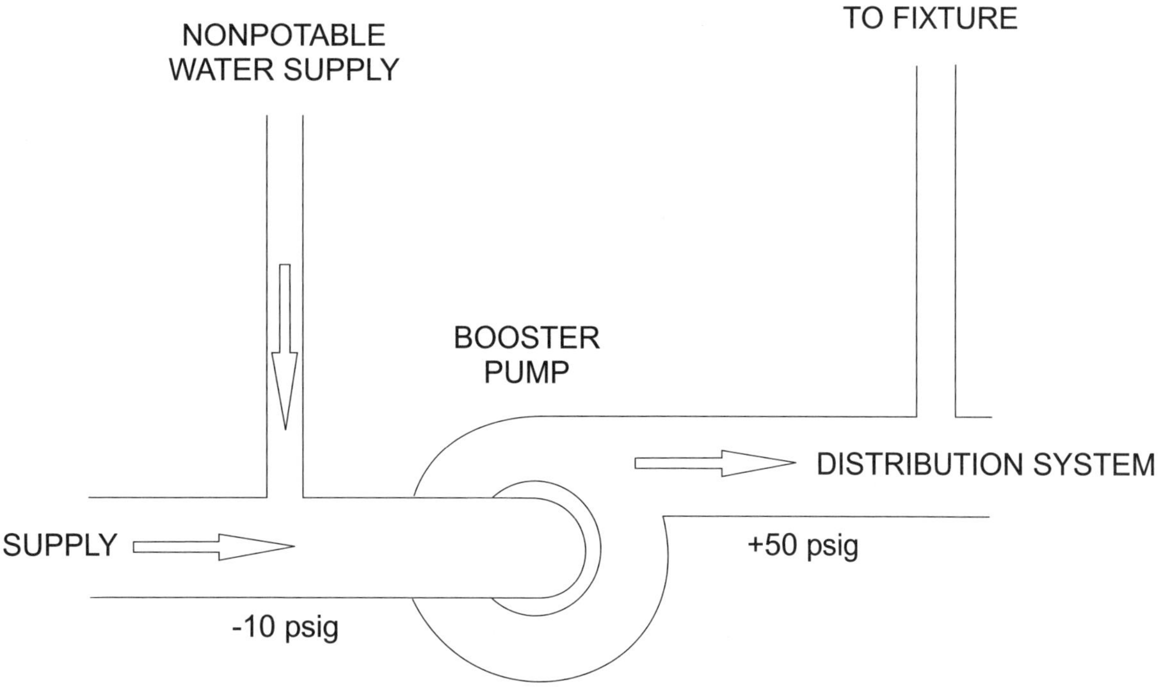

NONPOTABLE WATER SUPPLY

TO FIXTURE

BOOSTER PUMP

DISTRIBUTION SYSTEM

SUPPLY

+50 psig

-10 psig

For SI: 1 pound per square inch = 6.895 kPa.

Figure 608.1(10)
EXAMPLE OF PUMP CAUSING NEGATIVE PRESSURE

TABLE 608.1
APPLICATION OF BACKFLOW PREVENTERS

DEVICE	DEGREE OF HAZARD[a]	APPLICATION[b]	APPLICABLE STANDARDS
Air gap	High or low hazard	Backsiphonage or backpressure	ASME A112.1.2
Air gap fittings for use with plumbing fixtures, appliances and appurtenances	High or low hazard	Backsiphonage or backpressure	ASME A112.1.3
Antisiphon-type fill valves for gravity water closet flush tanks	High hazard	Backsiphonage only	ASSE 1002, CSA B125.3
Backflow preventer for carbonated beverage machines	Low hazard	Backpressure or backsiphonage Sizes $^1/_4$″- $^3/_8$″	ASSE 1022
Backflow preventer with intermediate atmospheric vents	Low hazard	Backpressure or backsiphonage Sizes $^1/_4$″ - $^3/_4$″	ASSE 1012, CAN/CSA B64.3
Barometric loop	High or low hazard	Backsiphonage only	(See Section 608.13.4)
Double check backflow prevention assembly and double check fire protection backflow prevention assembly	Low hazard	Backpressure or backsiphonage Sizes $^3/_8$″ - 16″	ASSE 1015, AWWA C510, CSA B64.5, CSA B64.5.1
Double check detector fire protection backflow prevention assemblies	Low hazard	Backpressure or backsiphonage (Fire sprinkler systems) Sizes 2″ - 16″	ASSE 1048
Dual-check-valve-type backflow preventer	Low hazard	Backpressure or backsiphonage Sizes $^1/_4$″ - 1″	ASSE 1024, CSA B64.6
Hose connection backflow preventer	High or low hazard	Low head backpressure, rated working pressure, backpressure or backsiphonage Sizes $^1/_2$″-1″	ASSE 1052, CSA B64.2.1.1
Hose connection vacuum breaker	High or low hazard	Low head backpressure or backsiphonage Sizes $^1/_2$″, $^3/_4$″, 1″	ASSE 1011, CAN/CSA B64.2, CSA B64.2.1
Laboratory faucet backflow preventer	High or low hazard	Low head backpressure and backsiphonage	ASSE 1035, CSA B64.7
Pipe-applied atmospheric-type vacuum breaker	High or low hazard	Backsiphonage only Sizes $^1/_4$″ - 4″	ASSE 1001, CAN/CSA B64.1.1
Pressure vacuum breaker assembly	High or low hazard	Backsiphonage only Sizes $^1/_2$″ - 2″	ASSE 1020, CSA B64.1.2
Reduced pressure principle backflow preventer and reduced pressure principle fire protection backflow preventer	High or low hazard	Backpressure or backsiphonage Sizes $^3/_8$″- 16″	ASSE 1013, AWWA C511, CAN/CSA B64.4, CSA B64.4.1
Reduced pressure detector fire protection backflow prevention assemblies	High or low hazard	Backsiphonage or backpressure (Fire sprinkler systems)	ASSE 1047
Spillproof vacuum breaker	High or low hazard	Backsiphonage only Sizes $^1/_4$″-2″	ASSE 1056
Vacuum breaker wall hydrants, frost-resistant, automatic draining type	High or low hazard	Low head backpressure or backsiphonage Sizes $^3/_4$″, 1″	ASSE 1019, CAN/CSA B64.2.2

For SI: 1 inch = 25.4 mm.

a. Low hazard–See Pollution (Section 202).
High hazard–See Contamination (Section 202).

b. See Backpressure (Section 202).
See Backpressure, low head (Section 202).
See Backsiphonage (Section 202).

608.3 Devices, appurtenances, appliances and apparatus. All devices, appurtenances, appliances and apparatus intended to serve some special function, such as sterilization, distillation, processing, cooling, or storage of ice or foods, and that connect to the water supply system, shall be provided with protection against backflow and contamination of the water supply system. Water pumps, filters, softeners, tanks and all other appliances and devices that handle or treat potable water shall be protected against contamination.

❖ This section requires backflow protection for fixtures, appliances and equipment that serve some special function and are connected to the potable water supply system. The specialty equipment being regulated is typically associated with sanitation, food service and health care facilities. This section combines the intents of Sections 608, 609 and 802.

Generally, one backflow prevention device is required for each fixture, appliance and appurtenance. The water downstream of a backflow prevention device is considered to be potable water until a backflow event occurs to cause the water to be nonpotable. Where a single fixture, piece of equipment or individual appliance is connected downstream of the backflow preventer, a malfunction of that appurtenance that causes the water to become nonpotable only affects the appurtenance causing the contamination. If multiple fixtures, appliances or appliances are supplied with water from a single backflow prevention device, the water contamination caused by one fixture could cause the other fixture to be supplied with nonpotable water. Thus, a backflow protection device is required for each item to prevent "cross contamination" between items.

Only where all of the connected fixtures do not require potable water could one backflow preventer serve all of those items. One example is a warehouse having multiple standard frostproof yard hydrants to supply water for floor and lot cleaning. Because a standard frostproof yard hydrant has a stop-and-waste valve below grade, the potable water supply to the hydrant must be protected from possible contamination from any one of those yard hydrants. One backflow preventer could serve this multiple hydrant system as long as the piping identification requirements of Section 608.8 are met (noting that the riser of the yard hydrant is part of the piping). Another example might be a series of machines requiring water (but not necessarily potable water) for cooling purposes. One backflow preventer could supply water for this equipment line. Again, note the piping identification requirements of Section 608.8.

608.3.1 Special equipment, water supply protection. The water supply for hospital fixtures shall be protected against backflow with a reduced pressure principle backflow preventer, an atmospheric or spill-proof vacuum breaker, or an *air gap*. Vacuum breakers for bedpan washer hoses shall not be located less than 5 feet (1524 mm) above the floor. Vacuum breakers for hose connections in health care or laboratory areas shall not be less than 6 feet (1829 mm) above the floor.

❖ This section establishes the minimum level of protection required for the water supply to various health care fixtures. In health care facilities, the risk of contamination to the potable water supply is greater because of the biological hazards present in many of the procedures conducted. Therefore, to reduce the potential for contamination, procedures or processes that use a hose must have their vacuum breakers installed at an elevation higher than that where the procedure is taking place.

608.4 Water service piping. Water service piping shall be protected in accordance with Sections 603.2 and 603.2.1.

❖ To reduce the risk of contaminating the potable water supply, water service piping must be installed in accordance with Sections 603.2 and 603.2.1.

608.5 Chemicals and other substances. Chemicals and other substances that produce either toxic conditions, taste, odor or discoloration in a potable water system shall not be introduced into, or utilized in, such systems.

❖ The prohibition of chemicals, regardless of hazard classification, includes the use of substances that clean the inside of the piping system or that de-lime various components, all of which could affect the potability of the water and therefore must never be added to the system. The generalization can be made that nothing should be introduced into a potable water supply after it leaves the water treatment facility.

608.6 Cross-connection control. Cross connections shall be prohibited, except where *approved* protective devices are installed.

❖ A cross connection is an attachment or potential attachment of the potable water supply to any other source or system where it is possible to contaminate the potable water supply with nonpotable water, industrial fluid, gas or other substances. Cross connections to a nonpotable source or system can include the following:

- Bypass arrangements,
- Jumper connections,
- Removable sections,
- Swivel or changeover devices,
- Direct connections and
- Submerged inlets.

Where a cross connection is unavoidable, the water supply must be protected by an approved means of backflow prevention. The intent of this section is to prevent cross connections from occurring except where such connection is absolutely necessary. Cross connections are inherently hazardous and must be avoided where reasonably possible.

608.6.1 Private water supplies. Cross connections between a private water supply and a potable public supply shall be prohibited.

❖ A private water supply (auxiliary water supply) is a supply or source not under control or management of a public water purveyor. These sources typically include wells, springs and surface supply sources, such as lakes and reservoirs. These sources are subject to contamination by industrial or animal wastes, or other natural or man-made materials (see Commentary Figure 608.6.1).

608.7 Valves and outlets prohibited below grade. Potable water outlets and combination stop-and-waste valves shall not be installed underground or below grade. Freezeproof yard hydrants that drain the riser into the ground are considered to be stop-and-waste valves.

Exception: Freezeproof yard hydrants that drain the riser into the ground shall be permitted to be installed, provided that the potable water supply to such hydrants is protected upstream of the hydrants in accordance with Section 608 and the hydrants are permanently identified as nonpotable outlets by *approved* signage that reads as follows: "Caution, Nonpotable Water. Do Not Drink."

❖ Any fixture or device that incorporates a stop-and-waste feature is prohibited if the waste opening is underground or in any location where waste water or water-borne contaminants could enter the device or water supply from the ground or other source by reversal of flow.

Many water utilities routinely use stop-and-waste valves on curb stops and fire hydrants. Another common location for stop-and-waste valves is in yard hydrants equipped with a below-grade self-draining weep hole [see Commentary Figures 608.7(1) and 608.7(2)].

Sanitary yard hydrants are available that protect against ground-water contamination by eliminating the waste opening below grade. These devices use a sealed reservoir to collect water draining from the hydrant head and riser and then empty the reservoir upon opening of the valve the next time the hydrant is used.

The exception allows for installations of frostproof yard hydrants (having a combination stop-and-waste valve below grade) for applications where nonpotable water is acceptable for the use intended. Examples of

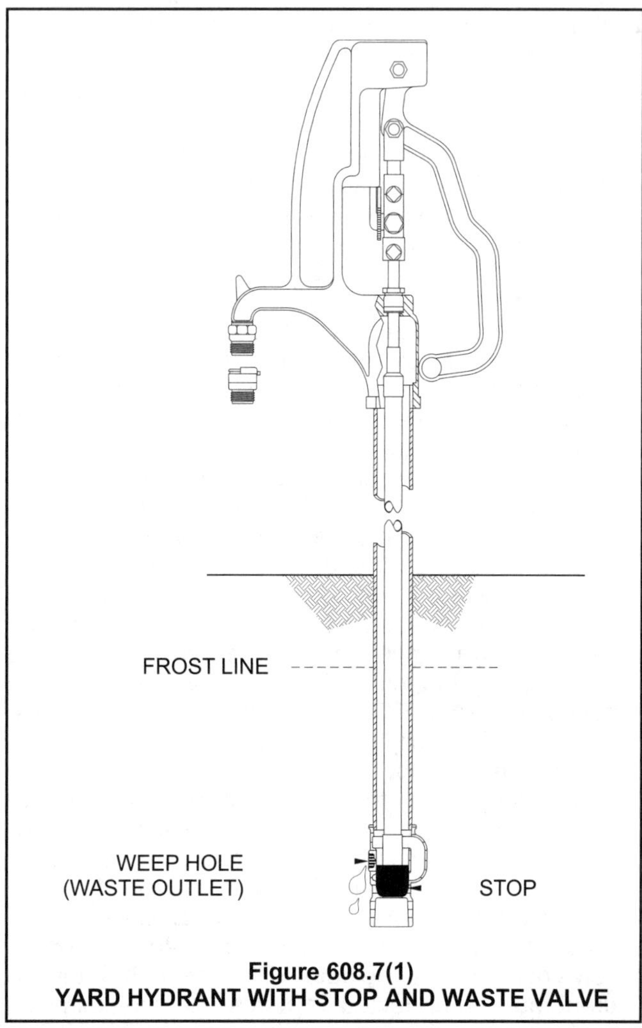

Figure 608.7(1)
YARD HYDRANT WITH STOP AND WASTE VALVE

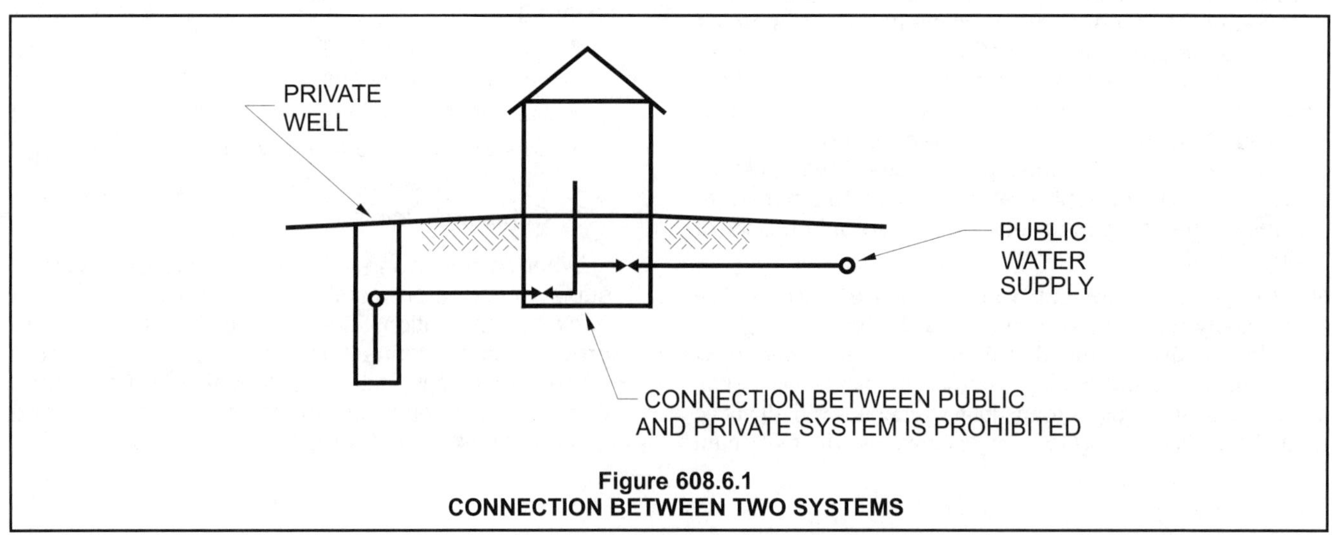

Figure 608.6.1
CONNECTION BETWEEN TWO SYSTEMS

nonpotable water use applications are hydrants used for farm area wash down, public park area cleaning and garbage area cleanup. Yard hydrants having stop-and-waste valves below grade must have the potable water supply supplying the hydrant protected with backflow protection as specified in Section 608. In addition, the hydrants must be marked permanently with a sign having the code-specified words. The sign must meet the approval of the code official.

608.8 Identification of nonpotable water. In buildings where nonpotable water systems are installed, the piping conveying the nonpotable water shall be identified either by color marking or metal tags in accordance with Sections 608.8.1 through 608.8.3. All nonpotable water outlets such as hose connections, open ended pipes, and faucets shall be identified at the point of use for each outlet with the words, "Nonpotable—not safe for drinking." The words shall be indelibly printed on a tag or sign constructed of corrosion-resistant waterproof material or shall be indelibly printed on the fixture. The letters of the words shall be not less than 0.5 inches in height and color in contrast to the background on which they are applied.

❖ Where nonpotable water supply systems are installed in a building, the piping of the nonpotable system must be identified so as to prevent inadvertent cross connection between different systems, especially the potable water system, during repairs or renovations. For example, a building could be provided with a recycled gray-water supply system that supplies nonpotable water for flushing urinals and water closets. Without identification of the recycled gray water piping, the piping system might be assumed to convey potable water since it is commonplace for urinals and water closets to be supplied with potable water. A connection to an unmarked nonpotable piping system for the purposes of obtaining potable water could lead to significant health risks for the building's occupants.

Some nonpotable water systems have outlets such as hose connections, open-ended pipes and faucets that could be mistaken for potable water outlets. Therefore, at these locations, the code requires signage to warn the building's occupants that water obtained from these outlets is not safe for drinking.

608.8.1 Information. Pipe identification shall include the contents of the piping system and an arrow indicating the direction of flow. Hazardous piping systems shall also contain information addressing the nature of the hazard. Pipe identification shall be repeated at maximum intervals of 25 feet (7620 mm) and at each point where the piping passes through a wall, floor or roof. Lettering shall be readily observable within the room or space where the piping is located.

❖ In addition to the piping identification indicating the contents and the direction of flow, any hazardous nature of the flow such as temperature, pressure, flammability and toxicity must also be indicated. The

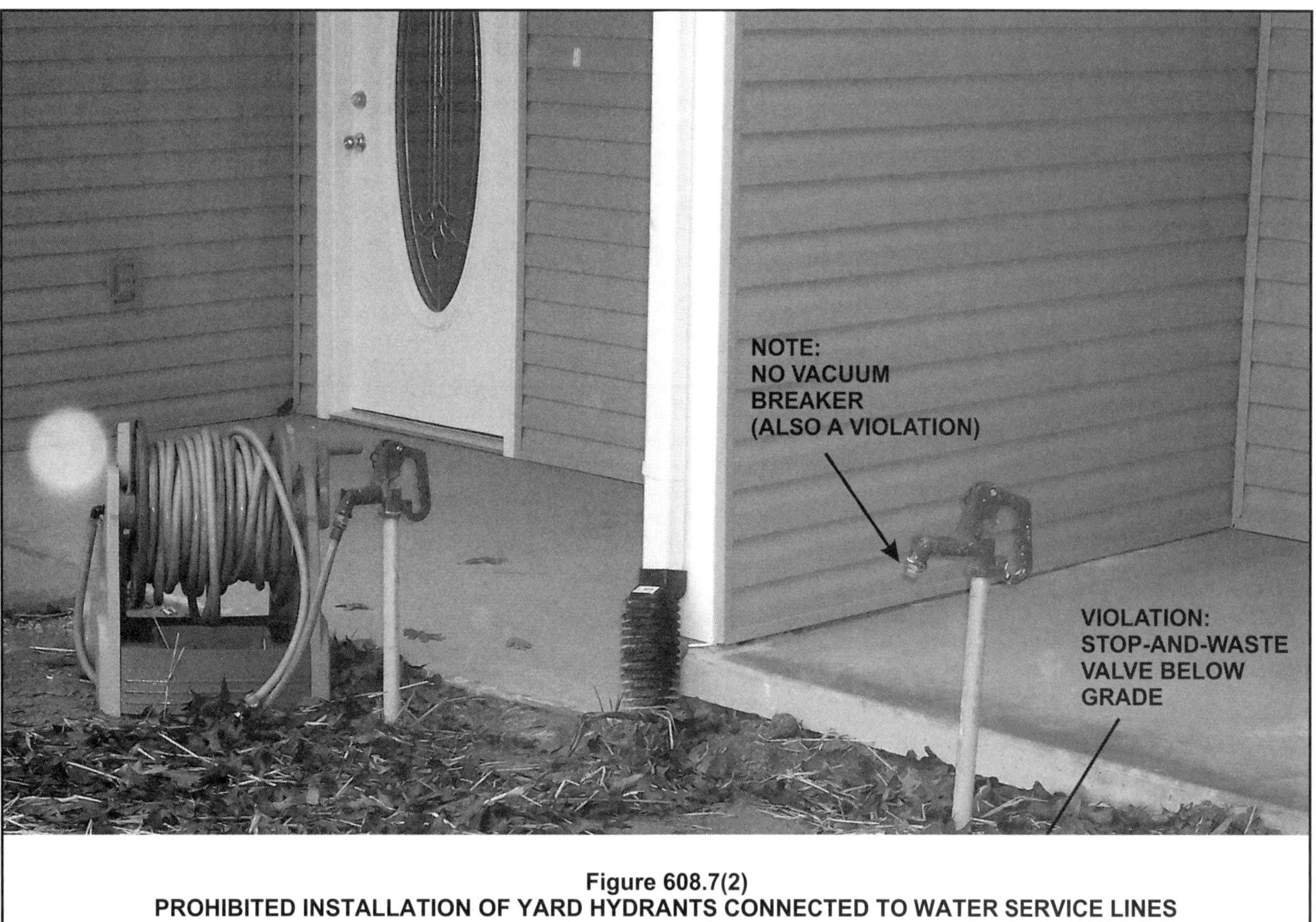

Figure 608.7(2)
PROHIBITED INSTALLATION OF YARD HYDRANTS CONNECTED TO WATER SERVICE LINES

repetition of the identification locations is necessary so that those performing maintenance and modifications can easily find the pipe identification to determine the contents of the piping. Obviously, the pipe identification must be in a position such that a person within the room or space can easily see the identification markings.

608.8.2 Color. The color of the pipe identification shall be discernable and consistent throughout the building. The color purple shall be used to identify reclaimed, rain and gray water distribution systems.

❖ Where color is used for pipe identification, it must be clearly recognizable from any other pipe identification colors and be the same color for the same type of piping in the entire building. Where identifying reclaimed, rain and gray water distribution systems with color, the color purple must be used.

608.8.3 Size. The size of the background color field and lettering shall comply with Table 608.8.3.

❖ See Table 608.8.3 for size of pipe identification.

TABLE 608.8.3
SIZE OF PIPE IDENTIFICATION

PIPE DIAMETER (inches)	LENGTH BACKGROUND COLOR FIELD (inches)	SIZE OF LETTERS (inches)
$^3/_4$ to $1^1/_4$	8	0.5
$1^1/_2$ to 2	8	0.75
$2^1/_2$ to 6	12	1.25
8 to 10	24	2.5
over 10	32	3.5

For SI: 1 inch = 25.4 mm.

608.9 Reutilization prohibited. Water utilized for the cooling of equipment or other processes shall not be returned to the potable water system. Such water shall be discharged into a drainage system through an *air gap* or shall be utilized for nonpotable purposes.

❖ Water used for cooling or processing could become contaminated and therefore must not be returned to the potable water system. Water usage that has no effect on the potability of the water could be acceptable in accordance with Section 501.2 or under the alternative approval provisions of Section 105.2. Systems affecting the potability of the water are prohibited.

It is not the intent of this section to prohibit potable water heaters from supplying hot water to heat exchangers that are constructed of materials approved for potable water distribution piping. Such systems are used to provide space heating and are designed to have no effect on the potability of the water (see Section 501.2).

608.10 Reuse of piping. Piping that has been utilized for any purpose other than conveying potable water shall not be utilized for conveying potable water.

❖ When a fluid or gas other than potable water is conveyed through piping, it can react with the interior pipe wall. The interior of the pipe could develop deposits or a coating that could contaminate the potable water if that pipe is used in a potable water supply system. Therefore, piping used to convey potable water must either be new (unused) or have been previously used only to convey potable water.

608.11 Painting of water tanks. The interior surface of a potable water tank shall not be lined, painted or repaired with any material that changes the taste, odor, color or potability of the water supply when the tank is placed in, or returned to, service.

❖ Only coatings that do not affect the potability of the water can be applied to a potable-water tank.

608.12 Pumps and other appliances. Water pumps, filters, softeners, tanks and all other devices that handle or treat potable water shall be protected against contamination.

❖ Pumps, filters, tanks and apparatus for use with potable water in a water supply system must be designed

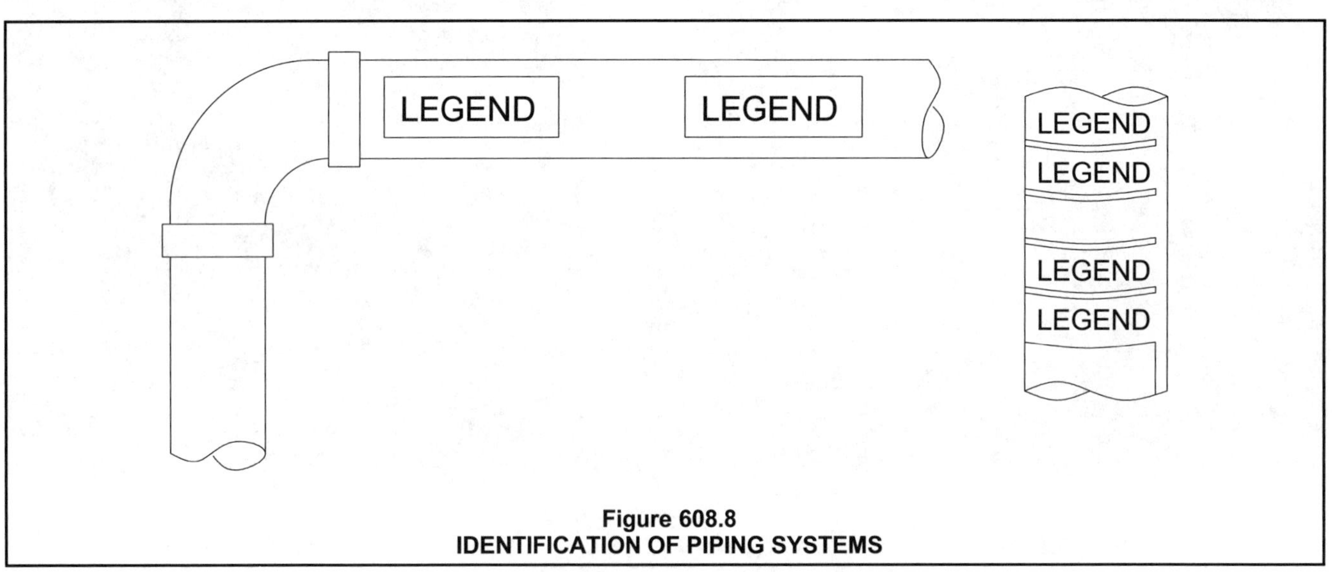

Figure 608.8
IDENTIFICATION OF PIPING SYSTEMS

and approved for this specific use. Such devices and equipment must be installed to prevent contamination. For example, a water softener regeneration cycle drain must be protected against backflow.

608.13 Backflow protection. Means of protection against backflow shall be provided in accordance with Sections 608.13.1 through 608.13.9.

❖ Sections 608.13.1 through 608.13.9 list the eight methods of protection against backflow that are recognized by the code. Any other method would require compliance with Table 608.1 or an alternative approval in accordance with Section 105.2.

608.13.1 Air gap. The minimum required *air gap* shall be measured vertically from the lowest end of a potable water outlet to the *flood level rim* of the fixture or receptacle into which such potable water outlet discharges. Air gaps shall comply with ASME A112.1.2 and *air gap* fittings shall comply with ASME A112.1.3.

❖ ASME A112.1.2 is the standard for air gaps, which require specific dimensions as illustrated in Table 608.15.1. An air gap is not a device and has no moving parts. It is, therefore, the most effective and dependable method of preventing backflow and should be used where feasible. However, when this method of separation is used, there must be an assurance or at least a level of awareness by the user that the air gap cannot be bypassed or defeated by adding a hose or extension from the end of the potable water outlet to the source of contamination. The potable water opening or outlet must terminate at an elevation above the level of the source of contamination. The potable water could be contaminated only if the entire area or room becomes flooded to a depth that would submerge the potable water opening or outlet. An air gap installation must always be constructed so that the air gap remains permanently fixed. Air gaps can be fabricated from commercially available plumbing components or obtained as separate units and integrated into plumbing and piping systems.

This section requires air gaps to comply with Table 608.15.1 based on two variables:

- The size of the effective opening of the outlet or supply, and
- The distance between the outlet and a wall or similar obstruction (see Table 608.15.1).

ASME A112.1.3 is the standard for air gap fittings for use with plumbing fixtures, appliances and appurtenances. This standard was developed for applications where an air gap fitting is manufactured for appliance applications.

608.13.2 Reduced pressure principle backflow preventers. Reduced pressure principle backflow preventers shall conform to ASSE 1013, AWWA C511, CAN/CSA B64.4 or CSA B64.4.1. Reduced pressure detector assembly backflow preventers shall conform to ASSE 1047. These devices shall be permitted to be installed where subject to continuous pressure conditions. The relief opening shall discharge by *air gap* and shall be prevented from being submerged.

❖ A reduced pressure principle backflow preventer is a device considered to be the most reliable mechanical method to prevent backflow. These devices consist of dual independently acting, spring-loaded check valves that are separated by a chamber or "zone" equipped with a relief valve. The pressure downstream of the device and in the central chamber between the check valves is maintained at a minimum of 2 pounds psi (13.8 kPa) less than the potable water supply pressure at the device inlet, hence the name "reduced pressure principle." The relief valve located in the central chamber is held closed by the pressure differential between the inlet supply pressure and the central chamber pressure. If the inlet supply pressure decreases, the device will discharge water from the central chamber, thus maintaining a lower pressure in the chamber. In the event of backpressure or negative supply pressure, the relief valve will open to the atmosphere and drain any backflow that has leaked through the check valve. The relief vent will also allow air to enter to prevent any siphonage.

During a static or no-flow condition, both check valves and the relief vent remain closed. When there is normal flow, the two check valves open and the relief vent port remains closed. The supply pressure drops across the valve because of the force required to open the spring-loaded check valves. If backsiphonage or backpressure occurs, both check valves close, and the relief vent opens to the atmosphere, discharging the water in the intermediate zone. If both check valves are fouled and held in the open position, and there is either a backsiphonage or backpressure condition or both, the relief vent port opens, discharging the possibly contaminated water to the atmosphere and allowing air to enter to break the siphon. This gives a visual indication of a malfunction of one or both of the check valves or relief valve.

ASSE 1013 is the standard for two types of backflow prevention assemblies, identified as "reduced pressure principle backflow preventers" (RP) and "reduced pressure principle fire protection backflow preventers" (RPF). An RP device is designed for a working pressure of 150 psi (1024 kPa), and an RPF device is designed for a working pressure of 175 psi (1207 kPa). The RP and RPF are identical in their backflow protection. The RPF has specific performance requirements relating to its use on fire protection systems.

ASSE 1047 is the standard for devices known as "reduced pressure detector fire protection backflow prevention assemblies" (RPDF). This device is primarily used to protect the public water supply from contaminants found in fire sprinkler systems and to detect low rates of flow up to 0.126 L/s (2 gpm) within the sprinkler system caused by leakage or unauthorized use. This standard also allows provisions for alarm signaling devices to be included in the assembly. The application and installation of a reduced pressure principle backflow preventer is as follows:

- Provides protection against low- or high-hazard contamination.
- Provides protection against backsiphonage and backpressure backflow.
- The device can be used where it is under continuous pressure from the water supply.
- The relief vent opening must discharge through an air gap.
- The device must be accessible for testing and inspection.
- The device cannot be installed below grade where it may be subject to submersion.

Provisions must be made at or near the location of the installation to prevent drainage from the relief vent opening causing damage to the structure when the device is installed in a building or structure.

608.13.3 Backflow preventer with intermediate atmospheric vent. Backflow preventers with intermediate atmospheric vents shall conform to ASSE 1012 or CAN/CSA B64.3. These devices shall be permitted to be installed where subject to continuous pressure conditions. The relief opening shall discharge by *air gap* and shall be prevented from being submerged.

❖ A backflow preventer with intermediate atmospheric vent is a nontestable device consisting of two spring-loaded check valves in the closed position and a vent to atmosphere between the two check valves. The design and principle of operation of these backflow preventers are similar to those of a reduced pressure principle backflow preventer. Supply pressure keeps the atmospheric vent closed, but zero supply pressure or backsiphonage will open the inner chamber to the atmosphere.

The application and installation of a backflow preventer with intermediate atmospheric vent is as follows:

- Provides protection against low-hazard contamination.
- Provides protection against backsiphonage and backpressure backflow.
- The relief vent opening must discharge through an air gap.
- The device can be used where it is under continuous pressure from the water supply.
- The device cannot be installed below grade where it may be subject to submersion.
- The device must be accessible for inspection and replacement.

Provisions must be made at or near the location of the installation to prevent drainage from the relief vent opening causing damage to the structure when the device is installed in a building or structure.

608.13.4 Barometric loop. Barometric loops shall precede the point of connection and shall extend vertically to a height of 35 feet (10 668 mm). A barometric loop shall only be utilized as an atmospheric-type or pressure-type vacuum breaker.

❖ The code considers a barometric loop to be equivalent to a vacuum breaker. A barometric loop may be installed in any location where the code would otherwise require a vacuum breaker.

The principle of a barometric loop is that siphon action is limited to the height that a column of water can be drawn. The 35-foot (10 668 mm) dimension is greater than the theoretical head of a pump suction. A perfect vacuum is equivalent to zero psi (0 kPa) absolute (gauge pressure minus atmospheric pressure). Atmospheric pressure cannot support a water column in excess of 34 feet (10 363 mm) in height. Because siphonage is dependent on atmospheric pressure, siphonage cannot occur if the vertical distance exceeds 34 feet (10 363 mm).

A barometric loop would have control valves located in the vertical pipe or downstream of the connection (see Commentary Figure 608.13.4). A barometric loop cannot prevent backflow caused by backpressure.

608.13.5 Pressure-type vacuum breakers. Pressure-type vacuum breakers shall conform to ASSE 1020 or CSA B64.1.2 and spillproof vacuum breakers shall comply with ASSE 1056. These devices are designed for installation under continuous pressure conditions when the critical level is installed at the required height. Pressure-type vacuum breakers shall not be installed in locations where spillage could cause damage to the structure.

❖ A pressure-type vacuum breaker contains one or two independently operating spring-loaded check valves and an independently operating spring-loaded air inlet valve located on the discharge side of the check valve. This device is testable because it has a shutoff valve and test cock at each end (see Commentary Figure 608.13.5). While a pressure-type vacuum breaker operates by the same principle as an atmospheric vacuum breaker, a pressure-type vacuum breaker is spring activated. The water pressure acting against the spring closes off the air opening. Because of the spring activation, a valve can be located downstream of a pressure vacuum breaker. A spillproof-type vacuum breaker is considered a pressure-type vacuum breaker (see Section 608.13.8).

The application and installation of a pressure-type vacuum breaker is as follows:

- Provides protection against low-hazard and high-hazard contamination.
- Provides protection against backsiphonage only.
- Can be used where it is under continuous pressure from the water supply (valving permitted downstream).

- The device has a critical level of installation of 6 inches (152 mm) above flood level rim.
- The device must be accessible for testing and inspection.
- Provisions must be made at or near the location of the installation to prevent leakage from the air vent opening causing damage to the structure when the device is installed in a building or structure.

ASSE 1056 and CSA B64.1.1 were developed to specifically address indoor applications offering the same vacuum breaker capabilities as ASSE 1020; however, water discharges each time the valve is pressurized. As with the ASSE 1020 valves, backflow protection is achieved against backsiphonage by a check valve backed up with an air inlet vent that opens in response to a loss of supply pressure. Such valves are not for use in any system where backpressure is applied to the device. Where the system is pressurized, the vent closes to prevent flow through the upstream check valve and to eliminate vent spillage (see Commentary Section 608.13.8).

608.13.6 Atmospheric-type vacuum breakers. Pipe-applied atmospheric-type vacuum breakers shall conform to ASSE 1001 or CAN/CSA B64.1.1. Hose-connection vacuum breakers shall conform to ASSE 1011, ASSE 1019, ASSE 1035, ASSE 1052, CAN/CSA B64.2, CSA B64.2.1, CSA B64.2.1.1, CAN/CSA B64.2.2 or CSA B64.7. These devices shall operate under normal atmospheric pressure when the critical level is installed at the required height.

❖ An atmospheric vacuum breaker is designed to prevent siphonic action from occurring downstream of the device. When siphon action or a vacuum is applied to the water supply, the vacuum breaker opens to the atmosphere, allowing air to enter the piping system, thus breaking the vacuum. This prevents contamination of the potable water system by stopping the backsiphonage backflow. When a siphon occurs on the supply inlet of the vacuum breaker, the disc float drops over the opening. Air enters the atmospheric port opening, allowing the remaining water in the piping downstream from the vacuum breaker to drain.

These devices are not effective in preventing backflow resulting from backpressure. The outlet of atmospheric vacuum breakers must remain open to the atmosphere by terminating with a pipe, spout or similar unobstructed opening. Valves must not be installed downstream of this device because this would subject

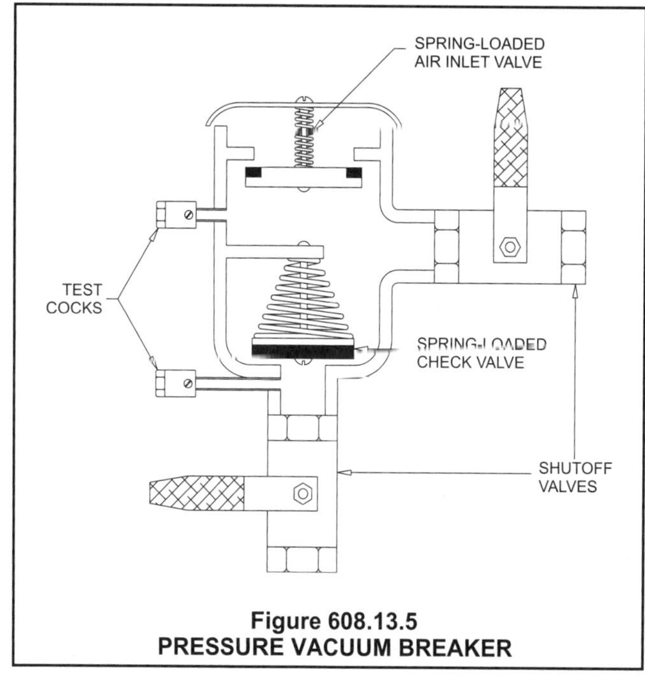

Figure 608.13.5
PRESSURE VACUUM BREAKER

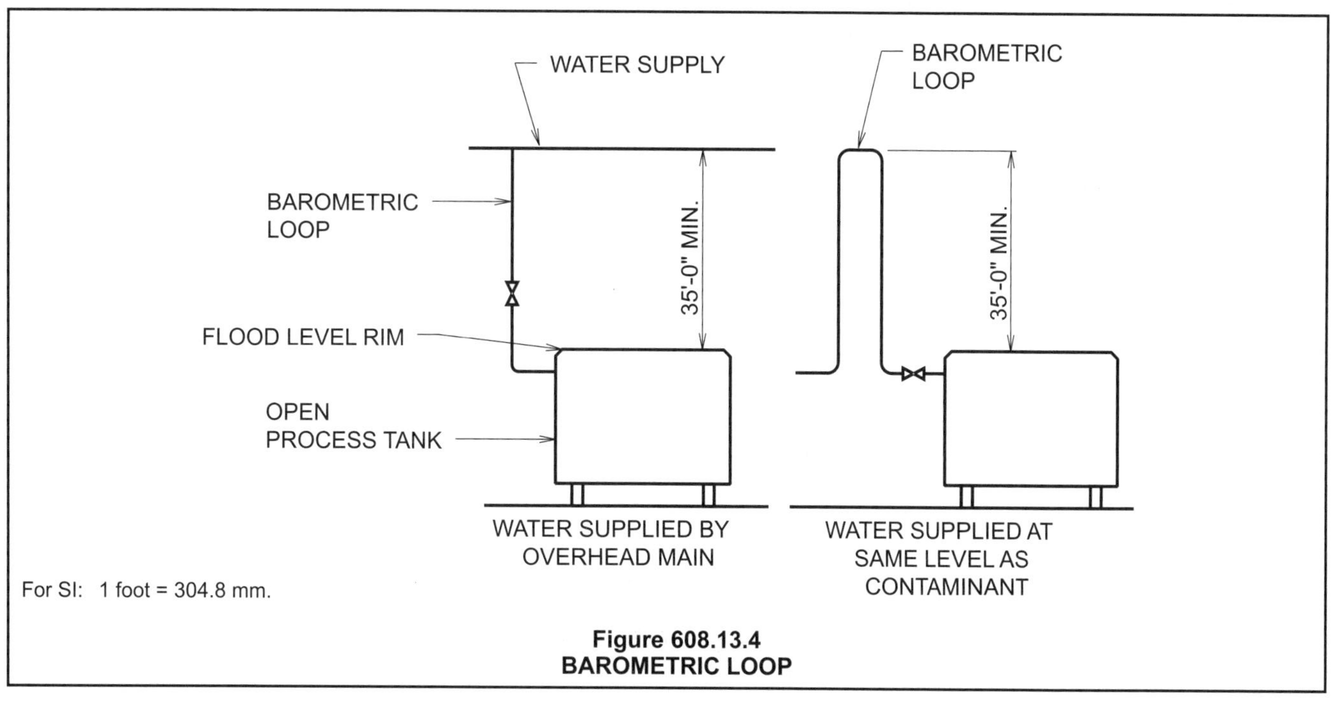

Figure 608.13.4
BAROMETRIC LOOP

the device to supply pressure, thereby rendering it inoperative.

Application and installation of an atmospheric-type vacuum breaker is as follows:

- Provides protection against low-hazard or high-hazard contamination.
- Provides protection against backsiphonage only.
- The device cannot be installed where it is under continuous pressure from the water supply (12-hour or less intervals).
- Critical level installed 6 inches (152 mm) above highest point of use downstream.
- The device must be accessible for inspection and replacement.

Hose connection backflow preventers are designed to be installed on the discharge side of a hose-threaded outlet on a potable water system. This two-check device protects against backflow resulting from backsiphonage and low-head backpressure, under the high-hazard conditions present at a hose-threaded outlet. This device must be used only on systems where sources of backpressure are not introduced and where low-head backpressure does not exceed that generated by an elevated hose equal to or less than 10 feet (3048 mm) in height.

608.13.7 Double check-valve assemblies. Double check-valve assemblies shall conform to ASSE 1015, CSA B64.5, CSA B64.5.1 or AWWA C510. Double-detector check-valve assemblies shall conform to ASSE 1048. These devices shall be capable of operating under continuous pressure conditions.

❖ These devices are designed for low-hazard applications subject to backpressure and backsiphonage. The devices consist of two independent spring-loaded check valves in series. Test cocks are provided to permit testing of the devices (see Commentary Figure 608.13.7). Note that these devices must not be confused with dual check-valve devices or two single check valves placed in series.

Double-detector check-valve assemblies are always permitted as an option where double check-valve assemblies are required because they offer identical protection. Double-detector devices provide the ability to monitor water flow through the device to determine whether there is a system leak or water is being intentionally withdrawn from the system downstream of the device.

608.13.8 Spillproof vacuum breakers. Spillproof vacuum breakers (SVB) shall conform to ASSE 1056. These devices are designed for installation under continuous-pressure conditions when the critical level is installed at the required height.

❖ A spillproof vacuum breaker is otherwise referred to, and is actually identified in ASSE 1056, as a backsiphonage backflow vacuum breaker. A spillproof or backsiphonage backflow vacuum breaker can be installed in any location where the code would otherwise require a pressure-type vacuum breaker. A spillproof vacuum breaker is identical to a pressure vacuum breaker except for an additional feature that solves the problem of water discharging through the air inlet vent each time the device is pressurized. A spillproof vacuum breaker is therefore permitted in locations where a pressure vacuum breaker would otherwise be prohibited because of damage caused by leakage.

A spillproof vacuum breaker is a testable device consisting of one check valve, spring loaded to the closed position, and an air inlet vent valve, spring loaded to the open position. The air inlet valve is located downstream from the check valve. The check valve and air inlet vent are located between two shutoff valves and two test cocks.

The application and installation of a spillproof vacuum breaker is as follows:

- Protects against low-hazard and high-hazard contamination.
- Protects against backsiphonage backflow only.
- Can be subjected to continuous pressure (valving downstream is permitted).
- The critical level must be installed 6 inches (152 mm) above the highest outlet served.
- Must be accessible for testing and inspection.

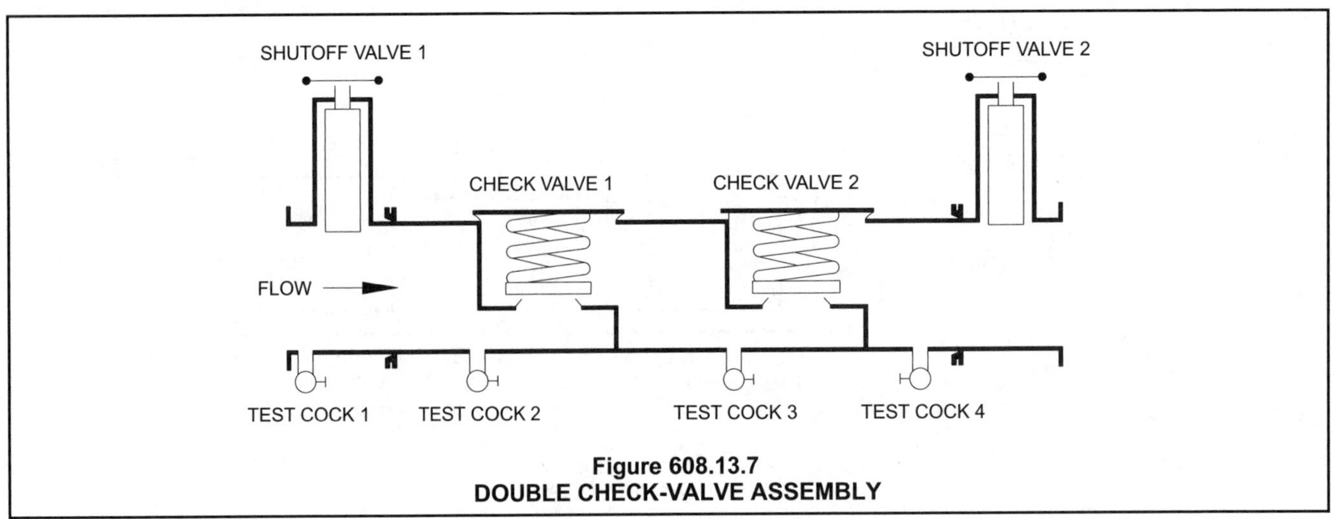

Figure 608.13.7
DOUBLE CHECK-VALVE ASSEMBLY

608.13.9 Chemical dispenser backflow devices. Back- flow devices for chemical dispensers shall comply with ASSE 1055 or shall be equipped with an *air gap* fitting.

❖ Chemical dispensing systems mix potable water with chemicals to provide the user with a ready-to-use chemical solution. These systems must be equipped with a backflow prevention device complying with ASSE 1055 or an air gap fitting, which is a device with an integral air gap. ASSE 1055 applies to devices classified as chemical dispensing systems having self-contained means of backflow protection, as follows:

Type A: These devices have the chemical(s) pressurized above atmospheric pressure.

Type B: These devices do not pressurize the chemical(s) above atmospheric pressure, with the only source of backpressure coming from an elevated hose.

608.14 Location of backflow preventers. *Access* shall be provided to backflow preventers as specified by the installation instructions of the *approved* manufacturer.

❖ Every backflow preventer must be located so that it can be reached for inspection, testing, repair and replacement without removing a wall, floor, ceiling or other permanent portion of the building structure. It can be located behind access panels or in utility-type spaces. Backflow preventer devices, other than double-check-valve assemblies, have an atmospheric opening, and the location must take this into consideration. The devices must be located not only in ventilated spaces to allow air to enter them, but also where water discharge from them will not cause damage or create a hazard.

Backflow preventers must not be installed in locations subject to flooding, such as underground vaults. If the atmospheric opening on a backflow preventer is submerged, the protection provided by the device is lost.

608.14.1 Outdoor enclosures for backflow prevention devices. Outdoor enclosures for backflow prevention devices shall comply with ASSE 1060.

❖ ASSE 1060 is the standard that details the requirements of an outside enclosure for various types of backflow prevention assemblies. It includes enclosure types for freezing and nonfreezing locations. The enclosures incorporate features to provide for freeze protection, positive drainage to prevent submergence of the assembly, security and accessibility for testing and repair.

Freeze protection enclosures are divided into the following classes:

1. Class I enclosures are designed for all devices other than pressure and atmospheric vacuum breakers.
2. Class I-V enclosures are designed for pressure and atmospheric vacuum breakers.

Freeze-retardant enclosures are divided into the following classes:

1. Class II enclosures are designed for all devices other than pressure and atmospheric vacuum breakers.
2. Class II-V enclosures are designed for pressure and atmospheric vacuum breakers.

Enclosures without freeze protection are divided into the following classes:

1. Class III enclosures are designed for all devices other than pressure and atmospheric vacuum breakers.
2. Class III-V enclosures are designed for pressure and atmospheric vacuum breakers.

Freeze-protection enclosures are designed and constructed to maintain a minimum internal temperature of 40°F (4°C) when the external temperature is as low as -30°F (-34°C).

608.14.2 Protection of backflow preventers. Backflow preventers shall not be located in areas subject to freezing except where they can be removed by means of unions or are protected from freezing by heat, insulation or both.

❖ Frequently, backflow prevention valves are required to be located outside of buildings for a variety of reasons. If a valve will be exposed to freezing temperatures, the valve must be either protected during the freezing condition or removed during freezing weather. Where valves are necessary to function during freezing weather, removal is not possible, thus requiring that insulation or heat or both be used to keep the valve above freezing temperatures.

There is usually no question as to the need for an insulated enclosure and heat source for backflow preventers exposed to severe freezing weather. Valves exposed to milder winter conditions might only need insulation or, where freezing conditions rarely exist, no protection may be an appropriate solution. The code is silent on the level of protection needed for these milder climates. Enclosure vendors and code officials could be sources for information to decide on the degree of freeze protection required.

For those backflow prevention valve applications where use of the valve is seasonal, such as for lawn sprinklers, or swimming pool water service, removal is usually the easiest solution. If removal is intended, unions must be provided to enable the valve to be easily removed without disturbing permanent piping. Even though the valve is removed, removal of the water in the remaining permanent piping should not be overlooked in order to prevent freeze damage to the piping.

608.14.2.1 Relief port piping. The termination of the piping from the relief port or *air gap* fitting of a backflow preventer shall discharge to an *approved* indirect waste receptor or to the outdoors where it will not cause damage or create a nuisance.

❖ In particular, reduced pressure zone (RPZ) backflow prevention valves will discharge water during a backflow event, a drop in supply pressure or where an obstruction (dirt) fouls the check valves. If the valve is located outdoors, an air gap fitting is usually not needed as the discharge dumps to the ground. If the valve is indoors or the discharge in an outdoor location would cause damage or a nuisance, the valve must be located over an appropriate waste receptor (floor drain, sump with drain, etc.) to capture the flow from the relief port. If the relief port discharge must be piped to a suitable discharge point (either outdoors or to a nearby waste receptor), the relief port must be fitted with an air gap fitting (as supplied by the valve manufacturer) for connection of the relief discharge piping. Where the piping discharges to a waste receptor, it must connect indirectly.

In most situations, relief port discharges are a trickle of water that is easily accommodated by the pipe size dictated by the air gap connection. However, large obstructions (dirt, rocks, etc.) or sudden large drops in supply pressure may cause an extreme and prolonged discharge event such that the air gap fitting and relief pipe size will be ineffective in capturing the discharge. Designers should consider this type of event and provide for a means to capture and direct the flow from an extreme event to a safe location in order to avoid nearby equipment and property damage.

608.15 Protection of potable water outlets. All potable water openings and outlets shall be protected against backflow in accordance with Section 608.15.1, 608.15.2, 608.15.3, 608.15.4, 608.15.4.1 or 608.15.4.2.

❖ An opening or outlet is where the potable water system comes in contact with the atmosphere.This includes, but is not limited to, faucets, hose bibbs, plumbing fixtures, plumbing appliances and other points where water is withdrawn from the potable water supply system. The openings should be protected by one of four methods: air gap, backflow preventer with intermediate atmospheric vent, vacuum breaker or reduced pressure principle backflow preventer.

There are two types of cross-connection control philosophies: containment and isolation.

Containment: The practice of confining potential contamination within the facility where it arises by installing a backflow preventer at the meter or curbstop.

Isolation: The practice of confining potential contamination by installing a backflow preventer at each outlet or connection to a fixture or nonpotable system.

The code requires isolation only for cross connections. The practice of containment is optional and widely used in waste-water treatment plants and marinas where potable water is supplied to boats and buildings with numerous nonpotable piping systems.

Many water supplies to commercial or industrial facilities have containment protection due in part to the complexity or diversity of use of the potable water supply within the facility. Although containment alone will protect the public water system, it does not protect the potable water within the facility.

608.15.1 Protection by air gap. Openings and outlets shall be protected by an *air gap* between the opening and the fixture *flood level rim* as specified in Table 608.15.1. Openings and outlets equipped for hose connection shall be protected by means other than an *air gap*.

❖ An air gap is the simplest, most common and most reliable form of backflow protection for openings and outlets. Most sink and lavatory faucets are designed with a built-in air gap. A hose connection outlet cannot rely on an air gap to provide backflow protection. Once a hose is added to the outlet, the air gap no longer exists. Therefore, an alternative method of backflow prevention is required for any hose outlet.

Although an air gap is an extremely effective backflow preventer, it does interrupt the piping flow with corresponding loss of pressure for subsequent use.

Consequently, air gaps are used primarily at end-of-the-line service where reservoirs or storage tanks are desired.

Additionally, it exposes the potable water supply to the surrounding air with its inherent bacteria, dust particles and other air-borne pollutants or contaminants. This presents a backflow potential and must be considered when selecting appropriate backflow methods.

An air gap must be inspected as frequently as mechanical backflow preventers in accordance with Section 312.9 [see Commentary Figures 608.15.1(1) and 608.15.1(2)].

TABLE 608.15.1. See page 6-69.

❖ This table has been a mainstay in the plumbing industry since 1940.The values specified have not changed in more than 50 years.

The middle column of the table shows minimum air gap dimensions for outlets located away from walls or obstructions. The third column contains more stringent air gap dimensions when the outlet is located near a wall or obstruction [see Commentary Figures 608.15.1(3) and 608.15.1(4)].

608.15.2 Protection by a reduced pressure principle backflow preventer. Openings and outlets shall be protected by a reduced pressure principle backflow preventer.

❖ A reduced pressure principle backflow preventer can also be used to protect outlets and openings, although it is rarely used for this purpose (see Commentary Section 608.13.2).

TABLE 608.15.1
MINIMUM REQUIRED AIR GAPS

FIXTURE	MINIMUM AIR GAP	
	Away from a wall[a] (inches)	Close to a wall (inches)
Lavatories and other fixtures with effective opening not greater than $^1/_2$ inch in diameter	1	$1^1/_2$
Sink, laundry trays, gooseneck back faucets and other fixtures with effective openings not greater than $^3/_4$ inch in diameter	$1^1/_2$	$2^1/_2$
Over-rim bath fillers and other fixtures with effective openings not greater than 1 inch in diameter	2	3
Drinking water fountains, single orifice not greater than $^7/_{16}$ inch in diameter or multiple orifices with a total area of 0.150 square inch (area of circle $^7/_{16}$ inch in diameter)	1	$1^1/_2$
Effective openings greater than 1 inch	Two times the diameter of the effective opening	Three times the diameter of the effective opening

For SI: 1 inch = 25.4 mm.

a. Applicable where walls or obstructions are spaced from the nearest inside-edge of the spout opening a distance greater than three times the diameter of the effective opening for a single wall, or a distance greater than four times the diameter of the effective opening for two intersecting walls.

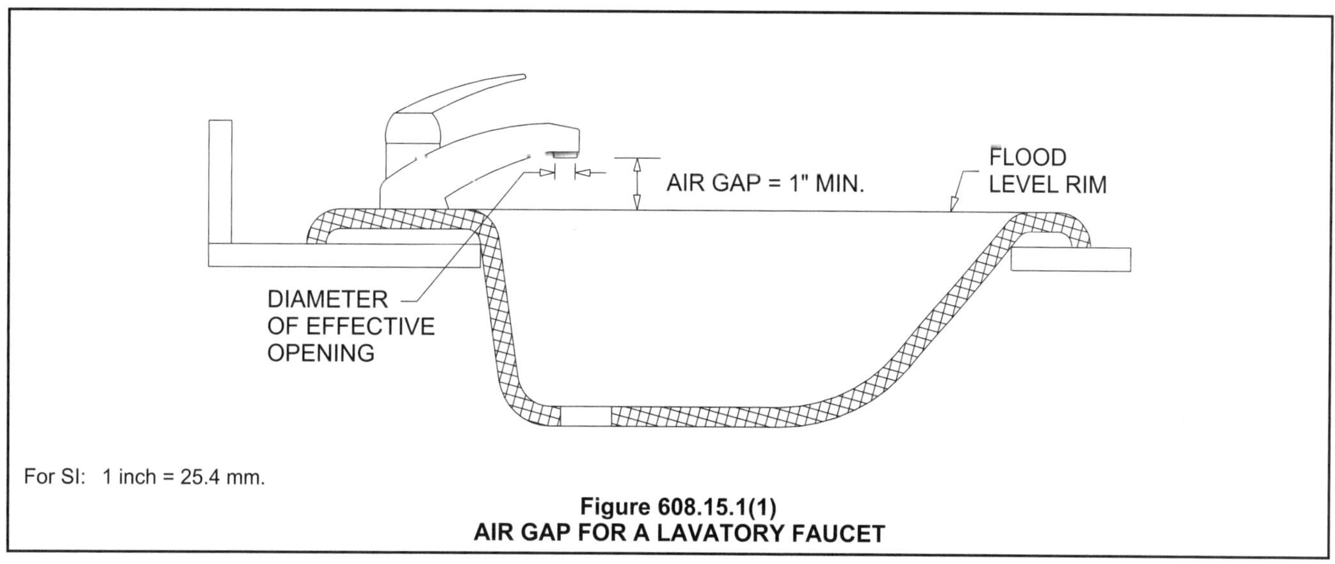

For SI: 1 inch = 25.4 mm.

Figure 608.15.1(1)
AIR GAP FOR A LAVATORY FAUCET

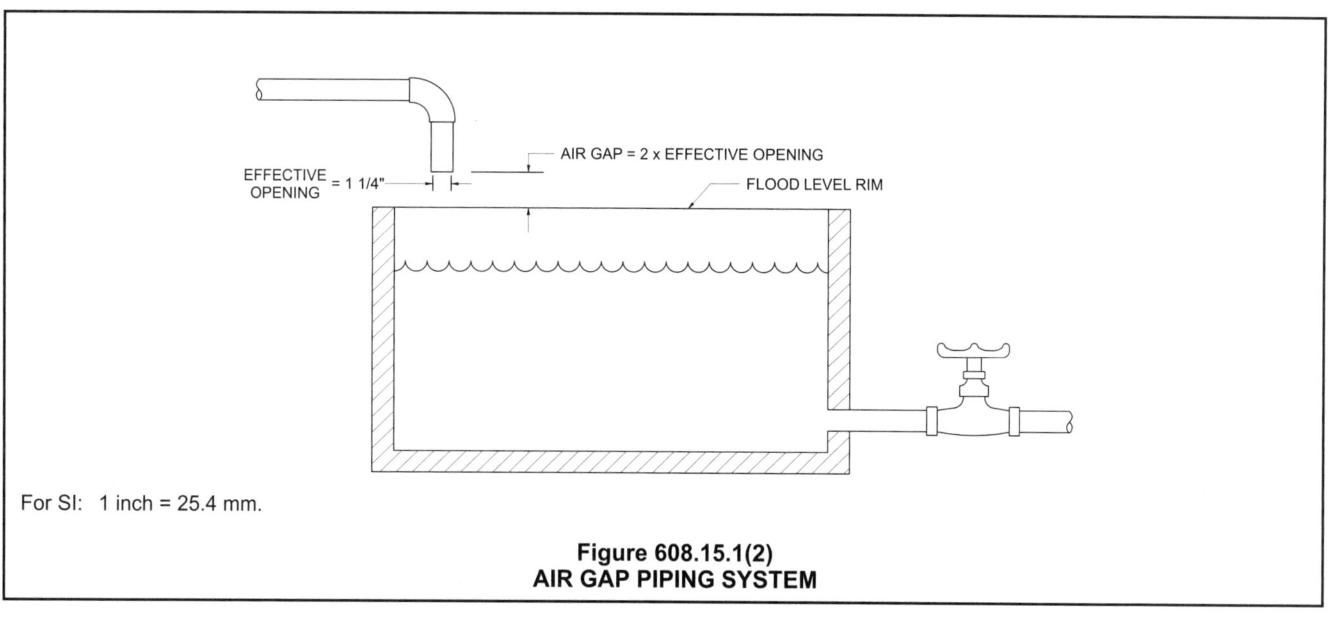

For SI: 1 inch = 25.4 mm.

Figure 608.15.1(2)
AIR GAP PIPING SYSTEM

608.15.3 Protection by a backflow preventer with intermediate atmospheric vent. Openings and outlets shall be protected by a backflow preventer with an intermediate atmospheric vent.

❖ The devices addressed in Section 608.13.3 are acceptable means of backflow prevention for openings and outlets.

608.15.4 Protection by a vacuum breaker. Openings and outlets shall be protected by atmospheric-type or pressure-type vacuum breakers. The critical level of the vacuum breaker shall be set a minimum of 6 inches (152 mm) above the *flood level rim* of the fixture or device. Fill valves shall be set in accordance with Section 425.3.1. Vacuum breakers shall not be installed under exhaust hoods or similar locations that will contain toxic fumes or vapors. Pipe-applied vacuum breakers shall be installed not less than 6 inches (152 mm) above the *flood level rim* of the fixture, receptor or device served.

❖ For a vacuum breaker to operate properly, it must be located above the flood level rim of the fixture or device that it is protecting.

If the vacuum breaker is below the flood level rim, it is subject to pressure from the fixture or outlet piping be-

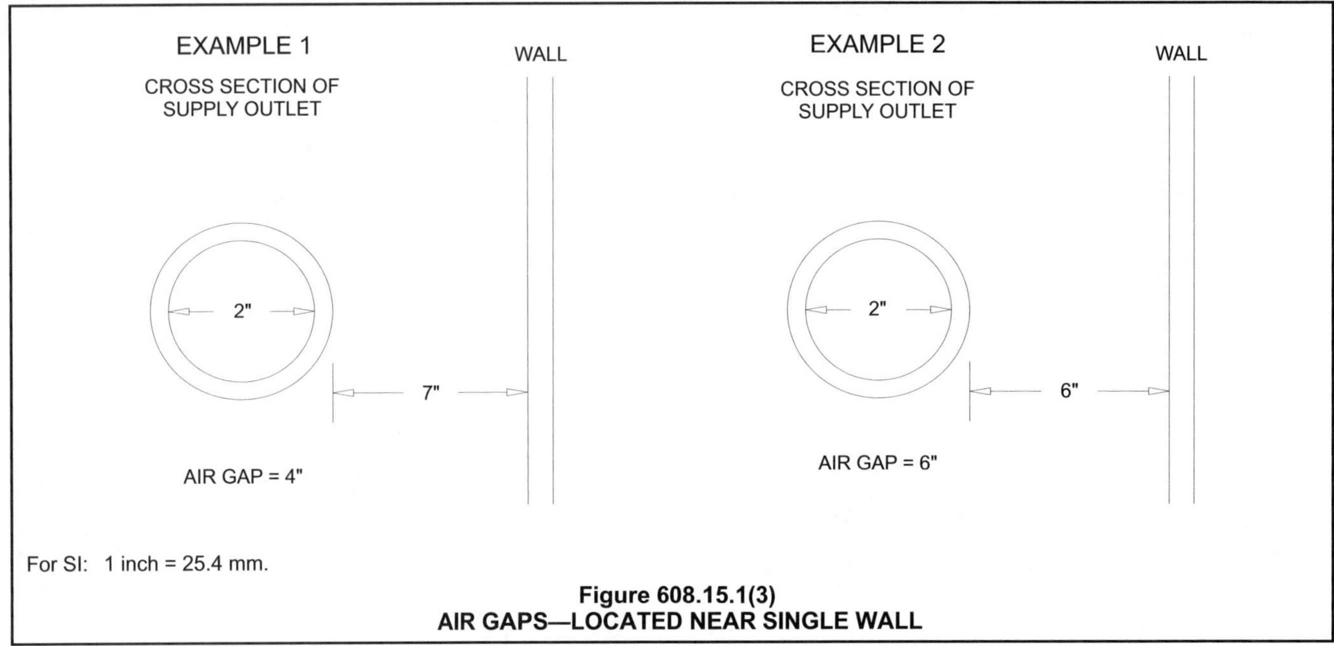

Figure 608.15.1(3)
AIR GAPS—LOCATED NEAR SINGLE WALL

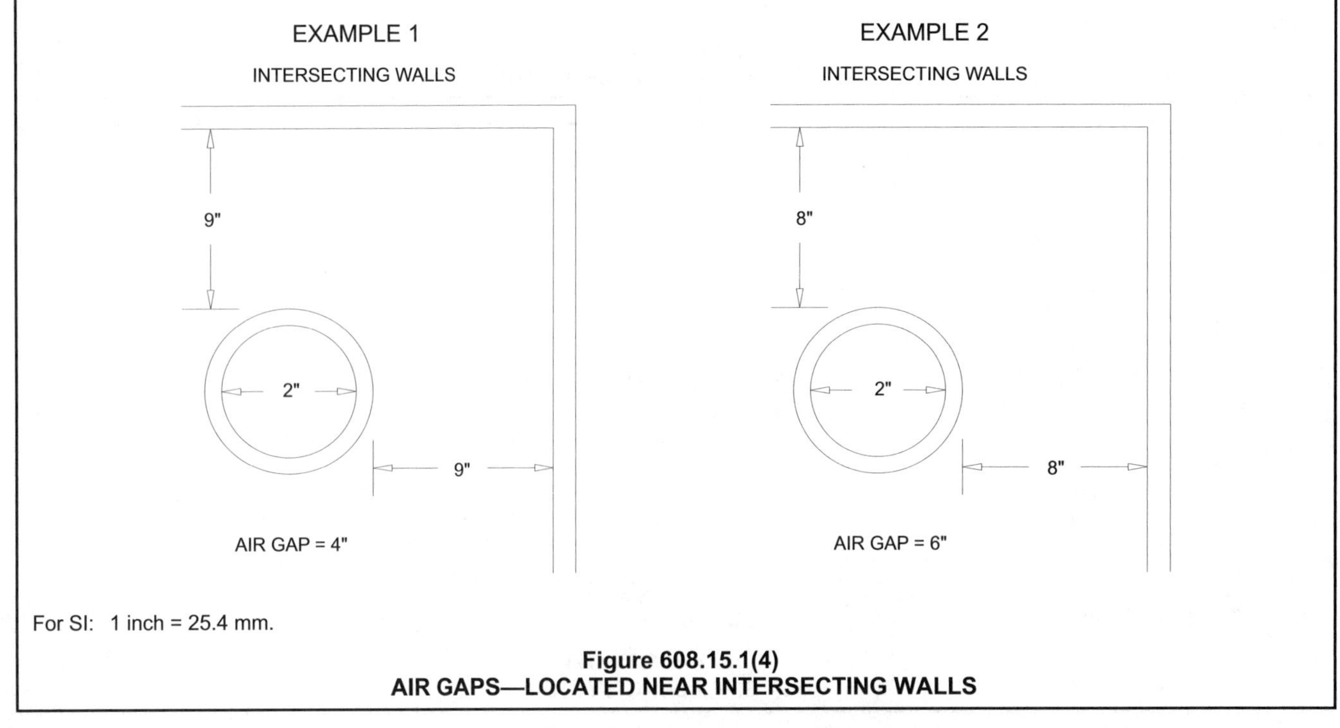

Figure 608.15.1(4)
AIR GAPS—LOCATED NEAR INTERSECTING WALLS

cause of the static head (see commentary, Section 608.13.6). Valves cannot be located downstream from an atmospheric vacuum breaker. A downstream valve will subject the device to backpressure, thereby causing damage or impaired operation. All valves or controlling devices must be located upstream of the device. Valves may be downstream only where a pressure vacuum breaker is installed (see Commentary Figure 608.15.4).

608.15.4.1 Deck-mounted and integral vacuum breakers. *Approved* deck-mounted or equipment-mounted vacuum breakers and faucets with integral atmospheric or spillproof vacuum breakers shall be installed in accordance with the manufacturer's instructions and the requirements for labeling with the critical level not less than 1 inch (25 mm) above the *flood level rim*.

❖ This section recognizes a type of vacuum breaker that is installed on the deck (horizontal surface at rim) of plumbing fixtures such as bathtubs, whirlpools, bidets and sinks. Such devices must be tested for installation with lesser distances between the critical level and the fixture flood level rim. Such devices are intended to be more aesthetically pleasing and would be built into fixture trim hardware, such as fill spouts or diverter controls.

608.15.4.2 Hose connections. Sillcocks, hose bibbs, wall hydrants and other openings with a hose connection shall be protected by an atmospheric-type or pressure-type vacuum breaker or a permanently attached hose connection vacuum breaker.

Exceptions:

1. This section shall not apply to water heater and boiler drain valves that are provided with hose connection threads and that are intended only for tank or vessel draining.
2. This section shall not apply to water supply valves intended for connection of clothes washing machines where backflow prevention is otherwise provided or is integral with the machine.

❖ In accordance with this section, outlets and openings to which a hose can be connected must be protected by some type of vacuum breaker device. Commentary Figure 608.15.4.2(1) shows a hose connection in violation of this requirement. Commentary Figure 608.15.4.2(2) shows one type of hose faucet that has one type of integral vacuum breaker. Atmospheric vacuum breakers screwed onto the threads of the hose faucet may also be acceptable as long as the connection is permanent (locking type) or the threads between the hose faucet and the vacuum breaker are nonstandard (i.e., neither pipe threads nor garden hose threads). The devices addressed in Sections 608.13.2 and 608.13.3 would also be permitted because they would afford a greater level of protection. Where a hose can be elevated above the critical level of a vacuum breaker, the vacuum breaker must be designed to function under such conditions. For example, hose connection vacuum breakers incorporate a check valve in conjunction with an atmospheric port. Such devices can prevent siphonage and can also resist backflow caused by the head pressure developed in an elevated hose. In accordance with Table 608.1 and the definition for "Backflow, low head backpressure" in Section 202, this backpressure is limited to 10 feet (3048 mm) of water column or 4.33 psi (29.9 kPa). If the possibility exists for backpressure to exceed 10 feet (3048 mm) of water column [the equivalent of raising a hose filled with water 10 feet (3048 mm) above the vacuum breaker], another backflow prevention device, designed to handle greater backpressures, must be installed. If a hose is equipped with a control valve, an atmospheric vacuum breaker cannot be used to protect that outlet [see commentary, Section 608.15.4 and Commentary Figure 608.15.4.2(3)].

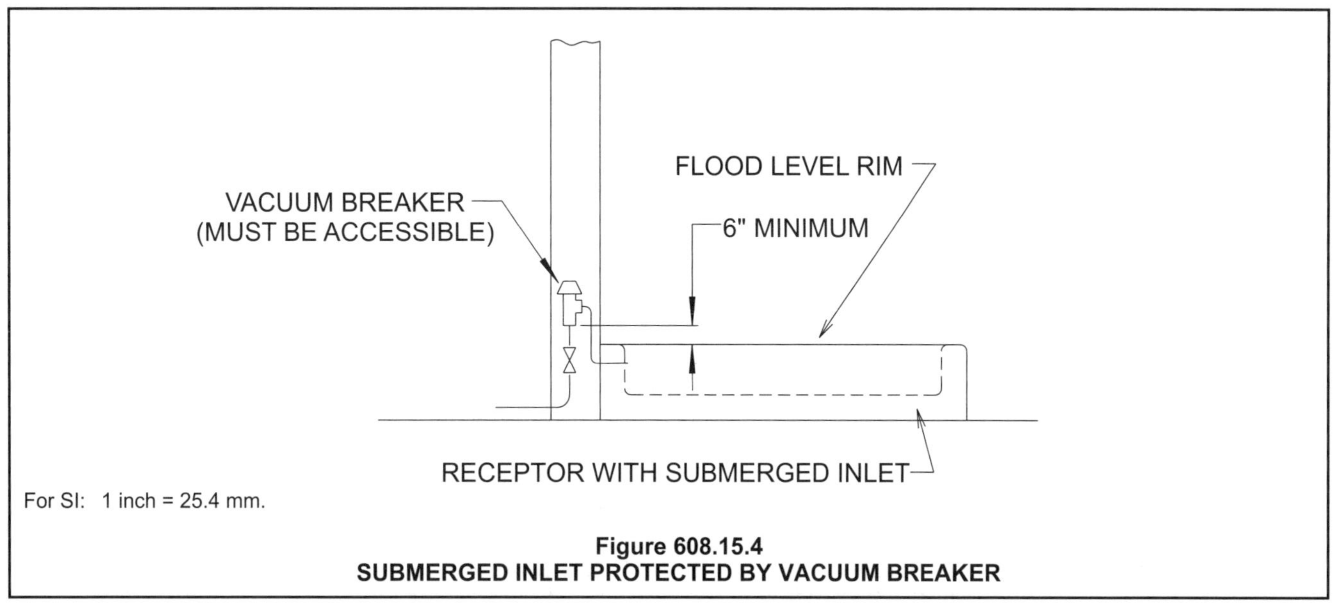

Figure 608.15.4
SUBMERGED INLET PROTECTED BY VACUUM BREAKER

In Commentary Figure 608.15.4.2(3), the supply valve shutoff disc (1) seals against the diaphragm (2) when there is no flow in the pipe. The atmospheric port (3) is open. As flow occurs in the pipe, the atmospheric ports are closed before the supply valve disc opens. Note that this type of vacuum breaker can be acceptable only if the installation is made permanent (e.g., a locking break-off set screw).

The exceptions are intended to address situations where hose threads are commonly provided but do not require the installation of a vacuum breaker. The two exceptions are not exempting these outlets from any form of backflow prevention. In Exception 1, water heater and boiler drain valves (with or without hose threads) must be provided with an air gap, and in Exception 2, the clothes washing machine must be provided with an integral form of backflow prevention (typically an air gap between the fill pipe and the overflow rim of the washing machine tub).

Figure 608.15.4.2(1)
HOSE (SILL) FAUCET WITHOUT VACUUM BREAKER

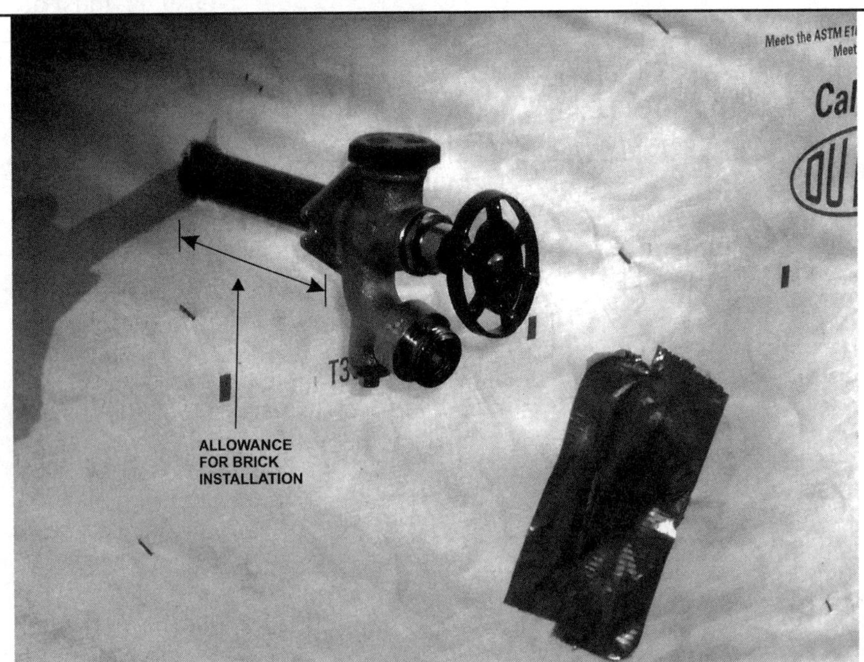

Figure 608.15.4.2(2)
HOSE (SILL) FREEZEPROOF FAUCET WITH INTEGRAL ANTISIPHON DEVICES

Table 608.1 contains a reference to a relatively new device called a "hose connection backflow preventer." The device is testable and consists of two check valves spring loaded in the closed position with an atmospheric vent between them. The backflow preventer can either be an add-on device to or integral with a wall hydrant or hose bibb. Because the add-on device is designed to attach to hose threads, it would be appropriate for these devices to serve as the required method of backflow protection for hose connections.

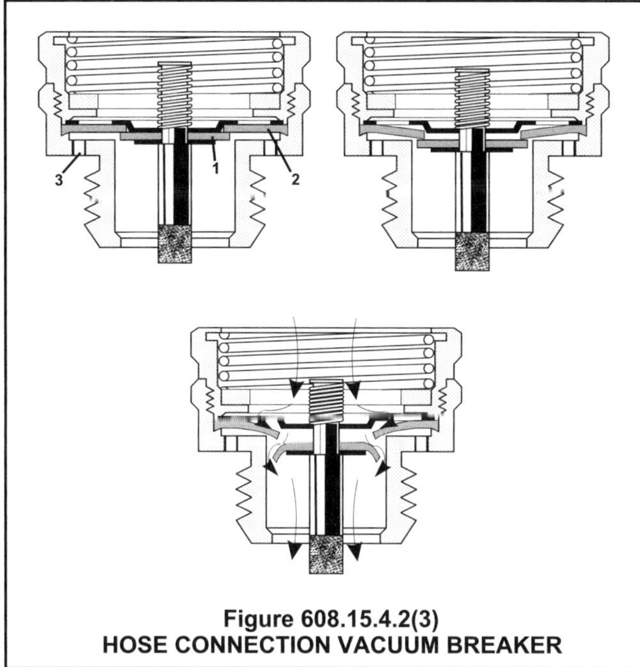

Figure 608.15.4.2(3)
HOSE CONNECTION VACUUM BREAKER

608.16 Connections to the potable water system. Connections to the potable water system shall conform to Sections 608.16.1 through 608.16.10.

❖ Sections 608.16.1 through 608.16.10 regulate backflow prevention for direct connections between a potable and nonpotable system. A cross connection exists where there is no atmospheric opening (air gap) between the potable water supply and the source of contamination.

This section addresses cross connections that result from the direct connection of the potable water supply to a closed system such as boilers, vessels, tanks, heat exchangers and nonpotable piping systems.

608.16.1 Beverage dispensers. The water supply connection to beverage dispensers shall be protected against backflow by a backflow preventer conforming to ASSE 1022 or by an *air gap*. The portion of the backflow preventer device downstream from the second check valve and the piping downstream therefrom shall not be affected by carbon dioxide gas.

❖ The potable water supply to a carbonated beverage dispenser must be protected by a backflow preventer with an intermediate atmospheric vent conforming to ASSE 1022 or by an air gap. Concerns have been raised that a dual check valve alone will not show any visible indication of failure. The backflow preventer for beverage dispensing equipment includes two check valves and an atmospheric vent. If there is failure of the downstream check and the backpressure exceeds the supply pressure, the vent will discharge, giving a visual indication of the check valve's failure. Piping downstream of the backflow preventer with intermediate atmospheric vent must not be copper. These requirements are intended to prevent CO^2 backflow into copper water supply lines, which causes leaching of copper into the potable water supply (see Commentary Figure 608.16.1).

Prior to the 2009 edition of the code, a carbonated beverage dispenser backflow preventer that met the CSA B64.3.1 standard was acceptable. An in-depth analysis comparing the test protocols between the ASSE and CSA standards suggested that the CSA standard was not equal to the requirements of ASSE. Because the use of ASSE 1022 devices has a long history of providing safe conditions, and the majority of beverage dispenser manufacturers only specify the ASSE 1022 device for use with their equipment, the CSA standard was removed from this section.

608.16.2 Connections to boilers. The potable supply to the boiler shall be equipped with a backflow preventer with an intermediate atmospheric vent complying with ASSE 1012 or CAN/CSA B64.3. Where conditioning chemicals are introduced into the system, the potable water connection shall be protected by an *air gap* or a reduced pressure principle backflow preventer, complying with ASSE 1013, CAN/CSA B64.4 or AWWA C511.

❖ Because boilers are pressurized vessels, the potential for backflow caused by backpressure is quite high. Any time the potable water supply pressure drops below boiler pressure, backflow can occur.

If a boiler system contains only untreated, unconditioned water supplied from a potable source, a backflow preventer with an intermediate atmospheric vent can be installed (see Commentary Figure 608.16.2).

If any chemicals are added to the boiler system, the potable water supply connection must be protected with a reduced pressure principle backflow preventer.

608.16.3 Heat exchangers. Heat exchangers utilizing an essentially toxic transfer fluid shall be separated from the potable water by double-wall construction. An *air gap* open to the atmosphere shall be provided between the two walls. Heat exchangers utilizing an essentially nontoxic transfer fluid shall be permitted to be of single-wall construction.

❖ The extent of isolation required for a heat exchanger depends on the type of fluid used on the exchanger's nonpotable side. From the definition of "Essentially nontoxic transfer fluids," the Gosselin rating of nonpotable fluid must be evaluated. If the fluid has a Gosselin rating of 1, a single-wall heat exchanger is permitted.

The Gosselin rating is a measure of the toxicity of a substance. The name originates from one of the prime developers of the rating system, Dr. Robert E. Gosselin, professor of Pharmacology at Dartmouth Medical School in New Hampshire.

Gosselin toxicity ratings are the values used by medical personnel to analyze poison victims. The ratings are based on the probable lethal dose for a human. The six levels of Gosselin ratings are outlined in Commentary Table 608.16.3(1).

Some of the commercially available transfer fluids with a Gosselin rating of 1 are identified in Commentary Table 608.16.3(2).

If the heat transfer fluid has a Gosselin rating of 2 or more, a double-wall heat exchanger is required. The double-wall heat exchanger must have an intermediate space between the walls that is open to the atmosphere. This type of construction would allow any leakage of fluid through the walls of the heat exchanger to discharge externally to the heat exchanger where it would be observable [see Commentary Figure 608.16.3(3)].

Table 608.16.3(1)
GOSSELIN RATINGS

TOXICITY RATING OR CLASS	PROBABLE ORAL LETHAL DOSE (HUMAN)	
	Dose	For 70 kg person (150 pounds)
6 Super toxic	Less than 5 mg/kg	A taste (less than 7 drops)
5 Extremely toxic	5-50 mg/kg	Between 7 drops and 1 teaspoon
4 Very toxic	50-500 mg/kg	Between 1 teaspoon and 1 ounce
3 Moderately toxic	0.5-5 gm/kg	Between 1 ounce and 1 pint (pound)
2 Slightly toxic	5-15 gm/kg	Between 1 pint and 1 quart
1 Practically nontoxic	Above 15 gm/kg	More than 1 quart (2.2 pounds)

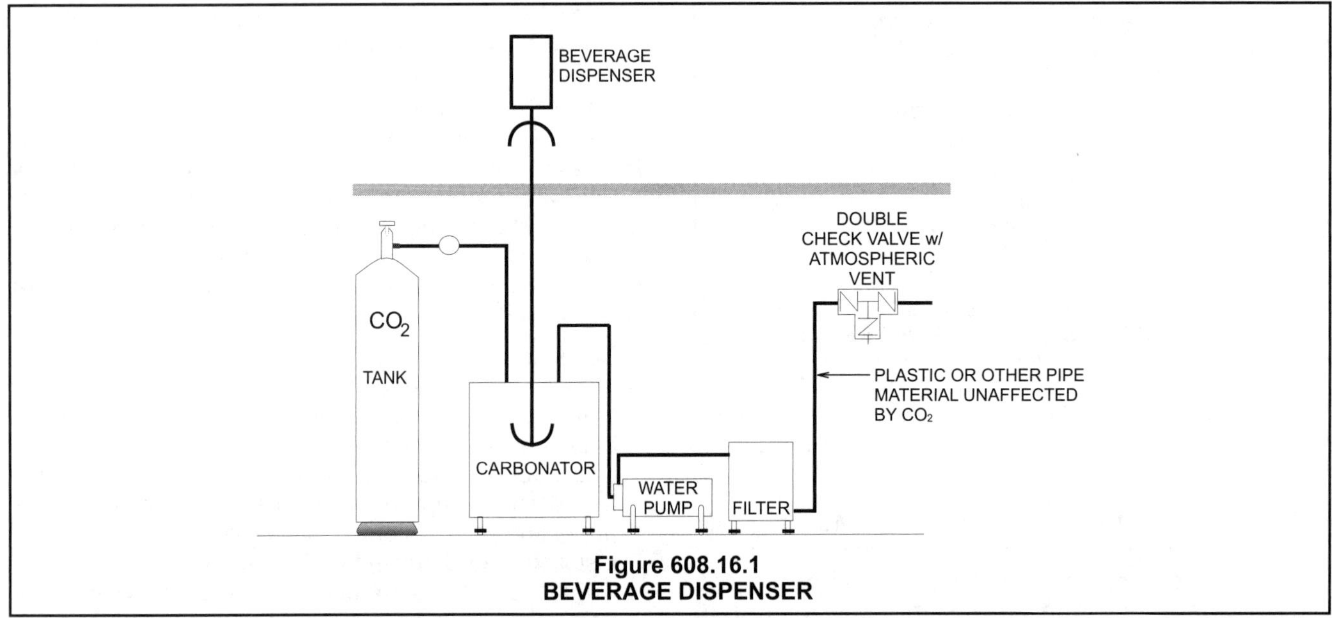

Figure 608.16.1
BEVERAGE DISPENSER

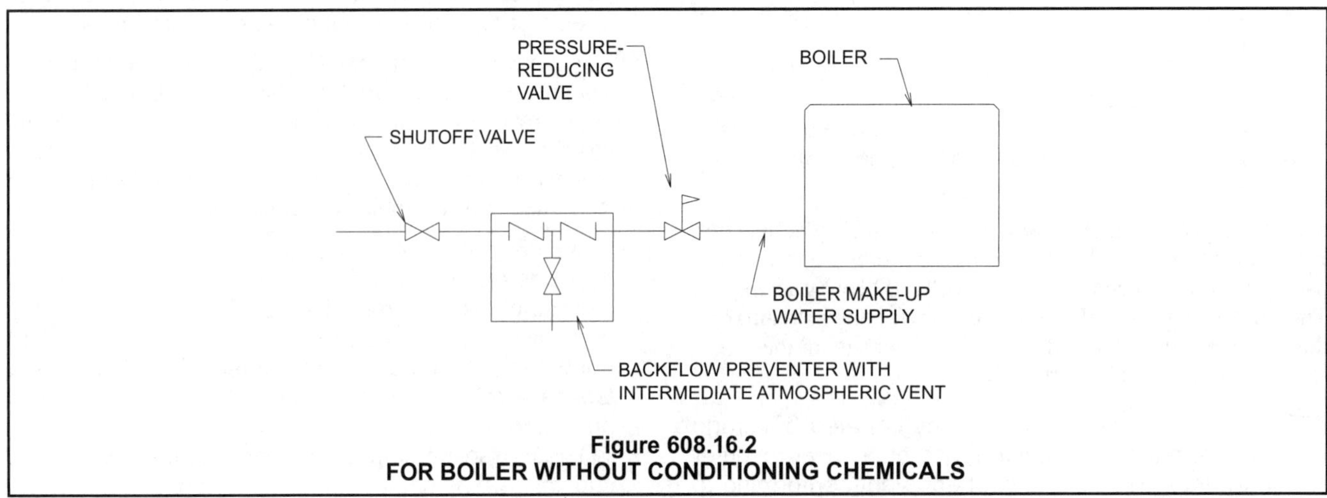

Figure 608.16.2
FOR BOILER WITHOUT CONDITIONING CHEMICALS

Table 608.16.3(2)
HEAT TRANSFER FLUIDS (ESSENTIALLY NONTOXIC) GOSSELIN RATING OF 1

TRADE NAME	MANUFACTURER
Caloria Ht-43	Exxon Co.
DowFrost	Dow
Drewsol	Drew Chemical
Freeze Proof	Commonwealth
Mobiltherm 603	Mobile Oil Co.
Mobiltherm Light	Mobile Oil Co.
Nutek 835	Nuclear Technology
Nutek 876	Nuclear Technology
Process Oil 3029	Exxon Co.
Propylene Glycol U.S.P.	Union Carbide
Solargard G.	Daystar Corp.
Solar Winter Ban	Solar Alternative Inc.
Sunsol 60	Sunworks
Suntemp	Resource Technology Corp.
Syltherm 444	Dow Corning Corp.
Thorminol 00	Monsanto Co.
UCAR Food Freeze 35	Union Carbide
FDA Approved Boiler Additives	
Freon 12	
Freon 22	
Freon 112	
Freon 114	

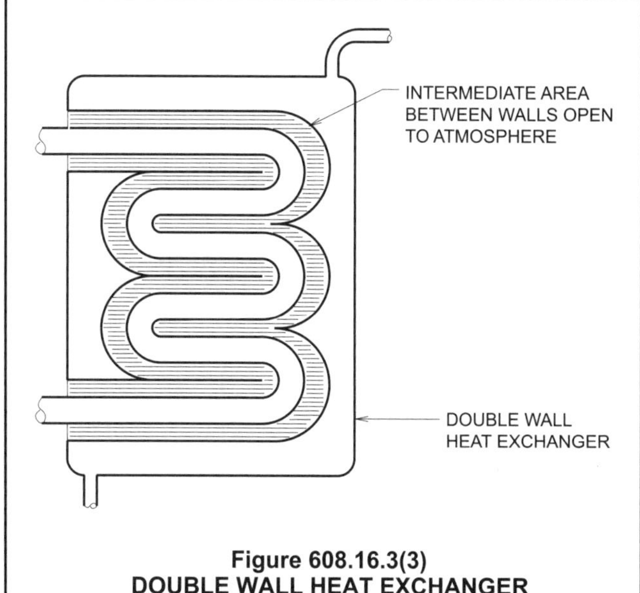

Figure 608.16.3(3)
DOUBLE WALL HEAT EXCHANGER

608.16.4 Connections to automatic fire sprinkler systems and standpipe systems. The potable water supply to automatic fire sprinkler and standpipe systems shall be protected against backflow by a double check-valve assembly or a reduced pressure principle backflow preventer.

Exceptions:

1. Where systems are installed as a portion of the water distribution system in accordance with the requirements of this code and are not provided with a fire department connection, isolation of the water supply system shall not be required.
2. Isolation of the water distribution system is not required for deluge, preaction or dry pipe systems.

❖ A double-check-valve assembly is the minimum form of backflow prevention required between the potable water supply and an automatic fire sprinkler system or standpipe system. Protection by a double-check-valve assembly is permitted only where the sprinkler or standpipe system is filled from a potable source.

A double-check-valve assembly is tested and rated for low-hazard backpressure applications typically associated with automatic sprinkler systems.

If antifreeze or other chemicals are added to a sprinkler or standpipe system or if a nonpotable hazardous secondary supply system is involved, such as water of unknown quality entering from the fire department connection during fire fighting, the potable water supply must be protected with a reduced pressure principle backflow preventer (see Commentary Section 609.16.4.1).

Exception 1 addresses the installation of dual-purpose (combination) water distribution and fire sprinkler piping systems. Where only spot protection is provided with limited-area sprinkler systems and in residential applications, it is not uncommon for fire sprinklers to be supplied from the domestic water service and water distribution system. In such systems, there is no fire department connection through which water can be pumped into the system, and the entire piping system, including sprinkler piping, is constructed of materials that are approved for potable water distribution. Because there is no potential source of backflow contamination in such systems, a backflow preventer is not required [see Commentary Figure 608.16.4(1)].

Exception 1 is not intended to apply to separate (independent) water distribution systems and fire sprinkler systems that share only a common water service. Where the fire sprinklers and water distribution system share no piping in common other than the main supply (water service) and both systems are constructed of materials approved for water distribution, isolation of the fire sprinkler system would still be required [see Commentary Figure 608.16.4(2)].

Exception 2 allows installation of dry-piped sprinkler systems without a backflow preventer. However, this does not supercede the need to provide a double check valve or reduced pressure principle backflow preventer where low- or high-hazard contaminants can be introduced through the fire department connection or auxiliary source of water supply. The following four cases are provided to assist in determining the minimum level of backflow prevention at connections to fire protection systems.

Case 1—A reduced pressure principle backflow preventer is required at the connection of the water supply to a fire protection system if any of the following conditions exist:

- The fire protection system contains underground piping in close proximity to a sanitary sewer, septic tank, seepage pit, privy, cesspool or other piping system carrying high-hazard materials. Table 608.17.1 and Section 703.1 can be used to establish acceptable protection methods, such as vertical and horizontal separations from sources of contamination.
- The fire protection system (wet or dry) is equipped with a fire department connection where the potability of the water entering the fire department connection cannot be guaranteed. This includes water from the tanks of fire trucks containing corrosion inhibitors, antifreeze, extinguishing agents or other chemicals; surface water from lakes, ponds, rivers and reservoirs; and other nonpotable auxiliary water sources such as wells, mills or industrial sources. The above-mentioned water sources have to be close enough to the fire department connection to be used for fire pump suctioning. AWWA Manual 14 identifies 1,700 feet (518 m) as the maximum distance for an auxiliary water source to be considered a potential suction supply.
- The fire protection system is directly connected to a nonpotable secondary water supply, such as any of the examples mentioned above.
- The fire protection system contains chemical additives, antifreeze or liquid foam concentrates used for fighting certain types of fires.

Case 2—A double check-valve assembly is required as the minimum form of backflow protection at the connection of the water supply to a wet fire protection system where none of the conditions in Case 1 exist.

Case 3—No backflow protection is required at the connection of a dry fire protection system (includes dry, preaction and deluge systems), regardless of the type of pipe material, where none of the conditions of Case 1 exist.

Case 4—No backflow protection is required at the connections of sprinkler head branches to multipurpose piping where the sprinkler head branches are constructed of potable water materials. Case 4 addresses only residential sprinklers and limited-area sprinkler systems [see Commentary Figure 608.16.4(1)]. Multipurpose piping is considered to be dual-purpose piping that delivers water for domestic purposes as well as supplying sprinkler heads.

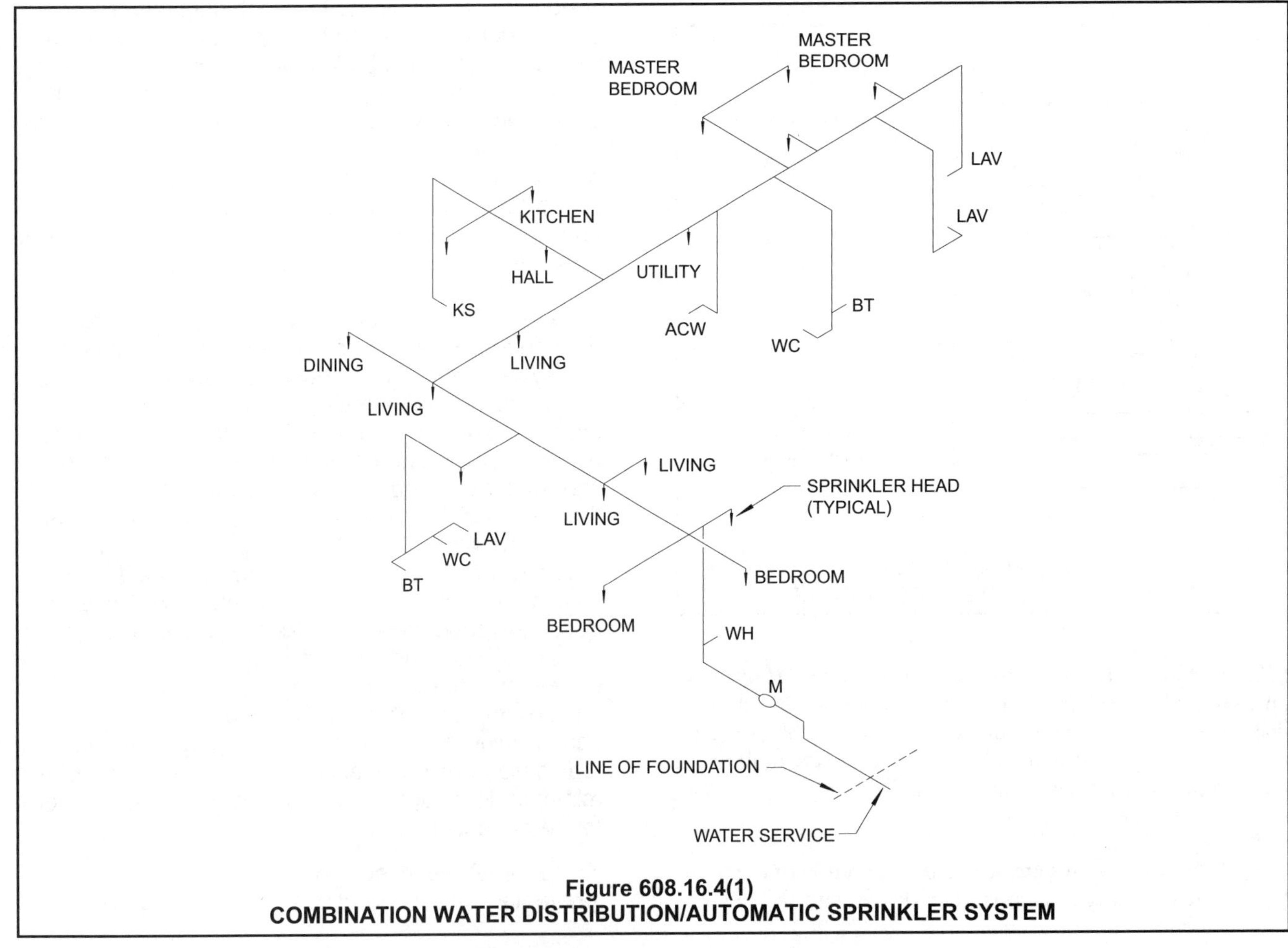

Figure 608.16.4(1)
COMBINATION WATER DISTRIBUTION/AUTOMATIC SPRINKLER SYSTEM

608.16.4.1 Additives or nonpotable source. Where systems under continuous pressure contain chemical additives or antifreeze, or where systems are connected to a nonpotable secondary water supply, the potable water supply shall be protected against backflow by a reduced pressure principle backflow preventer. Where chemical additives or antifreeze are added to only a portion of an automatic fire sprinkler or standpipe system, the reduced pressure principle backflow preventer shall be permitted to be located so as to isolate that portion of the system. Where systems are not under continuous pressure, the potable water supply shall be protected against backflow by an *air gap* or a pipe applied atmospheric vacuum breaker conforming to ASSE 1001 or CAN/CSA B64.1.1.

❖ A nonpotable secondary water supply could include above- or below-ground tanks, private wells, ponds, reservoirs and lakes. In the event of loss or inadequacy of the primary potable supply, the secondary nonpotable water source is pumped into the system, thereby contaminating the potable supply in the event backflow occurs. In some localities, fire departments will use mobile equipment to pump water from nonpotable sources through the fire department connection into the sprinkler system. Where chemicals, additives, antifreeze or nonpotable water are introduced into a sprinkler system, a reduced pressure principle backflow preventer is required. It is not uncommon for a reduced pressure principle backflow preventer to be installed on a branch of the sprinkler system to isolate only that portion of the system containing antifreeze.

Fire suppression systems that are not under continuous pressure are classified as indirect cross connections. When subject to only backsiphonage, these systems may be protected by an atmospheric-type vacuum breaker conforming to ASSE 1001 or CSA B64.1.1. These devices protect the potable water supply against pollutants or contaminants that enter the system as a result of backsiphonage through the outlet. Under backsiphonage conditions, a small amount of water may exit through the air ports. These devices must be installed downstream of the last control valve, not subject to backpressure nor more than 12 hours of continuous water pressure.

608.16.5 Connections to lawn irrigation systems. The potable water supply to lawn irrigation systems shall be protected against backflow by an atmospheric-type vacuum breaker, a pressure-type vacuum breaker or a reduced pressure principle backflow preventer. A valve shall not be installed downstream from an atmospheric vacuum breaker. Where chemicals are introduced into the system, the potable water supply shall be protected against backflow by a reduced pressure principle backflow preventer.

❖ Lawn irrigation systems are not considered plumbing systems and begin at the point of connection to the potable water distribution (the point of connection being the backflow preventer interface). Irrigation system outlets can be below grade level, exposing them to contamination. The piping in irrigation systems is not regulated by the code. For these reasons, the minimum form of backflow prevention is an atmospheric vacuum breaker.

However, most irrigation systems use valved sprinklers and zone valve controls that could render atmospheric vacuum breakers inoperative; therefore, in most cases atmospheric vacuum breakers will not be permitted. Pressure-type vacuum breakers are designed to operate under pressure at the device outlet. These devices are permitted for all irrigation systems except those having chemical injection. Where fertilizers, herbicides, pesticides or any other chemicals can be introduced into an irrigation system, the potable water supply must be protected by a reduced pressure principle backflow preventer.

It is not uncommon for lawn maintenance personnel to pump chemicals directly into an irrigation system, thereby creating a hazardous cross connection subject to backpressure. Installation of pumps, connections for pumping equipment and auxiliary air tanks on

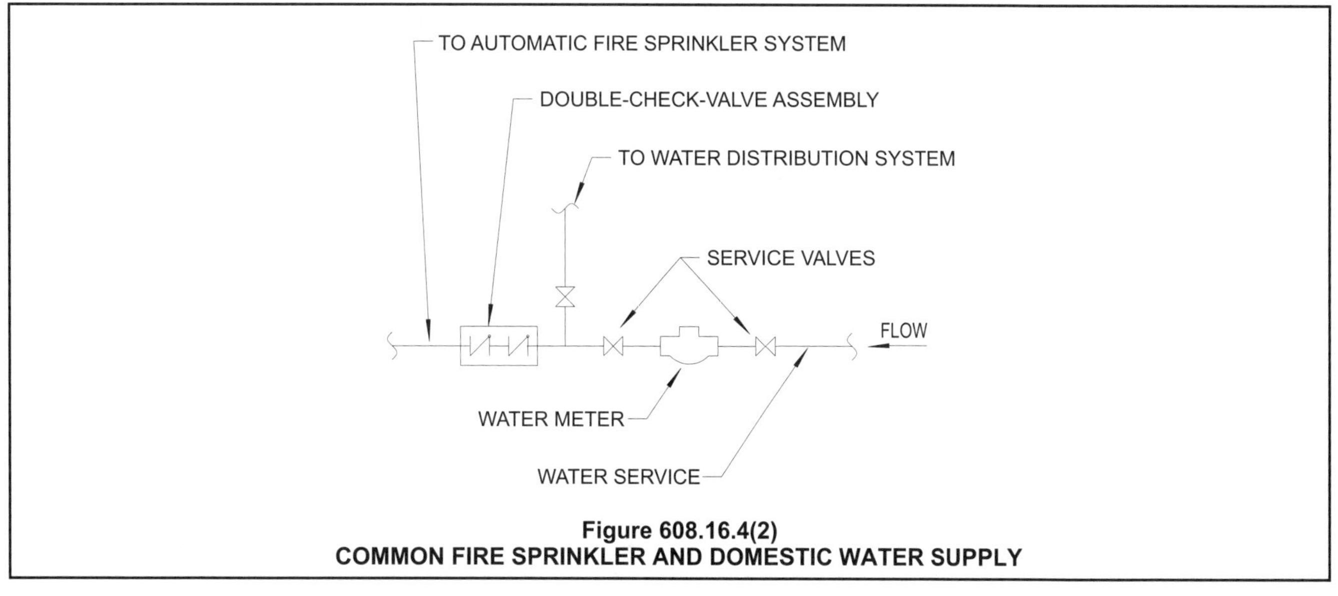

Figure 608.16.4(2)
COMMON FIRE SPRINKLER AND DOMESTIC WATER SUPPLY

irrigation systems is also common to improve pressure. These types of installations produce backpressure within the system. Under these circumstances, the only acceptable means of backflow prevention is a reduced pressure principle backflow preventer.

Because of the varying nature and unpredictable usage of irrigation systems, most installations are equipped with pressure-type vacuum breakers or reduced pressure principle backflow preventers.

608.16.6 Connections subject to backpressure. Where a potable water connection is made to a nonpotable line, fixture, tank, vat, pump or other equipment subject to backpressure, the potable water connection shall be protected by a reduced pressure principle backflow preventer.

❖ A reduced pressure principle backflow preventer is the only acceptable means of protecting a direct connection between the potable water supply and a source of high-hazard contamination that can create a back-pressure on the potable water supply. The backpressure can be created by mechanical means, such as a pump, or by static pressure head, including an elevated tank or pipe arrangement.

A direct connection to a pressurized source of contamination is the most dangerous form of cross connection; therefore, the most stringent means of backflow prevention is required (see Commentary Figure 608.16.6).

This section is not intended to supersede the requirements of Section 608.16.4, which permits the use of a double-check-valve assembly for automatic fire sprinkler systems. Although a connection by the fire department during a fire will likely elevate the automatic sprinkler pressure beyond the domestic water supply pressure (resulting in backpressure), the double check-valve backflow prevention assembly is tested and rated for backpressure applications typically associated with automatic sprinkler systems. However, a double-check-valve assembly is not rated for high-hazard applications. Therefore, the potential sources of contamination must be analyzed before a backflow prevention device is selected (see Commentary Section 608.16.4).

608.16.7 Chemical dispensers. Where chemical dispensers connect to the potable water distribution system, the water supply system shall be protected against backflow in accordance with Section 608.13.1, 608.13.2, 608.13.5, 608.13.6, 608.13.8 or 608.13.9.

❖ Chemical dispensers are typically cleaning-related devices having multiple chambers that hold sanitizing solvents. They are stationary self-contained units that, when connected directly to the potable water supply, allow the user to mix the chemicals it contains with water, producing the desired type and strength of cleaning agent. Chemical dispensers are commonly found in a janitor closet or cleaning room in large commercial operations that require a higher-than-average level of sanitation, such as a school or hospital. A connection between the potable water supply and a chemical dispenser creates a situation that is susceptible to backsiphonage with the potential for backpressure. Because of these sources for contamination, chemical dispensers must be protected using an air gap, a reduced principle backflow preventer, a backflow preventer with intermediate vents, a pressure-type vacuum breaker or an atmospheric-type vacuum breaker. After identifying the degree of hazard of the potential contaminants and establishing whether backsiphonage or backpressure can occur, the appropriate device can be selected using Table 608.1.

608.16.8 Portable cleaning equipment. Where the portable cleaning equipment connects to the water distribution system, the water supply system shall be protected against backflow in accordance with Section 608.13.1, 608.13.2, 608.13.3, 608.13.7 or 608.13.8.

❖ Portable cleaning equipment, similar to the chemical dispensers described in Section 608.16.7, are typically cleaning-related devices having multiple chambers that hold sanitizing solvents. These are self-contained mobile units that transport the chemicals with the unit. They can be connected directly to the potable

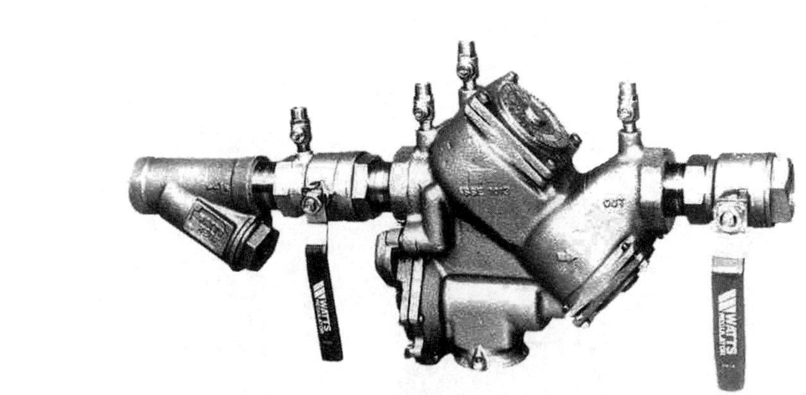

Figure 608.16.6
REDUCED PRESSURE PRINCIPLE BACKFLOW PREVENTER
(Photo courtesy of Watts Regulator Company)

water supply with a hose, which allows the user to mix the chemicals with water, producing the desired type and strength of cleaning agent. Because the location (elevation) of the hose is not restricted, any connection between the potable water supply and portable cleaning equipment is susceptible to backsiphonage with the potential for backpressure. Because of the sources for contamination, portable cleaning equipment must be protected using an air gap, a reduced pressure principle backflow preventer, a backflow preventer with intermediate vents or a double-check-valve assembly.

Because this cleaning equipment is portable, it is imperative that potential points of connection to the potable water supply be properly protected. Some manufacturers of portable cleaning equipment have tried to address this issue by installing a backflow preventer integral with the equipment. When evaluating this equipment, it is important to verify that the integral devices are rated for the type of hazard and installed in a fixed position with the critical level above the equipment outlet. See Section 608.13 for additional discussion on the proper selection of backflow prevention methods.

608.16.9 Dental pump equipment. Where dental pumping equipment connects to the water distribution system, the water supply system shall be protected against backflow in accordance with Section 608.13.1, 608.13.2, 608.13.5, 608.13.6 or 608.13.8.

❖ Dental equipment that is connected to the potable water supply and operates through the use of a pump is susceptible to backsiphonage with the potential for backpressure and therefore must be protected using an air gap, a pressure-type vacuum breaker or an atmospheric-type vacuum breaker.

This equipment could introduce pathogens or become reservoirs of water-borne disease because of contact with the microorganisms in a patient's mouth. These devices include water-flushed cuspidors, water/air syringes, water-powered aspirators/suction systems, water-flushed ultrasonic cleaning apparatus and water-cooled handpieces.

Some of this equipment has built-in air gaps and does not require external protection or is not provided with water power. However, water pumped into water/air syringes and drill handpieces connected to waterlines are still a significant backflow concern. Given the high-hazard nature of dental unit water lines, a backflow preventer must be installed to isolate them from the rest of the plumbing system.

The code requirements of Section 608.16.9 can be found to conflict with dental industry literature; therefore, the code official must consider the intent of these requirements. Biohazard contamination of the potable water supply has some of the most severe implications, which justifies thorough consideration and the necessity for protection against actual or potential sources of contamination. Therefore, each application of water supply use in conjunction with dental equipment must be scrutinized carefully, considering the design of the equipment and how the water supply integrates into the function of the equipment. The slightest potential for cross contamination must be addressed where potable water supply use is intergrated into the function of any medical equipment.

608.16.10 Coffee machines and noncarbonated beverage dispensers. The water supply connection to coffee machines and noncarbonated beverage dispensers shall be protected against backflow by a backflow preventer conforming to ASSE 1022 or by an *air gap*.

❖ The same protection must be provided for these types of devices as one must provide for carbonated drink dispensers.

608.17 Protection of individual water supplies. An individual water supply shall be located and constructed so as to be safeguarded against contamination in accordance with Sections 608.17.1 through 608.17.8.

❖ The important concern with an individual water supply is that the water remains potable and free from any contamination.

608.17.1 Well locations. A potable ground water source or pump suction line shall not be located closer to potential sources of contamination than the distances shown in Table 608.17.1. In the event the underlying rock structure is limestone or fragmented shale, the local or state health department shall be consulted on well site location. The distances in Table 608.17.1 constitute minimum separation and shall be increased in areas of creviced rock or limestone, or where the direction of movement of the ground water is from sources of contamination toward the well.

❖ A well must be separated from sources of contamination to reduce the possibility of affecting the potability of the water. An analysis of the well location must also include an evaluation of soil and drainage conditions of the area.

TABLE 608.17.1
DISTANCE FROM CONTAMINATION TO PRIVATE WATER SUPPLIES AND PUMP SUCTION LINES

SOURCE OF CONTAMINATION	DISTANCE (feet)
Barnyard	100
Farm silo	25
Pasture	100
Pumphouse floor drain of cast iron draining to ground surface	2
Seepage pits	50
Septic tank	25
Sewer	10
Subsurface disposal fields	50
Subsurface pits	50

For SI: 1 foot = 304.8 mm.

❖ The distances specified in the table have been established by soil scientists and engineers. They take into consideration the variety of soil conditions that may exist and have a built-in factor of safety to drastically reduce the possibility of a well becoming contaminated.

608.17.2 Elevation. Well sites shall be positively drained and shall be at higher elevations than potential sources of contamination.

❖ It is always preferable to have a well located at a higher elevation than any potential source of contamination. This is not always possible, however, especially when the source of contamination is located on an adjacent lot.

608.17.3 Depth. Private potable well supplies shall not be developed from a water table less than 10 feet (3048 mm) below the ground surface.

❖ So the water will remain potable, the well must be at least 10 feet (3048 mm) below grade. It is preferable, however, to have a well a minimum of 20 feet (6096 mm) deep. High water tables and shallow wells are more susceptible to contamination because of their close proximity to the ground surface (see Commentary Figure 608.17.3).

608.17.4 Water-tight casings. Each well shall be provided with a water-tight casing to a minimum distance of 10 feet (3048 mm) below the ground surface. All casings shall extend at least 6 inches (152 mm) above the well platform. The casing shall be large enough to permit installation of a separate drop pipe. Casings shall be sealed at the bottom in an impermeable stratum or extend several feet into the water-bearing stratum.

❖ Any contaminant could enter the well with relative ease if the well does not have a proper casing. High water tables are more susceptible to contamination because of their close proximity to the ground surface (see Commentary Figure 608.17.4).

608.17.5 Drilled or driven well casings. Drilled or driven well casings shall be of steel or other *approved* material. Where drilled wells extend into a rock formation, the well casing shall extend to and set firmly in the formation. The annular space between the earth and the outside of the casing shall be filled with cement grout to a minimum distance of 10 feet (3048 mm) below the ground surface. In an instance of casing to rock installation, the grout shall extend to the rock surface.

❖ The most common type of well, a drilled well, can be drilled through rock. This is considered the best all-around method for constructing a well. A driven well is most practical when the water table is relatively shallow [not deeper than 50 feet (15 240 mm)].

The well casing must be installed to prohibit the entrance of contaminants. The grout between the earth

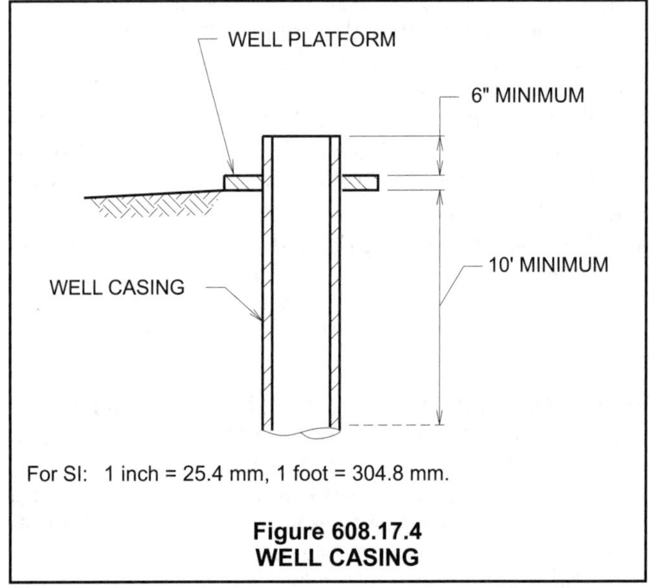

For SI: 1 inch = 25.4 mm, 1 foot = 304.8 mm.

Figure 608.17.4
WELL CASING

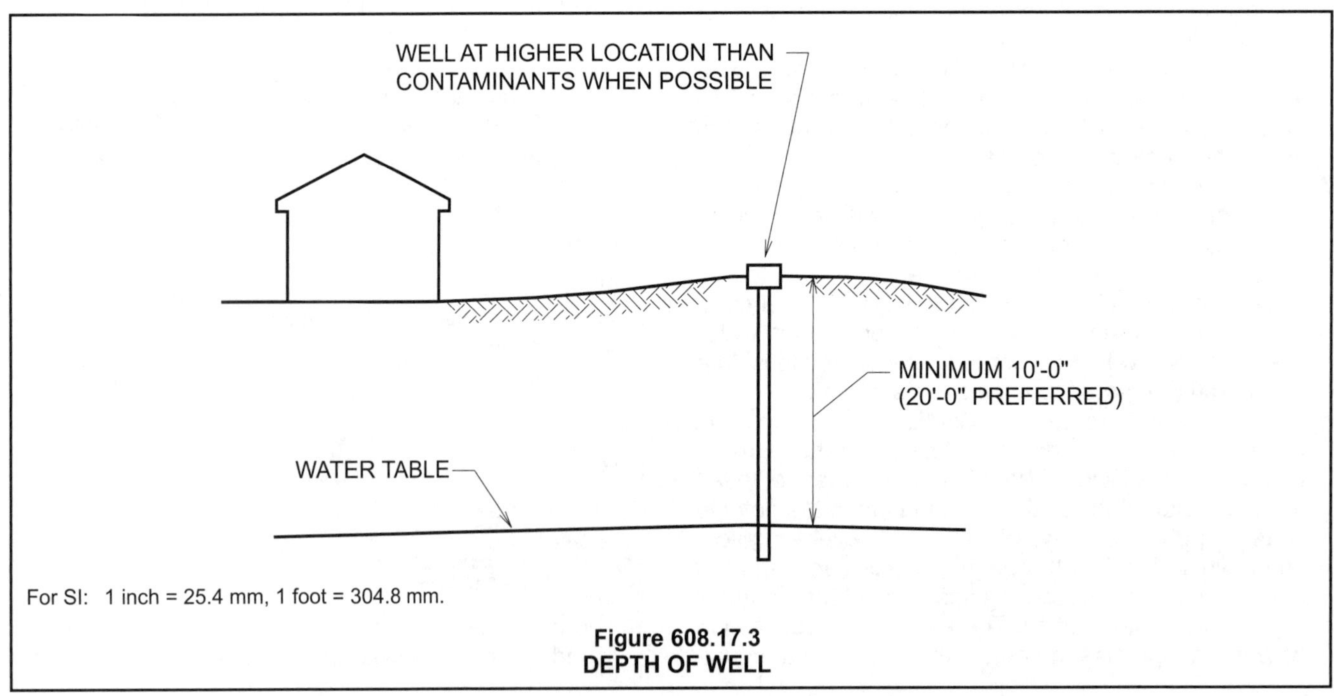

For SI: 1 inch = 25.4 mm, 1 foot = 304.8 mm.

Figure 608.17.3
DEPTH OF WELL

and the well casing provides the well with an additional barrier of protection. Use of a drive shoe is recommended to prevent casing distortion (see Commentary Figure 608.17.5).

608.17.6 Dug or bored well casings. Dug or bored well casings shall be of water-tight concrete, tile, or galvanized or corrugated metal pipe to a minimum distance of 10 feet (3048 mm) below the ground surface. Where the water table is more than 10 feet (3048 mm) below the ground surface, the water-tight casing shall extend below the table surface. Well casings for dug wells or bored wells constructed with sections of concrete, tile, or galvanized or corrugated metal pipe shall be surrounded by 6 inches (152 mm) of grout poured into the hole between the outside of the casing and the ground to a minimum depth of 10 feet (3048 mm).

❖ The opening is much larger with a dug or bored well than with a drilled well. A water-tight casing designed to resist cave-in is required to protect such a well from the entrance of a contaminant.

608.17.7 Cover. Every potable water well shall be equipped with an overlapping water-tight cover at the top of the well casing or pipe sleeve such that contaminated water or other substances are prevented from entering the well through the annular opening at the top of the well casing, wall or pipe sleeve. Covers shall extend downward at least 2 inches (51 mm) over the outside of the well casing or wall. A dug well cover shall be provided with a pipe sleeve permitting the withdrawal of the pump suction pipe, cylinder or jet body without disturbing the cover. Where pump sections or discharge pipes enter or leave a well through the side of the casing, the circle of contact shall be water tight.

❖ A cover is important to protect the well water from contamination from an outside source. It must fit firmly to the casing with cracks or openings filled tightly to create a water-tight seal (see Commentary Figure 608.17.7).

608.17.8 Drainage. All potable water wells and springs shall be constructed such that surface drainage will be diverted away from the well or spring.

❖ One of the greatest threats to a well is the risk of contamination from surface drainage. To protect against this, surface drainage must be prevented from draining to the well area and must be diverted away from the casing. In flood-hazard and high-hazard zones, well casings must extend at least 1 foot (305 mm) above the base flood elevation, protecting the well from being contaminated by floodwaters.

SECTION 609
HEALTH CARE PLUMBING

609.1 Scope. This section shall govern those aspects of health care plumbing systems that differ from plumbing systems in other structures. Health care plumbing systems shall conform to the requirements of this section in addition to the other requirements of this code. The provisions of this section shall apply to the special devices and equipment installed and maintained in the following occupancies: nursing homes, homes for the aged, orphanages, infirmaries, first aid stations, psychiatric facilities, clinics, professional offices of dentists and doctors, mortuaries, educational facilities, surgery, dentistry, research and testing laboratories, establishments manufacturing pharmaceutical drugs and medicines, and other structures with similar apparatus and equipment classified as plumbing.

❖ This section governs the water supply to health care facilities and fixtures. Hospitals are required to have dual water services so that in the event of a water main or water service pipe break, the hospital will not experience a complete loss of water supply. In health care facilities there are unique plumbing fixtures with special

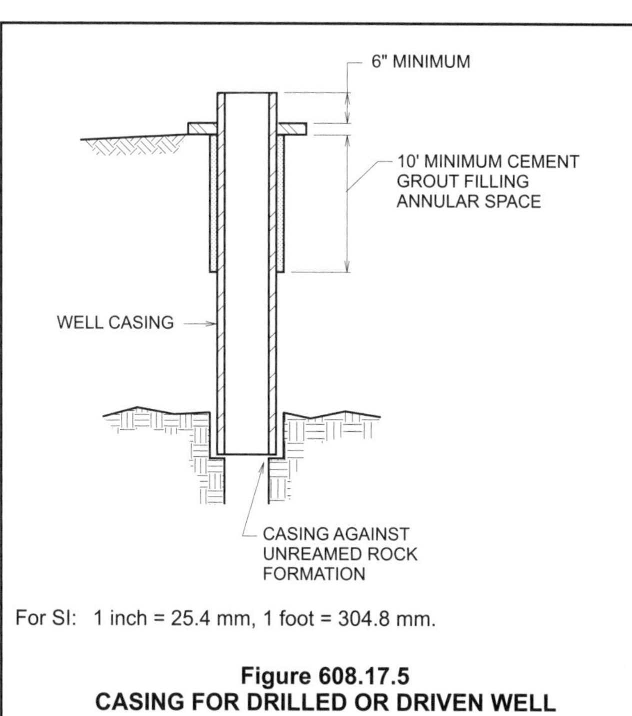

For SI: 1 inch = 25.4 mm, 1 foot = 304.8 mm.

Figure 608.17.5
CASING FOR DRILLED OR DRIVEN WELL

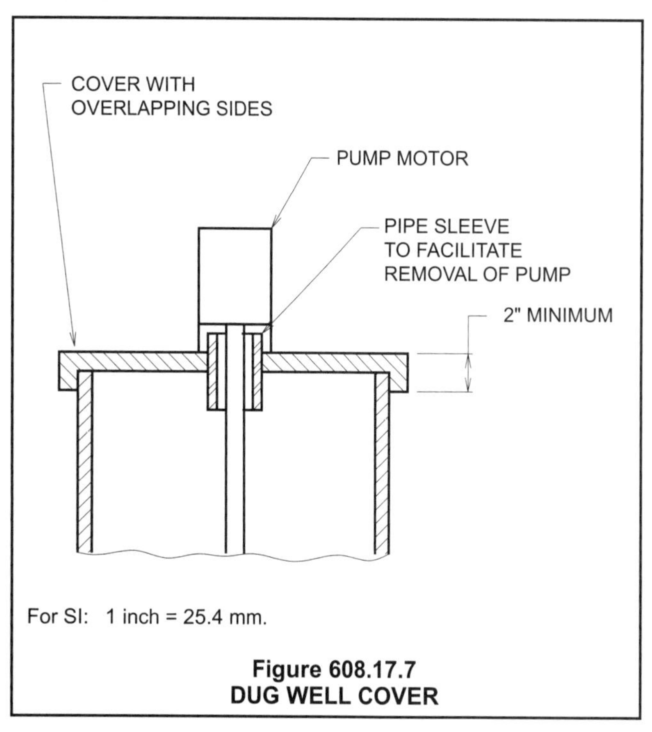

For SI: 1 inch = 25.4 mm.

Figure 608.17.7
DUG WELL COVER

requirements. Although typically found in hospitals, these fixtures could be located in other health care facilities. Note that the list of occupancies is not all-inclusive; it is intended only to be representative of the occupancies in which special plumbing fixtures are regulated by this chapter.

609.2 Water service. All hospitals shall have two water service pipes installed in such a manner so as to minimize the potential for an interruption of the supply of water in the event of a water main or water service pipe failure.

❖ This section applies only to hospitals. It is not applicable to other health care facilities, such as urgent care centers or doctors' offices. The intent of the section is to require two remote water services; however, they can connect to the same water main. Where a single water main is provided, it is desirable to connect the water services on opposite sides of the water main control valve. In the event that the water main control valve is closed, one of the two water service taps will remain in service (see Commentary Figure 609.2).

609.3 Hot water. *Hot water* shall be provided to supply all of the hospital fixture, kitchen and laundry requirements. Special fixtures and equipment shall have hot water supplied at a temperature specified by the manufacturer. The hot water system shall be installed in accordance with Section 607.

❖ The hot water demands of a hospital vary greatly depending on its type of equipment and the services provided. When a laundry is installed, the newer systems have a separate hot water supply system independent of the hospital system. The systems are designed for uninterrupted laundry service.

Kitchen equipment demands also vary depending on the type of kitchen and the accompanying equipment. The design professional must determine the needs of hospital systems when designing the hot water supply system.

609.4 Vacuum breaker installation. Vacuum breakers shall be installed a minimum of 6 inches (152 mm) above the *flood level rim* of the fixture or device in accordance with Section 608. The *flood level rim* of hose connections shall be the maximum height at which any hose is utilized.

❖ Because a vacuum breaker has a critical level and must not be subjected to any backpressure, vacuum breakers with hose connections must be elevated above the highest level that the hose can be raised (see commentary, Section 608.13). Vacuum breakers serving hose connections are typically located 7 feet (2134 mm) or more above the floor.

609.5 Prohibited water closet and clinical sink supply. Jet- or water-supplied orifices, except those supplied by the flush connections, shall not be located in or connected with a water closet bowl or clinical sink. This section shall not prohibit an *approved* bidet installation.

❖ The concern with water closets and clinical sinks is that any submerged water supply outlet must be protected against backflow. If the water supply is through the flush connection, it is protected by the vacuum breaker of the flushing device.

609.6 Clinical, hydrotherapeutic and radiological equipment. All clinical, hydrotherapeutic, radiological or any equipment that is supplied with water or that discharges to the waste system shall conform to the requirements of this section and Section 608.

❖ This section parallels Section 423 in that it addresses any and all specialty-type fixtures, equipment and ap-

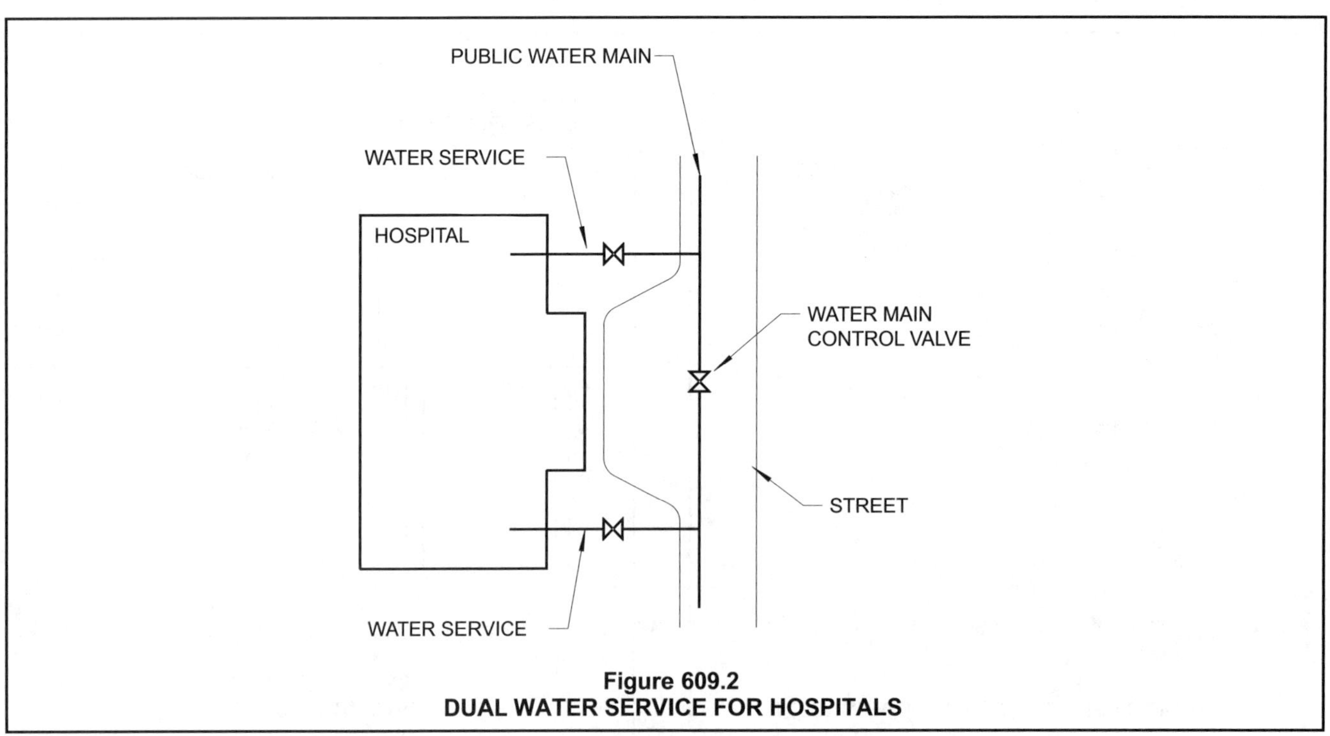

Figure 609.2
DUAL WATER SERVICE FOR HOSPITALS

pliances that would not generally be classified as plumbing fixtures. The intent is to protect the potable water supply from backflow contamination that could be caused by the variety of special health-care-related equipment supplied by the potable water distribution system. All such fixtures and equipment must comply with the provisions in Section 609 and with the water supply protection requirements of Section 608.

Because such fixtures and equipment are not normally viewed as plumbing fixtures, they are commonly overlooked during the design or plumbing inspection stages. Specialty fixtures and equipment could pose a backflow hazard equal to or greater than any typical plumbing fixture (see Commentary Section 608).

609.7 Condensate drain trap seal. A water supply shall be provided for cleaning, flushing and resealing the condensate trap, and the trap shall discharge through an *air gap* in accordance with Section 608.

❖ An indirect waste receptor or floor drain receiving the discharge of condensate in a health care facility must have a water supply for flushing the receptor and resealing the trap should the sealing water evaporate. There must be an air gap between the water supply opening and the waste receptor to protect against backflow.

609.8 Valve leakage diverter. Each water sterilizer filled with water through directly connected piping shall be equipped with an *approved* leakage diverter or bleed line on the water supply control valve to indicate and conduct any leakage of unsterile water away from the sterile zone.

❖ Sterilized water must be protected against leakage from the unsterile potable water supply.

SECTION 610 DISINFECTION OF POTABLE WATER SYSTEM

610.1 General. New or repaired potable water systems shall be purged of deleterious matter and disinfected prior to utilization. The method to be followed shall be that prescribed by the health authority or water purveyor having jurisdiction or, in the absence of a prescribed method, the procedure described in either AWWA C651 or AWWA C652, or as described in this section. This requirement shall apply to "on-site" or "in-plant" fabrication of a system or to a modular portion of a system.

1. The pipe system shall be flushed with clean, potable water until dirty water does not appear at the points of outlet.
2. The system or part thereof shall be filled with a water/chlorine solution containing at least 50 parts per million (50 mg/L) of chlorine, and the system or part thereof shall be valved off and allowed to stand for 24 hours; or the system or part thereof shall be filled with a water/chlorine solution containing at least 200 parts per million (200 mg/L) of chlorine and allowed to stand for 3 hours.
3. Following the required standing time, the system shall be flushed with clean potable water until the chlorine is purged from the system.
4. The procedure shall be repeated where shown by a bacteriological examination that contamination remains present in the system.

❖ Every potable water system must be disinfected before being placed in service. Disinfection methods require more than just flushing the system with potable water. As outlined in this section, the disinfection of a system usually includes the use of some form of chlorine solution.

Before a system is chlorinated, the main must be filled to eliminate air pockets and flushed to remove any particulates. Calcium hypochlorite granules are placed in pipe sections in accordance with AWWA C651. This procedure provides a strong chlorine concentration to remove any residual contamination. This is an important procedure for the protection of public health. Newly installed or repaired potable water piping systems can contain dirt, debris, solder fluxes, chemicals and disease-causing organisms.

SECTION 611 DRINKING WATER TREATMENT UNITS

611.1 Design. Drinking water treatment units shall meet the requirements of NSF 42, NSF 44, NSF 53 or NSF 62.

❖ This section contains the requirements for point of use and point of entry drinking water treatment units. These units are intended to augment the treatment of the drinking water by the municipality and to provide user treatment for private wells. Water delivered to a building is expected to be within established guidelines for health, odor and chemical content. These guidelines are established by the U.S. Department of Health and are applicable to potable water sources, whether they are public municipal water treatment facilities or individual private wells. Even though the guidelines make certain that the water is safe to consume, there are times when the water supply has an objectionable odor or dissolved materials within it are noticeable to the occupants of the building, particularly in drinking water.

NSF 42 contains requirements for units that are designed to improve the perceived quality of drinking water supplied to the occupants of the building. These units are typically installed either at the final distribution point of the drinking water, referred to as "point of use" units, or at a central location where several fixtures can be served from the same treatment unit, referred to as "point of entry" units.

Regardless of the location of the treatment units, their function is the same. These units are designed to reduce the amount of specific contaminants present in the water supply. The contaminants affect the aesthetic quality of the water, causing occupants to smell an offensive odor or to notice some chemical or particulate in the water. Most water treatment units are designed to remove certain odors or dissolved materials in the drinking water supply.

The user or installer of the treatment unit is responsible for the selection of the proper unit based on the offensive characteristics of the water and the unit's reported performance capabilities. Treatment units require periodic maintenance so they will continue to function as designed. Because of this requirement, the treatment unit must be installed in a location where it is accessible for maintenance.

NSF 53, similar to the NSF 42 standard discussed above, contains the standards for units designed to improve the quality of drinking water by removing dissolved materials or substances that might prove to have a negative effect on the health of the building occupants. These materials or substances could be microbiological, chemical or particulate and are often considered either an established or potential health hazard. Although the exact source of the potentially harmful material or substances is not known, the intent of these treatment units is to remove these materials or substances from the water just prior to its consumption by the occupants.

611.2 Reverse osmosis systems. The discharge from a reverse osmosis drinking water treatment unit shall enter the drainage system through an *air gap* or an *air gap* device that meets the requirements of NSF 58.

❖ Similar to the drinking water treatment units discussed in Section 611.1, the reverse osmosis systems manufactured in accordance with NSF 58 are intended to remove specific contaminants from public or private drinking water supplies considered to be micro-biologically safe. The treatment units designed in conformance with NSF 58 are intended to reduce the total dissolved solids (TDS) and other contaminants present in the drinking water, many of which are considered to be either established or potential health hazards.

A reverse osmosis system reduces the objectionable dissolved materials by the use of pressure to force the water through a filter membrane or medium in such a manner that the concentration level of the discharged water is less than the concentration of the incoming water. Reverse osmosis treatment systems are intended for installation as a point of use system. This means that the treatment unit serves the drinking water fixture located immediately adjacent to the treatment system.

Treatment units manufactured in accordance with this standard are to be clearly identified as conforming to this standard, including the specific performance claims of the unit as verified and substantiated by test data. The waste discharge from a reverse osmosis drinking water treatment unit must be designed, constructed and located so that the discharge orifice is directed downward and the lower edge of the outlet must be at an elevation not less than 2 inches (51 mm) above the flood rim of the waste receptacle.

611.3 Connection tubing. The tubing to and from drinking water treatment units shall be of a size and material as recommended by the manufacturer. The tubing shall comply with NSF 14, NSF 42, NSF 44, NSF 53, NSF 58 or NSF 61.

❖ The tubing used for drinking water treatment units must be of a size and material recommended by the manufacturer. The tubing materials are certified during the evaluation of the equipment for compliance with materials or products that come into contact with drinking water.

SECTION 612
SOLAR SYSTEMS

612.1 Solar systems. The construction, installation, alterations and repair of systems, equipment and appliances intended to utilize solar energy for space heating or cooling, domestic hot water heating, swimming pool heating or process heating shall be in accordance with the *International Mechanical Code*.

❖ Solar energy can be used for a variety of purposes including space heating, space cooling, domestic water heating, swimming pool heating and process heating. Therefore, the scope encompasses all of these potential solar energy applications and includes the construction, installation, alteration and repair of all systems and equipment.

SECTION 613
TEMPERATURE CONTROL DEVICES AND VALVES

613.1 Temperature-actuated mixing valves. Temperature-actuated mixing valves, which are installed to reduce water temperatures to defined limits, shall comply with ASSE 1017.

❖ ASSE 1017 is the standard for temperature-actuated mixing valves for hot water distribution systems. Water mixing (also defined as "tempering" or "blending") valves are used extensively in applications for domestic service to mix hot and cold water to reduce high service water temperature to the building distribution piping system.

These devices are not intended for final temperature control at the fixtures and appliances (see ASSE 1016). Final control must be provided by suitable individual mixing devices, which should be supplemented by appropriate high-limit alarms or temperature-limiting devices at the point of use. These valves are designed for automatic control of hot water temperature and to maintain this temperature within a reasonable degree of uniformity.

Bibliography

The following resource materials are referenced in this chapter or are relevant to the subject matter addressed in this chapter.

ASME A112.1.2-1991 (R2002), *Air Gaps in Plumbing Systems*. New York, NY: American Society of Mechanical Engineers, 2002.

ASME A112.1.3-00, *Air Gap Fittings for Use with Plumbing Fixtures, Appliances and Appurtenances*. New York, NY: American Society of Mechanical Engineers, 2000.

ASME A112.18.1-03, *Plumbing Fixture Fittings*. New York, NY: American Society of Mechanical Engineers, 2003.

ASME A112.18.3M-02, *Performance Requirements for Backflow Protection Devices and Systems in Plumbing Fixture Fittings*. New York, NY: American Society of Mechanical Engineers, 2002.

ASME A112.18.6-03, *Flexible Water Connectors*. New York, NY: American Society of Mechanical Engineers, 2003.

ASME A112.19.2M-03, *Vitreous China Plumbing Fixtures with 1996 Errata*. New York, NY: American Society of Mechanical Engineers, 2003.

ASME A112.19.6-95, *Hydraulic Performance Requirements for Water Closets and Urinals*. New York, NY: American Society of Mechanical Engineers, 1995.

ASSE 1003-01, *Performance Requirements for Water Pressure Reducing Valves*. Westlake, OH: American Society of Sanitary Engineering, 2001.

ASSE 1013-99, *Performance Requirements for Reduced Pressure Principle Backflow Preventers and Reduced Pressure Fire Protection Principle Backflow Preventers*. Westlake, OH: American Society of Sanitary Engineering, 1999.

ASSE 1016-96, *Performance Requirements for Individual Thermostatic, Pressure Balancing and Combination Control Valves for Bathing Facilities*. Westlake, OH: American Society of Sanitary Engineering, 1996.

ASSE 1020-98, *Performance Requirements for Pressure Vacuum Breaker Assembly*. Westlake, OH: American Society of Sanitary Engineering, 1998.

ASSE 1047-99, *Performance Requirements for Reduced Pressure Detector Fire Protection Backflow Prevention Assemblies*. Westlake, OH: American Society of Sanitary Engineering, 1999.

ASSE 1055-97, *Performance Requirements for Backflow Devices for Chemical Dispensing Systems*. Westlake, OH: American Society of Sanitary Engineering, 1997.

ASSE 1056-01, *Performance Requirements for Back Siphonage Vacuum Breaker*. Westlake, OH: American Society of Sanitary Engineering, 2001.

ASSE 1060-96, *Performance Requirements for Outdoor Enclosures for Backflow Prevention Assemblies*. Westlake, OH: American Society of Sanitary Engineering, 1996.

ASSE 1066-97, *Performance Requirements for Individual Pressure Balancing Valves for Individual Fixture Fittings*. Westlake, OH: American Society of Sanitary Engineering, 1997.

ASTM A 53/A 53M-02, *Specification for Pipe, Steel, Black and Hot-dipped, Zinc-coated Welded and Seamless*. West Conshohocken, PA: ASTM International, 2002.

ASTM A 733-03, *Specification for Welded and Seamless Carbon Steel and Austenitic Stainless Steel Pipe Nipples*. West Conshohocken, PA: ASTM International, 2003.

ASTM B 32-03, *Specification for Solder, Metal*. West Conshohocken, PA: ASTM International, 2003.

ASTM B 42-02e01, *Specification for Seamless Copper Pipe, Standard Sizes*. West Conshohocken, PA: ASTM International, 2002.

ASTM B 43-04, *Specification for Seamless Red Brass Pipe, Standard Sizes*. West Conshohocken, PA: ASTM International, 2004.

ASTM B 75-02, *Specification for Seamless Copper Tube*. West Conshohocken, PA: ASTM International, 2002.

ASTM B 88-03, *Specification for Seamless Copper Water Tube*. West Conshohocken, PA: ASTM International, 2003.

ASTM B 302-02, *Specification for Threadless Copper Pipe, Standard Sizes*. West Conshohocken, PA: ASTM International, 2002.

ASTM B 447-03, *Specification for Welded Copper Tube*. West Conshohocken, PA: ASTM International, 2003.

ASTM B 687-99, *Specification for Brass, Copper, and Chromium-plated Pipe Nipples*. West Conshohocken, PA: ASTM International, 1999.

ASTM B 813-00e01, *Specification for Liquid and Paste Fluxes for Soldering of Copper and Copper Alloy Tube*. West Conshohocken, PA: ASTM International, 1993.

ASTM B 828-02, *Practices for Making Capillary Joints by Soldering of Copper and Copper Alloy Tube and Fittings*. West Conshohocken, PA: ASTM International, 2002.

ASTM C 296-00, *Specification for Asbestos-Cement Pressure Pipe*. West Conshohocken, PA: ASTM International, 2000.

ASTM D 1527-99e01, *Specification for Acrylonitrile-Butadiene-Styrene (ABS) Plastic Pipe, Schedules 40 and 80*. West Conshohocken, PA: ASTM International, 1999.

ASTM D 1785-04, *Specification for Poly (Vinyl Chloride)(PVC) Plastic Pipe, Schedules 40, 80 and 120*. West Conshohocken, PA: ASTM International, 2004.

ASTM D 1869-1995(2000), *Specification for Rubber Rings for Asbestos-Cement Pipe*. West Conshohocken, PA: ASTM International, 2000.

ASTM D 2235-01, *Specification for Solvent Cement for Acrylonitrile-butadiene-styrene (ABS) Plastic Pipe and Fittings*. West Conshohocken, PA: ASTM International, 2001.

ASTM D 2239-03, *Specification for Polyethylene (PE) Plastic Pipe (SIDR-PR) Based on Controlled Inside Diameter*. West Conshohocken, PA: ASTM International, 2003.

ASTM D 2241-04a, *Specification for Poly (Vinyl Chloride) (PVC) Pressure-Rated Pipe (SDR-Series)*. West Conshohocken, PA: ASTM International, 2004.

ASTM D 2282-99e01, *Specification for Acrylonitrile-Butadiene-Styrene (ABS) Plastic Pipe (SDR-PR)*. West Conshohocken, PA: ASTM International, 1999.

ASTM D 2468-96a, *Specification for Acrylonitrile-Butadiene-Styrene (ABS) Plastic Pipe and Fittings Schedule 40*. West Conshohocken, PA: ASTM International, 1996.

ASTM D 2662-96a, *Specification for Polybutylene (PB) Plastic Pipe (SDR-PR) Based on Controlled Inside Diameter*. West Conshohocken, PA: ASTM International, 1996.

ASTM D 2666-96a, *Specification for Polybutylene (PB) Plastic Tubing*. West Conshohocken, PA: ASTM International, 1996.

ASTM D 2672-96a (2003), *Specification for Joints for IPS PVC Pipe Using Solvent Cement*. West Conshohocken, PA: ASTM International, 2003.

ASTM D 2737-03, *Specification for Polyethylene (PE) Plastic Tubing*. West Conshohocken, PA: ASTM International, 2003.

ASTM D 2846/D 2846M-99, *Specification for Chlorinated Poly (Vinyl Chloride) (CPVC) Plastic Hot and Cold Water Distribution Systems*. West Conshohocken, PA: ASTM International, 1999.

ASTM D 2855-96 (2002), *Standard Practice for Making Solvent-Cemented Joints with Poly (Vinyl Chloride) (PVC) Pipe and Fittings*. West Conshohocken, PA: ASTM International, 2002.

ASTM D 3309-96a (2002), *Specification for Polybutylene (PB) Plastic Hot and Cold Water Distribution Systems*. West Conshohocken, PA: ASTM International, 2002.

ASTM F 441/F 441 M-02, *Specification for Chlorinated Poly (Vinyl Chloride) (CPVC) Plastic Pipe, Schedules 40 and 80*. West Conshohocken, PA: ASTM International, 2002.

ASTM F 442/F 442M-99, *Specification for Chlorinated Poly (Vinyl Chloride) (CPVC) Plastic Pipe (SDR-PR)*. West Conshohocken, PA: ASTM International, 1999.

ASTM F 493-04, *Specification for Solvent Cements for Chlorinated Poly (Vinyl Chloride) (CPVC) Plastic Pipe and Fittings*. West Conshohocken, PA: ASTM International, 2004.

ASTM F 656-02, *Specification for Primers for Use in Solvent Cement Joints of Poly (Vinyl Chloride) (PVC) Plastic Pipe and Fittings*. West Conshohocken, PA: ASTM International, 2002.

ASTM F 877-02e01, *Specification for Cross-linked Polyethylene (PEX) Plastic Hot- and Cold-water Distribution Systems*. West Conshohocken, PA: ASTM International, 2002.

ASTM F 1807-04, *Specification for Metal Insert Fittings Utilizing a Copper Crimp Ring SDR9 Cross-linked Polyethylene (PEX) Tubing*. West Conshohocken, PA: ASTM International, 2004.

AWWA C104-98, *Standard for Cement-mortar Lining for Ductile-iron Pipe and Fittings for Water*. Denver, CO: American Water Works Association, 1998.

AWWA C111-00, *Standard for Rubber-gasket Joints for Ductile-iron Pressure Pipe and Fittings*. Denver, CO: American Water Works Association, 2000.

AWWA C115-99, *Standard for Flanged Ductile-iron Pipe with Threaded Flanges*. Denver, CO: American Water Works Association, 1999.

AWWA C151-02, *Standard for Ductile-iron Pipe, Centrifugally Cast for Water*. Denver, CO: American Water Works Association, 2002.

AWWA C651-99, *Disinfecting Water Mains*. Denver, CO: American Water Works Association, 1999.

CSA B125-01, *Plumbing Fittings*. Rexdale (Toronto), Ontario, Canada: Canadian Standards Association, 2001.

CSA B137.6-02, *CPVC Pipe, Tubing and Fittings for Hot and Cold Water Distribution Systems–with Revisions through May 1986*. Rexdale (Toronto), Ontario, Canada: Canadian Standards Association, 2002.

Hunter, Roy B. *Water-distribution Systems for Buildings*. Washington, DC: National Bureau of Standards, NBS BMS, 1941.

IECC-09, *International Energy Conservation Code*. Washington, DC: International Code Council, Inc., 2009.

IFGC-09, *International Fuel Gas Code*. Washington, DC: International Code Council, Inc., 2009.

IMC-09, *International Mechanical Code*. Washington, DC: International Code Council, Inc., 2009.

IPSDC-09, *International Private Sewage Disposal Code*. Washington, DC: International Code Council, Inc., 2009.

NSF 14-03, *Plastic Piping Components and Related Materials*. Ann Arbor, MI: National Sanitation Foundation, 2003.

NSF 42-02e, *Drinking Water Treatment Units—Aesthetic Effects*. Ann Arbor, MI: National Sanitation Foundation, 2002.

NSF 53-02e, *Drinking Water Treatment Units—Health Effects*. Ann Arbor, MI: National Sanitation Foundation, 2002.

NSF 58-04, *Reverse Osmosis Drinking Water Treatment Systems*. Ann Arbor, MI: National Sanitation Foundation, 2004.

NSF 61-03e, *Drinking Water System Components—Health Effects*. Ann Arbor, MI: National Sanitation Foundation, 2003.

NSF 62-04, *Drinking Water Distillation Systems*. Ann Arbor, MI: National Sanitation Foundation, 2004.

Chapter 7: Sanitary Drainage

General Comments

The conventional method of sizing a sanitary drainage system is by drainage fixture unit (dfu) load values. The drainage fixture unit approach takes into consideration the probability of a load on a drainage system.

The probability method of sizing drainage systems was developed largely by Dr. Roy Hunter. Through his research, Dr. Hunter attempted to standardize and simplify design principles, while reducing the cost of plumbing systems.

Fixture unit values were determined based on the average rate of discharge by a fixture, the time of a single operation and the frequency of use or interval between operations. The theoretical approach considers a large group of fixtures being connected to the plumbing system with only a small fraction of the total number of fixtures in use simultaneously. The probability method has also been effective in the design of smaller plumbing systems because of the excessive design factors added by Dr. Hunter.

Dr. Hunter sought an adequate design methodology to provide satisfactory service without interruption or inconvenience to the user.

Because the fixture unit values have a built-in probability factor, they cannot be directly translated into flow or discharge rates. For example, one dfu is equivalent to a discharge rate of 1 cubic foot per minute (0.4719 L/s) [approximately 7.5 gallons per minute (gpm) (28 L/min)]. Two independent fixtures, however, each having a value of one dfu, cannot be considered as having a combined discharge rate of 15 gpm (57 L/min) because the dfu value incorporates the element of probability.

Purpose

The purpose of Chapter 7 is to regulate the materials, design and installation of sanitary drainage piping systems as well as the connections made to the system. The intent is to design and install sanitary drainage systems that will function reliably, are neither undersized nor oversized and are constructed from materials, fittings and connections whose quality is regulated by this section.

In the nineteenth century, typhoid fever, cholera and dysentery were a threat to survival. The modern plumbing system, with proper drainage piping, has been one of the main reasons for the elimination of these diseases.

Medical professionals give much of the credit to the plumbing profession for improvements in health and longevity. Medicine alone would have only a marginal effect without improved sanitation practices.

SECTION 701 GENERAL

701.1 Scope. The provisions of this chapter shall govern the materials, design, construction and installation of sanitary drainage systems.

❖ This section establishes that design and installation provisions for sanitary drainage systems are located in Chapters 3, 7 and 10.

Chapter 7 regulates the design, sizing, components and installation of a sanitary drainage system. Many other chapters relate to Chapter 7 because they regulate components of sanitary drainage systems. For example, Chapters 3, 9 and 10 regulate pipe support, venting and traps, respectively, all of which are combined to create a sanitary drainage system.

Piping in connection with swimming pool water recirculation and filtering is not covered by this code.

701.2 Sewer required. Every building in which plumbing fixtures are installed and all premises having drainage piping shall be connected to a *public sewer*, where available, or an *approved private* sewage disposal system in accordance with the *International Private Sewage Disposal Code.*

❖ Plumbing fixtures installed in a building or structure must be directly connected to a drainage system. The sanitary drainage systems must connect to the public systems provided by the local jurisdiction or public utility if they are available. If a public system is not available, the drainage system must connect to an approved private sewage disposal system in accordance with the *International Private Sewage Disposal Code®* (IPSDC®), which contains requirements for the design and installation of private sewage disposal systems. This section must not be construed to prevent the indirect waste systems required by Chapter 8.

701.3 Separate sewer connection. Every building having plumbing fixtures installed and intended for human habitation, occupancy or use on premises abutting on a street, alley or easement in which there is a *public sewer* shall have a separate connection with the *sewer*. Where located on the same lot, multiple buildings shall not be prohibited from connecting to a common *building sewer* that connects to the *public sewer.*

❖ This requirement intends to prohibit the combining of sewers serving different buildings prior to connection to the public sewer. The only exception is where the sewers to be combined are serving buildings on the same lot or parcel of land. This section does not prohibit the use of adjoining properties that have been included in a dedicated easement approved by the administrative authority. Commentary Figure 701.3(1) illustrates the use of a dedicated public sewer easement to accommodate Buildings A and B. The common building sewer is actually an extension of the public sewer and under control of the public authority.

When the sanitary discharge from more than one building is connected to the public sewer by means of a shared or common sewer line, the responsibility for the maintenance of the common sewer line can become a problem. Should a stoppage develop in the common portion of the drain, all parties connected to the drain could disclaim responsibility for having the stoppage cleared. This presents a problem because there is no one party to hold responsible for the maintenance of the drain. Also, replacement or repairs of a sewer are the responsibility of all parties sharing a common drain connection, and the process of determining the point of origin of the wastes into the common drain is more difficult. The first sentence of this section is intended to apply to the connection of more than one building where each building is under separate ownership, and the second sentence applies in instances where connection of more than one structure is proposed where all the buildings are under the same ownership. The commercial property owner, such as a shopping complex owner, may want to build another building on his or her property, and connecting the new building sewer to existing sewers on the same property would be convenient and less expensive than making a new connection to the public sewer [see Commentary Figure 701.3.(2)].

701.4 Sewage treatment. Sewage or other waste from a plumbing system that is deleterious to surface or subsurface waters shall not be discharged into the ground or into any waterway unless it has first been rendered innocuous through subjection to an *approved* form of treatment.

❖ The discharge of untreated waste and sewage from a sanitary drainage system creates an environmental and human health hazard. This section requires treatment of sewage that poses a hazard before discharging to ponds, rivers, streams, lakes, aquifers or similar bodies of water. The form of treatment for the waste must be approved by the code official and is often governed by the local health department or state or U.S. Environmental Protection Agency (EPA) having jurisdiction. These requirements are also applicable to gray-water recycling systems, as stated in Appendix C of the code.

701.5 Damage to drainage system or public sewer. Wastes detrimental to the *public sewer* system or to the functioning of the sewage-treatment plant shall be treated and disposed of in accordance with Section 1003 as directed by the code official.

❖ Materials that are detrimental to the drainage system must not discharge into such system. Any substances that will clog, produce explosive mixtures, degrade the pipes or their joints or any wastes that interfere with the sewage disposal or treatment process must not be allowed to enter the building drainage system. Sections 803.2 and 1003.1 contain requirements for collecting harmful materials and the neutralization of corrosive wastes (see commentary, Sections 803.2 and 1003.1).

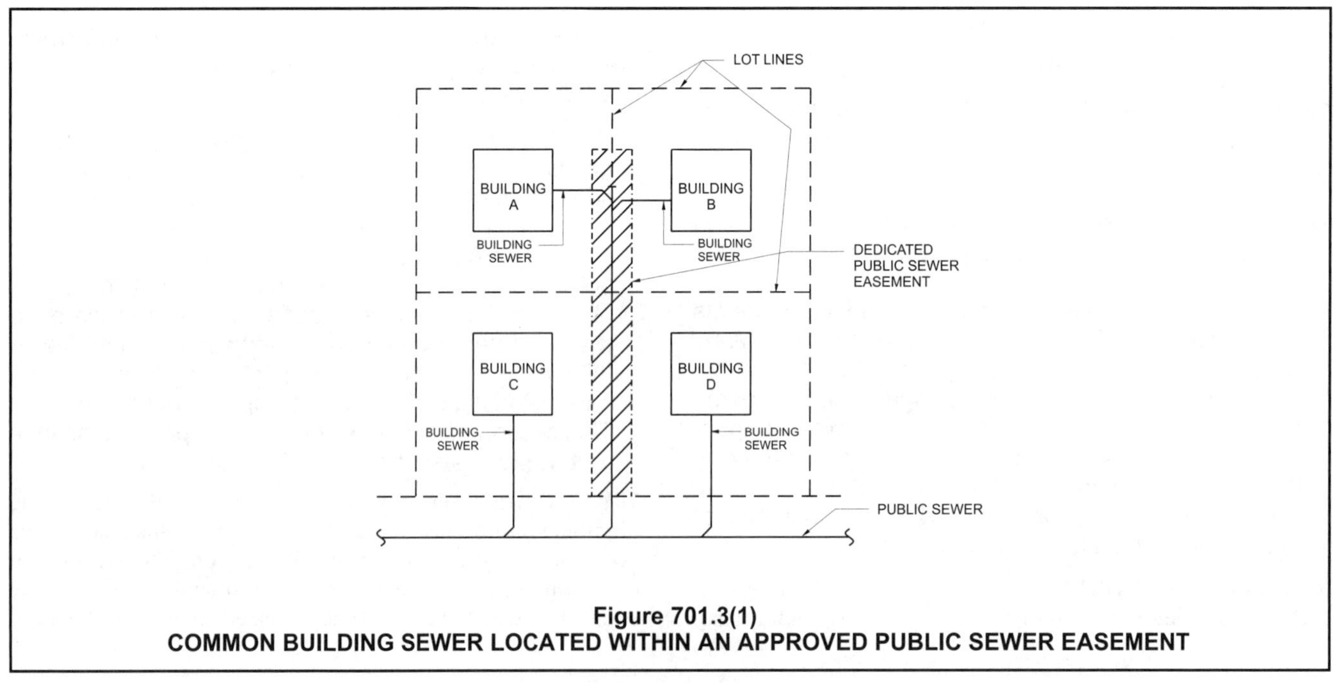

Figure 701.3(1)
COMMON BUILDING SEWER LOCATED WITHIN AN APPROVED PUBLIC SEWER EASEMENT

701.6 Tests. The sanitary drainage system shall be tested in accordance with Section 312.

❖ Requirements for sanitary drainage and vent system testing are described in Section 312.

701.7 Connections. Direct connection of a steam exhaust, blowoff or drip pipe shall not be made with the building drainage system. Wastewater when discharged into the building drainage system shall be at a temperature not higher than 140°F (60°C). When higher temperatures exist, *approved* cooling methods shall be provided.

❖ The temperature of wastewater is limited to a maximum of 140°F (60°C) to protect the sanitary drainage piping. Excessive high-temperature wastes are detrimental to the piping system because of expansion and contraction resulting therefrom. For example, pipe joints may become disturbed or pulled apart, or solidly bedded piping can be strained or broken. Section 803.1 also provides limitations on the temperature for special waste discharges and prohibits the connection of steam piping to the drainage system (see commentary, Section 803.1). Such waste must be cooled, and material and installation methods must be approved by the code official. Typically, this is accomplished by piping the high-temperature discharge to a sump or cooling tank. The high-temperature waste may also be used to preheat the cold water supply to water heaters, thus accomplishing two things at once and conserving energy at the same time. Another method after cooling is to discharge the waste into a branch of the building drain where it can mix with the total sanitary sewage flow from the building, and no temperature effects may be transmitted into the building's piping system in the event the cooling process shuts down or is inoperative at times.

701.8 Engineered systems. Engineered sanitary drainage systems shall conform to the provisions of Sections 105.4 and 714.

❖ Engineered sanitary drainage systems are designed using approved computer program methods and may not comply with all the provisions prescribed in this chapter. The system design must be approved by the code official in accordance with Section 105.4 and must conform to the provisions of Section 714.

701.9 Drainage piping in food service areas. Exposed soil or waste piping shall not be installed above any working, storage or eating surfaces in food service establishments.

❖ A potential for contamination exists in areas where piping is exposed. This protection is necessary for health, safety and sanitation of food service areas. Many design alternatives are available, such as secondary containment, ceilings and perimeter installations that avoid these areas.

SECTION 702 MATERIALS

702.1 Above-ground sanitary drainage and vent pipe. Above-ground soil, waste and vent pipe shall conform to one of the standards listed in Table 702.1.

❖ This section contains material requirements for all sanitary drainage, waste and vent (DWV) systems. The referenced table specifies allowable piping materials for above-ground DWV systems. In selecting pipe and fittings, it is important to consider their reliability for installation in a drainage system. The corrosion resistance of the pipe and fittings is a measure of the ability of the materials used to resist both the internal corrosive effects of the effluent likely to flow through them

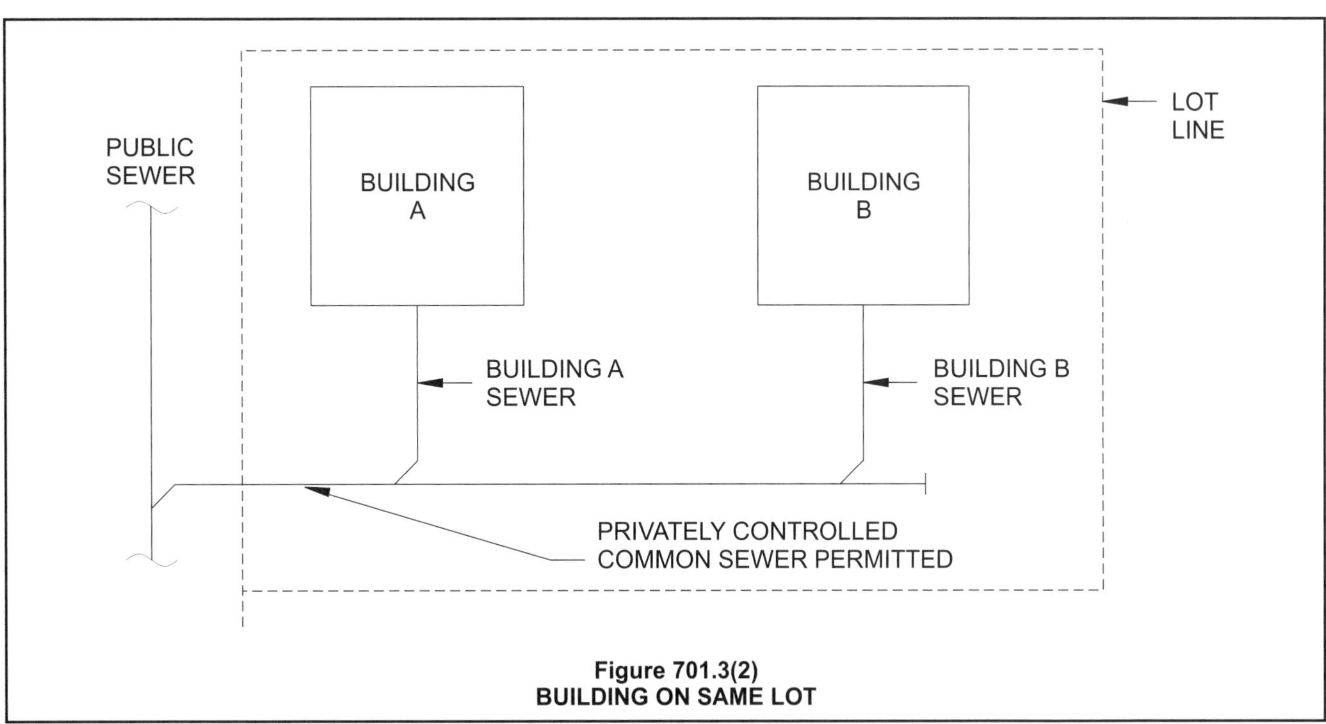

Figure 701.3(2)
BUILDING ON SAME LOT

and the effects of soils on their exterior. Corrosion can be reduced or eliminated by the application of coating, lining and cathodic protection. Physical strength of the pipe and fittings is another consideration because the physical damage may occur either during installation or after being placed in service. Fire-resistance rating of the pipe and fittings is a measure of their ability to remain intact and not fail, or to carry water, during a fire. Additionally, pipe, fittings, joints, hangers and supports should be selected taking this consideration into account.

TABLE 702.1
ABOVE-GROUND DRAINAGE AND VENT PIPE

MATERIAL	STANDARD
Acrylonitrile butadiene styrene (ABS) plastic pipe in IPS diameters, including Schedule 40, DR 22 (PS 200) and DR 24 (PS 140); with a solid, cellular core or composite wall	ASTM D 2661; ASTM F 628; ASTM F 1488; CSA B181.1
Brass pipe	ASTM B 43
Cast-iron pipe	ASTM A 74; ASTM A 888; CISPI 301
Copper or copper-alloy pipe	ASTM B 42; ASTM B 302
Copper or copper-alloy tubing (Type K, L, M or DWV)	ASTM B 75; ASTM B 88; ASTM B 251; ASTM B 306
Galvanized steel pipe	ASTM A 53
Glass pipe	ASTM C 1053
Polyolefin pipe	ASTM F 1412; CAN/CSA B181.3
Polyvinyl chloride (PVC) plastic pipe in IPS diameters, including schedule 40, DR 22 (PS 200), and DR 24 (PS 140); with a solid, cellular core or composite wall	ASTM D 2665; ASTM F 891; ASTM F 1488; CSA B181.2
Polyvinyl chloride (PVC) plastic pipe with a 3.25-inch O.D. and a solid, cellular core or composite wall	ASTM D 2949, ASTM F 1488
Polyvinylidene fluoride (PVDF) plastic pipe	ASTM F 1673; CAN/CSA B181.3
Stainless steel drainage systems, Types 304 and 316L	ASME A112.3.1

❖ Table 702.1 identifies the materials that are approved and the standards with which these materials must comply.

ABS plastic pipe.

There are two types of acrylonitrile butadiene styrene (ABS) plastic pipe used for drainage, waste and vent. One is a solid-wall pipe (ASTM D 2661 or CSA B181.1), and the other has a cellular core (ASTM F 628). The solid-wall pipe is produced by either a single extrusion or a simultaneous multiple coextrusion. ABS plastic pipe with a cellular core is produced by a coextrusion process with concentric inner and outer solid ABS layers and the core consisting of closed-cell framed ABS plastic.

Schedule 40 ABS is black and is marked on two sides 180 degrees (3.14 rad) apart at least every 2 feet (610 mm) with the following information:

- The manufacturer's name.
- The designation ASTM D 2661, ASTM F 628 or CSA B181.1.
- Nominal pipe size.
- Single Extrusion Pipe "ABS DWV."
- Coextruded Pipe "COEX ABS DWV."
- Coextruded ABS Pipe with Cellular Core "COEX ABS CELLULAR CORE DWV."

ASTM D 2661 and CSA B181.1 are the referenced standards for both pipe and drainage fittings. ASTM F 628 is the referenced standard for cellular core ABS plastic.

Brass pipe.

See the commentary to Table 605.3, Brass pipe.

Cast-iron pipe.

It is manufactured in three classifications: service (standard) weight, extra-heavy weight and hubless. This pipe is usually lined internally with cement or coal-tar enamel and is coated externally with a variety of materials to reduce corrosion by soils.Two types of pipe ends are manufactured: hub-and-spigot and hubless. ASTM A 74 is the referenced standard for hub-and-spigot and is available in diameters from 3 to 15 inches (76 to 381 mm) and manufactured in 5-and 10-foot (1524 mm and 3048 mm) lengths. CISPI 301 and ASTM A 888 are the referenced standards for hubless pipe, which is available in diameters from $1^1/_2$ to 15 inches (38 to 381 mm) and manufactured in 5-, 10- and 18-foot (1524, 3048 and 5486 mm) lengths. The hub and spigot ends may be joined by either caulking or an elastomeric compression gasket. Hubless ends can be joined only by external compression couplings.

Cast iron is marked continuously on the barrel with the following information:

- The manufacturer's name or registered trademark.
- The country of origin.
- Letters to indicate the proper classification, extra heavy (XH) or service weight (SV).

Copper or copper-alloy pipe.

See the commentary to Table 605.3, Copper or copper-alloy pipe.

Copper or copper-alloy tubing (Type K, L, M or DWV).

ASTM B 306 is the referenced standard for Type DWV copper tubing, which has a thinner pipe wall than Type M copper. The tubing is produced only in a drawn (hard) temper (see commentary, Table 605.3, Copper or copper-alloy tubing, Type K, L, M or DWV). Type DWV copper tubing is identified with a continuous yellow marking with the following information:

- The manufacturer's name or registered trademark and "DWV."
- The country of origin.

Galvanized steel pipe.

See the commentary to Table 605.3, Galvanized steel pipe.

Glass pipe.

Borosilicate glass pipe is low-expansion, low-alkali-content glass pipe that is extremely chemical resistant. Glass pipe is used primarily for above-ground chemical waste drainage systems. ASTM C 1053 is the referenced standard for borosilicate glass pipe and fittings used for drainage, waste and vent systems and govern the chemical requirements, operating temperatures, pressure rating and dimensions for glass pipe and fittings. Glass piping has a minimum rated operating temperature of -40°F (-40°C) and a maximum operating temperature of 212°F (100°C). The maximum rated working pressure in DWV piping systems is the pressure caused by gravity flow. Glass pipe is manufactured in $1^{1}/_{2}$-, 2-, 3-, 4- and 6-inch (38, 51, 76, 102 and 152 mm) diameters and standard lengths of 5 and 10 feet (1524 and 3048 mm).

Polyolefin pipe.

Polyolefin is a term used to describe a family of plastics that includes polyethylene (PE) and polypropylene (PP). ASTM F1412 and CSA B181.3 are the referenced standards for polyolefin pipe and fittings used in laboratory drainage systems for corrosive applications.This material may not be compatible with all chemicals, making consultation with the manufacturer necessary to determine the suitability of the pipe for a particular application (see commentary, Section 702.5). Polyolefin is identified with the following information:

- The manufacturer's name or trademark.
- The designation CSA B181.3 or the wording "Lab Drainage."
- Nominal pipe size.
- Raw material designation.

Polyvinyl chloride (PVC) plastic pipe.

PVC plastic pipe for drainage and vent systems is available in solid wall, coextruded with either a cellular core or composite pipe.The pipe is either white or a milky white (cream) and is continuously marked with the manufacturer's name, the ASTM or CSA standard and Type DWV. Cellular core pipe is also marked COEX CELLULAR CORE PVC PIPE, and composite pipe is marked with the applicable dimension ratio (DR) and pipe stiffness (PS), for example, PS DR35/PS50 or sewer and drain series DR35/PS50. If the pipe also conforms to the requirements for water service pipe, it will be double marked. ASTM D 2665 and CAN/CSA-B181.2 are the referenced standards for PVC plastic pipe and fittings. The pipe has the same dimensions as standard-weight (Schedule 40) steel pipe. For this reason, it is often called "Schedule 40 PVC drain pipe." ASTM F 891 is the referenced standard for coextruded cellular core PVC pipe. The standard specifies various grades of pipe. Only Schedule 40 pipe, also designated as Type DWV, is approved for above-ground use. ASTM F 1488 is the referenced standard for coextruded composite PVC pipe. The standard specifies various grades of pipe based on different DR and PS applications. IPS Schedule 40 and IPS DR-PS Series are used for above- or below-ground installation for drain, waste and vent pipe. Sewer and Drain DR-PS Series is used for gravity flow sewer and drain pipe. ASTM D 2949 is the referenced standard for 3-inch (76 mm) PVC plastic pipe and fittings only. The pipe has an outside diameter of $3^{1}/_{4}$ inches (83 mm), as compared to $3^{1}/_{2}$ inches (89 mm) for other 3-inch (76 mm) PVC plastic pipe. The inside diameter is the same as a 3-inch (76 mm) pipe conforming to ASTM D 2665. The resulting pipe wall is $^{1}/_{8}$ inch (3.2 mm) less in thickness.

The $3^{1}/_{4}$-inch (83 mm) outside diameter pipe was developed to permit installation of PVC plastic pipe in a nominal 2-inch by 4-inch (51 mm by 102 mm) stud wall cavity. The pipe is often incorrectly identified as Schedule 30 PVC plastic pipe or thin wall pipe (see commentary, Table 605.3, Polyvinyl chloride plastic pipe).

Stainless steel pipe.

Stainless steel pipe for drainage and vent systems is made of a stainless steel core alloy, types 304 and 316L. Stainless steel pipe is used for both DWV and chemical waste systems. ASME A112.3.1 is the referenced standard for stainless steel pipe and fittings. For building drainage and vent piping, Type 304 is approved for use only in above-ground applications and Type 316L is approved for use only in below-ground applications. Either type may be used for building sewers. Stainless steel pipe is manufactured in 2-, 3-, 4- and 6-inch (51, 76, 102 and 152 mm) diameters and in lengths of 3.3, 4.9, 6.6, 9.8 and 16.4 feet (1006, 1494, 2012, 2987 and 4999 mm). Stainless steel pipe is identified with the following information:

- Manufacturer's name or trademark.
- Designation ASME A112.3.1.
- Nominal pipe size.
- Date of manufacture.

702.2 Underground building sanitary drainage and vent pipe. Underground building sanitary drainage and vent pipe shall conform to one of the standards listed in Table 702.2.

❖ The referenced table specifies the acceptable piping materials for underground drainage and vent systems.

TABLE 702.2. See below.

❖ The materials listed in this table are for installation underground within a building. A more detailed description of each material is found in the commentary to Tables 702.1 and 702.3.

TABLE 702.3. See next column.

❖ Table 702.3 identifies the materials that are permitted and the standards with which these materials must comply.

702.3 Building sewer pipe. *Building sewer* pipe shall conform to one of the standards listed in Table 702.3.

❖ The referenced table specifies the acceptable piping materials for sanitary sewer systems.

TABLE 702.2
UNDERGROUND BUILDING DRAINAGE AND VENT PIPE

MATERIAL	STANDARD
Acrylonitrile butadiene styrene (ABS) plastic pipe in IPS diameters, including schedule 40, DR 22 (PS 200) and DR 24 (PS 140); with a solid, cellular core, or composite wall	ASTM D 2661; ASTM F 628; ASTM F 1488; CSA B181.1
Asbestos-cement pipe	ASTM C 428
Cast-iron pipe	ASTM A 74; ASTM A 888; CISPI 301
Copper or copper-alloy tubing (Type K, L, M or DWV)	ASTM B 75; ASTM B 88; ASTM B 251; ASTM B 306
Polyolefin pipe	ASTM F 1412; CAN/CSA B181.3
Polyvinyl chloride (PVC) plastic pipe in IPS diameters, including schedule 40, DR 22 (PS 200) and DR 24 (PS 140); with a solid, cellular core, or composite wall	ASTM D 2665; ASTM F 891; ASTM F 1488; CSA B181.2
Polyvinyl chloride (PVC) plastic pipe with a 3.25-inch O.D. and a solid, cellular core, or composite wall	ASTM D 2949, ASTM F 1488
Polyvinylidene fluoride (PVDF) plastic pipe	ASTM F 1673; CAN/CSA B181.3
Stainless steel drainage systems, Type 316L	ASME A 112.3.1

TABLE 702.3
BUILDING SEWER PIPE

MATERIAL	STANDARD
Acrylonitrile butadiene styrene (ABS) plastic pipe in IPS diameters, including schedule 40, DR 22 (PS 200) and DR 24 (PS 140); with a solid, cellular core or composite wall	ASTM D 2661; ASTM F 628; ASTM F 1488; CSA B181.1
Acrylonitrile butadiene styrene (ABS) plastic pipe in sewer and drain diameters, including SDR 42 (PS 20), PS 35, SDR 35 (PS 45), PS 50, PS 100, PS 140, SDR 23.5 (PS 150) and PS 200; with a solid, cellular core or composite wall	ASTM F 1488; ASTM D 2751
Asbestos-cement pipe	ASTM C 428
Cast-iron pipe	ASTM A 74; ASTM A 888; CISPI 301
Concrete pipe	ASTM C14; ASTM C76; CAN/CSA A257.1M; CAN/CSA A257.2M
Copper or copper-alloy tubing (Type K or L)	ASTM B 75; ASTM B 88; ASTM B 251
Polyethylene (PE) plastic pipe (SDR-PR)	ASTM F 714
Polyvinyl chloride (PVC) plastic pipe in IPS diameters, including schedule 40, DR 22 (PS 200) and DR 24 (PS 140); with a solid, cellular core or composite wall	ASTM D 2665; ASTM F 891; ASTM F 1488
Polyvinyl chloride (PVC) plastic pipe in sewer and drain diameters, including PS 25, SDR 41 (PS 28), PS 35, SDR 35 (PS 46), PS 50, PS 100, SDR 26 (PS 115), PS 140 and PS 200; with a solid, cellular core or composite wall	ASTM F 891; ASTM F 1488; ASTM D 3034; CSA B182.2; CSA B182.4
Polyvinyl chloride (PVC) plastic pipe with a 3.25-inch O.D. and a solid, cellular core or composite wall.	ASTM D 2949, ASTM F 1488
Polyvinylidene fluoride (PVDF) plastic pipe	ASTM F 1673; CAN/CSA B181.3
Stainless steel drainage systems, Types 304 and 316L	ASME A112.3.1
Vitrified clay pipe	ASTM C 4; ASTM C 700

702.4 Fittings. Pipe fittings shall be *approved* for installation with the piping material installed and shall comply with the applicable standards listed in Table 702.4.

❖ Fittings are designed to be installed in a particular system with a given material or combination of materials.

Drainage pattern fittings must be installed in drainage systems. Vent fittings are limited to the venting system. Many fittings are intended for use only for water distribution systems. There are also fittings available that may be used in any type of plumbing system. Fittings must be of the same material as, or compatible with, the pipe. This prevents any chemical or corrosive action between dissimilar materials. Many pipe standards also include fittings. There are a number of other standards that strictly regulate pipe fittings. For example, ASTM D 3311 addresses fitting patterns for ABS and PVC complying with ASTM D 2661, D 2665, F 628 and F 891. It does not address material or performance requirements, only fitting geometric and laying lengths. So in some instances, several referenced standards may regulate a single fitting.

TABLE 702.4. See next column.

❖ The table specifies acceptable materials for pipe fittings, along with the appropriate standards with which the various fittings must comply.

702.5 Chemical waste system. A chemical waste system shall be completely separated from the sanitary drainage system. The chemical waste shall be treated in accordance with Section 803.2 before discharging to the sanitary drainage system. Separate drainage systems for chemical wastes and vent pipes shall be of an *approved* material that is resistant to corrosion and degradation for the concentrations of chemicals involved.

❖ A measure of the degree of acidity or alkalinity of a liquid is its pH value, which defines the concentration of free hydrogen ions on a scale of 0 to 14. A pH value of 7 is considered neutral. As the value decreases, the liquid becomes more acidic; as the value increases, it becomes more alkaline. Mildly alkaline liquids are not harmful to a piping system. When the pH of a liquid is less than 6.8, it is considered to be corrosive. A drainage and vent system that handles waste with a pH lower than 5.5 is called an "acid waste system." This type of system is usually found in hospitals, research laboratories, food processing plants, photo labs, battery rooms and similar facilities.

There is no universal material available to resist all chemical action. Each pipe has qualities that make it resistant to certain chemical waste systems. An individual analysis of compatibility is required for the installation of a chemical waste and vent system. Piping material such as borosilicate glass and polyolefin plastic is often used for special waste applications and is listed in Table 702.1; however, not all materials commonly used for chemical waste and vent systems are listed in the code. Materials such as high-silicon iron pipe, lead pipe and ceramic glazed or unglazed vitrified clay are commonly used. Some of those materials are proprietary products and cannot be listed in the code.

Acid wastes in the vapor state are more corrosive than in the liquid state; therefore, it is important to provide material that resists chemical action for the vent piping as well as for the drainage piping. The code official must evaluate the material or rely on an evaluation by authoritative sources, and approve the product as an alternative material. Section 803 governs the requirements for discharge of special wastes into the sanitary drainage system.

**TABLE 702.4
PIPE FITTINGS**

MATERIAL	STANDARD
Acrylonitrile butadiene styrene (ABS) plastic pipe in IPS diameters	ASTM D 2661; ASTM F 628; CSA B181.1
Acrylonotrile butadiene styrene (ABS) plastic pipe in sewer and drain diameters	ASTM D 2751
Asbestos cement	ASTM C 428
Cast iron	ASME B 16.4; ASME B 16.12; ASTM A 74; ASTM A 888; CISPI 301
Copper or copper alloy	ASME B 16.15; ASME B 16.18; ASME B 16.22; ASME B 16.23; ASME B 16.26; ASME B 16.29
Glass	ASTM C 1053
Gray iron and ductile iron	AWWA C 110
Malleable iron	ASME B 16.3
Polyolefin	ASTM F 1412; CAN/CSA B181.3
Polyvinyl chloride (PVC) plastic in IPS diameters	ASTM D 2665; ASTM F 1866
Polyvinyl chloride (PVC) plastic pipe in sewer and drain diameters	ASTM D 3034
Polyvinyl chloride (PVC) plastic pipe with a 3.25-inch O.D.	ASTM D 2949
Polyvinylidene fluoride (PVDF) plastic pipe	ASTM F 1673; CAN/CSA B181.3
Stainless steel drainage systems, Types 304 and 316L	ASME A 112.3.1
Steel	ASME B 16.9; ASME B 16.11; ASME B 16.28

ABS plastic pipe.

ASTM D 2661, D 2751 and F 628 are the referenced standards for ABS sewer pipe and fittings. ASTM D 2751 is the referenced standard for pipe that is produced in standard dimension ratios (SDRs) of 23.5, 35 and 42. SDR is the ratio between the average outside diameter of the pipe and the minimum wall thickness (see commentary, Table 702.1).

Asbestos-cement pipe.

The pipe is rated by type of material and class. Classes are based on the minimum crush load in pound-force per square foot (Pa). The different classes include 1,500, 2,400, 3,300, 4,000, 5,000, 6,000 and 7,000. There are two types of pipe material: Types I and II. Both types correspond to the chemical requirements for the material. Asbestos cement is made from portland cement and asbestos fibers. The pipe is produced in diameters of 4 through 42 inches (102 through 1067 mm).

Concrete pipe.

The pipe is made from portland cement, blast-furnace slag cement or portland-pozzolan cement-and-aggregate mixture. The pipe is produced both reinforced and nonreinforced. ASTM C 14 and CAN/CSA 257.1 are the referenced standards for nonreinforced concrete pipe. ASTM C 14 concrete pipe is available in three classes: 1, 2 or 3. The pipe classification relates to the concrete strength based on diameter: Class 1 has the lowest strength requirements and Class 3 has the highest.The pipe is manufactured in diameters of 4 through 36 inches (102 through 914 mm). ASTM C 76 and CAN/CSA 257.2 are the referenced standards for reinforced concrete pipe, which is strengthened with steel wire, hot rod or welded wire fabric. ASTM C 76 pipe is produced in five classes: 1, 2, 3, 4 or 5. The class determines the strength of the concrete, pipe and reinforcement geometry: Class I has the lowest strength requirements and Class 5 has the highest. The pipe is manufactured in sizes 12 inches (305 mm) and larger.

Polyethylene (PE) plastic pipe (SDR-PR).

ASTM F 714 is the referenced standard for polyethylene plastic pipe based on outside diameters of 3.5 inches (90 mm) and larger. This piping material is intended for new construction and insertion renewal of old piping systems used for the transport of municipal sewage, domestic sewage, industrial process liquids and effluents in both pressure and nonpressure systems. All pipes produced under ASTM F 714 are pressure rated. Polyethylene plastics are defined as plastics or resins prepared by the polymerization of no less than 85-percent ethylene and no less than 95 percent of total olefins with additional compounding ingredients.

PVC plastic pipe (Type DWV, SDR 26, SDR 35, SDR 41, PS50 or PS 100).

ASTM F 891 specifies sewer-grade pipe by a PS designation, in addition to Schedule 40 (Type DWV). PS indicates pipe stiffness. The acceptable grades for sewers are PS 50 and PS 100. The numerical designation indicates pipe stiffness in pounds per square inch (psi). The higher value indicates the thicker pipe wall. ASTM D 2665 and D 3034 cover sewer-grade PVC pipe and fittings. The materials may be used interchangeably. Both pipes are produced in an SDR of 35 and 41, with a hub end. The PVC material used to produce the pipe is classified as 12454-B, 12454-C or 13364-B. The pipe is produced in diameters of 4 through 15 inches (102 through 381 mm). CSA B182.2 is the referenced standard for SDR 28, 35 and 41 PVC sewer pipe and fittings, and CAN/CSA-B182.4 is the referenced standard for PVC ribbed sewer pipe and fittings (see Commentary Table 702.1).

Vitrified clay pipe.

The pipe material is surface clay, fire clay, shale or a combination of these materials. The pipe is classified as standard strength, extra strength, extra quality and heavy duty. The classification designates the strength of the pipe. Vitrified clay pipe is produced in diameters of 3 inches (76 mm) and larger. Each length of pipe is marked with the manufacturer's name and the classification of the pipe.

702.6 Lead bends and traps. Lead bends and traps shall not be less than $^1/_8$ inch (3.2 mm) wall thickness.

❖ The walls of lead bends and traps must be at least 0.125 inch (3.2 mm). They are typically installed in chemical waste systems.

SECTION 703 BUILDING SEWER

703.1 Building sewer pipe near the water service. Where the *building sewer* is installed within 5 feet (1524 mm) of the water service, the installation shall comply with the provisions of Section 603.2.

❖ This section contains general requirements that apply to the installation and size of building sewers. Where the building sewer is installed in close proximity to the water service, there is a concern that the water service can be subjected to contamination from a sewer leak or failure. To minimize the possibility of contamination, only materials with high reliability are permitted to be installed in a common trench. See Section 603.2 for additional requirements for water service and building sewer separation.

703.2 Drainage pipe in filled ground. Where a *building sewer* or *building drain* is installed on filled or unstable ground, the drainage pipe shall conform to one of the standards for ABS plastic pipe, cast-iron pipe, copper or copper-alloy tubing, or PVC plastic pipe listed in Table 702.3.

❖ Filled or unstable ground is considered to be a marshy or landfill area. Because of the increased possibility of movement, only the strongest materials with the greatest ability to maintain pipe alignment are permitted to be installed for the building sewer.

703.3 Sanitary and storm sewers. Where separate systems of sanitary drainage and storm drainage are installed in the same property, the sanitary and storm building sewers or drains shall be permitted to be laid side by side in one trench.

❖ The concern for contamination of the storm sewer from a sewer leak or failure is not as critical as that of contamination of the water service; therefore, it is permissible to place the storm and sanitary sewers side by side in the same trench.

703.4 Existing building sewers and drains. Existing building sewers and drains shall connect with new *building sewer* and drainage systems only where found by examination and test to conform to the new system in quality of material. The code official shall notify the owner to make the changes necessary to conform to this code.

❖ Use of an existing drainage system may continue to be used if the piping material is found to be in satisfactory condition. Some existing materials are no longer manufactured; however, the piping may be more than adequate. Materials of this type that may be encountered include wrought-iron pipe, cast-iron medium-weight pipe and cast-iron "victory" pipe.

703.5 Cleanouts on building sewers. Cleanouts on building sewers shall be located as set forth in Section 708.

❖ Improperly located cleanouts may create difficulties in the cleaning of stoppages in the drainage system. Cleanout locations and spacing requirements are stated in Section 708.

SECTION 704 DRAINAGE PIPING INSTALLATION

704.1 Slope of horizontal drainage piping. Horizontal drainage piping shall be installed in uniform alignment at uniform slopes. The minimum slope of a horizontal drainage pipe shall be in accordance with Table 704.1.

❖ This section contains installation requirements for sanitary drainage systems. The minimum desired velocity in a horizontal drain pipe is approximately 2 feet per second (0.61 m/s). This velocity is often referred to as the "scouring velocity," even though it may not always clean the walls of the drainage pipe. The scouring velocity is intended to keep solids in suspension. For example, if the velocity is too low where a drain pipe is excessively oversized, the solids tend to drop out of suspension, settling to the bottom of the pipe. This may eventually result in a drain stoppage. A greater velocity is required to move solids at rest than is required to keep moving solids in suspension. A true scouring velocity would then be higher than the velocity initially used to transport the solid.

TABLE 704.1
SLOPE OF HORIZONTAL DRAINAGE PIPE

SIZE (inches)	MINIMUM SLOPE (inch per foot)
$2^1/_2$ or less	$^1/_4$
3 to 6	$^1/_8$
8 or larger	$^1/_{16}$

For SI: 1 inch = 25.4 mm, 1 inch per foot = 83.3 mm/m.

❖ A simplified method of determining the minimum slope of a drainage pipe is provided. The slope listed for each given pipe size is intended to produce the minimum required drainage flow velocity. A drainage pipe can be installed with greater slopes because a maximum slope is not prescribed. The slope is commonly expressed in percentages as total fall per length of pipe. For example, a 1-percent slope is approximately equivalent to $^1/_8$ inch per foot (10 mm per m) and a 2-percent slope is approximately equal to $^1/_4$ inch per foot (20 mm per m).

704.2 Change in size. The size of the drainage piping shall not be reduced in size in the direction of the flow. A 4-inch by 3-inch (102 mm by 76 mm) water closet connection shall not be considered as a reduction in size.

❖ One of the fundamental requirements of a drainage system is that the size of piping cannot be reduced in the direction of drainage flow. A size reduction would create an obstruction to flow, possibly resulting in a backup of flow, an interruption of service in the drainage system or stoppage in the pipe. A 4-inch by 3-inch (102 mm by 76 mm) water closet connection is not considered to be a reduction in pipe size.

704.3 Connections to offsets and bases of stacks. Horizontal branches shall connect to the bases of stacks at a point located not less than 10 times the diameter of the drainage *stack* downstream from the *stack*. Except as prohibited by Section 711.2, horizontal branches shall connect to horizontal *stack* offsets at a point located not less than 10 times the diameter of the drainage *stack* downstream from the upper *stack*.

❖ At the base of every stack, a phenomenon of flow may occur that is commonly called "hydraulic jump." Hydraulic jump is the rising in the depth of flow above a half full pipe. A horizontal drain designed in accordance with the code is expected to have a normal flow depth of not greater than half full. The rise in flow may be great enough to close off the opening of the pipe, thus creating pressure fluctuations in the system that may affect trap seals. Hydraulic jump occurs within a distance of 10 times the diameter of the drainage stack downstream of the stack connection. To avoid flow interference, backup of flow and extreme pressure fluctuations, horizontal branch connections are prohibited in this area. Note that Section 711.2 prohibits all connections to and within 2 feet (610 mm) above or below a horizontal stack offset that has more than four branch intervals above it. Hydraulic jump also results from the high-velocity vertical flow in the stack making a transition to a slower horizontal flow in the building drain. The initial velocity remains high in the horizontal building drain until flow levels off, achieving a uniform flow condition (see Commentary Figure 704.3). A horizontal stack offset creates flow conditions similar to that created at the base of any other stack. The stack above the offset connects to horizontal piping, thereby causing hydraulic jump.

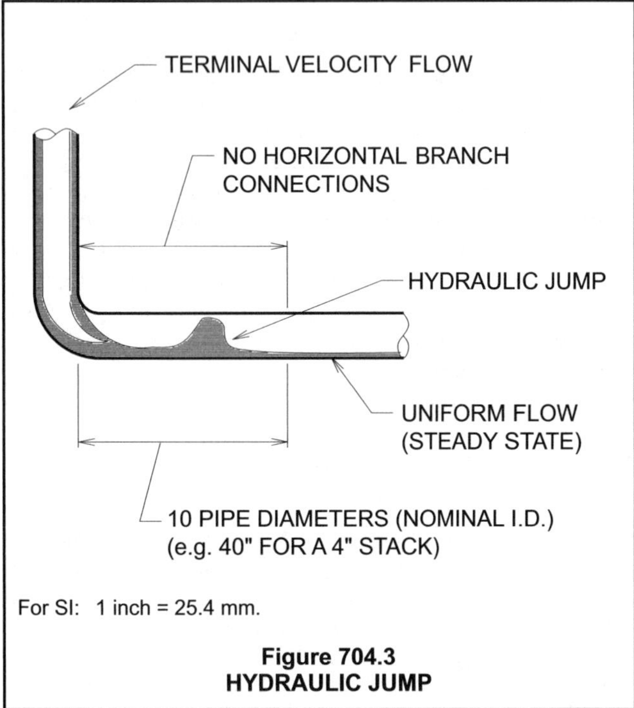

Figure 704.3
HYDRAULIC JUMP

704.4 Future fixtures. Drainage piping for future fixtures shall terminate with an *approved* cap or plug.

❖ If piping is installed for future fixtures, it must be closed with caps or plugs to prevent the escape of sewer gas and waste and the entrance of foreign substances. Section 905.6 also requires a vent to be roughed in for drainage pipe that will serve fixtures to be installed in the future.

SECTION 705 JOINTS

705.1 General. This section contains provisions applicable to joints specific to sanitary drainage piping.

❖ This section contains requirements that apply to materials used for the joining of pipes and the installation of those materials. Acceptable joints and connections for the sanitary drainage piping materials addressed in Section 702 are listed in alphabetical order in this chapter. The main section title identifies the piping material to which the section applies. Subsequent sections list acceptable joints or connections for each particular material. Joints and connections between dissimilar materials are regulated in Section 705.19.

705.2 ABS plastic. Joints between ABS plastic pipe or fittings shall comply with Sections 705.2.1 through 705.2.3.

❖ Acceptable joints for ABS plastic pipe are identified in the referenced sections.

705.2.1 Mechanical joints. Mechanical joints on drainage pipes shall be made with an elastomeric seal conforming to ASTM C 1173, ASTM D 3212 or CSA B602. Mechanical joints shall be installed only in underground systems unless otherwise *approved*. Joints shall be installed in accordance with the manufacturer's instructions.

❖ Mechanical joints are predominantly used in underground applications (e.g., bell or hub compression gasket joints and elastomeric transition couplings). Mechanical joints are typically of the unrestrained type. Above-ground mechanical joints are limited primarily to rehabilitation installations and connections to existing systems when it is impractical to make another type of joint. Mechanical joints are sometimes used where other joining means cannot be used. For example, underground PVC pipe may be joined with an elastomeric coupling where, because of water in the trench, solvent cementing is not possible. The referenced standards are addressed only with respect to the elastomeric material used to make the seal in a mechanical joint, not the entire coupling or fitting assembly.

705.2.2 Solvent cementing. Joint surfaces shall be clean and free from moisture. Solvent cement that conforms to ASTM D 2235 or CSA B181.1 shall be applied to all joint surfaces. The joint shall be made while the cement is wet. Joints shall be made in accordance with ASTM D 2235, ASTM D 2661, ASTM F 628 or CSA B181.1. Solvent-cement joints shall be permitted above or below ground.

❖ Solvent cementing is the most common method of joining ABS plastic pipe and fittings. This section clarifies that solvent cementing is an acceptable joining method for both above-ground and below-ground installations. Although it is the most common method, solvent cementing is one of the most misunderstood methods of joining pipe.

The ASTM and CSA referenced standards indicate how to handle solvent cement safely and recommend a practice for making joints. ABS solvent cement is a flammable liquid and a source of flammable vapors. It must be kept a safe distance away from any source of ignition. Solvent cement should also be used in a well-ventilated area. Prolonged breathing of solvent cement vapors can have adverse health effects. Solvent cement is also an eye and skin irritant. If contact with the eyes is possible, protective goggles should be worn. When frequent contact with the skin is likely, impervious gloves should be worn.

The pipe and fittings should be approximately the same temperature prior to solvent cementing. The pipe should be cut square and chamfered. The pipe end and socket fitting must be clean, dry and free from grease, oil and other foreign materials. The solvent cement must be uniformly applied to the socket of the fitting and to the end of the pipe to the depth of the socket. Excessive solvent cement should be avoided to prevent a puddle of cement from forming on the inside of the pipe joint, which may deform or structurally weaken the fitting or pipe. The joint is made by twisting the pipe into the fitting socket immediately after applying the cement. The pipe will tend to back out unless it is restrained; therefore, it must be held in place for a few seconds until the solvent cement begins to set.

A small bead of solvent cement forms between the pipe exterior and fitting. The bead must be removed with a clean cloth to avoid the possibility of weakening the pipe wall. There should be no rough handling of the pipe for at least 1 hour after the joint is made. A solvent-cement joint should remain for a period of 24 hours before being exposed to working pressure. Because each type of solvent cement is specifically designed for a given piping material, all-purpose solvent cement or universal solvent cement cannot be used to join ABS pipe or fittings unless it conforms to ASTM D 2235 or CSA B181.1.

Solvent cement for each plastic pipe material requires a unique mixture or combination of solvents and dissolved plastic resins to conform to the applicable standard. Although all-purpose cement may dissolve the surfaces of ABS, PVC and CPVC, it may be too aggressive or not aggressive enough to make a joint strong enough to meet the minimum strength requirements of the referenced standards. Solvent cement must be identified on its container as complying with the applicable ASTM or CSA standard.

705.2.3 Threaded joints. Threads shall conform to ASME B1.20.1. Schedule 80 or heavier pipe shall be permitted to be threaded with dies specifically designed for plastic pipe. *Approved* thread lubricant or tape shall be applied on the male threads only.

❖ Pipe thread dimensions can be historically traced to Robert Briggs, mentioned in the commentary to Table 605.3 for galvanized steel pipe. ASME B1.20.1 identifies pipe dimensions as "NPT," which is often thought of as an abbreviation. Some of the terms from which NPT has been shortened are: National Pipe Thread, Nominal Pipe Thread and National Pipe Tapered Thread. NPT is a coded designation: N stands for USA Standard, P indicates pipe and T means that the threads are tapered.

Thread lubricant (pipe-joint compound) must be specifically designed for ABS plastic pipe. It is not the same lubricant that is commonly used for metallic pipe, such as steel. Thread lubricant and tape are designed for two purposes: to lubricate the joint for proper thread mating and to fill in small imperfections on the surfaces of the threads. Field threading of ABS is limited to solid-core Schedule 80 pipe because the wall thickness is sufficient to accommodate threads without affecting pipe strength.

705.3 Asbestos cement. Joints between asbestos-cement pipe or fittings shall be made with a sleeve coupling of the same composition as the pipe, sealed with an elastomeric ring conforming to ASTM D 1869.

❖ The only joining method recognized for asbestos-cement pipe is a mechanical coupling, which makes the seal with an elastomeric O-ring. The ASTM referenced standard regulates the material and dimensional requirements for the O-ring (see Commentary Figure 705.3).

705.4 Brass. Joints between brass pipe or fittings shall comply with Sections 705.4.1 through 705.4.4.

❖ Acceptable joints for brass pipe or fittings are identified in the referenced sections.

705.4.1 Brazed joints. All joint surfaces shall be cleaned. An *approved* flux shall be applied where required. The joint shall be brazed with a filler metal conforming to AWS A5.8.

❖ Although brazed joints are similar to soldered joints, brazed joints are joined at a higher temperature [in excess of 1,000°F (538°C)]. Because of the brazed joint's inherent resistance to elevated temperatures, it is much stronger than a soldered joint. Brazing is often referred to as "silver soldering." Silver soldering is more accurately described as silver brazing and uses high silver-bearing alloys primarily composed of silver, copper and zinc. Silver soldering (brazing) typically re-

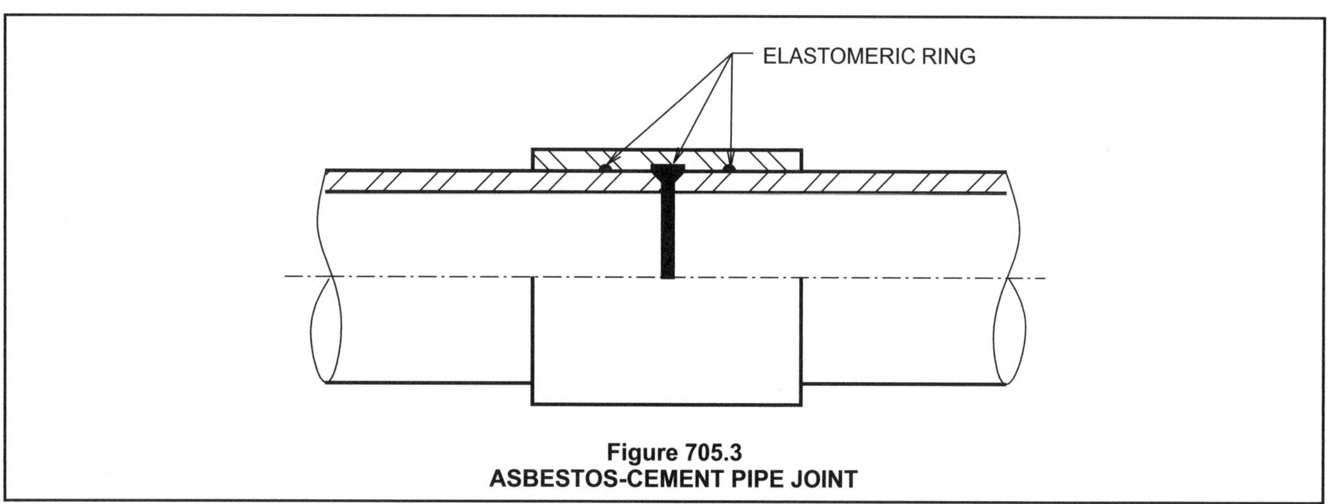

Figure 705.3
ASBESTOS-CEMENT PIPE JOINT

quires temperatures in excess of 1,000°F (538°C), and such solders are classified as "hard" solders.

There has always been confusion over the distinction between "silver solder" and "silver-bearing solder." Silver solders are unique and may be further subdivided into soft and hard categories, which are determined by the percentages of silver and the other component elements of the particular alloy. The distinction must be made that silver-bearing solders [with a melting point less than 600°F (316°C)] are used in soft-soldered joints, whereas silver solders [with a melting point above 1,000°F (538°C)] are used in silver-brazed joints. The two common series of filler metals used for brazing are BCuP and BAg. BCuP contains phosphorus and copper alloys. BAg contains 30- to 60-percent silver with zinc and copper alloys. Some of the BAg filler metals also contain cadmium.

During brazing, such filler metal produces toxic fumes. Brazing should always be done in well-ventilated areas, and protective eye gear must also be worn. The joint surfaces to be brazed must be cleaned, which is typically accomplished by polishing the pipe with an emery cloth. The socket fitting is cleaned with a specially designed brush. Brazing flux is applied to the joint surfaces. Certain types of BCuP filler metals do not require flux for certain types of joints. The manufacturer's recommendations should be consulted to determine when flux is necessary. The joints are typically heated with an oxyacetylene torch, though an air acetylene torch may be used on a smaller-diameter pipe. The pipe and fitting must be at approximately the same temperature when making the joint. The filler metal flows to the heat with the aid of the flux. The pipe must first be heated above the fitting. When the flux becomes clear, the heat must be directed to the base of the fitting socket. When the flux becomes clear and quiet, the filler metal is applied to the joint. The heat draws the filler metal into the socket. Any remaining flux residue must be removed from the pipe and joint surfaces (see Commentary Figure 705.4.1).

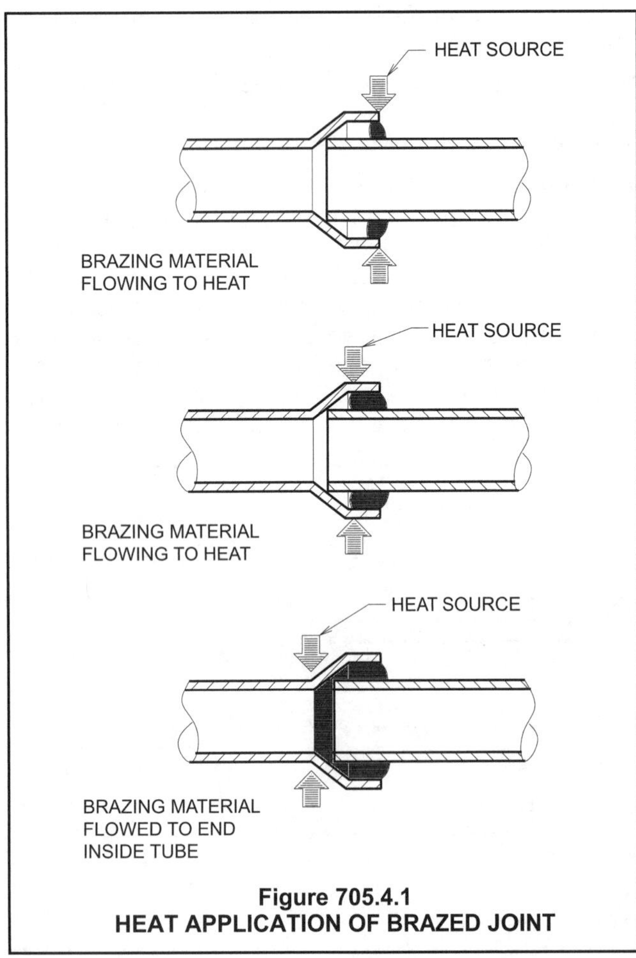

Figure 705.4.1
HEAT APPLICATION OF BRAZED JOINT

705.4.2 Mechanical joints. Mechanical joints shall be installed in accordance with the manufacturer's instructions.

❖ Mechanical joints are unique in that there are many acceptable types available. Most mechanical joints use an elastomeric material to form a seal. Compression fittings that use a compression ring (ferrule) which forms a metal-to-metal seal are also considered mechanical joints. In all cases, the joint must be carefully installed in accordance with the manufacturer's instructions.

705.4.3 Threaded joints. Threads shall conform to ASME B1.20.1. Pipe-joint compound or tape shall be applied on the male threads only.

❖ The pipe-joint compound is typically designed for all metallic pipe. Generally, the pipe-joint compound used for brass pipe may also be applied to copper and steel pipe (see Commentary Section 705.2.3).

705.4.4 Welded joints. All joint surfaces shall be cleaned. The joint shall be welded with an *approved* filler metal.

❖ A welded joint is similar to a brazed joint. The primary differences are the temperature at which the joint is made, the type of filler metal used and the fact that welding reaches the melting point of the base metal, whereas brazing temperatures are well below the melting point of the pipe (see Commentary Section 705.4.1).

705.5 Cast iron. Joints between cast-iron pipe or fittings shall comply with Sections 705.5.1 through 705.5.3.

❖ Acceptable joints for cast-iron pipe or fittings are identified in the referenced sections.

705.5.1 Caulked joints. Joints for hub and spigot pipe shall be firmly packed with oakum or hemp. Molten lead shall be poured in one operation to a depth of not less than 1 inch (25 mm). The lead shall not recede more than $^1/_8$ inch (3.2 mm) below the rim of the hub and shall be caulked tight. Paint, varnish or other coatings shall not be permitted on the jointing material until after the joint has been tested and *approved*. Lead shall be run in one pouring and shall be caulked tight. Acid-resistant rope and acidproof cement shall be permitted.

❖ A caulked joint is the oldest form of joining cast-iron pipe (see Commentary Figure 705.5.1). The joint is sealed with tightly packed oakum or hemp. The oakum or hemp expands the first time it becomes wet, thus sealing the joint. Lead is poured in place behind the oakum to prevent the oakum from expanding out of the joint. Quite often, a caulked joint will leak when it is first tested. This is a result of the oakum not yet expanding in place. Once the oakum has expanded, the joint will not leak.

The joint is made by packing the oakum or hemp into the hub fitting around the pipe to be joined. The surfaces must be dry to prevent the oakum from expanding prematurely. Once in place, the oakum is packed tight with a packing iron and hammer. Molten lead is poured in one continuous operation, securing the oakum. If the joint surfaces or oakum is wet when pouring the lead, rapidly expanding steam causes the lead to pop or explode. After the lead solidifies, it must be packed tight. A properly made caulked joint will bear marks on the lead caused by a finishing iron used to pack the lead.

Because caulked joints are labor intensive and involve the risk of worker injury, and because lead is a health hazard, such joints are becoming increasingly rare. The code does not recognize a substitute for lead in caulked joints. In the case of remodeling and repair work where one or more caulked joints must be made to allow transitions to different types of piping, code officials had on occasion permitted the use of approved lead substitutes, such as cement-based compounds. Elastomeric seals are also available for making transition joints with hubbed cast-iron pipe (see Commentary Section 705.5.2).

705.5.2 Compression gasket joints. Compression gaskets for hub and spigot pipe and fittings shall conform to ASTM C 564 and shall be tested to ASTM C 1563. Gaskets shall be compressed when the pipe is fully inserted.

❖ A mechanical compression joint is used with hubbed cast-iron drainage pipe. The elastomeric gasket is often called a "rubber gasket" or "rubber joint." The gasket is inserted into the hub of the pipe. A lubricant is used to coat the end of the joining pipe and the gasket. The pipe is inserted into the hub using a special tool that pulls the pipe and hub together, compressing the gasket in the annular space between the pipe and the hub. The pipes must be completely pulled together to make the joint properly. A typical compression joint has two main compression points in the seal (see Commentary Figure 705.5.2).

705.5.3 Mechanical joint coupling. Mechanical joint couplings for hubless pipe and fittings shall comply with CISPI 310, ASTM C 1277 or ASTM C 1540. The elastomeric sealing sleeve shall conform to ASTM C 564 or CAN/CSA B602 and shall be provided with a center stop. Mechanical joint couplings shall be installed in accordance with the manufacturer's installation instructions.

❖ Mechanical joint couplings are devices used with hubless cast-iron drainage, waste and vent pipe and fittings. Mechanical joint couplings are manufactured in a variety of types, all of which include a metallic shield or sleeve elastomeric seal and a clamp assembly that compresses the seal around the outside circumference of the pipe or fitting. Mechanical joint couplings are typically designed to resist the shear forces that can cause misalignment of pipe and fittings. ASTM C 564 and CAN/CSA-B602 apply only to the elastomeric seal component of mechanical joint couplings. CISPI 310 and ASTM C 1277 address the entire coupling assembly. The center stop prevents the pipe or fitting from penetrating too far into the coupling. The center stop is designed to ensure proper penetration depth for both of the pipes or fittings being joined.

The code regulates the design and quality of the elastomeric seals and the overall design of mechanical joint coupling assemblies used to join cast-iron DWV pipe and fittings. CISPI 310 and ASTM C 1277 address couplings that use a metal shield. The shield protects the elastomeric seal and contributes to the shear force resistance and rigidity to the assembled joint. The use of coupling designs that do not use a metal shield may have to be approved in accordance with Section 105.2. Mechanical joints and mechanical joint couplings must be installed in accordance with the manufacturer's instructions (see Commentary Figure 705.5.3).

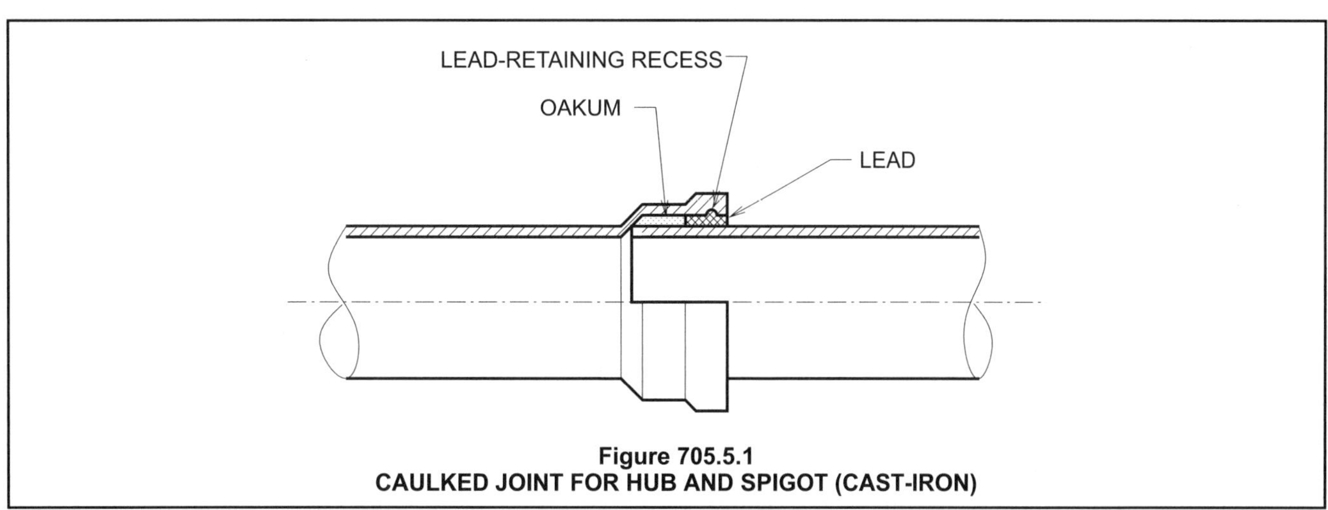

Figure 705.5.1
CAULKED JOINT FOR HUB AND SPIGOT (CAST-IRON)

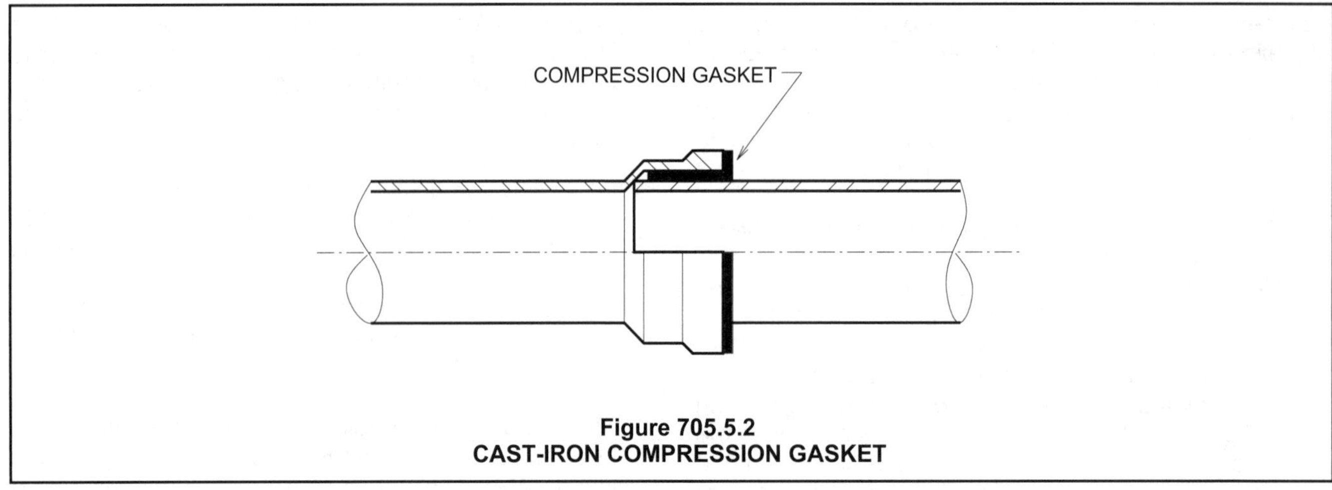

Figure 705.5.2
CAST-IRON COMPRESSION GASKET

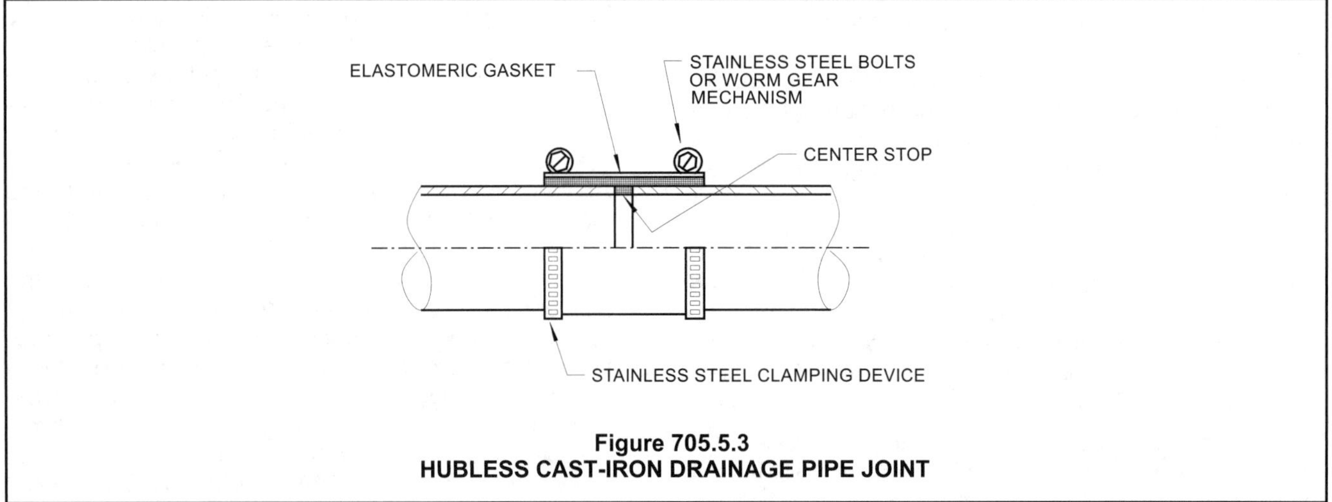

Figure 705.5.3
HUBLESS CAST-IRON DRAINAGE PIPE JOINT

705.6 Concrete joints. Joints between concrete pipe and fittings shall be made with an elastomeric seal conforming to ASTM C 443, ASTM C 1173, CAN/CSA A257.3M or CAN/CSA B602.

❖ The only acceptable method of joining concrete pipe is an elastomeric gasket that fits in the hub of the pipe. The gasket is compressed between the hub and the pipe to form a seal. The ASTM and CSA standards regulate the design, material and performance of the elastomeric gasket (see Commentary Figure 705.6).

705.7 Coextruded composite ABS pipe, joints. Joints between coextruded composite pipe with an ABS outer layer or ABS fittings shall comply with Sections 705.7.1 and 705.7.2.

❖ Acceptable joints for ABS plastic pipe are identified in the referenced sections.

705.7.1 Mechanical joints. Mechanical joints on drainage pipe shall be made with an elastomeric seal conforming to ASTM C1173, ASTM D 3212 or CSA B602. Mechanical joints shall not be installed in above-ground systems, unless otherwise *approved*. Joints shall be installed in accordance with the manufacturer's instructions.

❖ The installation technique used to make a mechanical joint for coextruded composite ABS pipe and fittings is identical to that of a mechanical joint for ABS plastic pipe and fittings (see Commentary Section 705.2.1).

705.7.2 Solvent cementing. Joint surfaces shall be clean and free from moisture. Solvent cement that conforms to ASTM D 2235 or CSA B181.1 shall be applied to all joint surfaces. The joint shall be made while the cement is wet. Joints shall be made in accordance with ASTM D 2235, ASTM D 2661, ASTM F 628 or CSA B181.1. Solvent-cement joints shall be permitted above or below ground.

❖ The installation technique used to make a solvent cement joint for coextruded composite ABS pipe and fittings is identical to that of a solvent cement joint for ABS plastic pipe and fittings (see Commentary Section 705.2.2).

705.8 Coextruded composite PVC pipe. Joints between coextruded composite pipe with a PVC outer layer or PVC fittings shall comply with Sections 705.8.1 and 705.8.2.

❖ Acceptable joints for coextruded composite PVC plastic pipe or fittings are identified in the referenced sections.

705.8.1 Mechanical joints. Mechanical joints on drainage pipe shall be made with an elastomeric seal conforming to ASTM D 3212. Mechanical joints shall not be installed in above-ground systems, unless otherwise *approved*. Joints shall be installed in accordance with the manufacturer's instructions.

❖ The installation technique used to make a mechanical joint for coextruded composite PVC plastic pipe and fittings is identical to that of a mechanical joint for PVC plastic pipe and fittings (see Commentary Section 705.14.1).

705.8.2 Solvent cementing. Joint surfaces shall be clean and free from moisture. A purple primer that conforms to ASTM F 656 shall be applied. Solvent cement not purple in color and conforming to ASTM D 2564, CSA B137.3, CSA B181.2 or CSA B182.1 shall be applied to all joint surfaces. The joint shall be made while the cement is wet and shall be in accordance with ASTM D 2855. Solvent-cement joints shall be permitted above or below ground.

❖ The installation technique used to make a solvent cement joint for coextruded composite PVC plastic pipe and fittings is identical to that of solvent cementing for PVC plastic pipe and fittings (see Commentary Section 705.14.2).

705.9 Copper pipe. Joints between copper or copper-alloy pipe or fittings shall comply with Sections 705.9.1 through 705.9.5.

❖ Acceptable joints for copper or copper-alloy pipe or fittings are identified in the referenced sections.

705.9.1 Brazed joints. All joint surfaces shall be cleaned. An *approved* flux shall be applied where required. The joint shall be brazed with a filler metal conforming to AWS A5.8.

❖ The installation technique and materials used to make a brazed joint for copper or copper-alloy pipe are identical to that of brass pipe and copper tubing (see Commentary Section 705.4.1).

705.9.2 Mechanical joints. Mechanical joints shall be installed in accordance with the manufacturer's instructions.

❖ Mechanical joints for copper or copper alloy pipe are similar to those used to join brass pipe (see Commentary Section 705.4.2). Because many different types of mechanical joints are available and most couplings are unique to one manufacturer, the joining method must be in accordance with the manufacturer's installation instructions. Such joints are highly resistant to being pulled apart.

705.9.3 Soldered joints. Solder joints shall be made in accordance with the methods of ASTM B 828. All cut tube ends shall be reamed to the full inside diameter of the tube end. All joint surfaces shall be cleaned. A flux conforming to ASTM B 813 shall be applied. The joint shall be soldered with a solder conforming to ASTM B 32.

❖ Soldering is the most common method of joining copper pipe and tubing. ASTM B 828 governs the procedures for making capillary joints by soldering copper and copper alloy tube and fittings. To consistently make satisfactory joints, the following sequence of joint preparation and operations must be followed: measuring and cutting, reaming, cleaning, fluxing, assembly and support, heating, applying the solder and cooling and cleaning. The joint surfaces must be accurately measured because if the tube is too short, it will not reach all the way into the fitting and a proper joint cannot be made. If the tube is too long, the system strain may affect service life.

Tubing may be cut in a number of different ways to produce a square end. The tubing may be cut with a tubing cutter, a hacksaw, an abrasive wheel or a band saw. Regardless of the method, the cut must be square with the run of the tube so it will seat properly in the fitting. The tubing ends, outside of the tube ends to the inside diameter of the tube, must be reamed to remove the burrs created by the cutting operation (see Commentary Figure 705.9.3).

The joint surfaces for a soldered joint must be cleaned to remove all oxides and surface soil from the tube ends and fittings. Cleaning may be accomplished using a sand cloth, an abrasive pad and a fitting brush. The space between the tube and the fitting is where the solder metal fills this gap by capillary action. The spacing is critical for the solder metal to flow into the gap and form a strong joint. Uniformity of capillary space will ensure metal capillary flow.

Once the joint surfaces are cleaned, do not touch the cleaned area with bare hands or oily gloves. Skin

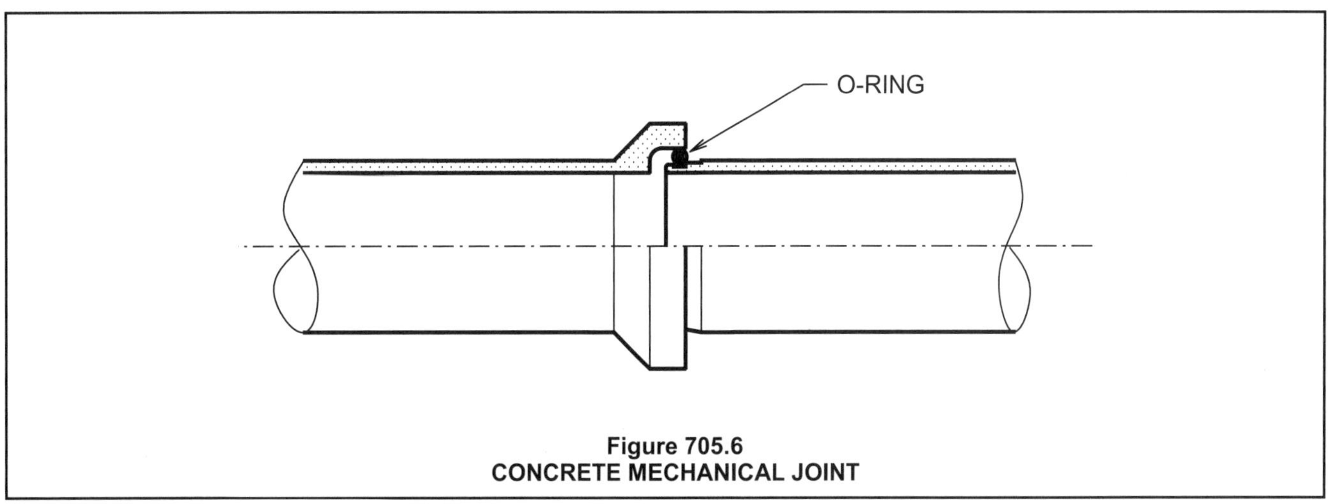

Figure 705.6
CONCRETE MECHANICAL JOINT

oils, lubricating oils or grease could interfere with the process. Apply a thin, even coating of flux to both the tubing and fitting that will dissolve and remove traces of oxide from the cleaned surfaces. Care must be taken in applying flux because excessive amounts will cause corrosion that may perforate the wall of the tubing, the fitting or both. The flux material must conform to ASTM B 813, which establishes minimum performance criteria. This also limits the maximum lead content to 0.2 percent.

Begin heating with the flame alternating from the fitting cup back onto the tube. Touch the solder to the joint and if the solder does not melt, remove it and continue the heating process. Once the melting temperatures are reached, apply heat to the base of the fitting to aid capillary action in drawing the molten solder into the cup toward the heat source. Molten solder metal is drawn into the joint by capillary action regardless of the direction that the metal is being fed (horizontal or vertical). If excessive heat is used, the flux will burn out and char the pipe, and where the temperature is too uneven, the solder will run down either the inside or the outside of the pipe. Either condition will prevent capillary action.

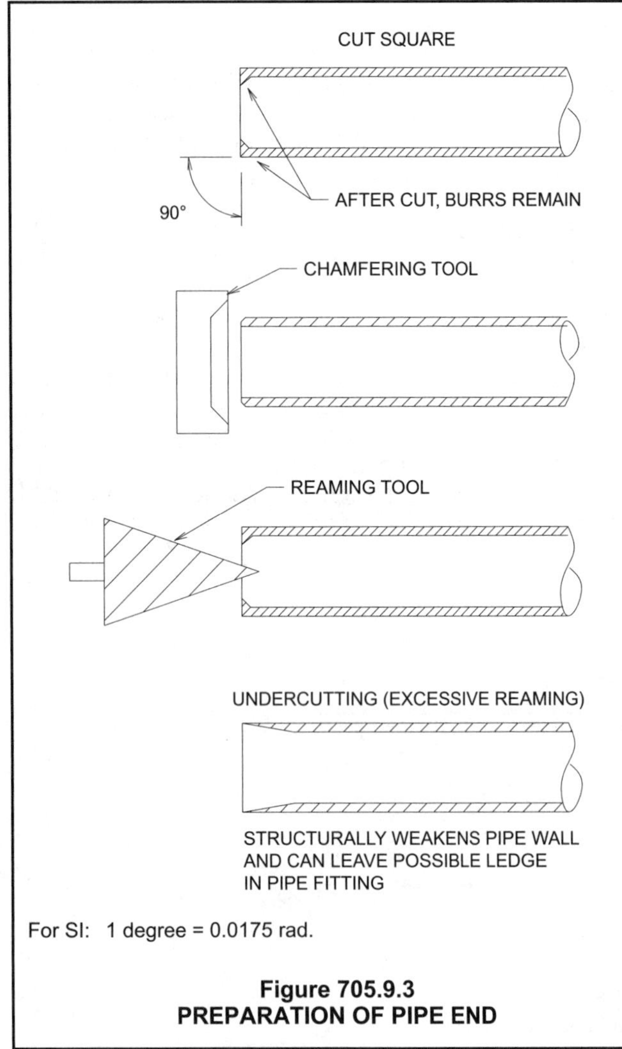

Figure 705.9.3
PREPARATION OF PIPE END

Allow the completed joint to cool naturally; cooling it with water will cause unnecessary shock and stress on the joint. When the joint is cool, clean off any remaining flux.

705.9.4 Threaded joints. Threads shall conform to ASME B1.20.1. Pipe-joint compound or tape shall be applied on the male threads only.

❖ Before assembly, the threads must be cleaned with a wire brush to remove burrs or chips resulting from the cutting process. Pipe-joint compound or Teflon tape is applied on the male thread to ensure a tight seal. If pipe-joint compound is applied to the female thread, it will enter the system. The male thread is placed inside the female thread and the joint is tightened using a pipe wrench. Threaded joints are generally limited to smaller pipe diameters because of the amount of effort required to turn a pipe of larger size in making the joint.

705.9.5 Welded joints. All joint surfaces shall be cleaned. The joint shall be welded with an *approved* filler metal.

❖ The installation technique and materials used to make a welded joint for copper or copper-alloy tubing are similar to that of brass pipe (see Commentary Section 705.4.4).

705.10 Copper tubing. Joints between copper or copper-alloy tubing or fittings shall comply with Sections 705.10.1 through 705.10.3.

❖ Acceptable joints for copper or copper-alloy tubing or fittings are identified in the referenced sections.

705.10.1 Brazed joints. All joint surfaces shall be cleaned. An *approved* flux shall be applied where required. The joint shall be brazed with a filler metal conforming to AWS A5.8.

❖ The brazing of a joint for copper or copper-alloy tubing is identical to that of copper, copper-alloy and brass pipe (see Commentary Section 705.4.1).

705.10.2 Mechanical joints. Mechanical joints shall be installed in accordance with the manufacturer's instructions.

❖ There is a variety of joining methods for mechanical joints for copper or copper-alloy tubing. The most common is the compression-type joint. This type of joint is made by placing a compression nut over the outside of the pipe followed by a compression ring. The pipe is inserted into the fitting and tightened. As the compression nut is tightened, the pipe and fitting are brought together, thereby forming a leakproof joint. Mechanical joints have some flexibility, allowing the pipes to be installed with an angle of deflection. Such joints are highly resistant to being pulled apart.

705.10.3 Soldered joints. Solder joints shall be made in accordance with the methods of ASTM B 828. All cut tube ends shall be reamed to the full inside diameter of the tube end. All joint surfaces shall be cleaned. A flux conforming to ASTM B 813 shall be applied. The joint shall be soldered with a solder conforming to ASTM B 32.

❖ The requirements and procedures for a soldered joint for copper and copper-alloy tubing are identical to those for copper and copper-alloy pipe (see commentary, Section 705.9.3 and Commentary Figure 705.9.3).

705.11 Borosilicate glass joints. Glass-to-glass connections shall be made with a bolted compression-type stainless steel (300 series) coupling with contoured acid-resistant elastomeric compression ring and a fluorocarbon polymer inner seal ring; or with caulked joints in accordance with Section 705.11.1.

❖ Two methods are recognized for joining glass pipe: one is a mechanical compression joint that is used to join bead-end or plain-end glass pipe; the other is a caulked joint that is used to join bead-end glass pipe to hubbed pipe. The compression-type mechanical joint is available in two standard types: one type is used to join beaded-end glass pipe to beaded-end glass pipe, and the other one is used to join beaded-end glass pipe to plain-end glass pipe.

Both couplings consist of an elastomeric compression ring (typically rubber) with a chemically inert tetrafluoroethylene (TFE) inner seal ring and a stainless steel outer band with bolt. The bead-end to bead-end pipe connection is made by slipping the compression liner and inner seal ring over one section of pipe, aligning the pipes to be joined and sliding the compression liner over both bead ends. The stainless steel outer band is then slid over the compression liner and the joint is secured by tightening the bolt on the outer band. The bead-end to plain-end joint is made in a similar manner. The methods for joining glass pipe to other pipe materials are listed in Section 705.19.6.

705.11.1 Caulked joints. Every lead-caulked joint for hub and spigot soil pipe shall be firmly packed with oakum or hemp and filled with molten lead not less than 1 inch (25 mm) deep and not to extend more than $^1/_8$ inch (3.2 mm) below the rim of the hub. Paint, varnish or other coatings shall not be permitted on the jointing material until after the joint has been tested and *approved*. Lead shall be run in one pouring and shall be caulked tight. Acid-resistant rope and acidproof cement shall be permitted.

❖ The method recognized for joining glass pipe to hubbed pipe is a poured-lead caulk joint. This joint method is used to join beaded-end glass pipe to glass or metal hubbed pipe. The bead end of the glass pipe is inserted into the bell of the other pipe and oakum, hemp or acid-resistant rope is then packed around the glass pipe to prevent direct contact between the two pipes. The installation of the joint is identical to that of cast-iron pipe (see Commentary Section 705.5.1).

705.12 Steel. Joints between galvanized steel pipe or fittings shall comply with Sections 705.12.1 and 705.12.2.

❖ Acceptable joints for galvanized steel pipe or fittings are identified in the referenced sections.

705.12.1 Threaded joints. Threads shall conform to ASME B1.20.1. Pipe-joint compound or tape shall be applied on the male threads only.

❖ The pipe joint compound used for steel pipe will also be used for both copper and brass pipe (see Commentary Section 705.2.3). The application of pipe-joint compound or Teflon tape to only male threads prevents the compound from entering the piping system.

705.12.2 Mechanical joints. Joints shall be made with an *approved* elastomeric seal. Mechanical joints shall be installed in accordance with the manufacturer's instructions.

❖ A wide variety of mechanical joints are available to join steel pipe, including plain or beveled ends, cut or rolled grooves. Each requires different pipe-end preparation in making a joint. Rolled grooves (instead of cut grooves) are used where the wall of the pipe is not thick enough to have a groove formed around the pipe near the ends. The coupling assembly for both types of grooving is the same. The joint consists of an inner elastomeric gasket and an outer two-piece (split) collar with integral bolts used for tightening. The assembly is placed over both ends and set into the grooves in the pipe. The bolts are tightened to the torque requirements listed in the manufacturer's installation instructions.

705.13 Lead. Joints between lead pipe or fittings shall comply with Sections 705.13.1 and 705.13.2.

❖ Acceptable joints for lead pipe or fittings are identified in the referenced sections.

705.13.1 Burned. Burned joints shall be uniformly fused together into one continuous piece. The thickness of the joint shall be at least as thick as the lead being joined. The filler metal shall be of the same material as the pipe.

❖ A burned joint is formed by fitting the end of one lead pipe into the flared end of another lead pipe. Heat is then applied evenly around the joint, melting the overlapping edges and fusing them together. Lead burning is a method used for joining sheet lead or making lead pans.

705.13.2 Wiped. Joints shall be fully wiped, with an exposed surface on each side of the joint not less than $^3/_4$ inch (19.1 mm). The joint shall be at least 0.325 inch (9.5 mm) thick at the thickest point.

❖ A wiped joint is formed by fitting the end of one pipe into the flared end of the other. Molten lead is poured onto the joint and wiped with a hand-held pad, providing a minimum dimension of $^3/_4$ inch (19.1 mm) and a maximum thickness of $^3/_8$ inch (9.5 mm) (see Commentary Figure 705.13.2). With the aid of modern technology and introduction of new materials, lead pipe is rarely used in drainage systems.

705.14 PVC plastic. Joints between PVC plastic pipe or fittings shall comply with Sections 705.14.1 through 705.14.3.

❖ Acceptable joints for PVC plastic pipe or fittings are identified in the referenced sections.

705.14.1 Mechanical joints. Mechanical joints on drainage pipe shall be made with an elastomeric seal conforming to ASTM C 1173, ASTM D 3212 or CAN/CSA B602. Mechanical joints shall not be installed in above-ground systems, unless otherwise *approved*. Joints shall be installed in accordance with the manufacturer's instructions.

❖ A typical underground sewer line is plastic hub pipe with an elastomeric compression gasket inside the hub. The pipe must be inserted to the full depth of the hub to make a proper joint (see commentary, Section 705.2.1). The typical mechanical joint must be restrained to prevent separation; therefore, such joints are limited to underground installations where restraint is accomplished by burial.

705.14.2 Solvent cementing. Joint surfaces shall be clean and free from moisture. A purple primer that conforms to ASTM F 656 shall be applied. Solvent cement not purple in color and conforming to ASTM D 2564, CSA B137.3, CSA B181.2 or CSA B182.1 shall be applied to all joint surfaces. The joint shall be made while the cement is wet and shall be in accordance with ASTM D 2855. Solvent-cement joints shall be permitted above or below ground.

❖ PVC plastic pipe requires a primer prior to solvent cementing. Although ASTM F 656 only recommends that PVC primer be purple, this is a code requirement. The purple coloring is intended to allow the installer and the inspector to verify that the required primer has been applied. To distinguish the difference between primer and solvent cement, the solvent cement must be a color other than purple. Solvent cement for PVC is typically clear (see commentary, Sections 705.2.2 and 605.22.2). A PVC primer is first applied to both the pipe and the socket fitting. The PVC solvent cement is then applied to the pipe and socket fitting. The joint is made with a slight twisting until the pipe reaches the full depth of the socket fitting. It is then held in place until the solvent cement begins to set. It is wise to wear impervious gloves to prevent solvent cement from coming into contact with the skin. The solvent-cementing procedure must be done in accordance with the referenced standard.

To minimize the health and fire hazards associated with solvent cements and primers, the manufacturers' instructions for the use and handling of these materials must be followed, in addition to the requirements found in ASTM D 2855.

705.14.3 Threaded joints. Threads shall conform to ASME B1.20.1. Schedule 80 or heavier pipe shall be permitted to be threaded with dies specifically designed for plastic pipe. *Approved* thread lubricant or tape shall be applied on the male threads only.

❖ The requirements for threaded joints in PVC plastic pipe are identical to those of ABS and CPVC plastic pipe (see commentary, Section 705.2.3).

705.15 Vitrified clay. Joints between vitrified clay pipe or fittings shall be made with an elastomeric seal conforming to ASTM C 425, ASTM C 1173 or CAN/CSA B602.

❖ An acceptable joint for vitrified clay pipe is a mechanical joint sealed with an elastomeric material. This is a type of compression joint similar to that used with other hubbed pipe and fittings.

705.16 Polyethylene plastic pipe. Joints between polyethylene plastic pipe and fittings shall be underground and shall comply with Section 705.16.1 or 705.16.2.

❖ Acceptable joints for polyethylene plastic pipe or fittings are identified in the referenced sections. Note that joints can be located only underground.

705.16.1 Heat-fusion joints. Joint surfaces shall be clean and free from moisture. All joint surfaces shall be cut, heated to melting temperature and joined using tools specifically designed for the operation. Joints shall be undisturbed until cool. Joints shall be made in accordance with ASTM D 2657 and the manufacturer's instructions.

❖ Three types of processes are used: (1) hot plate welding for butt welding of pipes, (2) heat fusion for joining pipe to fittings and (3) electro-fusion welding for joining pipe to fittings. Special clamping fixtures are used to hold the pipe ends and fittings in alignment. Some fixtures for small pipe diameters are small enough to hand carry; larger pipes require fixtures to be on a wheeled cart or even a motorized/tracked machine.

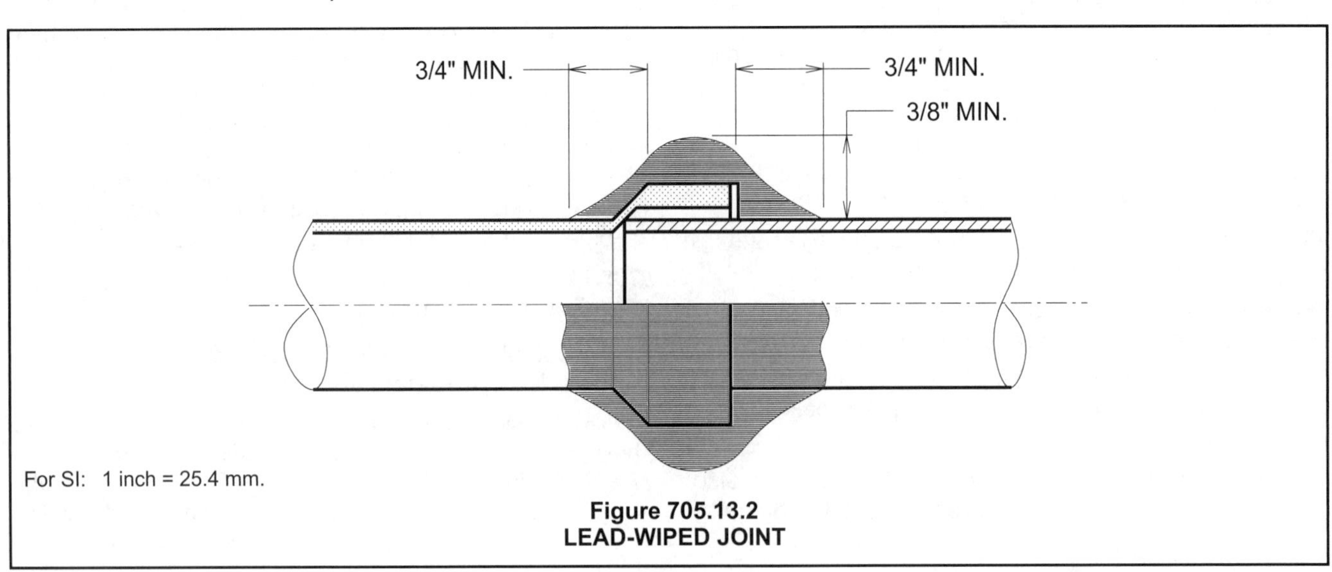

Figure 705.13.2
LEAD-WIPED JOINT

In the hot plate welding process, the ends of pipe are faced and squared with the centerline of the pipe using a rotating-blade planer block that accurately rotates around the pipe end while planing.

After cleaning debris from the pipes, the two pipe ends are clamped in precise alignment in the fixture so that no perceptible gap will exist between the pipe ends once they are pushed together. A plate with internal heating elements and a nonstick surface is inserted between and in contact with the two pipe ends. After sufficient heating to soften the ends of the pipe, the heating plate is removed and the ends of the pipe are mated with a specified force, "fusing" the softened pipe ends together. After an appropriate cooling time, the pipe is removed from the fixture and is ready for installation. A completed quality butt weld has (1) complete and uniform beads, (2) bead rollback onto pipe and (3) proper alignment between the two pipes with centerlines parallel and, more importantly, outside wall surfaces in line (no offset).

In the heat fusion and electrofusion processes, a pipe end is joined to a fitting such as coupling, tee or valve having a recessed socket.

In the heat fusion method, the exterior of the pipe and interior of the fitting socket are electrically heated independently with specialized electric heating devices. Once the materials are sufficiently softened, the heating devices are removed and the pipe is precisely inserted in the socket fitting and allowed to cool.

In the electrofusion method, the socket fitting has a heating element "implant" that is connected to an appropriate power source. The pipe is inserted into the fitting and the heating element implant is energized for a specific period of time, causing the pipe and fitting interior to be softened and fused together. The joint is allowed to cool for an appropriate amount of time before the joint is disturbed.

All three processes require trained and qualified personnel as well as strict adherence to procedures to make fusion joints with integrity. Cleanliness and protection against weather elements during the process is essential to produce high quality joints.

705.16.2 Mechanical joints. Mechanical joints in drainage piping shall be made with an elastomeric seal conforming to ASTM C 1173, ASTM D 3212 or CAN/CSA B602. Mechanical joints shall be installed in accordance with the manufacturer's instructions.

❖ Three types of mechanical joints can be used depending on the size of the pipe: (1) compression nut fittings [small diameters up to 2 inches (51 mm)], (2) stab type fittings [small diameters up to 2 inches (51 mm)], and (3) bolt type compression fittings [from $1^1/_4$ inches (32 mm) and up], and a variety of flanged ends. Each requires following the manufacturer's instructions for proper application and installation. Note that only joints using an elastomeric seal conforming to ASTM C 1173, D 3212 or CAN/CSA B602 comply with the code.

705.17 Polyolefin plastic. Joints between polyolefin plastic pipe and fittings shall comply with Sections 705.17.1 and 705.17.2.

❖ Acceptable joints for polyethylene plastic pipe or fittings are identified in the referenced sections.

705.17.1 Heat-fusion joints. Heat-fusion joints for polyolefin pipe and tubing joints shall be installed with socket-type heat-fused polyolefin fittings or electrofusion polyolefin fittings. Joint surfaces shall be clean and free from moisture. The joint shall be undisturbed until cool. Joints shall be made in accordance with ASTM F 1412 or CAN/CSA B181.3.

❖ Only the heat fusion and electrofusion process using socket type fitting is allowed (see Commentary Section 705.16.1).

705.17.2 Mechanical and compression sleeve joints. Mechanical and compression sleeve joints shall be installed in accordance with the manufacturer's instructions.

❖ Three types of mechanical joints can be used depending upon the size of the pipe: (1) compression nut fittings [small diameters up to 2 inches (51 mm)], (2) stab type fittings [small diameters up to 2 inches (51 mm)], and (3) bolt type compression fittings [from $1^1/_4$ inches (32 mm) and up], and a variety of flanged ends. Each requires following the manufacturer's instructions for proper application and installation.

705.18 Polyvinylidene fluoride plastic. Joints between polyvinylidene plastic pipe and fittings shall comply with Sections 705.18.1 and 705.18.2.

❖ This pipe material (pronounced poly-vy-NIL-ih-DEAN floor-IDE) is a highly nonreactive and pure thermoplastic fluoropolymer. Several different companies manufacture this material under trademarked trade names but the material is the same. Because of the multiple strong carbon-fluorine molecular bonds, the material has high purity, high strength and an exceptional resistance to solvents, acids and bases, making it an excellent material for conveying corrosive wastes. It has a melting point of about 351°F (177°C) and is a low generator of heat and smoke in a fire event.

705.18.1 Heat-fusion joints. Heat-fusion joints for polyvinylidene fluoride pipe and tubing joints shall be installed with socket-type heat-fused polyvinylidene fluoride fittings or electrofusion polyvinylidene fittings and couplings. Joint surfaces shall be clean and free from moisture. The joint shall be undisturbed until cool. Joints shall be made in accordance with ASTM F 1673.

❖ Polyvinylidene fluoride pipe and tubing joints can be heat fused or electrofused in a manner similar to that used for polyolefin pipe.

705.18.2 Mechanical and compression sleeve joints. Mechanical and compression sleeve joints shall be installed in accordance with the manufacturer's instructions.

❖ Polyvinylidene fluoride pipe and tubing can be joined with compression fittings or mechanical joints that are specifically intended for use with the piping material. The manufacturer's instructions for installation of the fittings must be followed.

705.19 Joints between different materials. Joints between different piping materials shall be made with a mechanical joint of the compression or mechanical-sealing type conforming to ASTM C 1173, ASTM C 1460 or ASTM C 1461. Connectors and adapters shall be *approved* for the application and such joints shall have an elastomeric seal conforming to ASTM C 425, ASTM C 443, ASTM C 564, ASTM C 1440, ASTM D 1869, ASTM F 477, CAN/CSA A257.3M or CAN/CSA B602, or as required in Sections 705.19.1 through 705.19.7. Joints between glass pipe and other types of materials shall be made with adapters having a TFE seal. Joints shall be installed in accordance with the manufacturer's instructions.

❖ Mechanical joints are typically used to join dissimilar piping materials. Because many piping materials have the same outside diameters, many fittings are designed by manufacturers for use with a number of different piping materials. Some mechanical joints are manufactured specifically for joining pipes having different outside diameters. Transition couplings use bushings (inserts) to match the outside diameters of the pipes or are simply designed to fit over different outside diameter pipe ends. ASTM C 1173, ASTM C 1460 or ASTM C 1461 are the standards for the complete coupling assembly, which includes an elastomeric gasket, a clamp assembly or both.

705.19.1 Copper or copper-alloy tubing to cast-iron hub pipe. Joints between copper or copper-alloy tubing and cast-iron hub pipe shall be made with a brass ferrule or compression joint. The copper or copper-alloy tubing shall be soldered to the ferrule in an *approved* manner, and the ferrule shall be joined to the cast-iron hub by a caulked joint or a mechanical compression joint.

❖ An adapter must be used when joining copper tubing to a cast-iron hub. The adapter is called by a wide variety of names, the most common being "soil adapter" or "caulk adapter." The adapter creates a spigot end similar to that of cast-iron pipe and fittings.

705.19.2 Copper or copper-alloy tubing to galvanized steel pipe. Joints between copper or copper-alloy tubing and galvanized steel pipe shall be made with a brass converter fitting or dielectric fitting. The copper tubing shall be soldered to the fitting in an *approved* manner, and the fitting shall be screwed to the threaded pipe.

❖ When joining copper or copper-alloy tubing with galvanized steel pipe, a method of protecting against galvanic corrosion is required. Galvanic corrosion occurs when two different metals come in contact in the presence of an electrolyte, such as water. Galvanic corrosion accelerates the natural corrosion process that occurs in metals, and, because certain metals corrode faster than others, they have been placed in a hierarchy in order of rate of corrosion. The more reactive metal at a connection is called the "anode" (galvanized steel pipe), and the less reactive metal is called the "cathode" (copper tubing). When the copper tubing and galvanized steel pipe are coupled, the galvanized steel pipe will tend to dissolve in the electrolyte, thereby generating electric current flow between the metals. This section prescribes two methods of protection against galvanic corrosion: dielectric fittings and brass fittings. The fittings are designed to provide a barrier or buffer between the two materials.

705.19.3 Cast-iron pipe to galvanized steel or brass pipe. Joints between cast-iron and galvanized steel or brass pipe shall be made by either caulked or threaded joints or with an *approved* adapter fitting.

❖ Galvanized steel pipe and brass pipe may be caulked directly into the hub of cast-iron pipe without an adapter fitting. Certain cast-iron fittings have a female tapped inlet to permit a direct threaded connection (see Commentary Figure 705.18.3).

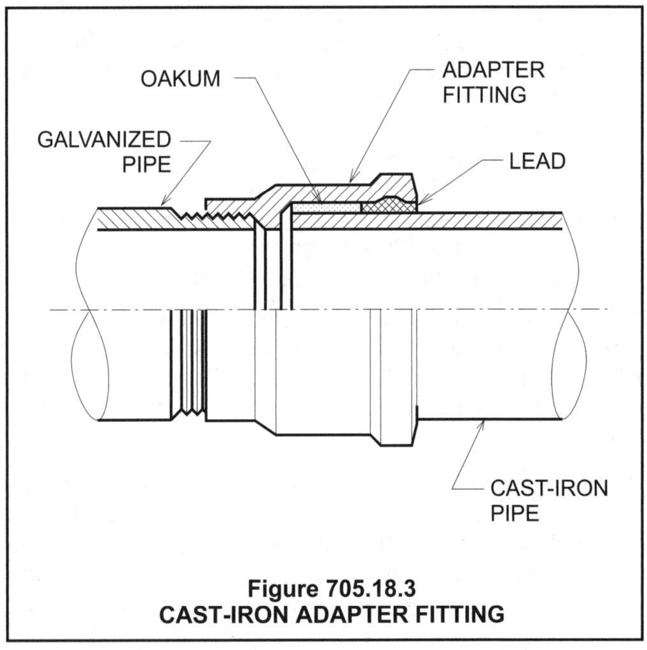

Figure 705.18.3
CAST-IRON ADAPTER FITTING

705.19.4 Plastic pipe or tubing to other piping material. Joints between different types of plastic pipe or between plastic pipe and other piping material shall be made with an *approved* adapter fitting. Joints between plastic pipe and cast-iron hub pipe shall be made by a caulked joint or a mechanical compression joint.

❖ ABS and PVC plastic pipe may connect directly into the hub of cast-iron pipe with a caulked joint or by using a compression gasket without the use of an adapter fitting. Special caulk-adapter plastic fittings are available that provide a spigot end for use with hubbed pipe and fittings. Where different grades of plastic pipe are joined or where plastic pipe and cast-iron spigot end pipe, Schedule 40 steel pipe or copper pipe are joined, an adapter fitting must be used.

705.19.5 Lead pipe to other piping material. Joints between lead pipe and other piping material shall be made by a wiped joint to a caulking ferrule, soldering nipple, or bushing or shall be made with an *approved* adapter fitting.

❖ The common method of adapting to lead is by a wiped joint to a ferrule, nipple or bushing; however, a mechanical adapter fitting using an elastomeric material may also be used.

705.19.6 Borosilicate glass to other materials. Joints between glass pipe and other types of materials shall be made with adapters having a TFE seal and shall be installed in accordance with the manufacturer's instructions.

❖ Glass pipe may be joined to plain-end or threaded-end metal or plastic pipe with adapter couplings that incorporate a TFE (tetrafluoroethylene) seal. This seal isolates the glass pipe from contact with the pipe it is being joined to and helps provide a positive seal around the perimeter of the glass pipe. Typically, a manufacturer of glass pipe also makes the couplings that are designed for use with the pipe; therefore, the coupling manufacturer's installation instructions must be followed. The TFE seal is resistant to chemical attack, as is glass pipe. TFE is commonly known by the registered trademark name "Teflon."

705.19.7 Stainless steel drainage systems to other materials. Joints between stainless steel drainage systems and other piping materials shall be made with *approved* mechanical couplings.

❖ Joints between stainless steel drainage systems must be made with a mechanical coupling and have an elastomeric seal conforming to ASTM C 425, ASTM C 443, ASTM C 564, ASTM C 1173, ASTM D 1869, ASTM F 477, CAN/CSA A257.3M or CAN/CSA B602. Because many different types of mechanical joints are available and most couplings are unique to each manufacturer, the joining method must be in accordance with the manufacturer's installation instructions.

705.20 Drainage slip joints. Slip joints shall comply with Section 405.8.

❖ An elastomeric gasket must be used to make the seal for drainage slip joints [see Commentary Figures 705.20(1) and 705.20(2) and the definition of "Slip joint" in Chapter 2]. Slip joints cannot be concealed and are limited to those connections that occur between a fixture waste outlet and its connection to the fixture drain, including the trap adapter on the trap outlet arm. In locations where a slip joint connection is concealed, an access panel must be provided so that the connection can be serviced. This would include tub waste and overflow connections (see Commentary Section 405.8).

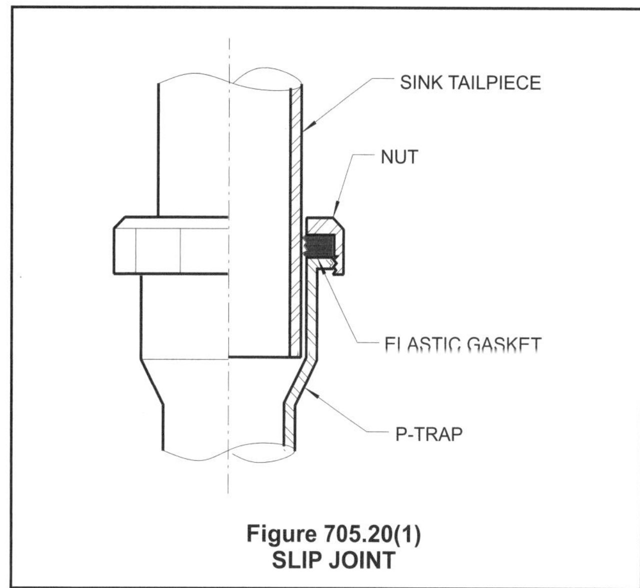

Figure 705.20(1)
SLIP JOINT

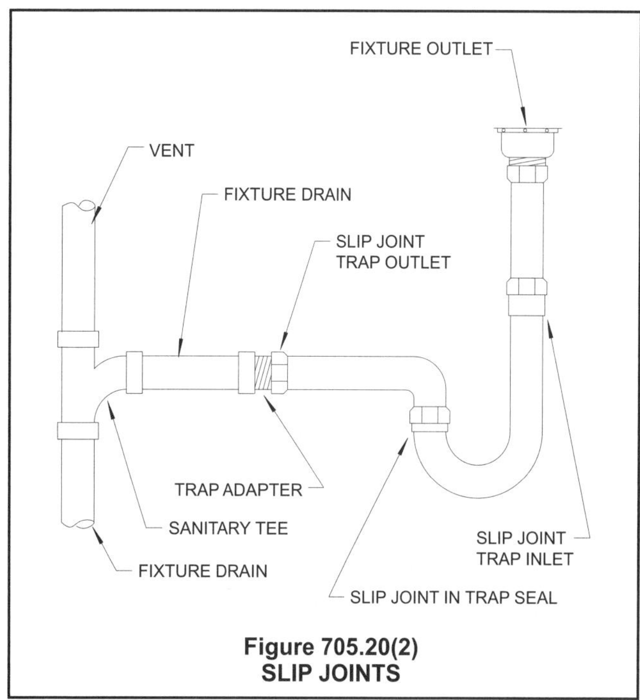

Figure 705.20(2)
SLIP JOINTS

705.21 Caulking ferrules. Ferrules shall be of red brass and shall be in accordance with Table 705.21.

❖ Caulking ferrules are used to adapt cast-iron hub pipe to either copper tubing or threaded pipe. The ferrule is installed in the hub of the cast-iron pipe by either a caulked joint or an elastomeric gasket compression joint. The copper tubing is soldered to, or the threaded pipe is screwed into, the ferrule.

TABLE 705.21
CAULKING FERRULE SPECIFICATIONS

PIPE SIZES (inches)	INSIDE DIAMETER (inches)	LENGTH (inches)	MINIMUM WEIGHT EACH
2	$2^1/_4$	$4^1/_2$	1 pound
3	$3^1/_4$	$4^1/_2$	1 pound 12 ounces
4	$4^1/_4$	$4^1/_2$	2 pounds 8 ounces

For SI: 1 inch = 25.4 mm, 1 ounce = 28.35 g, 1 pound = 0.454 kg.

❖ The table specifies the acceptable dimensions and weights for caulking ferrules.

705.22 Soldering bushings. Soldering bushings shall be of red brass and shall be in accordance with Table 705.22.

❖ A soldering bushing is used to adapt lead pipe to other piping materials.

TABLE 705.22
SOLDERING BUSHING SPECIFICATIONS

PIPE SIZES (inches)	MINIMUM WEIGHT EACH
$1^1/_4$	6 ounces
$1^1/_2$	8 ounces
2	14 ounces
$2^1/_2$	1 pound 6 ounces
3	2 pounds
4	3 pounds 8 ounces

For SI: 1 inch = 25.4 mm, 1 ounce = 28.35 g, 1 pound = 0.454 kg.

❖ The table specifies the acceptable weights for soldering bushing sizes.

705.23 Stainless steel drainage systems. O-ring joints for stainless steel drainage systems shall be made with an *approved* elastomeric seal.

❖ O-ring joints for stainless steel drainage systems are identical to those between stainless steel drainage systems and other piping materials (see Commentary Section 705.19.7).

SECTION 706
CONNECTIONS BETWEEN DRAINAGE PIPING AND FITTINGS

706.1 Connections and changes in direction. All connections and changes in direction of the sanitary drainage system shall be made with *approved* drainage fittings. Connections between drainage piping and fixtures shall conform to Section 405.

❖ This section contains the requirements for the proper use of fittings and connections used to accomplish changes in direction. The design of a drainage system is based on uniform flow in the piping. Special drainage fittings have been developed with design patterns providing the least resistance to flow. All types of connections and drainage fittings are subject to the approval of the code official. Section 405 regulates the installation of fixtures connected to the sanitary drainage system.

706.2 Obstructions. The fittings shall not have ledges, shoulders or reductions capable of retarding or obstructing flow in the piping. Threaded drainage pipe fittings shall be of the recessed drainage type.

❖ Because the flow of waste within the drainage system is slow and induced by gravity, any shoulder edge or obstruction on the inside surface of the pipe may inhibit or retard flow and likely cause a stoppage. Therefore, joints, connections and fittings must be of the drainage type, free from flow-obstructing elements. Recessed drainage-type threaded fittings are designed so that the inside surface of the pipe will be flush with the inside surface of the fitting. An offset closet flange is not considered an obstruction.

706.3 Installation of fittings. Fittings shall be installed to guide sewage and waste in the direction of flow. Change in direction shall be made by fittings installed in accordance with Table 706.3. Change in direction by combination fittings, side inlets or increasers shall be installed in accordance with Table 706.3 based on the pattern of flow created by the fitting. Double sanitary tee patterns shall not receive the discharge of back-to-back water closets and fixtures or appliances with pumping action discharge.

> **Exception:** Back-to-back water closet connections to double sanitary tees shall be permitted where the horizontal *developed length* between the outlet of the water closet and the connection to the double sanitary tee pattern is 18 inches (457 mm) or greater.

❖ Drainage fittings and connections must provide a smooth transition of flow without creating obstructions or causing interference. The use of proper fittings helps to maintain the required flow velocities and reduces the possibility of stoppage in the drainage system. Combination fittings are commonly used in drainage systems and must be evaluated for their pattern of flow. Combination fittings include two or more fittings, such as combination wye and eighth bends or tee-wyes. A double sanitary tee (cross) cannot be used for connections to fixtures and appliances with pumping action because it has a short pattern for change of direction. In addition to blowout-type water closets, 1.6-gallon-per-flush (6.1 L/flush) flushometer tank (pressure assisted) and gravity tank-type fixtures may also produce crossover flow into the opposite fixture connection. The high-velocity flow may discharge through the opposing openings of a double sanitary tee, causing an interruption or blowback at the other fixture. Back-to-back water closets are fixtures that are installed directly opposite of each other and discharge through a short distance into a double pattern fitting. If back-to-back fixtures discharge through several feet of pipe or multiple changes of direction, the flow velocity in the fixture drains would decrease to the point where cross flow would not occur. The exception provides for back-to-back water closet connections to double sanitary tee patterns where the horizontal developed length between the outlet of the water closet and connection to the double sanitary tee pattern is 18 inches (457 mm) or more. The critical distance at which there is potential for crossover is a horizontal developed length of less than 10 times the diameter of the horizontal pipe, for example, 30 inches (762 mm) for a 3-inch-diameter (76 mm) pipe. Once the developed length is over 30 inches (762 mm), the surging flow conditions will no longer exist, and the frictional resistance of the pipe retards the velocity to that of uniform flow conditions.

It is mistakenly thought that the use of a double combination fitting as shown in Commentary Figure 706.3(5) is not permitted for the water closet arrangement because it will not permit the vent to connect above the trap weir. However, this arrangement is permitted because the vent connection limitation is designed to prevent self-siphonage, and water closets must self-siphon to operate properly. Use of fittings such as those having side inlets, heel inlets and specialty opening configurations is permitted only in applications that comply with the intent of Table 706.3 [see Commentary Figures 706.3(1) through (5)]

TABLE 706.3
FITTINGS FOR CHANGE IN DIRECTION

TYPE OF FITTING PATTERN	CHANGE IN DIRECTION		
	Horizontal to vertical	Vertical to horizontal	Horizontal to horizontal
Sixteenth bend	X	X	X
Eighth bend	X	X	X
Sixth bend	X	X	X
Quarter bend	X	X[a]	X[a]
Short sweep	X	X[a,b]	X[a]
Long sweep	X	X	X
Sanitary tee	X[c]	—	—
Wye	X	X	X
Combination wye and eighth bend	X	X	X

For SI: 1 inch = 25.4 mm.

a. The fittings shall only be permitted for a 2-inch or smaller fixture drain.

b. Three inches or larger.

c. For a limitation on double sanitary tees, see Section 706.3.

❖ When the flow is in the horizontal plane only, long-pattern fittings are necessary to prevent excessive reduction in flow velocity. When flow is from the horizontal to the vertical or from the vertical to the horizontal, a shorter pattern fitting may be used because the acceleration of flow caused by gravity will maintain higher velocities.

An "X" in a column indicates that a particular fitting is acceptable for the described change in direction. For example, the only "X" shown for a sanitary tee is in the column for horizontal to vertical change in direction. The table has been developed to identify where use of a particular drainage pattern fitting is allowed.

The terminology used to identify drainage pattern fittings was originally developed by the cast-iron soil pipe industry. The pattern of the fittings is regulated by the fitting standards referenced in Table 702.4. A fitting of one material will not necessarily look the same as a

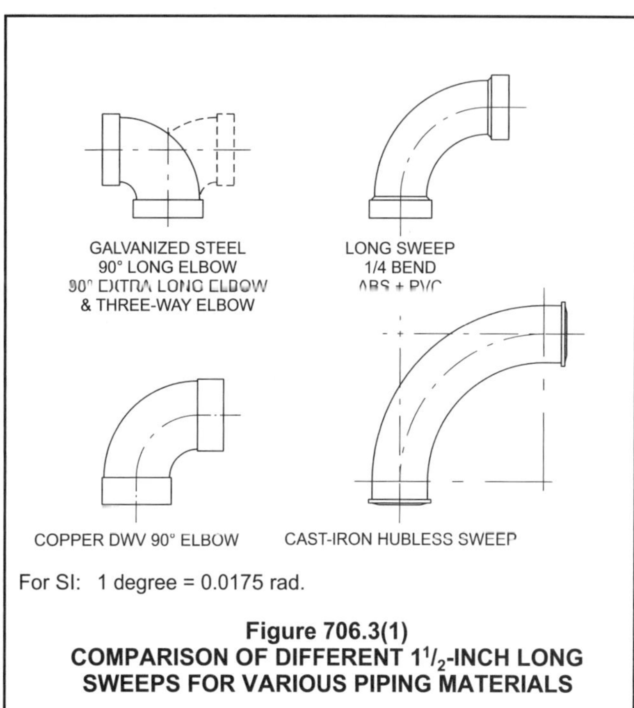

For SI: 1 degree = 0.0175 rad.

Figure 706.3(1)
COMPARISON OF DIFFERENT $1^1/_2$-INCH LONG SWEEPS FOR VARIOUS PIPING MATERIALS

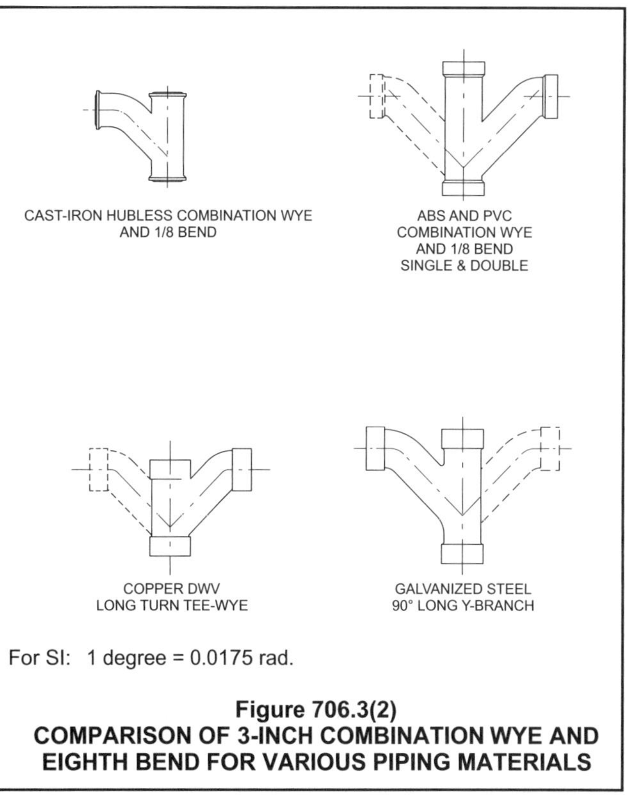

For SI: 1 degree = 0.0175 rad.

Figure 706.3(2)
COMPARISON OF 3-INCH COMBINATION WYE AND EIGHTH BEND FOR VARIOUS PIPING MATERIALS

fitting of another material with the same name. Only cast-iron pipe fittings have a separate quarter-bend fitting and short-sweep fitting. For other drainage piping materials, a quarter bend and a short sweep are the same fitting.

Copper tubing fittings are unique by having only one sweep-type fitting identified as a DWV 90-degree (1.6 rad) elbow. This is considered a long-sweep drainage fitting. Some manufacturers make a copper fitting identified as a DWV 90-degree (1.6 rad) long radius elbow. This is also not included in the fitting standards, but could be used as a long sweep.

Note b permits the use of short-sweep fittings of 3 inches (76 mm) in diameter or larger where the change in direction is from the vertical to the horizontal.

Note a permits the use of 2-inch (51 mm) or smaller short-sweep and quarter-bend fittings for drains serving a single fixture where the change in direction is from the vertical to the horizontal or from the horizontal to the horizontal.

The intent of Note c is to allow short-pattern fittings for fixture drains, which are generally difficult to install using long-pattern fittings in frame construction. For example, a long-sweep PVC 90-degree (1.6 rad) elbow would not fit within a nominal 2-inch by 4-inch stud wall cavity where used to make a horizontal-to-horizontal change in direction. A quarter bend with a very short radius, less than that of a short sweep, is identified as a vent elbow or vent quarter bend. These particular fittings are designed for vent system applications only.

Double sanitary tees must not be installed where water closets or appliances with pump action are back to back. Appliances with pump action are typically washing machines or dishwashers (see commentary, Section 706.3).

Side inlet fittings are acceptable for drainage connections. The typical application for side inlet fittings is horizontal to vertical changes in direction involving side inlet sanitary tees, side inlet elbows and side inlet wyes. When used to connect a vent, however, the requirements for venting as specified in Section 905 must be met [see Commentary Figures 706.3(1) through 706.3(8)].

Low-heel and high-heel inlet elbows are acceptable depending on the application. A low-heel branch inlet must not be used for a drainage connection where solid waste enters the run of the elbow from the vertical. In such a case, solid waste could enter the branch inlet opening, causing a blockage. Because there is no waste flow in vent piping, and the airflow in vent piping

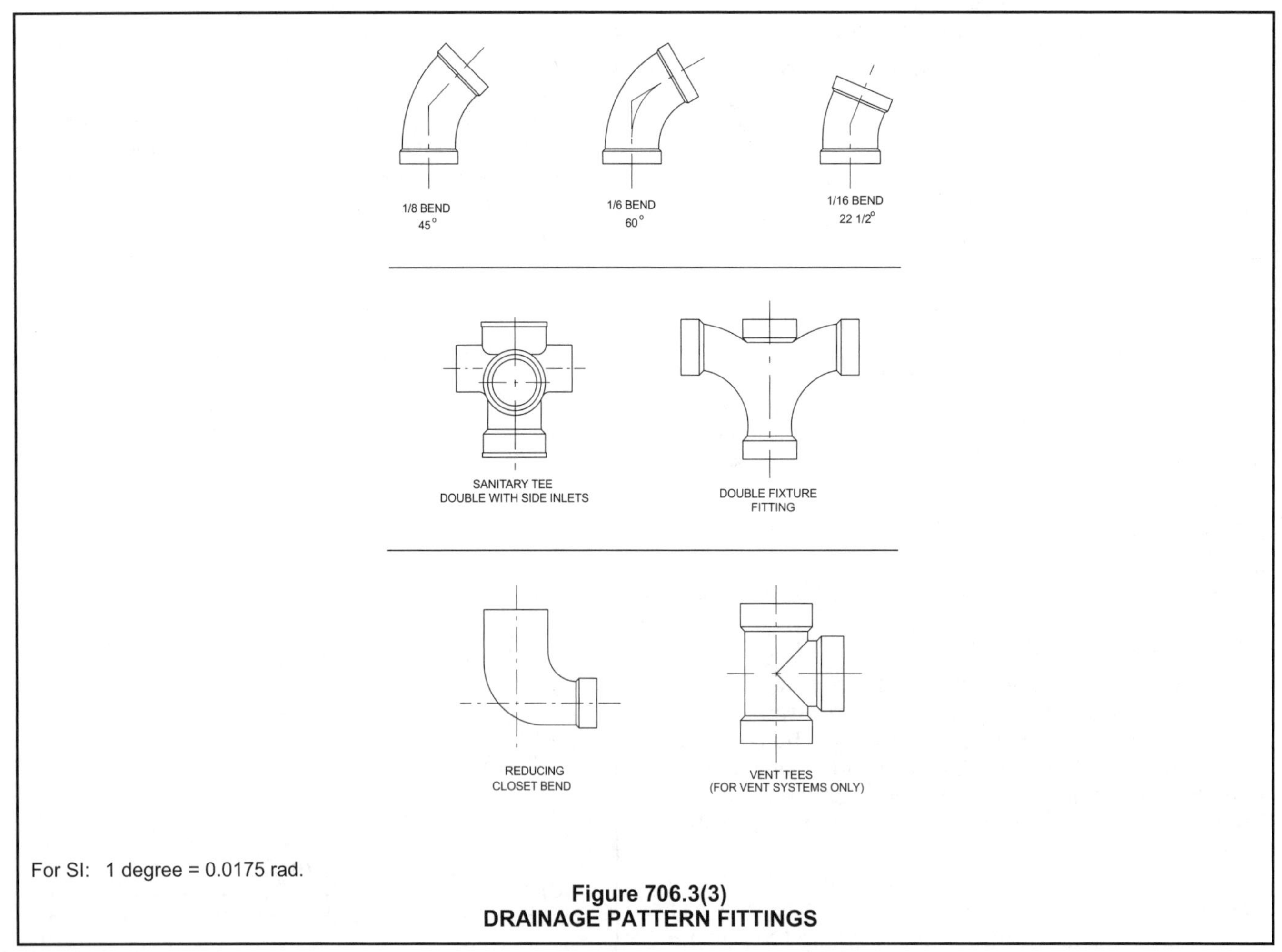

For SI: 1 degree = 0.0175 rad.

Figure 706.3(3)
DRAINAGE PATTERN FITTINGS

may be in either direction, the types of fittings used in vent piping are not regulated by this section. For example, a sanitary tee placed on its back with the branch inlet connected to a vertical dry vent riser is a common installation practice and is not prohibited. In this application, the only waste flow is through the straight run of the fitting, and the branch inlet conducts only air into or out of the horizontal drain pipe. Note that the tee must be oriented to allow compliance with Section 905.3. The restrictions placed on the use of a sanitary tee are based on its inability to guide waste flow entering through the branch inlet. Such restrictions do not apply where the branch inlet is used for a dry vent connection in compliance with Section 905. In all cases, it is the intent of the code to require that fittings be installed in a manner that will encourage high-velocity drainage flow and minimize the chances of blockage.

The details of the patterns for drainage, waste and vent fittings made of PVC and ABS are covered by ASTM D 3311. This standard requires 90 degree [1.6 rad] fittings (except vent patterns) to have sockets aligned in the fitting to cause horizontal drainage pipes to have a slope of $^1/_4$ inch per foot (21 mm per m) [see Commentary Figure 706.3(9)]. In some applications, the use of these quarter-bend fittings, having an angle of 91.2 degrees, results in an improper alignment of the connecting pipes resulting in a poor joint. Figure 706.3.(10) shows the correct orientation for the sanitary tee to prevent this piping alignment problem. Many plumber apprentices learned to install sanitary tees in vent piping in this manner after being told that the orientation promoted better airflow or allowed the condensate to flow better back to the drain; however, there is no technical substantiation for these explanations.

Sanitary tees may be used for horizontal drainage flow through the run of the tee where the branch of the tee is for venting only and is oriented within 45 degrees of the vertical. A sanitary tee can be installed on its back for venting applications in sanitary drainage.

Other bend patterns for DWV fittings of PVC and ABS do not have this special socket alignment; thus, an $^1/_8$ bend fitting is 45 degrees (0.79 rad), a $^1/_6$ bend fitting is 60 degrees (1 rad) and so forth. In some applications, use of these fittings might a create pipe slope problem. Although the code is silent on the use of a close combination of fittings to serve the purpose of one fitting, full and square engagement of the pipes into a close combination "made up" 90 degree (1.6 rad) bend might produce undesirable results.

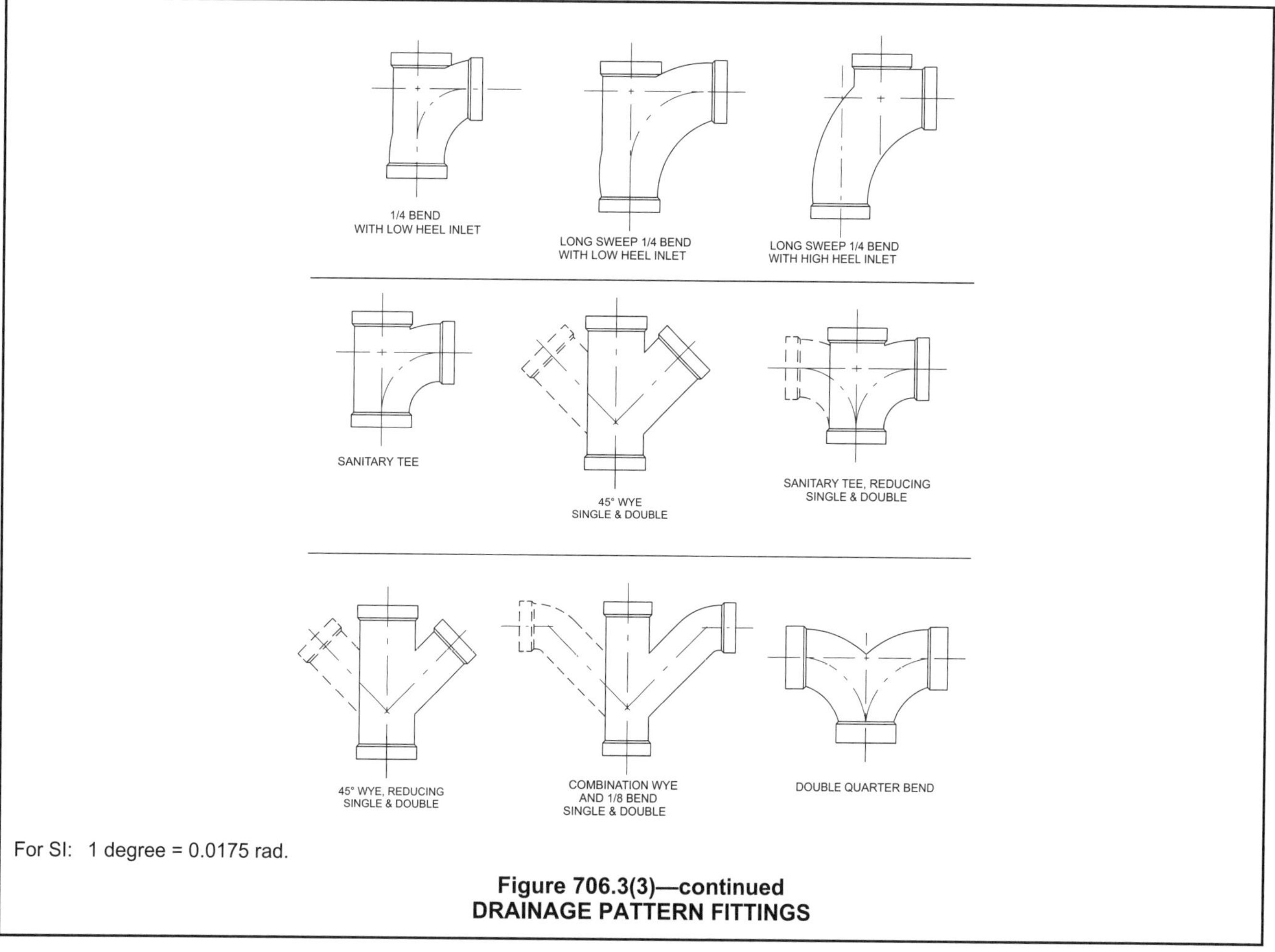

For SI: 1 degree = 0.0175 rad.

Figure 706.3(3)—continued
DRAINAGE PATTERN FITTINGS

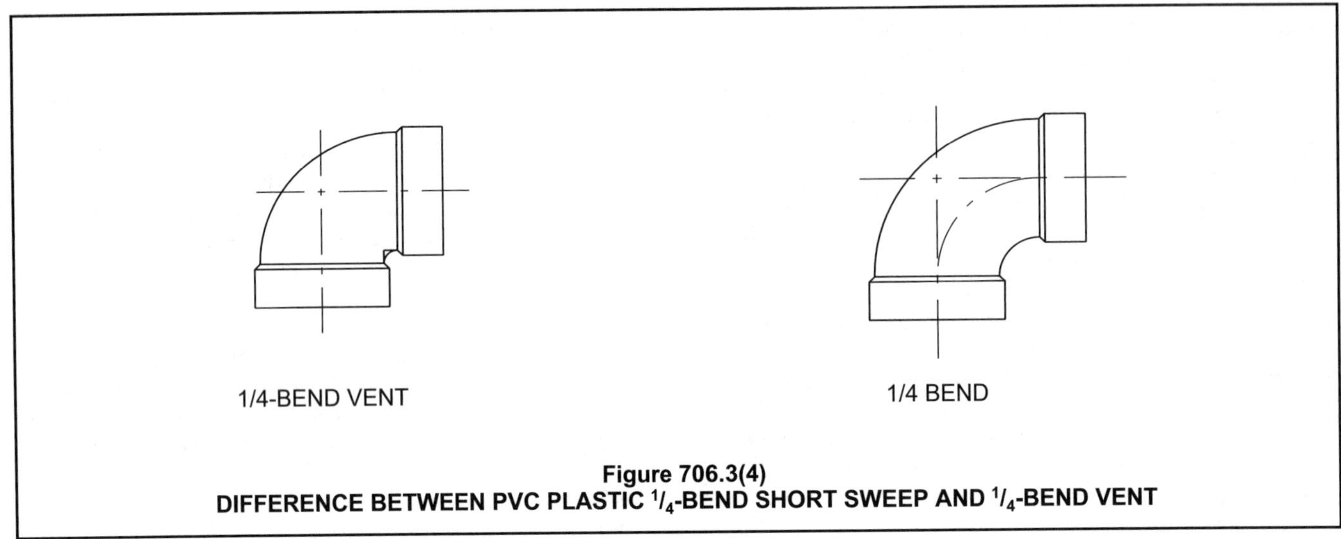

Figure 706.3(4)
DIFFERENCE BETWEEN PVC PLASTIC 1/4-BEND SHORT SWEEP AND 1/4-BEND VENT

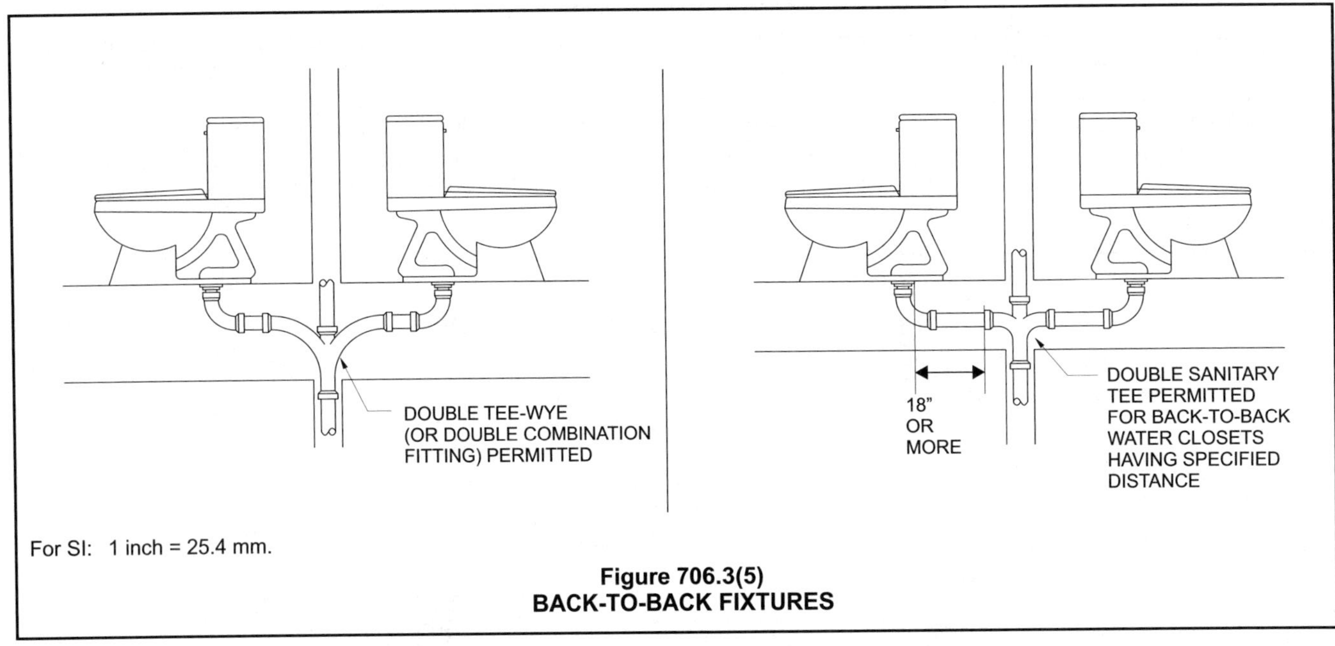

For SI: 1 inch = 25.4 mm.

Figure 706.3(5)
BACK-TO-BACK FIXTURES

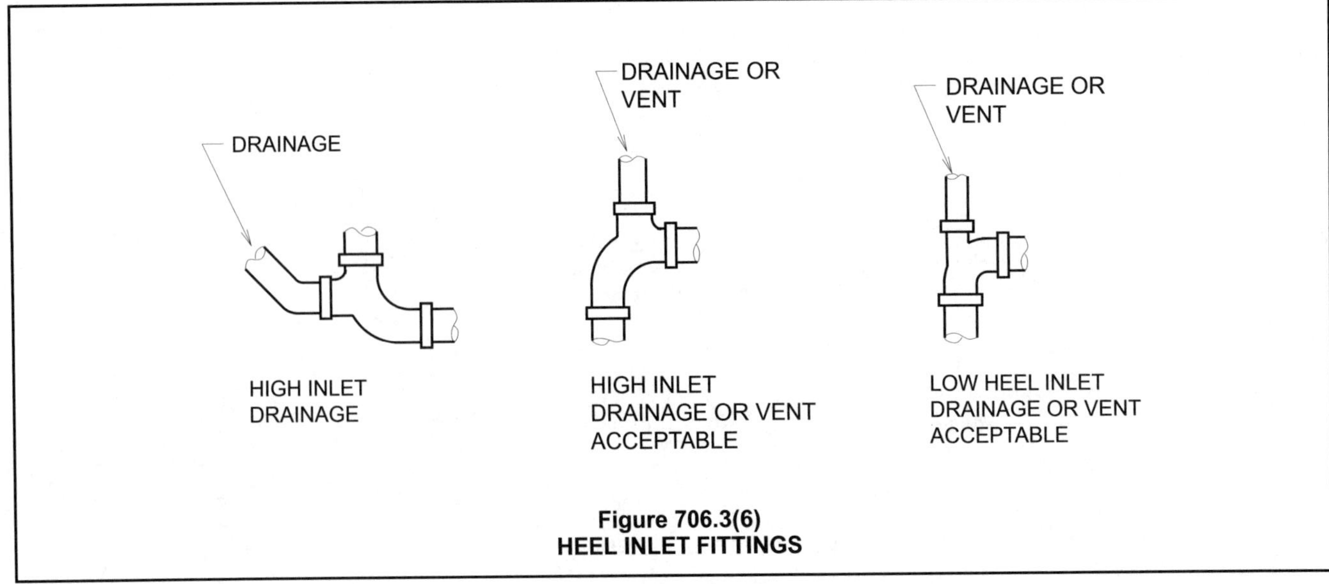

Figure 706.3(6)
HEEL INLET FITTINGS

DOUBLE SANITARY TEE
WITH SIDE INLETS
TWIN TEE
COTTAGE TEE

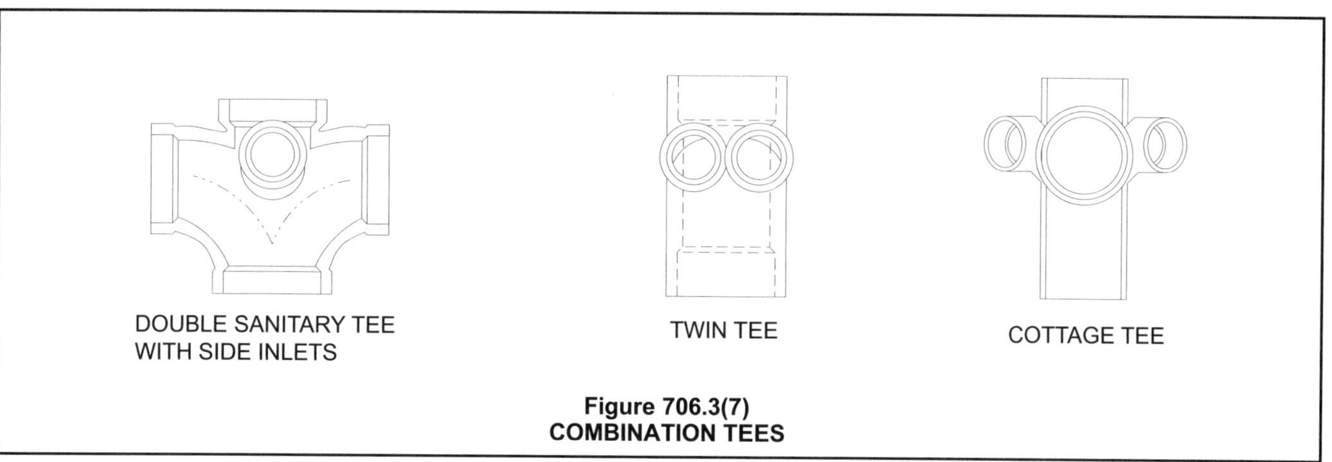

Figure 706.3(7)
COMBINATION TEES

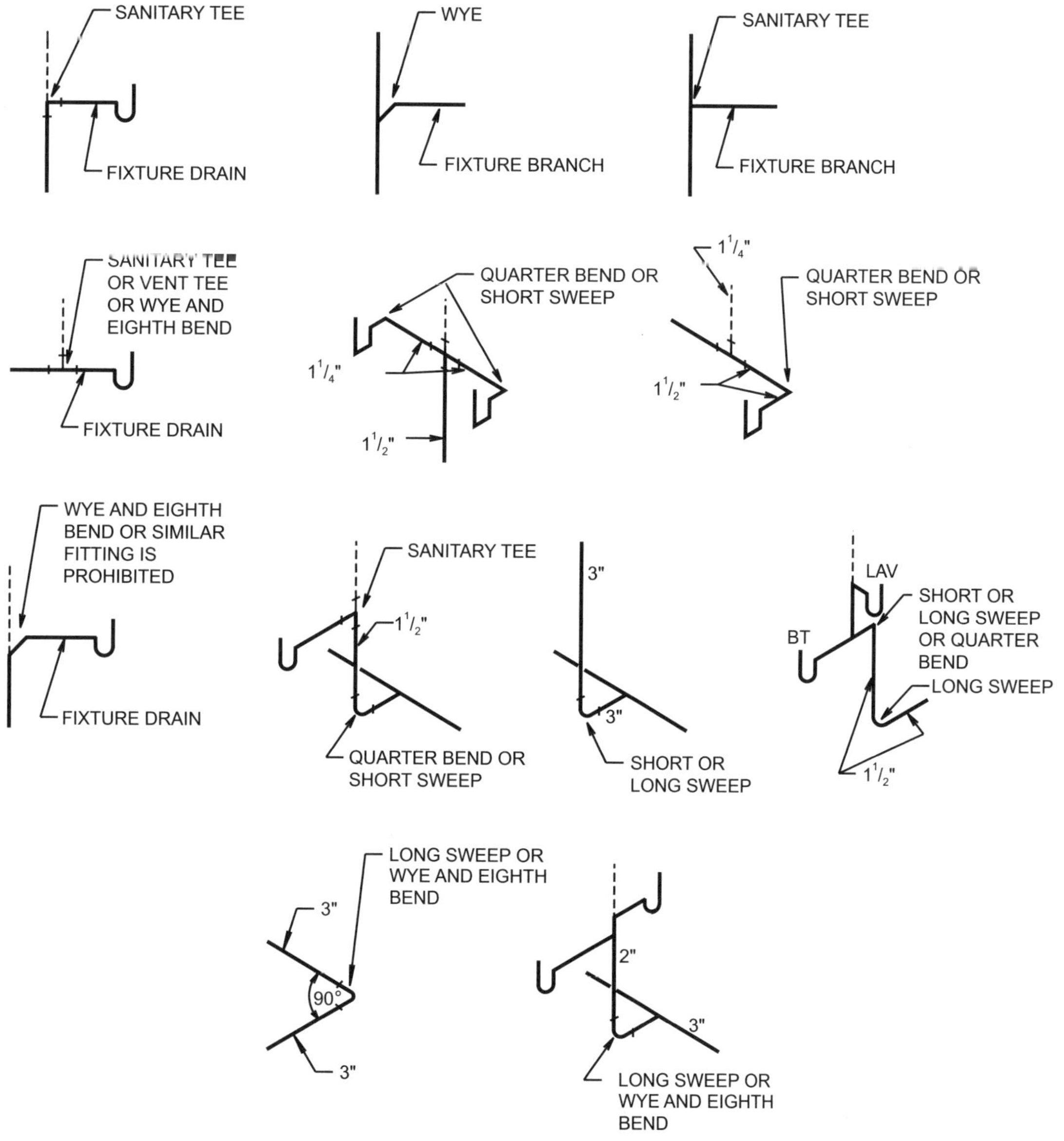

SANITARY TEE
FIXTURE DRAIN
WYE
FIXTURE BRANCH
SANITARY TEE
FIXTURE BRANCH
SANITARY TEE
OR VENT TEE
OR WYE AND
EIGHTH BEND
FIXTURE DRAIN
QUARTER BEND OR
SHORT SWEEP
1 1/4"
1 1/2"
1 1/4"
QUARTER BEND OR
SHORT SWEEP
1 1/2"
WYE AND EIGHTH
BEND OR SIMILAR
FITTING IS
PROHIBITED
FIXTURE DRAIN
SANITARY TEE
1 1/2"
QUARTER BEND OR
SHORT SWEEP
3"
3"
SHORT OR
LONG SWEEP
LAV
BT
SHORT OR
LONG SWEEP
OR QUARTER
BEND
LONG SWEEP
1 1/2"
LONG SWEEP OR
WYE AND EIGHTH
BEND
3"
90°
3"
2"
3"
LONG SWEEP OR
WYE AND EIGHTH
BEND

For SI: 1 inch = 25.4 mm, 1 degree = 0.0175 rad.

Figure 706.3(8)
FITTING USES

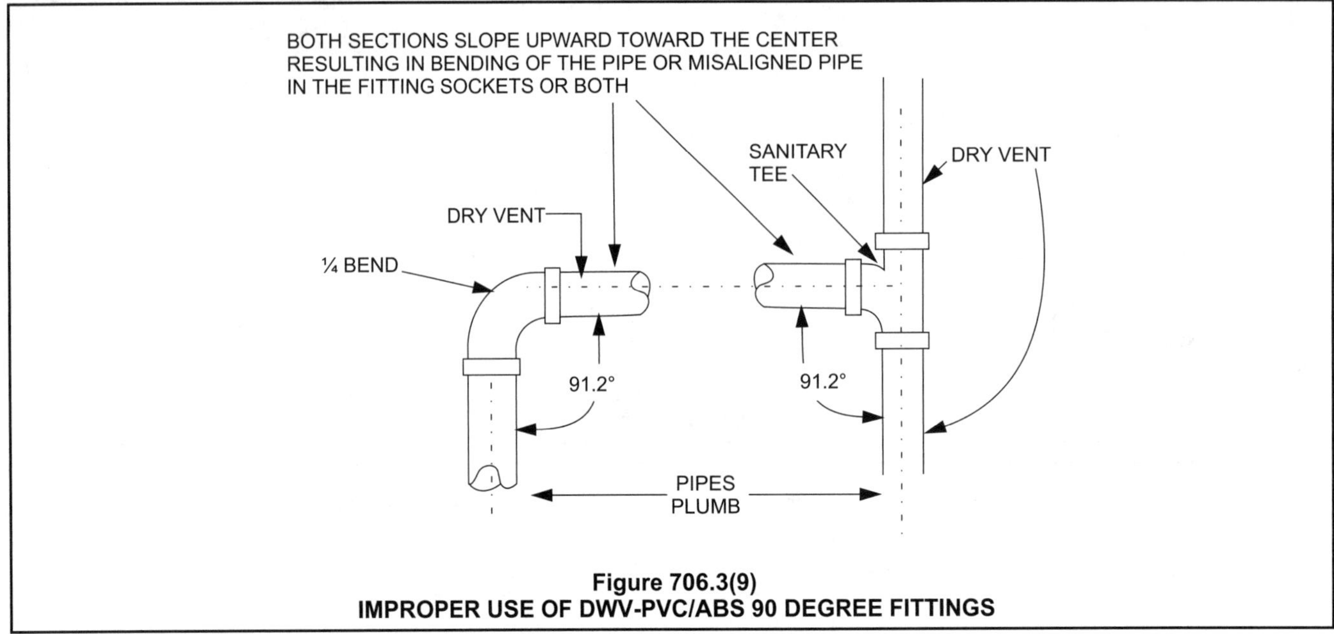

Figure 706.3(9)
IMPROPER USE OF DWV-PVC/ABS 90 DEGREE FITTINGS

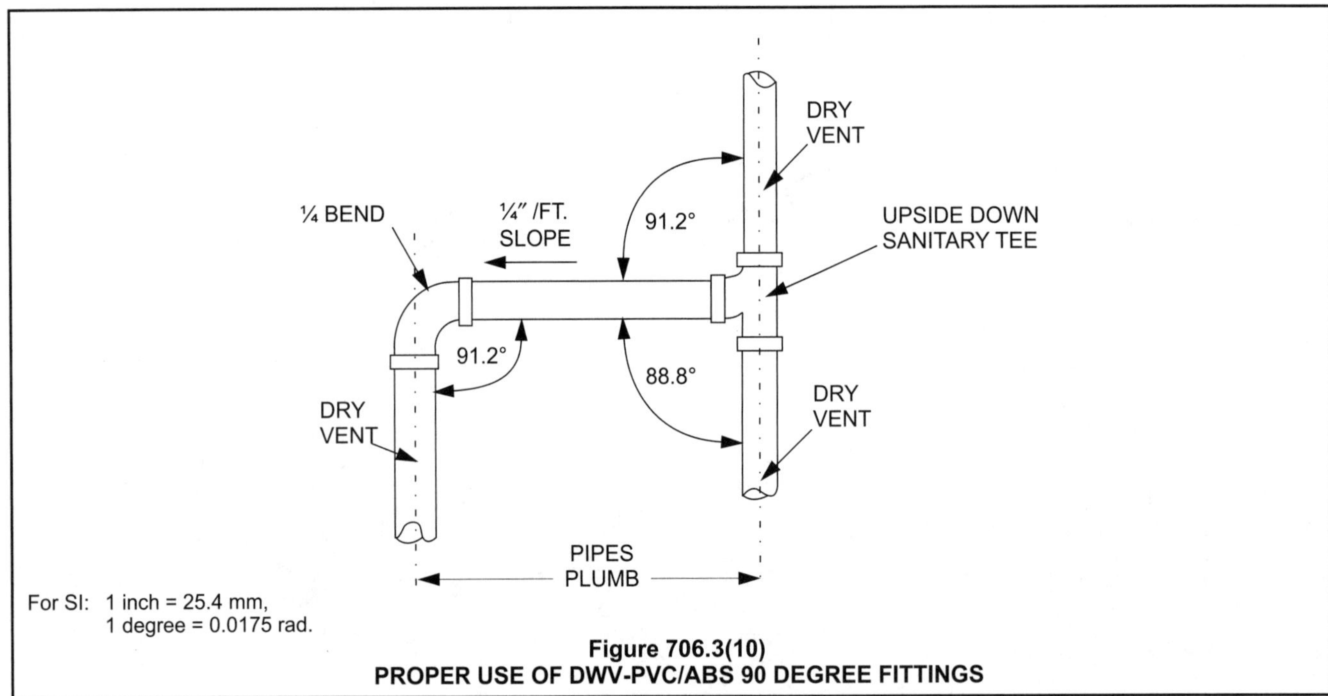

For SI: 1 inch = 25.4 mm,
1 degree = 0.0175 rad.

Figure 706.3(10)
PROPER USE OF DWV-PVC/ABS 90 DEGREE FITTINGS

706.4 Heel- or side-inlet quarter bends. Heel-inlet quarter bends shall be an acceptable means of connection, except where the quarter bend serves a water closet. A low-heel inlet shall not be used as a wet-vented connection. Side-inlet quarter bends shall be an acceptable means of connection for drainage, wet venting and *stack* venting arrangements.

❖ Because of the possibility of waste from a water closet clogging a heel-inlet (either high or low heel) quarter bend, this fitting must not be used in a closet bend application. Similarly, low-heel-inlet quarter bends must not receive flow from drainage pipes that serve as wet vents for a fixture. Only side-inlet quarter bends can be used for these applications [see Commentary Figures 706.4(1), (2) and 706.3(6)]. Note the obvious cut floor joist violation in Commentary Figure 706.4(2).

SECTION 707
PROHIBITED JOINTS AND CONNECTIONS

707.1 Prohibited joints. The following types of joints and connections shall be prohibited:

1. Cement or concrete joints.
2. Mastic or hot-pour bituminous joints.
3. Joints made with fittings not *approved* for the specific installation.

4. Joints between different diameter pipes made with elastomeric rolling O-rings.
5. Solvent-cement joints between different types of plastic pipe.
6. Saddle-type fittings.

❖ This section contains requirements for joints between pipe and fittings of different materials.

Cement and concrete are not effective in sealing a pipe joint because such materials are inflexible and susceptible to cracking and displacement. A rolling O-ring has no resistance to being pushed or rolled out of a joint when exposed to pressure or the movement of pipe caused by expansion and contraction. A solvent-cement joint is a homogeneous chemical bond made between a pipe and a fitting. The bond is accomplished because the chemical composition of the joint surfaces is the same. If the materials are different, the joint may not form a proper chemical bond, and the strength and integrity of the joint will be adversely affected. For example, solvent cementing is prohibited between ABS and PVC, PVC and CPVC or ABS and CPVC (see Commentary Section 705.2.2). Saddle-type fittings that are moved out of alignment can weaken the pipe because of the pipe wall penetration and typically do not form a drainage pattern connection. "Saddle-type" fittings are commonly used for connecting a building sewer to an existing public sewer.

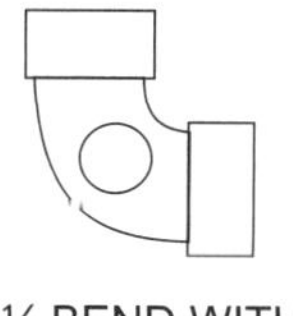

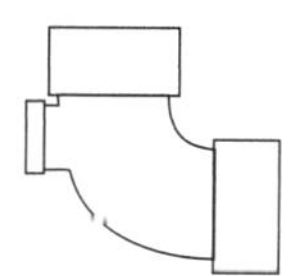

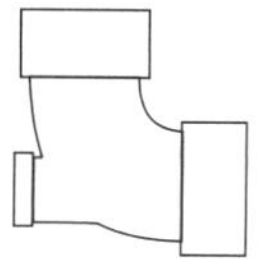

Figure 706.4(1)
HEEL- AND SIDE-INLET QUARTER BEND APPLICATION RESTRICTIONS

Figure 706.4(2)
SIDE-INLET QUARTER BEND FOR WATER CLOSET

SECTION 708
CLEANOUTS

708.1 Scope. This section shall govern the size, location, installation and maintenance of drainage pipe cleanouts.

❖ Sanitary drainage pipe cleanouts must provide access to the drainage system for clearing stoppages. Although sanitary drainage systems are designed to function without stoppages under all conditions, stoppages do occur. Cleanouts are intended to make the interior of the drainage system accessible for the clearing of stoppages with relative convenience and without the costs and hardship involved with the cutting or dismantling of pipe and fittings. The typical cleanout is an accessible opening in a drainage system installed to facilitate inspection and provide a convenient place for inserting drain-cleaning equipment for the removal of obstructions.

The definition of "Cleanout" in Section 202 specifies what is considered to be a means of access to the drainage system. Note that removal of a fixture, such as a water closet, to gain access to a drain system provides an acceptable cleanout point, although inconvenient. The most recognized cleanouts are cleanout plugs. These cleanouts are usually located at changes of direction in a drain or at the base of stacks.

The code also recognizes other acceptable types of cleanouts, such as a removable fixture trap where the "U" bend and trap arm can be readily removed after loosening slip-joint connections. A fixture drain may involve vent connections and multiple changes in direction, which could make drain cleaning difficult from a fixture trap point of access.

In most cases, a removed "P" trap will provide suitable access to the fixture drain; however, there are circumstances where a properly placed cleanout plug would provide greater convenience and ease of drain cleaning and avoid the disassembly of drainage system components.

The code also considers the removal of a plumbing fixture such as a water closet as an acceptable means of accessing the drainage system. A water closet may be removed without disturbing any of the drainage piping system. The removal and reinstallation of a water closet requires a considerable amount of time and skilled labor. The replacement of the flange-sealing ring, replacement of the closet bolts, replacement of the water supply line, repair of the flange, repair of the tank-to-bowl seals and leveling and stabilizing the fixture are some of the procedures that may be involved in the removal and replacement of a water closet.

A well-placed cleanout plug would be considerably more convenient than removing a plumbing fixture having an integral trap, such as a water closet or urinal. Cleanouts must be maintained gas tight and water tight to prevent leakage of waste or sewer gases.

A cleanout may be any opening to the drainage system that is designed to allow ready access to the interior of the piping (see Commentary Figure 708.1).

708.2 Cleanout plugs. Cleanout plugs shall be brass or plastic, or other *approved* materials. Brass cleanout plugs shall be utilized with metallic drain, waste and vent piping only, and shall conform to ASTM A 74, ASME A112.3.1 or ASME A112.36.2M. Cleanouts with plate-style *access* covers shall be fitted with corrosion-resisting fasteners. Plastic cleanout plugs shall conform to the requirements of Section 702.4. Plugs shall

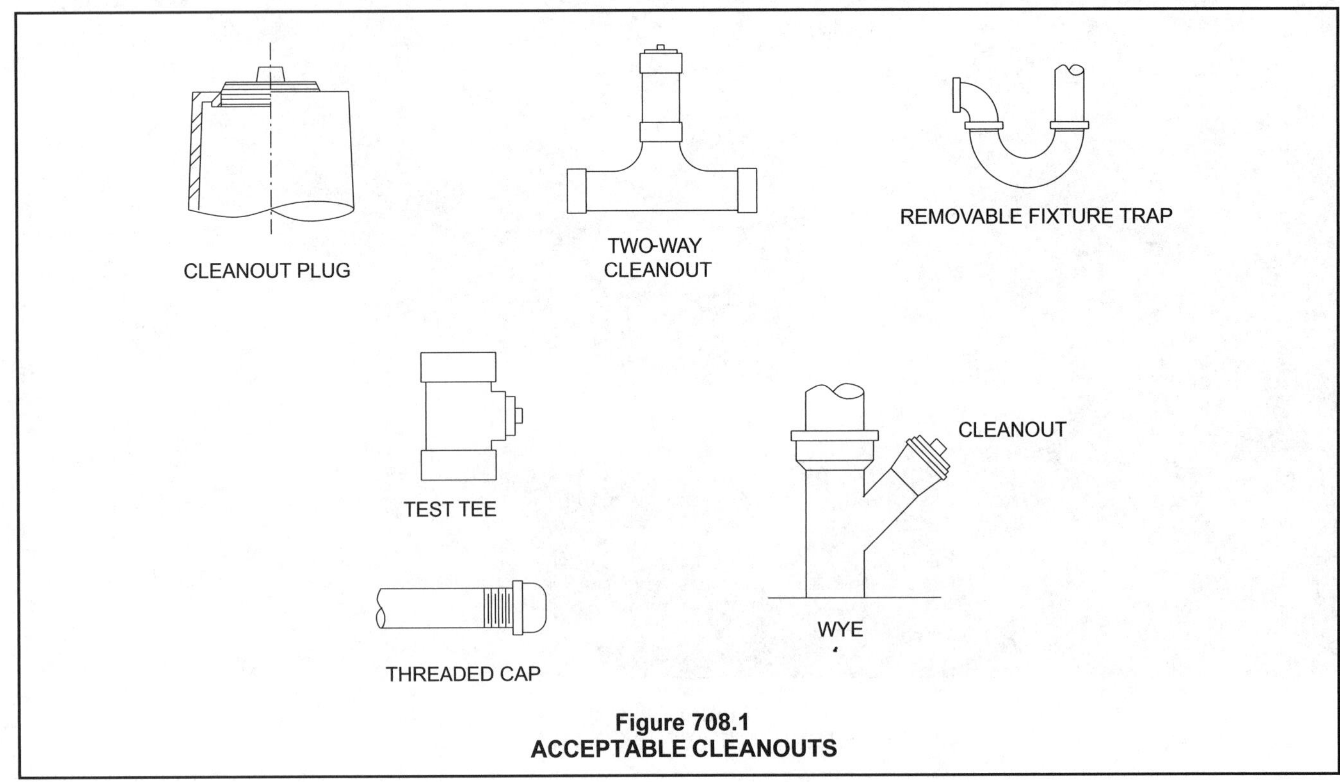

Figure 708.1
ACCEPTABLE CLEANOUTS

have raised square or countersunk square heads. Countersunk heads shall be installed where raised heads are a trip hazard. Cleanout plugs with borosilicate glass systems shall be of borosilicate glass.

❖ Metallic cleanout plugs must be brass to provide for easy removal. If the cleanout plug is corroded in place, the brass plug is a soft enough material to be chiseled out. Brass cleanout plugs are limited to metallic fittings because the metal plug threads may damage the softer plastic threads of a fitting. Plastic plugs are intended for use with plastic fittings; however, this section does not prohibit the use of plastic plugs with metallic fittings. Like brass plugs, plastic plugs are less likely to seize in metallic fittings. The cleanout plug must have a square turning surface to allow for ease of removal while minimizing the possibility of stripping the surface during removal.

Because a cleanout is designed to provide access into the drainage system, the cleanout itself must be accessible. The cleanout must extend up to and be flush with the finished floor level or grade. A cleanout plug located on a walking surface must be installed so that it does not create a tripping hazard; these cleanouts typically are countersunk to minimize such hazard and protect the cleanout from damage (see Commentary Figure 708.2).

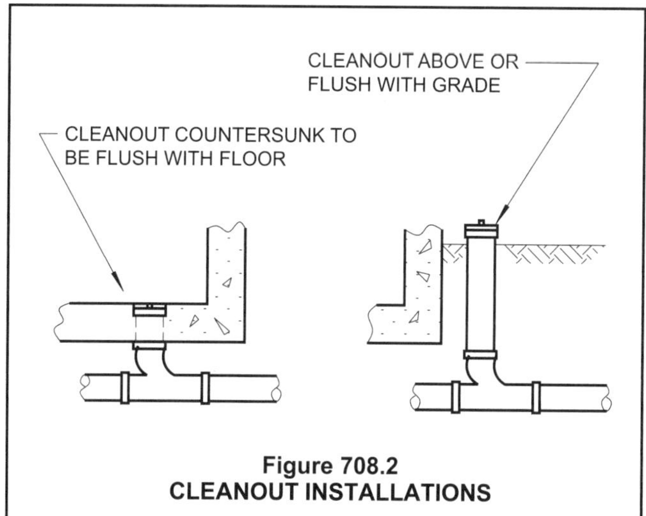

Figure 708.2
CLEANOUT INSTALLATIONS

The first sentence of this section refers to other approved materials for the construction of cleanout plugs; however, the intent is to recognize other types of fittings that may be used as cleanouts, such as hubless cast-iron blind plugs and removable caps. The intent of the code is to require that threaded cleanout plugs be constructed of either brass or plastic. Specialized borosilicate glass drainage piping requires the use of a compatible borosilicate glass cleanout plug to maintain corrosion resistance.

708.3 Where required. Cleanouts shall be located in accordance with Sections 708.3.1 through 708.3.6.

❖ This section contains the requirements for the location of cleanouts as described in Sections 708.3.1 through 708.3.6.

708.3.1 Horizontal drains within buildings. All horizontal drains shall be provided with cleanouts located not more than 100 feet (30 480 mm) apart.

❖ Cleanouts must be spaced a reasonable distance apart to facilitate cleaning any portion of the drainage system. This distance is based on the use of modern-day cleaning equipment, and attempts to minimize the inconvenience and health hazard associated with creating a mess inside the building as a result of removing a blockage. Inadequate cleanout spacing may require the use of excessively long rodding cables, which complicates the task and increases the likelihood that the rodding cable will jam or break inside the pipe. Every horizontal drain must be able to be rodded from a cleanout or from a fixture trap that serves as a cleanout, regardless of the length (see Commentary Figure 708.3.1).

708.3.2 Building sewers. Building sewers shall be provided with cleanouts located not more than 100 feet (30 480 mm) apart measured from the upstream entrance of the cleanout. For building sewers 8 inches (203 mm) and larger, manholes shall be provided and located not more than 200 feet (60 960 mm)

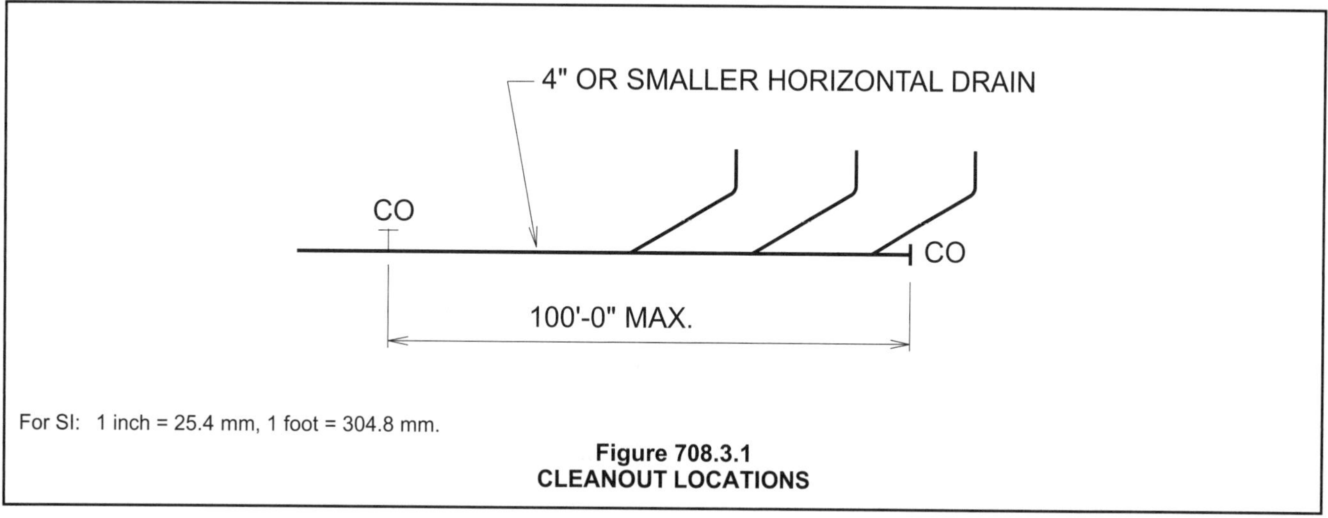

Figure 708.3.1
CLEANOUT LOCATIONS

from the junction of the *building drain* and *building sewer*, at each change in direction and at intervals of not more than 400 feet (122 m) apart. Manholes and manhole covers shall be of an *approved* type.

❖ A distance of up to 100 feet (30 480 mm) is permitted between cleanouts on building sewers because drain-cleaning equipment may be used outdoors where this distance can be easily accommodated. The use of outdoor cleanouts does not involve the same concerns as indoor cleanouts relative to health hazards and protection of property. Outdoor cleanouts for sewers must be brought up to grade for access, and the length of piping between the actual cleanout access opening and the sewer must be included in the overall developed length of piping between cleanouts [see Commentary Figure 708.3.2(1)].

When a building sewer is 8 inches (203 mm) or larger, manholes are required as cleanouts, and a distance of 200 feet (60 960 mm) is required from the junction of the building drain and the building sewer. Because of the size and maneuverability afforded by a manhole, the spacing between cleanout points is increased to a maximum of 400 feet (121.9 m). A manhole is required at each change of direction of building sewers of 8 inches (203 mm) in diameter and larger, regardless of the distance between manholes [see Commentary Figure 708.3.2(2)]. The designer specifying the installation of manholes in areas subject to flooding must consider the protection of manholes from the effects of floodwaters. Floodwaters may damage or completely displace manholes and covers if they are not properly sealed and secured. Additionally, submerged manhole openings (covers) may allow floodwaters to enter the drainage system. The manhole cover may be located above the base flood level elevation or designed to resist hydrostatic forces when submerged in water and dynamic forces resulting from wave action or high-velocity water.

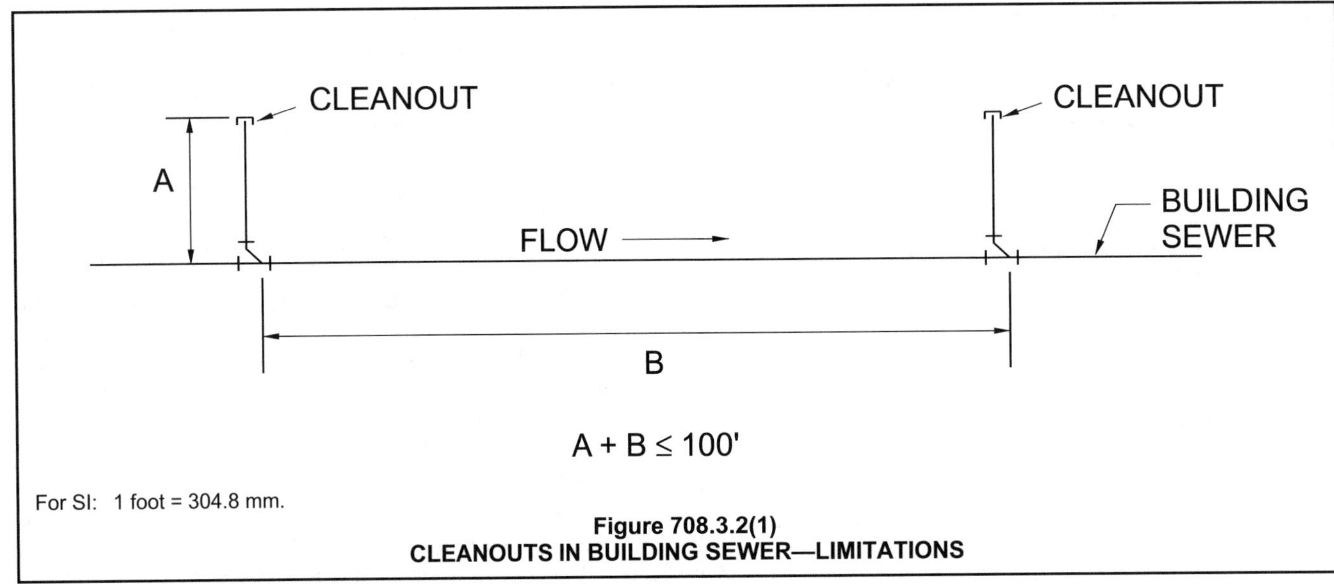

Figure 708.3.2(1)
CLEANOUTS IN BUILDING SEWER—LIMITATIONS

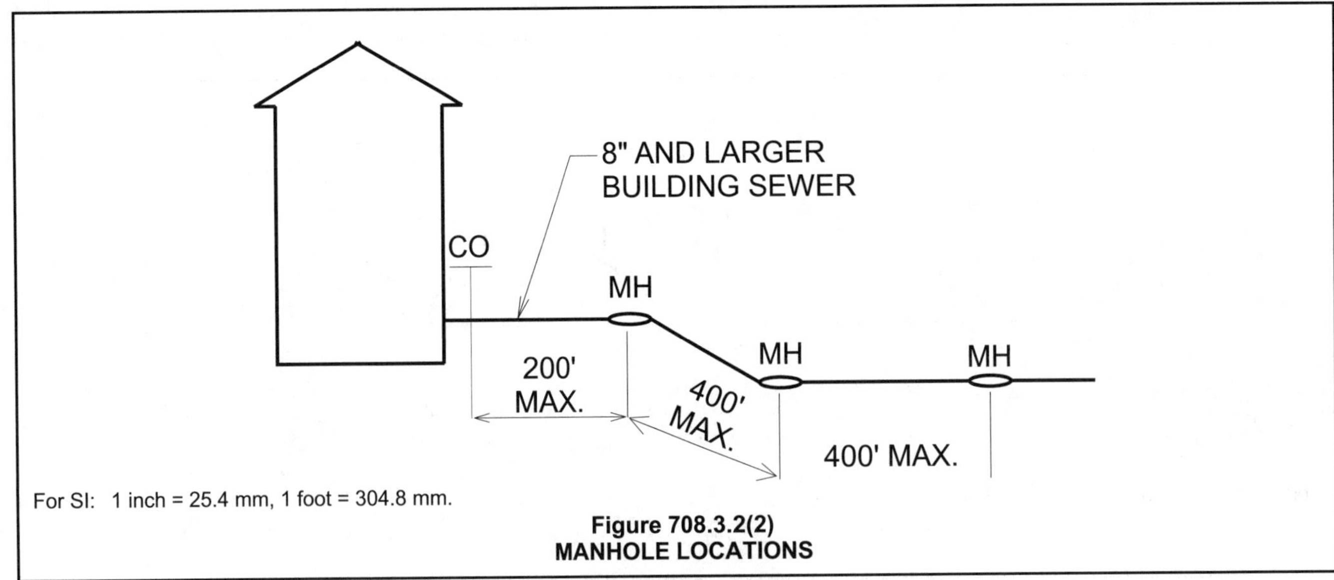

Figure 708.3.2(2)
MANHOLE LOCATIONS

708.3.3 Changes of direction. Cleanouts shall be installed at each change of direction greater than 45 degrees (0.79 rad) in the *building sewer*, *building drain* and horizontal waste or soil lines. Where more than one change of direction occurs in a run of piping, only one cleanout shall be required for each 40 feet (12 192 mm) of *developed length* of the drainage piping.

❖ The requirement for a cleanout at a change in direction greater than 45 degrees (0.79 rad) is for an individual fitting, not a combination of fittings. The building sewer as well as all horizontal drain lines within the building is covered by this requirement. If a 90-degree (1.6 rad) change in direction is accomplished with a single fitting, such as a sweep or combination tee-wye, a cleanout is required. If the same change in direction is accomplished with two one-eighth bends, a cleanout is not required.

Although this requirement appears to be inconsistent, it is logical based on the limitations of cleaning equipment. Rodding equipment will easily pass through fittings having a change in direction of 45 degrees (0.79 rad) or less, even if multiple fittings are connected. A fitting with a greater change in direction, however, could inhibit the rodding equipment from passing through the fitting.

A cleanout is intended to provide access to the drainage system for cleaning; therefore, cleanouts should be located near where potential blockages will most likely occur in the drainage system. Cleanout locations at or adjacent to abrupt changes in direction allow such access. A cleanout is not required at changes in direction less than or equal to 45 degrees (0.79 rad) or where changes in direction are accomplished with multiple fittings, none of which exceed 45 degrees (0.79 rad) (see Commentary Figure 708.3.3).

Earlier editions of the code required installation of a cleanout at every change in direction. This section permits a single cleanout to serve up to 40 feet (12 192 mm) of developed length of drainage pipe, regardless of the number of changes in direction that occur within that developed length. This provision takes into consideration that modern drain-cleaning equipment can easily pass through fittings over this distance and that technological advances in the manufacturing of pipe and fittings have improved flow characteristics and decreased the likelihood of stoppages.

This section also requires that cleanouts be reachable or accessible without having to remove a permanent portion of the structure (see Commentary Section 708.4).

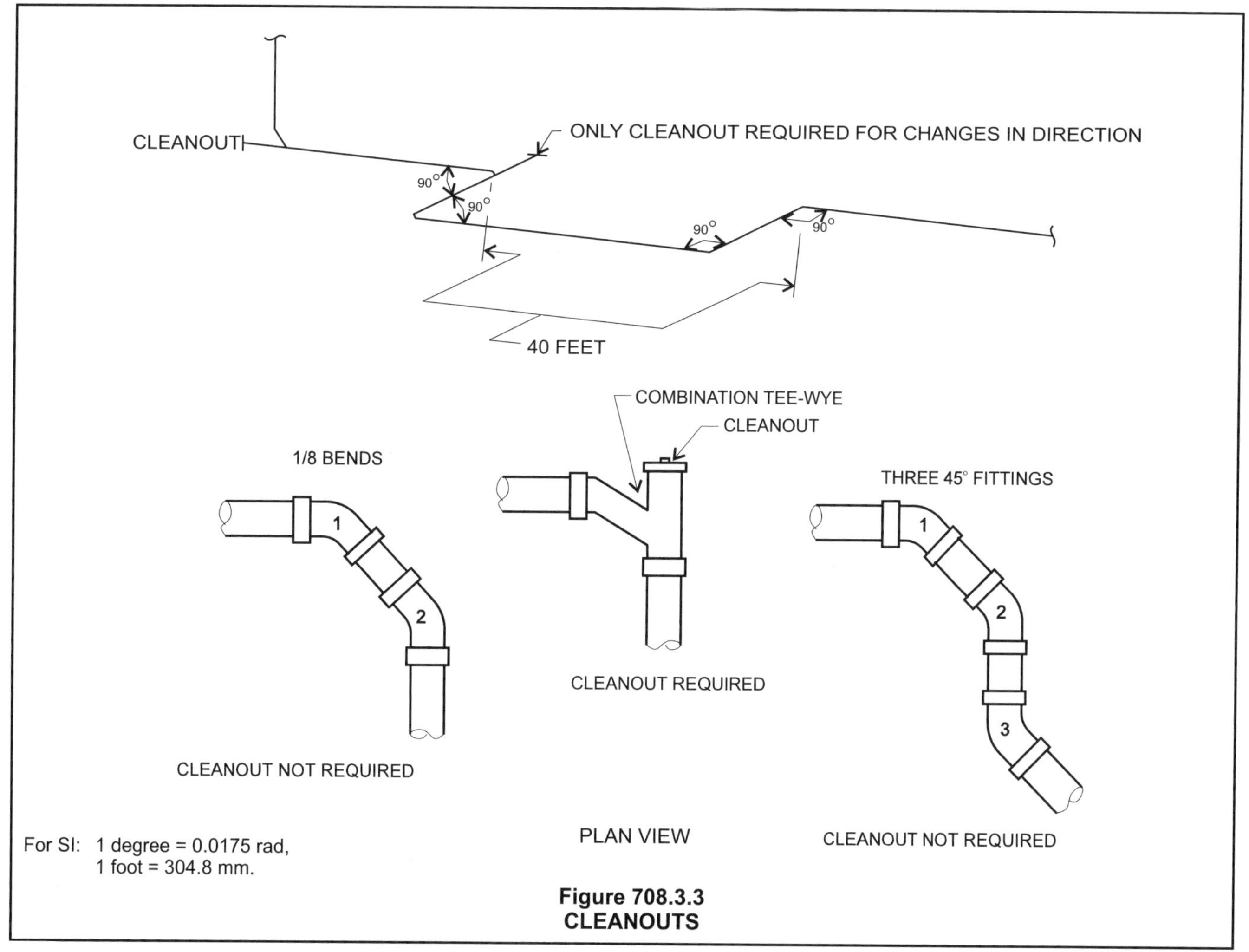

Figure 708.3.3
CLEANOUTS

708.3.4 Base of stack. A cleanout shall be provided at the base of each waste or soil *stack*.

❖ The characteristics of the drainage flow in horizontal pipe increase the probability of a stoppage. Additionally, it is possible for solids to collect at the change of direction from vertical to horizontal; therefore, a cleanout is required at the base of every drainage stack to provide access to the horizontal piping that serves the stack (see Commentary Figure 708.3.4).

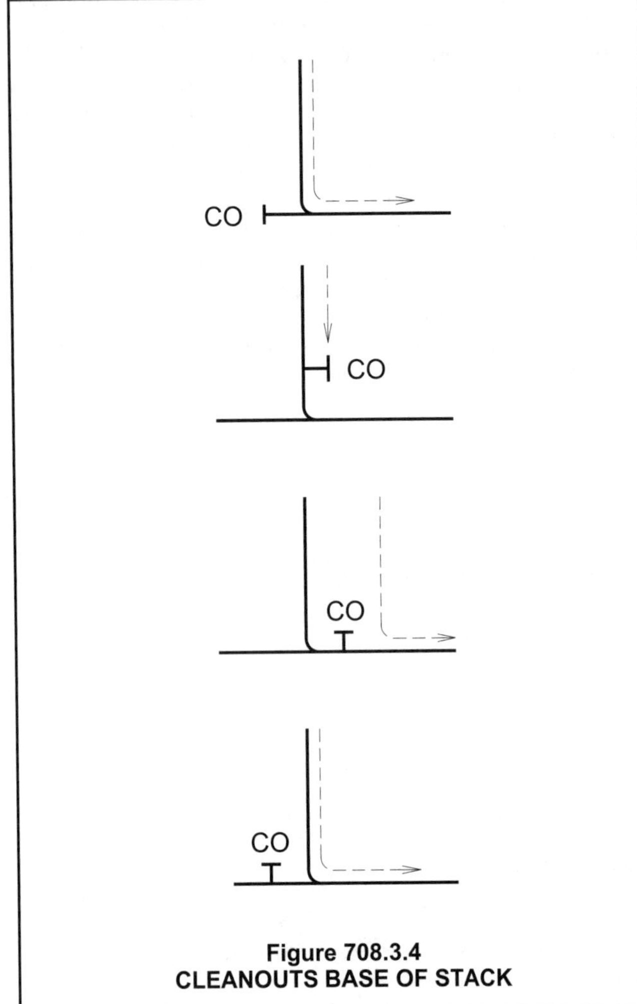

Figure 708.3.4
CLEANOUTS BASE OF STACK

708.3.5 Building drain and building sewer junction. There shall be a cleanout near the junction of the *building drain* and the *building sewer*. The cleanout shall be either inside or outside the building wall and shall be brought up to the finished ground level or to the basement floor level. An *approved* two-way cleanout is allowed to be used at this location to serve as a required cleanout for both the *building drain* and *building sewer*. The cleanout at the junction of the *building drain* and *building sewer* shall not be required if the cleanout on a 3-inch (76 mm) or larger diameter soil *stack* is located within a *developed length* of 10 feet (3048 mm) of the *building drain* and *building sewer* connection. The minimum size of the cleanout at the junction of the *building drain* and *building sewer* shall comply with Section 708.7.

❖ When a stoppage is known to be outside of the building within the building sewer, it is most convenient to have a cleanout at the junction of the building drain and the sewer. The code permits locating the cleanout either inside or outside the building, but as noted in the commentary to Section 708.3.2, locating the cleanout on the exterior is preferred, when possible. Note that the code permits a cleanout that services a 3-inch (76 mm) or larger-diameter soil stack if it is within a developed length of 10 feet (3048 mm). The majority of drain stoppages occur in the building sewer because of low-flow velocities and tree roots that penetrate the sewer [see Commentary Figure 708.3.5(1)].

A two-way cleanout uses a specialized fitting constructed with a branch throat that curves toward both the upstream and downstream directions of the run of the fitting [see Commentary Figure 708.3.5(2)].

A two-way cleanout would be permitted to satisfy the requirements of both this section and Section 708.3.1 if the building drain is not more than 100 feet (30 480 mm) long.

708.3.6 Manholes. Manholes serving a *building drain* shall have secured gas-tight covers and shall be located in accordance with Section 708.3.2.

❖ Where a manhole is provided to serve as a cleanout for a building drain, the manhole must have a secured, gas-tight cover. This requirement prevents the escape of sewer gas and reduces the possibility of unauthorized access to the manhole (see Commentary Figure 708.3.6). See Section 708.3.2 for additional requirements and concerns related to the installation of manholes.

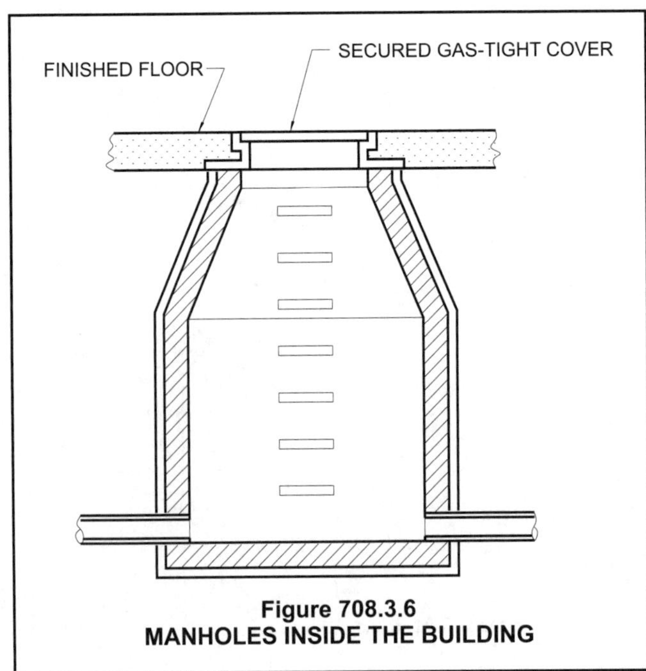

Figure 708.3.6
MANHOLES INSIDE THE BUILDING

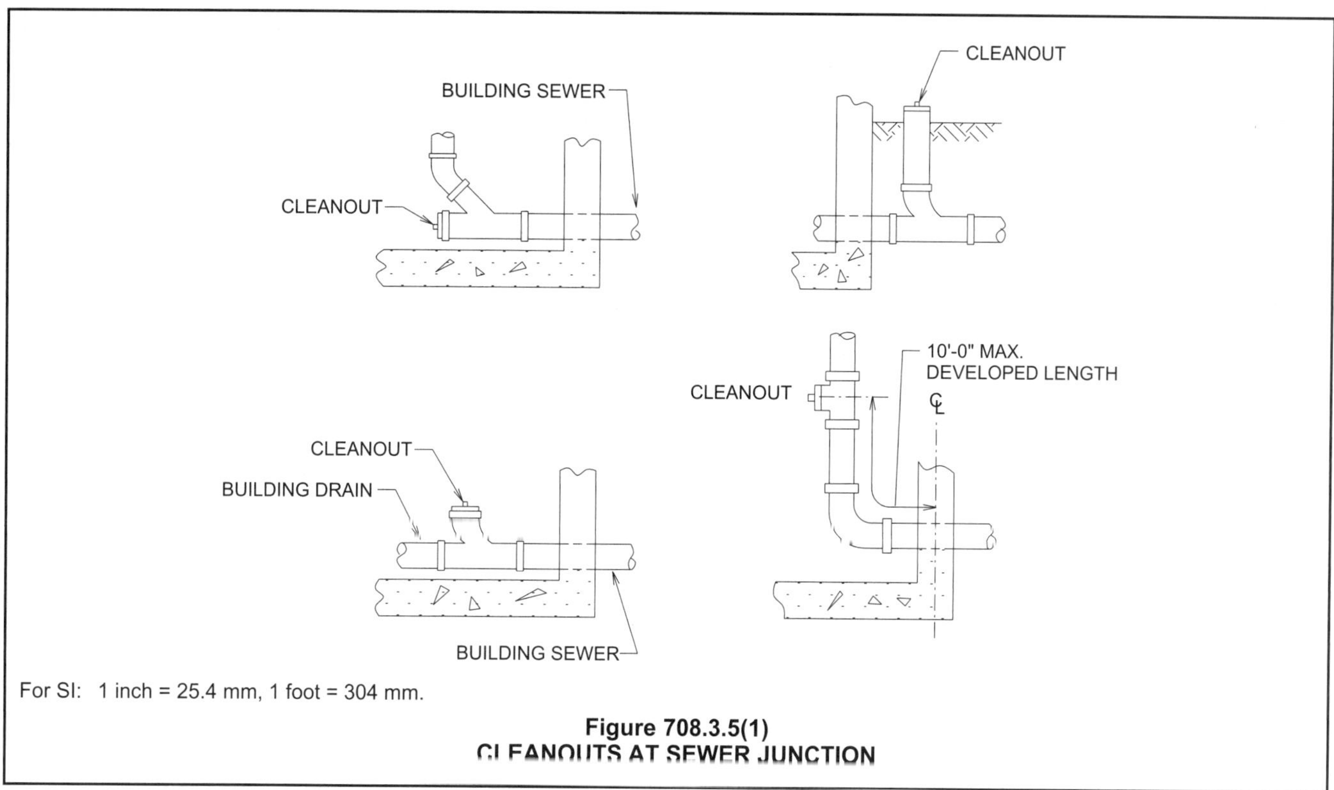

Figure 708.3.5(1)
CLEANOUTS AT SEWER JUNCTION

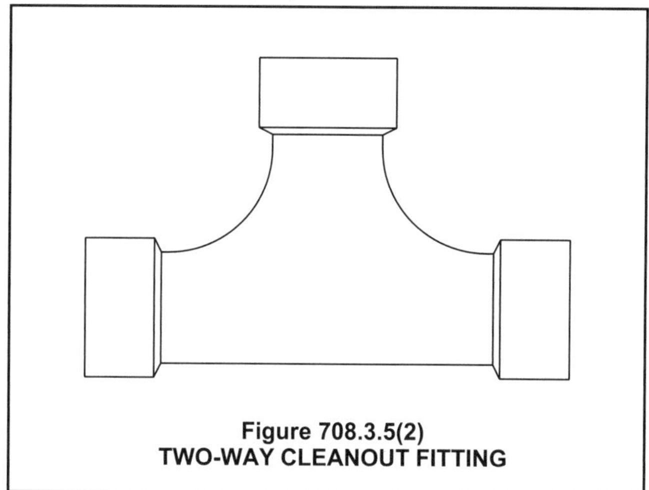

Figure 708.3.5(2)
TWO-WAY CLEANOUT FITTING

708.4 Concealed piping. Cleanouts on concealed piping or piping under a floor slab or in a crawl space of less than 24 inches (610 mm) in height or a plenum shall be extended through and terminate flush with the finished wall, floor or ground surface or shall be extended to the outside of the building. Cleanout plugs shall not be covered with cement, plaster or any other permanent finish material. Where it is necessary to conceal a cleanout or to terminate a cleanout in an area subject to vehicular traffic, the covering plate, *access* door or cleanout shall be of an *approved* type designed and installed for this purpose.

❖ Because a cleanout is designed to provide access into the drainage system, the cleanout itself must be accessible, regardless of whether the piping it serves is concealed or in a location not readily accessed. The cleanout must extend up to and be flush with the finished floor level or outside grade or to a location that is flush with a finished wall. Cleanouts located on walking surfaces must be countersunk to minimize the tripping hazard and to protect the cleanout from damage (see Commentary Figure 708.4). A cleanout must be accessible for use wherever it is located in the drainage system and must not be covered by any permanent, nonaccessible construction, such as plaster. If a cleanout is concealed in a floor or wall assembly, an approved access cover must be provided. The access opening must be of suitable size to allow for the removal of the plug and for the drain-cleaning procedure. Cleanouts must terminate flush with the wall or floor surface where an access pit or chase opening is not provided (see Commentary Figure 708.4). When a cleanout is located in an area subject to vehicular traffic, or even excessive pedestrian traffic, the designer needs to evaluate the cover plate's structural capacity.

708.5 Opening direction. Every cleanout shall be installed to open to allow cleaning in the direction of the flow of the drainage pipe or at right angles thereto.

❖ The normal practice for rodding or cleaning a drainage pipe is in the direction of flow. The cleaning rod (cable) or instrument will follow in the normal direction of flow to remove the obstruction at the point of stoppage. If a cleaning instrument is inserted opposite to the direction of the flow, the instrument may go in either direction at branch connections, thus the travel of the instrument cannot be controlled. There is also a possibility of damaging fixtures if the instrument goes in the wrong direction into a fixture branch. For example, a floor-mounted water closet may easily be damaged by a cable connected to an electric drain-cleaning machine. Additionally, cleaning a drain to remove an upstream stoppage can result in spillage when the stoppage is cleared and the backed-up waste suddenly flows toward the open cleanout.

This section does not prohibit the use of two-way cleanouts; however, such two-way cleanouts are considered as cleanouts only for the direction of flow. For example, a test tee opens at right angles to the pipe and is, therefore, permitted. Though not required by the code, some permanent indication of flow should be indicated at two-way cleanouts, except those where the direction of flow is obvious, such as on a vertical pipe.

708.6 Prohibited installation. Cleanout openings shall not be utilized for the installation of new fixtures, except where *approved* and where another cleanout of equal *access* and capacity is provided.

❖ If a cleanout opening is eliminated by a permanent piping connection, a new approved cleanout of equal access and capacity must be installed. In existing structures, cleanouts provide a convenient opening for the connection of new piping in remodeling, addition and alteration work. The cleanout fitting is commonly removed to allow a new connection; however, a substitute cleanout must be provided to serve in the same capacity as the cleanout that was eliminated. Many threaded cleanout openings have only a few threads and, therefore, are not intended to receive threaded pipe or male adapters.

It is not uncommon in existing structures to find floor cleanouts or cleanouts integral with floor drains that have been opened for the purpose of using the cleanout as a floor drain. Because such openings are not protected by a trap, sewer gases will enter the building interior; therefore, this practice is prohibited.

708.7 Minimum size. Cleanouts shall be the same nominal size as the pipe they serve up to 4 inches (102 mm). For pipes larger than 4 inches (102 mm) nominal size, the minimum size of the cleanout shall be 4 inches (102 mm).

Exceptions:

1. "P" trap connections with slip joints or ground joint connections, or *stack* cleanouts that are not more than one pipe diameter smaller than the drain served, shall be permitted.
2. Cast-iron cleanout sizing shall be in accordance with referenced standards in Table 702.4, ASTM A 74 for hub and spigot fittings or ASTM A 888 or CISPI 301 for hubless fittings.

❖ This section contains the requirements and criteria for sizing cleanouts for drainage piping.

Cleanout openings for $1^1/_4$-, $1^1/_2$-, 2-, $2^1/_2$-, 3- and 4-inch (32, 38, 51, 64, 76 and 102 mm) pipes must not be smaller than the pipe served, thus allowing use of

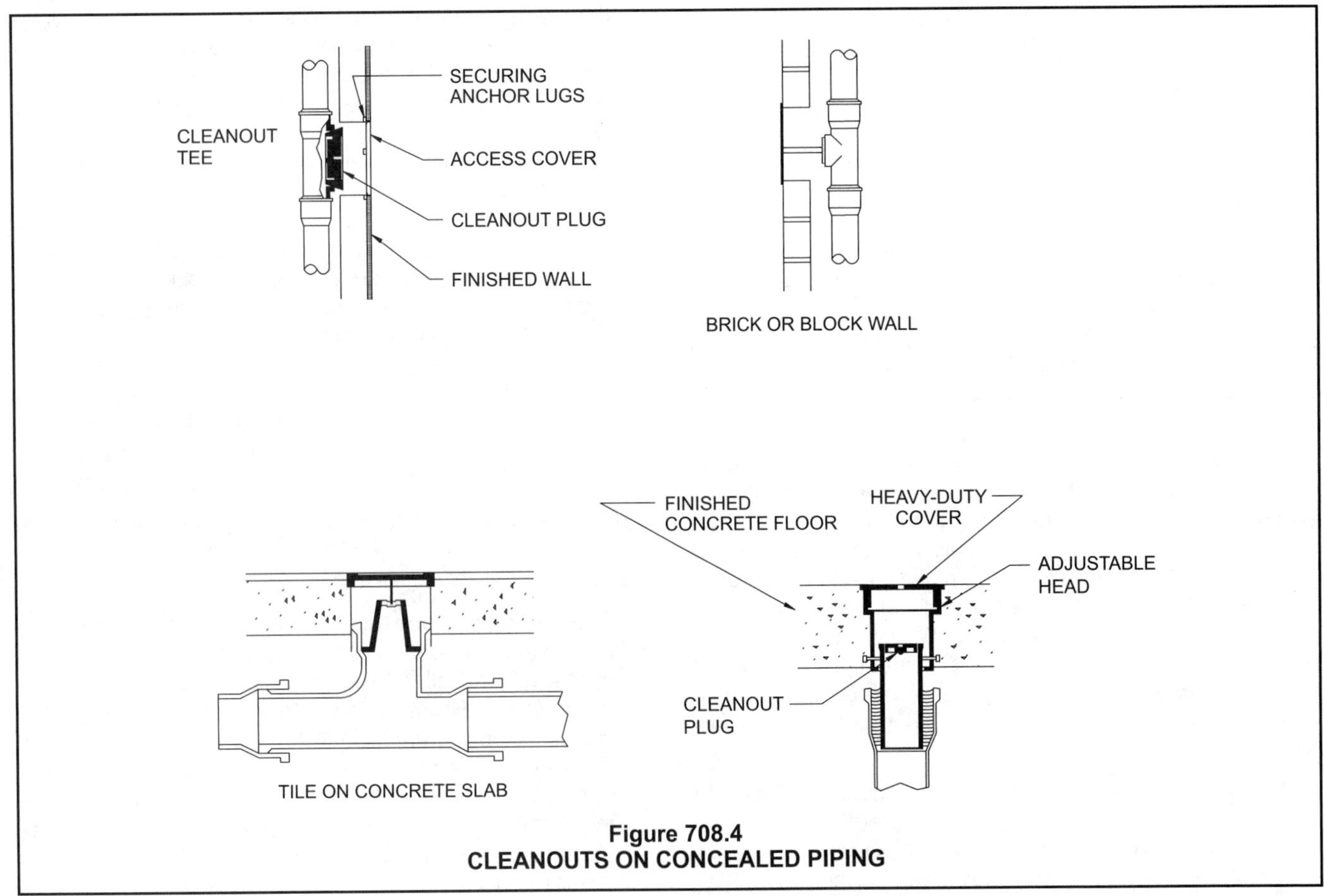

Figure 708.4
CLEANOUTS ON CONCEALED PIPING

full-bore cleaning equipment. For larger piping, a 4-inch (102 mm) opening is considered adequate for today's cleaning equipment, and full-bore cleaning equipment is typically not used for such piping.

Manufactured cleanout fittings, like pipe, are sized by nominal dimensions. An industry rule of thumb for pipes larger than 4 inches (102 mm) is that the cleanout must be sized in accordance with Commentary Table 708.7. The requirements contained in Commentary Table 708.7, although not required by the code, are intended to facilitate the use of drain-cleaning attachments (heads) and equipment of the proper size to accomplish full-bore or near full-bore pipe cleaning.

Commentary Table 708.7 contains minimum cleanout sizes for larger-diameter drains and sewers. The cleanout sizing is based solely on the need to provide a large enough access opening into the drain pipe for adequate service with larger drain-cleaning equipment. The table provides sizing for drainage pipe that is 10 inches (254 mm) or larger in diameter. The 8-inch (203 mm) minimum cleanout size for a 10-inch (254 mm) or larger drain, however, applies only to cleanouts serving drains other than building sewers because Section 708.3.2 requires a manhole for all building sewers 8 inches (203 mm) or larger.

Exception 1 shown in Section 708.7 allows a smaller nominal size fitting to serve as a cleanout for a larger nominal pipe size. This exception is very specific and limited to removable "P" traps and cleanout fittings, both of which must be not less than one pipe size smaller than the drain pipe served. The first exception is intended to permit the removal of fixture traps to provide access to the connecting drains. The second exception references ASTM A 74, ASTM A 888 or CISPI 301 for cleanout sizing under Table 702.4 because the standards require a different trap size than the above requirements.

Table 708.7
CLEANOUT SIZES

NOMINAL PIPE SIZE (inches)	NOMINAL SIZE CLEANOUT (inches)
5	4
6	4
8	6
10 and above	8

For SI: 1 inch = 25.4 mm.

708.8 Clearances. Cleanouts on 6-inch (153 mm) and smaller pipes shall be provided with a clearance of not less than 18 inches (457 mm) for rodding. Cleanouts on 8-inch (203 mm) and larger pipes shall be provided with a clearance of not less than 36 inches (914 mm) for rodding.

❖ This section states minimum access clearances for cleanouts. The code is very specific on the requirements related to providing drainage piping cleanouts. The usability of these cleanouts is not to be compromised by their placement in the building. The clearances given herein are viewed as the minimum conditions that will allow the cleanout to serve its purpose. For this reason, the designer must attempt to provide additional clearance whenever possible and practical.

For a cleanout to be used for cleaning a drain, adequate clear space must be provided to facilitate the cleaning operation. Although the instrument used for cleaning is flexible, there still must be room in front of the cleanout to allow it to be inserted and guided into the pipe (see Commentary Figure 708.8). The clearances are to be measured perpendicular to the face of the cleanout opening.

On smaller drains, smaller equipment and smaller, more flexible cable is used, requiring less room to insert and guide it into the cleanout. These clearances are viewed as the minimum, and whenever possible, more space should be provided.

A cleanout is not permitted to be permanently obstructed or concealed by permanent finish materials. There must always be a way to gain access to a cleanout without disturbing building finishes. Just as attention is being given to the cleanout, so must attention be given to the equipment to do the drain cleaning. The best located, most accessible cleanout will be of little use if the cleaning equipment cannot get close to it, or if excessive damage is done to the surrounding area.

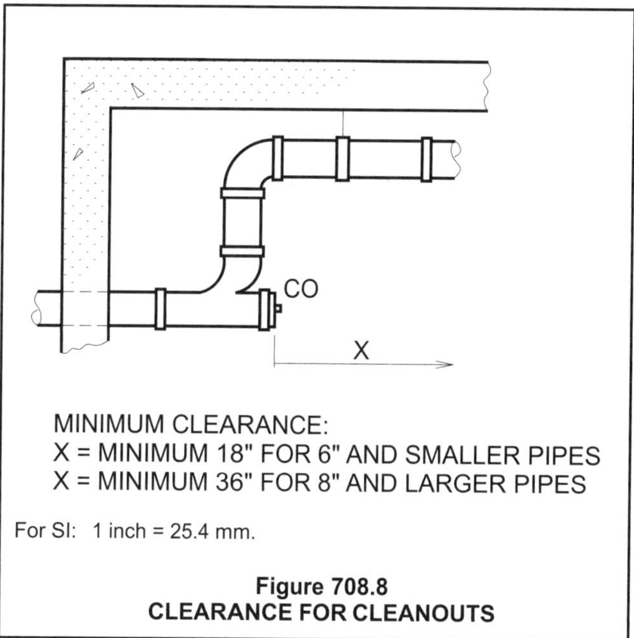

For SI: 1 inch = 25.4 mm.

Figure 708.8
CLEARANCE FOR CLEANOUTS

708.9 Access. *Access* shall be provided to all cleanouts.

❖ Cleanouts concealed behind permanent construction without access are of no value and could not be located if needed. Access may be provided to concealed cleanouts by doors and removable panels, covers and plates.

SECTION 709
FIXTURE UNITS

709.1 Values for fixtures. *Drainage fixture unit* values as given in Table 709.1 designate the relative load weight of different kinds of fixtures that shall be employed in estimating the total load carried by a soil or waste pipe, and shall be used in connection with Tables 710.1(1) and 710.1(2) of sizes for soil, waste and vent pipes for which the permissible load is given in terms of fixture units.

❖ This section contains the requirements to establish fixture unit amounts for plumbing fixtures for use in determining drainage system piping size. For a discussion on the determination of fixture units, see the General Comments at the beginning of this chapter. The term "fixture unit" refers to "drainage fixture units." Refer to the commentary on the definitions of "Fixture units" and "Drainage fixture units" in Chapter 2.

The method of sizing a sanitary drainage system is based on the dfu values listed in Table 709.1.

TABLE 709.1. See page 7-39.

❖ This table identifies common plumbing fixtures with their corresponding dfu values. The dfu values are used to determine the load on a drain, which, in turn, is used to determine the minimum drainage pipe size requirement for the load served. In determining the fixture unit values, each fixture was evaluated for its impact on the sanitary drainage system. The original development of fixture unit values was based on an arbitrary scale established by Dr. Hunter. Dr. Hunter conceived the idea of assigning a fixture unit value to represent the degree to which a fixture loads a system when used at the maximum assumed frequency. The scale was based on two fixtures: the lavatory and water closet (public). Dr. Hunter determined that a lavatory load on a drainage system was one-sixth of that for a water closet. Hence, the lavatory was assigned the value of 1 while the water closet was 6.

The sole purpose of the fixture unit concept is to make it possible to calculate the design load on the system directly when the system is composed of different kinds of fixtures, each having a different loading characteristic. The last column of the table lists the minimum trap size required for the corresponding fixture. Minimum trap size is also the minimum drainage pipe size because the size of a drainage pipe cannot be reduced in the direction of flow (see Commentary Section 1002.5). Although the code does not regulate the maximum size of a fixture trap, traps should not be oversized because the reduced flow volume and velocity will impair the trap's self-scouring ability.

Note a references Table 709.2 for determining dfu values for commercial washing machines with traps larger than 3 inches (76 mm) in diameter.

Note b allows for shower heads and whirlpool attachments to a bathtub fixture without increasing the fixture unit value.

Note c references Sections 709.2 and 709.4 to determine fixture unit values of plumbing fixtures not found in the table and fixtures with intermittent flows.

Note d requires consistency between the fixture outlet size and trap size of urinals and water closets. Both the trap size and the design of a water closet are regulated by ANSI Z124.4 and ASME A112.19.2.

Note e permits assigning a water closet or urinal a lower dfu value than that listed in the table when appropriate testing justifies the reduction. This can be significant when installing 1.6-gallon-per-flush (6.1 L per flush) water closets because some manufacturers have conducted tests and determined that such fixtures have a dfu value of 2. An emergency floor drain is installed in areas such as toilet rooms, boiler rooms and laundry rooms and is used only in cases of fixture overflow, water pipe failure and vandalism (see Commentary Section 412.4). Such floor drains do not add to the normal load of the drainage system.

Note f permits adding fixtures to a dwelling unit bathroom group to the dfu value of the bathroom group fixture count.

Note g references Section 406.3 for sizing requirements for the trap and fixture drain for an automatic clothes washer and the branch drain or drainage stack.

Note h addresses waste receptors and the appropriate determination of dfu loads based upon what is being discharged to the waste receptor. Section 709.4 requires a minimum dfu load to be assumed for a trap size if the sum of indirectly connected fixtures to the waste receptor is less than the minimum dfu load for the trap size. Section 709.4.1 allows for the special application of condensate drains for refrigerated cases and a reduction of dfu load (see Commentary Section 709.4.1).

The 2009 edition expanded the shower fixture row by adding the required trap sizes and dfu values for shower drain flows greater than 5.7 gpm (21.6 lpm). This change was precipitated by the increasing prevalence of residential building owners wanting multiple shower heads and body massage sprays installed in residential showers to provide users with a "luxury showering experience." Although the code limits the flow rate of an individual showerhead to not greater than 2.5 gpm (9.5 lpm), there is no restriction against the simultaneous operation of multiple showerheads in one shower compartment. Because the $1^1/_2$ inch (38 mm) minimum size shower drain is not capable of accommodating the flow from more than two shower heads operating at the same time, the proponent of this change calculated the allowable flows for larger drain sizes. The allowable flows were based on the drain pipe installed at the minimum allowable slope for the size of pipe and a friction factor of 0.010 (in the familiar Darcy-Weisbach equation).

TABLE 709.1
DRAINAGE FIXTURE UNITS FOR FIXTURES AND GROUPS

FIXTURE TYPE	DRAINAGE FIXTURE UNIT VALUE AS LOAD FACTORS	MINIMUM SIZE OF TRAP (inches)
Automatic clothes washers, commercial[a,g]	3	2
Automatic clothes washers, residential[g]	2	2
Bathroom group as defined in Section 202 (1.6 gpf water closet)[f]	5	—
Bathroom group as defined in Section 202 (water closet flushing greater than 1.6 gpf)[f]	6	—
Bathtub[b] (with or without overhead shower or whirpool attachments)	2	$1^1/_2$
Bidet	1	$1^1/_4$
Combination sink and tray	2	$1^1/_2$
Dental lavatory	1	$1^1/_4$
Dental unit or cuspidor	1	$1^1/_4$
Dishwashing machine,[c] domestic	2	$1^1/_2$
Drinking fountain	$^1/_2$	$1^1/_4$
Emergency floor drain	0	2
Floor drains[h]	2[h]	2
Floor sinks	Note h	2
Kitchen sink, domestic	2	$1^1/_2$
Kitchen sink, domestic with food waste grinder and/or dishwasher	2	$1^1/_2$
Laundry tray (1 or 2 compartments)	2	$1^1/_2$
Lavatory	1	$1^1/_4$
Shower (based on the total flow rate through showerheads and body sprays) Flow rate: 5.7 gpm or less Greater than 5.7 gpm to 12.3 gpm Greater than 12.3 gpm to 25.8 gpm Greater than 25.8 gpm to 55.6 gpm	 2 3 5 6	 $1^1/_2$ 2 3 4
Service sink	2	$1^1/_2$
Sink	2	$1^1/_2$
Urinal	4	Note d
Urinal, 1 gallon per flush or less	2[e]	Note d
Urinal, nonwater supplied	$^1/_2$	Note d
Wash sink (circular or multiple) each set of faucets	2	$1^1/_2$
Water closet, flushometer tank, public or private	4[e]	Note d
Water closet, private (1.6 gpf)	3[e]	Note d
Water closet, private (flushing greater than 1.6 gpf)	4[e]	Note d
Water closet, public (1.6 gpf)	4[e]	Note d
Water closet, public (flushing greater than 1.6 gpf)	6[e]	Note d

For SI: 1 inch = 25.4 mm, 1 gallon = 3.785 L, gpf = gallon per flushing cycle, gpm = gallon per minute.

a. For traps larger than 3 inches, use Table 709.2.

b. A showerhead over a bathtub or whirlpool bathtub attachment does not increase the drainage fixture unit value.

c. See Sections 709.2 through 709.4.1 for methods of computing unit value of fixtures not listed in this table or for rating of devices with intermittent flows.

d. Trap size shall be consistent with the fixture outlet size.

e. For the purpose of computing loads on building drains and sewers, water closets and urinals shall not be rated at a lower drainage fixture unit unless the lower values are confirmed by testing.

f. For fixtures added to a dwelling unit bathroom group, add the dfu value of those additional fixtures to the bathroom group fixture count.

g. See Section 406.3 for sizing requirements for fixture drain, branch drain, and drainage stack for an automatic clothes washer standpipe.

709.2 Fixtures not listed in Table 709.1. Fixtures not listed in Table 709.1 shall have a *drainage fixture unit* load based on the outlet size of the fixture in accordance with Table 709.2. The minimum trap size for unlisted fixtures shall be the size of the drainage outlet but not less than $1^1/_4$ inches (32 mm).

❖ When a specific plumbing fixture is not listed in Table 709.1, the dfu value is based on the outlet size of the fixture. For example, floor sinks (receptors) are not listed in Table 709.1; therefore, Table 709.2 would be used. A $1^1/_4$-inch (32 mm) drain is the minimum acceptable size to permit proper open channel flow in the pipe for sanitary drainage. The minimum trap size is based on this pipe size (see Commentary Section 709.4).

TABLE 709.2
DRAINAGE FIXTURE UNITS FOR FIXTURE DRAINS OR TRAPS

FIXTURE DRAIN OR TRAP SIZE (inches)	DRAINAGE FIXTURE UNIT VALUE
$1^1/_4$	1
$1^1/_2$	2
2	3
$2^1/_2$	4
3	5
4	6

For SI: 1 inch = 25.4 mm.

❖ If a fixture is not listed in the table, the dfu value is based on the outlet or trap size of the fixture. This method of assigning fixture unit values is consistent with the historical development of sizing methodology because the outlet size restricts the discharge flow rate of the fixture. Special consideration should be given to the sizing of receptors, such as floor sinks that receive the discharge of multiple indirectly connected fixtures. For example, the table may be inappropriate for determining the fixture unit value for an indirect waste receptor with a 4-inch (102 mm) trap that serves five or six $1^1/_2$-inch-diameter (38 mm) culinary fixture drains (see Commentary Section 709.4).

709.3 Values for continuous and semicontinuous flow. *Drainage fixture unit* values for continuous and semicontinuous flow into a drainage system shall be computed on the basis that 1 gpm (0.06 L/s) of flow is equivalent to two fixture units.

❖ Equipment that discharges either continuously or semicontinuously, such as pumps and ejectors, is assigned two fixture units for each gpm of discharge rate. For example, a continuously operating pump with a discharge rate of 20 gpm (78 L/min) would have a dfu value of 40. This conversion applies only to the type of continuous flow addressed in this section. Because a dfu is based on probability, only whole numbers are used to express fixture unit values. A fractional fixture unit value is always rounded up to the next whole number. This calculation does not work in reverse. The relationship of dfu to gpm is not a constant ratio that allows direct conversion of units. Thus, it cannot be determined that a dfu value of 40 would yield a continuous flow of 20 gpm (78 L/min).

709.4 Values for indirect waste receptor. The *drainage fixture unit* load of an indirect waste receptor receiving the discharge of indirectly connected fixtures shall be the sum of the *drainage fixture unit* values of the fixtures that discharge to the receptor, but not less than the *drainage fixture unit* value given for the indirect waste receptor in Table 709.1 or 709.2.

❖ Waste receptors such as hub drains and floor sinks are typically used to receive waste from multiple indirectly connected fixtures (see Section 802). The simultaneous discharge of indirectly connected fixtures may result in a greater load on the drainage system than accounted for by Table 709.2. For example, a 3-inch (76 mm) hub drain could be serving four $1^1/_2$-inch (38 mm) indirect wastes, which may represent a total load of 8 dfu (4 by 2 dfu); however, Table 709.2 would assign a total load of only 5 dfu. If the total dfu load determined by this section is less than that given by Table 709.1 or 709.2, as applicable, the table values must prevail.

709.4.1 Clear-water waste receptors. Where waste receptors such as floor drains, floor sinks and hub drains receive only clear-water waste from display cases, refrigerated display cases, ice bins, coolers and freezers, such receptors shall have a *drainage fixture unit* value of one-half.

❖ In large food stores and warehouses having many refrigerated display cases and freezer units, many traps are installed with either hub drains or waste receptors to collect the condensate from the units. Even though the condensate flow from each unit is only a trickle, the trap sizes dictated the size of the drain system connecting all the traps. For example, a 3-inch (76 mm) trap with hub drain has a drainage fixture unit value (derived from Table 709.2) of 5 dfu. In situations with many traps, the total dfu load results in unnecessarily large pipe sizes compared to the actual flow in the pipe. This section allows the designer to use a smaller, more realistic dfu loading for these applications.

SECTION 710
DRAINAGE SYSTEM SIZING

710.1 Maximum fixture unit load. The maximum number of drainage fixture units connected to a given size of *building sewer*, *building drain* or horizontal *branch* of the *building drain* shall be determined using Table 710.1(1). The maximum number of drainage fixture units connected to a given size of horizontal *branch* or vertical soil or waste *stack* shall be determined using Table 710.1(2).

❖ This section contains the requirements for the sizing of gravity drainage piping.

The maximum number of dfu that can be connected to a drainage pipe is given in Tables 710.1(1) and 710.1(2) and is dependent on whether the pipe is a building drain, sewer, stack or horizontal branch.

TABLE 710.1(1). See below.

❖ This table gives the maximum fixture units that can be connected to building drains, sewers and every horizontal branch connecting to a building drain. Such drains are the lowest portions of the system that drain horizontally by gravity (see the definition of "Building drain" in Chapter 2). The pipe diameters indicated are the nominal dimension designations for piping material. These dimensions are regulated by the piping material standards referenced in this chapter.

The minimum size of each pipe section is based on the slope of the pipe (pitch or grade) and the number of dfu that connect to that section. Drainage fixture units are additive in the direction of flow. The dfu values listed in the table are the maximum value for the given pipe size and slope.

The principle of a building drain flowing approximately half full at a maximum capacity design load was used by Dr. Hunter to develop a sizing method. The sizing is based further on a steady-state flow condition in the pipe (level flow).

The building drain actually has a combination of a steady-state and surging (highly fluctuating) flow. Surging flow results from the high entrance velocity from the discharge of a fixture, horizontal branch or stack. The surging flow also allows an increase in the capacity of the drain; however, by sizing on a steady flow, an added factor of safety results from the effects of surging flow.

The dfu load values in this table are conservative, and half-full flow will rarely, if ever, occur [see Commentary Figure 710.1(1)].

TABLE 710.1(2). See page 7-42.

❖ The table may be considered as four separate tables: one for sizing horizontal branches connecting to a stack, one for limiting the discharge per branch interval for any height stack, one for sizing stacks three branch intervals or less in height and one for sizing stacks more than three branch intervals in height. The pipe dimensions listed in the first column are again the nominal dimension designations identified in the pipe standards.

To size a horizontal branch, only the first two columns of the table are required [see Commentary Table 710.1(1)]. The second column relates only to horizontal branches; it does not affect the sizing of stacks. A horizontal branch, however, may have a vertical section of pipe.

The table lists the maximum number of fixture units that can be discharged into a given size pipe. As the number of dfu increases in the direction of flow, the pipe may also have to be increased in size.

The dfu load value for a horizontal branch is typically lower than the permitted load for a building drain, a branch of a building drain or a sewer the same size as listed in Table 710.1(1). When sizing a horizontal branch, consideration must be given to the flow entering a stack. If there is simultaneous flow in the stack and horizontal branch, the flow in the horizontal branch may adversely affect the flow in the stack, and the flow in the stack may impede the flow in the horizontal branch. A significant flow in the stack may cause a backup of flow in the horizontal branch connecting to the stack. The table accounts for the effect on the stack by limiting the load capacity of the horizontal branch [see Commentary Figure 710.1(2)].

TABLE 710.1(1)
BUILDING DRAINS AND SEWERS

DIAMETER OF PIPE (inches)	MAXIMUM NUMBER OF DRAINAGE FIXTURE UNITS CONNECTED TO ANY PORTION OF THE BUILDING DRAIN OR THE BUILDING SEWER, INCLUDING BRANCHES OF THE BUILDING DRAIN[a]			
	Slope per foot			
	$^1/_{16}$ inch	$^1/_8$ inch	$^1/_4$ inch	$^1/_2$ inch
$1^1/_4$	—	—	1	1
$1^1/_2$	—	—	3	3
2	—	—	21	26
$2^1/_2$	—	—	24	31
3	—	36	42	50
4	—	180	216	250
5	—	390	480	575
6	—	700	840	1,000
8	1,400	1,600	1,920	2,300
10	2,500	2,900	3,500	4,200
12	3,900	4,600	5,600	6,700
15	7,000	8,300	10,000	12,000

For SI: 1 inch = 25.4 mm, 1 inch per foot = 83.3 mm/m.

a. The minimum size of any building drain serving a water closet shall be 3 inches.

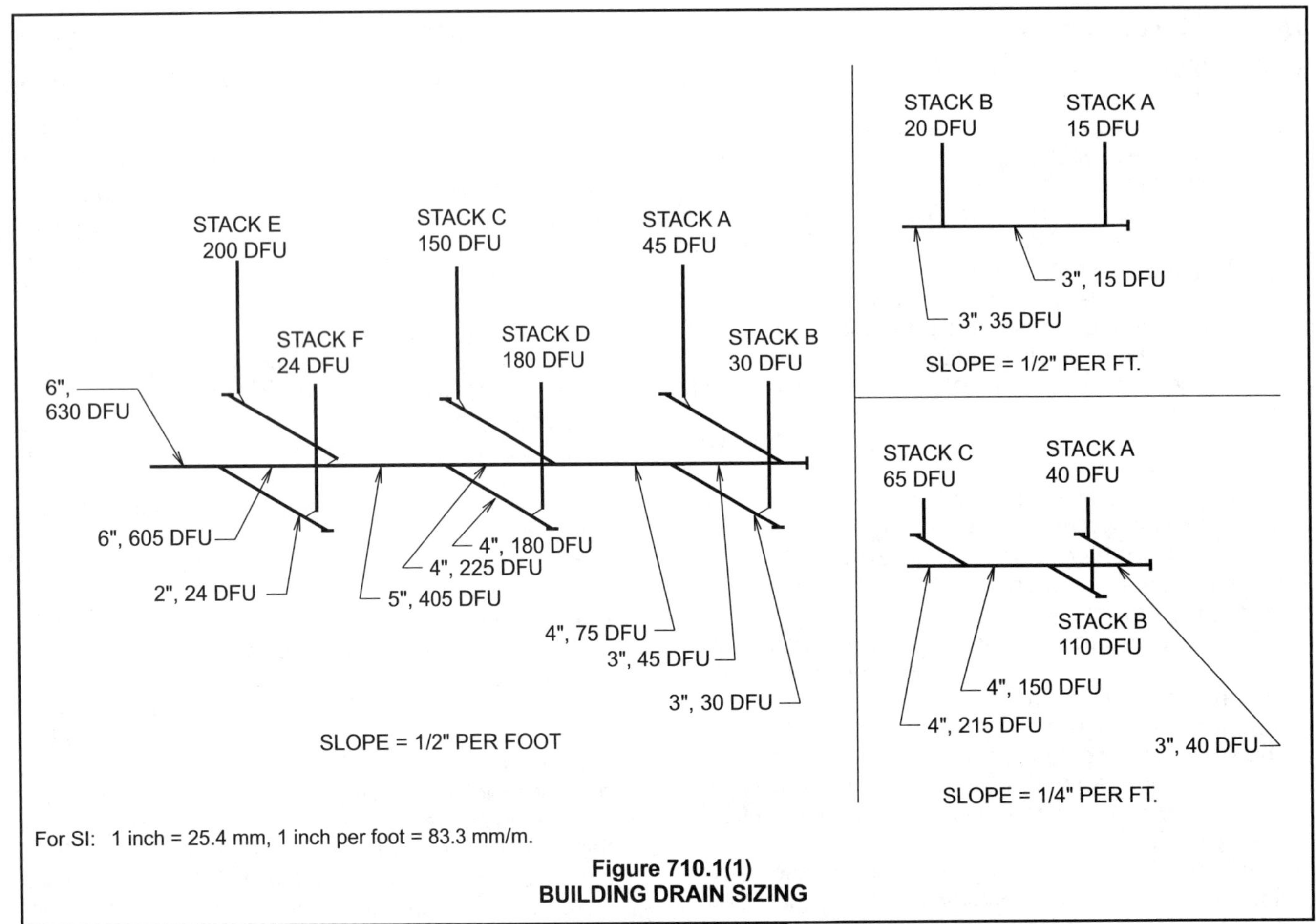

For SI: 1 inch = 25.4 mm, 1 inch per foot = 83.3 mm/m.

Figure 710.1(1)
BUILDING DRAIN SIZING

TABLE 710.1(2)
HORIZONTAL FIXTURE BRANCHES AND STACKS[a]

DIAMETER OF PIPE (inches)	MAXIMUM NUMBER OF DRAINAGE FIXTURE UNITS (dfu)			
	Total for horizontal branch	Stacks[b]		
		Total discharge into one branch interval	Total for stack of three branch Intervals or less	Total for stack greater than three branch intervals
$1^1/_2$	3	2	4	8
2	6	6	10	24
$2^1/_2$	12	9	20	42
3	20	20	48	72
4	160	90	240	500
5	360	200	540	1,100
6	620	350	960	1,900
8	1,400	600	2,200	3,600
10	2,500	1,000	3,800	5,600
12	3,900	1,500	6,000	8,400
15	7,000	Note c	Note c	Note c

For SI: 1 inch = 25.4 mm.

a. Does not include branches of the building drain. Refer to Table 710.1(1).

b. Stacks shall be sized based on the total accumulated connected load at each story or branch interval. As the total accumulated connected load decreases, stacks are permitted to be reduced in size. Stack diameters shall not be reduced to less than one-half of the diameter of the largest stack size required.

c. Sizing load based on design criteria.

TABLE 710.1(2). See below.

❖ The table does not have an entry for a $1^1/_4$-inch-diameter (32 mm) pipe, and the most upstream lavatory in Commentary Figure 710.1(2) is served by a $1^1/_4$-inch (32 mm) horizontal fixture branch. This is intended to illustrate that the omission of a $1^1/_4$-inch (32 mm) pipe from the table does not prohibit its use as a horizontal drain. A $1^1/_4$-inch (32 mm) pipe is limited to 1 dfu.

When sizing a stack, there are three main criteria that must be considered: the discharge into a branch interval of a stack, the total discharge into a stack and the total number of branch intervals on the stack.

For sizing a stack of three branch intervals or less, the third and fourth columns of the table are used [see Commentary Table 710.1(3)].Within any given branch interval, the maximum amount of discharge a stack may receive is limited. If the maximum discharge per branch interval is exceeded, the stack would have to be increased in size. The discharge may be received from more than one horizontal branch.

Commentary Figure 710.1(3), Example 1, depicts two 3-inch (76 mm) horizontal branches connecting within one branch interval. The total fixture unit load is 30, meaning that the stack diameter must be a minimum of 4 inches (102 mm). The example demonstrates a situation where the branches must be only 3 inches (76 mm) and, because of the location of their

Table 710.1(1)
HORIZONTAL BRANCH CAPACITY

DIAMETER OF PIPE (inches)	TOTAL FOR A HORIZONTAL BRANCH
$1^1/_2$	3
2	6
$2^1/_2$	12
3	20
4	160
5	360
6	620
8	1,400
10	2,500
12	2,900
15	7,000

For SI: 1 inch = 25.4 mm.

Table 710.1(2)
STACK CAPACITIES

DIAMETER OF PIPE (inches)	TOTAL DISCHARGE INTO ONE BRANCH INTERVAL	TOTAL FOR STACK OF THREE BRANCH INTERVALS OR LESS
$1^1/_2$	2	4
2	6	10
$2^1/_2$	9	20
3	20	48
4	90	240
5	2000	540
6	350	960
8	600	2,200
10	1,000	3,800
12	1,500	6,000
15	—	—

For SI: 1 inch = 25.4 mm.

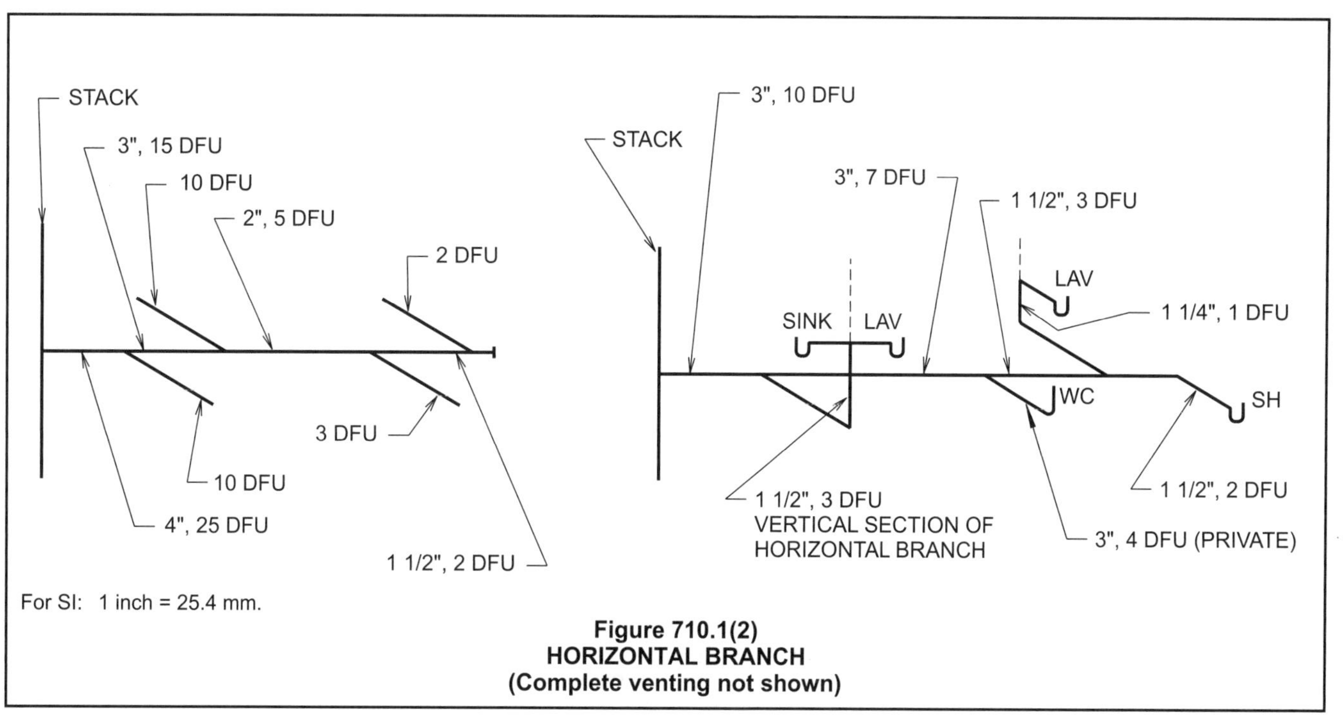

Figure 710.1(2)
HORIZONTAL BRANCH
(Complete venting not shown)

connection to the stack, a larger-size stack is required. In this example, the total permitted capacity of a 3-inch (76 mm) stack was not exceeded; only the limitation for a given branch interval was exceeded.

The flow, discharging within a branch interval, is limited to avoid interference with the flow in the stack. The collision of horizontal flow in a branch with the vertical flow in a stack may disrupt the annular flow pattern in the stack. This disruption may cause the cross-sectional area of the stack to be partially or completely filled with waste, which in turn causes the formation of slugs or pistons of waste in the stack. As stated previously, the flow from the horizontal branches may greatly affect the pressure differentials in the drainage system.

Another example of the limitations for a given branch interval, as shown in Commentary Figure 710.1(3), Example 2, is a single 4-inch (102 mm) pipe connecting to a stack.

If a total of 110 dfu discharged into a horizontal branch, the pipe size is a minimum of 4 inches (102 mm). The stack receiving the discharge of the branch must be a minimum of 5 inches (127 mm). The horizontal branch exceeds the 90 dfu limitation for a single branch interval of a 4-inch (102 mm) stack.

The stack is also sized based on the total dfu discharge to the stack. The discharge of each horizontal branch is added to compute the stack's total dfu load.

For sizing stacks over three branch intervals, the third and fifth columns are used [see Commentary Table 710.1(3)]. The difference in sizing the taller stacks is the total dfu load permitted in the stack. Taller stacks are allowed a much greater capacity than stacks three branch intervals or less [see Commentary Figures 710.1(4) and 710.1(5)]. The flow in the stack, based on the table, will occupy approximately one-quarter of the cross-sectional area of the pipe under the maximum design condition. Dr. Hunter discovered in his early research that this capacity of flow results in the drainage spiraling down the pipe wall with a center core of air. Only small volumes of waste tend to fall down the inner core of the stack. As the volume increases, any slug of water is forced by air frictional factors against the pipe wall [see Commentary Figure 710.1(6)].

After a short distance of fall [approximately 15 feet (4572 mm)], terminal velocity is reached by the drainage flow in the stack. Terminal velocity is the maximum speed of drainage flow in the stack, resulting from a

Table 710.1(3)
ALLOWABLE STACK DRAINAGE FIXTURE UNITS

DIAMETER OF PIPE (inches)	TOTAL DISCHARGE INTO ONE BRANCH INTERVAL	TOTAL FOR STACK GREATER THAN THREE BRANCH INTERVALS
$1^1/_2$	2	8
2	6	24
$2^1/_2$	9	42
3	20	72
4	90	500
5	2000	1,100
6	350	1,900
8	600	3,600
10	1,000	5,600
12	1,500	8,400
15	—	—

For SI: 1 inch = 25.4 mm.

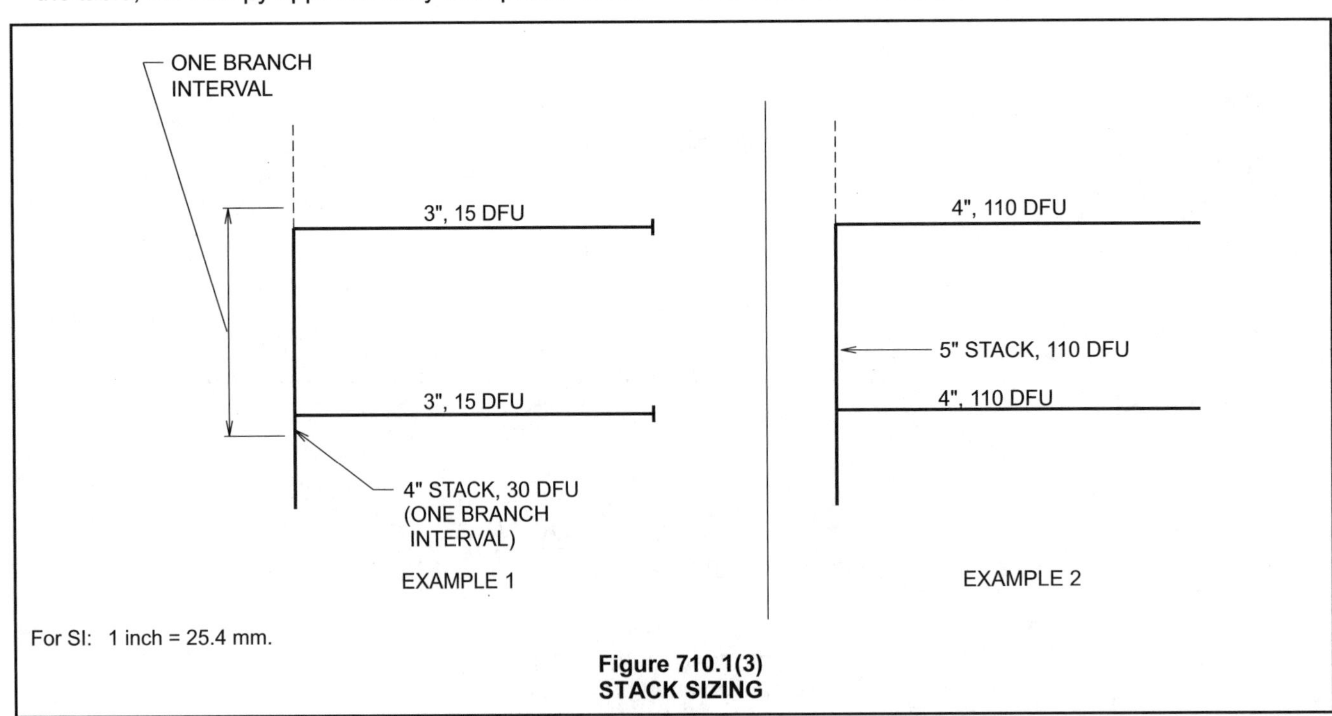

Figure 710.1(3)
STACK SIZING

balancing of air and pipe wall friction with gravitational force.

Although terminal velocity always occurs in a tall stack, it rarely occurs in shorter stacks of three branch intervals or less. This phenomenon is reflected in the table by permitting a greater capacity of flow in stacks with more than three branch intervals.

Note a of the table is a reminder that Table 710.1(1) is to be used for sizing horizontal branches of the building drain.

Note b permits reducing the size of the stack in the direction opposite of flow as the dfu load decreases. This is commonly referred to as "telescoping the stack." A stack cannot be reduced to less than half the diameter of its base where it terminates as a stack vent [see Commentary Figure 710.1(7)].

710.1.1 Horizontal stack offsets. Horizontal *stack* offsets shall be sized as required for building drains in accordance with Table 710.1(1), except as required by Section 711.4.

❖ Horizontal offsets (see the definition of "Horizontal pipe" in Chapter 2) cause a disruption of flow in the stack, and horizontal piping has less flow capacity than vertical piping. The vertical flow must convert to a horizontal flow in horizontal offsets; therefore, the increased turbulence and decreased flow capacity of the horizontal offset must be considered.

In many cases, Table 710.1(1) will require that the diameter of the offset be larger than the stack piping above it. The reference to Section 711.4 in the last phrase of this section is of no consequence because Section 711.4 would require the same sizing as this section.

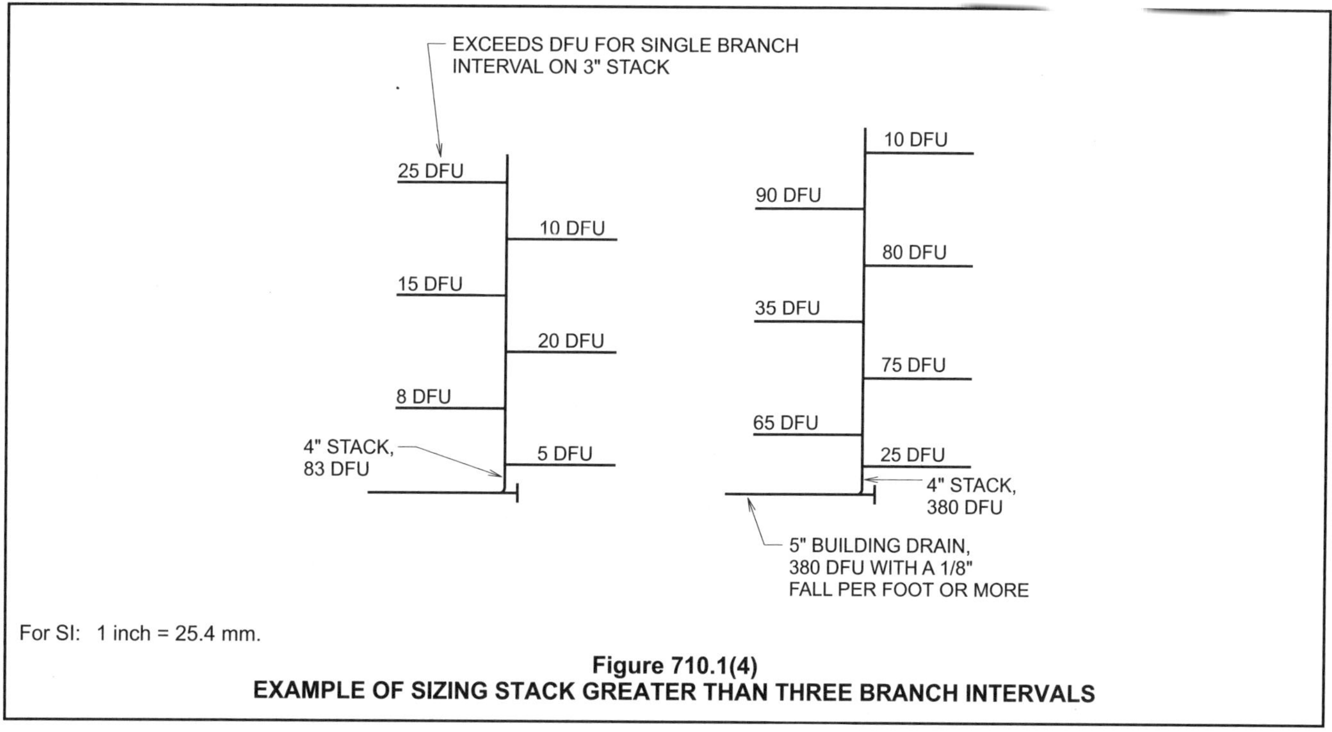

For SI: 1 inch = 25.4 mm.

Figure 710.1(4)
EXAMPLE OF SIZING STACK GREATER THAN THREE BRANCH INTERVALS

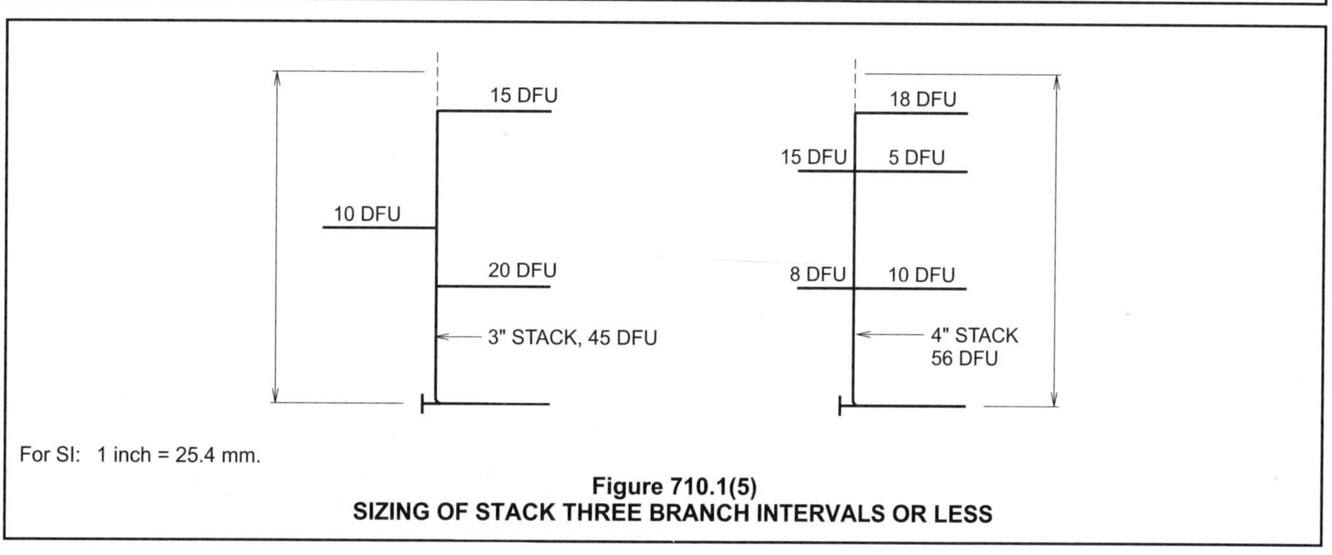

For SI: 1 inch = 25.4 mm.

Figure 710.1(5)
SIZING OF STACK THREE BRANCH INTERVALS OR LESS

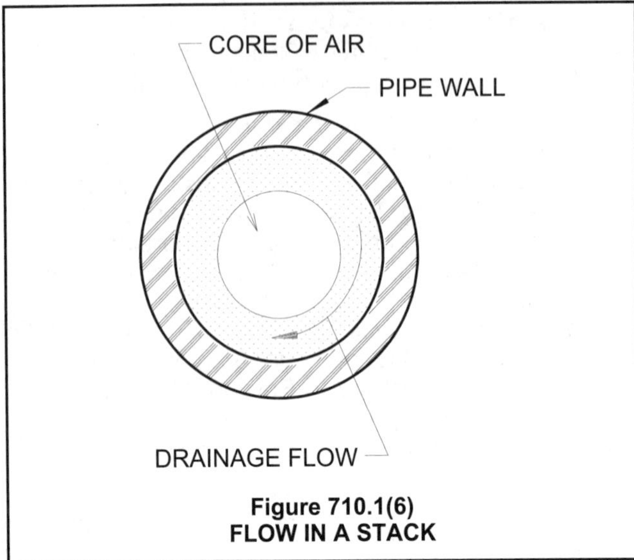

Figure 710.1(6)
FLOW IN A STACK

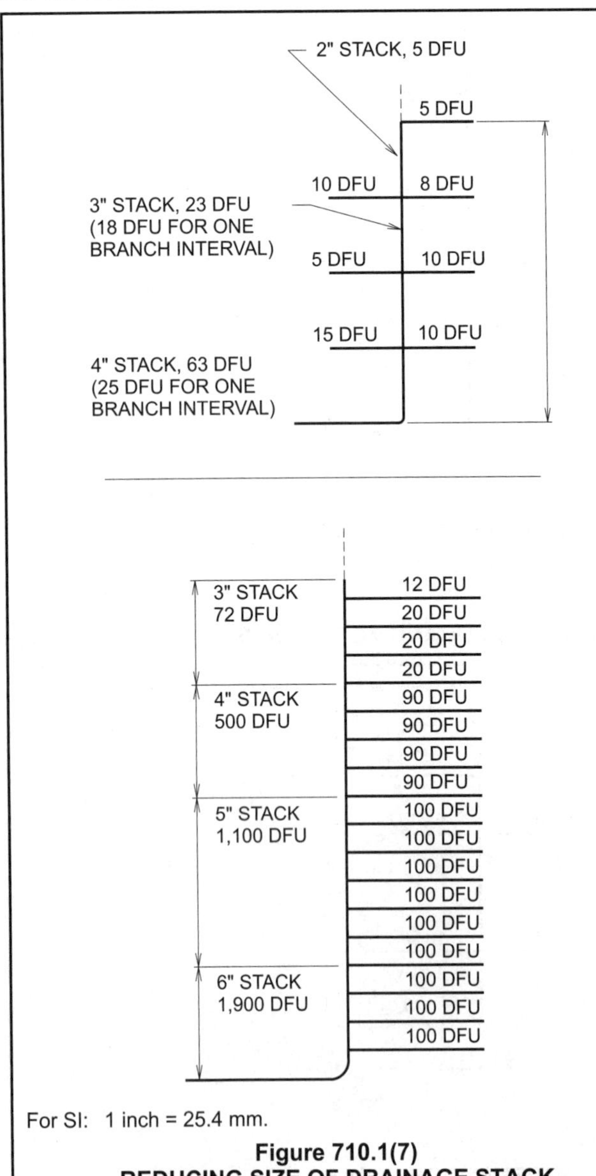

Figure 710.1(7)
REDUCING SIZE OF DRAINAGE STACK

710.1.2 Vertical stack offsets. Vertical *stack* offsets shall be sized as required for straight stacks in accordance with Table 710.1(2), except where required to be sized as a *building drain* in accordance with Section 711.1.1.

❖ Because vertical offsets (see the definition of "Vertical pipe" in Chapter 2) do not have as much impact on flow in the stack, such offsets are sized no differently than the stack itself (see commentary, Section 711.1.1).

710.2 Future fixtures. Where provision is made for the future installation of fixtures, those provided for shall be considered in determining the required sizes of drain pipes.

❖ Anticipated dfu loads for future plumbing fixtures must be considered in the design of the spacing. Piping installed for future plumbing fixtures must be closed to prevent the escape of sewer gas and waste and to prevent the entrance of foreign substances. Section 905.6 also requires that a vent be roughed in for drainage pipe that will serve fixtures to be installed in the future.

SECTION 711
OFFSETS IN DRAINAGE PIPING IN BUILDINGS OF FIVE STORIES OR MORE

711.1 Horizontal branch connections above or below vertical stack offsets. If a horizontal *branch* connects to the *stack* within 2 feet (610 mm) above or below a vertical *stack* offset, and the offset is located more than four *branch intervals* below the top of the *stack*, the offset shall be vented in accordance with Section 916.

❖ Offsets in stacks, particularly those having multiple branch intervals above, produce turbulent flow, thus creating pressure differentials that can adversely affect traps connected to branches that connect to the stack. Points within 2 feet (610 mm) of an offset, above or below, would be subject to pressure fluctuations. Offset venting in accordance with Section 915 is intended to neutralize such pressure fluctuations (see Commentary Figure 711.1).

711.1.1 Omission of vents for vertical stack offsets. Vents for vertical offsets required by Section 711.1 shall not be required where the *stack* and its offset are sized as a *building drain* [see Table 710.1(1)].

❖ This section presents an alternative to the venting requirements of Section 915. Comparing the last column of Table 710.1(2) with any of the columns of Table 710.1(1) shows that sizing the stack and offset as if they were a building drain will result in larger piping being required in most cases. Another way to view this is that the stack and offset piping may not need to be increased if the loading on the stack is reduced. In all cases, the result is that pressure fluctuations are controlled by limiting the flow in the stack (see Commentary Figure 711.1).

711.2 Horizontal branch connections to horizontal stack offsets. Where a horizontal *stack* offset is located more than four *branch intervals* below the top of the *stack*, a horizontal *branch* shall not connect within the horizontal *stack* offset or within 2 feet (610 mm) above or below such offset.

❖ This section addresses horizontal stack offsets and prohibits branch connections within the offset and within 2 feet above or below it (see commentary, Section 711.1). The turbulence and pressure fluctuations in horizontal offsets are worse than in vertical offsets; therefore, greater restrictions are necessary (see Commentary Figure 711.2).

711.3 Horizontal stack offsets. A *stack* with a horizontal offset located more than four *branch intervals* below the top of the *stack* shall be vented in accordance with Section 915 and sized as follows:

1. The portion of the *stack* above the offset shall be sized as for a vertical *stack* based on the total number of drainage fixture units above the offset.
2. The offset shall be sized in accordance with Section 710.1.1.
3. The portion of the *stack* below the offset shall be sized as for the offset or based on the total number of drainage fixture units on the entire *stack*, whichever is larger [see Table 710.1(2), Column 5].

❖ When sizing a horizontal offset, there are three portions of the drainage stack system that must be sized. The section of the stack above the offset is simply sized as if it were a separate stack. The horizontal or near horizontal section of the offset is sized as a building drain using Table 710.1(1). The lower section of the stack below the offset must be at least the same size as the offset. Depending on the total dfu load on the stack, the portion below the offset could be required by Table 710.1(2) to be larger than the offset

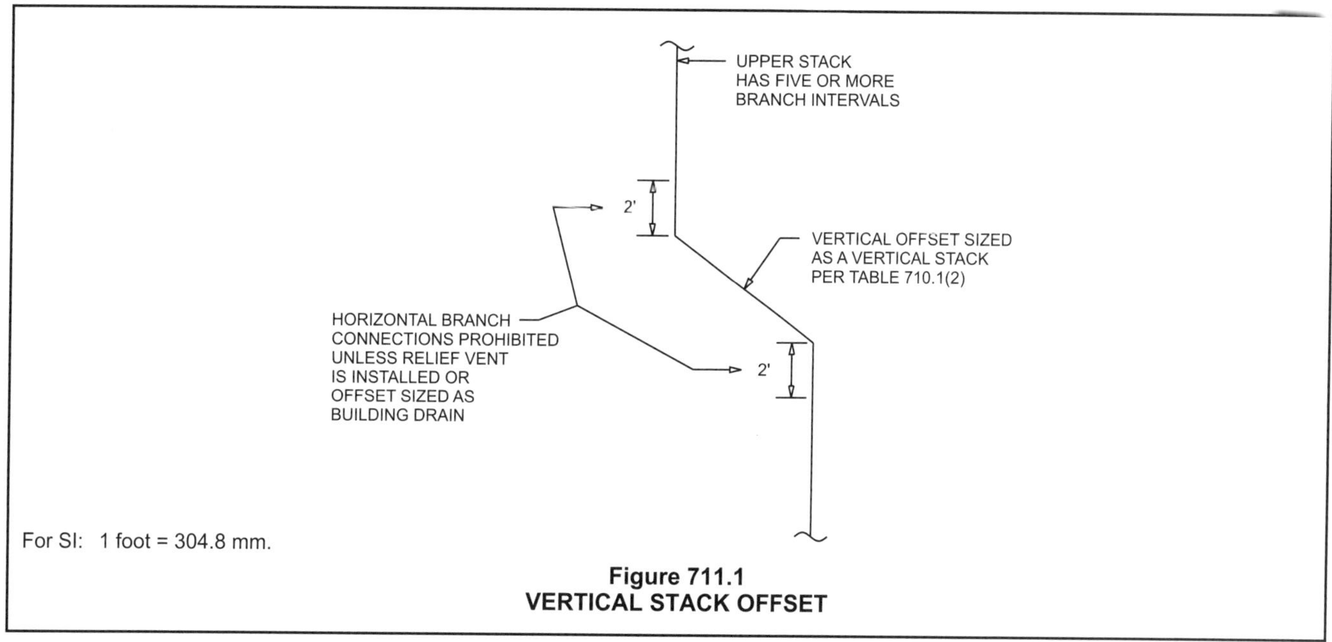

Figure 711.1
VERTICAL STACK OFFSET

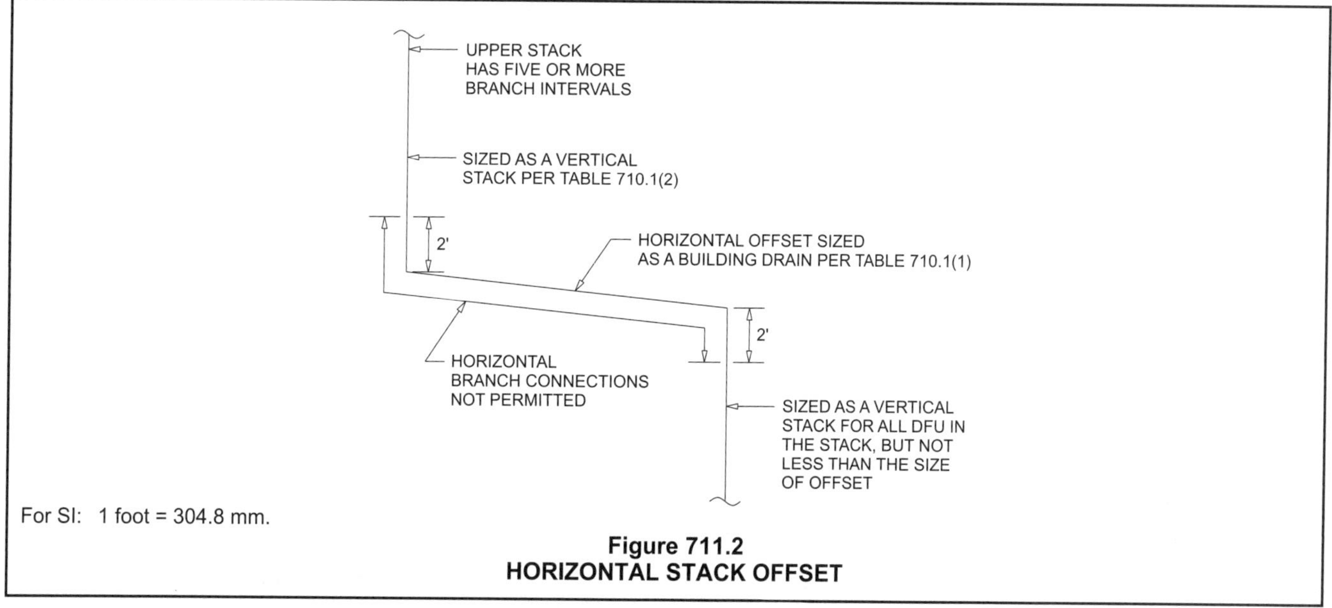

Figure 711.2
HORIZONTAL STACK OFFSET

(see Commentary Figure 711.3). The dfu load discharging through the offset would be added to the total dfu load on the stack below the offset.

The size of drainage pipe may never be reduced in the direction of flow. A relief vent is required for the offset to neutralize pressure differences that result from the turbulent flow in the offset (see Section 915 for vent requirements).

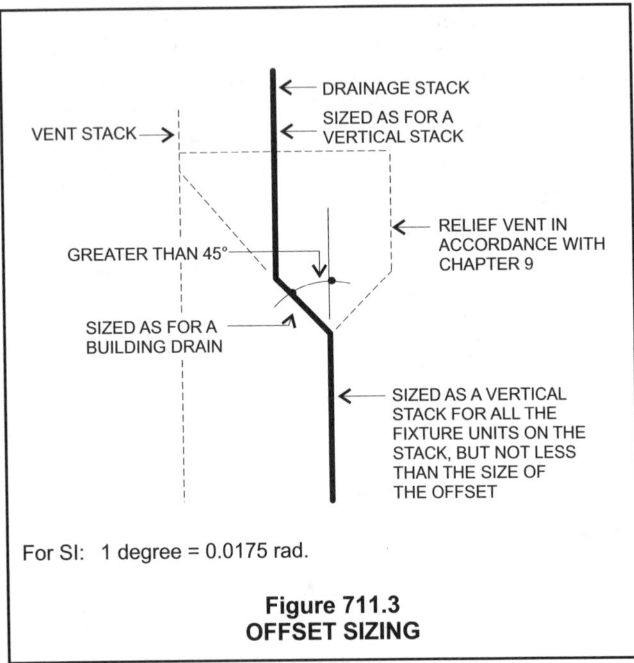

Figure 711.3
OFFSET SIZING

711.3.1 Omission of vents for horizontal stack offsets. Vents for horizontal *stack* offsets required by Section 711.3 shall not be required where the *stack* and its offset are one pipe size larger than required for a *building drain* [see Table 710.1(1)] and the entire *stack* and offset are not less in cross-sectional area than that required for a straight *stack* plus the area of an offset vent as provided for in Section 915. Omission of offset vents in accordance with this section shall not constitute approval of horizontal *branch* connections within the offset or within 2 feet (610 mm) above or below the offset.

❖ Venting is not required for horizontal offsets when the stack and offset are sized one pipe size larger than that required for a vent building drain, as provided for in Section 915. The last column in Table 710.1(2), contains the size requirements for straight stacks (considering that Section 711.3 applies only to offsets more than four branch intervals below the top of the stack). Section 915 contains the size requirements for the offset relief vent. With this increase in cross-sectional area, the stack will act as both a waste and a vent by allowing more air into the system, which will moderate the pressure fluctuations caused by the offset.

711.4 Offsets below lowest branch. Where a vertical offset occurs in a soil or waste *stack* below the lowest horizontal *branch*, a change in diameter of the *stack* because of the offset shall not be required. If a horizontal offset occurs in a soil or waste *stack* below the lowest horizontal *branch*, the required diameter of the offset and the *stack* below it shall be determined as for a *building drain* in accordance with Table 710.1(1).

❖ No size increase is required in vertical offsets that are below the lowest horizontal branch connection to the stack [see Commentary Figure 711.4(1)]. The pressure differences created by such offsets mainly affect the behavior of horizontal branches downstream of the offset, and if there are no horizontal branches downstream, there is nothing to protect by sizing increases. Horizontal offsets that are below the lowest horizontal branch connection must be sized as a building drain in accordance with Table 710.1(1) [see Commentary Figure 711.4(2)]. Because of the greater offset angle, oversizing is required to minimize pressure fluctuations; however, offset venting is not required.

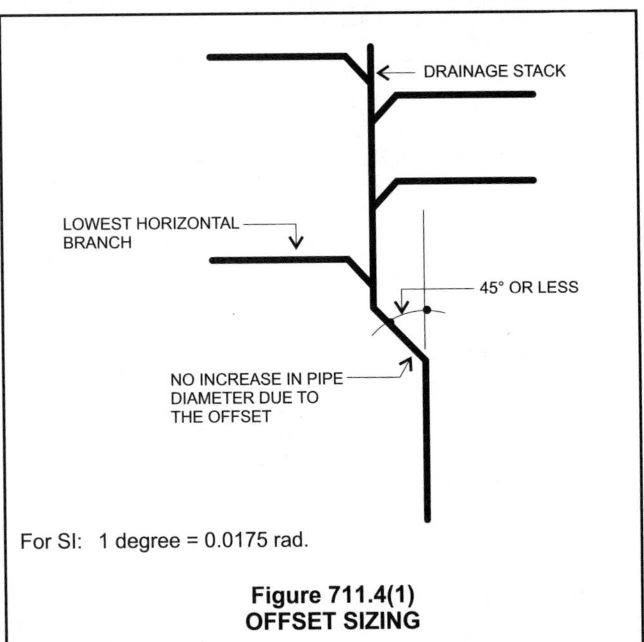

Figure 711.4(1)
OFFSET SIZING

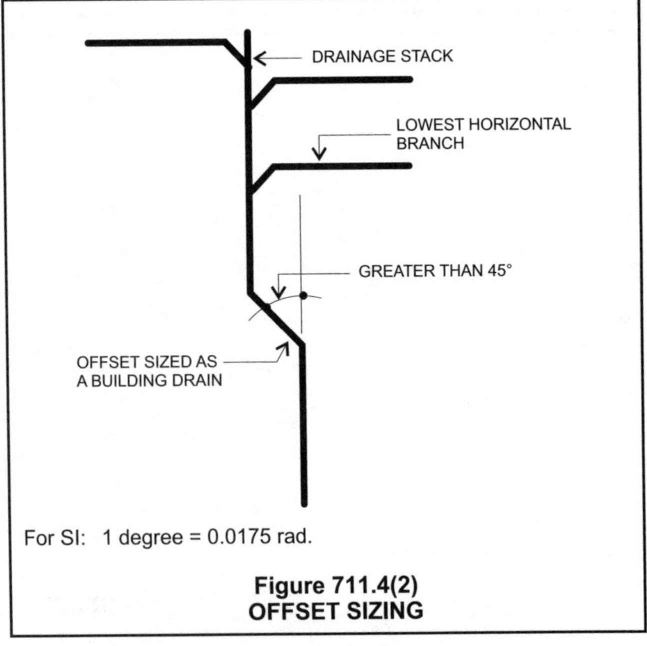

Figure 711.4(2)
OFFSET SIZING

SECTION 712
SUMPS AND EJECTORS

712.1 Building subdrains. Building subdrains that cannot be discharged to the *sewer* by gravity flow shall be discharged into a tightly covered and vented sump from which the liquid shall be lifted and discharged into the building gravity drainage system by automatic pumping equipment or other *approved* method. In other than existing structures, the sump shall not receive drainage from any piping within the building capable of being discharged by gravity to the *building sewer*.

❖ Where the drainage system or portions of it cannot discharge by gravity to the sewer, the drainage is collected in a tightly sealed and vented sump and pumped to a gravity sewer or drain. The sump must be sized to provide adequate holding capacity and to limit the retention period of the waste. Though not specifically required by the code, the intent of this section is that the capacity of the sump must not exceed one-half of a day's (12 hours) discharge load from the piping system connected to the sump under normal use. The intent is to keep the waste retention period short to prevent the sump from acting as a waste decomposition (septic) tank.

The minimum capacity of the sump must be such that the pumping equipment operates for at least 15 seconds per pumping cycle to prevent short cycling, thereby extending the life of the equipment.

The cover for the sump must be gas tight to prevent the escape of sewer gas into the building (see Commentary Section 916.5). Sumps, other than pneumatic ejectors, must be vented in accordance with Section 916.5.1.

Sumps receiving water closet and urinal discharge use a sewage ejector to pump the soil and waste to a gravity drain. Quite often, the pumping equipment is referred to as a sewage pump by technical standards. A pneumatic ejector is a special type of sewage "pump" that operates by air pressure, which forces the sewage from the pressurized receiver instead of mechanically pumping it. Pneumatic ejectors require special relief vents to instantaneously relieve the pressure in the receiver [see commentary, Section 916.5.2 as well as Commentary Figures 712.1(1) and 712.1(2) for two types of pumps].

712.2 Valves required. A check valve and a full open valve located on the discharge side of the check valve shall be installed in the pump or ejector discharge piping between the pump or ejector and the gravity drainage system. *Access* shall be provided to such valves. Such valves shall be located above the sump cover required by Section 712.1 or, where the discharge pipe from the ejector is below grade, the valves shall be accessibly located outside the sump below grade in an *access* pit with a removable *access* cover.

❖ The discharge piping must have a check valve to prevent the previously pumped waste from returning to the sump pit when the pump or ejector shuts off. This prolongs the life of the pumping equipment and conserves energy (see Commentary Figure 712.2). The check valve will also prevent the gravity drainage system from backing up into the sump or receiver.

The requirement for installation of a full-open valve on the discharge side of the check valve indicates a similar concern. There will be some point over the service life of the sump or check valve when replacement or repairs are necessary. This full-open valve is there to prevent waste from running back into the discharge piping when maintenance is performed on the check valve or pump/ejector. It also allows isolation of the drainage system to prevent backups and sewer gas leakage during pump replacement and maintenance. Both valves must be accessible for maintenance on the pump and be installed on the discharge pipe between the pump or ejector and the gravity drainage system.

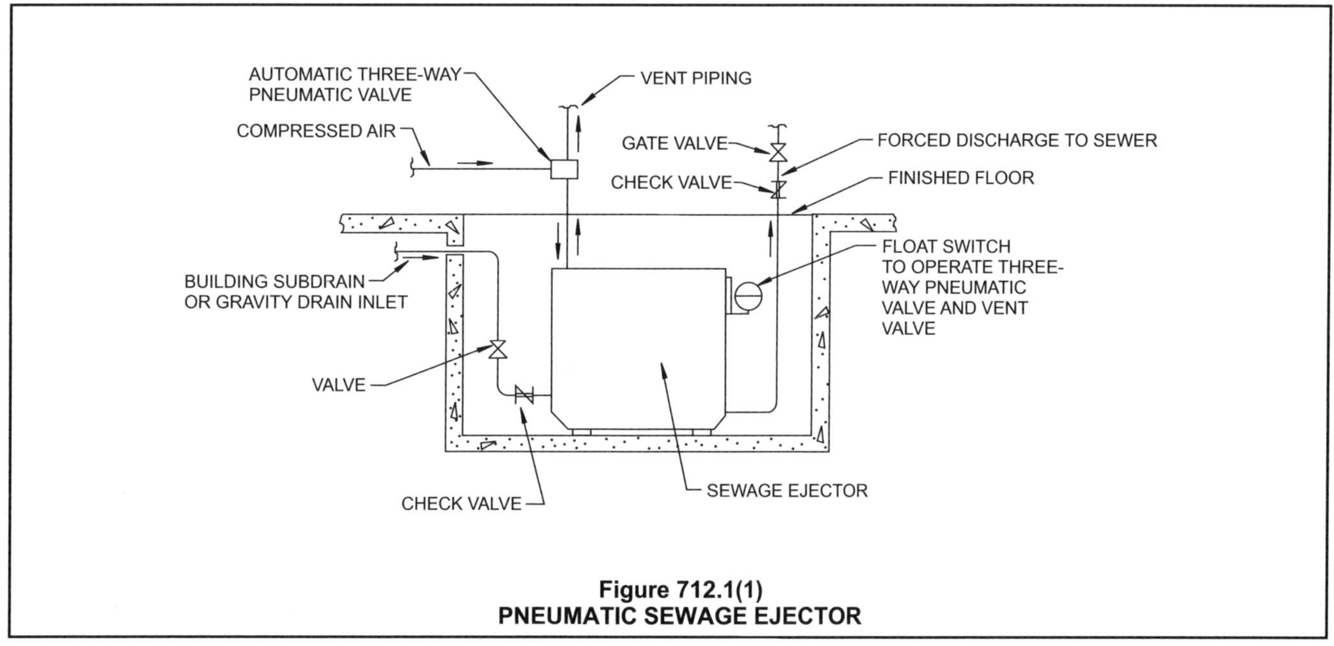

Figure 712.1(1)
PNEUMATIC SEWAGE EJECTOR

OUTLET
MOTOR
CONTROL
SUMP VENT
GAS-TIGHT COVER
FROM BUILDING SUBDRAIN
WATER-TIGHT DRAINAGE SUMP
INLET
FLOAT

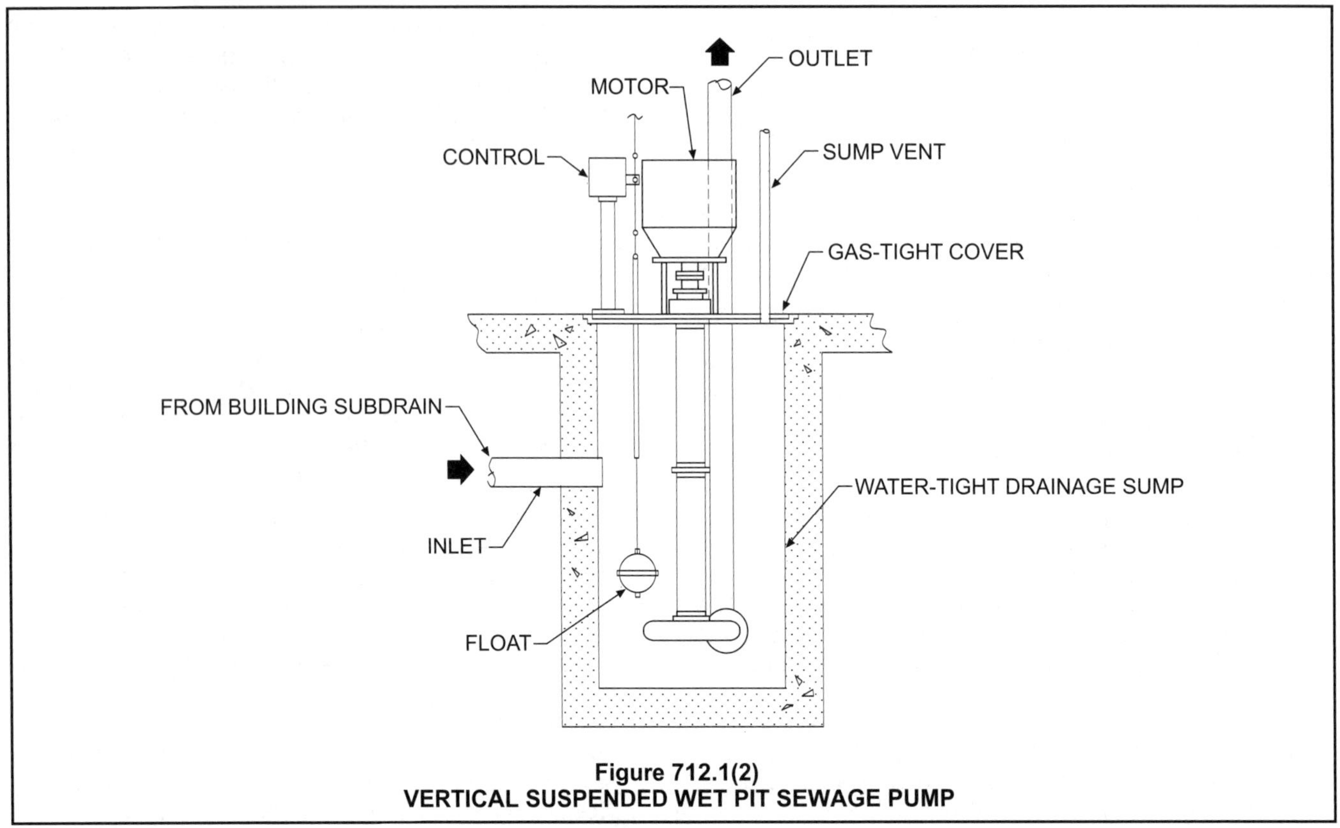

Figure 712.1(2)
VERTICAL SUSPENDED WET PIT SEWAGE PUMP

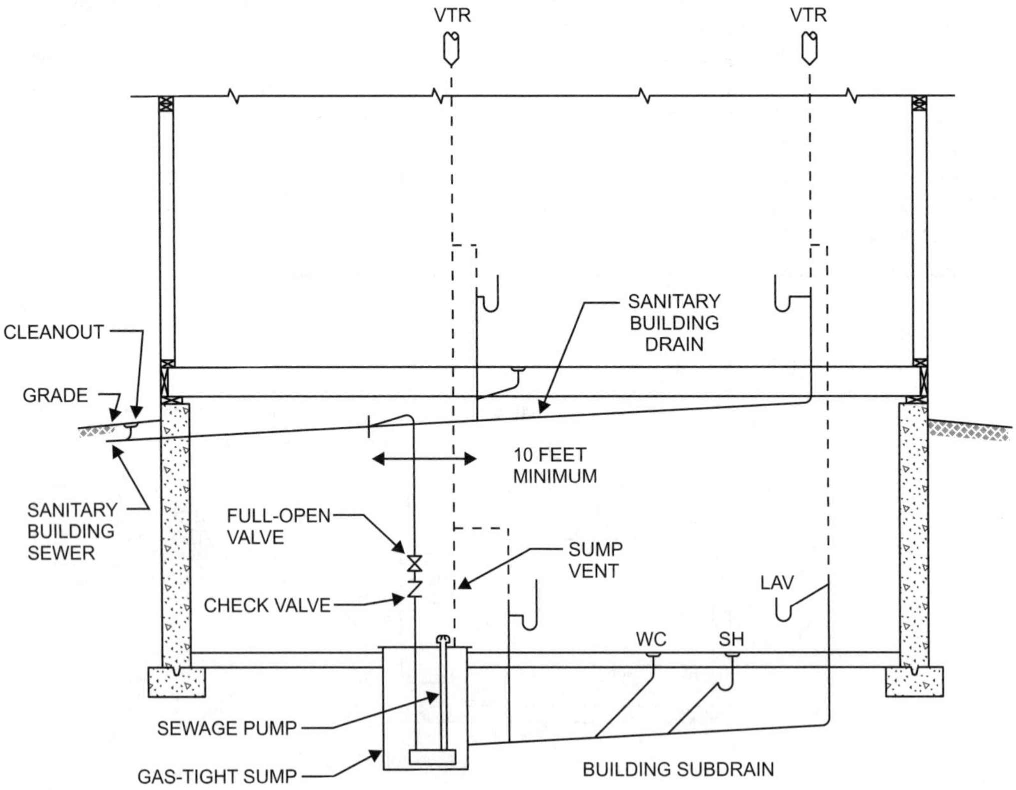

VTR
VTR
CLEANOUT
SANITARY BUILDING DRAIN
GRADE
10 FEET MINIMUM
SANITARY BUILDING SEWER
FULL-OPEN VALVE
SUMP VENT
LAV
CHECK VALVE
WC
SH
SEWAGE PUMP
GAS-TIGHT SUMP
BUILDING SUBDRAIN

For SI: 1 foot = 304.8 mm.

Figure 712.2
SEWAGE EJECTOR INSTALLATION

712.3 Sump design. The sump pump, pit and discharge piping shall conform to the requirements of Sections 712.3.1 through 712.3.5.

❖ This is the main section, under which Sections 712.3.1 to 712.3.5 will give the specific requirements for the construction of the sump pump, the pit and the discharge piping related to sumps and ejectors. Section 712.3.1 relates to sump pump construction. Section 712.3.2 relates to pit construction. Section 71 2.3.3 relates to discharge piping. Section 712.3.4 contains the amount of space to be maintained between the inlet piping into the sump pit and the top of the effluent level. Section 712.3.5 contains requirements for the connection of the discharge piping to the drainage system.

712.3.1 Sump pump. The sump pump capacity and head shall be appropriate to anticipated use requirements.

❖ The sewage pump or pneumatic ejector must be sized properly to accommodate the peak flow into the receiver or pit, to provide the head pressure required to lift or eject waste, to prevent excessive cycling of the pumping equipment and to handle the type of waste that discharges to the sump.

Sewage pumps and ejectors must be able to handle, without creating blockages, the types of solid waste discharged into the sump pit or receiver. Pumps and ejectors serving water closets must be able to handle the solids associated with soil drainage.

Grinder pumps and grinder ejectors that serve water closets are exempt from the 2-inch (51 mm) spherical solids-handling requirements of this section. These pumps and ejectors reduce solids to a near-liquid state and pump the slurry to the drainage system. The liquidized solids pass readily through the pump and into the drainage system, thus reducing the possibility of creating a stoppage in either the pump or the receiving drainage line.

In certain instances, the design professional may choose to install duplex pumping equipment. Duplex pumping installations consist of two pumps with special controls installed to alternate pump duty. If the inflow into the sump is at a high rate or one pump fails to operate, the second pump provides assistance.

The intent behind duplex pumping equipment is to reduce the possibility of disruption of service of the plumbing system. If a pump fails, it may be repaired while the drainage system functions normally with the remaining pump. Duplex pumping equipment is not required by the code.

The sewage pump or ejector must have a minimum capacity that is based on the size of the discharge pipe. The required capacity provides a full-flow velocity of at least 2 feet per second (0.61 m/s) in the discharge pipe (see Table 712.4.2).

712.3.2 Sump pit. The sump pit shall be not less than 18 inches (457 mm) in diameter and 24 inches (610 mm) deep, unless otherwise *approved*. The pit shall be accessible and located such that all drainage flows into the pit by gravity. The sump pit shall be constructed of tile, concrete, steel, plastic or other *approved* materials. The pit bottom shall be solid and provide permanent support for the pump. The sump pit shall be fitted with a gas-tight removable cover adequate to support anticipated loads in the area of use. The sump pit shall be vented in accordance with Chapter 9.

❖ As discussed in the commentary to Section 712.1, the sump pit must be properly sized to receive waste and sewage from the plumbing fixtures that it will serve, as well as proper operation of the pump.

In this section, the minimum dimensions of 18 inches (457 mm) in diameter and 24 inches (610 mm) in depth are given, unless the designer or manufacturer presents calculations and other supporting data that will allow the code official to determine that other dimensions work for the situation being addressed. This approval is strictly on a case-by-case basis by the code official. This section states that the sump pit is to be accessible for maintenance and requires that all drainage flow into the pit must be by the force of gravity.

The sump pit must be constructed of a durable material, such as tile, concrete, steel, plastic or other approved materials. The bottom of the sump pit must be solid and structurally capable of supporting the sump pump. The sump pit must have a gas-tight removable cover to prevent the escape of sewer gas. This cover is to be structurally capable of supporting the weight of loads it will receive based on the location of the sump pit. The sump pit must be vented in accordance with Chapter 9 (see Commentary Section 916.5). The sump pit cover must be readily removable; therefore, the method used to connect the vent to the sump pit must be designed for disassembly.

712.3.3 Discharge piping. Discharge piping and fittings shall be constructed of *approved* materials.

❖ The pump discharge piping and fittings must be suitable for the pressure and conditions of service. The materials must be approved by the code official. See commentary for Section 712.2. Previous editions of the code were silent on the allowable materials.

Refer to Commentary Figures 712.1(1) and 712.1(2) for illustrations of sumps and related components.

712.3.4 Maximum effluent level. The effluent level control shall be adjusted and maintained to at all times prevent the effluent in the sump from rising to within 2 inches (51 mm) of the invert of the gravity drain inlet into the sump.

❖ This section states that the control mechanism that starts and stops the pump must be maintained and adjusted to limit the level of effluent in the sump to not higher than 2 inches (51 mm) below the invert level of the gravity drainage pipe(s) entering the sump. This requirement is to reduce the chance that the gravity drainage piping will become flooded or clogged because of standing effluent in the sump that may back up into the drain piping.

712.3.5 Ejector connection to the drainage system. Pumps connected to the drainage system shall connect to the *building sewer* or shall connect to a wye fitting in the *building drain* a minimum of 10 feet (3048 mm) from the base of any soil *stack*, waste *stack* or *fixture drain*. Where the discharge line connects into horizontal drainage piping, the connector shall be made through a wye fitting into the top of the drainage piping.

❖ This is part of the main section, titled "Sumps and ejectors." As such, it may be interpreted to be applicable to both sewage pumps (including the "grinder type") and pneumatic sewage ejectors. The terms "sewage pump" and "ejector" are sometimes used interchangeably to describe the same device. This section imposes certain restrictions on how and where sewage pumps and pneumatic sewage ejectors may connect to the gravity drainage system.

The connection of the discharge piping is to occur either directly to the building sewer or to a wye fitting in the building drain. If connected to the building drain, the wye fitting must be a minimum of 10 feet (3048 mm) away from other pipe connections or fixtures to reduce the likelihood that discharge from the pump/ejector will interfere with the gravity flow in the drainage system. The 10-foot (3048 mm) distance will allow the pumped waste flow to settle in the invert of the building drain or sewer without creating backups and pressure surges. In a related requirement, where discharge piping connects to a horizontal drainage pipe, this connection must be made with a wye fitting with the fitting wye branch oriented toward the top of the drainage piping. This requirement is to help prevent disruption of the flow of waste inside the drainage piping because the normal flow is on the bottom of the pipe. Therefore, introducing discharge flow into the top of the pipe will not cause blockages of the pipe or disruption of other flow inside the pipe.

712.4 Sewage pumps and sewage ejectors. A sewage pump or sewage ejector shall automatically discharge the contents of the sump to the building drainage system.

❖ This section is redundant with Section 712.1 and requires such equipment to operate automatically.

712.4.1 Macerating toilet systems. Macerating toilet systems shall comply with CSA B45.9 or ASME A112.3.4 and shall be installed in accordance with the manufacturer's installation instructions.

❖ A macerating toilet system is a factory assembled unit having a water-tight housing, an electrically-powered macerating pump assembly and an automatic switch control unit. One configuration is a tank, without a water closet, containing the pump and control unit to which the discharge of a water closet is connected. Another configuration has the macerating pump and control unit integral to a specially designed water closet. Most stand alone tank units are intended to be mounted on the same floor as the fixtures are mounted so that under floor piping is not required. This is especially convenient where the floor is a concrete slab such as in a basement [Commentary Figure 714.4.1(1)]. Some tank units are configured to allow a lavatory and a shower or bathtub to be connected to the tank. Because the discharge from showers, tub and lavatories has minimal and very small solids, those flows are directed to the pump and are not macerated.

Figure 712.4.1(1)
MACERATING TOILET SYSTEMS
(Photo Courtesy of SFA Saniflo)

The macerating pump for both types emulsifies the waste from the water closet to enable a small pump to eject the waste to a sanitary drainage system. The discharge pipe size of these pumps is typically $^{3}/_{4}$ inch (19 mm). One design of a macerating pump assembly is shown in Commentary Figure 712.4.1(2).

712.4.2 Capacity. A sewage pump or sewage ejector shall have the capacity and head for the application requirements. Pumps or ejectors that receive the discharge of water closets shall be capable of handling spherical solids with a diameter of up to and including 2 inches (51 mm). Other pumps or ejectors shall be capable of handling spherical solids with a diameter of up to and including 1 inch (25.4 mm). The minimum capacity of a pump or ejector based on the diameter of the discharge pipe shall be in accordance with Table 712.4.2.

Exceptions:

1. Grinder pumps or grinder ejectors that receive the discharge of water closets shall have a minimum discharge opening of $1^{1}/_{4}$ inches (32 mm).
2. Macerating toilet assemblies that serve single water closets shall have a minimum discharge opening of $^{3}/_{4}$ inch (19 mm).

❖ The sewage pump or ejector must be sized properly to accommodate the peak flow into the receiver, to provide the head pressure required for the elevation of lift, to prevent excessive cycling of the pumping equipment and to handle the type of waste that discharges to the sump. Sewage pumps and ejectors must be able to handle, without creating blockages, the types of solid waste discharged into the sump or receiver. Pumps and ejectors serving water closets must be able to handle the solids associated with soil drainage. Grinder pumps/ejectors and macerating toilet assemblies that serve water closets are exempt from the 2-inch (51 mm) spherical solids-handling requirements of this section. These pumps and ejectors pulverize solids to a near liquid state and pump the slurry to the drainage system. The $1^{1}/_{4}$-inch (32 mm) minimum-size discharge for grinder pumps and ejectors and $^{3}/_{4}$-inch (19.1 mm) minimum-size discharge for macerating grinder/pump systems will permit solids that are completely reduced to pass without creating a stoppage.

TABLE 712.4.2
MINIMUM CAPACITY OF SEWAGE PUMP OR SEWAGE EJECTOR

DIAMETER OF THE DISCHARGE PIPE (inches)	CAPACITY OF PUMP OR EJECTOR (gpm)
2	21
$2^{1}/_{2}$	30
3	46

For SI: 1 inch = 25.4 mm, 1 gallon per minute = 3.785 L/m.

❖ Table 712.4.2 specifies sewage pump capacity in gallons per minute with respect to the pump's discharge

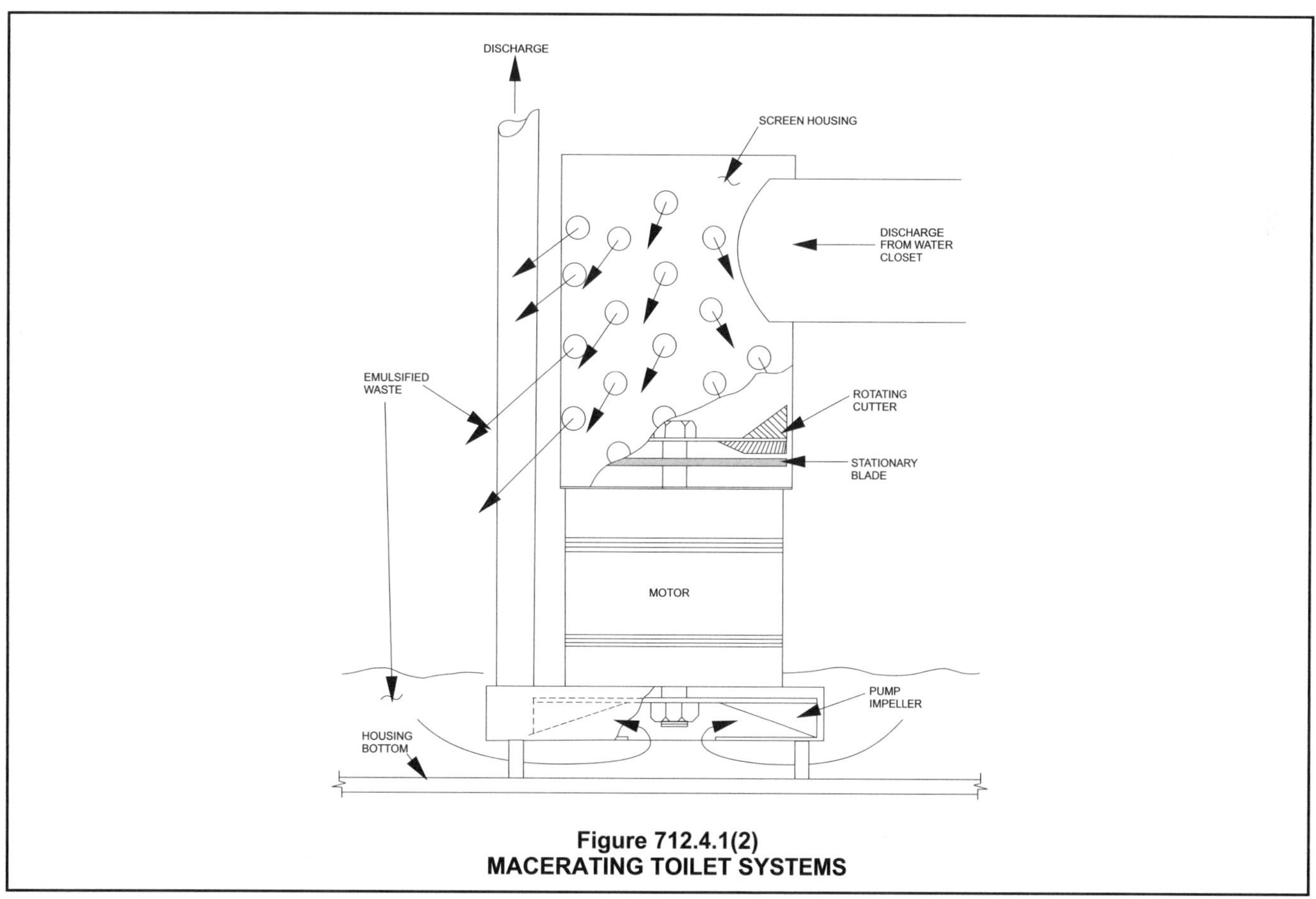

Figure 712.4.1(2)
MACERATING TOILET SYSTEMS

opening size. This relationship between pipe size and pump capacity will maintain flow velocity in the piping to help prevent blockage and restriction.

SECTION 713
HEALTH CARE PLUMBING

713.1 Scope. This section shall govern those aspects of health care plumbing systems that differ from plumbing systems in other structures. Health care plumbing systems shall conform to this section in addition to the other requirements of this code. The provisions of this section shall apply to the special devices and equipment installed and maintained in the following occupancies: nursing homes; homes for the aged; orphanages; infirmaries; first aid stations; psychiatric facilities; clinics; professional offices of dentists and doctors; mortuaries; educational facilities; surgery, dentistry, research and testing laboratories; establishments manufacturing pharmaceutical drugs and medicines; and other structures with similar apparatus and equipment classified as plumbing.

❖ Section 713 regulates the installation of special fixtures and the related waste and vent connections found in health care facilities. The special plumbing fixtures addressed in this section include clinical sinks, bedpan washers, bedpan steamers, vacuum fluid suction systems, sterilizers and hydrotherapeutic baths.

Section 713.1 defines the scope of this chapter and describes the occupancies involved.

Section 202 contains the definitions of key terms and phrases used throughout this section.

Section 422 regulates the use of special fixtures in the health care industry.

Section 609 governs the water supply to health care facilities and fixtures. Hospitals are required to have dual water services so that in the event of a water main or water service pipe break, the hospital will not experience a complete loss of water supply.

The unique plumbing fixtures in health care facilities are regulated by the special requirements of this section as well as those previously mentioned such as Sections 202, 422 and 609. Health care fixtures must be installed to maintain the same level of sanitation and protection against backflow as other plumbing fixtures.

Some of the requirements for health care fixtures in this section may appear archaic. As technology improves, the fixtures change. Requirements addressing older fixtures will no longer be applicable to certain newer fixtures.

This is most evident with bedpan washers and sterilizers. Water closets are often equipped with an add-on flushing mechanism that may function as a bedpan washer. Installation of a separate fixture may not be needed.

Sterilizers may have a local or vapor vent connection. The new units are designed to condense the vapor rather than discharge it through a vent.

Although the code lists many applications, some are bound to be missing. The requirements for plumbing installations in health care facilities apply to facilities whether or not they are specifically listed. These requirements are to be considered the minimum special attention given to health care plumbing. The fixture installer must carefully follow any other requirements contained in the manufacturer's instructions.

This section regulates the waste and vent connections for health care plumbing fixtures, appliances and equipment. The requirements of this section are in addition to other applicable code requirements (see Commentary Sections 202, 422 and 609).

713.2 Bedpan washers and clinical sinks. Bedpan washers and clinical sinks shall connect to the drainage and vent system in accordance with the requirements for a water closet. Bedpan washers shall also connect to a local vent.

❖ These fixtures are to be vented and connected to the drainage system like a water closet because they receive similar human waste and function in a similar manner. A local vent is required because a bedpan washer uses high-temperature hot water to clean the bedpan. A clinical sink is not equipped with a local vent (see Commentary Section 713.3).

713.3 Indirect waste. All sterilizers, steamers and condensers shall discharge to the drainage through an indirect waste pipe by means of an *air gap*. Where a battery of not more than three sterilizers discharges to an individual receptor, the distance between the receptor and a sterilizer shall not exceed 8 feet (2438 mm). The indirect waste pipe on a bedpan steamer shall be trapped.

❖ An indirect waste pipe protects sterilizers, steamers and condensers from contamination from the drainage system through the use of the required air gap. An indirect waste pipe is typically piped individually to the indirect waste receptor. This section permits a common indirect waste pipe for two or three sterilizers installed together (see Commentary Figure 713.3). Additionally, the indirect waste pipe on the bedpan steamer must be trapped to prevent the escape of steam vapors and odor.

713.4 Vacuum system station. Ready *access* shall be provided to vacuum system station receptacles. Such receptacles shall be built into cabinets or recesses and shall be visible.

❖ Vacuum system stations are typically installed in recesses in locations where medical professionals have ready access to them. The person locating the station must take into consideration the movements of personnel during an operation or emergency procedure.

These stations are to be visible to allow for ready supervision of the equipment to make sure it is functioning as required.

713.5 Bottle system. Vacuum (fluid suction) systems intended for collecting, removing and disposing of blood, pus or other fluids by the bottle system shall be provided with receptacles equipped with an overflow prevention device at each vacuum outlet station.

❖ A vacuum or fluid suction system is a central system of piping in a health care facility, with terminals located throughout the building. The terminals are typically found in operating rooms, emergency rooms, delivery rooms, recovery rooms, intensive care rooms and similar locations.

Typically, a bottle collector connects to the terminal when the suction system is used. Fluids collected must be discharged to the bottle while the vacuum system remains free of fluids; therefore, a means of preventing the overflow of the bottle is required. This is commonly done by having a separate vacuum trap with a shutoff installed between the suction collection bottle and the vacuum system terminal inlet.

713.6 Central disposal system equipment. All central vacuum (fluid suction) systems shall provide continuous service. Systems equipped with collecting or control tanks shall provide for draining and cleaning of the tanks while the system is in operation. In hospitals, the system shall be connected to the emergency power system. The exhausts from a vacuum pump serving a vacuum (fluid suction) system shall discharge separately to open air above the roof.

❖ The vacuum system is essential for a health care facility and must operate under all conditions. To provide continuous service, at least two vacuum pumps are installed. These systems must also connect to the building's emergency power system (see Commentary Figure 713.6).

Many systems operate with a receiver tank to prevent the constant cycling of the vacuum pumps. The tanks must be equipped with a drain and must be arranged for the cleaning of the tank while the system

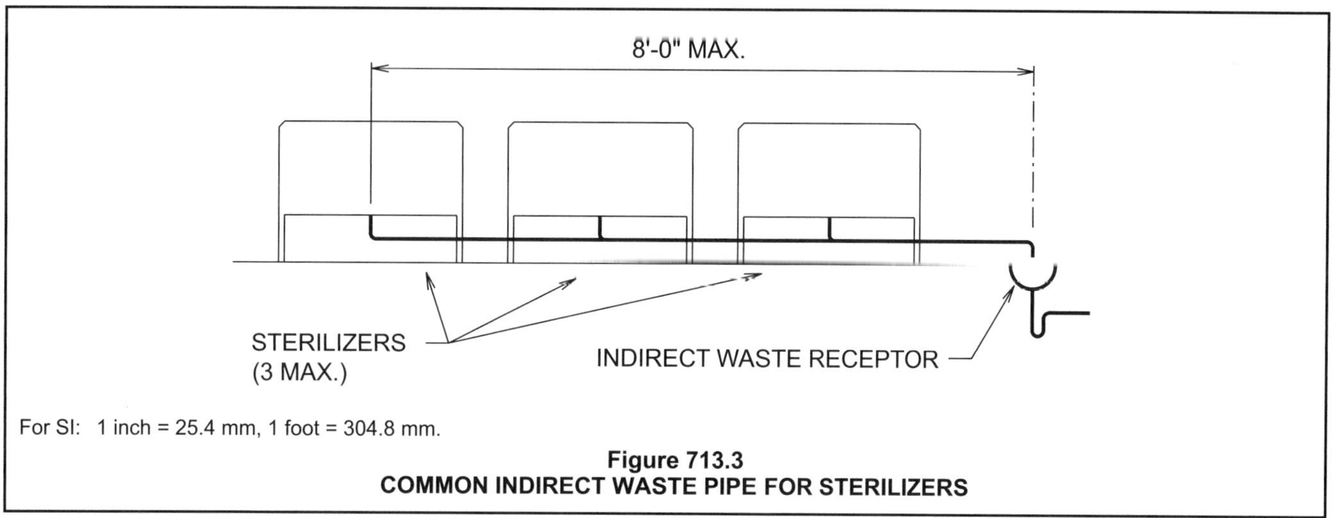

Figure 713.3
COMMON INDIRECT WASTE PIPE FOR STERILIZERS

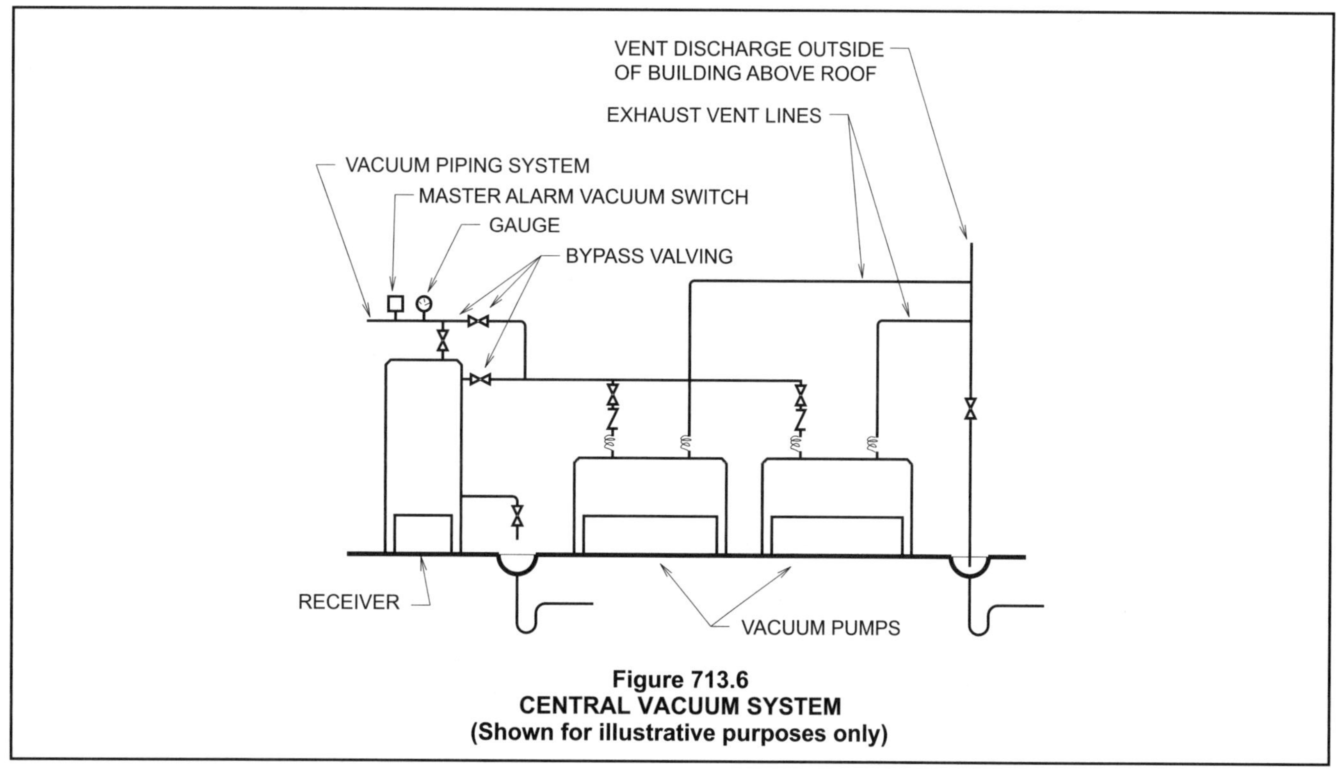

Figure 713.6
CENTRAL VACUUM SYSTEM
(Shown for illustrative purposes only)

is still in service. This is accomplished with a bypass piping arrangement located around the receiver tank. This section also requires the exhaust from the vacuum pump to discharge to the outdoors at an elevation that is above the roof of the building in which the equipment is located. The exhaust may be considered a biohazard and must be discharged in a manner similar to plumbing vent terminals.

713.7 Central vacuum or disposal systems. Where the waste from a central vacuum (fluid suction) system of the barometric-lag, collection-tank or bottle-disposal type is connected to the drainage system, the waste shall be directly connected to the sanitary drainage system through a trapped waste.

❖ Certain vacuum systems are designed to connect directly to the drainage system. These types of installations cannot be connected by an indirect waste pipe. Rather, they must be directly connected. This is intended to minimize possible contamination by preventing the waste from being exposed to the atmosphere. The connection must be trapped to protect the system from sewer gases.

713.7.1 Piping. The piping of a central vacuum (fluid suction) system shall be of corrosion-resistant material with a smooth interior surface. A *branch* shall not be less than $^1/_2$ inch (12.7 mm) nominal pipe size for one outlet and shall be sized in accordance with the number of vacuum outlets. A main shall not be less than 1-inch (25 mm) nominal pipe size. The pipe sizing shall be increased in accordance with the manufacturer's instructions as stations are increased.

❖ The common piping materials used in vacuum systems are copper tubing, galvanized steel pipe and stainless steel pipe. These materials are corrosion resistant and have smooth interior surfaces. The sizing, cleanout and access requirements are considered the very minimum acceptable.

713.7.2 Velocity. The velocity of airflow in a central vacuum (fluid suction) system shall be less than 5,000 feet per minute (25 m/s).

❖ The air velocity of a vacuum system is limited to 5,000 feet per minute (25 m/s) to reduce the noise factor and erosion in the piping.

713.8 Vent connections prohibited. Connections between local vents serving bedpan washers or sterilizer vents serving sterilizing apparatus and normal sanitary plumbing systems are prohibited. Only one type of apparatus shall be served by a local vent.

❖ Both a local vent and a sterilizer vent are used for the removal of condensate and odor. Both vents function differently than a vent system designed to protect the trap seals in a drainage system. Local vents must be completely independent of the sanitary drain and waste venting system. An interconnection between local vents and sanitary drain and waste vents would allow sewer gases to enter the local vents, thereby resulting in equipment contamination and the escape of sewer gases into the building interior.

Each local or sterilizer vent must be designed to serve only one type of equipment to prevent possible cross contamination. A local vent connecting to equipment used to clean or sterilize bedpans must not connect to any other sterilizer vents. Although the equipment might be sterilizing the bedpan, there could be bacterial growth in the local vent originating from human waste. The installer must carefully read and follow the fixture manufacturer's instructions.

713.9 Local vents and stacks for bedpan washers. Bedpan washers shall be vented to open air above the roof by means of one or more local vents. The local vent for a bedpan washer shall not be less than a 2-inch-diameter (51 mm) pipe. A local vent serving a single bedpan washer is permitted to drain to the fixture served.

❖ A local vent is occasionally referred to as a "vapor vent." It connects to the fixture or equipment and is designed to vent steam or hot water vapor and foul odors. The local vent prevents water vapor from escaping into the room or space. Besides increasing the humidity or forming condensation, escaping water vapor may also contain biological contaminants. Local vents are disappearing with the use of modern equipment designed to contain and condense the water vapor in the equipment. This type of equipment may not provide a connection for, nor does it require, a local vent. This requirement applies only to bedpan washers having a connection for a local vent.

713.9.1 Multiple installations. Where bedpan washers are located above each other on more than one floor, a local vent *stack* is permitted to be installed to receive the local vent on the various floors. Not more than three bedpan washers shall be connected to a 2-inch (51 mm) local vent *stack*, not more than six to a 3-inch (76 mm) local vent *stack* and not more than 12 to a 4-inch (102 mm) local vent *stack*. In multiple installations, the connections between a bedpan washer local vent and a local vent *stack* shall be made with tee or tee-wye sanitary pattern drainage fittings installed in an upright position.

❖ The requirements for this section may be simplified into a table form (see Commentary Table 713.9.1). The stack may be increased in size as additional bedpan washers connect to the local vent stack (see Commentary Figure 713.9.1).

Table 713.9.1
MULTIPLE CONNECTION FOR BEDPAN WASHERS TO LOCAL VENT STACK

LOCAL VENT STACK SIZE (inches	MAXIMUM NUMBER OF BEDPAN WASHERS
2	3
3	6
4	12

For SI: 1 inch = 25.4 mm.

713.9.2 Trap required. The bottom of the local vent *stack*, except where serving only one bedpan washer, shall be drained by means of a trapped and vented waste connection to the sanitary drainage system. The trap and waste shall be the same size as the local vent *stack*.

❖ Because a local vent is designed to remove water vapor, there will be a certain amount of water condensing inside the pipe. The condensate must be drained to the sanitary drainage system to prevent the local vent stack from accumulating water. The connection to the drainage system must be by means of a trapped and vented waste connection (see Commentary Figure 713.9.2).

713.9.3 Trap seal maintenance. A water supply pipe not less than $^1/_4$ inch (6.4 mm) in diameter shall be taken from the flush supply of each bedpan washer on the discharge or fixture side of the vacuum breaker, shall be trapped to form not less than a 3-inch (76 mm) water seal, and shall be connected to the local vent *stack* on each floor. The water supply shall be installed so as to provide a supply of water to the local vent *stack* for cleansing and drain trap seal maintenance each time a bedpan washer is flushed.

❖ A $^1/_4$-inch (6.4 mm) tube is looped to form a trap and is connected to the local vent. With each use of the bedpan washer, water is discharged to maintain the trap seal of the drainage connection. The looped tube must be connected downstream of the fixture's vacuum breaker to prevent a cross connection.

713.10 Sterilizer vents and stacks. Multiple installations of pressure and nonpressure sterilizers shall have the vent connections to the sterilizer vent *stack* made by means of inverted wye fittings. *Access* shall be provided to vent connections for the purpose of inspection and maintenance.

❖ A sterilizer vent is similar to a local vent. The vent is designed to remove the steam vapor developed in sterilizer equipment (see Commentary Figure 713.10).

Like local vents, sterilizer vents are not used because current equipment condenses the steam vapor in the equipment.

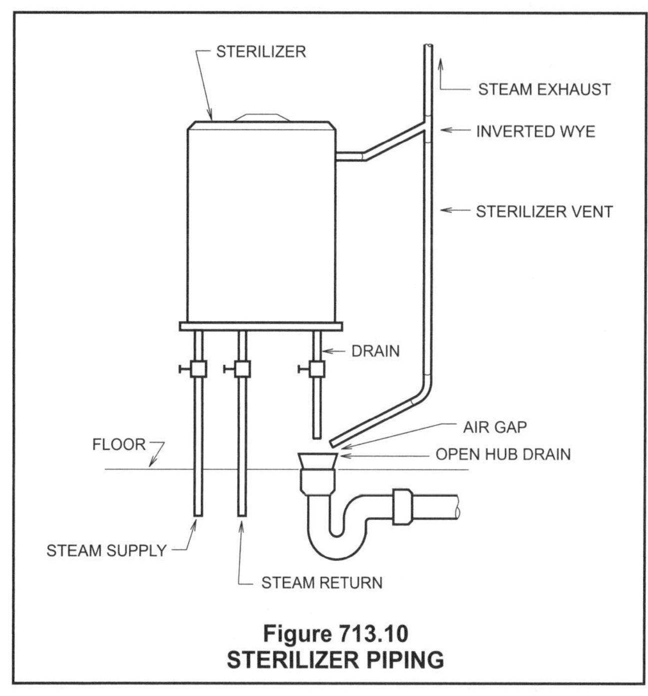

Figure 713.10
STERILIZER PIPING

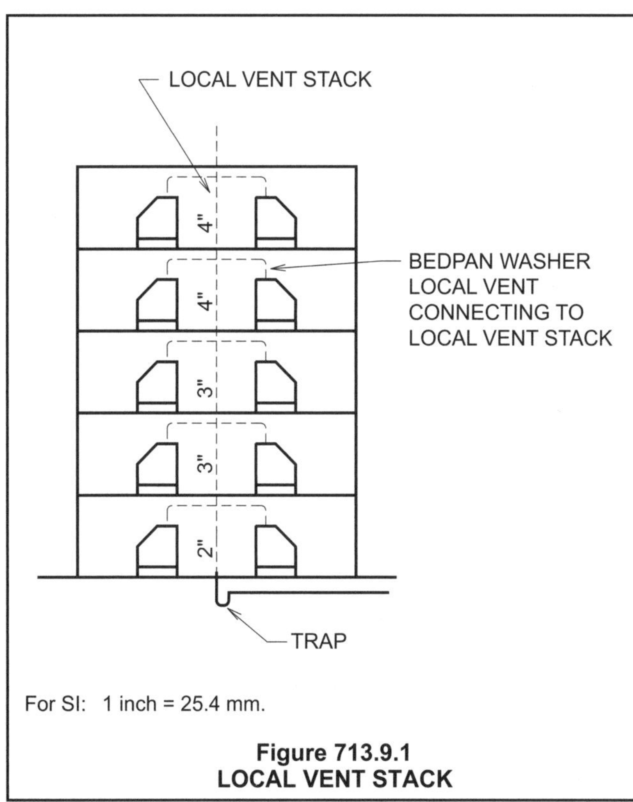

Figure 713.9.1
LOCAL VENT STACK

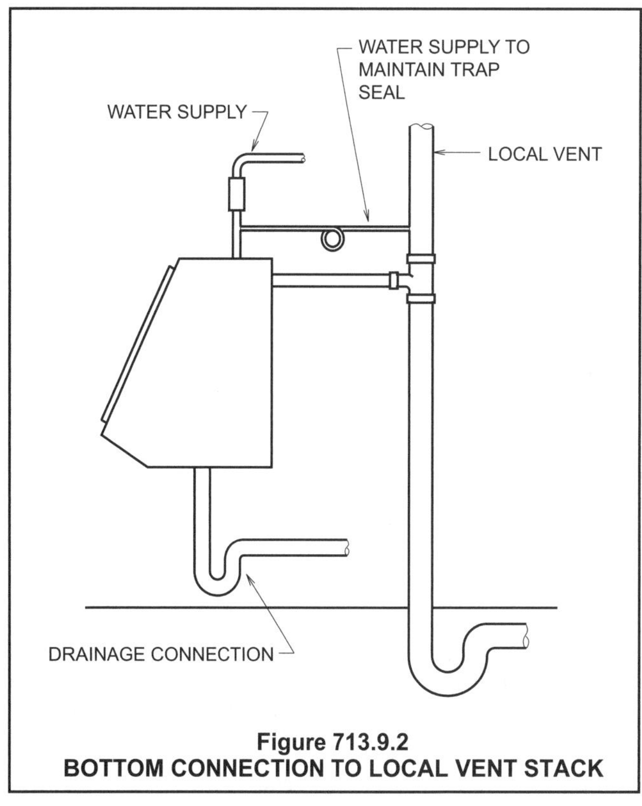

Figure 713.9.2
BOTTOM CONNECTION TO LOCAL VENT STACK

713.10.1 Drainage. The connection between sterilizer vent or exhaust openings and the sterilizer vent *stack* shall be designed and installed to drain to the funnel or basket-type waste fitting. In multiple installations, the sterilizer vent *stack* shall be drained separately to the lowest sterilizer funnel or basket-type waste fitting or receptor.

❖ The sterilizer vent exhausts the steam created in the sterilizer. As a result, condensation forms inside the piping. To maintain proper operation of the system, the sterilizer vent must be designed to drain the condensate to an indirect waste receptor at the base of the vent.

713.11 Sterilizer vent stack sizes. Sterilizer vent *stack* sizes shall comply with Sections 713.11.1 through 713.11.4.

❖ The requirements for sterilizer vent stack sizes are described in Sections 713.11.1 through 713.11.4.

713.11.1 Bedpan steamers. The minimum size of a sterilizer vent serving a bedpan steamer shall be $1^1/_2$ inches (38 mm) in diameter. Multiple installations shall be sized in accordance with Table 713.11.1.

❖ The minimum size sterilizer vent is $1^1/_2$ inches (38 mm). When multiple sterilizer vents are interconnected, the piping must be sized in accordance with Table 713.11.1.

TABLE 713.11.1
STACK SIZES FOR BEDPAN STEAMERS AND BOILING-TYPE STERILIZERS
(Number of Connections of Various Sizes Permitted to Various-sized Sterilizer Vent Stacks)

STACK SIZE (inches)	CONNECTION SIZE		
	$1^1/_2$″		2″
$1^1/_2$[a]	1	or	0
2[a]	2	or	1
2[b]	1	and	1
3[a]	4	or	2
3[b]	2	and	2
4[a]	8	or	4
4[b]	4	and	4

For SI: 1 inch = 25.4 mm.
a. Total of each size.
b. Combination of sizes.

❖ This table often appears confusing because of the use of the terms "or" and "and." When the term "or" is used under the permitted connection size, it means that either one or the other listed connection is permitted. When the term "and" is used, it means that both of the listed connections are permitted together.

713.11.2 Boiling-type sterilizers. The minimum size of a sterilizer vent *stack* shall be 2 inches (51 mm) in diameter where serving a utensil sterilizer and $1^1/_2$ inches (38 mm) in diameter where serving an instrument sterilizer. Combinations of boiling-type sterilizer vent connections shall be sized in accordance with Table 713.11.1.

❖ Table 713.11.1 is also used for sizing boiling-type sterilizer vent stacks. Refer to Section 713.11.1 for commentary on use of the table.

713.11.3 Pressure sterilizers. Pressure sterilizer vent stacks shall be $2^1/_2$ inches (64 mm) minimum. Those serving combinations of pressure sterilizer exhaust connections shall be sized in accordance with Table 713.11.3.

❖ Pressure sterilizer vent stacks must not be less than $2^1/_2$ inches (64 mm) in diameter; however, where such stacks serve multiple pressure sterilizers, the vent stacks must be sized in accordance with Table 713.11.3.

TABLE 713.11.3
STACK SIZES FOR PRESSURE STERILIZERS
(Number of Connections of Various Sizes Permitted To Various-sized Vent Stacks)

STACK SIZE (inches)	CONNECTION SIZE			
	$^3/_4$″	1″	$1^1/_4$″	$1^1/_2$″
$1^1/_2$[a]	3 or	2 or	1	—
$1^1/_2$[b]	2 and	1	—	—
2[a]	6 or	3 or	2 or	1
2[b]	3 and	2	—	—
2[b]	2 and	1 and	1	—
2[b]	1 and	1 and	—	1
3[a]	15 or	7 or	5 or	3
3[b]	1 and	1 and 5 and	2 and	2 1

For SI: 1 inch = 25.4 mm.
a. Total of each size.
b. Combination of sizes.

❖ The table dictates the allowable number of individual sterilizer vent connections to a vent stack. Connections are listed as numbers and combinations of various sizes (see Commentary Table 713.11.1). The rows designated by Note a allow one or more connections of a single sterilizer exhaust connection size. Note a rows do not permit commingling of different sterilizer exhaust connection sizes. The rows designated by Note b allow multiple connections of sterilizer exhausts that are combinations of sizes.

713.11.4 Pressure instrument washer sterilizer sizes. The minimum diameter of a sterilizer vent *stack* serving an instrument washer sterilizer shall be 2 inches (51 mm). Not more than two sterilizers shall be installed on a 2-inch (51 mm) *stack*, and not more than four sterilizers shall be installed on a 3-inch (76 mm) *stack*.

❖ The requirements of this section are illustrated in Commentary Table 713.11.4.

Table 713.11.4
PRESSURE INSTRUMENT WASHER STERILIZER SIZES

STACK SIZE (inches)	NUMBER OF STERILIZERS CONNECTED
2	2
3	4

For SI: 1 inch = 25.4 mm.

SECTION 714
COMPUTERIZED DRAINAGE DESIGN

714.1 Design of drainage system. The sizing, design and layout of the drainage system shall be permitted to be designed by *approved* computer design methods.

❖ This section establishes a parameter for sizing a drainage system using a computerized method. Various computer programs have been developed using the same technical guidelines used to develop the sizing tables in Sections 703, 706, 710 and 711. One of the major differences is that the computerized methods are more exact, knowing the exact discharge considerations or load on the drainage system fixture profiles, material specifications and slope of drainage piping.

714.2 Load on drainage system. The load shall be computed from the simultaneous or sequential discharge conditions from fixtures, appurtenances and appliances or the peak usage design condition.

❖ When sizing to exact parameters, the discharge conditions for the particular system must be known. This is similar to determining the peak demand of the water distribution system, as indicated in the commentary to Section 604.3.

714.2.1 Fixture discharge profiles. The discharge profiles for flow rates versus time from fixtures and appliances shall be in accordance with the manufacturer's specifications.

❖ Each fixture has a specific discharge profile, plotting flow rates versus time. When designing by a computerized method, the actual discharge profile of the individual fixtures being installed may be used. The table in Section 604.3 applies a conservative, general discharge profile for each category of fixtures.

714.3 Selections of drainage pipe sizes. Pipe shall be sized to prevent full-bore flow.

❖ The drainage system must be designed to maintain open channel flow (gravity). If full-bore flow is achieved, the piping may have pressure flow occurring. This may be detrimental to the operation of the system because the venting design provisions are based on the hydraulics of a gravity system.

714.3.1 Selecting pipe wall roughness. Pipe size calculations shall be conducted with the pipe wall roughness factor (ks), in accordance with the manufacturer's specifications and as modified for aging roughness factors with deposits and corrosion.

❖ Exact sizing uses the parameters of the given material to be installed. The roughness coefficient of the interior pipe wall is used in sizing. Pipe wall roughness is a factor used to determine the flow losses attributable to friction. These factors must contain modifying factors that reflect that changes occur as the pipe ages and develops deposits and corrosion.

714.3.2 Slope of horizontal drainage piping. Horizontal drainage piping shall be designed and installed at slopes in accordance with Table 704.1.

❖ Piping sized by a computerized method is subject to the same slope requirements as conventionally sized piping. The minimum slope must be as specified in Table 704.1. Refer to that section and its commentary for related issues, requirements and concerns.

SECTION 715
BACKWATER VALVES

715.1 Sewage backflow. Where the flood level rims of plumbing fixtures are below the elevation of the manhole cover of the next upstream manhole in the *public sewer*, such fixtures shall be protected by a backwater valve installed in the *building drain*, *branch* of the *building drain* or horizontal *branch* serving such fixtures. Plumbing fixtures having flood level rims above the elevation of the manhole cover of the next upstream manhole in the *public sewer* shall not discharge through a backwater valve.

❖ A backwater valve is required in areas where the public sewer might back up into the building through the sanitary drainage system [see Commentary Figure 715.1(1)]. When plumbing fixtures are located above the next upstream manhole cover from the building sewer connection to the public sewer, the sewer will back up through the street manhole before entering the building.

Public sewers might become blocked or overloaded, which will result in sewage backing up into the manholes and any laterals (taps) connected to the sewer system. The point of overflow for the public sewer will be the top of the manholes in the backed-up portion of the system.

Fixtures or drains located at an elevation below that of the tops of the manholes for the relative portion of the sewer system are subject to backflow and must be protected by backwater valves [see Commentary Figure 715.1(2)]. Plumbing fixtures that are not subject to backflow are not permitted to discharge through a backwater valve. In theory, limiting the fixtures that discharge through a backwater valve will prevent waste from upstream fixtures from backing up through downstream fixtures because it cannot pass through the backwater valve when the public sewer is blocked or overloaded. Additionally, the valve will be protected from excess wear and potential failure resulting from debris and accumulations.

The last sentence of this section, if taken literally, would apply to fixtures having a flood level rim a fraction of an inch above the upstream manhole cover. Such fixtures might still be subject to the backup of sewage, which may be insanitary and present the risk of sewage entering the structure through fixture connections, traps and other "weak links" in the piping system.

715.2 Material. All bearing parts of backwater valves shall be of corrosion-resistant material. Backwater valves shall comply with ASME A112.14.1, CSA B181.1 or CSA B181.2.

❖ This section requires that bearing parts of the backwater valve be made of corrosion-resistant materials so that the valve remains operable for its expected life, thus protecting the building it serves. This section also gives the industry standards to which the backwater

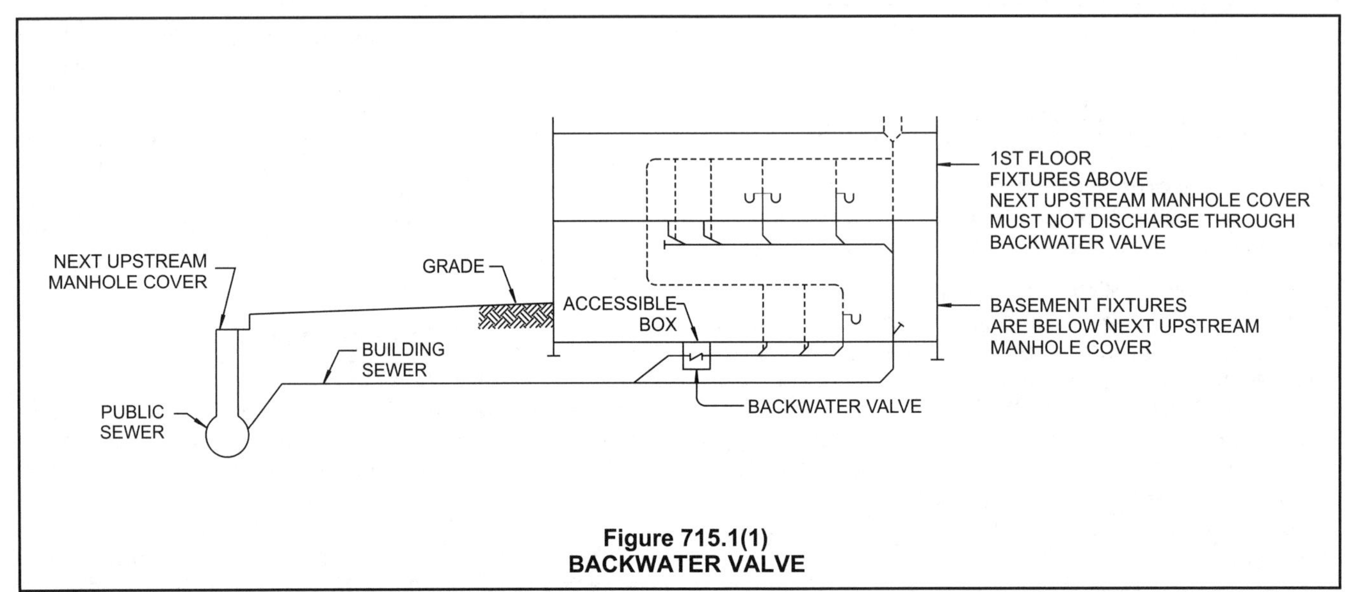

Figure 715.1(1)
BACKWATER VALVE

ACCESS PLUG
FLOOR DRAIN
SWING CHECK VALVE
SPIGOT OUTLET CONNECTION
COMBINATION FLOOR DRAIN BACKWATER VALVE
BOLTED GASKETED COVER
SWING CHECK VALVE
INLET
HORIZONTAL BACKWATER VALVE
OUTLET
INLET
BACKWATER VALVE WITH MANUAL GATE VALVE

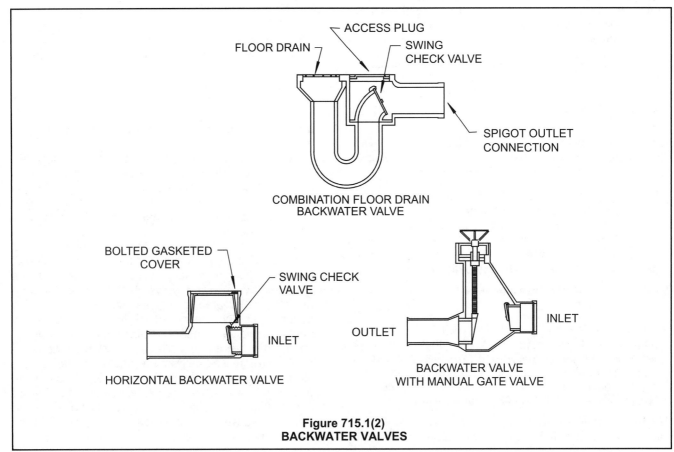

Figure 715.1(2)
BACKWATER VALVES

valve must be manufactured. As required in Section 107, these backwater valves are to be inspected and verified by an independent third-party quality-assurance facility.

715.3 Seal. Backwater valves shall be so constructed as to provide a mechanical seal against backflow.

❖ A backwater valve is designed to prevent the backflow of drainage in a piping system. The valve incorporates a swing check on the inlet side of the body. This section requires that the swing check have a positive seal against the backflow of drainage. Some backwater valves are also equipped with a gate valve that is manually operated to close the outlet side of the device. Backwater valves equipped with gate valves are typically used in areas subject to severe flooding conditions [see Commentary Figure 715.1(2)].

715.4 Diameter. Backwater valves, when fully opened, shall have a capacity not less than that of the pipes in which they are installed.

❖ This section simply requires that the capacity of the pipe in which the backwater valve is located not be reduced by the valve. This would include components of the backwater valve, including any optional manual gate valves. Backwater valves are designed so as to not create a restriction in the piping.

715.5 Location. Backwater valves shall be installed so that *access* is provided to the working parts for service and repair.

❖ Because a backwater valve has movable parts, there is a possibility of stoppage or malfunction. This section requires that backwater valves be located to be accessible to permit the necessary maintenance or repairs.

Bibliography

The following resource materials are referenced in this chapter or are relevant to the subject matter addressed in this chapter.

ANSI Z 124.4-96, *Plastic Water Closet Bowls and Tanks*. New York, NY: American National Standards Institute, 1996.

ASME A 112.3.1-93, *Performance Standard and Installation Procedures for Stainless Steel Drainage Systems for Sanitary, Storm and Chemical Applications, Above and Below Ground*. New York, NY: American Society of Mechanical Engineers, 1993.

ASME A 112.3.4-00 (Reaffirmed 2004), *Macerating Toilet Systems and Related Components*. New York, NY: American Society of Mechanical Engineers, 2000.

ASME A 112.19.2M-2003, *Vitreous China Plumbing*. New York, NY: American Society of Mechanical Engineers, 2003.

ASME B1.20.1-83 (R2006), *Pipe Threads, General Purpose (inch)*. New York, NY: American Society of Mechanical Engineers, 2006.

ASTM A 74-04a, *Specification for Cast Iron Soil Pipe and Fittings*. West Conshohocken, PA: ASTM International, 2004.

ASTM A 888-04a, *Specification for Hubless Cast Iron Soil Pipe and Fittings for Sanitary and Storm Drain, Waste, and Vent Piping Application*. West Conshohocken, PA: ASTM International, 2004.

ASTM B 306-02, *Specification for Copper Drainage Tube (DWV)*. West Conshohocken, PA: ASTM International, 2002.

ASTM B 813-00e01, *Standard Specification for Liquid and Paste Fluxes for Soldering Applications of Copper and Copper Alloy Tube*. West Conshohocken, PA: ASTM International, 2001.

ASTM B 828-02, *Practice for Making Capillary Joints by Soldering of Copper and Copper Alloy Tube and Fittings*. West Conshohocken, PA: ASTM International, 2002.

ASTM C 14-03, *Specification for Concrete Sewer, Storm Drain and Culvert Pipe*. West Conshohocken, PA: ASTM International, 2003.

ASTM C 76-04a, *Specification for Reinforced Concrete Culvert, Storm Drain and Sewer Pipe*. West Conshohocken, PA: ASTM International, 2004.

ASTM C 425-04, *Specification for Compression Joints for Vitrified Clay Pipe and Fittings*. West Conshohocken, PA: ASTM International, 2004.

ASTM C 443-03, *Specification for Joints for Concrete Pipe and Manholes, Using Rubber Gaskets*. West Conshohocken, PA: ASTM International, 2003.

ASTM C 564-04a, *Specification for Rubber Gaskets for Cast Iron Soil Pipe and Fittings*. West Conshohocken, PA: ASTM International, 2004.

ASTM C 1053-00, *Specification for Borosilicate Glass Pipe and Fittings for Drain, Waste and Vent (DWV) Applications*. West Conshohocken, PA: ASTM International, 2000.

ASTM C 1173-02, *Specification for Flexible Transition Couplings for Underground Piping Systems*. West Conshohocken, PA: ASTM International, 2002.

ASTM C 1277-04, *Specification for Shielded Coupling Joining Hubless Cast Iron Soil Pipe and Fittings*. West Conshohocken, PA: ASTM International, 2004.

ASTM C 1460-04, *Specification for Shielded Transition Couplings for Use with Dissimilar DWV Pipe and Fittings Above Ground P3003.18*. West Conshohocken, PA: ASTM International, 2004.

ASTM C 1461-06, *Specification for Mechanical Couplings Using Thermoplastic Elastomeric (TPE) Gaskets for Joining Drain, Waste and Vent (DWV) Sewer, Sanitary and Storm Plumbing Systems for Above and Below Ground Use*. West Conshohocken, PA: ASTM International, 2006.

ASTM D 1869-95 (2000), *Specification for Rubber Rings for Asbestos-cement Pipe*. West Conshohocken, PA: ASTM International, 2000.

ASTM D 2235-01, *Specification for Solvent Cement for Acrylonitrile-Butadiene-Styrene (ABS) Plastic Pipe and Fittings*. West Conshohocken, PA: ASTM International, 2001.

ASTM D 2661-02, *Specification for Acrylonitrile-Butadiene-Styrene (ABS) Schedule 40 Plastic Drain, Waste and Vent Pipe and Fittings*. West Conshohocken, PA: ASTM International, 2002.

ASTM D 2665-04ae01, *Specification for Poly (VinylChloride) (PVC) Plastic Drain, Waste and Vent Pipe and Fittings*. West Conshohocken, PA: ASTM International, 2004.

ASTM D 2751-96a, *Specification for Acrylonitrile Butadiene-Styrene (ABS) Sewer Pipe and Fittings*. West Conshohocken, PA: ASTM International, 1996.

ASTM D 2855-96 (2002), *Standard Practice for Making Solvent-cemented Joints with Poly (Vinyl Chloride) (PVC) Pipe and Fittings*. West Conshohocken, PA: ASTM International, 2002.

ASTM D 2949-01a, *Specification for 3.25-in Outside Diameter Poly (Vinyl Chloride) (PVC) Plastic Drain, Waste and Vent Pipe and Fittings*. West Conshohocken, PA: ASTM International, 2001.

ASTM D 3034-04, *Specification for Type PSM Poly (Vinyl Chloride) (PVC) Sewer Pipe and Fittings*. West Conshohocken, PA: ASTM International, 2004.

ASTM D 3212-96a (2003)e01, *Specification for Joints for Drain and Sewer Plastic Pipes Using Flexible Elastomeric Seals*. West Conshohocken, PA: ASTM International, 2003.

ASTM D 3311-02, *Specification for Drain, Waste and Vent (DWV) Plastic Fitting Patterns*. West Conshohocken, PA: ASTM International, 2002.

ASTM F 477-02e01, *Specification for Elastomeric Seals (Gaskets) for Joining Plastic Pipe*. West Conshohocken, PA: ASTM International, 2002.

ASTM F 628-01, *Specification for Acrylonitrile-Butadiene-Styrene (ABS) Schedule 40 Plastic Drain, Waste and Vent Pipe with a Cellular Core*. West Conshohocken, PA: ASTM International, 2001.

ASTM F 656-02, *Specification for Primers for Use In Solvent Cement Joints of Poly (Vinyl Chloride) PVC Plastic Pipe and Fittings*. West Conshohocken, PA: ASTM International, 2002.

ASTM F 714-06a, *Specification for Polyethylene (PE) Plastic Pipe (SDR-PR) Based on Outside Diameter*. West Conshohocken, PA: ASTM International, 2006.

ASTM F 891-00e01, *Specification for Coextruded Poly (Vinyl Chloride) (PVC) Plastic Pipe with a Cellular Core*. West Conshohocken, PA: ASTM International, 2000.

ASTM F 1412-01e01, *Specification for Polyolefin Pipe and Fittings for Corrosive Waste Drainage*. West Conshohocken, PA: ASTM International, 2001.

ASTM F 1488-03, *Standard Specification for Coextruded Composite Pipe*. West Conshohocken, PA: ASTM International, 2003.

CISPI 301-04a, *Specification for Hubless Cast-Iron Soil Pipe and Fittings for Sanitary and Storm Drain, Waste and Vent Piping Applications*. Chattanooga, TN: Cast Iron Soil Pipe Institute, 2004.

CISPI 310-04, *Specification for Coupling for Use in Connection with Hubless Cast-Iron Soil Pipe and Fittings for Sanitary and Storm Drain Waste and Vent Piping Applications*. Chattanooga, TN: Cast Iron Soil Pipe Institute, 2004.

CSA A 257.1-92, *Circular Concrete Culvert, Storm Drain, Sewer Pipe and Fittings*. Rexdale, Ontario, Canada: Canadian Standards Association, 1992.

CSA A 257.2-92, *Reinforced Circular Concrete Culvert, Storm Drain, Sewer Pipe and Fittings*. Rexdale, Ontario, Canada: Canadian Standards Association, 1992.

CSA A 257.3-92, *Joints for Circular Concrete Sewer and Culvert Pipe, Manhole Sections, and Fittings Using Rubber Gaskets*. Rexdale, Ontario, Canada: Canadian Standards Association, 1992.

CSA B45.9-02, *Macerating Systems and Related Components*. Rexdale, Ontario, Canada: Canadian Standards Association, 2002.

CSA B 181.1-02, *ABS Drain, Waste and Vent Pipe and Pipe Fittings*. Rexdale, Ontario, Canada: Canadian Standards Association, 2002.

CSA B181.2-02, *PVC Drain, Waste, and Vent Pipe and Pipe Fittings—with Revisions Through December 1993*. Rexdale, Ontario, Canada: Canadian Standards Association, 2002.

CSA B 181.3-02, *Polyolefin Laboratory Drainage System*. Rexdale, Ontario, Canada: Canadian Standards Association, 2002.

CSA B 182.2-02, *PVC Sewer Pipe and Fittings (PSM Type)*. Rexdale, Ontario, Canada: Canadian Standards Association, 2002.

CSA B 182.4-02, *Profile PVC Sewer Pipe and Fittings*. Rexdale, Ontario, Canada: Canadian Standards Association, 2002.

CSA B 602-02, *Mechanical Couplings for Drain, Waste and Vent Pipe and Sewer Pipe*. Rexdale, Ontario, Canada: Canadian Standards Association, 2002.

IPSDC-09, *International Private Sewage Disposal Code*. Washington, DC: International Code Council, 2009.

Chapter 8:
Indirect/Special Waste

General Comments

An indirect waste pipe is a method of discharging waste from a fixture or appliance into the sanitary drainage system to prevent backflow and possible contamination. The waste pipe discharges through an air gap or air break into an open receptor or standpipe. Section 801 outlines the scope of the chapter, which is to indicate how to protect fixtures from possible contamination and to indicate how to dispose of special wastes properly. Section 802 requires most plumbing fixtures and appliances in food-handling establishments to discharge through an indirect waste. This eliminates a possible cause of food contamination from the sanitary drainage system. Section 802 also regulates the location of floor drains in food storage areas and waste water from swimming pools, pool filters and deck drains. Section 803 regulates special wastes that require treatment before being discharged to the sanitary drainage system, such as waste from a chemical laboratory.

Commonly referred to as "open site" or "safe waste," an indirect waste pipe is a means of protecting a fixture or appliance in the event of a backup of the sanitary drainage system. Sewage cannot back up into a fixture or appliance if a stoppage occurs downstream of the waste receptor receiving the discharge of an indirect waste pipe. Instead, sewage will spill onto the surrounding floor area. This protection is necessary when the contamination presents a health hazard or creates a potential cross connection with the potable water system. Note that an indirect waste pipe conveying waste from a fixture such as a sink, floor drain or swimming pool is part of the drainage system and, therefore, must conform to other applicable provisions of the code, including those found in Chapters 3 and 7. Condensate disposal piping must comply with Section 314 of this code which is identical to and controlled by Section 307 of the *International Mechanical Code*® (IMC®). This section requires condensate pipe to be cast iron, galvanized steel, copper, cross-linked polyethylene (PEX), polyethylene (PE), polybutylene (PB), acrylonitrile butadiene styrene (ABS), chlorinated polyvinyl (CPVC) or polyvinyl chloride (PVC). Condensate piping must be not less than $^3/_4$ inch (19.1 mm) internal diameter and sized in accordance with an approved method. Where condensate is generated by multiple refrigeration units, the minimum size of the piping carrying the combined flows must be in accordance with Table 314.2.2 of this code. A condensate drain is required to have a trap installed if required by the equipment or appliance manufacturer. Condensate drain traps are not required to be vented as is required for sanitary drainage traps.

Purpose

Chapter 8 regulates drainage installations requiring an indirect connection to the sanitary drainage system. Fixtures and plumbing appliances, such as those associated with food preparation or handling, health care facilities and potable liquids, must be protected from contamination that can result from connection to the drainage system. An indirect connection prevents sewage from backing up into a fixture or appliance, thus providing protection against potential health hazards. The chapter also regulates special wastes containing hazardous chemicals. The waste must be treated to prevent any damage to the piping or sewage treatment process.

SECTION 801
GENERAL

801.1 Scope. This chapter shall govern matters concerning indirect waste piping and special wastes. This chapter shall further control matters concerning food-handling establishments, sterilizers, clear-water wastes, swimming pools, methods of providing air breaks or air gaps, and neutralizing devices for corrosive wastes.

❖ This section outlines the scope of the chapter, which is to protect fixtures and appliances from possible contamination and to dispose of special wastes properly.

Indirect waste connections are necessary where a higher degree of health protection from sewage backup into the fixture or appliance is required because of the nature of its use. The chapter also has performance requirements for the neutralizing of special waste.

801.2 Protection. All devices, appurtenances, appliances and apparatus intended to serve some special function, such as sterilization, distillation, processing, cooling, or storage of ice or foods, and that discharge to the drainage system, shall be provided with protection against backflow, flooding, fouling, contamination and stoppage of the drain.

❖ This section requires protection of appliances and specialized equipment against contamination resulting from backflow. This section combines the intent of Sections 608, 609 and 802.

SECTION 802
INDIRECT WASTES

802.1 Where required. Food-handling equipment and clear-water waste shall discharge through an indirect waste pipe as specified in Sections 802.1.1 through 802.1.8. All health-care related fixtures, devices and equipment shall discharge to the drainage system through an indirect waste pipe by means of an *air gap* in accordance with this chapter and Section 713.3. Fixtures not required by this section to be indirectly connected shall be directly connected to the plumbing system in accordance with Chapter 7.

❖ This section establishes where an indirect waste pipe is required, as well as appropriate indirect waste methods and installation procedures. The requirements for indirect wastes are described in Sections 802.1.1 through 802.1.8. Health care plumbing fixtures, appliances and equipment that must discharge through an indirect waste pipe are included in Section 713.3. Fixtures that need not be indirectly connected by Chapter 8 must be directly connected to the plumbing system in accordance with Section 301.3.

802.1.1 Food handling. Equipment and fixtures utilized for the storage, preparation and handling of food shall discharge through an indirect waste pipe by means of an *air gap*.

❖ All food must be protected from possible contamination caused by the sanitary drainage system. The requirement for indirect waste connections extends to all storage, cooking and preparation equipment, including vegetable sinks, food washing sinks, refrigerated cases and cabinets, ice boxes, ice-making machines, steam kettles, steam tables, potato peelers, egg boilers, coffee urns and brewers, drink dispensers and similar types of equipment and fixtures [see Commentary Figures 802.1.1(1), 802.1.1(2) and 802.1.1(3)]. Food stored in walk-in coolers and freezers is specifically addressed in Section 802.1.2. Dishwashing sinks do not require an indirect waste connection because they are designed for use with contaminated water and do not come in contact with food. Furthermore, many of these fixtures in a food-handling establishment connect to a grease interceptor. If the discharge were through an indirect waste connection, the grease would tend to collect in the waste receptor, creating a health hazard. It is common, however, for local health officials to require such sinks to discharge through an indirect waste connection because sinks designated for dish, pot, pan and utensil washing are often used also for food preparation/washing. However, domestic and commercial dishwashing machines must discharge indirectly through an air gap or air break into a standpipe or waste receptor or as allowed by Section 802.1.6 (see Commentary Figure 802.1.6 or 802.1.7).

Occasionally, commercial kitchen designers will desire to mount a food waste grinder to a food preparation sink for ease of disposal of food prep wastes such as leaves, skins, spoiled produce, bones and meat trimmings. This creates several code conflicts. A food waste grinder is not a plumbing "fixture" that is required to be indirectly connected; therefore, it must be directly connected to the drain system. However, because the discharge of the food waste grinder is also the outlet for the food prep sink, this section requires the drain to be indirectly connected. The concern of the code for indirectly connecting drains of food prep sinks is to avoid contamination of food from a backup of the drainage system. Therefore, the code's intent is

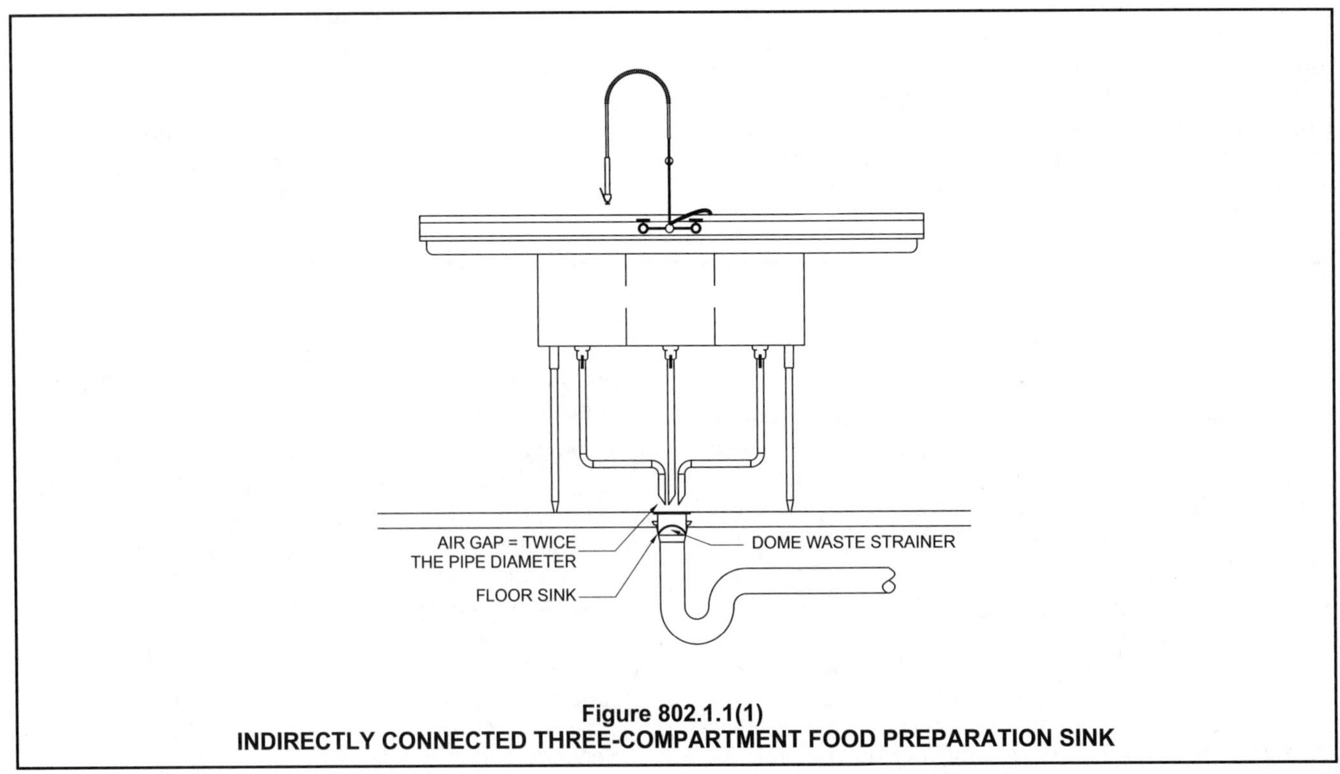

Figure 802.1.1(1)
INDIRECTLY CONNECTED THREE-COMPARTMENT FOOD PREPARATION SINK

to protect food from bacterial contamination. Without question, the grinding chamber of a food waste grinder will always contain significant bacteria regardless of any cleaning or sanitation practices. Although not directly stated in the code, the requirement for food waste grinders to be directly connected and the code's intent to protect food from contamination suggests that food waste grinders should not be installed on food prep sinks.

802.1.2 Floor drains in food storage areas. Floor drains located within walk-in refrigerators or freezers in food service and food establishments shall be indirectly connected to the sanitary drainage system by means of an *air gap*. Where a floor drain is located within an area subject to freezing, the waste line serving the floor drain shall not be trapped and shall indirectly discharge into a waste receptor located outside of the area subject to freezing.

Exception: Where protected against backflow by a backwater valve, such floor drains shall be indirectly connected to the sanitary drainage system by means of an *air break* or an *air gap*.

❖ This section is consistent with the intent of Section 802.1.1, which is to protect food stored in walk-in refrigerators and freezers from possible contamination caused by the sanitary drainage system. Floor drains located in these areas must be indirectly connected to the sanitary drainage system through an air gap. The exception allows for floor drains in refrigerators and

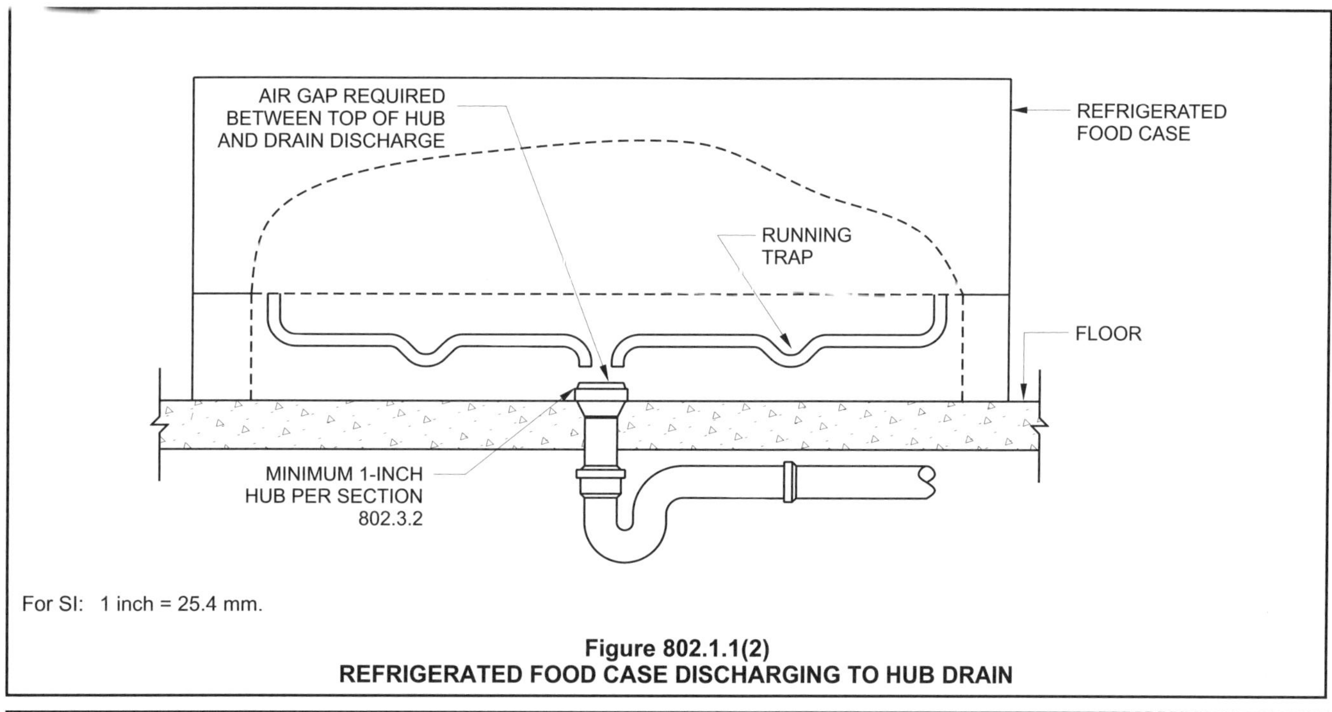

Figure 802.1.1(2)
REFRIGERATED FOOD CASE DISCHARGING TO HUB DRAIN

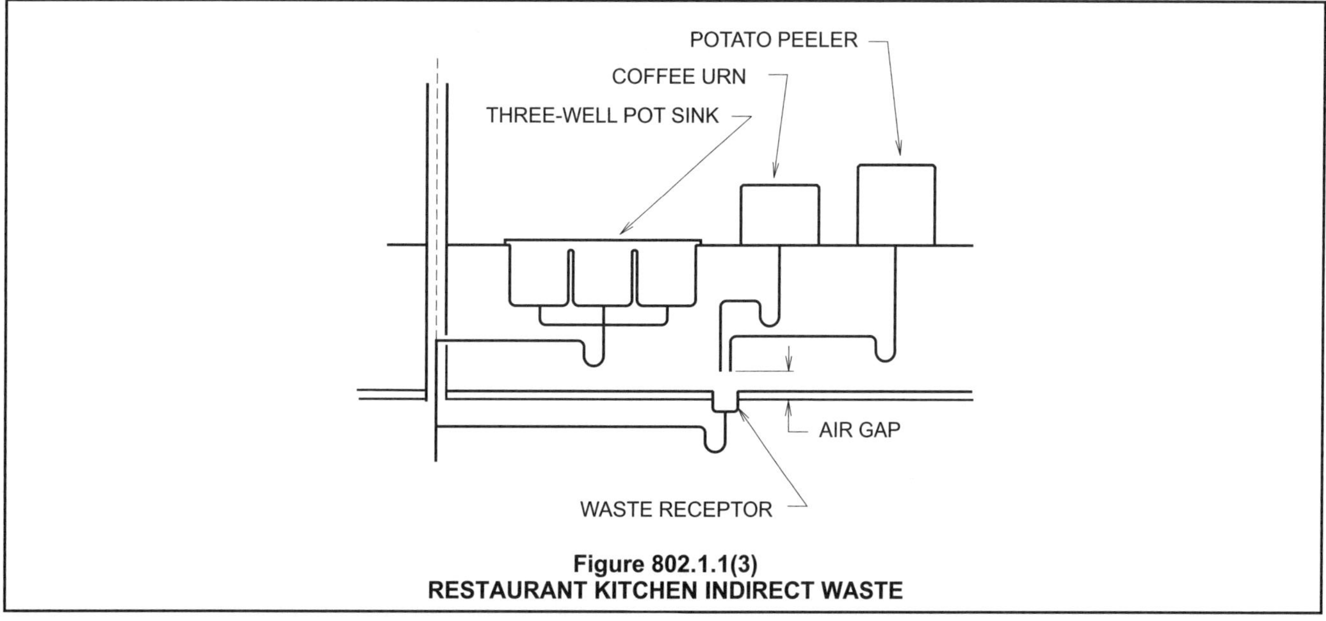

Figure 802.1.1(3)
RESTAURANT KITCHEN INDIRECT WASTE

freezers to connect by means of an air break. This maintains backflow protection while providing an indirect waste connection option for large food storage areas where it is impractical to elevate the cooler/freezer floor to a level necessary to achieve an air gap (see Commentary Figure 802.1.2). Section 305.6 prohibits locating traps in areas subject to freezing, and Section 802.2 requires indirect waste piping to be trapped where the length limit is exceeded.

802.1.3 Potable clear-water waste. Where devices and equipment, such as sterilizers and relief valves, discharge potable water to the building drainage system, the discharge shall be through an indirect waste pipe by means of an *air gap*.

❖ Clear-water waste includes all of the various condensate drains plus the other devices and systems listed. Clear-water wastes are not required to discharge to the sanitary drainage system. However, where they do, an indirect waste connection by means of an air gap is required. An air gap prevents possible cross connection between a potable water source and the drain system (see Commentary Figure 802.1.3). Indirect waste pipes from potable water sources should not be connected together (manifolded) as the discharge from one source might contaminate the other source of potable water. For example, the connection of two separate ice storage bin indirect waste pipes would be prohibited. Each must discharge independently to the waste receptor.

802.1.4 Swimming pools. Where wastewater from swimming pools, backwash from filters and water from pool deck drains

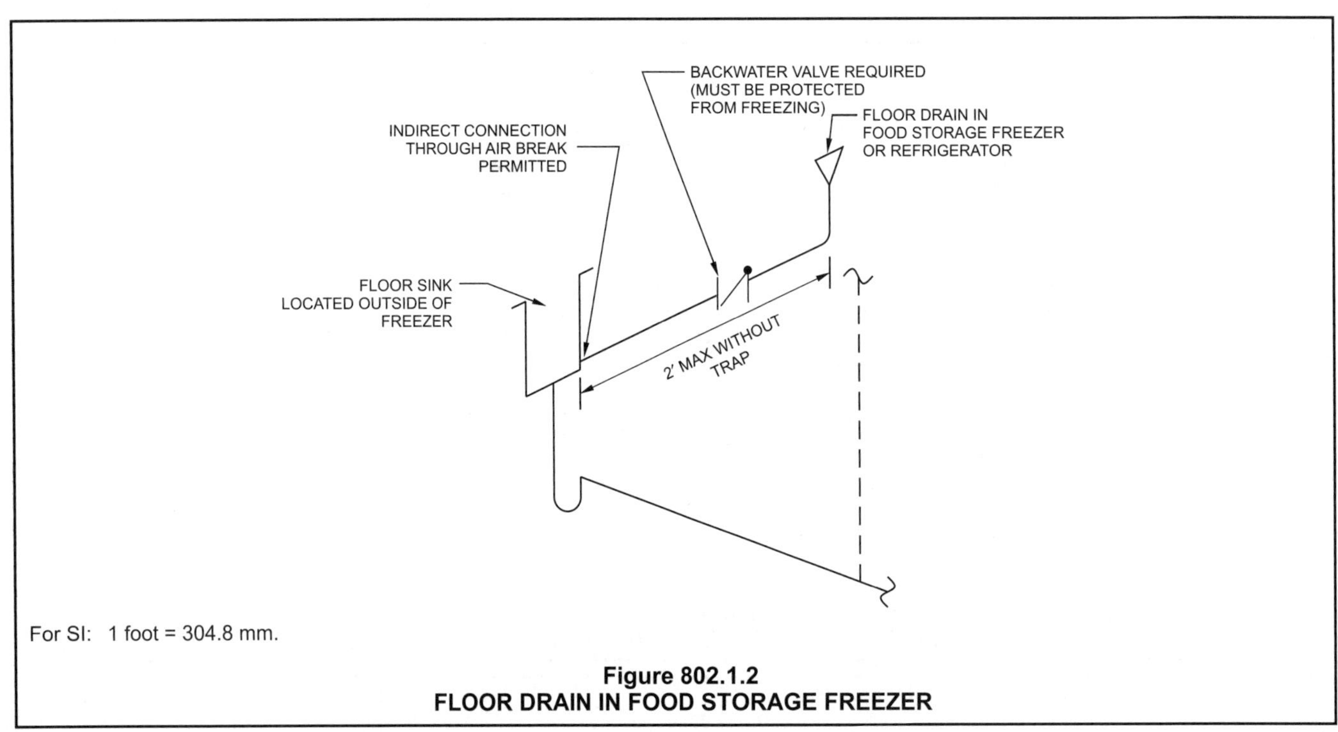

Figure 802.1.2
FLOOR DRAIN IN FOOD STORAGE FREEZER

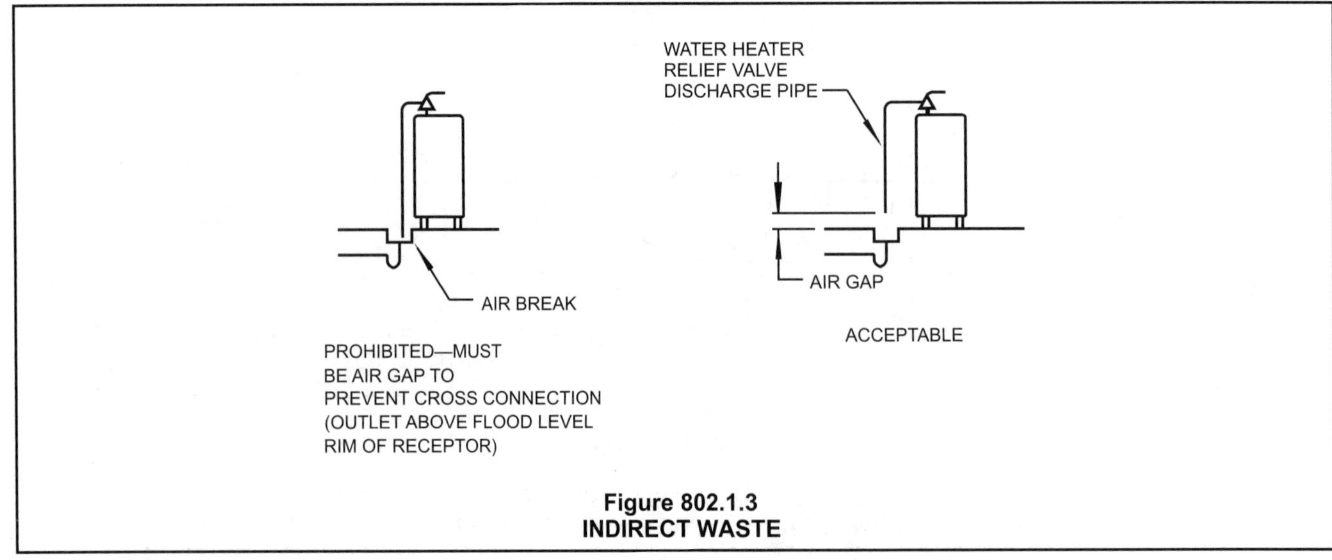

Figure 802.1.3
INDIRECT WASTE

discharge to the building drainage system, the discharge shall be through an indirect waste pipe by means of an *air gap*.

❖ This section addresses the method of discharge where a swimming pool drain connects to the drainage system. An indirect waste connection by means of an air gap is required.

802.1.5 Nonpotable clear-water waste. Where devices and equipment such as process tanks, filters, drips and boilers discharge nonpotable water to the building drainage system, the discharge shall be through an indirect waste pipe by means of an *air break* or an *air gap*.

❖ Where there is no possibility of contaminating the potable water (nonpotable discharge), the indirect waste may be by means of an air gap or air break. An air break is often preferred to reduce any splashing that may occur. This section does not require the devices and equipment listed to discharge to the drainage system; it only indicates the method of discharge if they do connect (see Commentary Figure 802.1.5).

802.1.6 Domestic dishwashing machines. Domestic dishwashing machines shall discharge indirectly through an *air gap* or *air break* into a standpipe or waste receptor in accordance with Section 802.2, or discharge into a wye-branch fitting on the tailpiece of the kitchen sink or the dishwasher connection of a food waste grinder. The waste line of a domestic dishwashing machine discharging into a kitchen sink tailpiece or food waste grinder shall connect to a deck-mounted *air gap* or the waste line shall rise and be securely fastened to the underside of the sink rim or counter.

❖ Dishwashing machines must discharge indirectly to the drainage system. An indirect connection by air gap or air break is required. The waste line must be looped as high as possible and securely fastened to the underside of the sink rim or countertop to minimize the potential for waste backflow into the dishwasher cabinet (see Commentary Figure 802.1.6).

802.1.7 Commercial dishwashing machines. The discharge from a commercial dishwashing machine shall be through an *air gap* or *air break* into a standpipe or waste receptor in accordance with Section 802.2.

❖ The waste from a commercial dishwashing machine must discharge by means of an air gap or air break to prevent sewage from backing up into the dishwashing compartment in the event of a stoppage in the drainage system (see Commentary Figure 802.1.7). In accordance with ASSE 1004, the minimum drain connection is $1^1/_2$ inches (38 mm) nominal size.

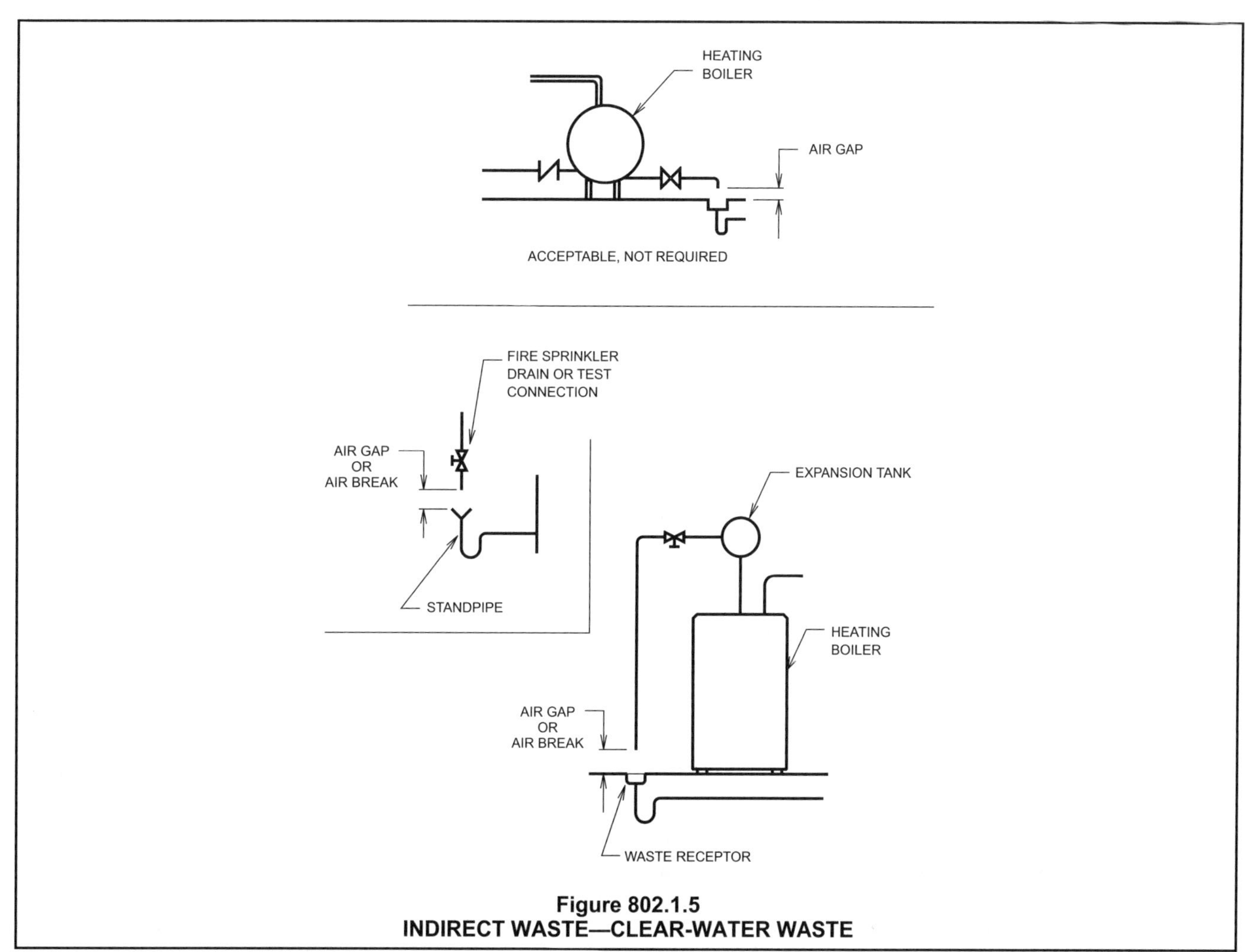

Figure 802.1.5
INDIRECT WASTE—CLEAR-WATER WASTE

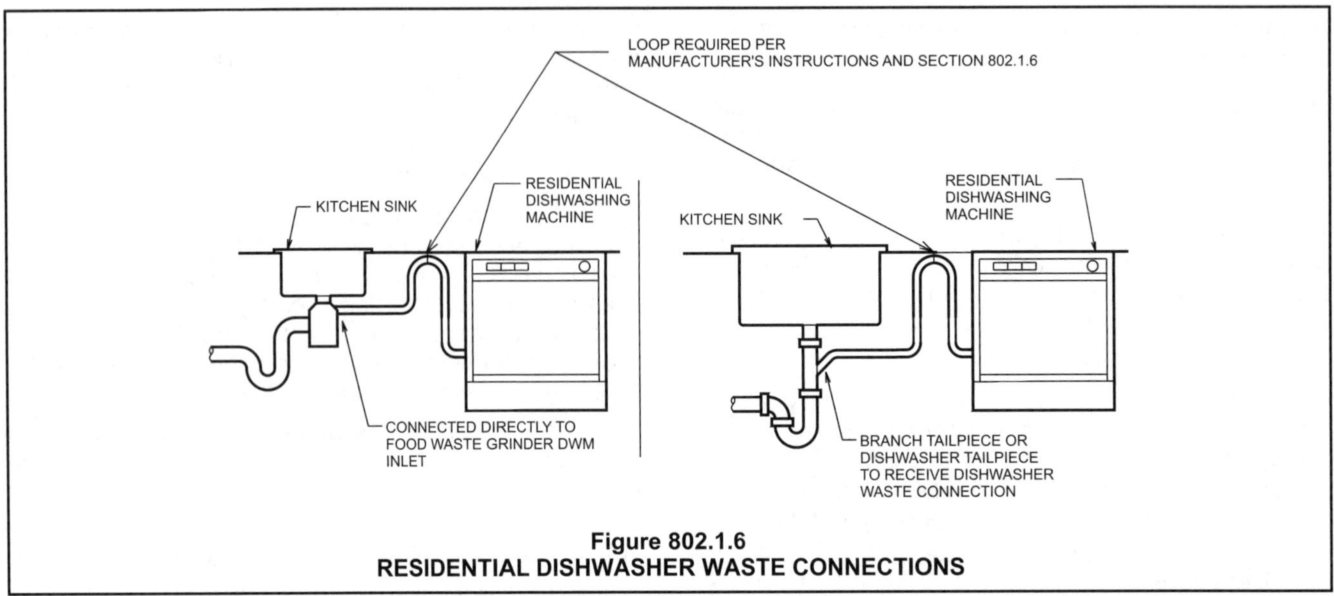

Figure 802.1.6
RESIDENTIAL DISHWASHER WASTE CONNECTIONS

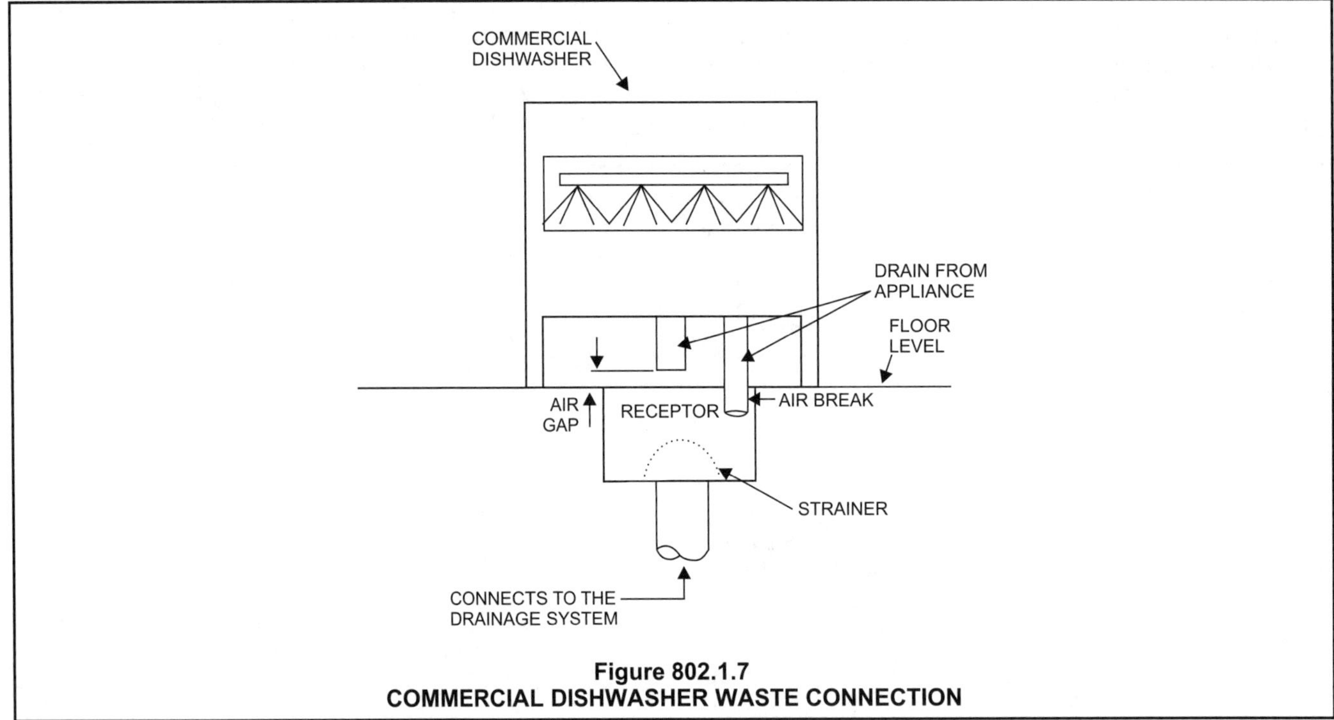

Figure 802.1.7
COMMERCIAL DISHWASHER WASTE CONNECTION

802.1.8 Food utensils, dishes, pots and pans sinks. Sinks used for the washing, rinsing or sanitizing of utensils, dishes, pots, pans or serviceware used in the preparation, serving or eating of food shall discharge indirectly through an *air gap* or an *air break* or directly connect to the drainage system.

❖ It is not uncommon for commercial kitchen pot and pan sinks to indirectly connect to the sanitary drainage system in order to meet local health department regulations. However, prior to the 2009 edition, if the sink was not a food preparation sink, the sink was required to be directly connected to the sanitary drainage system. The addition to this section allows for pot, pan and serviceware washing sinks to be indirectly connected through an air gap or air break. Indirect connection of such sinks offers protection from contamination where the sinks are used for both warewashing and food preparation.

802.2 Installation. All indirect waste piping shall discharge through an *air gap* or *air break* into a waste receptor or standpipe. Waste receptors and standpipes shall be trapped and vented and shall connect to the building drainage system. All indirect waste piping that exceeds 2 feet (610 mm) in *developed length* measured horizontally, or 4 feet (1219 mm) in total *developed length*, shall be trapped.

❖ To minimize bacterial growth and odor in indirect waste piping, a trap is required when the piping exceeds 2 feet (610 mm) in developed length measured horizontally or 4 feet (1219 mm) in total developed length. The trap requirement is intended to apply only to fixtures or devices that are open to the room atmosphere. It does not apply to an indirect waste pipe making a direct connection to a device, such as a water heater relief valve. A trapped relief valve discharge pipe is prohibited because it creates a restriction, causes backpressure to develop and could possibly freeze, creating a blockage. The trap prescribed by this section does not require a vent because such trap does not serve the same purpose as other traps in the drainage system (see Commentary Figure 802.2).

802.2.1 Air gap. The *air gap* between the indirect waste pipe and the *flood level rim* of the waste receptor shall be a minimum of twice the *effective opening* of the indirect waste pipe.

❖ An air gap is a drainage pipe that terminates above the flood level rim of the waste receptor. Any stoppage downstream of the waste receptor will result in waste overflowing the receptor without submerging the indirect waste pipe outlet. This type of indirect waste connection offers the highest level of protection by not allowing any possibility for backsiphoning of waste or sewage [see Commentary Figures 802.2.1]. An air gap is always acceptable where an air break is specified.

802.2.2 Air break. An *air break* shall be provided between the indirect waste pipe and the trap seal of the waste receptor or standpipe.

❖ The indirect waste pipe is allowed to extend below the flood level rim, but it must terminate above the seal of the trap serving the waste receptor [see Commentary Figures 802.2.2(1) and 802.2.2(2)]. An air gap is always acceptable where an air break is specified.

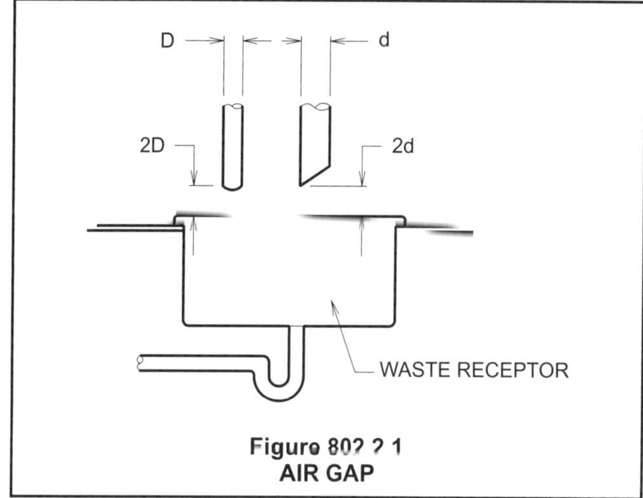

Figure 802.2.1
AIR GAP

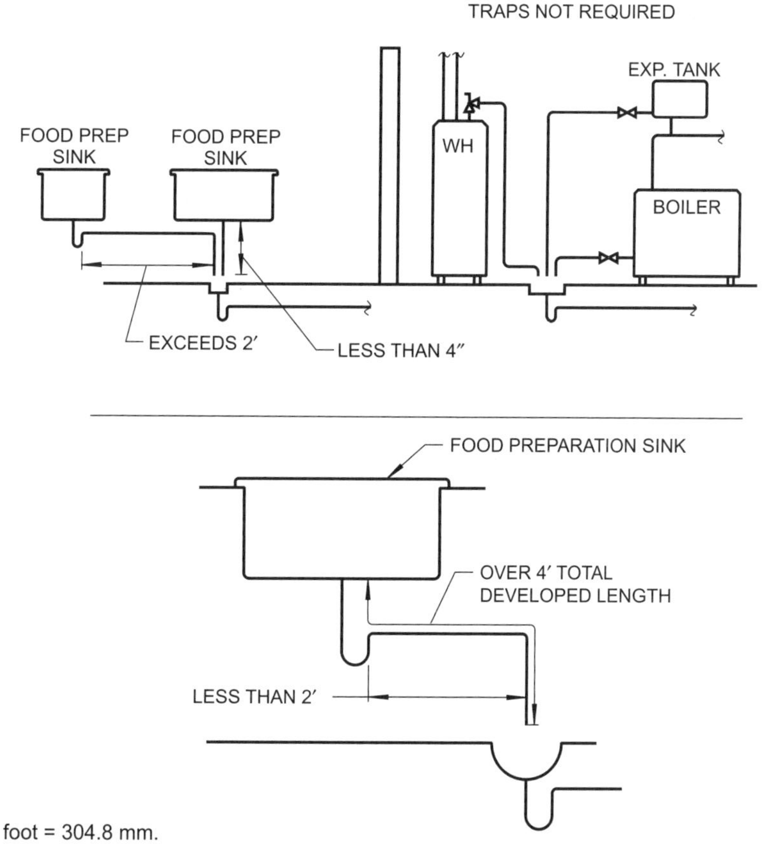

For SI: 1 inch = 25.4 mm, 1 foot = 304.8 mm.

Figure 802.2
TRAPS FOR INDIRECT WASTE PIPES

802.3 Waste receptors. Every waste receptor shall be of an *approved* type. A removable strainer or basket shall cover the waste outlet of waste receptors. Waste receptors shall be installed in ventilated spaces. Waste receptors shall not be installed in bathrooms or toilet rooms or in any inaccessible or unventilated space such as a closet or storeroom. Ready *access* shall be provided to waste receptors.

❖ There are no specific standards regulating waste receptors. They may be identified by various names, including open hub drain, floor sink and waste sink. A floor drain can be considered to be a waste receptor where an indirect waste pipe is directed into it or where it is fitted with a means (such as a funnel attachment) to receive an indirect waste pipe. Otherwise, a floor drain is a plumbing fixture intended to drain a floor surface and is not subject to the code requirements for waste receptors.

Floor sinks are specialized sinks designed to be installed in the floor. Manufacturers produce a special line of fixtures designed to be used as waste receptors, some of which look like kitchen sinks. A laundry sink is commonly used as a waste receptor to receive the discharge of an automatic clothes washer [see Commentary Figures 802.3(1) and 802.3(2)]. Waste receptors must be located to allow adequate ventilation and air circulation, minimize misuse, prevent a tripping or walking hazard and allow maintenance.

802.3.1 Size of receptors. A waste receptor shall be sized for the maximum discharge of all indirect waste pipes served by the receptor. Receptors shall be installed to prevent splashing or flooding.

❖ Sizing a waste receptor is as difficult as sizing a drainage system. There must be a determination of the probability of simultaneous discharge of the various fixtures connecting to the same waste receptor. The receptor must be capable of accepting all of the discharge without overflowing. The code does not specify a method for sizing receptors; however, it would be logical to size the drain for the receptor based on the total drainage fixture units (dfu's) that discharge to the receptor. While accepting the discharge of an indirect waste pipe, the receptor is simultaneously discharging to the drainage system at a rate based on the size of its drain pipe.

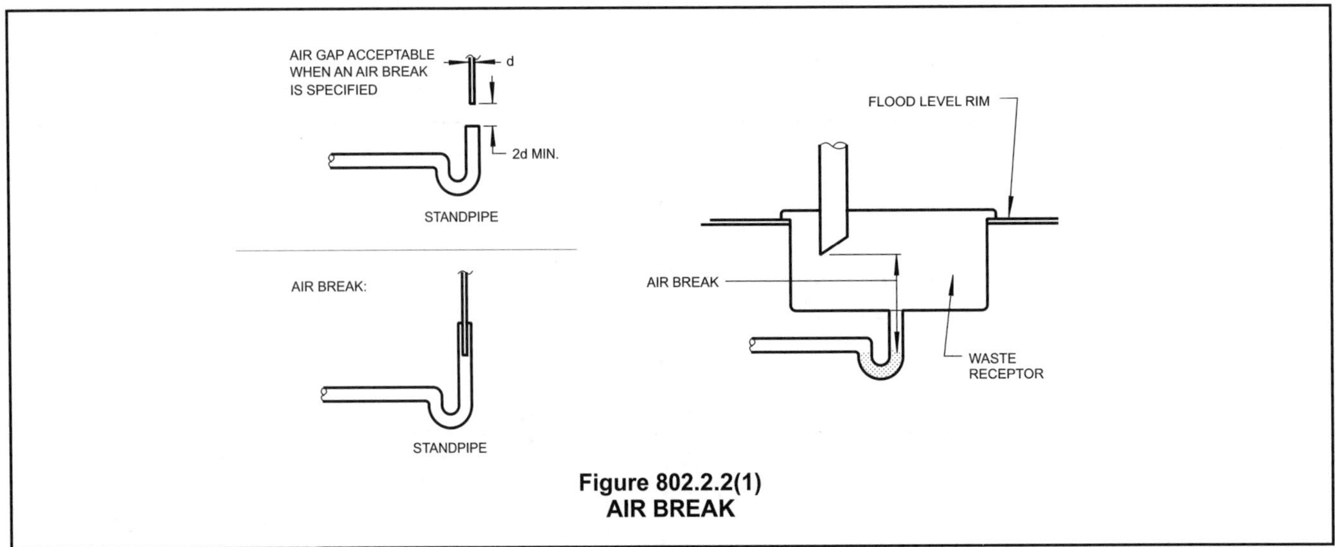

Figure 802.2.2(1)
AIR BREAK

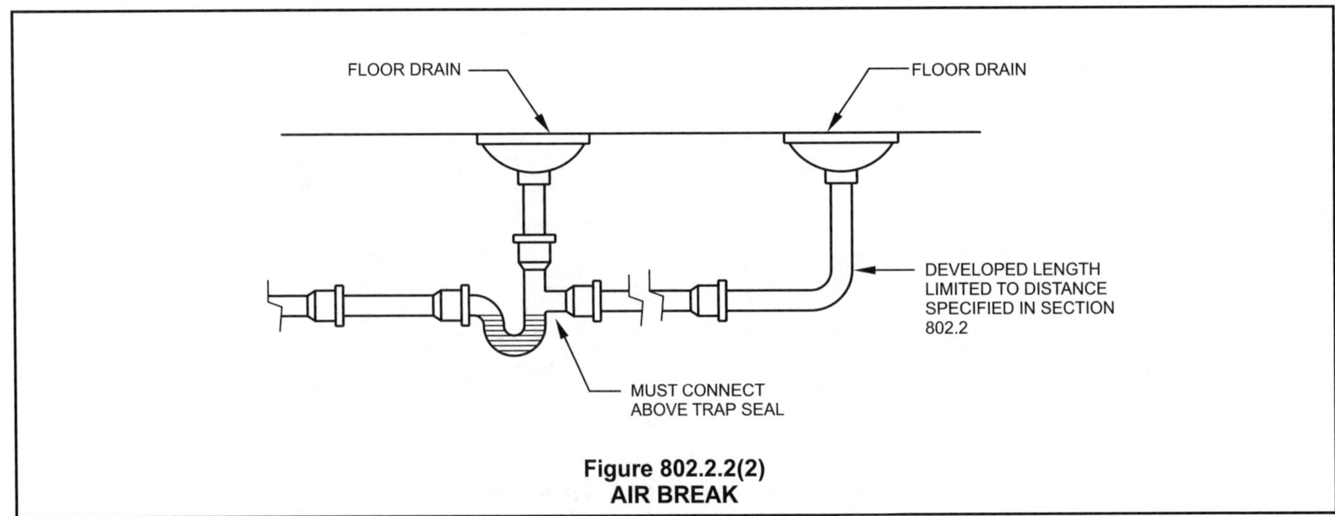

Figure 802.2.2(2)
AIR BREAK

Two major problems in the design and sizing of a waste receptor are splashing and flooding. Both must be minimized to avoid creating an insanitary condition. Both the shape of the waste receptor and the discharge rate of the indirect waste pipe affect the amount of splashing and flooding involved. A shallow, flat bottom receptor receiving a high rate of discharge will splash to the exterior of the receptor.

The depth and bowl shape may have to be selected to reduce splashing. Another means of reducing splashing into a waste receptor is by terminating the indirect waste pipe with a diagonally cut end. The lower section of the diagonal cut is faced to the closest exterior wall of the waste receptor. If the waste receptor receives only the discharge of emergency drains, such as a temperature and pressure relief valve, splashing is of little concern. If the receptor is used only during an emergency, not on an everyday basis, such limited use will not create any major health hazard [see Commentary Figures 802.3.1(1) and 802.3.1(2)].

A high rate of discharge also increases a waste receptor's susceptibility to flooding. The designer should pay particular attention to receptor depth, trap sizing and the free area of the drain's protective grating. As is often the case, the waste receptor trap size restricts the discharge flow rate of the fixture. A close coordination of the sizing of the trap, rate of discharge, receptor depth and drain selection will circumvent an insanitary flooding condition.

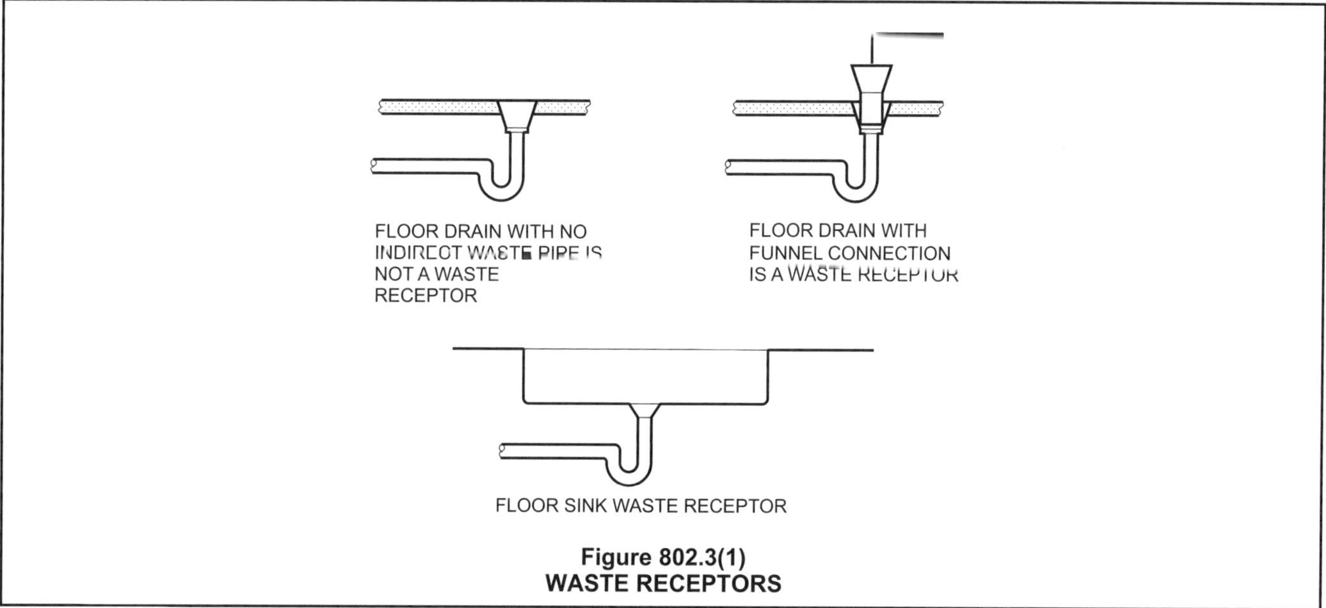

Figure 802.3(1)
WASTE RECEPTORS

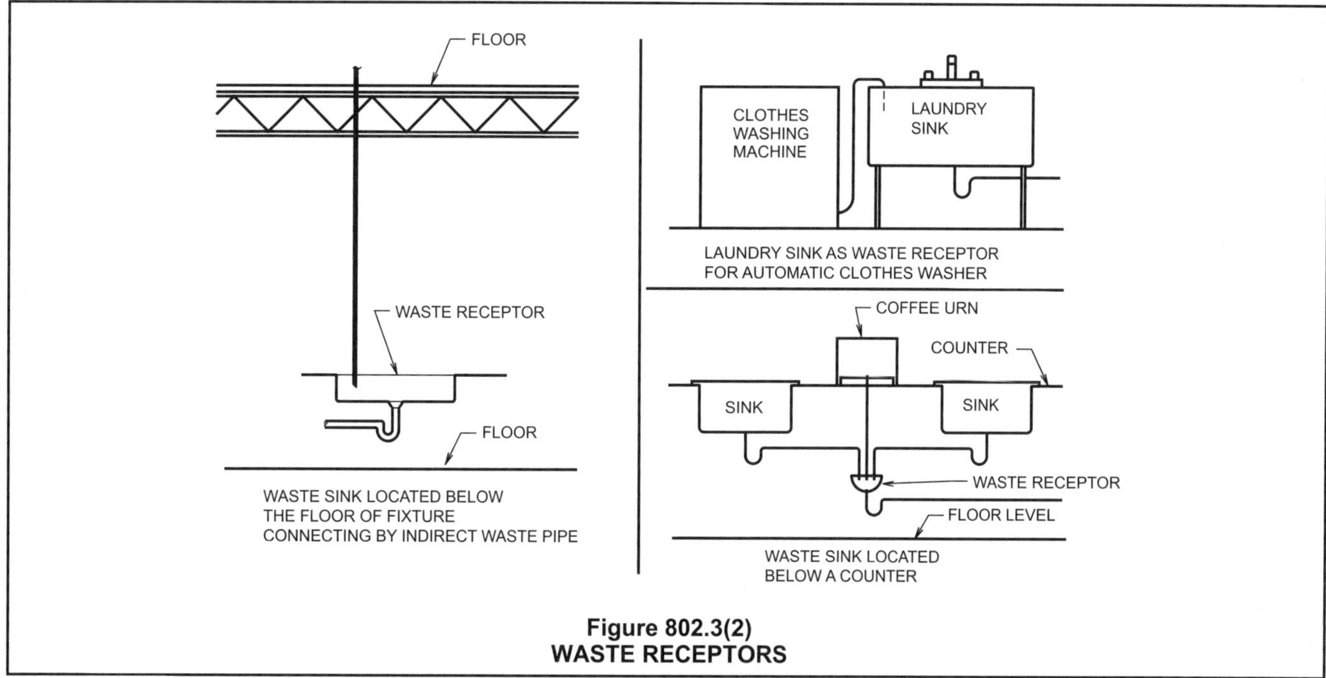

Figure 802.3(2)
WASTE RECEPTORS

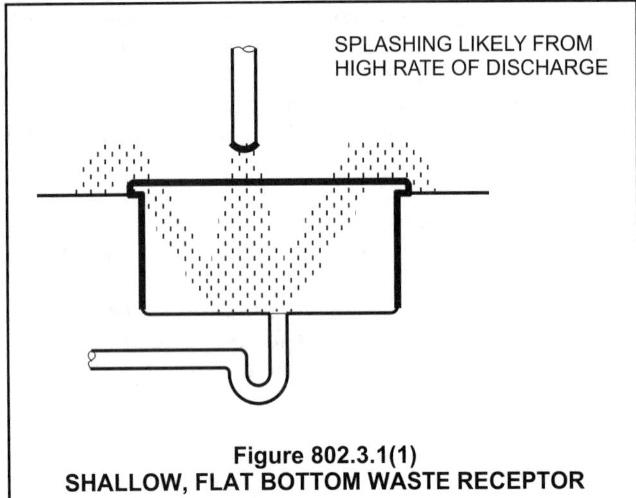

Figure 802.3.1(1)
SHALLOW, FLAT BOTTOM WASTE RECEPTOR

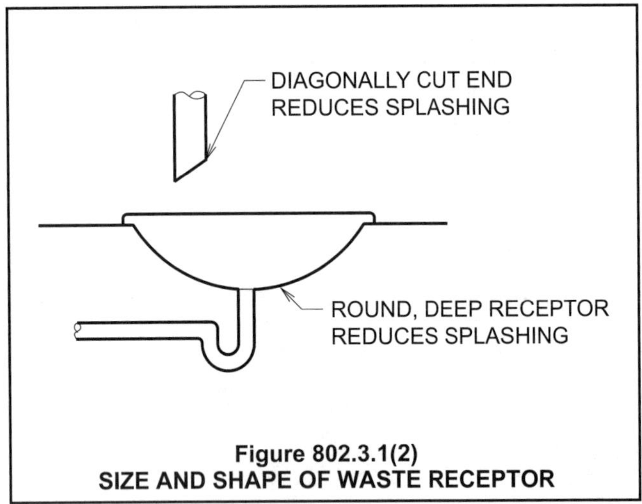

Figure 802.3.1(2)
SIZE AND SHAPE OF WASTE RECEPTOR

802.3.2 Open hub waste receptors. Waste receptors shall be permitted in the form of a hub or pipe extending not less than 1 inch (25.4 mm) above a water-impervious floor and are not required to have a strainer.

❖ An open hub drain is a common method of providing a waste receptor in water-impervious floors. It eliminates the need to provide a separate fixture with a strainer. A hubbed drain pipe or plain-end section of pipe serves as the receptor for receiving indirect waste. Because no strainer is required, an open hub drain is restricted to clear-water waste without solids. Because a hub drain does have the ability to exclude solids and debris, the pipe or hub is required to be a minimum of 1 inch (25 mm) above the floor to prevent the drain from serving as a floor drain (see Commentary Figure 802.3.2).

802.4 Standpipes. Standpipes shall be individually trapped. Standpipes shall extend a minimum of 18 inches (457 mm) and a maximum of 42 inches (1066 mm) above the trap weir. *Access* shall be provided to all standpipes and drains for rodding.

❖ A standpipe is a type of indirect waste receptor. The minimum dimension for the height of the receptor is in-

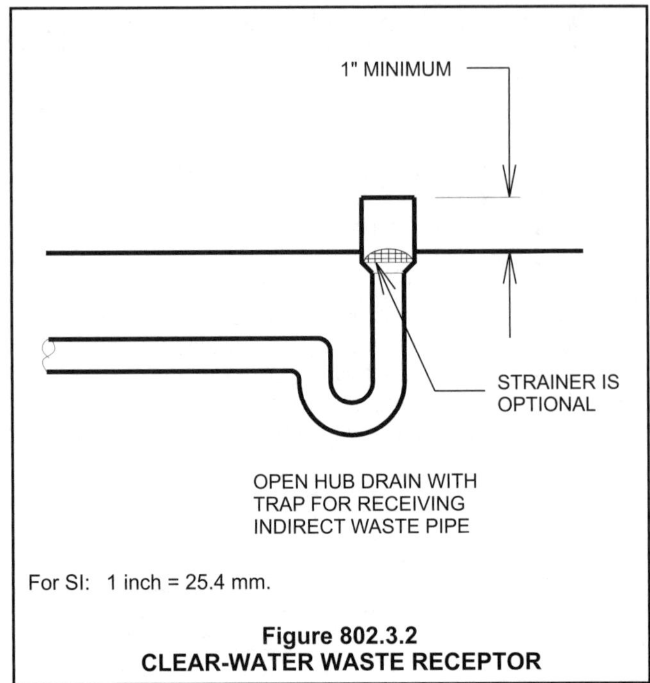

Figure 802.3.2
CLEAR-WATER WASTE RECEPTOR

tended to provide minimal retention capacity and head pressure to prevent an overflow. This is especially necessary when an indirect waste pipe receives a high rate of discharge, such as the pumped discharge of a clothes washer (see Commentary Figure 802.4). The maximum height requirement is intended to control the waste flow velocity at the trap inlet. Excessive inlet velocity can promote trap siphonage.

SECTION 803
SPECIAL WASTES

803.1 Wastewater temperature. Steam pipes shall not connect to any part of a drainage or plumbing system and water above 140°F (60°C) shall not be discharged into any part of a drainage system. Such pipes shall discharge into an indirect waste receptor connected to the drainage system.

❖ Special wastes contain higher concentrations of corrosive, radioactive or toxic materials than ordinarily encountered in the building drainage system. Waste water with a temperature in excess of 140°F (60°C) is also considered as a special waste. This section establishes where special waste systems are required and how they are to be designed.

The temperature of wastewater is limited to a maximum of 140°F (60°C) to protect the sanitary drainage piping. Such elevated temperatures may increase erosion for all pipe materials. For example, ABS and PVC plastic pipe may soften. Dishwashing machines are sometimes singled out as having elevated wastewater temperature. Quite often, the hot water (sanitizing rinse) is supplied at a temperature of 180°F (82°C). Although initially above the threshold temperature of 140°F(60°C), after the washing or rinsing cycle, the water is usually reduced to a temperature that will not

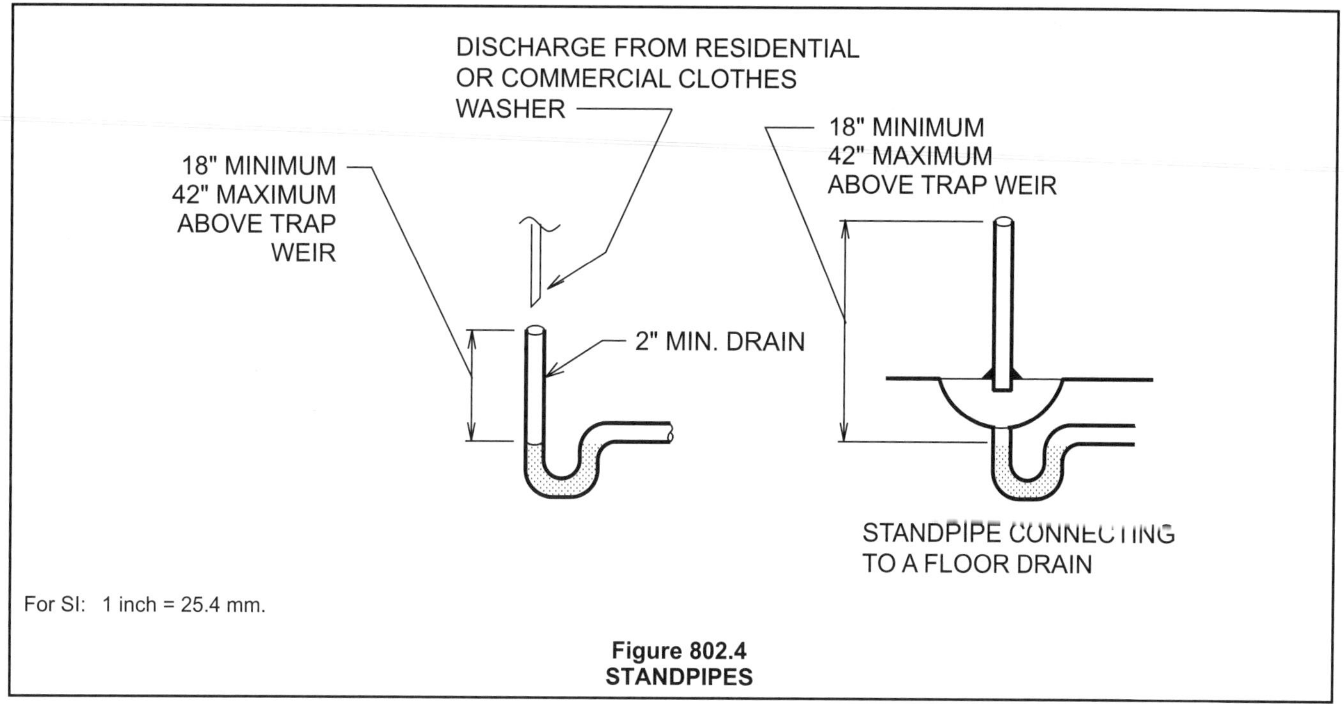

Figure 802.4
STANDPIPES

adversely affect the system piping. If waste temperatures are too high [above 140°F (60°C)], the waste must be cooled or special piping material must be installed to receive the waste. The special piping material installation must be approved as an alternative material by the code official. Typically, waste retention tanks are used as the means of cooling waste.

803.2 Neutralizing device required for corrosive wastes. Corrosive liquids, spent acids or other harmful chemicals that destroy or injure a drain, *sewer*, soil or waste pipe, or create noxious or toxic fumes or interfere with sewage treatment processes shall not be discharged into the plumbing system without being thoroughly diluted, neutralized or treated by passing through an *approved* dilution or neutralizing device. Such devices shall be automatically provided with a sufficient supply of diluting water or neutralizing medium so as to make the contents noninjurious before discharge into the drainage system. The nature of the corrosive or harmful waste and the method of its treatment or dilution shall be *approved* prior to installation.

❖ When deleterious waste is suspected, it must be completely analyzed for any detrimental effects on the sanitary drainage system, public sewer or sewage treatment facility. A special waste system must treat any deleterious waste and reduce the threat to an acceptable level before it is discharged to the sanitary drainage system. The code does not specify what an "acceptable level" is because it varies by pipe material and type of waste. Special waste systems are associated with atypical plumbing systems, including laboratories, chemical plants, processing plants and hospitals. They may be necessary, however, when emergency shower systems are installed for use by persons exposed to harmful chemicals or nuclear material.

803.3 System design. A chemical drainage and vent system shall be designed and installed in accordance with this code. Chemical drainage and vent systems shall be completely separated from the sanitary systems. Chemical waste shall not discharge to a sanitary drainage system until such waste has been treated in accordance with Section 803.2.

❖ When a chemical drainage system is installed, the drainage and vent piping must still conform to all code requirements. The pipe sizing and installation would have to conform to the requirements of Chapters 7 and 9 and other chapters as applicable. Both the drainage and vent piping must be completely separate systems. Although it is understood that the drainage piping must be separate until the waste is treated or neutralized, it is always questionable whether a separate system is necessary for a vent. The reason for a completely separate vent system is the uncertainty of the nature of fumes and vapors in a chemical waste system. The gases in the chemical drain, waste and vent (DWV) system could be incompatible with the piping materials and gases in the normal DWV system. Because the effect of combining the systems is unknown, such venting systems are kept separate.

SECTION 804
MATERIALS, JOINTS AND CONNECTIONS

804.1 General. The materials and methods utilized for the construction and installation of indirect waste pipes and systems shall comply with the applicable provisions of Chapter 7.

❖ This section references all of the acceptable materials and methods described in Chapter 7 that are used specifically for the installation and joining of indirect waste pipes and systems.

Bibliography

The following resource material is referenced in this chapter or is relevant to the subject matter addressed in this chapter.

IMC-09, *International Mechanical Code*. Washington, DC: International Code Council, Inc. 2009.

Chapter 9: Vents

General Comments

In the most common method of venting, each trap or trapped fixture is provided with a separate or individual vent, which then connects to the building's main venting system. The minimum size of any vent is one-half of the required size of the drain pipe, but not less than $1^{1}/_{4}$ inches (32 mm) (see commentary, Section 916.2).

Other methods of venting include common venting (Section 908), wet venting (Section 909), waste stack venting (Section 910), circuit venting (Section 911), combination drain and vent (Section 912) and island fixture venting (Section 913).

A common vent is a single vent pipe that vents more than one fixture. It functions as an individual vent for each fixture. Wet venting is a method of venting single or double bathroom groups or combinations thereof. One vent pipe may function as a vent for all of the fixtures connected to the wet vent. Circuit venting is the venting of a battery of up to eight fixtures with a single vent pipe. A combination drain and vent system is restricted to floor drains, sinks, drinking fountains and lavatories. The system relies on the oversizing of the drain pipe. Island fixture venting is a method of individual or common venting in which the vent is installed below the flood level rim of the fixture before rising to connect to another vent or vent terminal. A waste stack vent is a method of venting individual fixtures through a drainage stack. The oversized stack functions as a vent.

Other areas addressed in Chapter 9 include stack requirements (Section 903), vent terminals and frost closure (Section 904), vent grade (Section 905), trap to vent distance limitations (Section 906) and relief vents (Section 914).

Some have related a vent system to the open/closed straw phenomenon. For example, imagine that a soda straw is inserted into a glass of water. A thumb is placed over the open end of the straw, and the straw is raised above the glass. Water remains in the straw until the thumb is removed. The claim is made that without a vent, the drainage system would not properly operate—just as the straw did not with the thumb over the top. (This analogy is not completely accurate because a drainage system is not flowing completely full as is the soda straw; therefore, a column of water could not be suspended in this manner.) Drainage systems are designed to flow partially filled.

The vent system reduces the pressure differentials in the drainage system, thus protecting the trap seals from positive pressures, self-siphonage and induced siphonage that may result in complete or partial loss of the seals.

To better understand venting methods and terminology, all venting methods may be conceptually categorized as either dry vents or wet vents. Vent stacks, stack vents, branch vents, relief vents, island vents and individual vents are dry vents. Wet vents, circuit vents, combination drain and vents, waste stack vents and different level common vents are versions of "wet venting," in which the vent is wetted by drainage flow.

Purpose

The purpose of venting is to protect the trap seal of each trap. The vents are designed to maintain maximum differential pressures at each trap of not more than 1 inch of water column (249 Pa). Pressure fluctuations are created by waste flow in the drainage system. Venting is not intended to provide for the circulation of air within the drainage system. If there were no traps in a drainage system, venting would not be required. The system would function adequately because it would be open to the atmosphere at the fixture connections, thereby allowing airflow.

Chapter 9 covers the requirements for vents and venting. Knowing why venting is required makes it easier to understand the intent of this chapter. Venting protects every trap against the loss of its seal. Provisions set forth in this chapter are geared toward limiting the pressure differentials in the drainage system to a maximum of 1 inch of water column (249 Pa) above or below atmospheric pressure (i.e., positive or negative pressures)

SECTION 901 GENERAL

901.1 Scope. The provisions of this chapter shall govern the materials, design, construction and installation of vent systems.

❖ Chapter 9 describes a variety of methods to vent plumbing fixtures and traps. The methods have been laboratory tested to determine sizing and installation requirements that provide proper venting to a drainage system. The venting methods have also been field tested, establishing a long history of satisfactory service.

901.2 Trap seal protection. The plumbing system shall be provided with a system of vent piping that will permit the admission or emission of air so that the seal of any fixture trap shall not be subjected to a pneumatic pressure differential of more than 1 inch of water column (249 Pa).

❖ Protection of the trap seal means that during the normal operation of the plumbing system, the water seal remains in the trap. Each trap has a minimum of a 2-inch (51 mm) trap seal depth translating to a hydrostatic pressure equal to a 2-inch water column (498 Pa). If exposed to a 1-inch water column (249 Pa) pressure differential, a 1-inch (25 mm) water seal remains in the trap. The vent methods identified in this chapter are intended to limit the pressure differential at trap seals to 1 inch of water column (249 Pa) or less.

Pressure fluctuations downstream of the trap caused by certain flow conditions can occur. High positive pressures could force the trap seal back into the fixture, possibly resulting in water spilling onto the floor from fixtures such as floor drains. This will also result in sewer gases being pushed through the trap seal. For example, bubbles rising through the trap seal of a water closet are an indication of excessive positive pressure on the drainage side of the trap.

When a trap seal is exposed to a lower pressure on the drainage side of the trap, the water seal will rise and flow over the trap weir and into the drainage system. This will result in a complete or partial loss of the trap water seal. The lower pressure on the drainage side of the trap (also called "negative pressure" or "partial vacuum") causes loss of trap seal that is proportional to the amount of negative pressure. Such a pressure differential must never exceed 1 inch of water column (249 Pa) [see Commentary Figures 901.2(1) and 901.2(2)].

Section 1002.4 contains additional provisions to protect a trap seal that is subject to loss by evaporation, which might occur where trap seals do not regularly receive waste.

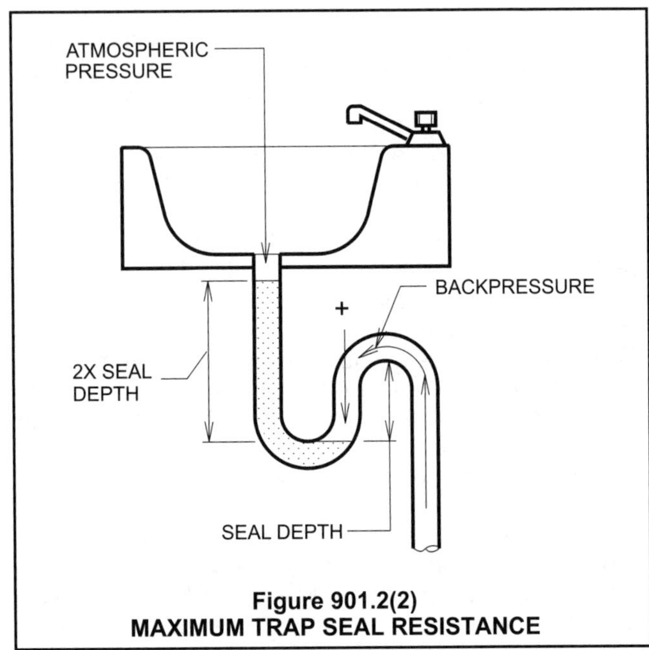

Figure 901.2(2)
MAXIMUM TRAP SEAL RESISTANCE

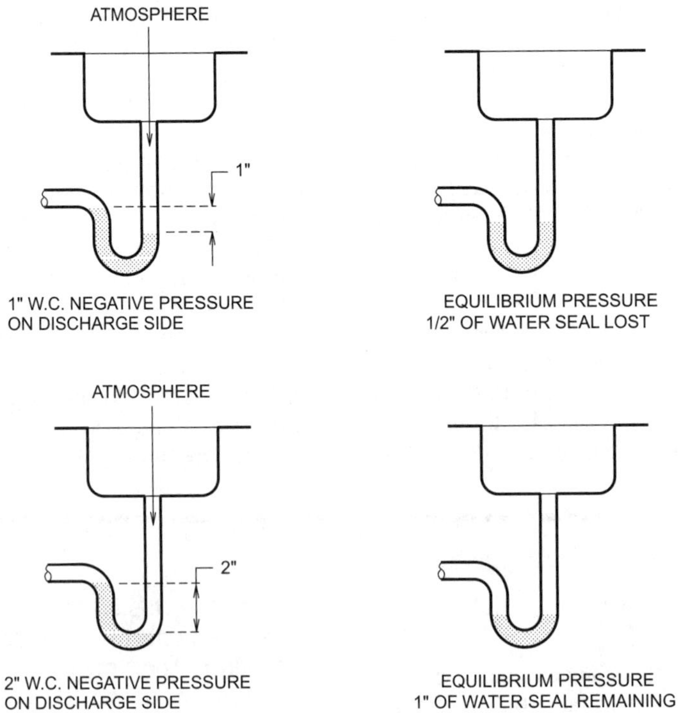

ATMOSPHERE
1"
1" W.C. NEGATIVE PRESSURE ON DISCHARGE SIDE
EQUILIBRIUM PRESSURE 1/2" OF WATER SEAL LOST
ATMOSPHERE
2"
2" W.C. NEGATIVE PRESSURE ON DISCHARGE SIDE
EQUILIBRIUM PRESSURE 1" OF WATER SEAL REMAINING

For SI: 1 inch = 25.4 mm.

Figure 901.2(1)
LOSS OF TRAP SEAL

901.2.1 Venting required. Every trap and trapped fixture shall be vented in accordance with one of the venting methods specified in this chapter.

❖ This section establishes that traps and trapped fixtures must be vented. This section also indicates that the method of venting can be any one of the applicable methods described in Chapter 9. Proper application of these venting options can substantially reduce the amount of pipe and fittings while still providing proper venting. Inspectors and plumbers often overlook the opportunities afforded by the different venting methods. For example, an particular installation might not be in compliance with the code for a certain type of vent; however, that same installation might comply if it were considered as a different type of vent system or slightly modified to comply with the rules of a different type of venting.

901.3 Chemical waste vent system. The vent system for a chemical waste system shall be independent of the sanitary vent system and shall terminate separately through the roof to the open air.

❖ Where a chemical waste system is installed, it is understood that the drainage system must be separate from the sanitary system (see commentary, Section 803.3). The vent system is also required to be separate. The fumes and vapors in a chemical vent system may adversely affect a sanitary vent system. In accordance with Section 702.5, vent materials must be resistant to corrosion and degradation from the chemical waste and vapors involved.

901.4 Use limitations. The plumbing vent system shall not be utilized for purposes other than the venting of the plumbing system.

❖ The venting system is designed for a specific function and cannot be used for any other purpose because that might interfere with the proper operation of the plumbing system or create an extreme hazard. For example, vent piping must not serve as a drain for condensate from heating, ventilating and air-conditioning (HVAC) equipment. Such use introduces waste into piping that is not intended to convey waste and might allow sewer gases to enter the buildings via the condensate drain connection.

901.5 Tests. The vent system shall be tested in accordance with Section 312.

❖ This section requires that vent systems be tested in accordance with Section 312, which describes the tests and inspections necessary to verify a vent system performs as required by the code. Although visual inspections play a vital part in the integrity of the system, it is impossible for the code official to identify defects without testing.

901.6 Engineered systems. Engineered venting systems shall conform to the provisions of Section 918.

❖ This section requires that the design of alternative vent systems be in accordance with Section 918. Section 918 describes an alternative sizing method for individual fixture vents, which may not comply with the requirements of this chapter. Specific requirements are applicable, however, with respect to design criteria, submittals, technical data, construction documents, design approval and inspection and tests.

SECTION 902 MATERIALS

902.1 Vents. The materials and methods utilized for the construction and installation of venting systems shall comply with the applicable provisions of Section 702.

❖ This section requires the construction and installation of venting systems, which are a necessary part of the complete sanitary drainage system, to comply with Section 702, which includes requirements for materials used in the sanitary drainage system (see commentary, Section 702).

902.2 Sheet copper. Sheet copper for vent pipe flashings shall conform to ASTM B 152 and shall weigh not less than 8 ounces per square foot (2.5 kg/m^2).

❖ Sheet copper was a predominant material used as a lining in plumbing fixtures. Today, it is rarely used because of the high cost of material and labor. Often in today's roofing systems, copper is replaced with rubber or asphaltic materials because they are less costly and labor intensive. The ASTM standard regulates cold-rolled tempered sheet copper, hot-rolled tempered sheet copper and annealed sheet copper. When sheet copper is used for flashing around vents, the sheeting must not weigh less than 8 ounces per square foot (2.5 kg/m^2). This provides better durability and protection for the area being penetrated by the vent. Sheet copper is still used for flashing on long-life roof systems, such as slate or tile. Copper can cause corrosion to occur when used with or contacting other metals; therefore, the designer who specifies copper flashing must use only copper or carefully isolate it from other metals.

902.3 Sheet lead. Sheet lead for vent pipe flashings shall weigh not less than 3 pounds per square foot (15 kg/m^2) for field-constructed flashings and not less than $2^1/_2$ pounds per square foot (12 kg/m^2) for prefabricated flashings.

❖ Where sheet lead is used in flashing installations around vent pipe, it must weigh at least 3 pounds per square foot (15 kg/m^2) where constructed at the job site and $2^1/_2$ pounds per square foot (12 kg/m^2) for prefabricated units.

Lead flashing is usually wrapped over the exposed rim of the vent terminal. If the flashing is not properly formed around the internal surface of the vent pipe, it will reduce the effective cross-sectional area of the vent terminal and restrict the flow of air into and out of the drainage system. This may compromise the effectiveness of the vent system to protect trap seals and may promote frost closure at the vent terminal.

SECTION 903
OUTDOOR VENT EXTENSION

903.1 Required vent extension. The vent system serving each *building drain* shall have at least one vent pipe that extends to the outdoors.

❖ The intent of this section of the code is to make sure that the building drain is vented to the outdoors. The building drain must have not less than one vent pipe routed to the exterior of the building. The required vent connecting to the outdoors can be satisfied in a number of ways including using a (1) soil stack vent, (2) waste stack vent, (3) individual fixture vent, (4) common vent, (5) vent stack or (6) relief vent.

The code does not specify where the required vent must be located or the required routing except that it must eventually connect to the building drain either directly or through an extension of a drain. Therefore, any vent connecting to a soil or waste stack, any vent stack, any relief vent or any fixture vent may be considered as the required vent extension to outdoors [see Section 903.1.1 and Commentary Figures 903.1(1) through (3)]. However, in no case can the required vent extension to outdoors be less than one-half of the required size of the building drain (see Sections 903.1.2 and 916.1).

Note that the requirement for a stack vent "running undiminished in size" to the outdoors was deleted from this section because it is understood that no vent can be reduced (diminished) to less than its required minimum size. Also, this deleted text dates back to the days when full-size (i.e., same size as building drain) vent stacks and stack vents were required for each building drain. The following examples illustrate the required vent extension for building drains with a soil stack:

Example A:

A 3-inch (76 mm) soil stack with a stack vent serving as the required vent extension to the outdoors, connecting to a 3-inch (76 mm) building drain, must have at least a $1^1/_2$-inch (38 mm) stack vent [maximum of 102 fixture units served and a maximum 25-foot (7620 mm) developed length] in accordance with Section 916.1 and Table 916.1. This $1^1/_2$-inch (38 mm) stack vent is the minimum size required to comply with this section because it is at least one-half the size of the building drain.

Example B:

A 3-inch (76 mm) soil stack with a stack vent serving as the required vent extension to the outdoors, connecting to a 4-inch building drain must have at least a 2-inch (51 mm) stack vent [maximum of 102 fixture units served and a maximum 86-foot (26 213 mm) developed length] in accordance with Section 916.1 and Table 916.1. The 2-inch (51 mm) stack vent is the minimum size required for the purpose of complying with this section because the 4-inch building drain requires

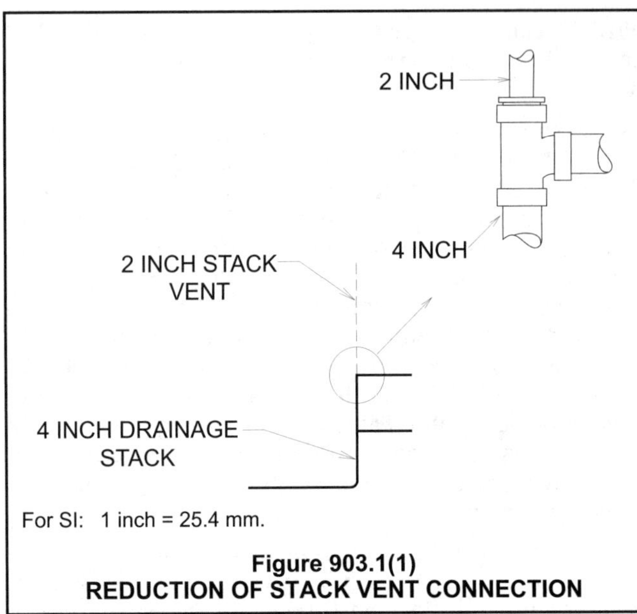

For SI: 1 inch = 25.4 mm.

Figure 903.1(1)
REDUCTION OF STACK VENT CONNECTION

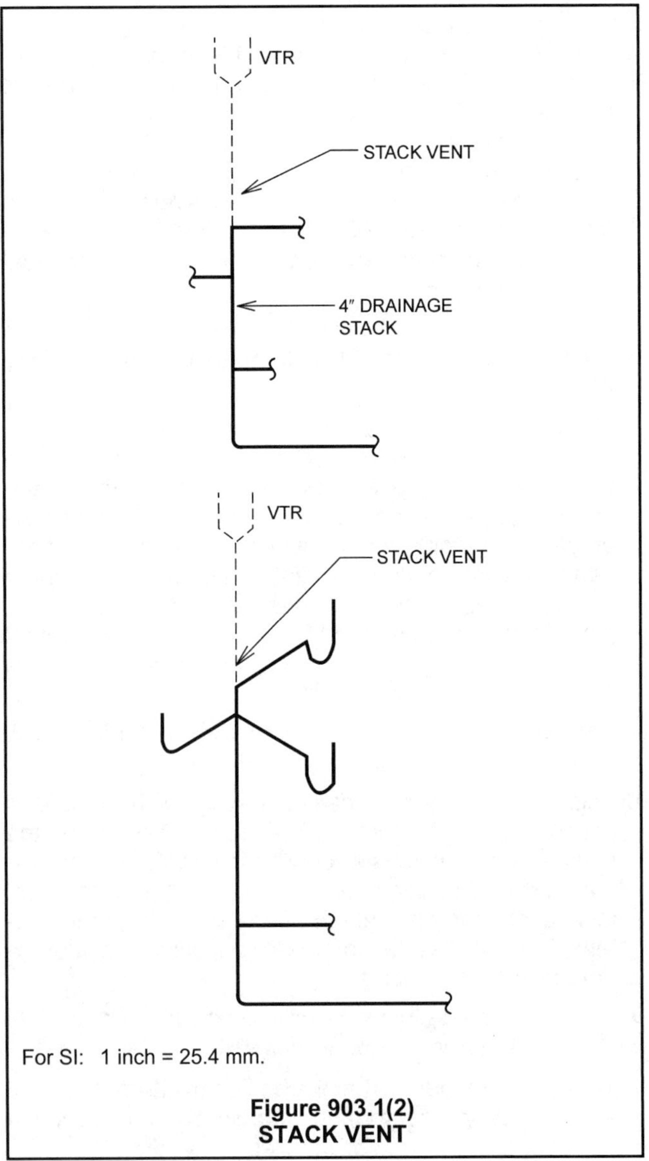

For SI: 1 inch = 25.4 mm.

Figure 903.1(2)
STACK VENT

a vent extension to outdoors of not less than one-half the building drain size.

Example C:

A $1^1/_2$-inch (38 mm) waste stack with a $1^1/_2$-inch (38 mm) stack vent serving as the required vent extension to the outdoors, connecting to a 4-inch building drain must have at least a 2-inch (51 mm) stack vent in accordance with Section 916.1. Therefore, the waste stack must also be at least 2-inch (51 mm) from the dry vent connection down to the building drain. If the waste stack with stack vent was not serving as the required vent extension to the outdoors, the stack could remain at $1^1/_2$ inches (32 mm) and the stack vent $1^1/_2$ inches (32 mm) [maximum of two fixture units served (in accordance with Table 910.4) and a maximum 150-foot (45 720 mm) developed length (in accordance with Table 916.1)].

This section was revised to remove the reference to "stack" which implied that all buildings were required to have a stack. Not all buildings have stacks. For example, in slab-on-grade single-story construction, there are no soil or waste stacks because all fixtures drain to a horizontal building drain below the slab. The following examples illustrate the required vent extension for building drains without stacks.

Example D:

A 3-inch (76 mm) building drain with a $1^1/_2$ inch (38 mm) vented individual lavatory drain serving as the required vent extension to the outdoors must have at least a $1^1/_2$-inch (38 mm) vent [up to 40 feet (12 192 mm) developed length] in accordance with Section 916.2 because the vent must be at least one half the size of the drain that it serves. If this fixture drain with vent was not serving as the required vent extension to the outdoors, the vent size minimum would be $1^1/_4$ inches (32 mm) [up to 40 feet (12 192 mm) developed length].

Example E:

A 4-inch (102 mm) building drain with a $1^1/_2$ inch (38 mm) individual vented lavatory drain serving as the required vent extension to the outdoors must have at least a 2-inch (51 mm) vent [up to 40 feet (12 192 mm) developed length] in accordance with Section 916.2 because the vent must be at least one-half the size of the drain it serves. Therefore, any portion of the fixture drain downstream of the vent connection must also be at least 2 inches (51 mm). If this fixture drain with vent was not serving as the required vent extension to the outdoors, the vent size minimum would be $1^1/_4$ inches (32 mm) [up to 40 feet (12 192 mm) developed length to outdoors].

The wording "directly as possible" was also eliminated in this section because it was not enforceable and was not clear as to where the required stack had to connect to the building drain or if the stack could have offsets on its way to the outdoors. Therefore, the required vent extension to outdoors can have offsets (with the exception for waste stack vents in accordance with Section 910.2) and can connect to other vents that connect to the outdoors.

Waste and soil stacks need not terminate with stack vents leading to the outdoors (see Commentary Section 903.3).

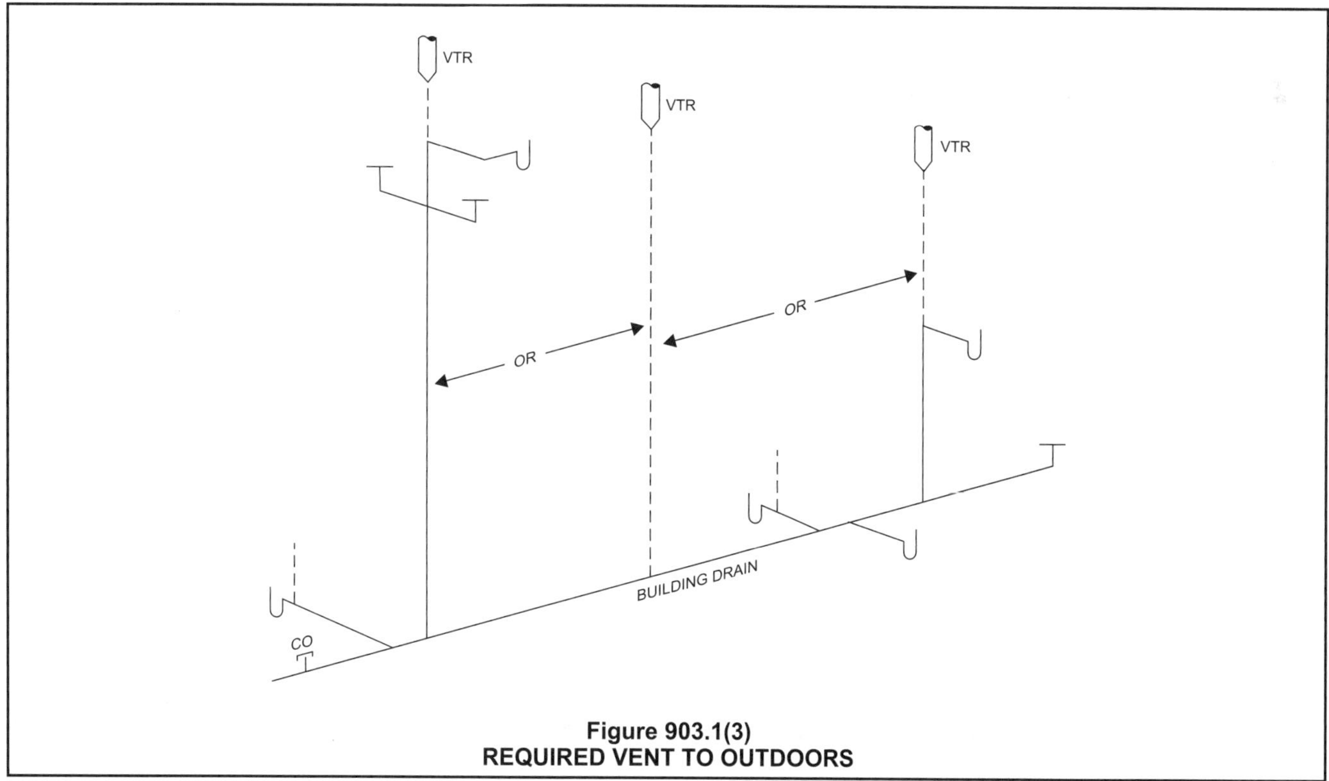

Figure 903.1(3)
REQUIRED VENT TO OUTDOORS

Historically, in buildings having soil stacks, stack vents were once considered to be the primary protection for trap seals often supplemented by individual vents called "revents" or "backvents." The previously required stack with stack vent ("main vent") was a large diameter stack vent (same size as the soil stack) that served as a convenient point to connect branch vents. Present venting methodology depends on the vents provided for fixtures and traps for the necessary airflow.

903.1.1 Installation. The required vent shall be a dry vent that connects to the *building drain* or an extension of a drain that connects to the *building drain*. Such vent shall not be an island fixture vent as allowed by Section 913.

❖ A vent leading to the outdoors which connects directly to the building drain, or to a drain connecting to the building drain, can be used for the required vent extension. Island fixture vents cannot be used for this function because of the high potential for clogging of the vent piping that has horizontal components below the flood rim level of the island fixture (see Commentary Section 903.1).

903.1.2 Size. The required vent shall be sized in accordance with Section 916.2 based on the required size of the *building drain*.

❖ The minimum size is based on Section 916.2 unless required to have a 3-inch (76 mm) minimum diameter vent terminal as required by Section 904.2 (see Commentary Section 903.1).

903.2 Vent stack required. A vent *stack* shall be required for every drainage *stack* that has five *branch intervals* or more.

Exception: Drainage stacks installed in accordance with Section 910.

❖ If a drainage stack is five or more branch intervals in height, a vent stack is required. If the drainage stack is less than five branch intervals in height, a vent stack is not required because the pressures in the drainage stack are not likely to create a pressure differential at the trap seals in excess of 1 inch of water column (249 Pa). Vent stacks are dry and required to connect at or near the base of the stack served to act as a relief vent for the pressures that develop in the lowest portion of stacks (see commentary, Section 903.4). The code does not require stack vents for drainage stacks except as required by Section 910.3 because the minimum required venting of each fixture has been accomplished when the system complies with the venting methods outlined in this chapter. A stack vent is typically used as a collection point for vent pipes so that a single roof penetration can be made. Keep in mind that "vent stacks" and "stack vents" are distinct (see Section 202).

The exception allows waste-stack vented systems not to have a vent stack when the stack has five or more branch intervals connected. Waste-stack vented systems (Section 910) are already oversized to provide for adequate venting without the need for additional venting.

903.3 Vent termination. Vent stacks or stack vents shall terminate outdoors to the open air or to a stack-type air admittance valve in accordance with Section 917.

❖ No vent is allowed to terminate as an open-ended pipe in an attic, cupola, open parking structure, screened-in porch or any other space within a building envelope, regardless of how well such space is ventilated. The vent must extend to the outdoors if it is the required vent extension for the building drain. If the vent is not the required vent extension for the building drain, it may terminate outdoors or terminate with an air admittance valve in accordance with Section 917. Code editions of 2003 and earlier required that every vent stack or stack vent terminate to open air. The 2006 edition changed this section to allow for termination with an air admittance valve (AAV). However, as worded in the 2006 edition, this section suggested that all vent stacks or stack vents could terminate with an AAV with no system vent to outdoors. This would be in conflict with Sections 903.1 and 917.7 requiring at least one vent to outdoors. The deletion of the word "every" was intended to clarify the intent.

903.4 Vent connection at base. Every vent *stack* shall connect to the base of the drainage *stack*. The vent *stack* shall connect at or below the lowest horizontal *branch*. Where the vent *stack* connects to the *building drain*, the connection shall be located downstream of the drainage *stack* and within a distance of 10 times the diameter of the drainage *stack*.

❖ A drainage stack of five or more branch intervals must be vented at or below the lowest branch connection to relieve the positive pressure developed in the stack. The connection can be to the stack or to the building drain downstream of the stack and within a distance of 10 times the drainage stack diameter.

The size of the vent connection to the drainage system must not be less than the size required for the vent stack. Vent stacks are sized in accordance with Section 916.1, based on the size of the drainage stack at its base. Where the vent stack is connected to the building drain at the base of the drainage stack, the point of connection must be within the section of horizontal drain where the greatest pressure fluctuations and turbulence occur (see Commentary Section 704.3 and Commentary Figure 903.4).

903.5 Vent headers. *Stack vents* and vent stacks connected into a common vent header at the top of the stacks and extending to the open air at one point shall be sized in accordance with the requirements of Section 916.1. The number of fixture units shall be the sum of all fixture units on all stacks connected thereto, and the *developed length* shall be the longest vent length from the intersection at the base of the most distant *stack* to the vent terminal in the open air, as a direct extension of one *stack*.

❖ It is quite often economically or aesthetically desirable to connect a number of vent stacks or stack vents to a common header to minimize the number of penetrations of the roof. The header must be adequately sized to both supply and relieve air in every vent stack simultaneously.

Section 916.1 requires that the header be sized to carry the total fixture unit load served by the vent stacks connected to the header. In determining the developed length, the longest vent lengths connected thereto are added to the overall length of the header.

Example:

Commentary Figure 903.5 illustrates how to determine the size of a vent header connecting four vent stacks serving soil stacks more than three branch intervals in height.

The fixture unit served by each vent stack is added as it connects to the header. Each section of the header is sized as a vent stack serving the accumulated load being vented by the individual vent stacks.

The section of the vent header between vent stack A and vent stack B is a horizontal offset of vent stack A. Therefore, it is the same size as vent stack A. The developed length for vent stack A is 240 feet (220 feet + 20 feet) (73 152 mm). The number of drainage fixture units (dfu) served by vent stack A is 200.

The total fixture unit load served by the vent header between stacks B and C is 350 dfu (200 dfu + 150 dfu). The section of the header between vent stacks B and C is sized based on the accumulated load of drainage stacks A and B and the longest developed

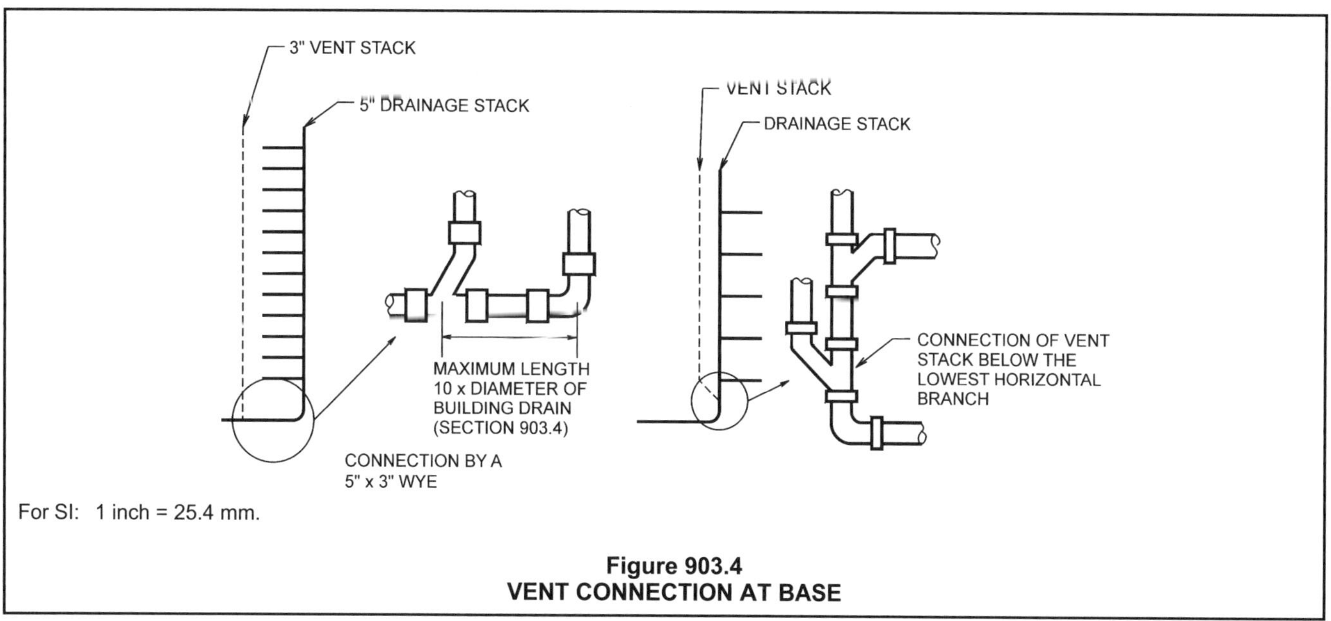

For SI: 1 inch = 25.4 mm.

Figure 903.4
VENT CONNECTION AT BASE

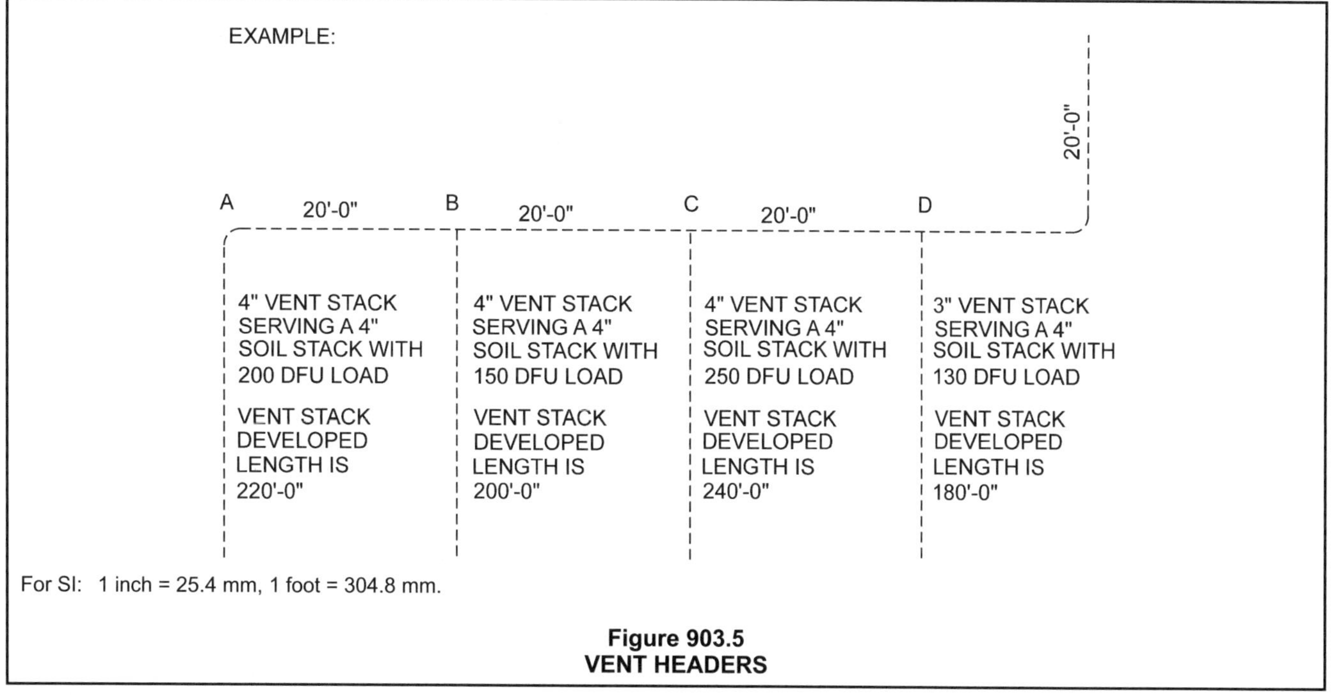

For SI: 1 inch = 25.4 mm, 1 foot = 304.8 mm.

Figure 903.5
VENT HEADERS

length of either stack A or B. The developed length is the longest vent length from the base of the most distant stack (vent stack A) to the vent terminal in the open air. In this example, the developed length is 220 feet + 20 feet + 20 feet + 20 feet + 20 feet = 300 feet (91 440 mm).

From Table 916.1, the stack and the vent must be 4 inches (102 mm), which can serve up to 540 fixture units at a maximum developed length of 580 feet (176.8 m); therefore, the size required for the vent stack header is 4 inches (102 mm).

The total fixture unit load served by the vent header between vent stacks C and D is 600 dfu, which exceeds 540 dfu listed as the maximum total for a 4-inch (102 mm) drainage stack; therefore, a 5-inch (127 mm) drainage stack and vent would be required, which may serve up to 1,400 dfu at a maximum developed length of 590 feet (179.8 m). Thus, the required size of the vent stack header is 5 inches (127 mm).

The total fixture unit load between vent stack D and the termination is 730 dfu. This value is still within the capacity of a 5-inch (127 mm) drainage stack and vent stack; therefore, the size of the vent stack header between stack D and the termination is still 5 inches (127 mm).

SECTION 904
VENT TERMINALS

904.1 Roof extension. All open vent pipes that extend through a roof shall be terminated at least [NUMBER] inches (mm) above the roof, except that where a roof is to be used for any purpose other than weather protection, the vent extensions shall be run at least 7 feet (2134 mm) above the roof.

❖ This section contains requirements for the location, installation and protection of vent terminals.

Where the roof is occupied for any purpose, such as a sunning or promenade deck, plumbing vents must extend at least 7 feet (2134 mm) above the roof to prevent harmful sewer gases from polluting the area. The sewer gases will tend to disperse into the air, rather than accumulate near the roof surface. The minimum termination height must be determined by the local jurisdiction based on its snowfall rates. The termination height should be no less than 6 inches (152 mm) above the roof to protect the terminal from being blocked by snow and to allow sufficient length of pipe for proper roof flashing.

The vent terminal through the roof must be designed to tolerate movement resulting from expansion and contraction of the piping material. This is extremely important where the vent is a piping material with a high rate of expansion and contraction, such as acrylonitrile butadiene styrene (ABS) and polyvinyl chloride (PVC) plastic pipe. Vent terminals extending higher than the distances specified in Table 308.5 for vertical pipe hanger spacing must have intermediate supports (see Commentary Figure 904.1).

904.2 Frost closure. Where the 97.5-percent value for outside design temperature is 0°F (-18°C) or less, every vent extension through a roof or wall shall be a minimum of 3 inches (76 mm) in diameter. Any increase in the size of the vent shall be made inside the structure a minimum of 1 foot (305 mm) below the roof or inside the wall.

❖ The possibility of frost closure occurs only in areas of the country having cold climates with winter outside $97^1/_2$-percent design temperatures of less than 0°F (-18°C). Temperatures for many cities in the United States are shown in Appendix D. For example, Pueblo, Colorado, has a winter $97^1/_2$-percent temperature of 0°F (-18°C) and may not require a minimum of 3 inches (76 mm) in diameter vent, but in Concord, New Hampshire, a 3-inch-diameter (76 mm) termination may be required.

A vent 3 inches (76 mm) or larger in size is not likely to completely close up from frost. However, vents

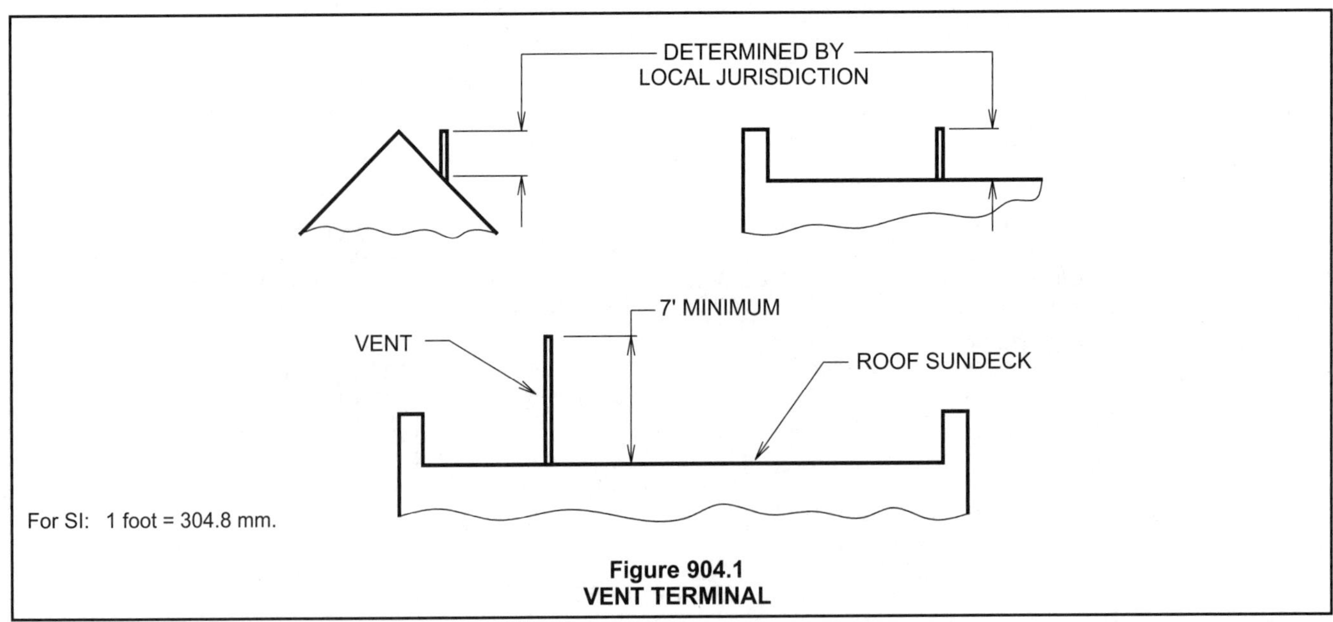

Figure 904.1
VENT TERMINAL

smaller than 3 inches (76 mm) in diameter must be increased in size where frost closure is possible. Any increase must occur within the structure to protect the smaller-diameter pipe from direct exposure to the outside atmosphere.

See the "Degree Day and Design Temperature" table (located in Appendix D) and the ASHRAE *Fundamentals Handbook* for temperature information.

904.3 Flashings. The juncture of each vent pipe with the roof line shall be made water-tight by an *approved* flashing.

❖ Water that is allowed to enter between the intersections of roofing and vent piping will cause decay of the wood and degradation of other building materials; therefore, flashings, including those of materials described in Sections 902.2 and 902.3, are required at these intersections.

904.4 Prohibited use. Vent terminals shall not be used as a flag pole or to support flag poles, television aerials or similar items, except when the piping has been anchored in an *approved* manner.

❖ The use of the vent terminal to support imposed loads is prohibited unless the vent terminal is properly anchored in a manner that is approved by the code official. Vent terminals must be anchored or supported at intervals determined by Table 308.5 for vertical pipe.

904.5 Location of vent terminal. An open vent terminal from a drainage system shall not be located directly beneath any door, openable window, or other air intake opening of the building or of an adjacent building, and any such vent terminal shall not be within 10 feet (3048 mm) horizontally of such an opening unless it is at least 2 feet (610 mm) above the top of such opening.

❖ The vent terminal must be located away from any building air intake opening (gravity or mechanical) to reduce the possibility of sewer gases entering a building (see Commentary Figure 904.5).

904.6 Extension through the wall. Vent terminals extending through the wall shall terminate a minimum of 10 feet (3048 mm) from the lot line and 10 feet (3048 mm) above average ground level. Vent terminals shall not terminate under the overhang of a structure with soffit vents. Side wall vent terminals shall be protected to prevent birds or rodents from entering or blocking the vent opening.

❖ Sidewall vent terminations are an alternative to roof penetrations that may result in significant cost savings and a more aesthetically pleasing installation. For example, a sidewall vent may be preferred to penetrating membrane built-up, slate or tile roofs. Such roof penetrations are difficult to make leakproof, may be expensive and are often considered unsightly.

In multistory buildings where remodeling work is being done, a sidewall vent termination can serve as a welcome alternative to running vent piping through finished or occupied stories above.

Where a sidewall vent is installed, the vent opening must be protected with a screen or louver to prevent birds from building a nest in the pipe. This also prevents rodents from entering the vent system (see Commentary Figure 904.6). Such vents must not ter-

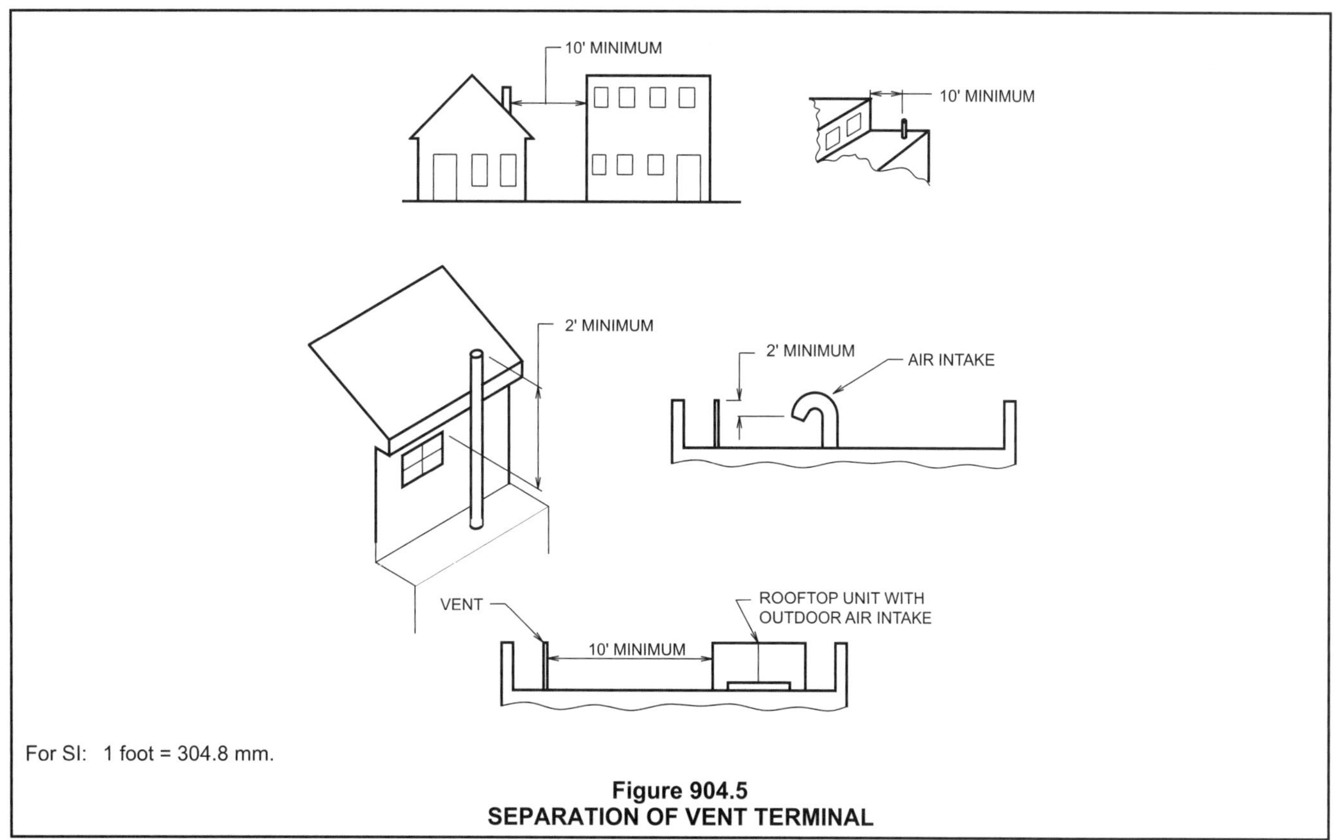

For SI: 1 foot = 304.8 mm.

Figure 904.5
SEPARATION OF VENT TERMINAL

minate where the emissions from the vent can either cause structural damage or enter the building envelope through soffit vents. Such vents must also comply with the location requirements of Section 904.5.

904.7 Extension outside a structure. In climates where the 97.5-percent value for outside design temperature is less than 0°F (-18°C), vent pipes installed on the exterior of the structure shall be protected against freezing by insulation, heat or both.

❖ A vent pipe located outside a building increases the likelihood of frost closure in colder climates. To prevent a buildup of frost on the inside of the vent pipe, it must be protected. If provided with adequate insulation, the heat of the vent gases and vapors may be relied upon to prevent frost from developing in the vent pipe.

SECTION 905 VENT CONNECTIONS AND GRADES

905.1 Connection. All individual, *branch* and circuit vents shall connect to a vent *stack*, *stack vent*, air admittance valve or extend to the open air.

❖ This section further emphasizes the requirement that vents are to extend to the outdoors except where allowed to terminate with an air admittance valve. This might be accomplished by connecting to a vent stack or a stack vent that extends outdoors and terminates to the open air (see commentary, Section 903.3) or by extending these vents independently to the outdoors. This prevents sewer gases from escaping into the building. The use of air admittance valves must comply with the provisions of Section 917.

905.2 Grade. All vent and *branch* vent pipes shall be so graded and connected as to drain back to the drainage pipe by gravity.

❖ Vents convey water vapor from the drainage system that might condense in the vent piping. Also, rainwater can enter a vent system at the vent termination; therefore, the vent must be graded (sloped) to the drainage system, thus preventing any accumulation of condensate or rainwater. The slope of a vent pipe does not affect the movement of air inside the pipe. It makes no difference whether the vent piping is sloped to drain back to the drain piping served by the vent or the vent piping slopes to drain to a vent stack or stack vent, provided that water will not stand in any portion of the venting system (see Commentary Figure 905.2).

Vent piping must be supported according to Table 308.5 to maintain slope and to prevent sags where condensate could collect and restrict air flow.

905.3 Vent connection to drainage system. Every dry vent connecting to a horizontal drain shall connect above the centerline of the horizontal drain pipe.

❖ When drainage enters a pipe, it is naturally assumed that the liquid proceeds down the pipe in the direction of flow. Being a liquid, however, the drainage seeks its own level and may move against the direction of flow before reversing and draining down the pipe. When this occurs, the solids may drop out of suspension to the bottom of the pipe. In a drain, the solids will again move down the pipe with the next discharge of liquid into the drain.

If the horizontal pipe serves only as a dry vent, the solids have no possibility of being brought back into suspension and, over a period of time, the horizontal vent pipe may be completely obstructed by waste. To avoid these blockages, the vent must connect to horizontal drains above the centerline (see Commentary Figure 905.3). The intent is to prevent waste from entering a dry vent by connecting the vent above the flow line of the drain. Such connections will result in vent piping that forms an angle of 45-degree (0.79 rad) or more with the horizontal.

905.4 Vertical rise of vent. Every dry vent shall rise vertically to a minimum of 6 inches (152 mm) above the *flood level rim* of the highest trap or trapped fixture being vented.

Exception: Vents for interceptors located outdoors.

❖ This code provision is very important for the proper design, longevity and maintenance-free operation of a plumbing system. Surprisingly, this is also one of the most frequently violated code provisions.

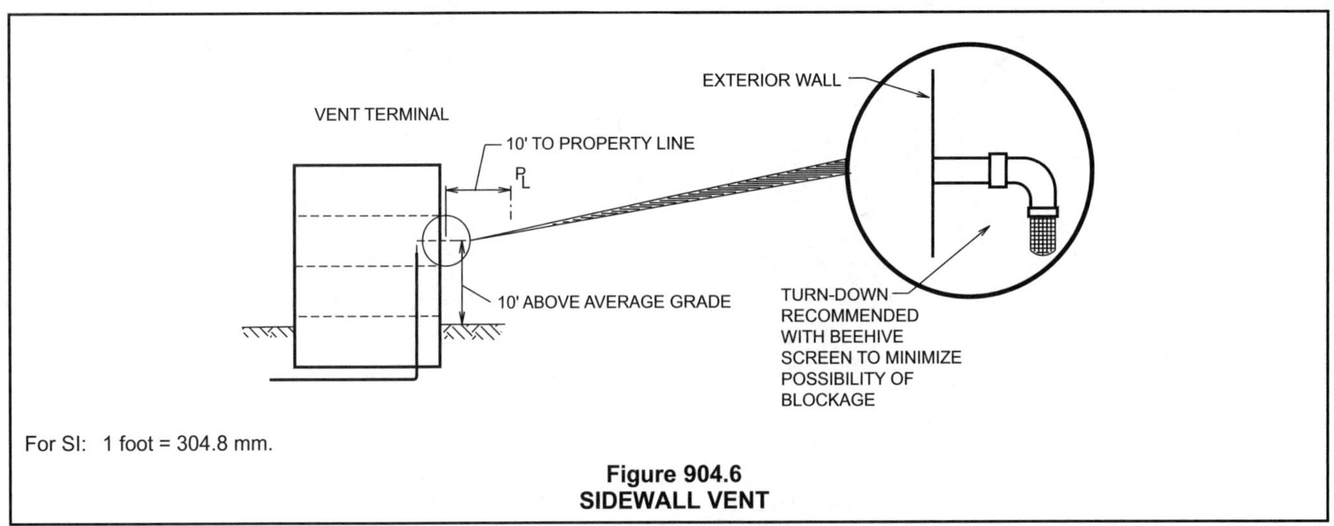

Figure 904.6 SIDEWALL VENT

The intent of this section is to prohibit horizontal vent piping from occurring where it will be subjected to waste flow. A vertical rise of the vent pipe substantially reduces the possibility of it having a blockage or stoppage. Most pipe stoppages occur in horizontal pipes. The vertical rise of the vent to a point above the flood level rim is necessary to protect the vent from a backup of waste in the event of blockage in the drainage system. A blockage causes the drainage to rise in the vent pipe and in the plumbing fixture. Any drainage system backup will flow over the fixture's flood level rim, thereby preventing it from entering the horizontal vent pipe.

A stoppage in the drain may be easily cleared by rodding through a cleanout; however, this is not the case with a vent blockage, which is seldom, if ever, accessible (see Commentary Figure 905.4).

A vent blockage will rarely produce any noticeable or recognizable symptoms of a problem; therefore, extraordinary precautions are taken to protect vents. Note that a drain obstruction will always be apparent, but a vent obstruction will almost never be apparent to the building occupants. It is often argued that horizontal vent runs below the fixture rim are necessary in order to place the vent connection within the distance limits of Section 906.1. This argument may be valid if the code specified individual fixture vents as the only venting option; however, this is not the case. The code allows many venting options; thus; there is always an option to avoid the installation of horizontal vents below the fixture flood rim.

The exception provides a measure of design flexibility and acknowledges where a grease interceptor is located in an open parking area, it is impossible to install a vent that rises 6 inches (152 mm) above the top of the interceptor.

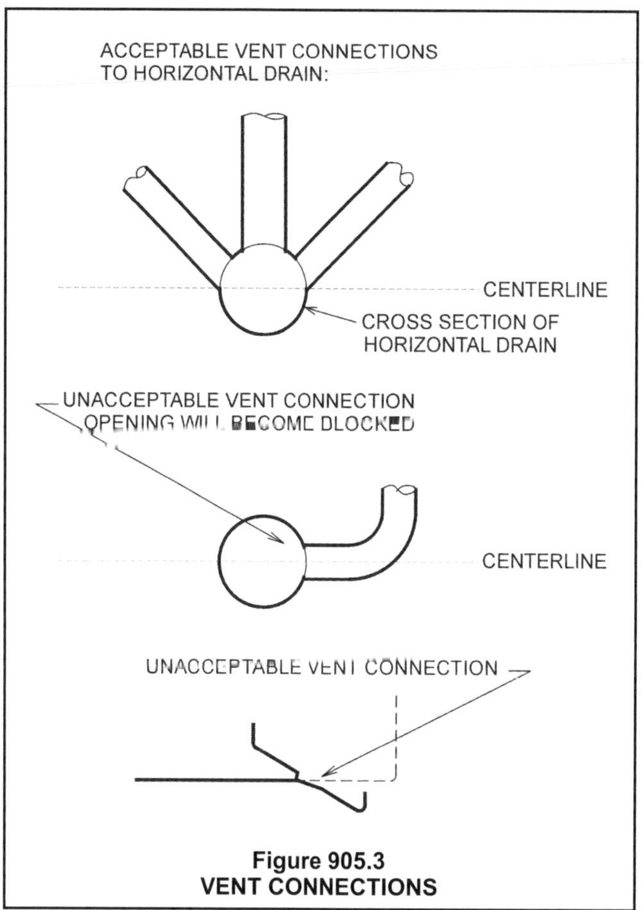

Figure 905.3
VENT CONNECTIONS

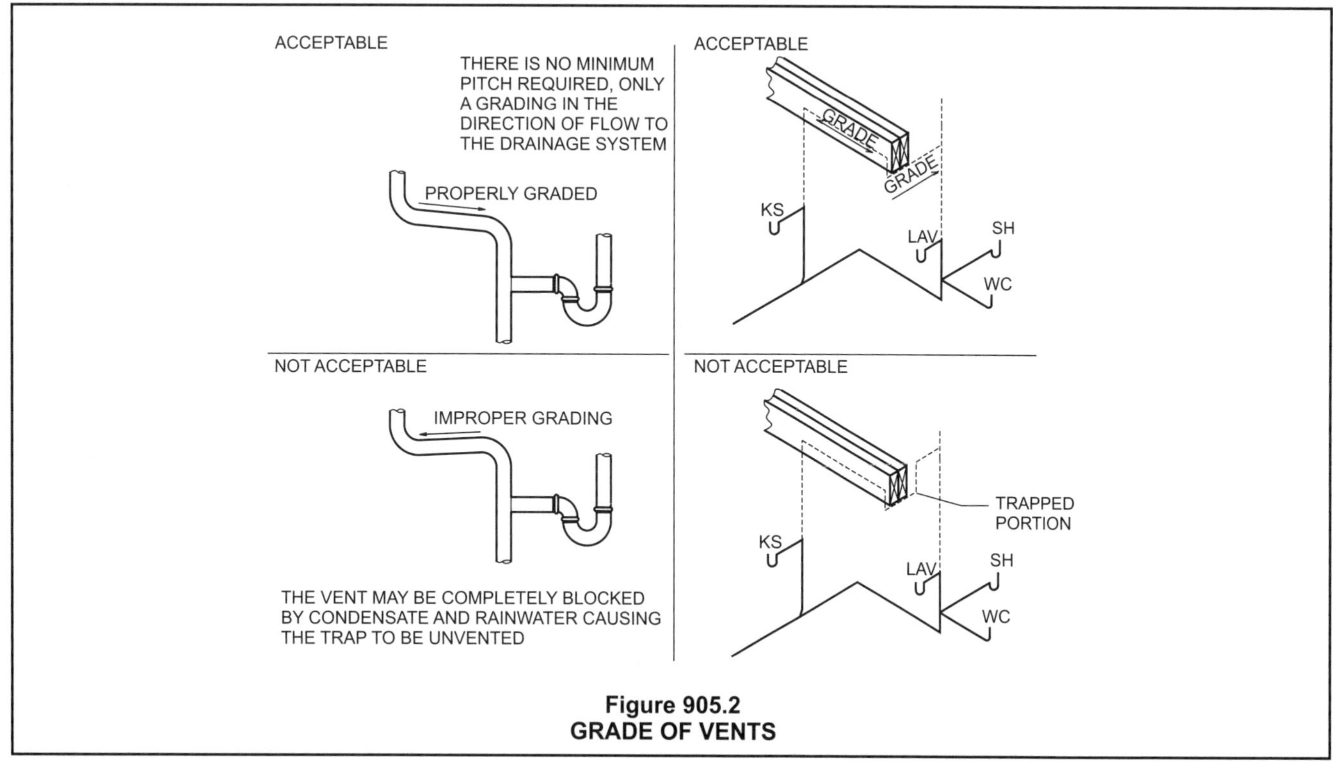

Figure 905.2
GRADE OF VENTS

ACCEPTABLE:

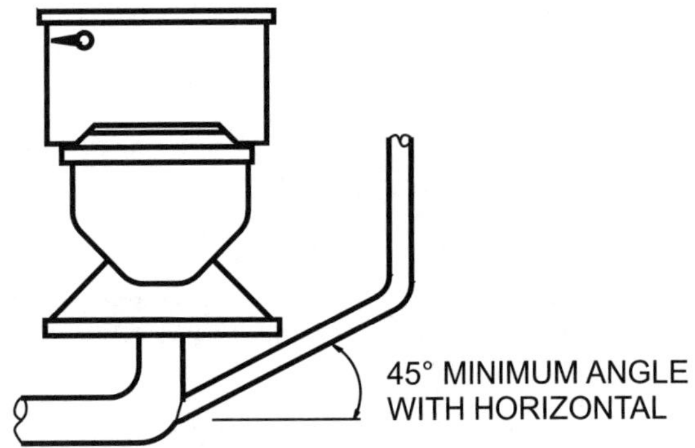

ACCEPTABLE:

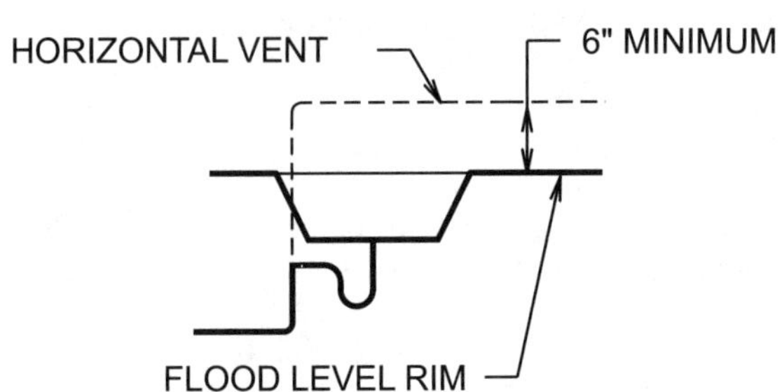

ACCEPTABLE:

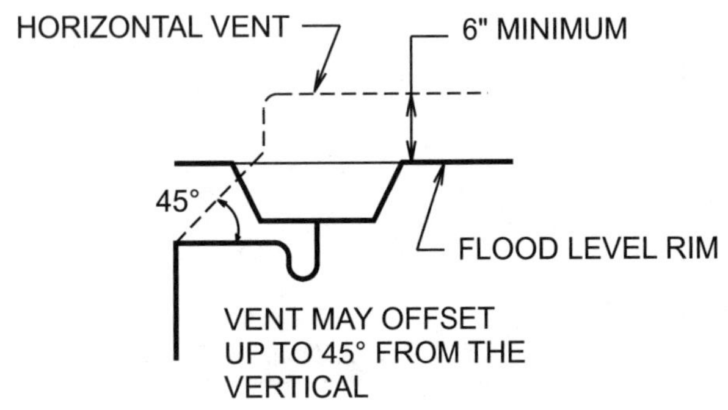

ACCEPTABLE:

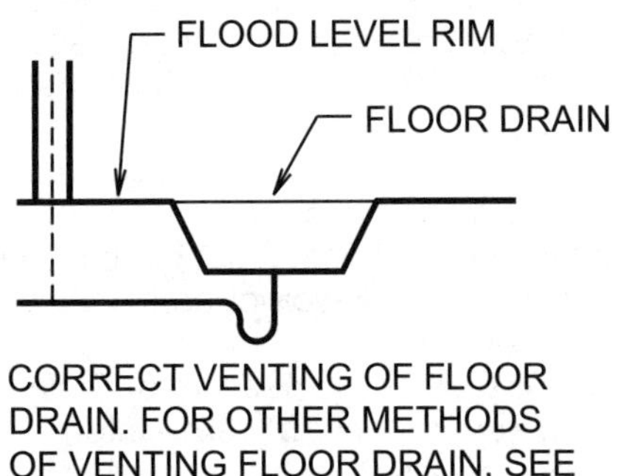

CORRECT VENTING OF FLOOR DRAIN. FOR OTHER METHODS OF VENTING FLOOR DRAIN, SEE SECTIONS 912 AND 913.

UNACCEPTABLE:

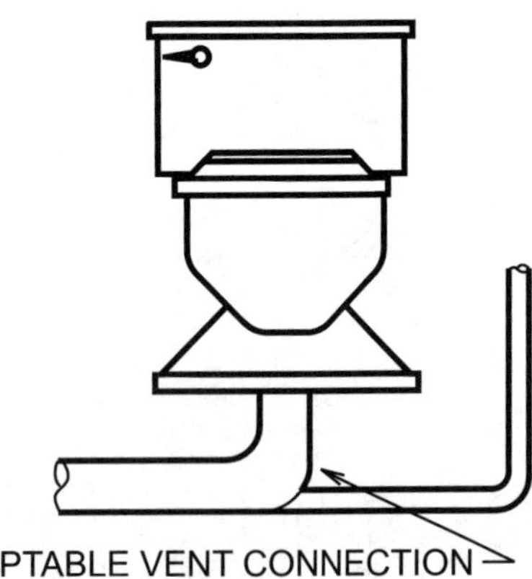

UNACCEPTABLE:

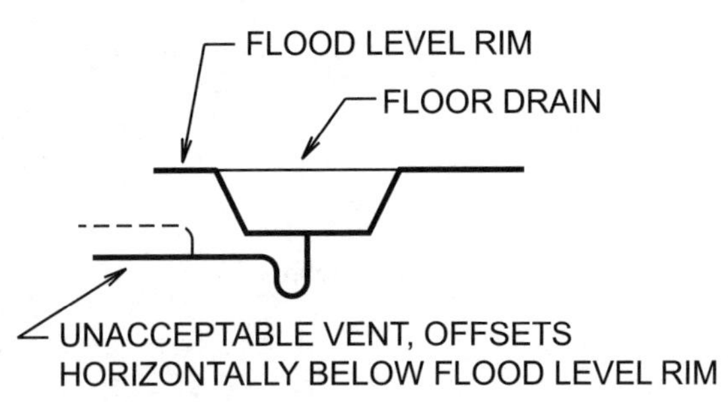

For SI: 1 inch = 25.4 mm, 1 degree = 0.0175 rad.

Figure 905.4
VERTICAL RISE OF VENT

905.5 Height above fixtures. A connection between a vent pipe and a vent *stack* or *stack vent* shall be made at least 6 inches (152 mm) above the *flood level rim* of the highest fixture served by the vent. Horizontal vent pipes forming *branch* vents, relief vents or loop vents shall be at least 6 inches (152 mm) above the *flood level rim* of the highest fixture served.

❖ This section expresses the same concern as the previous section. In the event of drain stoppage, waste could rise into the vent piping. Over a period of time, the vent piping may become blocked with solids that have settled out of the waste liquids. Additionally, individual, common or branch vents that improperly connect to vent stacks or stack vents may become drains if the fixture drain or drains served are blocked or restricted. It is entirely possible for a vent to serve unintentionally as a drain, and this condition may go unnoticed for a long period of time [see Commentary Figures 905.5(1) and 905.5(2)].

A properly installed horizontal dry vent pipe will never contain waste liquids because the fixture or fixtures served will overflow before the waste can enter the vent piping [see Commentary Figure 905.5(3) and commentary, Section 905.4].

905.6 Vent for future fixtures. Where the drainage piping has been roughed-in for future fixtures, a rough-in connection for a vent shall be installed. The vent size shall be not less than one-half the diameter of the rough-in drain to be served. The vent rough-in shall connect to the vent system, or shall be vented by

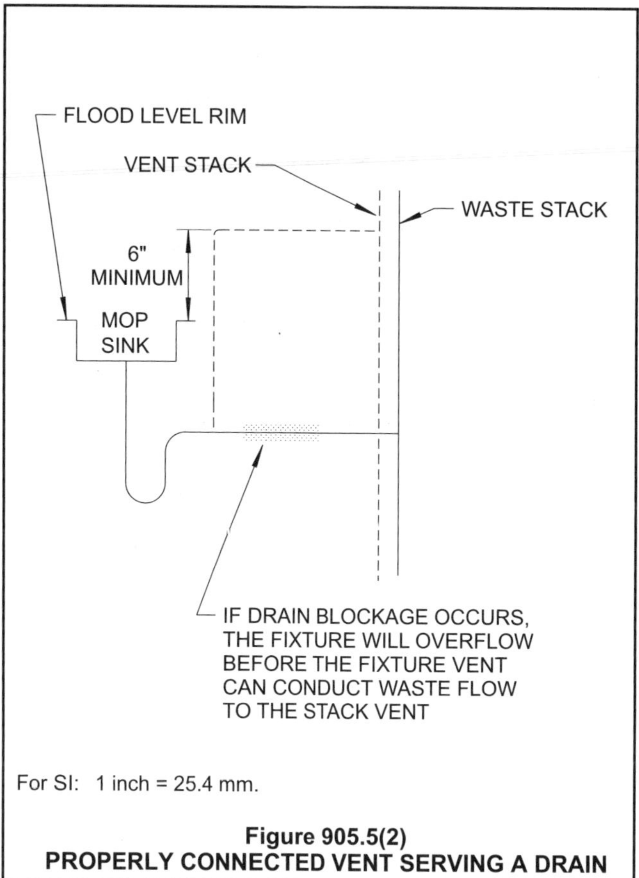

For SI: 1 inch = 25.4 mm.

Figure 905.5(2)
PROPERLY CONNECTED VENT SERVING A DRAIN

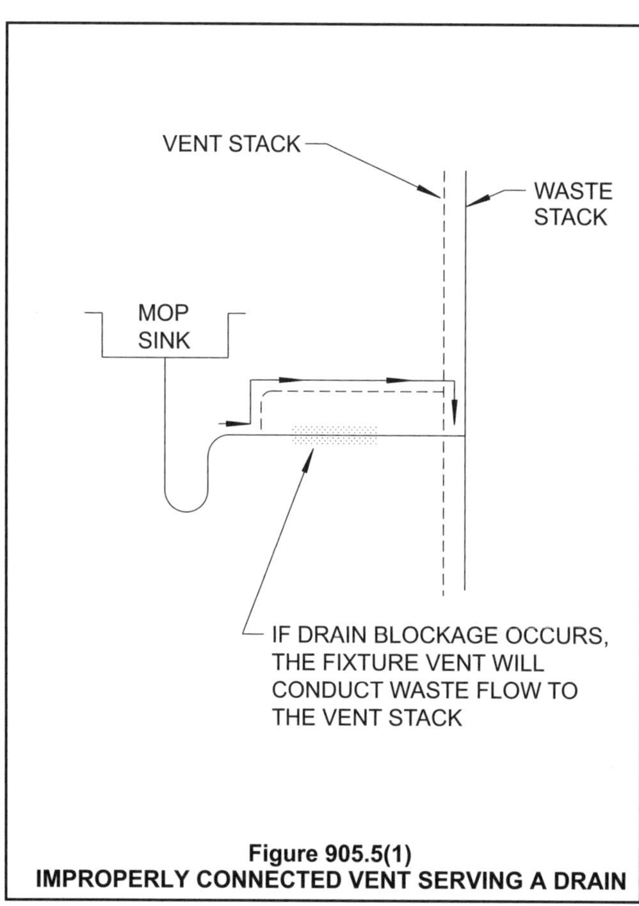

Figure 905.5(1)
IMPROPERLY CONNECTED VENT SERVING A DRAIN

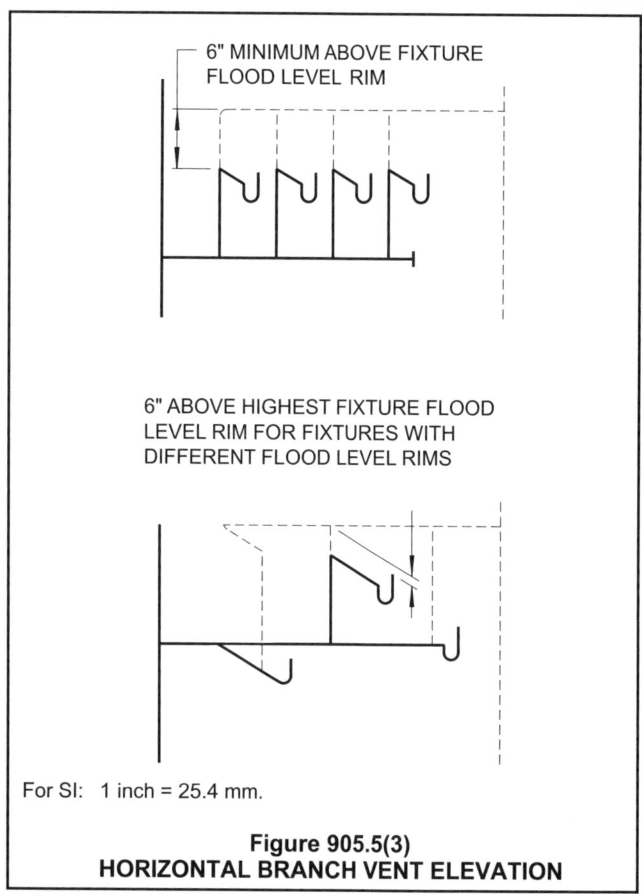

For SI: 1 inch = 25.4 mm.

Figure 905.5(3)
HORIZONTAL BRANCH VENT ELEVATION

other means as provided for in this chapter. The connection shall be identified to indicate that it is a vent.

❖ Section 710.2 addresses the installation of drainage pipe for future plumbing fixtures. Where future fixture rough-in piping occurs, vent piping must also be installed. This section does not necessarily require complete connection of the rough-in vent to the future fixture drainage pipe. The vent pipe is simply roughed-in in a manner to remain accessible for future connection when plumbing fixtures are installed. The vent rough-in must be tied in to the vent system, as required by Section 905.1, or must extend to a vent terminal in the outside air.

The vent rough-in requirement of this section is intended to make sure that a venting means will be available in advance of the future fixture installation. It also eliminates the cost and hardship associated with the installation of vent piping at the time fixtures are installed. It is common to find that fixtures have been installed without vents because a vent was not provided for future use.

Note that a future fixture rough-in may be served by a wet vent, common vent, waste stack vent, circuit vent, combination drain and vent or an air admittance valve; therefore, a dry vent rough-in may not be required. Venting configurations must be sized and installed to accommodate future fixture installations.

Where the installation of a future fixture will require making a connection to the vent rough-in, such connection must be identified as a vent so that the purpose of the original installation is evident when the future fixture is installed.

SECTION 906
FIXTURE VENTS

906.1 Distance of trap from vent. Each fixture trap shall have a protecting vent located so that the slope and the *developed length* in the *fixture drain* from the trap weir to the vent fitting are within the requirements set forth in Table 906.1.

Exception: The *developed length* of the *fixture drain* from the trap weir to the vent fitting for self-siphoning fixtures, such as water closets, shall not be limited.

❖ This section contains the location and installation requirements for vent pipes serving as fixture vents. The distance from every trap to its vent is limited to reduce the possibility of the trap self-siphoning. Self-siphoning of a trap is the siphoning caused by the discharge from the fixture the trap serves. The intent is to locate the vent within the hydraulic gradient of the fixture drain so as to prevent the opening to the vent from being occluded by waste flow [see Commentary Section 906.2 and Commentary Figure 906.1(1)].

Self-siphoning fixtures, such as water closets and siphon action urinals, can have an unlimited distance from the trap weir to the vent. The maximum distance between the trap weir and the vent is to protect the trap against self-siphoning. Fixtures that depend on siphon action for proper operation will operate properly no matter how far the vent is located from the fixture.

Studies have been done to address the trap-to-vent distance for a water closet. Because the water closet relies on self-siphonage to operate properly, and the trap is resealed after each use, there is no need for limiting the distance from the water closet to a vent. The water closet fixture drain must be vented in all cases, but limiting the distance between the vent connection and the water closet appears to serve no purpose. A water closet or other self-siphoning fixture drain can have a total fall greater than the diameter of the fixture drain. This is reinforced by Section 906.2.

Commentary Figure 906.1(2) illustrates the slope required to achieve the maximum length of a $1^1/_2$ inch (38 mm) fixture drain between the trap and the vent serving the trap. For shorter distances between the trap weir, the fixture drain could be sloped greater that $^1/_4$ inch per foot (20.8 mm/m) as long as the elevation of the trap weir is not higher than the elevation of the top of the vent opening. Because some drainage fittings might not accommodate the connection of a fixture drain having a significant slope, it would be prudent to set all fixture drain pipes from the trap weir to the vent at a slope of $^1/_4$ inch per foot (20.8 mm/m). Note that the illustration shows the measurement of the pipe length to be from the back of each fitting's hub and not necessarily from the trap weir to the top of the opening of the fitting in the stack. There are two reasons for this: (1) the depth of flow slightly decreases as the liquid flows towards the vent opening because the flow velocity increases slightly due to the slope of the fixture drain and (2) the indicated points are easily identified in the field whereas the exact points of the trap weir and the vent opening locations are difficult to determine.

TABLE 906.1
MAXIMUM DISTANCE OF FIXTURE TRAP FROM VENT

SIZE OF TRAP (inches)	SLOPE (inch per foot)	DISTANCE FROM TRAP (feet)
$1^1/_4$	$^1/_4$	5
$1^1/_2$	$^1/_4$	6
2	$^1/_4$	8
3	$^1/_8$	12
4	$^1/_8$	16

For SI: 1 inch = 25.4 mm, 1 foot = 304.8 mm,
1 inch per foot = 83.3 mm/m.

❖ The distances listed in Table 906.1 are intended to prevent the weir of the trap from being located above the highest inlet to the vent. As the total change in elevation of a fixture drain (resulting from the fixture drain's slope) approaches the inside diameter of the drain, the weir of the trap approaches the highest level of the vent connection and self-siphoning is more likely.

When the trap weir rises above the highest inlet of the vent opening, it is commonly referred to as either a full or half S trap. This slang expression is derived from the concept that such a piping arrangement creates a self-siphoning potential similar to that of an S trap [see Commentary Figures 906.1(1) and 906.1(2)]. The intent of Section 906.2 is to prevent creating an S-trap effect, which increases the possibility of self-siphonage.

906.2 Venting of fixture drains. The total fall in a *fixture drain* due to pipe slope shall not exceed the diameter of the *fixture drain*, nor shall the vent connection to a *fixture drain*, except for water closets, be below the weir of the trap.

❖ This section reinforces the requirement in Section 906.1. The trap weir must stay below the highest inlet to the vent except for fixtures with integral traps (see commentary, Section 906.1). The fixture drain cannot offset vertically or drop vertically, other than the required slope, between the trap and its vent connection.

The type of fitting used to connect a fixture drain to the fixture branch is also limited. Where the fixture drain pipe runs horizontally and then turns vertically at the point of connection to the vent, a short-pattern fitting is necessary to keep the trap weir below the highest inlet to the vent. The type of transition fitting will necessarily be a sanitary tee except where the vent connects to the horizontal portion of the fixture drain. Although a long-pattern fitting (including a tee-wye or a combination wye-eighth bend) is recommended for creating the proper flow into the vertical drain, such a fitting greatly increases the possibility of self-siphonage of the trap seal and is, therefore, prohibited by this section [see Commentary Figures 906.2(1) and (2)].

Water closets are not regulated by this section. This section specifies a requirement to reduce the possibility of self-siphoning of the trap seal for a given fixture. Water closets (and siphon-action urinals), however, are designed for siphon action of the fixture. The fixture traps must self-siphon during each flushing cycle to accomplish total evacuation of the fixture contents. The flushing device is further designed to refill the trap seal at the completion of the flushing cycle. This operation may be observed during each flushing cycle of a water closet when the water in the bowl is almost completely removed. This self-siphoning action is what the requirements of Sections 906.1 and 906.2 attempt to prevent in other fixtures [see Commentary Figure 906.2(2)].

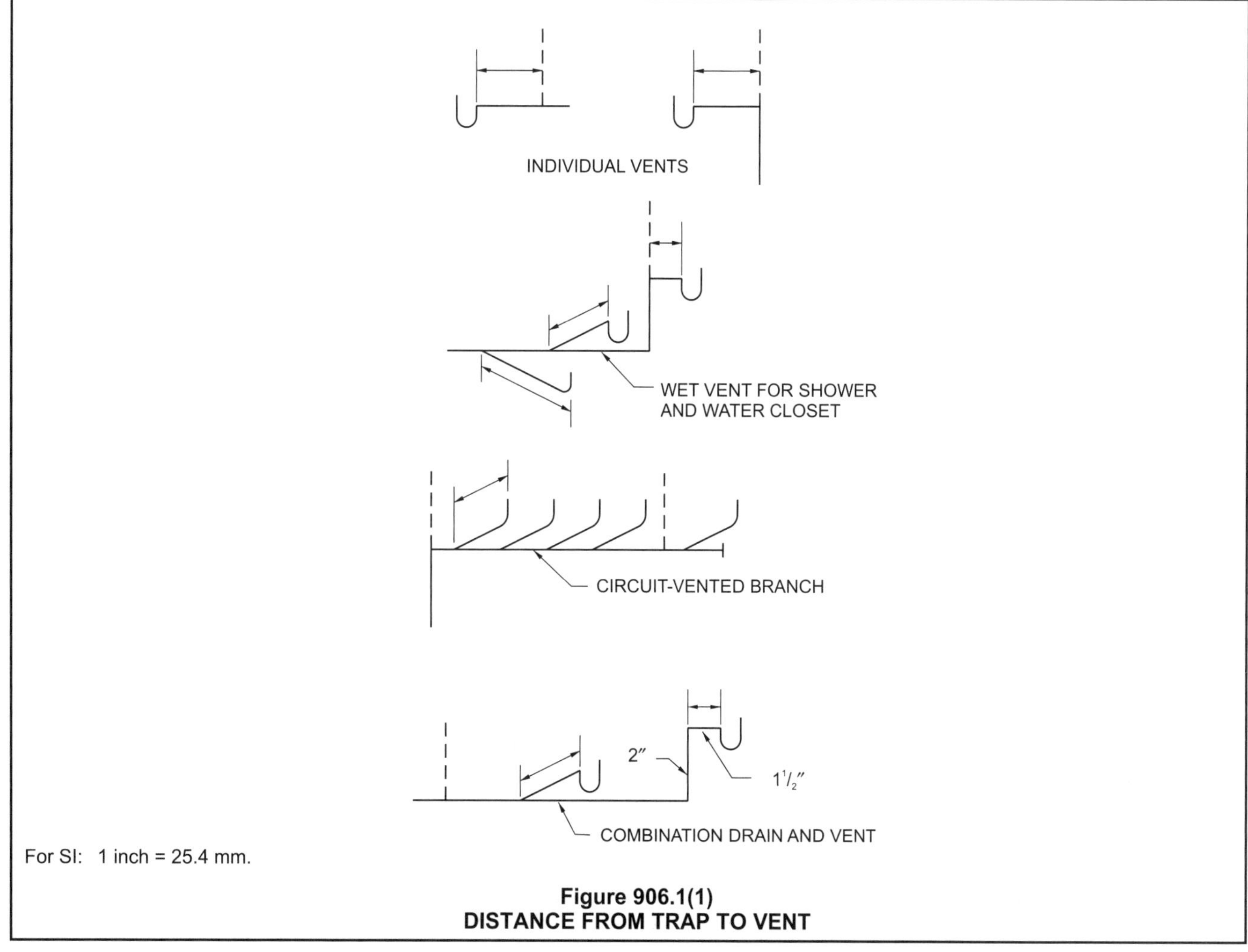

Figure 906.1(1)
DISTANCE FROM TRAP TO VENT

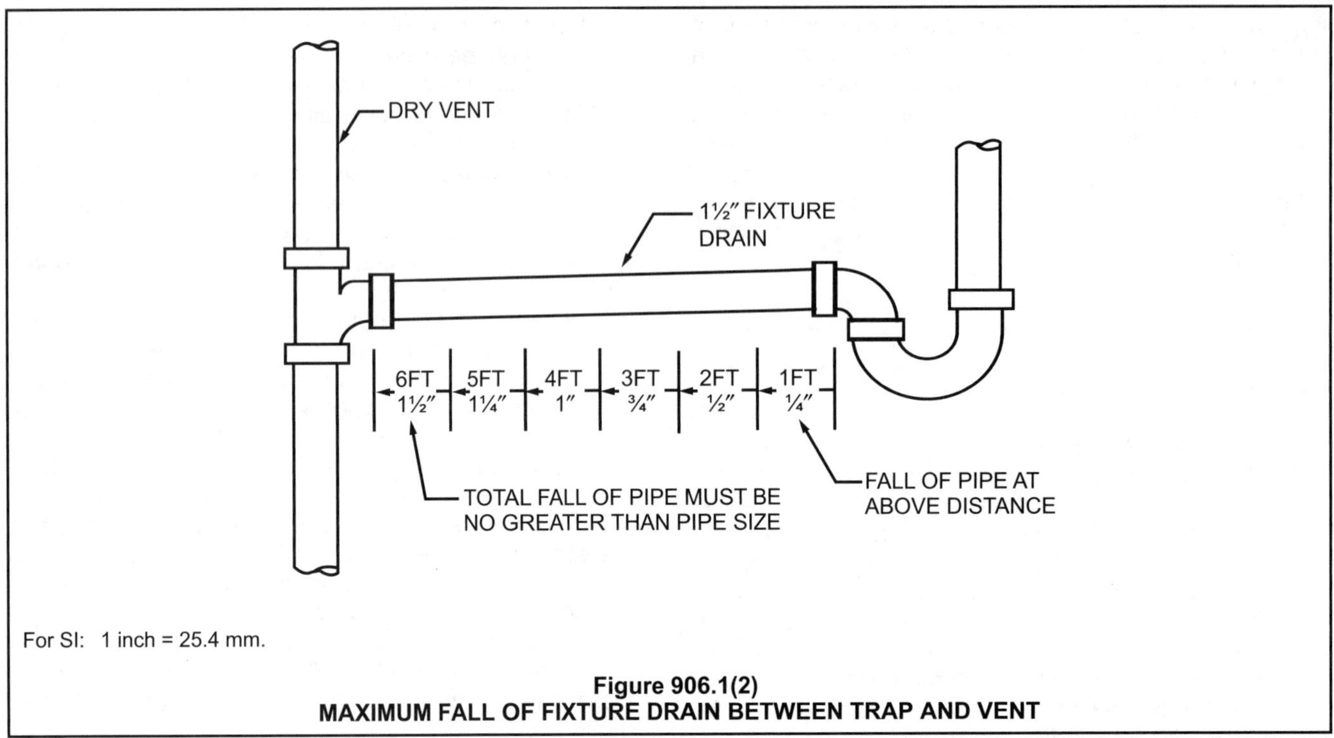

For SI: 1 inch = 25.4 mm.

Figure 906.1(2)
MAXIMUM FALL OF FIXTURE DRAIN BETWEEN TRAP AND VENT

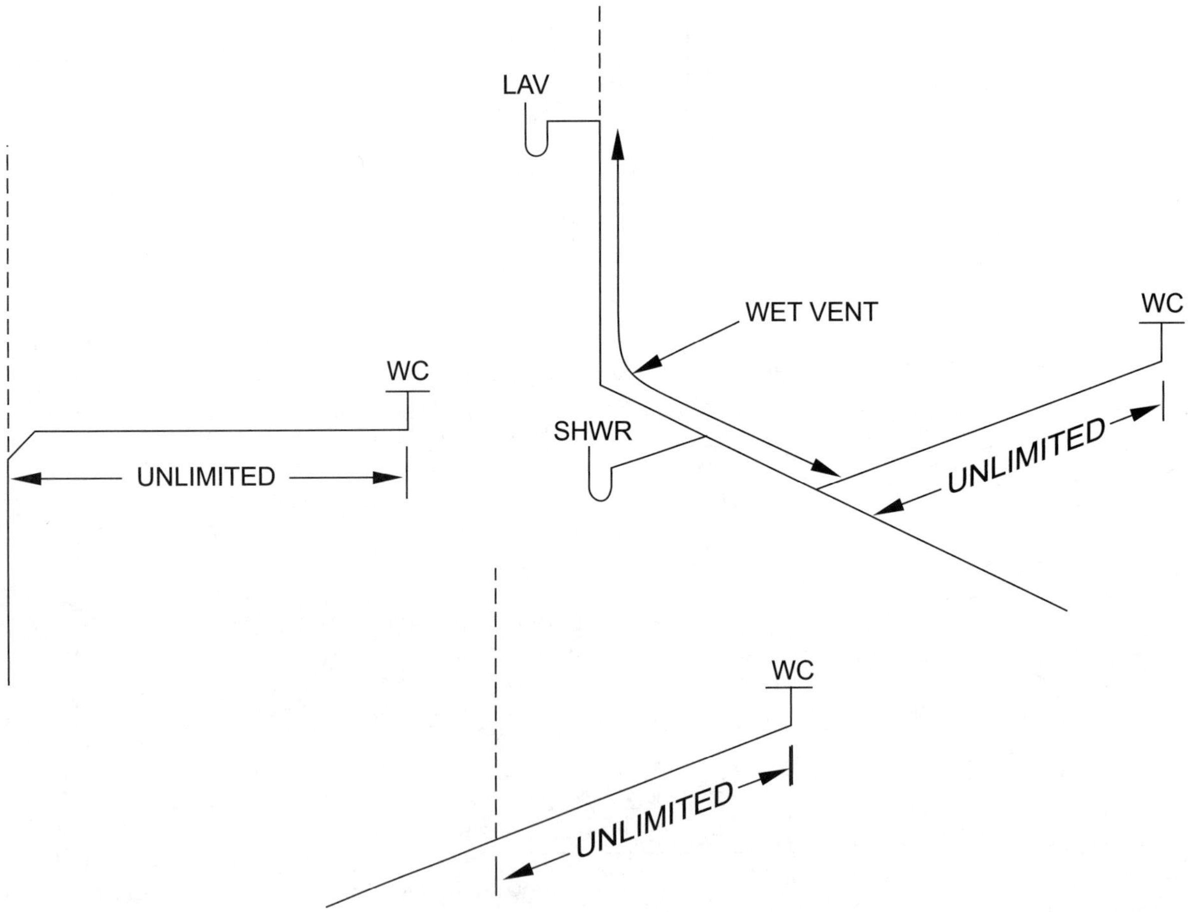

LAV
WET VENT
WC
WC
UNLIMITED
SHWR
UNLIMITED
WC
UNLIMITED

Figure 906.1(3)
TRAP TO VENT DISTANCE FOR WATER CLOSETS

B
A
LONG-PATTERN FITTING NOT ACCEPTABLE. THE TRAP WEIR (A) RISES ABOVE THE VENT INLET (B) WHICH INCREASES THE POSSIBILITY OF SELF-SIPHONAGE.
B
A
TRAP WEIR (A) IS ABOVE THE VENT INLET (B) WHICH IS UNACCEPTABLE. SUCH AN ARRANGEMENT IS SIMILAR TO CREATING AN "S" TRAP BY USE OF PIPE FITTINGS.
$^{1}/_{4}$" PER FOOT PITCH
$1^{1}/_{2}$"
(1) HIGHEST VENT OPENING
(2) TRAP WEIR
VENT IS PERMITTED TO CONNECT BELOW THE TRAP WEIR FOR FIXTURES THAT DEPEND ON SELF-SIPHONING FOR TRAP CONTENT EVACUATION.
THE TRAP WEIR (2) MUST BE PLACED BELOW THE HIGHEST OPENING TO THE VENT (1).

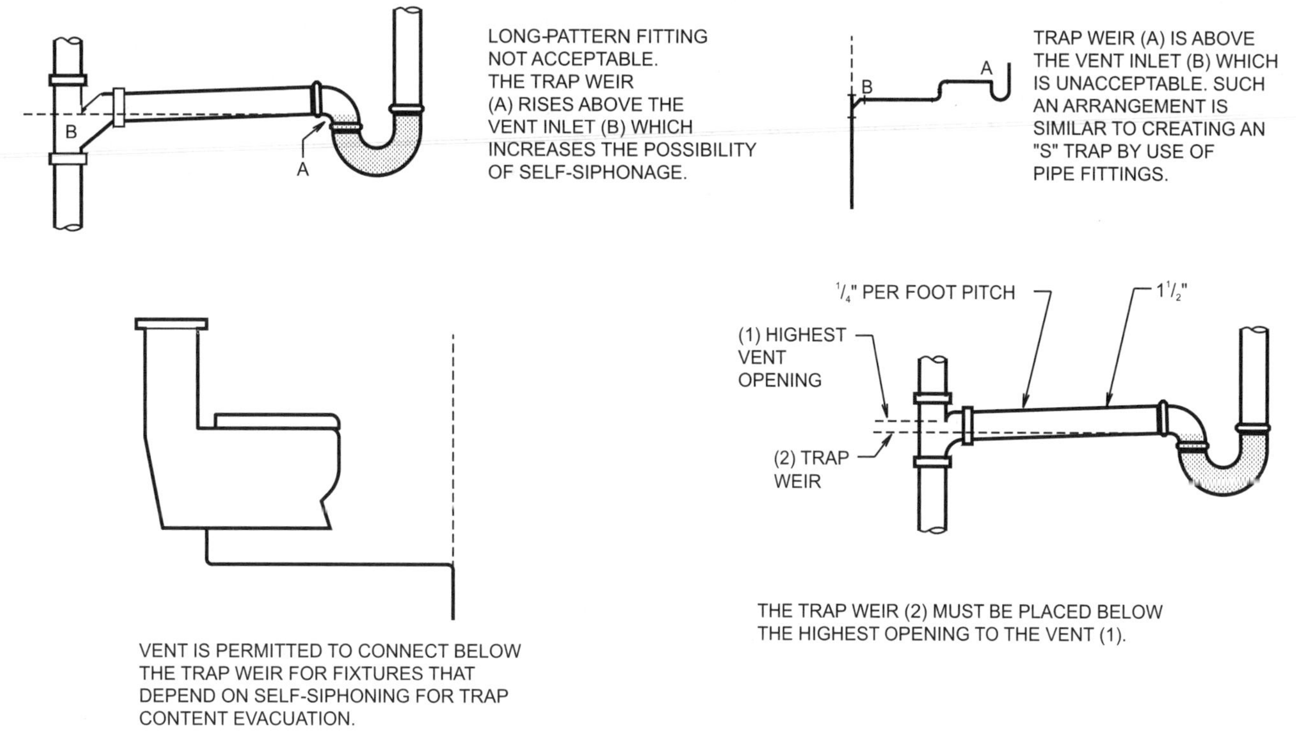

For SI: 1 inch = 25.4 mm, 1 inch per foot = 83.33 mm/m.

Figure 906.2(1)
VENT CONNECTION

TRAP FULL OF WATER
AIR
TRAP WATER
COMPRESSED AIR
TRAP WATER LEAVING TRAP
BOWL WATER FLOWING OUT
PRESSURE OF ATMOSPHERE AND MOMENTUM FORCING WATER OUT
GRAVITY BREAKING WATER SURFACE ALLOWING AIR TO ENTER
AIR
SIPHON BROKEN
WATER IN LEG RUNNING BACK BY GRAVITY
LOSS OF SEAL
WATER LEFT NOT SUFFICIENT TO FILL TRAP

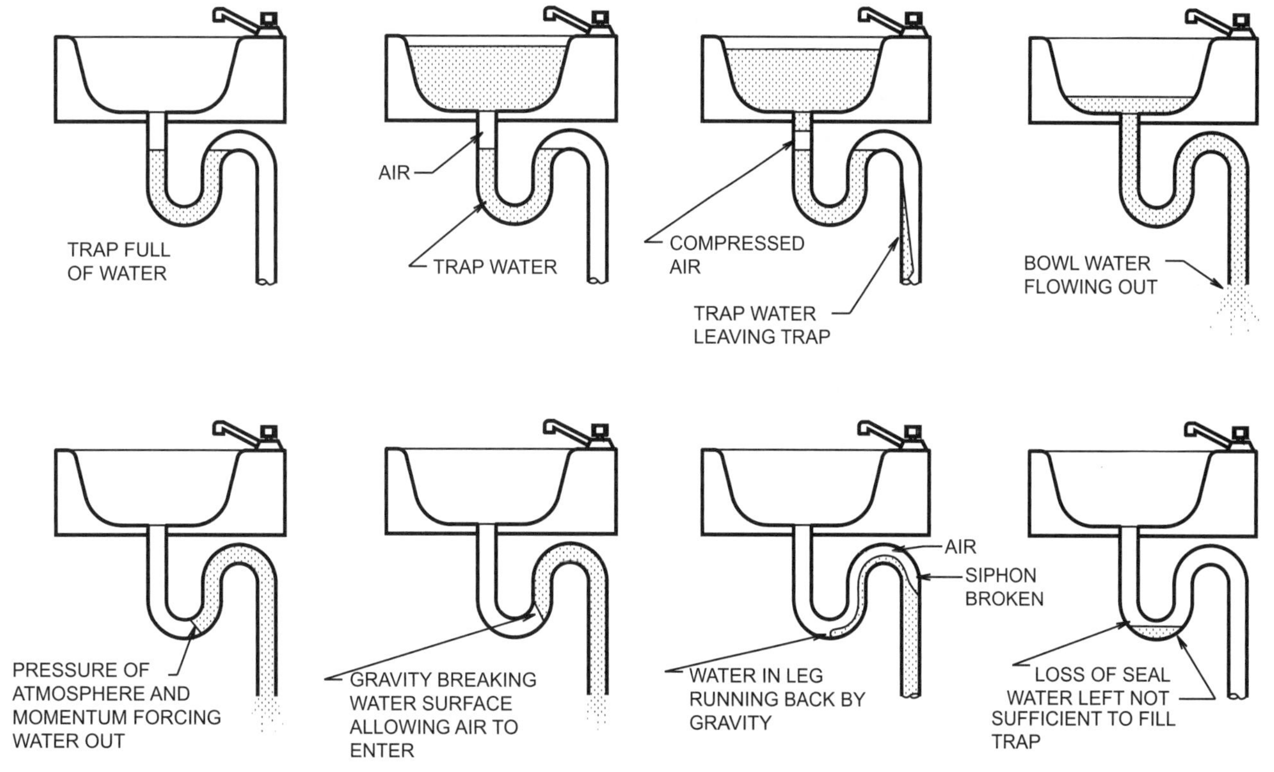

Figure 906.2(2)
SELF-SIPHONING PHENOMENON

906.3 Crown vent. A vent shall not be installed within two pipe diameters of the trap weir.

❖ Crown venting is any arrangement in which a vent connects at the top of the weir (crown) of a trap. Crown vented "P" and "S" traps used to be manufactured with vent connections as an integral part of the trap.

The problem with this type of connection is that the vent opening becomes blocked, thus closing the vent. The blockage is a result of the action of the drainage flowing through the trap. The flow direction and velocity will force waste up into the vent connection, eventually clogging it with debris. The vent connection must therefore be a minimum of two pipe diameters downstream from the trap weir to prevent the vent from becoming blocked [see Commentary Figures 906.3(1) and 906.3(2)].

SECTION 907
INDIVIDUAL VENT

907.1 Individual vent permitted. Each trap and trapped fixture is permitted to be provided with an individual vent. The individual vent shall connect to the *fixture drain* of the trap or trapped fixture being vented.

❖ The simplest form of venting any trap or trapped fixture is an individual vent for each trap. A single vent pipe is connected between the trap or fixture and the branch connection to the drainage system. The protecting vent must be between the fixture trap and the source of pressure fluctuations. If drainage flows from another fixture between the vent connection and the fixture trap, the vent is not installed as an individual vent. The only drainage flowing past a properly installed individual vent is from the fixture being vented.

An individual vent was previously called a "revent." This expression dates back to the original concepts in venting when it was believed that the only vent necessary was an extension of the drainage stack through the roof. It was later determined that a stack vent was inadequate for protecting the trap seals. Individual vents were added to assist the stack vent in protecting the trap seals. The individual vent was called a "revent" because it was considered to be additional venting to an already-vented system. The definition of the prefix "re-" is "repetition of a previous action." The name was changed to "individual vent" when it was determined that a stack vent was unnecessary if each trap was protected by a vent [see Commentary Figures 907.1(1) and 907.1(2)].

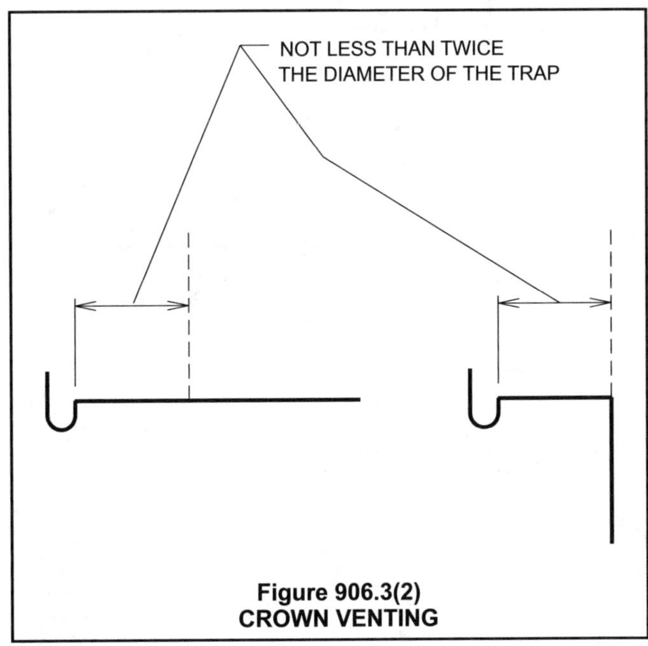

Figure 906.3(2)
CROWN VENTING

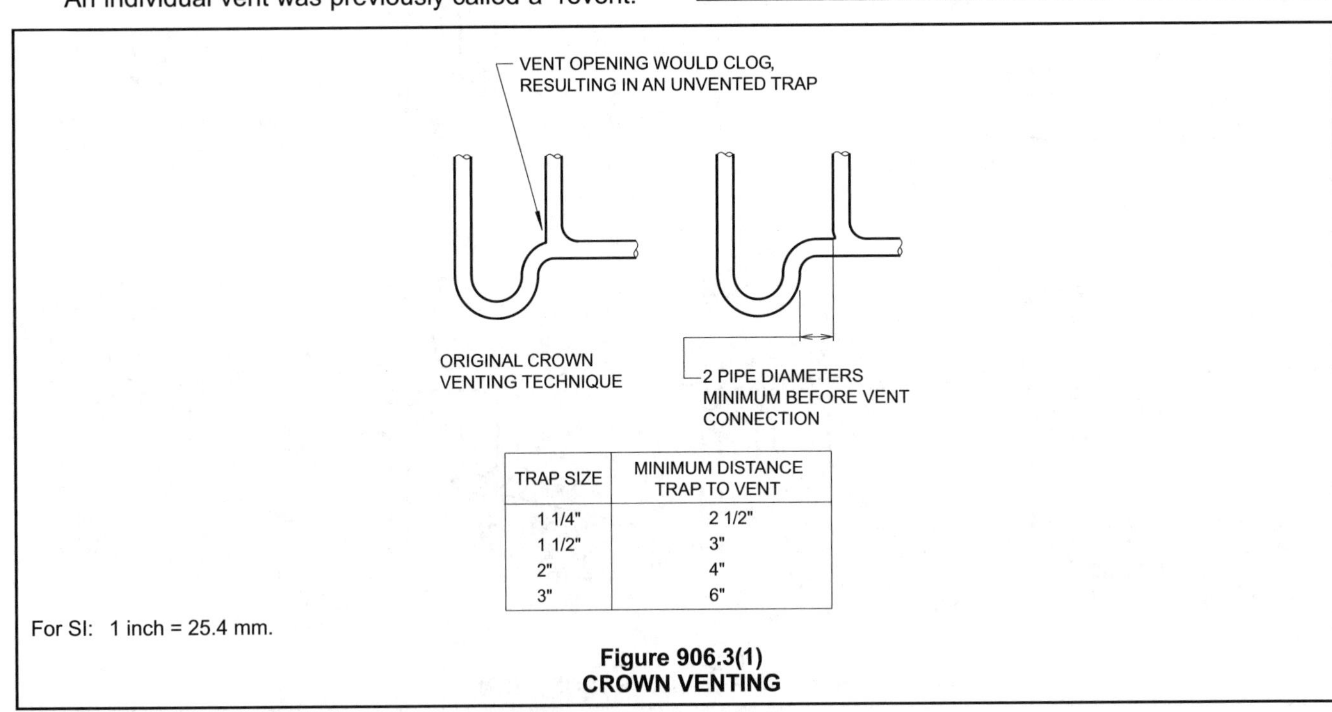

TRAP SIZE	MINIMUM DISTANCE TRAP TO VENT
1 1/4"	2 1/2"
1 1/2"	3"
2"	4"
3"	6"

Figure 906.3(1)
CROWN VENTING

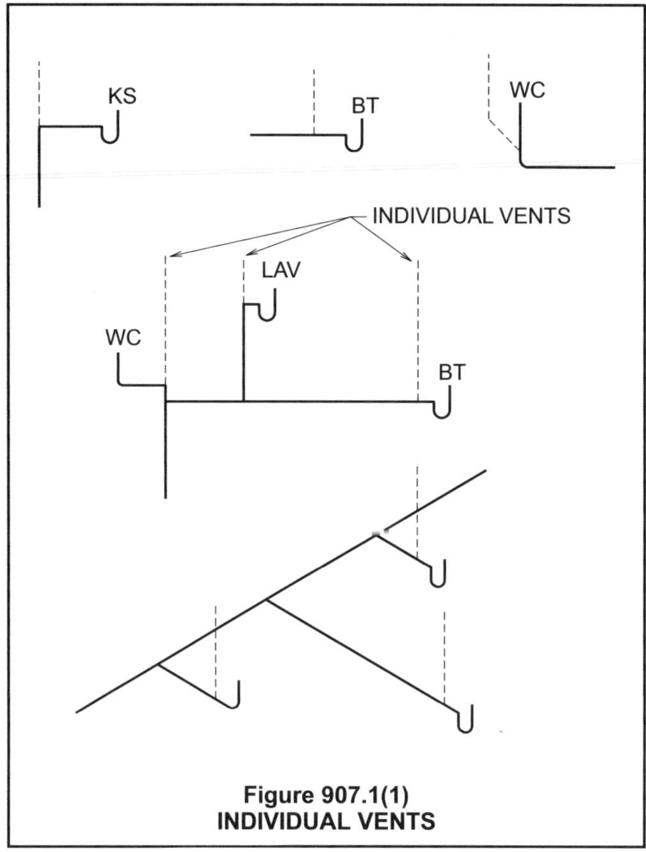

Figure 907.1(1)
INDIVIDUAL VENTS

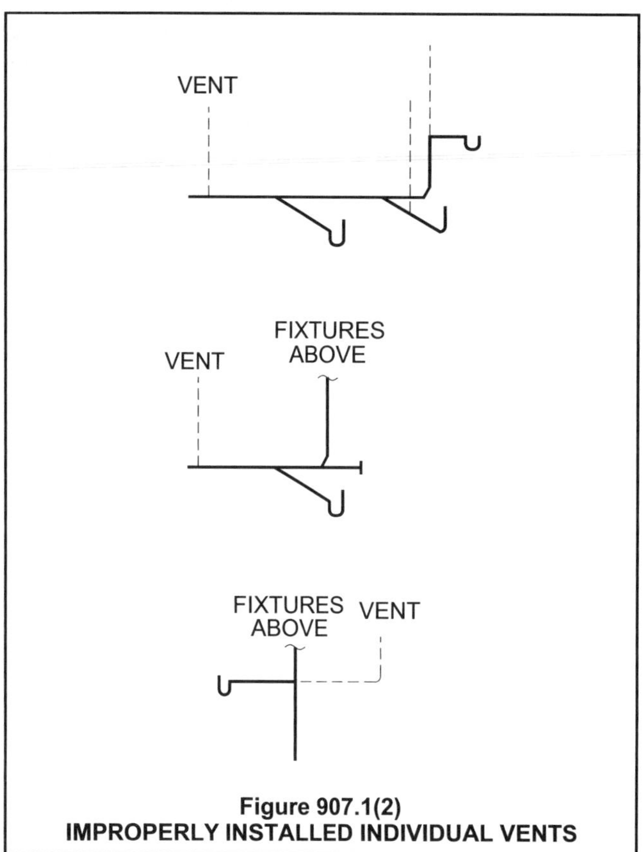

Figure 907.1(2)
IMPROPERLY INSTALLED INDIVIDUAL VENTS

SECTION 908
COMMON VENT

908.1 Individual vent as common vent. An individual vent is permitted to vent two traps or trapped fixtures as a common vent. The traps or trapped fixtures being common vented shall be located on the same floor level.

❖ For common venting, the vent is classified and sized as an individual vent. The fixture drains being common vented are allowed to connect to either a vertical or horizontal drainage pipe. Any two fixtures can be common vented. Where pump-action discharge fixtures or water closets are common vented, Section 706.3 requires directional drainage fittings to prevent the discharge action of one fixture from interfering with the other fixture or causing a blowback into another fixture.

If two water closets are installed back-to-back at the same level and are common vented, the connection to the vertical drainage branch must be through a double combination wye and eighth bend. An exception for the installation of back-to-back water closet connections allows the use of a double sanitary tee where the distance between the outlet of the water closet and the connection to the double sanitary tee pattern is 18 inches (457 mm) or more. The fixtures being common vented must be located on the same floor level so that the flow velocities are kept low by the limiting distance of the vertical drop of waste [see Commentary Figures 908.1 and 706.3(5)].

908.2 Connection at the same level. Where the fixture drains being common vented connect at the same level, the vent connection shall be at the interconnection of the fixture drains or downstream of the interconnection.

❖ The most popular form of common venting is connecting two fixtures at the same level. A typical installation is two horizontal drains connecting to a vertical drain through a double-pattern fitting. The extension of the vertical pipe serves as the vent; however, two drains connecting horizontally to a horizontal drain through a double-pattern fitting is also a form of common venting. The vent is allowed to connect at the interconnection of the fixture drains or downstream along the horizontal drain.

The requirements of Section 905 regarding vent connection would apply, as well as those of Section 906. The trap-to-vent distances for each of the fixtures must comply with Table 906.1 (see Commentary Section 908.1 and Commentary Figure 908.2).

908.3 Connection at different levels. Where the fixture drains connect at different levels, the vent shall connect as a vertical extension of the vertical drain. The vertical drain pipe connecting the two fixture drains shall be considered the vent for the lower *fixture drain*, and shall be sized in accordance with Table 908.3. The upper fixture shall not be a water closet.

❖ This section applies only to vertical drain applications. Two fixtures within a single story are allowed to connect at different levels to a vertical drain and still be common vented. The vent remains an individual vent serving as a common vent. The drain pipe between the upper fixture and the lower fixture, however, must meet the oversizing requirement specified in Table 908.3. Although drainage flows in the pipe between the upper and lower fixture connections, the section of pipe is not classified as a wet vent despite the fact that it functions in the same manner as a wet vent. The pipe is wet and does serve as a vent; however, it is classified as a common vent for purposes of the code (see Commentary Figure 908.3).

The fixture at the higher level is not allowed to be a water closet because the pressures created by the high-volume rapid discharge of a water closet could adversely affect the venting of lower fixtures.

TABLE 908.3
COMMON VENT SIZES

PIPE SIZE (inches)	MAXIMUM DISCHARGE FROM UPPER FIXTURE DRAIN (dfu)
$1^1/_2$	1
2	4
$2^1/_2$ to 3	6

For SI: 1 inch = 25.4 mm.

❖ The sizing of the common vent serving as a drain for the upper fixture and vent for the lower fixture is the same as the sizing for a wet vent. The same concepts of venting apply, hence the same sizing. The dry vent extension above the upper fixture connection is sized as an individual vent and is based on the largest fixture drain served. For example, the size of a dry vent serving a lavatory and water closet connecting at different levels is $1^1/_2$ inches (38 mm) because the water closet fixture drain cannot be less than 3 inches (76 mm).

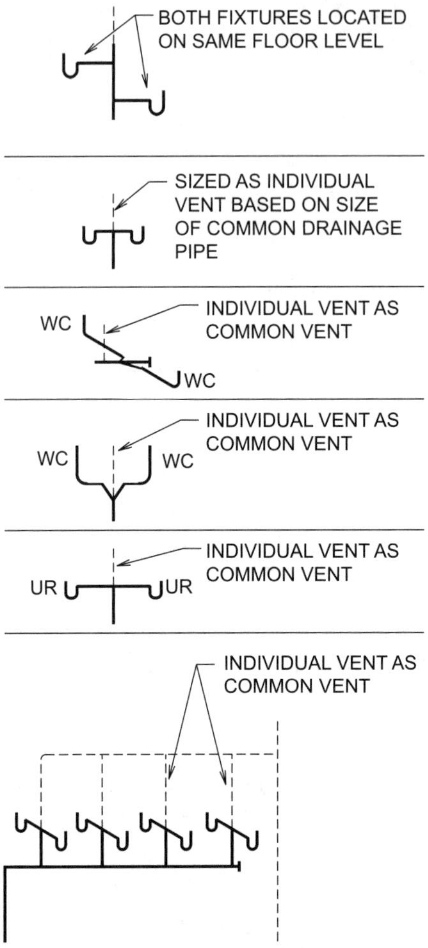

BOTH FIXTURES LOCATED ON SAME FLOOR LEVEL

SIZED AS INDIVIDUAL VENT BASED ON SIZE OF COMMON DRAINAGE PIPE

WC
INDIVIDUAL VENT AS COMMON VENT
WC

INDIVIDUAL VENT AS COMMON VENT
WC WC

INDIVIDUAL VENT AS COMMON VENT
UR UR

INDIVIDUAL VENT AS COMMON VENT

Figure 908.1
COMMON VENTS

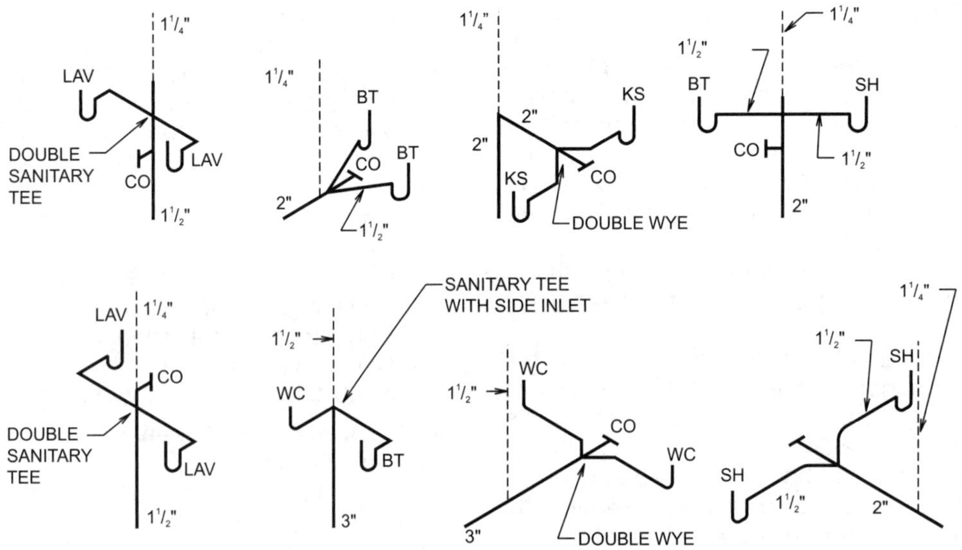

LAV
$1^1/_4$"
DOUBLE SANITARY TEE
CO
LAV
$1^1/_2$"
$1^1/_4$"
BT
CO
BT
2"
$1^1/_2$"
$1^1/_4$"
2"
2"
KS
KS
CO
DOUBLE WYE
$1^1/_4$"
$1^1/_2$"
BT
SH
CO
$1^1/_2$"
2"
LAV
$1^1/_4$"
CO
DOUBLE SANITARY TEE
LAV
$1^1/_2$"
SANITARY TEE WITH SIDE INLET
$1^1/_2$"
WC
BT
3"
$1^1/_2$"
WC
CO
WC
3"
DOUBLE WYE
$1^1/_4$"
$1^1/_2$"
SH
SH
$1^1/_2$"
2"

For SI: 1 inch = 25.4 mm.

Figure 908.2
COMMON VENTS WITH FIXTURE DRAINS CONNECTING AT SAME LEVEL

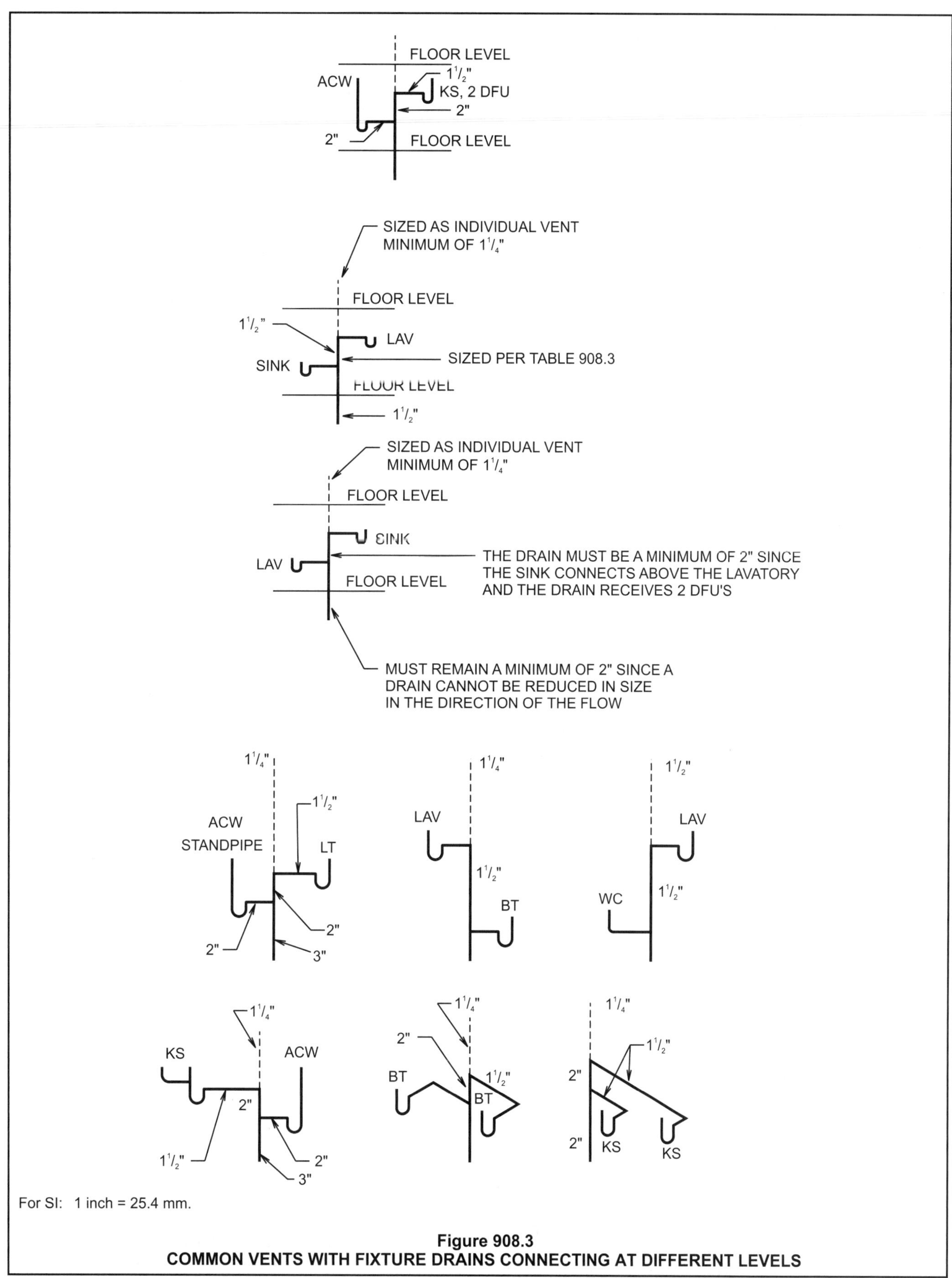

Figure 908.3
COMMON VENTS WITH FIXTURE DRAINS CONNECTING AT DIFFERENT LEVELS

SECTION 909
WET VENTING

909.1 Horizontal wet vent permitted. Any combination of fixtures within two bathroom groups located on the same floor level is permitted to be vented by a horizontal wet vent. The wet vent shall be considered the vent for the fixtures and shall extend from the connection of the dry vent along the direction of the flow in the drain pipe to the most downstream *fixture drain* connection to the *horizontal branch drain*. Each wet-vented *fixture drain* shall connect independently to the horizontal wet vent. Only the fixtures within the bathroom groups shall connect to the wet-vented *horizontal branch drain*. Any additional fixtures shall discharge downstream of the horizontal wet vent.

❖ A horizontal wet vent is a horizontal branch drain pipe that is sized larger than is required by Section 710. The oversizing of the pipe allows for a large air space above the maximum probable waste flow level in the pipe so as to provide for adequate venting airflow in the same pipe as the waste flow. In other words, the horizontal wet vent is special type of vent pipe that is "wet" because it carries waste flow. Commentary Figure 909.1(1) shows a commonly installed horizontal wet vent for a single bathroom group. A horizontal wet vent can be used only for a limited number and type of fixtures contained in a maximum of two bathroom groups (see definition of "Bathroom group" in Section 202).

The acceptance of horizontal wet venting for bathroom groups is based on the low probability of simultaneous discharge from the types of fixtures within the group, significant over-sizing of the drain pipe, and the limited number of drainage fixtures units allowed. Years of satisfactory performance of horizontal wet venting of bathroom groups in single family residential dwellings [as covered by the *International Residential Code®* (IRC®)] has proven this method of venting. This method is similar to the long-proven methods of circuit venting and combination drain and venting arrangements which also use air space above the waste flow level for venting fixtures connected to a horizontal drain pipe.

The maximum number and type of fixtures that can be accommodated by a horizontal wet vent are: two water closets, two lavatories, two bidets, two emergency floor drains (see definition of "Emergency floor drain" in Section 202), and two bathing fixtures (either bathtubs or showers). Note that a urinal is not in the list of bathroom group fixtures that can be served by a horizontal wet vent. A horizontal wet vent could have as few as two fixtures or as many as ten fixtures but not more than two fixtures of any type can be connected to the system. For example, three lavatories or three water closets cannot be connected to a horizontal wet vent. No other types of fixtures are allowed to discharge into the horizontal wet vent system. All fixtures connecting to a horizontal wet vent must be on the same floor level to avoid high flow velocities which could cause excessive turbulence that might block the air space (venting area) above the waste flow level.

As with other horizontal wet venting methods (e.g., combination drain and vent, and circuit vent), the main section of the horizontal wet vent pipe is intended to be installed horizontally without vertical offsets with each fixture drain connected "independently" to the horizontal drain in order to keep flow turbulence at a minimum. Note that fixtures connecting though a double pattern fitting are considered to be independently connected to the horizontal wet vent. The individual fixture drains vented by a horizontal wet vent must enter the horizontal wet vent in the horizontal plane so that the trap weir of the fixture trap is not above the connection. Water closets, having integral traps that depend on self-siphoning, are exempt from this requirement. Other fixtures that are not vented by the horizontal wet vent (e.g., vented by a dry vent) can enter the horizontal wet vent horizontally, vertically or at any angle in between.

The slope of the piping and the required fittings must be in accordance with Chapter 7. While there is no specified maximum slope for the horizontal sections of horizontal wet vent piping, the intent is to avoid high flow velocities which might disrupt the air space for venting. Thus, the horizontal wet vent pipe is intended to be installed at or near the minimum slope requirements for horizontal drains.

Section 909.2.1 regulates the dry vent connection to the horizontal wet vent. The dry vent for a horizontal wet vent must connect to a vent stack, stack vent, air admittance valve or terminate outdoors.

The required vent could be a dry vent connected directly to the main section of the horizontal wet vent pipe downstream of the most upstream fixture. See Commentary Figure 909.1(2). In this arrangement, the dry vent is also the individual vent for the most upstream fixture. The required vent could be a dry vent for common vented fixture. See Commentary Figure 909.1.(3).

The last sentence of Section 909.2.1 reinforces the definition of a horizontal wet vent so that improper fixture connections are avoided. A horizontal wet vent begins at the dry vent connection and extends down-

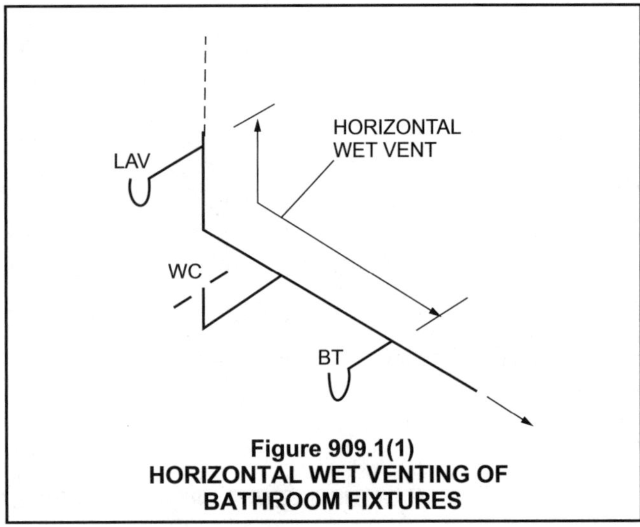

Figure 909.1(1)
HORIZONTAL WET VENTING OF BATHROOM FIXTURES

stream to the last fixture being wet vented. The horizontal wet vent does not extend upstream of the dry vented fixture connection. Thus, multiple interconnected fixture drains that connect upstream of the dry vented fixture connection are not vented by the horizontal wet vent. Commentary Figures 909.1(4) and (5) are examples of prohibited arrangements.

Not all fixtures connected to a horizontal wet vent are required to be vented by the horizontal wet vent. See Commentary Figure 909.1(6). In this arrangement the shower trap was required to be located at a distance that exceeded the maximum trap-to-vent distance requirements of Table 906.1, thus requiring the trap to be independently vented. Any number of individually vented fixtures of the types allowed for a bathroom group can connect to the horizontal wet vent as long as the total number of fixtures allowed for the bathroom groups is not exceeded. [See Commentary Figure 909.1(7)].

The sizing for both the horizontal and vertical "wet" sections of a horizontal wet vent is regulated by Section 909.3 and Table 909.3. A unique feature of a horizontal wet vent that is different from other horizontal "wet type" venting methods described in this chapter is that the pipe does not have to be of uniform size throughout. The horizontal wet vent pipe size need only increase in size to accommodate the connected load and provide adequate air space above the flow line for venting. Note that the last fixture downstream, vented by the horizontal wet vent, does not contribute to the dfu loading and thus does not affect the sizing of the wet vent. Where a water closet is connected to a horizontal wet vent, the piping must be at least 3 inches (76 mm) in diameter from that point onward downstream. Pipe sizes larger than the minimum required size might also be necessary in order to satisfy clean-out access provisions as required by Chapter 7. The code does not specify a maximum length for a horizontal wet vent. However, since this method was originally developed for applications where two bathroom groups were adjacent or back-to-back, the intent for the use of this method is that the fixtures to be horizontally wet vented are in the same vicinity and not in remote groups. However, the code is silent as to how

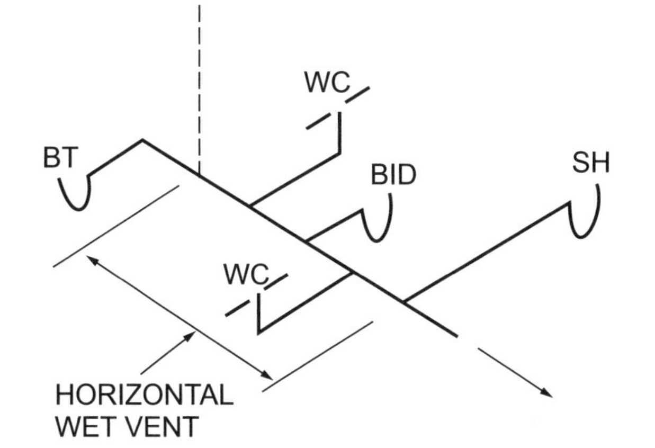

Figure 909.1(2)
HORIZONTAL WET VENT WITH INDIVIDUAL VENT

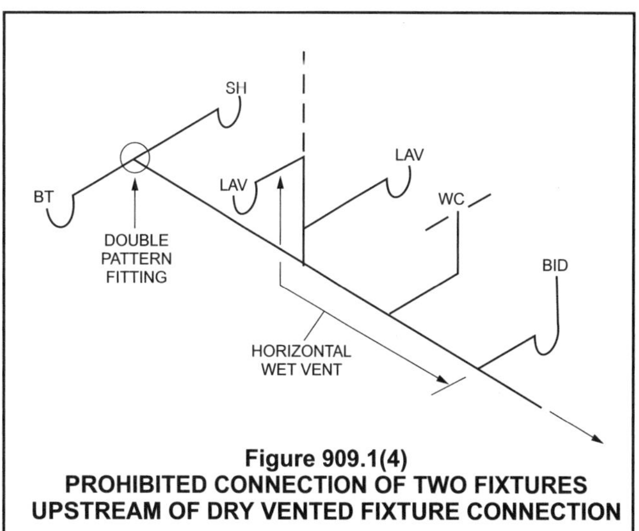

Figure 909.1(4)
PROHIBITED CONNECTION OF TWO FIXTURES UPSTREAM OF DRY VENTED FIXTURE CONNECTION

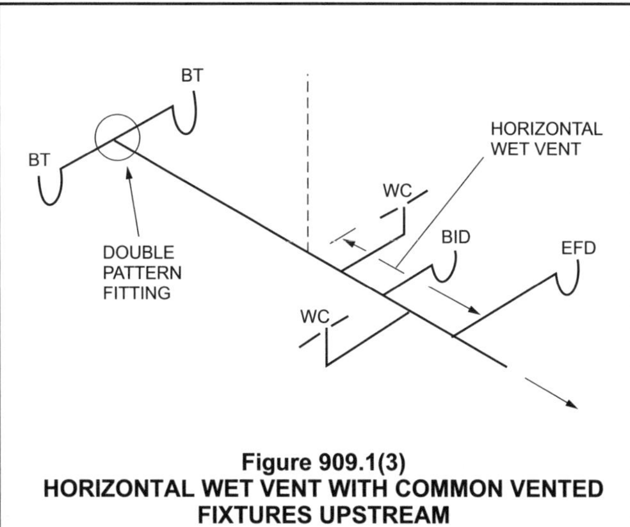

Figure 909.1(3)
HORIZONTAL WET VENT WITH COMMON VENTED FIXTURES UPSTREAM

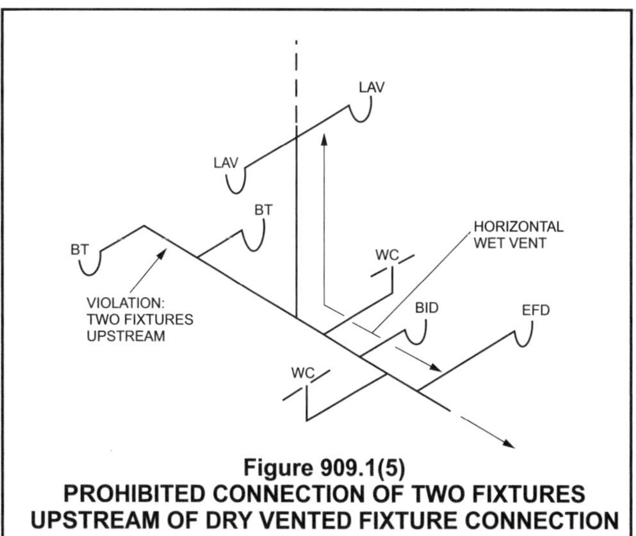

Figure 909.1(5)
PROHIBITED CONNECTION OF TWO FIXTURES UPSTREAM OF DRY VENTED FIXTURE CONNECTION

long a horizontal wet vent can be.

Various acceptable horizontal wet vent arrangements, complete with the required pipe sizes, are shown in Commentary Figures 909.1(8) through (12). Horizontal offsets are shown, as well as pipe size increases as required by the increase in dfu loading in accordance with Table 909.3. Using pipe sizes greater than those required by Table 909.3 is acceptable. However, as with any horizontal drain installed at near minimum slope, excessive oversizing could result in sluggish flow and clogging problems.

Commentary Figure 909.1(11) is a good example to illustrate the dry vent sizing requirements for horizontal wet vents. The size of the required dry vent for a horizontal wet vent is regulated by Section 909.3. For dry vents that are not the "required" dry vent for the horizontal wet vent, sizing is to be in accordance with the appropriate sections covering those venting methods. For example, if the dry vent of an individually vented lavatory fixture is not serving as the required vent for the horizontal wet vent, then the minimum size of the dry vent is $1^1/_4$ inches (32 mm) in accordance with Section 916.2. However, if this dry vent is the "required" dry vent for a horizontal wet vent that serves a 3-inch drain (76 mm), then the required dry vent must be at least $1^1/_2$ inches (38 mm) as the required vent must be at least one-half the size of the "largest required diameter of pipe within the wet-vented system." This is the reason why the dry vent shown in Commentary Figure 909.1 (11) is $1^1/_2$ inches (38 mm) and not $1^1/_4$ inches (32 mm).

Prohibited horizontal wet vented arrangements are illustrated in Commentary Figures 909.1(13) through (18). Notes in each figure indicate the reason for noncompliance.

Commentary Figures 909.1(19) through (24) show examples of proper horizontal wet venting.

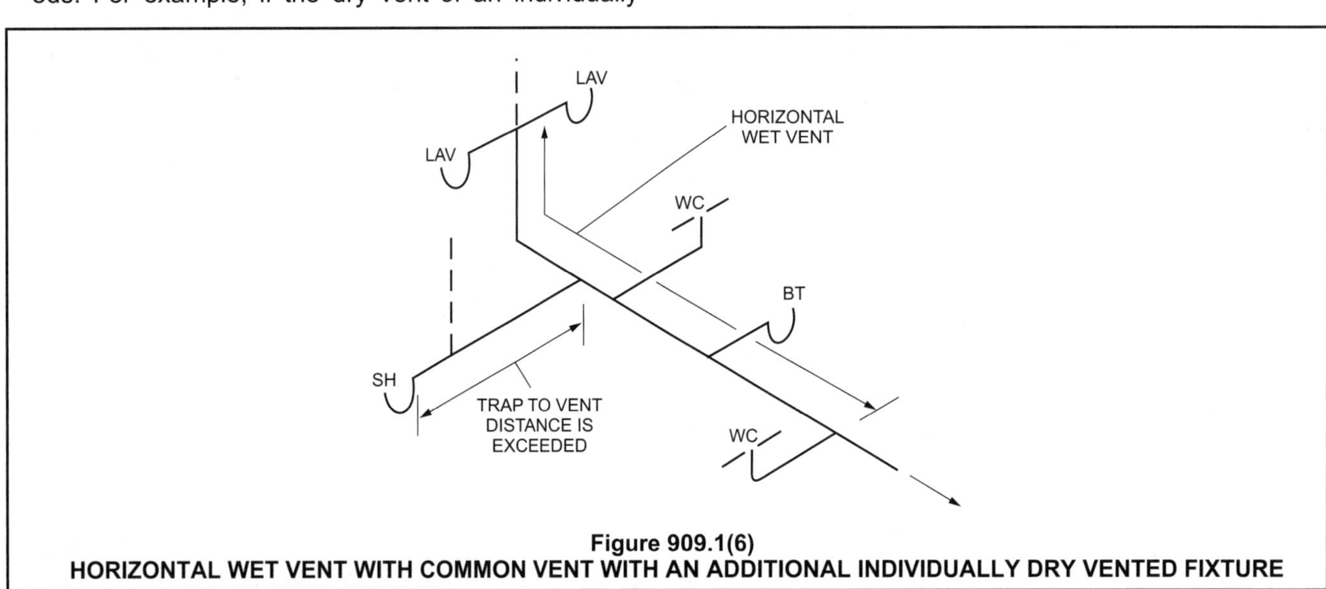

Figure 909.1(6)
HORIZONTAL WET VENT WITH COMMON VENT WITH AN ADDITIONAL INDIVIDUALLY DRY VENTED FIXTURE

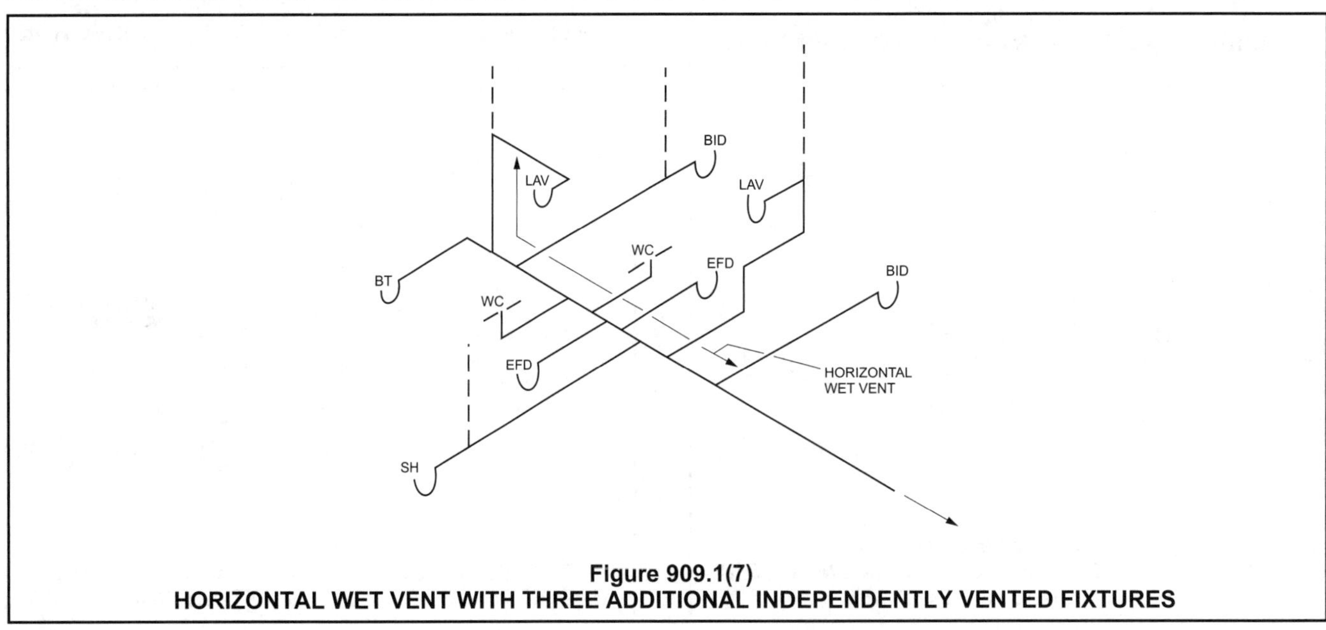

Figure 909.1(7)
HORIZONTAL WET VENT WITH THREE ADDITIONAL INDEPENDENTLY VENTED FIXTURES

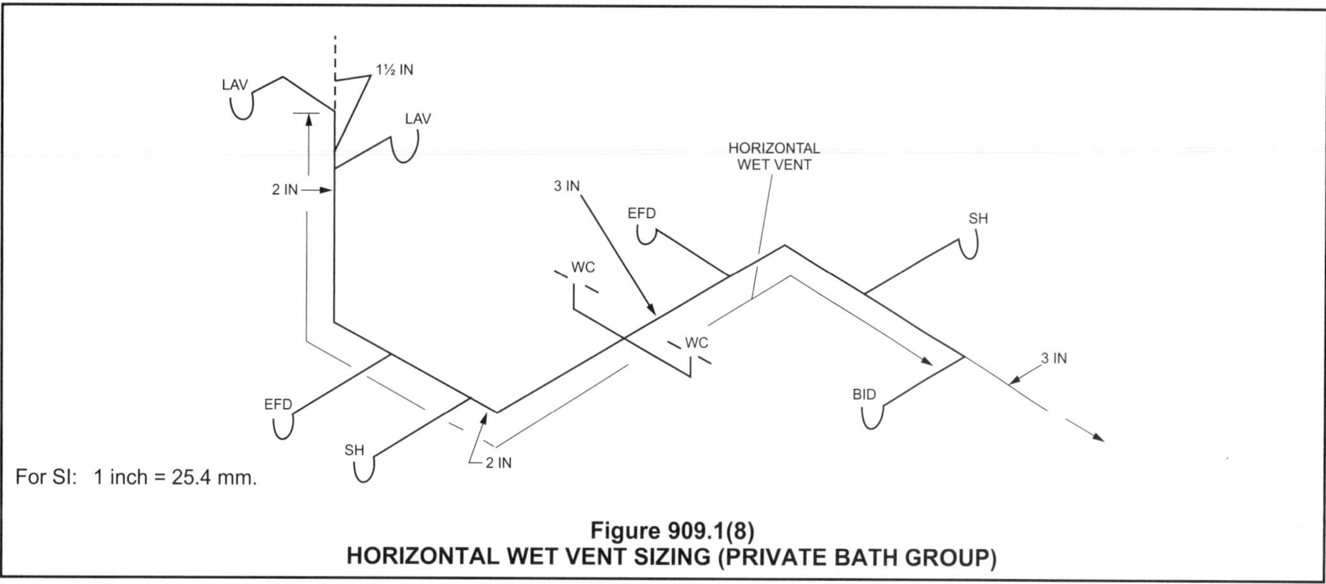

For SI: 1 inch = 25.4 mm.

Figure 909.1(8)
HORIZONTAL WET VENT SIZING (PRIVATE BATH GROUP)

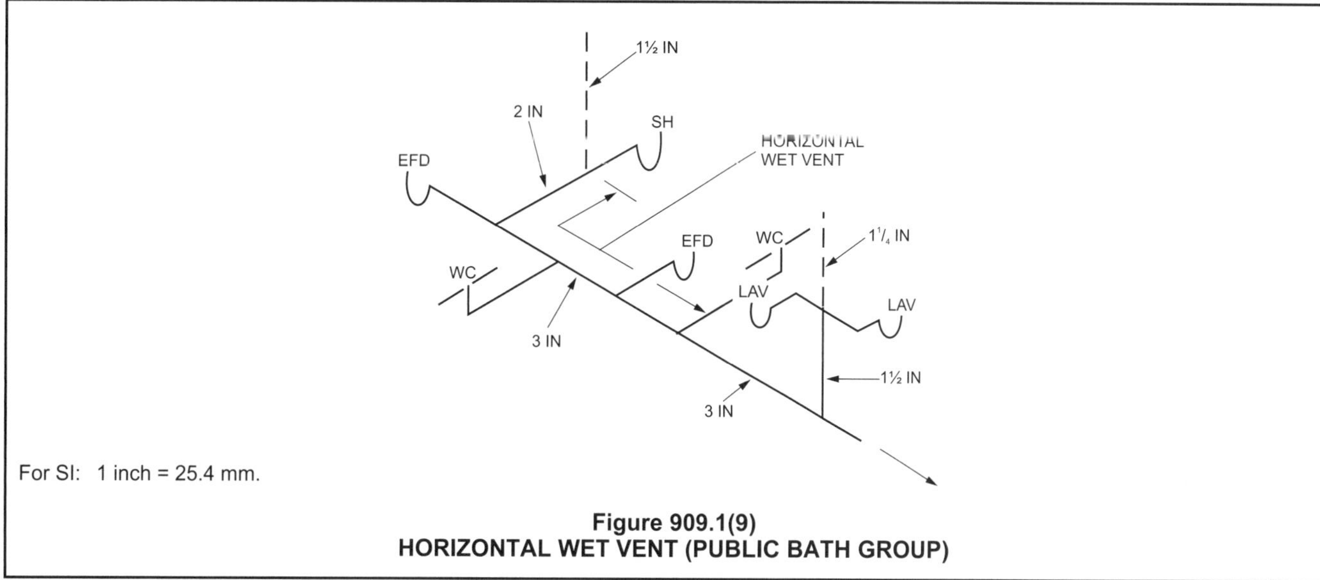

For SI: 1 inch = 25.4 mm.

Figure 909.1(9)
HORIZONTAL WET VENT (PUBLIC BATH GROUP)

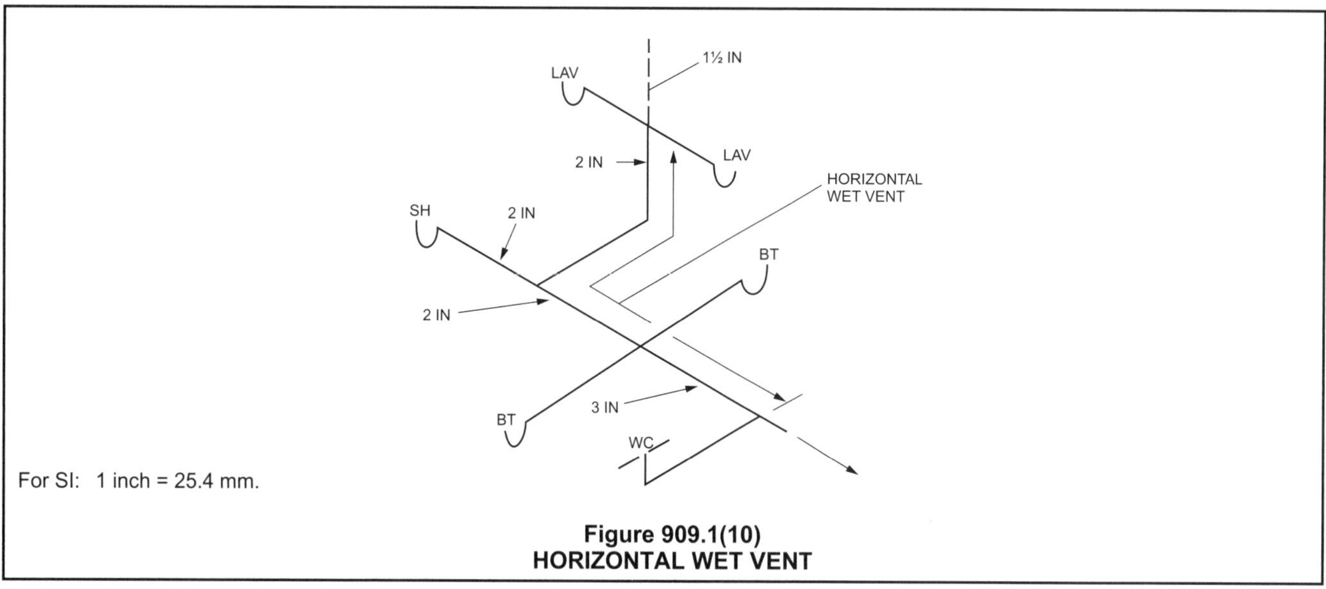

For SI: 1 inch = 25.4 mm.

Figure 909.1(10)
HORIZONTAL WET VENT

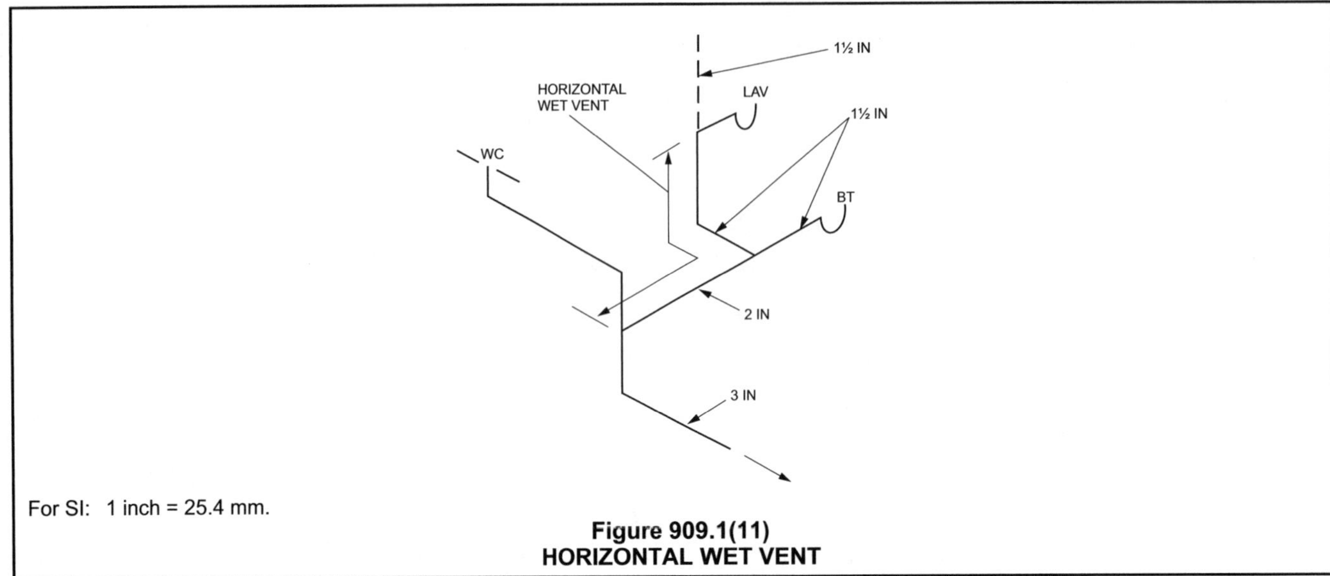

Figure 909.1(11)
HORIZONTAL WET VENT

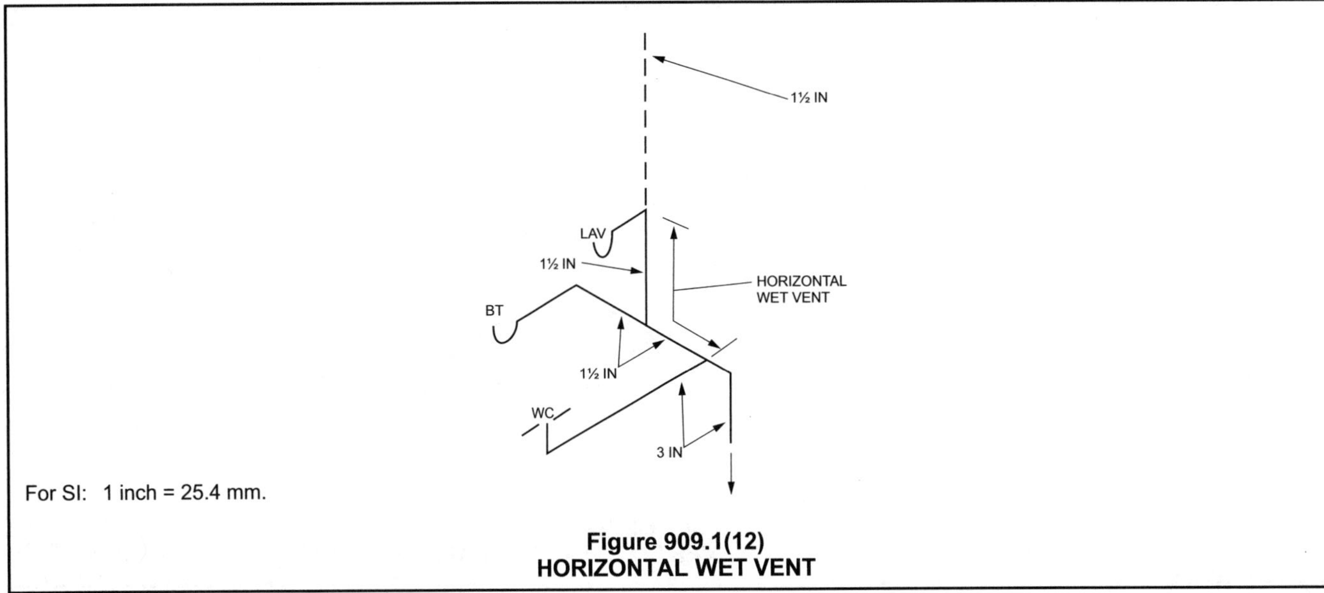

Figure 909.1(12)
HORIZONTAL WET VENT

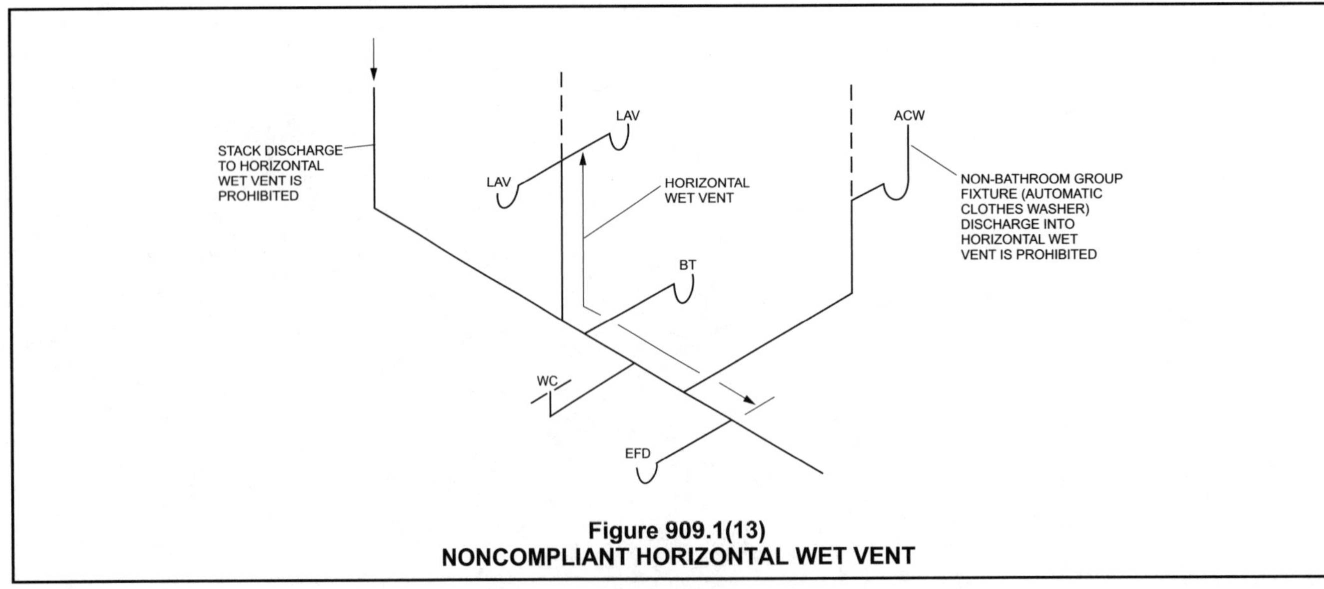

Figure 909.1(13)
NONCOMPLIANT HORIZONTAL WET VENT

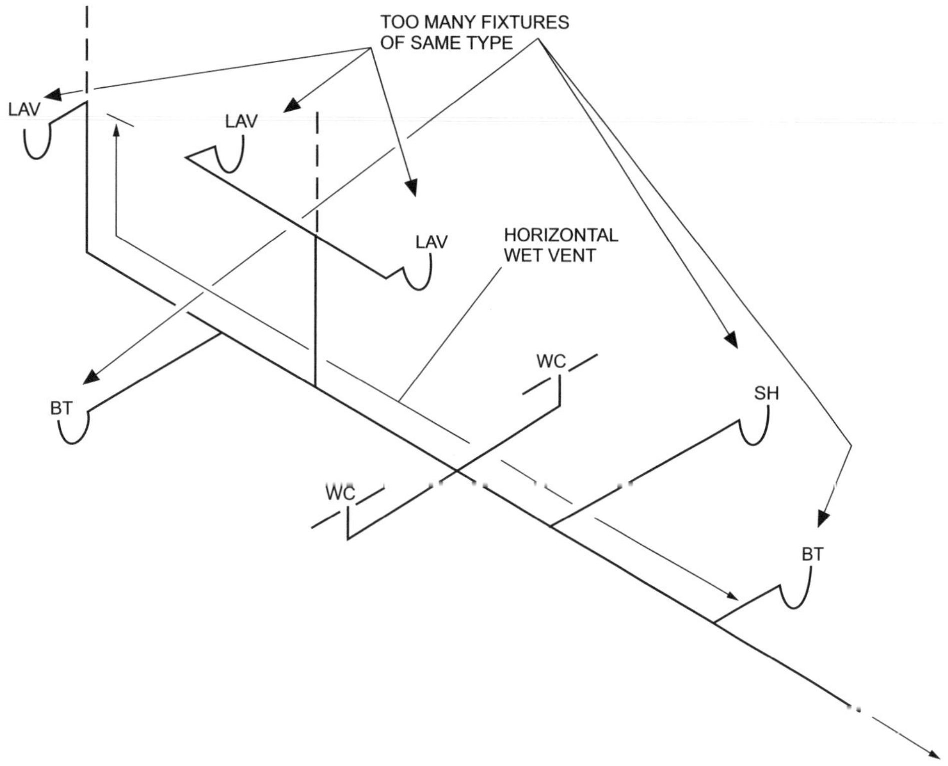

Figure 909.1(14)
NONCOMPLIANT HORIZONTAL WET VENT

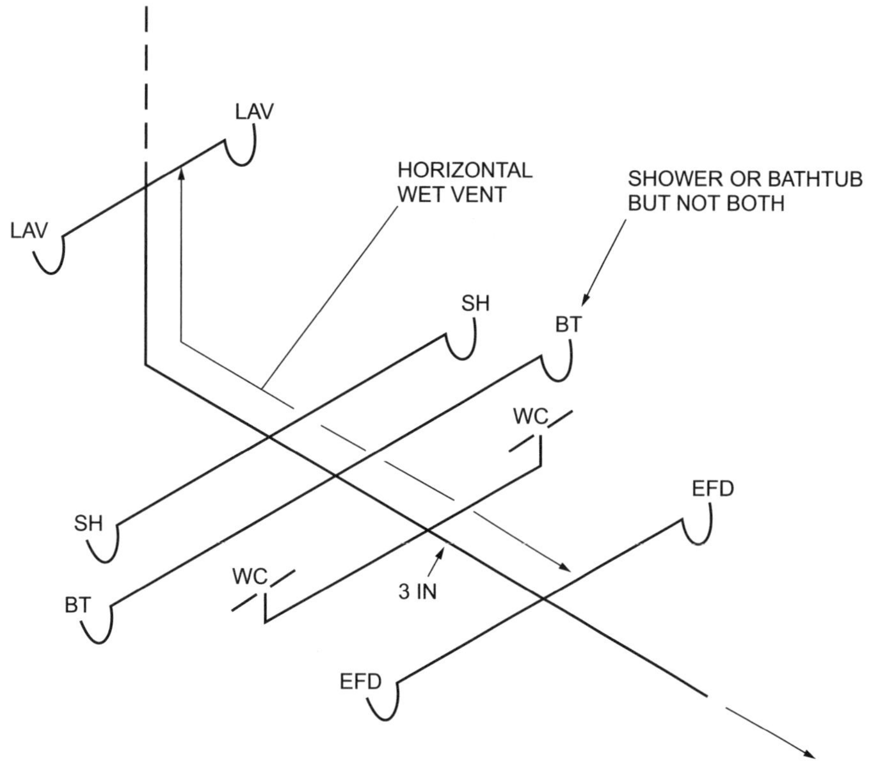

Figure 909.1(15)
NONCOMPLIANT HORIZONTAL WET VENT

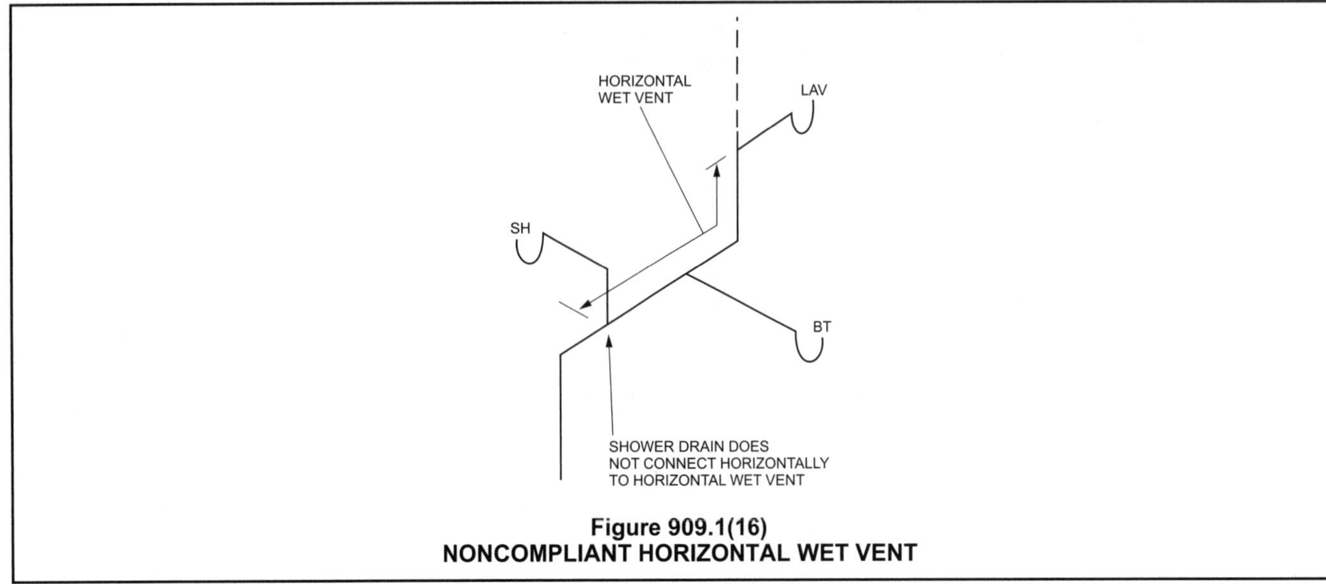

Figure 909.1(16)
NONCOMPLIANT HORIZONTAL WET VENT

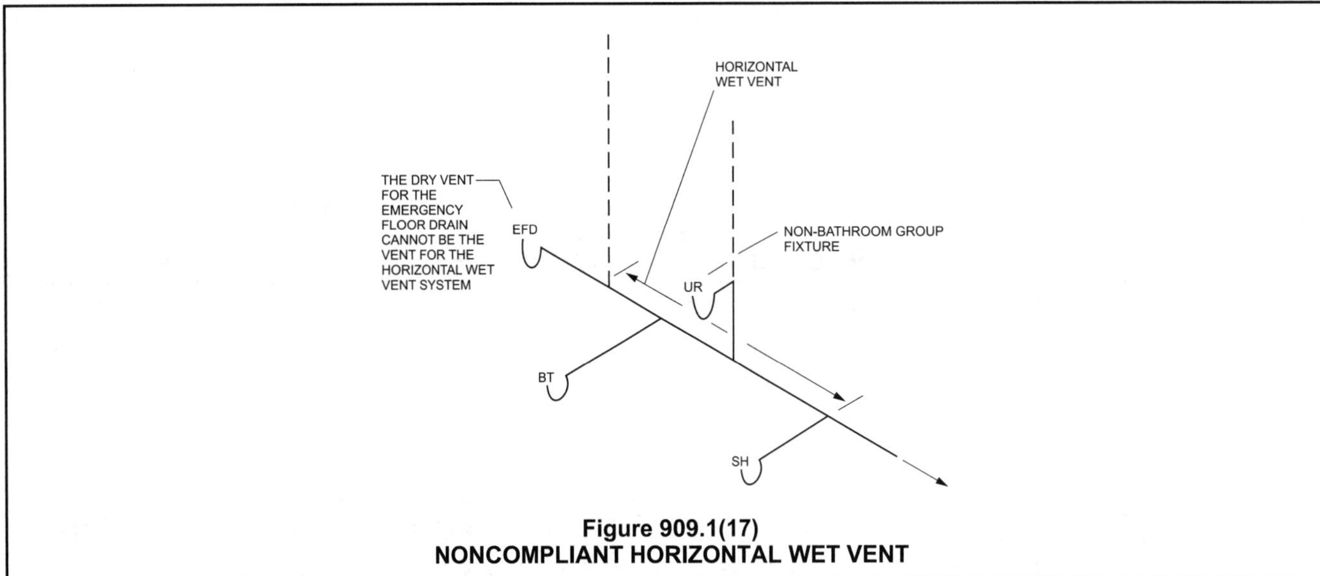

Figure 909.1(17)
NONCOMPLIANT HORIZONTAL WET VENT

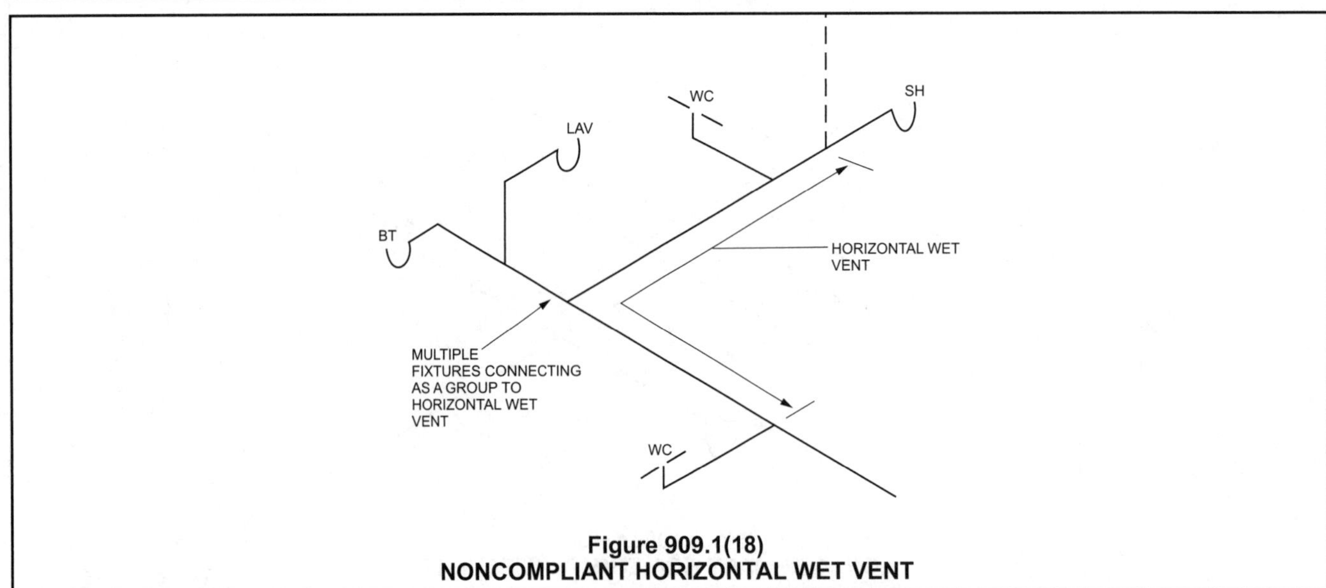

Figure 909.1(18)
NONCOMPLIANT HORIZONTAL WET VENT

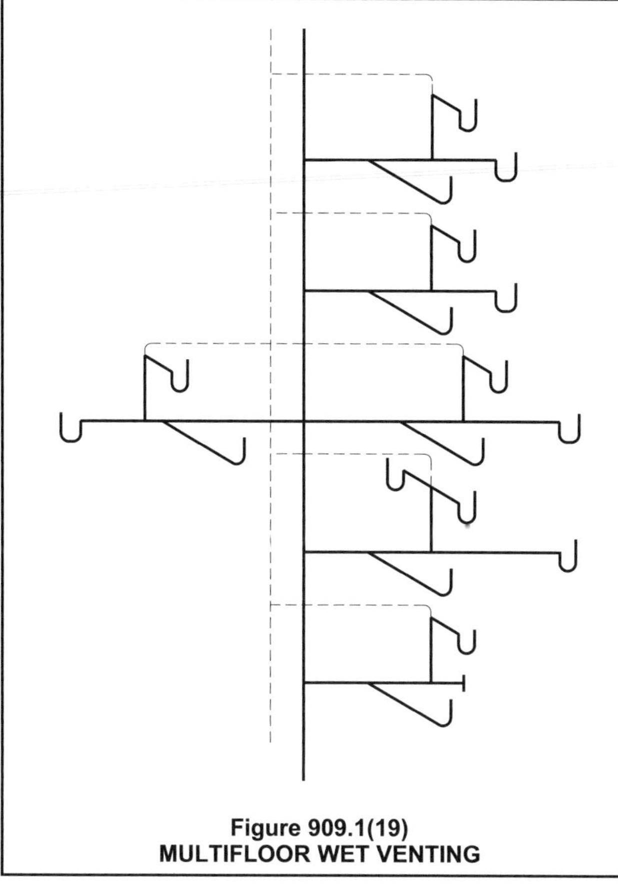

Figure 909.1(19)
MULTIFLOOR WET VENTING

DRAIN COMPONENT	MIN. SIZE	DFU (TABLE 709.1)
LAV FIXTURE DRAIN	1 1/4"	1
SH FIXTURE DRAIN	1 1/2"	2
WC FIXTURE DRAIN	3"	3
A-B[a]	1 1/2"[a]	1
B-C[a]	2"[a]	3
A-X	1 1/2"	NA

a. FROM TABLE 909.3
NA. NOT APPLICABLE

THE WET VENT EXTENDS FROM C TO A

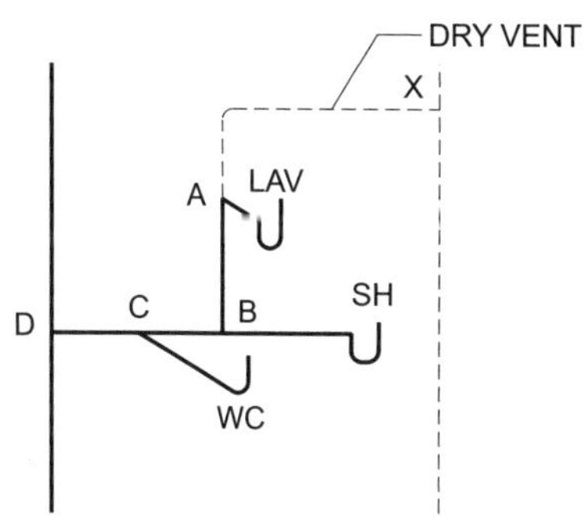

SIZING OF THE SYSTEM IS ACCORDING TO TABLE

For SI: 1 inch = 25.4 mm.

Figure 909.1(20)
SINGLE BATHROOM GROUP (PRIVATE) WET VENT

THE DISTANCE FROM EACH TUB TO POINT C MUST BE WITHIN THE DISTANCE SPECIFIED IN TABLE 906.1

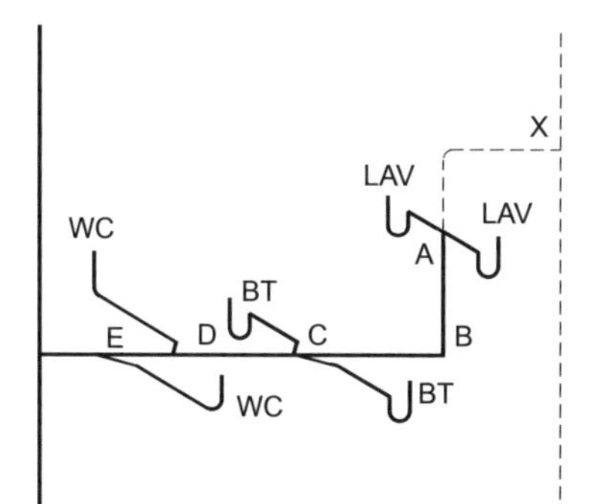

DRAIN COMPONENT	MIN. SIZE	DFU (TABLE 709.1)
LAV DRAIN	1 1/4"	1
BT DRAIN	1 1/2"	2
WC DRAIN	3"	3
A-B[a]	2"[a]	2
B-C[a]	2"[a]	2
C-D[a]	2 1/2"[a]	6
D-E[a]	3"[a]	9
A-X	1 1/2"	NA

a. FROM TABLE 909.3
NA. NOT APPLICABLE

For SI: 1 inch = 25.4 mm.

Figure 909.1(21)
DOUBLE BATHROOM GROUP (PRIVATE) WET VENT

SIZE

SECTION	WHEN WET VENT STARTS AT A	WHEN WET VENT STARTS AT C
A-B	1 1/2"	1 1/4"
B-C	1 1/2"	2"
B-D	2"	2"

WHEN THE BATHTUB FIXTURE DRAIN EXCEEDS THE DISTANCE FROM TRAP TO VENT REQUIREMENTS, AN ADDITIONAL INDIVIDUAL VENT IS ADDED FOR THE FIXTURE. THE WET VENT MAY BE CONSIDERED AS EITHER SECTION A-D BY USE OF THE LAV CONNECTION, OR SECTION C-D BY USE OF THE BATHTUB CONNECTION.

For SI: 1 inch = 25.4 mm.

Figure 909.1(22)
WET VENT VARIATIONS FOR A SINGLE BATHROOM GROUP

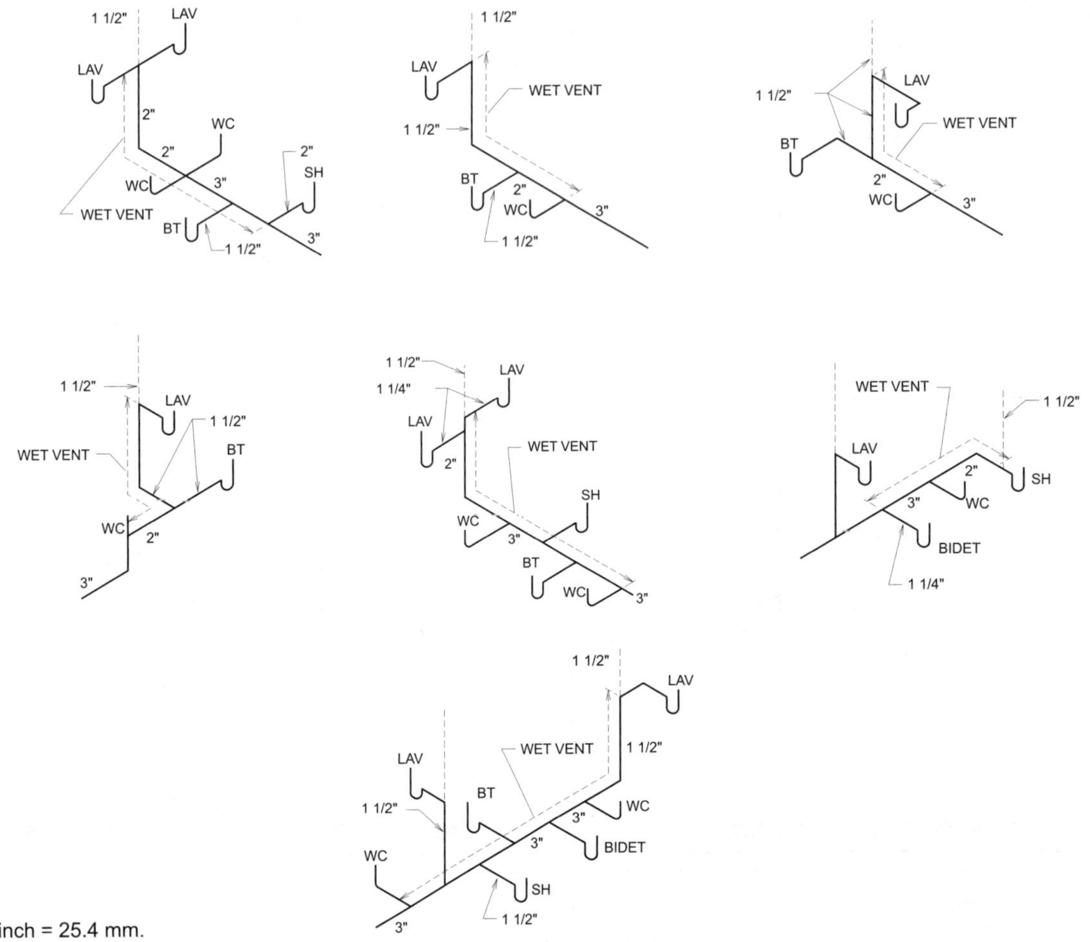

For SI: 1 inch = 25.4 mm.

Figure 909.1(23)
HORIZONTAL WET VENTING

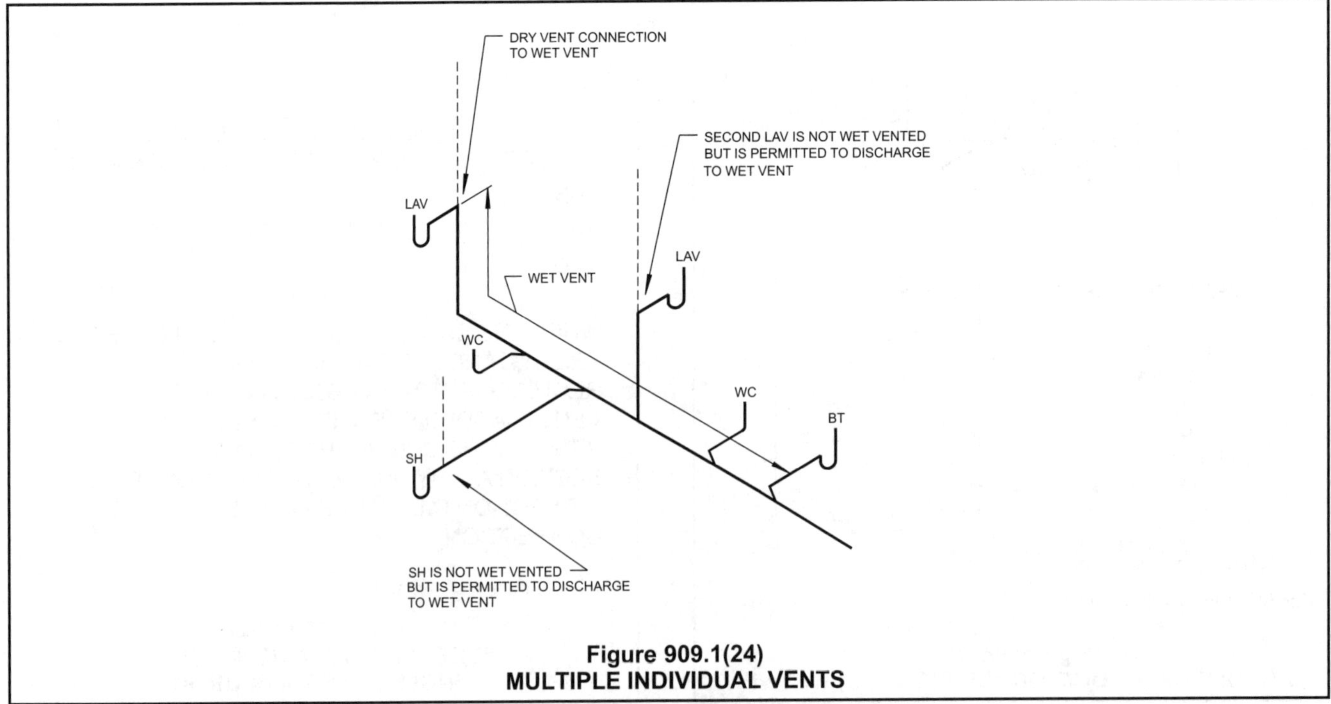

Figure 909.1(24)
MULTIPLE INDIVIDUAL VENTS

909.1.1 Vertical wet vent permitted. Any combination of fixtures within two bathroom groups located on the same floor level is permitted to be vented by a vertical wet vent. The vertical wet vent shall be considered the vent for the fixtures and shall extend from the connection of the dry vent down to the lowest *fixture drain* connection. Each wet-vented fixture shall connect independently to the vertical wet vent. Water closet drains shall connect at the same elevation. Other fixture drains shall connect above or at the same elevation as the water closet fixture drains. The dry vent connection to the vertical wet vent shall be an individual or common vent serving one or two fixtures.

❖ Vertical wet vents used to be referred to as "stack venting." The vertical wet vent extends from the connection to the dry vent down to the lowest fixture drain connection. A common example of vertical wet venting of a single bathroom group is shown in Commentary Figure 909.1.1(1). The two most important requirements of vertical wet venting are that all fixtures must connect above or at the water closet connection elevation and all fixtures must connect independently to the vertical wet vent. Note that fixtures connected through a double pattern fitting are considered to be independently connected. If more than one water closet is to be connected, both must connect at the same elevation [see Commentary Figure 909.1.1(2)]. Note that in this figure, a double sanitary tee is only acceptable where the water closet drain connection to the sanitary tee is 18 inches (457 mm) or more in horizontal developed length distance from the outlet of the water closet (see Section 706.3, Exception). Otherwise, the water closets must connect through double wye and eighth-bend fitting (also known as a "double T-wye") [see Commentary Figure 909.1.1(3)].

For SI: 1 inch = 25.4 mm.

Figure 909.1.1(1)
VERTICAL WET VENT OF SINGLE BATHROOM GROUP

Any combination of fixtures within two bathroom groups located on the same floor level can be installed without individual fixture vents (see definition of "Bathroom group" in Section 202). However, each fixture drain must connect to the wet vent within the maximum length for trap-to-vent distance as specified in Table 906.1. Note that a bathroom group does not include a urinal fixture. Without being specifically stated, the vertical wet vent must be vertical without vertical or horizontal offsets. Commentary Figure 909.1.1(4) illustrates various vertical wet vent arrangements.

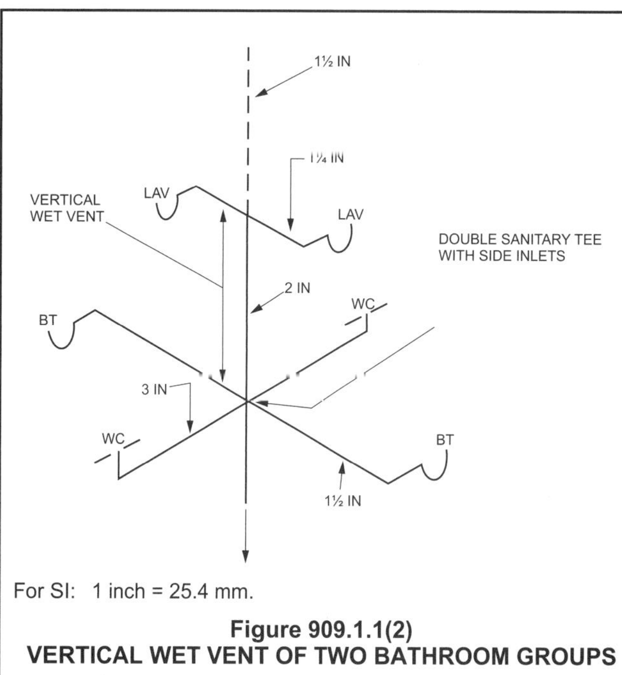

For SI: 1 inch = 25.4 mm.

Figure 909.1.1(2)
VERTICAL WET VENT OF TWO BATHROOM GROUPS

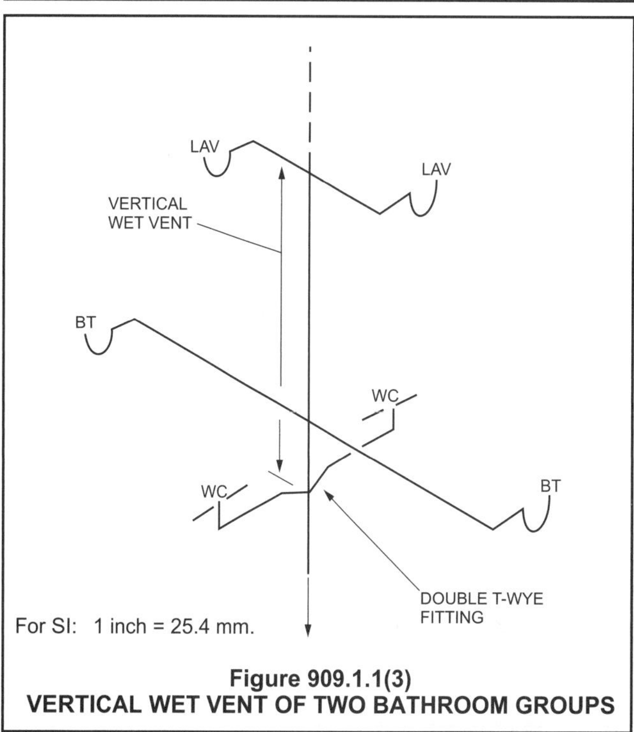

For SI: 1 inch = 25.4 mm.

Figure 909.1.1(3)
VERTICAL WET VENT OF TWO BATHROOM GROUPS

Section 909.2.2 regulates how the dry vent connection is made to the vertical wet vent and Section 909.3 regulates the size of the dry vent. Both are discussed in this commentary section for continuity of understanding. The dry vent is an extension of the vertical wet vent and is connected to a vent stack, stack vent, air admittance valve or point of termination outside of the building. The dry vent size must be at least one-half of the largest required pipe size served by the wet vent. For example, in Commentary Figure 909.1.1(2), the largest pipe served by the vertical wet vent is 3 inches (76 mm), requiring the dry vent to be a minimum of $1^1/_2$ inches (38 mm). If the run of the dry vent from the start of the dry vent fixture connection to the termination of the vent (to a vent stack, stack vent, air admittance valve or termination outdoors) exceeds 40 feet (12 mm) developed length, Section 916.2 requires the dry vent size to be increased by one nominal pipe size for its entire length.

909.2 Dry vent connection. The required dry-vent connection for wet-vented systems shall comply with Sections 909.2.1 and 909.2.2.

❖ See the commentary for Sections 909.2.1 and 909.2.2.

909.2.1 Horizontal wet vent. The dry-vent connection for a horizontal wet-vent system shall be an individual vent or a common vent for any *bathroom group* fixture, except an *emergency floor drain*. Where the dry-vent connects to a water closet *fixture drain*, the drain shall connect horizontally to the horizontal wet-vent system. Not more than one wet-vented *fixture drain* shall discharge upstream of the dry-vented *fixture drain* connection.

❖ See the commentary for Sections 909.1. Previous code editions did not allow a water closet to be the upstream dry vented fixture drain connection. Because a water closet can be dry vented in a non wet vented system, there is not any technical reason why this venting arrangement will not work in a wet vented system. Thus, this edition of the code allows a dry vented water closet to be the dry vented fixture that vents the horizontal wet vent.

Another change for this edition is that two (or more) fixture drains cannot connect upstream of the dry vented fixture drain connection for a horizontal wet vent system. Commentary Figure 909.1(5) illustrates one common violation. As the figure shows, the wet vent does not continue horizontally upstream towards the bathtub. The section of pipe between the first bathtub upstream and the dry vented fixture connection is simply an individual fixture drain the connects to the wet vent. Because the code requires that each fixture trap be provided with its own vent (see Section 906.1), (i.e. each fixture does not connect individually to the horizontal wet vent) the most upstream bathtub fixture drain in Commentary Figure 909.1(5) does not have a vent. The violation illustrated in Commentary Figure 909.1(4) follows the same reasoning.

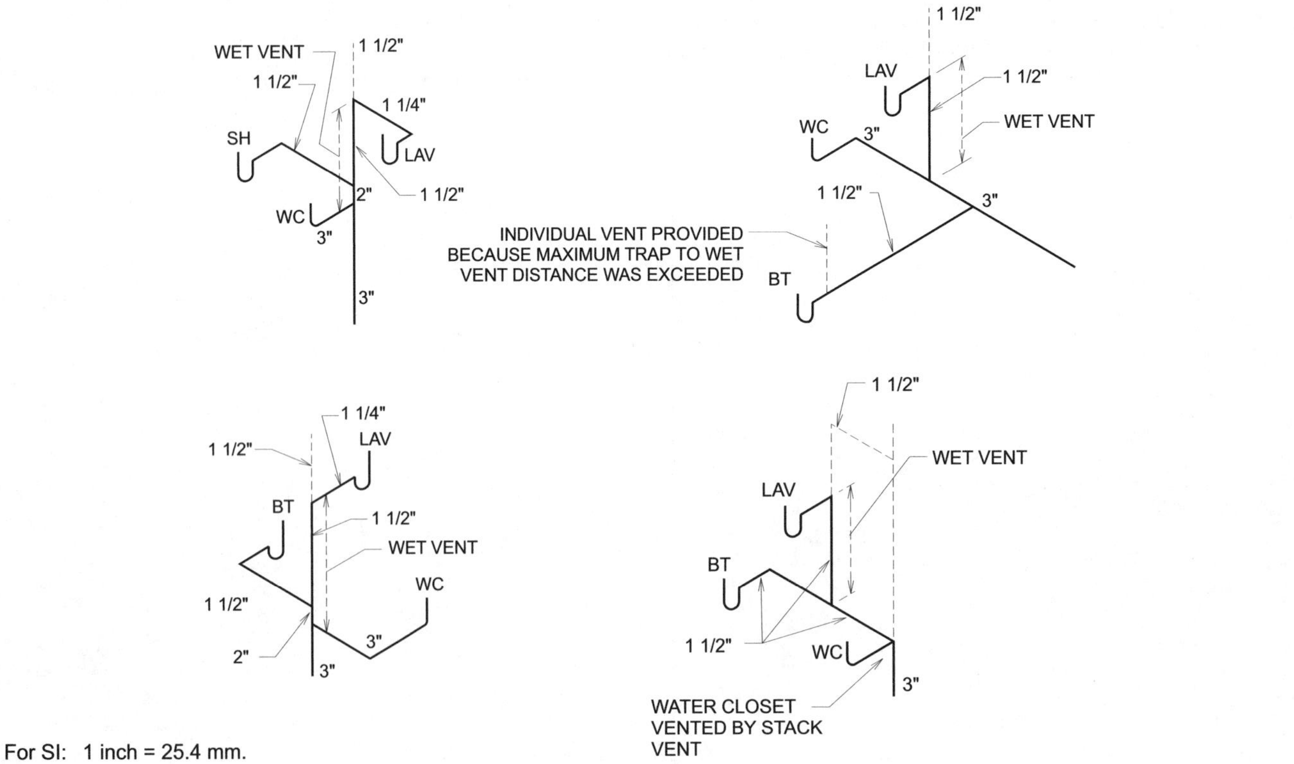

For SI: 1 inch = 25.4 mm.

Figure 909.1.1(4)
VERTICAL WET VENTING

909.2.2 Vertical wet vent. The dry-vent connection for a vertical wet-vent system shall be an individual vent or common vent for the most upstream *fixture drain*.

❖ See commentary for Section 909.1.1.

909.3 Size. The dry vent serving the wet vent shall be sized based on the largest required diameter of pipe within the wet-vent system served by the dry vent. The wet vent shall be of a minimum size as specified in Table 909.3, based on the fixture unit discharge to the wet vent.

❖ See commentary for Sections 909.1 and 909.1.1 which specifically address this section for horizontal wet vents and vertical wet vents, respectively.

Each section of the wet-vented system is individually sized, depending on the dfu load carried by that section. In other words, wet vent piping size increases as the load increases. As opposed to circuit venting, a wet-vented horizontal branch drain need not be uniformly sized. The fixture unit loads shown in Table 909.3 range from 1 dfu (single lavatory or bidet) to 12 dfu (two bathroom groups) (see the definition of "Bathroom group" in Section 202).

TABLE 909.3
WET VENT SIZE

WET VENT PIPE SIZE (inches)	DRAINAGE FIXTURE UNIT LOAD (dfu)
$1^1/_2$	1
2	4
$2^1/_2$	6
3	12

For SI: 1 inch = 25.4 mm.

❖ The table states the drain pipe oversizing that is fundamental to wet venting. The sizing of wet vent piping is based on the fixture unit load discharging through such piping. Comparing Table 710.1(2) to Table 909.3, the degree of wet vent oversizing is evident. The individual fixture drains are sized in accordance with Section 710.1.

The developed length of a fixture drain is measured from the fixture to the point of connection to the wet vent. The length of the fixture drain is limited to the distance specified in Table 906.1 (see Commentary Sections 909.1 and 909.1.1).

SECTION 910
WASTE STACK VENT

910.1 Waste stack vent permitted. A waste *stack* shall be considered a vent for all of the fixtures discharging to the *stack* where installed in accordance with the requirements of this section.

❖ A waste stack vent uses the waste stack as the vent for fixtures other than urinals and water closets. The principles of use are based on some of the original research that was done in development of the first plumbing codes. The system has been identified by a variety of names, including vertical wet vent, Philadelphia single-stack and multifloor stack venting. Note that due to the oversized design of a waste stack vent system, no relief vents are needed for stacks greater than 10 branch intervals.

910.2 Stack installation. The waste *stack* shall be vertical, and both horizontal and vertical offsets shall be prohibited between the lowest *fixture drain* connection and the highest *fixture drain* connection. Every *fixture drain* shall connect separately to the waste *stack*. The *stack* shall not receive the discharge of water closets or urinals.

❖ Because the drainage stack serves as the vent, there are certain limitations on the design to prevent pressures in the system from exceeding plus or minus 1 inch of water column (249 Pa). The system is identified as a "waste" stack vent because it prohibits the connection of water closets and urinals. Only "waste" may discharge to the stack.

Water closets and some urinals have a surging discharge that results in too great a pressure fluctuation in the system. The system can serve sinks, lavatories, bathtubs, bidets, showers, floor drains, drinking fountains and standpipes.

To preserve the desirable annular flow in the stack, offsets (of any degree) are prohibited and the stack must be vertical [90 degrees (1.57 rad) from horizontal] from the lowest fixture connection to the highest fixture connection. Offsets are permitted in the drainage stack below the lowest branch connection to the waste stack vent and within the dry stack vent portion (see Commentary Section 910.3).

Each fixture must connect independently through a single or double sanitary tee to the drainage stack. Independent connections will prevent large discharges into the stack at any one point and will direct the flow into the stack with minimum interference with any vertical flow already occurring, thus minimizing pressure fluctuations [see Commentary Figures 910.2(1) and 910.2(2)].

910.3 Stack vent. A *stack vent* shall be provided for the waste *stack*. The size of the *stack vent* shall be not less than the size of the waste *stack*. Offsets shall be permitted in the *stack vent*, shall be located at least 6 inches (152 mm) above the flood level of the highest fixture and shall be in accordance with Section 905.2. The *stack vent* shall be permitted to connect with other *stack vents* and vent stacks in accordance with Section 903.5.

❖ A full-size stack vent provides the vent opening to the outdoors, allowing the stack to remain at neutral pressures. This is the only type of stack addressed in the code that is required to have a stack vent. Offsets are allowed in the dry stack vent because such offsets are dry and have no effect on the annular flow in the waste stack. As always, dry vents must not run horizontally below the flood level rim of fixtures because this may allow waste and solids to enter the vent and impair its function. The stack vent can be terminated at a stack-type air admittance valve, extended to the outdoors, or tied into another vent stack, stack vent or vent header (see Commentary Figure 910.3).

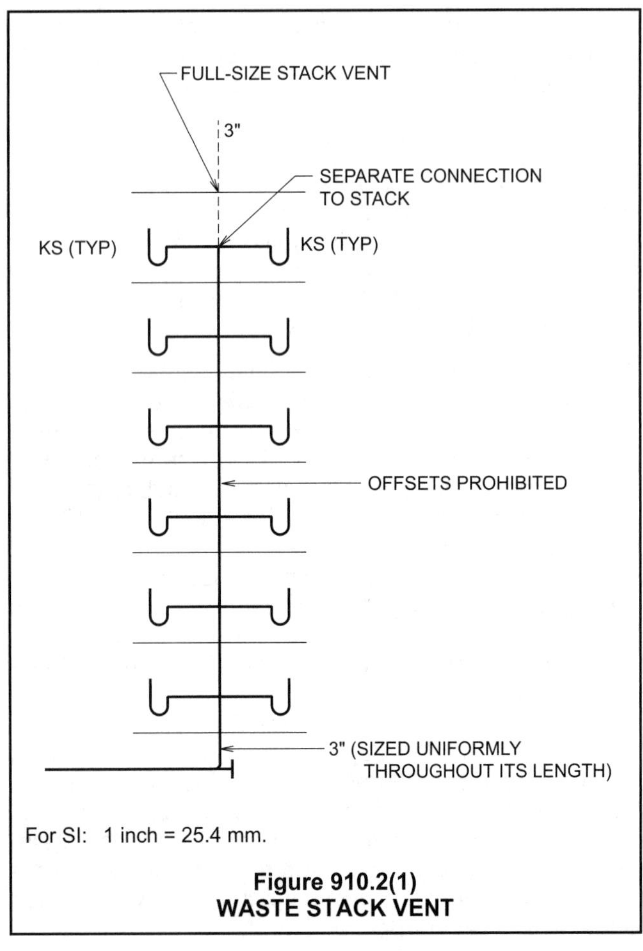

For SI: 1 inch = 25.4 mm.

Figure 910.2(1)
WASTE STACK VENT

910.4 Waste stack size. The waste *stack* shall be sized based on the total discharge to the *stack* and the discharge within a *branch* interval in accordance with Table 910.4. The waste *stack* shall be the same size throughout its length.

❖ Two factors govern sizing of the waste stack vent: the total discharge into the stack and the total discharge within any one branch interval. The size of the stack must be uniform from the base of the stack to the stack vent terminal to allow unimpeded air movement to control pressure fluctuations.

TABLE 910.4
WASTE STACK VENT SIZE

STACK SIZE (inches)	MAXIMUM NUMBER OF DRAINAGE FIXTURE UNITS (dfu)	
	Total discharge into one branch interval	Total discharge for stack
$1^1/_2$	1	2
2	2	4
$2^1/_2$	No limit	8
3	No limit	24
4	No limit	50
5	No limit	75
6	No limit	100

For SI: 1 inch = 25.4 mm.

❖ To reduce the possibility of exceeding plus or minus 1 inch of water column (249 Pa) pressure, the stack is

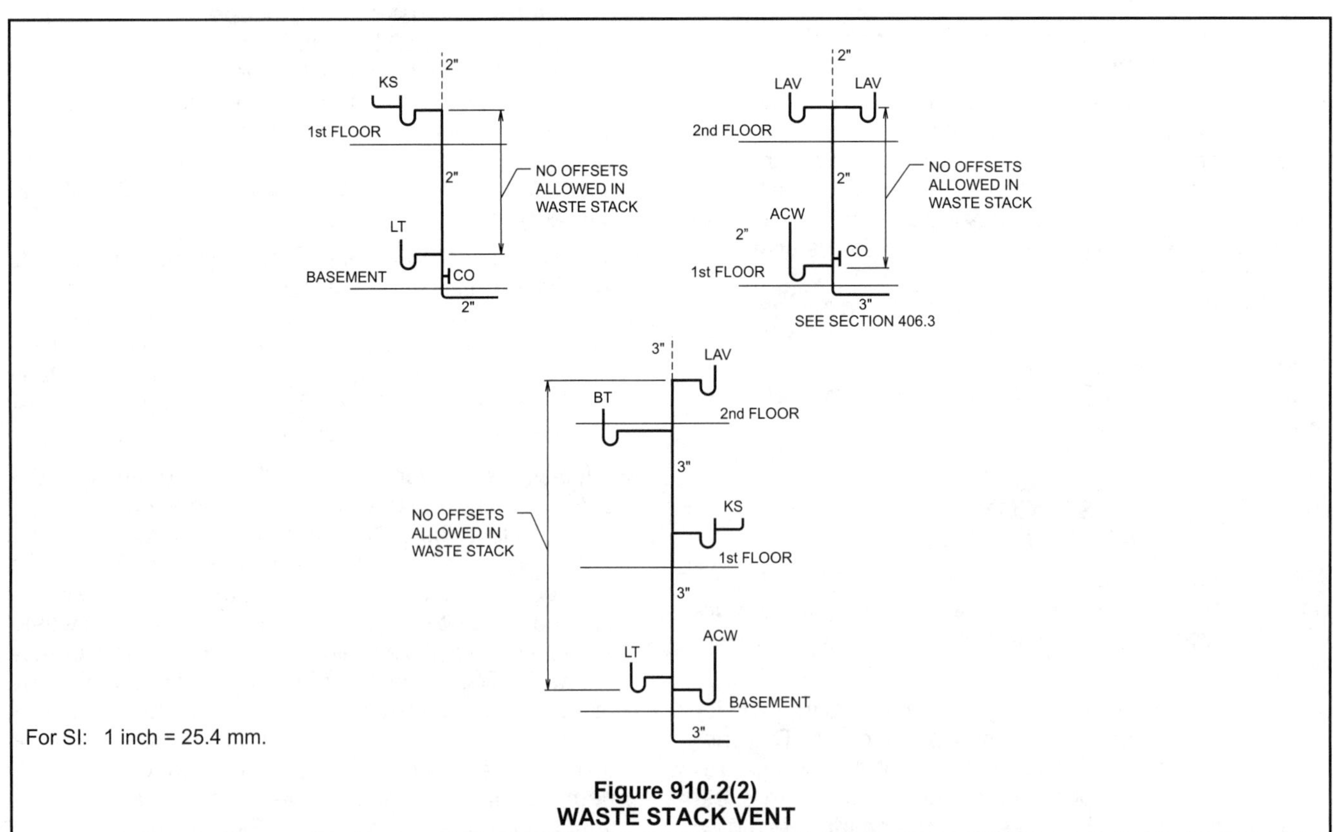

For SI: 1 inch = 25.4 mm.

Figure 910.2(2)
WASTE STACK VENT

greatly oversized. Although Chapter 7 permits a drainage stack to be fully loaded, a waste stack vent may be loaded only to a fraction of its total capacity. Full loading of a stack will result in pressure fluctuations that will necessitate additional venting for fixtures connecting to the stack.

SECTION 911 CIRCUIT VENTING

911.1 Circuit vent permitted. A maximum of eight fixtures connected to a horizontal *branch* drain shall be permitted to be circuit vented. Each *fixture drain* shall connect horizontally to the horizontal *branch* being circuit vented. The horizontal *branch* drain shall be classified as a vent from the most downstream *fixture drain* connection to the most upstream *fixture drain* connection to the horizontal *branch.*

❖ Circuit venting involves up to eight fixtures and a single vent. A dry vent connects between the two most upstream fixtures on the horizontal branch. The horizontal branch serves as a "wet vent" for the battery of fixtures. The piping arrangement has the appearance of multiple fixtures connecting to a horizontal branch, with only the most upstream fixture being individually vented.

The principle of circuit venting is that the flow of drainage never exceeds a half-full flow condition. The air for venting the fixtures circulates in the top half of the horizontal branch drain pipe. The flow velocity in the horizontal branch is slow and nonturbulent, thereby preventing pressure differentials from affect-

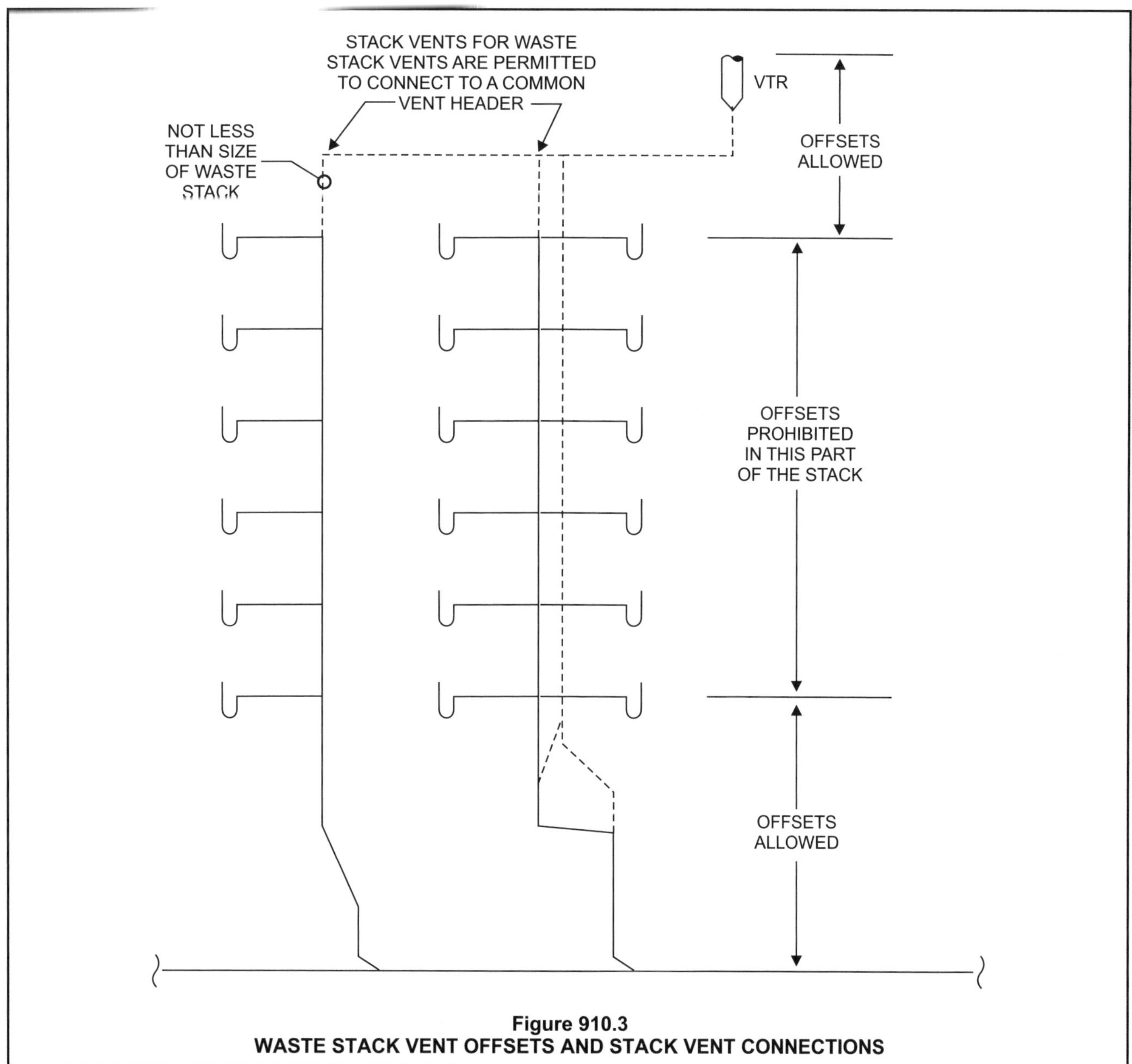

Figure 910.3
WASTE STACK VENT OFFSETS AND STACK VENT CONNECTIONS

ing the connecting fixtures [see Commentary Figure 911.1(1)]. The circuit-vented fixtures must connect to the circuit-vented branch in the horizontal plane to limit the amount of turbulence created by fixture discharge [see Commentary Figure 911.1(2)].

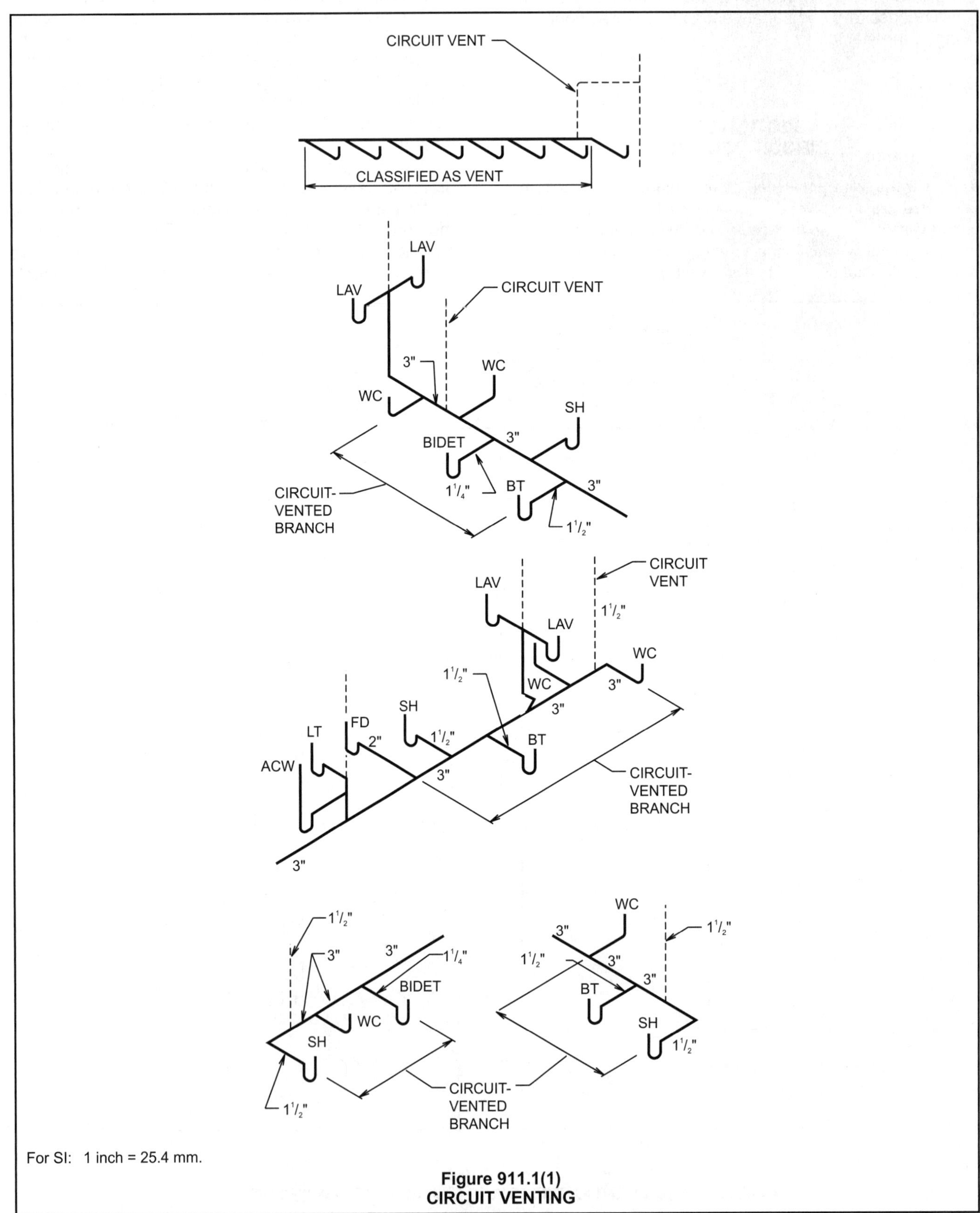

For SI: 1 inch = 25.4 mm.

Figure 911.1(1)
CIRCUIT VENTING

Figure 911.1(2)
ROUGH-IN FOR EIGHT CIRCUIT VENTED FIXTURES
(Photo courtesy of B&I Contractors)

911.1.1 Multiple circuit-vented branches. Circuit-vented horizontal *branch* drains are permitted to be connected together. Each group of a maximum of eight fixtures shall be considered a separate circuit vent and shall conform to the requirements of this section.

❖ Circuit-vented branches may connect either in series or in parallel. Each group of up to eight fixtures must be considered as a separate circuit-vented branch and be installed accordingly.

The concept of multiple circuit-vented branches is that horizontal flow in a drain pipe does not require a large quantity of air to balance the pressure. The slower flow velocity in a horizontal drain also prevents pressure fluctuations that may affect the traps of the connecting fixtures (see Commentary Figure 911.1.1).

911.2 Vent connection. The circuit vent connection shall be located between the two most upstream fixture drains. The vent shall connect to the horizontal *branch* and shall be installed in accordance with Section 905. The circuit vent pipe shall not receive the discharge of any soil or waste.

❖ The circuit vent must connect between the two most upstream fixture drains to allow proper circulation of air. The vent is located so that the most upstream fixture discharges past the vent connection, thereby washing that section of pipe and preventing the buildup of solids. For this reason, the most upstream fixture should not be a fixture that is seldom used (such as a floor drain). When back-to-back fixtures are installed, the vent is connected between the most upstream pairs (see commentary, Section 911.1 and Commentary Figure 911.2).

Section 905 requires that the vent connect vertically to the drain and rise vertically to 6 inches (152 mm) above the flood level rim before offsetting horizontally.

SERIES:

CIRCUIT VENT

CIRCUIT VENT

SECOND CIRCUIT-VENTED BRANCH

ONE CIRCUIT-VENTED BRANCH

PARALLEL:

VENT STACK

CIRCUIT VENT

CIRCUIT-VENTED BRANCH

CIRCUIT VENT

CIRCUIT-VENTED BRANCH

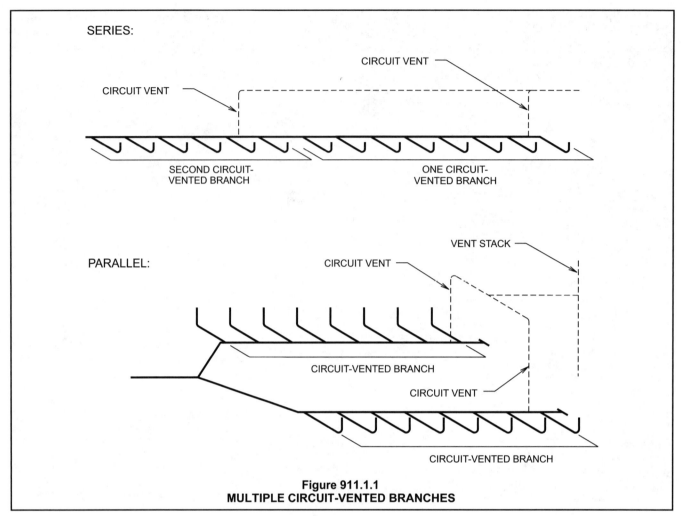

Figure 911.1.1
MULTIPLE CIRCUIT-VENTED BRANCHES

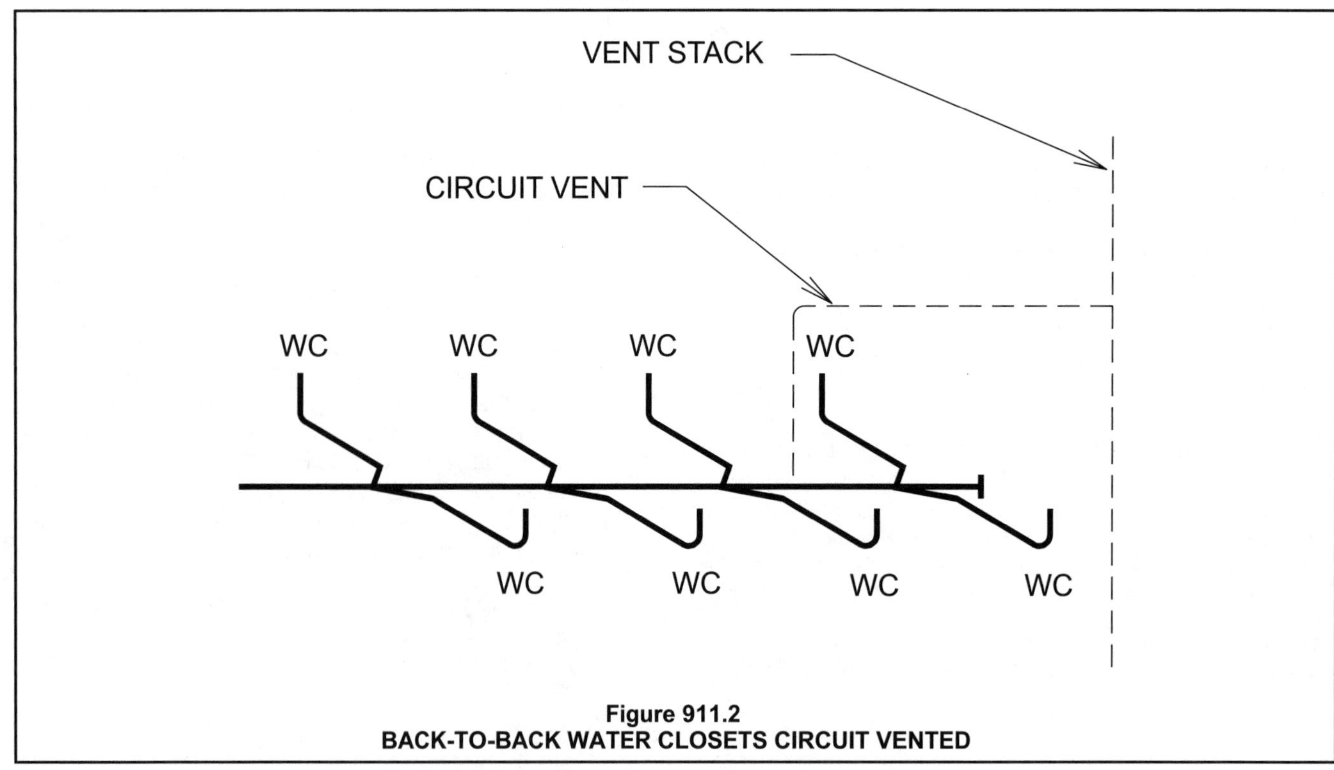

Figure 911.2
BACK-TO-BACK WATER CLOSETS CIRCUIT VENTED

911.3 Slope and size of horizontal branch. The maximum slope of the vent section of the *horizontal branch drain* shall be one unit vertical in 12 units horizontal (8-percent slope). The entire length of the vent section of the *horizontal branch drain* shall be sized for the total drainage discharge to the *branch*.

❖ The principle of a circuit vent is that the drainage branch is entirely horizontal. A maximum pitch of 1 inch per foot (83 mm/m) ensures that the branch remains horizontal without any vertical offsets.

The horizontal drainage branch must also be uniformly sized to establish consistent flow characteristics and the free movement of air. Because it is uniformly sized for its full length, the horizontal branch will be oversized for all but the most downstream portion. Table 710.1(2) is used to size the horizontal branch (see Commentary Figure 911.3).

911.3.1 Size of multiple circuit vent. Each separate circuit-vented horizontal *branch* that is interconnected shall be sized independently in accordance with Section 911.3. The downstream circuit-vented horizontal *branch* shall be sized for the total discharge into the *branch*, including the upstream branches and the fixtures within the *branch*.

❖ Each circuit-vented branch must be uniformly sized based on the total drainage load on that branch, including any drainage load from upstream branches. If an increase in pipe size is required, the location must be at the interconnection of the circuit-vented branches (see Commentary Figure 911.3.1).

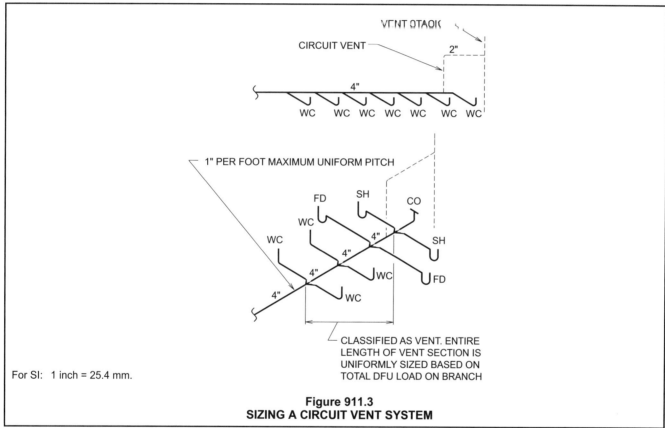

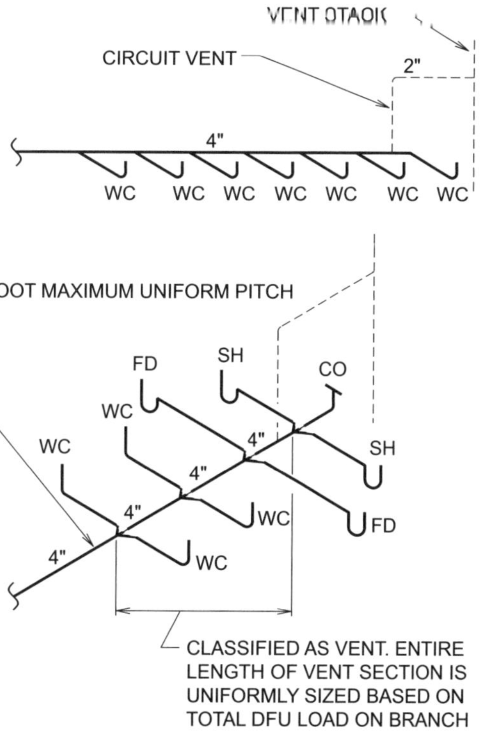

For SI: 1 inch = 25.4 mm.

Figure 911.3
SIZING A CIRCUIT VENT SYSTEM

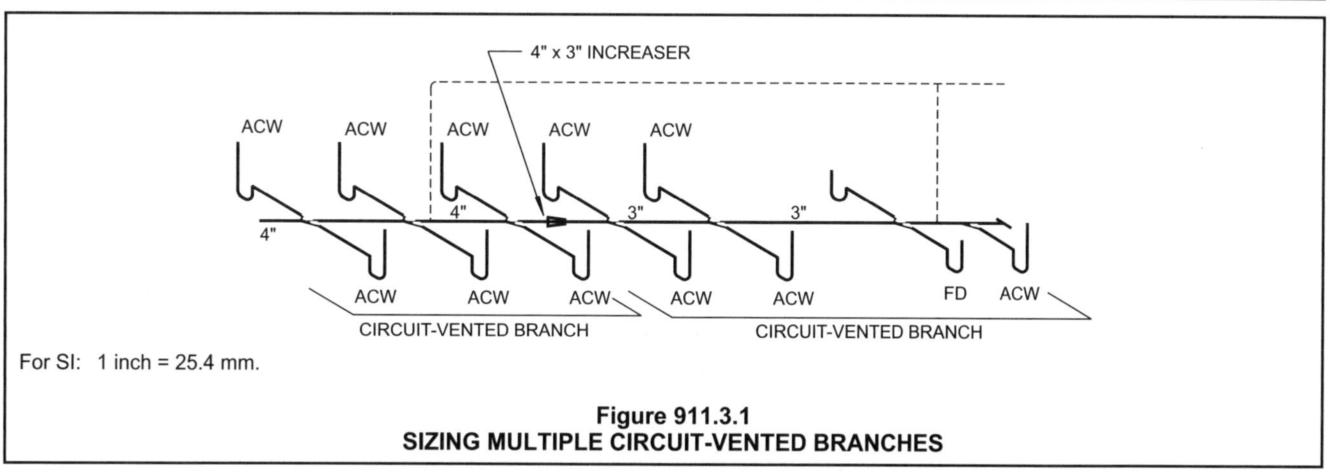

For SI: 1 inch = 25.4 mm.

Figure 911.3.1
SIZING MULTIPLE CIRCUIT-VENTED BRANCHES

911.4 Relief vent. A relief vent shall be provided for circuit-vented horizontal branches receiving the discharge of four or more water closets and connecting to a drainage *stack* that receives the discharge of soil or waste from upper horizontal branches.

❖ If a circuit-vented horizontal branch connects to a drainage stack that is receiving drainage discharge from floors above, a relief vent is required if more than three water closets connect to the circuit-vented horizontal branch. The relief vent will prevent any pressure differential in the drainage stack from affecting the horizontal branch by relieving the pressures.

911.4.1 Connection and installation. The relief vent shall connect to the horizontal *branch* drain between the *stack* and the most downstream *fixture drain* of the circuit vent. The relief vent shall be installed in accordance with Section 905.

❖ The relief vent must connect between the stack and the most downstream fixture on the horizontal branch. The relief vent is sized in accordance with Section 916.2 (see Commentary Figure 911.4.1).

911.4.2 Fixture drain or branch. The relief vent is permitted to be a *fixture drain* or fixture *branch* for fixtures located within the same *branch interval* as the circuit-vented horizontal *branch*. The maximum discharge to a relief vent shall be four fixture units.

❖ Unlike the circuit vent, which must be dry, the relief vent is permitted to serve as a drain for fixtures installed within the same circuit-vented branch. When a relief vent serves as a fixture branch, that vent is subject to the same sizing requirements as a common vent or wet vent. A $1^1/_2$-inch (38 mm) relief vent may receive 1 dfu and a 2-inch (51 mm) or larger relief vent is limited to 4 dfu.

911.5 Additional fixtures. Fixtures, other than the circuit-vented fixtures, are permitted to discharge to the horizontal *branch* drain. Such fixtures shall be located on the same floor as the circuit-vented fixtures and shall be either individually or common vented.

❖ Where additional fixtures (other than the circuit-vented fixtures) are located on the same floor as a circuit-vented branch, such fixtures are allowed to connect to the circuit-vented branch. The additional fixtures must be separately vented, and the additional drainage load of those fixtures must be considered in sizing the circuit-vented branch [see Commentary Figures 911.5(1) and 911.5(2)]. This provision is intended to allow the connection of lavatories, urinals and similar fixtures that are located in the same area as the circuit-vented fixtures. This is common practice for toilet room installations that use circuit vents.

SECTION 912 COMBINATION DRAIN AND VENT SYSTEM

912.1 Type of fixtures. A combination drain and vent system shall not serve fixtures other than floor drains, sinks, lavatories and drinking fountains. Combination drain and vent systems shall not receive the discharge from a food waste grinder or clinical sink.

❖ This section contains requirements for a venting method serving floor drains, sinks, lavatories and drinking fountains. Like wet vents and circuit vents, a combination drain and vent system is another variation of "wet" venting. Although the specific requirements for each are unique, all of the "wet" venting means (circuit, wet and combination) utilize the same basic principle of oversized drains with ample space for airflow above the waste flow.

The number of fixtures connecting to a combination drain and vent system is unlimited, provided that the fixtures are floor drains, sinks, lavatories or drinking fountains. The system is commonly used where floor drains are installed in large open areas that cannot ac-

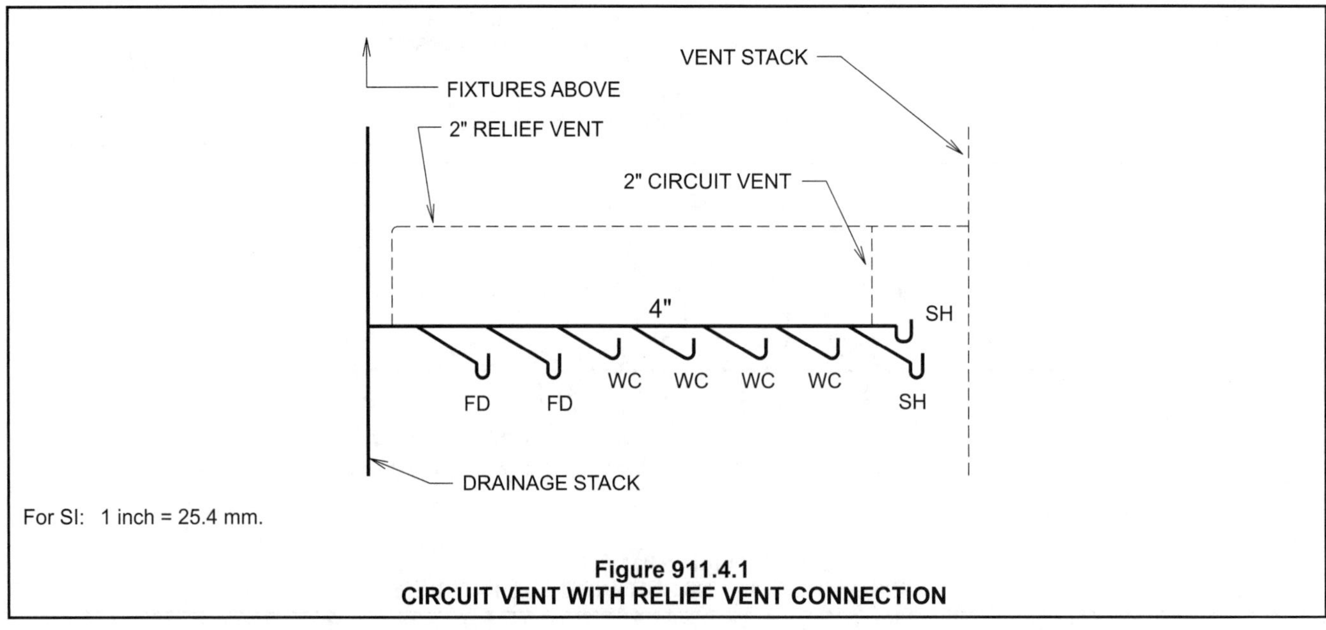

Figure 911.4.1
CIRCUIT VENT WITH RELIEF VENT CONNECTION

commodate vertical vent risers from the floor drains. A combination drain and vent is also practical to use where island sinks are installed in dwelling units, laboratories and classrooms.

To prevent blockages, such systems are intended to receive only clear and gray-water wastes. Because of the low-flow velocities and volumes that occur in these systems, any solids introduced would accumulate, causing blockages and reduction of the internal cross-sectional area of the piping. Therefore, water closets, sinks equipped with food waste grinders, clinical sinks and fixtures that discharge solid waste are not allowed to be served by a combination drain and vent system. Ground food waste might collect in the drain, causing blockages. Combination drain and vent systems are oversized drains; therefore, the velocity of the flow will be low, thereby reducing normally available scouring action. Note that the definition of this system in Chapter 2 has not been revised to coincide with the previous revisions to Section 912 specifically; lavatories and drinking fountains are allowed on combination drain and vent systems.

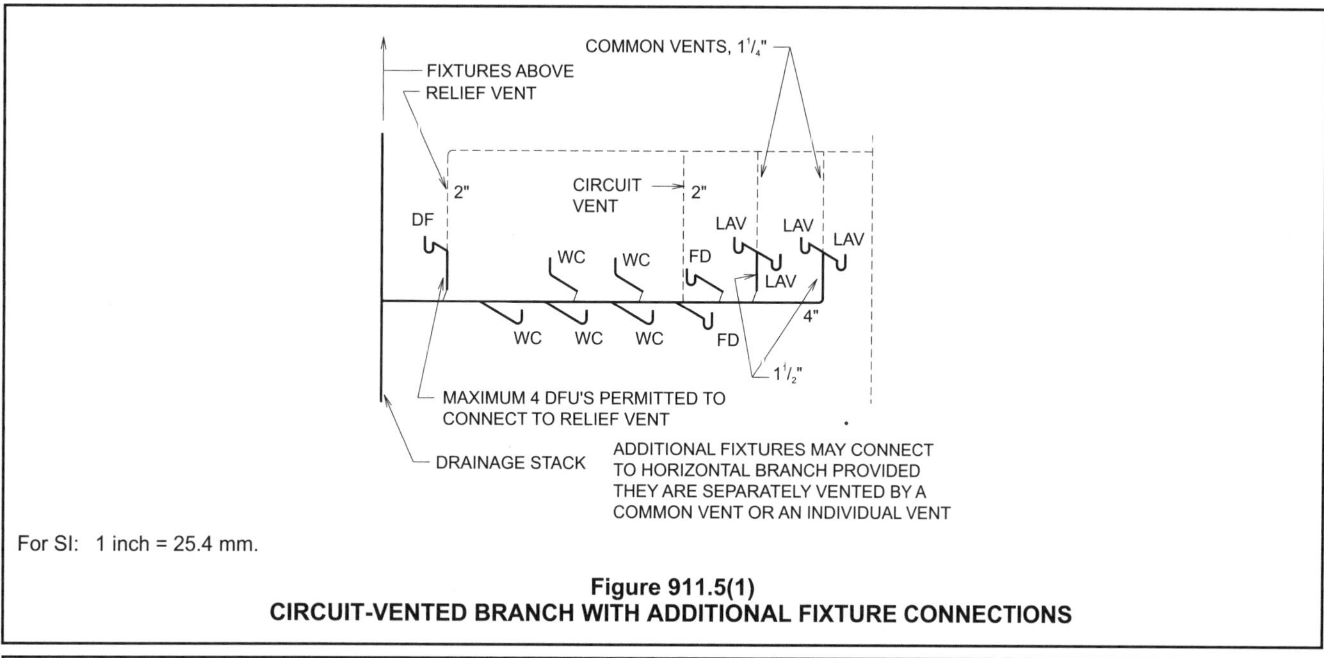

For SI: 1 inch = 25.4 mm.

Figure 911.5(1)
CIRCUIT-VENTED BRANCH WITH ADDITIONAL FIXTURE CONNECTIONS

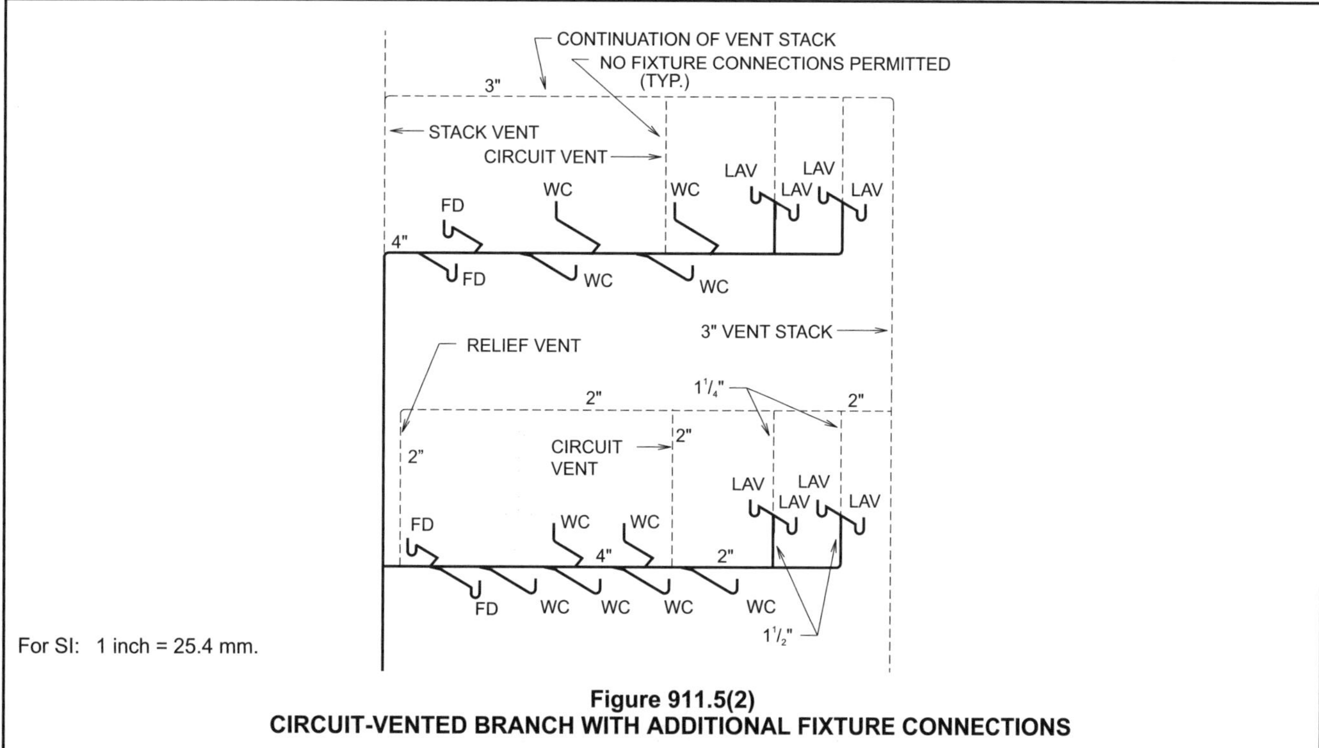

For SI: 1 inch = 25.4 mm.

Figure 911.5(2)
CIRCUIT-VENTED BRANCH WITH ADDITIONAL FIXTURE CONNECTIONS

912.2 Installation. The only vertical pipe of a combination drain and vent system shall be the connection between the *fixture drain* and the horizontal combination drain and vent pipe. The maximum vertical distance shall be 8 feet (2438 mm).

❖ In a combination drain and vent system, the drain serves as the vent for the fixture. The system is intended to be a horizontal piping system, with the only vertical piping [limited to 8 feet (2438 mm) in height] being the vertical portion of a fixture drain located above the level of the combination drain and vent [see Commentary Figure 912.2(5)].

Each fixture drain connected to a combination drain and vent system is sized in accordance with Chapter 7. Depending on the method by which it is sized, the fixture drain may or may not be considered as a portion of the combination drain and vent system. Where the fixture drain is sized as a combination drain and vent in accordance with Table 912.3, the trap-to-vent distances listed in Table 906.1 do not apply because the drain is also a vent for its entire length.

Additional installation requirements for the combination drain and vent and the fixture branch or drain connections are contained in Sections 912.2.1 through 912.2.4 [see Commentary Figures 912.2(1) through 912.2(4)].

912.2.1 Slope. The horizontal combination drain and vent pipe shall have a maximum slope of one-half unit vertical in 12 units horizontal (4-percent slope). The minimum slope shall be in accordance with Table 704.1.

❖ This section specifies the maximum and minimum slopes for the combination drain and vent to regulate the flow velocity of the waste. The intent of this section is to provide a minimum slope to facilitate drainage by gravity while providing a maximum slope to limit flow velocity to a range that minimizes adverse pressure fluctuations. This section and Section 912.2 act together to make sure such systems maintain a low flow velocity by incorporating height limitations on vertical piping.

912.2.2 Connection. The combination drain and vent system shall be provided with a dry vent connected at any point within the system or the system shall connect to a horizontal drain that is vented in accordance with one of the venting methods specified in this chapter. Combination drain and vent systems connecting to building drains receiving only the discharge from a *stack* or stacks shall be provided with a dry vent. The vent connection to the combination drain and vent pipe shall extend vertically a minimum of 6 inches (152 mm) above the *flood level rim* of the highest fixture being vented before offsetting horizontally.

❖ A dry vent pipe must connect somewhere within the combination drain and vent system, or the system must connect to a horizontal drain that is vented. A horizontal drain that is vented is any horizontal drain that permits the free flow of air above the liquid flow level in the drain pipe. The code does not define "horizontal drain that is vented," and this has been and still is the subject of much debate. This section implies that any horizontal drain is vented if it is serving fixtures within the same story that are vented by a method in this chapter [see Commentary Figures 912.2.2(1) through 912.2.2(5)]. For example, a branch of a building drain that serves only an individually vented laundry sink would be considered a horizontal drain that is vented. Keep in mind that the sink drain would have to be oversized to allow for both the simultaneous flow of waste and air and the connection of a 2-inch (51 mm) or larger combination drain and vent pipe. Building

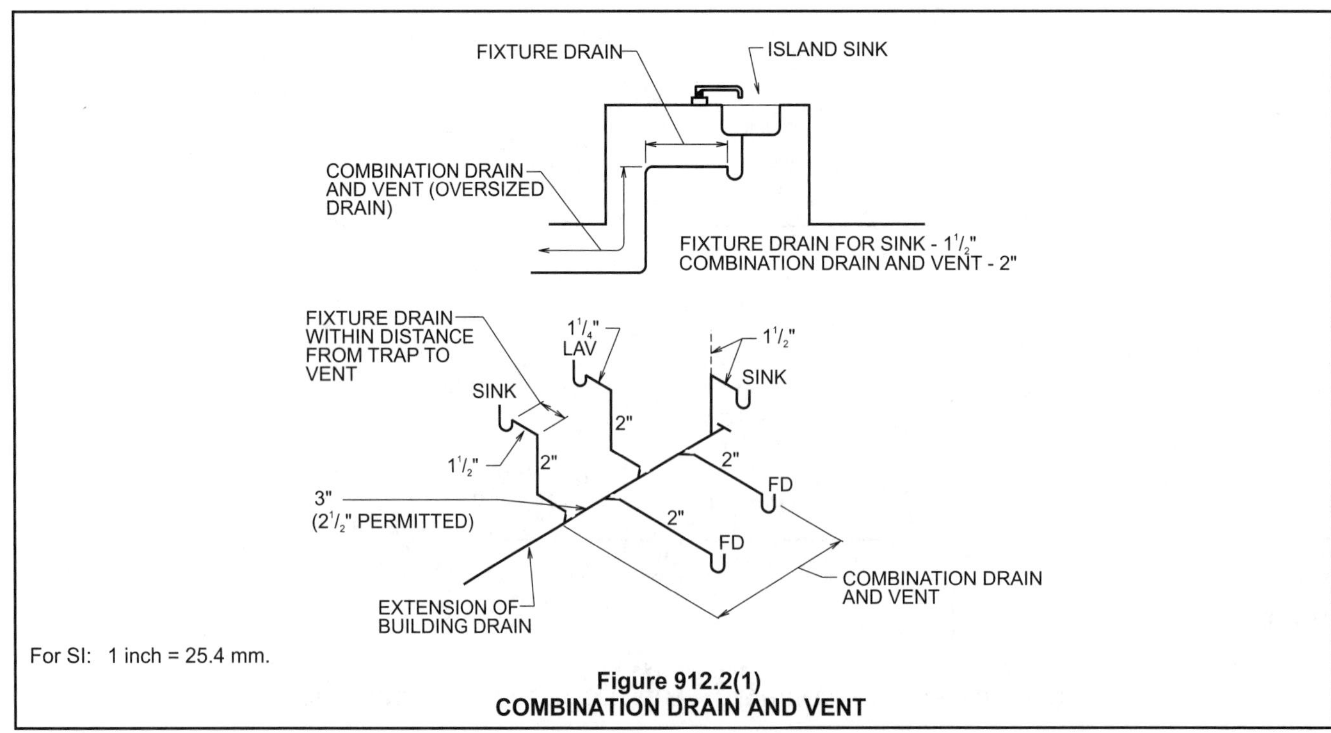

Figure 912.2(1)
COMBINATION DRAIN AND VENT

drains and branches of building drains that receive only the discharge of stacks are not considered to be horizontal drains that are vented. Clearly, by their nature, wet vent systems, circuit vent systems and combination drain and vent systems are "horizontal drains that are vented." Section 704.3 prohibits the connection of horizontal branches at or near waste and soil stacks unless those connections are 10 pipe diameters downstream from the stack.

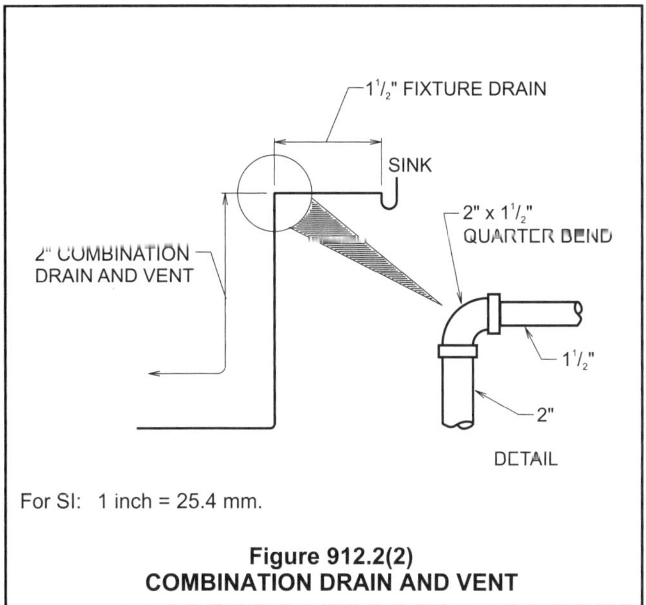

For SI: 1 inch = 25.4 mm.

Figure 912.2(2)
COMBINATION DRAIN AND VENT

If a building drain receives only the discharge of stacks, it cannot be considered as a "horizontal drain that is vented" because of the positive pressures expected to occur in such piping. Because the combination drain and vent system is both a drain and a vent for its entire length, it is not important where the dry vent connects to the system. In fact, where these systems connect to a horizontal drain that is vented, the combination drain and vent may be considered as "end-vented" because the vent source is at the termination of the combination system. If a combination drain and vent system connects to a stack, a dry vent must connect directly to the combination drain and vent system (see the column titled "Connecting to a horizontal branch or stack" in Table 912.3).

This section expresses the same concern as Section 905.4 regarding vertical rise of the dry vent above the flood level rim. The vertical rise of the vent to a point above the flood level rim is necessary to protect the vent against clogging (see Commentary Sections 905.4 and 905.5).

912.2.3 Vent size. The vent shall be sized for the total *drainage fixture unit* load in accordance with Section 916.2.

❖ Where a dry vent connects directly to the combination drain and vent, the vent must be sized in accordance with Section 916.2. The dry vent sizing need not be based on the installed size of the combination drain and vent pipe, but rather on the size of a drain pipe required to carry the dfu load served by the combination drain and vent system.

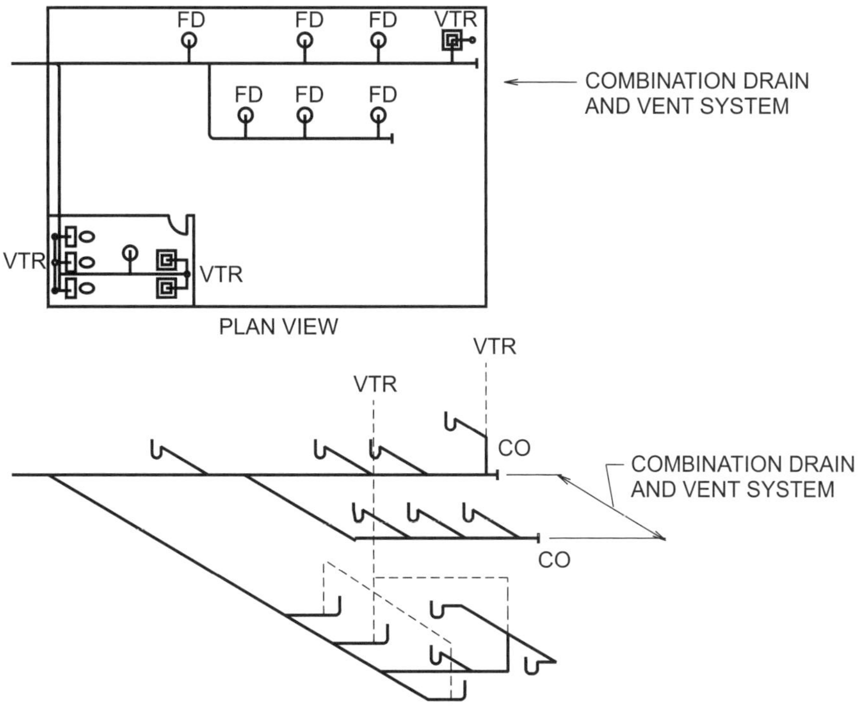

Figure 912.2(3)
COMBINATION DRAIN AND VENT

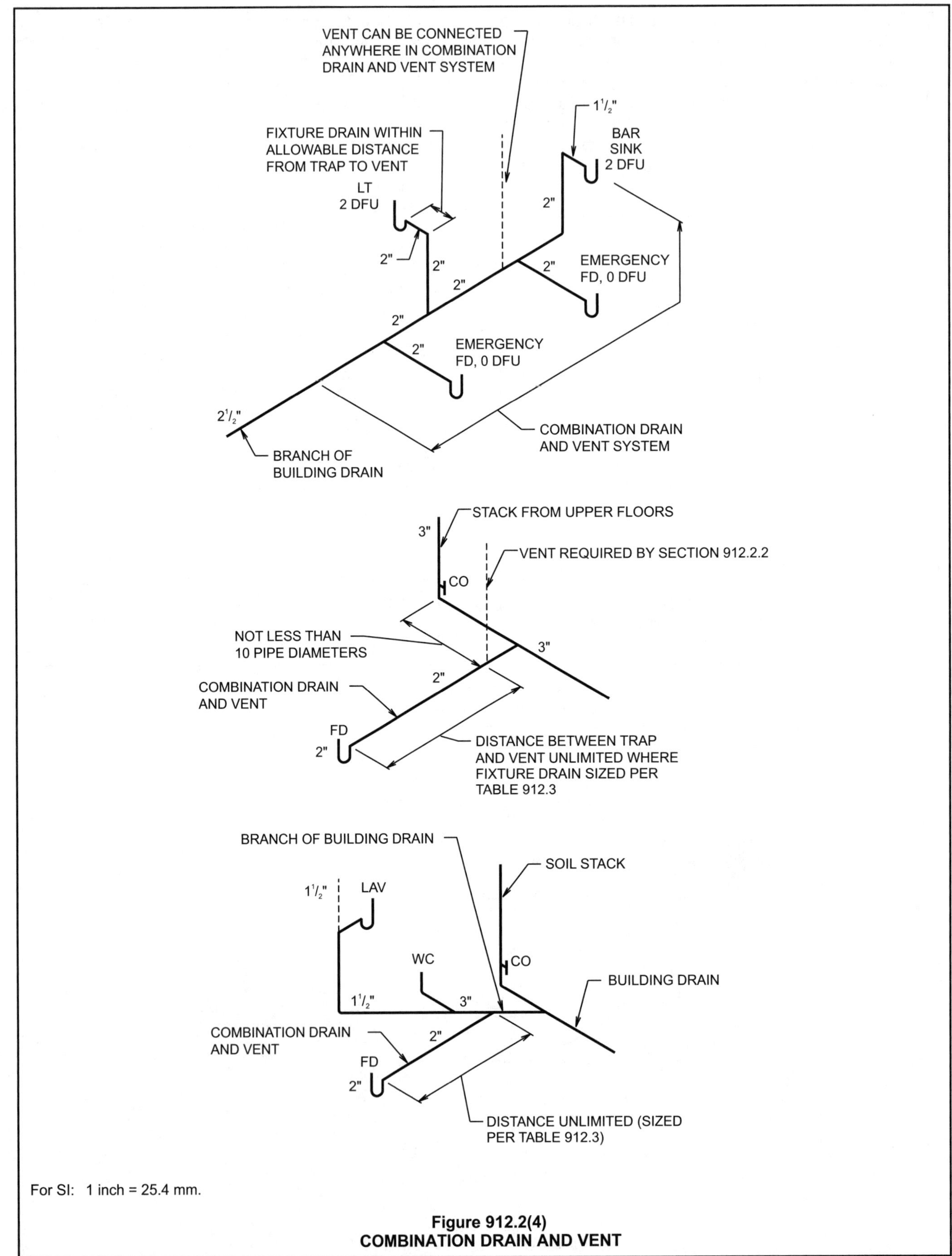

For SI: 1 inch = 25.4 mm.

Figure 912.2(4)
COMBINATION DRAIN AND VENT

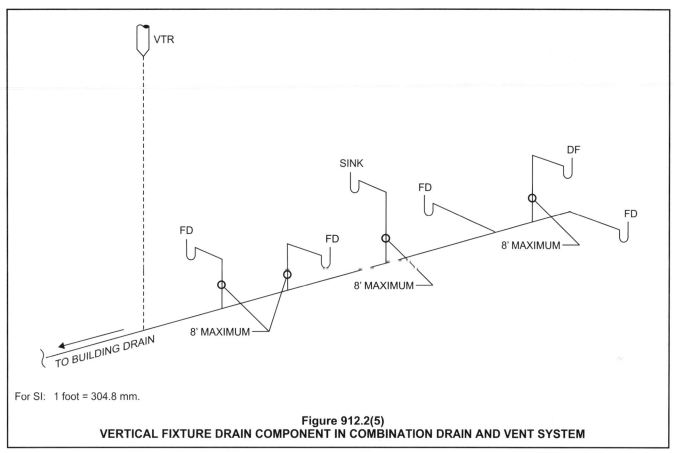

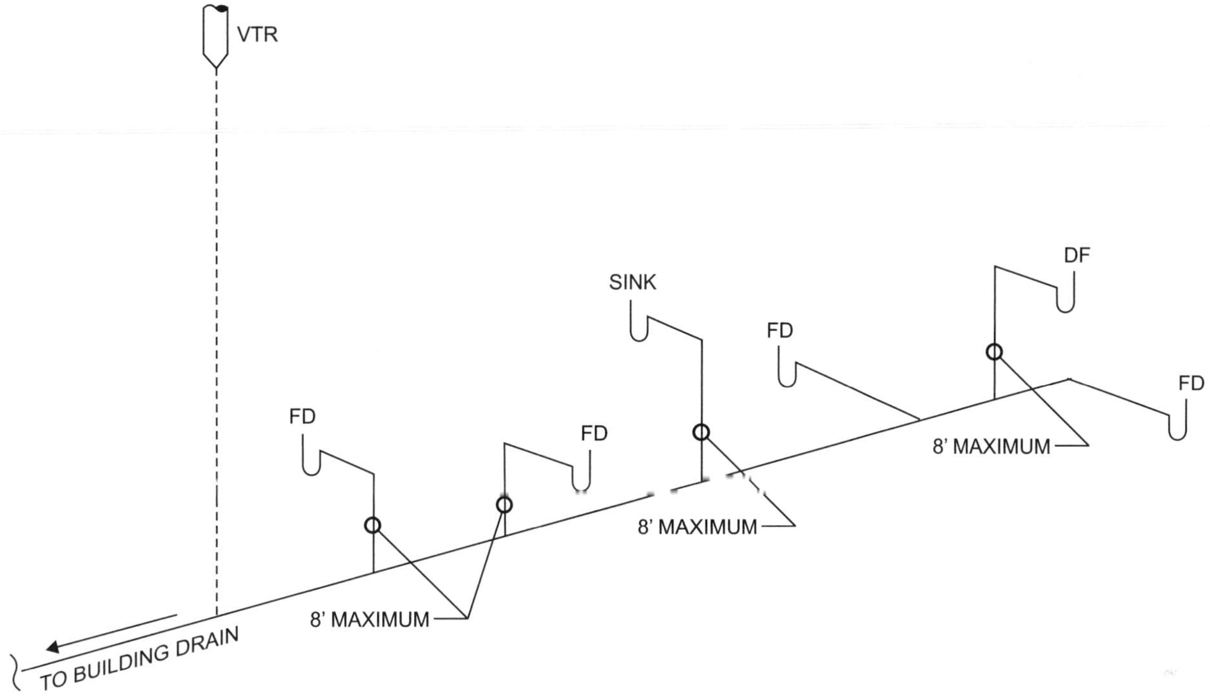

For SI: 1 foot = 304.8 mm.

Figure 912.2(5)
VERTICAL FIXTURE DRAIN COMPONENT IN COMBINATION DRAIN AND VENT SYSTEM

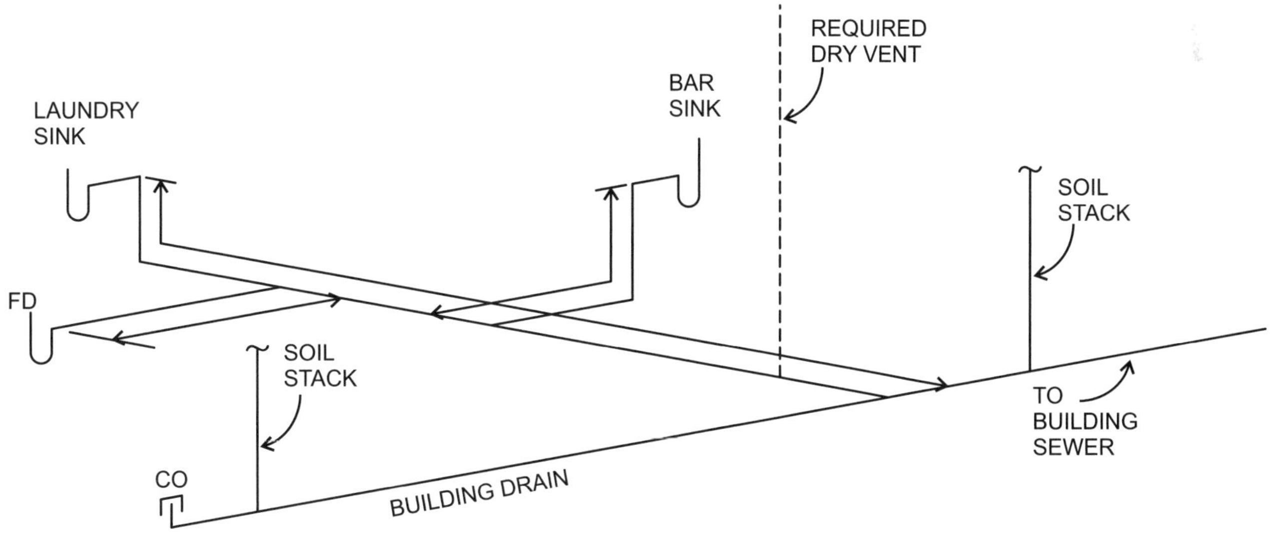

Figure 912.2.2(1)
COMBINATION DRAIN AND VENT

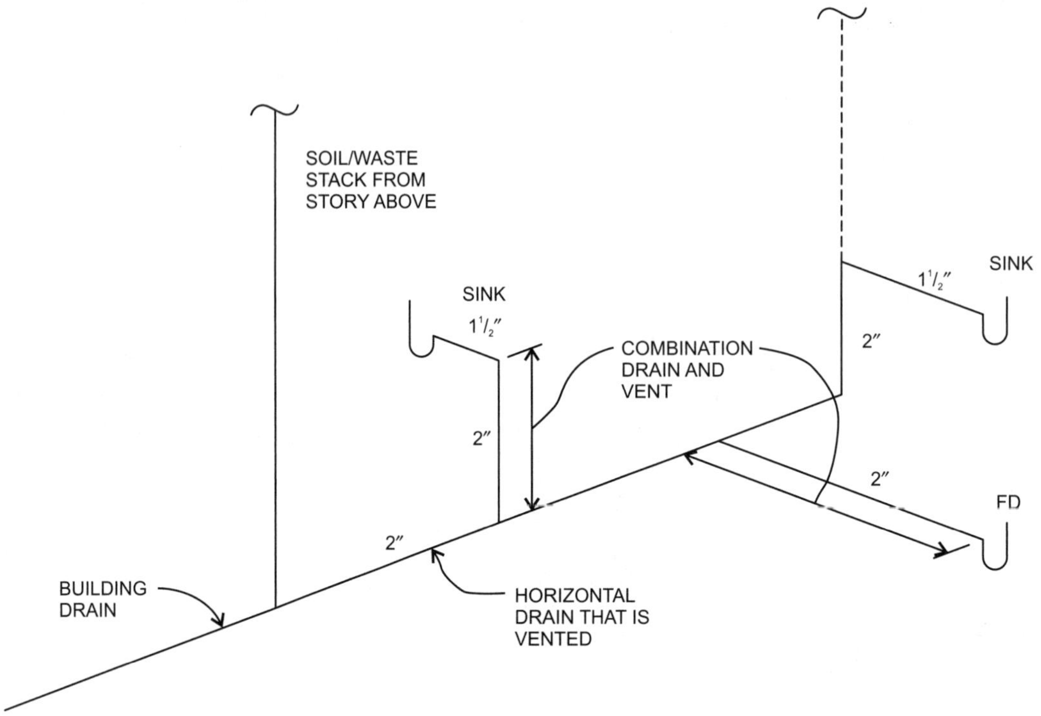

SOIL/WASTE STACK FROM STORY ABOVE

SINK

1$^1/_2$″

SINK

1$^1/_2$″

COMBINATION DRAIN AND VENT

2″

2″

2″

2″

FD

BUILDING DRAIN

HORIZONTAL DRAIN THAT IS VENTED

For SI: 1 inch = 25.4 mm.

Figure 912.2.2(2)
COMBINATION DRAIN AND VENT

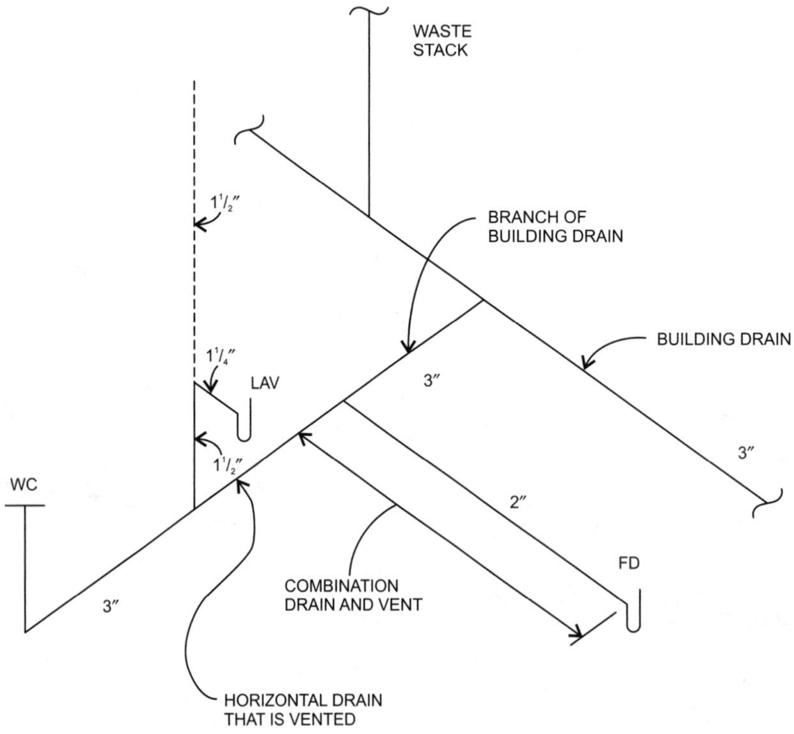

WASTE STACK

1$^1/_2$″

BRANCH OF BUILDING DRAIN

BUILDING DRAIN

1$^1/_4$″

LAV

3″

1$^1/_2$″

3″

WC

2″

FD

3″

COMBINATION DRAIN AND VENT

HORIZONTAL DRAIN THAT IS VENTED

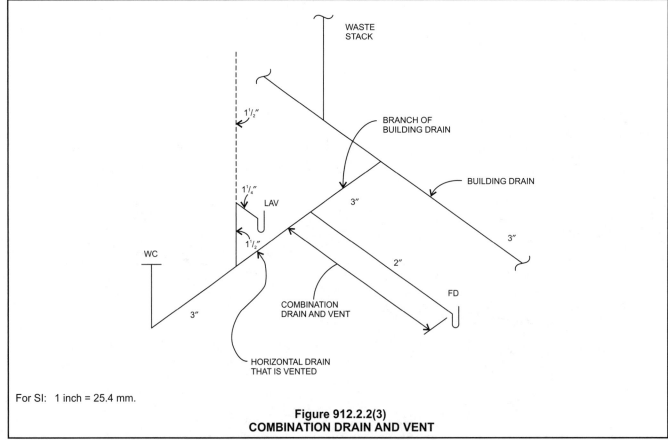

For SI: 1 inch = 25.4 mm.

Figure 912.2.2(3)
COMBINATION DRAIN AND VENT

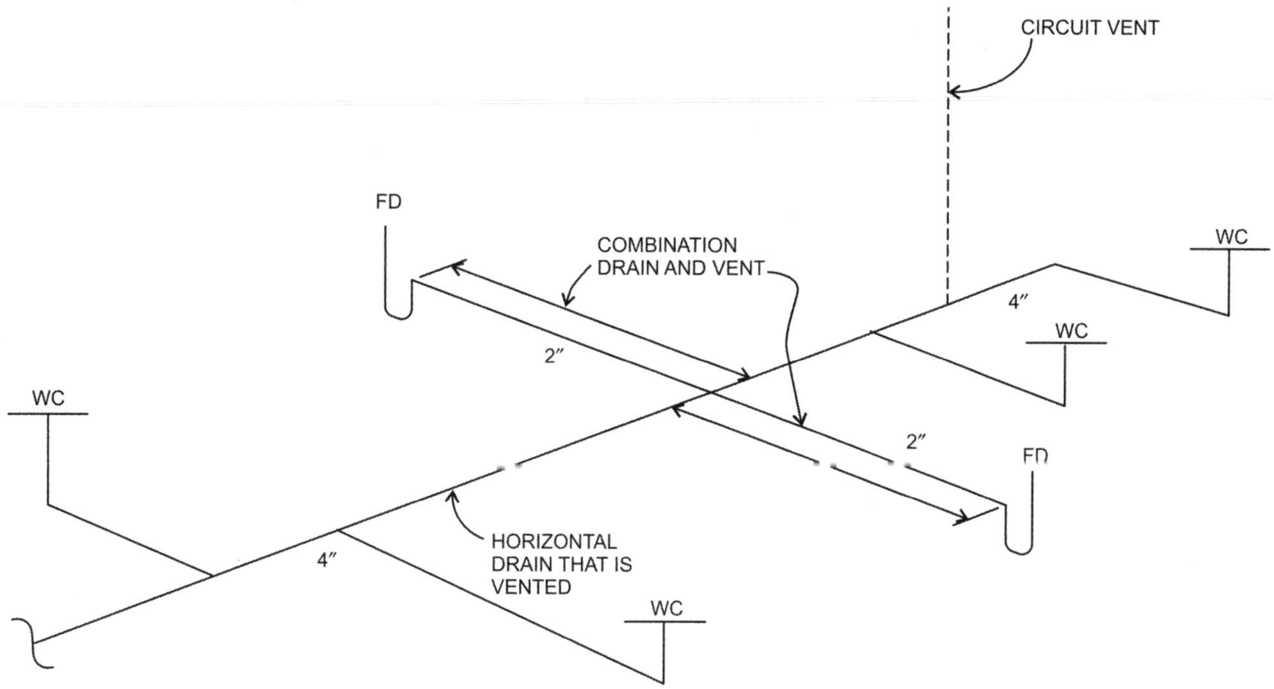

For SI: 1 inch = 25.4 mm.

Figure 912.2.2(4)
COMBINATION DRAIN AND VENT

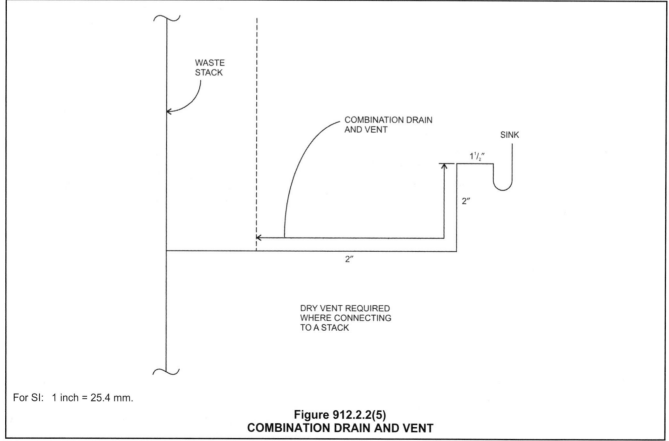

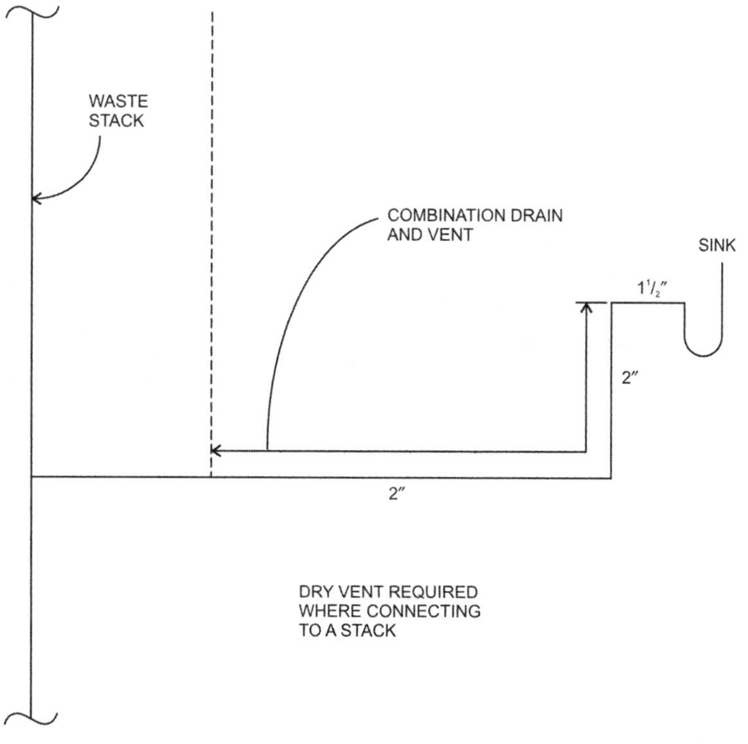

For SI: 1 inch = 25.4 mm.

Figure 912.2.2(5)
COMBINATION DRAIN AND VENT

912.2.4 Fixture branch or drain. The fixture *branch* or *fixture drain* shall connect to the combination drain and vent within a distance specified in Table 906.1. The combination drain and vent pipe shall be considered the vent for the fixture.

❖ The oversized horizontal drain branch is considered to be the vent for any individual fixture it serves. The distance from the fixture trap to the connection to the combination drain and vent system is limited to the length specified in Table 906.1 (see Commentary Section 906.1). Note that the length of a fixture drain sized in accordance with Table 912.3 can be unlimited because it, too, is considered a combination drain and vent.

912.3 Size. The minimum size of a combination drain and vent pipe shall be in accordance with Table 912.3.

❖ The referenced table sizes the drain serving as a combination drain and vent. Because of the degree of oversizing and the low slope for such piping, the flow velocity and depth will be low. These facts indicate that such systems cannot tolerate waste discharge containing solids, hence the reason for excluding the discharge from water closets, food waste grinders, clinical sinks and other fixtures that might discharge solids.

TABLE 912.3
SIZE OF COMBINATION DRAIN AND VENT PIPE

DIAMETER PIPE (inches)	MAXIMUM NUMBER OF DRAINAGE FIXTURE UNITS (dfu)	
	Connecting to a horizontal branch or stack	Connecting to a building drain or building subdrain
2	3	4
$2^1/_2$	6	26
3	12	31
4	20	50
5	160	250
6	360	575

For SI: 1 inch = 25.4 mm.

❖ The system requires that the combination drain and vent pipe be oversized drainage for two reasons. First, the velocity of the drainage flowing in the pipe is greatly reduced by oversizing the pipe, which reduces the possibility of creating a siphon or blowback action. Second, the cross-sectional area of waste flow (or depth of drainage) in the pipe is reduced by increasing the drainage pipe size, which results in a greater area of the pipe above the flow of drainage to permit the movement of air for venting.

Note that when connecting to a stack or an unvented horizontal drain, a combination drain and vent system must have a dry vent connection (see commentary, Sections 912.2.2 and 912.2.3).

SECTION 913
ISLAND FIXTURE VENTING

913.1 Limitation. Island fixture venting shall not be permitted for fixtures other than sinks and lavatories. Residential kitchen sinks with a dishwasher waste connection, a food waste grinder, or both, in combination with the kitchen sink waste, shall be permitted to be vented in accordance with this section.

❖ This section contains a venting option for plumbing fixtures located where an individual vent cannot be installed without horizontal sections of piping below the flood level rim of the fixture served.

Island fixture venting is another method of venting island sinks and lavatories.The other options are a combination drain and vent regulated by Section 912 and air admittance valves in accordance with Section 917. This section also clarifies that a food waste grinder, a dishwasher or both may be connected to a kitchen sink drain that is served by an island vent where the installation occurs in a residential occupancy.

913.2 Vent connection. The island fixture vent shall connect to the *fixture drain* as required for an individual or common vent. The vent shall rise vertically to above the drainage outlet of the fixture being vented before offsetting horizontally or vertically downward. The vent or *branch* vent for multiple island fixture vents shall extend to a minimum of 6 inches (152 mm) above the highest island fixture being vented before connecting to the outside vent terminal.

❖ For an island fixture vent, the installation is similar to a conventional drainage system that is either individually or common vented. The only difference is that the vent cannot rise 6 inches (152 mm) above the flood level rim before offsetting horizontally. The vent offsets horizontally below the flood level rim so that the piping can be contained under the cabinet or other fixture support. The concern with a vent offsetting horizontally below the flood level rim of the fixture is the possibility of waste backing up into the vent. To cause a stoppage in the drain to be noticeable to the occupants, the vent must rise above the bottom of the fixture bowl before offsetting. This will result in the water rising in the fixture if there is a stoppage (see Commentary Figure 913.2).

The island vent must rise to at least 6 inches (152 mm) above the fixture flood level rim before connecting to any other vent or vent terminal. In the event of drain stoppage, waste may be discharged to other vents or the vent terminal if this installation precaution was not taken.

913.3 Vent installation below the fixture flood level rim. The vent located below the *flood level rim* of the fixture being vented shall be installed as required for drainage piping in accordance with Chapter 7, except for sizing. The vent shall be sized in accordance with Section 916.2. The lowest point of the island fixture vent shall connect full size to the drainage system. The connection shall be to a vertical drain pipe or to the top half of a horizontal drain pipe. Cleanouts shall be provided in the island fixture vent to permit rodding of all vent piping located below the *flood level rim* of the fixtures. Rodding in both directions shall be permitted through a cleanout.

❖ Because it is possible to have the vent fill with waste when it is located below the flood level rim, the vent must be installed as a drain. The installation will require drainage pattern fittings and a cleanout. The cleanout may be used to rod in both directions because the piping is a vent. A connection to the drainage piping is required at the lowest point of the vent to prevent it from retaining water and becoming blocked by water or waste (see Commentary Section 913.2). The vent will have to be at a higher elevation than the drain pipe to allow the vent to drain any waste water that has collected therein.

SECTION 914 RELIEF VENTS—STACKS OF MORE THAN 10 BRANCH INTERVALS

914.1 Where required. Soil and waste stacks in buildings having more than 10 *branch intervals* shall be provided with a relief vent at each tenth interval installed, beginning with the top floor.

❖ This section requires a method of relieving pressure conditions in drainage stacks more than 10 branch intervals in height.

The flow in a drainage stack creates both negative and positive pressures. The vent system is designed to equalize the air pressures at the trap seal. A relief vent is required to assist the fixture venting and the venting at the base of the stack by providing midpoint connections to the stack. The relief vent prevents excessive pressure from being created by emitting or admitting air at specified points within the stack. A relief vent must be located every 10 branch intervals, measured from the highest horizontal drainage branch, which is typically the top floor. The interval breaks are then calculated downward to the base of the stack (see Commentary Figure 914.1).

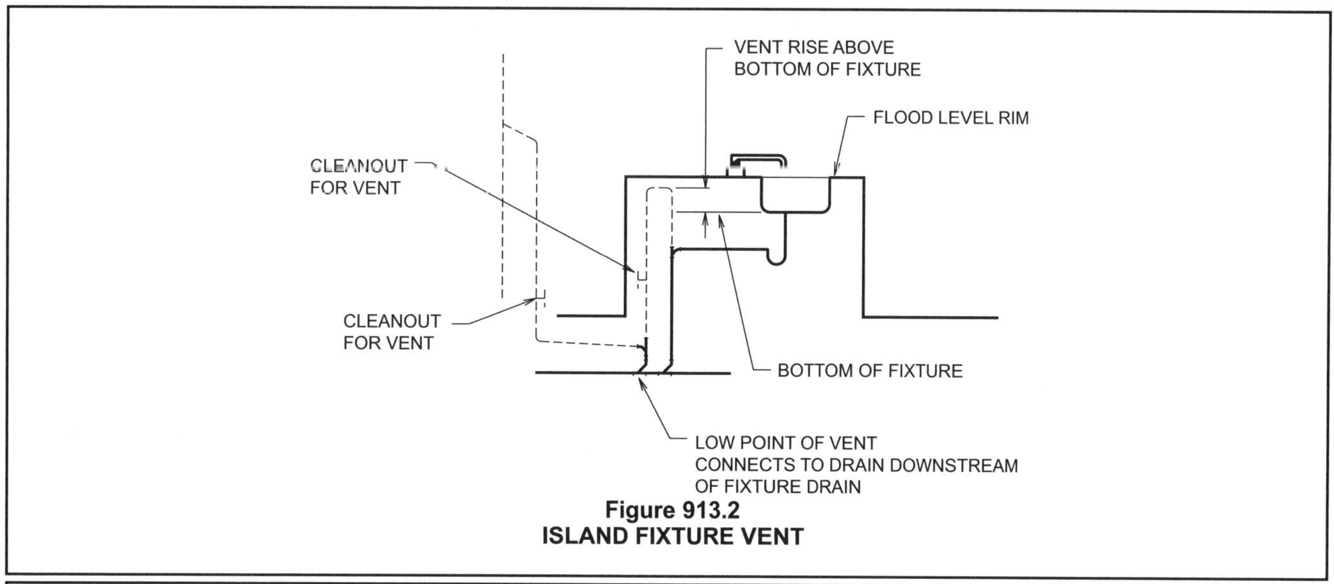

Figure 913.2
ISLAND FIXTURE VENT

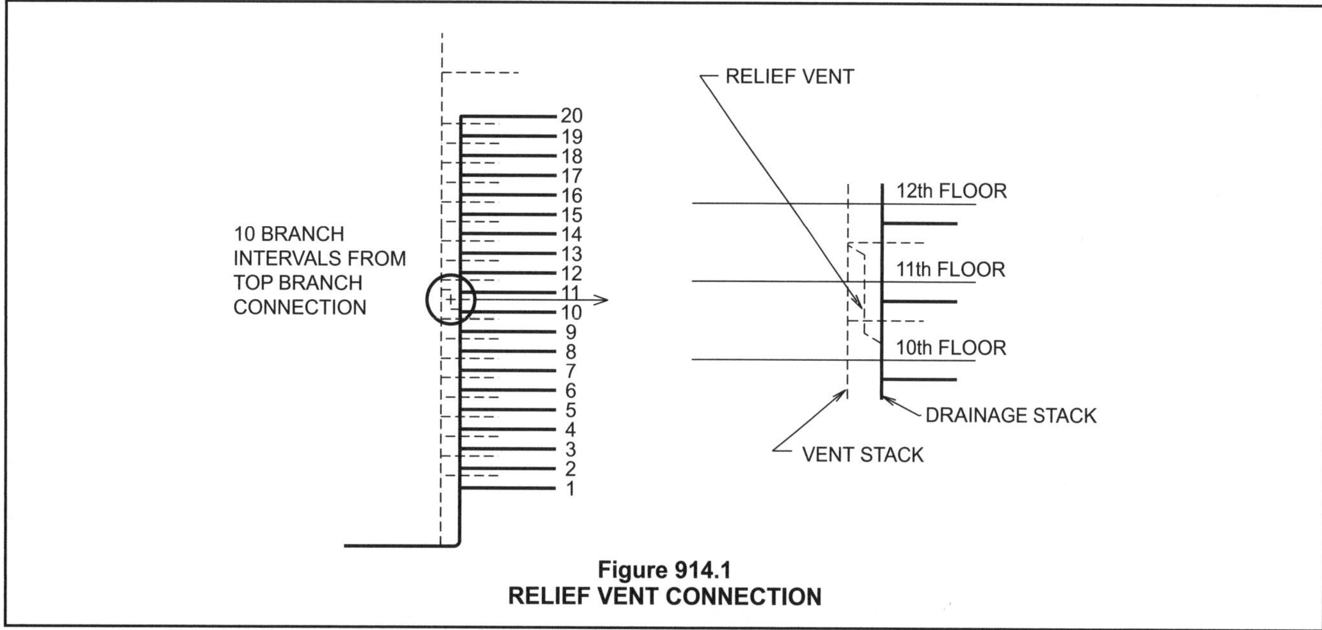

Figure 914.1
RELIEF VENT CONNECTION

914.2 Size and connection. The size of the relief vent shall be equal to the size of the vent *stack* to which it connects. The lower end of each relief vent shall connect to the soil or waste *stack* through a wye below the horizontal *branch* serving the floor, and the upper end shall connect to the vent *stack* through a wye not less than 3 feet (914 mm) above the floor.

❖ This section requires the size of the relief vent to be at least equal to the size of the vent stack to which it connects. This is considered the minimum necessary to provide sufficient venting capacity in both the relief vent and the stack vent to which it is connected.

The lower end of the relief vent is connected to the soil or waste stack below the level of the horizontal branch that serves the floor level within the branch interval required to have the relief vent. The location of this connection is intended to allow waste that might get into the relief vent, including condensation, to reach a waste line. This connection is made using a wye fitting installed as a drainage fitting in order not to impair the flow. The upper connection of the relief vent is made to the vent stack and is to be located a minimum of 3 feet (914 mm) above the floor level of the same horizontal branch. This connection is made using a wye fitting installed in an inverted position. The 3-foot (914 mm) minimum height required is a common theme in this chapter and is intended to prevent waste flow from entering the vent stack.

SECTION 915
VENTS FOR STACK OFFSETS

915.1 Vent for horizontal offset of drainage stack. Horizontal offsets of drainage stacks shall be vented where five or more *branch intervals* are located above the offset. The offset shall be vented by venting the upper section of the drainage *stack* and the lower section of the drainage *stack.*

❖ This section requires venting offsets to reduce pressure differentials. Offsets involve changes in the direction of flow from true vertical to greater than 45 degrees (0.79 rad) from vertical. Such changes upset the nonturbulent annular flow in the stack, thus creating turbulent flow with the possibility for full bore flow in sections of the offset. Significant pressure fluctuations may result, which are stabilized by the offset vents. Refer to Section 704.3 for information on hydraulic jump.

An offset is vented to reduce the pressure differentials that occur in the drainage stack and offset (see the commentary to Section 704.3 for flow in an offset). Venting is necessary when the drainage stack above the offset is five or more branch intervals in height.

An offset in a drainage stack is vented as two separate stacks with a vent connection for both the upper stack and the lower stack (see Commentary Sections 915.2 and 915.3). The offset vents must be no less than the size of the vent stack, which is determined in accordance with Section 916.

915.2 Upper section. The upper section of the drainage *stack* shall be vented as a separate *stack* with a vent *stack* connection installed in accordance with Section 903.4. The offset shall be considered the base of the *stack*.

❖ The vent for the upper section of the stack must connect to the stack in accordance with Section 903.4. The venting would be equivalent to venting the upper section as if it were a separate stack. If the vent stack for the upper section does not connect to the vent stack for the lower section (see commentary, Section 915.3), the vent need be sized only for the upper section (see Commentary Figure 915.2).

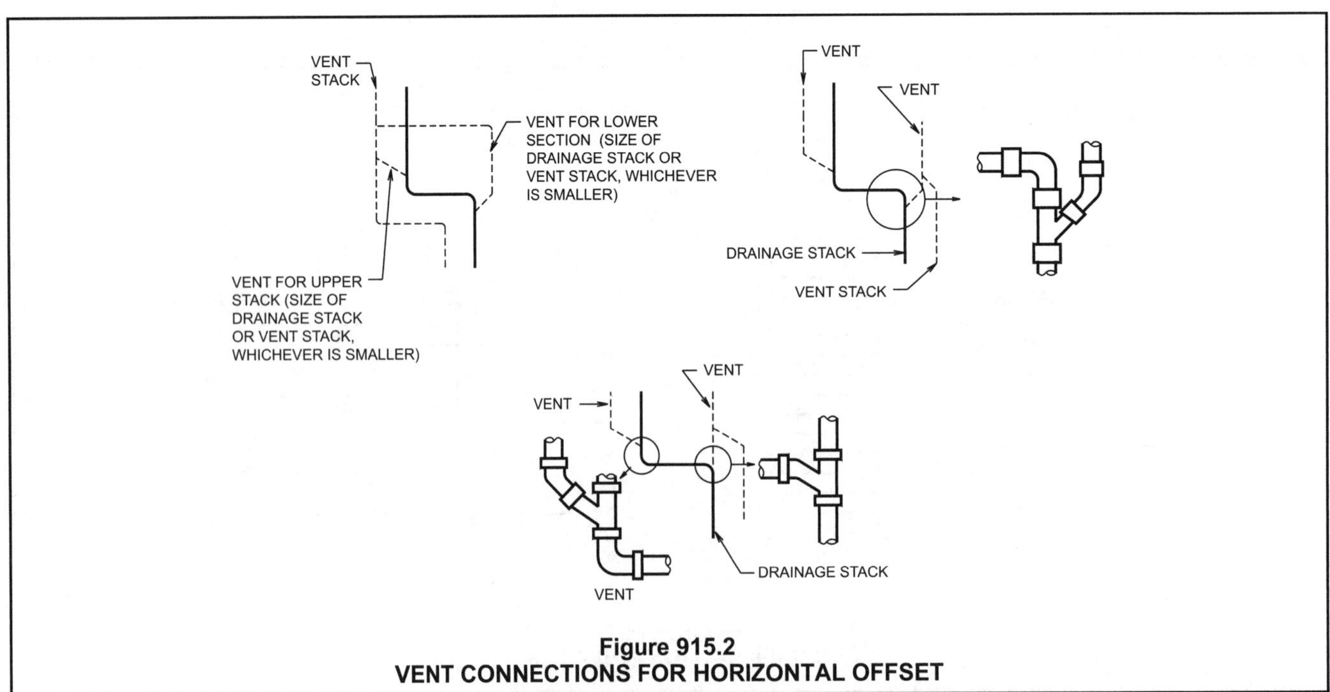

Figure 915.2
VENT CONNECTIONS FOR HORIZONTAL OFFSET

915.3 Lower section. The lower section of the drainage *stack* shall be vented by a yoke vent connecting between the offset and the next lower horizontal *branch*. The yoke vent connection shall be permitted to be a vertical extension of the drainage *stack*. The size of the yoke vent and connection shall be a minimum of the size required for the vent *stack* of the drainage *stack*.

❖ The vent for the lower section of the stack may connect as an extension of the drainage stack (similar to a stack vent) or as a yoke vent. A yoke vent is an upright (vertical) wye connection to the vertical stack. The vent system for the lower section must be sized for the total load on the drainage stack. The size of the vent connection must not be less than the size of the vent stack that serves the drainage stack (see Commentary Figure 915.2).

SECTION 916 VENT PIPE SIZING

916.1 Size of stack vents and vent stacks. The minimum required diameter of *stack vents* and vent stacks shall be determined from the *developed length* and the total of drainage fixture units connected thereto in accordance with Table 916.1, but in no case shall the diameter be less than one-half the diameter of the drain served or less than $1^1/_4$ inches (32 mm).

❖ The sizing criteria for stack vents and vent stacks are based on three variables: the developed length of the vent, the size of the stack served by the vent and the total dfu connected to the stack.

The vent is sized in accordance with Table 916.1, but in no case is the vent pipe diameter to be less than one-half the required diameter of the drain served or less than $1^1/_4$ inches (32 mm).

The developed length of the vent pipe must also be considered when sizing the vent system. The volume of air that a given pipe size can convey with a limited friction loss depends on the pipe length. The longer the pipe, the greater the effect friction will have on the flow of air. If the resistance to airflow is too high, the pressure drop will be excessive, and the vent will be unable to prevent pressure differentials from exceeding 1 inch of water column (249 Pa). For example, if the pressure loss in a vent caused by pipe friction exceeds 2 inches of water column (498 Pa), pressures in the drainage system would exceed 2 inches of water column (498 Pa) before any air would move through the vent to neutralize pressure fluctuations. The vent pipe is often increased in size to compensate for this effect. Refer to the description of "Developed length" in Section 916.3. Commentary Figures 916.1(1), 916.1(2) and 916.1(3) contain examples for proper sizing of vents.

TABLE 916.1. See page 9-52.

❖ This table contains the requirements for sizing the piping serving as a stack vent or a vent stack. Included in this table are the size of the soil or waste stack to be vented, the number of (drainage) fixture units connected to the drain line being vented and the developed length of vent pipe. Since the sizing limitations are in one table, the user can easily determine that the requirements for stack vents and vent stacks have been satisfied.

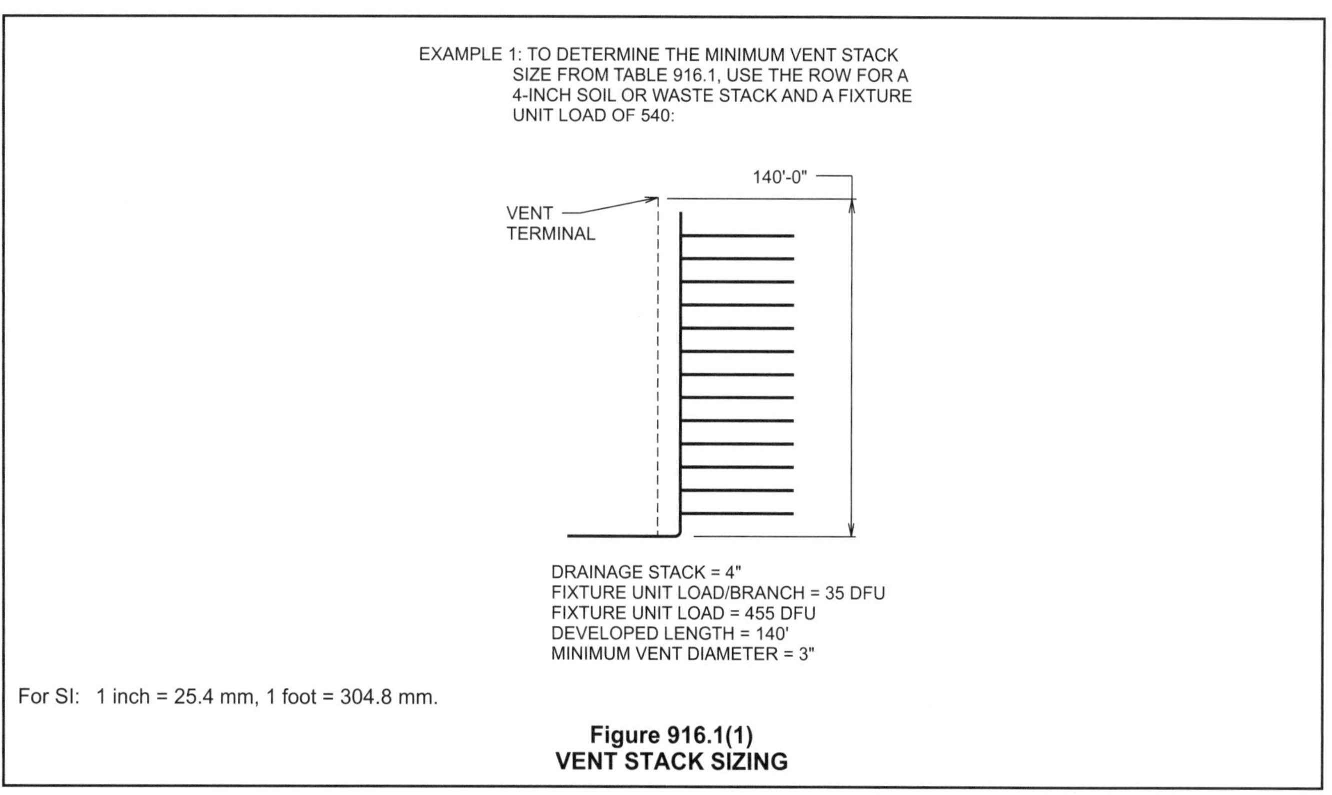

EXAMPLE 1: TO DETERMINE THE MINIMUM VENT STACK SIZE FROM TABLE 916.1, USE THE ROW FOR A 4-INCH SOIL OR WASTE STACK AND A FIXTURE UNIT LOAD OF 540:
140'-0"
VENT TERMINAL
DRAINAGE STACK = 4"
FIXTURE UNIT LOAD/BRANCH = 35 DFU
FIXTURE UNIT LOAD = 455 DFU
DEVELOPED LENGTH = 140'
MINIMUM VENT DIAMETER = 3"
For SI: 1 inch = 25.4 mm, 1 foot = 304.8 mm.

Figure 916.1(1)
VENT STACK SIZING

TABLE 916.1
SIZE AND DEVELOPED LENGTH OF STACK VENTS AND VENT STACKS

DIAMETER OF SOIL OR WASTE STACK (inches)	TOTAL FIXTURE UNITS BEING VENTED (dfu)	MAXIMUM DEVELOPED LENGTH OF VENT (feet)[a] DIAMETER OF VENT (inches)										
		$1^1/_4$	$1^1/_2$	2	$2^1/_2$	3	4	5	6	8	10	12
$1^1/_4$	2	30										
$1^1/_2$	8	50	150	—	—	—	—	—	—	—	—	—
$1^1/_2$	10	30	100									
2	12	30	75	200								
2	20	26	50	150		—	—	—	—	—	—	—
$2^1/_2$	42		30	100	300							
3	10		42	150	360	1,040						
3	21	—	32	110	270	810	—	—	—	—	—	—
3	53		27	94	230	680						
3	102		25	86	210	620						
4	43	—		35	85	250	980	—	—	—	—	—
4	140			27	65	200	750					
4	320			23	55	170	640					
4	540	—	—	21	50	150	580		—	—	—	—
5	190				28	82	320	990				
5	490				21	63	250	760				
5	940	—	—	—	18	53	210	670	—	—	—	—
5	1,400				16	49	190	590				
6	500					33	130	400	1,000			
6	1,100	—	—	—	—	26	100	310	780	—	—	—
6	2,000					22	84	260	660			
6	2,900					20	77	240	600			
8	1,800	—	—	—	—		31	95	240	940	—	—
8	3,400						24	73	190	720		
8	5,600						20	62	160	610		
8	7,600	—	—	—	—	—	18	56	140	560		—
10	4,000							31	78	310	960	
10	7,200							24	60	240	740	
10	11,000	—	—	—	—	—		20	51	200	630	—
10	15,000							18	46	180	570	
12	7,300								31	120	380	940
12	13,000	—	—	—	—	—	—	—	24	94	300	720
12	20,000								20	79	250	610
12	26,000								18	72	230	500
15	15,000	—	—	—	—	—	—	—		40	130	310
15	25,000									31	96	240
15	38,000	—	—	—	—	—	—	—	—	26	81	200
15	50,000									24	74	180

For SI: 1 inch = 25.4 mm, 1 foot = 304.8 mm.

a. The developed length shall be measured from the vent connection to the open air.

916.2 Vents other than stack vents or vent stacks. The diameter of individual vents, *branch* vents, circuit vents and relief vents shall be at least one-half the required diameter of the drain served. The required size of the drain shall be determined in accordance with Table 710.1(2). Vent pipes shall not be less than $1^1/_4$ inches (32 mm) in diameter. Vents exceeding 40 feet (12 192 mm) in *developed length* shall be increased by one nominal pipe size for the entire *developed length* of the vent pipe. Relief vents for soil and waste stacks in buildings having more than 10 *branch intervals* shall be sized in accordance with Section 914.2.

❖ This section contains the sizing criteria for dry portions of a conventional venting system, including individual vents, branch vents, circuit vents, relief vents, common vents and dry vent connections to wet vents. The size of the vent is based on two variables: the required size of the drain served by the vent and the developed length of the vent.

EXAMPLE 2: TO DETERMINE THE MINIMUM VENT STACK SIZE FROM TABLE 916.1, USE THE ROW FOR A 6-INCH SOIL OR WASTE STACK AND A FIXTURE UNIT LOAD OF 1,100:
0" DRAINAGE STACK
VENT STACK
DRAINAGE STACK = 6"
FIXTURE UNIT LOAD = 750 DFU
DEVELOPED LENGTH = 280'
MIN. VENT DIAMETER = 5"

For SI: 1 inch = 25.4 mm, 1 foot = 304.8 mm.

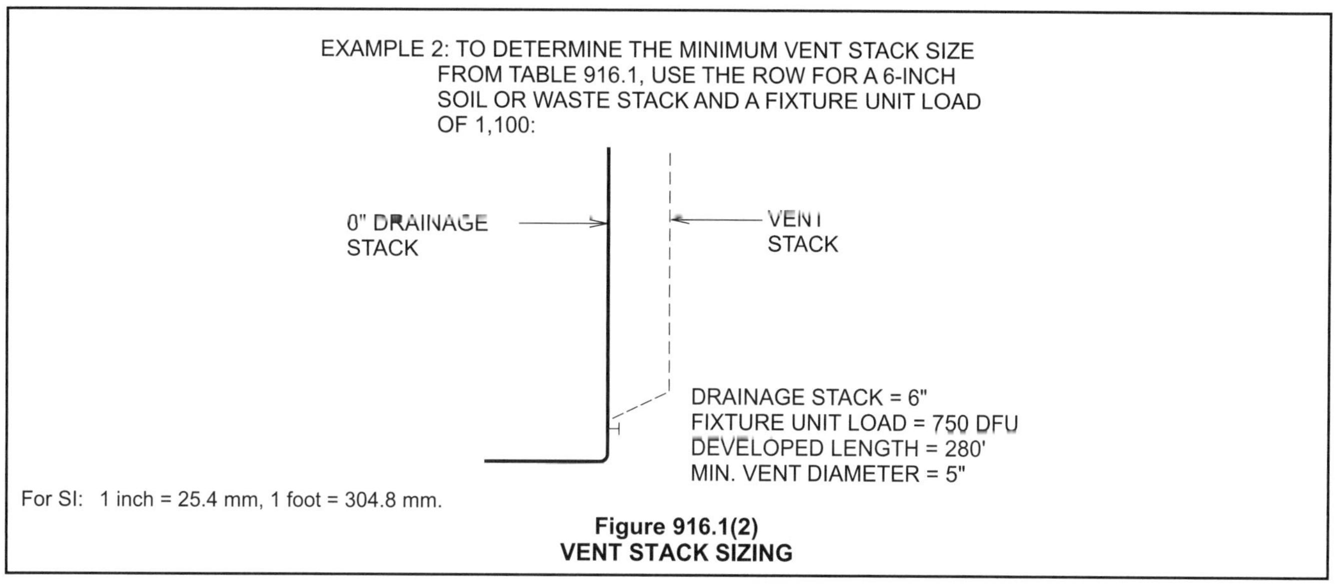

Figure 916.1(2)
VENT STACK SIZING

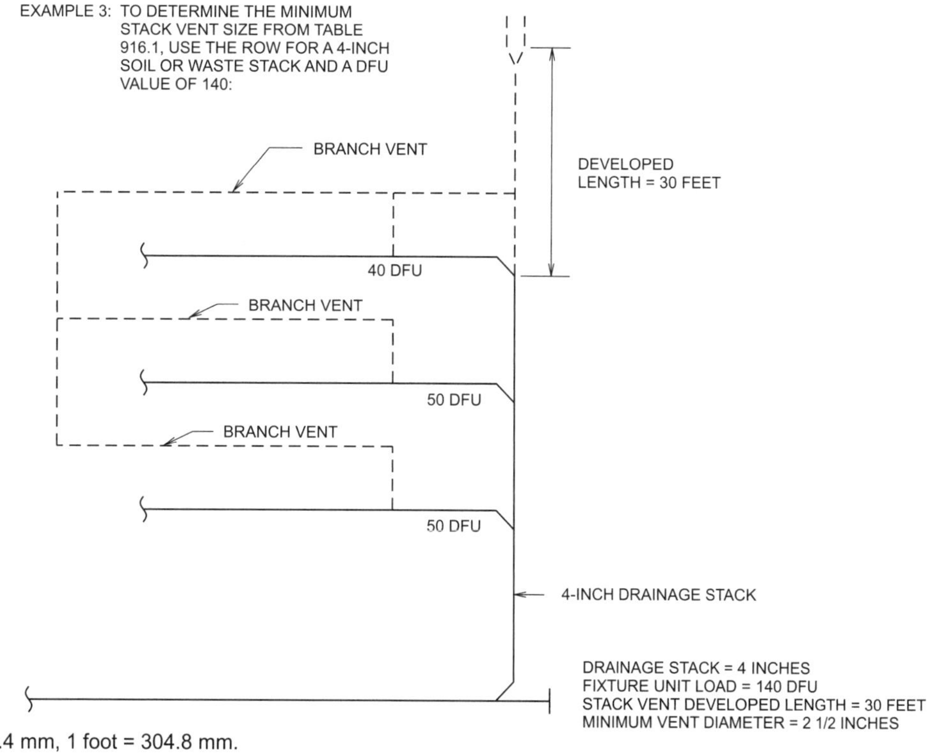

EXAMPLE 3: TO DETERMINE THE MINIMUM STACK VENT SIZE FROM TABLE 916.1, USE THE ROW FOR A 4-INCH SOIL OR WASTE STACK AND A DFU VALUE OF 140:
BRANCH VENT
DEVELOPED LENGTH = 30 FEET
40 DFU
BRANCH VENT
50 DFU
BRANCH VENT
50 DFU
4-INCH DRAINAGE STACK
DRAINAGE STACK = 4 INCHES
FIXTURE UNIT LOAD = 140 DFU
STACK VENT DEVELOPED LENGTH = 30 FEET
MINIMUM VENT DIAMETER = 2 1/2 INCHES

For SI: 1 inch = 25.4 mm, 1 foot = 304.8 mm.

Figure 916.1(3)
STACK VENT SIZING

The sizing of vent pipes other than vent stacks and stack vents is very similar to the method used to size the vent stacks and the stack vents discussed in Section 916.1. The drainage piping is sized in accordance with Table 710.1(2) based on the drainage fixture load determined by Tables 709.1 and 709.2 and Sections 709.3 and 709.4. The vent size is then determined to be a minimum of one-half the required diameter of the drain served, but not less than $1^1/_4$ inches (32 mm).

Note that this section ties vent sizing to the drain sizing criteria of Table 710.1(2) for horizontal fixture branches, which results in more conservative vent sizing. For example, a circuit-vented branch serving 30 dfu may consist of a 3-inch (76 mm) drain and a $1^1/_2$-inch (38 mm) circuit vent in accordance with Table 710.1(1), but would have to consist of a 4-inch (102 mm) drain and a 2-inch (51 mm) circuit vent in accordance with Table 710.1(2). The developed length of the vent pipe must also be considered when sizing the vent system. The amount of air that a given pipe size can convey depends on its length. The effect of friction increases as pipe length increases and has a greater effect on the flow of air. If the friction increases to such a level that the vent is not able to prevent pressure differentials from exceeding 1 inch of water column (249 Pa), the vent pipe size must be increased to compensate for this effect. Refer to the definition of "Developed length" in Section 916.3. Commentary Figures 916.2(1) through 916.2(4) contain several examples for proper sizing of vents. This section makes the assumption that a developed length exceeding 40 feet (12 192 mm) will result in excessive friction losses; therefore, such vents must be increased by one nominal pipe size. A vent increased in size in accordance with this section would be allowed to have an unlimited developed length; however, there will always be some length at which the friction loss would again be excessive. It is assumed that the natural constraints of construction practice will prevent vent lengths that would exceed the friction losses of a vent increased by one nominal pipe size. Note that the increase in pipe size caused by the 40-foot (12 192 mm) length limitation will increase the size of the entire developed length of the vent, not only the portion that extends beyond 40 feet (12 192 mm).

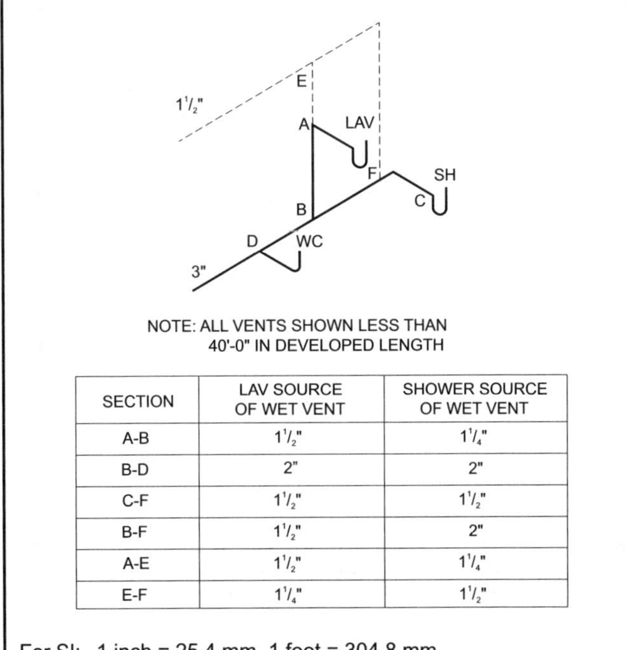

SECTION	LAV SOURCE OF WET VENT	SHOWER SOURCE OF WET VENT
A-B	$1^1/_2$"	$1^1/_4$"
B-D	2"	2"
C-F	$1^1/_2$"	$1^1/_2$"
B-F	$1^1/_2$"	2"
A-E	$1^1/_2$"	$1^1/_4$"
E-F	$1^1/_4$"	$1^1/_2$"

For SI: 1 inch = 25.4 mm, 1 foot = 304.8 mm.

Figure 916.2(2)
VENT SIZING

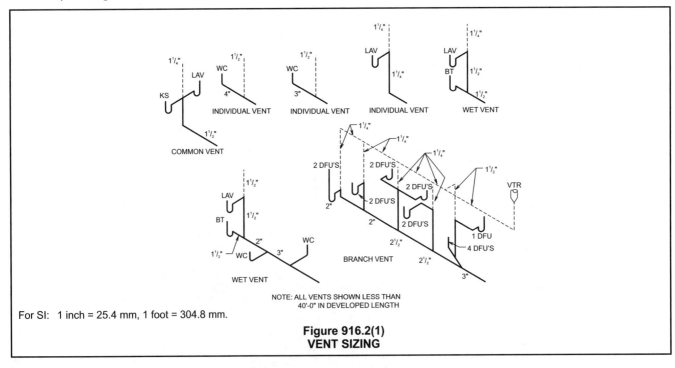

NOTE: ALL VENTS SHOWN LESS THAN 40'-0" IN DEVELOPED LENGTH

For SI: 1 inch = 25.4 mm, 1 foot = 304.8 mm.

Figure 916.2(1)
VENT SIZING

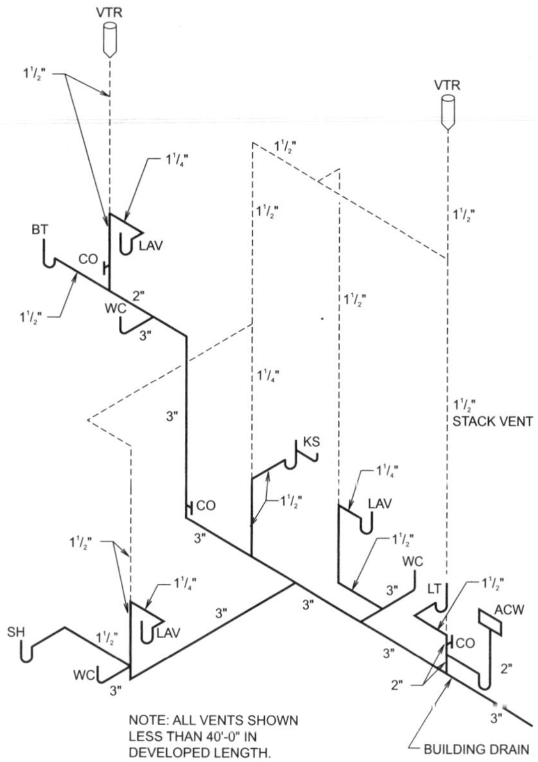

NOTE: ALL VENTS SHOWN LESS THAN 40'-0" IN DEVELOPED LENGTH.

For SI: 1 inch = 25.4 mm, 1 foot = 304.8 mm.

Figure 916.2(3)
VENT SIZING

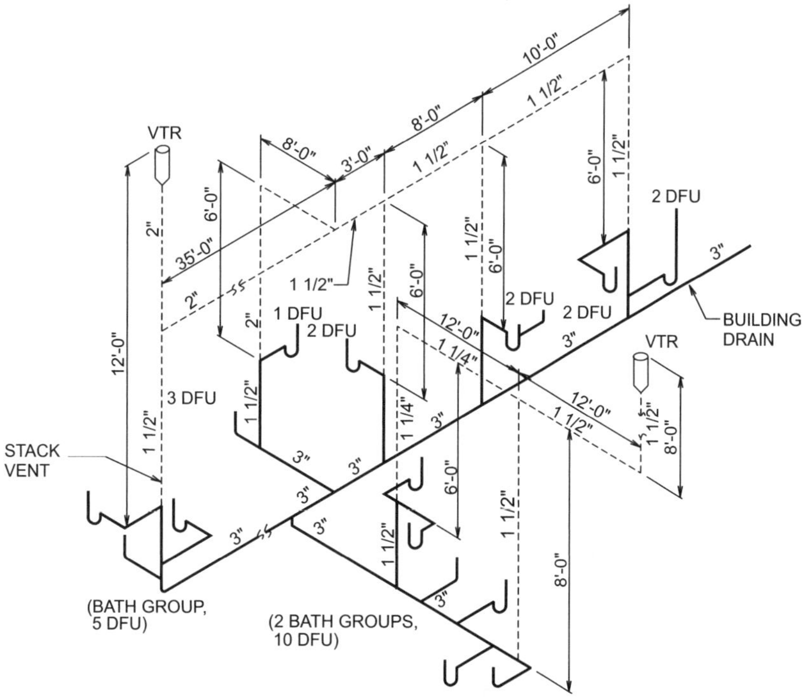

For SI: 1 inch = 25.4 mm, 1 foot = 304.8 mm.

Figure 916.2(4)
VENT SIZING

916.3 Developed length. The *developed length* of individual, *branch*, circuit and relief vents shall be measured from the farthest point of vent connection to the drainage system to the point of connection to the vent *stack*, *stack vent* or termination outside of the building.

❖ Section 916.2 requires that a vent pipe be increased one nominal pipe size when its developed length exceeds 40 feet (12 192 mm). This section establishes how to determine the developed length for individual, branch, circuit and relief vents [see Commentary Figures 916.3(1) and 916.3(2)].

The developed length of a vent is measured from the farthest point of connection to the drainage system to a vent stack, stack vent or termination outdoors. Where connecting to a vent stack or stack vent, either one must be sized to supply adequate air to each vent connection. For example, the circumstance may arise where the size of a stack vent or vent stack would have to be increased to serve a connected, individual, branch, circuit or relief vent. Consideration must be given to the fact that the developed length of a stack vent will also contribute to the overall friction loss of any vent connected to the stack vent. For example, a 2-inch (51 mm) branch vent connecting to a 2-inch (51 mm) stack vent will have an overall friction loss that is based on the developed length of the branch vent plus the developed length of the stack vent.

If a vent does not connect to a vent stack or stack vent, the developed length is measured to the point of termination in the outdoor air.

916.4 Multiple branch vents. Where multiple *branch* vents are connected to a common *branch* vent, the common *branch* vent shall be sized in accordance with this section based on the size of the common horizontal drainage *branch* that is or would be required to serve the total *drainage fixture unit* (*dfu*) load being vented.

❖ A common branch vent with multiple branch vent connections must be sized as though it were serving a branch drain sized to carry the total fixture load of all drains served by the branch vents.

Example:

Determine the proper sizes for the vent system shown in Commentary Figure 916.4. Note that vent developed lengths are assumed to be less than 40 feet (12 192 mm) up to point Z.

The total dfu load of the horizontal drain branches served by branch vents A, B and C at point W is 18 dfu. In accordance with Table 710.1(2), the minimum required size of a horizontal drain branch with that load is 3 inches (76 mm); therefore, the size of the common branch vent between points W and X is $1^{1}/_{2}$ inches (38 mm).

The horizontal drain branch served by the common branch vent at point X is 22 dfu. Table 710.1(2) requires the horizontal branch drain to be 4 inches (102 mm). The common branch vent between points X and Y is then required to be 2 inches (51 mm). Because the horizontal branch drain size is now 4 inches (102 mm), the common branch vent between points Y and Z must also be 2 inches (51 mm) in diameter. Note that the vertical vent at point Z must be at least 2 inches (51 mm) in diameter. If the developed length of the vertical vent from point Z to the vent terminal plus the developed length of any branch vent exceeds 40 feet (12 192 mm), the vertical vent and such branch vent must be increased in diameter by one pipe size.

916.4.1 Branch vents exceeding 40 feet in developed length. *Branch* vents exceeding 40 feet (12 192 mm) in *developed length* shall be increased by one nominal size for the entire *developed length* of the vent pipe.

❖ This section addresses the length of common branch vent piping that serves more than one branch vent. This section expresses the same intent as Section 916.2, except that it specifically covers common branch piping that serves more than one branch vent. If the common portion of vent piping exceeds the length limitation, it is increased in size to compensate for the friction loss within the common vent piping.

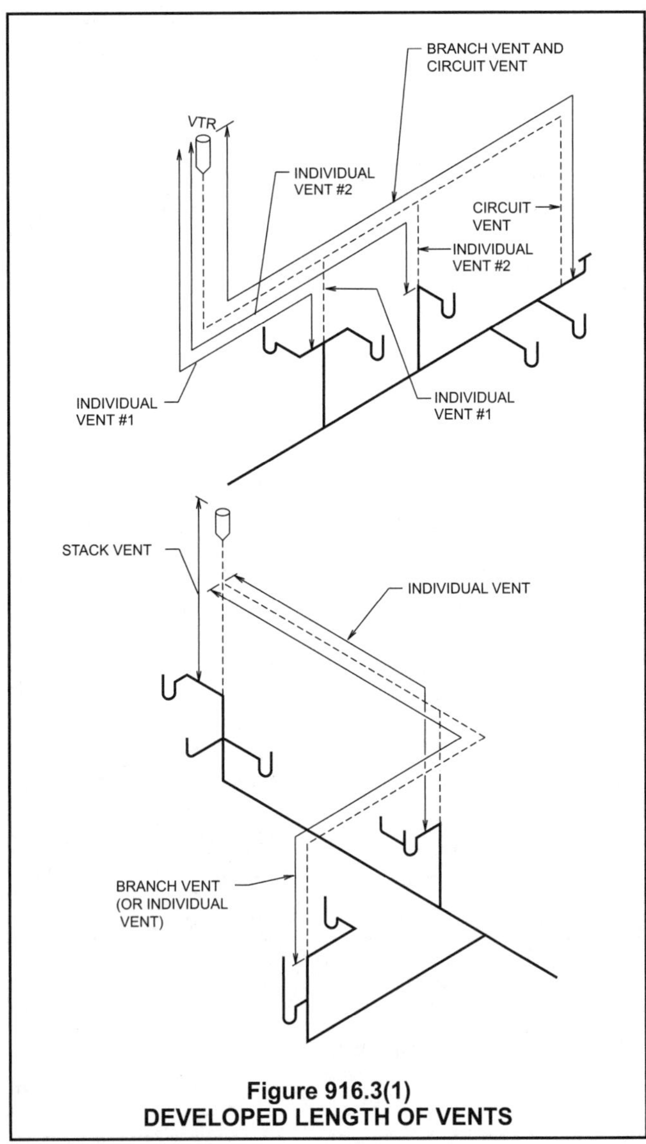

Figure 916.3(1)
DEVELOPED LENGTH OF VENTS

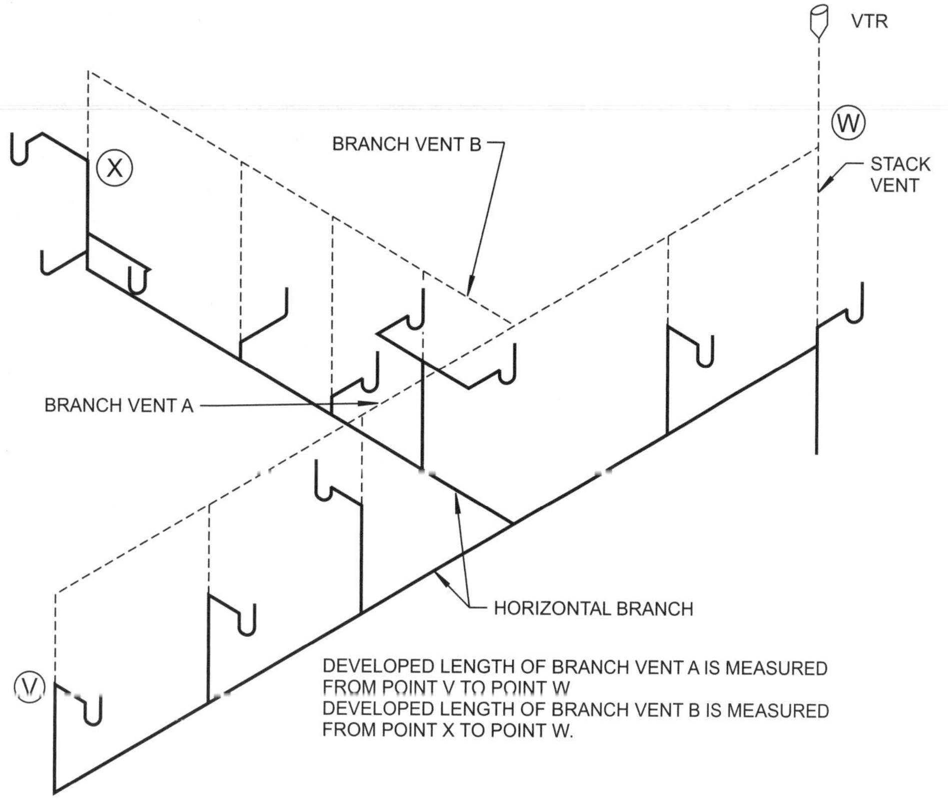

VTR

W

STACK VENT

BRANCH VENT B

X

BRANCH VENT A

HORIZONTAL BRANCH

V

DEVELOPED LENGTH OF BRANCH VENT A IS MEASURED FROM POINT V TO POINT W
DEVELOPED LENGTH OF BRANCH VENT B IS MEASURED FROM POINT X TO POINT W.

Figure 916.3(2)
DEVELOPED LENGTH OF VENTS

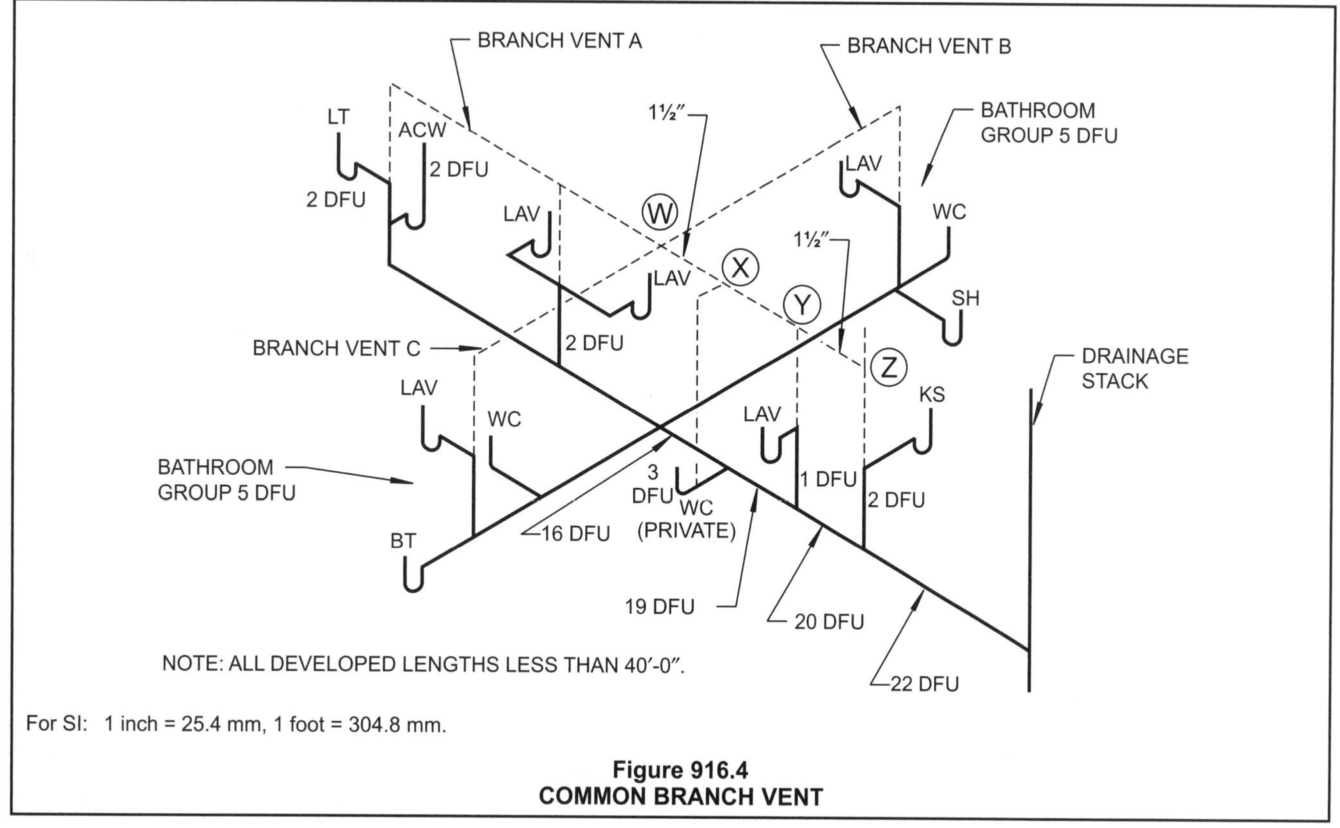

For SI: 1 inch = 25.4 mm, 1 foot = 304.8 mm.

Figure 916.4
COMMON BRANCH VENT

916.5 Sump vents. Sump vent sizes shall be determined in accordance with Sections 916.5.1 and 916.5.2.

❖ The following sections specify the method of sizing sump vents and sewage ejector relief pipes.

916.5.1 Sewage pumps and sewage ejectors other than pneumatic. Drainage piping below *sewer* level shall be vented in a similar manner to that of a gravity system. Building sump vent sizes for sumps with sewage pumps or sewage ejectors, other than pneumatic, shall be determined in accordance with Table 916.5.1.

❖ Sumps receiving sanitary drainage must be vented. Where a sewage pump is used, the vent is allowed to connect to the building venting system. Table 916.5.1 specifies the minimum vent pipe size based on the discharge capacity of the pump and the developed length of vent pipe. The sump vent must be a minimum of $1^1/_4$ inches (32 mm) (see Commentary Figure 916.5.1).

Building subdrain systems, including their venting systems, are designed no differently than gravity drainage systems, except that a sewage pump or ejector is required. The vents for both gravity and subdrain systems may be combined if pneumatic ejector vents are independently terminated in the outside air (see Section 916.5.2).

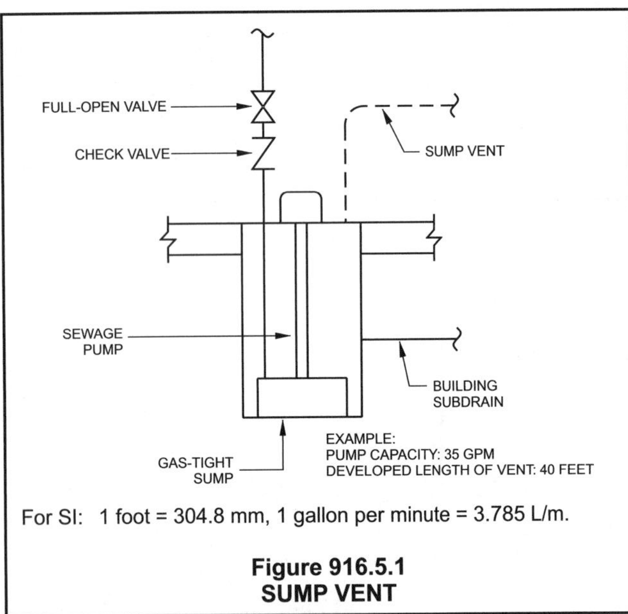

For SI: 1 foot = 304.8 mm, 1 gallon per minute = 3.785 L/m.

Figure 916.5.1
SUMP VENT

916.5.2 Pneumatic sewage ejectors. The air pressure relief pipe from a pneumatic sewage ejector shall be connected to an independent vent *stack* terminating as required for vent extensions through the roof. The relief pipe shall be sized to relieve air pressure inside the ejector to atmospheric pressure, but shall not be less than $1^1/_4$ inches (32 mm) in size.

❖ Pneumatic sewage ejectors operate at high air pressures, forcing the sewage from the ejector sump to the gravity drainage system. An independent vent is necessary to relieve the pressure in the sump at the completion of the ejector cycle. If such a vent is connected to the building venting system, tremendous pressure differentials would be created, disrupting the normal operation of the drainage system.

TABLE 916.5.1. See page 9-59.

❖ The vent for a sump provides a supply of air into the sump while the pumping equipment is operating. The supply of air prevents a pressure differential from occurring on the drainage supply pipe.

The sump vent also allows air to be displaced when drainage is filling the sump and air to be admitted when the sump is emptied. Because the air requirements are minimal, the resulting pipe size is small.

The developed length of a sump vent is measured from the sump connection to a vent stack or stack vent connection. If the sump vent does not connect to a vent stack, the developed length is measured to the point of termination outside. In accordance with Note a, the overall (equivalent) length of a vent is the actual developed length plus an allowance for the equivalent length of fittings. For example, using the rule-of-thumb 50-percent allowance, a sump vent with an actual developed length of 90 feet (27 432 mm) would have an overall developed length of 135 feet (41 148 mm) because of the addition of the equivalent lengths of fittings.

SECTION 917
AIR ADMITTANCE VALVES

917.1 General. Vent systems utilizing air admittance valves shall comply with this section. Stack-type air admittance valves shall conform to ASSE 1050. Individual and branch-type air admittance valves shall conform to ASSE 1051.

❖ This section contains the requirements for air admittance valves. Air admittance valves must comply with the provisions of ASSE 1051, which requires marking each device with the name of the manufacturer, model number or device description and the device classification.

The standard divides the classification of air admittance valves into two types: ASSE 1051—Type A and ASSE 1051—Type B. Type A air admittance valves are permitted to serve only as a vent for an individual fixture, and Type B air admittance valves are permitted to serve one or more individual vents (see the definition of "Air admittance valve" in Section 202).

917.2 Installation. The valves shall be installed in accordance with the requirements of this section and the manufacturer's installation instructions. Air admittance valves shall be installed after the DWV testing required by Section 312.2 or 312.3 has been performed.

❖ The installation of air admittance valves must conform to the requirements of Sections 917.3 through 917.8 and the manufacturer's instructions. Where differences occur between the provisions of the code and the manufacturer's instructions, the most restrictive provisions must be applied.

Air admittance valves are designed to withstand limited positive pressures and because extreme pressures may damage the devices, they must not be subjected to the pressures of system testing. Drain, waste and vent testing involves pressures far in excess of the pressures of normal system operation.

917.3 Where permitted. Individual, *branch* and circuit vents shall be permitted to terminate with a connection to an individual or branch-type air admittance valve. *Stack vents* and vent stacks shall be permitted to terminate to stack-type air admittance valves. Individual and branch-type air admittance valves shall vent only fixtures that are on the same floor level and connect to a *horizontal branch drain*. The *horizontal branch drain* having individual and branch-type air admittance valves shall conform to Section 917.3.1 or 917.3.2. Stack-type air admittance valves shall conform to Section 917.3.3.

❖ This section establishes the specific locations within the vent system where an air admittance valve may be installed. ASSE 1050 devices may be used only for stack venting. ASSE 1051—Type A devices maybe used only as an individual vent for a single fixture drain [see Commentary Figure 917.3(1)]. ASSE 1051—Type B devices may serve as branch vents for one or more individual vents, including the dry vent connecting to a common vent, wet vent or circuit vent [see Commentary Figures 917.3(2) through 917.3(5)].

ASSE 1051 air admittance valves are not permitted to serve as the vent for multiple fixtures located on more than one floor [see Commentary Figure 917.3(6)]. The ASSE 1051 devices are not permitted to serve fixtures that do not connect into a horizontal drain branch on the same floor level. Refer to Section 917.8 for other prohibited air admittance valve installations.

Because air admittance valves relieve only negative pressures (less than atmospheric) in the drainage system, special provisions are necessary to reduce the likelihood of pressure differentials occurring in portions of drainage systems served by air admittance valves. Section 917.3.2 contains measures for pressure relief by requiring the installation of a relief vent. Section 917.3.1 requires a horizontal branch drain to connect into a drainage stack at a point where adverse pressure conditions are not likely to occur.

Note that Sections 917.3.1 and 917.3.2 describe two distinct methods of protecting horizontal branches from excessive pressure fluctuations. An installation need comply with only one of the sections.

917.3.1 Location of branch. The *horizontal branch drain* shall connect to the drainage *stack* or *building drain* a maximum of four *branch* intervals from the top of the *stack*.

❖ As drainage flows down a stack, it tends to draw air downward with it. Air located in the drainage stack ahead of the flow also tends to be pushed downward by the flowing drainage—similar to the action of a piston in a cylinder. Measuring the air pressure in the stack with respect to atmospheric pressure will indicate that the flow action creates negative pressures behind the flow and positive pressures ahead of the flow [see Commentary Figure 917.3.1(1)]. A horizontal branch drain served by an air admittance valve must be located within the top

TABLE 916.5.1
SIZE AND LENGTH OF SUMP VENTS

DISCHARGE CAPACITY OF PUMP (gpm)	MAXIMUM DEVELOPED LENGTH OF VENT (feet)[a]					
	Diameter of vent (inches)					
	$1^1/_4$	$1^1/_2$	2	$2^1/_2$	3	4
10	No limit[b]	No limit	No limit	No limit	No limit	No limit
20	270	No limit	No limit	No limit	No limit	No limit
40	72	160	No limit	No limit	No limit	No limit
60	31	75	270	No limit	No limit	No limit
80	16	41	150	380	No limit	No limit
100	10[c]	25	97	250	No limit	No limit
150	Not permitted	10[c]	44	110	370	No limit
200	Not permitted	Not permitted	20	60	210	No limit
250	Not permitted	Not permitted	10	36	132	No limit
300	Not permitted	Not permitted	10[c]	22	88	380
400	Not permitted	Not permitted	Not permitted	10[c]	44	210
500	Not permitted	Not permitted	Not permitted	Not permitted	24	130

For SI: 1 inch = 25.4 mm, 1 foot = 304.8 mm, 1 gallon per minute = 3.785 L/m.

a. Developed length plus an appropriate allowance for entrance losses and friction due to fittings, changes in direction and diameter. Suggested allowances shall be obtained from NSB Monograph 31 or other approved sources. An allowance of 50 percent of the developed length shall be assumed if a more precise value is not available.

b. Actual values greater than 500 feet.

c. Less than 10 feet.

four branch intervals of the stack where pressure differentials would be tolerable. Otherwise, a method of relieving pressure is required because of the greater pressure differentials expected in lower portions of a stack [see commentary, Section 917.3.2 and Commentary Figure 917.3.1(2)].

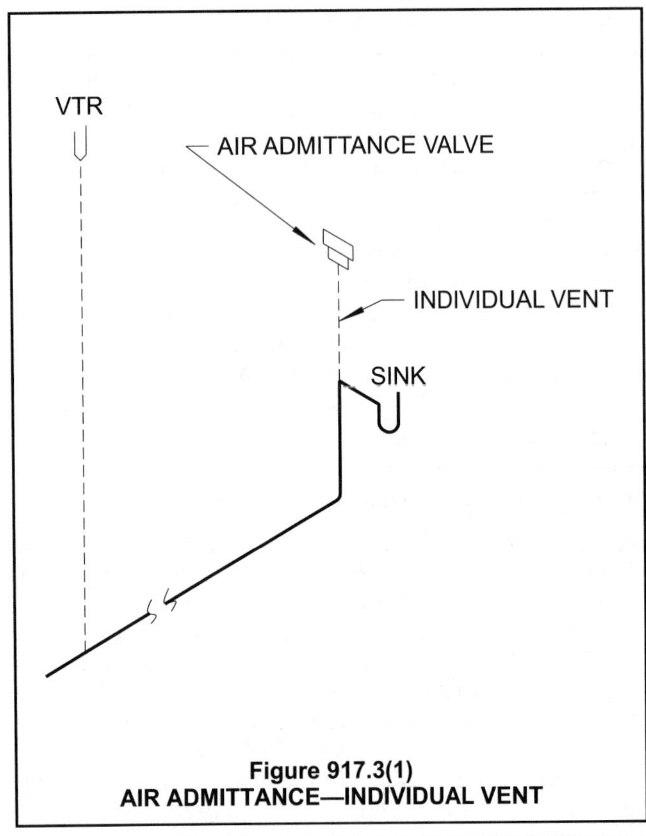

Figure 917.3(1)
AIR ADMITTANCE—INDIVIDUAL VENT

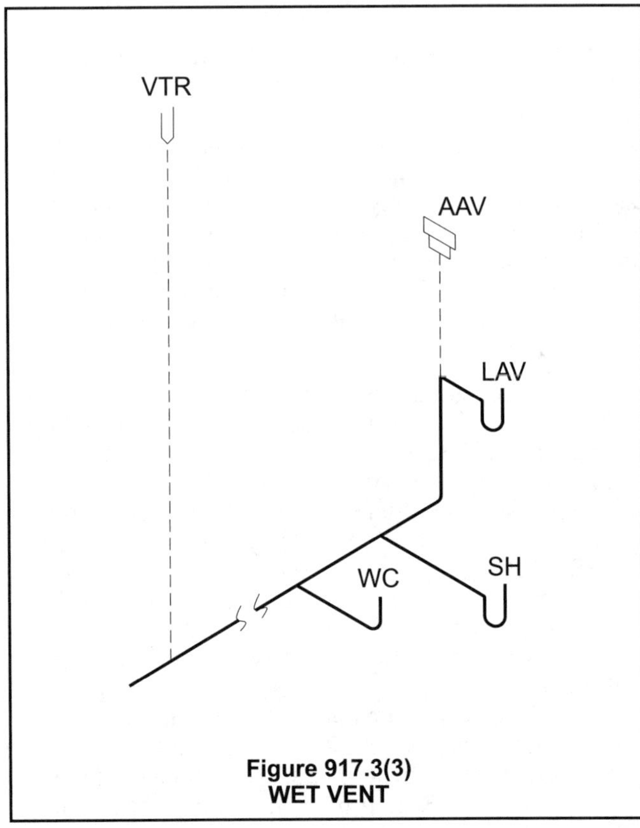

Figure 917.3(3)
WET VENT

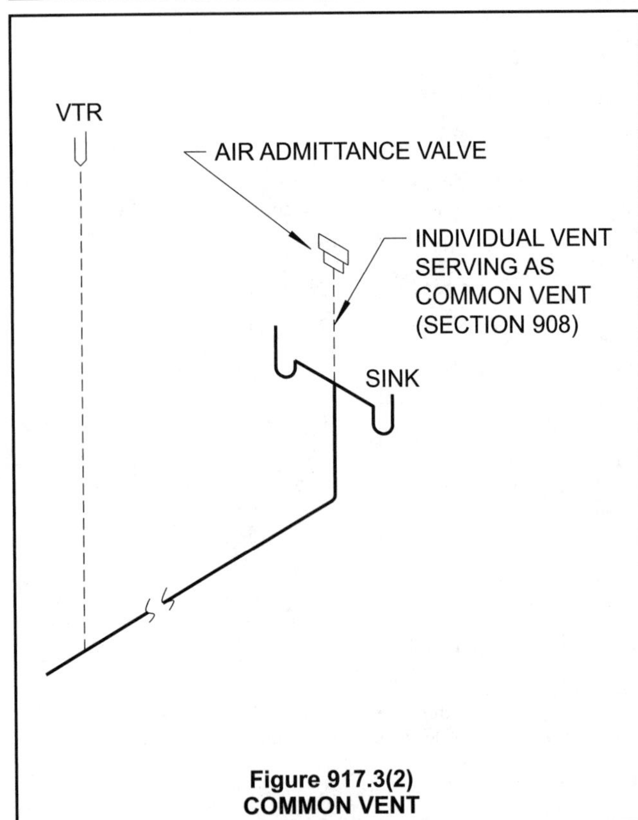

Figure 917.3(2)
COMMON VENT

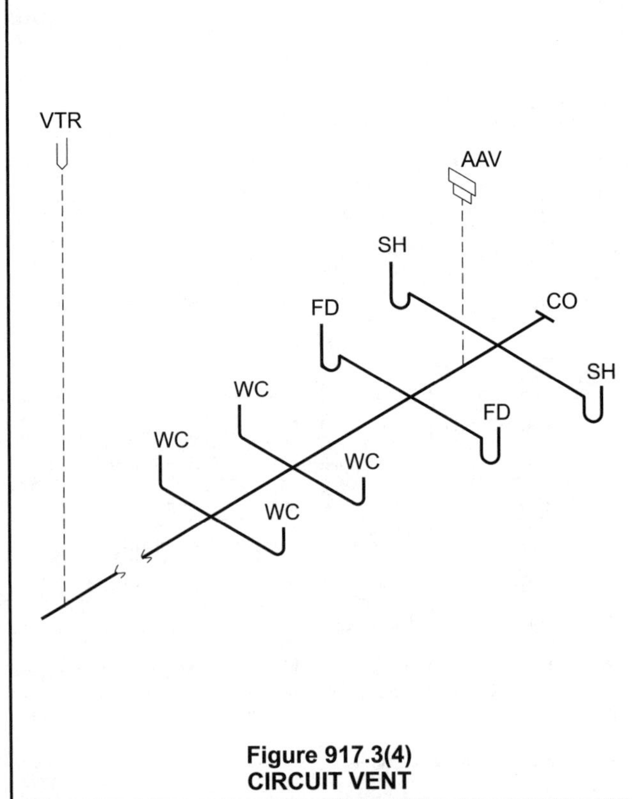

Figure 917.3(4)
CIRCUIT VENT

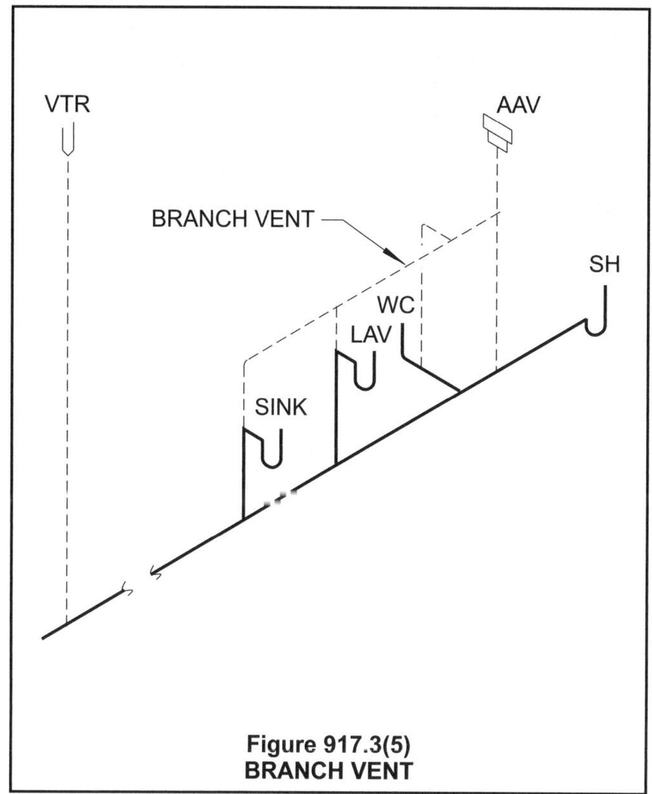

Figure 917.3(5)
BRANCH VENT

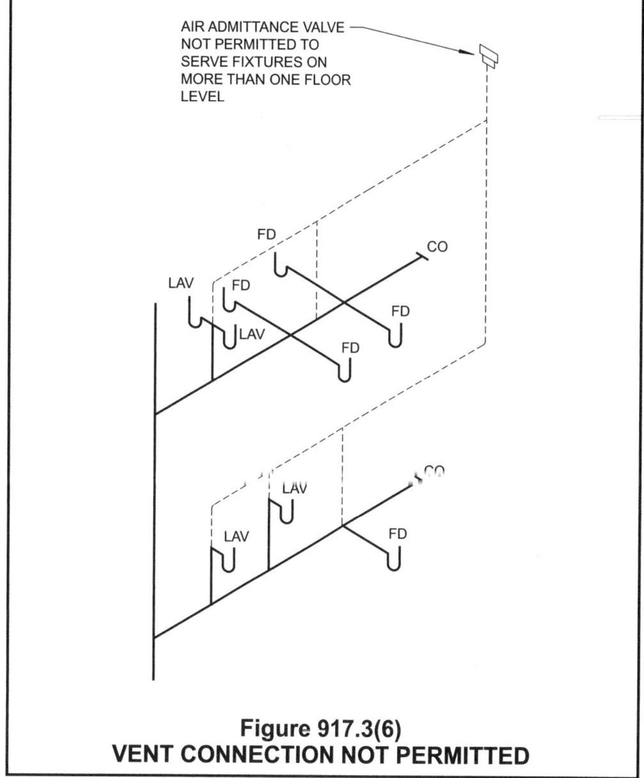

Figure 917.3(6)
VENT CONNECTION NOT PERMITTED

917.3.2 Relief vent. Where the horizontal *branch* is located more than four *branch intervals* from the top of the *stack*, the horizontal *branch* shall be provided with a relief vent that shall connect to a vent *stack* or *stack vent*, or extend outdoors to the open air. The relief vent shall connect to the horizontal *branch* drain between the *stack* and the most downstream *fixture drain* connected to the horizontal *branch* drain. The relief vent shall be sized in accordance with Section 916.2 and installed in accordance with Section 905. The relief vent shall be permitted to serve as the vent for other fixtures.

❖ The intent of this section is similar to that of Sections 911.4 and 914. A relief vent connected to horizontal branch drains assists in neutralizing excessive pressures before they become detrimental to the trap seal of fixtures served by an air admittance valve. A relief vent is not required for horizontal branch drains that conform to Section 917.3.1.

917.3.3 Stack. Stack-type air admittance valves shall not serve as the vent terminal for vent stacks or *stack vents* that serve drainage stacks having more than six *branch intervals*.

❖ When a building has a drainage stack that has more than six branch intervals, stack type air admittance valves must not be used to vent these drains. The pressure fluctuations in stacks with more than six branch intervals can be excessive, and could cause the loss of trap water seals, thereby allowing sewer gases into the building.

917.4 Location. Individual and branch-type air admittance valves shall be located a minimum of 4 inches (102 mm) above the *horizontal branch drain* or *fixture drain* being vented. Stack-type air admittance valves shall be located not less than 6 inches (152 mm) above the *flood level rim* of the highest fixture being vented. The air admittance valve shall be located within the maximum *developed length* permitted for the vent. The air admittance valve shall be installed a minimum of 6 inches (152 mm) above insulation materials.

❖ An air admittance valve has one moving part (a seal), which must be maintained a safe distance above the drain served. In the event of a drain stoppage, the seal may become inoperable or operate improperly if waste is permitted to rise into the air admittance valve assembly. Note that the air admittance valve need not extend above the flood level rim of the fixture served (see Commentary Figure 917.4) because in the event of a drain blockage, such device will trap air between it and the rising waste, thereby protecting the device from contamination.

The maximum developed length of a vent pipe equipped with an air admittance valve is regulated by Sections 916.2 and 916.3. Section 916.2 requires increasing the size of a vent exceeding 40 feet (12 192 mm) in developed length (measured from the air admittance valve to the point of connection to the drain pipe) by one nominal pipe size.

The distance between a trap and its protecting vent must also meet the maximum developed length requirements of Table 906.1.

An air admittance valve must be located a safe distance above insulation materials that may block air inlets or otherwise impair the operation of the device.

917.5 Access and ventilation. *Access* shall be provided to all air admittance valves. The valve shall be located within a ventilated space that allows air to enter the valve.

❖ Because an air admittance valve is a device with a moving part, it must be accessible for inspection, service, repair or replacement. The code defines "Access (to)" as being able to be reached but which first may require the removal of a panel, door or similar obstruction. Access to an air admittance valve is not achieved if it is necessary to remove any portion of the structure's permanent finish materials such as drywall, plaster, paneling, built-in furniture or cabinets, or any other similar permanently affixed building component.

Where air admittance valves are located in confined spaces, ventilation openings are required. Because these devices admit air into the drainage system, an unimpeded air supply must be available at all times. Additionally, access and ventilation requirements are addressed in the manufacturer's installation instructions.

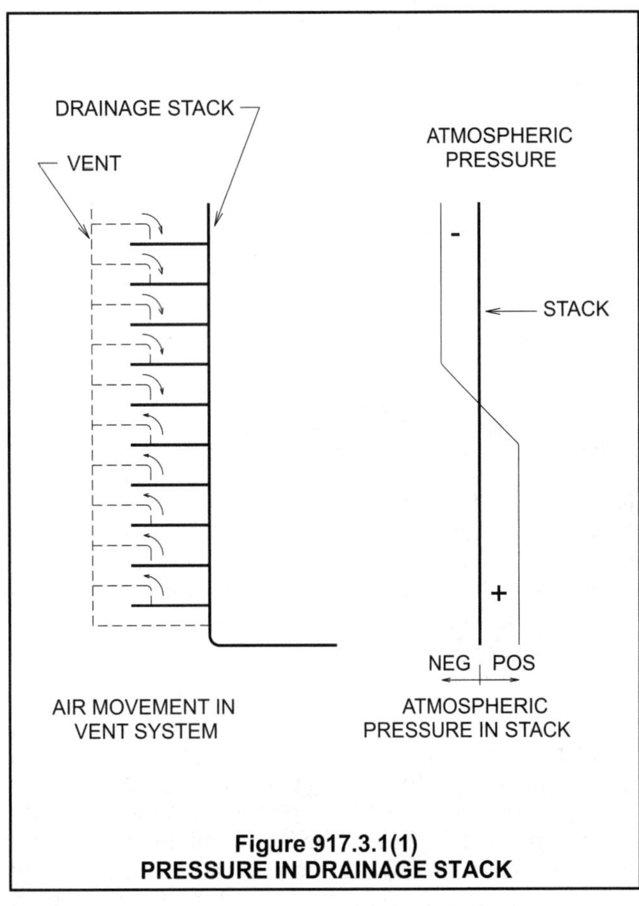

Figure 917.3.1(1)
PRESSURE IN DRAINAGE STACK

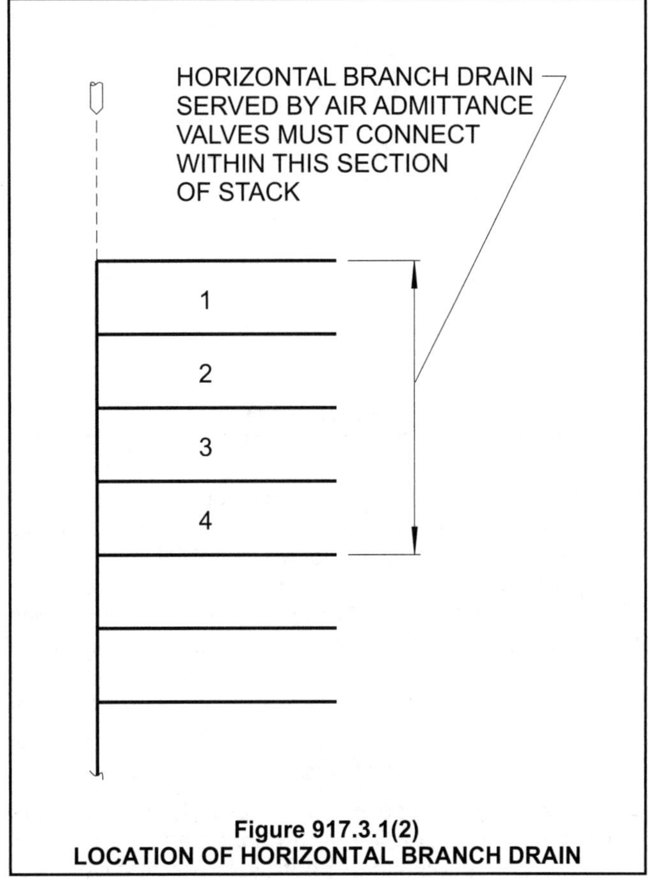

Figure 917.3.1(2)
LOCATION OF HORIZONTAL BRANCH DRAIN

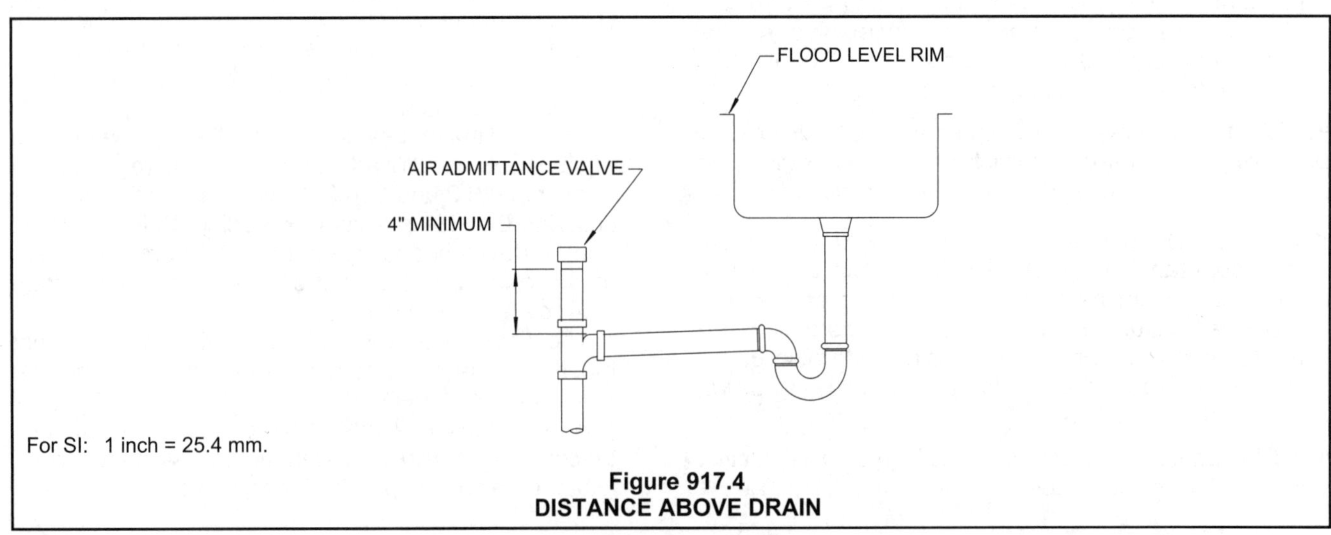

For SI: 1 inch = 25.4 mm.

Figure 917.4
DISTANCE ABOVE DRAIN

917.6 Size. The air admittance valve shall be rated in accordance with the standard for the size of the vent to which the valve is connected.

❖ The size of the vent pipe to which the air admittance valve is connected is regulated by Section 916. The air admittance valve is sized based on the required size of the vent pipe and the dfu load limitations found in the manufacturer's installation instructions. The dfu load rating is based on testing performed in accordance with ASSE 1051. The air admittance valves now available have dfu limitations that are consistent with the maximum number permitted on a horizontal branch as found in Table 710.1(2).

917.7 Vent required. Within each plumbing system, a minimum of one *stack vent* or vent *stack* shall extend outdoors to the open air.

❖ Conventional venting methods are intended to relieve both positive and negative pressures that occur in the drainage system. The primary function of an air admittance valve is relieving negative pressures. The intent of this section is to require at least one conventional vent pipe per plumbing system to terminate outside the building. The vent to the open air serves both to enhance the positive pressure neutralization in the drainage system and to protect the plumbing system from pressure excursions caused by the use of public sewer cleaning equipment (see Commentary Figure 917.7). This section does not limit the size of the plumbing system or the distances to the required stack vent or vent stack and states no required size for the stack vent or vent stack.

917.8 Prohibited installations. Air admittance valves shall not be installed in nonneutralized special waste systems as described in Chapter 8. Air admittance valves shall not be located in spaces utilized as supply or return air plenums. Air admittance valves without an engineered design shall not be utilized to vent sumps or tanks of any type.

❖ Because of the deleterious nature of the material and vapors associated with corrosive wastes, an air admittance valve is prohibited for use with such systems. The valves are also not permitted in spaces where pressure conditions adversely affect the valve's operation. Concealed wall and ceiling spaces used as plenums as part of an air distribution system are under negative or positive pressure. An air admittance valve located in this environment will be affected and may not function properly. For example, if located in a return plenum, greater negative pressure will be allowed to develop in the drain before the device will open and admit air to the drainage system. If located in a supply plenum, the device may be held open continuously because the pressure at the device inlet will be greater than the internal pressure of the drainage system; this is the case when the device functions normally to neutralize negative pressures.

Air admittance valves must not be used to vent sumps or tanks as there is positive pressure created when the tank or sump is filling. Because air admittance valves cannot relieve positive pressure within the sump or tank, the positive pressure might be relieved through the traps of the fixtures connected to the sump or tank. However, where air admittance valves are used as part of an engineered design that allows for positive pressure relief in other ways, AAVs can be used to allow air into the tank or sump (see Commentary Figures 917.8(1) and (2)].

SECTION 918
ENGINEERED VENT SYSTEMS

918.1 General. Engineered vent systems shall comply with this section and the design, submittal, approval, inspection and testing requirements of Section 105.4.

❖ This section contains an alternative sizing method for individual vents. An engineered vent system is considered to be an alternative engineered design. As such, it must comply with both the requirements specifically for this engineered vent system (see Commentary Section 918.1) as well as the general design, documentation, inspection, testing and approval of an alternative engineered design (see commentary, Section 105.4). Note that the engineered vent system must be designed, signed and sealed by a registered design professional (see commentary, Section 105.4.4).

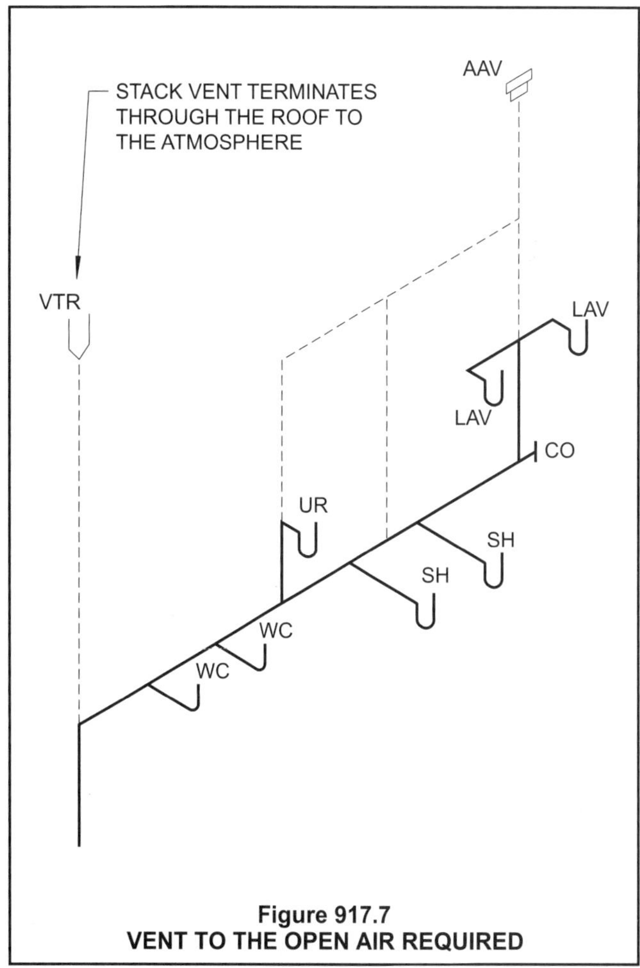

Figure 917.7
VENT TO THE OPEN AIR REQUIRED

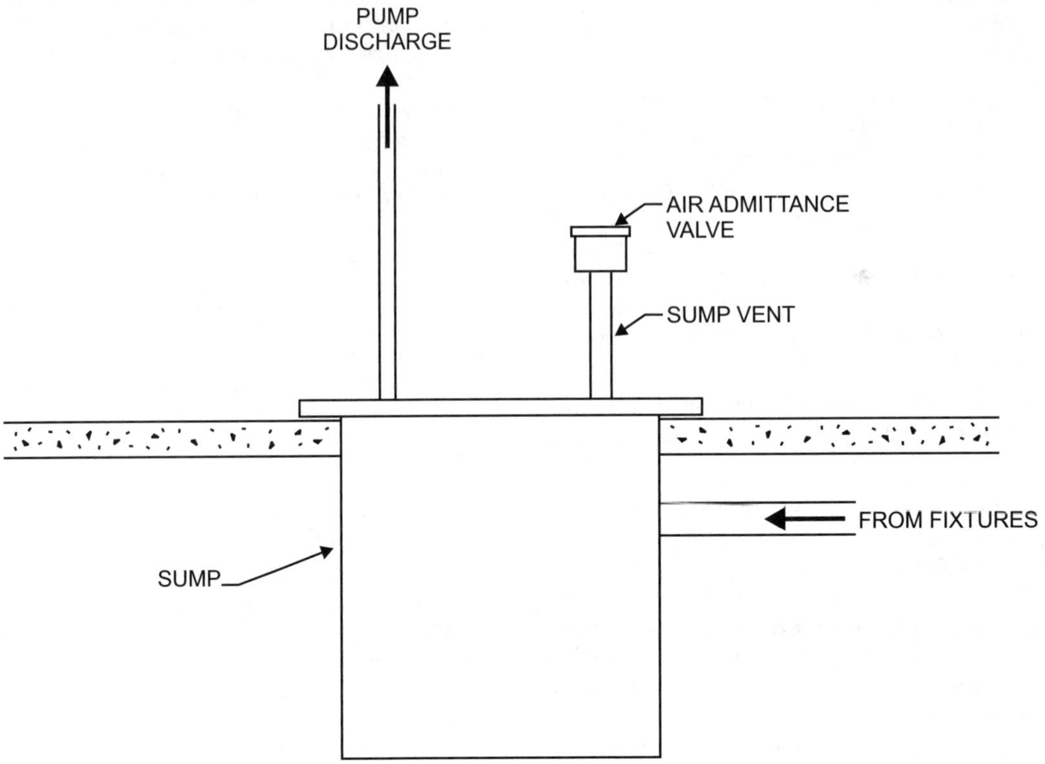

Figure 917.8(1)
PROHIBITED USE OF AIR ADMITTANCE VALVE

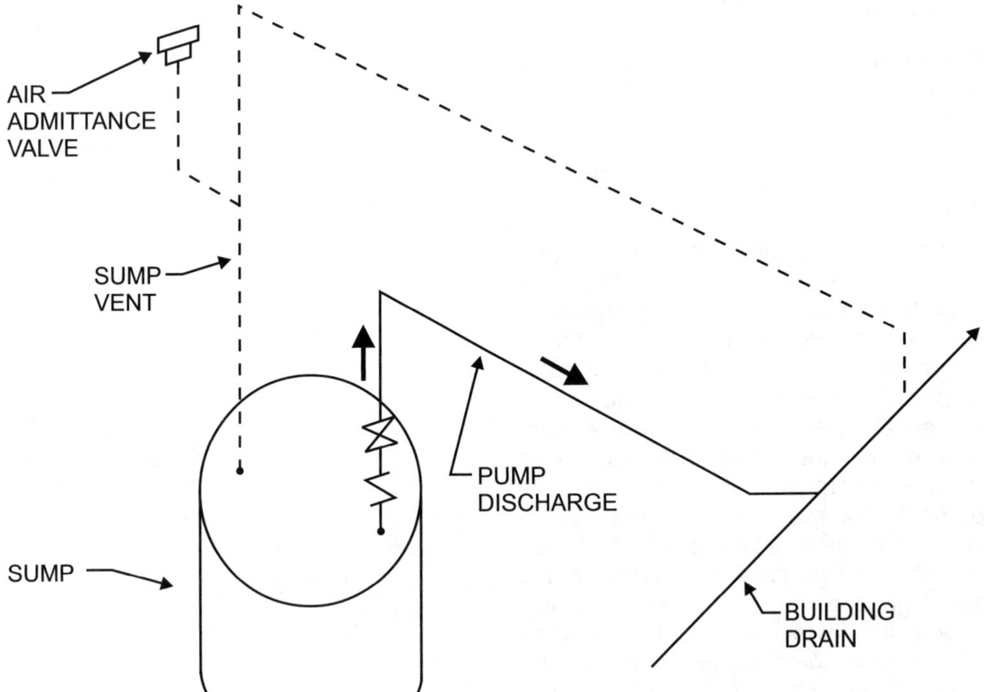

Figure 917.8(2)
ONE MANUFACTURER'S ENGINEERED DESIGN FOR USE OF AN AIR ADMITTANCE VALVE ON A SUMP

918.2 Individual branch fixture and individual fixture header vents. The maximum *developed length* of individual fixture vents to vent branches and vent headers shall be determined in accordance with Table 918.2 for the minimum pipe diameters at the indicated vent airflow rates.

The individual vent airflow rate shall be determined in accordance with the following:

$$Q_{h,b} = N_{n,b}\,Q_v \qquad \textbf{(Equation 9-1)}$$

For SI: $Q_{h,b} = N_{n,b}\,Q_v$ (0.4719 L/s)

where:

$N_{n,b}$ = Number of fixtures per header (or vent *branch*) ÷ total number of fixtures connected to vent *stack*.

$Q_{h,b}$ = Vent *branch* or vent header airflow rate (cfm).

Q_v = Total vent *stack* airflow rate (cfm).

Q_v (gpm) = $27.8\, r_s^{2/3} (1 - r_s)\, D^{8/3}$

Q_v (cfm) = $0.134\, Q_v$ (gpm)

where:

D = Drainage *stack* diameter (inches).

Q_w = Design discharge load (gpm).

r_s = Waste water flow area to total area.

$$= \frac{Q_w}{27.8\, D^{8/3}}$$

Individual vent airflow rates are obtained by equally distributing $Q_{h,b}$ into one-half the total number of fixtures on the *branch* or header for more than two fixtures; for an odd number of total fixtures, decrease by one; for one fixture, apply the full value of $Q_{h,b}$.

Individual vent *developed length* shall be increased by 20 percent of the distance from the vent *stack* to the fixture vent connection on the vent *branch* or header.

❖ The individual vent pipe size may be reduced to as small as 1/2 inch (13 mm) in diameter by calculating the vent airflow rate. The sizing method is based on every fixture being individually vented and connected to a vent header or vent stack. The reduced size applies only to the length of the individual vent up to the point where it connects to either the branch vent or vent stack. This distance is also the individual vent developed length.

Each fixture or group of fixtures must be individually calculated to determine the individual vent airflow rate that is applied to Section 918.2 to determine the individual vent minimum pipe size. For example purposes, the fifth floor of a 10-story building will be analyzed to determine the minimum individual vent pipe size [see Commentary Figures 918.2(1) and 918.2(2)].

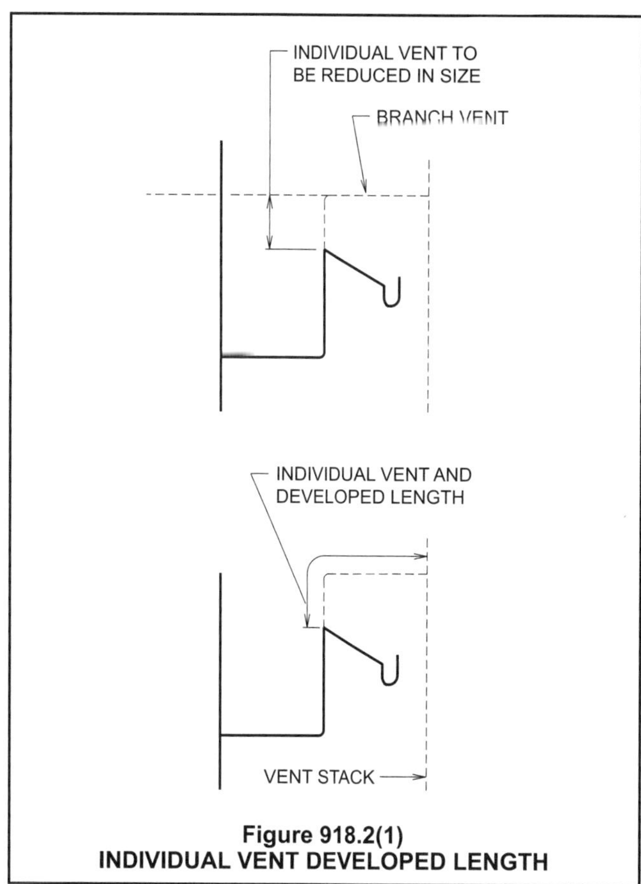

Figure 918.2(1)
INDIVIDUAL VENT DEVELOPED LENGTH

TABLE 918.2
MINIMUM DIAMETER AND MAXIMUM LENGTH OF INDIVIDUAL BRANCH FIXTURE VENTS AND INDIVIDUAL FIXTURE HEADER VENTS FOR SMOOTH PIPES

DIAMETER OF VENT PIPE (inches)	INDIVIDUAL VENT AIRFLOW RATE (cubic feet per minute)																			
	Maximum developed length of vent (feet)																			
	1	2	3	4	5	6	7	8	9	10	11	12	13	14	15	16	17	18	19	20
1/2	95	25	13	8	5	4	3	2	1	1	1	1	1	1	1	1	1	1	1	1
3/4	100	88	47	30	20	15	10	9	7	6	5	4	3	3	3	2	2	2	2	1
1	—	—	100	94	65	48	37	29	24	20	17	14	12	11	9	8	7	7	6	6
1 1/4	—	—	—	—	—	—	—	100	87	73	62	53	46	40	36	32	29	26	23	21
1 1/2	—	—	—	—	—	—	—	—	—	—	—	100	96	84	75	65	60	54	49	45
2	—	—	—	—	—	—	—	—	—	—	—	—	—	—	—	—	—	—	—	100

For SI: 1 inch = 25.4 mm, 1 cubic foot per minute = 0.4719 L/s, 1 foot = 304.8 mm.

Example:

Building statistics:

Building height 10 stories

Total drainage fixture units (dfu) to drainage stack 320

Size of drainage stack [Table 710.1(2)] 4 inches

Size of vent stack with developed length exceeding 40 feet (Section 916.1) $2^1/_2$ inches

Total number of plumbing fixtures 108

Fifth Floor

Number of fixtures (1 sink, 3 water closets, 5 lavatories, 3 showers) 12

Total dfu for horizontal branch 31

Size of horizontal branch[Table 710.1(2)] 4 inches

Size of branch vent (Section 916.2) 2 inches

Maximum developed length of individual vent 25 feet

The *Q* (gpm) must first be calculated using the equation: $Q \text{ (gpm)} = 27.8 r_s^{2/3}(1\text{-}r_s)D^{8/3}$

The diameter of the drainage stack (*D*) is 4 inches. The r_s may be determined by calculation or by use of the values in Commentary Table 918.2 (an excerpt from National Bureau of Standards Monograph 31).

Commentary Table 918.2 lists an r_s of 0.25 for a 4-inch stack with a total fixture unit load of 320. Substituting the values into the equation, it reads:

$$Q \text{ (gpm)} = 27.8\ (0.25)^{2/3}\ (1\text{-}0.25)(4)^{8/3}$$

$$Q \text{ (gpm)} = 333.6 \text{ gpm}$$

The value *Q* (gpm) must be converted to *Q* (cfm) by the following equation:

$$Q \text{ (cfm)} = 0.134\ Q \text{ (gpm)}$$

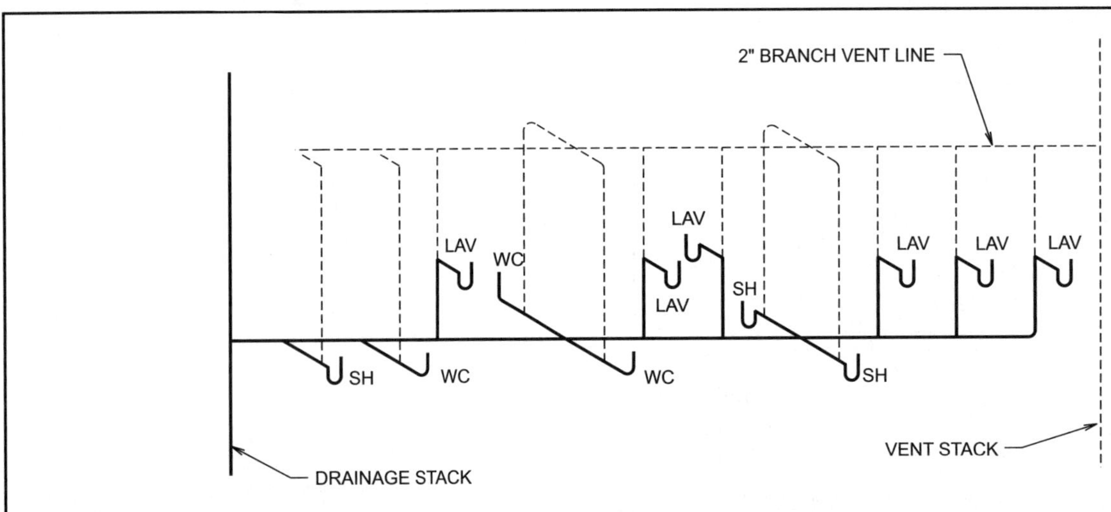

For SI: 1 inch = 25.4 mm.

Figure 918.2(1)
EXAMPLE OF ENGINEERED REDUCED-SIZE VENT

Table 918.2(2)
MAXIMUM LOADS ON MULTIPLE-STORY DRAINAGE STACKS

STACK DIAMETER (inches)	EQUIVALENT LOAD (fixture units)	PROPORTION OF CROSS SECTION OCCUPIED BY FALLING WATER	WATER DISCHARGE RATE (gpm)
4	43	0.15	47.5
4	140	0.20	76.7
4	320	0.25	111.0
4	530	0.29	144.0
4	580	0.30	151.0
4	850	0.33	180.0

For SI: 1 inch = 25.4 mm, 1 gallon per minute = 3.785 L/m.

Substituting the value for Q (gpm):

$$Q\text{ (cfm)} = 0.134\ (333.6\text{ gpm})$$

$$Q\text{ (cfm)} = 44.7\text{ cfm}$$

The vent header airflow rate is calculated using the equation:

$$Q_{h,b} = N_{n,b}Q$$

The value of $N_{n,b}$ for the fifth floor is equal to the number of fixture vents on the fifth floor divided by the total number of fixtures connected to the drainage stack.

Substituting the values into the vent header equation:

$$Q_{h,b} = (0.11)(44.7\text{ cfm})$$

$$Q_{h,b} = 4.9\text{ cfm}$$

The individual vent airflow rate is calculated by dividing the branch airflow rate by half the number of fixtures connected to the branch. The fifth floor has 12 fixtures, half of which equals 6. The individual vent airflow rate becomes:

Individual vent airflow rate = 4.9 cfm/6 = 0.82

The value used for Table 918.2 is rounded up to 1. With a maximum developed length of 25 feet (7620 mm), each individual vent on the fifth floor may be a minimum size of $^1/_2$ inch (13 mm).

For SI: 1 inch = 25.4 mm, 1 foot = 304.8 mm,
1 gallon per minute = 3.785 L/m,
1 cubic foot per minute = 0.4719 L/s.

Note that the example listed here would be solved in a similar manner for a design using metric values or "System International" (SI). The information shown in the code text as "For SI" includes the necessary conversion factor to change the resultant value from cubic feet per minute (cfm) to liters per second (L/s).

When the individual vent airflow rate is determined by the equation in Section 918.2, the size and developed length are determined by Table 918.2. The values in the table indicate the maximum developed length for a given pipe size and the individual vent airflow rate. Note that as a precaution against vent blockage, it may be desirable to use larger diameter vent piping from the connection to the fixture drain, to the point 6 inches (152 mm) above the flood level rim of the fixture served.

SECTION 919
COMPUTERIZED VENT DESIGN

919.1 Design of vent system. The sizing, design and layout of the vent system shall be permitted to be determined by *approved* computer program design methods.

❖ This section provides an alternative sizing method for venting systems, similar to Section 918.

This section is the logical complement to Section 714, which addresses computerized drainage system design (see Commentary Section 714).

919.2 System capacity. The vent system shall be based on the air capacity requirements of the drainage system under a peak load condition.

❖ This section says to design for the worst-case scenario. The airflow capacity of the venting system must be equal to or greater than the capacity demanded by a drainage system under its maximum anticipated load condition.

Bibliography

The following resource materials are referenced in this chapter or are relevant to the subject matter addressed in this chapter.

ASHRAE, *Fundamentals Handbook 2001*. Atlanta, GA: American Society of Heating, Refrigerating and Air-Conditioning Engineers, Inc., 2001.

ASPE Data Book. *"Vents and Venting" (Chapter 17)*. Westlake, CA: American Society of Plumbing Engineers, 1991.

ASSE 1051-02, *Performance Requirements for Air Admittance Valves for Plumbing Drainage Systems, Fixtures and Branch Devices*. Westlake, OH: American Society of Sanitary Engineers, 2002.

IRC-09, *International Residential Code*. Washington, DC: International Code Council, Inc., 2009.

Chapter 10: Traps, Interceptors and Separators

General Comments

The plumbing trap is, perhaps, the single most important device in a sanitary plumbing system as it prevents exposure of a building's occupants to sewer gases and vapors contained within the drainage system. Many requirements in this code are based upon protecting the "liquid seal" created by traps located at every plumbing fixture. Many more complicated designs of trap devices have been developed, used and abandoned throughout plumbing history in favor of the simplicity and reliability of the "U"-shaped trap.

If only "domestic" wastewater generated by humans for bathing, household cooking and cleaning, and the flushing of waste elimination was to be conveyed and treated, there would be no need for interceptors and separators. However, our society also generates commercial and industrial wastewaters from manufacturing processes, mass food production, vehicle cleaning, chemical production and product recycling which have substances that are detrimental to the piping systems and wastewater treatment processes. These substances must be prevented from entering waste water systems by the use of interceptors and separators.

Purpose

This chapter contains design requirements and installation limitations for traps. Prohibited types of traps are specifically identified. Where fixtures do not frequently replenish the liquid in traps, a method is provided to ensure that the trap seal of the trap will be replenished.

Requirements for the design and location of various types of interceptors and separators are provided. Specific venting requirements are given for separators and interceptors as those requirements are not addressed in Chapter 9.

SECTION 1001 GENERAL

1001.1 Scope. This chapter shall govern the material and installation of traps, interceptors and separators.

❖ This section contains the scope of the chapter, which is to provide and maintain a physical barrier between a building and its drainage system, protect the drainage system from prohibited materials and prevent the emission of sewer gases into the building.

SECTION 1002 TRAP REQUIREMENTS

1002.1 Fixture traps. Each plumbing fixture shall be separately trapped by a liquid-seal trap, except as otherwise permitted by this code. The vertical distance from the fixture outlet to the trap weir shall not exceed 24 inches (610 mm), and the horizontal distance shall not exceed 30 inches (610 mm) measured from the centerline of the fixture outlet to the centerline of the inlet of the trap. The height of a clothes washer standpipe above a trap shall conform to Section 802.4. A fixture shall not be double trapped.

Exceptions:

1. This section shall not apply to fixtures with integral traps.
2. A combination plumbing fixture is permitted to be installed on one trap, provided that one compartment is not more than 6 inches (152 mm) deeper than the other compartment and the waste outlets are not more than 30 inches (762 mm) apart.
3. A grease interceptor intended to serve as a fixture trap in accordance with the manufacturer's installation instructions shall be permitted to serve as the trap for a single fixture or a combination sink of not more than three compart2ments where the vertical distance from the fixture outlet to the inlet of the interceptor does not exceed 30 inches (762 mm) and the *developed length* of the waste pipe from the most upstream fixture outlet to the inlet of the interceptor does not exceed 60 inches (1524 mm).

❖ This section provides design and installation criteria for fixture traps, which are needed to protect building occupants from the hazards associated with the contents of drainage systems. A trap is the method used to keep sewer gases from emanating out of the drainage system. The liquid seal prevents sewer gases and aerosol-borne bacteria from entering the building space. Sewer gases often contain methane gas and could cause an explosion when exposed to an ignition source.

The configuration of a trap interferes with the flow of drainage; however, the interference is minimal because of the construction of the trap and the relatively high inlet velocity of the waste flow [see Commentary Figure 1002.1(1), Exception 1]. Double trapping of a fixture is prohibited because of the additional obstruction of flow and potential for stoppages. Double trapping will cause air to be trapped between two trap seals, and the "air-bound" drain will impede the flow.

The primary purpose of the horizontal distance of 30 inches (762 mm) from a fixture outlet to the trap weir is to limit the amount of bacterial growth and resulting odor. It is desirable to locate the trap as close as possible to the fixture. Buildup on the wall of the fixture outlet pipe will breed bacteria and cause odors to develop. The vertical distance is also limited to control the velocity of the drainage flow. If the trap has a long vertical separation distance from the fixture, the velocity of flow at the trap inlet can create a self-siphoning of the trap [see Commentary Figure 1002.1(1), Exception 2].

Exception 1 addresses fixtures such as water closets and urinals that have integral traps and, therefore, do not require a field-installed trap [see Commentary Figure 1002.1(1), Exception 1].

Exception 2 allows a single trap to serve multiple compartment sinks and other combination fixtures. A difference in elevation of sink wells (compartments) in combination fixtures will affect the vertical distance between the fixture outlet and the trap. The head pressure difference caused by elevation could affect the drainage rate of the fixture or cause waste from one well of a sink to discharge into a lower well. The limitation on distance between outlets is intended to minimize the amount of waste piping on the inlet side of the trap [see Commentary Figure 1002.1(1), Exception 2].

Exception 3 in Commentary Figure 1002.1(2) intends to permit the common practice of allowing a grease interceptor to act as the fixture trap where recommended by the grease interceptor manufacturer. Because grease interceptors function like a trap, double trapping will result where the typically designed grease interceptor is preceded by a fixture trap. Grease interceptors of the submerged inlet design will not allow the gases in the holding tank to enter the fixture drain; therefore, the source of odor would be limited to the section of piping between the fixture outlet and the interceptor inlet [see Commentary Figure 1002.1(2)].

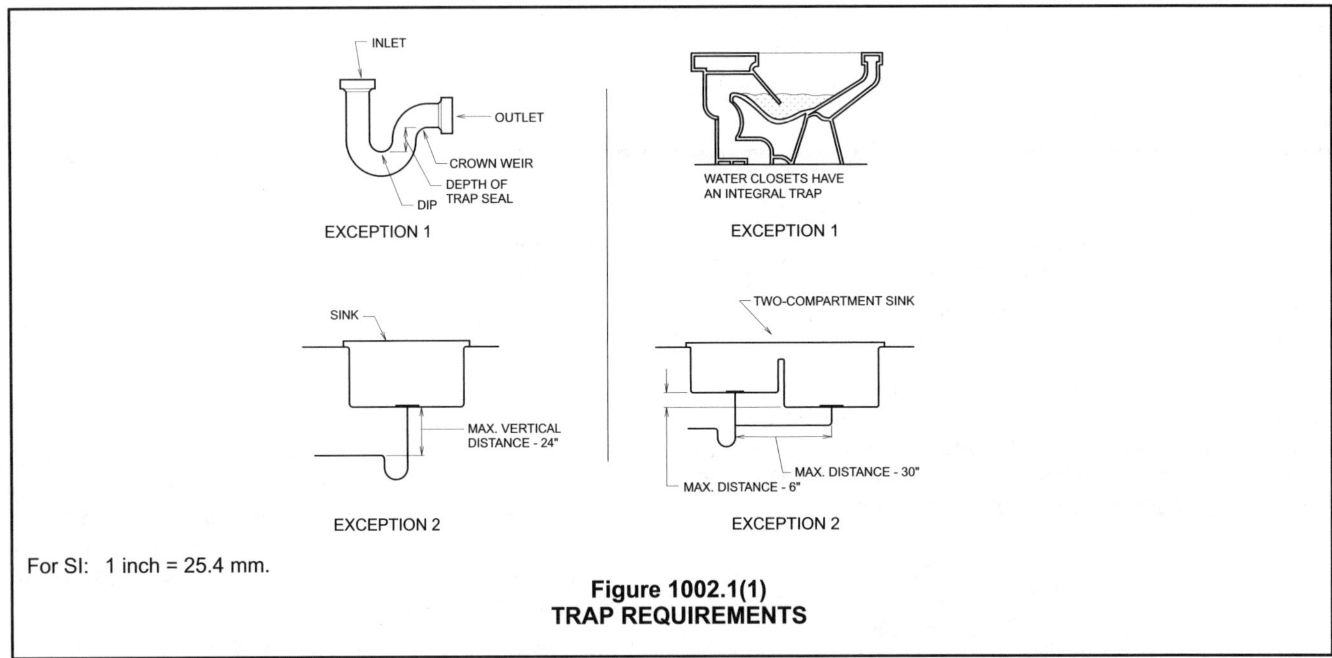

For SI: 1 inch = 25.4 mm.

Figure 1002.1(1)
TRAP REQUIREMENTS

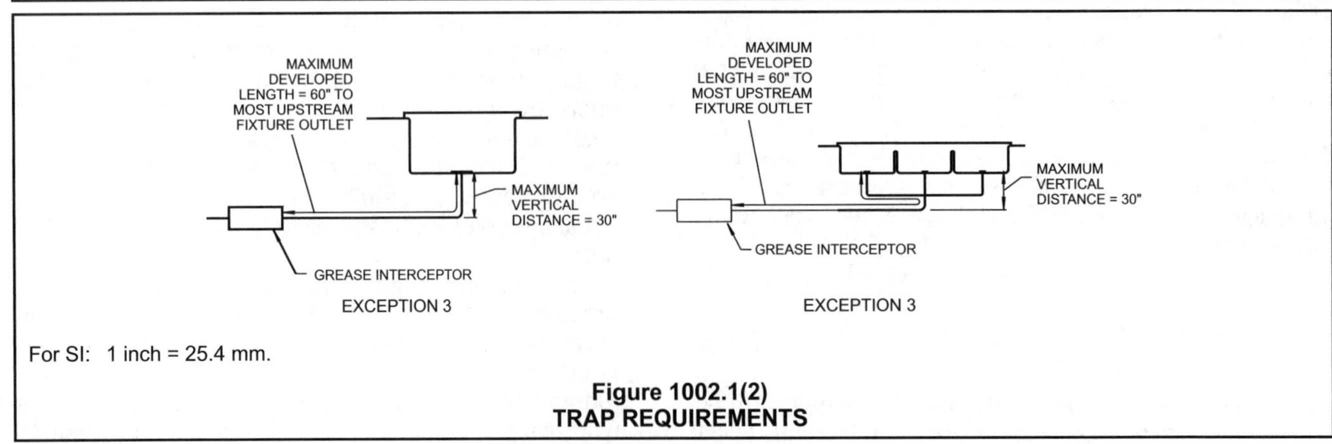

For SI: 1 inch = 25.4 mm.

Figure 1002.1(2)
TRAP REQUIREMENTS

1002.2 Design of traps. Fixture traps shall be self-scouring. Fixture traps shall not have interior partitions, except where such traps are integral with the fixture or where such traps are constructed of an *approved* material that is resistant to corrosion and degradation. Slip joints shall be made with an *approved* elastomeric gasket and shall be installed only on the trap inlet, trap outlet and within the trap seal.

❖ A trap must have a pattern allowing unobstructed flow to the drain. Interior partitions are allowed where constructed of a material resistant to corrosion and degradation. Bottle traps, having an interior partition, are commonly used in areas where the tailpiece of the fixture is very close to the wall such as is sometimes encountered with pedestal lavatory installation. However, the use of bottle traps is not limited only to close-quarter installations as they are also used where a more "refined" appearance is desired for an exposed trap application.

Most traps are premanufactured; however, a trap can be field fabricated with pipe and fittings (see commentary, Section 705.16 and Commentary Figure 1002.2).

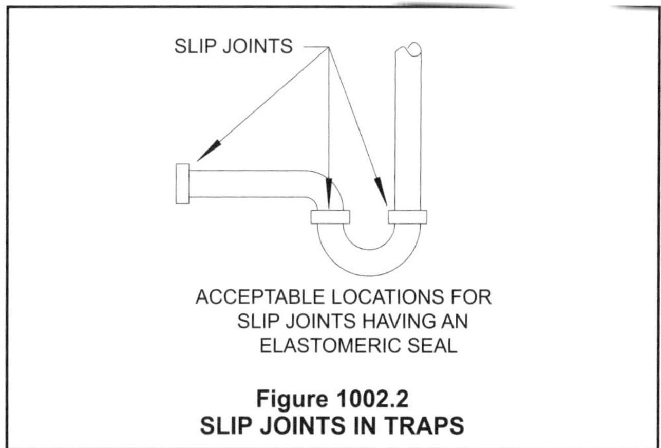

Figure 1002.2
SLIP JOINTS IN TRAPS

1002.3 Prohibited traps. The following types of traps are prohibited:

1. Traps that depend on moving parts to maintain the seal.
2. Bell traps.
3. Crown-vented traps.
4. Traps not integral with a fixture and that depend on interior partitions for the seal, except those traps constructed of an *approved* material that is resistant to corrosion and degradation.
5. "S" traps.
6. Drum traps.

Exception: Drum traps used as solids interceptors and drum traps serving chemical waste systems shall not be prohibited.

❖ A trap is intended to be a simple U-shaped piping arrangement that offers minimal resistance to flow. Based on their design or configuration, prohibited traps are not a simple U-shape piping design. Such traps typically impede drainage flow by moving parts, design (tends to clog with debris), flow pattern and ability to lose a trap seal (see Commentary Figure 1002.3).

Item 1 refers to mechanical traps that use moving parts such as floats or flappers. Such designs are not dependable because of corrosion, clogging and waste deposits that interfere with the operation of moving components and seals.

Item 2 prohibits bell traps that, because of their design, tend to clog with debris, and their trap seal has a larger exposed surface area that accelerates evaporation.

Item 3 more accurately describes a venting arrangement than a type of trap. A crown-vented trap is a factory-made or field-constructed "P" or "U" trap that has a vent opening located in the crown of the outlet weir of the trap. The flow pattern and momentum of waste flow will cause waste to enter and eventually clog the vent connection.

Item 4 refers to independent traps that are not integral with fixtures. For example, water closet traps are partition traps but, of course, are not prohibited because they are regulated as part of the fixture.

Partition traps are single-wall traps that rely on a single partition to separate the house side of the trap from the sewer side of the trap. Partition traps made of materials that are not corrosion resistant can structurally fail without any leakage that would indicate a failure of the partition; therefore, the inability to maintain a trap seal would not be indicated by a partition trap.

Item 5 prohibits factory-made and field-constructed "S" configuration traps because of their inherent ability to self-siphon, which results in a loss of trap seal.

Item 6 prohibits drum traps because they are not self-cleaning and therefore clog easily. An exception provides for the use of such traps where installed as a solids interceptor or serving chemical waste systems. Many drum traps are used to prevent solids from entering the drainage system. Hair traps, metal traps, plaster traps and single-fixture acid neutralizers are variations of the drum-trap design.

1002.4 Trap seals. Each fixture trap shall have a liquid seal of not less than 2 inches (51 mm) and not more than 4 inches (102 mm), or deeper for special designs relating to accessible fixtures. Where a trap seal is subject to loss by evaporation, a trap seal primer valve shall be installed. Trap seal primer valves shall connect to the trap at a point above the level of the trap seal. A trap seal primer valve shall conform to ASSE 1018 or ASSE 1044.

❖ A liquid seal of 2 inches (51 mm) is standard for most traps. Some larger pipes, 3 through 6 inches (76 through 152 mm), have a greater seal of up to 4 inches (102 mm) to construct a smooth pattern of flow for the given pipe size [see Commentary Figure 1002.1(1), Exception 1].

A trap seal must be deep enough to resist the pressures that can develop in a properly vented drainage system, but not deep enough to promote the retention of solids or the growth of bacteria.

Traps that do not periodically receive waste discharge will eventually lose their seal as a result of evaporation. The rate of trap seal evaporation is somewhat dependent on the location of the trap. For example, water in fixture traps in environments with high ambient temperatures or high-volume air movement will evaporate rapidly. Fixtures, such as floor drains, typically have trap seals that are subject to loss by evaporation and, therefore, must be protected by trap seal primers [see Commentary Figure 1002.4(1)]. A deep seal will not prevent the loss of a trap seal, but will simply lengthen the time it will take for it to evaporate [see Commentary Figure 1002.4(2)]. Trap seal primers are either water-supply-fed or drainage types. These devices cause a small quantity of water to be discharged to the trap served at each use of a particular plumbing fixture. For example, a floor drain in a toilet room can be maintained by a trap seal primer valve that is connected to a water supply pipe for a lavatory located in the same room as the floor drain.

The connection of the trap seal primer piping to the trap must be above the trap seal water line. A submerged inlet from the trap primer could cause blockage, reducing the effectiveness of the priming system. The standards listed for trap primer valves require that the valves incorporate an air gap to prevention cross connection of the potable water supply with the drain system.

1002.5 Size of fixture traps. Fixture trap size shall be sufficient to drain the fixture rapidly and not less than the size indicated in Table 709.1. A trap shall not be larger than the drainage pipe into which the trap discharges.

❖ The minimum fixture trap sizes are listed in Table 709.1. The minimum trap sizes for fixtures not listed in Table 709.1 must be the size of the fixture outlet, but in no case less than $1^1/_4$ inches (32 mm). The code does not prescribe a maximum trap size; however, if a trap is oversized, it will not scour (cleanse) itself and is therefore prone to clogging. A trap that is larger than the drainage pipe into which it discharges is also subject to frequent clogging because the reduced size outlet pipe will not allow a waste flow velocity that is adequate to scour and clean the trap (see commentary, Table 709.1 and Section 704.2).

1002.6 Building traps. Building (house) traps shall be prohibited, except where local conditions necessitate such traps. Building traps shall be provided with a cleanout and a relief vent or fresh air intake on the inlet side of the trap. The size of the relief vent or fresh air intake shall not be less than one-half the diameter of the drain to which the relief vent or air intake connects. Such relief vent or fresh air intake shall be carried above grade and shall be terminated in a screened outlet located outside the building.

❖ Building traps were originally installed on the building drain before connection to the building sewer as a form of rat control. In the major cities, the rats would breed in the city sewer system and enter a building

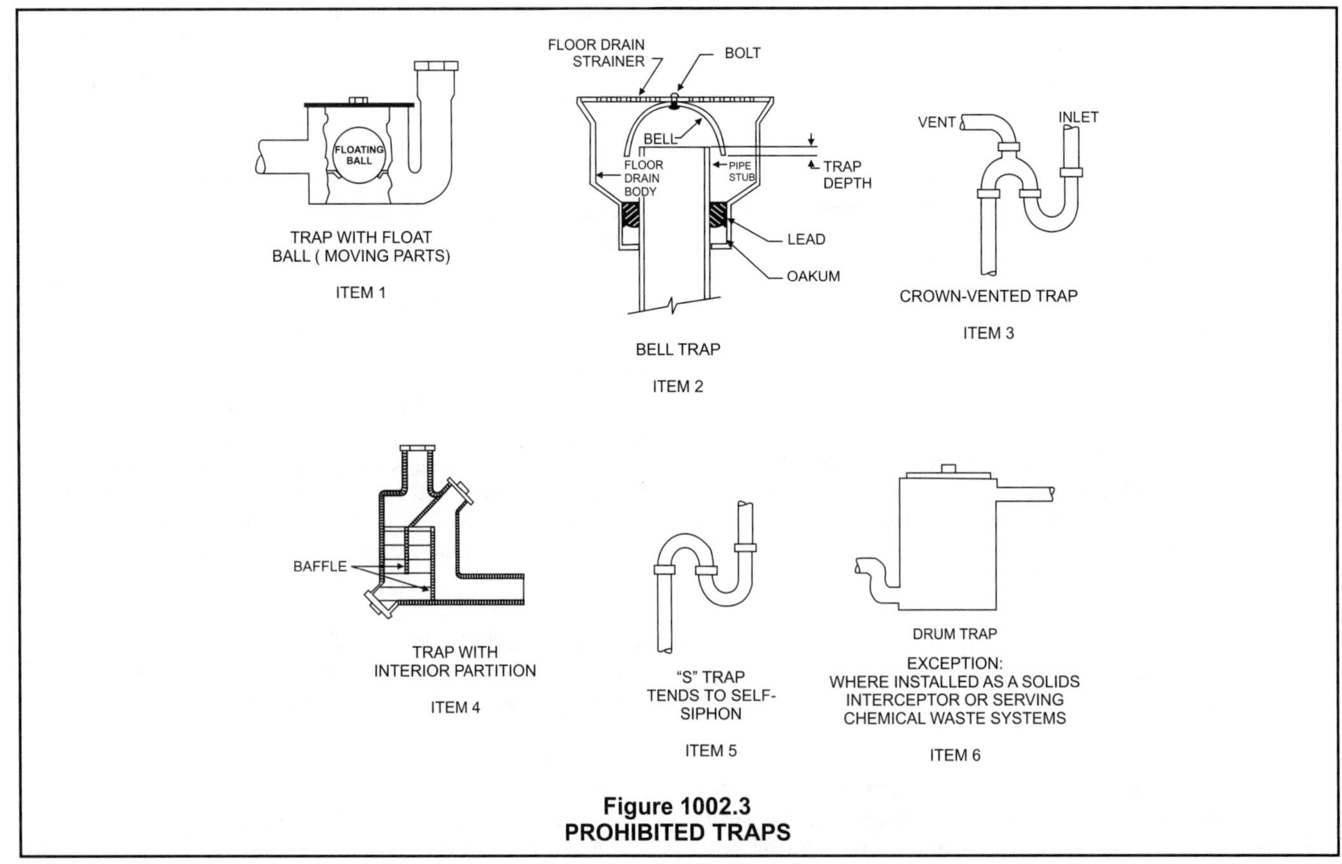

Figure 1002.3
PROHIBITED TRAPS

through the sewer connection. It was believed that a water trap seal would prevent the rats from entering the building.

In rare cases, building traps could be necessary where the public sewer exerts positive backpressure on the connected building sewers. This backpressure can cause the building vent terminations to emit strong sewer gases, thereby creating a serious odor problem around the building.

A building trap creates a problem because it is an obstruction to flow, possibly causing a complete stoppage in the drainage system. Newer building traps added two cleanouts, one on either side of the trap to facilitate rodding. Another problem was that the building trap eliminated the possibility of ventilating the public sewer through the building sewer connection. To solve this problem, some major cities required a building trap for every building except the last one on the block. That building was prohibited from having a building trap to provide ventilation for the public sewer. Building traps were never intended to eliminate individual fixture traps and are prohibited except where necessitated by local conditions (see Commentary Figure 1002.6).

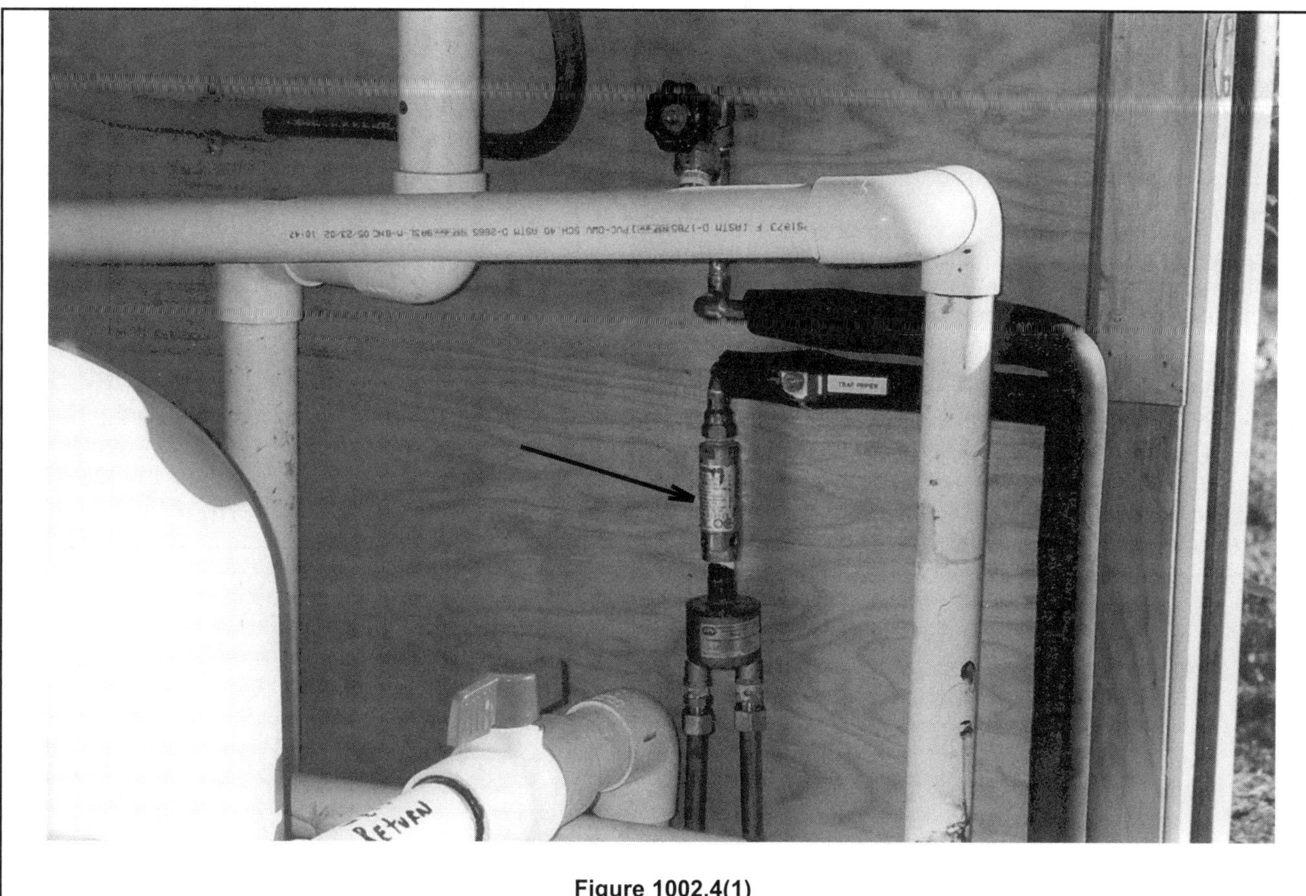

Figure 1002.4(1)
TRAP PRIMER FOR TWO FLOOR DRAINS

TRAP SEAL PRIMER VALVE
FLOOR DRAIN
SINK
DRAINAGE-TYPE TRAP SEAL PRIMER
FLOOR DRAIN

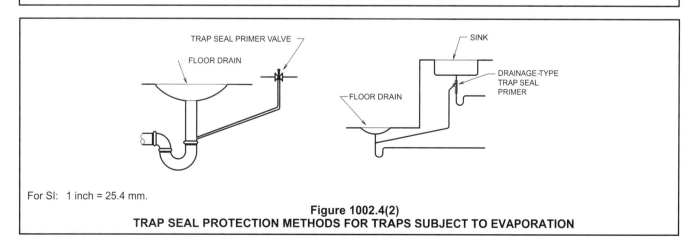

For SI: 1 inch = 25.4 mm.

Figure 1002.4(2)
TRAP SEAL PROTECTION METHODS FOR TRAPS SUBJECT TO EVAPORATION

1002.7 Trap setting and protection. Traps shall be set level with respect to the trap seal and, where necessary, shall be protected from freezing.

❖ Traps must be set level to maintain the trap seal depth and reduce the possibility of self-siphoning (see Commentary Figure 1002.7).

1002.8 Recess for trap connection. A recess provided for connection of the underground trap, such as one serving a bathtub in slab-type construction, shall have sides and a bottom of corrosion-resistant, insect- and verminproof construction.

❖ Where a trap is installed below the building, such as in a recess below a slab on grade, requirements similar to those in the *International Building Code*® (IBC®) for waterproofing, ratproofing and termite protection apply (see Commentary Figure 1002.6).

1002.9 Acid-resisting traps. Where a vitrified clay or other brittleware, acid-resisting trap is installed underground, such trap shall be embedded in concrete extending 6 inches (152 mm) beyond the bottom and sides of the trap.

❖ Because of their brittle nature, earthenware-type traps, including vitrified clay, must be embedded in concrete where installed below grade (see Section 702.5).

1002.10 Plumbing in mental health centers. In mental health centers, pipes and traps shall not be exposed.

❖ In certain mental health centers, patients are considered a danger to themselves. As a result of this concern, special care is required to secure the plumbing fixtures and piping in these facilities.

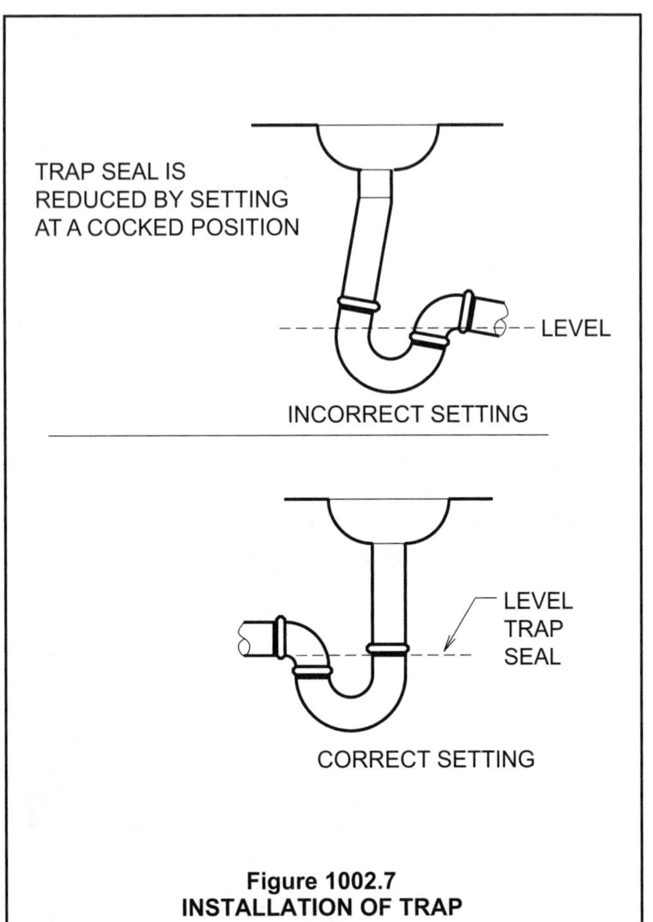

Figure 1002.7
INSTALLATION OF TRAP

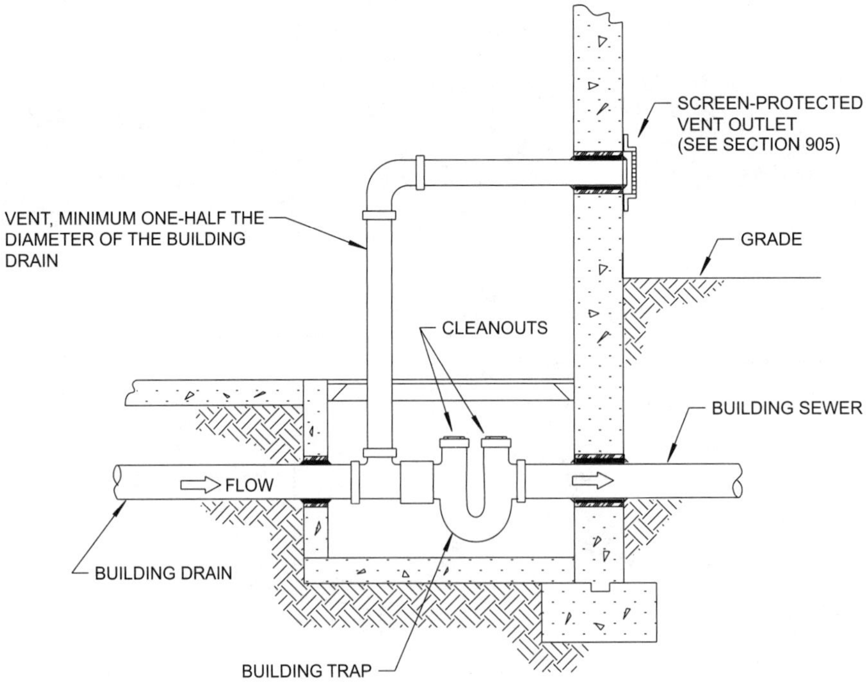

Figure 1002.6
BUILDING TRAP

SECTION 1003 INTERCEPTORS AND SEPARATORS

1003.1 Where required. Interceptors and separators shall be provided to prevent the discharge of oil, grease, sand and other substances harmful or hazardous to the building drainage system, the *public sewer,* the private sewage disposal system or the sewage treatment plant or processes.

❖ The design and installation of sanitary drain and sewer systems and the waste treatment systems for disposal of wastewater are based upon the known general characteristics of "domestic" wastewater. Domestic wastewater is largely that which is generated by the use of potable water by humans for bathing, household cooking and cleaning, and for the flushing of waste eliminations from humans. Wastewaters generated from commercial and industrial sources such as manufacturing processes, mass food production, vehicle cleaning, chemical production, and product recycling can have detrimental effects to the materials and methods used to convey the wastewater as well as the mechanical, biological and chemical processes used to make the wastewater safe to discharge into the environment. Therefore, these "commercial/industrial" wastewaters must have the detrimental components removed in order for the wastewater to be safe for drainage systems and less problematic for domestic wastewater treatment biological and chemical processes.

The terms "separator" and "interceptor" are used interchangeably in the code, this commentary and by many involved in the separator/interceptor industry. Both terms mean a device that causes one or more entrained materials in a wastewater flow to divert from the flow for retention within the device or follow a different flow path for capture and retention outside of the device. The retained materials require periodic removal from the devices or storage vessels in order to maintain the efficiency and function of the separating device. The removal and disposal of these detrimental wastes from the devices or storage vessels is typically performed by specialized service companies that are state licensed to transport and dispose of the wastes in an environmentally safe manner. Note that separators/interceptors do not chemically alter commercial/industrial wastewater but only mechanically separate out physical components from wastewater flows. Section 803 covers applications where wastewater flows require adjustment of pH, temperature reduction, noxious odor abatement or removal of hazardous chemicals.

This section identifies grease, oil and sand as specific substances of concern that require removal from wastewater before discharge to the sanitary drain system. The congealing nature of grease can cause severe clogging problems in building drainage systems, sewers and public sewer systems, resulting in wastewater overflows in the building or to the environment. Oil can be detrimental to sewage treatment processes as it tends to float on water surfaces, making it difficult to capture and remove in the treatment process. Sand can be abrasive to most pipe line materials and the pumps used to move wastewater. Sand can also build up and cause clogging problems in manholes located where turns occur in gravity sewer lines. Sections 1003.3 and 1003.4 indicate what occupancy uses specifically require oil and grease interceptors, respectively. Sections 1003.6, 1003.7 and 1003.8 require separators for commercial laundries, bottling establishments and slaughterhouses, respectively. The code does not require sand interceptors for a specific occupancy use; however, where sand interceptors are installed, Section 1003.5 has design requirements for those interceptors. Other occupancies that might require separators include hospitals; fowl-, fish- or meat-packaging plants; canneries; cement producers; beauty salons; refineries and machine shops. The code official is responsible for identifying any other occupancy uses where similar discharges might require a separator.

There is some debate over where interceptors/separators should be located. Read literally, the section text indicates that oil, grease and sand are to be prevented from discharging into building drainage systems. The code intends to prevent the discharge of substances that are detrimental but does not clearly state what piping must be protected. This seems to imply that those materials are not allowed to enter any fixture because a fixture drain is part of the building drainage system. However, because interceptors/separators are designed for pipe connection on the inlet and outlet of the unit, interceptors/separators are intended to be located in the building drainage system at some point downstream of the fixture. Ideally, each fixture producing wastewater requiring an interceptor would have the unit located in the fixture drain immediately downstream of the fixture. While such an arrangement would protect the greatest amount of piping in the building drainage system, it is not always feasible or cost effective to provide interceptors at every fixture. Locating a single interceptor far downstream of multiple fixtures might require increased maintenance (cleaning and repair) of the piping upstream of the interceptor. For example, placing the only grease interceptor outdoors will not protect the building drainage system from the accumulation of grease; it can only protect the sewer system and sewage treatment system. The designer of the building or tenant space is responsible for locating interceptors with respect to the occupant's expectations for frequency and degree of building drainage system maintenance, ease of cleaning and floor space allocation for equipment.

Where wastewater flows are destined for a private sewage disposal system that ultimately discharges to the environment through various ground infiltration methods, state and county groundwater environmental control departments might require additional separators, special local treatment methods, holding tanks or monitoring stations. Those requirements are not covered by the code and can vary from state to state.

Every fixture connecting to an interceptor or separator must be individually trapped, except for the specific case of a grease interceptor for a single fixture or combination sink (see Exception 3 of Section 1002.1). Although many interceptor designs have the appearance of having a "trap function," they might not be designed in accordance with the requirements of Section 1002 in order to provide the same protection as a fixture trap.

1003.2 Approval. The size, type and location of each interceptor and of each separator shall be designed and installed in accordance with the manufacturer's instructions and the requirements of this section based on the anticipated conditions of use. Wastes that do not require treatment or separation shall not be discharged into any interceptor or separator.

❖ Interceptors are devices that are designed to remove a specific substance from a wastewater stream flowing at a maximum rate of flow. Although the code has a few requirements for the design of certain types of interceptors, the interceptor designer/manufacturer is responsible for all other details of construction, such as size and materials as well as the installation alignments, elevations and access requirements for maintenance. Because the size and, therefore, the cost of interceptors is directly proportional to the flow rate, and some wastewater flows not having substances that require removal might have substances that will cause problems in separation, only the wastewater flows having the substances that need to be removed are allowed to be routed through an interceptor (see Commentary Figure 1003.2). The commentary for Section 1003.1 provides information regarding the location of interceptors.

1003.3 Grease interceptors. Grease interceptors shall comply with the requirements of Sections 1003.3.1 through 1003.3.5.

❖ The need for and the performance of grease interceptors has received increased visibility over the years due to costly sewer overflows as a result of congealed grease clogging public sewer lines. Although grease problems in public sewer systems have existed since the first commercial kitchens discharged to public sewers, the extreme attention to preventing grease discharges into public sewers is due largely to the severe fines levied by state environmental health departments for wastewater overflows which are in violation of the Federal Clean Air and Water Act. Once grease enters the sewer system, it cools and congeals in the pipes, creating blockages for wastewater flow. Even though sewer system operators are accustomed to routine maintenance cleaning of sewer lines, it is nearly impossible for any reasonable cleaning program to keep sewers free of grease so that overflows don't occur. Therefore, more and more sewer system operators are establishing severe limits for grease concentration in discharges, extraordinary inspection/reporting requirements and strong enforcement procedures for the producers of grease-laden wastewaters in order to attack the problem at the source.

The code does not prescribe grease concentration limits for wastewater discharge or inspection/maintenance requirements for grease interceptors. The only code requirement for the performance of grease interceptors is provided in the applicable referenced standards in Section 1003.3.4. The performance requirements provided in the standards are not a guarantee that any grease interceptor covered by those standards will limit the grease concentration in the wastewater to the levels required by local sewer system operators or wastewater discharge enforcement agencies. The designer/manufacturer is responsible for providing a device design that meets customer-specified removal rates for the concentration of grease-laden wastewater as specified by the customer.

There are two types of grease interceptors that are commonly used for grease-laden wastewater applica-

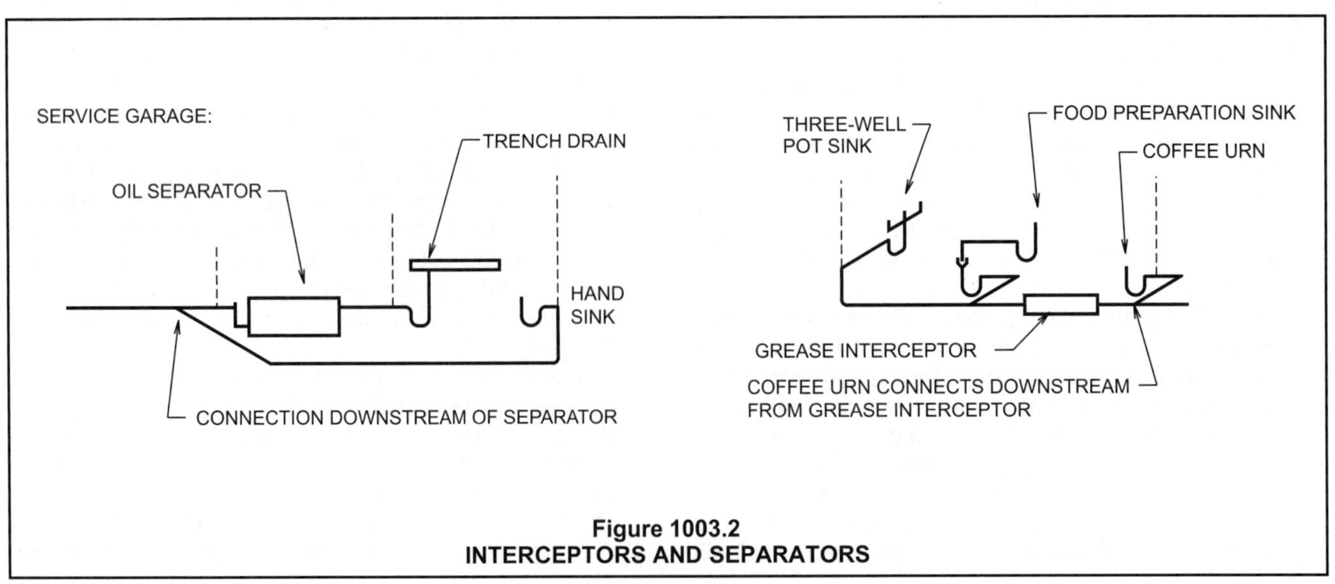

Figure 1003.2
INTERCEPTORS AND SEPARATORS

tions: (1) gravity and (2) hydromechanical. The terms for these two types of grease interceptors have recently come into use because the code-defined term for one of these devices was changed in the 2003 edition. Prior to that edition, gravity-type grease interceptors were called grease interceptors and hydromechanical-type grease interceptors were called grease traps. The term "trap" was changed to interceptor because not all grease trap designs provide a "trap-like" function insofar as preventing sewer gases from backflowing through the device and escaping from untrapped fixture drains. Changing the term "trap" to "interceptor" did not recognize that in doing so, the distinction between the two types of interceptors was blurred. Even though the current edition of the code does not provide a clear distinction between the two types, this commentary necessarily uses the gravity and hydromechanical definitions as provided in PDI G101 in order to clearly discuss code sections concerning grease interceptors.

The gravity-type grease interceptor is large in size (e.g., at least 500 gallons in volume), usually of concrete or polymer construction and provides for extremely low flow velocity through the unit to allow the grease to collect in a compartment [see Commentary Figure 1003.3(1)]. The hydromechanical-type grease interceptor is smaller in size (e.g., typically less than 250 gallons in volume), typically of fabricated steel or polymer construction and utilizes changes in flow direction along with air-entrainment to cause grease to be retained in the device [see Commentary Figure 1003.3(2)]. Both rely upon gravity as the means of separation. As the code requires all grease interceptors to comply with the referenced standards as stated in Section 1003.3.4, this automatically excludes gravity-type grease interceptors from complying with the code. The code does not reference any standards for gravity-type grease interceptors.

However, gravity-type grease interceptors are sometimes required by jurisdictions instead of what is required by the code, simply due to local experience, familiarity and comfort with gravity-type units. In some areas, grease interceptor installations are controlled not by building code officials but by departments of health, wastewater system utility operators or environmental discharge departments, none of which are bound by the code. There are few published "standards" for the design of gravity-type grease interceptors. Decades of design experience by credible engineering concerns have resulted in design philosophies and construction criteria that have proven that gravity-type grease interceptors can provide for adequate grease removal from most wastewater flows. However, due to the variety of fats, oils and greases (FOG) produced by some applications and tighter limitations by certain jurisdictions on discharges of FOG, gravity-type grease interceptors alone might not produce the desired results. Hydromechanical devices fitted to the discharge of upstream grease-producing fixtures and equipment might be necessary in addition to gravity-type interceptors in order to attain compliance with local discharge limits of FOG.

Note that an automatic grease removal device is not a third type of grease interceptor but only a variation of the hydromechanical type that offers automatic removal of collected grease to an external container.

1003.3.1 Grease interceptors and automatic grease removal devices required. A grease interceptor or automatic grease removal device shall be required to receive the drainage from fixtures and equipment with grease-laden waste located in food preparation areas, such as in restaurants, hotel kitchens, hospitals, school kitchens, bars, factory cafeterias and clubs. Fixtures and equipment shall include pot sinks, prerinse sinks; soup kettles or similar devices; wok stations; floor drains or sinks into which kettles are drained; automatic hood wash units and dishwashers without prerinse sinks. Grease interceptors and automatic grease removal devices shall receive waste only from fixtures and equipment that allow fats, oils or grease to be discharged.

❖ This section requires grease interceptors for specific fixtures, equipment and certain floor drains in food preparation areas for occupancy uses where food is commercially prepared. Although the text provides a list of occupancy uses, this list is not all-inclusive, but an indication as to the types of facilities that might have a food preparation area. For example, church kitchens, catering kitchens, delis, meat markets and recreation center kitchens could be occupancy uses that could be generators of grease-laden wastewaters. The code official has the responsibility to define the fixtures and equipment and their intended uses that constitute the need for grease interception. For example, a bar that only serves beverages in glasses and uses a dishwasher or three-compartment sinks to clean only the glasses would not require a grease interceptor. However, a bar that also serves food on reusable plates with silverware might require a grease interceptor for the three-compartment sink or dishwasher that washes those items.

Only the wastewater flows that require grease interception are allowed to be connected to a grease interceptor. Although this requirement is generally stated in Section 1003.2, it is even a more critical requirement for grease interceptors as cooler wastewater and wastewater flows with high concentrations of cleaning chemicals, sugars or acids can affect the separation efficiency of a grease interceptor (see Commentary Figure 1003.2).

1003.3.2 Food waste grinders. Where food waste grinders connect to grease interceptors, a solids interceptor shall separate the discharge before connecting to the grease interceptor. Solids interceptors and grease interceptors shall be sized and rated for the discharge of the food waste grinder. Emulsifiers, chemicals, enzymes and bacteria shall not discharge into the food waste grinder.

❖ As the code only recognizes hydromechanical grease interceptors, the intent of this section is to prevent large particles of ground-up food discharged by the food waste grinder from clogging or rapidly filling these types of grease interceptors. Hydromechanical-type grease interceptors have numerous baffles, sometimes with small openings for flow such that food particles could block the flow. Also, the storage capacity for solids in these relatively small units is limited. Without a solids interceptor on the discharge of the food waste grinder, a hydromechanical grease interceptor could become inefficient or blocked within hours of being cleaned. Therefore, a solids interceptor is required for the food waste grinder discharge before connection to a hydromechanical grease interceptor (see Commentary Figure 1003.3.2).

Even though a solids interceptor "filters out" large particles of ground-up food, there is still a grease load that must be accommodated by the grease intercep-

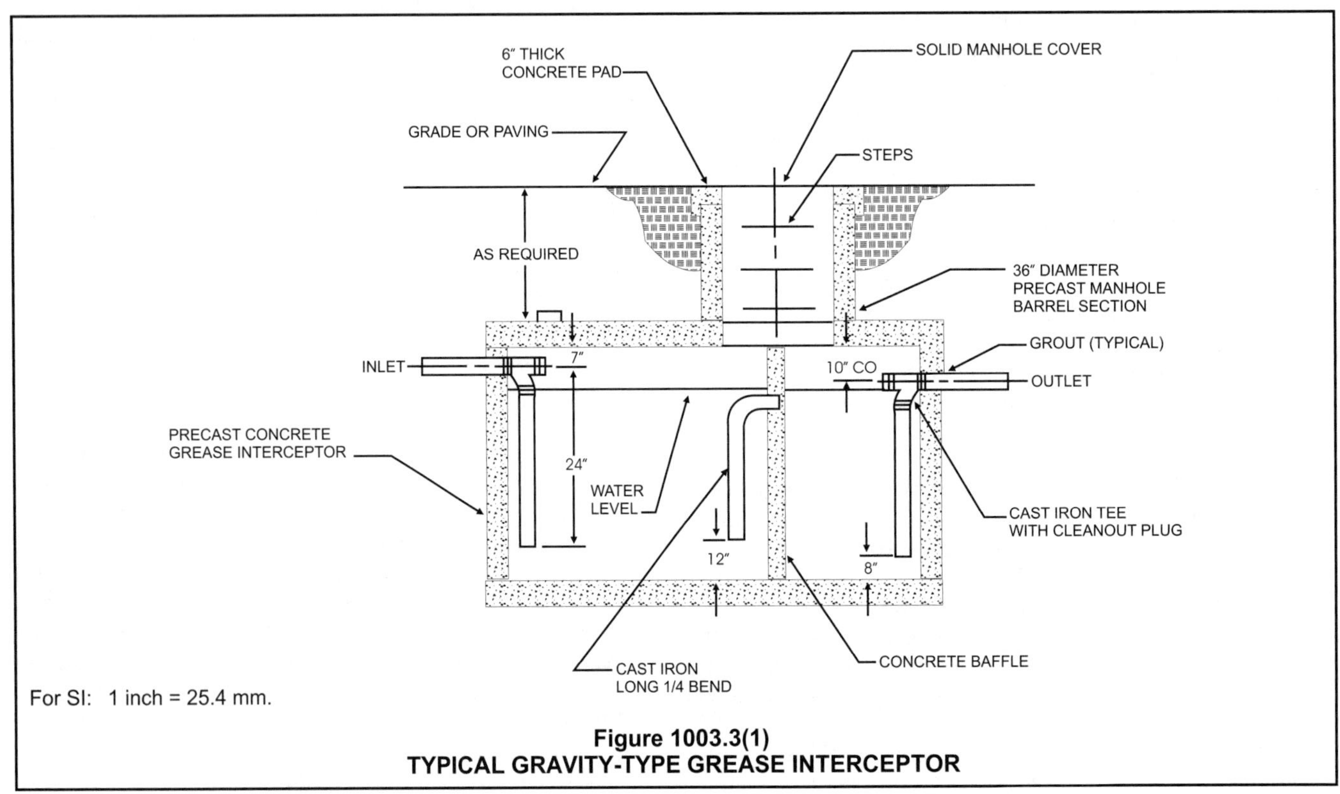

Figure 1003.3(1)
TYPICAL GRAVITY-TYPE GREASE INTERCEPTOR

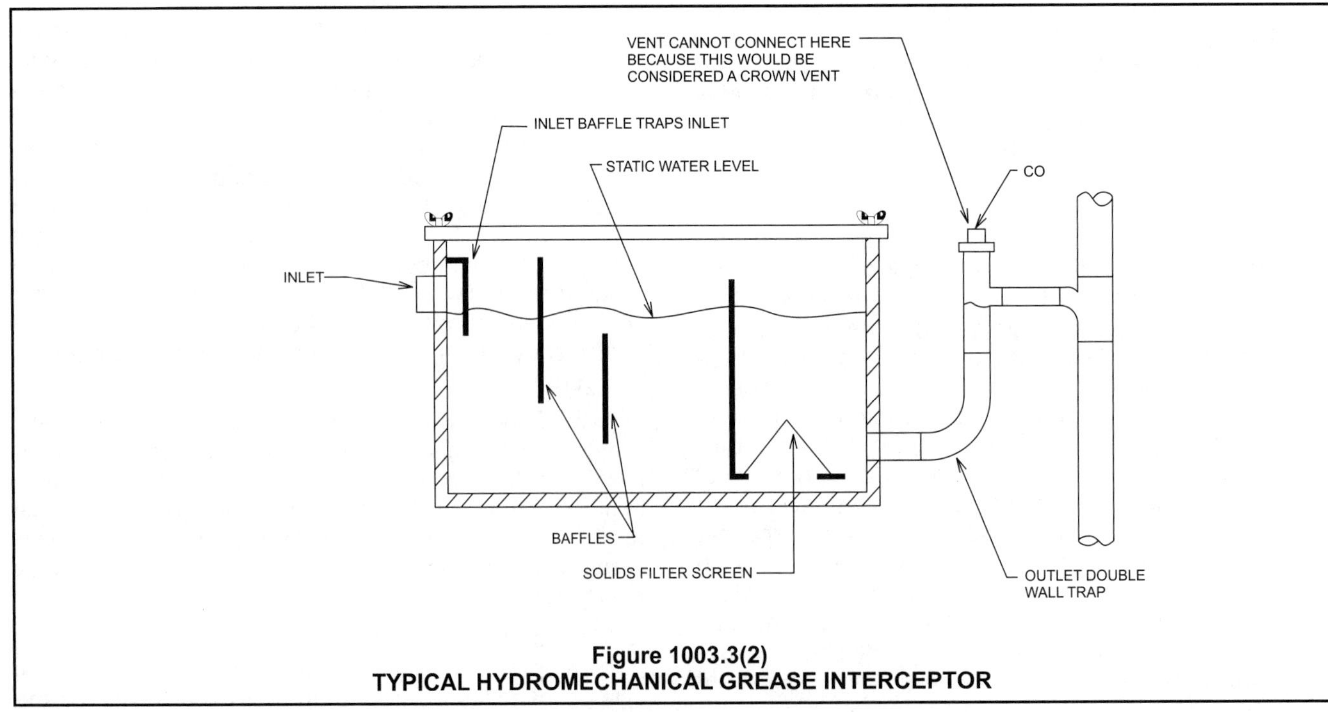

Figure 1003.3(2)
TYPICAL HYDROMECHANICAL GREASE INTERCEPTOR

tor. The specifier of the interceptor must include this information on the request for the interceptor or provide for an interceptor somewhat larger than the size selected to accommodate all other flows to provide for more storage capacity.

This section does not require that the discharge from a food waste grinder be routed through a grease interceptor. The code does not prohibit the discharge from food waste grinders from connecting downstream of a grease interceptor (either of the hydromechanical type or of the gravity type). There are mixed opinions concerning where food waste grinder discharges should be connected. Because studies have indicated that dishwasher prerinse (food scrap) sinks are a significant source of grease wastewater and that many food waste grinder installations receive the waste from prerinse sinks, the grinder discharge should go through the grease interceptor. However, since a food waste grinder operates best with a cold water flow and chops the food waste (and the grease contained within) into solidified particles, the particles will readily flow in the sewer system without the congealing problems associated with hot or warm grease-laden wastewater discharges. Manufacturers of food waste grinders understandably find the requirement for a solids interceptor for the grinder discharge illogical, because the food particles are transported from the grinder to the solids interceptor, from which the particles must be manually removed and disposed of again. Ultimately, the code official or the public sewer system operators will make the decision on where food waste grinders will be connected.

As mitigating problems associated with grease clogging in plumbing piping systems and the cost of interceptor cleaning is a marketable concept, many producers of emulsifiers, chemicals, enzymes and bacteria products offer products to aid in the reduction of grease. While the effectiveness of these products is beyond the discussion of this commentary, the manufacturers of hydromechanical grease interceptors are certain that the addition of these products to the flow stream and into the grease interceptor are detrimental to the operation of the interceptor. Thus, this code section, with the assumption that the grinder is discharging to a hydromechanical grease interceptor, prohibits the addition of these products to a food waste grinder. The intent of the prohibition has no relevance to the protection of food waste grinder, nor is it intended to apply to food waste grinders discharging only to gravity-type grease interceptors.

Note that for gravity-type interceptors with large pathways for internal flow, limited number of baffles and significant storage capacity, clogging and solids build up due to food waste grinder discharge is usually a nonissue unless maintenance cleaning (usually performed on a weekly or monthly basis) is ignored.

1003.3.3 Grease interceptors and automatic grease removal devices not required. A grease interceptor or an automatic grease removal device shall not be required for individual dwelling units or any private living quarters.

❖ The volume and concentration of grease in the wastewater discharge from a dwelling unit or private living quarters is usually not a significant source of grease discharged to public sewer systems. Therefore, the complications for installation of and the resulting expense (initially and ongoing) of a grease interceptor are not warranted. However, this does not mean that users and owners of individual or private living quarters should not be encouraged to "can the grease (cooking oils and fats)" rather than pour such waste down the drain. Many public sewer system operators realize that even small quantities of grease from many locations can result in sewer system prob-

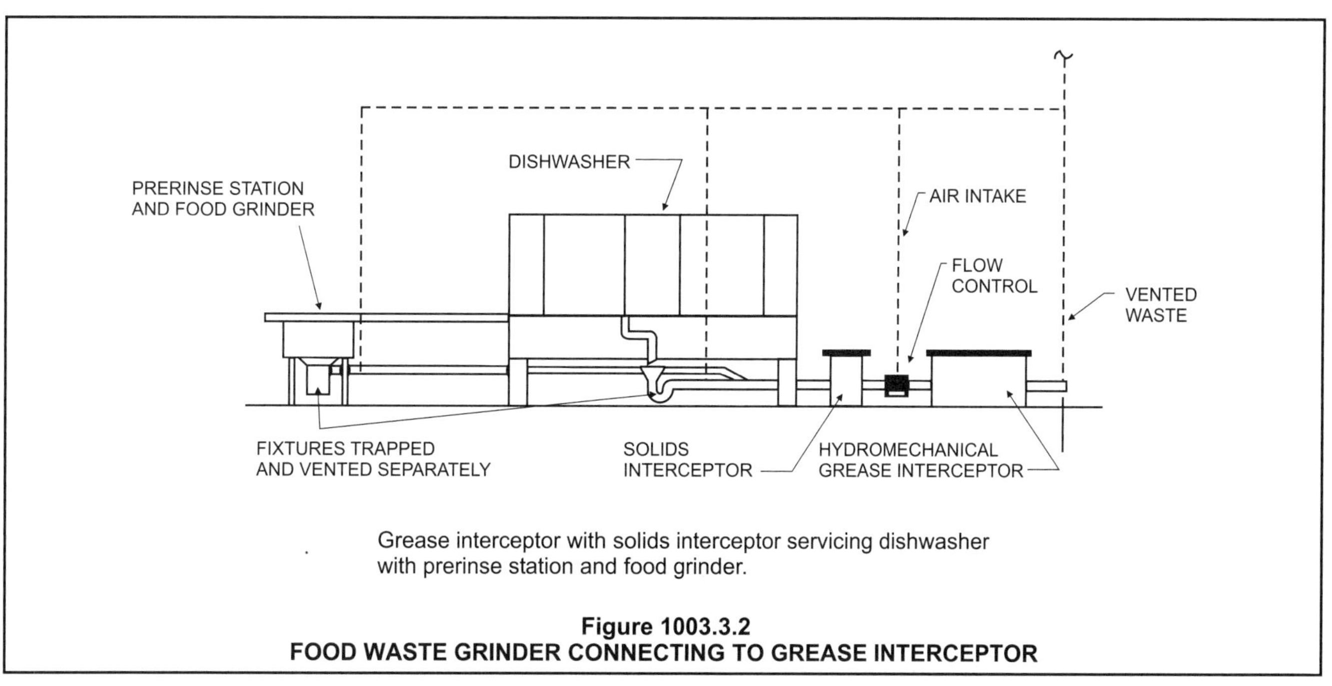

Figure 1003.3.2
FOOD WASTE GRINDER CONNECTING TO GREASE INTERCEPTOR

lems. With advance techniques in real-time grease monitoring and the ease of camera inspection of sewer lines, users who are chronic dumpers of fats, oils and greases into the public sewer system or even clandestine commercial kitchen operations might be identified for enforcement follow-up.

1003.3.4 Grease interceptors and automatic grease removal devices. Grease interceptors and automatic grease removal devices shall be sized in accordance with PDI G101, ASME A112.14.3 Appendix A, or ASME A112.14.4. Grease interceptors and automatic grease removal devices shall be designed and tested in accordance with PDI G101, ASME A112.14.3 or ASME A112.14.4. Grease interceptors and automatic grease removal devices shall be installed in accordance with the manufacturer's instructions.

Exception: Interceptors that have a volume of not less than 500 gallons (1893 L) and that are located outdoors shall not be required to meet the requirements of this section.

❖ The referenced standards PDI G101 and ASME A112.14.3 are nearly identical and describe the testing and certification ratings (flow-through and grease retention) required of a manufacturer's design of (hyrdromechanical) grease interceptor. The standards do not specify any construction requirements other than requiring the use of standard pipe threads for threaded connections. ASME A112.14.4 is the standard for automatic grease removal devices, an "automatic purging" variation of hydromechanical grease interceptors.

The standards do not provide any mandatory sizing methods for hydromechanical grease interceptors or automatic grease removal devices. The PDIG 101 and ASME 112.14.3 standards provide an identical sizing method in a nonmandatory appendix of each standard. Sizing methods for grease interceptors have been debated for decades. Although there is most likely a point at which too small of a grease interceptor size does reduce the removal efficiency under ideal conditions (i.e., just cleaned), there is no penalty (other than initial cost) for the selection of a unit larger than required. A larger unit than "required by the numbers" might cost only slightly more with the benefit of less frequent cleanings (cleaning being a cost as well). Ultimately, the performance of any grease interceptor is directly related to adherence to an appropriate maintenance schedule for actual loading conditions. Maintenance schedules are usually developed after monitoring grease collection amounts over numerous cleanings to determine an optimum cleaning frequency. However, local jurisdictions might require that cleanings occur at intervals not less than weekly or monthly (regardless of whether the cleaning is actually needed or not) based upon their experience with the type of facility operation. Although a discussion of a "best" method for sizing grease interceptors is not within the scope of this commentary, the following discussion of sizing methods is provided for informational purposes only. Grease interceptor manufacturers might also provide sizing methods for their interceptor design. The authority having jurisdiction is responsible for providing or approving the methods used for sizing grease interceptors.

Generally, there are four sizing methods that are used in the prefabricated grease interceptor industry (both gravity and hydromechanical types): (1) the PDI method, (2) the 1980 USEPA "seats" method, (3) the 1980 USEPA "meals" method and (4) the "dfu" method. Some jurisdictions use methods similar to these but have added other requirements or changed the "constants" used in the equations. Others have developed completely different methods which will not be discussed in this commentary. Grease interceptor designs can also be "engineered." There are no consensus standards for the sizing of grease interceptors. The authority having jurisdiction must provide the sizing method to be used or approve the sizing method that is used.

The PDI method

This method is published in the nonmandatory Appendix A of the Plumbing and Drainage Institute's standard G101. This method derives a gallon per minute flow-through rate for selecting a hydromechanical grease interceptor.

Sample Problem 1: A delicatessen has a three-compartment sink for washing utensils, pots and pans used for preparing meats and other deli products for sale. Each compartment of the sink is 24 inches wide by 24 inches long by 18 inches deep (maximum water depth). The three compartment sink will be directly connected through a flow control located at the same elevation as the inlet to interceptor. The highest water elevation in the sink will be at an elevation that is not greater than 6 feet above the inlet the grease interceptor. Determine the required grease interceptor size.

<u>Problem Approach</u>

Calculate the total water holding volume of the sink in gallons, assume a drain time for the sink and choose a "certified" size from Table 1003.3.4.1. Check with the manufacturer to determine if special flow control orifice size is required for the elevation difference between the water level and the interceptor inlet.

<u>Solution</u>

Total sink volume = [3 x (24 x 24 x 18)]/231 cubic inches per gallon

Total sink volume = 134 gallons

Consider the fact that the actual water capacity in the sink will be not greater than approximately 75 percent of the sink due to the volume taken up by the items being washed and the typical fill levels. Therefore,

Actual sink volume = (75/100) x 134 gallons

Actual sink volume = 101 gallons

Consider the fact that the time to drain the sink is not critical. Therefore, assume a 2-minute drain time. (To be more conservative, a 1-minute drain time could be assumed). Therefore,

Flow rate through
interceptor = 101 gallons/2 minutes
= 51 gallons per minute

Refer to Table 1003.3.4.1 to determine that a 75-gallon per minute flow through rating exceeds the required 51 gallons per minute. Although a 50 gpm unit is technically not large enough for the application, use of a 50 gpm-rated interceptor (with a 50 gpm flow control) would be suitable as the only drawback would be that the drain time for the sink would be slightly greater.

Sample Problem 2: Determine the required grease interceptor size for a hotel restaurant kitchen having the following fixtures. The grease interceptor(s) will be located in the floor of the kitchen:

One three-compartment sink for washing cooking utensils, pots and pans. Each compartment of the sink is 24 inches wide by 24 inches long by 18 inches deep (maximum water depth).

One two-compartment sink for food preparation. Each compartment of the sink is 24 inches wide by 24 inches long by 18 inches deep (maximum water depth).

One one-compartment sink for food preparation. The sink compartment is 24 inches wide by 24 inches long by 12 inches deep (maximum water depth).

One steam kettle with a 60-gallon capacity.

One dishwasher with a 30-gallon per minute rating for washing plates, bowls and silverware.

One glassware washer with a 20-gallon per minute rating for washing glasses and cups.

One prewash station (no food waste grinder): 1.6 gallons per minute

(All of the above will be individually indirectly connected to the drain system through floor sinks.)

Three hand sinks.
One mop (service) sink.
Three floor drains.

Problem Approach

Eliminate fixtures that do not produce a grease load from the calculations, calculate the total water holding volume of the sinks in gallons, assume a drain time for the sinks, add equipment discharge rates, add up all the discharge rates and choose a "certified" size from Table 1003.3.4.1.

Solution—Part I

The glassware washer, hand sinks, mop sink and floor drains are not significant grease contributors and therefore are eliminated from the computation. The drain piping from these fixtures must not discharge to the grease interceptor.

Solution—Part 2

Three-compartment sink:

Total sink volume = [3 x (24 x 24 x 18)]/231 cubic inches per gallon

Total sink volume = 134 gallons

Consider the fact that the actual water capacity in the sink will be not greater than approximately 75 percent of the sink due to the volume taken up by the items being washed. Therefore,

Actual sink volume = (75/100) x 134 gallons
= 101 gallons

Because the discharge of the sink compartments are individually indirectly connected into floor sinks, the draining of the sink will be rapid. Assuming a 1-minute drain time:

Flow rate from
entire sink = 101 gallons/1 minute
= 101 gallons per minute

Two-compartment sink:

Total sink volume = [2 compartments x (24 x 24 x 18)]/231 cubic inches per gallon

Total sink volume = 90 gallons

Consider the fact that the actual water capacity in the sink will be not greater than approximately 75 percent of the sink due to the volume taken up by the items being washed and necessary "freeboard" space above the waterline. Therefore,

Actual sink volume = (75/100) x 90 gallons

Actual sink volume = 68 gallons

Because the discharge of the sink compartments are individually indirectly connected into floor sinks, the draining of the sink will be rapid. Assuming a 1-minute drain time:

Flow rate through
interceptor = 68 gallons/1 minute
= 68 gallons per minute

One-compartment sink:

Total sink volume = [1 compartment x (24 x 24 x 18)]/231 cubic inches per gallon
= 45 gallons

Consider the fact that the actual water capacity in the sink will be no greater than approximately 75 percent of the sink due to the volume taken up by the items being washed and necessary "freeboard" space above the waterline. Therefore,

Actual sink volume = (75/100) x 45 gallons
= 34 gallons

Because the discharge of the sink compartments are individually indirectly connected into floor sinks, the draining of the sink will be rapid. Assuming a 1-minute drain time:

Flow rate through interceptor = 34 gallons/1 minute
= 34 gallons per minute

Solution—Part 3

The following equipment discharge rates are identified:

Dishwasher: 30 gpm

Steam kettle: 60 gallons (assume a drain time of 2 minutes) = 30 gpm

Prerinse station: 1.6 gpm

Total flow rate from all sinks and equipment = 101 + 68 + 34 + 30 + 30 + 1.6
= 265 gpm

Because the total flow rate exceeds 100 gpm (the maximum allowable flow rate for a certified grease interceptor), several grease interceptors will be required. The following configuration is one of many possible ways to configure the drain system for this kitchen.

Grease interceptor 1: 100 gpm rating with 100 gpm flow control to accommodate the three-compartment sink.

Grease interceptor 2: 100 gpm rating with 100 gpm flow control to accommodate the two-compartment sink, the dishwasher and the prerinse station.

Grease interceptor 3: 75 gpm rating with 75 gpm flow control to accommodate the one-compartment sink and the steam kettle.

The "seats" method for restaurants

In 1980, the United States Environmental Protection Agency released the *Design Manual for Onsite Wastewater Treatment and Disposal Systems* (USEPA document no. 625/1-80-012). The derived interceptor volume (in gallons) is intended for choosing a gravity-type grease interceptor. Note that the interceptor is not sized for a flow-through rating (gallons per minute). Section 8.2 of the manual offers the following formula for deriving the size of a grease interceptor for a restaurant:

Grease interceptor size, gallons[1] = *D* x *GL* x *ST* x (*hr*/2) x *LF*
where,

D = number of seats in dining area

GL = gallons of wastewater per meal, normally 5 gallons

ST = storage capacity factor, minimum value = 1.7
maximum value = 2.5

16 = number of hours open

LF = loading factor,
1.25 for interstate freeway locations
1.0 for other freeways
1.0 for recreational areas
0.8 main highways
0.8 other highways

Footnote 1: minimum volume is 750 gallons.

The "meals" method for hospitals, nursing homes, other types of commercial kitchens

This method was also published in the 1980 United States Environmental Protection Agency's *Design Manual for Onsite Wastewater Treatment and Disposal Systems*. The derived interceptor volume (in gallons) is intended for choosing a gravity-type grease interceptor. Note that the interceptor is not sized for a flow-through rating (gallons per minute). Section 8.2 of the manual offers the following formula for deriving the size of a grease interceptor for commercial kitchens other than a restaurant kitchen:

Grease interceptor size, gallons[1] =
M x *GL* x *ST* x 2.5 x *LF*
where,

M = number of meal served per day

G = gallons of wastewater per meal, normally 4.5 gallons

LF = loading factor,
1.25 for kitchens with food waste grinders and dishwashing
1.0 for kitchens without food waste grinders but with dishwashers
0.75 for kitchens with food waste grinders but without dishwash
0.5 for kitchens without food waste grinders and dishwashers

Footnote 1: minimum volume is 750 gallons.

The "dfu" method

This method provides a table that equates the number of drainage fixture units ("dfu") for the total drainage flow requiring grease interception to the required interceptor capacity. Separate tables are often provided to size hydromechanical- and gravity-type grease interceptors, respectively. Although some tables are developed by arbitrary assignments of sizes to flows, other tables are developed by converting the number of dfu (as assigned by the adopted plumbing code) to a gallon per minute flow. The choice of a hydromechanical grease interceptor is straightforward whereas, for a gravity-type interceptor, sizing is based upon providing at least a 30-minute retention time through the grease interceptor vessel. Because dfu assignments vary between model codes, the reader is cautioned about applying one code's dfu assignments to another model code's interceptor sizing tables.

1003.3.4.1 Grease interceptor capacity. Grease interceptors shall have the grease retention capacity indicated in Table 1003.3.4.1 for the flow-through rates indicated.

❖ Table 1003.3.4.1 comes directly from the PDI G101 and ASME A112.14.3 standards. Footnote a is not included in the standard's table but was added to the code in the 2006 edition. The minimum grease retention capacity for a hydromechanical grease interceptor provides for a "reasonable" interceptor cleaning interval for an "average" concentration of grease within the wastewater at the interceptor's rated flow. The minimum grease retention capacity was determined by extensive testing long ago and establishes a performance baseline for all manufacturers to comply with. Because the grease concentration in grease-laden wastewater varies widely, the actual cleaning interval necessary to prevent desired grease concentration at the outlet of the interceptor from being exceeded is determined by observation or testing.

The PDI G101 and ASME A112.14.3 standards only provide for testing and certification for flow-through ratings of up to 100 gpm. Because some interceptor manufacturers produce designs having flow-through ratings greater than 100 gpm, footnote a was added to Table 1003.3.4.1 to extend the same baseline of performance to these larger designs. However, as the standards do not provide for testing of these larger units, hydromechanical grease interceptor designs over 100 gpm must be submitted for approval under Section 105.2 as they cannot meet the testing and certification requirements of Section 1003.3.4.

TABLE 1003.3.4.1
CAPACITY OF GREASE INTERCEPTORS[a]

TOTAL FLOW-THROUGH RATING (gpm)	GREASE RETENTION CAPACITY (pounds)
4	8
6	12
7	14
9	18
10	20
12	24
14	28
15	30
18	36
20	40
25	50
35	70
50	100
75	150
100	200

For SI: 1 gallon per minute = 3.785 L/m, 1 pound = 0.454 kg.

a. For total flow-through ratings greater than 100 (gpm), double the flow-through rating to determine the grease retention capacity (pounds).

❖ Table 1003.3.4.1 indicates the standard "sizes" (flow-through ratings) of hydromechanical grease interceptors to which numerous manufacturers have designed, tested and certified their units to meet the requirements of PDI G101 and ASME A112.14.3. The use of standard sizes allows for manufacturers to produce a tested and certified unit without having to test and certify the actual unit being constructed for a customer. This allows for faster delivery times and allows manufacturers to build frequently purchased sizes in advance of customer orders. While it might be possible to order a "nonstandard" size, the delivery time and cost for special testing and certification of such a unit will most likely be substantially more than if the next larger size of standard unit was ordered. This is not to say that a nonstandard unit could not be ordered, but given that most projects have significant time and financial constraints, ordering standard sizes should offer the best economy.

The grease retention capacity for each of the standard interceptor sizes is the minimum required capacity. Larger grease retention capacities offer the advantage of decreased cleaning frequency; however, if the local jurisdiction requires a cleaning interval of not less than a certain frequency, the additional grease capacity might never be used.

1003.3.4.2 Rate of flow controls. Grease interceptors shall be equipped with devices to control the rate of water flow so that the water flow does not exceed the rated flow. The flow-control device shall be vented and terminate not less than 6 inches (152 mm) above the flood rim level or be installed in accordance with the manufacturer's instructions.

❖ A hydromechancial grease interceptor is designed for a maximum flow-through rate to allow for separation to occur and so that velocities within the unit do not cause excessive turbulence which might cause "carry over" of grease to the outlet. The rate of discharge from an individual plumbing fixture varies depending on the "head" (height of water) in each receptor (sink compartment), the number of sink compartments discharged at once and the configuration of the compartment outlets and the drain piping. The probability of simultaneous discharge of multiple plumbing fixtures along with other equipment (e.g., dishwashers, kettles) also varies. The selection of a grease interceptor size might be predicated only on certain combinations of fixtures discharging simultaneously in order to optimize interceptor sizing for the most probable flow condition. Note that the sample problems for the "PDI" sizing method provided in the commentary for Section 1003.3.4 include assumptions for the drain times of sinks. In reality, sinks could drain faster than the assumed time and more fixtures/equipment could discharge simultaneously than were considered in the selection of the interceptor size. Therefore, the interceptor must be provided with a flow-control device to limit flow rate to the interceptor (see Commentary Figure 1003.3.4.2).

Although an interceptor might be sized to accommodate the maximum possible flow rate of all plumbing fixtures and equipment (i.e., simultaneous discharge), the flow control is still required as it also serves to entrain air into the waste flow to aid in the grease separation process. Entrained air in the waste causes the grease, oils and fats to become more buoyant because of attachment of tiny air bubbles. The emhanced buoyancy causes the grease, fats and oils to more readily rise to the retention area within the interceptor.

The requirement in the second sentence of this section for the flow control to be "vented" ensures that the flow control will be installed with a means to readily obtain air for entrainment into the waste flow. The requirement for the air intake to terminate (within the same room as the fixture) not less than 6 inches (152 mm) above the flood level rim of the fixture applies only to a specific fixture/interceptor arrangement as follows: The grease interceptor must be capable of serving as a fixture trap for a single fixture or combination sink of no more than three compartments and must be within a specific proximity to the fixture (see Exception 3 of Section 1002.1). Terminating the flow control vent 6 inches (152 mm) above the fixture flood level rim prevents waste overflow through the vent should the grease interceptor become clogged. Otherwise, all fixture and equipment drains must be trapped and vented as required by Chapter 9 and the flow control air intake must be connected to the sanitary drain vent system or must terminate outdoors in accordance with the requirements for vent terminations in Chapter 9.

The manufacturer's installation instructions will indicate whether the grease interceptor can serve as a trap for a single fixture or combination sink and might require additional vent connections to the interceptor as well.

The code does not prohibit a grease interceptor from being indirectly connected to a sanitary drainage system through a waste receptor. For example, where a grease interceptor serves an adjacent untrapped single fixture or multicompartment combination sink (as allowed by Exception 3 of Section 1002.1), the interceptor could discharge, through an air break, into a waste receptor such as a floor sink. See Sections 802.1 and 802.1.8. This is especially useful where a change in kitchen configuration requires the addition of a floor-mounted interceptor for a new or relocated fixture. Note that in this arrangement, if a vent is required to prevent self-siphoning of the interceptor, it must vent to the room or to the outdoors, not to the sanitary vent system.

1003.3.5 Automatic grease removal devices. Where automatic grease removal devices are installed, such devices shall be located downstream of each fixture or multiple fixtures in accordance with the manufacturer's instructions. The automatic grease removal device shall be sized to pretreat the measured or calculated flows for all connected fixtures or equipment. Ready *access* shall be provided for inspection and maintenance.

❖ Automatic grease removal devices serve the same function as grease interceptors except that the unpleasant task of removing the grease is performed by automated internal means such as a pump or skimmer that discharges the collected grease to a container outside of the interceptor. The container, when full, is replaced with an empty one. Although ASME A112.14.4 is the standard for automatic grease removal devices, the standard requires that grease removal devices comply with the ratings and grease retention capacities required by ASME A112.14.3 for hydromechanical grease interceptors.

Note that this section requires ready access for grease removal devices so that maintenance and repairs can be properly performed.

1003.4 Oil separators required. At repair garages, car-washing facilities, at factories where oily and flammable liquid wastes are produced and in hydraulic elevator pits, separators

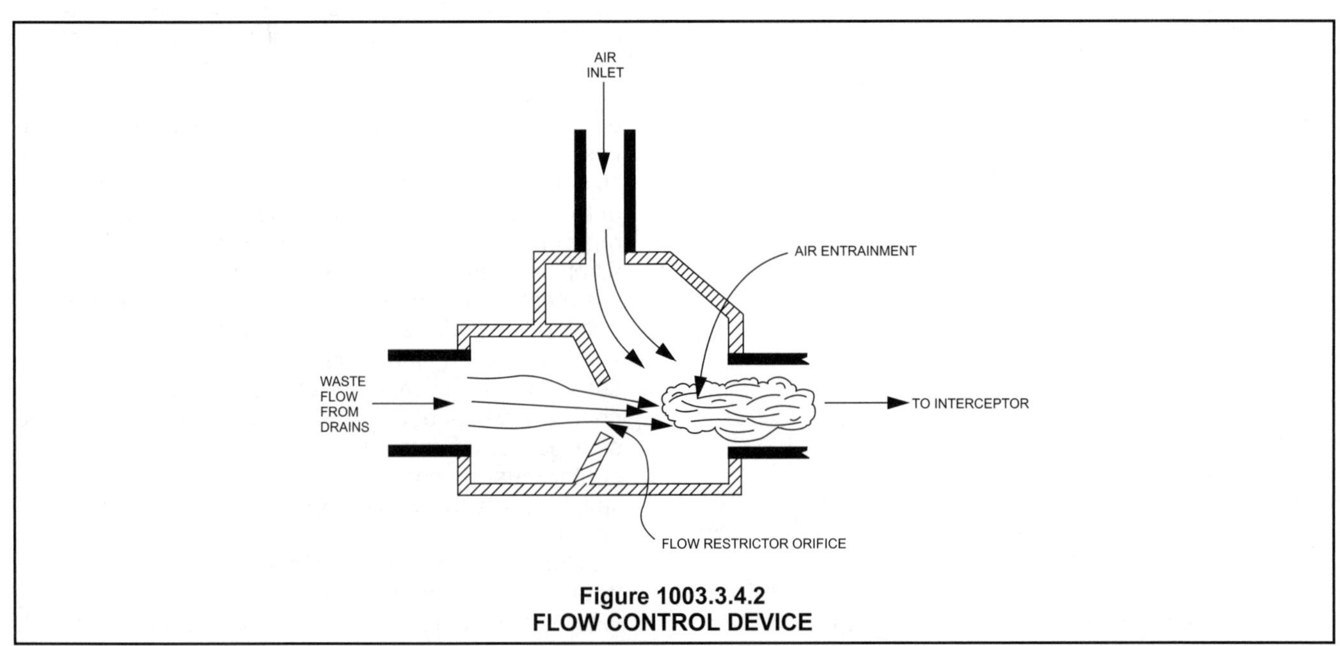

Figure 1003.3.4.2
FLOW CONTROL DEVICE

shall be installed into which all oil-bearing, grease-bearing or flammable wastes shall be discharged before emptying into the building drainage system or other point of disposal.

Exception: An oil separator is not required in hydraulic elevator pits where an *approved* alarm system is installed.

❖ The intent of this section is to require an oil separator only where oily, grease-bearing and flammable wastes are known to be generated. Floor and trench drains located in areas used for car-washing that have engine or undercarriage cleaning capabilities and in motor vehicle repair garages must be connected to an oil separator before connecting to the building drainage system. Where oily and flammable liquid wastes are produced in factories, floor and trench drains serving those areas must also be connected to an oil separator before connecting to a building drain.

Pits below hydraulic elevators collect hydraulic oil leaking from equipment connections and worn seals. The designer has a choice to drain the elevator pit through an oil separator or as stated by the exception, an approved oil detection alarm can be installed in the pit to warn occupants of an accumulating oil condition.

The code is silent concerning the details required of an approved oil alarm system. As the intent of this code section is to prevent an oil discharge to the sanitary drain system, an approved alarm system should have the means to alert maintenance personnel long before oil is discharged from the sump pump. Typically, hydraulic elevator pits are located below grade and, many times, below building drain level. Within the pit, a sump pump in sump basin is provided to remove any rainwater that enters the elevator pit. A number of manufacturers produce a specially fitted sump pump with an oil sensor and alarm unit. When oil enters the sump basin, the sensor will detect the presence of oil (floating on the surface of water in the sump basin), and sound an alarm before any oil is pumped out of the pit.

There could be other arrangements where the elevator pit has a floor drain. In that case, there could be an oil sensor connected to a control system that would sound an alarm if the presence of oil is detected. Because it would be difficult for the code to cover all possible equipment arrangements, the code simply leaves it up to the code official to determine if the oil alarm (and related equipment) meets the intent of the code, which is to prevent oil from entering the drain system.

Note that Section 301.6 prohibits any floor drain or sump pump discharge line coming from the base of an elevator shaft from being directly connected to the building drainage system. The connection must be indirect. For example, the sump pump discharge line could discharge through an air break or air gap into a standpipe or floor sink or other suitable waste receptor located outside of the elevator shaft or equipment room.

It is not the intent of this section to require that an oil separator serve all plumbing fixtures installed in the facilities as described above. While the occupants of these facilities might wash their hands or even shower at these facilities in order to aid in the complete removal of oil or grease, it is understood that the majority of oily and greasy substances have already been removed from the body using shop towels and absorbent cloths. The quantity of oil and grease from showering or hand washing is negligible such that the waste can be safely discharged directly to the sanitary drain system. However, service sinks, especially those used for washing/rinsing floor mops and the discharge of mop water should be considered for connection to the oil separator.

Where the discharge from an oil separator could contain materials that would be detrimental to a private sewage disposal system, the discharge from the oil separator might be required to be routed to a holding tank instead of to the private sewage disposal system. The local authority having jurisdiction for private sewage disposal systems should be consulted before connecting a separator discharge to a private sewage disposal system.

1003.4.1 Separation of liquids. A mixture of treated or untreated light and heavy liquids with various specific gravities shall be separated in an *approved* receptacle.

❖ There is a variety of designs for oil separators. Typically an oil separator is a large tank designed to retain oil by minimizing turbulent flow. Oil is removed through a draw-off pipe, and grease, sludge and solids settle to the bottom or collect on the filter screen. Oil separators function by relying on the differences in specific gravity (density) of the waste water and oil.

1003.4.2 Oil separator design. Oil separators shall be designed in accordance with Sections 1003.4.2.1 and 1003.4.2.2.

❖ The design criteria for effective performance depend on the minimum cross-sectional area and the minimum ratio of depth to width. In turn, design and shape of the separator depend on the character and quality of the oils to be separated. A properly sized separator allows the separation and prevents the evacuation of solids into the drainage system.

1003.4.2.1 General design requirements. Oil separators shall have a depth of not less than 2 feet (610 mm) below the invert of the discharge drain. The outlet opening of the separator shall have not less than an 18-inch (457 mm) water seal.

❖ The sizing specified in this section relates to the open-tank-type separator. The minimum depth is necessary to provide sludge and solids retention capacity and to obtain efficient retention of oil or other volatile liquid wastes by minimizing turbulent flow through the separator. The outlet must be designed to provide at least 18 inches (457 mm) of liquid seal to provide for the storage of oil to at least that depth.

1003.4.2.2 Garages and service stations. Where automobiles are serviced, greased, repaired or washed or where gasoline is dispensed, oil separators shall have a minimum capacity of 6 cubic feet (0.168 m^3) for the first 100 square feet (9.3 m^2) of area to be drained, plus 1 cubic foot (0.28 m^3) for each additional 100 square feet (9.3 m^2) of area to be drained into the separator. Parking garages in which servicing, repairing or washing is not conducted, and in which gasoline is not dispensed, shall not require a separator. Areas of commercial garages utilized only for storage of automobiles are not required to be drained through a separator.

❖ In addition to the general requirements of Section 1003.4.2.1, this section specifically applies to motor vehicle garages or service stations where lubrication, oil changing, fuel dispensing, repair work, or hand or mechanical washing takes place. The intent of this section is to require separators in locations where flammable or combustible liquids are to be discharged into the drainage system, typically in occupancies provided with floor or trench drains.

The requirement for a separator does not, in itself, create the requirement for floor or trench drains. Where a drainage system is provided in these occupancies, provisions must be made to prevent flammable and combustible liquids from entering the drainage system, sewers and waste treatment facilities.

The sizing of a separator can be reduced to two equations as follows:

Floor Area $\leq$ 100 square feet $C = 6$

Floor Area $>$ 100 square feet $C = 5 + (A/100)$

where:

C = Minimum capacity of separator (ft^3).

A = Floor area served (ft^2).

For SI: 1 ft^2 = 0.0929 m^2, 1 ft^3 = 0.0283 m^3.

Garages used exclusively for parking or storage of automobiles do not normally produce wastes requiring separation; therefore, separators are not required for drains in such locations.

1003.5 Sand interceptors in commercial establishments. Sand and similar interceptors for heavy solids shall be designed and located so as to be provided with ready *access* for cleaning, and shall have a water seal of not less than 6 inches (152 mm).

❖ Like grease interceptors, sand interceptors need periodic maintenance to function correctly. They must be installed to allow access for cleaning and servicing. Note that this section does not require installation of sand or heavy solids interceptors but only regulates their installation and design.

1003.6 Laundries. Laundry facilities not installed within an individual dwelling unit or intended for individual family use shall be equipped with an interceptor with a wire basket or similar device, removable for cleaning, that prevents passage into the drainage system of solids $^1/_2$ inch (12.7 mm) or larger in size, string, rags, buttons or other materials detrimental to the public sewage system.

❖ Commercial laundries and any laundries not for individual family use must be equipped with an interceptor that is capable of preventing string, lint and other solids from entering the sewage system. The filter, screen or basket must allow for cleaning and removing intercepted solids.

1003.7 Bottling establishments. Bottling plants shall discharge process wastes into an interceptor that will provide for the separation of broken glass or other solids before discharging waste into the drainage system.

❖ Drains from areas in both bottling and food processing plants where glass bottles and jars are used as containers present a greater problem from broken glass than generally recognized. Separators should be located to intercept and treat this waste separately from the remainder of the processing area.

1003.8 Slaughterhouses. Slaughtering room and dressing room drains shall be equipped with *approved* separators. The separator shall prevent the discharge into the drainage system of feathers, entrails and other materials that cause clogging.

❖ The type of interceptors necessary to protect slaughtering room drains depends to a great extent on the particular operation. For example, in a poultry processing plant, separators have various-size screens designed to prevent feathers from the killing and defeathering area from entering the drainage system. Additionally, the drainage system will need protection against entrails and other heavy solids that could gain entry from the cutting room. In such operations, a screen or basket-type interceptor is necessary to prevent the passage of feathers, entrails and similar animal matter.

1003.9 Venting of interceptors and separators. Interceptors and separators shall be designed so as not to become air bound where tight covers are utilized. Each interceptor or separator shall be vented where subject to a loss of trap seal.

❖ Most interceptors and separators require venting of the containment tank and the outlet trap if so equipped. Typically, a vent is required to allow the escape or admittance of air to compensate for the variable fluid level in the tank. Interceptors and separators with integral traps must be vented to prevent the tank from emptying its contents by siphon action. Oil and gasoline separators require vents independent of the sanitary drain venting system to help control and dissipate vapor buildup that may occur in the holding tank. Depending on the size, configuration and location of an interceptor or separator, however, a vent is not always necessary [see Commentary Figures 1003.9(1) through 1003.9(3)].

AIR INTAKE
FLOW CONTROL
FIXTURE NOT TRAPPED
AIR INTAKE
FLOW CONTROL
FIXTURE TRAPPED AND VENTED SEPARATELY

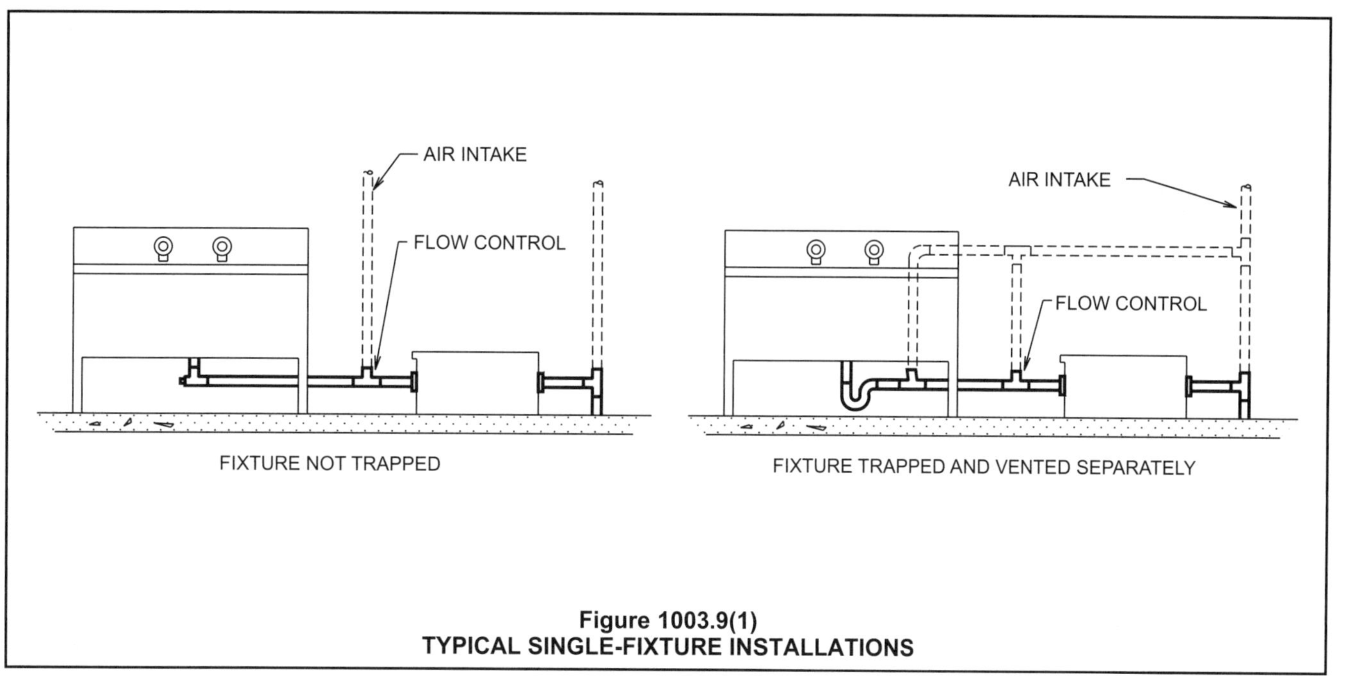

Figure 1003.9(1)
TYPICAL SINGLE-FIXTURE INSTALLATIONS

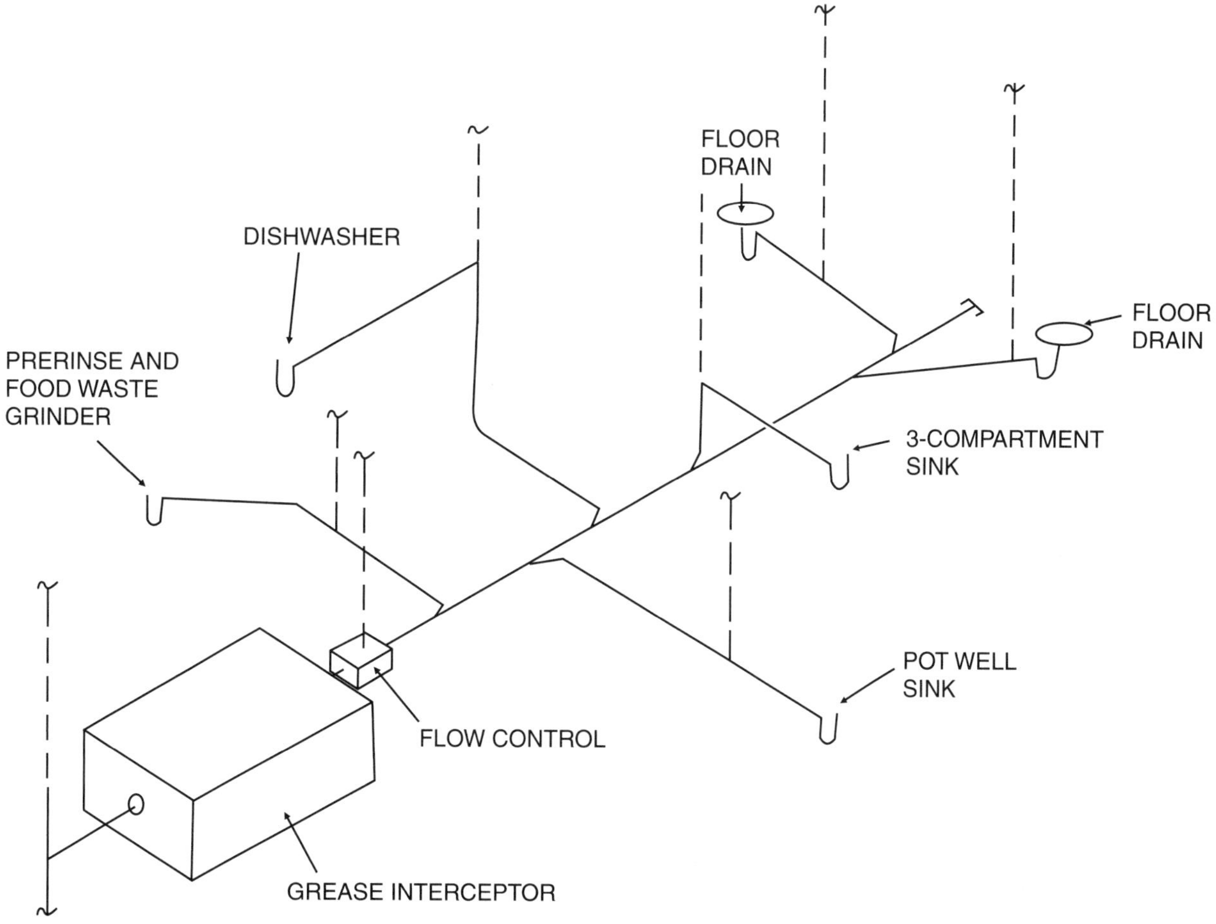

FLOOR DRAIN
DISHWASHER
FLOOR DRAIN
PRERINSE AND FOOD WASTE GRINDER
3-COMPARTMENT SINK
POT WELL SINK
FLOW CONTROL
GREASE INTERCEPTOR

Figure 1003.9(2)
VENTING OF GREASE INTERCEPTOR

1003.10 Access and maintenance of interceptors and separators. *Access* shall be provided to each interceptor and separator for service and maintenance. Interceptors and separators shall be maintained by periodic removal of accumulated grease, scum, oil, or other floating substances and solids deposited in the interceptor or separator.

❖ For an interceptor or separator to function properly, it must be cleaned periodically. The solids and liquids, such as grease, oil, large solids, feathers or glass, must be removed before they exceed the capacity of the interceptor or separator. An interceptor or separator that has not been cleaned will cause stoppage in the drainage system.

When a grease interceptor or oil separator reaches its maximum retention capacity, it will cease to function, which will allow oil and grease to pass through to the sewage system and possibly clog the drainage pipes downstream. An interceptor or separator that is no longer functioning properly will not provide any indication of malfunction; therefore, such devices must be periodically monitored to determine the volume of substances being retained. Oil separators are typically equipped with test cocks to determine the need for oil removal from the device.

Grease interceptors that have not been properly maintained have caused some jurisdictions and health authorities to prohibit the location of these devices in kitchens. Without proper maintenance, interceptors would become full of grease and eventually restrict or plug the interceptors and drain lines downstream, causing greasy waste to leak from the cover of the interceptor. The resulting mess is considered a health hazard.

Some manufacturers have addressed the problems associated with the lack of maintenance by providing various methods of removing or drawing off the grease from the grease interceptor semiautomatically or automatically. Additionally, the use of bacteria (bioremediation) traps and digests greases in the interceptor to convert them into the byproducts of digestion; however, these devices still require periodic maintenance and cleaning [see Commentary Figure 1003.10(1).

SECTION 1004 MATERIALS, JOINTS AND CONNECTIONS

1004.1 General. The materials and methods utilized for the construction and installation of traps, interceptors and separators shall comply with this chapter and the applicable provi-

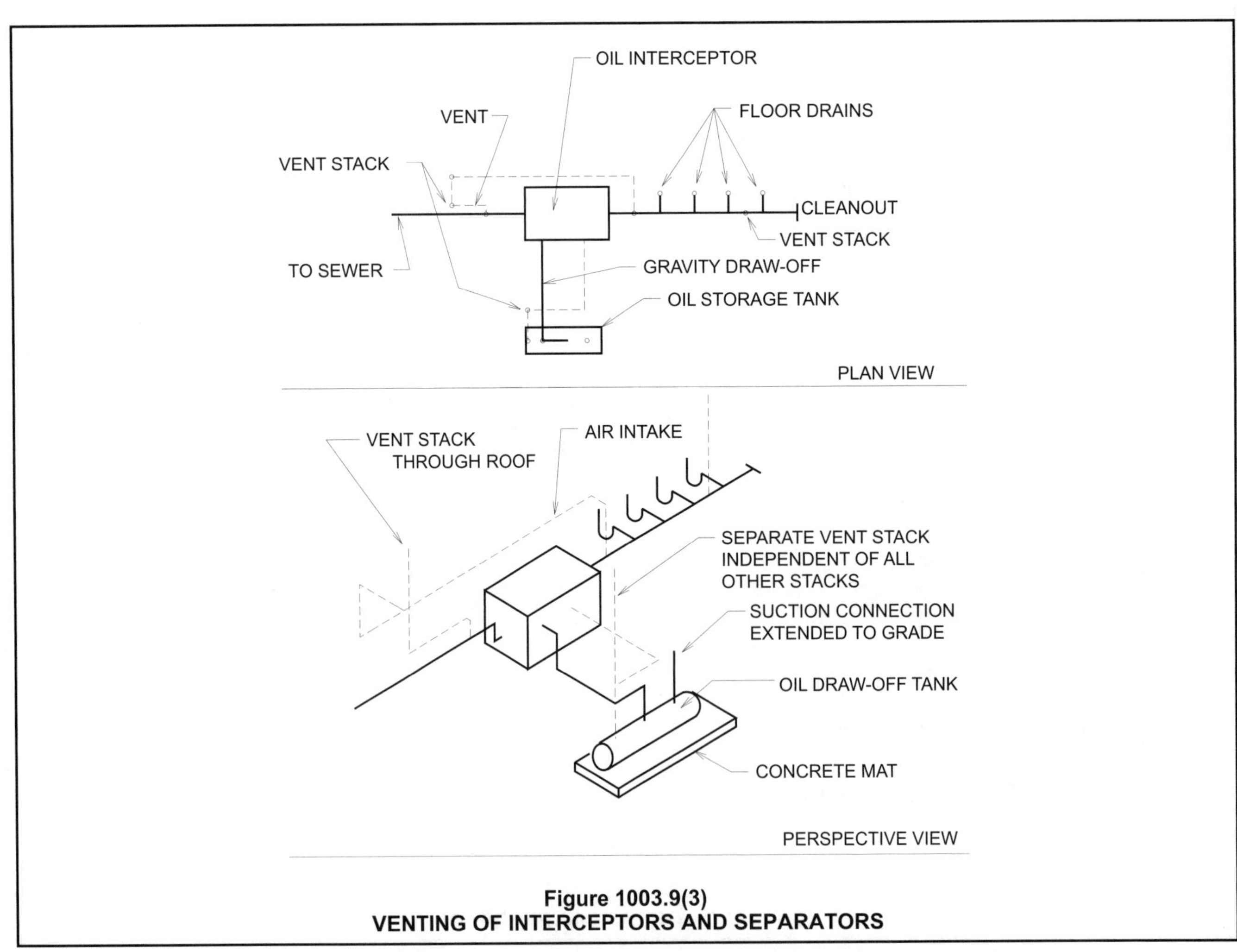

Figure 1003.9(3)
VENTING OF INTERCEPTORS AND SEPARATORS

sions of Chapters 4 and 7. The fittings shall not have ledges, shoulders or reductions capable of retarding or obstructing flow of the piping.

❖ The materials and methods used to install traps, interceptors and separators are regulated by different sections in Chapters 4 and 7. Section 405.8 requires that slip joint connections be made only on the trap inlet, trap outlet and within the trap seal. Additionally, fixtures with such connections must have an access opening at least 12 inches (305 mm) in diameter. Section 405.7.1 requires that the overflow from any fixture discharge into the drainage system on the inlet or fixture side of the trap (except water closets and urinals). Section 702 requires the material used to complete the sanitary system to conform to the standards listed in Table 702.1, 702.2, 702.3 or 702.4. Section 705 regulates the method of joining pipe and fittings. Section 707 states the joining methods that are prohibited. Section 708 governs the location, installation, maintenance, size and clearances of drainage pipe cleanouts. Section 709 regulates drain pipe sizing.

Bibliography

The following resource materials are referenced in this chapter or are relevant to the subject matter addressed in this chapter.

ASME A112.14.1-2003, *Backwater Valves*. New York, NY: American Society of Mechanical Engineers, 2003.

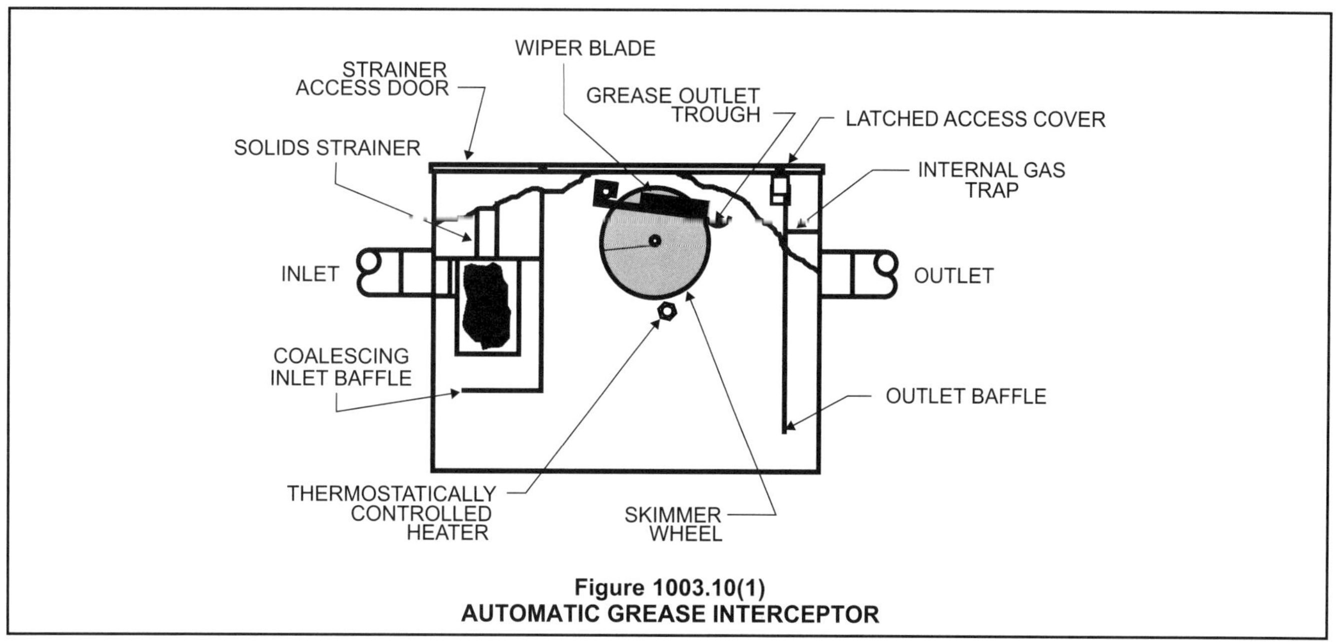

Figure 1003.10(1)
AUTOMATIC GREASE INTERCEPTOR

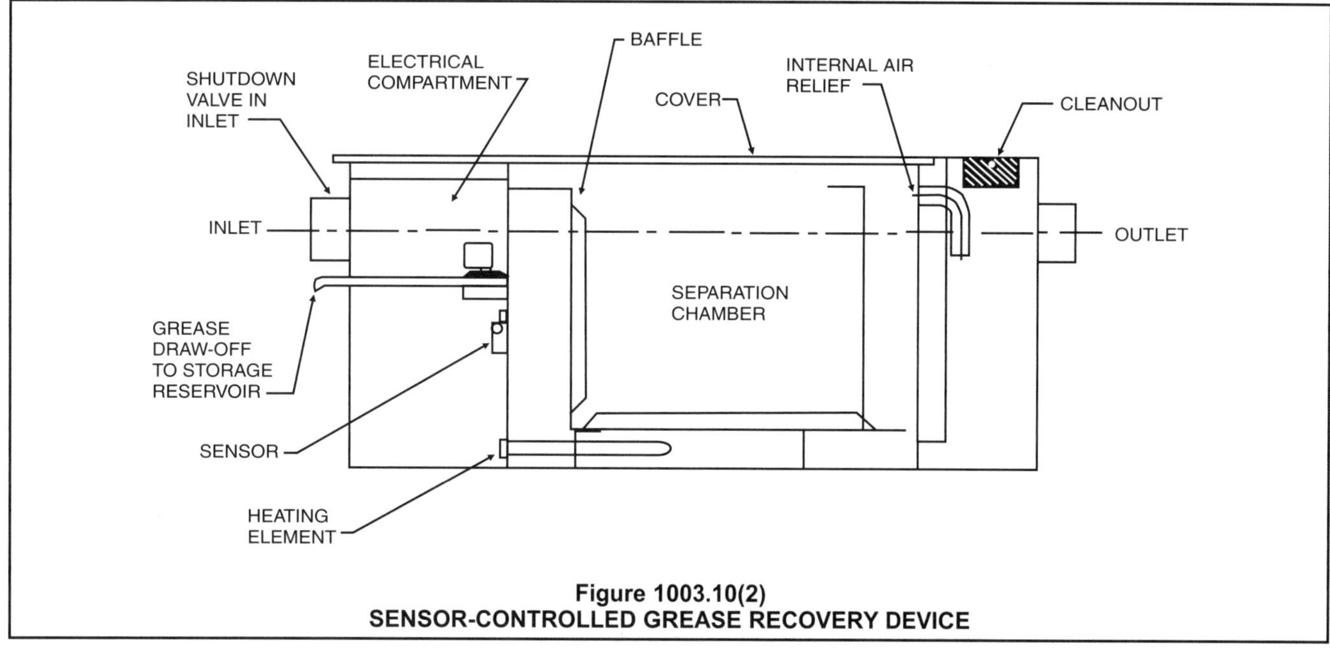

Figure 1003.10(2)
SENSOR-CONTROLLED GREASE RECOVERY DEVICE

ASME A112.14.3-00, *Grease Interceptors*. New York, NY: American Society of Mechanical Engineers, 2000.

ASME A112.14.4-01, *Grease Removal Devices*. New York, NY: American Society of Mechanical Engineers, 2001.

ASSE 1018-01, *Performance Requirements for Trap Seal Primer Valves—Potable Water Supply Fed*. Westlake, OH: American Society of Sanitary Engineering, 2001.

ASSE 1044-01, *Performance Requirements for Trap Seal Primer Valves; Drainage Type*. Westlake, OH: American Society of Sanitary Engineering, 2001.

CSA-B181.1-99, *ABS Drain, Waste, and Vent Pipe and Pipe Fittings*. Rexdale (Toronto), Ontario: Canadian Standards Association, 1999.

CSA-B181.2-02, *PVC Drain, Waste, and Vent Pipe and Pipe Fittings—with Revisions through December 1993*. Rexdale (Toronto), Ontario: Canadian Standards Association, 2002.

Design Manual for Onsite Wastewater Treatment and Disposal Systems.

Guide to Grease Interceptors. South Easton, MA: Plumbing and Drainage Institute, 1998.

IBC-09. *International Building Code*. Washington, DC: International Code Council, Inc., 2009.

PDI G101 (2003), *Testing and Rating Procedure for Grease Interceptors with Appendix of Sizing and Installation Data*. South Easton, MA: Plumbing and Drainage Institute, 2003.

US EPA document no. 625/1-80-012, *Design Manual for Onsite Wastewater Treatment and Disposal Systems.* Cincinnati, OH: United States Environmental Protection Agency, 1980.

USEPA-80, *Design Manual for Onsite Wastewater Treatment and Disposal Systems*. Washington, DC: United States Environmental Protection Agency, 1980.

Chapter 11: Storm Drainage

General Comments

Storm drainage systems are generally assumed to be systems for the collection and transport of rainwater and ground water away from a building. The final disposal of this water normally occurs at some remote location, typically a stream, river or lake. This discharged water is normally left untreated.

The code prohibits the connection of floor drains to the storm drainage system (see commentary, Section 1104.3). If the purpose of this prohibition is to prevent the connection of floor drains located within the building to the storm drainage system, it raises the question of whether floor drains that are part of an open parking garage are included in this limitation. When a parking garage does not contain repair and service operations or other similar activity, it can be considered equivalent to a parking lot that is commonly drained to the storm drainage system. The code is silent on this specific issue, and therefore, the code will be interpreted and applied by the code official, based on the facts related to the individual parking garage. If the parking garage will function as a parking lot, with no increased possibility that repair and service operations will occur, floor drains in the garage could be connected to the storm drainage system the same as for a parking lot.

This chapter also covers the collection and discharge of subsoil water at the exterior of the building. This subsoil water is to be prevented from entering the building and causing possible damage to the structure and its contents. This subsoil water collection and discharge requirement, while not specifically stated, needs to be coordinated with the *International Building Code*® (IBC®) and its requirement for foundation wall dampproofing. These two elements, when properly designed and installed, will work together to keep subsoil water out of the building.

Purpose

Chapter 11 regulates the removal of storm water typically associated with rainfall. The proper installation of a storm drainage system reduces the possibility of structural collapse on a flat roof, prevents the leakage of water through the roof, prevents damage to the footings and foundation of the building and prevents flooding of the lower levels of the building.

SECTION 1101 GENERAL

1101.1 Scope. The provisions of this chapter shall govern the materials, design, construction and installation of storm drainage.

❖ Chapter 11 is one of the few chapters in the code that contains requirements that are applicable beyond the building itself. The main concern is protection of the building from the effects of rainwater; therefore, the regulations apply to roofs and paved areas around the building. Once driveways and walkways are paved, the ground does not have the ability to absorb the rainwater. A storm water system is necessary to control the runoff from roofs and paved areas.

1101.2 Where required. All roofs, paved areas, yards, courts and courtyards shall drain into a separate storm *sewer* system, or a combined *sewer* system, or to an *approved* place of disposal. For one- and two-family dwellings, and where *approved*, storm water is permitted to discharge onto flat areas, such as streets or lawns, provided that the storm water flows away from the building.

❖ The intent of Section 1101.2 is simply to require the safe control and disposal of rainwater that has collected on roofs, paved areas, yards, courts and courtyards. Construction and the resulting subsequent change to the natural grade can greatly affect the existing runoff of rainwater from the undeveloped area. A storm drainage system is designed (as a minimum level of performance) to maintain the status quo of the area during a rainfall after completion of the construction.

The existing exposed soil, trees, shrubbery and grass absorb a percentage of rainwater, while the remaining unabsorbed rainwater runs off to low-lying areas, tributary streams, rivers or ponds. When a newly constructed building or paved area replaces the existing exposed ground surface, rainwater absorption is prevented; thus, the runoff is now 100 percent.

The storm drainage system must drain the rainwater to an acceptable point of disposal, such as a retention basin, river, stream, pond, lake or public storm sewer system. The discharge could be to the surrounding land if it does not adversely affect the building or any adjacent properties.

Storm water must be directed away from the building to prevent damage to the foundation system or leakage into the building.

Buildings with pitched roofs must also control and divert storm water to an approved location. Gutters and downspouts or other approved disposal methods must prevent rainwater from washing soil away from the building foundation and building up around and increasing ground water pressure on basement walls. These methods also minimize the amount of water running down exterior walls and causing leaks at windows and wall penetrations.

1101.3 Prohibited drainage. Storm water shall not be drained into sewers intended for sewage only.

❖ A sanitary sewer system will become overloaded during a rainstorm if storm water is discharged into it. Typically, sewage treatment plants are not designed to take the surcharge imposed on them from additional storm water discharge. Discharging storm water into the sanitary system could result in either flooding and backups or the polluting of nearby waterways that receive inadequately treated sanitary discharge. The intent is to discharge into sanitary sewers only that which requires treatment.

The only exception to the prohibition stated in this section occurs where combined sanitary and storm sewers exist as they do in many older cities (see Commentary, Section 1108), and the code official approves such discharge connection.

1101.4 Tests. The conductors and the building *storm drain* shall be tested in accordance with Section 312.

❖ This section references Section 312, which establishes testing and inspection requirements for the completed, installed storm drainage system. The required testing is the same as would be required for the sanitary drainage system. When the storm system handles the discharge from a large roof area, and the storm discharge piping is contained on the interior of the building, the potential for vast building damage exists if the storm drainage system fails. The volume of water, pressures and dynamic forces that are possible in these systems oftentimes greatly exceed that for sanitary drainage systems. Unlike sanitary systems, storm drainage systems can be designed for full-bore flow, meaning that the storm system would have to function under more severe service conditions.

1101.5 Change in size. The size of a drainage pipe shall not be reduced in the direction of flow.

❖ One of the fundamental requirements of any form of drainage system is that reduction of the piping size is not permitted in the direction of drainage flow. This is also the case with storm drainage systems. A size reduction creates an obstruction to the flow of discharge, possibly resulting in a backup, an interruption of service in the drainage system or a stoppage occurring in the piping.

1101.6 Fittings and connections. All connections and changes in direction of the storm drainage system shall be made with *approved* drainage-type fittings in accordance with Table 706.3. The fittings shall not obstruct or retard flow in the system.

❖ This section requires the use of approved drainage-type fittings in the storm drainage system to make necessary directional changes. Although the flow through the storm drainage system is significantly different from that of the sanitary drainage system, a smooth transition of flow is still necessary (see Commentary Section 706.3).

1101.7 Roof design. Roofs shall be designed for the maximum possible depth of water that will pond thereon as determined by the relative levels of roof deck and overflow weirs, scuppers, edges or serviceable drains in combination with the deflected structural elements. In determining the maximum possible depth of water, all primary roof drainage means shall be assumed to be blocked.

❖ This section contains structural design requirements for the roof of a building. Basically, the roof structure must be capable of supporting the maximum ponding of water that will occur when the primary roof drainage means are blocked. In Section 1107, the code requires the installation of a secondary (emergency) roof drainage system to limit the amount of rainwater that could be retained on top of the roof. This also limits the potential increased load to be structurally supported by the roof. Refer to Section 1107 to determine how elements such as scuppers, overflow weirs, serviceable drains and other similar openings, along with the roof edge itself, are addressed in the code.

Although not totally defined, the primary storm drainage system is generally interpreted to include roof drains, leaders, conductors and horizontal storm drains within the structure. Although not immune to failure, the storm sewer system is usually not considered to be part of the building's primary roof drainage means. On large roofs with multiple, independently drained areas or bays, the primary drainage means for each such area could be considered separately.

The maximum depth of the ponding water should be measured to the highest flow depth allowed by the design of the secondary (emergency) roof drainage system. The volume of flow into a scupper or roof drain is a function of water depth at the scupper or roof drain. A scupper or roof drain, for example, will typically not reach its maximum flow capacity until the water has reached a certain depth above the weir of the drain. This means that water may necessarily pond on a roof in order for the drain to reach design flow conditions for the drainage system. Commentary Table 1101.7 shows several roof drains and the depth of water (hydraulic heads) above the drain inlet.

Consideration must be given to roof structure deflection, which will increase the volume of water held by ponding. An improperly designed roof could collapse under the weight of the water held on it (see Commentary Figure 1101.7). The roof structure in Commentary Figure 1101.7 must be designed to support the weight of water at the maximum possible depth. The maximum depth is determined assuming that the primary drainage system is blocked and the water must rise above the scuppers to a height sufficient to cause flow. Because of the many variables, each roof drainage system design must be evaluated individually, taking into consideration such factors as roof drain capacity at ponding depth, roof type, roof configurations, mains and sewer sizes, drain capacities, locations of trees, climate and probability of system failures.

The secondary (emergency) roof drainage system must be completely independent of the primary roof drainage system so that a failure in the primary system does not influence the proper functioning of the secondary system. The secondary (emergency) roof drainage system does not conduct storm water under normal conditions; therefore, such systems are generally not connected to a storm sewer system and simply discharge to grade. This discharge must occur in a location above grade that will be noticed by either the users or service personnel of the building. The discharge point becomes a visual indication that the primary system has failed or become blocked.

Table 1101.7
FLOW RATE, IN GALLONS PER MINUTE, OF VARIOUS ROOF DRAINS AT VARIOUS WATER DEPTHS AT DRAIN INLETS (INCHES)

DRAINAGE SYSTEM	FLOW RATE (gpm) Depth of water above drain inlet (hydraulic head) (inches)									
	1	2	2.5	3	3.5	4	4.5	5	7	8
4-inch-diameter drain	80	170	180							
6-inch-diameter drain	100	190	270	380	540					
8-inch-diameter drain	125	230	340	560	850	1,100	1,170			
6-inch-wide, open-top scupper	18	50	*	90	*	140	*	194	321	303
24-inch-wide, open-top scupper	72	200	*	360	*	560	*	776	1,284	1,572
6-inch-wide, 4-inch-high, closed-top scupper	18	50	*	90	*	140	*	177	231	253
24-inch-wide, 4-inch-high, closed-top scupper	72	200	*	360	*	560	*	708	924	1,012
6-inch-wide, 6-inch-high, closed-top scupper	18	50	*	90	*	140	*	194	303	343
24-inch-wide, 6-inch-high, closed-top scupper	72	200	*	360	*	560	*	776	1,212	1,372

For SI: 1 inch = 25.4 mm, 1 gallon per minute = 3.785 L/m.
Source: Factory Mutual Engineering Corp. Loss Prevention Data 1-54.

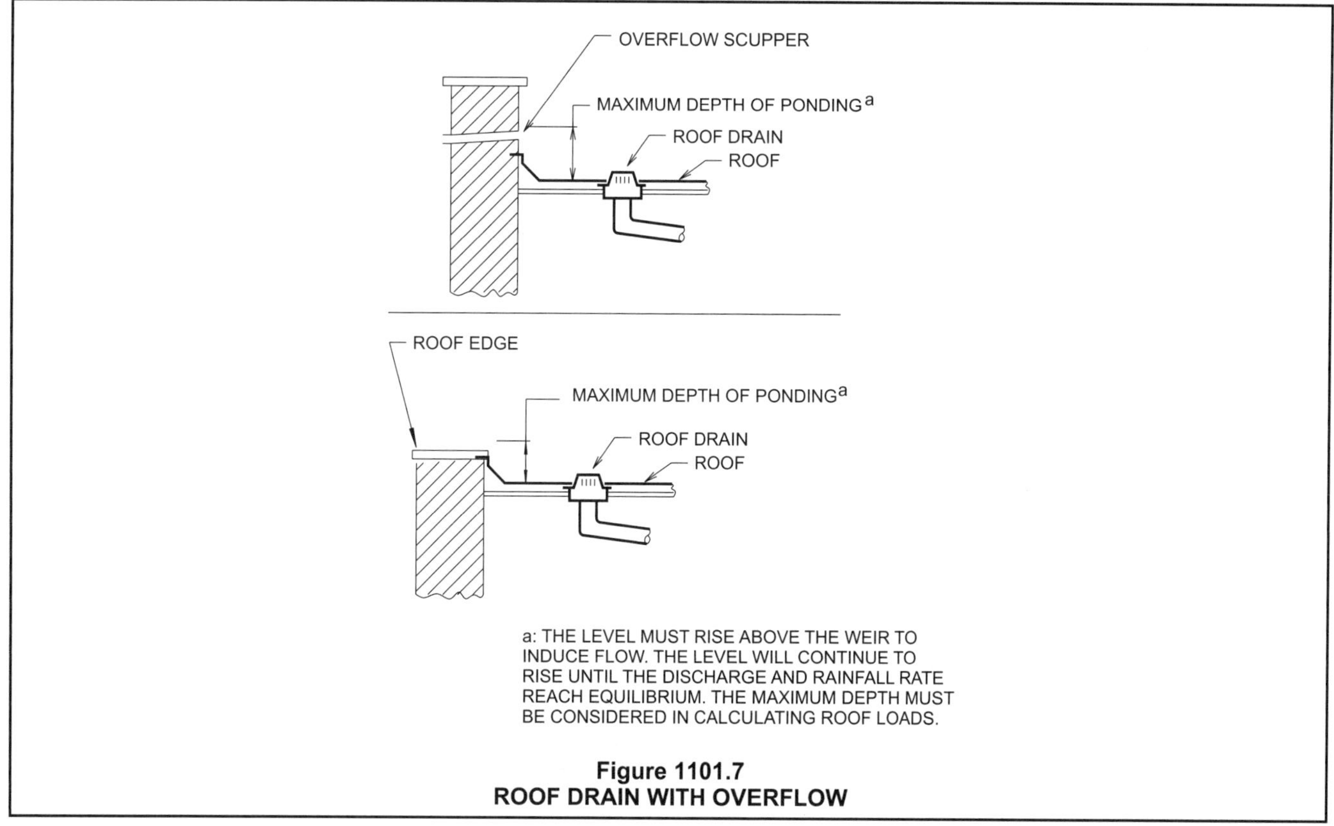
OVERFLOW SCUPPER
MAXIMUM DEPTH OF PONDING a
ROOF DRAIN
ROOF
ROOF EDGE
MAXIMUM DEPTH OF PONDING a
ROOF DRAIN
ROOF
a: THE LEVEL MUST RISE ABOVE THE WEIR TO INDUCE FLOW. THE LEVEL WILL CONTINUE TO RISE UNTIL THE DISCHARGE AND RAINFALL RATE REACH EQUILIBRIUM. THE MAXIMUM DEPTH MUST BE CONSIDERED IN CALCULATING ROOF LOADS.

Figure 1101.7
ROOF DRAIN WITH OVERFLOW

1101.8 Cleanouts required. Cleanouts shall be installed in the storm drainage system and shall comply with the provisions of this code for sanitary drainage pipe cleanouts.

Exception: Subsurface drainage system.

❖ The cleanout requirements apply to drainage systems without making a distinction between sanitary and storm with the exception of subsurface drainage systems. These systems serve to drain water away from foundations and use perforated pipe. A storm drain is susceptible to stoppage in the same way as a sanitary system; hence, cleanouts are necessary as required by Section 708.

1101.9 Backwater valves. Storm drainage systems shall be provided with backwater valves as required for sanitary drainage systems in accordance with Section 715.

❖ Although backwater valves are typically associated with the installation of the sanitary sewer and the connection between the building and the sanitary sewer, they are also installed on storm systems connected to combination storm and sanitary sewers. There is often a danger of backflow of storm water into a building using combination storm and sanitary drainage systems where the public sewer becomes overloaded, clogged or surcharged. To prevent damage from backflow, a backwater valve should be installed in accordance with Section 715 [see Commentary Figure 1111.1(1)].

SECTION 1102 MATERIALS

1102.1 General. The materials and methods utilized for the construction and installation of storm drainage systems shall comply with this section and the applicable provisions of Chapter 7.

❖ This section requires the use of approved sanitary drainage-type piping, fittings and any other components in the storm drainage system. Although the flow through the storm drainage system is significantly different from that of the sanitary drainage system (see Commentary Section 1101.6), the need to provide a system with a smooth flow of discharge is still a valid concern; therefore, this section requires compliance with the requirements in Chapter 7.

1102.2 Inside storm drainage conductors. Inside storm drainage conductors installed above ground shall conform to one of the standards listed in Table 702.1.

❖ Any of the materials acceptable for sanitary drainage are allowable for storm drainage; hence, the reference to Table 702.1, which is the table for above-ground sanitary drainage and vent pipe.

1102.3 Underground building storm drain pipe. Underground building *storm drain* pipe shall conform to one of the standards listed in Table 702.2.

❖ Similar to above-ground piping, as discussed in Section 1102.2, any of the piping materials listed in Table 702.2 for underground building drainage and vent pipe are acceptable for use as building storm drain piping underground.

1102.4 Building storm sewer pipe. Building storm *sewer* pipe shall conform to one of the standards listed in Table 1102.4.

❖ The referenced table specifies acceptable piping materials for storm sewers (see Commentary Table 1102.4).

TABLE 1102.4 BUILDING STORM SEWER PIPE

MATERIAL	STANDARD
Acrylonitrile butadiene styrene (ABS) plastic pipe	ASTM D 2661; ASTM D 2751; ASTM F 628; CAN/CSA B181.1; CAN/CSA B182.1
Asbestos-cement pipe	ASTM C 428
Cast-iron pipe	ASTM A 74; ASTM A 888; CISPI 301
Concrete pipe	ASTM C 14; ASTM C 76; CAN/CSA A257.1M; CAN/CSA A257.2M
Copper or copper-alloy tubing (Type K, L, M or DWV)	ASTM B 75; ASTM B 88; ASTM B 251; ASTM B 306
Polyethylene (PE) plastic pipe	ASTM F 2306/F 2306M
Polyvinyl chloride (PVC) plastic pipe (Type DWV, SDR26, SDR35, SDR41, PS50 or PS100)	ASTM D 2665; ASTM D 3034; ASTM F 891; CSA B182.4; CSA B181.2; CSA B182.2
Vitrified clay pipe	ASTM C 4; ASTM C 700
Stainless steel drainage systems, Type 316L	ASME A112.3.1

❖ The table identifies the materials that are permitted and the standards with which these materials must comply.

ASTM D 2751 is the referenced standard for typical acrylonitrile butadiene styrene (ABS) sewer pipe and fittings. The pipe may be produced from ABS material with a cell classification of 1-3-3, 3-2-2 or 2-2-3. This pipe is produced in standard dimension ratios (SDRs) of 23.5, 35 and 42. SDR is the ratio between the average outside diameter of the pipe and the minimum wall thickness. The other standards referenced are for Schedule 40 plastic pipe fittings and Schedule 40 drain, waste and vent (DWV) pipe with a cellular core.

Asbestos-cement pipe is rated by type of material and class. Classes are based on the minimum crush load in pounds-force per foot. The different classes include 1,500, 2,400, 3,300, 4,000, 5,000, 6,000 and 7,000. There are two types of pipe material: Type I and Type II. Both types correspond to the chemical requirements for the material. Asbestos cement is made from portland cement and asbestos fibers. The pipe is produced in diameters of 4 through 42 inches (102 through 1067 mm).

Cast-iron pipe is available in two pipe styles: hubbed

and hubless. Hubbed cast-iron pipe is the oldest type. It uses a hub at joints and connections; hubless pipe is joined using mechanical couplings. Hubbed cast iron is produced in two weights: service and extra heavy. Extra heavy is characterized by a thicker pipe wall.

Previously, other grades of cast-iron pipe were produced: medium weight and "victory pipe." Medium-weight pipe was a class between extra-heavy and service-weight pipe. A pipe referred to as "victory pipe" was manufactured during World War II—it was lighter and had a thinner pipe wall than service-weight pipe.

The wall thickness of hubless cast-iron pipe is approximately equal to service-weight hub pipe. Hubless pipe is produced both with and without a spigot bead. Cast-iron pipe is commonly referred to as "soil pipe" or "cast-iron soil pipe." The term "soil pipe" was used when cast iron was essentially the only material used for soil drainage pipes. The term is still in use today when differentiating between cast-iron pipe used for drainage and cast-iron pipe used for water. Cast iron is typically produced in 5-foot (1524 mm) and 10-foot (3048 mm) lengths. The pipe is marked with the manufacturer's name and designated as hubless, extra heavy (XH) or service (SV).

Concrete pipe is made from portland cement, blast-furnace slag cement or portland-pozzolan cement and aggregate mixture. The pipe is produced both reinforced and nonreinforced.

ASTM C 14 and CAN/CSA A257.1M are the referenced standards for unnreinforced concrete pipe. ASTM C 14 concrete pipe is available in three classes: 1, 2 or 3. The pipe classification relates to the concrete strength based on diameter. Class 1 has the lowest strength requirements and Class 3 has the highest. The pipe is manufactured in diameters of 4 through 36 inches (102 through 914 mm).

ASTM C 76 and CAN/CSA A257.2M are the referenced standards for reinforced concrete pipe, which is strengthened with either steel wire, hot rod or welded wire fabric. ASTM C 76 pipe is produced in five classes: 1, 2, 3, 4 or 5. The class determines the strength of the concrete, pipe and reinforcement geometry. The pipe is manufactured in sizes 12 inches (305 mm) and larger.

Copper or copper-alloy tubing (Type K, L, M or DWV) is manufactured in two different tempers: drawn (called hard copper with the designation "H") and annealed (called soft copper with the designation "O"). Drawn copper tubing typically comes in straight lengths, where annealed copper tubing typically comes in both straight lengths and coils. Annealed and tempered drawn copper can be formed by bending. Bending-tempered copper tubing is identified on the pipe with the designation "BT."

The four types of copper tubing are identified with a colored marking system for ease of identification, as follows:

- Green—Type K copper tubing
- Blue—Type L copper tubing
- Red—Type M copper tubing
- Yellow—Type DWV copper tubing

Corrugated double wall polyethylene (PE) pipe (ASTM F 2306) for storm water conveyance has been in widespread use by utilities and road building contractors for over 20 years. Single wall corrugated PE pipe has been approved for use in subsoil drain systems for many years but for reasons unexplained, the double wall version was never proposed for inclusion in the list of approved materials for building storm drains. The double wall version of PE pipe covered by the referenced ASTM number has a smooth, uniform bore with a corrugated exterior that imparts strength to resist crushing.

Polyvinyl chloride (PVC) plastic pipe (Type DWV, SDR 26, SDR 35, SDR 41, PS 50 or PS 100) is regulated by ASTM F 891 which specifies that sewer grade pipe be designated by a "PS" in addition to a Schedule 40 designation (Type DWV). PS indicates pipe stiffness. The grades of PVC plastic pipes used in sewers are PS 50 and 100. The numerical designation indicates the pipe stiffness in pounds per square inch (psi). The higher value indicates plastic pipe with a thicker pipe wall.

ASTM D 2665 and ASTM D 3034 are the standards regulating sewer-grade PVC pipe and fittings. The materials may be used interchangeably. Both pipes are produced in SDRs of 35 and 41, with a hub end. The PVC material used to produce the pipe is classified as 12454-B, 12454-C or 13364-B. The pipe is produced in diameters of 4 through 15 inches (102 mm through 381 mm). CSA B182.2 is the referenced standard for SDR 28, 35 and 41 PVC sewer pipe and fittings, and CSA B182.4 is the referenced standard for PVC ribbed sewer pipe and fittings.

Vitrified clay pipe is surface clay, fire clay, shale or a combination of the materials, and is classified as standard strength, extra strength, extra quality or heavy duty. The classification determines the strength of the pipe. It is produced in diameters of 3 inches (76 mm) and larger. Each length of pipe is marked with the manufacturer's name and the classification of the pipe.

Type 316L stainless steel provides excellent corrosion resistance when installed either above or below ground in a storm sewer system. Stainless steel pipe and fittings have been used for years worldwide to convey sanitary wastes because of the strength, durability and corrosion resistance of the material. The additional cost of installing stainless steel piping can often be offset by the piping's longer expected life.

1102.5 Subsoil drain pipe. Subsoil drains shall be open-jointed, horizontally split or perforated pipe conforming to one of the standards listed in Table 1102.5.

❖ The referenced table specifies acceptable piping materials for subsoil drainage systems (see Commentary Table 1102.5).

TABLE 1102.5
SUBSOIL DRAIN PIPE

MATERIAL	STANDARD
Asbestos-cement pipe	ASTM C 508
Cast-iron pipe	ASTM A 74; ASTM A 888; CISPI 301
Polyethylene (PE) plastic pipe	ASTM F 405; CAN/CSA B182.1; CSA B182.6; CSA B182.8
Polyvinyl chloride (PVC) Plastic pipe (type sewer pipe, PS25, PS50 or PS100)	ASTM D 2729; ASTM F 891; CSA B182.2; CAN/CSA B182.4
Stainless steel drainage systems, Type 316L	ASME A 112.3.1
Vitrified clay pipe	ASTM C 4; ASTM C 700

❖ The materials not described herein are described in detail in the commentary to Table 1102.4. Polyethylene (PE) plastic pipe: ASTM F 405 is the referenced standard for corrugated pipe and PE tubing. The pipe is available with perforations and is classified as either standard or heavy-duty tubing.

Polyvinyl chloride (PVC) plastic pipe (type sewer pipe, PS 25, PS 50 or PS 100): ASTM D 2729 regulates thin-wall drain pipe, which is available as both perforated and solid wall pipe.

1102.6 Roof drains. Roof drains shall conform to ASME A112.21.2M or ASME A112.3.1.

❖ ASME A112.21.2M regulates the design and construction of roof drains. Roof drains must have either a dome or grate covering the opening. This helps restrict leaves and debris from entering the storm drainage piping and creating a stoppage. Because roof drains are exposed to extreme temperatures, it is common practice to install an expansion joint between the roof drain and the storm drainage pipe.

ASME A112.21.2M permits roof drains to be constructed only of cast iron, red brass or leaded nickel bronze. ASME A112.3.1 provides details for Types 304 and 316L stainless steel roof drains. Roof drains constructed of other materials, such as ABS and PVC, must be evaluated and approved as alternative materials by the code official.

The last edition of the standard ASME A112.21.2 is 1983. A reorganization within ASME resulted in the creation of a new roof drain standard, ASME A112.6.4. The ASME website description for this new standard states "A112.6.4-2003 has incorporated into it A112.21.2M-1983 on Roof Drains." While the ASME A112.6.4-2003 standard does incorporate all the information and requirements that were in the A112.21.2 standard, the new standard now includes plastic roof drain materials and has requirements and testing for strength of the grates and top rims of all material types of roof drains. However, because the A112.6.4 standard is not listed in the code, use of material made to this standard will require alternate approval in accordance with Section 105.2.

1102.7 Fittings. Pipe fittings shall be *approved* for installation with the piping material installed, and shall conform to the respective pipe standards or one of the standards listed in Table 1102.7. The fittings shall not have ledges, shoulders or reductions capable of retarding or obstructing flow in the piping. Threaded drainage pipe fittings shall be of the recessed drainage type.

❖ Each fitting is designed to be installed in a particular system with a given material or combination of materials. Drainage pattern fittings must be installed in drainage systems. Vent fittings are limited to the venting system. Fittings are also available that could be used in any type of plumbing system.

Fittings must be the same material as, or compatible with, the pipe. This prevents any chemical or corrosive action between dissimilar materials.

Many pipe standards also include provisions regulating fittings. There are, however, a number of standards that strictly regulate pipe fittings.

TABLE 1102.7
PIPE FITTINGS

MATERIAL	STANDARD
Acrylonitrile butadiene styrene (ABS) plastic	ASTM D 2661; ASTM D 3311; CSA B181.1
Cast-iron	ASME B16.4; ASME B16.12; ASTM A 888; CISPI 301; ASTM A 74
Coextruded composite ABS sewer and drain DR-PS in PS35, PS50, PS100, PS140, PS200	ASTM D 2751
Coextruded composite ABS DWV Schedule 40 IPS pipe (solid or cellular core)	ASTM D 2661; ASTM D 3311; ASTM F 628
Coextruded composite PVC DWV Schedule 40 IPS-DR, PS140, PS200 (solid or cellular core)	ASTM D 2665 ASTM D 3311; ASTM F 891
Coextruded composite PVC sewer and drain DR-PS in PS35, PS50, PS100, PS140, PS200	ASTM D 3034
Copper or copper alloy	ASME B16.15; ASME B16.18; ASME B16.22; ASME B16.23; ASME B16.26; ASME B16.29
Gray iron and ductile iron	AWWA C110
Malleable iron	ASME B16.3
Plastic, general	ASTM F 409
Polyethylene (PE) plastic pipe	ASTM F 2306/F 2306M
Polyvinyl chloride (PVC) plastic	ASTM D 2665; ASTM D 3311; ASTM F 1866
Steel	ASME B16.9; ASME B16.11; ASME B16.28
Stainless steel drainage systems, Type 316L	ASME A112.3.1

❖ The table specifies acceptable materials for pipe fittings, along with the standard with which the various fittings must comply.

SECTION 1103 TRAPS

1103.1 Main trap. Leaders and storm drains connected to a combined *sewer* shall be trapped. Individual storm water traps shall be installed on the storm water drain *branch* serving each conductor, or a single trap shall be installed in the main *storm drain* just before its connection with the combined *building sewer* or the *public sewer*.

❖ This section regulates the installation of traps on combination sanitary and storm sewer systems (see Commentary Figure 1103.1).

The trap on the storm drain is intended to prevent sewer gas from entering the system and is required only where the storm drain discharges into the sanitary sewer system. Unlike a sanitary system, a storm drainage system is not designed for any precautions against sewer gas. Any storm drain connection to a combination sanitary and storm sewer must be trapped to prevent the escape of sewer gases. For example, an air intake opening for an air-handling unit may be located next to a roof drain. A properly located trap prevents sewer gas from escaping through a roof drain and entering the outdoor air intake.

1103.2 Material. Storm water traps shall be of the same material as the piping system to which they are attached.

❖ This section simply requires that the material for the trap on the storm sewer line connecting to the combination sanitary and storm sewer be the same material as the combination sanitary and storm sewer.

1103.3 Size. Traps for individual conductors shall be the same size as the horizontal drain to which they are connected.

❖ This section requires that the trap be the same size as the horizontal drain pipe to which it discharges to prevent any obstruction of flow in the piping.

1103.4 Cleanout. An accessible cleanout shall be installed on the building side of the trap.

❖ This section requires the installation of an accessible cleanout and also states where it should be located.

SECTION 1104 CONDUCTORS AND CONNECTIONS

1104.1 Prohibited use. Conductor pipes shall not be used as soil, waste or vent pipes, and soil, waste or vent pipes shall not be used as conductors.

❖ This section regulates the installation of conductors, leaders and connections.

A sanitary drainage stack and storm drainage conductors are designed differently and serve different purposes. They cannot be interconnected and the components of the two different systems cannot be interchanged (see Commentary Sections 1101.3 and 1104.3).

1104.2 Combining storm with sanitary drainage. The sanitary and storm drainage systems of a structure shall be entirely separate except where combined *sewer* systems are utilized. Where a combined *sewer* is utilized, the building *storm drain* shall be connected in the same horizontal plane through a single-wye fitting to the combined *sewer* at least 10 feet (3048 mm) downstream from any soil *stack*.

❖ This section emphasizes that separate storm and sanitary systems are required. These requirements are consistent with Sections 1101.3 and 1104.1.

In jurisdictions where a combined sewer is used, this section identifies the required method of connecting the sanitary and storm drainage pipes. The requirement for connecting on the same horizontal plane with a 10-foot (3048 mm) separation reduces the interference and pressure differentials created by the storm drainage system during a heavy rainfall.

If the jurisdiction has a combined system (sanitary and storm), each connection of a storm conductor must be through a single-wye pattern fitting located in accordance with this section. The fitting properly directs the full flow from the storm drainage system.

1104.3 Floor drains. Floor drains shall not be connected to a *storm drain*.

❖ To prevent contamination of the storm water discharge with chemicals, waste, sewage, etc., the code prohibits the connection of floor drains to a storm drain. A

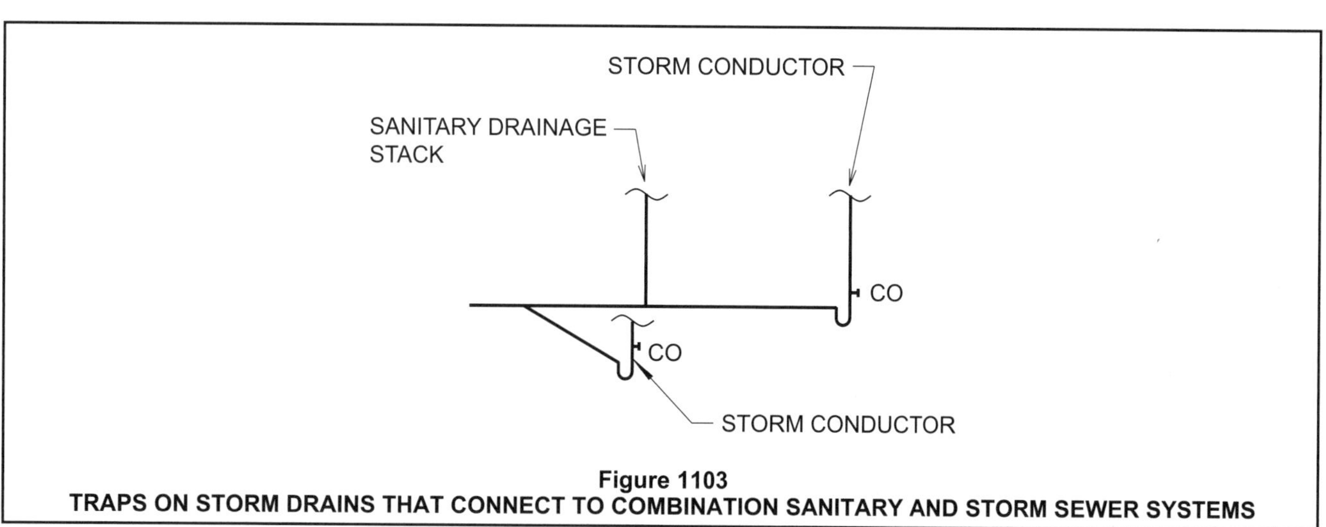

Figure 1103
TRAPS ON STORM DRAINS THAT CONNECT TO COMBINATION SANITARY AND STORM SEWER SYSTEMS

floor drain connected to a storm drainage system invites the discharge of sanitary waste into the storm system, which will create an environmental hazard, considering that storm drainage is discharged to the environment without treatment. Another concern is that under flow conditions, a floor drain connection will act as a relief opening, thereby allowing water flow into the building.

Areaway drains serving subsurface spaces outside of the building, such as window wells or basement or cellar entrance wells, are typically not considered to be floor drains when applying this section. Additionally, where approved by the code official, if a parking garage will function only as a parking lot, with no increased possibility that repair and service operations will occur, floor drains in the garage could be connected to the storm drainage system the same as for a parking lot.

SECTION 1105 ROOF DRAINS

1105.1 Strainers. Roof drains shall have strainers extending not less than 4 inches (102 mm) above the surface of the roof immediately adjacent to the roof drain. Strainers shall have an available inlet area, above roof level, of not less than one and one-half times the area of the conductor or leader to which the drain is connected.

❖ This section requires strainers on roof drains. A strainer, sometimes called a "beehive strainer," helps prevent leaves, debris and roof ballast materials from entering the storm drain. The beehive shape of the strainer helps prevent the entire opening of the strainer from being obstructed when there is an accumulation of debris. This section regulates how far the strainer should extend above the roof. The requirement for the inlet area to be at least one and one-half times (150 percent) the area of the pipe connected to the drain provides some assurance that the roof drain will continue to flow at full capacity when the strainer is partially obstructed.

1105.2 Flat decks. Roof drain strainers for use on sun decks, parking decks and similar areas that are normally serviced and maintained shall comply with Section 1105.1 or shall be of the flat-surface type, installed level with the deck, with an available inlet area not less than two times the area of the conductor or leader to which the drain is connected.

❖ Where a roof deck is used for parking, recreation and other purposes, the roof drain will be flat. It is assumed that any blockage would be noticed by individuals frequenting the roof (see commentary, Section 1105.1). If the designer elects to use this flat-surface-type drain, the amount of available inlet area must increase to a minimum of two times (200 percent) the area of the attached conductor or leader.

1105.3 Roof drain flashings. The connection between roofs and roof drains which pass through the roof and into the interior of the building shall be made water-tight by the use of *approved* flashing material.

❖ The roof drain must be installed with a proper flashing to prevent leakage between the roof deck and the roof drain. A flashing ring is used to attach the roof membrane to the roof drain body to maintain a water-tight roof.

SECTION 1106 SIZE OF CONDUCTORS, LEADERS AND STORM DRAINS

1106.1 General. The size of the vertical conductors and leaders, building storm drains, building storm sewers, and any horizontal branches of such drains or sewers shall be based on the 100-year hourly rainfall rate indicated in Figure 1106.1 or on other rainfall rates determined from *approved* local weather data.

❖ This section regulates the size of storm drainage systems as a function of a 60-minute duration, 100-year rainfall rate on the roof or paved area to be drained. Selection of the correct rainfall rate is critical for the proper sizing of the storm drain system.

FIGURE 1106.1. See page 11-9.

❖ Figure 1106.1 spans five consecutive pages of the code and is broken into major regions of the country. This figure indicates the rainfall rates for a storm of 1-hour duration and a 100-year return period. Rainfall rates indicate the maximum rate of rainfall within a given period of time occurring every so many years. For example, Figure 1106.1 of the code indicates a rainfall rate of 2.1 inches (53 mm) per hour for Burlington, Vermont. Thus, it is predicted that it will rain 2.1 inches (53 mm) within 1 hour once every 100 years.

The rainfall rates are calculated by a statistical analysis of weather records. Because the statistics are based on previous or historical weather conditions, it is conceivable to have two 100-year storms in one week's time. The probability of this occurring, however, is very low.

1106.2 Vertical conductors and leaders. Vertical conductors and leaders shall be sized for the maximum projected roof area, in accordance with Tables 1106.2(1) and 1106.2(2).

❖ Table 1106.2(1) applies to circular leaders and conductors. The sizing information, including the projected roof area, the diameter of the leader and the rainfall rates, listed in Table 1106.2(1), is based on the maximum probable capacity of a vertical pipe, which is approximately 29 percent of the pipe cross-sectional area.

Table 1106.2(2) applies to rectangular leaders and conductors. This table was developed based upon the area of an ellipse inscribed within the rectangular shape. The inscribed ellipse was converted into a circle of equivalent area and using the information in Table 1106.2(1), projected roof areas were determined for the rectangular shapes.

FIGURE 1106.1
100-YEAR, 1-HOUR RAINFALL (INCHES) EASTERN UNITED STATES

For SI: 1 inch = 25.4 mm.
Source: National Weather Service, National Oceanic and Atmospheric Administration, Washington D.C.

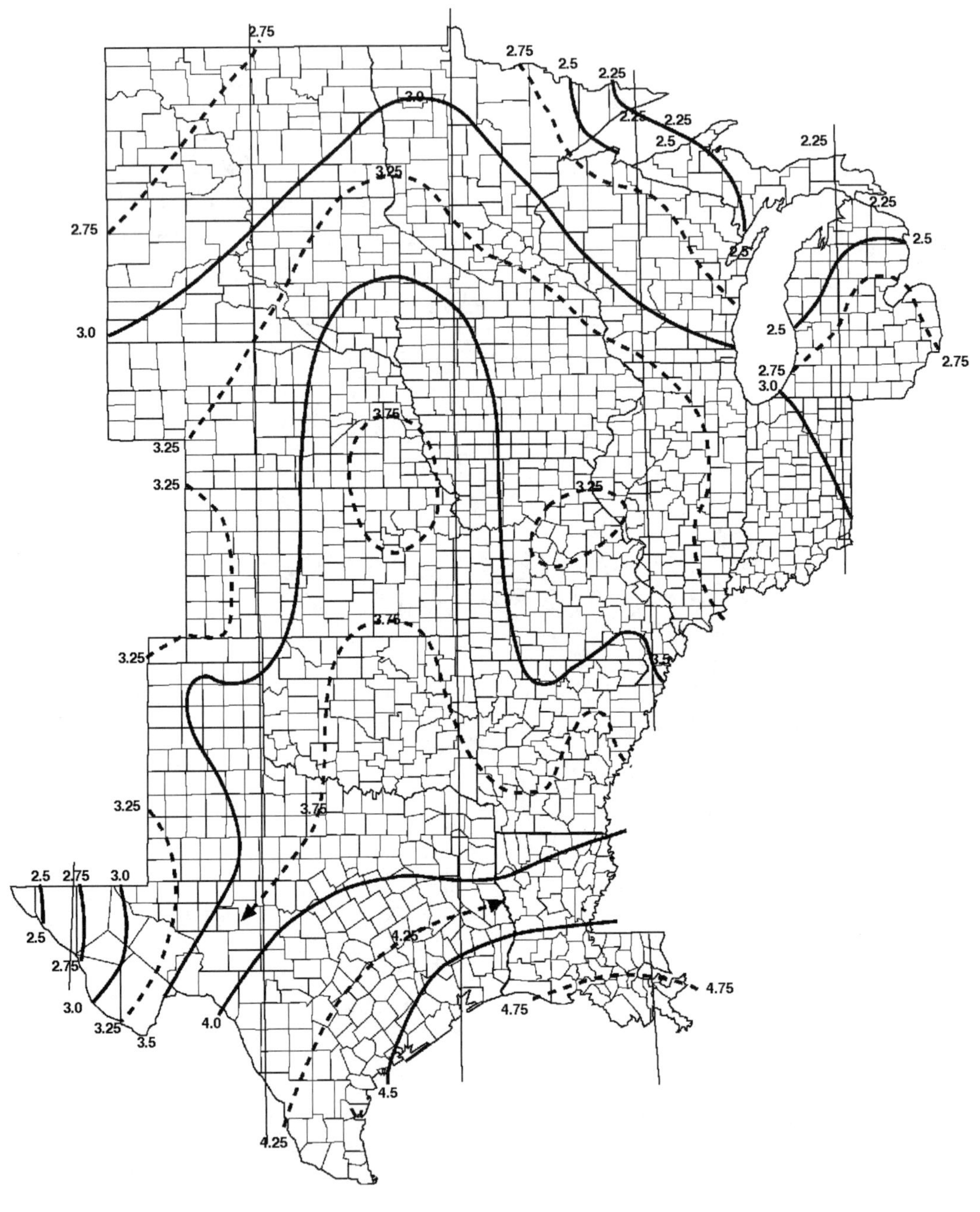

FIGURE 1106.1—continued
100-YEAR, 1-HOUR RAINFALL (INCHES) CENTRAL UNITED STATES

For SI: 1 inch = 25.4 mm.
Source: National Weather Service, National Oceanic and Atmospheric Administration, Washington D.C.

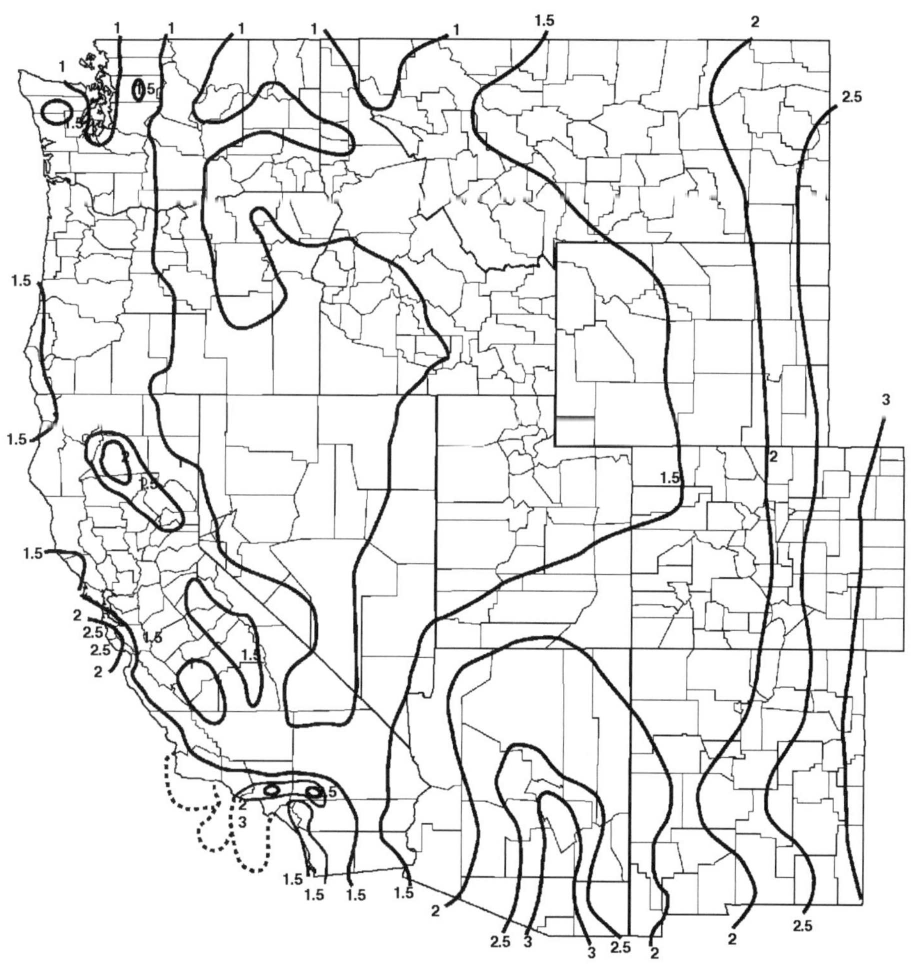

FIGURE 1106.1—continued
100-YEAR, 1-HOUR RAINFALL (INCHES) WESTERN UNITED STATES

For SI: 1 inch = 25.4 mm.
Source: National Weather Service, National Oceanic and Atmospheric Administration, Washington D.C.

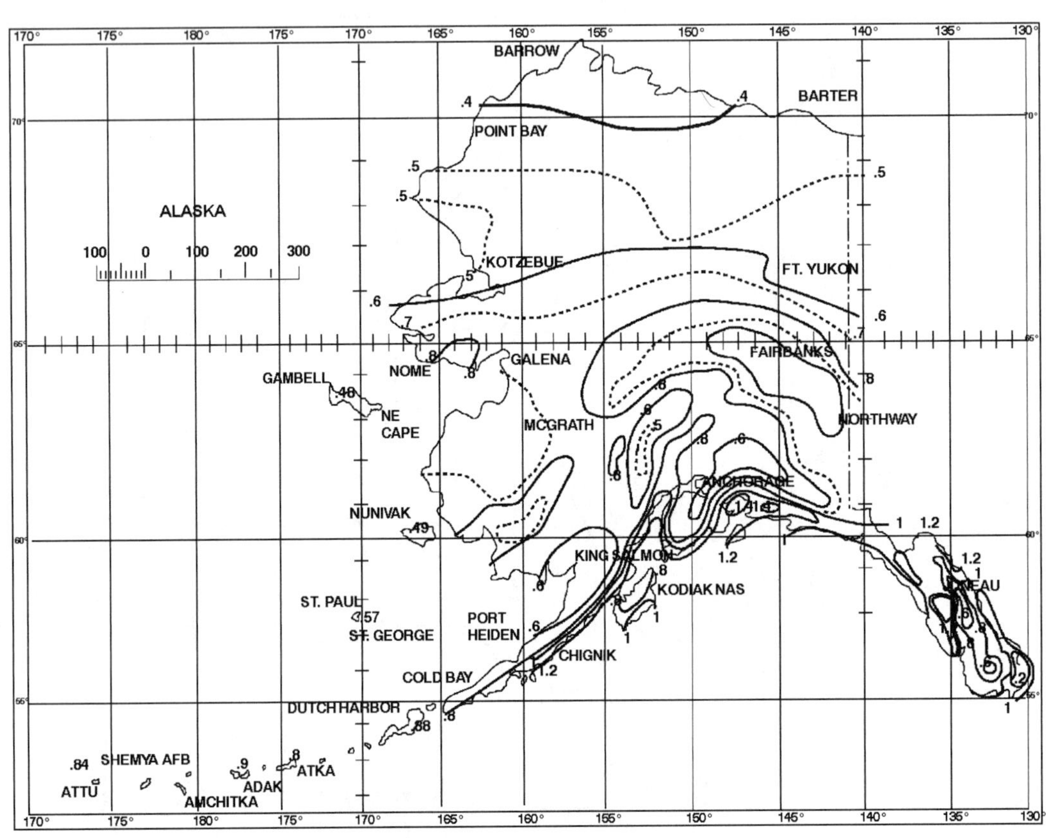

FIGURE 1106.1—continued
100-YEAR, 1-HOUR RAINFALL (INCHES) ALASKA

For SI: 1 inch = 25.4 mm.
Source: National Weather Service, National Oceanic and Atmospheric Administration, Washington D.C.

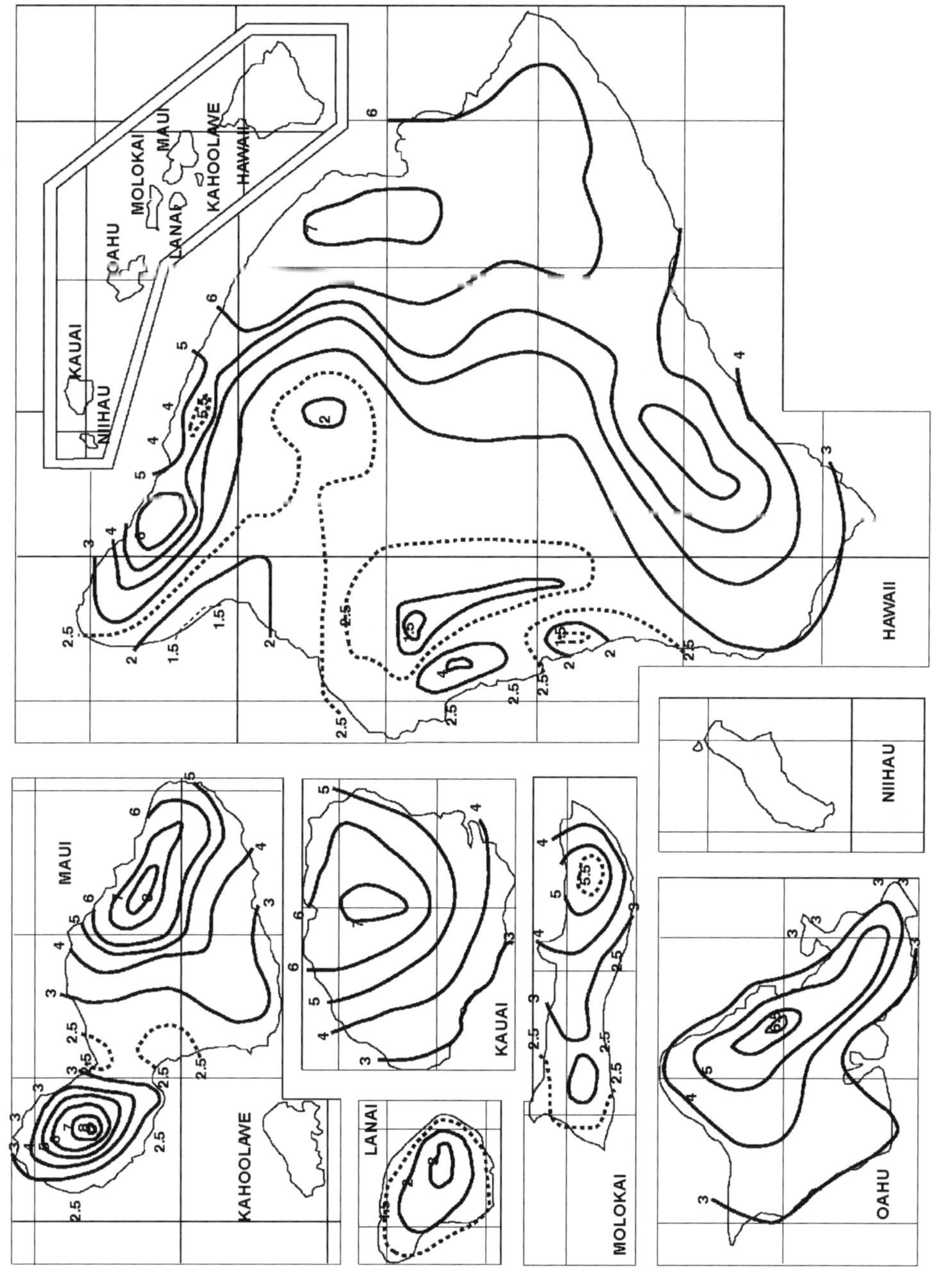

FIGURE 1106.1—continued
100-YEAR, 1-HOUR RAINFALL (INCHES) HAWAII

For SI: 1 inch = 25.4 mm.
Source: National Weather Service, National Oceanic and Atmospheric Administration, Washington D.C.

TABLE 1106.2(1). See below.

❖ To size a circular conductor or leader, the designer must determine the rainfall rate for the locality and the horizontally projected roof area served by each conductor or leader.

Example:

Building location: Oklahoma City, Oklahoma
Horizontal projected roof area: 10,000 square feet
Number of conductors: 2
Rainfall rate: 3.8 inches per hour (from Figure 1106.1 or Appendix B of the code)

For SI: 1 inch = 25.4 mm, 1 square foot = 0.0929 m^2.

Each conductor serves a horizontally projected roof area of 5,000 square feet. Although the table shows only whole numbers for the rainfall rates, it is acceptable to interpolate between rainfall rates to determine the maximum projected roof area. However, to be conservative in sizing, rainfall rates are typically rounded up to the next whole number. For this example, 3.8 inches per hour is rounded up to 4. Read down the "4" column until a value exceeding 5,000 square feet is found, then read to the left in the table to determine that each conductor must be a minimum of 5 inches in diameter.

TABLE 1106.2(2). See page 11-15.

❖ To size a rectangular conductor or leader, the designer must determine the rainfall rate for the locality and the horizontally projected roof area served by each conductor or leader.

Example:

Building location: Charleston, SC
Horizontal projected roof area: 50,000 square feet
Number of conductors: 5
Rainfall rate: 4.3 inches per hour (from Figure 1106.1 or Appendix B of the code)

For SI: 1 inch = 25.4 mm, 1 square foot = 0.0929 m^2.

Each conductor serves a horizontally projected roof area of 10,000 square feet. Although the table shows only whole numbers for the rainfall rates, it is acceptable to interpolate between rainfall rates to determine the maximum projected roof area. However, to be conservative in sizing, rainfall rates are typically rounded up to the next whole number. For this example, 4.3 inches per hour is rounded up to 5 so the allowable projected roof areas can be directly read from the table column. Read down the "5" column until a value exceeding 10,000 square feet is found, then read to the left in the table to determine that each conductor must have a minimum square dimension of $7^1/_2$ x $7^1/_2$ inches.

Where the size indicated by the table is not convenient for whatever reason, table Note b allows the designer to calculate an equivalent circular diameter so that other rectangular sizes can be selected. Using the example, the equivalent circular diameter of the $7^1/_2$ inch x $7^1/_2$ inch square section is;

$$D_e = (7.5 \times 7.5)^{½}$$
$$= 7.5 \text{ inches}$$

If one dimension of the rectangular conductor is required to be 9 inches, the equivalent circular diameter equation can be used to find the other required dimension, *L*.

$$D_e = 7.5 = (9 \times L)^{½}$$
$$L = (7.5^2)/9$$
$$= 6.25 \text{ inches}$$

1106.3 Building storm drains and sewers. The size of the building *storm drain*, building storm *sewer* and their horizontal branches having a slope of one-half unit or less vertical in 12 units horizontal (4-percent slope) shall be based on the maximum projected roof area in accordance with Table 1106.3. The minimum slope of horizontal branches shall be one-eighth unit vertical in 12 units horizontal (1-percent slope) unless otherwise *approved*.

❖ Unlike a sanitary drainage system, a horizontal storm drain or sewer is sized for a full-flow condition. The

TABLE 1106.2(1)
SIZE OF CIRCULAR VERTICAL CONDUCTORS AND LEADERS

DIAMETER OF LEADER (inches)[a]	HORIZONTALLY PROJECTED ROOF AREA (square feet)											
	Rainfall rate (inches per hour)											
	1	2	3	4	5	6	7	8	9	10	11	12
2	2,880	1,440	960	720	575	480	410	360	320	290	260	240
3	8,800	4,400	2,930	2,200	1,760	1,470	1,260	1,100	980	880	800	730
4	18,400	9,200	6,130	4,600	3,680	3,070	2,630	2,300	2,045	1,840	1,675	1,530
5	34,600	17,300	11,530	8,650	6,920	5,765	4,945	4,325	3,845	3,460	3,145	2,880
6	54,000	27,000	17,995	13,500	10,800	9,000	7,715	6,750	6,000	5,400	4,910	4,500
8	116,000	58,000	38,660	29,000	23,200	19,315	16,570	14,500	12,890	11,600	10,545	9,600

For SI: 1 inch = 25.4 mm, 1 square foot = 0.0929 m^2.

pipe may be filled to capacity under a worst-case condition, and is still sized for gravity flow. Under "normal" operation, the storm sewer is not full and functions more like a sanitary sewer.

When sizing a storm drainage system, the local rainfall rate must be used. Table 1106.3 and maps of the United States (which are shown in Figure 1106.1 of the code) indicate the rainfall rates for a storm of 1-hour duration and a 100-year return period (see Commentary Section 1106.1).

TABLE 1106.3. See page 11-16.

❖ The storm drainage system is sized for the horizontally projected roof area that it serves and the slope of the pipe. Knowing the roof area that discharges into a storm drain pipe and the rainfall rate, the designer determines the pipe size by reading the appropriate "slope table," finding the area that is at least as large as the horizontally projected roof area and then reading across to find the pipe diameter.

Note that a storm pipe of the same size has a greater capacity as its slope increases. The projected area in the tables can be converted to gallons per minute (gpm) flow by the following formula.

gpm = 0.0104 × R × A

where:

R = Rainfall rate (inches per hour).

A = Roof area (square feet) to be drained. Example: gpm

R = 1-inch rainfall rate.

A = 3,288-square-foot roof area to be drained. gpm = 0.0104 × 1 × 3,288 = 34 gpm

Example:

Building location: Burlington, Vermont

Building and parking lot area: 10,000 square feet (see Commentary Figure 1106.3)

Pitch of storm sewer: $^1/_4$ inch per foot

Rainfall rate: 2.1 inches per hour (determined in Appendix B of the code).

For SI: 1 inch = 25.4 mm, 1 square foot = 0.0929 m^2.

To determine the size of the horizontal storm drain shown in Commentary Figure 1106.3, locate the portion of the table that reflects the values for piping that is sloped at a rate of one-quarter unit vertical in 12 units horizontal. This is the center portion of the table. Next, look in the column under the rainfall rate of 3 (inches per hour) for the number for the horizontal projected roof area that was at least as large as the area you are designing for. Always round up to the next larger rate when selecting a rainfall rate to be conservative in the design. Next, read across the table to the left to find the diameter of the piping required for this application.

TABLE 1106.2(2)
SIZE OF RECTANGULAR VERTICAL CONDUCTORS AND LEADERS

DIMENSIONS OF COMMON LEADER SIZES width x length (inches)[a]	HORIZONTALLY PROJECTED ROOF AREA (square feet)											
	Rainfall rate (inches per hour)											
	1	2	3	4	5	6	7	8	9	10	11	12
$1^3/_4 \times 2^1/_2$	3,410	1,700	1,130	850	680	560	480	420	370	340	310	280
2 × 3	5,540	2,770	1,840	1,380	1,100	920	790	690	610	550	500	460
$2^3/_4 \times 4^1/_4$	12,830	6,410	4,270	3,200	2,560	2,130	1,830	1,600	1,420	1,280	1,160	1,060
3 × 4	13,210	6,600	4,400	3,300	2,640	2,200	1,880	1,650	1,460	1,320	1,200	1,100
$3^1/_2 \times 4$	15,900	7,950	5,300	3,970	3,180	2,650	2,270	1,980	1,760	1,590	1,440	1,320
$3^1/_2 \times 5$	21,310	10,650	7,100	5,320	4,260	3,550	3,040	2,660	2,360	2,130	1,930	1,770
$3^3/_4 \times 4^3/_4$	21,960	10,980	7,320	5,490	4,390	3,660	3,130	2,740	2,440	2,190	1,990	1,830
$3^3/_4 \times 5^1/_4$	25,520	12,760	8,500	6,380	5,100	4,250	3,640	3,190	2,830	2,550	2,320	2,120
$3^1/_2 \times 6$	27,790	13,890	9,260	6,940	5,550	4,630	3,970	3,470	3,080	2,770	2,520	2,310
4 × 6	32,980	16,490	10,990	8,240	6,590	5,490	4,710	4,120	3,660	3,290	2,990	2,740
$5^1/_2 \times 5^1/_2$	44,300	22,150	14,760	11,070	8,860	7,380	6,320	5,530	4,920	4,430	4,020	3,690
$7^1/_2 \times 7^1/_2$	100,500	50,250	33,500	25,120	20,100	16,750	14,350	12,560	11,160	10,050	9,130	8,370

a. Sizes indicated are nominal width × length of the opening for rectangular piping.

b. For shapes not included in this table, Equation 11-1 shall be used to determine the equivalent circular diameter, D_e, of rectangular piping for use in interpolation using the data from Table 1106.2(1).

$D_e = [\text{width} \times \text{length}]^{1/2}$ **(Equation 11-1)**

where:

D_e = equivalent circular diameter and D_e, width and length are in inches.

In this example, under the rainfall rate of 3, locate the area value of 10,066 (square feet), which then requires a pipe size of 6 (inches). Even if one failed to round the rainfall rate up, incorrectly using a rainfall rate of 2, the area value of 15,100 (square feet) still requires a pipe of 6 (inches).

1106.4 Vertical walls. In sizing roof drains and storm drainage piping, one-half of the area of any vertical wall that diverts rainwater to the roof shall be added to the projected roof area for inclusion in calculating the required size of vertical conductors, leaders and horizontal storm drainage piping.

❖ This section includes the requirement for the mandatory inclusion of half of the area of vertical walls, including parapet walls, that are adjacent to and above a roof area, into the calculation for projected horizontal roof area when sizing storm drainage components. This acknowledges that vertical walls can catch and divert rainwater onto the roof. This additional water must be accounted for in the sizing of the components that will catch and discharge this water (see commentary, Figure 1106.4).This additional water must also be accounted for in the design of a controlled flow system (see Commentary Section 1110).

The example in Commentary Figure 1106.4 illustrates how to compute the wall area that must be added to the lower projected roof area for sizing the storm drains.

1106.5 Parapet wall scupper location. Parapet wall roof drainage scupper and overflow scupper location shall comply with the requirements of the *International Building Code*.

❖ This section requires that the parapet wall scuppers be installed to satisfy the requirements of the IBC. This is a statement of mandatory coordination between two parts of the code; in this case, plumbing and building. There are other sections that require this coordination, but are not specifically referenced. Section 1101.7, for example, requires that a roof be designed structurally to support the maximum depth of rainwater that can pond on it, but does not require designing a roof in accordance with the requirements of the IBC, even though it would apply as well.

TABLE 1106.3
SIZE OF HORIZONTAL STORM DRAINGE PIPING

SIZE OF HORIZONTAL PIPING (inches)	HORIZONTALLY PROJECTED ROOF AREA (square feet)					
	Rainfall rate (inches per hour)					
	1	2	3	4	5	6
$^1/_8$ unit vertical in 12 units horizontal (1-percent slope)						
3	3,288	1,644	1,096	822	657	548
4	7,520	3,760	2,506	1,800	1,504	1,253
5	13,360	6,680	4,453	3,340	2,672	2,227
6	21,400	10,700	7,133	5,350	4,280	3,566
8	46,000	23,000	15,330	11,500	9,200	7,600
10	82,800	41,400	27,600	20,700	16,580	13,800
12	133,200	66,600	44,400	33,300	26,650	22,200
15	218,000	109,000	72,800	59,500	47,600	39,650
$^1/_4$ unit vertical in 12 units horizontal (2-percent slope)						
3	4,640	2,320	1,546	1,160	928	773
4	10,600	5,300	3,533	2,650	2,120	1,766
5	18,880	9,440	6,293	4,720	3,776	3,146
6	30,200	15,100	10,066	7,550	6,040	5,033
8	65,200	32,600	21,733	16,300	13,040	10,866
10	116,800	58,400	38,950	29,200	23,350	19,450
12	188,000	94,000	62,600	47,000	37,600	31,350
15	336,000	168,000	112,000	84,000	67,250	56,000
$^1/_2$ unit vertical in 12 units horizontal (4-percent slope)						
3	6,576	3,288	2,295	1,644	1,310	1,096
4	15,040	7,520	5,010	3,760	3,010	2,500
5	26,720	13,360	8,900	6,680	5,320	4,450
6	42,800	21,400	13,700	10,700	8,580	7,140
8	92,000	46,000	30,650	23,000	18,400	15,320
10	171,600	85,800	55,200	41,400	33,150	27,600
12	266,400	133,200	88,800	66,600	53,200	44,400
15	476,000	238,000	158,800	119,000	95,300	79,250

For SI: 1 inch = 25.4 mm, 1 square foot = 0.0929 m^2.

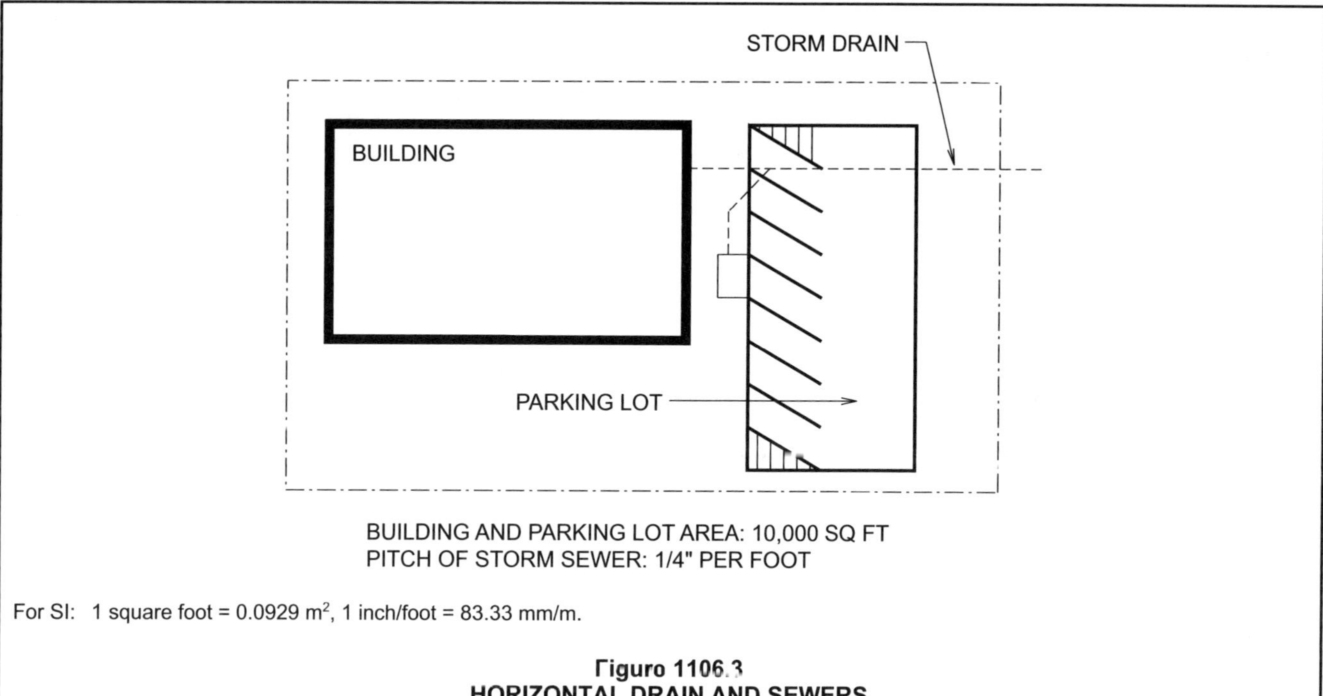

For SI: 1 square foot = 0.0929 m^2, 1 inch/foot = 83.33 mm/m.

Figure 1106.3
HORIZONTAL DRAIN AND SEWERS

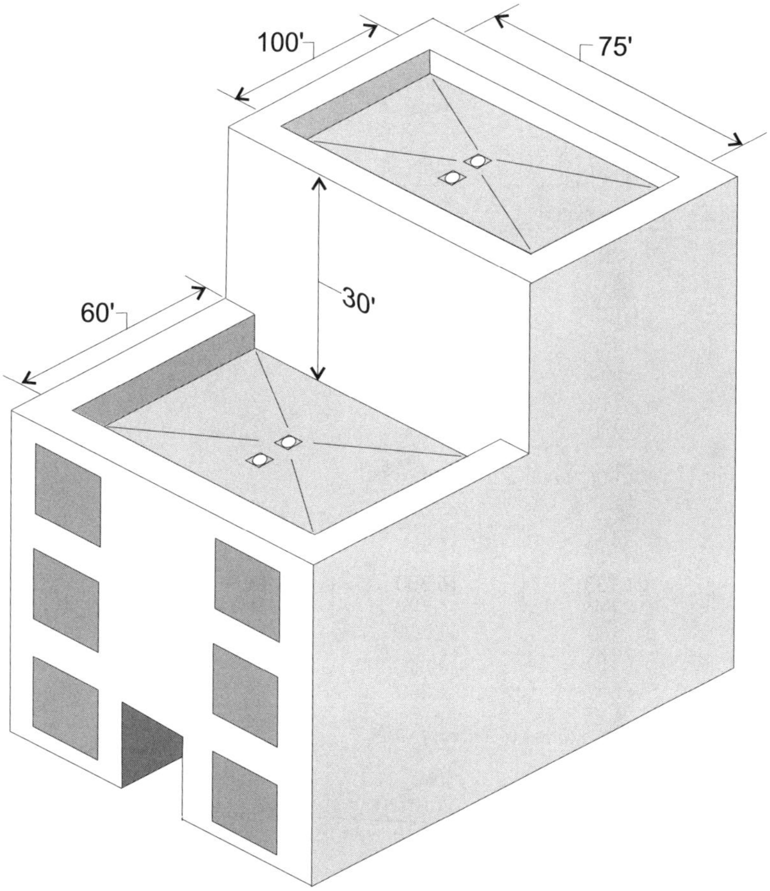

100'

75'

30'

60'

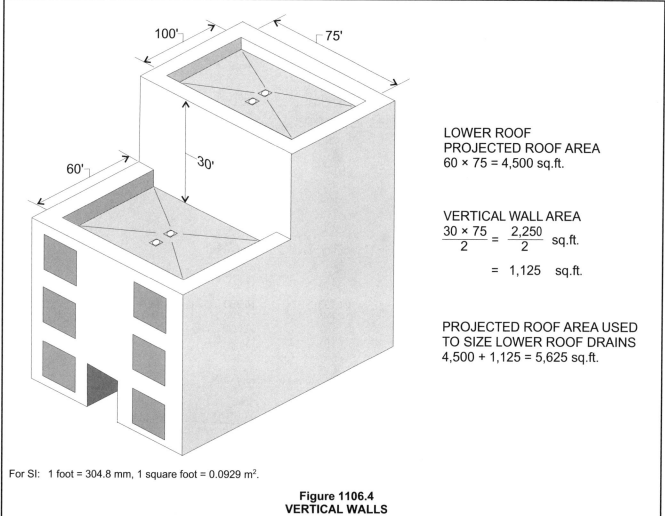

LOWER ROOF
PROJECTED ROOF AREA
60 × 75 = 4,500 sq.ft.

VERTICAL WALL AREA

$$\frac{30 \times 75}{2} = \frac{2,250}{2} \text{ sq.ft.}$$

$$= 1,125 \text{ sq.ft.}$$

PROJECTED ROOF AREA USED TO SIZE LOWER ROOF DRAINS
4,500 + 1,125 = 5,625 sq.ft.

For SI: 1 foot = 304.8 mm, 1 square foot = 0.0929 m^2.

Figure 1106.4
VERTICAL WALLS

1106.6 Size of roof gutters. The size of semicircular gutters shall be based on the maximum projected roof area in accordance with Table 1106.6.

❖ Roof gutters are designed for full-flow drainage, assuming the volume of rainwater associated with a 60-minute duration, 100-year rain. The sizing table is based on full capacity of the gutters. The size and number of conductors or leader connections to a roof gutter system must be adequate to prevent overflow of the gutters.

TABLE 1106.6. See below.

❖ The method of sizing roof gutters is similar to that for storm drains and sewers. Information necessary to use this table is the rainfall rate to be used and the desired slope of the installed gutter. With this information, the required gutter size for a given area can be determined. Note that the horizontal projected roof area must include at least 50 percent of the area of adjacent vertical walls according to Section 1106.4.

The table gives slopes from one-sixteenth to one-half unit vertical to 12 units horizontal. Any other slope as well as any profile other than semicircular needs to be approved by the code official.

SECTION 1107 SECONDARY (EMERGENCY) ROOF DRAINS

1107.1 Secondary drainage required. Secondary (emergency) roof drains or scuppers shall be provided where the roof perimeter construction extends above the roof in such a manner

TABLE 1106.6
SIZE OF SEMICIRCULAR ROOF GUTTERS

DIAMETER OF GUTTERS (inches)	HORIZONTALLY PROJECTED ROOF AREA (square feet)					
	Rainfall rate (inches per hour)					
	1	2	3	4	5	6
$^1/_{16}$ unit vertical in 12 units horizontal (0.5-percent slope)						
3	680	340	226	170	136	113
4	1,440	720	480	360	288	240
5	2,500	1,250	834	625	500	416
6	3,840	1,920	1,280	960	768	640
7	5,520	2,760	1,840	1,380	1,100	918
8	7,960	3,980	2,655	1,990	1,590	1,325
10	14,400	7,200	4,800	3,600	2,880	2,400
$^1/_8$ unit vertical 12 units horizontal (1-percent slope)						
3	960	480	320	240	192	160
4	2,040	1,020	681	510	408	340
5	3,520	1,760	1,172	880	704	587
6	5,440	2,720	1,815	1,360	1,085	905
7	7,800	3,900	2,600	1,950	1,560	1,300
8	11,200	5,600	3,740	2,800	2,240	1,870
10	20,400	10,200	6,800	5,100	4,080	3,400
$^1/_4$ unit vertical in 12 units horizontal (2-percent slope)						
3	1,360	680	454	340	272	226
4	2,880	1,440	960	720	576	480
5	5,000	2,500	1,668	1,250	1,000	834
6	7,680	3,840	2,560	1,920	1,536	1,280
7	11,040	5,520	3,860	2,760	2,205	1,840
8	15,920	7,960	5,310	3,980	3,180	2,655
10	28,800	14,400	9,600	7,200	5,750	4,800
$^1/_2$ unit vertical in 12 units horizontal (4-percent slope)						
3	1,920	960	640	480	384	320
4	4,080	2,040	1,360	1,020	816	680
5	7,080	3,540	2,360	1,770	1,415	1,180
6	11,080	5,540	3,695	2,770	2,220	1,850
7	15,600	7,800	5,200	3,900	3,120	2,600
8	22,400	11,200	7,460	5,600	4,480	3,730
10	40,000	20,000	13,330	10,000	8,000	6,660

For SI: 1 inch = 25.4 mm, 1 square foot = 0.0929 m^2.

that water will be entrapped if the primary drains allow buildup for any reason.

❖ This section requires all buildings to have some method for preventing the accumulation of unplanned excessive rainwater. A secondary drainage system is required where the building has parapet walls or other construction on the perimeter of the building that would cause ponding. See Section 1101.7 regarding roof loads resulting from ponding on roofs.

The intent here is to limit the amount of ponding water that will be placed on the roof by rainfall. If the building is designed so water cannot pond on the roof, such as roofs sloped toward the edge of the building, secondary drainage is not required (see Commentary Figure 1107.1). This simple concept should be carried to all portions of the roof, so that if any portion is designed so water can pond, secondary drains would be required.

Bear in mind that a roof could be intentionally designed to allow ponding to some design depth such as for a rainwater harvesting system.

1107.2 Separate systems required. Secondary roof drain systems shall have the end point of discharge separate from the primary system. Discharge shall be above grade, in a location that would normally be observed by the building occupants or maintenance personnel.

❖ This section requires the installation of a completely separate system for secondary roof drains. This eliminates past confusion over how far the secondary roof drain piping had to run before it could be connected to the primary drainage system. This section requires that the two systems be completely independent and that they have separate points of discharge. This requirement is consistent with the IBC as well as ASCE 7, which is referenced by the IBC.

This system is designed to function in an emergency situation; therefore, it is required to discharge above grade. By requiring this discharge above grade, the discharge will be a means of signaling that there is a problem with the primary drain that requires immediate attention. Additionally, because this discharge is not a constant means of relieving the roof rainwater, it would not normally constitute a runoff problem. Discharge into an open ditch is considered "above grade." Even though this section has been interpreted to prohibit open ditch discharge, that is incorrect. The intent is to prohibit hidden discharge points. Commentary Figure 1107.2(1) illustrates typical examples of separate secondary drains. Commentary Figure 1107.2(2) illustrates the secondary drain improperly connecting to the primary roof drain.

1107.3 Sizing of secondary drains. Secondary (emergency) roof drain systems shall be sized in accordance with Section 1106 based on the rainfall rate for which the primary system is sized in Tables 1106.2, 1106.3 and 1106.6. Scuppers shall be sized to prevent the depth of ponding water from exceeding that for which the roof was designed as determined by Section 1101.7. Scuppers shall not have an opening dimension of less than 4 inches (102 mm). The flow through the primary system shall not be considered when sizing the secondary roof drain system.

❖ This section gives the method for sizing the secondary (emergency) roof drainage system and its associated components. The sizing tables contained in Chapter 11 (Tables 1106.2, 1106.3 and 1106.6) are based on handling the anticipated rainfall rate (in inches per hour) for a particular horizontal projected roof area. This area is to include at least 50 percent of the area of adjacent vertical walls that can convey rainwater onto the roof.

The actual process of sizing the components of a secondary drainage system is identical to sizing the primary system. The same tables and rainfall rates are used.

Instead of installing a secondary system of roof drains and pipes, many designers install scuppers to allow rainwater to overflow the roof. The height and length of a scupper is critical in determining the maximum depth of ponding on a roof. Although the code states that the minimum scupper dimension is 4 inches (102 mm), it would be necessary to increase the length dimension to keep the ponding depth at or below the level for which the roof was designed (see Commentary Section 1101.7 and Commentary Table 1101.7). Commentary Figure 1107.3 contains a table and equation that can be used to design the scupper.

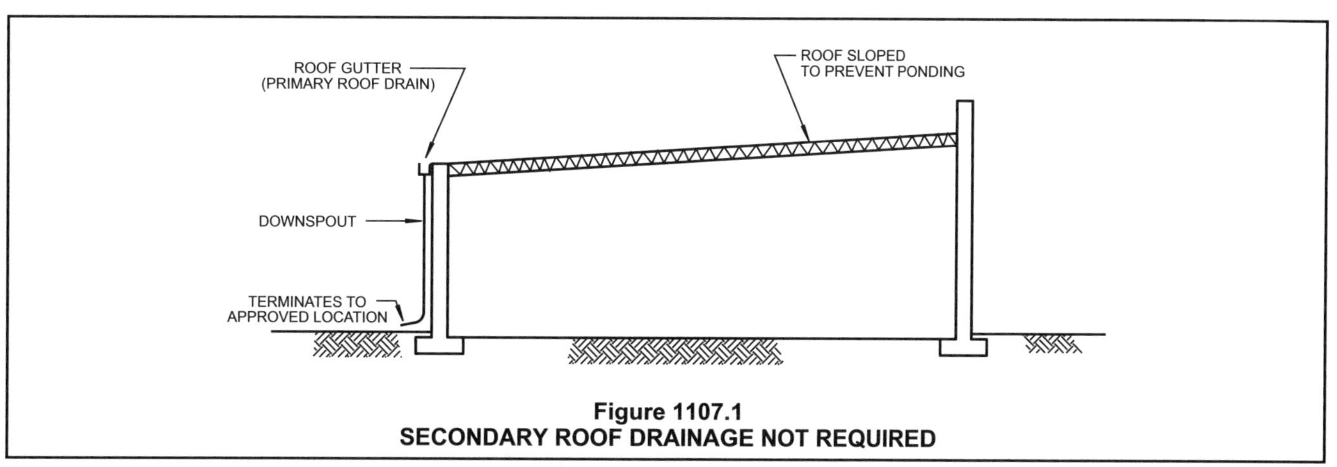

Figure 1107.1
SECONDARY ROOF DRAINAGE NOT REQUIRED

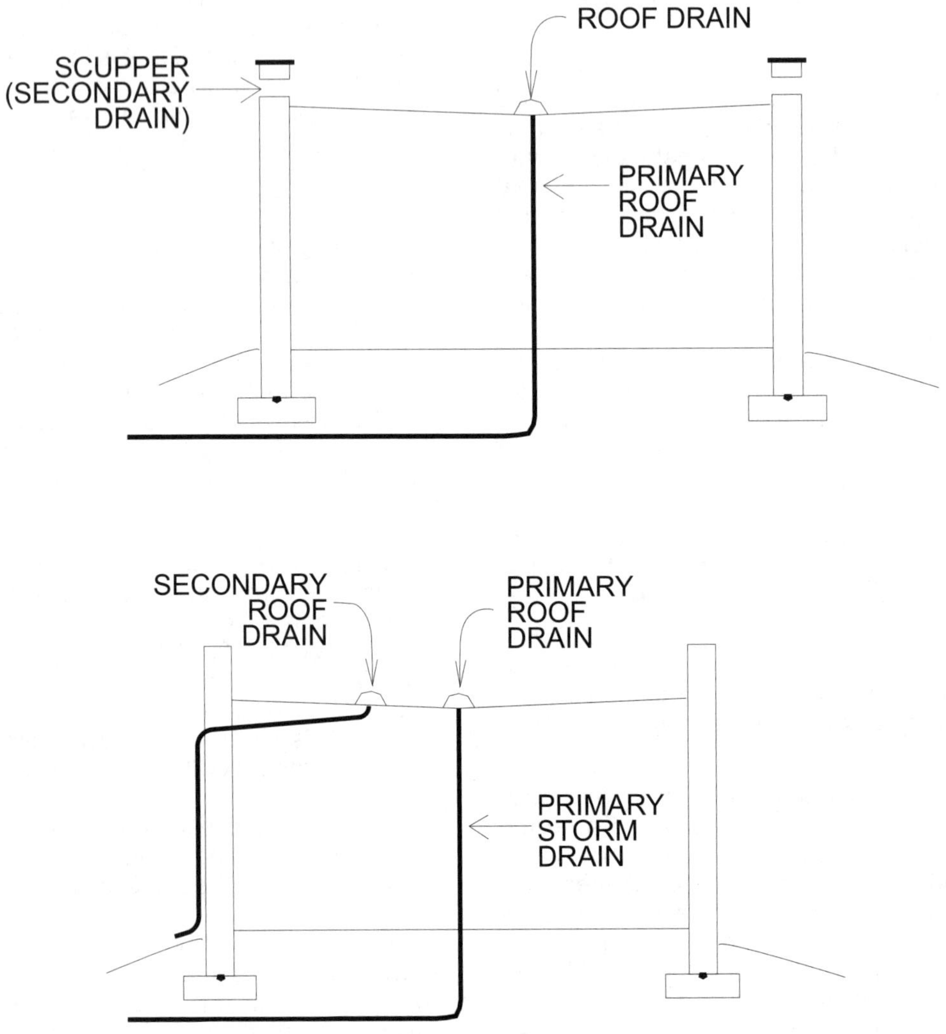

ROOF DRAIN

SCUPPER (SECONDARY DRAIN)

PRIMARY ROOF DRAIN

SECONDARY ROOF DRAIN

PRIMARY ROOF DRAIN

PRIMARY STORM DRAIN

Figure 1107.2(1)
SEPARATE PRIMARY AND SECONDARY ROOF DRAINS

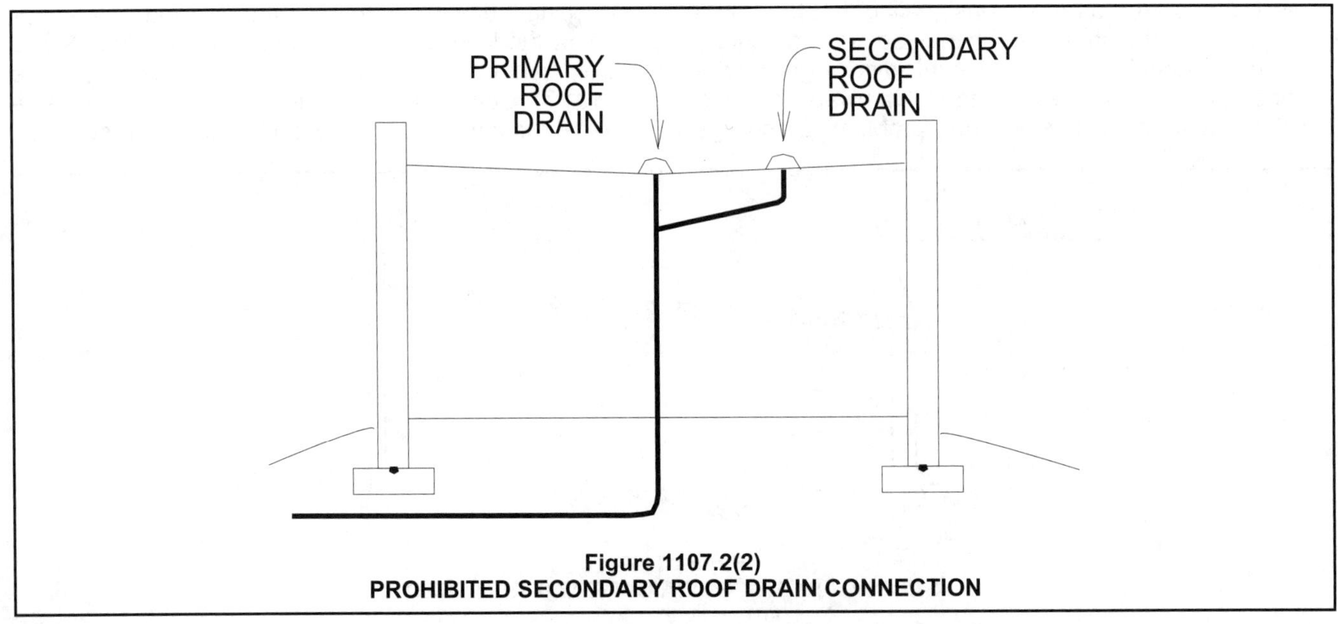

Figure 1107.2(2)
PROHIBITED SECONDARY ROOF DRAIN CONNECTION

SECTION 1108
COMBINED SANITARY AND STORM SYSTEM

1108.1 Size of combined drains and sewers. The size of a combination sanitary and *storm drain* or *sewer* shall be computed in accordance with the method in Section 1106.3. The fixture units shall be converted into an equivalent projected roof or paved area. Where the total fixture load on the combined drain is less than or equal to 256 fixture units, the equivalent drainage area in horizontal projection shall be taken as 4,000 square feet (372 m^2). Where the total fixture load exceeds 256 fixture units, each additional fixture unit shall be considered the equivalent of 15.6 square feet (1.5 m^2) of drainage area. These values are based on a rainfall rate of 1 inch (25 mm) per hour.

❖ This section acknowledges that this system still exists. Although not currently being installed as a new system, some existing combined systems require work done on them, which is regulated by this section.

In some older cities, combined sanitary and storm sewers still exist. In fact, approximately 1,100 cities use combination sanitary and storm sewers (see Commentary Figure 1108.1) that affect 42 million people nationwide. Many individual local jurisdictions have banned any increased load on this type of system. Because the storm drainage system normally results in the larger-size pipe, the (dfu) load is converted to a projected drainage area and is added to the storm sewer load. The factors used to perform this conversion are based on a rainfall rate of 1 inch (25 mm) per hour.

The combined drain or sewer size is computed as follows:

Step 1. Identify actual roof area and any adjacent contributing vertical wall areas to determine horizontal projected roof area for rainfall discharge.

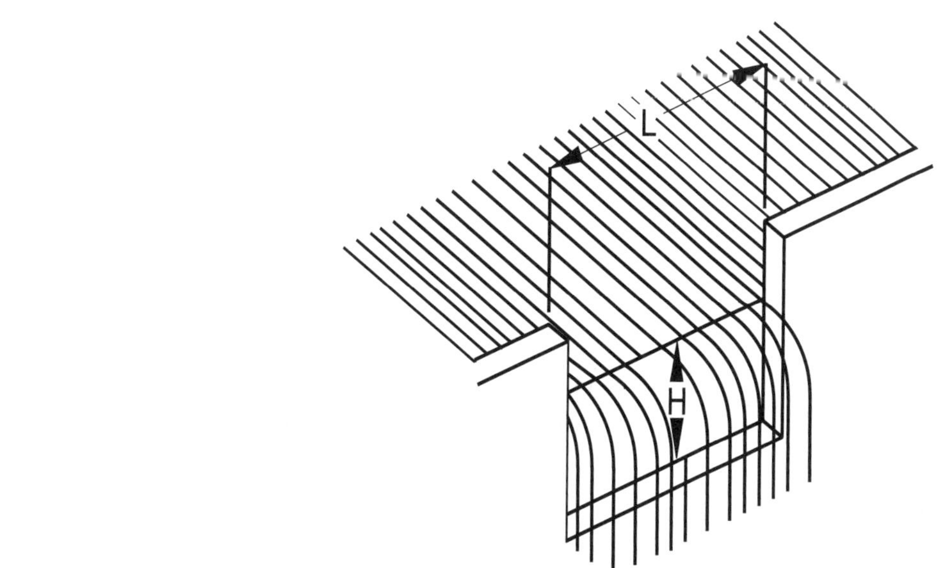

Head (H) (inches)	CAPACITY OF SCUPPER (gallons per minute)									
	Length (*L*) of scupper (inches)									
	4	6	8	10	12	18	24	30	36	48
1	10.7	17.4	23.4	29.3	35.4	53.4	71.5	89.5	107.5	143.7
2	30.5	47.5	64.4	81.4	98.5	149.4	200.3	251.1	302.1	404.0
3	52.9	84.1	115.2	146.3	177.8	271.4	364.9	458.5	552.0	739.0
4	76.7	124.6	172.6	220.5	269.0	413.3	557.5	701.8	846.0	1135.0
6	123.3	211.4	299.5	387.5	476.5	741.1	1005.8	1270.4	1535.0	2067.5

For SI: 1 inch = 25.4 mm, 1 foot = 304.8 mm, 1 gallon per minute = 3.785 L/m.
Based on the Francis formula:

$Q = 3.33 (L - 0.2H) H^{1.5}$

where:

Q = Flow rate (cubic feet per second).

L = Length of scupper opening (feet).

H = Head on scupper [feet (measured 6 feet back from opening)].

Figure 1107.3
SIZE OF SCUPPERS

Step 2: Convert the dfu load for sanitary discharge to an equivalent drainage area.

Equivalent horizontal projected roof area = 4,000 where total fixture load is less than 256 dfu

OR

Equivalent horizontal projected roof area = 4,000 + [(dfu total - 256) × 15.6] where total fixture load is greater than 256 dfu.

Step 3: Add the areas from Steps 1 and 2 above to find the total combined load expressed in square feet.

Step 4: Go to Table 1106.3 and locate the column for 1-inch (25 mm) rainfall rate and the row with the appropriate pipe slope. Divide the horizontal projected roof areas in the 1-inch (25 mm) rainfall rate column by the design rainfall rate. Determine the required pipe size using the new horizontal projected roof area values and the equivalent horizontal projected roof area found in Step 3 above.

Example:

Building location: Hartford, Connecticut
Actual roof drainage area: 10,000 square feet
Total drainage fixture units: 300 dfu
Rainfall rate: 2.7 inches per hour
Pitch of sewer: $^{1}/_{4}$ inch per foot

For SI: 1 inch = 25.4 mm, 1 square foot = 0.0929 m^2, 1 inch/foot = 83.33 mm/m.

Step 1: The actual drainage area is 10,000 sq ft

Step 2: Drainage fixture unit equivalent projected area = 4,000 + [(300 - 256) × 15.6] = 4,686 square feet.

Step 3: 10,000 + 4,686 dfu equivalent area = 14,686 square feet.

Step 4: Divide the horizontal projected roof area values in the 1-inch rainfall rate column and $^{1}/_{4}$-inch per foot slope row by the actual rainfall rate of 2.7 in Table 1106.3 (see commentary, Table 1108.1).

The equivalent projected roof area of 14,686 square feet exceeds the 11,185 square foot area that can be served by a 6-inch pipe; therefore, the next larger pipe size, 8 inches, which can handle up to 24,148 square feet, is required.

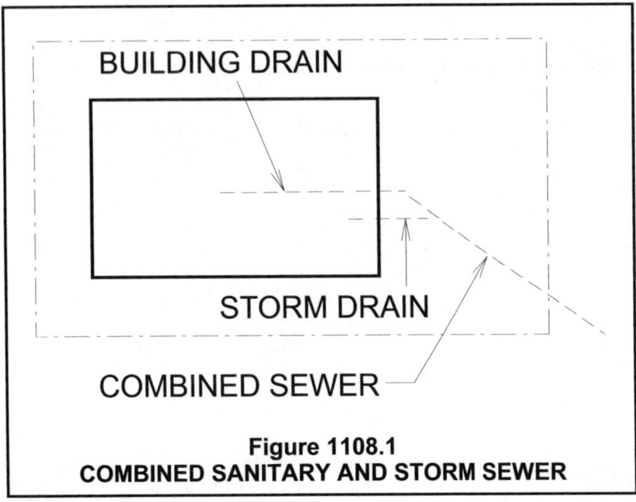

Figure 1108.1
COMBINED SANITARY AND STORM SEWER

Table 1108.1
EQUIVALENT ROOF AREA CONVERSIONS

SIZE OF HORIZONTAL PIPING (inches)	EQUIVALENT ROOF AREA BASED ON 2.7-INCH RAINFALL RATE (square feet)
3	1,719
4	3,926
5	6,993
6	11,185
8	24,148
10	43,259
12	69,630
15	124,444

For SI: 1 inch = 25.4 mm, 1 square foot = 0.0929 m^2.

SECTION 1109
VALUES FOR CONTINUOUS FLOW

1109.1 Equivalent roof area. Where there is a continuous or semicontinuous discharge into the building *storm drain* or building storm *sewer*, such as from a pump, ejector, air conditioning plant or similar device, each gallon per minute (L/m) of such discharge shall be computed as being equivalent to 96 square feet (9 m^2) of roof area, based on a rainfall rate of 1 inch (25.4 mm) per hour.

❖ This section regulates sizing of storm drain and storm sewer piping that receives a continuous or semicontinuous discharge from a pump, ejector, air conditioning plant or other similar device.

Continuous flow is not converted to the local rainfall rate because of the "constant nature" of its discharge and is considered in the same manner as the dfu area in combined sewers (see Commentary Section 1108.1).

Pump discharge capacity (in gpm) × 96 = Equivalent projected roof area in square feet, or metric conversion.

Pump discharge (in liters per minute) × 2.356 = Equivalent projected roof area in square meters.

SECTION 1110
CONTROLLED FLOW ROOF DRAIN SYSTEMS

1110.1 General. The roof of a structure shall be designed for the storage of water where the storm drainage system is engineered for controlled flow. The controlled flow roof drain system shall be an engineered system in accordance with this section and the design, submittal, approval, inspection and testing requirements of Section 105.4. The controlled flow system shall be designed based on the required rainfall rate in accordance with Section 1106.1.

❖ This section contains requirements for a storm drainage system that is designed as a controlled flow system. Often such a system is used to maintain a small level of water on the roof to help with cooling. Section 1107 also contains a requirement for emergency systems.

Controlled-flow roof drainage systems have smaller diameter storm drainage pipes relying on the roof to serve as a temporary storage or retention basin. Rainwater collects on the roof while the vertical conductors discharge at an assumed rate.

The principles of design for a controlled-flow roof drainage system are the same as for a retention basin or a dam for flood control. The water discharges at a gradual rate and continues after the storm has subsided.

Many engineers will size such a system for a maximum period of storage, usually 24 hours. The storm used for sizing is typically a 25-year frequency storm. (When listed as a 25-year storm without a period of intensity, it includes the total amount of rainfall for the duration of the storm.) The engineer may further design for a maximum depth of water of 3 inches (76 mm).

A controlled-flow system is an alternative engineering design practice and is not required by the code. In accordance with Section 1101.7, the structural engineer must design for the roof loads imposed by any depth of ponding. Such a system replaces the primary system of Section 1105 and does not replace the emergency system of Section 1107.

This section also requires the testing, inspecting and approval requirements contained in Section 105.4 (see Commentary Section 105.4).

1110.2 Control devices. The control devices shall be installed so that the rate of discharge of water per minute shall not exceed the values for continuous flow as indicated in Section 1109.1.

❖ A controlled flow system relies on a gradual discharge to the drainage system. A controlling device, usually a properly sized roof drain, is necessary to control the discharge to the system.

The flow capacities listed in Tables 1106.2 and 1106.3 are the maximum allowable values for these components. The controlled-flow system must be designed based on those discharge rates being maximum values so that the discharge rate of water per minute does not exceed the values shown in Section 1109.1.

1110.3 Installation. Runoff control shall be by control devices. Control devices shall be protected by strainers.

❖ Control devices are usually roof drains that are properly sized (see Commentary Section 1110.2). This section requires that whatever control device is used must be protected by a strainer against blockage.

1110.4 Minimum number of roof drains. Not less than two roof drains shall be installed in roof areas 10,000 square feet (929 m^2) or less and not less than four roof drains shall be installed in roofs over 10,000 square feet (929 m^2) in area.

❖ For controlled-flow systems, a system redundancy safety precaution is built into the design by the requirement for a minimum of two drains for roof areas of 10,000 square feet (929 m^2) or less. The objective is that if one drain becomes clogged, the other would allow water to drain. For larger buildings in excess of 10,000 square feet (929 m^2) of roof area, a minimum of four roof drains is required (see Commentary Figure 1110.4). Although no specific requirements exist in this section covering design of the controlled-flow system, the potential for partial clogging of part of the system must be evaluated so that the contained water will drain off the roof without any structural or other problems.

SECTION 1111
SUBSOIL DRAINS

1111.1 Subsoil drains. Subsoil drains shall be open-jointed, horizontally split or perforated pipe conforming to one of the standards listed in Table 1102.5. Such drains shall not be less than 4 inches (102 mm) in diameter. Where the building is subject to backwater, the subsoil drain shall be protected by an accessibly located backwater valve. Subsoil drains shall discharge to a trapped area drain, sump, dry well or *approved* location above ground. The subsoil sump shall not be required to have either a gas-tight cover or a vent. The sump and pumping system shall comply with Section 1113.1.

❖ This section requires that subsoil drain material conform to the requirements for subsoil drain pipe contained in Table 1102.5. The collection of water into these subsoil drain pipes should be either open jointed, horizontally split or perforated. The minimum allowable size for a subsoil drain pipe is 4 inches (102 mm).

As with other below-grade piping, subsoil drain piping must have an accessible backwater valve if the area being drained is subject to backwater conditions [see commentary, Figure 1111.1(1)].

The discharge of the subsoil drain system is regulated in this section. Discharge to either another storm water discharge area drain, a storm water sump, a storm water dry well or an approved above-ground dis-

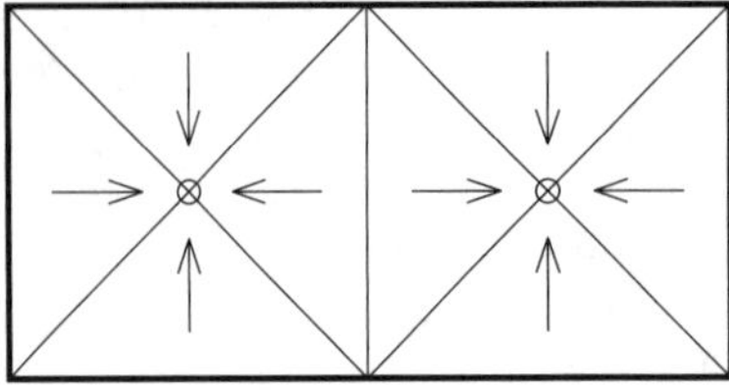

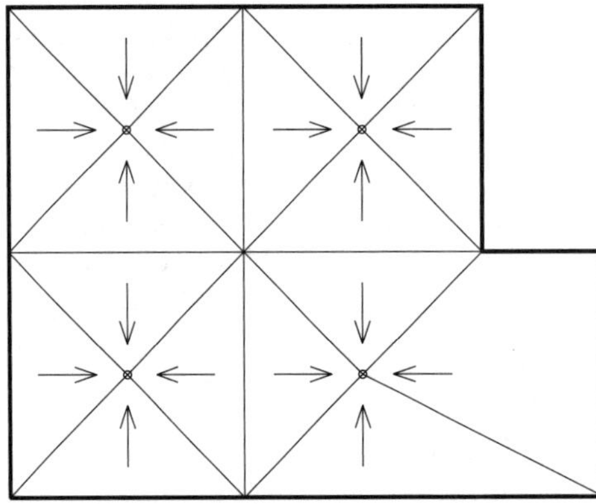

For SI: 1 square foot = 0.0929 m^2.

Figure 1110.4
MINIMUM NUMBER OF ROOF DRAINS FOR CONTROLLED FLOW

charge point in the adjacent grading of the property is allowed. When a subsoil sump is used, a gas-tight cover is not required on top of the sump or to vent the sump because it is not connected to a sanitary sewer; thus, there are no sewer gases to be concerned about [see Commentary Figure 1111.1(2)]. If the discharge is to a sump, the sump and pumping systems must be installed in accordance with Section 1113.1.

SECTION 1112
BUILDING SUBDRAINS

1112.1 Building subdrains. Building subdrains located below the *public sewer* level shall discharge into a sump or receiving tank, the contents of which shall be automatically lifted and discharged into the drainage system as required for building sumps. The sump and pumping equipment shall comply with Section 1113.1.

❖ This section regulates the installation of building subdrains located below the public sewer.

The building subdrains identified include subsoil drainage around the perimeter of the building (foundation drains) and other clear-water storm drains, such as basement window well and outdoor stairwell drains.

Storm rainwater systems that discharge into sumps and sump pump systems need not conform to Sections 712 and 916.5 for sanitary sump drainage systems. The sump and pumping systems must comply with Section 1113.1.

This section requires that the collected storm water that is fed into the sump automatically be discharged from the sump. This requirement is intended to protect against flooding and potential damage to the building should someone fail to turn on the pump when necessary. Additionally, this will help minimize damage to the pump caused by unnecessary extended operation.

The discharge from the sump can be to grade in the surrounding area of the building if the discharge does not create a nuisance to the surrounding area and is approved by the code official. The sump pump system receiving subsoil drainage can also discharge to the storm drainage system.

Subsoil drainage should not be discharged into the sanitary drainage system because it places a substantial burden on the sanitary treatment facility (see commentary, Section 1108). This is also a concern in areas that still use combined sanitary and storm sewer systems.

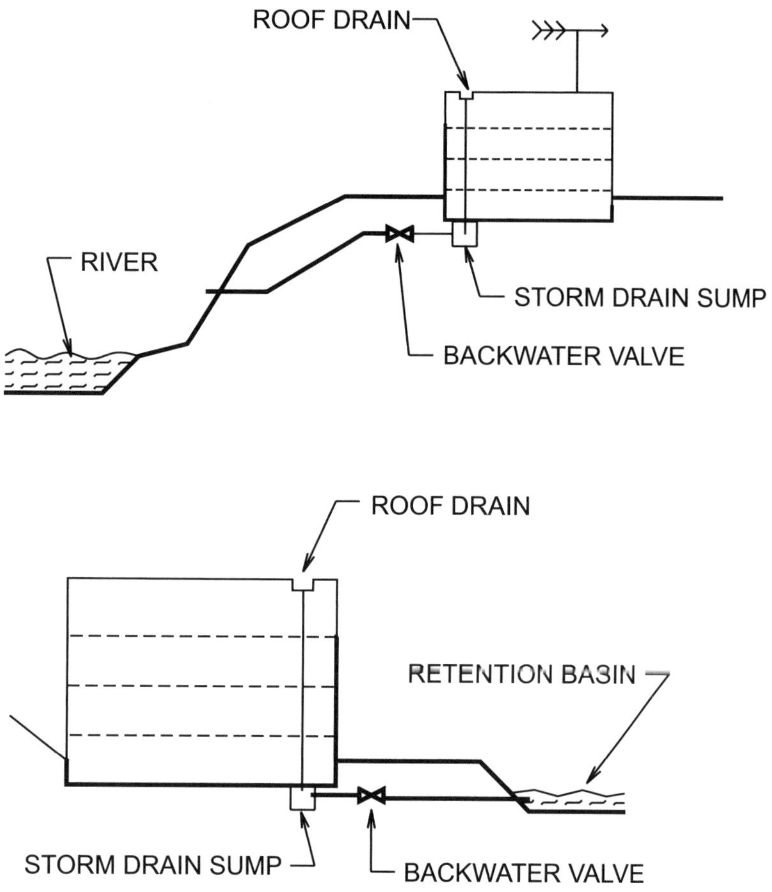

ROOF DRAIN
RIVER
STORM DRAIN SUMP
BACKWATER VALVE

ROOF DRAIN
RETENTION BASIN
STORM DRAIN SUMP
BACKWATER VALVE

Figure 1111.1(1)
SUBSOIL DRAIN SUBJECT TO BACKFLOW

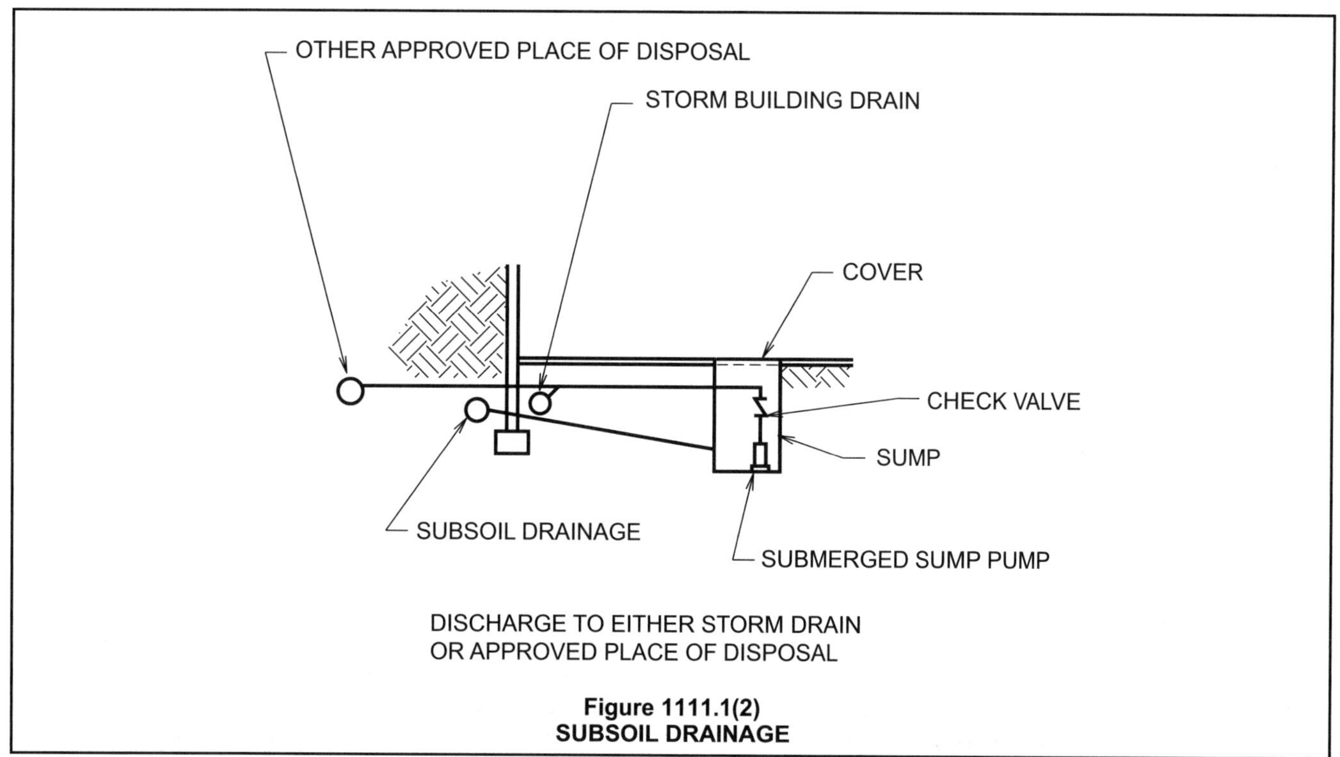

Figure 1111.1(2)
SUBSOIL DRAINAGE

SECTION 1113 SUMPS AND PUMPING SYSTEMS

1113.1 Pumping system. The sump pump, pit and discharge piping shall conform to Sections 1113.1.1 through 1113.1.4

❖ This section regulates the sumps and pumping systems associated with subsoil drains and building subdrains.

This section contains requirements for the construction of a subdrain storm system, including:

- Pump capacity and head pressure (see Commentary Section 1113.1.1),
- Sump pit construction (see Commentary Section 1113.1.2),
- Electrical service outlets (see Commentary Section 1113.1.3) and
- Discharge piping material (see Commentary Section 1113.1.4).

1113.1.1 Pump capacity and head. The sump pump shall be of a capacity and head appropriate to anticipated use requirements.

❖ This section requires that the pump be sized properly to function under the anticipated load. The sump pump system must be designed to remove the subsoil drainage at the anticipated inflow rate under a worst-case condition. The sizing is determined by evaluating the local ground conditions, the length of piping and the weather in the area. A worst-case condition for subsoil drainage usually happens in the spring when heavy rains occur and any snow accumulation melts. The selection of a sump pump should include consideration for the on-off cycle time and the duration of operation of the pump in the "on" phase.

1113.1.2 Sump pit. The sump pit shall not be less than 18 inches (457 mm) in diameter and 24 inches (610 mm) deep, unless otherwise *approved.* The pit shall be accessible and located such that all drainage flows into the pit by gravity. The sump pit shall be constructed of tile, steel, plastic, cast-iron, concrete or other *approved* material, with a removable cover adequate to support anticipated loads in the area of use. The pit floor shall be solid and provide permanent support for the pump.

❖ This section requires that the minimum dimensions of the sump pit be 18 inches (457 mm) in diameter and 24 inches (610 mm) deep. The pit size is very important in the design of the system because it determines the amount of water that is stored in the system prior to discharge. It also determines the operation time of the pump for each "on" cycle (see Commentary Section 712.3.2).

The pit must be constructed of tile, steel, plastic, cast iron, concrete or any material that performs like these and is approved by the code official. The pit must have a removable cover to allow access to the pump and other related items, such as the check valve. The cover is to be structurally capable of holding anticipated loads. The pit must also have a floor that is capable of permanently supporting the pump, including piping that bears on the pump. This requirement basically eliminates the use of gravel as the bottom of the sump because gravel would not allow for the permanent support of the pump.

1113.1.3 Electrical. Electrical service outlets, when required, shall meet the requirements of NFPA 70.

❖ When the pump is powered by electricity, the receptacles are required to be installed in accordance with NFPA 70. The user of this commentary should consult that document for specific requirements.

1113.1.4 Piping. Discharge piping shall meet the requirements of Section 1102.2, 1102.3 or 1102.4 and shall include a gate valve and a full flow check valve. Pipe and fittings shall be the same size as, or larger than, pump discharge tapping.

> **Exception:** In one- and two-family dwellings, only a check valve shall be required, located on the discharge piping from the pump or ejector.

❖ The piping material used for the sump discharge must conform to the requirements contained in Sections 1102.2, 1102.3 and 1102.4, which refer to Tables 702.1, 702.2 and 1102.4, respectively. The piping in the sump system discharge line must have a check valve and a full-open valve installed to permit maintenance on the other components of the system. The requirement for a full-open valve does not apply to one- and two-family dwellings. This check valve should be placed as close as practical to the sump pump and is required to prevent discharge water from the pump from running back into the pit after the pump stops, thus preventing the constant cycling of the sump pump.

Access must be provided to such valves and located above the sump cover. Where the discharge pipe is below grade, the valve must be located outside the sump below grade in an access pit with a removeable cover.

Pipe and fittings of the discharge piping are not to be smaller than the size of the discharge outlet of the pump itself. Reduced-size discharge piping will affect pumping capacity, thereby causing the system not to function as designed.

Bibliography

The following resource materials are referenced in this chapter or are relevant to the subject matter addressed in this chapter.

ASCE 7-05, *Minimum Design Loads for Buildings and Other Structures.* New York, NY: American Society of Civil Engineers, 2005.

ASME A 112.3.1-93, *Performance Standard and Installation Procedures for Stainless Steel Drainage Systems or Sanitary, Storm and Chemical Applications, Above and Below Ground.* New York, NY: American Society of Mechanical Engineers, 1993.

ASME A112.6.4-03, *Roof, Deck, and Balcony Drains*. New York, NY: American Society of Mechanical Engineers, 2003.

ASME A112.21.2M-83, *Roof Drains*. New York, NY: American Society of Mechanical Engineers, 1983.

ASTM C 14-99, *Specification for Concrete Sewer, Storm Drain, and Culvert Pipe*. West Conshohocken, PA: American Society for Testing and Materials, 1999.

ASTM C 76-04a, *Specification for Reinforced Concrete Culvert, Storm Drain, and Sewer Pipe*. West Conshohocken, PA: American Society for Testing and Materials, 2004.

ASTM D 2665-04ae01, *Specification for Poly (Vinyl Chloride) (PVC) Plastic Drain, Waste, and Vent Pipe and Fittings*. West Conshohocken, PA: American Society for Testing and Materials, 2004.

ASTM D 2729-96a, *Specification for Poly (Vinyl Chloride) (PVC) Sewer Pipe and Fittings*. West Conshohocken, PA: American Society for Testing and Materials, 1996.

ASTM D 2751-96a, *Specification for Acrylonitrile-butadiene-styrene (ABS) Sewer Pipe and Fittings*. West Conshohocken, PA: American Society for Testing and Materials, 1996.

ASTM D 3034-04, *Specification for Type PSM Poly (Vinyl Chloride) (PVC) Sewer Pipe and Fittings*. West Conshohocken, PA: American Society for Testing and Materials, 2004.

ASTM F 405-97, *Specification for Corrugated Polyethylene (PE) Tubing and Fittings*. West Conshohocken, PA: American Society for Testing and Materials,1997.

ASTM F 891-00e01, *Specification for Coextruded Poly (Vinyl Chloride) (PVC) Plastic Pipe with a Cellular Core*. West Conshohocken, PA: American Society for Testing and Materials, 2001.

ASTM F 2306-05, *12 inch to 60 inch Annular Corrugated Profile-wall Polyethylene (PE) Pipe and Fittings for Gravity Flow Storm Sewer and Subsurface Drainage Applications*. West Conshohocken, PA: American Society for Testing and Materials, 2005.

CAN/CSA A257.1-92, *Circular Concrete Culvert, Storm Drain, Sewer Pipe and Fittings*. Rexdale (Toronto), Ontario, Canada: Canadian Standards Association, 1992.

CAN/CSA A257.2M-92, *Reinforced Circular Concrete Culvert, Storm Drain, Sewer Pipe and Fittings*. Rexdale (Toronto), Ontario, Canada: Canadian Standards Association, 1992.

CSA B182.2-02, *PVC Sewer Pipe and Fittings (PSM Type)*. Rexdale (Toronto), Ontario, Canada: Canadian Standards Association, 2002.

CAN/CSA-B182.4-02, *Profile PVC Sewer Pipe and Fittings*. Rexdale (Toronto), Ontario, Canada: Canadian Standards Association, 2002.

IBC-09, *International Building Code*. Washington DC: International Code Council, 2009.

NFPA 70-08, *National Electrical Code*. Quincy, MA: National Fire Protection Agency, 2008.

Rainfall Rates, *National Weather Service, National Oceanic and Atmospheric Administration*. Washington, DC.

Chapter 12: Special Piping and Storage Systems

General Comments

Even though the gases used in medical and inhalation anesthetic systems are nonflammable, they are frequently oxidizing gases and, as such, create an environment that supports combustion and where ignition of combustible materials might occur. Therefore, the proper design, installation and performance of these systems is essential to protect against the physical hazard they could pose. The portion of these systems not regulated by this chapter and the *International Fire Code*® (IFC®) is the mechanical exhaust of vacuum systems, which must be designed and installed in accordance with the requirements of the *International Mechanical Code*® (IMC®).

The nonmedical oxygen systems regulated by this chapter are bulk oxygen storage systems and oxygen-fuel-gas systems used for welding and cutting. Bulk oxygen storage systems contain oxygen in either a liquid or gaseous state, which though nonflammable, possesses properties that support combustion. In a pure oxygen environment, many materials that do not burn in air can burn under pressure. The proper selection of materials used as components of bulk oxygen systems as well as the location of bulk oxygen storage containers are essential to minimizing the potential fire hazards posed by an oxygen-enriched environment.

Welding and cutting operations using fuel gas and oxygen pose an inherent risk of fire or explosion because of the combined presence of flammable gases, flames and oxygen. Additionally, the generation of acetylene gas, the most common flammable gas used with oxygen for welding and cutting operations, and the storage and use of calcium carbide to produce acetylene gas and other hazardous processes, are regulated by this chapter. Calcium carbide, when exposed to water, generates acetylene gas, which poses an explosion hazard where not properly stored and handled.

Purpose

This chapter sets the requirements for the design, installation, storage, handling and use of nonflammable medical gas systems, including inhalation anesthetic and vacuum piping systems, bulk oxygen storage systems and oxygen-fuel gas systems used for welding and cutting operations. The intent of these requirements is to minimize the potential fire and explosion hazards associated with the gases used in these systems.

SECTION 1201 GENERAL

1201.1 Scope. The provisions of this chapter shall govern the design and installation of piping and storage systems for nonflammable medical gas systems and nonmedical oxygen systems. All maintenance and operations of such systems shall be in accordance with the *International Fire Code.*

❖ This chapter contains the requirements and referenced standards that regulate nonflammable medical gas, inhalation anesthetic, vacuum piping and nonmedical oxygen systems. The referenced standards contain the design, installation, system components, materials, testing, operation, maintenance and safety requirements for these systems. The IFC regulates the maintenance and operation of these systems.

SECTION 1202 MEDICAL GASES

[F] 1202.1 Nonflammable medical gases. Nonflammable medical gas systems, inhalation anesthetic systems and vacuum piping systems shall be designed and installed in accordance with NFPA 99C.

Exceptions:

1. This section shall not apply to portable systems or cylinder storage.
2. Vacuum system exhaust terminations shall comply with the *International Mechanical Code*.

❖ This section contains the requirements and referenced standards that apply to nonflammable medical gas, inhalation anesthetic and vacuum piping systems.

NFPA 99C applies to nonflammable medical gas, inhalation anesthetic and permanently installed vacuum piping systems, with the exception of portable systems and the storage of gas cylinders. The mechanical exhaust of vacuum piping systems is also exempt from NFPA 99C requirements and must comply with the requirements of the IMC.

Gases such as oxygen, nitrous oxide, compressed air, carbon dioxide, helium, nitrogen and various mixtures of these gases are typically used in medical gas-piping systems.

Oxygen and nitrous oxide are two of the most common gases used in medical gas and inhalation anesthetic systems. Even though these two gases are nonflammable, they are strong oxidizing gases, and their presence creates an environment where the potential for fire, chemical, mechanical and electrical hazards exists. Being a strong oxidizer, these gases, either by themselves or in combination, provide an environment that lowers the ignition energies of combustibles.

Because of the hazards that medical gases and inhalation anesthetics pose, NFPA 99C contains specific criteria for connection of the gas source to the distribution piping system, system identification and system alarm requirements. These requirements help reduce the potential for mistaken interconnection of a gas system to a vacuum system, which could present a fire and explosion hazard.

NFPA 99C requires that pipe and fittings used in a medical gas system must be of Type K or L (ASTM B 819) copper tubing, precleaned and capped. Additionally, pipe must be identified with one of the following markings: OXY, MED, OXY/MED, ACR/OXY or ACR/MED. Valves, fittings and other piping components must be cleaned for "oxygen service" by the manufacturer or an approved supplier, with the result a sealed, oil-free fitting enclosed in a dry nitrogen-purged container.

The piping system must be continuously purged prior to, during and after installation with oil-free dry nitrogen to prevent the formation of scale within the tubing. Openings must be plugged or capped to prevent debris or contamination on the inside of the tube.

The soldered or brazed joint is to be visually checked after it has been washed with hot water and a stainless steel wire brush has been used to remove scale. The following conditions are considered unacceptable:

1. Flux or flux residue present.
2. Excessive oxidation of the joint.
3. Cracks in the tube or component.
4. Cracks in the braze filler metal.
5. Presence of unmelted filler metal.
6. Presence of filler metal clearly visible around the entire joint and at the interface between the socket and tube.
7. Failure of the joint to hold the test pressure.

Vacuum system piping must be corrosion-resistant metal such as seamless copper water tube (ASTM B 88, Types K, L and M), copper ACR tube (ASTM B 280), copper medical gas tube (ASTM B 819), stainless steel tube or galvanized steel pipe (ASTM A 53). Joints in copper must be soldered or brazed. Joints in stainless steel tube must be brazed or welded. Joints in galvanized piping must be flanged or threaded. The vacuum piping must be marked or labeled prior to installation with one of the following markings: Medical-Surgical Vacuum or MED/SURG VAC. They must be marked not more than 20 feet (6096 mm) apart. Vacuum systems must never be converted for use as gas systems.

Where flux is used for soldered or brazed joints, it must be used sparingly to avoid excess flux inside the joint. If flux has been used, it must be washed with hot water.

NFPA 99C contains requirements regarding flammable inhalation anesthetic gases; however, use of flammable inhalation gases is not permitted in a piped distribution system. According to B.R. Klein, in 1950 the NFPA Committee on Gases prohibited piped distribution of flammable anesthetic gases to eliminate one possible source of explosion and fire in health care facilities. The IFC also contains requirements applicable to the storage, use and handling of compressed gases used in connection with nonflammable medical gases and inhalation anesthetic systems. The mechanical exhaust discharge and discharge termination of vacuum piping systems must comply with the IMC.

SECTION 1203
OXYGEN SYSTEMS

[F] 1203.1 Design and installation. Nonmedical oxygen systems shall be designed and installed in accordance with NFPA 50 and NFPA 51.

❖ This section contains the general requirements and referenced standards that apply to nonmedical oxygen systems.

NFPA 50 applies to bulk oxygen systems that are located at a consumer site and have a storage capacity exceeding 20,000 cubic feet (566 m^3) of liquid or gaseous oxygen, including unconnected reserves. The bulk oxygen originates off site, and the supply is delivered to the premises by mobile delivery equipment. This standard does not apply to oxygen manufacturing plants, oxygen suppliers or systems having capacities less than 20,000 cubic feet (566 m^3).

Oxygen is an odorless, tasteless, colorless, nontoxic nonflammable gas. Although oxygen gas is nonflammable, it is an oxidizer, and as such creates an environment where the ignition of combustibles could occur.

The location of bulk storage systems is critical to protect them from becoming involved in fires that originate from sources adjacent to the system. NFPA 50 contains the minimum clearance requirements between bulk oxygen storage containers and adjacent structures, combustible materials stored on site, places of public assembly, public sidewalks, parked vehicles, adjoining property lines and other storage containers containing flammable and combustible liquids or gases.

The standard requires that bulk oxygen storage containers comply with Section VIII of the ASME *Boiler and Pressure Vessel Code* and be corrosion resistant, or designed, constructed, tested and maintained in accordance with U.S. Department of Transportation (DOT) specifications and regulations for 4L containers. The containers must also be equipped with safety relief valves. Piping, tubing and fittings used for bulk oxygen systems must be suitable for oxygen service and must conform to the requirements of ASME B 31.3 for chemical plant and petroleum refinery piping joints in piping, and tubing must be either welded or brazed, flanged, threaded, socket, slip or compression fittings. The use of gaskets and threaded sealants must be suitable for oxygen service. Brazing materials must have a melting point above 1,000°F (538°C). Field-assembled piping must be tested prior to being placed in service, and the bulk oxygen storage location must be permanently identified with signage stating "Oxygen–No Smoking–No Open Flames" or an equivalent warning.

The IFC also contains requirements applicable to the storage, handling and transportation of liquid oxygen.

NFPA 51 applies to oxygen-fuel-gas systems for welding, cutting and allied processes, the use and storage of fuels, acetylene generation and calcium carbide storage. The standard does not apply to systems consisting of regulators, hoses, a torch and a single tank each of oxygen and fuel gas. Additionally, in systems where oxygen is not used with fuel gases, the manufacture of gases and filling of cylinders, the storage of empty cylinders and compressed air-fuel systems are not governed by this standard.

Oxygen-fuel gas systems for welding and cutting consist of flammable fuel gases such as acetylene, hydrogen, LP-gas, natural gas or stabilized methylacetylene-propadiene used with oxygen to create a high-temperature flame. Although NFPA 51 applies to fuel gases used with oxygen, very specific regulations governing the generation of acetylene and the storage of calcium carbide are stated because acetylene is the most commonly used gas for welding and cutting.

Acetylene is a colorless, flammable gas produced by the direct combination of hydrogen and carbon. The generation of acetylene gas is limited by NFPA51 to the carbide-to-water process, in which pieces of calcium carbide are placed in water. The calcium carbide, a gray, stone-like material produced by the smelting of lime and coke, reacts with water to produce gas bubbles that are the acetylene gas.

Acetylene generators must be listed and are not permitted to generate acetylene at pressures exceeding 15 pounds per square inch (103 kPa) gauge. Acetylene generated at higher pressures is unstable and will, under certain circumstances, decompose into hydrogen and carbon with explosive violence.

Acetylene generators must be installed on a level foundation, protected from freezing and have sufficient area around them for maintenance, operation, adjustment and charging. Rooms or buildings in which acetylene generators are located must have explosion relief vents at not less than 1 square foot per 50 cubic feet (0.0929 m^2/1.4 m^3) of room volume.

Cylinders used for the storage of fuel gases or oxygen are required to conform to U.S. DOT or Canada Transport Commission regulations.

Pipe and fittings must comply with ASME B31.3. Pipe must be Schedule 40 steel, brass, copper or stainless steel, and fittings must be standard weight in sizes up to and including 6 inches (152 mm) in diameter. Copper tubing must be Type K or L in accordance with ASTM B 88. Piping for acetylene and methylacetylene-propadiene systems must be steel. Joints in steel piping must be welded, threaded or flanged. Joints in brass or copper pipe must be welded, brazed, threaded or flanged. Where brazed, joints must be made with a silver-brazing alloy or similar high-melting-point filler metal.

So that no contaminants are present in the piping system, piping must be visually inspected for contaminants prior to assembly and the assembled must be system blown out with compressed air. Prior to being placed in service, the entire system must be pressure tested to a minimum 1.5 times the maximum operating pressure.

Finally, piping systems must be provided with a protective equipment device designed to prevent not only backflow of oxygen into the fuel gas supply but also a flashback of flame into the gas supply. The device must also prevent the development of an oxygen-fuel gas mixture at pressures that if ignited will achieve combustion pressure that will cause the device to fail and thus not prevent both of the previous items from occurring. Pressure-relief devices are also required, and listed shutoff valves and backflow check valves must be installed at every fuel gas and oxygen outlet station.

The IFC contains requirements applicable to welding and cutting operations and equipment, calcium carbide and acetylene generators and related piping manifolds, hose systems and other system components for fuel gas and oxygen systems. It also contains provisions for the storage, use and handling of both compressed gases used in cutting and welding operations and calcium carbide.

Bibliography

The following resource materials are referenced in this chapter or are relevant to the subject matter addressed in this chapter.

ASME B31.3-02, *Process Piping*. New York, NY: American Society of Mechanical Engineers, 2002.

ASME-07, *Boiler and Pressure Vessel Code*. New York, NY: American Society of Mechanical Engineers, 2007.

ASTM A 53/A 53M-02, *Specification for Pipe, Steel, Black and Hot-dipped, Zinc-coated, Welded and Seamless*. West Conshohocken, PA: American Society for Testing and Materials, 2002.

ASTM B 88-03, *Specification for Seamless Copper Water Tube*. West Conshohocken, PA: American Society for Testing and Materials, 2003.

ASTM B 280-02, *Specification for Seamless Copper Tube for Air-conditioning and Refrigeration Field Service*. West Conshohocken, PA: American Society for Testing and Materials, 2002.

ASTM B 819-00, *Specification for Seamless Copper Tube for Medical Gas Systems*. West Conshohocken, PA: American Society for Testing and Materials, 2000.

IBC-09, *International Building Code*. Washington DC: International Code Council, 2009.

IFC-09, *International Fire Code*. Washington DC: International Code Council, 2009.

IMC-09, *International Mechanical Code*. Washington DC: International Code Council, 2009.

NFPA 50-01, *Bulk Oxygen Systems at Consumer Sites*. Quincy, MA: National Fire Protection Association, 2001.

NFPA 51-02, *Design and Installation of Oxygen-fuel Gas Systems for Welding, Cutting, and Allied Processes*. Quincy, MA: National Fire Protection Association, 2002.

NFPA 99C-02, *Gas and Vacuum Systems*. Quincy, MA: National Fire Protection Association, 2002.

Chapter 13: Referenced Standards

General Comments

This chapter lists the standards that are referenced in various sections of this document. The standards are listed herein by the promulgating agency of the standard, the standard identification, the date and title, and the section or sections of this document that reference the standard. The application of the referenced standards is as specified in Section 102.8.

It is important to understand that not every document related to building design and construction is qualified to be a "referenced standard." The International Code Council® (ICC®) has adopted a criterion that standards referenced in the *International Codes*® (I-Codes®) and standards intended for adoption into the I-Codes must meet to qualify as a referenced standard. The policy is summarized as follows:

- Code references: The scope and application of the standard must be clearly identified in the code text.
- Standard content: The standard must be written in mandatory language and be appropriate for the subject covered. The standard shall not have the effect of requiring proprietary materials or prescribing a proprietary testing agency.
- Standard promulgation: The standard must be readily available and developed and maintained in a consensus process such as ASTM or ANSI.

The ICC Code Development Procedures, of which the standards policy is a part, are updated periodically. A copy of the latest version can be obtained from the ICC offices.

Once a standard is incorporated into the code through the code development process, it becomes an enforceable part of the code. When the code is adopted by a jurisdiction, the standard also is part of that jurisdiction's adopted code. It is for this reason that the criteria were developed. Compliance with this policy means that documents incorporated into the code are developed through the use of a consensus process, written in mandatory language, and do not mandate the use of proprietary materials or agencies. The requirement for a standard to be developed through a consensus process is vital because it means that the standard will be representative of the most current body of available knowledge on the subject as determined by a broad spectrum of interested or affected parties without dominance by any single interest group. A true consensus process has many attributes, including but not limited to:

- An open process that has formal (published) procedures that allow for the consideration of all viewpoints,
- A definitive review period that allows for the standard to be updated or revised,
- A process of notification to all interested parties and
- An appeals process.

Many available documents related to design, installation and construction, though useful, are not "standards" and are not appropriate for reference in the code. Often, these documents are developed or written with the intention of being used for regulatory purposes and are unsuitable for use as a regulation because of extensive use of recommendations, advisory comments and nonmandatory terms. Typical examples of such documents include installation instructions, guidelines and practices.

The objective of ICC's standards policy is to provide regulations that are clear, concise and enforceable—thus the requirement for standards to be written in mandatory language. This requirement is not intended to mean that a standard cannot contain informational or explanatory material that will aid the user of the standard in its application. When it is the desire of the standard's promulgating agency for such material to be included, however, the information must appear in a nonmandatory location, such as an annex or appendix, and be clearly identified as not being part of the standard.

Overall, standards referenced by the code must be authoritative, relevant, up to date and, most important, reasonable and enforceable. Standards that comply with the ICC standards policy fulfill these expectations.

Purpose

As a performance-oriented code, the code contains numerous references to documents that are used to regulate materials and methods of construction. The references to these documents within the code text consist of the promulgating agency's acronym and its publication designation (e.g., ASME A112.1.3) and a further indication that the document being referenced is the one that is listed in Chapter 13. Chapter 13 contains all of the information that is necessary to identify the specific referenced document. Included is the following information on a document's promulgating agency (see Figure 13):

- The promulgating agency (i.e., the agency's title),
- The promulgating agency's acronym and
- The promulgating agency's address. For example, a reference to an ASME standard within the code indicates that the document is promulgated by the American Society of Mechanical

Engineers (ASME), which is located in New York City. Chapter 13 lists the standards agencies alphabetically for ease of identification.

Chapter 13 also includes the following information on the referenced document itself (see Figure 13):

- The document's publication designation,
- The document's edition year,
- The document's title,
- Any addenda or revisions to the document that are applicable and
- Every section of the code in which the document is referenced.

For example, a reference to ASME A112.1.3 indicates that this document can be found in Chapter 13 under the heading ASME. The specific standards designation is A112.1.3. For convenience, these designations are listed in alphanumeric order. Chapter 13 identifies that ASME A112.1.3 is titled *Air Gap Fittings for Use with Plumbing Fixtures, Appliances and Appurtenances;* the applicable edition (i.e., its year of publication) is 2000; and it is referenced in two sections of the code.

The key aspect of the manner in which standards are referenced by the code is that a specific edition of a specific standard is clearly identified. In this manner, the requirements necessary for compliance can be readily determined. The basis for code compliance is, therefore, established and available on an equal basis to the code official, the contractor, the designer and the owner.

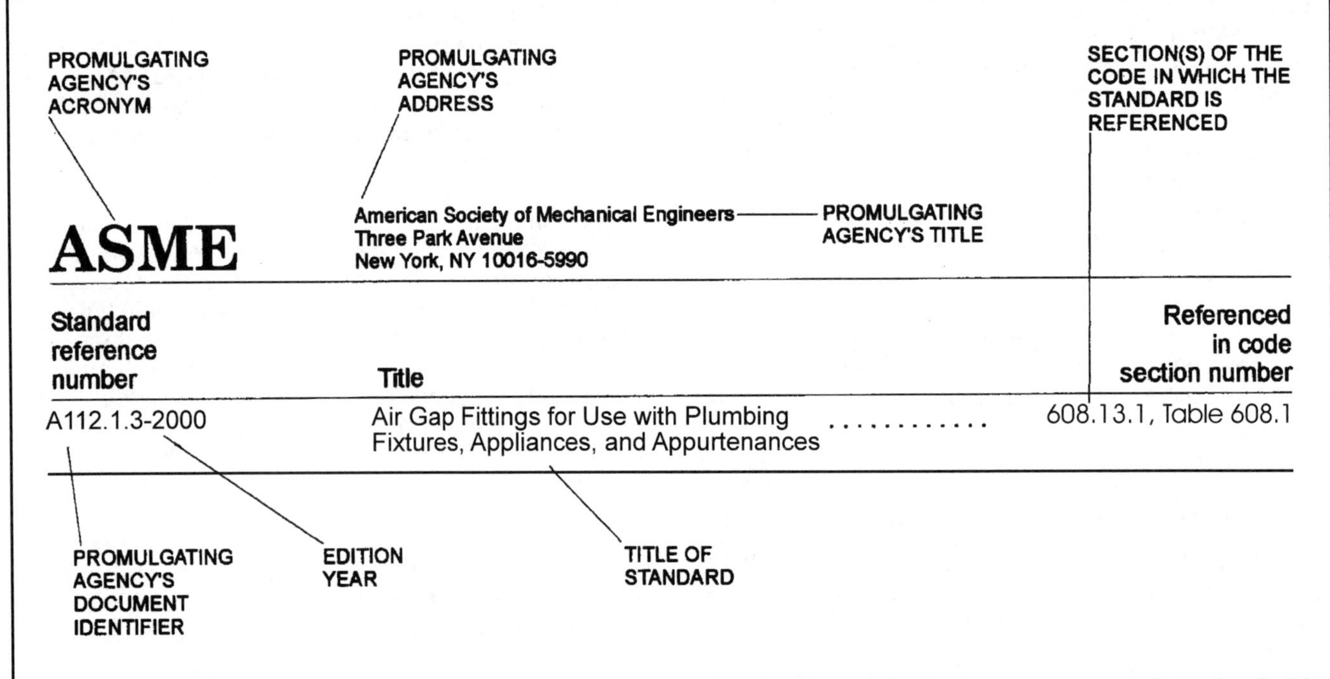

This chapter lists the standards that are referenced in various sections of this document. The standards are listed herein by the promulgating agency of the standard, the standard identification, the effective date and title, and the section or sections of this document that reference the standard. The application of the referenced standards shall be as specified in Section 102.8.

ANSI

American National Standards Institute
25 West 43rd Street, Fourth Floor
New York, NY 10036

Standard Reference Number	Title	Referenced in code section number
A118.10—99	Specifications for Load Bearing, Bonded, Waterproof Membranes for Thin Set Ceramic Tile and Dimension Stone Installation	417.5.2.5
Z4.3—95	Minimum Requirements for Nonsewered Waste-disposal Systems	311.1
Z21.22—99 (R2003)	Relief Valves for Hot Water Supply Systems with Addenda Z21.22a-2000 (R2003) and Z21.22b-2001 (R2003)	501.2, 501.4
Z124.1—95	Plastic Bathtub Units	407.1
Z124.2—95	Plastic Shower Receptors and Shower Stalls	417.1
Z124.3—95	Plastic Lavatories	416.1, 416.2, 417.1
Z124.4—96	Plastic Water Closet Bowls and Tanks	420.1
Z124.6—97	Plastic Sinks	415.1, 418.1
Z124.9—94	Plastic Urinal Fixtures	419.1

AHRI

Air-Conditioning, Heating, & Refrigeration Institute
4100 North Fairfax Drive, Suite 200
Arlington, VA 22203

Standard Reference Number	Title	Referenced in code section number
1010—02	Self-contained, Mechanically Refrigerated Drinking-Water Coolers	410.1

ASME

American Society of Mechanical Engineers
Three Park Avenue
New York, NY 10016-5990

Standard Reference Number	Title	Referenced in code section number
A112.1.2—2004	Air Gaps in Plumbing Systems	Table 608.1, 608.13.1
A112.1.3—2000 Reaffirmed 2005	Air Gap Fittings for Use with Plumbing Fixtures, Appliances and Appurtenances	Table 608.1, 608.13.1
A112.3.1—2007	Stainless Steel Drainage Systems for Sanitary, DWV, Storm and Vacuum Applications Above and Below Ground	412.1, Table 702.1, Table 702.2, Table 702.3, Table 702.4, 708.2, Table 1102.4, Table 1102.5, 1102.6, Table 1102.7
A112.3.4—2000 (Reaffirmed 2004)	Macerating Toilet Systems and Related Components	712.4.1
A112.4.1—1993 (R2002)	Water Heater Relief Valve Drain Tubes	504.6
A112.4.3—1999 (Reaffirmed 2004)	Plastic Fittings for Connecting Water Closets to the Sanitary Drainage System	405.4
A112.6.1M—1997 (R2002)	Floor-affixed Supports for Off-the-floor Plumbing Fixtures for Public Use	405.4.3
A112.6.2—2000 (Reaffirmed 2004)	Framing-affixed Supports for Off-the-floor Water Closets with Concealed Tanks	405.4.3
A112.6.3—2001 (Reaffirmed 2007)	2001 Floor and Trench Drains	412.1
A112.6.7—2001 (Reaffirmed 2007)	Enameled and Epoxy-coated Cast-iron and PVC Plastic Sanitary Floor Sinks	427.1
A112.14.1—2003	Backwater Valves	715.2
A112.14.3—2000	Grease Interceptors	1003.3.4
A112.14.4—2001 (Reaffirmed 2007)	Grease Removal Devices	1003.3.4
A112.18.1-2005/ CSA B125.1-2005	Plumbing Supply Fittings	424.1, 424.2, 424.3, 607.4, 608.2

ASME—continued

A112.18.2-2005/ CSA B125.2-2005	Plumbing Waste Fittings	424.1.2
A112.18.3—2002	Performance Requirements for Backflow Protection Devices and Systems in Plumbing Fixture Fittings	424.2, 424.6
A112.18.6—2003	Flexible Water Connectors	605.6
A112.18.7—1999 (Reaffirmed 2004)	Deck mounted Bath/Shower Transfer Valves with Integral Backflow Protection	424.8
A112.19.1M—2004 (Reaffirmed 2004)	Enameled Cast Iron Plumbing Fixtures	407.1, 410.1, 415.1, 416.1, 418.1
A112.19.2—2003	Vitreous China Plumbing Fixtures and Hydraulic Requirements for Water Closets and Urinals	401.2, 405.9, 408.1, 410.1, 416.1, 418.1, 419.1, 420.1
A112.19.3M—2000 (Reaffirmed 2007)	Stainless Steel Plumbing Fixtures (Designed for Residential Use)	405.9, 415.1, 416.1, 418.1
A112.19.4M—1994 (Reaffirmed 2004)	Porcelain Enameled Formed Steel Plumbing Fixtures	407.1, 416.1, 418.1
A112.19.5—2005	Trim for Water-closet Bowls, Tanks and Urinals	425.4
A112.19.6—1995	Hydraulic Performance Requirements for Water Closets and Urinals	419.1, 420.1
A112.19.7M—2006	Hydromassage Bathtub Appliances	421.1
A112.19.8M—2007	Suction Fittings for Use in Swimming Pools, Wading Pools, Spas, Hot Tubs	421.4
A112.19.9M—1991(R2002)	Nonvitreous Ceramic Plumbing Fixtures with 2002 Supplement	407.1, 408.1, 410.1, 415.1, 416.1, 417.1, 418.1, 420.1
A112.19.12—2006	Wall Mounted and Pedestal Mounted, Adjustable, Elevating, Tilting and Pivoting Lavatory, Sink and Shampoo Bowl Carrier Systems and Drain Systems	416.4, 418.3
A112.19.13—2001 (Reaffirmed 2007)	Electrohydraulic Water Closets	420.1
A112.19.15— 2005	Bathtub/Whirlpool Bathtubs with Pressure Sealed Doors	407.4, 421.5
A112.19.19—2006	Vitreous China Nonwater Urinals	419.1
A112.21.2M—1983	Roof Drains	1102.6
A112.36.2M—1991(R2002)	Cleanouts	708.2
B1.20.1—1983(R2006)	Pipe Threads, General Purpose (inch)	605.10.3, 605.12.3, 605.14.4, 605.16.3, 605.18.1, 705.2.3, 705.4.3, 705.9.4, 705.12.1, 705.14.3
B16.3—2006	Malleable Iron Threaded Fittings Classes 150 and 300	Table 605.5, Table 702.4, Table 1102.7
B16.4—2006	Gray Iron Threaded Fittings Classes 125 and 250	Table 605.5, Table 702.4, Table 1102.7
B16.9—2003	Factory-made Wrought Steel Buttwelding Fittings	Table 605.5, Table 702.4, Table 1102.7
B16.11—2005	Forged Fittings, Socket-welding and Threaded	Table 605.5, Table 702.4, Table 1102.7
B16.12—1998 (Reaffirmed 2006)	Cast-iron Threaded Drainage Fittings	Table 605.5, Table 702.4, Table 1102.7
B16.15—2006	Cast Bronze Threaded Fittings	Table 605.5, Table 702.4, Table 1102.7
B16.18—2001 (Reaffirmed 2005)	Cast Copper Alloy Solder Joint Pressure Fittings	Table 605.5, Table 702.4, Table 1102.7
B16.22—2001 (Reaffirmed 2005)	Wrought Copper and Copper Alloy Solder Joint Pressure Fittings	Table 605.5, Table 702.4, Table 1102.7
B16.23—2002 (Reaffirmed 2006)	Cast Copper Alloy Solder Joint Drainage Fittings DWV	Table 605.5, Table 702.4, Table 1102.7
B16.26—2006	Cast Copper Alloy Fittings for Flared Copper Tubes	Table 605.5, Table 702.4, Table 1102.7
B16.28—1994	Wrought Steel Buttwelding Short Radius Elbows and Returns	Table 605.5, Table 702.4, Table 1102.7
B16.29—2001	Wrought Copper and Wrought Copper Alloy Solder Joint Drainage Fittings (DWV)	Table 605.5, Table 702.4, Table 1102.7

ASSE

American Society of Sanitary Engineering
901 Canterbury Road, Suite A
Westlake, OH 44145

Standard Reference Number	Title	Referenced in code section number
1001—02	Performance Requirements for Atmospheric Type Vacuum Breakers	425.2, Table 608.1, 608.13.6, 608.16.4.1
1002—99	Performance Requirements for Antisiphon Fill Valves (Ballcocks) for Gravity Water Closet Flush Tanks	425.3.1, Table 608.1
1003—01	Performance Requirements for Water Pressure Reducing Valves	604.8
1004—90	Performance Requirements for Backflow Prevention Requirements for Commercial Dishwashing Machines	409.1
1005—99	Performance Requirements for Water Heater Drain Valves	501.3
1006—89	Performance Requirements for Residential Use Dishwashers	409.1
1007—92	Performance Requirements for Home Laundry Equipment	406.1, 406.2
1008—89	Performance Requirements for Household Food Waste Disposer Units	413.1

ASSE—continued

1009—90	Performance Requirements for Commercial Food Waste Grinder Units	413.1
1010—04	Performance Requirements for Water Hammer Arresters	604.9
1011—04	Performance Requirements for Hose Connection Vacuum Breakers	Table 608.1, 608.13.6
1012—02	Performance Requirements for Backflow Preventers with Intermediate Atmospheric Vent	Table 608.1, 608.13.3, 608.16.2
1013—05	Performance Requirements for Reduced Pressure Principle Backflow Preventers and Reduced Pressure Fire Protection Principle Backflow Preventers	Table 608.1, 608.13.2, 608.16.2
1015—05	Performance Requirements for Double Check Backflow Prevention Assemblies and Double Check Fire Protection Backflow Prevention Assemblies	Table 608.1, 608.13.7
1016—96	Performance Requirements for Individual Thermostatic, Pressure Balancing and Combination Control Valves for Individual Fixture Fittings	424.3, 424.4, 607.4
1017—03	Performance Requirements for Temperature Actuated Mixing Valves for Hot Water Distribution Systems	501.2, 613.1
1018—01	Performance Requirements for Trap Seal Primer Valves; Potable Water Supplied	1002.4
1019—04	Performance Requirements for Vacuum Breaker Wall Hydrants, Freeze Resistant, Automatic Draining Type	Table 608.1, 608.13.6
1020—04	Performance Requirements for Pressure Vacuum Breaker Assembly	Table 608.1, 608.13.5
1022—03	Performance Requirements for Backflow Preventer for Beverage Dispensing Equipment	Table 608.1, 608.16.1, 608.16.10
1024—04	Performance Requirements for Dual Check Valve Type Backflow Preventers (for Residential Supply Service or Individual Outlets)	605.3.1, Table 608.1
1035—02	Performance Requirements for Laboratory Faucet Backflow Preventers	Table 608.1, 608.13.6
1037—90	Performance Requirements for Pressurized Flushing Devices for Plumbing Fixtures	425.2
1044—01	Performance Requirements for Trap Seal Primer Devices Drainage Types and Electronic Design Types	1002.4
1047—05	Performance Requirements for Reduced Pressure Detector Fire Protection Backflow Prevention Assemblies	Table 608.1, 608.13.2
1048—05	Performance Requirements for Double Check Detector Fire Protection Backflow Prevention Assemblies	Table 608.1, 608.13.7
1050—02	Performance Requirements for Stack Air Admittance Valves for Sanitary Drainage Systems	917.1
1051—02	Performance Requirements for Individual and Branch Type Air Admittance Valves for Sanitary Drainage Systems-fixture and Branch Devices	917.1
1052—04	Performance Requirements for Hose Connection Backflow Preventers	Table 608.1, 608.13.6
1055—97	Performance Requirements for Chemical Dispensing Systems	608.13.9
1056—01	Performance Requirements for Spill Resistant Vacuum Breaker	Table 608.1, 608.13.5, 608.13.8
1060—96	Performance Requirements for Outdoor Enclosures for Backflow Prevention Assemblies	608.14.1
1061—06	Performance Requirements for Removable and Nonremovable Push Fit Fittings	Table 605.5
1062—97	Performance Requirements for Temperature Actuated, Flow Reduction Valves to Individual Fixture Fittings	424.7
1066—97	Performance Requirements for Individual Pressure Balancing In-line Valves for Individual Fixture Fittings	604.11
1069—05	Performance Requirements for Automatic Temperature Control Mixing Valves	424.4
1070—04	Performance Requirements for Water-temperature Limiting Devices	408.3, 416.5, 424.5, 607.1
1079—2005	Dielectric Pipe Unions	605.24.1, 605.24.3
5013—98	Performance Requirements for Testing Reduced Pressure Principle Backflow Prevention Assembly (RPA) and Reduced Pressure Fire Protection Principle Backflow Preventers (RFP)	312.10.2
5015—98	Performance Requirements for Testing Double Check Valve Backflow Prevention Assembly (DCVA)	312.10.2
5020—98	Performance Requirements for Testing Pressure Vacuum Breaker Assembly (PVBA)	312.10.2
5047—98	Performance Requirements for Testing Reduced Pressure Detector Fire Protection Backflow Prevention Assemblies (RPDA)	312.10.2
5048—98	Performance Requirements for Testing Double Check Valve Detector Assembly (DCDA)	312.10.2
5052—98	Performance Requirements for Testing Hose Connection Backflow Preventers	312.10.2
5056—98	Performance Requirements for Testing Spill Resistant Vacuum Breaker	312.10.2

ASTM

ASTM International
100 Barr Harbor Drive
West Conshohocken, PA 19428-2959

Standard Reference Number	Title	Referenced in code section number
A 53/A 53M—06a	Specification for Pipe, Steel, Black and Hot-dipped, Zinc-coated Welded and Seamless	Table 605.3, Table 605.4, Table 702.1
A 74—06	Specification for Cast-iron Soil Pipe and Fittings	Table 702.1, Table 702.2, Table 702.3, Table 702.4, 708.2, 708.7, Table 1102.4, Table 1102.5, Table 1102.7
A 312/A 312M—06	Specification for Seamless and Welded Austenitic Stainless Steel Pipes	Table 605.3, Table 605.4, Table 605.5, 605.23.2

ASTM—continued

A 733—03 Specification for Welded and Seamless Carbon Steel and Austenitic Stainless Steel Pipe Nipples . . . Table 605.8

A 778—01 Specification for Welded Unannealed Austenitic Stainless Steel Tubular Products . . . Table 605.3, Table 605.4, Table 605.5

A 888—07a Specification for Hubless Cast-iron Soil Pipe and Fittings for Sanitary and Storm Drain, Waste, and Vent Piping Application . . . Table 702.1, Table 702.2, Table 702.3, Table 702.4, 708.7, Table 1102.4, Table 1102.5, Table 1102.7

B 32—04 Specification for Solder Metal . . . 605.14.3, 605.15.4, 705.9.3, 705.10.3

B 42—02e01 Specification for Seamless Copper Pipe, Standard Sizes . . . Table 605.3, Table 605.4, Table 702.1

B 43—98(2004) Specification for Seamless Red Brass Pipe, Standard Sizes . . . Table 605.3, Table 605.4, Table 702.1

B 75—02 Specification for Seamless Copper Tube . . . Table 605.3, Table 605.4, Table 702.1, Table 702.2, Table 702.3, Table 1102.4

B 88—03 Specification for Seamless Copper Water Tube . . . Table 605.3, Table 605.4, Table 702.1, Table 702.2, Table 702.3, Table 1102.4

B 152/B 152M—06a Specification for Copper Sheet, Strip Plate and Rolled Bar . . . 402.3, , 417.5.2.4, 425.3.3, 902.2

B 251—02e01 Specification for General Requirements for Wrought Seamless Copper and Copper-alloy Tube . . . Table 605.3, Table 605.4, Table 702.1, Table 702.2, Table 702.3, Table 1102.4

B 302—02 Specification for Threadless Copper Pipe, Standard Sizes . . . Table 605.3, Table 605.4, Table 702.1

B 306—02 Specification for Copper Drainage Tube (DWV) . . . Table 702.1, Table 702.2, Table 1102.4

B 447—07 Specification for Welded Copper Tube . . . Table 605.3, Table 605.4

B 687—99(2005)e01 Specification for Brass, Copper and Chromium-plated Pipe Nipples . . . Table 605.8

B 813—00e01 Specification for Liquid and Paste Fluxes for Soldering of Copper and Copper Alloy Tube . . . 605.14.3, 605.15.4, 705.9.3, 705.10.3

B 828—02 Practice for Making Capillary Joints by Soldering of Copper and Copper Alloy Tube and Fittings . . . 605.14.3, 605.15.4, 705.9.3, 705.10.3

C 4—04e01 Specification for Clay Drain Tile and Perforated Clay Drain Tile . . . Table 702.3, Table 1102.4, Table 1102.5

C 14—07 Specification for Nonreinforced Concrete Sewer, Storm Drain and Culvert Pipe . . . Table 702.3, Table 1102.4

C 76—07 Specification for Reinforced Concrete Culvert, Storm Drain and Sewer Pipe . . . Table 702.3, Table 1102.4

C 296—(2004)e01 Specification for Asbestos-cement Pressure Pipe . . . Table 605.3

C 425—04 Specification for Compression Joints for Vitrified Clay Pipe and Fittings . . . 705.15, 705.19

C 428—97(2006) Specification for Asbestos-cement Nonpressure Sewer Pipe . . . Table 702.2, Table 702.3, Table 702.4, Table 1102.4

C 443—05a Specification for Joints for Concrete Pipe and Manholes, Using Rubber Gaskets . . . 705.6, 705.19

C 508—00(2004) Specification for Asbestos-cement Underdrain Pipe . . . Table 1102.5

C 564—04a Specification for Rubber Gaskets for Cast-iron Soil Pipe and Fittings . . . 705.5.2, 705.5.3, 705.19, Table 1102.4

C 700—07 Specification for Vitrified Clay Pipe, Extra Strength, Standard Strength, and Perforated . . . Table 702.3, 702.4, Table 1102.4, Table 1102.5

C 1053—00(2005) Specification for Borosilicate Glass Pipe and Fittings for Drain, Waste, and Vent (DWV) Applications . . . Table 702.1, Table 702.4

C 1173—06 Specification for Flexible Transition Couplings for Underground Piping System . . . 705.2.1, 705.7.1, 705.14.1, 705.15, 705.16.1, 705.19

C 1277—06 Specification for Shielded Coupling Joining Hubless Cast-iron Soil Pipe and Fittings . . . 705.5.3

C 1440—03 Specification for Thermoplastic Elastomeric (TPE) Gasket Materials for Drain, Waste, and Vent (DWV), Sewer, Sanitary and Storm Plumbing Systems . . . 705.19

C 1460—04 Specification for Shielded Transition Couplings for Use with Dissimilar DWV Pipe and Fittings Above Ground . . . 705.19

C 1461—06 Specification for Mechanical Couplings Using Thermoplastic Elastomeric (TPE) Gaskets for Joining Drain, Waste and Vent (DWV) Sewer, Sanitary and Storm Plumbing Systems for Above and Below Ground Use . . . 705.19

C 1540—04 Specification for Heavy Duty Shielded Couplings Joining Hubless Cast-iron Soil Pipe and Fittings . . . 705.5.3

C 1563—04 Standard Test Method for Gaskets for Use in Connection with Hub and Spigot Cast Iron Soil Pipe and Fittings for Sanitary Drain, Waste, Vent and Storm Piping Applications. . . . 705.5.2

D 1527—99(2005) Specification for Acrylonitrile-Butadiene-Styrene (ABS) Plastic Pipe, Schedules 40 and 80 . . . Table 605.3

D 1785—06 Specification for Poly (Vinyl Chloride) (PVC) Plastic Pipe, Schedules 40, 80 and 120 . . . Table 605.3

D 1869—95(2005) Specification for Rubber Rings for Asbestos-cement Pipe . . . 605.11, 605.24, 705.3, 705.19

D 2235—04 Specification for Solvent Cement for Acrylonitrile-Butadiene-Styrene (ABS) Plastic Pipe and Fittings . . . 605.10.2, 705.2.2, 705.7.2

D 2239—03 Specification for Polyethylene (PE) Plastic Pipe (SIDR-PR) Based on Controlled Inside Diameter . . . Table 605.3

D 2241—05 Specification for Poly (Vinyl Chloride) (PVC) Pressure-rated Pipe (SDR-Series) . . . Table 605.3

D 2282—(2005)99e01 Specification for Acrylonitrile-Butadiene-Styrene (ABS) Plastic Pipe (SDR-PR) . . . Table 605.3

D 2464—06 Specification for Threaded Poly (Vinyl Chloride) (PVC) Plastic Pipe Fittings, Schedule 80 . . . Table 605.5, Table 1102.7

D 2466—06 Specification for Poly (Vinyl Chloride) (PVC) Plastic Pipe Fittings, Schedule 40 . . . Table 605.5, Table 1102.7

D 2467—06 Specification for Poly (Vinyl Chloride) (PVC) Plastic Pipe Fittings, Schedule 80 . . . Table 605.5, Table 1102.7

D 2468—96a Specification for Acrylonitrile-Butadiene-Styrene (ABS) Plastic Pipe Fittings, Schedule 40 . . . Table 605.5, Table 1102.7

D 2564—04e01 Specification for Solvent Cements for Poly (Vinyl Chloride) (PVC) Plastic Piping Systems . . . 605.21.2, 705.8.2, 705.14.2

ASTM—continued

Standard reference number	Title	Referenced in code section number
D 2609—02	Specification for Plastic Insert Fittings for Polyethylene (PE) Plastic Pipe	Table 605.5, Table 1102.7
D 2657—07	Practice for Heat Fusion-joining of Polyolefin Pipe and Fitting	605.19.2, 705.16.1
D 2661—06	Specification for Acrylonitrile-Butadiene-Styrene (ABS) Schedule 40 Plastic Drain, Waste, and Vent Pipe and Fittings	Table 702.1, Table 702.2, Table 702.3, Table 702.4, 705.2.2, 705.7.2, Table 1102.4, Table 1102.7
D 2665—07	Specification for Poly (Vinyl Chloride) (PVC) Plastic Drain, Waste, and Vent Pipe and Fittings	Table 702.1, Table 702.2, Table 702.3, Table 702.4, Table 1102.4, Table 1102.7
D 2672—96a(2003)	Specification for Joints for IPS PVC Pipe Using Solvent Cement	Table 605.3
D 2683—04	Standard Specification for Socket-type Polyethylene fittings for Outside Diameter-controlled Polyethylene Pipe and Tubing	Table 605.5
D 2729—04e01	Specification for Poly (Vinyl Chloride) (PVC) Sewer Pipe and Fittings	Table 1102.5
D 2737—03	Specification for Polyethylene (PE) Plastic Tubing	Table 605.3
D 2751—05	Specification for Acrylonitrile-Butadiene-Styrene (ABS) Sewer Pipe and Fittings	Table 702.3, Table 702.4, Table 1102.7
D 2846/D 2846M—06	Specification for Chlorinated Poly (Vinyl Chloride) (CPVC) Plastic Hot and Cold Water Distribution Systems	Table 605.3, Table 605.4, Table 605.5, 605.16.2
D 2855—96(2002)	Standard Practice for Making Solvent-cemented Joints with Poly (Vinyl Chloride) (PVC) Pipe and Fittings	605.22.2, 705.8.2, 705.14.2
D 2949—01ae01	Specification for 3.25-in Outside Diameter Poly (Vinyl Chloride) (PVC) Plastic Drain, Waste, and Vent Pipe and Fittings	Table 702.1, Table 702.2, Table 702.3, Table 702.4
D 3034—06	Specification for Type PSM Poly (Vinyl Chloride) (PVC) Sewer Pipe and Fittings	Table 702.3, Table 702.4, Table 1102.7, Table 1102.4
D 3035-03	Standard Specification for Polyethylene (PE) Plastic Pipe (DR-PR) Based on Controlled Outside Diameter	Table 605.3
D 3139—98(2005)	Specification for Joints for Plastic Pressure Pipes Using Flexible Elastomeric Seals	605.10.1, 605.22.1
D 3212—96a(2003)e01	Specification for Joints for Drain and Sewer Plastic Pipes Using Flexible Elastomeric Seals	705.2.1, 705.8.1, 705.14.1, 705.16.2
D 3261—03	Standard Specification for Butt Heat Fusion Polyethylene (PE) Plastic fittings for Polyethylene (PE) Plastic Pipe and Tubing	Table 605.5
D 3311—06a	Specification for Drain, Waste and Vent (DWV) Plastic Fittings Patterns	Table 1102.7
D 4068—01	Specification for Chlorinated Polyethlene (CPE) Sheeting for Concealed Water-containment Membrane	417.5.2.2
D 4551—96(2001)	Specification for Poly (Vinyl Chloride) (PVC) Plastic Flexible Concealed Water-containment Membrane	417.5.2.1
F 405—05	Specification for Corrugated Polyethylene (PE) Tubing and Fittings	Table 1102.5
F 409—02	Specification for Thermoplastic Accessible and Replaceable Plastic Tube and Tubular Fittings	424.1.2, Table 1102.7
F 437—06	Specification for Threaded Chlorinated Poly (Vinyl Chloride) (CPVC) Plastic Pipe Fittings, Schedule 80	Table 605.5
F 438—04	Specification for Socket-type Chlorinated Poly (Vinyl Chloride) (CPVC) Plastic Pipe Fittings, Schedule 40	Table 605.5
F 439—06	Standard Specification for Chlorinated Poly (Vinyl Chloride) (CPVC) Plastic Pipe Fittings, Schedule 80	Table 605.5
F 441/F 441M—02	Specification for Chlorinated Poly (Vinyl Chloride) (CPVC) Plastic Pipe, Schedules 40 and 80	Table 605.3, Table 605.4
F 442/F 442M—99(2005)	Specification for Chlorinated Poly (Vinyl Chloride) (CPVC) Plastic Pipe (SDR-PR)	Table 605.3, Table 605.4
F 477—07	Specification for Elastomeric Seals (Gaskets) for Joining Plastic Pipe	605.24, 705.19
F 493—04	Specification for Solvent Cements for Chlorinated Poly (Vinyl Chloride) (CPVC) Plastic Pipe and Fittings	605.16.2
F 628—06e01	Specification for Acrylonitrile-Butadiene-Styrene (ABS) Schedule 40 Plastic Drain, Waste, and Vent Pipe with a Cellular Core	Table 702.1, Table 702.2, Table 702.3, Table 702.4, 705.2.2, 705.7.2, Table 1102.4, Table 1102.7
F 656—02	Specification for Primers for Use in Solvent Cement Joints of Poly (Vinyl Chloride) (PVC) Plastic Pipe and Fittings	605.22.2, 705.8.2, 705.14.2
F 714—06a	Specification for Polyethylene (PE) Plastic Pipe (SDR-PR) Based on Outside Diameter	Table 702.3
F 876—06	Specification for Cross-linked Polyethylene (PEX) Tubing	Table 605.3, Table 605.4
F 877—07	Specification for Cross-linked Polyethylene (PEX) Plastic Hot and Cold Water Distribution Systems	Table 605.3, Table 605.4, Table 605.5
F 891—04	Specification for Coextruded Poly (Vinyl Chloride) (PVC) Plastic Pipe with a Cellular Core	Table 702.1 Table 702.2, Table 702.3, Table 1102.4, Table 1102.5, Table 1102.7
F 1055—98(2006)	Standard Specification for Electrofusion Type Polyethylene Fittings for Outside Diameter Controlled Polyethylene Pipe and Tubing	Table 605.5
F 1281—07	Specification for Cross-linked Polyethylene/Aluminum/ Cross-linked Polyethylene (PEX-AL-PEX) Pressure Pipe	Table 605.3, Table 605.4, Table 605.5, 605.21.1
F 1282—06	Specification for Polyethylene/Aluminum/Polyethylene (PE-AL-PE) Composite Pressure Pipe	Table 605.3, Table 605.4, Table 605.5, 605.21.1

ASTM—continued

F 1412—01e01	Specification for Polyolefin Pipe and Fittings for Corrosive Waste Drainage	Table 702.1, Table 702.2, Table 702.4, 705.17.1
F 1488—03	Specification for Coextruded Composite Pipe	Table 702.1, Table 702.2, Table 702.3
F 1673—04	Polyvinylidene Fluoride (PVDF) Corrosive Waste Drainage Systems	Table 702.1, Table 702.2, Table 702.3, Table 702.4, 705.18.1
F 1807—07	Specification for Metal Insert Fittings Utilizing a Copper Crimp Ring for SDR9 Cross-linked Polyethylene (PEX) Tubing	Table 605.5
F 1866—07	Specification for Poly (Vinyl Chloride) (PVC) Plastic Schedule 40 Drainage and DWV Fabricated Fittings	Table 702.4, Table 1102.7
F 1960—07	Specification for Cold Expansion Fittings with PEX Reinforcing Rings for use with Cross-linked Polyethylene (PEX) Tubing	Table 605.5
F 1974—04	Specification for Metal Insert Fittings for Polyethylene/Aluminum/Polyethylene and Cross-linked Polyethylene/Aluminum/Cross-linked Polyethylene Composite Pressure Pipe	Table 605.5, 605.21.1
F 1986—01(2006)	Specification for Multilayer Pipe, Type 2, Compression Fittings and Compression Joints for Hot and Cold Drinking Water Systems	Table 605.3, Table 605.4, Table 605.5
F 2080—05	Specifications for Cold-expansion Fittings with Metal Compression-sleeves for Cross-linked Polyethylene (PEX) Pipe	Table 605.5
F 2098—04e01	Standard specification for Stainless Steel Clamps for Securing SDR9 Cross-linked Polyethylene (PEX) Tubing to Metal Insert Fittings	Table 605.5
F 2159—05	Specification for Plastic Insert Fittings Utilizing a Copper Crimp Ring for SDR9 Cross-linked Polyethylene (PEX) Tubing	Table 605.5
F 2262—05	Specification for Cross-linked Polyethylene/Aluminum/Cross-linked Polyethylene Tubing OD Controlled SDR9	Table 605.3, Table 605.4
F 2306/F 2306M-05	12" to 60" Annular Corrugated Profile-wall Polyethylene (PE) Pipe and Fittings for Gravity Flow Storm Sewer and Subsurface Drainage Applications	Table 1102.4, Table 1102.7
F 2389—06	Specification for Pressure-rated Polypropylene (PP) Piping Systems	Table 605.3, Table 605.4, Table 605.5, 605.20.1
F 2434—05	Standard Specification for Metal Insert Fittings Utilizing a Copper Crimp Ring for SDR9 Cross-linked Polyethylene (PEX) Tubing and SDR9 Cross-linked Polyethylene/Aluminum/Cross-linked Polyethylene (PEX AL-PEX) Tubing	Table 605.5

AWS

American Welding Society
550 N.W. LeJeune Road
Miami, FL 33126

Standard Reference Number	Title	Referenced in code section number
A5.8—04	Specifications for Filler Metals for Brazing and Braze Welding	605.12.1, 605.14.1, 605.15.1, 705.4.1, 705.9.1, 705.10.1

AWWA

American Water Works Association
6666 West Quincy Avenue
Denver, CO 80235

Standard Reference Number	Title	Referenced in code section number
C104—98	Standard for Cement-mortar Lining for Ductile-iron Pipe and Fittings for Water	605.3, 605.5
C110—/A21.10—03	Standard for Ductile-iron and Gray-iron Fittings, 3 Inches through 48 Inches, for Water	Table 605.5, Table 702.4, Table 1102.7
C111—00	Standard for Rubber-gasket Joints for Ductile-iron Pressure Pipe and Fittings	605.13
AWWA—continued		
C115/A21.15—99	Standard for Flanged Ductile-iron Pipe with Ductile-iron or Gray-iron Threaded Flanges	Table 605.3, Table 605.4
C151/A21.51—02	Standard for Ductile-iron Pipe, Centrifugally Cast for Water	Table 605.3, Table 605.4
C153—00/A21.53—00	Standard for Ductile-iron Compact Fittings for Water Service	Table 605.5
C510—00	Double Check Valve Backflow Prevention Assembly	Table 608.1, 608.13.7
C511—00	Reduced-pressure Principle Backflow Prevention Assembly	Table 608.1, 608.13.2, 608.16.2
C651—99	Disinfecting Water Mains	610.1
C652—02	Disinfection of Water-storage Facilities	610.1

CISPI

Cast Iron Soil Pipe Institute
5959 Shallowford Road, Suite 419
Chattanooga, TN 37421

Standard Reference Number	Title	Referenced in code section number
301—04a	Specification for Hubless Cast-iron Soil Pipe and Fittings for Sanitary and Storm Drain, Waste and Vent Piping Applications	Table 702.1, Table 702.2, Table 702.3, Table 702.4, 708.7, Table 1102.4, Table 1102.5, Table 1102.7
310—04	Specification for Coupling for Use in Connection with Hubless Cast-iron Soil Pipe and Fittings for Sanitary and Storm Drain, Waste and Vent Piping Applications	705.5.3

CSA

Canadian Standards Association
5060 Spectrum Way.
Mississauga, Ontario, Canada L4W 5N6

Standard Reference Number	Title	Referenced in code section number
B45.1—02	Ceramic Plumbing Fixtures	408.1, 416.1, 418.1, 419.1, 420.1
B45.2—02	Enameled Cast-iron Plumbing Fixtures	407.1, 415.1, 416.1, 418.1
B45.3—02	Porcelain Enameled Steel Plumbing Fixtures	407.1, 416.1, 418.1
B45.4—02	Stainless-steel Plumbing Fixtures	415.1, 416.1, 418.1, 420.1
B45.5—02	Plastic Plumbing Fixtures	407.1, 416.2, 417.1, 419.1, 420.1, 421.1
B45.9—99	Macerating Systems and Related Components	712.4.1
B64.1.2—01	Vacuum Breakers, Pressure Type (PVB)	Table 608.1, 608.13.5
B64.2.1—01	Vacuum Breakers, Hose Connection Type (HCVB) with Manual Draining Feature	Table 608.1, 608.13.6
B64.2.1.1—01	Vacuum Breakers, Hose Connection Dual Check Type (HCDVB)	Table 608.1, 608.13.6
B64.4.1—01	Backflow Preventers, Reduced Pressure Principle Type for Fire Sprinklers (RPF)	Table 608.1, 608.13.2
B64.5—01	Backflow Preventers, Double Check Type (DCVA)	Table 608.1, 608.13.7
B64.5.1—01	Backflow Preventers, Double Check Type for Fire Systems (DCVAF)	Table 608.1 608.13.7
B64.6—01	Backflow Preventers, Dual Check Valve Type (DuC)	605.3.1, Table 608.1
B64.7—94	Vacuum Breakers, Laboratory Faucet Type (LFVB	Table 608.1, 608.13.6
B64.10/B64.10.1—01	Manual for the Selection and Installation of Backflow Prevention Devices/Manual for the Maintenance and Field Testing of Backflow Prevention Devices	312.10.2
B79—94(2000)	Floor, Area and Shower Drains, and Cleanouts for Residential Construction	412.1
B125—01	Plumbing Fittings	424.4, 424.6, 425.4
B125.1/ASME A112.18.1—05	Plumbing Supply Fittings	424.1, 424.2, 424.3, 607.4, 608.2
B125.2/ASME A112.18.2—05	Plumbing Waste Fittings	424.1.2
B125.3—2005	Plumbing Fittings	416.5, 424.5, 425.3.1, Table 608.1
B137.1—02	Polyethylene Pipe, Tubing and Fittings for Cold Water Pressure Services	Table 605.3
B137.2—02	PVC Injection-moulded Gasketed Fittings for Pressure Applications	Table 605.5, Table 1102.7
B137.3—02	Rigid Poly (Vinyl Chloride) (PVC) Pipe for Pressure Applications	Table 605.3, Table 605.4, Table 605.5, 605.22.2, 705.8.2, 705.14.2
B137.5—02	Cross-linked Polyethylene (PEX) Tubing Systems for Pressure Applications—with Revisions through September 1992	Table 605.3, Table 605.4, Table 605.5
B137.6—02	CPVC Pipe, Tubing and Fittings for Hot and Cold Water Distribution Systems—with Revisions through May 1986	Table 605.3, Table 605.4
B137.11—02	Polypropylene (PP-R) Pipe and Fittings for Pressure Applications	Table 605.3, Table 605.4, Table 605.5
B181.1—02	ABS Drain, Waste and Vent Pipe and Pipe Fittings	Table 702.1, Table 702.2, Table 702.3, Table 702.4, 705.2.2, 705.7.2, 715.2, Table 1102.4, Table 1102.7
B181.2—02	PVC Drain, Waste, and Vent Pipe and Pipe Fittings—with Revisions through December 1993	Table 702.1 Table 702.2, 705.8.2, 705.14.2, 715.2
B182.1—02	Plastic Drain and Sewer Pipe and Pipe Fittings	705.8.2, 705.14.2, Table 1102.4
B182.2—02	PVC Sewer Pipe and Fittings (PSM Type)	Table 702.3, Table 1102.4, Table 1102.5
B182.4—02	Profile PVC Sewer Pipe and Fittings	Table 702.3, Table 1102.4, Table 1102.5
B182.6—02	Profile Polyethylene Sewer Pipe and Fittings for Leak-proof Sewer Applications	Table 1102.5
B182.8—02	Profile Polyethylene Storm Sewer and Drainage Pipe and Fittings	Table 1102.5
CAN/CSA-A257.1M—92	Circular Concrete Culvert, Storm Drain, Sewer Pipe and Fittings	Table 702.3, Table 1102.4
CAN/CSA-A257.2M—92	Reinforced Circular Concrete Culvert, Storm Drain, Sewer Pipe and Fittings	Table 702.3, Table 1102.4
CAN/CSA-A257.3M—92	Joints for Circular Concrete Sewer and Culvert Pipe, Manhole Sections and Fittings Using Rubber Gaskets	705.6, 705.19
CAN/CSA-B64.1.1—01	Vacuum Breakers, Atmospheric Type (AVB)	425.2, Table 608.1, 608.13.6
CAN/CSA-B64.2—01	Vacuum Breakers, Hose Connection Type (HCVB)	Table 608.1, 608.13.6

CSA—continued

CAN/CSA-B64.2.2—01	Vacuum Breakers, Hose Connection Type (HCVB) with Automatic Draining Feature	Table 608.1, 608.13.6
CAN/CSA-B64.3—01	Backflow Preventers, Dual Check Valve Type with Atmospheric Port (DCAP)	Table 608.1, 608.13.3, 608.16.2
CAN/CSA-B64.4—01	Backflow Preventers, Reduced Pressure Principle Type (RP)	Table 608.1, 608.13.2, 608.16.2
CAN/CSA-B64.10—01	Manual for the Selection, Installation, Maintenance and Field Testing of Backflow Prevention Devices	312.10.2
CAN/CSA-B137.9—02	Polyethylene/Aluminum/Polyethylene Composite Pressure Pipe Systems	Table 605.3, Table 605.5, 605.21.1
CAN/CSA-B137.10M—02	Cross-linked Polyethylene/Aluminum/Polyethylene Composite Pressure Pipe Systems	Table 605.3, Table 605.4, Table 605.5, 605.21.1
CAN/CSA-B181.3—02	Polyolefin Laboratory Drainage Systems	Table 702.1, Table 702.2, Table 702.4, 705.17.1
CAN/CSA-B182.4—02	Profile PVC Sewer Pipe and Fittings	Table 702.3, Table 1102.4, Table 1102.5
CAN/CSA-B602—02	Mechanical Couplings for Drain, Waste and Vent Pipe and Sewer Pipe	705.2.1, 705.5.3, 705.6, 705.7.1, 705.14.1, 705.15, 705.16.2, 705.19

ICC

International Code Council, Inc.
500 New Jersey Ave, NW
6th Floor
Washington, DC 20001

Standard Reference Number	Title	Referenced in code section number
IBC—09	International Building Code®	201.3, 305.4, 307.1, 307.2, 307.3, 308.2, 309.1,310.1, 310.3, 403.1, Table 403.1, 404.1, 407.3, 417.6, 502.4, 606.5.2, 1106.5
IEBC—09	International Existing Building Code	101.2
IECC—09	International Energy Conservation Code	313.1, 607.2, 607.2.1
IFC—09	International Fire Code®	201.3, 1201.1
IFGC—09	International Fuel Gas Code®	101.2, 201.3, 502.1
IMC—09	International Mechanical Code®	201.3, 307.6, 310.1, 422.9, 502.1, 612.1, 1202.1
IPSDC—09	International Private Sewage Disposal Code®	701.2
IRC—09	International Residential Code	101.2

ISEA

International Safety Equipment Association
1901 N. Moore Street, Suite 808
Arlington, VA 22209

Standard Reference Number	Title	Referenced in code section number
Z358.1—98	Emergency Eyewash and Shower Equipment	411.1

NFPA

National Fire Protection Association
1 Batterymarch Park
Quincy, MA 02169-7471

Standard Reference Number	Title	Referenced in code section number
50—01	Bulk Oxygen Systems at Consumer Sites	1203.1
51—07	Design and Installation of Oxygen-fuel Gas Systems for Welding, Cutting and Allied Processes	1203.1
70—08	National Electric Code	502.1, 504.3, 1113.1.3
99C—05	Gas and Vacuum Systems	1202.1

NSF

NSF International
789 Dixboro Road
Ann Arbor, MI 48105

Standard Reference Number	Title	Referenced in code section number
3—2007	Commercial Warewashing Equipment	409.1
14—2007	Plastic Piping System Components and Related Materials	303.3, 611.3
18—2007	Manual Food and Beverage Dispensing Equipment	426.1
42—2007e	Drinking Water Treatment Units—Aesthetic Effects	611.1, 611.3

NSF—continued

44—2004	Residential Cation Exchange Water Softeners	611.1, 611.3
53—2007	Drinking Water Treatment Units—Health Effects	611.1, 611.3
58—2006	Reverse Osmosis Drinking Water Treatment Systems	611.2
61—2007a	Drinking Water System Components—Health Effects	410.1, 424.1, 605.3, 605.4, 605.5, 605.7, 611.3, 611.3
62—2004	Drinking Water Distillation Systems	611.1

PDI

Plumbing and Drainage Institute
800 Turnpike Street, Suite 300
North Andover, MA 01845

Standard Reference Number	Title	Referenced in code section number
G101(2003)	Testing and Rating Procedure for Grease Interceptors with Appendix of Sizing and Installation Data	1003.3.4

UL

Underwriters Laboratories, Inc.
333 Pfingsten Road
Northbrook, IL 60062-2096

Standard Reference Number	Title	Referenced in code section number
UL508—99	Industrial Control Equipment with Revision through July 2005	314.2.3

Appendix A: Plumbing Permit Fee Schedule

General Comments

Appendix A is a sample, guide or prototype of a permit fee schedule. Once established, the schedule should be made available to all plumbing contractors to aid them in preparing bids and applying for permits. This sample fee schedule will more than likely have to be modified to account for local contractor licensing laws, local practices and the typical work seen in the particular jurisdiction. The ultimate goal is to develop a fee schedule that is fair to all permit applicants. The permit fee schedule needs to balance the revenues necessary for department funding without becoming a financial burden to the contractors and building owners purchasing the plumbing permits.

Purpose

Section 106.6 requires that fees be collected for the issuance of plumbing work permits. These fees support the department of plumbing inspection. Many department costs must be funded, including inspector salaries, clerical worker salaries, legal assistance costs, office space costs, equipment costs, insurance costs, education costs and transportation costs. The majority of plumbing inspection departments are either totally or partially funded by permit fees.

Section 106.6.2 requires the creation of a permit fee schedule for permit issuance in the jurisdiction. This schedule sets the fees for permits based on the total cost of the proposed work or the total number of fixtures, devices, water and drainage outlets and systems or some combination of the above. Some jurisdictions simply charge a fee for the number of each type of fixture; others charge a fee based on a percentage of the total plumbing job cost. For example, a $10 fee per $1,000 of job cost would be a 1-percent fee. In all cases, the fees must be determined at the local level based on the needs of the plumbing department. The fees will depend on the amount of funding, if any, that the plumbing department receives from sources other than permit fees.

The provisions contained in this appendix are not mandatory unless specifically referenced in the adopting ordinance.

Permit Issuance

1. For issuing each permit . . . . $______
2. For issuing each supplemental permit . . . . ______

Unit Fee Schedule

1. For each plumbing fixture or trap or set of fixtures on one trap (including water, drainage piping and backflow protection thereof) . . . . ______
2. For each building sewer and each trailer park sewer . . . . ______
3. Rainwater systems—per drain (inside building) . . . . ______
4. For each cesspool (where permitted) . . . . ______
5. For each private sewage disposal system . . . . ______
6. For each water heater and/or vent . . . . ______
7. For each industrial waste pretreatment interceptor including its trap and vent, excepting kitchen-type grease interceptors functioning as fixture traps . . . . ______
8. For installation, alteration or repair of water-piping and/or water-treating equipment, each . . . . ______
9. For repair or alteration of drainage or vent piping, each fixture . . . . ______
10. For each lawn sprinkler system on any one meter including backflow protection devices therefor . . . . ______
11. For atmospheric-type vacuum breakers not included in Item 2:

    1 to 5 . . . . ______

    over 5, each . . . . ______

12. For each backflow protective device other than atmospheric-type vacuum breakers:
 2 inches (51 mm) and smaller . . . . ______
 Over 2 inches (51 mm) . . . . ______

Other Inspections and Fees

1. Inspections outside of normal business hours . . . . ______ per hour
 (minimum charge two hours)
2. Reinspection fee assessed under provisions of Section 107.4.3 . . . . ______ each
3. Inspections for which no fee is specifically indicated . . . . ______ per hour
 (minimum charge one-half hour)
4. Additional plan review required by changes, additions or revisions to approved
 plans (minimum charge one-half hour) . . . . ______ per hour

Appendix B: Rates of Rainfall for Various Cities

General Comments

Gutters, leaders, conductors and downspouts are all components of the storm drainage system and are sized based on the rainfall rate. The rainfall rates, in inches per hour, are based on a storm of 1-hour duration and a 100-year return period. The rainfall rates shown in this appendix are derived from Figure 1106.1. Rainfall rates indicate the maximum rate of rainfall within a given period of time. The rainfall rates are calculated from a statistical analysis of weather records. The roof area listed must be matched to the local rainfall rate.

Purpose

Appendix B provides the hourly rainfall rates indicated in Figure 1106.1.

This appendix is informative and is not part of the code.

Rainfall rates, in inches per hour, are based on a storm of 1-hour duration and a 100-year return period. The rainfall rates shown in the appendix are derived from Figure 1106.1.

Alabama:
- Birmingham 3.8
- Huntsville 3.6
- Mobile 4.6
- Montgomery 4.2

Alaska:
- Fairbanks 1.0
- Juneau 0.6

Arizona:
- Flagstaff 2.4
- Nogales 3.1
- Phoenix 2.5
- Yuma 1.6

Arkansas:
- Fort Smith 3.6
- Little Rock 3.7
- Texarkana 3.8

California:
- Barstow 1.4
- Crescent City 1.5
- Fresno 1.1
- Los Angeles 2.1
- Needles 1.6
- Placerville 1.5
- San Fernando 2.3
- San Francisco 1.5
- Yreka 1.4

Colorado:
- Craig 1.5
- Denver 2.4
- Durango 1.8
- Grand Junction 1.7
- Lamar 3.0
- Pueblo 2.5

Connecticut:
- Hartford 2.7
- New Haven 2.8
- Putnam 2.6

Delaware:
- Georgetown 3.0
- Wilmington 3.1

District of Columbia:
- Washington 3.2

Florida:
- Jacksonville 4.3
- Key West 4.3
- Miami 4.7
- Pensacola 4.6
- Tampa 4.5

Georgia:
- Atlanta 3.7
- Dalton 3.4
- Macon 3.9
- Savannah 4.3
- Thomasville 4.3

Hawaii:
- Hilo 6.2
- Honolulu 3.0
- Wailuku 3.0

Idaho:
- Boise 0.9
- Lewiston 1.1
- Pocatello 1.2

Illinois:
- Cairo 3.3
- Chicago 3.0
- Peoria 3.3
- Rockford 3.2
- Springfield 3.3

Indiana:
- Evansville 3.2
- Fort Wayne 2.9
- Indianapolis 3.1

Iowa:
- Davenport 3.3
- Des Moines 3.4
- Dubuque 3.3
- Sioux City 3.6

Kansas:
- Atwood 3.3
- Dodge City 3.3
- Topeka 3.7
- Wichita 3.7

Kentucky:
- Ashland 3.0
- Lexington 3.1
- Louisville 3.2
- Middlesboro 3.2
- Paducah 3.3

Louisiana:
- Alexandria 4.2
- Lake Providence 4.0
- New Orleans 4.8
- Shreveport 3.9

Maine:
- Bangor 2.2
- Houlton 2.1
- Portland 2.4

Maryland:
- Baltimore 3.2
- Hagerstown 2.8
- Oakland 2.7
- Salisbury 3.1

Massachusetts:
- Boston 2.5
- Pittsfield 2.8
- Worcester 2.7

Michigan:
- Alpena 2.5
- Detroit 2.7
- Grand Rapids 2.6
- Lansing 2.8
- Marquette 2.4
- Sault Ste. Marie 2.2

Minnesota:
- Duluth 2.8
- Grand Marais 2.3
- Minneapolis 3.1
- Moorhead 3.2

Worthington 3.5

Mississippi:
Biloxi 4.7
Columbus 3.9
Corinth 3.6
Natchez 4.4
Vicksburg 4.1

Missouri:
Columbia 3.2
Kansas City 3.6
Springfield 3.4
St. Louis 3.2

Montana:
Ekalaka 2.5
Havre 1.6
Helena 1.5
Kalispell 1.2
Missoula 1.3

Nebraska:
North Platte 3.3
Omaha 3.8
Scottsbluff 3.1
Valentine 3.2

Nevada:
Elko 1.0
Ely 1.1
Las Vegas 1.4
Reno 1.1

New Hampshire:
Berlin 2.5
Concord 2.5
Keene 2.4

New Jersey:
Atlantic City 2.9
Newark 3.1
Trenton 3.1

New Mexico:
Albuquerque 2.0
Hobbs 3.0
Raton 2.5
Roswell 2.6
Silver City 1.9

New York:
Albany 2.5
Binghamton 2.3
Buffalo 2.3
Kingston 2.7
New York 3.0
Rochester 2.2

North Carolina:
Asheville 4.1
Charlotte 3.7
Greensboro 3.4
Wilmington 4.2

North Dakota:
Bismarck 2.8
Devils Lake 2.9
Fargo 3.1
Williston 2.6

Ohio:
Cincinnati 2.9
Cleveland 2.6
Columbus 2.8
Toledo 2.8

Oklahoma:
Altus 3.7
Boise City 3.3
Durant 3.8
Oklahoma City 3.8

Oregon:
Baker 0.9
Coos Bay 1.5
Eugene 1.3
Portland 1.2

Pennsylvania:
Erie 2.6
Harrisburg 2.8
Philadelphia 3.1
Pittsburgh 2.6
Scranton 2.7

Rhode Island:
Block Island 2.75
Providence 2.6

South Carolina:
Charleston 4.3
Columbia 4.0
Greenville 4.1

South Dakota:
Buffalo 2.8
Huron 3.3
Pierre 3.1
Rapid City 2.9
Yankton 3.6

Tennessee:
Chattanooga 3.5
Knoxville 3.2
Memphis 3.7
Nashville 3.3

Texas:
Abilene 3.6
Amarillo 3.5
Brownsville 4.5
Dallas 4.0
Del Rio 4.0
El Paso 2.3
Houston 4.6
Lubbock 3.3
Odessa 3.2
Pecos 3.0
San Antonio 4.2

Utah:
Brigham City 1.2
Roosevelt 1.3
Salt Lake City 1.3
St. George 1.7

Vermont:
Barre 2.3
Bratteboro 2.7
Burlington 2.1
Rutland 2.5

Virginia:
Bristol 2.7
Charlottesville 2.8
Lynchburg 3.2
Norfolk 3.4
Richmond 3.3

Washington:
Omak 1.1
Port Angeles 1.1
Seattle 1.4
Spokane 1.0
Yakima 1.1

West Virginia:
Charleston 2.8
Morgantown 2.7

Wisconsin:
Ashland 2.5
Eau Claire 2.9
Green Bay 2.6
La Crosse 3.1
Madison 3.0
Milwaukee 3.0

Wyoming:
Cheyenne 2.2
Fort Bridger 1.3
Lander 1.5
New Castle 2.5
Sheridan 1.7
Yellowstone Park 1.4

Appendix C:
Gray Water Recycling Systems

General Comments

Gray-water recycling, like all recycling of resources, is an environmentally friendly "green" initiative. This appendix and the water conservation provisions of Section 604.4 can work together to reduce greatly the amount of fresh water we consume and, at the same time, reduce the amount of energy expended using potable water and treating waste water. Water use equates to energy use. Gray-water recycling is not common today because of the expense of the additional required plumbing, the additional maintenance requirements and the general reluctance of people to reuse waste water. In a world of limited natural resources, however, gray-water recycling is an excellent step toward water conservation.

Jurisdictions adopting the *International Plumbing Code®* (IPC®) should also adopt Appendix C and modify Section 301.3 as suggested therein. The modification to Section 301.3 will permit collection and reuse of specified waste water for flushing water closets and urinals. Unlike most plumbing fixtures, water closets and urinals have never needed potable water to function properly and safely. Gray water recycling is a big component of "green" building technology.

Purpose

The provisions of Appendix C allow the construction of gray-water recycling systems that will function as intended with minimal maintenance and negligible health risk.

The provisions contained in this appendix are not mandatory unless specifically referenced in the adopting ordinance.

Note: Section 301.3 of this code requires all plumbing fixtures that receive water or waste to discharge to the sanitary drainage system of the structure. In order to allow for the utilization of a gray water system, Section 301.3 should be revised to read as follows:

301.3 Connections to drainage system. All plumbing fixtures, drains, appurtenances and appliances used to receive or discharge liquid wastes or sewage shall be directly connected to the sanitary drainage system of the building or premises, in accordance with the requirements of this code. This section shall not be construed to prevent indirect waste systems required by Chapter 8.

Exception: Bathtubs, showers, lavatories, clothes washers and laundry trays shall not be required to discharge to the sanitary drainage system where such fixtures discharge to an *approved* gray water system for flushing of water closets and urinals or for subsurface landscape irrigation.

SECTION C101
GENERAL

C101.1 Scope. The provisions of this appendix shall govern the materials, design, construction and installation of gray water systems for flushing of water closets and urinals and for subsurface landscape irrigation (see Figures 1 and 2).

❖ This section contains the requirements for design of, materials for and installation of gray water systems.

C101.2 Definition. The following term shall have the meaning shown herein.

GRAY WATER. Waste discharged from lavatories, bathtubs, showers, clothes washers and laundry trays.

❖ Gray water is the waste discharge from those fixtures that do not or are not intended to receive human bodily waste (urine, feces) or food waste. Water containing human bodily waste is commonly referred to as "black water." Although gray water is certainly not potable, it does not have nearly the disease-causing potential of "black water." Fixtures that can use gray water for flushing are water closets and urinals because neither require potable water to function properly. Raw gray water to be used in fixtures must be filtered, disinfected and identified by color, effectively reducing the potential health risks and any objectionable odors or appearances. Gray water can also be used for subsurface irrigation of yards, lawns, gardens, plantings, etc., but does not have to be disinfected or dyed when used for this purpose.

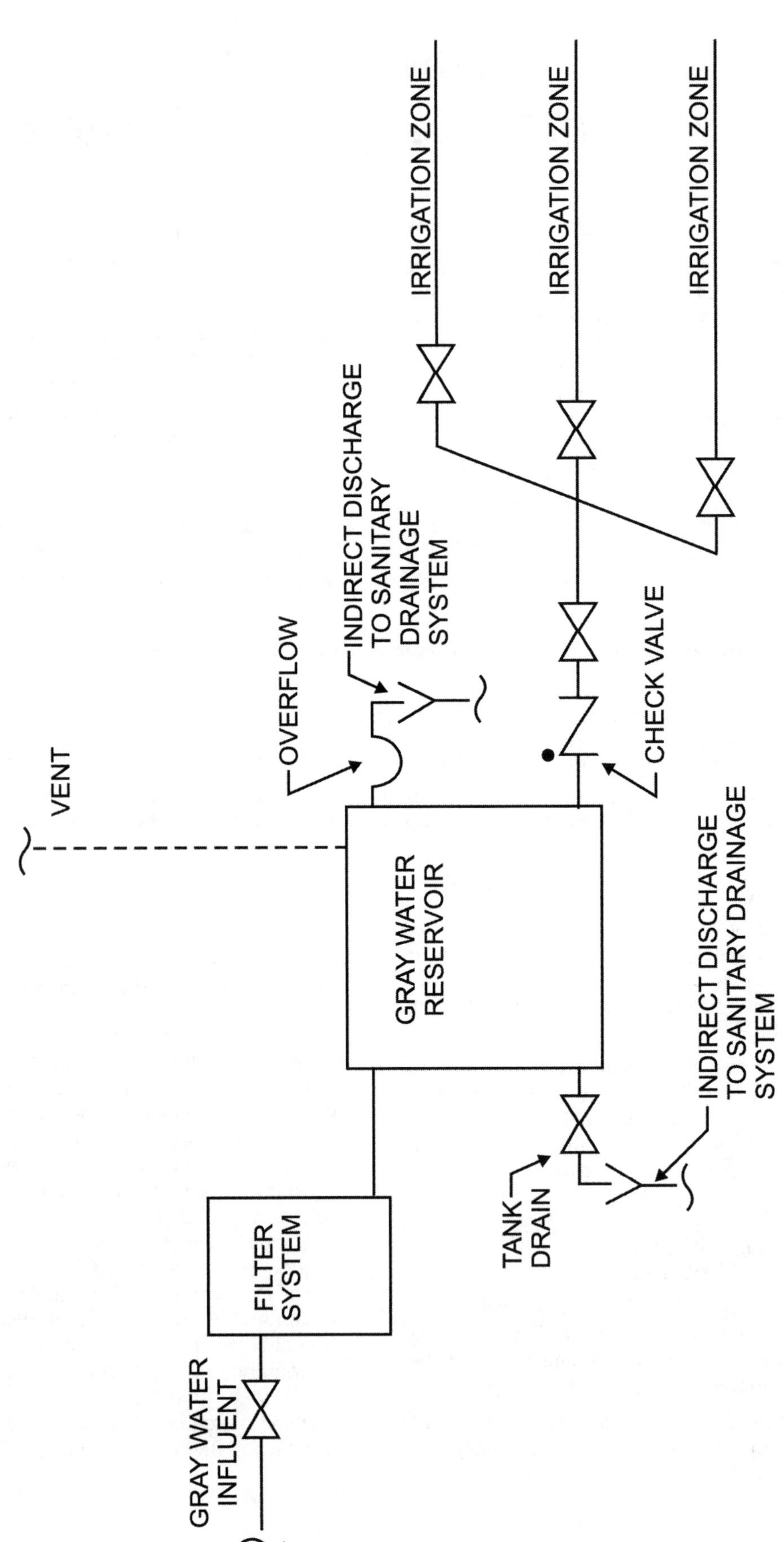

FIGURE 1
GRAY WATER RECYCLING SYSTEM FOR SUBSURFACE LANDSCAPE IRRIGATION

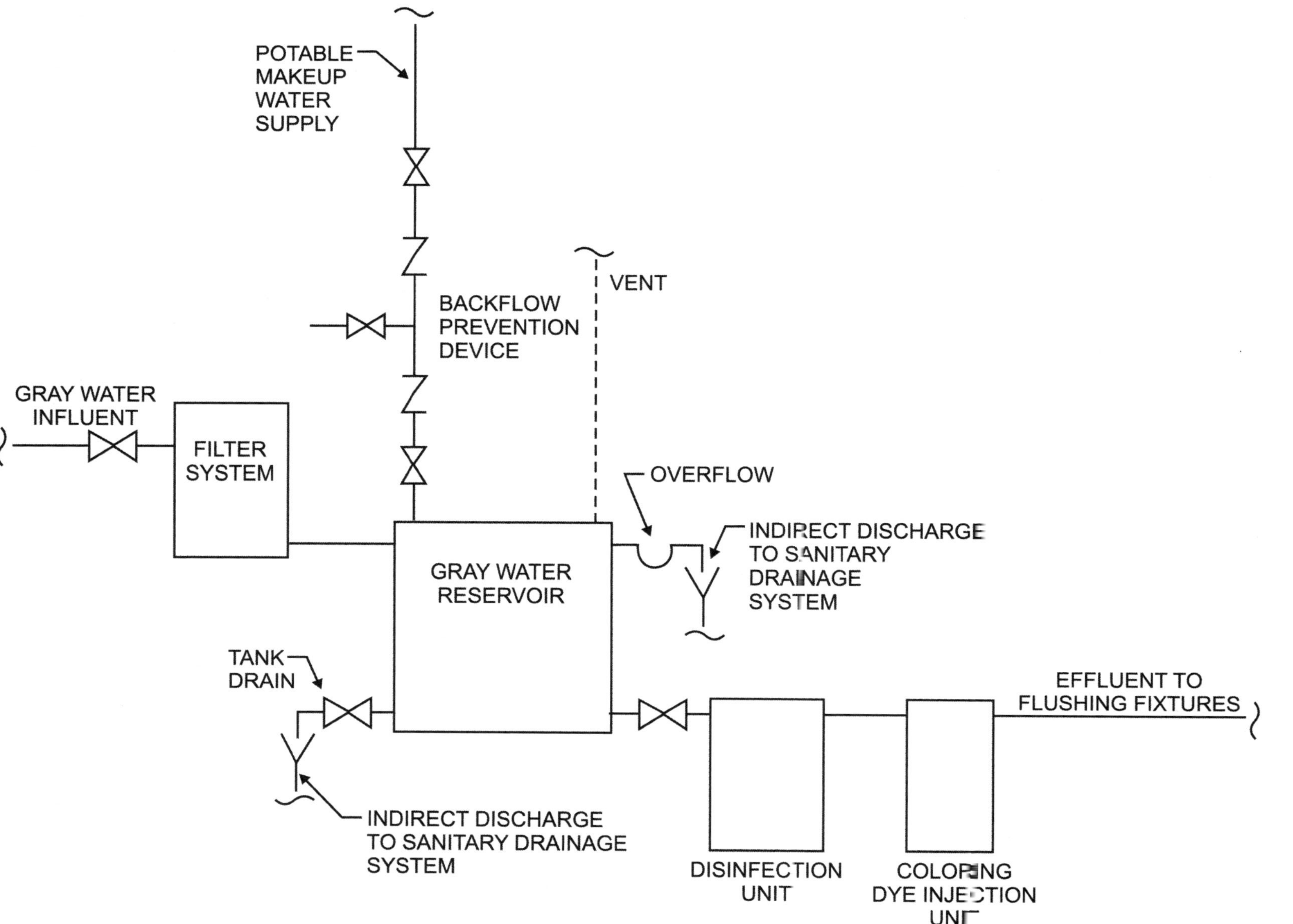

FIGURE 2
GRAY WATER RECYCLING SYSTEM FOR FLUSHING WATER CLOSETS AND URINALS

C101.3 Permits. Permits shall be required in accordance with Section 106.

❖ The installation and repair of any piping, fixtures and equipment associated with gray water recycling systems requires that a permit be obtained just as would be necessary for sanitary drainage plumbing system installation or repair.

C101.4 Installation. In addition to the provisions of Section C101, systems for flushing of water closets and urinals shall comply with Section C102 and systems for subsurface landscape irrigation shall comply with Section C103. Except as provided for in Appendix C, all systems shall comply with the provisions of the *International Plumbing Code*.

❖ The piping and hardware used to construct a gray-water recycling system must be installed no differently than any other drain, waste and vent (DWV) piping. The application and design concerns are identical for gray-water systems and traditional DWV systems. A building designed for gray-water recycling will have two independent fixture drainage collection systems. The nongray-water drainage system will discharge directly to the building sewer and the gray-water drainage system will discharge to a reservoir/treatment system. The venting systems can be independent or combined. The building will have two independent water distribution systems, one for potable water and one for treated gray water. Gray-water drain and supply piping should never be interconnected with any nongray-water piping.

C101.5 Materials. Above-ground drain, waste and vent piping for gray water systems shall conform to one of the standards listed in Table 702.1. Gray water underground building drainage and vent pipe shall conform to one of the standards listed in Table 702.2.

❖ Gray water systems are installed using the same materials for drain waste and vent piping as used in sanitary drainage systems.

C101.6 Tests. Drain, waste and vent piping for gray water systems shall be tested in accordance with Section 312.

❖ Gray water systems must be tested in the same manner as drain waste and vent piping used in sanitary drainage systems.

C101.7 Inspections. Gray water systems shall be inspected in accordance with Section 107.

❖ Gray water systems are inspected in a similar manner as sanitary drainage and potable water systems.

C101.8 Potable water connections. Only connections in accordance with Section C102.3 shall be made between a gray water recycling system and a potable water system.

❖ In some gray water applications, the addition of potable makeup water may be needed to allow fixtures using gray water to continue operation even when the gray water volume is exhausted. Section C102.3 specifies the requirements for backflow protection.

C101.9 Waste water connections. Gray water recycling systems shall receive only the waste discharge of bathtubs, showers, lavatories, clothes washers or laundry trays.

❖ By definition, gray water is the discharge from bathtubs, showers, lavatories, clothes washers and laundry sinks used for bathing and cleaning. Therefore, only these fixtures can be connected to the gray-water collection system.

The discharge from water closets, clinical sinks, urinals, bidets, kitchen sinks and dishwashers cannot be connected to the gray-water collection system.

C101.10 Collection reservoir. Gray water shall be collected in an *approved* reservoir constructed of durable, nonabsorbent and corrosion-resistant materials. The reservoir shall be a closed and gas-tight vessel. *Access* openings shall be provided to allow inspection and cleaning of the reservoir interior.

❖ The gray-water collection reservoir must be a water- and gas-tight tank with a volume capacity that will be adequate for the anticipated fixture flushing water demand with a specified reserve capacity. The reserve capacity is intended to cover the worst-case demand conditions, as well as account for conditions when the use of gray-water discharging fixtures is lower than anticipated. When the discharge from gray-water discharging fixtures does not fulfill the demand for fixture flushing water, potable water must be provided as makeup water (see Section C102.3).

The storage capacity of the reservoir tank must also be limited so that the gray water is not retained in the tank too long. Long retention times would cause the human skin cells and other matter to undergo digestion by anaerobic bacteria. Bacterial reduction of the matter in gray water would produce strong odors.

C101.11 Filtration. Gray water entering the reservoir shall pass through an *approved* filter such as a media, sand or diatomaceous earth filter.

❖ Gray water must be filtered to remove clothes fibers, food particles, hair and other solids that would otherwise collect in the reservoir and enter the gray-water distribution piping and plumbing fixtures. Without filtering, the entire system would be contaminated with debris that could cause, among other things, fixture flush valve and fill valve damage, pipe blockages, pump damage, increased maintenance needs, bacteria growth, odors and unsightly plumbing fixtures. Unfiltered gray water would also be more difficult to disinfect effectively. Diatomaceous earth is a siliceous earth composed of the shells of algae plants known as diatoms and is commonly used as a filter medium in swimming pools. Disinfection methods will be greatly affected by the effectiveness of filtering. To simplify maintenance and avoid a health risk to maintenance personnel, the filtering system should be designed for backwashing the filtered debris directly to the sanitary drainage system.

C101.11.1 Required valve. A full-open valve shall be installed downstream of the last fixture connection to the gray water discharge pipe before entering the required filter.

❖ Maintenance of a gray water reservoir, filter, disinfection and pump system requires that gray water flow be stopped from entering. Although installation of a full open valve on a gravity drainage line seems out of the ordinary, it is necessary for the safety of maintenance personnel. Those fixtures producing gray water should be "tagged out" to prevent use during maintenance of the gray water system.

C101.12 Overflow. The collection reservoir shall be equipped with an overflow pipe having the same or larger diameter as the influent pipe for the gray water. The overflow pipe shall be trapped and shall be indirectly connected to the sanitary drainage system.

❖ If the gray-water influent exceeds the demand for fixture flushing water, the reservoir could be filled completely. The excess gray water must be allowed to overflow into the building's sanitary drainage system. An indirect connection is required to eliminate the possibility of sewage backflow into the gray water system. The reservoir is part of a closed system similar to any other portion of the DWV system downstream of the fixture traps. Note that connecting the overflow pipe indirectly to the sanitary drainage system creates a new problem as it eliminates another. The gray water reservoir will now be open to the building allowing gases and vapors from the reservoir and the required vent to escape. This could be corrected by trapping the overflow pipe and providing a means of maintaining the trap seal. See the second sentence of Section C101.10.

C101.13 Drain. A drain shall be located at the lowest point of the collection reservoir and shall be indirectly connected to the sanitary drainage system. The drain shall be the same diameter as the overflow pipe required in Section C101.12.

❖ A drain will facilitate reservoir maintenance and also be used in the event of failure of the overflow piping. Indirect connection to the sanitary drainage system is required because of the concern for possible backflow of sewage into the gray water system.

C101.14 Vent required. The reservoir shall be provided with a vent sized in accordance with Chapter 9 and based on the diameter of the reservoir influent pipe.

❖ A vent is necessary just as it is for a sewage sump. Air must be admitted or expelled as the water level rises and falls in the reservoir.

SECTION C102
SYSTEMS FOR FLUSHING WATER CLOSETS AND URINALS

C102.1 Collection reservoir. The holding capacity of the reservoir shall be a minimum of twice the volume of water required to meet the daily flushing requirements of the fixtures supplied with gray water, but not less than 50 gallons (189 L). The reservoir shall be sized to limit the retention time of gray water to a maximum of 72 hours.

❖ Sizing of the collection reservoir is critical to obtaining the best performance and therefore best cost advantage of a gray water system. Too large a holding tank causes collected gray water to stagnate, allowing anaerobic bacteria to generate strong odors. Too small a holding tank causes a shortage of gray water for flushing fixtures, resulting in potable makeup water being added to the reservoir (defeating the overall point of the gray water system saving potable water).

C102.2 Disinfection. Gray water shall be disinfected by an *approved* method that employs one or more disinfectants such as chlorine, iodine or ozone that are recommended for use with the pipes, fittings and equipment by the manufacturer of the pipes, fittings and equipment.

❖ Disinfection will reduce the health risk to people and animals that come into contact with gray water. Chlorination, ozone generation and ultraviolet radiation are commonly used methods of killing microorganisms in wastewater. The disinfection process should occur after filtering to improve effectiveness.

C102.3 Makeup water. Potable water shall be supplied as a source of makeup water for the gray water system. The potable water supply shall be protected against backflow in accordance with Section 608. There shall be a full-open valve located on the makeup water supply line to the collection reservoir.

❖ Potable water must be provided to make up for any shortage of gray water. Depending on the circumstances, the flushing water demand could exceed the production of gray water, and makeup water would be necessary. It is imperative that the potable water supply be protected from contamination. Connecting potable water piping to a gray-water reservoir is an extreme health risk requiring dependable protection against backflow. An air gap would be ideal, but might be difficult to install because the reservoir must be a gas-tight vessel. It might be feasible to install an air gap arrangement where the potable water enters the vessel through an open trapped inlet pipe or a gray-water discharging fixture. Vacuum breaker devices and barometric loops might not be suitable either, because of the air-tight nature of the reservoir and the possibility of an overflow pipe failure and the gray water backing up to elevations higher than the reservoir. Section 608 does not specifically address the connection of potable water piping to an air-tight vessel containing nonpotable waste water. An acceptable method of backflow protection for this situation is the installation of a reduced pressure principle device.

C102.4 Coloring. The gray water shall be dyed blue or green with a food grade vegetable dye before such water is supplied to the fixtures.

❖ The gray water must be dyed to make it readily distinguishable from potable water. Blue or green water in a water closet or urinal will alert the occupants that the water is recycled gray water and definitely not potable. Blue and green are considered to be ideal identification colors because they are distinct from the colors of any typical waste associated with such fixtures. A food-grade dye is required to prevent the introduction of toxic chemical dyes that could have a detrimental effect on the drainage system, gray-water system, plumbing fixtures and waste treatment facilities.

C102.5 Materials. Distribution piping shall conform to one of the standards listed in Table 605.3.

❖ Gray water distribution systems are installed using the same materials as used for potable water distribution systems.

C102.6 Identification. Distribution piping and reservoirs shall be identified as containing nonpotable water. Piping identification shall be in accordance with Section 608.8.

❖ Gray-water distribution piping and reservoirs must be identified as nonpotable to prevent accidental connections to piping mistakenly believed to be potable water piping (see Commentary Sections 608.8 through 608.8.3).

SECTION C103 SUBSURFACE LANDSCAPE IRRIGATION SYSTEMS

C103.1 Collection reservoir. Reservoirs shall be sized to limit the retention time of gray water to a maximum of 24 hours.

❖ The gray-water collection reservoir must be a water- and gas-tight tank with a volume capacity that will be adequate for the anticipated irrigation water demand. because the demand for irrigation is usually based on timed watering events, there is no need for a reserve capacity or minimum size of a holding tank. If less gray water is produced, the frequency/duration of watering can be reduced. If more gray water is produced, the gray water can be wasted to the sanitary drainage system and/or the frequency/duration of watering can be increased. In either case, the gray water must not be held longer than 24 hours because disinfection of the gray water is not required and odors could be generated quickly.

C103.1.1 Identification. The reservoir shall be identified as containing nonpotable water.

❖ Gray-water distribution piping and reservoirs must be identified as nonpotable to prevent accidental connections to piping mistakenly believed to be potable water piping (see Commentary Sections 608.8 through 608.8.3).

C103.2 Valves required. A check valve and a full-open valve located on the discharge side of the check valve shall be installed on the effluent pipe of the collection reservoir.

❖ Subsurface irrigation systems might be installed at an elevation above the gray water reservoir. The check valve prevents backflow of water from irrigation lines. To perform maintenance on the reservoir or check valve, a full-open shut-off valve is required in the discharge line from the reservoir.

C103.3 Makeup water. Makeup water shall not be required for subsurface landscape irrigation systems. Where makeup water is provided, the installation shall be in accordance with Section C102.3.

❖ Because irrigation is not a demand like a fixture that requires flushing water, a makeup water system is not required. However, if makeup water is desired, the connections must be made in accordance with Section C102.3.

C103.4 Disinfection. Disinfection shall not be required for gray water used for subsurface landscape irrigation systems.

❖ Subsurface irrigation means that the discharge of the gray water will be underground. Because there will be no human contact and the water will not be exposed to the air, disinfection is not necessary.

C103.5 Coloring. Gray water used for subsurface landscape irrigation systems shall not be required to be dyed.

❖ Coloring of gray water for subsurface irrigation is not necessary because the water cannot be seen by humans; aesthetics and identification as nonpotable are not issues.

C103.6 Estimating gray water discharge. The system shall be sized in accordance with the gallons-per-day-per-occupant number based on the type of fixtures connected to the gray water system. The discharge shall be calculated by the following equation:

$C = A \times B$

A = Number of occupants:

Residential—Number of occupants shall be determined by the actual number of occupants, but not less than two occupants for one bedroom and one occupant for each additional bedroom.

Commercial—Number of occupants shall be determined by the *International Building Code*®.

B = Estimated flow demands for each occupant:

Residential—25 gallons per day (94.6 Lpd) per occupant for showers, bathtubs and lavatories and 15 gallons per day (56.7 Lpd) per occupant for clothes washers or laundry trays.

Commercial—Based on type of fixture or water use records minus the discharge of fixtures other than those discharging gray water.

C = Estimated gray water discharge based on the total number of occupants.

❖ To size a gray water reservoir and determine the discharge volume, a calculation must be performed using the number of occupants and the type of fixtures producing gray water. Commentary Table C103.6 summarizes the gray water discharge volumes.

C103.7 Percolation tests. The permeability of the soil in the proposed absorption system shall be determined by percolation tests or permeability evaluation.

❖ Before a subsurface irrigation system can be designed, the rate of absorption of the soil must be known. This test is similar to that for determining sanitary septic system drain fields.

C103.7.1 Percolation tests and procedures. At least three percolation tests in each system area shall be conducted. The holes shall be spaced uniformly in relation to the bottom depth of the proposed absorption system. More percolation tests shall be made where necessary, depending on system design.

❖ Three test holes are located in the area that is to receive the gray water discharge. The bottom of the holes needs to be close to the same level as the proposed subsurface irrigation system to provide as accurate a percolation test as possible.

C103.7.1.1 Percolation test hole. The test hole shall be dug or bored. The test hole shall have vertical sides and a horizontal dimension of 4 inches to 8 inches (102 mm to 203 mm). The bottom and sides of the hole shall be scratched with a sharp-pointed instrument to expose the natural soil. All loose material shall be removed from the hole and the bottom shall be covered with 2 inches (51 mm) of gravel or coarse sand.

❖ The measurement that is being performed in a percolation test is the loss of a specific volume of water over time. Therefore, all test holes need to be dug or bored to the same dimension to facilitate averaging of the results from three holes and comparison of the percolation rate to a performance rating scale. Because digging or boring can cause the hole surfaces to be smoothed and compacted, the surfaces require scoring (scratching) to bring those surfaces back to a condition which simulates undisturbed earth. Removal of the loose material in the hole reduces the possibility that the test water will become muddy and not percolate as intended. The addition of coarse sand or gravel in the bottom of the hole simulates the conditions of an actual installation of subsurface piping.

C103.7.1.2 Test procedure, sandy soils. The hole shall be filled with clear water to a minimum of 12 inches (305 mm) above the bottom of the hole for tests in sandy soils. The time for this amount of water to seep away shall be determined, and this procedure shall be repeated if the water from the second filling of the hole seeps away in 10 minutes or less. The test shall proceed as follows: Water shall be added to a point not more than 6 inches (152 mm) above the gravel or coarse sand. Thereupon, from a fixed reference point, water levels shall be measured at 10-minute intervals for a period of 1 hour. Where 6 inches (152 mm) of water seeps away in less than 10 minutes, a shorter interval between measurements shall be used, but in no case shall the water depth exceed 6 inches (152 mm). Where 6 inches (152 mm) of water seeps away in less than 2 minutes, the test shall be stopped and a rate of less than 3 minutes per inch (7.2 s/mm) shall be reported. The final water level drop shall be used to calculate the percolation rate. Soils not meeting the above requirements shall be tested in accordance with Section C103.7.1.3.

❖ Clear water is added to the hole to a depth 12 inches (305 mm) above the bottom of the hole (not to the top of the gravel or sand) and the time for it to seep away completely is recorded. The hole is filled again to 12 inches (305 mm) above the bottom of the hole and the time for the water to seep away completely is recorded. If upon the second filling, the water seeps away in 10 minutes or less, the hole is filled a third time to a depth of 12 inches (305 mm) above the bottom of the hole and the time for it to seep completely away is recorded. This part of the test is to provide some saturation of the soil in the test hole to simulate a saturation level comparable to that found in an actual system.

The test consists of filling the hole to a depth of 6 inches (152 mm) above the coarse sand or gravel and measuring the drop in water level from a fixed reference point every 10 minutes. The measurements are recorded. If the 6 inches (152 mm) of water takes 2 minutes or more to seep completely away, the time for the water level drop occurring in the last time interval will be the final result of the test in minutes per inch.

If the 6 inches (152 mm) of water seeps away in less than 2 minutes, the test is ended with the result of the test being recorded at 3 minutes per inch.

C103.7.1.3 Test procedure, other soils. The hole shall be filled with clear water, and a minimum water depth of 12 inches

Table C103.6
GRAY WATER DISCHARGE VOLUMES

	RESIDENTIAL	COMMERCIAL
A. Number of occupants	Actual number of occupants but not less than 2 for one bedroom and 1 for each additional bedroom	As per *International Building Code*© (IBC©)
B. Gallons of gray water produced per occupant	25 gallons per day per occupant[a] plus 15 gallons per day per occupant[b]	25 gallons per day per occupant[a] plus 15 gallons per day per occupant[b]
C = A × B	Gallons per day	Gallons per day[c]

a. This is the total volume produced by each occupant's use of bathtubs, lavatories and showers combined.
b. This is the total volume produced by each occupant's portion of water used by clothes washers and laundry sinks, combined.
c. Alternatively, if approved by the code official, water use records could be used to determine gray water discharge with the nongray-water-producing fixtures subtracted from the total.

(305 mm) shall be maintained above the bottom of the hole for a 4-hour period by refilling whenever necessary or by use of an automatic siphon. Water remaining in the hole after 4 hours shall not be removed. Thereafter, the soil shall be allowed to swell not less than 16 hours or more than 30 hours. Immediately after the soil swelling period, the measurements for determining the percolation rate shall be made as follows: Any soil sloughed into the hole shall be removed and the water level shall be adjusted to 6 inches (152 mm) above the gravel or coarse sand. Thereupon, from a fixed reference point, the water level shall be measured at 30-minute intervals for a period of 4 hours, unless two successive water level drops do not vary by more than $^1/_{16}$ inch (1.59 mm). At least three water level drops shall be observed and recorded. The hole shall be filled with clear water to a point not more than 6 inches (152 mm) above the gravel or coarse sand whenever it becomes nearly empty. Adjustments of the water level shall not be made during the three measurement periods except to the limits of the last measured water level drop. When the first 6 inches (152 mm) of water seeps away in less than 30 minutes, the time interval between measurements shall be 10 minutes and the test run for 1 hour. The water depth shall not exceed 5 inches (127 mm) at any time during the measurement period. The drop that occurs during the final measurement period shall be used in calculating the percolation rate.

❖ Clear water is added to the hole to a depth 12 inches (305 mm) above the bottom of the hole (not to the top of the gravel or sand) and is replenished to that level over a period of 4 hours. After 4 hours, the hole, with the water in it, is left undisturbed for at least 16 hours but for no more than 30 hours. Within the 16 to 30 hour time period, the hole is carefully cleared of any loose soil that might have fallen into the hole and the water level adjusted to a depth of 6 inches (152 mm) above the coarse sand or gravel. At 30 minute intervals, the depth of the water is measured from a fixed reference point for 4 hours. If, after the first time interval, the water level drops no more than $^1/_{16}$ inch (1.16 mm) over the next two time intervals, the test is ended. The water level drop measurement in the last time interval will be the final result of the test in minutes per inch.

If the water level drops to the coarse sand or gravel during the 4 hour test period, the water must be replenished to a point 6 inches (152 mm) above the coarse sand or gravel. At least three time intervals must occur before the water is replenished again

If the first 6 inches (152 mm) of water seeps away in less than 30 minutes, measurements must be taken at 10 minute intervals for an hour. At no time should the water depth over the coarse sand or gravel be more than 5 inches (127 mm). The water level drop measurement in the last time interval will be the final result of the test in minutes per inch.

C103.7.1.4 Mechanical test equipment. Mechanical percolation test equipment shall be of an *approved* type.

❖ Because the preparation of percolation test holes and the administering of a conventional test can take considerable time and labor, alternative methods have been devised to reduce time and labor. One method to arrive at similar results is called a soil permeability test. The device used is called a permeameter and requires boring only a 2 inch (51 mm) diameter hole down to the level at which the surface irrigation piping will be installed. Drilling of the test hole is a one-man operation with hand-held equipment. Measurements take no more than 2 hours. However, because percolation tests have been performed in the same manner for so many years, code officials may be reluctant to try new methods such as permeability over the standard percolation tests.

C103.7.2 Permeability evaluation. Soil shall be evaluated for estimated percolation based on structure and texture in accordance with accepted soil evaluation practices. Borings shall be made in accordance with Section C103.7.1 for evaluating the soil.

❖ This section allows for permeability testing instead of the standard percolation test to determine the estimated percolation rate.

C103.8 Subsurface landscape irrigation site location. The surface grade of all soil absorption systems shall be located at a point lower than the surface grade of any water well or reservoir on the same or adjoining property. Where this is not possible, the site shall be located so surface water drainage from the site is not directed toward a well or reservoir. The soil absorption system shall be located with a minimum horizontal distance between various elements as indicated in Table C103.8. Private sewage disposal systems in compacted areas, such as parking lots and driveways, are prohibited. Surface water shall be diverted away from any soil absorption site on the same or neighboring lots.

❖ Protection of ground water sources as well as any water reservoirs is necessary when planning for the placement of a subsurface irrigation system. When on the same or adjoining property, the elevation of the top of the ground over the irrigation site must be lower than the top of the ground at a nearby water well or reservoir. This allows the rain water to be flow toward the irrigation site as opposed to away from the site toward the water sources, possibly causing contamination of those water sources. The site cannot be located under parking lots or driveways (paved or unpaved) or any ground area where the soil can become compacted from repeated use such as fairgrounds, ball fields, and overflow parking fields. In addition to these restrictions, there are minimum horizontal distance separations for both the gray-water holding tanks and the irrigation disposal field as indicated by Table C103.8.

C103.9 Installation. Absorption systems shall be installed in accordance with Sections C103.9.1 through C103.9.5 to provide landscape irrigation without surfacing of gray water.

❖ To prevent the gray water from coming up through the surface of the ground that covers the system, specific details given in Sections C103.9.1 through C103.9.5 must be followed.

TABLE C103.8
LOCATION OF GRAY WATER SYSTEM

ELEMENT	MINIMUM HORIZONTAL DISTANCE	
	HOLDING TANK (feet)	IRRIGATION DISPOSAL FIELD (feet)
Buildings	5	2
Property line adjoining private property	5	5
Water wells	50	100
Streams and lakes	50	50
Seepage pits	5	5
Septic tanks	0	5
Water service	5	5
Public water main	10	10

For SI: 1 foot = 304.8 mm.

C103.9.1 Absorption area. The total absorption area required shall be computed from the estimated daily gray water discharge and the design-loading rate based on the percolation rate for the site. The required absorption area equals the estimated gray water discharge divided by the design-loading rate from Table C103.9.1.

❖ The percolation rate of the soil determines the design loading factor for the irrigation site. Table C103.9.1 shows the design loading factors for ranges of percolation rates. The number of gallons per day of gray water discharge (estimated) divided by the design loading factor equals the area required. For example, if the percolation rate is 3 minutes per inch, the design loading factor is 1.2 gpsf/day. Given an estimated gray water discharge of, say, 400 gallons per day, the computation is

Area required = 400/1.2 = 333.3 square feet

If the percolation rate was longer, say 35 minutes, the design loading factor would be 0.72 gpsf/day.

For the same gray water discharge of 400 gallons per day, the computation is

Area required = 400/0.72 = 555.5 square feet

Note that the areas computed are seepage areas of the trenches or beds and not an overall general area for the site. Refer to Sections C103.9.2 and C103.9.3 for details on calculating the seepage areas of excavation.

TABLE C 103.9.1
DESIGN LOADING RATE

PERCOLATION RATE (minutes per inch)	DESIGN LOADING FACTOR (gallons per square foot per day)
0 to less than 10	1.2
10 to less than 30	0.8
30 to less than 45	0.72
45 to 60	0.4

For SI: 1 minute per inch = min/25.4 mm,
1 gallon per square foot = 40.7 L/m^2.

C103.9.2 Seepage trench excavations. Seepage trench excavations shall be a minimum of 1 foot (304 mm) to a maximum of 5 feet (1524 mm) wide. Trench excavations shall be spaced a minimum of 2 feet (610 mm) apart. The soil absorption area of a seepage trench shall be computed by using the bottom of the trench area (width) multiplied by the length of pipe. Individual seepage trenches shall be a maximum of 100 feet (30 480 mm) in *developed length*.

❖ There are two types of seepage excavations: trench type and bed type. The trench type consists of relatively narrow trenches, 1 to 5 feet wide (304 to 1524 mm) with a single length of seepage pipe in each trench. The bed method allows for areas wider than 5 feet (1524 mm) to be excavated with multiple seepage pipes installed in the excavation. Which method is selected depends on the amount of area needed, the lay of the land and the equipment available for excavating.

In the trench method, excavations are made from 1 foot to 5 feet wide (304 mm to 1524 mm) by a maximum of 100 feet (30 480 mm) long with the space between adjacent trenches being at least 2 feet (610 mm) . The bottom of the excavation must be level. Although the code does not indicate what degree of level is required, in practice, rotary laser level devices are typically used allowing for an accuracy of $^{1}/_{16}$ inch (1.16 mm) at a 100-foot (30 480 mm) radius. Seasoned excavators can usually hold level within plus or minus $^{1}/_{4}$ inch (6.4 mm) over the length of a 100-foot (30 480 mm) excavation.

The seepage area for the trench method is the bottom width of the trench by the length of the seepage pipe to be installed in the trench. For example, if a 2 foot (610 mm) wide trench is dug for a 100-foot (30 480 mm) long pipe, the seepage area for the trench is 200 square feet (18.6 m^2).

C103.9.3 Seepage bed excavations. Seepage bed excavations shall be a minimum of 5 feet (1524 mm) wide and have more than one distribution pipe. The absorption area of a seepage bed shall be computed by using the bottom of the trench area. Distribution piping in a seepage bed shall be uniformly spaced a maximum of 5 feet (1524 mm) and a minimum of 3 feet (914 mm) apart, and a maximum of 3 feet (914 mm) and a minimum of 1 foot (305 mm) from the sidewall or headwall.

❖ The excavation must be at least 5 feet (1524 mm) wide and have more than one distribution pipe evenly spaced between 3 feet and 5 feet (914 and 1524 mm) apart from the adjacent distribution pipe. The pipes closest to the side walls of the excavation can be no closer than 1 foot (305 mm) but no more than three feet (914 mm) from the sides of the excavation. For example, an excavation that is 16 feet (4877 mm) wide could have three distribution pipes, the outside pipes being 3 feet (914 mm) from the side walls and the spacing between the pipes being 5 feet (1524 mm) . This would result in maximizing the seepage area for this three-pipe system. Although the code does not specify a maximum length for each distribution pipe for a bed type installation, given the similar nature of design and operation to the trench method, the distribution pipes may perform best if held to lengths no more than 100 feet (30 480 mm).

The seepage area for the bed method is the bottom width of the trench by the length of the excavation. The length of the one pipe would be equal to the length of the excavation. Given one pipe length of, say, 100 feet (30 480 mm), the seepage area is 16 feet x 100 feet = 1600 square feet (4877 mm x 30 480 mm = 148 644 m^2).

C103.9.4 Excavation and construction. The bottom of a trench or bed excavation shall be level. Seepage trenches or beds shall not be excavated where the soil is so wet that such material rolled between the hands forms a soil wire. All smeared or compacted soil surfaces in the sidewalls or bottom of seepage trench or bed excavations shall be scarified to the depth of smearing or compaction and the loose material removed. Where rain falls on an open excavation, the soil shall be left until sufficiently dry so a soil wire will not form when soil from the excavation bottom is rolled between the hands. The bottom area shall then be scarified and loose material removed.

❖ The bottom of the excavation must be level (see Commentary, Section C103.9.2). An excavation cannot be made in soil wet enough to enable a person to form a "soil wire" with their hand using soil from near the bottom of the excavation. A "soil wire" is formed by picking up soil in one hand, squeezing it with that hand to compact it and then rolling the clump between flattened hands to try to roll it into a pencil-shaped form and then into a diameter as small as $^1/_8$ inch (3 mm) in diameter.

Soil smearing and compaction can occur in the excavated areas and must be scarified (scratched/scored) with straight rakes or similar devices to allow for adequate seepage. Any loose material must be removed from the bottom of the excavation. If the excavation is rained upon, the excavation must be allowed to dry out so that a soil wire cannot be formed from soil material from the bottom of the excavation. Once the excavation has dried out, the bottom of the excavation must be scarified, followed by removal of the loose material.

C103.9.5 Aggregate and backfill. A minimum of 6 inches (152 mm) of aggregate ranging in size from $^1/_2$ to $2^1/_2$ inches (12.7 mm to 64 mm) shall be laid into the trench below the distribution piping elevation. The aggregate shall be evenly distributed a minimum of 2 inches (51 mm) over the top of the distribution pipe. The aggregate shall be covered with *approved* synthetic materials or 9 inches (229 mm) of uncompacted marsh hay or straw. Building paper shall not be used to cover the aggregate. A minimum of 9 inches (229 mm) of soil backfill shall be provided above the covering.

❖ A 6-inch (152 mm) layer of aggregate, of a sieve size from $^1/_2$ inch to $2^1/_2$ inches (12.7 to 64 mm), is placed evenly in the bottom of the excavation. The distribution piping is located on top of this placed aggregate and then additional aggregate is placed on the sides and top of the pipe. At least 2 inches (51 mm) of aggregate must cover the pipe. An approved synthetic filter membrane designed to keep the soil fines of the backfill out of the aggregate is placed over the entire aggregate surface. Alternatively, 9 inches (229 mm) of "marsh" hay or straw can be used instead of the filter membrane. '"Marsh hay" is a shortened term for salt marsh hay. It grows in salt marshes in the coastal regions of the United States and is known for its strong fibers and resistance to rot. Marsh straw is a threatened plant species and is rarely used. Building paper is not to be used for this purpose.

Finally, the excavation is backfilled with soil and leveled but not compacted. There must be at least 9 inches (229 mm) of soil cover on top of the filter membrane or hay layer.

C103.10 Distribution piping. Distribution piping shall be not less than 3 inches (76 mm) in diameter. Materials shall comply with Table C103.10. The top of the distribution pipe shall be not less than 8 inches (203 mm) below the original surface. The slope of the distribution pipes shall be a minimum of 2 inches (51 mm) and a maximum of 4 inches (102 mm) per 100 feet (30 480 mm).

❖ Three-inch (76 mm) pipe is the minimum size that can be used. Pipe materials must be one of those listed in Table C103.10. The installed pipe must have a slight slope away from the distribution header or source of gray water. This slope must be no less than 2 inches (51 mm) in 100 feet (30 480 mm) but no more than 4 inches (102 mm) in 100 feet (30 480 mm). This slope is nearly level but is just enough to allow the distribution pipe to be evenly loaded by the gray water throughout its entire length.

TABLE C103.10
DISTRIBUTION PIPE

MATERIAL	STANDARD
Polyethylene (PE) plastic pipe	ASTM F 405
Polyvinyl chloride (PVC) plastic pipe	ASTM D 2729
Polyvinyl chloride (PVC) plastic pipe with pipe stiffness of PS 35 and PS 50	ASTM F 1488

C103.11 Joints. Joints in distribution pipe shall be made in accordance with Section 705 of this code.

❖ The piping, including distribution headers and required fittings, must be connected in the manner prescribed by Section 705 for sanitary drainage piping.

Appendix D: Degree Day and Design Temperatures

General Comments

Appendix D specifies outdoor design conditions. Although the data were extracted from the 1985 ASHRAE *Handbook of Fundamentals*, updated values for design conditions can be obtained from the 1987, 1993, 1997 and 2001 editions of the ASHRAE handbook. The exterior design conditions are described below in three groups. The groups are established according to how the data are used.

Purpose

Appendix D is reprinted for informational purposes and provides climate conditions for 223 locations throughout the United States. These summaries include sources for heating degree days and values of dry-bulb and wet-bulb temperatures to be used for evaluating the protection of vent terminals or the thickness of insulation required for recirculating hot water distribution systems as required by the code. The appendix also assigns the outdoor design conditions required for the calculations called for in the *International Property Maintenance Code*® (IPMC®) for minimum space heating requirements.

Load Calculations

Outdoor design temperatures are used in the load calculations required by the IPMC for determining minimum heating facilities for residential structures (see Sections 602.2 and 602.3 of the IPMC).

Winter Design Dry-Bulb (97$^1/_2$ percent). This is the outdoor dry-bulb temperature used to calculate heating loads. The value means that, on average, the outdoor temperature will be above this temperature for 97½ percent of the hours during the heating season. For the northern hemisphere, the heating season in the United States is the three-month winter period from December through February.

Summer Design Dry-Bulb (2$^1/_2$ percent). This is the outdoor dry-bulb temperature used to calculate cooling loads. The value means that, on average, the outdoor temperature will be above this temperature for 2½ percent of the hours during the cooling season. For the northern hemisphere, the cooling season in the United States is the four-month summer period from May through August.

Mean Coincident Wet-Bulb (2$^1/_2$ percent). This is the outdoor wet-bulb temperature used to calculate cooling loads. The value means that, on average, the outdoor temperature will be above this temperature for 2$^1/_2$ percent of the hours during the cooling season.

Piping Insulation

Distribution losses impact building energy use both in the energy required to make up for the lost heat and in the additional load that can be placed on the space cooling system if the heat is released to the air-conditioned space. In circulating systems, hot water is exposed to loss throughout the entire distribution system as long as the water is circulating. These losses can be limited by insulating the entire hot water supply and return piping system to the requirements of the *International Energy Conservation Code*® (IECC®).

Protection of Vent Terminals

The possibility of frost closure of vent terminals predominantly occurs in areas of the country having cold climates, as determined where the winter design dry-bulb (97$^1/_2$ percent) temperature is less than 0°F (-18°C) (see Sections 904.2 and 904.7).

This appendix is informative and is not part of the code.

TABLE D101
DEGREE DAY AND DESIGN TEMPERATURES[a] FOR CITIES IN THE UNITED STATES

STATE	STATION[b]	HEATING DEGREE DAYS (yearly total)	DESIGN TEMPERATURES			DEGREES NORTH LATITUDE[c]
			Winter	Summer		
			97 1/2%	Dry bulb 2 1/2%	Wet bulb 2 1/2%	
AL	Birmingham	2,551	21	94	77	33°30′
	Huntsville	3,070	16	96	77	34°40′
	Mobile	1,560	29	93	79	30°40′
	Montgomery	2,291	25	95	79	32°20′
AK	Anchorage	10,864	-18	68	59	61°10′
	Fairbanks	14,279	-47	78	62	64°50′
	Juneau	9,075	1	70	59	58°20′
	Nome	14,171	-27	62	56	64°30′
AZ	Flagstaff	7,152	4	82	60	35°10′
	Phoenix	1,765	34	107	75	33°30′
	Tuscon	1,800	32	102	71	33°10′
	Yuma	974	39	109	78	32°40′
AR	Fort Smith	3,292	17	98	79	35°20′
	Little Rock	3,219	20	96	79	34°40′
	Texarkana	2,533	23	96	79	33°30′
CA	Fresno	2,611	30	100	71	36°50′
	Long Beach	1,803	43	80	69	33°50′
	Los Angeles	2,061	43	80	69	34°00′
	Los Angeles[d]	1,349	40	89	71	34°00′
	Oakland	2,870	36	80	64	37°40′
	Sacramento	2,502	32	98	71	38°30′
	San Diego	1,458	44	80	70	32°40′
	San Francisco	3,015	38	77	64	37°40′
	San Francisco[d]	3,001	40	71	62	37°50′
CO	Alamosa	8,529	-16	82	61	37°30′
	Colorado Springs	6,423	2	88	62	38°50′
	Denver	6,283	1	91	63	39°50′
	Grand Junction	5,641	7	94	63	39°10′
	Pueblo	5,462	0	95	66	38°20′
CT	Bridgeport	5,617	9	84	74	41°10′
	Hartford	6,235	7	88	75	41°50′
	New Haven	5,897	7	84	75	41°20′
DE	Wilmington	4,930	14	89	76	39°40′
DC	Washington	4,224	17	91	77	38°50′
FL	Daytona	879	35	90	79	29°10′
	Fort Myers	442	44	92	79	26°40′
	Jacksonville	1,239	32	94	79	30°30′
	Key West	108	57	90	79	24°30′
	Miami	214	47	90	79	25°50′
	Orlando	766	38	93	78	28°30′
	Pensacola	1,463	29	93	79	30°30′
	Tallahassee	1,485	30	92	78	30°20′
	Tampa	683	40	91	79	28°00′
	West Palm Beach	253	45	91	79	26°40′

(continued)

TABLE D101—continued
DEGREE DAY AND DESIGN TEMPERATURES[a] FOR CITIES IN THE UNITED STATES

STATE	STATION[b]	HEATING DEGREE DAYS (yearly total)	DESIGN TEMPERATURES			DEGREES NORTH LATITUDE[c]
			Winter	Summer		
			$97^1/_2$%	Dry bulb $2^1/_2$%	Wet bulb $2^1/_2$%	
GA	Athens	2,929	22	92	77	34°00′
	Atlanta	2,961	22	92	76	33°40′
	Augusta	2,397	23	95	79	33°20′
	Columbus	2,383	24	93	78	32°30′
	Macon	2,136	25	93	78	32°40′
	Rome	3,326	22	93	78	34°20′
	Savannah	1,819	27	93	79	32°10′
HI	Hilo	0	62	83	74	19°40′
	Honolulu	0	63	86	75	21°20′
ID	Boise	5,809	10	94	66	43°30′
	Lewiston	5,542	6	93	66	46°20′
	Pocatello	7,033	-1	91	63	43°00′
IL	Chicago (Midway)	6,155	0	91	75	41°50′
	Chicago (O'Hare)	6,639	-4	89	76	42°00′
	Chicago[d]	5,882	2	91	77	41°50′
	Moline	6,408	-4	91	77	41°30′
	Peoria	6,025	-4	89	76	40°40′
	Rockford	6,830	-4	89	76	42°10′
	Springfield	5,429	2	92	77	39°50′
IN	Evansville	4,435	9	93	78	38°00′
	Fort Wayne	6,205	1	89	75	41°00′
	Indianapolis	5,699	2	90	76	39°40′
	South Bend	6,439	1	89	75	41°40′
IA	Burlington	6,114	-3	91	77	40°50′
	Des Moines	6,588	-5	91	77	41°30′
	Dubuque	7,376	-7	88	75	42°20′
	Sioux City	6,951	-7	92	77	42°20′
	Waterloo	7,320	-10	89	77	42°30′
KS	Dodge City	4,986	5	97	73	37°50′
	Goodland	6,141	0	96	70	39°20′
	Topeka	5,182	4	96	78	39°00′
	Wichita	4,620	7	98	76	37°40′
KY	Covington	5,265	6	90	75	39°00′
	Lexington	4,683	8	91	76	38°00′
	Louisville	4,660	10	93	77	38°10′
LA	Alexandria	1,921	27	94	79	31°20′
	Baton Rouge	1,560	29	93	80	30°30′
	Lake Charles	1,459	31	93	79	30°10′
	New Orleans	1,385	33	92	80	30°00′
	Shreveport	2,184	25	96	79	32°30′
ME	Caribou	9,767	-13	81	69	46°50′
	Portland	7,511	-1	84	72	43°40′
MD	Baltimore	4,654	13	91	77	39°10′
	Baltimore[d]	4,111	17	89	78	39°20′
	Frederick	5,087	12	91	77	39°20′

(continued)

TABLE D101—continued
DEGREE DAY AND DESIGN TEMPERATURES[a] FOR CITIES IN THE UNITED STATES

STATE	STATION[b]	HEATING DEGREE DAYS (yearly total)	DESIGN TEMPERATURES			DEGREES NORTH LATITUDE[c]
			Winter	Summer		
			97 1/2%	Dry bulb 2 1/2%	Wet bulb 2 1/2%	
MA	Boston	5,634	9	88	74	42°20′
	Pittsfield	7,578	-3	84	72	42°30′
	Worcester	6,969	4	84	72	42°20′
MI	Alpena	8,506	-6	85	72	45°00′
	Detroit (City)	6,232	6	88	74	42°20′
	Escanaba[d]	8,481	-7	83	71	45°40′
	Flint	7,377	1	87	74	43°00′
	Grand Rapids	6,894	5	88	74	42°50′
	Lansing	6,909	1	87	74	42°50′
	Marquette[d]	8,393	-8	81	70	46°30′
	Muskegon	6,696	6	84	73	43°10′
	Sault Ste. Marie	9,048	-8	81	70	46°30′
MN	Duluth	10,000	-16	82	70	46°50′
	Minneapolis	8,382	-12	89	5	44°50′
	Rochester	8,295	-12	87	75	44°00′
MS	Jackson	2,239	25	95	78	32°20′
	Meridian	2,289	23	95	79	32°20′
	Vicksburg[d]	2,041	26	95	80	32°20′
MO	Columbia	5,046	4	94	77	39°00′
	Kansas City	4,711	6	96	77	39°10′
	St. Joseph	5,484	2	93	79	39°50′
	St. Louis	4,900	6	94	77	38°50′
	St. Louis[d]	4,484	8	94	77	38°40′
	Springfield	4,900	9	93	77	37°10′
MT	Billings	7,049	-10	91	66	45°50′
	Great Falls	7,750	-15	88	62	47°30′
	Helena	8,129	-16	88	62	46°40′
	Missoula	8,125	-6	88	63	46°50′
NE	Grand Island	6,530	-3	94	74	41°00′
	Lincoln[d]	5,864	-2	95	77	40°50′
	Norfolk	6,979	-4	93	77	42°00′
	North Platte	6,684	-4	94	72	41°10′
	Omaha	6,612	-3	91	77	41°20′
	Scottsbluff	6,673	-3	92	68	41°50′
NV	Elko	7,433	-2	92	62	40°50′
	Ely	7,733	-4	87	59	39°10′
	Las Vegas	2,709	28	106	70	36°10′
	Reno	6,332	10	92	62	39°30′
	Winnemucca	6,761	3	94	62	40°50′
NH	Concord	7,383	-3	87	73	43°10′
NJ	Atlantic City	4,812	13	89	77	39°30′
	Newark	4,589	14	91	76	40°40′
	Trenton[d]	4,980	14	88	76	40°10′
NM	Albuquerque	4,348	16	94	65	35°00′
	Raton	6,228	1	89	64	36°50′
	Roswell	3,793	18	98	70	33°20′
	Silver City	3,705	10	94	64	32°40′

(continued)

TABLE D101—continued
DEGREE DAY AND DESIGN TEMPERATURES[a] FOR CITIES IN THE UNITED STATES

STATE	STATION[b]	HEATING DEGREE DAYS (yearly total)	DESIGN TEMPERATURES			DEGREES NORTH LATITUDE[c]
			Winter	Summer		
			$97^1/_2$%	Dry bulb $2^1/_2$%	Wet bulb $2^1/_2$%	
NY	Albany	6,875	-1	88	74	42°50′
	Albany[d]	6,201	1	88	74	42°50′
	Binghamton	7,286	1	83	72	42°10′
	Buffalo	7,062	6	85	73	43°00′
	NY (Central Park)[d]	4,871	15	89	75	40°50′
	NY (Kennedy)	5,219	15	87	75	40°40′
	NY(LaGuardia)	4,811	15	89	75	40°50′
	Rochester	6,748	5	88	73	43°10′
	Schenectady[d]	6,650	1	87	74	42°50′
	Syracuse	6,756	2	87	73	43°10′
NC	Charlotte	3,181	22	93	76	35°10′
	Greensboro	3,805	18	91	76	36°10′
	Raleigh	3,393	20	92	77	35°50′
	Winston-Salem	3,595	20	91	75	36°10′
ND	Bismarck	8,851	-19	91	71	46°50′
	Devils Lake[d]	9,901	-21	88	71	48°10′
	Fargo	9,226	-18	89	74	46°50′
	Williston	9,243	-21	88	70	48°10′
OH	Akron-Canton	6,037	6	86	73	41°00′
	Cincinnati[d]	4,410	6	90	75	39°10′
	Cleveland	6,351	5	88	74	41°20′
	Columbus	5,660	5	90	75	40°00′
	Dayton	5,622	4	89	75	39°50′
	Mansfield	6,403	5	87	74	40°50′
	Sandusky[d]	5,796	6	91	74	41°30′
	Toledo	6,494	1	88	75	41°40′
	Youngstown	6,417	4	86	73	41°20′
OK	Oklahoma City	3,725	13	97	77	35°20′
	Tulsa	3,860	13	98	78	36°10′
OR	Eugene	4,726	22	89	67	44°10′
	Medford	5,008	23	94	68	42°20′
	Portland	4,635	23	85	67	45°40′
	Portland[d]	4,109	24	86	67	45°30′
	Salem	4,754	23	88	68	45°00′
PA	Allentown	5,810	9	88	75	40°40′
	Erie	6,451	9	85	74	42°10′
	Harrisburg	5,251	11	91	76	40°10′
	Philadelphia	5,144	14	90	76	39°50′
	Pittsburgh	5,987	5	86	73	40°30′
	Pittsburgh[d]	5,053	7	88	73	40°30′
	Reading[d]	4,945	13	89	75	40°20′
	Scranton	6,254	5	87	73	41°20′
	Williamsport	5,934	7	89	74	41°10′
RI	Providence	5,954	9	86	74	41°40′
SC	Charleston	2,033	27	91	80	32°50′
	Charleston[d]	1,794	28	92	80	32°50′
	Columbia	2,484	24	95	78	34°00′

(continued)

TABLE D101—continued
DEGREE DAY AND DESIGN TEMPERATURES[a] FOR CITIES IN THE UNITED STATES

STATE	STATION[b]	HEATING DEGREE DAYS (yearly total)	DESIGN TEMPERATURES			DEGREES NORTH LATITUDE[c]
			Winter	Summer		
			$97^1/_2\%$	Dry bulb $2^1/_2\%$	Wet bulb $2^1/_2\%$	
SD	Huron	8,223	-14	93	75	44°30′
	Rapid City	7,345	-7	92	69	44°00′
	Sioux Falls	7,839	-11	91	75	43°40′
TN	Bristol	4,143	14	89	75	36°30′
	Chattanooga	3,254	18	93	77	35°00′
	Knoxville	3,494	19	92	76	35°50′
	Memphis	3,232	18	95	79	35°00′
	Nashville	3,578	14	94	77	36°10′
TX	Abilene	2,624	20	99	74	32°30′
	Austin	1,711	28	98	77	30°20′
	Dallas	2,363	22	100	78	32°50′
	El Paso	2,700	24	98	68	31°50′
	Houston	1,396	32	94	79	29°40′
	Midland	2,591	21	98	72	32°00′
	San Angelo	2,255	22	99	74	31°20′
	San Antonio	1,546	30	97	76	29°30′
	Waco	2,030	26	99	78	31°40′
	Wichita Falls	2,832	18	101	76	34°00′
UT	Salt Lake City	6,052	8	95	65	40°50′
VT	Burlington	8,269	-7	85	72	44°30′
VA	Lynchburg	4,166	16	90	76	37°20′
	Norfolk	3,421	22	91	78	36°50′
	Richmond	3,865	17	92	78	37°30′
	Roanoke	4,150	16	91	74	37°20′
WA	Olympia	5,236	22	83	66	47°00′
	Seattle-Tacoma	5,145	26	80	64	47°30′
	Seattle[d]	4,424	27	82	67	47°40′
	Spokane	6,655	2	90	64	47°40′
WV	Charleston	4,476	11	90	75	38°20′
	Elkins	5,675	6	84	72	38°50′
	Huntington	4,446	10	91	77	38°20′
	Parkersburg[d]	4,754	11	90	76	39°20′
WI	Green Bay	8,029	-9	85	74	44°30′
	La Crosse	7,589	-9	88	75	43°50′
	Madison	7,863	-7	88	75	43°10′
	Milwaukee	7,635	-4	87	74	43°00′
WY	Casper	7,410	-5	90	61	42°50′
	Cheyenne	7,381	-1	86	62	41°10′
	Lander	7,870	-11	88	63	42°50′
	Sheridan	7,680	-8	91	65	44°50′

a. All data were extracted from the 1985 ASHRAE Handbook, Fundamentals Volume.
b. Design data developed from airport temperature observations unless noted.
c. Latitude is given to the nearest 10 minutes. For example, the latitude for Miami, Florida, is given as 25°50′, or 25 degrees 50 minutes.
d. Design data developed from office locations within an urban area, not from airport temperature observations.

Bibliography

The following resource materials are referenced in this chapter or are relevant to the subject matter addressed in this chapter.

ASHRAE, Handbook of Fundamentals. Atlanta, GA: American Society of Heating, Refrigerating and Air-Conditioning Engineers, Inc., 2001.

IECC-09, *International Energy Conservation Code.* Washington, DC: International Code Council, 2009.

IPMC-09, *International Property Maintenance Code.* Washington, DC: International Code Council, 2009.

Owenby, J.R., D.S. Ezell, and R.R. Heim, Jr. *Annual Degree Days to Selected Bases Derived from the 1961 to 1990 Normals: Climatography of the United States No. 81 - Supplement No. 2.* Asheville, N.C.: U.S. Department of Commerce, National Oceanic and Atmospheric Administration, National Climatic Data Center, 1992.

U.S. Department of Energy (DOE). *90.1 Code Compliance Manual.* Prepared by Eley Associates under contract to the Building Standards and Guidelines Program. Richland, WA: Pacific Northwest National Laboratory, 1995.

Appendix E: Sizing of Water Piping System

General Comments

The design of a water supply system ensures adequate water supply to all fixtures at all times and achieves economic sizing of the piping. To do this, the maximum probable rate of flow or demand for which provisions should be made to every portion of the system is used. The demand cannot be determined exactly because most fixtures are used intermittently, and the probability of simultaneous use of such fixtures cannot be established. To be acceptable for estimating the demand of the water supply system, a method must produce the estimated demand for all fixtures on the system during periods of peak demand, avoid oversizing of the piping and be adaptable for estimating the demand of like fixtures as well as different types of fixtures and building occupancy classifications. This procedure takes into consideration the pressure, velocity limitations, materials, characteristics of the water source and demand.

Purpose

Appendix E describes two methods of sizing a water supply system. These methods provide plumbing fixtures and equipment with water in sufficient volume and adequate pressures to enable them to function properly without excessive noise under normal conditions of use. There are many other sizing or design methods conforming to good engineering practices.

The provisions contained in this appendix are not mandatory unless specifically referenced in the adopting ordinance.

SECTION E101 GENERAL

E101.1 Scope.

E101.1.1 This appendix outlines two procedures for sizing a water piping system (see Sections E103.3 and E201.1). The design procedures are based on the minimum static pressure available from the supply source, the head changes in the system caused by friction and elevation, and the rates of flow necessary for operation of various fixtures.

E101.1.2 Because of the variable conditions encountered in hydraulic design, it is impractical to specify definite and detailed rules for sizing of the water piping system. Accordingly, other sizing or design methods conforming to good engineering practice standards are acceptable alternatives to those presented herein.

SECTION E102 INFORMATION REQUIRED

E102.1 Preliminary. Obtain the necessary information regarding the minimum daily static service pressure in the area where the building is to be located. If the building supply is to be metered, obtain information regarding friction loss relative to the rate of flow for meters in the range of sizes likely to be used. Friction loss data can be obtained from most manufacturers of water meters.

E102.2 Demand load.

E102.2.1 Estimate the supply demand of the building main and the principal branches and risers of the system by totaling the corresponding demand from the applicable part of Table E103.3(3).

E102.2.2 Estimate continuous supply demands in gallons per minute (L/m) for lawn sprinklers, air conditioners, etc., and add the sum to the total demand for fixtures. The result is the estimated supply demand for the building supply.

❖ The determination of water demand by the occupants of a building can be very difficult. Studies by Roy B. Hunter starting in 1923 resulted in a method of weighting plumbing fixtures with their water supply load producing effects. Based on experience of using this method for the plumbing design in large federal buildings, Hunter revised the method slightly and then published it in 1940. In 1968, Elmer Jones of the Agricultural Research Service further revised the method using data from small housing developments. This method of estimating demand has been widely recognized as the generally accepted method to produce reliable water demand for all types of occupancies.

Determining occupant water demand involves the probability of use the rate of use and the length of use. Not all fixtures will be used simultaneously (although halftime at spectator-viewed sporting events may come close), the fixtures are not necessarily operated in the same manner, and the time spent using a fixture varies from person to person. Although sizing a system for continuous operation of all fixtures would be the easiest solution, the resulting pipe sizes might be unrealistically large and thus costly to install. However, on the same note, undersize piping could result in user complaints or even dangerous conditions caused by low water pressure. Special occupancies such as health care, schools, restaurants and stadiums usually require significant engineering expertise to properly design water systems for these applications.

SECTION E103
SELECTION OF PIPE SIZE

E103.1 General. Decide from Table 604.3 what is the desirable minimum residual pressure that should be maintained at the highest fixture in the supply system. If the highest group of fixtures contains flush valves, the pressure for the group should not be less than 15 pounds per square inch (psi) (103.4 kPa) flowing. For flush tank supplies, the available pressure should not be less than 8 psi (55.2 kPa) flowing, except blowout action fixtures must not be less than 25 psi (172.4 kPa) flowing.

E103.2 Pipe sizing.

E103.2.1 Pipe sizes can be selected according to the following procedure or by other design methods conforming to acceptable engineering practice and *approved* by the administrative authority. The sizes selected must not be less than the minimum required by this code.

❖ In general, plumbing fixtures require at least 8 psig (55 kPa) of flowing pressure to perform adequately. There are special fixtures that require higher pressures under flowing conditions such as flush valve-type water closets and temperature-controlling (e.g., balanced-pressure/thermostatic control valves) shower controls. The fixture that has the highest flowing pressure demand in the system controls the limit for pressure loss used for system sizing.

E103.2.2 Water pipe sizing procedures are based on a system of pressure requirements and losses, the sum of which must not exceed the minimum pressure available at the supply source. These pressures are as follows:

1. Pressure required at fixture to produce required flow. See Sections 604.3 and 604.5.
2. Static pressure loss or gain (due to head) is computed at 0.433 psi per foot (9.8 kPa/m) of elevation change.

   Example: Assume that the highest fixture supply outlet is 20 feet (6096 mm) above or below the supply source. This produces a static pressure differential of 20 feet by 0.433 psi/foot (2096 mm by 9.8 kPa/m) and an 8.66 psi (59.8 kPa) loss.
3. Loss through water meter. The friction or pressure loss can be obtained from meter manufacturers.
4. Loss through taps in water main.
5. Losses through special devices such as filters, softeners, backflow prevention devices and pressure regulators. These values must be obtained from the manufacturers.
6. Loss through valves and fittings. Losses for these items are calculated by converting to equivalent length of piping and adding to the total pipe length.
7. Loss due to pipe friction can be calculated when the pipe size, the pipe length and the flow through the pipe are known. With these three items, the friction loss can be determined. For piping flow charts not included, use manufacturers' tables and velocity recommendations.

Note: For the purposes of all examples, the following metric conversions are applicable:

1 cubic foot per minute = 0.4719 L/s

1 square foot = 0.0929 m^2

1 degree = 0.0175 rad

1 pound per square inch = 6.895 kPa

1 inch = 25.4 mm

1 foot = 304.8 mm

1 gallon per minute = 3.785 L/m

❖ Frictional losses resulting from the flow of fluids in piping, fittings and standard components are well established from equations of fluid mechanics and testing. Textbooks, handbooks and other reputable publications as well as specific manufacturer's data are resources for frictional loss values. The most important aspect of the information gathering process is to make sure that no component is missed or considered to be "negligible" so that it is not reflected on the computation sheet. As projects come to life, actual component data sheet information should be compared to that used for calculation purposes to assure that all losses are accounted for.

E103.3 Segmented loss method. The size of water service mains, *branch* mains and risers by the segmented loss method, must be determined according to water supply demand [gpm (L/m)], available water pressure [psi (kPa)] and friction loss caused by the water meter and *developed length* of pipe [feet (m)], including equivalent length of fittings. This design procedure is based on the following parameters:

- Calculates the friction loss through each length of the pipe.
- Based on a system of pressure losses, the sum of which must not exceed the minimum pressure available at the street main or other source of supply.
- Pipe sizing based on estimated peak demand, total pressure losses caused by difference in elevation, equipment, *developed length* and pressure required at most remote fixture, loss through taps in water main, losses through fittings, filters, backflow prevention devices, valves and pipe friction.

Because of the variable conditions encountered in hydraulic design, it is impractical to specify definite and detailed rules for sizing of the water piping system. Current sizing methods do not address the differences in the probability of use and flow characteristics of fixtures between types of occupancies. Creating an exact model of predicting the demand for a building is impossible and final studies assessing the impact of water conservation on demand are not yet complete. The following steps are necessary for the segmented loss method.

1. **Preliminary.** Obtain the necessary information regarding the minimum daily static service pressure in the area where the building is to be located. If the building supply is to be metered, obtain information regarding friction loss relative to the rate of flow for meters in the range of sizes to be used. Friction loss data can be obtained from manufacturers of water meters. It is essential that enough pressure be available to overcome all system losses caused by friction and elevation so that plumbing fixtures operate properly. Section 604.6 requires the water distribution system to be designed for the minimum pressure available taking into consideration pressure fluctuations. The lowest pressure must be selected to guarantee a continuous, adequate supply of water. The lowest pressure in the public main usually occurs in the summer because of lawn sprinkling and supplying water for air-conditioning cooling towers. Future demands placed on the public main as a result of large growth or expansion should also be considered. The available pressure will decrease as additional loads are placed on the public system.

2. **Demand load.** Estimate the supply demand of the building main and the principal branches and risers of the system by totaling the corresponding demand from the applicable part of Table E103.3(3). When estimating peak demand sizing methods typically use water supply fixture units (w.s.f.u.) [see Table E103.3(2)]. This numerical factor measures the load-producing effect of a single plumbing fixture of a given kind. The use of such fixture units can be applied to a single basic probability curve (or table), found in the various sizing methods [Table E103.3(3)]. The fixture units are then converted into gallons per minute (L/m) flow rate for estimating demand.

    2.1. Estimate continuous supply demand in gallons per minute (L/m) for lawn sprinklers, air conditioners, etc., and add the sum to the total demand for fixtures. The result is the estimated supply demand for the building supply. Fixture units cannot be applied to constant use fixtures such as hose bibbs, lawn sprinklers and air conditioners. These types of fixtures must be assigned the gallon per minute (L/m) value.

❖ The code is silent on how to select and apply the values from Table E103.3(2) for an application. One approach to selection would be to use the individual cold and hot water supply fixture unit values for each fixture with no consideration given to the column headed "Total." This would be a conservative method that yields higher overall water demand from the system and may result in pipe sizes larger than if values from the "Total" column were used with appropriate guidance. Note that the "Total" column is not the sum of the fixture hot and cold values, but is a weighted total representing the "load producing effect" of that fixture's combined hot and cold water demand on the water distribution system. A fixture "total" water demand value could be used when the water flow through the pipe section is the result of both hot and cold water being used at the fixture. But if the pipe is carrying water flow as a result of only hot water flow, individual hot and cold fixture water supply fixture units need to be used for the demand loading.

3. **Selection of pipe size.** This water pipe sizing procedure is based on a system of pressure requirements and losses, the sum of which must not exceed the minimum pressure available at the supply source. These pressures are as follows:

    3.1. Pressure required at the fixture to produce required flow. See Section 604.3 and Section 604.5.

    3.2. Static pressure loss or gain (because of head) is computed at 0.433 psi per foot (9.8 kPa/m) of elevation change.

    3.3. Loss through a water meter. The friction or pressure loss can be obtained from the manufacturer.

    3.4. Loss through taps in water main [see Table E103.3(4)].

    3.5. Losses through special devices such as filters, softeners, backflow prevention devices and pressure regulators. These values must be obtained from the manufacturers.

    3.6. Loss through valves and fittings [see Tables E103.3(5) and E103.3(6)]. Losses for these items are calculated by converting to equivalent length of piping and adding to the total pipe length.

    3.7. Loss caused by pipe friction can be calculated when the pipe size, the pipe length and the flow through the pipe are known. With these three items, the friction loss can be determined using Figures E103.3(2) through E103.3(7). When using charts, use pipe inside diameters. For piping flow charts not included, use manufacturers' tables and velocity recommendations. Before attempting to size any water supply system, it is necessary to gather preliminary information which includes available pressure, piping material, select design velocity, elevation differences and *developed length* to most remote fixture. The water supply system is divided into sections at major changes in elevation or where branches lead to fixture groups. The peak demand must be determined in each part of the hot and cold water supply system which includes the corresponding water supply fixture unit and conversion to gallons per minute (L/m) flow rate to be expected through each section. Sizing methods require the determination of the "most hydraulically remote" fixture to compute the pressure loss caused by pipe and fittings. The hydraulically remote fixture represents the most downstream fixture along the circuit of piping requiring the most available pressure to operate properly. Consideration must be given to all pressure demands and losses, such as friction caused by pipe, fittings and equipment, elevation and the residual pressure required by Table 604.3. The two most common and frequent complaints about the water supply system operation are lack of adequate pressure and noise.

❖ The "Segmented Loss Method" is also known as the "Velocity-Pressure Drop" method. Once flow through a pipe is known in gallons per minute (L/m) for any particular pipe section under consideration, a pipe size can be selected based on a recommended maximum velocity of the water through a specific pipe material.

Experience and extensive research on all types of pipe materials have produced general guidelines for recommended maximum velocities in domestic water piping. For cold water piping, maximum velocities in the range of 5 to 8 feet per second (1.5 to 2.4 m/s) are typical and for hot water piping, especially for temperatures above 150°F (66°C), maximum velocities of 4 feet per second (1.2 m/s) are typical. However, the type of pipe material has a significant impact on the maximum recommended velocities. For example, in general, manufacturers of cross-linked polyethylene tubing (PEX) allow velocities to approach 12 feet per second (3.7 m/s) in both hot [maximum 140°F (60°F)] and cold piping, whereas, copper piping manufacturers recommend using the lower velocity general guidelines as mentioned above.

Control of velocity in piping is necessary for a variety of reasons. High velocities create flowing "noise" and increased possibility for erosion of the pipe material. High velocities also contribute to water hammer problems because rapidly flowing fluid has very high energy. When high energy fluid is stopped abruptly, that energy dissipates in a shock wave moving upstream from the valve that closed, causing noise and instantaneous high pressures (and possibly damaging forces) within the pipe. But most importantly, high fluid velocities cause a loss in flowing pressure. The inside wall of new building water distribution pipe, while visually appearing smooth, has a roughness that fluid moving though the pipe tends to "drag" on. The higher the velocity is in the pipe, the greater the drag (friction) forces are on the fluid. For any fluid flowing full in a pipe, given an initial flowing pressure and velocity at the beginning of a section of pipe, at any downstream point in the pipe, there will be a decrease in the flowing pressure in the pipe. The friction takes energy away from the fluid. This loss of energy (i.e., pressure) caused by fluid flow dragging on the inside walls of a pipe is called pressure drop or pressure loss.

Friction loss (or pressure drop) graphs are available for a variety of pipe materials. Figures E103.3(2) through (4) are for new copper water piping (types K, L and M, respectively) with recessed, streamline soldered joints (see graph note a). Note that these graphs show a velocity lines only up to 10 feet per second (3 m/s). The code does not provide specific friction loss graphs for any other types of water pipe as listed in Tables 605.3 and 605.4. but provides only the graphs in Figures E103.3(5) through (7). Figure E103.3(5), although not specifically labeled as such, is the friction loss graph for schedule 40 new steel pipe with inside walls of "fairly smooth" condition. New galvanized steel piping, rarely used in modern construction for water distribution, is considered to have a "fairly smooth" inside wall condition when new. However, once galvanized steel pipe has been in service for an extended length of time, the inside wall condition usually degrades to what is considered to be a "fairly rough or rough" condition as a result of corrosion and deposit build up. The pipe material and wall thickness used as the basis for Figures E103.3(6) and (7) is not specified, only the condition of the pipe interior. Figure E103.3(7) also has an apparent discrepancy in specifying the inside wall condition at the top of the graph as "rough." The caption of the graph says "fairly rough" and Note a says "very rough."

Perhaps of most importance is not what the friction losses are when piping is new but what the friction losses will be after extended service. Friction loss graphs for a range of inside wall surface conditions are available from many sources to aid the system designer in determining accurate values for expected pressure drop.

Problem: What size Type L copper water pipe, service and distribution will be required to serve a two-story factory building having on each floor, back-to-back, two toilet rooms each equipped with hot and cold water? The highest fixture is 21 feet (6401 mm) above the street main, which is tapped with a 2-inch (51 mm) corporation cock at which point the minimum pressure is 55 psi (379.2 kPa). In the building basement, a 2-inch (51 mm) meter with a maximum pressure drop of 11 psi (75.8 kPa) and 3-inch (76 mm) reduced pressure principle backflow preventer with a maximum pressure drop of 9 psi (621 kPa) are to be installed. The system is shown by Figure E103.3(1). To be determined are the pipe sizes for the service main and the cold and hot water distribution pipes.

Solution: A tabular arrangement such as shown in Table E103.3(1) should first be constructed. The steps to be followed are indicated by the tabular arrangement itself as they are in sequence, columns 1 through 10 and lines A through L.

Step 1

Columns 1 and 2: Divide the system into sections breaking at major changes in elevation or where branches lead to fixture groups. After point B [see Figure E103.3(1)], separate consideration will be given to the hot and cold water piping. Enter the sections to be considered in the service and cold water piping in Column 1 of the tabular arrangement. Column 1 of Table E103.3(1) provides a line-by-line recommended tabular arrangement for use in solving pipe sizing.

The objective in designing the water supply system is to ensure an adequate water supply and pressure to all fixtures and equipment. Column 2 provides the pounds per square inch (psi) to be considered separately from the minimum pressure available at the main. Losses to take into consideration are the following: the differences in elevations between the water supply source and the highest water supply outlet, meter pressure losses, the tap in main loss, special fixture devices such as water softeners and backflow prevention devices and the pressure required at the most remote fixture outlet. The difference in elevation can result in an increase or decrease in available pressure at the main. Where the water supply outlet is located above the source, this results in a loss in the available pressure and is subtracted from the pressure at the water source. Where the highest water supply outlet is located below the water supply source, there will be an increase in pressure that is added to the available pressure of the water source.

Column 3: According to Table E103.3(3), determine the gpm (L/m) of flow to be expected in each section of the system. These flows range from 28.6 to 108 gpm. Load values for fixtures must be determined as water supply fixture units and then converted to a gallon-per-minute (gpm) rating to determine peak demand. When calculating peak demands, the water supply fixture units are added and then converted to the gallon-per-minute rating. For continuous flow fixtures such as hose bibbs and lawn sprinkler systems, add the gallon-per-minute demand to the intermittent demand of fixtures. For example, a total of 120 water supply fixture units is converted to a demand of 48 gallons per minute. Two hose bibbs × 5 gpm demand = 10 gpm. Total gpm rating = 48.0 gpm + 10 gpm = 58.0 gpm demand.

Step 2

Line A: Enter the minimum pressure available at the main source of supply in Column 2. This is 55 psi (379.2 kPa). The local water authorities generally keep records of pressures at different times of day and year. The available pressure can also be checked from nearby buildings or from fire department hydrant checks.

Line B: Determine from Table 604.3 the highest pressure required for the fixtures on the system, which is 15 psi (103.4 kPa), to operate a flushometer valve. The most remote fixture outlet is necessary to compute the pressure loss caused by pipe and fittings, and represents the most downstream fixture along the circuit of piping requiring the available pressure to operate properly as indicated by Table 604.3.

Line C: Determine the pressure loss for the meter size given or assumed. The total water flow from the main through the service as determined in Step 1 will serve to aid in the meter selected. There are three common types of water meters; the pressure losses are determined by the American Water Works Association Standards for displacement type, compound type and turbine type. The maximum pressure loss of such devices takes into consideration the meter size, safe operating capacity (gpm) and maximum rates for continuous operations (gpm). Typically, equipment imparts greater pressure losses than piping.

Line D: Select from Table E103.3(4) and enter the pressure loss for the tap size given or assumed. The loss of pressure through taps and tees in pounds per square inch (psi) are based on the total gallon-per-minute flow rate and size of the tap.

Line E: Determine the difference in elevation between the main and source of supply and the highest fixture on the system. Multiply this figure, expressed in feet, by 0.43 psi (2.9 kPa). Enter the resulting psi loss on Line E. The difference in elevation between the water supply source and the highest water supply outlet has a significant impact on the sizing of the water supply system. The difference in elevation usually results in a loss in the available pressure because the water supply outlet is generally located above the water supply source. The loss is caused by the pressure required to lift the water to the outlet. The pressure loss is subtracted from the pressure at the water source. Where the highest water supply outlet is located below the water source, there will be an increase in pressure which is added to the available pressure of the water source.

Lines F, G and H: The pressure losses through filters, backflow prevention devices or other special fixtures must be obtained from the manufacturer or estimated and entered on these lines. Equipment such as backflow prevention devices, check valves, water softeners, instantaneous or tankless water heaters, filters and strainers can impart a much greater pressure loss than the piping. The pressure losses can range from 8 psi to 30 psi.

Step 3

Line I: The sum of the pressure requirements and losses that affect the overall system (Lines B through H) is entered on this line. Summarizing the steps, all of the system losses are subtracted from the minimum water pressure. The remainder is the pressure available for friction, defined as the energy available to push the water through the pipes to each fixture. This force can be used as an average pressure loss, as long as the pressure available for friction is not exceeded. Saving a certain amount for available water supply pressures as an area incurs growth, or because of aging of the pipe or equipment added to the system is recommended.

Step 4

Line J: Subtract Line I from Line A. This gives the pressure that remains available from overcoming friction losses in the system. This figure is a guide to the pipe size that is chosen for each section, incorporating the total friction losses to the most remote outlet (measured length is called *developed length*).

> **Exception:** When the main is above the highest fixture, the resulting psi must be considered a pressure gain (static head gain) and omitted from the sums of Lines B through H and added to Line J.

The maximum friction head loss that can be tolerated in the system during peak demand is the difference between the static pressure at the highest and most remote outlet at no-flow conditions and the minimum flow pressure required at that outlet. If the losses are within the required limits, then every run of pipe will also be within the required friction head loss. Static pressure loss is the most remote outlet in feet × 0.433 = loss in psi caused by elevation differences.

Step 5

Column 4: Enter the length of each section from the main to the most remote outlet (at Point E). Divide the water supply system into sections breaking at major changes in elevation or where branches lead to fixture groups.

Step 6

Column 5: When selecting a trial pipe size, the length from the water service or meter to the most remote fixture outlet must be measured to determine the *developed length*. However, in systems having a flush valve or temperature controlled shower at the topmost floors the *developed length* would be from the water meter to the most remote flush valve on the system. A rule of thumb is that size will become progressively smaller as the system extends farther from the main source of supply. Trial pipe size may be arrived at by the following formula:

Line J: (Pressure available to overcome pipe friction) × 100/equivalent length of run total *developed length* to most remote fixture × percentage factor of 1.5 (note: a percentage factor is used only as an estimate for friction losses imposed for fittings for initial trial pipe size) = psi (average pressure drops per 100 feet of pipe).

For trial pipe size see Figure E 103.3(3) (Type L copper) based on 2.77 psi and a 108 gpm = $2^1/_2$ inches. To determine the equivalent length of run to the most remote outlet, the *developed length* is determined and added to the friction losses for fittings and valves. The developed lengths of the designated pipe sections are as follows:

A - B 54 ft

B - C 8 ft

C - D 13 ft

D - E 150 ft

Total developed length = 225 ft

The equivalent length of the friction loss in fittings and valves must be added to the *developed length* (most remote outlet). Where the size of fittings and valves is not known, the added friction loss should be approximated. A general rule that has been used is to add 50 percent of the *developed length* to allow for fittings and valves. For example, the equivalent length of run equals the *developed length* of run (225 ft × 1.5 = 338 ft). The total equivalent length of run for determining a trial pipe size is 338 feet.

Example: 9.36 (pressure available to overcome pipe friction) × 100/ 338 (equivalent length of run = 225 × 1.5) = 2.77 psi (average pressure drop per 100 feet of pipe).

Step 7

Column 6: Select from Table E103.3(6) the equivalent lengths for the trial pipe size of fittings and valves on each pipe section. Enter the sum for each section in Column 6. (The number of fittings to be used in this example must be an estimate.) The equivalent length of piping is the *developed length* plus the equivalent lengths of pipe corresponding to friction head losses for fittings and valves. Where the size of fittings and valves is not known, the added friction head losses must be approximated. An estimate for this example is found in Table E.1.

Step 8

Column 7: Add the figures from Column 4 and Column 6, and enter in Column 7. Express the sum in hundreds of feet.

Step 9

Column 8: Select from Figure E103.3(3) the friction loss per 100 feet (30 480 mm) of pipe for the gallon-per-minute flow in a section (Column 3) and trial pipe size (Column 5). Maximum friction head loss per 100 feet is determined on the basis of total pressure available for friction head loss and the longest equivalent length of run. The selection is based on the gallon-per-minute demand, the uniform friction head loss, and the maximum design velocity. Where the size indicated by hydraulic table indicates a velocity in excess of the selected velocity, a size must be selected which produces the required velocity.

Step 10

Column 9: Multiply the figures in Columns 7 and 8 for each section and enter in Column 9.

Total friction loss is determined by multiplying the friction loss per 100 feet (30 480 mm) for each pipe section in the total *developed length* by the pressure loss in fittings expressed as equivalent length in feet. Note: Section C-F should be considered in the total pipe friction losses only if greater loss occurs in Section C-F than in pipe section D-E. Section C-F is not considered in the total *developed length*. Total friction loss in equivalent length is determined in Table E.2.

Step 11

Line K: Enter the sum of the values in Column 9. The value is the total friction loss in equivalent length for each designated pipe section.

Step 12

Line L: Subtract Line J from Line K and enter in Column 10.

The result should always be a positive or plus figure. If it is not, repeat the operation using Columns 5, 6, 8 and 9 until a balance or near balance is obtained. If the difference between Lines J and K is a high positive number, it is an indication that the pipe sizes are too large and should be reduced, thus saving materials. In such a case, the operations using Columns 5, 6, 8 and 9 should again be repeated.

The total friction losses are determined and subtracted

from the pressure available to overcome pipe friction for trial pipe size. This number is critical as it provides a guide to whether the pipe size selected is too large and the process should be repeated to obtain an economically designed system.

Answer: The final figures entered in Column 5 become the design pipe size for the respective sections. Repeating this operation a second time using the same sketch but considering the demand for hot water, it is possible to size the hot water distribution piping. This has been worked up as a part of the overall problem in the tabular arrangement used for sizing the service and water distribution piping. Note that consideration must be given to the pressure losses from the street main to the water heater (Section A-B) in determining the hot water pipe sizes.

TABLE E.1

COLD WATER PIPE SECTION	FITTINGS/VALVES	PRESSURE LOSS EXPRESSED AS EQUIVALENT LENGTH OF TUBE (feet)	HOT WATER PIPE SECTION	FITTINGS/VALVES	PRESSURE LOSS EXPRESSED AS EQUIVALENT OF TUBE (feet)
A-B	3-$2^{1}/_{2}$″ Gate valves	3	A-B	3-$2^{1}/_{2}$″ Gate valves	3
	1-$2^{1}/_{2}$″ Side branch tee	12		1-$2^{1}/_{2}$″ Side branch tee	12
B-C	1-$2^{1}/_{2}$″ Straight run tee	0.5	B-C	1-2″ Straight run tee	7
				1-2″ 90-degree ell	0.5
C-F	1-$2^{1}/_{2}$″ Side branch tee	12	C-F	1-$1^{1}/_{2}$″ Side branch tee	7
C-D	1-$2^{1}/_{2}$″ 90-degree ell	7	C-D	1-$^{1}/_{2}$″ 90-degree ell	4
D-E	1-$2^{1}/_{2}$″ Side branch tee	12	D-E	1-$1^{1}/_{2}$″ Side branch tee	7

TABLE E.2

PIPE SECTIONS	FRICTION LOSS EQUIVALENT LENGTH (feet)	
	Cold Water	Hot Water
A-B	0.69 × 3.2 = 2.21	0.69 × 3.2 = 2.21
B-C	0.085 × 3.1 = 0.26	0.16 × 1.4 = 0.22
C-D	0.20 × 1.9 = 0.38	0.17 × 3.2 = 0.54
D-E	1.62 × 1.9 = 3.08	1.57 × 3.2 = 5.02
Total pipe friction losses (Line K)	5.93	7.99

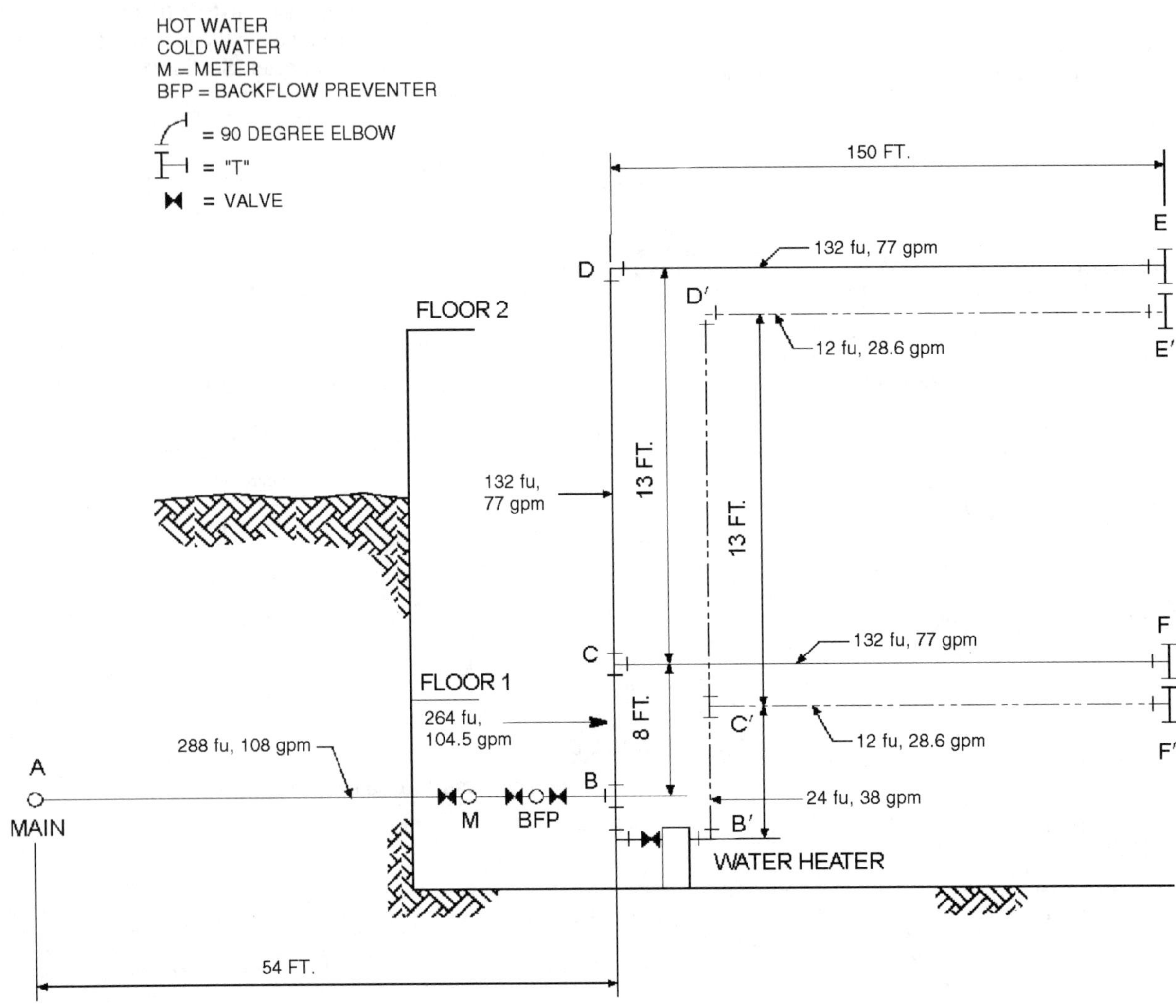

For SI: 1 foot = 304.8 mm, 1 gpm = 3.785 L/m.

FIGURE E103.3(1)
EXAMPLE-SIZING

TABLE E103.3(1)
RECOMMENDED TABULAR ARRANGEMENT FOR USE IN SOLVING PIPE SIZING PROBLEMS

COLUMN	1		2	3	4	5	6	7	8	9	10
Line	Description		Lb per square inch (psi)	Gal. per min through section	Length of section (feet)	Trial pipe size (inches)	Equivalent length of fittings and valves (feet)	Total equivalent length col. 4 and col. 6 (100 feet)	Friction loss per 100 feet of trial size pipe (psi)	Friction loss in equivalent length col. 8 x col. 7 (psi)	Excess pressure over friction losses (psi)
A	Service and cold water distribution piping[a]	Minimum pressure available at main	55.00								
B		Highest pressure required at a fixture (Table 604.3)	15.00								
C		Meter loss 2″ meter	11.00								
D		Tap in main loss 2″ tap (Table E103A)	1.61								
E		Static head loss 21 × 43 psi	9.03								
F		Special fixture loss backflow preventer	9.00								
G		Special fixture loss—Filter	0.00								
H		Special fixture loss—Other	0.00								
I		Total overall losses and requirements (Sum of Lines B through H)	45.64								
J		Pressure available to overcome pipe friction (Line A minus Lines B to H)	9.36								
	DESIGNATION Pipe section (from diagram) Cold water Distribution piping		FU								
		AB	288	108.0	54	$2^1/_2$	15.00	0.69	3.2	2.21	—
		BC	264	104.5	8	$2^1/_2$	0.5	0.85	3.1	0.26	—
		CD	132	77.0	13	$2^1/_2$	7.00	0.20	1.9	0.38	—
		CF[b]	132	77.0	150	$2^1/_2$	12.00	1.62	1.9	3.08	—
		DE[b]	132	77.0	150	$2^1/_2$	12.00	1.62	1.9	3.08	—
K	Total pipe friction losses (cold)			—	—	—	—	—	—	5.93	—
L	Difference (Line J minus Line K)			—	—	—	—	—	—	—	3.43
	Pipe section (from diagram) Diagram Hot water Distribution Piping	A′B′	288	108.0	54	$2^1/_2$	12.00	0.69	3.3	2.21	—
		B′C′	24	38.0	8	2	7.5	0.16	1.4	0.22	—
		C′D′	12	28.6	13	$1^1/_2$	4.0	0.17	3.2	0.54	—
		C′F′[b]	12	28.6	150	$1^1/_2$	7.00	1.57	3.2	5.02	—
		D′E′[b]	12	28.6	150	$1^1/_2$	7.00	1.57	3.2	5.02	—
K	Total pipe friction losses (hot)			—	—	—	—	—	—	7.99	—
L	Difference (Line J minus Line K)			—	—	—	—	—	—	—	1.37

For SI: 1 inch = 25.4 mm, 1 foot = 304.8 mm, 1 psi = 6.895 kPa, 1 gpm = 3.785 L/m.

a. To be considered as pressure gain for fixtures below main (to consider separately, omit from "I" and add to "J").

b. To consider separately, in K use C-F only if greater loss than above.

TABLE E103.3(2)
LOAD VALUES ASSIGNED TO FIXTURES[a]

FIXTURE	OCCUPANCY	TYPE OF SUPPLY CONTROL	LOAD VALUES, IN WATER SUPPLY FIXTURE UNITS (wsfu)		
			Cold	Hot	Total
Bathroom group	Private	Flush tank	2.7	1.5	3.6
Bathroom group	Private	Flush valve	6.0	3.0	8.0
Bathtub	Private	Faucet	1.0	1.0	1.4
Bathtub	Public	Faucet	3.0	3.0	4.0
Bidet	Private	Faucet	1.5	1.5	2.0
Combination fixture	Private	Faucet	2.25	2.25	3.0
Dishwashing machine	Private	Automatic	—	1.4	1.4
Drinking fountain	Offices, etc.	$^{3}/_{8}$″ valve	0.25	—	0.25
Kitchen sink	Private	Faucet	1.0	1.0	1.4
Kitchen sink	Hotel, restaurant	Faucet	3.0	3.0	4.0
Laundry trays (1 to 3)	Private	Faucet	1.0	1.0	1.4
Lavatory	Private	Faucet	0.5	0.5	0.7
Lavatory	Public	Faucet	1.5	1.5	2.0
Service sink	Offices, etc.	Faucet	2.25	2.25	3.0
Shower head	Public	Mixing valve	3.0	3.0	4.0
Shower head	Private	Mixing valve	1.0	1.0	1.4
Urinal	Public	1″ flush valve	10.0	—	10.0
Urinal	Public	$^{3}/_{4}$″ flush valve	5.0	—	5.0
Urinal	Public	Flush tank	3.0	—	3.0
Washing machine (8 lb)	Private	Automatic	1.0	1.0	1.4
Washing machine (8 lb)	Public	Automatic	2.25	2.25	3.0
Washing machine (15 lb)	Public	Automatic	3.0	3.0	4.0
Water closet	Private	Flush valve	6.0	—	6.0
Water closet	Private	Flush tank	2.2	—	2.2
Water closet	Public	Flush valve	10.0	—	10.0
Water closet	Public	Flush tank	5.0	—	5.0
Water closet	Public or private	Flushometer tank	2.0	—	2.0

For SI: 1 inch = 25.4 mm, 1 pound = 0.454 kg.

a. For fixtures not listed , loads should be assumed by comparing the fixture to one listed using water in similar quantities and at similar rates. The assigned loads for fixtures with both hot and cold water supplies are given for separate hot and cold water loads and for total load. The separate hot and cold water loads being three-fourths of the total load for the fixture in each case.

TABLE E103.3(3)
TABLE FOR ESTIMATING DEMAND

SUPPLY SYSTEMS PREDOMINANTLY FOR FLUSH TANKS			SUPPLY SYSTEMS PREDOMINANTLY FOR FLUSH VALVES		
Load	Demand		Load	Demand	
(Water supply fixture units)	(Gallons per minute)	(Cubic feet per minute)	(Water supply fixture units)	(Gallons per minute)	(Cubic feet per minute)
1	3.0	0.04104	—	—	—
2	5.0	0.0684	—	—	—
3	6.5	0.86892	—	—	—
4	8.0	1.06944	—	—	—
5	9.4	1.256592	5	15.0	2.0052
6	10.7	1.430376	6	17.4	2.326032
7	11.8	1.577424	7	19.8	2.646364
8	12.8	1.711104	8	22.2	2.967696
9	13.7	1.831416	9	24.6	3.288528
10	14.6	1.951728	10	27.0	3.60936
11	15.4	2.058672	11	27.8	3.716304
12	16.0	2.13888	12	28.6	3.823248
13	16.5	2.20572	13	29.4	3.930192
14	17.0	2.27256	14	30.2	4.037136
15	17.5	2.3394	15	31.0	4.14408
16	18.0	2.90624	16	31.8	4.241024
17	18.4	2.459712	17	32.6	4.357968
18	18.8	2.513184	18	33.4	4.464912
19	19.2	2.566656	19	34.2	4.571856
20	19.6	2.620128	20	35.0	4.6788
25	21.5	2.87412	25	38.0	5.07984
30	23.3	3.114744	30	42.0	5.61356
35	24.9	3.328632	35	44.0	5.88192
40	26.3	3.515784	40	46.0	6.14928
45	27.7	3.702936	45	48.0	6.41664
50	29.1	3.890088	50	50.0	6.684
60	32.0	4.27776	60	54.0	7.21872
70	35.0	4.6788	70	58.0	7.75344
80	38.0	5.07984	80	61.2	8.181216
90	41.0	5.48088	90	64.3	8.595624
100	43.5	5.81508	100	67.5	9.0234
120	48.0	6.41664	120	73.0	9.75864
140	52.5	7.0182	140	77.0	10.29336
160	57.0	7.61976	160	81.0	10.82808
180	61.0	8.15448	180	85.5	11.42964
200	65.0	8.6892	200	90.0	12.0312
225	70.0	9.3576	225	95.5	12.76644
250	75.0	10.026	250	101.0	13.50168

(continued)

TABLE E103.3(3)—continued
TABLE FOR ESTIMATING DEMAND

SUPPLY SYSTEMS PREDOMINANTLY FOR FLUSH TANKS			SUPPLY SYSTEMS PREDOMINANTLY FOR FLUSH VALVES		
Load	Demand		Load	Demand	
(Water supply fixture units)	(Gallons per minute)	(Cubic feet per minute)	(Water supply fixture units)	(Gallons per minute)	(Cubic feet per minute)
275	80.0	10.6944	275	104.5	13.96956
300	85.0	11.3628	300	108.0	14.43744
400	105.0	14.0364	400	127.0	16.97736
500	124.0	16.57632	500	143.0	19.11624
750	170.0	22.7256	750	177.0	23.66136
1,000	208.0	27.80544	1,000	208.0	27.80544
1,250	239.0	31.94952	1,250	239.0	31.94952
1,500	269.0	35.95992	1,500	269.0	35.95992
1,750	297.0	39.70296	1,750	297.0	39.70296
2,000	325.0	43.446	2,000	325.0	43.446
2,500	380.0	50.7984	2,500	380.0	50.7984
3,000	433.0	57.88344	3,000	433.0	57.88344
4,000	525.0	70.182	4,000	525.0	70.182
5,000	593.0	79.27224	5,000	593.0	79.27224

TABLE E103.3(4)
LOSS OF PRESSURE THROUGH TAPS AND TEES IN POUNDS PER SQUARE INCH (psi)

GALLONS PER MINUTE	SIZE OF TAP OR TEE (inches)						
	$^5/_8$	$^3/_4$	1	$1^1/_4$	$1^1/_2$	2	3
10	1.35	0.64	0.18	0.08	—	—	—
20	5.38	2.54	0.77	0.31	0.14	—	—
30	12.10	5.72	1.62	0.69	0.33	0.10	—
40	—	10.20	3.07	1.23	0.58	0.18	—
50	—	15.90	4.49	1.92	0.91	0.28	—
60	—	—	6.46	2.76	1.31	0.40	—
70	—	—	8.79	3.76	1.78	0.55	0.10
80	—	—	11.50	4.90	2.32	0.72	0.13
90	—	—	14.50	6.21	2.94	0.91	0.16
100	—	—	17.94	7.67	3.63	1.12	0.21
120	—	—	25.80	11.00	5.23	1.61	0.30
140	—	—	35.20	15.00	7.12	2.20	0.41
150	—	—	—	17.20	8.16	2.52	0.47
160	—	—	—	19.60	9.30	2.92	0.54
180	—	—	—	24.80	11.80	3.62	0.68
200	—	—	—	30.70	14.50	4.48	0.84
225	—	—	—	38.80	18.40	5.60	1.06
250	—	—	—	47.90	22.70	7.00	1.31
275	—	—	—	—	27.40	7.70	1.59
300	—	—	—	—	32.60	10.10	1.88

For SI: 1 inch = 25.4 mm, 1 pound per square inch = 6.895 kpa, 1 gallon per minute = 3.785 L/m.

TABLE E103.3(5)
ALLOWANCE IN EQUIVALENT LENGTHS OF PIPE FOR FRICTION LOSS IN VALVES AND THREADED FITTINGS (feet)

FITTING OR VALVE	PIPE SIZE (inches)							
	1/2	3/4	1	1 1/4	1 1/2	2	2 1/2	3
45-degree elbow	1.2	1.5	1.8	2.4	3.0	4.0	5.0	6.0
90-degree elbow	2.0	2.5	3.0	4.0	5.0	7.0	8.0	10.0
Tee, run	0.6	0.8	0.9	1.2	1.5	2.0	2.5	3.0
Tee, branch	3.0	4.0	5.0	6.0	7.0	10.0	12.0	15.0
Gate valve	0.4	0.5	0.6	0.8	1.0	1.3	1.6	2.0
Balancing valve	0.8	1.1	1.5	1.9	2.2	3.0	3.7	4.5
Plug-type cock	0.8	1.1	1.5	1.9	2.2	3.0	3.7	4.5
Check valve, swing	5.6	8.4	11.2	14.0	16.8	22.4	28.0	33.6
Globe valve	15.0	20.0	25.0	35.0	45.0	55.0	65.0	80.0
Angle valve	8.0	12.0	15.0	18.0	22.0	28.0	34.0	40.0

For SI: 1 inch = 25.4 mm, 1 foot = 304.8 mm, 1 degree = 0.0175 rad.

TABLE E103.3(6)
PRESSURE LOSS IN FITTINGS AND VALVES EXPRESSED AS EQUIVALENT LENGTH OF TUBE[a] (feet)

NOMINAL OR STANDARD SIZE (inches)	FITTINGS					VALVES			
	Standard Ell		90-Degree Tee						
	90 Degree	45 Degree	Side Branch	Straight Run	Coupling	Ball	Gate	Butterfly	Check
3/8	0.5	—	1.5	—	—	—	—	—	1.5
1/2	1	0.5	2	—	—	—	—	—	2
5/8	1.5	0.5	2	—	—	—	—	—	2.5
3/4	2	0.5	3	—	—	—	—	—	3
1	2.5	1	4.5	—	—	0.5	—	—	4.5
1 1/4	3	1	5.5	0.5	0.5	0.5	—	—	5.5
1 1/2	4	1.5	7	0.5	0.5	0.5	—	—	6.5
2	5.5	2	9	0.5	0.5	0.5	0.5	7.5	9
2 1/2	7	2.5	12	0.5	0.5	—	1	10	11.5
3	9	3.5	15	1	1	—	1.5	15.5	14.5
3 1/2	9	3.5	14	1	1	—	2	—	12.5
4	12.5	5	21	1	1	—	2	16	18.5
5	16	6	27	1.5	1.5	—	3	11.5	23.5
6	19	7	34	2	2	—	3.5	13.5	26.5
8	29	11	50	3	3	—	5	12.5	39

For SI: 1 inch = 25.4 mm, 1 foot = 304.8 mm, 1 degree = 0.01745 rad.

a. Allowances are for streamlined soldered fittings and recessed threaded fittings. For threaded fittings, double the allowances shown in the table. The equivalent lengths presented above are based on a C factor of 150 in the Hazen-Williams friction loss formula. The lengths shown are rounded to the nearest half-foot.

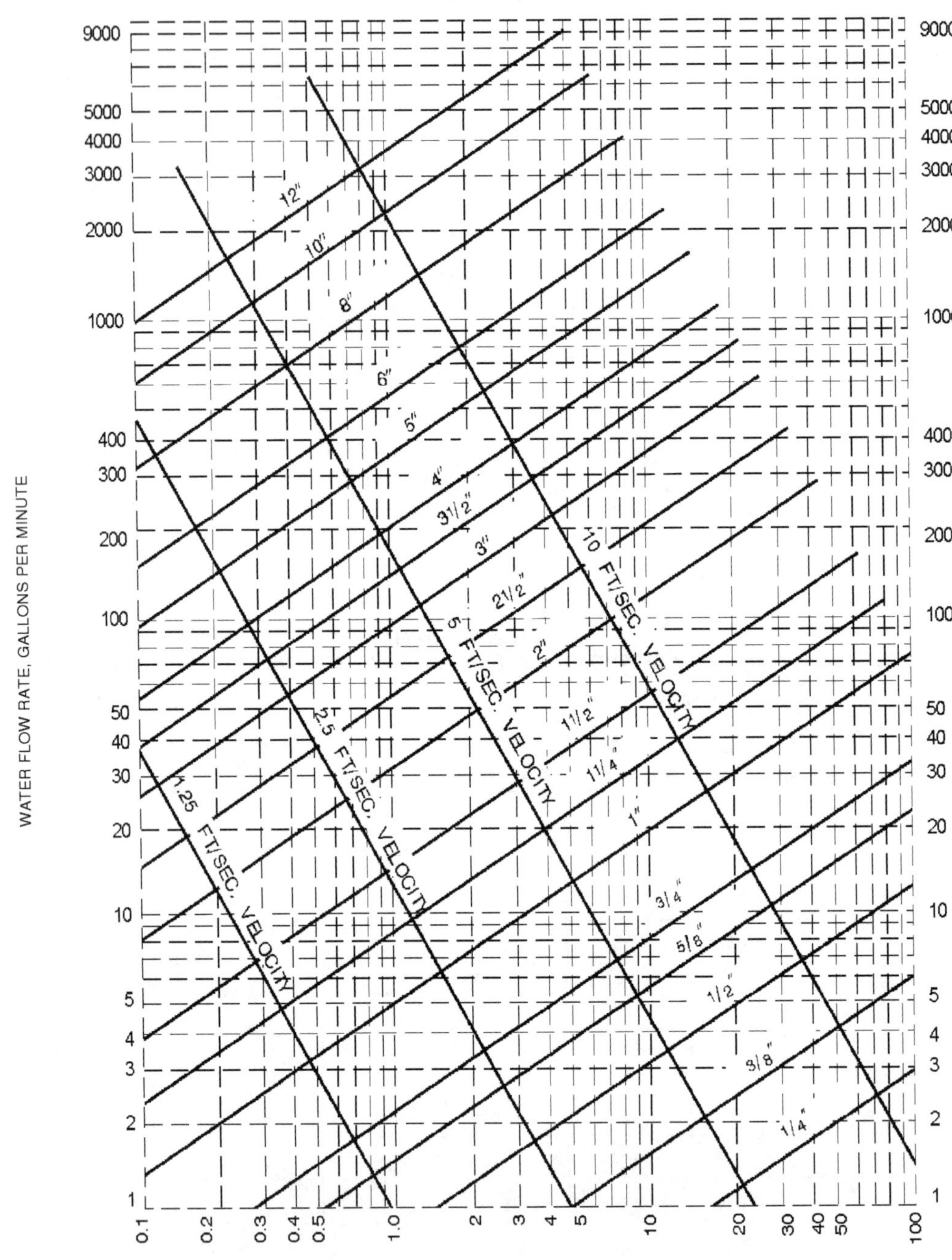

Note: Fluid velocities in excess of 5 to 8 feet/second are not usually recommended.

FIGURE E103.3(2)
FRICTION LOSS IN SMOOTH PIPE[a] (TYPE K, ASTM B 88 COPPER TUBING)

For SI: 1 inch = 25.4 mm, 1 foot = 304.8 mm, 1 gpm = 3.785 L/m, 1 psi = 6.895 kPa,
1 foot per second = 0.305 m/s.

a. This chart applies to smooth new copper tubing with recessed (streamline) soldered joints and to the actual sizes of types indicated on the diagram.

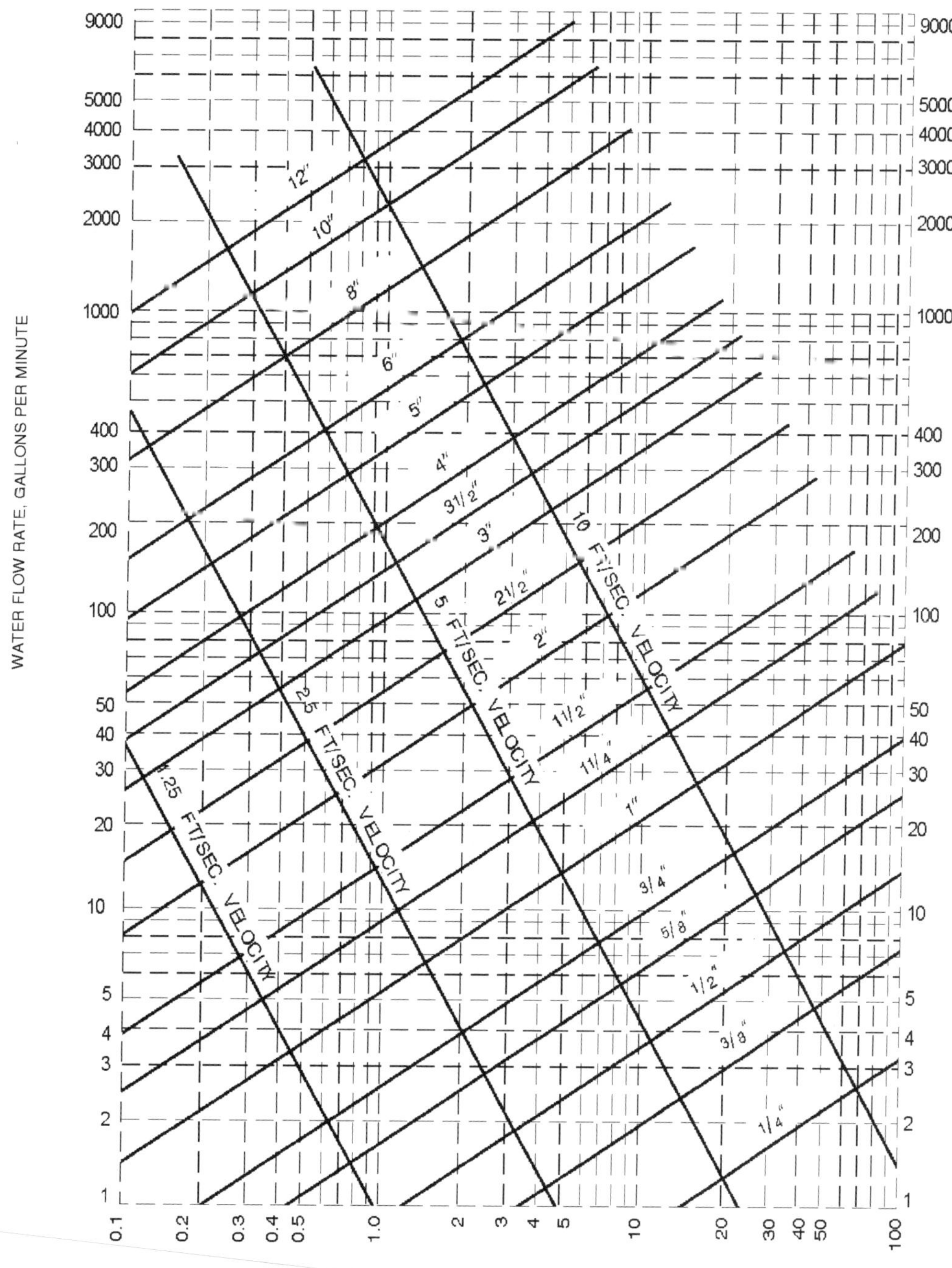

Note: Fluid velocities in excess of 5 to 8 feet/second are not usually recommended.

FIGURE E103.3(3)
FRICTION LOSS IN SMOOTH PIPE[a] (TYPE L, ASTM B 88 COPPER TUBING)

For SI: 1 inch = 25.4 mm, 1 foot = 304.8 mm, 1 gpm = 3.785 L/m, 1 psi = 6.895 kPa,
1 foot per second = 0.305 m/s.

a. This chart applies to smooth new copper tubing with recessed (streamline) soldered joints and to the actual sizes of types indicated on the diagram.

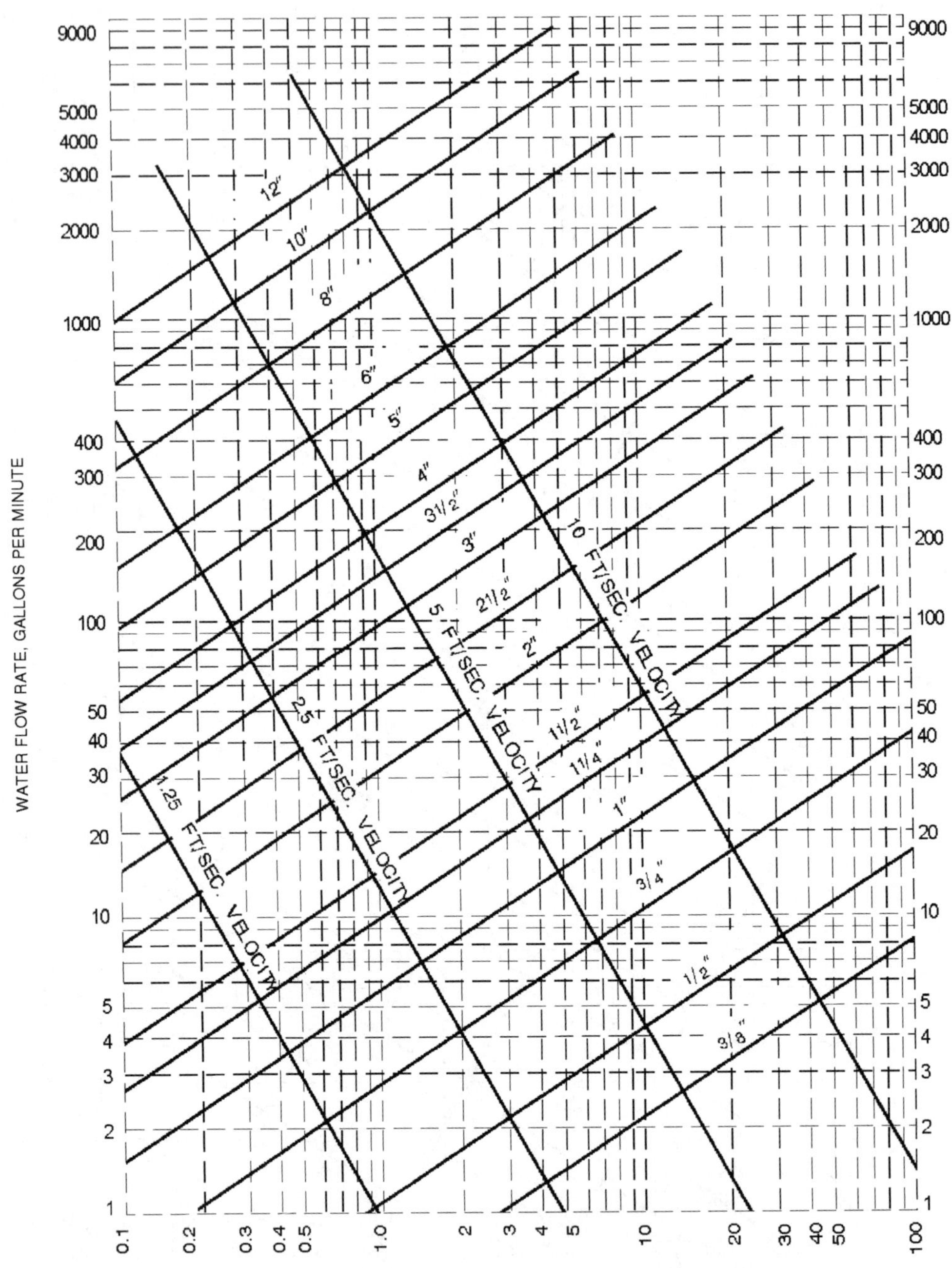

Note: Fluid velocities in excess of 5 to 8 feet/second are not usually recommended.

FIGURE E103.3(4)
FRICTION LOSS IN SMOOTH PIPE[a] (TYPE M, ASTM B 88 COPPER TUBING)

For SI: 1 inch = 25.4 mm, 1 foot = 304.8 mm, 1 gpm = 3.785 L/m, 1 psi = 6.895 kPa,
1 foot per second = 0.305 m/s.

a. This chart applies to smooth new copper tubing with recessed (streamline) soldered joints and to the actual sizes of types indicated on the diagram.

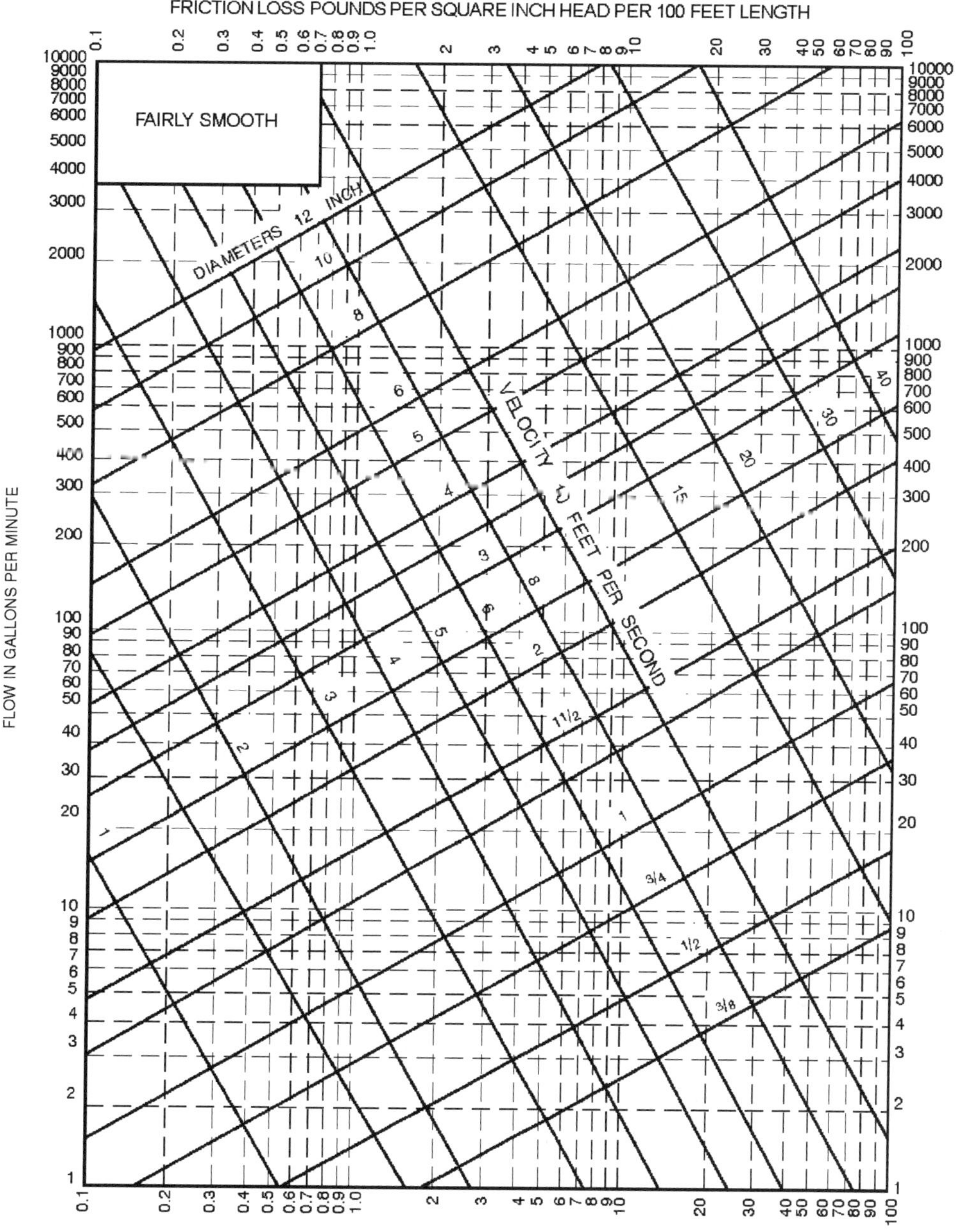

FIGURE E103.3(5)
FRICTION LOSS IN FAIRLY SMOOTH PIPE[a]

For SI: 1 inch = 25.4 mm, 1 foot = 304.8 mm, 1 gpm = 3.785 L/m, 1 psi = 6.895 kPa,
1 foot per second = 0.305 m/s.

a. This chart applies to smooth new steel (fairly smooth) pipe and to actual diameters of standard-weight pipe.

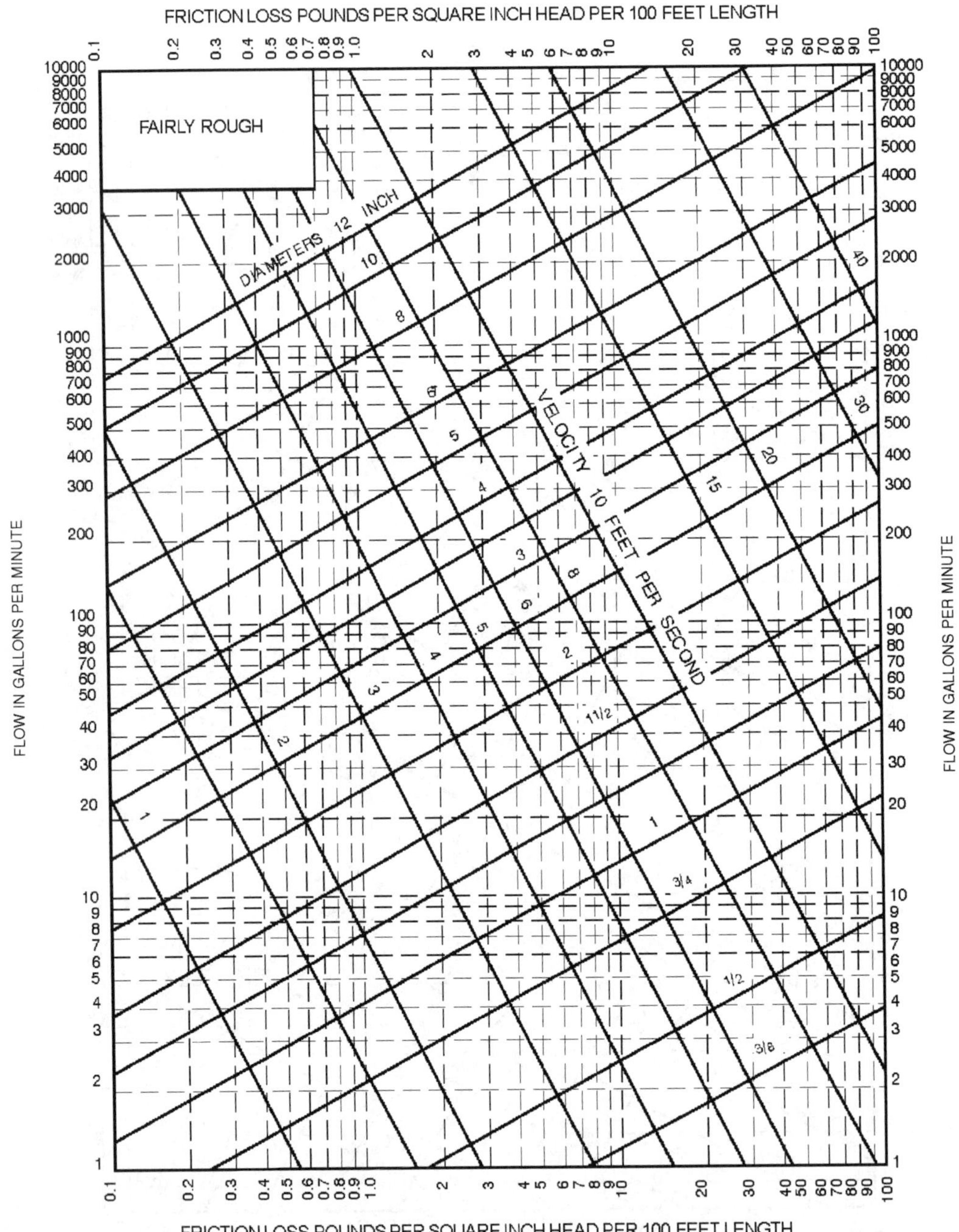

FIGURE E103.3(6)
FRICTION LOSS IN FAIRLY ROUGH PIPE[a]

For SI: 1 inch = 25.4 mm, 1 foot = 304.8 mm, 1 gpm = 3.785 L/m, 1 psi = 6.895 kPa,
1 foot per second = 0.305 m/s.

a. This chart applies to fairly rough pipe and to actual diameters which in general will be less than the actual diameters of the new pipe of the same kind.

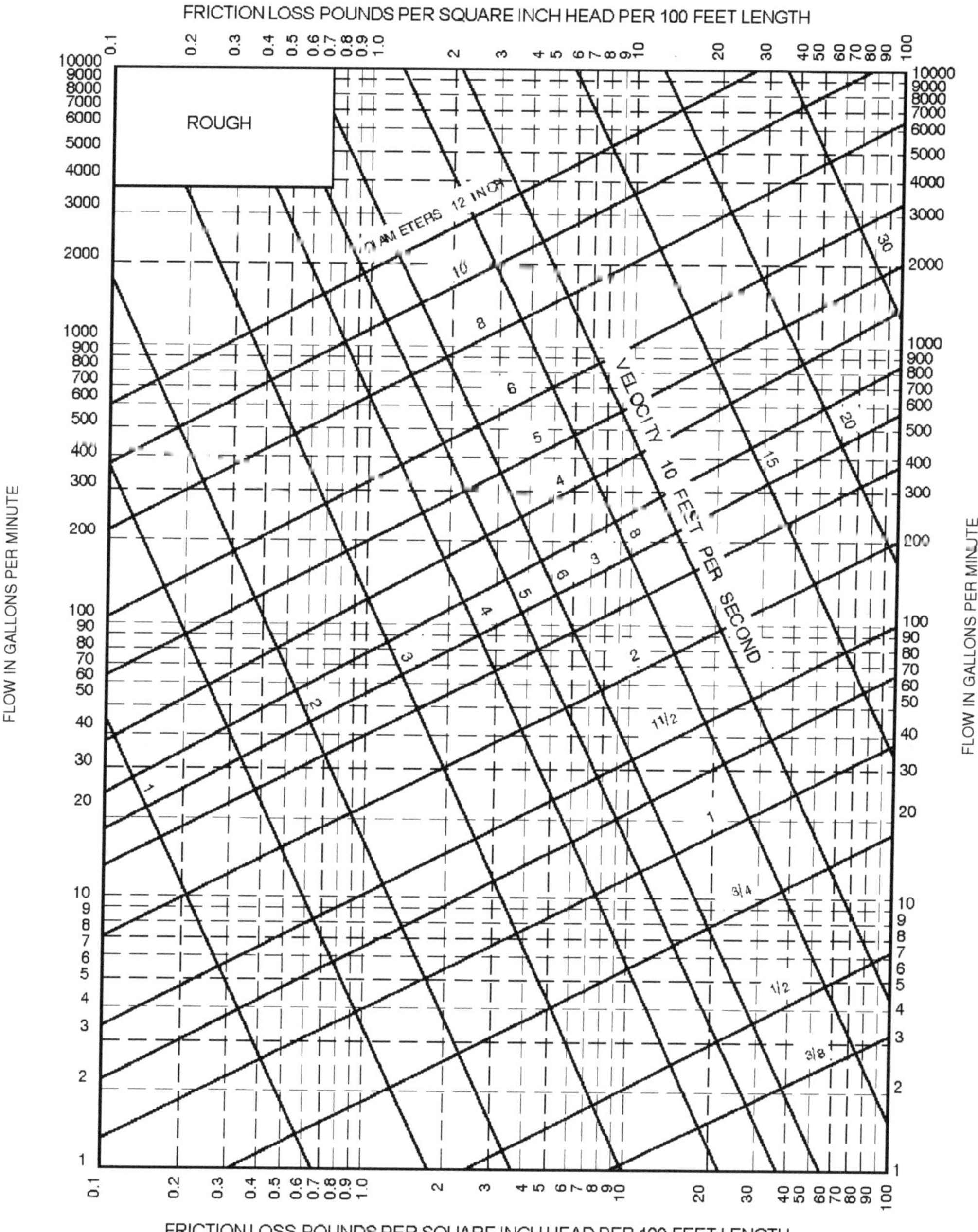

FIGURE E103.3(7)
FRICTION LOSS IN FAIRLY ROUGH PIPE[a]

For SI: 1 inch = 25.4 mm, 1 foot = 304.8 mm, 1 gpm = 3.785 L/m, 1 psi = 6.895 kPa,
1 foot per second = 0.305 m/s.

a. This chart applies to very rough pipe and existing pipe and to their actual diameters.

SECTION E201
SELECTION OF PIPE SIZE

E201.1 Size of water-service mains, branch mains and risers. The minimum size water service pipe shall be $^3/_4$ inch (19.1 mm). The size of water service mains, *branch* mains and risers shall be determined according to water supply demand [gpm (L/m)], available water pressure [psi (kPa)] and friction loss due to the water meter and *developed length* of pipe [feet (m)], including equivalent length of fittings. The size of each water distribution system shall be determined according to the procedure outlined in this section or by other design methods conforming to acceptable engineering practice and *approved* by the code official:

1. Supply load in the building water-distribution system shall be determined by total load on the pipe being sized, in terms of water-supply fixture units (w.s.f.u.), as shown in Table E103.3(2). For fixtures not listed, choose a w.s.f.u. value of a fixture with similar flow characteristics.
2. Obtain the minimum daily static service pressure [psi (kPa)] available (as determined by the local water authority) at the water meter or other source of supply at the installation location. Adjust this minimum daily static pressure [psi (kPa)] for the following conditions:
   - 2.1. Determine the difference in elevation between the source of supply and the highest water supply outlet. Where the highest water supply outlet is located above the source of supply, deduct 0.5 psi (3.4 kPa) for each foot (0.3 m) of difference in elevation. Where the highest water supply outlet is located below the source of supply, add 0.5 psi (3.4 kPa) for each foot (0.3 m) of difference in elevation.
   - 2.2. Where a water pressure reducing valve is installed in the water distribution system, the minimum daily static water pressure available is 80 percent of the minimum daily static water pressure at the source of supply or the set pressure downstream of the pressure reducing valve, whichever is smaller.
   - 2.3. Deduct all pressure losses due to special equipment such as a backflow preventer, water filter and water softener. Pressure loss data for each piece of equipment shall be obtained through the manufacturer of such devices.
   - 2.4. Deduct the pressure in excess of 8 psi (55 kPa) due to installation of the special plumbing fixture, such as temperature controlled shower and flushometer tank water closet.

     Using the resulting minimum available pressure, find the corresponding pressure range in Table E201.1.
3. The maximum *developed length* for water piping is the actual length of pipe between the source of supply and the most remote fixture, including either hot (through the water heater) or cold water branches multiplied by a factor of 1.2 to compensate for pressure loss through fittings.

   Select the appropriate column in Table E201.1 equal to or greater than the calculated maximum *developed length.*
4. To determine the size of water service pipe, meter and main distribution pipe to the building using the appropriate table, follow down the selected "maximum *developed length*" column to a fixture unit equal to, or greater than the total installation demand calculated by using the "combined" water supply fixture unit column of Table E103.3(2). Read the water service pipe and meter sizes in the first left-hand column and the main distribution pipe to the building in the second left-hand column on the same row.
5. To determine the size of each water distribution pipe, start at the most remote outlet on each *branch* (either hot or cold *branch*) and, working back toward the main distribution pipe to the building, add up the water supply fixture unit demand passing through each segment of the distribution system using the related hot or cold column of Table E103.3(2). Knowing demand, the size of each segment shall be read from the second left-hand column of the same table and maximum *developed length* column selected in Steps 1 and 2, under the same or next smaller size meter row. In no case does the size of any *branch* or main need to be larger that the size of the main distribution pipe to the building established in Step 4.

SECTION E202
DETERMINATION OF PIPE VOLUMES

E202.1 Determining volume of piping systems. Where required for engineering design purposes, Table E202.1 shall be used to determine the approximate internal volume of water distribution piping.

TABLE E201.1
MINIMUM SIZE OF WATER METERS, MAINS AND DISTRIBUTION PIPING BASED ON WATER SUPPLY FIXTURE UNIT VALUES (w.s.f.u.)

METER AND SERVICE PIPE (inches)	DISTRIBUTION PIPE (inches)	MAXIMUM DEVELOPMENT LENGTH (feet)									
Pressure Range 30 to 39 psi		40	60	80	100	150	200	250	300	400	500
3/4	1/2[a]	2.5	2	1.5	1.5	1	1	0.5	0.5	0	0
3/4	3/4	9.5	7.5	6	5.5	4	3.5	3	2.5	2	1.5
3/4	1	32	25	20	16.5	11	9	7.8	6.5	5.5	4.5
1	1	32	32	27	21	13.5	10	8	7	5.5	5
3/4	1 1/4	32	32	32	32	30	24	20	17	13	10.5
1	1 1/4	80	80	70	61	45	34	27	22	16	12
1 1/2	1 1/4	80	80	80	75	54	40	31	25	17.5	13
1	1 1/2	87	87	87	87	84	73	64	56	45	36
1 1/2	1 1/2	151	151	151	151	117	92	79	69	54	43
2	1 1/2	151	151	151	151	128	99	83	72	56	45
1	2	87	87	87	87	87	87	87	87	87	86
1 1/2	2	275	275	275	275	258	223	196	174	144	122
2	2	365	365	365	365	318	266	229	201	160	134
2	2 1/2	533	533	533	533	533	495	448	409	353	311

METER AND SERVICE PIPE (inches)	DISTRIBUTION PIPE (inches)	MAXIMUM DEVELOPMENT LENGTH (feet)									
Pressure Range 40 to 49 psi		40	60	80	100	150	200	250	300	400	500
3/4	1/2[a]	3	2.5	2	1.5	1.5	1	1	0.5	0.5	0.5
3/4	3/4	9.5	9.5	8.5	7	5.5	4.5	3.5	3	2.5	2
3/4	1	32	32	32	26	18	13.5	10.5	9	7.5	6
1	1	32	32	32	32	21	15	11.5	9.5	7.5	6.5
3/4	1 1/4	32	32	32	32	32	32	32	27	21	16.5
1	1 1/4	80	80	80	80	65	52	42	35	26	20
1 1/2	1 1/4	80	80	80	80	75	59	48	39	28	21
1	1 1/2	87	87	87	87	87	87	87	78	65	55
1 1/2	1 1/2	151	151	151	151	151	130	109	93	75	63
2	1 1/2	151	151	151	151	151	139	115	98	77	64
1	2	87	87	87	87	87	87	87	87	87	87
1 1/2	2	275	275	275	275	275	275	264	238	198	169
2	2	365	365	365	365	365	349	304	270	220	185
2	2 1/2	533	533	533	533	533	533	533	528	456	403

(continued)

TABLE E201.1—continued
MINIMUM SIZE OF WATER METERS, MAINS AND DISTRIBUTION PIPING BASED ON WATER SUPPLY FIXTURE UNIT VALUES (w.s.f.u.)

METER AND SERVICE PIPE (inches)	DISTRIBUTION PIPE (inches)	MAXIMUM DEVELOPMENT LENGTH (feet)									
Pressure Range 50 to 60 psi		40	60	80	100	150	200	250	300	400	500
$^3/_4$	$^1/_2$[a]	3	3	2.5	2	1.5	1	1	1	0.5	0.5
$^3/_4$	$^3/_4$	9.5	9.5	9.5	8.5	6.5	5	4.5	4	3	2.5
$^3/_4$	1	32	32	32	32	25	18.5	14.5	12	9.5	8
1	1	32	32	32	32	30	22	16.5	13	10	8
$^3/_4$	$1^1/_4$	32	32	32	32	32	32	32	32	29	24
1	$1^1/_4$	80	80	80	80	80	68	57	48	35	28
$1^1/_2$	$1^1/_4$	80	80	80	80	80	75	63	53	39	29
1	$1^1/_2$	87	87	87	87	87	87	87	87	82	70
$1^1/_2$	$1^1/_2$	151	151	151	151	151	151	139	120	94	79
2	$1^1/_2$	151	151	151	151	151	151	146	126	97	81
1	2	87	87	87	87	87	87	87	87	87	87
$1^1/_2$	2	275	275	275	275	275	275	275	275	247	213
2	2	365	365	365	365	365	365	365	329	272	232
2	$2^1/_2$	533	533	533	533	533	533	533	533	353	486

METER AND SERVICE PIPE (inches)	DISTRIBUTION PIPE (inches)	MAXIMUM DEVELOPMENT LENGTH (feet)									
Pressure Range Over 60		40	60	80	100	150	200	250	300	400	500
$^3/_4$	$^1/_2$[a]	3	3	3	2.5	2	1.5	1.5	1	1	0.5
$^3/_4$	$^3/_4$	9.5	9.5	9.5	9.5	7.5	6	5	4.5	3.5	3
$^3/_4$	1	32	32	32	32	32	24	19.5	15.5	11.5	9.5
1	1	32	32	32	32	32	28	28	17	12	9.5
$^3/_4$	$1^1/_4$	32	32	32	32	32	32	32	32	32	30
1	$1^1/_4$	80	80	80	80	80	80	69	60	46	36
$1^1/_2$	$1^1/_4$	80	80	80	80	80	80	76	65	50	38
1	$1^1/_2$	87	87	87	87	87	87	87	87	87	84
$1^1/_2$	$1^1/_2$	151	151	151	151	151	151	151	144	114	94
2	$1^1/_2$	151	151	151	151	151	151	151	151	118	97
1	2	87	87	87	87	87	87	87	87	87	87
$1^1/_2$	2	275	275	275	275	275	275	275	275	275	252
2	2	365	368	368	368	368	368	368	368	318	273
2	$2^1/_2$	533	533	533	533	533	533	533	533	533	533

For SI: 1 inch = 25.4, 1 foot = 304.8 mm.
a. Minimum size for building supply is $^3/_4$-inch pipe.

TABLE E202.1
INTERNAL VOLUME OF VARIOUS WATER DISTRIBUTION TUBING

	OUNCES OF WATER PER FOOT OF TUBE						
Size Nominal, Inch	Copper Type M	Copper Type L	Copper Type K	CPVC CTS SDR 11	CPVC SCH 40	Composite ASTM F 1281	PEX CTS SDR 9
$^3/_8$	1.06	0.97	0.84	N/A	1.17	0.63	0.64
$^1/_2$	1.69	1.55	1.45	1.25	1.89	1.31	1.18
$^3/_4$	3.43	3.22	2.90	2.67	3.38	3.39	2.35
1	5.81	5.49	5.17	4.43	5.53	5.56	3.91
$1^1/_4$	8.70	8.36	8.09	6.61	9.66	8.49	5.81
$1^1/_2$	12.18	11.83	11.45	9.22	13.20	13.88	8.09
2	21.08	20.58	20.04	15.79	21.88	21.48	13.86

For SI: 1 ounce = 0.030 liter.

Appendix F: Structural Safety

General Comments

Although the title of Section F101 is "Cutting, Notching and Boring in Wood Members," it also includes some regulations pertaining to structural steel and cold-formed steel structural members.

Purpose

This appendix contains criteria limiting the cutting away of a structural member during the installation of a plumbing system. These provisions are needed for the safety of the joists, studs, beams, columns or other structural members that support the building.

The provisions contained in this appendix are not mandatory unless specifically referenced in the adopting ordinance.

SECTION F101 CUTTING, NOTCHING AND BORING IN WOOD MEMBERS

[B] F101.1 Joist notching. Notches on the ends of joists shall not exceed one-fourth the joist depth. Holes bored in joists shall not be within 2 inches (51 mm) of the top or bottom of the joist, and the diameter of any such hole shall not exceed one-third the depth of the joist. Notches in the top or bottom of joists shall not exceed one sixth the depth and shall not be located in the middle third of the span.

❖ Wood is comprised of relatively small elongated cells oriented parallel to each other and bound together physically and chemically by lignin. The cells are continuous from end to end of a piece of lumber. When a beam or joist is notched or cut, the cell strands are interrupted. In notched beams, the effective depth is reduced equal to the depth of the notch (see Commentary Figure F101.1).

Some designs and installation practices require that limited notching and cutting occur. Notching should be avoided when possible. Holes bored in beams and joists create the same problems as notches. When necessary, the holes should be located in areas with the least stress concentration, generally along the neutral axis of the joist. Beams subject to high horizontal shear stress (short span/heavy load) should never be cut. Wood members having a thickness greater than 4 inches (102 mm) should never be notched, except at the ends of the members.

[B] F101.2 Stud cutting and notching. In exterior walls and bearing partitions, any wood stud is permitted to be cut or notched to a depth not exceeding 25 percent of its width. Cutting or notching of studs to a depth not greater than 40 percent of the width of the stud is permitted in nonbearing partitions supporting no loads other than the weight of the partition.

❖ To maintain the cross-sectional bearing area, studs should not be cut, notched or bored. When the alteration exceeds that specified in this section, the studs should be doubled or otherwise reinforced to provide the required strength (see Commentary Figure F101.2).

[B] F101.3 Bored holes. A hole not greater in diameter than 40 percent of the stud width is permitted to be bored in any wood stud. Bored holes not greater than 60 percent of the width of the stud are permitted in nonbearing partitions or in any wall where each bored stud is doubled, provided not more than two such successive doubled studs are so bored. In no case shall the edge of the bored hole be nearer than 0.625 inch (15.9 mm) to the edge of the stud. Bored holes shall not be located at the same section of stud as a cut or notch.

❖ Because of the redundancy of wood construction, limited hole boring is encouraged. This section describes the limits for bored holes in load-bearing and nonload-bearing wood studs. Commentary Figure F101.2 illustrates the requirements of this section.

[B] F101.4 Cutting, notching and boring holes in structural steel framing. The cutting, notching and boring of holes in structural steel framing members shall be as prescribed by the registered design professional.

❖ This section does not allow the cutting, notching or boring of structural steel girders, beams, joists, columns or other structural steel members unless it is approved first by the registered design professional responsible for the structural design of the building. The code does not provide any prescriptive limits as it does for wood framing. Since there are so many conditions that could be encountered, it is best for the design professional to consider them on a case-by-case basis.

[B] F101.5 Cutting, notching and boring holes in cold-formed steel framing. Flanges and lips of load-bearing cold-formed steel framing members shall not be cut or notched. Holes in webs of load-bearing cold-formed steel framing members shall be permitted along the centerline of the web of the framing member and shall not exceed the dimensional limitations, penetration spacing or minimum hole edge distance as prescribed by the registered design professional. Cutting, notching and boring holes of steel floor/roof decking shall be as prescribed by the registered design professional.

❖ No cutting, notching or boring is allowed at all in the flanges of the cold-formed lips of load-bearing studs, beams, columns or other load-carrying members, as that would result in a significant reduction of the load-carrying capacity.

This section requires the approval of the registered design professional prior to the boring of any holes in webs of any cold-formed steel member that is load bearing. The code does not provide any prescriptive limits as it does for wood framing.

[B] F101.6 Cutting, notching and boring holes in nonstructural cold-formed steel wall framing. Flanges and lips of nonstructural cold-formed steel wall studs shall not be cut or notched. Holes in webs of nonstructural cold-formed steel wall studs shall be permitted along the centerline of the web of the framing member, shall not exceed $1^1/_2$ inches (38 mm) in width or 4 inches (102 mm) in length, and the holes shall not be spaced less than 24 inches (610 mm) center to center from another hole or less than 10 inches (254 mm) from the bearing end.

❖ This section applies to nonload-bearing studs in partition walls. The holes in the studs are limited to a series of holes in the web, as described in this section. Some cold-formed studs are manufactured with these web openings already cut out for the installation of piping and electrical wiring.

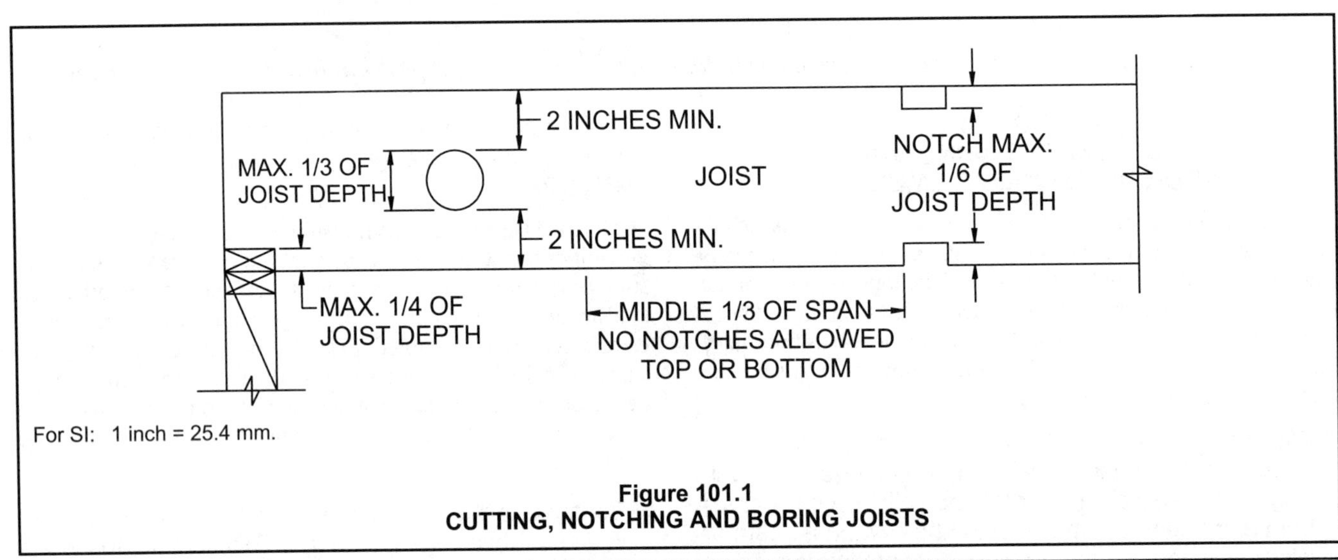

Figure 101.1
CUTTING, NOTCHING AND BORING JOISTS

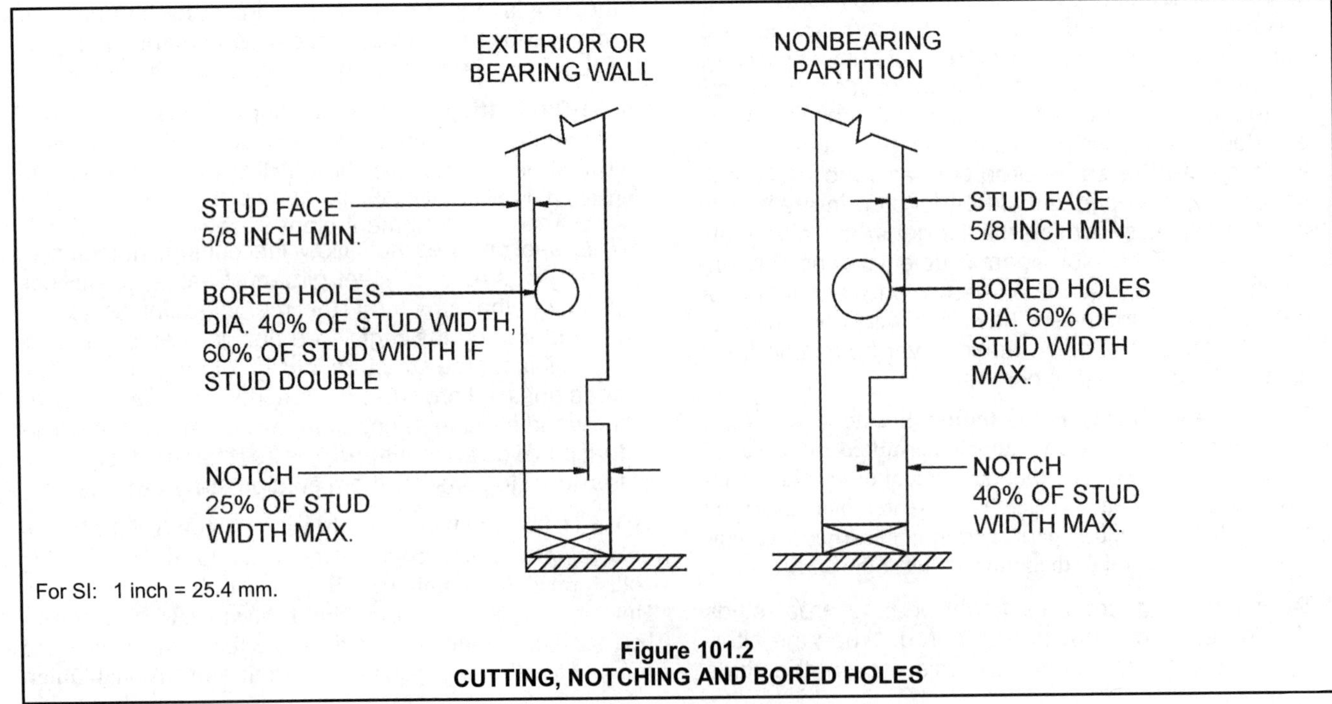

Figure 101.2
CUTTING, NOTCHING AND BORED HOLES

Appendix G: Vacuum Drainage System

General Comments

New designs and technology have evolved with the introduction of vacuum technology. This new technology is leading the way for drainage systems to transport sanitary waste and various other fluids and effluents. Although vacuum drainage systems have been around for many years in the transportation, marine and aviation industries, the technology is now being used in land-based projects.

Purpose

Appendix G regulates the design and installation of vacuum drainage systems.

The provisions contained in this appendix are not mandatory unless specifically referenced in the adopting ordinance.

SECTION G101 VACUUM DRAINAGE SYSTEM

G101.1 Scope. This appendix provides general guidelines for the requirements for vacuum drainage systems.

❖ This section regulates the design and installation methods for vacuum drainage systems. Plumbing systems are designed to drain by gravity whenever possible; however, there are many instances where the discharge from plumbing fixtures must be lifted by a vacuum to an approved point of disposal. This type of system is commonly specified in railcars, aircraft, buses, marine vessels, recreational vehicles and land-based projects.

G101.2 General requirements.

❖ Requirements for vacuum drainage systems are contained in this section. These requirements apply to system design, installation and other components utilized in such systems.

G101.2.1 System design. Vacuum drainage systems shall be designed in accordance with manufacturer's recommendations. The system layout, including piping layout, tank assemblies, vacuum pump assembly and other components/designs necessary for proper function of the system shall be per manufacturer's recommendations. Plans, specifications and other data for such systems shall be submitted to the local administrative authority for review and approval prior to installation.

❖ The vacuum system connections are made on the inlet rather than on the outlet as is the case with sewage ejectors or pumps. The wastes in these types of systems are being pulled, not pushed, through the piping system. Typically, the waste is introduced into the system from plumbing fixtures, conveyed to a collection tank and then to an approved point of disposal. A grinding pump that macerates waste is often installed prior to the collection tank, thereby providing additional holding capacity. The negative pressure is maintained in the system by a vacuum pump. A control panel links the system and typically includes a timing valve, discharge valve and water valve. For example, when the flush mechanism is activated for a water closet, the discharge valve opens, allowing atmospheric pressure to force waste through the piping to the vacuum tank. At the same time, the water valve injects water to rinse the bowl. A lavatory or sink waste flows by gravity to an interface valve, which then opens and allows the waste to enter the vacuum piping. The discharge from showers or floor drains is typically collected in a holding tank, and when the tank is filled, a pressure sensor triggers an actuator to open a valve, allowing the waste to move into the vacuum piping. A control system automatically regulates the operation of the vacuum and sewage ejector pumps. The vacuum pump creates a vacuum in the system by pumping air from the vacuum tank and piping. Once a predetermined level of sewage is reached in the vacuum tank, the sewage is automatically pumped out by sewage ejector pumps to a treatment plant or sewer main. Manufacturers must provide detailed installation and servicing instructions with each type of system [see Commentary Figures G101.2.1(1) and (2)].

G101.2.2 Fixtures. Gravity-type fixtures used in vacuum drainage systems shall comply with Chapter 4 of this code.

❖ Fixtures utilized in a vacuum system must adhere to the same standards for plumbing fixtures in Chapter 4 of the code. Plumbing fixtures regulated in this type of system must be free from deficiencies that affect the fixture's intended purpose. The referenced standards specify quality, dimensions, construction requirements and the testing of such fixtures (see Commentary Figure G101.2.2).

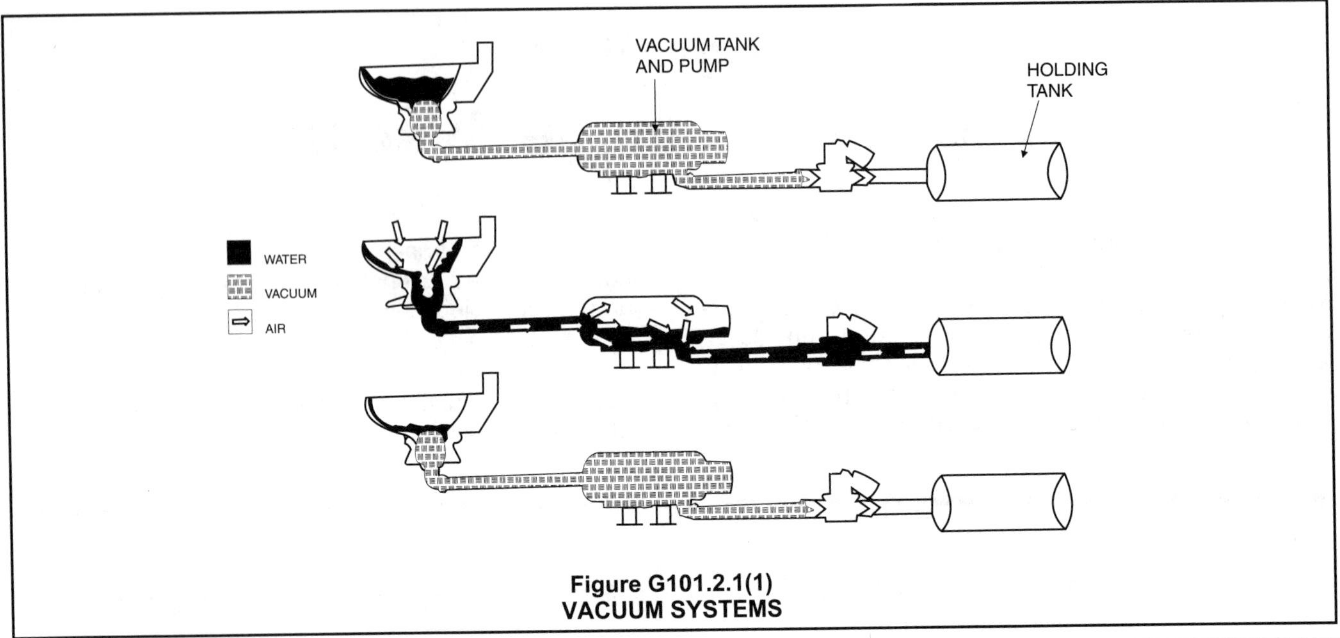

Figure G101.2.1(1)
VACUUM SYSTEMS

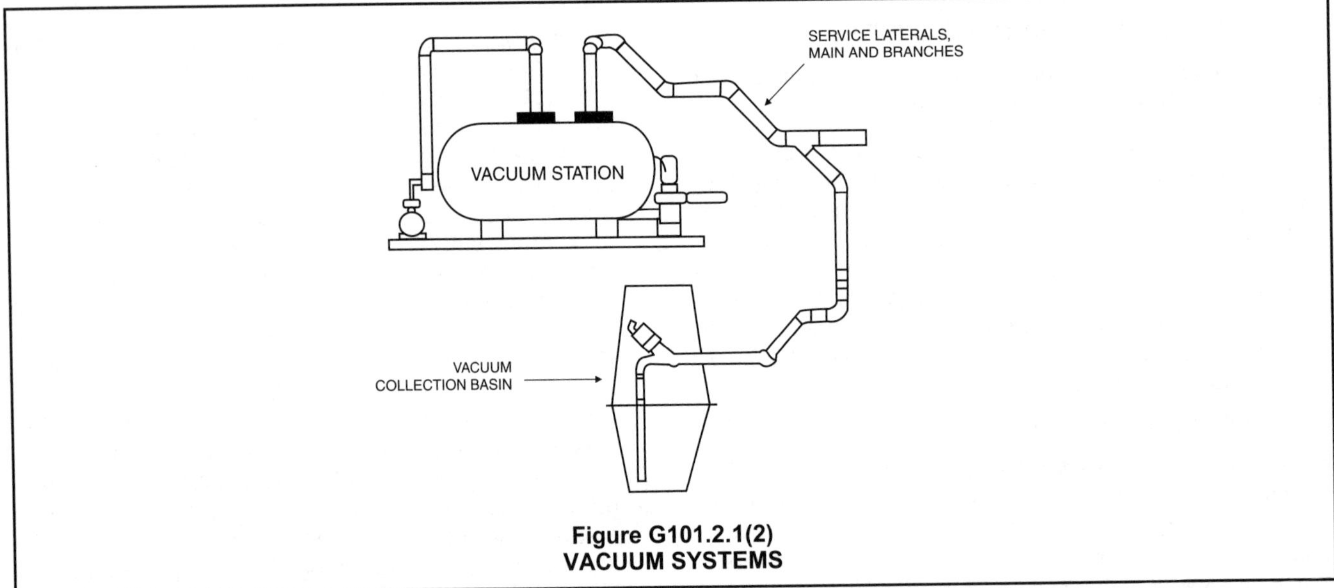

Figure G101.2.1(2)
VACUUM SYSTEMS

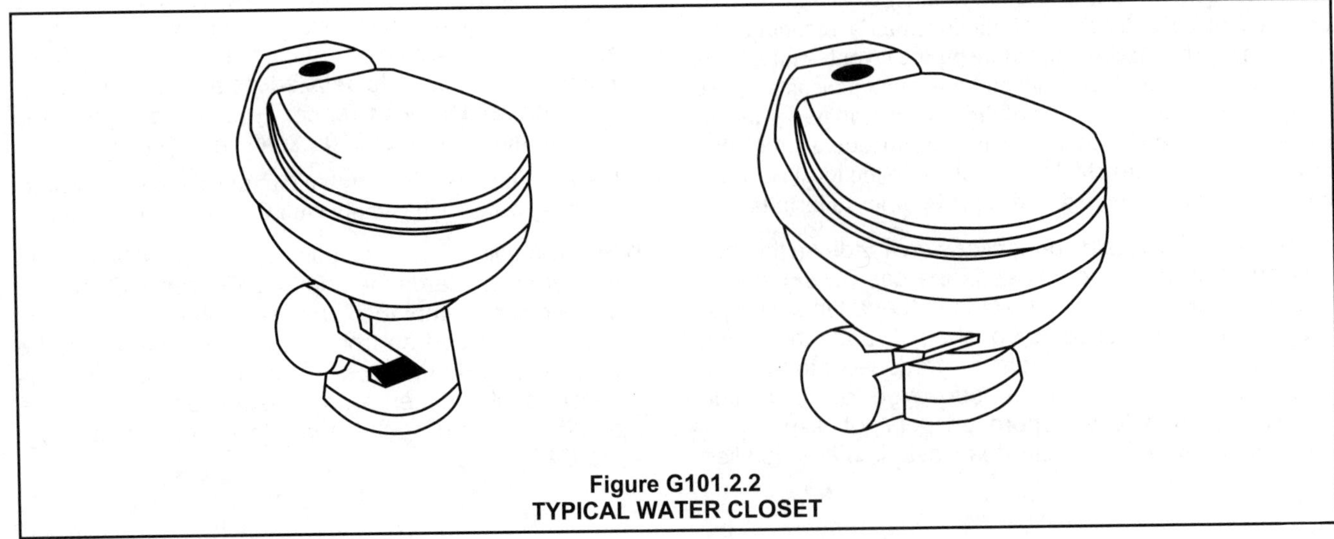

Figure G101.2.2
TYPICAL WATER CLOSET

G101.2.3 Drainage fixture units. Fixture units for gravity drainage systems which discharge into or receive discharge from vacuum drainage systems shall be based on values in Chapter 7 of this code.

❖ Table 709.1 of the code provides the drainage fixture unit (dfu) value for various type of fixtures. A water closet installed in a vacuum system typically discharges 1.1 liters or 1 quart of water per flush. Note e in this table indicates that water closets and urinals may have a lower dfu value confirmed by test results. Many vacuum-type water closets have a dfu value lower than the same fixture type indicated in Table 709.1.

G101.2.4 Water supply fixture units. Water supply fixture units shall be based on values in Chapter 6 of this code with the addition that the fixture unit of a vacuum-type water closet shall be "1."

❖ Appendix E contains water supply fixture units for various types of fixtures. For fixtures not listed, loads should be assumed by comparing the fixture to one listed using water in similar quantities and rates. Manufacturers' recommendations include the amount of water and pressure required to operate properly various types of fixtures utilized in a vacuum drainage system.

G101.2.5 Traps and cleanouts. Gravity-type fixtures shall be provided with traps and cleanouts in accordance with Chapters 7 and 10 of this code.

❖ Drainage piping in a vacuum system must have cleanouts installed in accordance with Section 708. Fixture traps must be provided in accordance with Section 1002.1. Many vacuum systems can eliminate fixture venting because of the positive seal between discharge waste water and the fixture; hence, the necessity for traps. The code official must evaluate each system individually based on the manufacturer's recommended installation instructions and the requirements of the code.

G101.2.6 Materials. Vacuum drainage pipe, fitting and valve materials shall be as recommended by the vacuum drainage system manufacturer and as permitted by this code.

❖ Before installation of drainage pipes and fittings, the manufacturer's recommendations must be consulted in order to evaluate the type of material to be used. Table 702.1 provides the acceptable material to be used in vacuum systems.

G101.3 Testing and demonstrations. After completion of the entire system installation, the system shall be subjected to a vacuum test of 19 inches (483 mm) of mercury and shall be operated to function as required by the administrative authority and the manufacturer. Recorded proof of all tests shall be submitted to the administrative authority.

❖ Vacuum systems must be subjected to testing to determine potential defects and leaks. Vacuum pumps maintain an operating range of 16 to 20 inches (406 to 508 mm) of mercury vacuum. Differential air pressure propels the sewage at velocities of 15 to 18 feet per second (4.5 to 5.5 m/s). Such systems are subjected to a vacuum test of 19 inches (483 mm) of mercury, which equals 9.3 pounds per square inch (64 kPa) of air pressure.

G101.4 Written instructions. Written instructions for the operations, maintenance, safety and emergency procedures shall be provided by the building owner as verified by the administrative authority.

❖ Manufacturer's installation instructions must be provided because they are an enforceable extension of the code for determining whether a system has been properly installed. A detailed description of operating maintenance, safety and emergency procedures is required to reflect the scope of information needed to determine compliance.

INDEX

D

E

F

G

H

I

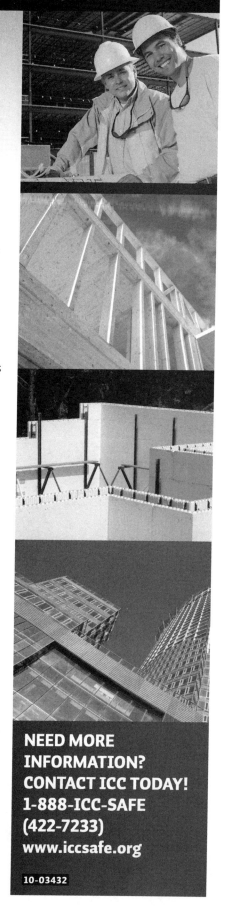